WILLIAM F. MAAG LIBRARY
YOUNGSTOWN STATE UNIVERSITY

McGraw-Hill Yearbook of Science & Technology

1982
1983

McGraw-Hill Yearbook of Science & Technology

1982
1983

COMPREHENSIVE COVERAGE OF RECENT EVENTS AND RESEARCH AS COMPILED BY THE STAFF OF THE McGRAW-HILL ENCYCLOPEDIA OF SCIENCE AND TECHNOLOGY

McGRAW-HILL BOOK COMPANY

New York St. Louis San Francisco
Auckland Bogotá Guatemala Hamburg
Johannesburg Lisbon London Madrid Mexico
Montreal New Delhi Panama Paris San Juan
São Paulo Singapore Sydney Tokyo Toronto

McGRAW-HILL YEARBOOK
OF SCIENCE & TECHNOLOGY
Copyright © 1982 by McGraw-Hill, Inc.
All rights reserved. Printed in the
United States of America. Except as permitted
under the United States Copyright Act
of 1976, no part of this publication may be
reproduced or distributed in any form or by any
means, or stored in a data base or retrieval
system, without the prior written permission of
the publisher. Philippines Copyright, 1982,
by McGraw-Hill, Inc.

1234567890 DODO 898765432

The Library of Congress has cataloged this serial
publication as follows:

McGraw-Hill yearbook of science and technology.
1962– . New York, McGraw-Hill Book Co.

v. illus. 26 cm.
Vols. for 1962– compiled by the staff of the
McGraw-Hill encyclopedia of science and
technology.

1. Science—Yearbooks. 2. Technology—
Yearbooks. I. McGraw-Hill encyclopedia of
science and technology.
Q1.M13 505.8 62-12028
Library of Congress (10)

ISBN 0-07-045489-2
ISSN 0076-2016

Oversize
Q
121
.M312
1982/83

Table of Contents

WILLIAM F. MAAG LIBRARY
YOUNGSTOWN STATE UNIVERSITY

International Editorial Advisory Board

Dr. Neil Bartlett
Professor of Chemistry
University of California, Berkeley

Dr. Richard H. Dalitz
Department of Theoretical Physics
University of Oxford

Dr. Freeman J. Dyson
The Institute for Advanced Study
Princeton, NJ

Dr. George R. Harrison
Dean Emeritus, School of Science
Massachusetts Institute of Technology

Dr. Leon Knopoff
Institute of Geophysics and
Planetary Physics
University of California, Los Angeles

Dr. H. C. Longuet-Higgins
Royal Society Research Professor,
Experimental Psychology
University of Sussex

Dr. Alfred E. Ringwood
Director, Research School of
Earth Sciences
Australian National University

Dr. Arthur L. Schawlow
Professor of Physics
Stanford University

Dr. Koichi Shimoda
Department of Physics
Keio University

Dr. A. E. Siegman
Director, Edward L. Ginzton Laboratory
Professor of Electrical Engineering
Stanford University

Prof. N. S. Sutherland
Director, Centre for Research
on Perception and Cognition
University of Sussex

Dr. Hugo Theorell
The Nobel Institute
Stockholm

Lord Todd of Trumpington
University Chemical Laboratory
Cambridge University

Dr. George W. Wetherill
Director, Department of Terrestrial
Magnetism
Carnegie Institution of Washington

Dr. E. O. Wilson
Professor of Zoology
Harvard University

Dr. Arnold M. Zwicky
Professor of Linguistics
Ohio State University

Editorial Staff

Sybil P. Parker, Editor in Chief

Jonathan Weil, Editor
Betty Richman, Editor

Edward J. Fox, Art director

Richard Roth, Art editor
Ann D. Bonardi, Art production supervisor
Cynthia M. Kelly, Art/traffic

Joe Faulk, Editing manager

Thomas M. Siracusa, Editing supervisor
Ruth L. Williams, Editing supervisor

Patricia W. Albers, Senior editing assistant
Judith Alberts, Editing assistant
Barbara Begg, Editing assistant

Art suppliers: Eric G. Hieber,
EH Technical Services, New York, New York;
James R. Humphrey, Nanuet, New York;
E. T. Steadman, New York, New York.

Typesetting and composition
by The Clarinda Company, Clarinda, Iowa.

Printed and bound by R. R. Donnelley
& Sons Company, the Lakeside Press
at Willard, Ohio, and Crawfordsville, Indiana.

Consulting Editors

Prof. George O. Abell. *Department of Astronomy, University of California, Los Angeles*. ASTRONOMY.

Vincent M. Altamuro. *President, Management Research Consultants, Yonkers, NY*. INDUSTRIAL ENGINEERING.

Prof. P. M. Anderson. *Department of Electrical and Computer Engineering, Arizona State University*. ELECTRICAL POWER ENGINEERING.

Prof. Eugene A. Avallone. *Department of Mechanical Engineering, City University of New York*. MECHANICAL POWER ENGINEERING AND PRODUCTION ENGINEERING.

Prof. B. Austin Barry. *Civil Engineering Department, Manhattan College*. CIVIL ENGINEERING.

Dr. Alexander Baumgarten. *Yale–New Haven Hospital*. IMMUNOLOGY AND VIROLOGY.

Dr. Salomon Bochner. *Edgar Odell Lovett Professor of Mathematics, Rice University*. MATHEMATICS.

Dr. Walter Bock. *Professor of Evolutionary Biology, Department of Biological Sciences, Columbia University*. ANIMAL ANATOMY; ANIMAL SYSTEMATICS; VERTEBRATE ZOOLOGY.

Robert E. Bower. *Director for Aeronautics, Langley Research Center, National Aeronautics and Space Administration*. AERONAUTICAL ENGINEERING; PROPULSION.

Edgar H. Bristol. *Corporate Research, Foxboro Company, Foxboro, MA*. CONTROL SYSTEMS.

Prof. D. Allan Bromley. *Henry Ford II Professor and Director, A. W. Wright Nuclear Structure Laboratory, Yale University*. ATOMIC, MOLECULAR, AND NUCLEAR PHYSICS.

Michael H. Bruno. *Graphic Arts Consultant, Nashua, NH*. GRAPHIC ARTS.

S. J. Buchsbaum. *Vice President, Network Planning and Customer Services, Bell Laboratories, Holmdel, NJ*. TELECOMMUNICATIONS.

Dr. John F. Clark. *Director, Space Application and Technology, RCA Laboratories, Princeton, NJ*. SPACE TECHNOLOGY.

Dr. Richard B. Couch. *Ship Hydrodynamics Laboratory, University of Michigan*. NAVAL ARCHITECTURE AND MARINE ENGINEERING.

Prof. David L. Cowan. *Department of Physics, University of Missouri*. CLASSICAL MECHANICS AND HEAT.

Dr. James Deese. *Department of Psychology, University of Virginia*. PHYSIOLOGICAL AND EXPERIMENTAL PSYCHOLOGY.

Prof. Todd M. Doscher. *Department of Petroleum Engineering, University of Southern California*. PETROLEUM CHEMISTRY; PETROLEUM ENGINEERING.

Dr. H. Fernandez-Moran. *A. N. Pritzker Divisional Professor of Biophysics, Division of Biological Sciences and the Pritzker School of Medicine, University of Chicago*. BIOPHYSICS.

Dr. John K. Galt. *Vice President, Sandia Laboratories, Albuquerque*. PHYSICAL ELECTRONICS.

Prof. M. Charles Gilbert. *Department of Geological Sciences, Virginia Polytechnic Institute and State University*. GEOLOGY (MINERALOGY AND PETROLOGY).

Prof. Roland H. Good, Jr. *Department of Physics, Pennsylvania State University*. THEORETICAL PHYSICS.

Dr. Alexander von Graevenitz. *Department of Medical Microbiology, University of Zurich*. MEDICAL BACTERIOLOGY.

Prof. David L. Grunes. *U.S. Plant, Soil and Nutrition Laboratory, U.S. Department of Agriculture*. SOILS.

Dr. Carl Hammer. *Research Consulting Services, Washington, DC*. COMPUTERS.

Prof. Dennis R. Heldman. *Department of Food Science and Human Nutrition, Michigan State University*. FOOD ENGINEERING.

Consulting Editors (continued)

Prof. Gertrude W. Hinsch. *Department of Biology, University of South Florida.* GROWTH AND MORPHOGENESIS.

Dr. R. P. Hudson. *Bureau International des Poids et Mesures, France.* LOW-TEMPERATURE PHYSICS.

Prof. Stephen F. Jacobs. *Professor of Optical Sciences, University of Arizona.* ELECTROMAGNETIC RADIATION AND OPTICS.

Prof. Richard C. Jarnagin. *Department of Chemistry, University of North Carolina, Chapel Hill.* PHYSICAL CHEMISTRY.

Dr. Gary Judd. *Vice Provost, Academic Programs and Budget, and Dean of the Graduate School, Rensselaer Polytechnic Institute.* METALLURGICAL ENGINEERING.

Dr. Donald R. Kaplan. *Miller Professor, Department of Botany, University of California, Berkeley.* PLANT ANATOMY.

Dr. Joseph J. Katz. *Senior Chemist, Argonne National Laboratory.* INORGANIC CHEMISTRY.

Dr. Peter B. Kaufman. *Division of Biological Sciences, University of Michigan.* PLANT PHYSIOLOGY.

Dr. George deVries Klein. *Professor of Sedimentology, University of Illinois at Urbana-Champaign.* GEOLOGY (SURFICIAL AND HISTORICAL); PHYSICAL GEOGRAPHY.

Prof. Kenneth D. Kopple. *Department of Chemistry, Illinois Institute of Technology.* ORGANIC CHEMISTRY.

Charles E. Lapple. *Consultant, Fluid and Particle Technology, Air Pollution and Chemical Engineering.* FLUID MECHANICS.

Prof. R. Bruce Lindsay. *Hazard Professor of Physics, Emeritus, Brown University.* ACOUSTICS.

Dr. Howard E. Moore. *Professor of Chemistry and Chairman, Department of Physical Sciences, Florida International University.* GEOCHEMISTRY.

Dr. N. Karle Mottet. *Professor of Pathology, University of Washington School of Medicine.* MEDICINE AND PATHOLOGY.

Dr. Royce W. Murray. *Kenan Professor of Chemistry, University of North Carolina.* ANALYTICAL CHEMISTRY.

Prof. Vjekoslav Pavelic. *Systems-Design Department, University of Wisconsin.* DESIGN ENGINEERING.

Dr. Guido Pontecorvo. *Imperial Cancer Research Fund Laboratories, London.* GENETICS AND EVOLUTION.

Prof. D. A. Roberts. *Plant Pathology Department, Institute of Food and Agricultural Sciences, University of Florida.* PLANT PATHOLOGY.

Prof. W. D. Russell-Hunter. *Professor of Zoology, Department of Biology, Syracuse University.* INVERTEBRATE ZOOLOGY.

Brig. Gen. Peter C. Sandretto. *(Retired) Director, Engineering Management, International Telephone and Telegraph Corporation.* NAVIGATION.

Dr. Bradley T. Scheer. *Professor Emeritus, Biology, University of Oregon.* GENERAL PHYSIOLOGY.

Prof. Thomas J. M. Schopf. *Department of Geophysical Sciences, University of Chicago.* PALEOBOTANY AND PALEONTOLOGY.

Prof. Frederick Seitz. *Formerly, President, The Rockefeller University.* SOLID-STATE PHYSICS.

Dr. Richard G. Wiegert. *Department of Zoology, University of Georgia.* ANIMAL ECOLOGY.

Dr. W. A. Williams. *Department of Agronomy and Range Science, University of California, Davis.* AGRICULTURE.

Contributors

A list of contributors, their affiliations, and the articles they wrote will be found on pages 479–482.

Preface

The 1982–1983 *McGraw-Hill Yearbook of Science and Technology*, continuing in the tradition of its 20 predecessors, presents the outstanding recent achievements in science and technology. Thus it serves as an annual review and also as a supplement to the *McGraw-Hill Encyclopedia of Science and Technology*, updating the basic information in the fifth edition (1982) of the Encyclopedia. Because this Yearbook is the first to appear on the new schedule entailing publication each fall, it is exceptional in carrying a double-year designation.

The Yearbook contains articles reporting on those topics that were judged by the 51 consulting editors and the editorial staff as being among the most significant recent developments. Each article is written by one or more authorities who are actively pursuing research or are specialists on the subject being discussed.

The Yearbook is organized in two independent sections. The first section includes seven feature articles, providing comprehensive, expanded coverage of subjects that have broad current interest and possible future significance. The second section comprises 152 alphabetically arranged articles on such topics as toxic shock syndrome, ape language, laser imaging, electronic cameras, solar-powered aircraft, geotextiles, and the space shuttle.

The *McGraw-Hill Yearbook of Science and Technology* provides librarians, students, teachers, the scientific community, and the general public with information needed to keep pace with scientific and technological progress throughout the world. The Yearbook has successfully served this need for the past 21 years through the ideas and efforts of the consulting editors and the contributions of eminent international specialists.

SYBIL P. PARKER
EDITOR IN CHIEF

McGraw-Hill Yearbook of Science & Technology

1982
1983

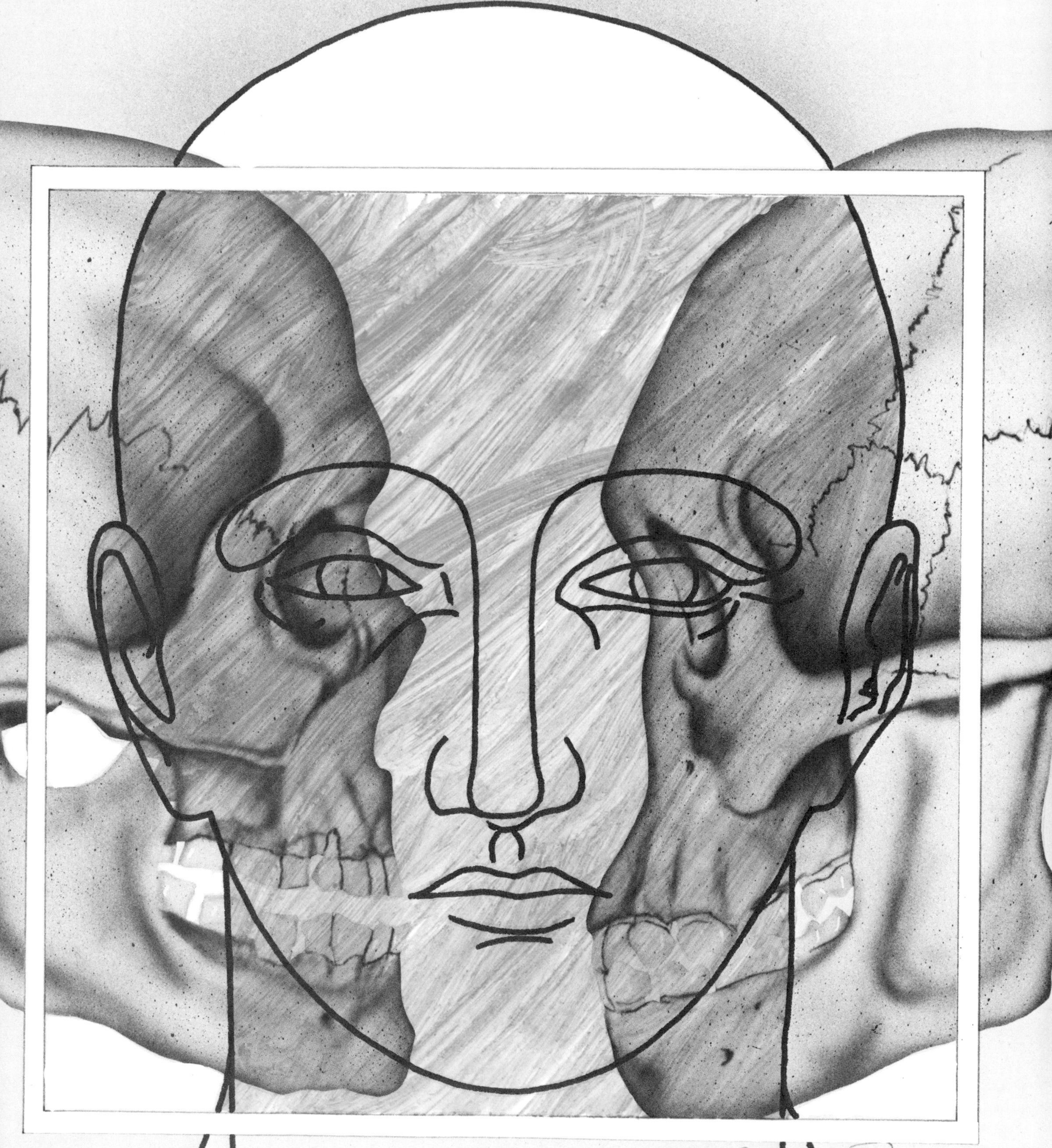
E.T. Steadman

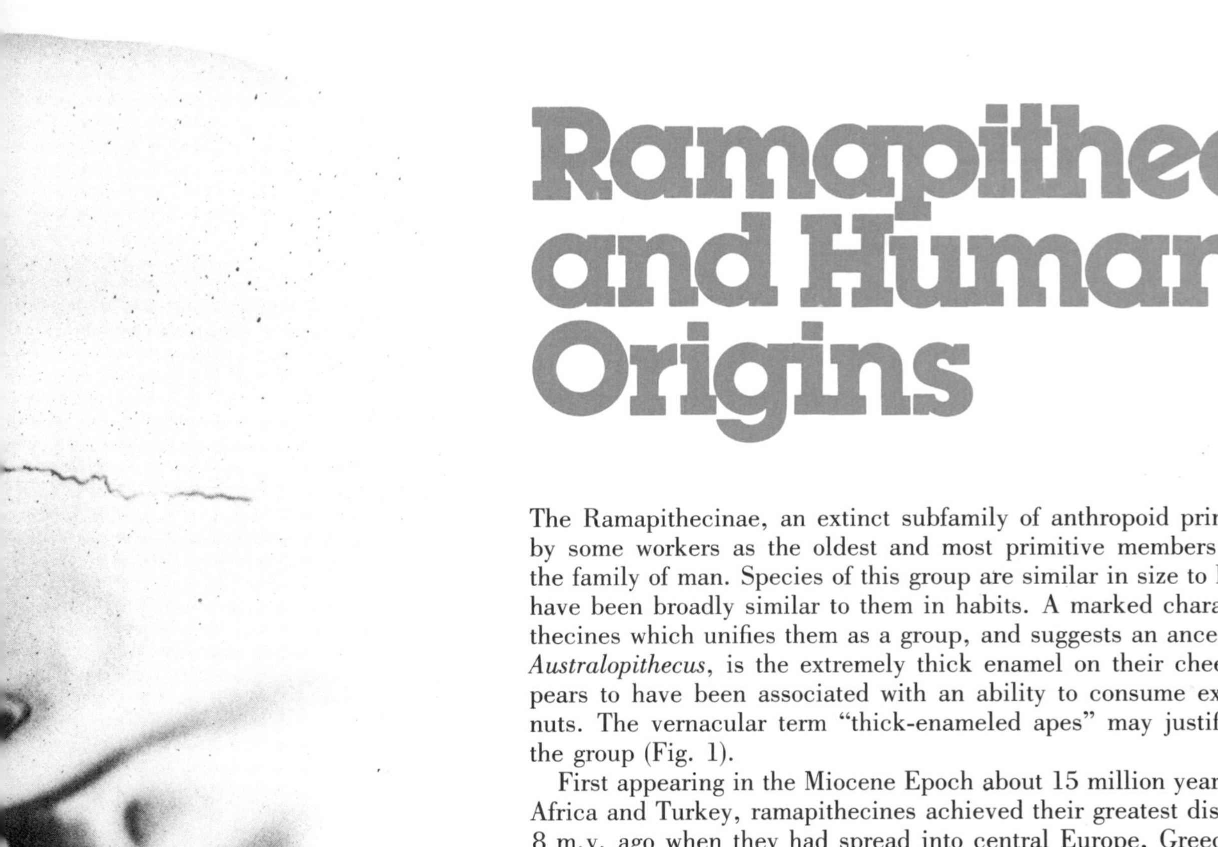

Ramapithecines and Human Origins

The Ramapithecinae, an extinct subfamily of anthropoid primates, are regarded by some workers as the oldest and most primitive members of the Hominidae, the family of man. Species of this group are similar in size to living apes and may have been broadly similar to them in habits. A marked characteristic of ramapithecines which unifies them as a group, and suggests an ancestral relationship to *Australopithecus*, is the extremely thick enamel on their cheek teeth, which appears to have been associated with an ability to consume extremely hard forest nuts. The vernacular term "thick-enameled apes" may justifiably be applied to the group (Fig. 1).

First appearing in the Miocene Epoch about 15 million years (m.y.) ago in East Africa and Turkey, ramapithecines achieved their greatest distribution about 10–8 m.y. ago when they had spread into central Europe, Greece, Pakistan, India, and China and were represented by no fewer than seven species. *Gigantopithecus*, a giant ape bigger than the mountain gorilla, from India, Pakistan, and China survived as a contemporary of early *Homo* until perhaps 1 m.y. ago. Other, less specialized branches of ramapithecines gave rise to *Australopithecus*, the earliest bipedal hominid, the oldest known occurrence of which is about 5.5 m.y. ago.

Among ramapithecines two genera are generally recognized, *Sivapithecus* and

Richard F. Kay, Associate Professor of Anatomy and Anthropology at Duke University, is a specialist in the reconstruction of adaptations of extinct primates, especially monkeys and apes. Since 1979 he has conducted field expeditions to recover the remains of the earliest apes in Egypt. He has written widely on primate, ape, and human origins and adaptations.

Fig. 1. Hypothetical reconstruction of middle Miocene *Sivapithecus* as it would have appeared in life. (*Drawing by Donna Gregory*)

Gigantopithecus, each containing several species. Discoveries of many fossils over the past 10 years demonstrate the great similarities of known parts (mostly jaws and teeth) among all ramapithecines, similarities which obviate the need to further subdivide *Sivapithecus* as some authorities do by continuing to recognize *Ramapithecus* and *Ouranopithecus*. These finds also show that although ramapithecines should be regarded as hominids because of their unique genealogical affinity with humans, they are much more primitive, or apelike, than had been recognized hitherto. This article gives a historical outline of the principal discoveries and a summary of the reasons for thinking ramapithecines are hominids. It also provides an analysis of the fossils for clues which would shed light on their adaptations.

HISTORY OF STUDY

An account of a ramapithecine fossil was first published in 1837, describing an upper canine tooth obtained from the Siwalik Hills in North India. This and other fossils from the Siwaliks range in age from 13 to 8 m.y. The authors of the report felt that this tooth, now lost, was that of an orangutan. In 1879 the British naturalist R. Lydekker described a ramapithecine palate with most of the teeth on one side. He called this specimen *Paleopithecus sivalensis*, and this name would continue to have been the standard attached to the group as a whole, were it not that *Paleopithecus* had already been applied in a German publication to fossil footprints of dinosaurs which were mistakenly thought to be those of an extinct "colossal age." In 1910 Guy Pilgrim named *Sivapithecus indicus* and in later papers advanced in great detail the idea that this animal was intimately attached to the human lineage, an idea which had been considered briefly by Lydekker in 1879. Pilgrim also described several other genera and species of fossil apes and considered them to be related to various living apes. It is now known that all the ape specimens he described are parts of several closely related species which take the name *Sivapithecus* and *Gigantopithecus*. Since those early finds, India and Pakistan have continued to yield many ramapithecine fossils. A joint Yale University–Geological Survey of Pakistan expedition to the Potwar Plateau area, under way since 1973, continues to recover important new specimens, including recently a part of a face of *Sivapithecus* and several fragmentary limb bones, the first known which definitely belong to ramapithecines.

European finds. The jaw of a large European ape was found in 1856 near the village of Saint-Gaudens, France, from a site where clay was being mined to make bricks. Edouard Lartet described this as *Dryopithecus fontani*. Later, isolated teeth recovered from deposits of the Vienna Basin in Germany were thought to belong to *Dryopithecus*, but several of these may belong to *Sivapithecus*. These and other European fossil apes are between 15 and 10 m.y. old. In recent years the scene of European discovery has shifted to Greece, the Balkans, and Turkey. Of particular importance are the finds of more than a dozen jaw fragments in one stratigraphic level in Macedonian Greece by a joint Greek and French expedition (Fig. 2). In 1980 Turkish and British paleontologists described a new facial fragment of *Sivapithecus* from near Ankara, Turkey (Fig. 3). Additionally, M. Kretzoi, a Hungarian paleontologist, has made many discoveries of *Sivapithecus* from coal layers near Budapest.

African finds. Between 1932 and 1935 L. S. B. Leakey's East African archeological expeditions made collections of ape fossils, among which was an upper jaw fragment with three teeth from Kenya. It was not until 1950, however, that the English anatomist Wilfred E. Le Gros Clark with Leakey

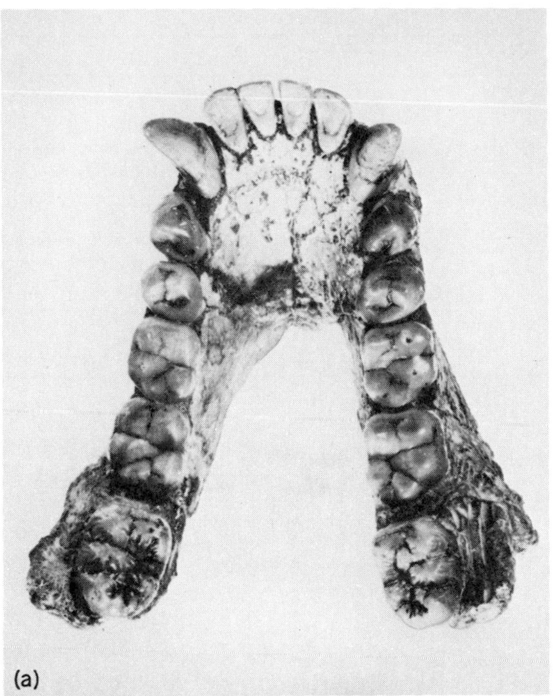
(a)

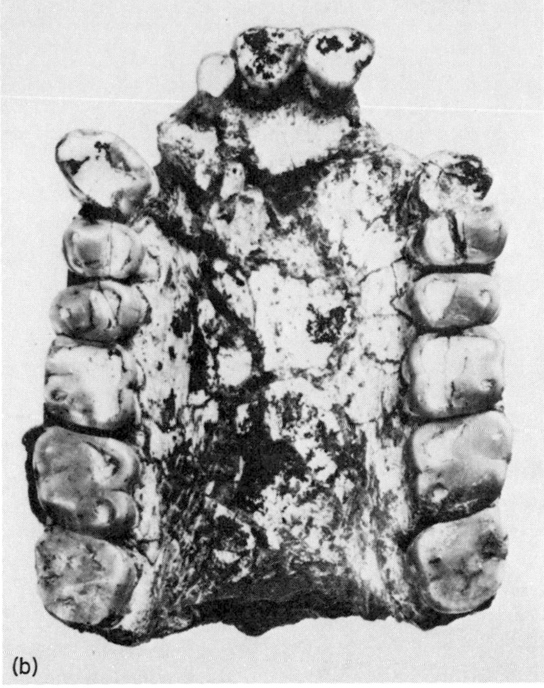
(b)

Fig. 2. Specimens of ramapithecine (*a*) lower jaw and (*b*) palate. (*Courtesy of L. de Bonis*)

recognized that this specimen was that of *Sivapithecus*, marking the first recognition of a ramapithecine from Africa. Since then, a handful of new specimens of ramapithecines have been discovered in Kenya at Fort Ternan and near Maboko Island (in Lake Victoria), most of which range in age between 15 and 13 m.y.

Chinese finds. Perhaps the most exciting finds of the last decade have been made in China. Until recently, Chinese *Sivapithecus* was known from a few isolated teeth described in the 1950s. However, expeditions starting in 1975 carried out by the Academia Sinica and the Yunnan Provincial Museum have recovered hundreds of new specimens aged roughly between 10 and 8 m.y. from coal fields close to Lufeng in Yunnan Province. Several beautifully preserved lower jaws have been described so far; pictures and accounts in the Chinese foreign language press indicate that as many as six crushed skulls (Fig. 4) are now known. As this material becomes better documented scientifically, it may revolutionize the understanding of ramapithecine adaptations and relationships.

SOME APE–HUMAN CONTRASTS

To set the stage for an assessment of details of ramapithecine anatomy, the anatomy and behavior of living great apes must first be considered in contrast with that of the earliest prehuman, *Australopithecus*, from Africa. The reason is that it has long been assumed, and rightly, that humans passed through a stage in their evolution in which they resembled living apes in many respects. Therefore, knowledge about living apes gives insight about early stages of human evolution. Furthermore, the spectrum of ape–human contrasts can be used as a heuristic device to visualize the closeness of approach of ramapithecines to the human condition.

The great apes are man's closest living relations; smaller "lesser apes," the gibbons, are more distantly related. There are four species of great apes (Table 1 and Fig. 5): the common chimpanzee and bonobo chimpanzee (*Pan troglodytes* and *P. paniscus*, respectively) and the gorilla (*Gorilla gorilla*)

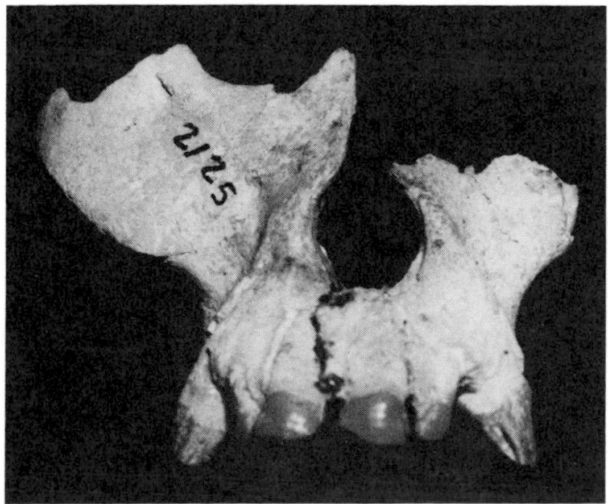

Fig. 3. Front view of a snout and face of *Sivapithecus* from the late Miocene of Turkey. Note the large size of canine teeth. The bottom of the right orbit is preserved at the top. (*Courtesy of P. Andrews*)

Table 1. Taxonomic groups of apes, humans, and prehumans

Family	Living forms	Extinct forms
Pongidae (all great apes)	*Pongo* (orangutan of Borneo and Sumatra) *Pan* (chimpanzee and bonobo of Africa) *Gorilla* (gorilla of Africa)	*Proconsul* (22–14 m.y.; several species from Africa and Saudi Arabia) *Dryopithecus* (15–10 m.y.; several species from Europe, southern Russia, and ?Africa)
Hominidae (humans and prehumans)	*Homo* (human beings)	*Australopithecus* (5.5–1.5 m.y.; from Africa) *Sivapithecus* (includes *Ouranopithecus* and *Ramapithecus*; 15–8 m.y.; from Africa and Eurasia) *Gigantopithecus* (9–?1 m.y.; from Asia)

live in Africa; and the orangutan (*Pongo pygmaeus*) lives in Borneo and Sumatra. Orangutans formerly inhabited parts of Southeast Asia, China, and Java but are now extinct there. Except for some fossil orangutan teeth of recent age, there is no fossil record of any of these living species.

Details of the anatomy and biochemistry show that African apes (gorilla and chimpanzee) are more closely allied to humans than is the orangutan. Some biochemists would go further and claim, based on their molecular studies, to place rather precise times for the branching of these lineages from one another. However, the validity of their approach has been challenged for a variety of sound reasons. It appears that study of molecular structure provides useful information for assessing relationships, as do details of the gross and microscopic anatomy of animals, but none can precisely determine the timing of branching of lineages owing to changing rates of evolution through time. For this, only the fossil record is useful.

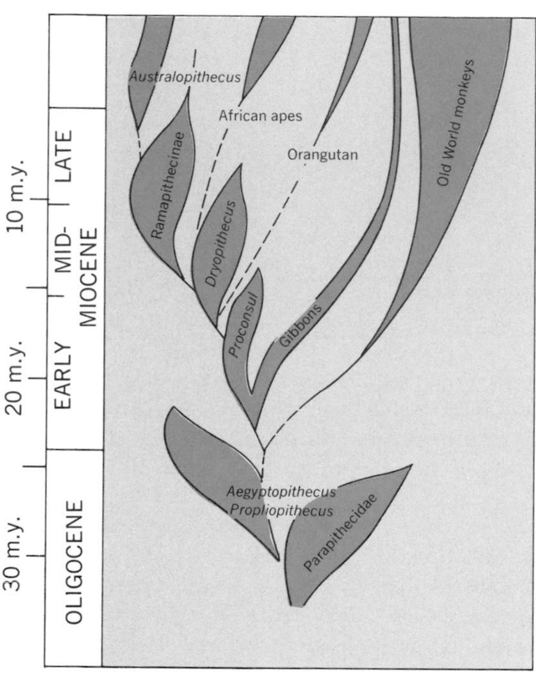

Fig. 5. Tentative phylogeny of monkeys and apes of Africa, Europe, and Asia.

Fig. 4. Badly crushed skull of *Sivapithecus* from the late Miocene of Lufeng, China. Front teeth are at the bottom. Orbits, nasal opening, and flaring, jug-handlelike zygomatic arches are also visible.

Orangutans, chimpanzees, and gorillas differ anatomically and behaviorally from modern humans in a variety of ways. The fossils document the antiquity of some of the anatomical differences and by inference the behavioral ones. For those characters for which there is a historical record, the process of hominization was extremely gradual.

Extremities and locomotion. Living apes are primarily forest-dwelling arboreal creatures. In the trees they are agile climbers and commonly suspend themselves by their arms under tree limbs while feeding and during locomotion. In conjunction with their arboreality, they have an opposable big toe on their foot, allowing them to grasp tree limbs for support. A short, broad thorax brings the elongate arms to the side of the body. These and modifications of the wrist, elbow, and shoulder joints greatly enhance reach and mobility of the arms, a requirement for suspensory activities. Also, the fingers of apes are long and curved, allowing the hand to operate somewhat like a hook during forelimb hanging. Orangutans are the most specialized of the apes for the arboreal mode of life. They are deliberate climbers. Their hindfoot is greatly modified as a hooklike appendage by reduction of the big toe; the range of movement at the hip joint is greater than the other apes, perhaps enhanced by loss of the round ligament of the femur. Their arms are even more elongate and their fingers more curved than those of the African apes, allowing greater reach and improved grasping ability. When they do come to the ground, living apes walk on all fours. The long, curved fingers are often or habitually brought into a fist or are turned so that the weight of the body is transmitted through the knuckles.

Human beings are of course terrestrial creatures walking bipedally. Modification of the human foot such that the big toe is no longer opposable, reorganization of the structure of knee and hip, and shortening of the fingers and arms are hallmarks of bipedalism. Based on these anatomical characters, bipedal locomotion had been achieved in *Australopithecus* at least by about 3.5 m.y. ago.

Feeding: dental and jaw structure. The living apes are almost completely herbivorous, although there is a great deal of difference among the species as to which plant parts are preferred. Gorillas prefer to eat more fibrous plant foods such as leaves, bark, buds, and pith. Chimpanzees, in marked contrast, specialize more in fruits which tend to be lower in total fiber. Orangutans have a mixed frugivorous-folivorous diet, showing great swings in preference in the wet and dry seasons. They prefer to eat fruit when available. More than any other living ape, orangutans will eat very hard-shelled fruits and nuts. The adaptations for these different herbivorous diets are reflected in the structure of the cheek teeth (molars and premolars). Compared with other apes, gorillas have more crown relief on their teeth, reflecting the comparatively well-developed cutting edges. A similar adaptive design is seen among most mammals which primarily eat fibrous foods. Chimpanzees and orangutans have comparatively low relief on their molar crowns; these flattened crown surfaces are more effective in crushing and pulping fruits with less fibrous cell walls. Orangutans have somewhat thicker cheek-tooth enamel than chimpanzees or gorillas, reflecting their abilities to open much harder-shelled fruits and nuts.

The upper incisors of the apes are spatulate-shaped devices used for separating pieces of food before chewing. This activity is accomplished either by a stripping action or by cutting against the lower incisors. The relative size of the incisors reflects the role they play in the different diets of the living apes. The more fruit-eating species, chimpanzee and orangutan, have relatively much larger incisors than does the gorilla.

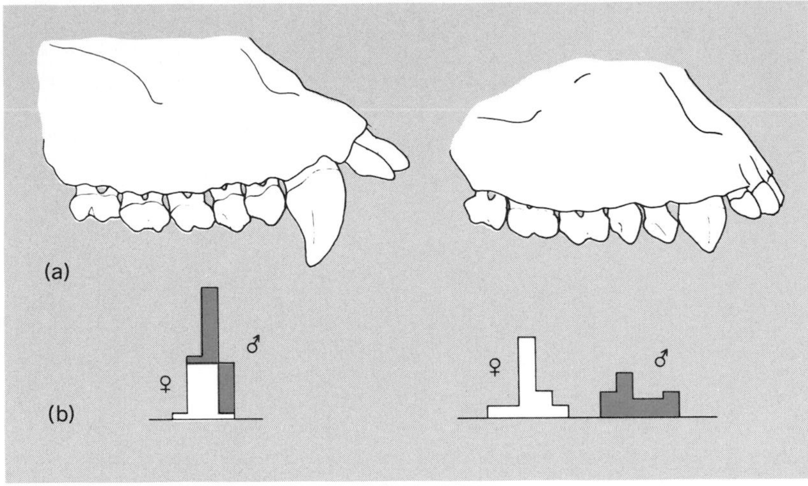

Fig. 6. Sexual dimorphism as seen in *Gorilla*. (*a*) Side views of the upper jaws of a male (left) and female (right) showing the larger canine tooth of the male. (*b*) Histograms showing the differences of molar size (left) and canine size (right). Sexually nondimorphic male molars (shaded part of histograms) overlap those of females in size, whereas sexually dimorphic male canines are always smaller.

All the great apes have large projecting canine teeth which they use to defend themselves and for threat and aggression in a sexual context. The canines are "sexually dimorphic"; that is, those of males are much larger than those of females (Fig. 6) of the species, which goes along with a higher degree of sexual competition among males. The canines of both sexes are also important tools for feed-

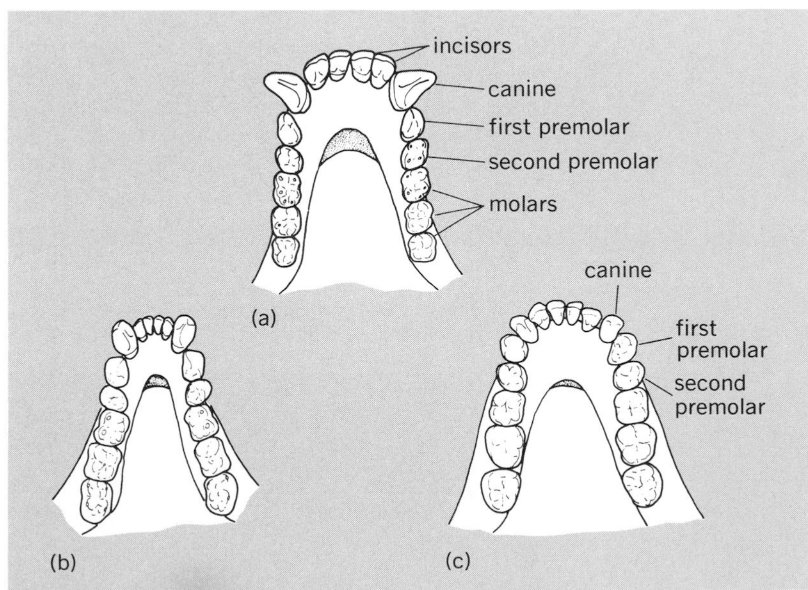

Fig. 7. Comparison of the lower jaws and teeth of (*a*) chimpanzee, (*b*) *Sivapithecus*, and (*c*) *Australopithecus*. The tooth rows of chimpanzees are straight-sided and U-shaped, whereas those of *Sivapithecus* are divergent and V-shaped and those of *Australopithecus* are slightly divergent and U-shaped. The incisors of chimpanzees are relatively much larger than either *Sivapithecus* or *Australopithecus*. Note also that the canines of *Australopithecus* are small and that chimpanzees have differently shaped premolars whereas in *Australopithecus* they are more similarly shaped.

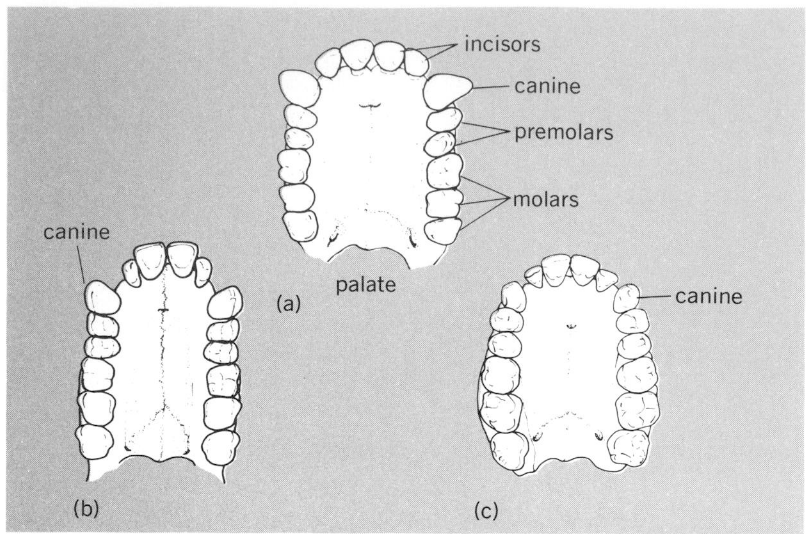

Fig. 8. Comparison of the upper jaws and teeth of (*a*) chimpanzee, (*b*) *Sivapithecus*, and (*c*) *Australopithecus*. Note that chimpanzees and *Sivapithecus* have large canines and straight-sided postcanine tooth rows, whereas *Australopithecus* has small canines and the postcanine teeth are bowed outward.

ing. In this context the upper canine, which is oval in cross section from front to back, contacts the front and outside face of the first of the lower premolars, thereby forming a cutting edge and sharpening device. Thus, the first and second lower premolars are very differently shaped, or heteromorphic, among apes: the first is elongate usually with a single cusp; the second is two-cusped (Fig. 7).

The shape of the palate of the living apes reflects the large size of the canines (Fig. 8). The canine, premolar, and molar teeth of apes are roughly parallel to one another on opposite sides of the palate; a space often occurs in front of the upper canine for the receipt of the large projecting lower canine. Incisors complete the U-shaped arrangement of the teeth in the palate.

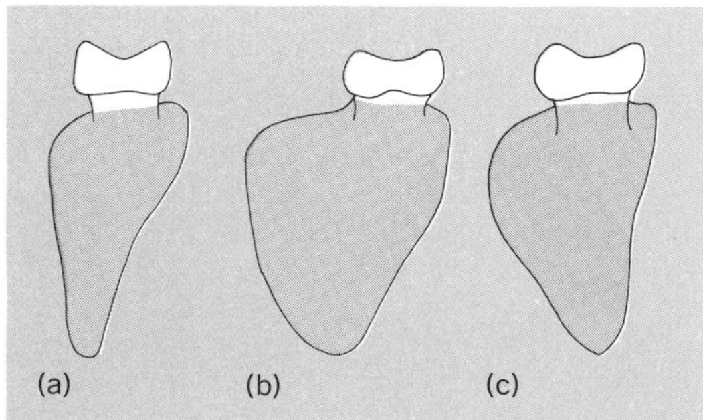

Fig. 9. Cross sections of the lower jaws of (*a*) *Pan troglodytes* (chimpanzee), (*b*) *Australopithecus* from Ethiopia, and (*c*) *Sivapithecus indicus* from Pakistan. *Sivapithecus* more nearly resembles *Australopithecus* in having a relatively thicker mandible when jaw depth is held constant. Sections are made through the jaws between the first and second molar teeth.

The lower jaws of the apes are deep or narrow under the molars; in cross section the distance between the base of the teeth and the bottom of the jaw is much greater than the dimension at right angles to this (Fig. 9). The midline region of the lower jaw, called the mandibular symphysis, is very deep; the dimension from the incisor margin to the back of the symphysis is much greater than the dimension at right angles to this (Fig. 10).

The dental and jaw structure of Plio/Pleistocene *Australopithecus* presents a sharp contrast with that of living apes. *Australopithecus* has extremely thick molar enamel, much thicker than any ape, and low

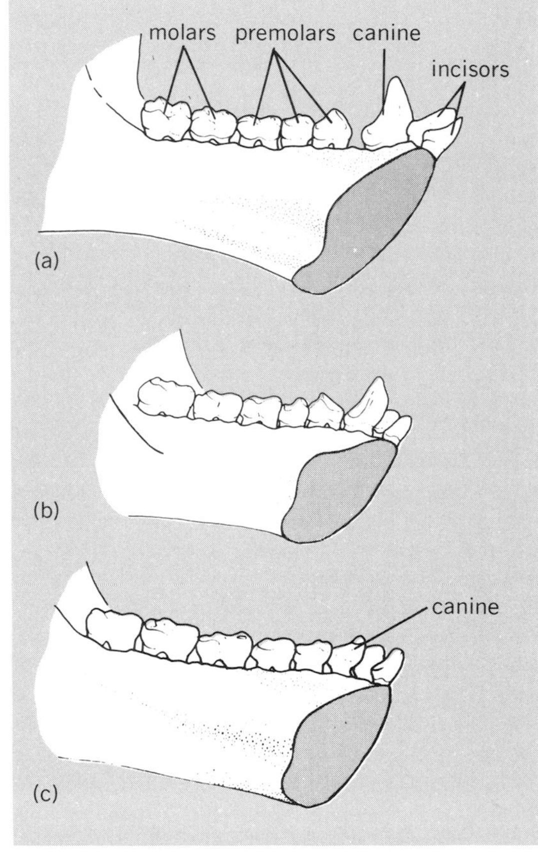

Fig. 10. Internal views of the mandibles of (*a*) *Pan troglodytes* (chimpanzee), male, (*b*) *Sivapithecus sivalensis* from India, and (*c*) early *Australopithecus* from Ethiopia. Each jaw has been cut in the midline to show a cross section of the mandibular symphysis (shaded area). Note the longer, narrower symphyseal cross section of the chimpanzee as well as the small size of the canine in *Australopithecus*.

crown-surface relief. The canines of this early hominoid are vary small (Figs. 7, 8, 10, and 11), and there is little discernible difference between males and females. The structure of the front lower premolar is much more similar to that of the back premolar (Fig. 7) because it is no longer designed to fit against and sharpen a very large upper canine (a second cusp is generally present on the front premolar as on the second one). With the great reduc-

Table 2. Geographic and temporal distribution of large apes and prehumans over the past 20 m.y. up until about 3 m.y. before present

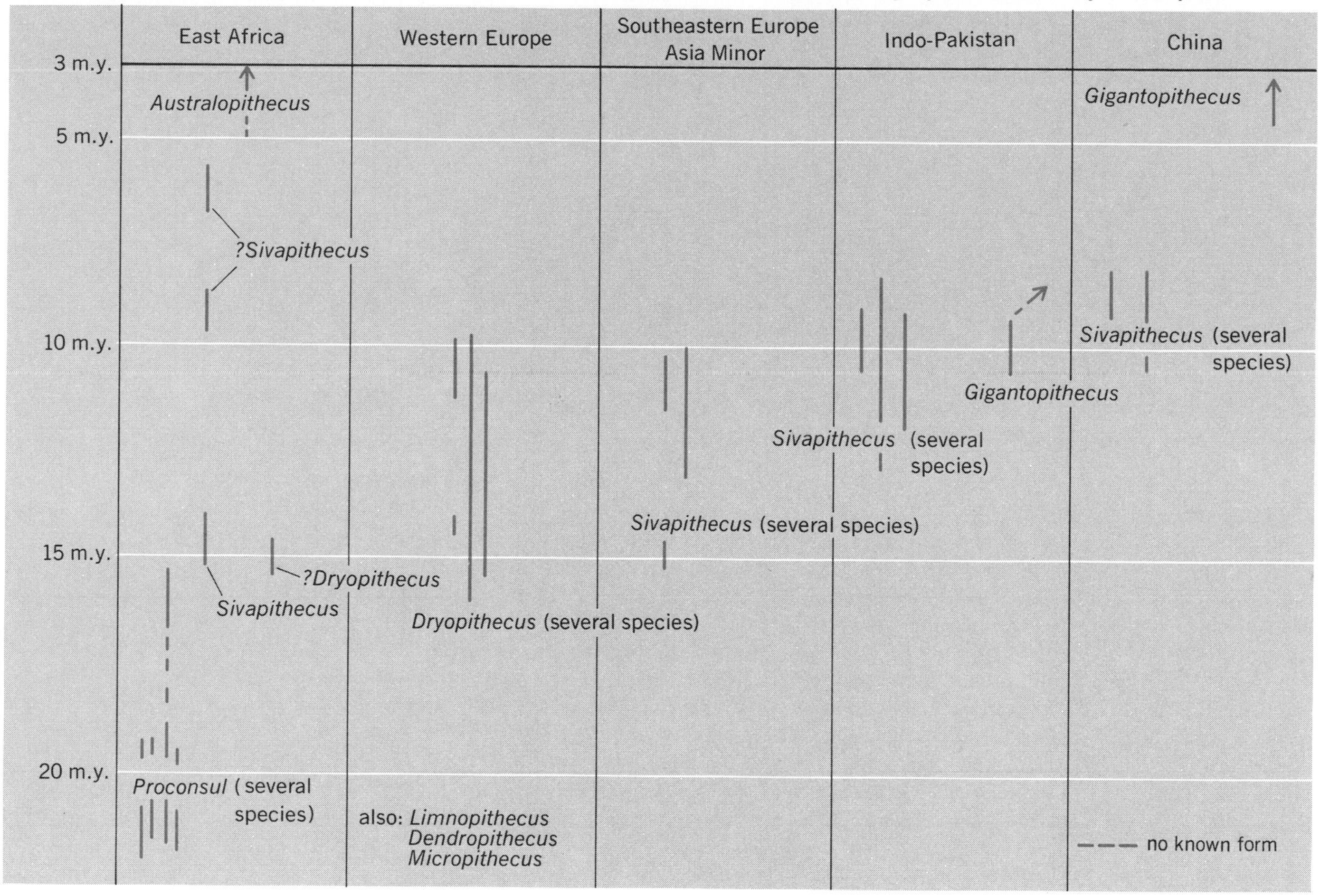

	East Africa	Western Europe	Southeastern Europe Asia Minor	Indo-Pakistan	China
3 m.y.	*Australopithecus*				*Gigantopithecus*
5 m.y.					
	?Sivapithecus				
10 m.y.				*Gigantopithecus*	*Sivapithecus* (several species)
				Sivapithecus (several species)	
			Sivapithecus (several species)		
15 m.y.	*?Dryopithecus* *Sivapithecus*	*Dryopithecus* (several species)			
20 m.y.	*Proconsul* (several species)	also: *Limnopithecus* *Dendropithecus* *Micropithecus*			

--- no known form

tion in size of the canines the palate converges more toward the front than in the apes (Figs. 8 and 11). The lower jaw under the molars is comparatively shallower or thicker in cross section than in apes, and the symphysis is short and thicker as well (Figs. 9 and 10).

EARLY APE HISTORY

The earliest record of ape and human ancestors is found in Egypt in Oligocene times, about 30 m.y. ago. Forms like *Aegyptopithecus* resemble apes in the structure of the teeth, but are very primitive in their limb structure and most closely resemble some of the living New World primates like the howling monkey. *Aegyptopithecus* may be close to the ancestry of both Old World monkeys and apes (Fig. 5). By the early Miocene period, about 23–17 m.y. ago, the earliest Old World monkeys and contemporary apes are known. These come mainly from Kenya and Uganda. The apes are given many names, including *Proconsul*, *Limnopithecus*, *Dendropithecus*, and *Micropithecus* (Table 2). They are still quite primitive in their limb structure, but are very like modern apes in tooth structure. These African apes are best represented by specimens clus-

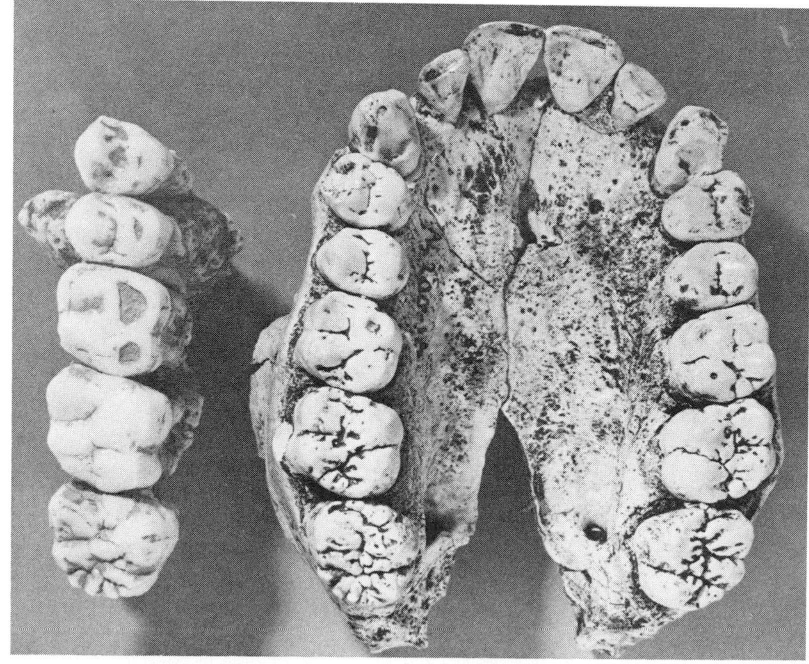

Fig. 11. Palate of an early *Australopithecus* from Ethiopia beside five cheek teeth of a *Sivapithecus* from Greece. Although separated from one another by 10 m.y., the two show remarkable anatomical similarities, implying an ancestral-descendant relationship. Both photographs, to the same scale, are of casts.

WILLIAM F. MAAG LIBRARY
YOUNGSTOWN STATE UNIVERSITY

tering around 20 m.y. old; thereafter the record of their occurrence becomes increasingly scrappy and inconclusive. However, *Proconsul*, *Limnopithecus*, and their allies survived in Africa until at least as recently as 14 m.y. ago.

At around 17 m.y. ago a land link was established between Africa and Eurasia, and this allowed several ape groups to reach Europe. One descendant of the *Proconsul* group which reached Europe at least 15 m.y. ago was *Dryopithecus* (Fig. 12). *Dryopithecus* survived in Europe and southern Russia until about 10 m.y. ago. Available bones of *Dryopithecus* show considerable advances over those of *Proconsul* toward the condition seen in modern apes, especially in the structure of the jaw, face, and arm bones. Although known *Dryopithecus* are contemporaries of earliest ramapithecines, the latter are probably descendants of a form like *Dryopithecus*. Ramapithecines, which are similar to *Dryopithecus* in many ways but differ in possessing very thick tooth enamel, first appeared in Africa and Turkey about 15 m.y. ago (Table 2 and Fig. 13). By about 10–8 m.y. ago they had spread widely in Eurasia, reaching Austria, Hungary, Greece, India, Pakistan, Nepal, and China, as well as continuing to exist in Turkey and Africa. Thereafter the spread of more open country with the decline of the forest regions, as well as climatic cooling, led to the virtual extinction of Eurasian *Dryopithecus* and ramapithecines; after about 8 m.y. ago, only *Gigantopithecus* survived in this once diverse larger ape community. *Gigantopithecus* has been recovered in caves dated as recently as 5 m.y. ago.

The poor quality of the African fossil record after about 17 m.y. ago has been alluded to in the above discussion. This is particularly true for the period between 10 and 6 m.y. ago, a time which is critical for clarifying the course of human evolution because *Australopithecus*, the first bipedal hominid, evolved in Africa in this period. In this time interval only two isolated ramapithecine teeth are known from the whole continent of Africa. These are sufficient only to establish continuity between thick-enameled ramapithecines and the first thick-enameled *Australopithecus*, which was found at Lothagam in Kenya from sediments dated at about 5 m.y. ago.

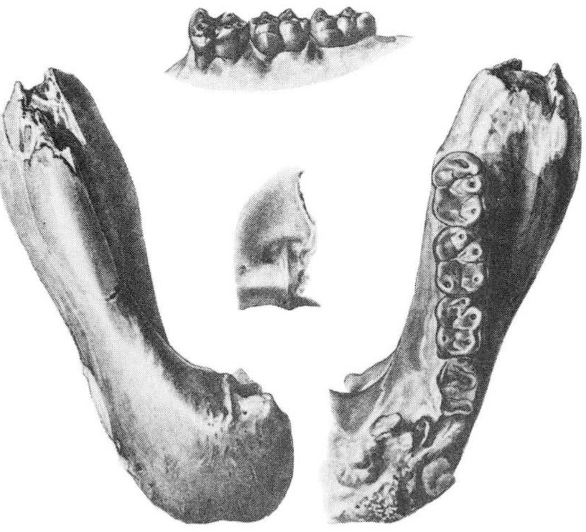

Fig. 12. Several views of a lower jaw of *Dryopithecus* from northern Spain. Left, a bottom view of the jaw; right, a top view with three molars shown; top, a side view of the cheek teeth; middle, a detail of the inside of the mandibular symphysis. (*From A. S. Woodward, 1914*)

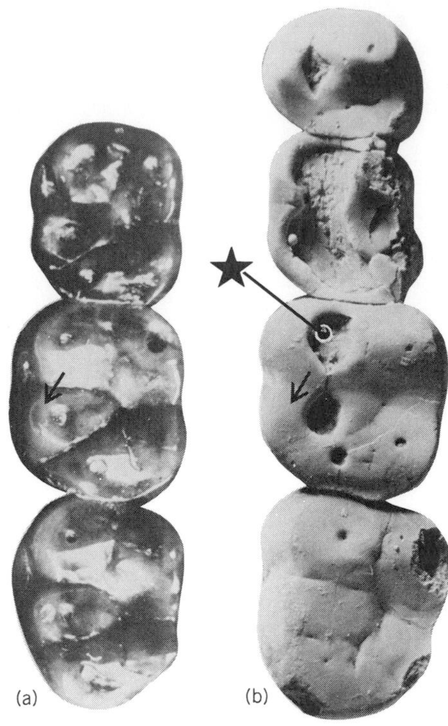

Fig. 13. Lower cheek teeth of (*a*) *Dryopithecus* from Spain with three molars, and (*b*) *Sivapithecus* from India with one premolar and three molars, the first of which is broken. The front of the jaw is at the top. In both specimens, the enamel has been worn while the animal was alive, so that the dentin shows in several places. A star identifies one such place. The three molars developed and erupted in sequence from front to back. The gradient of wear is much more in the *Sivapithecus* jaw than in *Dryopithecus*. Note that enamel is much thicker in *Sivapithecus* than *Dryopithecus* at the points indicated by the arrows.

RAMAPITHECINES IN HOMINID EVOLUTION

Currently there are three views as to the place of ramapithecines in human and ape evolution. Some researchers believe ramapithecines to be the ancestors of great apes and man. Others have emphasized some similarities of ramapithecines to the orangutan and argue for a close phyletic relationship between the two. The last and strongest documented view is that ramapithecines are more closely related to man than to any of the apes.

Ancestors of apes and humans. The first view, that ramapithecines are the ancestors of all living greater apes and humans, is currently propounded most strongly by the American anthropologist L. O. Greenfield. Greenfield has correctly noted that supposed resemblances between ramapithecines and *Australopithecus*, the earliest undoubted hominid, have been greatly exaggerated (and in some instances are purely fanciful). For example, there is no evidence that ramapithecines were bipedal or had small canines like *Australopithecus*, as claimed by some (Fig. 8). At the same time he views ramapithecines as bearing a great adaptive and phenotypic resemblance to living apes. From this, and his view that humans passed through an evolutionary stage where they looked like living great apes, Greenfield believes that ramapithecines may be ancestral to both apes and humans. Greenfield is right that ramapithecines show several resemblances to living great apes no longer found in modern humans. For example, the structure of the ramapithecine foot is more like that of living apes than humans, as is the overall shape of the facial bones. However, such similarities are most likely holdovers from the last common ancestor of great apes and man, and it is to be expected that the earliest member of a human lineage would resemble its ape ancestors in most ways. However, the important point is that, as will be discussed below, ramapithecines show certain key changes which point to the beginnings of human trends and suggest that the human lineage began with the ramapithecines.

Orangutan ancestry. The second view, currently propounded most strongly by the British paleontologist Peter Andrews, is that many features of the anatomy of *Sivapithecus* greatly resemble those of the orangutan, so *Sivapithecus* may be closer to orangutan ancestry than to that of man. Andrews mentions a number of details of the structure and shape of the face of *Sivapithecus*, particularly of specimens from Turkey (Fig. 3) and India, which resemble orangutans. Both have tall, oval-shaped orbits with a narrow bony bridge between them, and weak brow ridges above and between the orbits. In both, the upper face is flattened and flares outward; both have a much larger middle than outer upper incisor and thickened enamel on the molar teeth. At first glance, these resemblances might appear to establish this relationship, but study of the facial structure of early Miocene apes, *Australopithecus*, and living apes gives several plausible reasons why they may be insubstantial. The weak brow ridges of *Sivapithecus* and the orangutan are possibly what one might expect in the common ancestor of man and all great apes, given that early Miocene *Proconsul* has weakly developed brow ridges as well. A number of other *Sivapithecus*–orangutan similarities mentioned by Andrews are seen also in early *Australopithecus* and could be used with equal vigor to indicate a *Sivapithecus*–*Australopithecus* phyletic link. This includes the flattened, outwardly flaring upper facial bones, the much larger central than outer upper incisor, and the thickened enamel on the molar teeth. Finally, *Sivapithecus* and *Australopithecus* have much thicker enamel than do orangutans, suggesting that the latter evolved thick enamel independently in parallel with *Sivapithecus*.

This leaves the shape and disposition of the orbits as unique resemblances between *Sivapithecus* and organutans which are not found in *Australopithecus*. However, other impressive and unique similarities point to a *Sivapithecus*–*Australopithecus* link, making it more plausible to assume that the oval orbit shape and narrow interorbital bony bridge evolved convergently between *Sivapithecus* and *Australopithecus*.

Australopithecus and Homo links. The third view, advocated here, is that ramapithecines may be specially related to *Australopithecus* and *Homo* and not to other living or fossil groups. The similarities between *Sivapithecus* and *Australopithecus*, as distinct from modern apes, are illustrated in the accompanying figures. *Sivapithecus* resembles *Australopithecus* uniquely in the following ways (Figs. 7–11). (1) Each has a thick or shallow lower jaw under the molars and at the mandibular symphysis; in apes the jaws are narrower or deep. (2) Each has very thick enamel on the cheek teeth; apes, even orangutans, have thinner enamel. (3) There is a tendency to have two cusps on the foremost of the lower premolars; ape front premolars are usually single-cusped. (4) The upper canines are comparatively wider than long in cross section; in apes the greatest cross-sectional dimension is tangential to the dental arcade. (5) The canines are similar-sized in both sexes, whereas in apes they are much larger in males than in females. As will be reviewed below, these similarities show *Sivapithecus* to have shared two fundamental adaptations with early hominids—hard-object feeding and a possibly monogamous social structure. This makes the similarities more fundamental than the few resemblances of the shape and position of the orbits between orangutans and *Sivapithecus*, which are of uncertain functional or adaptive significance.

Thus, available evidence points to a link between ramapithecines and *Australopithecus*. It must always be recognized, however, that further knowledge about the details of ramapithecine anatomy could lead to a rejection of this hypothesis. Particularly vexing is the absence of informative fossils documenting the course of ape or human evolution between about 8 and 4 m.y.

RAMAPITHECINE ADAPTATIONS: LOCOMOTION

Locomotor differences are foremost among the ways by which humans and apes differ: Humans today are striding bipeds and facultative ground dwellers, while man's closest ape relatives move about on all fours in the trees. It is thought that humans passed through an evolutionary stage when they much more closely resembled chimpanzees in terms of locomotor capabilities. Precise information as to whether ramapithecines were primarily tree dwellers or spent some time on the ground, and as to their mode of locomotion, awaits further evidence about their limb structure. At present, from Pakistan there are some bits of the upper arm bones pos-

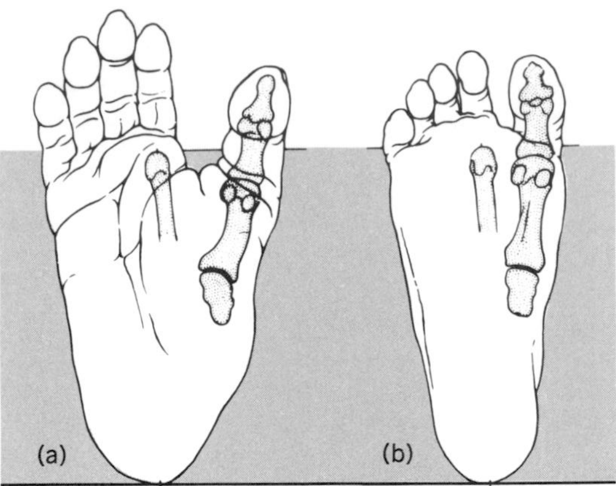

Fig. 14. Outlines of the sole of the foot of (*a*) a chimpanzee and (*b*) a human, showing the positions of the bones of the big toe and the head of the metatarsal bone of the second digit. Note that the ape big toe is a diverging grasping organ while the human big toe is in line with the other digit rays. (*From A. H. Schultz, The physical distinctions of man, Proc. Amer. Phil. Soc., 94:428–449, 1950*)

sibly of *Gigantopithecus* and of a large *Sivapithecus*. The latter is also known from fragments of the hand and a few pieces of the bones of the leg, ankle, and foot. The ankle and foot bones are the most informative so far. The joint surfaces give some information about the sorts of movement which could be made. *Sivapithecus* could turn its ankle inward freely, as can the living apes, and to a degree not allowed in the human ankle. Also the big-toe bones of *Sivapithecus* indicate that this digit most likely served a grasping role like in the chimpanzee foot but unlike the human, where the big toe is bound by ligaments in parallel to the other digit rays (Fig. 14). All of this points to an animal which used its hindfoot to grasp tree limbs, not as a propulsive strut in the manner of a human foot. Other known skeletal parts of ramapithecines also resemble those of great apes, suggesting that *Sivapithecus* was at least as tree-living as are chimpanzees today. One may speculate that the *Gorilla*-sized *Gigantopithecus* was somewhat more of a ground dweller.

RAMAPITHECINE ADAPTATIONS: DIET

Today, hunting and gathering societies like the Kung! Bushmen of Africa eat about 20% meat and 80% vegetable food. Chimpanzees, the most meat-eating of the apes, eat no more than 2% meat; thus, human evolution has seen a tenfold increase in the importance of animal food for energy. This appears to have been a relatively recent development, however, perhaps over the last 1–1.5 m.y. It is hard to imagine how prehuman hunters could have been efficient without highly developed tools, and there is nothing to suggest that earliest *Australopithecus*, the bipedal hominids of the Pliocene and early Pleistocene, were efficient hunters. Thus, factors other than hunting must be looked for to explain the origins of peculiar early hominid dental structures and bipedalism. In fact, if it is supposed that ramapithecines are the ancestors of *Australopithecus*, some hominid dental adaptations preceded hunting or meat eating by up to 10 m.y. As mentioned above, the cheek teeth of ramapithecines like *Australopithecus* have very low surface relief compared with those of apes, and are covered with a very thick layer of hard enamel, unlike those of most apes. Furthermore, ramapithecines resemble *Australopithecus* in that their jaws are very stout and thick, and the scars on these bones show a strong development of the muscles which provided masticatory force. All of these features point to a chewing system evolving early in human ancestry designed to produce and withstand extremely powerful chewing forces. Comparative study of the teeth of living primates and other mammals suggests an explanation for these features as adaptations for a specialized plant-eating diet.

Among herbivorous mammals a sharp adaptive dichotomy exists between browsers which primarily eat leaves and other plant parts high in fiber, and frugivores which eat fruits and other less fibrous plant parts. Browsers have specialized digestive tracts accommodating microorganisms which assist symbiotically in the digestion of plant fiber. Also browsers have well-developed cutting crests on their cheek teeth which can break up the fiber, increasing its surface area and improving digestibility. In contrast, fruit-eating mammals have relatively simple digestive systems and cheek teeth with large crushing surfaces and short, blunt, rounded cutting edges. Quantification of the development of the cutting systems on cheek teeth has been attempted. A shearing quotient (SQ) expresses the degree of enhancement of cutting edges on the unworn molars of browsing species compared to frugivorous ones. SQs for living apes serve as a model for the SQs of fossil apes and ramapithecines. SQs of chimpanzees and gibbons are near zero, as befits their frugivorous habits; SQs of the gorilla and siamang (a specialized browsing gibbon from Southeast Asia) are around 7–10, which goes along with their being browsers.

The SQs of ramapithecines cluster around zero, much like fruit-eating chimpanzees. However, it cannot be definitely concluded from this that *Sivapithecus* and *Gigantopithecus* had diets very similar to those of living chimpanzees. Certainly the relatively poor expression of cutting edges on the cheek teeth of the extinct forms suggests they ate very little fiber, like chimpanzees and unlike gorillas, but the added observation that ramapithecines have very thick enamel on their cheek teeth while chimps do not suggests some other fundamental dietary differences.

It has long been thought that the thick enamel of ramapithecines had something to do with their terrestrial habits: *Australopithecus* was a ground-dwelling biped with thick enamel on its cheek teeth. It is often assumed that ground dwelling and thick enamel evolved together so that having thick enamel could indicate terrestrial habits. In this view, human ancestors came to the ground and became bipedal to exploit new foods like roots, fibers, or cereal grains. These foods, because they are hard, tough, or grit-covered, might in turn have selected for thick enamel. It is now known that this view is incorrect for two reasons. First, *Sivapithecus* was probably a tree dweller, and certainly not a biped, but had thick enamel. So human ancestors probably evolved thick enamel before they assumed a ground-dwelling habitus. Second, thick cheek-tooth enamel is found today only in primates which eat very hard-shelled fruits or nuts. Two of the best examples of the adaptive significance of thick enamel are found in the feeding behavior of the cebus monkey from South America and Sumatran orangutans, both arboreal species. Each species has been observed to break open with its teeth nuts which are so hard that other species cannot eat them, and each has very thick enamel.

Orangutans do not always eat hard nuts, however. During the dry season they are primarily browsers. This may explain why their enamel, while relatively thicker than that of other apes, is not as thick as that of the cebus monkey. Field studies of the latter suggest that it is much more committed to nut eating on a yearly basis. From all of this it is reasonable to conclude that having thick enamel is a nut-eating adaptation. Such an adaptation involves very powerful muscles of mastication, as well as very thick enamel on the cheek teeth. Once the food has been broken open, the softer pith inside has very little fiber in it so molar cutting edges are not selected; living nut feeders have low SQs and thick enamel, precisely the pattern shown by ramapithecines. There seems to be no relationship with terrestriality.

In summary, it appears that the ramapithecines, although vegetarians like living apes, were specialized in particular for eating extremely hard foods like thick-shelled forest nuts.

SOCIAL STRUCTURE OF RAMAPITHECINES

One of the greatest distinctions between man and the apes is in the canine teeth. The canines of apes are large projecting teeth, larger in males than females (that is, sexually dimorphic; Fig. 5); human canines are not projecting and there is only minimal difference between the sexes. There are good reasons for believing that humans passed through an evolutionary stage when they had larger, sexually dimorphic canines; living great apes, most monkeys, and most fossil apes have such an arrangement. Moreover, although the crowns of human canines are small, the roots are stout and long, possibly a hint that they were once larger. The canines of ramapithecines are still quite large (Figs. 7, 8, and 10), but the amount of sexual dimorphism was low by the standards of living apes. This gives valuable information about the possible social structure of ramapithecines.

Most living monkeys and apes have sexually dimorphic canines, but in a few, notably the gibbons of Southeast Asia and many New World monkeys, the canines of males and females are similarly sized. Sexually dimorphic anthropoids form polygynous social groups with one or several adult males, a larger number of females, and their offspring. Nondimorphic anthropoids, on the other hand, form family groups composed of a single male, a single female, and their immature progeny. Therefore, ramapithecines, which have little canine dimorphism, may have had a family-type social group. From this scenario man's early preramapithecine ape ancestors had large, sexually dimorphic canines like those of living great apes, implying a polygynous social structure. Later, in the ramapithecine stage, the canines were reduced in size, giving the modern human configuration.

CLASSIFICATION

The available evidence suggests that ramapithecines are broadly ancestral to the hominids (*Australopithecus* and living humans) and not to any living ape. But in terms of their overall grade of organization, ramapithecines show a mixture of ape and hominid adaptations which on balance would make them appear quite apelike. So should ramapithecines be called hominids (members of the Hominidae, the family of man)? A satisfactory solution to this problem is elusive because it relies as much on philosophy as on the evidence of relationship. The reason the question is raised here is that it focuses on a key fact of human evolution: peculiarly human features or characteristics were not all acquired simultaneously as part of a common adaptive package. *Homo* is distinct from living apes in a variety of ways which could leave evidence in the fossil record—bipedality, tool use, enlarged brains, relatively small canines, reduced canine sexual dimorphism, and thick molar enamel are a few of the features which characterize the human adaptive "grade." As the fossil record improves, scientists are finding that some of those differences arose millions of years before others in human evolution. Human ancestors got their feet (and possibly their knuckles) very wet when they crossed the ape/human adaptive Rubicon, and this makes a definition of the family Hominidae extremely arbitrary. Should only those human ancestors who were bipeds (or tool users, or large brained) be included in the human family? The alternative, which R. Kay leans toward, is to define the human family as containing all creatures, no matter how apelike, that are more closely related to *Homo* and *Australopithecus* than to any of the living apes. By this definition ramapithecines should be included in the human family.

[RICHARD F. KAY]

Bibliography: R. F. Kay, *Amer. J. Phys. Anthropol.*, 55:141–152, 1981; R. F. Kay, *Proc. Nat. Acad. Sci.*, 1982; R. F. Kay, and E. L. Simons, in R. L. Ciochon and R. F. Corruccini (eds.), *New Interpretations of Ape and Human Ancestry*, 1982; D. R. Pilbeam, *Annu. Rev. Anthropol.*, 8:333–352, 1979; E. L. Simons, *J. Hum. Evol.*, 5:511–528, 1976.

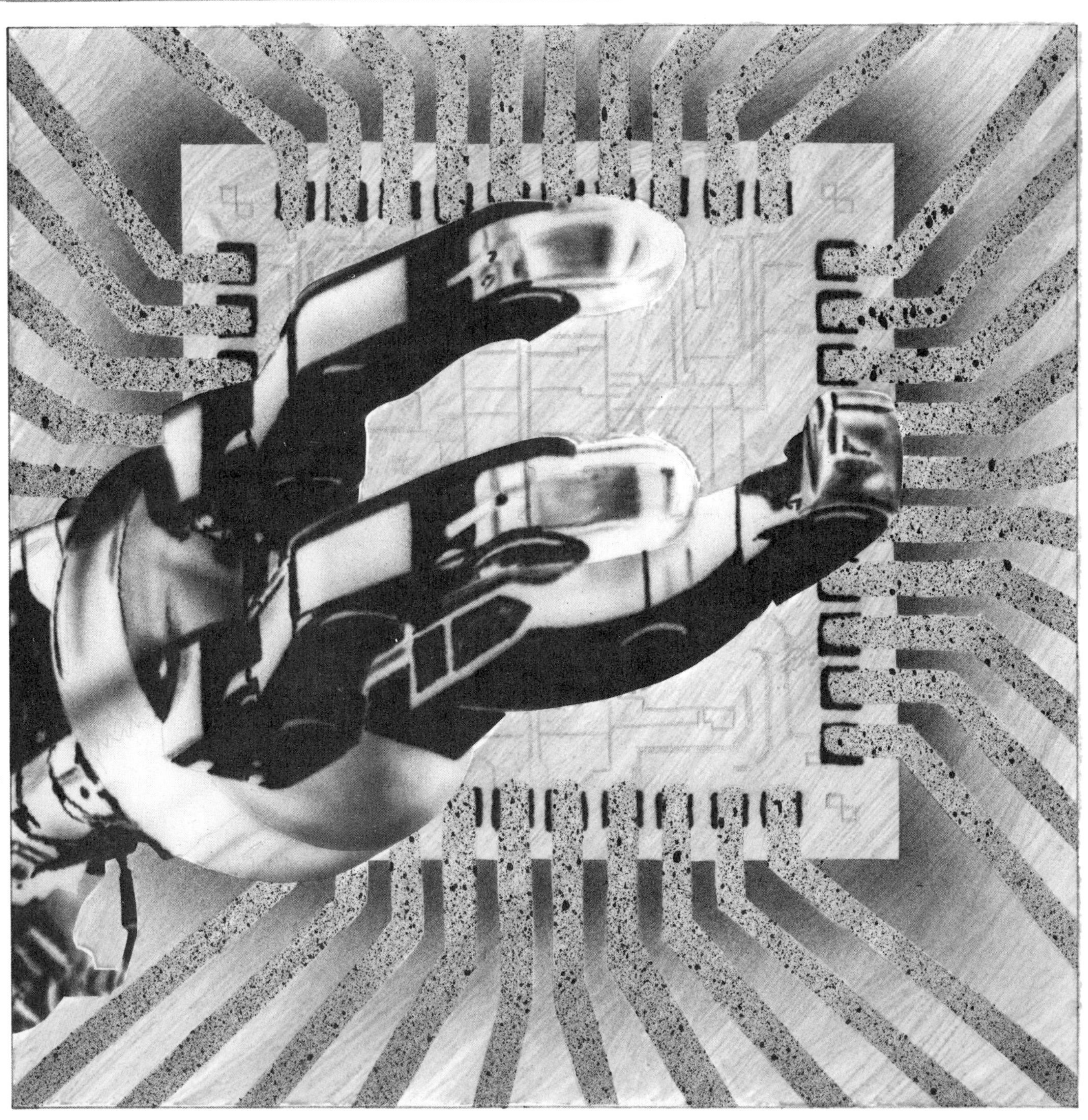
E.T. Steadman

Autofacturing

Autofacturing—automation in manufacturing—is a term to describe the overall activities of the automated factory. The 10 major components of a total autofacturing system are: product design for autofacturing, data-base construction and management, integrated computerized control system, automated materials-handling system, process and machine automation, automatic assembly machines, robots, automatic inspection and testing system, automatic maintenance system, and human interfaces with autofacturing.

Each of these components may be thought of as a subsystem of the total autofacturing system, and each, in turn, contains its own subsystems. The presence of all 10 is not required to have an autofacturing system. Each is presently in its own stage of development, and autofacturing may exist as a mixture of some advanced subsystems operating next to some which are less sophisticated. The ultimate goal should be to have all subsystems interconnected, complementary, and integrated into an overall autofacturing system in order to achieve the greatest possible benefits. It is expected, however, that advances will continue in each area, keeping the perfect total system a future target.

Autofacturing is not manufacturing—it is its technological successor. The term manufacturing was originally defined as the making of a product by hand; then it was expanded to include the making of a product by machinery. The word auto-

Vincent M. Altamuro became the president of Management Research Consultants in 1955. Formerly on the faculties of the graduate schools of engineering of Columbia University and The City University of New York, he has conducted seminars for Western Electric, the manufacturing subsidiary of AT&T, for over 20 years—one of the courses being in autofacturing, a term he originated.

mation is used to describe a wide range of activities, some not at all related to industrial plant operations, for example, airline reservations, banking, stock market transaction handling, traffic control, commercial data processing, word processing, and office automation. Recognizing the foregoing, V. M. Altamuro originated the word autofacturing to create a term which more accurately describes the activities of the factory of the future in which the products will not be made by hand but by automatic, self-actuated, computer-controlled machines, robots, and self-correcting feedback equipment.

The advantages of a fully operational autofacturing installation include greater output and lower per-unit costs, that is, increased productivity. Also to be expected are improvements in quality—more uniform, consistent output, with less handling damage, fewer rejects, and less scrap and rework because of continuous monitoring and rapid corrective feedback. It should also result in fewer personnel, reduced payroll expenses, fewer accidents, less material handling, less inventory, and reduced space requirements. One disadvantage is that its high output rate may be an all-or-nothing situation. That is, if the system breaks down, output could drop to zero. This problem can be lessened by designing the system to be compliant and flexible. The feature of compliability means that the system contains the ability to continue operating when a tool or work station breaks down (by using rerouting or redundancy) or varies in output rate (by using work-in-process accumulators). Flexibility refers to the ability of the system to adapt quickly to desired changes in products, output rates, mixes, and other specifications. The use of group technology—the coding and grouping of parts according to similarities in their sizes, shapes, materials, and production processes—may be an aid in this regard for products with shorter production runs. Another consideration is that the higher capital investment required for the installation raises the operation's break-even point. Also, while autofacturing requires fewer personnel, those remaining (engineers, computer specialists, and maintenance technicians) require a higher level of training, skill, and salary.

There are two basic ways in which autofacturing may be installed in a company. It may be installed "from scratch" in a new or remodeled facility or for a new product line in an existing facility. Or, it may be installed in sections as the technology, available capital, timing, and other circumstances allow. Then, at a later date and after the individual sections are debugged, these "islands" or "cells" may be connected and integrated into a total autofacturing system.

A brief description of the elements and state of the art of each of the 10 components of a total autofacturing system follows.

Product design for autofacturing. The first essential of an efficient autofacturing system is that the product—including its component piece parts and raw materials—be designed and specified so as to aid the system. That is, without detracting from the performance requirements, certain things can be done, at the product design stage, to permit a higher degree of efficient autofacturing.

Product. The product should be designed for modular assembly with simplified subassemblies. This means that rather than having it designed to be built by continuously adding part after part, it is designed to have a set of more simple subassemblies which, when brought together, become the end product. In addition, the subassemblies should be modular in such a way that various interchangeable combinations of them create different end products.

Piece parts. In autofacturing, the piece parts are handled by parts feeders, mechanisms, robots, and other nonhuman devices. They must be designed so that they are as simple, uniform, symmetrical, and sturdy as possible. The devices used to hold, sense, count, orient, feed, position, fixture, and assemble them must operate quickly and surely. The parts must not tangle, snag, nest, or shingle (overlap). They must not break or damage easily. They must have (or be given) features which make them easy to sense, move, and hold automatically and without error. An autofactured part or product may have features which have no function in the end product but exist merely to make it easier to handle the product as it flows through the autofacturing system. Also, the component piece parts should be designed so as to have as few at-rest orientations as possible. That is, a sphere has one orientation and is therefore always oriented properly; a coin has two (it may fall heads or tails). As parts become more asymmetric, irregular, and complex, they have an increasing number of possible orientations, making it much more difficult to get them into the orientation required at the next stage of the process.

Raw material. The raw material should be specified so as to be received from the suppliers in a form best suited for introduction into the autofacturing stream. The form, size, shape, packaging mode, and even the time and point of delivery can often be varied so as to best fit the requirements of the subsequent steps.

Data-base construction and management. For a true totally integrated autofacturing system, there must be a common data base. This means that all information used by and created by the system and its subsystems must be coded, structured, and organized in such a manner that it can be collected, stored, interpreted, processed, updated, retrieved, displayed, and communicated between and among the many elements of the overall system.

The data must be collected from the points where they originate and delivered to the points where needed. Some devices and media of data transmission are magnetic strip, bar code and alphanumeric scanners, keyboards, light pens, magnetic wands, touch-sensitive CRTs (where the displayed data touched are entered into the data base), voice, hand-held portable terminals, radio waves, line-of-sight light beams, lasers, fiber optic light guides, and the plant's standard 120-V electric power wiring used to carry data.

The data base could begin with the design engineer feeding a new product's specifications into the computer. The detail designer would retrieve them, specify exact dimensions, components, and hardware, and add those data to the base. The drafter would use that information to prepare the engineering drawings; then the information would be used for the preparation of parts lists, bills of material, exploded views, and so on. The engineering data would be an input to production information needed for the preparation of numerical control tapes and programs, work routings, machine loading, and so on. Subsequent design changes would be made by updating the original data, wherever located throughout the system. The principle is that one common set of data, once created, is kept current and used by all persons and functions of the organization.

The term distributed data-base management system is used to describe a data-base management system in which the data are collected from and distributed to different locations. The required data-base management system may be designed internally by the firm installing autofacturing, or it may be acquired from the vendors of computer program packages.

Integrated computerized control system. An autofacturing installation will not operate properly without an integrated computer system. The computer hardware can be any mixture of full-size mainframes, minicomputers, and microprocessors and microcomputers. Even the "intelligent" programmable controllers may be included. The firmware and software can be designed to perform the functions of computer-aided design and drafting, computer-aided design/computer-aided manufacturing (CAD/CAM), computer-aided engineering, computer-aided process planning, the planning and control of master schedules, material requirements planning, inventory control, tool control, machine controls, numerical controls, quality analysis and control, statistical analysis of data collected, data-base management systems, and management information systems and reports.

Automatic information feedback and control are essential to autofacturing. The system must operate at speeds too high to permit periodic measurements of conditions, human reactions, evaluations, decisions, and manual adjustments to bring conditions back to desired levels. In the time that the human or manual sequence would take, a multitude of substandard products could be made, an entire production batch ruined, or worse, the equipment seriously damaged. A closed-loop automatic control system continuously monitors key parameters, com-

Fig. 1. Microprocessor-based programmable controller. (*General Electric Co.*)

pares them to standards, computes a difference or correction signal, feeds it back to the input stage, and actuates adjustments to keep the process within preestablished limits—all so rapidly that no substandard products are made.

In autofacturing, most operations are under the complete direction of a computer—either from a remote location or right on the factory floor, possibly built into the machines. In addition, auxiliary devices such as programmable controllers (Fig. 1) and logic sequencers are hardy enough for the factory floor environment and are becoming sophisticated enough to perform computer functions. The programmable controller's place in the factory is that of an interface between the machine and the central computer. The programmable controller is used to direct the machine's operations and then to monitor and feed back the results to an information and control computer—which might have dozens of programmable controllers hooked into it.

The autofacturing computer system may be centralized (a large computer with many remote input/output terminals throughout the plant), or it may be a distributed system (microcomputers and programmable controllers at each location tied to supervisory minicomputers which, in turn, interface with the maxicomputer). Also, the system may be networked in a hierarchy, such that the computers (of all sizes) are connected so that information can flow from work stations to machine centers, to departments, from the plant shop floor to the functional offices, up to corporate headquarters. Remote computer CRT terminals, interactive graphics, and digital input and readout stations are all components of the integrated computerized control system. Artificial intelligence, voice communications with machines, and natural language programming of computers are developing and will find their place in future systems.

Automated materials-handling system. Beyond the conventional materials-handling equipment and systems of today's factories, the factory of the future operating in an autofacturing mode will also have programmable conveyors, computer-directed forklift trucks, automatic guided vehicle systems, and automatic storage and retrieval systems (AS/RS).

Conveyors. In autofacturing, conveyors will be programmed and instructed by control signals to switch paths, change speed and direction, accumulate work, release previously accumulated work, read bar codes, sense items, and react accordingly. The actions of the conveyors will be tied to inputs from a real-time information system and, in turn, will feed current data back to central control.

Forklift trucks. Computer-directed forklift trucks are already in operation and are growing in use. Their operation is not completely automatic in that a human operator is still required, but they do represent an additional step toward the automatic factory of the future. The installation consists of a central computerized base station and trucks equipped with an FM radio transceiver, a microprocessor data terminal with a liquid crystal or light-emitting diode display screen, an alphanumeric input keyboard, and in some cases, an electrostatic printer. This two-way real-time data flow reduces errors, time, and some personnel—although a truck operator is still required. In operation, coded signals are trans-

Fig. 2. Kenway Robocarrier automatic guided-vehicle system. (*Eaton Corp.*)

mitted from the base station to the truck receiver, whose microprocessor decodes it and converts it to human-readable messages on the display screen. After execution of the order, the operator enters an "order completed" message via the keyboard and transmits it to the central base station or, alternatively, may report the new location of the material moved so that the computer's data base can be updated.

Automatic guided vehicles. An automatic guided vehicle system provides the physical links to connect the factory's raw material storage, production machinery, and warehousing locations. This system is a fleet of operatorless, quiet, hard-rubber tire, low-profile, materials-handling vehicles (Fig. 2). They follow a path established by either a painted line or a tape on the floor or a wire buried a few inches into the floor. While their paths are fixed, some variation is possible via switches and path networks. They can be computer-controlled from a central location, as each vehicle can have a microprocessor or microcomputer on board. The central control console might contain a multicolor CRT display unit to show the location and load status of each cart in the network.

The vehicles can be self-loading and -unloading. As their loads are inventory, they are active interfaces to the plant's inventory-management system. They have applications in the office area as well, as mail and report distribution devices (Fig. 3). Expected future applications of the systems are uses in combination with other equipment and systems. They can be made to be an integral part of a network which includes conveyors, self-propelled monorail transporters, powered carrousels, automatic stacker cranes, the AS/RS, production machines, and assembly lines. Rather than merely transporting material to and from stores locations, they can be made to be active participants in the production process. For example, the work can be mounted on them and carried from machine to machine, or more imaginatively, the machine tools can be mounted on them and brought to the work. Or, both work and tooling can be mounted on the vehicle and the work processed "on the move." Further, that "tooling" could be a robot, such that a feature of the plant layout might involve stationary work with an automatic guided vehicle bringing a succession of robots, tools, parts, raw material, and other work in process to it.

Automatic storage and retrieval systems. An AS/RS has application at both ends of the production process in the automatic factory—at the beginning, to handle raw materials and purchased components, and at the end, to handle the finished products. Most plant layouts will have one physical AS/RS installation to handle both functions. Essentially, an AS/RS installation is a very-high-rise steel superstructure of racks designed to be addressable storage locations into which and from which material is moved. The storage and retrieval are accomplished by directed stacker-crane carriers which travel laterally on tracks to one of several aisles, horizontally down the aisle row to the proper bay, and then vertically, like elevators, to the proper spot (rack; Fig. 4). The carrier's load-holding section has its own lateral movement capability to deposit or fetch the load in or out of the rack position. Again, the AS/RS is integrated with the computer system such that once it stores material, the computer will keep track of its location and amount; when it is needed again, the AS/RS carrier will be automatically sent directly to that location.

Fig. 3. Mailmobile—an automatic guided vehicle for the office. (*Bell & Howell, Inc.*)

The AS/RS of the future will be more than merely an automated storage system; it will have a role in work-in-process handling. An installation placed among and between work centers can feed and receive work to and from the machines. The output of one machine can go into the AS/RS, and then be fed to the next machine, when desired—especially if the next machine is on a different factory level. The AS/RS can be interfaced with the automatic guided vehicle system, programmable conveyors, and assembly lines—all under the direction of a central computer—to form a linked system. The use of an AS/RS has the advantages of increasing the efficiency of space utilization (smaller aisles and higher stacking), speed of service, accuracy, control, and security.

The scope of the automated materials-handling system also covers the functions of order picking, packing, shipping, and physical distribution management. An example is a machine which automatically weighs, meters, fills, imprints, and counts bags of items preparatory to shipment. It can deposit the filled bags in cartons, ready for automatic seal-

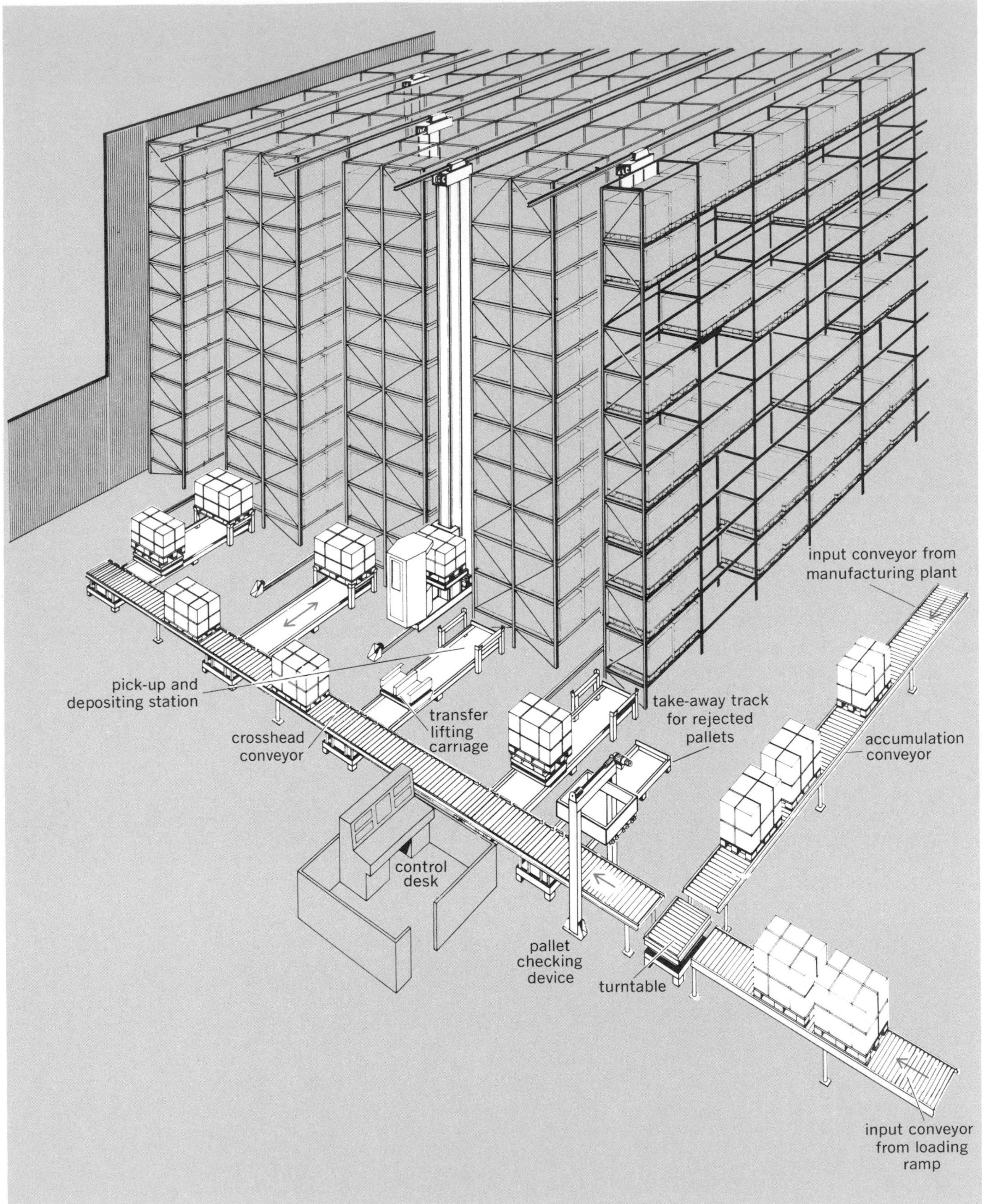

Fig. 4. Typical automatic storage and retrieval system; in this system there is an individual stacker crane for each aisle. (*Material Handling Division, Mannesmann Demag Corp.*)

Fig. 5. Transfer machine. (*a*) Transfer-matic machine—1 in a series of 41 in an installation capable of producing 246 automobile engine blocks per hour. (*b*) Close-up of an engine block in one work station. (*The Cross Company*)

ing and labeling with the customer's address. Further along, equipment is in existence that can lift full palletized loads of cartons and transport them on a film of air gently into large, standard-size unit containers which are interchangeably used on truck, railroad, airplane, or ship.

Equally important to the new equipment will be the concept that the functions of materials handling will no longer be thought of as merely the movement of things out of and into storage and between machines; it will also include the movements of, and temporary storages on and in, the production machinery. Therefore, in autofacturing, materials handling will be so integrated with production operations that it must be regarded as an integral and essential part of the total system.

Process and machine automation. Some products (such as gasoline, paper, plastic, and glass) are made by a continuous process which passes the raw materials through a series of connected stages which gradually alter their physical or chemical properties until the desired end product is obtained. These processes have long been automated, and any future advances are expected to come in the integration of more sophisticated on-line computers, sensors, and controls.

Other products are made in two phases—the conversion of raw material into piece parts (fabrication) and then the joining of sets of those piece parts into the desired end product (assembly). The newer machines that will accomplish the first phase will be discussed in this section, and those related to the second phase in the next section.

The evolution of machine tools includes their progression through operator-run machines, automatic machines in which the operator merely loaded and unloaded the work, a series of machines with both automatic cycles and automatic loading and unload-

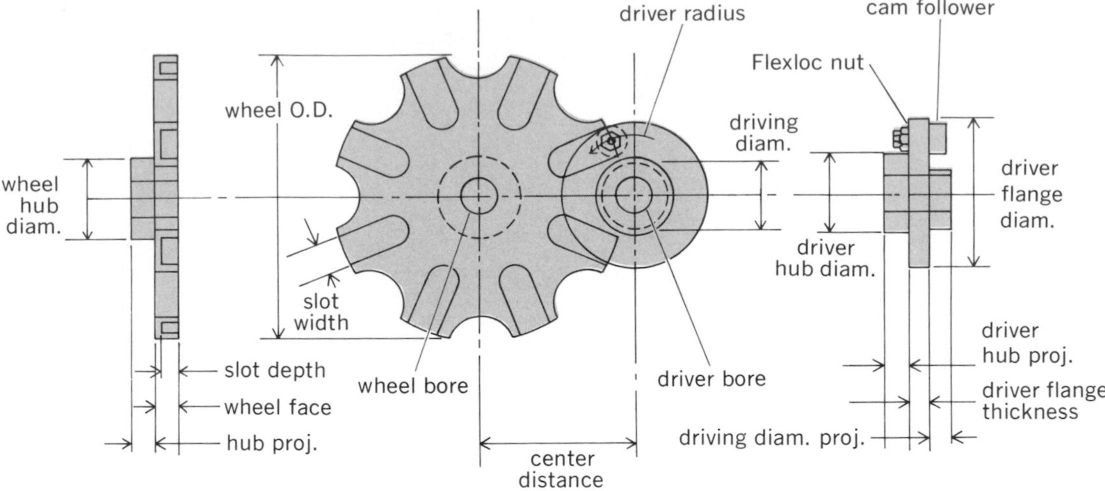

Fig. 6. Eight-position Geneva wheel drive mechanism. (*Geneva Mechanisms Corp.*)

ing, and then the foregoing with automatic materials handling between them. In autofacturing, the transfer machine dominates. It is a large complex of many machining stations, tied together and operating under a central control. It includes built-in materials-handling capabilities to transfer the work from station to station. The work stations typically perform the operations of mill, drill, ream, bore, tap, grind, and so on, as well as many inspections, but they can do other things too. The work stations may be in-line (Fig. 5), rotary-dial, or carrousel in configuration. The work may be mounted on a platen or fixture which is moved along, or it may be unmounted. The movement may be continuous or at periodic intervals (indexed) and synchronous (each work unit moves at the same time) or nonsynchronous (work can be moved independently to "surge" or accumulate). These are large and expensive machines, but they increase productivity dramatically. Large installations of nearly 1000 work stations which perform thousands of automatic operations cost millions of dollars but are becoming the only economical way to produce the products involved at high volume, high quality, and low cost.

Some other types of new machines are large, high-speed, complex pieces of equipment which are called turning centers and machining centers and which combine the capabilities of six or more conventional machine tools. Some of them have tool changer–tool holder magazines that contain up to 90 different tools, and can perform on multiple work faces (up to five), do several jobs simultaneously, and machine up to 15 different parts consecutively without stopping for setup changes. Most are numerical control, computer numerical control, or direct numerical control and have up to five axes of movement. Many have direct digital inputs and readout controls, microprocessors, and tie-ins to a central computer and the common data base. Some have adaptive controls which provide automatic adjustments in the machine's speed, feed, and depth of cut rates to accommodate the idiosyncrasies of differing raw material stock.

Numerical control is not new in plant operations, but it is an important component in the automatic factory of the future. Essentially, numerical control directs the motions of a machine tool by replacing the brain and hands of a skilled operator with a program (for example, a hole-punched or magnetic tape) and machine movement actuators (for example, stepping motors, servomechanisms, hydraulics, electronics). All machine and tool positions and paths are stated in sequential numerical codes so that once loaded with a blank workpiece, the machine tool can perform a uniform series of steps accurately and without operator intervention. The early—and most of the current—numerical control installations required only that the operator load and unload the machine. In future equipment that will also be done automatically, with simple mechanisms, transfer devices, or robots.

Numerical control requires the off-line preparation of a tape which must be brought to the machine tool, tested, and then run. With computer numerical control (CNC) the tape can be made, tested, edited, debugged, and updated on a computer before it is brought to the machine tool, or alternatively, the tape can be eliminated and the computer can control the machine directly. Using CAD/CAM, which is the on-line application of computers to design and manufacturing functions, the machines might be programmed by using interactive CRT graphics to establish the part's design and dimensions, and then converting that data to a numerical control–like instruction set and transmitting it directly to the machine's control unit. With direct numerical control (DNC) the tape is also eliminated and the computer directs the machine tool, but a larger computer is used so that it can control several machine tools, distribute and transfer jobs between machines to obtain a higher equipment utilization rate, optimize and balance work loads, and store and revise the library of numerical control tapes (programs).

In autofacturing, the new machines will have both internal transfer devices to transfer the work within the machine and external devices to transfer it to the next machine. They will have robots serving them and television cameras monitoring their output, operations, and condition. They will have sensors, microprocessors, and adaptive controls to measure, feed back, and optimize operations. If they get bad input material, get out of alignment, or experience broken or worn tools, they will not continue to turn out bad work, as do some of today's "automatic" machines. They will employ many of the newer "noiseless" and "chipless" processes, such as lasers, electron beams, ion beams, ultrasonics, and chemical and electrochemical machining. The machines will be integrated with the plant's materials-handling system, automatic assembly system, and information and control system.

Automatic assembly machines. Automatic assembly machines are somewhat similar to transfer machines, but their purpose is to assemble component piece parts into subassemblies or end products. They, too, can be in-line, rotary-dial, or carrousel. They are constructed to move either continuously or with an indexing motion—that is, move, stop while work is done, and move again. The indexing action is accomplished by using a barrel-cam mechanism or a Geneva wheel (Fig. 6) to move them quickly and yet smoothly and with controlled acceleration and deceleration. They also may be synchronous or nonsynchronous—that is, either all work stations, fixtures, or platens are linked together and move the same distance at the same time, or they are independent, permitting individual movement, work accumulation, and surges to allow for unequal work cycle times.

Usually several parts feeders serve the automatic assembly machines. Their operations include assembly, insertion, staking, riveting, screw fastening, crimping, soldering, and other joining methods. They can inspect their own work and reject substandard products. Advanced models with sensors, feedback, and integrated computer controls can adjust their operating parameters to keep quality within specified limits. They can operate at between 90 and 100% machine efficiency and produce hundreds of products per hour.

The assembly machine's material or piece parts-holding devices can be bins, hoppers, reels, spools, bobbins, trays, racks, or magazines. The characteristics of the material dictate the type of device used. The function of the devices is to hold and protect (and sometimes condition, for example, heat) the material prior to use. They should be equipped with sensors and signals to indicate when they are getting too low, and should be able to continuously dispense material from one end while being replenished at the other end. They are the supply side of the feeding and metering systems.

The function of feeders is to present the proper part or material to the operating work station, at the right time, in the correct orientation, and (if a continuous material, for example, adhesive) in the proper amount. Feeders can employ gravity, magnetics, hydraulics, pneumatics, centrifugal force, or mechanisms. In addition to orienting, metering, and feeding the parts, they can be made to inspect them and reject (eject) those not meeting established criteria. A widely used device is the vibratory bowl feeder. It operates by randomly selecting candidate parts to be fed to the operating work station. Through vibration, wiper blades, guides, rails, chutes, escapements, and so on, it attempts to align the parts in the desired orientation. Those properly aligned are advanced, and those not properly aligned are returned to the bowl to try again. The correctly oriented parts enter a track and are fed to the work station.

In addition to the parts' being oriented and fed to the work station, they must be positioned precisely where needed and then fixtured (held) while the operation is performed. The cycle of the work-holding fixture is (1) grip, (2) close, (3) lock, (4) dwell for time required to do operation, (5) open, and (6) release. As the assembly builds in size, it must be moved to successive work stations and manipulated as required. The work is advanced or rotated through the machine's positions until it is finished and unloaded (Fig. 7).

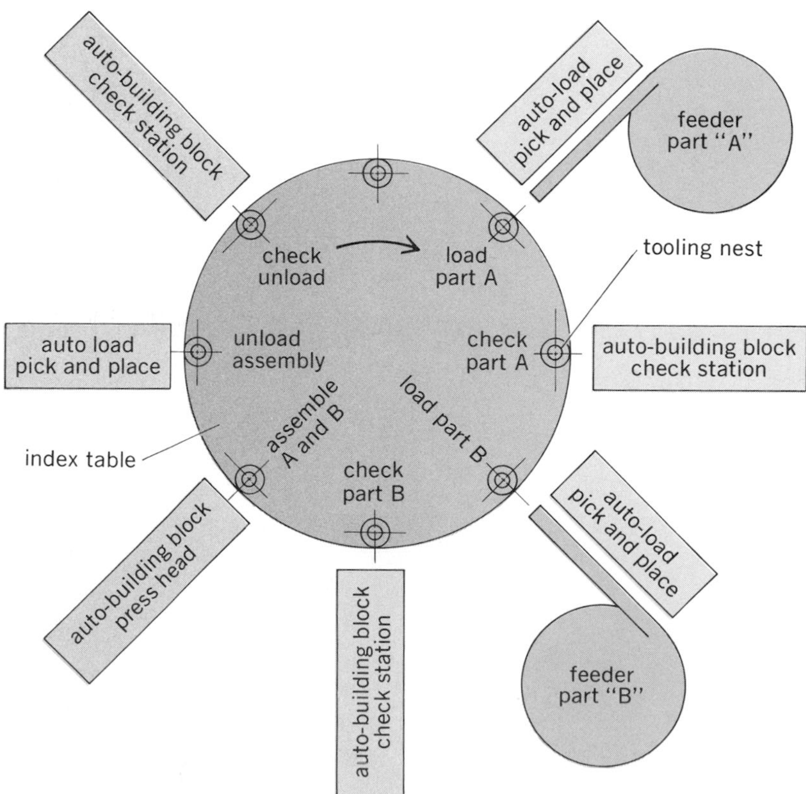

Fig. 7. Layout of the stations of a horizontal rotary-dial indexing assembly machine. (*Automation Designs, Inc.*)

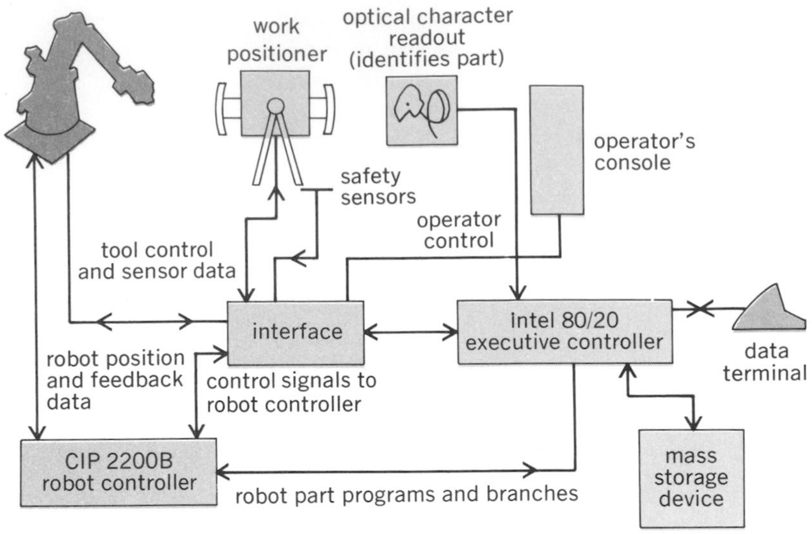

Fig. 8. Diagram of a robot work cell with all computer and controlling elements. (*K. Gettelman, Robots Drill and Profile, Mod. Machine Shop, pp. 129–136, April 1980*)

Robots. Robots are reprogrammable, multifunctional manipulators designed to move and do varied productive work, as directed (Figs. 8 and 9). They are especially valuable in autofacturing as they can often replace humans in the system. They can also serve as the flexible links required to connect a group of individual machines or work stations to form versatile work centers (or cells) and to connect the cells to form a single integrated system. They have a wide range of applications, from materials handling to production, and the growing number of suppliers offer them in a wide range of sizes and capabilities.

The simplest type is the mechanical manipulator or the pick-and-place robot. It is a simple, inexpensive transfer mechanism and is used merely to load or unload a machine or conveyor, or to orient, position, or hold an object. It may be classified as a nonservo-controlled robot in that its movements stop only at prefixed end points set along each axis of motion. Its motion path can be changed by resetting its microswitches and stops. Another class of robots continuously plays back its fixed instructions and repeats its cycle until stopped. The more sophisticated robots are considerably more complex and expensive than the early types, and their capabilities are extensive. Lightweight and heavyweight electric or hydraulic numerical control and servo-controlled robots are available in many sizes, shapes, and capacities. They can be programmed to move and stop at any point within their range of motion and with controlled acceleration and deceleration. They are being made to operate faster and faster to satisfy the needs of high-volume production. While strong, they can handle very small, delicate items gently and without damage. They can be made with dimensions similar to a human so that they and an operator can be interchangeable at a work station. They

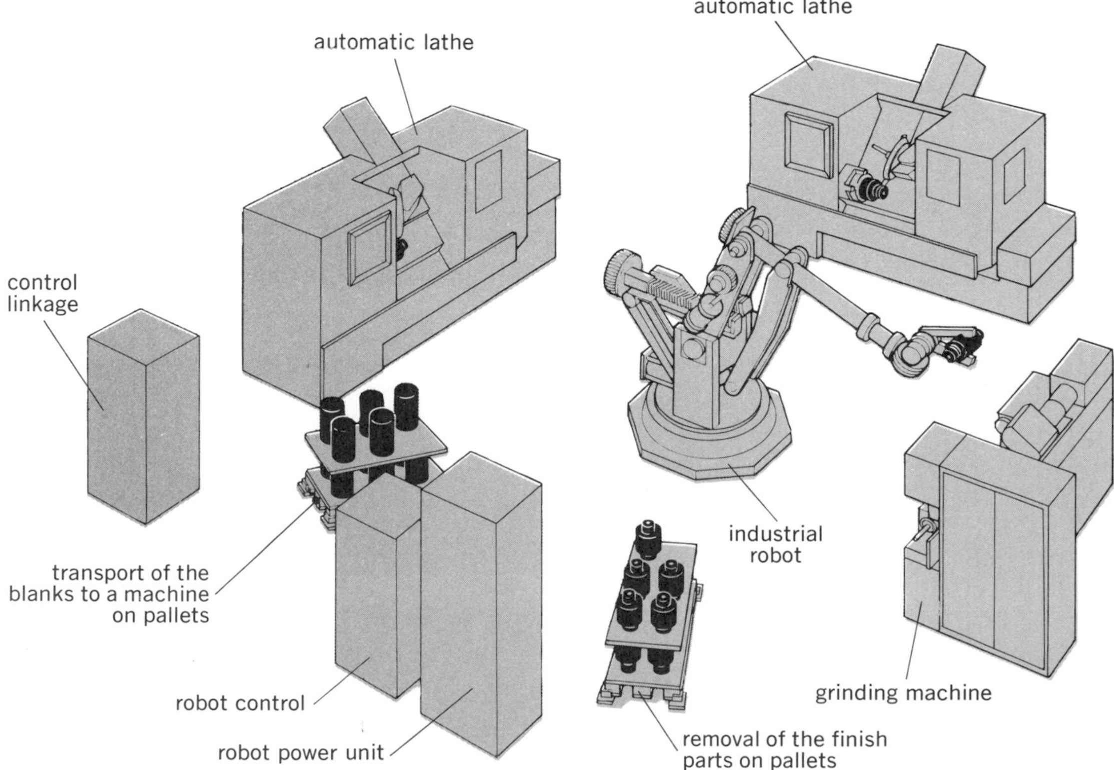

Fig. 9. Industrial robot servicing three machines. (*KUKA Augsberg*)

can have up to seven axes of motion (gripper, flange, wrist, elbow, shoulder, waist, and traverse), and the addition of audio, visual, and tactile senses is possible. The state of the art is that "intelligent" robots can recognize an unoriented part amid others, grasp it, and place it, correctly oriented, into position for assembly. Further, the robots can do the work operation. They can weld, rivet, paint, assemble, transfer work from one machine to the next, test, package, unpackage, stack, unstack, palletize, and so on. With the aid of a programmable controller or computer, they can stack successive pallets in different patterns, based upon the dimensions of the boxes coming to it for palletization.

A robot may be programmed in three basic ways. A prerecorded program may be loaded into it (via tape, disk, cams, stops, switches, and so on) which it will follow repeatedly once turned on. Next, it may be manually "walked through" the desired motions by human manipulation of its appendages. Or, it may be "led through" by a human using a control box (levers or buttons). The motions can be altered until the programmer is satisfied with them. In both of the last two methods, the debugged, finalized program is then recorded in the robot's memory, a switch is closed, and the robot will repeat the cycle tirelessly and accurately, until it is made to stop or reprogrammed. The old program may be removed and stored for future use. Remote reprogramming is also possible. When out of order or adjustment, the newer, computer-controlled robots will tell not only the location and cause of the problem but also how to correct it. Adaptive computerized robots will be able to "learn" as they go along—that is, they will be able to alter their programs themselves, based upon conditions their sensors observe.

Automatic inspection and testing system. A feature of autofacturing is that it is fast. It produces a great number of products rapidly. So as not to make many bad units, rapid identification and correction of substandard work are essential. Autofacturing usually uses 100% inspection and test, rather than the sampling common in manufacturing. Devices placed right in the equipment continuously monitor the input and output of each work station. Data are collected and analyzed in a real-time mode, statistical projections are made and compared to acceptance limits, and correction signals are fed back, and the parameters of the operation adjusted, so as to keep future output at the desired quality level (Fig. 10).

Autofacturing is characterized by the presence of sensors located throughout the system. They signal the presence or absence of material, parts, or products, or a particular characteristic of them, for quality control purposes. They can be used to help count, sort, time, measure, inspect, test, meter, position, refill, stop, restart, and so on. They can be of the contact or noncontact, mechanical, electromechanical, thermal, radio, fluidic (air), ultrasonic, capacitance, laser, photoelectric, or proximity type. Generally, they merely send a signal to a control device, but they can amplify, transduce, compare, and compute the data prior to transmitting a signal to an actuating mechanism. The more simple devices sense only one condition, but the more complex "cognitive" devices can sense multiple conditions and react to their combined effect.

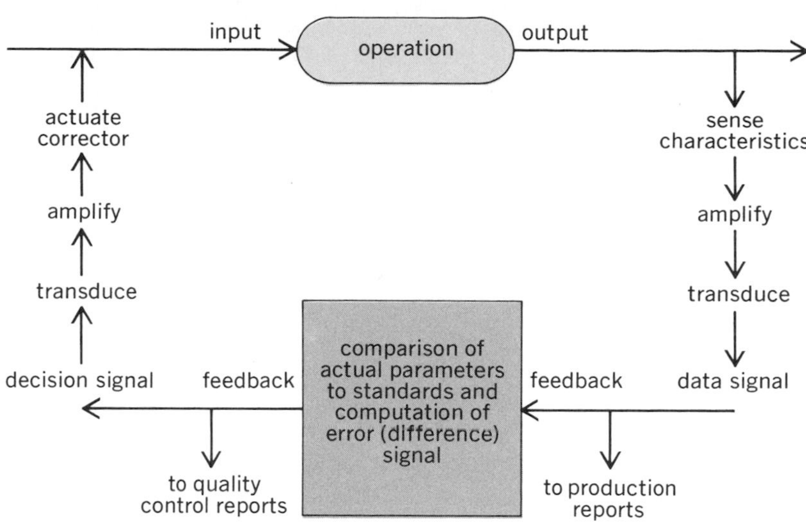

Fig. 10. Schematic of feedback flow of correction signals in a system which continuously measures and inspects output and automatically adjusts the input and internal operating variables so as to maintain output within preestablished limits.

Automatic maintenance system. It is extremely important that an autofacturing system operate at a smooth, steady, high speed. One of its major advantages is its high rate of output; but one of its serious limitations is that output can drop from 100% of capacity to zero during breakdowns. The reliability of the system must be high, therefore. Further, the more advanced and integrated the system, the more its overall operation may be dependent upon the performance of each subsystem—so subsystems, too, must have a high degree of reliability.

A three-pronged strategy can be employed to assure continued operation of the system: sensors, signals, and feedback controls; redundancy; and predictive or preventive maintenance.

Sensors can be placed throughout the system wherever there is a need for a continuous or remote monitoring of conditions. The sensors can feed into human-sensitive signals or can input an automatic feedback and correction control mechanism. Some human-sensitive signals are lights, bells, buzzers, alarms, chart pens, dials, and gages. The latest of these are voice-synthesizing integrated circuits which "speak." One linked to a bearing sensor might be programmed to say "The temperature of bearing #8 is now 140 degrees." In the automatic feedback mode, the sensor might send a signal to a device to release a drop of oil into a bearing at a certain sensed temperature.

Redundancy of design and construction merely

means that backup devices are built into the system at points of possible failure which could bring operations to a halt. A double tool holder is one example. If the first tool breaks, the tool holder automatically indexes 180° to present the backup tool to the work. The broken tool can then be replaced without stopping the machine.

Predictive or preventive maintenance anticipates trouble and "corrects" it before it occurs. The amount of the usage of a component (in hours, revolutions, output count, and so on) can be measured and the component replaced—even if it may have some life left in it—at a point at which the probability of a failure is judged to be significant.

Human interfaces with autofacturing. While a characteristic of the automatic factory is that it operates with fewer direct labor personnel, the complete elimination of people is not a requirement of autofacturing. Autofacturing is much more capital-intensive than the manufacturing it replaces, but it still requires some direct and indirect labor. More engineers, analysts, programmers, and maintenance people are required in autofacturing than in manufacturing. The direct laborers eliminated should be those doing dangerous, heavy, repetitive, tedious, boring, or low-skill-level work. The remaining people will represent those operations which are too difficult or uneconomical to automate. Their jobs must be redesigned to mesh with the new autofacturing operations.

Humans have been found to be better at picking out varied or complex signals against a high "noise" background, sensing and evaluating unusual or unexpected events, and reacting to them by drawing upon their large body of varied prior experience. Overall, the human brain and body are extremely versatile and adaptable, and to attempt to duplicate their capabilities with mechanisms and machines is not always successful from an engineering or economic viewpoint. Yet, machines are better than humans in exerting large, precise, consistent, repeated forces; sensing stimuli outside the human range; receiving, storing, and retrieving large amounts of coded data very rapidly and accurately; repeating a sequence or program for long periods of time without variation, fatigue, or diminution of performance; doing several things simultaneously; and working in hostile environments.

Where the equipment and the human interface, the devices should be designed so that people can use them most effectively, and the environment made so as to be suited to human comfort and safety. In effecting this human-machine interface, the influences to be considered include anthropometrics, biomechanics, ergonomics, cybernetics, human engineering factors, psychological and sociological factors, machine displays, control design, and the machine's effect (noise, fumes, radiation, and so on) on the workplace environment. The human-machine interfaces can be inputs or outputs of the equipment. Inputs are the buttons, switches, levers, and keyboards that control the machine. Outputs are the gages, dials, and CRTs and other displays through which the machine's conditions and information are presented to the operator. Both classes must be designed with the capabilities and limitations of a human in mind. The device's size, shape, location, color, method of activation, and force required must be selected carefully. Other human-machine interfaces involve the features of the equipment itself, for example, its color or seat design.

While not currently an important element in an autofacturing system, the audio human-machine interface may have a role in the future. It is possible to have devices monitor audio stimuli, either from humans or from other sources such as the product, machine, supporting equipment, or environment. The machine could "hear" a bottle crack, paper tear, a bearing squeak, a bell ring, and so on. The device would extract, categorize, and recognize certain predetermined properties (volume, frequency, pitch, sequence, pattern) of the audio signal and react as programmed. It could send a signal, activate a switch, type a message, or cause the machine to react. Making the voice-actuated control sophisticated enough to recognize a vocabulary of human spoken words is, obviously, more complex than having it respond only to a simple audio signal, but advances are being made in that area. The reading by a human into a voice-actuated control of handwritten numbers may prove to be more accurate than optical scanning. The applications in the more automatic storage of incoming packages, telephone inputs to the system, and control of robots are three obvious examples.

Together, the people and the equipment must be regarded as a single integrated system and must be designed for maximum mutual effectiveness. It must be recognized, however, that as companies move from manufacturing to autofacturing, they are going from a labor-intensive to a capital-intensive installation, and some workers will have to be retrained, relocated, retired, or released.

Conclusion. Manufacturing, as it is being performed today, will continue to exist (with expected improvements) in the majority of factories well into the future. But, as skilled labor and material become more scarce and expensive, products and processes become more complex, and international competition becomes more intense, autofacturing must replace manufacturing in many industries where mass production, consistently high quality, low unit costs, and high productivity are essential to their survival.

Most of the foregoing components of autofacturing already exist and are operating satisfactorily, but separately. What is needed now for the automatic factory of the future to become more of a reality is for those individual components to be combined and integrated into a total autofacturing system.

[VINCENT M. ALTAMURO]

Bibliography: "The Automatic Factory: New Wave of Productivity," *Mater. Handling Eng*. (Penton/IPC, Inc., Cleveland), vol. 36, no. 6, June 1981; C. Machover and R. E. Blauth (eds.), *The CAD/CAM Handbook,* Computervision Corporation, Bedford, MA, 1980; W. Tanner, *Industrial Robots,* vol. 1: *Fundamentals,* vol. 2: *Applications*, 2d ed., Society of Manufacturing Engineers, Dearborn, MI, 1981; K. Taraman (ed.), *CAD/CAM: Meeting Today's Productivity Challenge,* Computer and Automated Systems Association of the Society of Manufacturing Engineers, Dearborn, MI, 1980; K. R. Treer, *Automated Assembly,* Society of Manufacturing Engineers, Dearborn, MI, 1979.

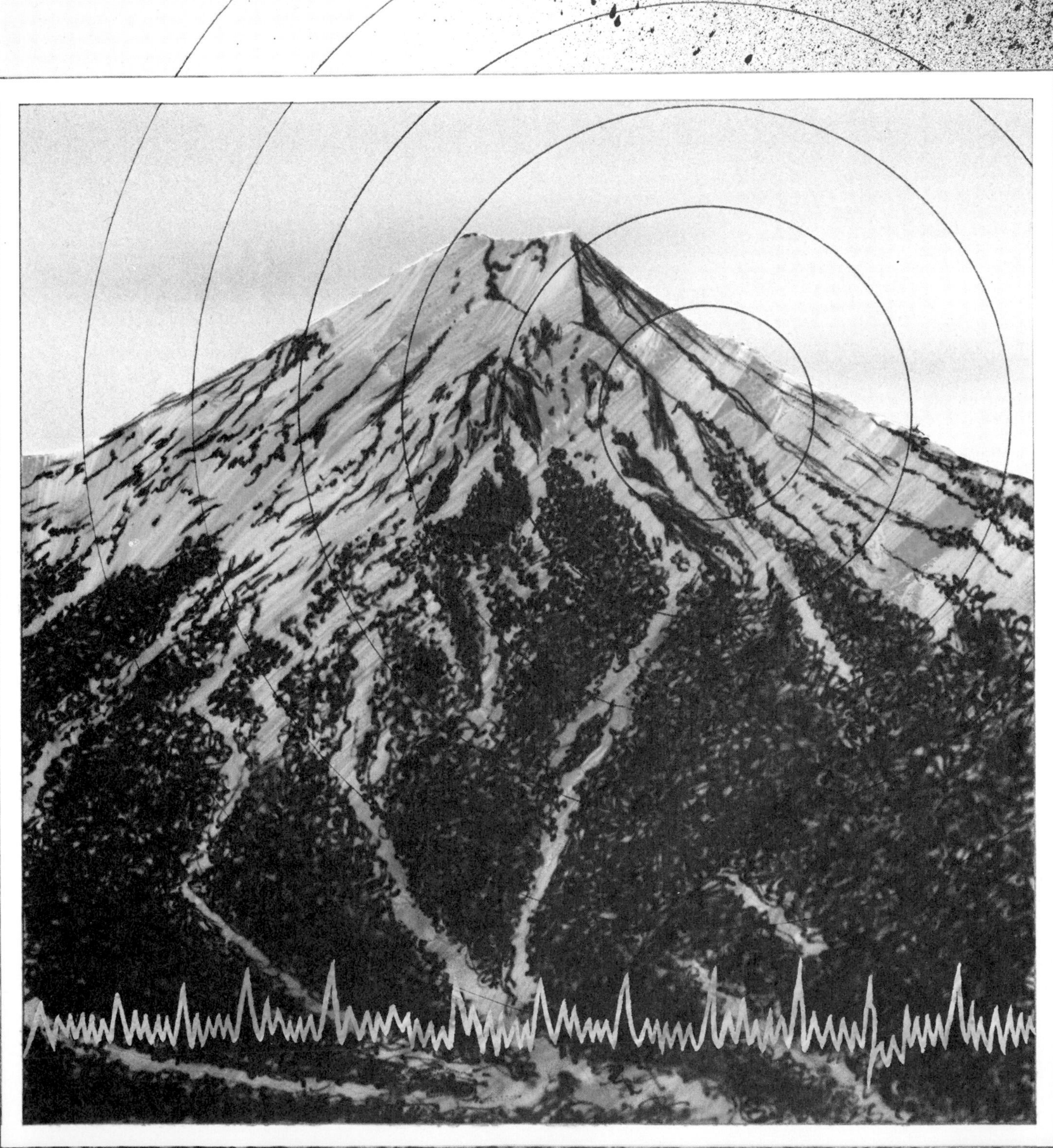

Volcano

The volcanic activity at Mount St. Helens that began March 20, 1980, and culminated in a violent eruption on May 18 followed by only a few years a comprehensive study of the eruptive history of the volcano. This history was inferred from the character, origin, and age of volcanic rocks and rock debris formed by repeated eruptions and widely distributed around the volcano. From this study it was concluded that Mount St. Helens was potentially the most dangerous volcano in the Cascade Range of the western United States and perhaps would erupt again before the end of this century. Thus the volcano's past behavior made it possible to anticipate eruptions such as those of 1980 and, when they occurred, to evaluate them in the context of the entire history of the volcano.

GEOLOGY OF MOUNT ST. HELENS

Mount St. Helens is one of the youngest major volcanoes in the Cascade Range, having appeared only about 40,000 years ago. Nevertheless, it has had a complex history and, during the last 2500 years, has erupted rocks that show a wide range of chemical composition. The record of eruptions is especially well preserved because the volcano was covered by large glaciers only once, and products of volcanism thus have been little affected by glacial erosion and deposition. Furthermore, the visible part of the pre-1980 cone was almost entirely formed by eruptions after large glaciers disappeared between 12,000 and 15,000 years ago.

Donal R. Mullineaux joined the U.S. Geological Survey in 1950. His areas of special interest include the geology of southeastern Puget Sound Lowland, the Cascade Range volcanoes, and the volcanic hazards of Cascade Range, Hawaii, and the western conterminous United States.

Dwight R. Crandell, a geologist with the U.S. Geological Survey, has specialized since 1967 in studies of the eruptive histories of volcanoes in the United States for the purpose of assessing the possible consequences of future eruptions. He served as volcanic-hazards coordinator for Mount St. Helens in 1980.

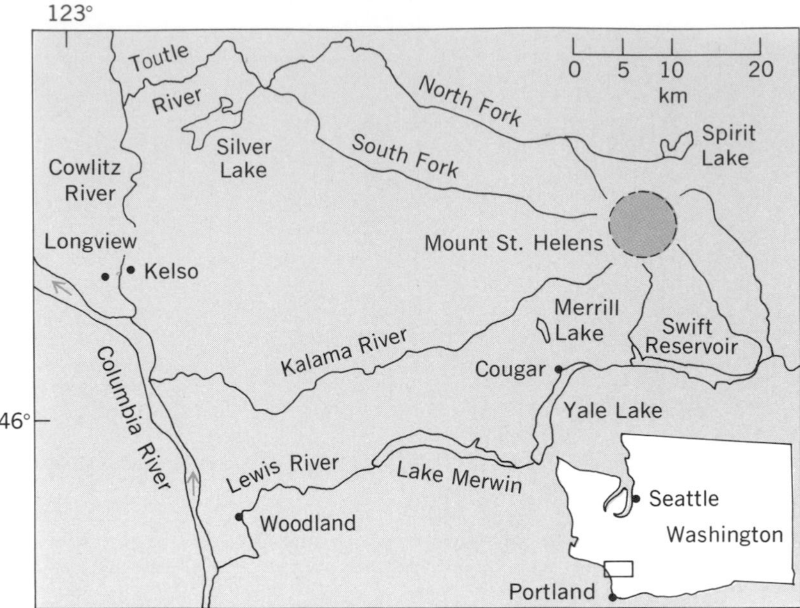

Fig. 1. Index map of the Mount St. Helens region, Washington. Circle shows a generalized outline of the base of the volcano.

In the past, vast amounts of rock debris originating at the volcano have moved eastward and southward into tributaries of the Lewis River, and westward and northward into the Toutle River and Kalama River drainage systems (Fig. 1). Studies of the resulting deposits provide clues as to the ways in which they were formed, and lead to inferences as to the kinds of eruptions that occurred at various times in the past. For example, a deposit consisting of a heterogeneous mixture of sand and rock fragments that contains logs of carbonized wood (charcoal) can be attributed to a pyroclastic flow. This is a rapidly moving mass of hot, dry rock debris mobilized chiefly by gravity and lubricated by hot air and other gases. Pyroclastic flows can be formed in several ways, but they are always directly or indirectly caused by eruptions. A complementary part of the eruptive record is represented by multiple layers of volcanic ash that mantle the surrounding region. Fragments of wood and charcoal preserved in volcanic deposits like these have been dated by the radiocarbon method, and the resulting ages place the eruptive history of the volcano into a framework of approximate calendar years.

ERUPTIVE HISTORY

The history of Mount St. Helens includes periods of a few years to a few thousand years during which intermittent eruptions occurred, separated by hundreds to thousands of years during which apparently no activity took place. The periods of volcanic activity are assigned informal names for the convenience of discussion (see the table); each period includes eruptions that are grouped because of a close association in time or because they produced mineralogically similar rocks, or both. Times during which no volcanic deposits were formed are believed to have been dormant intervals; during these times, weathering and erosion affected products of previous eruptions. During most eruptive periods, episodes

Summary of pre-1980 eruptive history of Mount St. Helens

Eruptive period	Approximate age, years	Kinds and products of eruptions
[Dormant interval of 123 years]		
Goat Rocks	180–120	Major dacite ash eruption A.D. 1800 was followed by andesite lava flow and dacite dome; many minor explosive eruptions occurred during 1831–1857.
[Dormant interval of about 200 years]		
Kalama	500–350	Major dacite ash eruption was followed by dome and pyroclastic flows of dacite, andesite lava flows, and another dome and pyroclastic flows of dacite.
[Dormant interval of about 700 years]		
Sugar Bowl	1150	Extrusion of dacite dome was accompanied by pyroclastic flows and a lateral explosive blast.
[Dormant interval of about 600 years]		
Castle Creek	>2200–1700	Initial eruption of andesite lava flow and ash was followed successively by dacite pyroclastic flows, basalt lava flows, andesite lava flows and pyroclastic flows, dacite ash, and basalt lava flows and ash.
[Dormant interval of less than 300 years]		
Pine Creek	3000–2500	Eruptions of domes and large pyroclastic flows of dacite were accompanied by a few eruptions of dacite ash.
[Dormant interval of no more than 300 years]		
Smith Creek	4000–3300	Several major eruptions of dacite ash alternated with pyroclastic flows probably derived from dacite domes.
[Dormant interval of more than 4000 years]		
Swift Creek	13,000–>8000	Several major eruptions of dacite ash preceded and followed formation of dacite domes and pyroclastic flows.
[Dormant interval of about 5000 years]		
Cougar	20,000–18,000	Several explosive eruptions formed dacite ash and large pyroclastic flows; nonexplosive dacite formed domes and a few lava flows.
[Dormant interval of about 15,000 years]		
Ape Canyon	>40,000–35,000	Repeated explosive eruptions of dacite ash were accompanied by pyroclastic flows, some of which probably were derived from domes.

of violent activity probably lasted only a few weeks or a few months rather than hundreds or thousands of years.

The following discussion refers to three kinds of volcanic rock—dacite, andesite, and basalt. These rocks differ in their chemical composition, and are defined by the percentage of silicon dioxide (SiO_2) they contain. The SiO_2 content of dacites erupted at Mount St. Helens ranges 62–69%, of andesites 56–62%, and of basalts 50–56%. Basalt lava is relatively fluid and tends to be erupted quietly to form lava flows; if explosive eruptions occur, they generally are of small volume. Dacite is relatively viscous, and gases do not readily escape from it. The initial eruption of gas-rich dacite commonly is explosive and forms ash and pyroclastic flows; at a later stage of an eruption, lava too stiff to flow laterally often will form a circular dome over the vent. Eruptions of andesite typically are intermediate in behavior between basalt and dacite. Some andesite eruptions at Mount St. Helens have been explosive and have formed ash; others have formed only lava flows.

Ape Canyon eruptive period. The first known eruptions of Mount St. Helens were explosive and produced thick deposits of dacite ash and pyroclastic flows. These deposits directly overlie mature soils developed on glacial deposits formed during the next-to-last glaciation of the surrounding mountains. Radiocarbon dates on these earliest volcanic deposits range from more than 40,000 to about 36,000 years. Although the time that Mount St. Helens appeared has not been dated closely, the initial eruptions are believed to have occurred shortly before 40,000 years ago. Immature soils preserved within the earliest ash deposits indicate that the Ape Canyon period included at least four episodes, perhaps brief, of volcanic activity separated by longer intervals of quiescence. The volume of one ash deposit produced during the period was as large as any subsequent ash erupted at Mount St. Helens. The Ape Canyon eruptive period was followed by a dormant interval of about 15,000 years.

Cougar eruptive period. Eruptions during the Cougar eruptive period produced lava flows, pyroclastic flows, and many thin deposits of ash. An andesite lava flow 180 m thick and at least 6 km long on the south side of the volcano is believed to be the oldest eruptive product of the period. It was followed by two even longer dacite pyroclastic flows that reached the Lewis River valley 14 km south of Mount St. Helens about 20,300 and 18,600 years ago, respectively, and by a pyroclastic flow that moved southeastward from the volcano about 19,000 years ago. Still other pyroclastic flows extended westward, southwestward, and eastward from the volcano. These large flows of hot rock debris must have devastated vegetation over a broad area around the volcano. Sand and gravel eroded from the resulting deposits and carried downstream by mudflows and floods produced a fill many tens of meters thick in the Lewis River valley, and probably also raised the bed of the Columbia River at the mouth of the Lewis.

Some thick deposits of rock debris of early Cougar age south and southwest of Mount St. Helens are thought to have resulted from mudflows generated by one or more massive landslides on the south flank of the volcano. These deposits may have had an origin similar to that of the deposits formed by the huge avalanche of rock debris on the north side of the volcano that was triggered by an earthquake on May 18, 1980.

No products of eruptions between about 18,000 and 13,000 years ago have been found at Mount St. Helens, so this apparently was a dormant interval. Alpine glaciers probably were at their maximum extents during the earlier part of this interval.

Swift Creek eruptive period. The Swift Creek eruptive period began about 13,000 years ago, when a series of explosive eruptions spread dacite ash at least as far as eastern Washington. About the same time, pyroclastic flows moved westward, southward, and eastward from the volcano. About 12,000 years ago many pyroclastic flows and mudflows extended into valleys on most sides of Mount St. Helens, and reached the Lewis River valley southeast of the volcano. Some of these pyroclastic flows and mudflows evidently originated when one or more dacite domes were erupted at the volcano.

This episode of activity was followed by a series of explosive eruptions that produced thick ash deposits about 11,500 years ago, between roughly 9000 and 11,000 years ago, and perhaps between 8300 and 9000 years ago. These last eruptions were followed by an apparent dormant interval of at least 4000 years. At the end of the eruptive period, outward-sloping fans of rock debris flanked one or more central domes and were bordered on the south by high remnants of the large lava flow of Cougar age.

Smith Creek eruptive period. Another series of explosive eruptions of dacite initiated the Smith Creek eruptive period about 4000 years ago. One of these eruptions, about 3500 years ago, produced one of the more voluminous ash deposits (estimated to be more than 3 km^3) of Mount St. Helens' history. It spread far northeast of the volcano and has been recognized 1100 km from its source, near Edmonton in Alberta. This eruption was soon followed by another major outburst which spread ash mostly eastward. Pyroclastic flows, probably caused by the explosive disruption of domes, were formed between 3500 and 3300 years ago, and mudflows possibly caused by melting of snow by hot pyroclastic flows extended at least 50 km down valleys west of the volcano. Spirit Lake may have first appeared at this time, when mudflows of volcanic rock debris blocked a valley north of the cone. The eruptive period was followed by a dormant interval of no more than about 300 years.

Pine Creek eruptive period. The Pine Creek period probably began about 3000 years ago and ended about 2500 years ago. Pine Creek eruptions seem to have consisted chiefly of dacite domes and pyroclastic flows, accompanied by small to moderate volumes of ash. Pyroclastic flows moved outward in nearly every direction from the volcanic center, and mudflows formed at the same time reached points many tens of kilometers downvalley. In the Cowlitz River valley downstream from the mouth of the Toutle River, deposits of mudflows and floods built a continuous fill across the valley floor about 6 m higher than the present river level. Near the mouth of the Lewis River, a similar fill about 7.5 m higher than the present river underlies the terrace on which Woodland is situated. Rock debris probably raised the bed of the Columbia River at the mouths of both the Lewis and the Cowlitz during the eruptions of Pine Creek time.

Some radiocarbon dates on deposits of Pine Creek age overlap those on deposits of the following Castle Creek eruptive period, and the two periods apparently were separated by only a brief dormant interval.

Castle Creek eruptive period. Castle Creek time marked the start of eruptions that built the modern volcano, and was characterized by a greater variety in eruptive behavior and type of rock than those of previous eruptive periods. These eruptions began soon after 2500 years ago and continued intermittently until about 1700 years ago. During this time, andesite and basalt were erupted, as well as dacite like that of preceding periods, and at times these rock types seem to have been erupted in quick succession. The earliest known eruptions of the period, between 2500 and 2200 years ago, formed andesite lava flows that were accompanied or followed by ash of similar composition. These were followed by explosive eruptions of dacite ash and, on the northwest side of the volcano, by dacite pyroclastic flows between 2200 and 2000 years ago. About 1900 years ago voluminous basalt lava flows poured southwestward and southward as far as the Lewis River valley.

The next recorded event was an andesite pyroclastic flow that moved down the east flank of the volcano, and this was followed shortly by an eruption of dacite ash about 1800 years ago. A dacite dome of Castle Creek age that underlies Dogs Head on the northeast flank of the volcano may have been formed at roughly the same time that the ash eruption took place. The dome is overlain by thin basalt lava flows that represent the latest eruptions of the period. These flows probably were accompanied by the eruption of a layer of basalt ash about 1700 years ago. The Castle Creek period was followed by a dormant period about 600 years long.

Sugar Bowl eruptive period. The only known eruption between about 1700 and 500 years ago was the extrusion of dacite lava about 1150 years ago to form Sugar Bowl dome low on the north flank of the volcano. A lateral explosive blast that occurred during dome extrusion threw rock fragments northeastward to a distance of 10 km or more across a sector at least 50° wide. Small pyroclastic flows were formed on the north side of the volcano at the same time that the lateral blast occurred, and mudflows moved into the Toutle River valley west of Spirit Lake. The deposits of the lateral blast are of special interest because in some ways they resemble those of the much larger blast of May 18, 1980.

Kalama eruptive period. Radiocarbon dates and ages of trees (determined by counting annual growth rings) on volcanic deposits of Kalama age suggest that the eruptive period extended from a little more than 500 to about 350 years ago, but the eruptions could have occurred during a shorter time span, perhaps less than a century. These eruptions affected nearly every side of the volcano and were largely responsible for the pre-1980 shape of Mount St. Helens.

The earliest known major eruption of Kalama time formed a deposit of dacite ash that extends into northeastern Washington and adjacent parts of Canada. As ash eruptions continued, dacite pyroclastic flows extended southwestward from the volcano. These were followed by andesite ash, andesite pyroclastic flows down the north and southwest flanks of the volcano, and andesite lava flows on the west, south, and east sides.

During the last phase of Kalama activity, a massive dacite dome rose into the summit crater, spilling avalanches and mudflows of hot rock debris down all sides of the volcano. Near the end of this phase, dacite pyroclastic flows swept down the northwest flank of the cone and locally covered deposits of the preceding mudflows. The Kalama period was followed by a dormant interval of about 200 years.

Goat Rocks eruptive period. The Goat Rocks eruptive period began with a major eruption of dacite ash about 1800 and ended in 1857. The principal vent of this period was on the north side of the cone at the site of Goat Rocks (Fig. 2; destroyed by the massive avalanche and eruption of May 18, 1980). Many minor eruptions were witnessed by inhabitants of the region between 1831 and 1857 and were reported in local newspapers. These accounts commonly included such terms as smoke, fire, and flame, and on at least one occasion a small amount of ash was deposited many tens of kilometers southeast of the volcano. In addition to activity at the site of Goat Rocks, vents were reported on the south side of the volcano (1842, 1843–1844, 1853) and on the northeast side (1850, 1853).

The dacite ash erupted at the beginning of the period had an estimated volume on the order of 0.1 km^3 and was carried by winds at least as far as northwestern Montana, roughly 575 km from the volcano. After this eruption, but before 1838, an andesite lava flow was extruded from a vent just west of the site of Goat Rocks and reached a length of about 5 km on the northwest flank of Mount St. Helens (Fig. 2). It was followed by the formation of the Goat Rocks dacite dome, probably during the 1840s. Several minor eruptions reported during the 1857–1980 dormant interval probably were small steam explosions.

Fig. 2. Aerial view of the northwest side of Mount St. Helens before the eruptions of 1980. Dogs Head Dome was erupted 1800 (?) years ago, Sugar Bowl Dome about 1150 years ago, Summit Dome 350–400 years ago, the lava flow between A.D. 1800 and 1838, and Goat Rocks Dome during the 1840s. The domes consist of dacite, and the lava flow is andesite. (*Courtesy of Delano Photographics*)

PERSPECTIVE

The catastrophic eruption of May 18, 1980, changed dramatically the landscape north of Mount St. Helens as well as the shape of the cone, just as major eruptions in the past changed other sides of the volcano, and affected areas tens of kilometers downvalley and hundreds of kilometers downwind. Past eruptions created lake basins such as those of Silver Lake and Spirit Lake (mudflows) and Merrill Lake (lava flow), filled or nearly filled with rock debris valleys heading at the volcano such as those of Pine Creek and the Kalama River, and repeatedly mantled broad areas beyond the volcano with thick layers of air-laid ash. From this perspective, the events of 1980 can be viewed as normal and expectable, even though they differed in some important details from eruptions of the past. Catastrophic events surely will recur in the future and will again endanger human life and property around the volcano. With the long explosive history of Mount St. Helens in mind, as well as the events of 1980, landowners, government officials, and scientists can now consider kinds of land use in threatened areas that are compatible with long-term risks.

Other volcanoes in the Cascade Range of the western United States also have long records of intermittent eruptions. Although their recent histories do not indicate that catastrophic eruptions are imminent, some of these volcanoes should be viewed as long-term threats to life and property. That threat can be mitigated to some extent by careful land-use practices, contingency plans for warning and evacuation, and monitoring of especially dangerous volcanoes. [DWIGHT R. CRANDELL]

1980 VOLCANIC ACTIVITY AT MOUNT ST. HELENS

The 1980 activity of Mount St. Helens, the first in more than a century, began on March 20 with an earthquake strong enough to be felt. A week of precursory earthquakes followed, and then nearly 2 months of minor, explosive eruptions that ejected only rock particles derived from the preexisting volcano. On May 18 a cataclysmic eruption that ejected molten rock (called magma below the surface, and lava aboveground) as well as solid rock fragments created large debris avalanches, a devastating laterally directed blast, mudflows, pumiceous pyroclastic flows (avalanches of hot fragments and gas), and an ash cloud that circled the Earth. Although the May 18 eruption was not as voluminous as several prehistoric Mount St. Helens eruptions, the 1980 debris avalanches and directed blast were much larger than any similar previous events at Mount St. Helens. Five more explosive magmatic eruptions as well as nonexplosive extrusion of several lava domes occurred after May 18; all the later explosive eruptions were much smaller than that of May 18.

Precursory earthquakes. The strong earthquake on March 20, which had a magnitude of about 4 on the Richter scale, was the first event of a rapidly developing seismic sequence under the volcano. Frequency increased to a peak on March 25, when more than 20 earthquakes of at least magnitude 4 were recorded. Observers in aircraft, however, saw no evidence of an eruption. Strong earthquakes became slightly less frequent on March 26 and 27, but the total energy release from them remained high.

Nonmagmatic steam and ash eruptions. Just after noon on March 27, a loud explosion marked the first eruption of 1980. Although clouds hid the volcano, an aerial observer reported a plume of ash rising above the clouds. Later, a small new crater about 60 m across could be seen within the old summit crater. In addition, two prominent systems of new cracks had formed; these systems trended east-west across the summit area, one just south and the other just north of the new crater. Between the two crack systems the surface had dropped slightly, and the upper northern slope of the volcano had been strongly fractured and pushed outward, upward, or both.

Additional explosive steam and ash eruptions, each lasting a few minutes to about an hour, began that night. These created a second crater (Fig. 3), and then enlarged both new craters until they merged into a single depression about 500 by 300 m across. Intermittent, similar eruptions continued until April 22, when they temporarily ceased. Winds carried most of the ash eastward, occasionally to distances of many tens of kilometers, but some ash was carried in all other directions. All the rock material erupted during that time was derived from the older rocks of the volcano.

Seismic activity continued to be high through late March and April. Earthquakes strong enough to be felt occurred each day, and the total energy they released remained at a high level. Continued shaking caused avalanches of snow, ice, and rock debris. A type of seismic activity called harmonic tremor, thought to indicate underground movement of magma, was detected in early April but was not associated with visible changes in activity.

Distention of the north flank of the volcano, first noticed on March 27, was continuous. By March 30, an elliptical bulge was visible on the upper north flank (Fig. 3), and by mid-April a much-enlarged bulge was conspicuous. Although no magma had been erupted at the surface, the bulge provided evidence that magma had been injected into the volcano. Surveying begun about mid-April measured the rate of swelling as about 2 m per day. This displacement of the north flank of the volcano was so continuous and so rapid that the possibility of a large landslide from the north flank was soon recognized and publicized as the most immediate hazard from the volcano.

No eruptions occurred between April 22 and May 7, although fumaroles steamed continuously in the new crater. Earthquakes had decreased somewhat in frequency by late April, but bulging of the north flank continued at approximately the same rate.

On May 7 another minor eruption marked a return to steam and ash outbursts similar to those of March and April. These eruptions continued during the next week, along with strong earthquake activity; two earthquakes of about magnitude 5 and two short periods of harmonic tremor were recorded during that period. A few new fumaroles and other warm spots appeared near the new crater and in the bulge. By May 12, bulging had displaced Goat Rocks on the north slope (Fig. 3) a total of about 100 m north of its pre-1980 position. Although a part of the bulge had moved slightly upward, the main displacement was outward. The bulge measured about 1.5 and 2 km across and down the slope, respectively.

On May 14 eruptions ceased again, but fumarolic activity and deformation continued.

Cataclysmic eruption. Mount St. Helens' first magmatic eruption of 1980, on Sunday, May 18, was sudden, violent, and disastrous. At 8:32 A.M., an earthquake of about magnitude 5 jolted the volcano. The earthquake caused a large part of the upper north flank of the volcano to break loose along a steep fracture and begin to slide downslope rapidly. The massive rockslide exposed a steep new rock face. Water from within the volcano flashed to steam—removal of the rockslide slab from the flank had depressurized or "uncorked" the underlying high-pressure system of magma and surrounding rocks. A laterally directed eruption cloud burst from the steep rock face exposed by the rockslide, and a vertical cloud rose from the summit crater. Within 30 s a second huge slide broke loose, and the laterally directed eruption expanded to involve and hide much of the north flank of the volcano.

More slides followed, and the succession of slides formed a series of huge avalanches of hot rock debris that swept into the North Fork Toutle River valley. One avalanche raced up and over the 300- to 400-m-high ridge north of the North Fork into the

Fig. 3. Mount St. Helens from the east on March 30, showing two new craters, the fracture system south (left) of the craters, and elliptical bulge (b) on the upper northeast slope. The bulge later expanded downslope to encompass Goat Rocks (G). (*Photograph by C. D. Miller, USGS*)

next valley. Another traveled more than 20 km down the North Fork valley, and still another plunged into Spirit Lake, temporarily forcing lake water up onto the sides of adjacent ridges. The deposits of the debris avalanches made a fill as much as 200 m thick in the North Fork Toutle valley and dammed many streams tributary to that valley.

As the debris avalanches swept off the volcano, a blast cloud of hot gas carrying large and small rock fragments erupted from the newly exposed face, expanded rapidly to the northwest, north, and northeast, and devastated an area of nearly 600 km^2. Moving at a speed of several hundred kilometers per hour, it quickly overtook the avalanches and, except toward the northwest, extended far beyond them. The blast cloud hugged the ground surface; viewed from a distance, it disappeared into valleys and reappeared as it overtopped ridges. Even large trees were picked up and carried away within about 10 km of the volcano. Between about 10 and 25 km, the blast cloud flattened trees but did not pick them up, and, as the cloud progressively slowed, it became increasingly deflected by peaks and valleys. Witnesses within but near the margin of the affected area reported that the cloud was hot but only for a few minutes. As the expanding blast cloud dropped the rock debris it was carrying, it eventually became lighter than the air it was approaching, and its front rose from the ground as if along a ramp. There, it scorched trees along a rising line until it rose above and merely dropped ash on otherwise unaffected green trees.

The cataclysmic eruption also caused small pyroclastic flows to spill down other flanks of the volcano. They melted snow and ice and started mudflows that streamed down the west, east, and south flanks of the volcano. Mudflows that reached Swift Reservoir in the Lewis River valley south of the volcano caused a 2-m-high wave, but the dam was not overtopped because the reservoir had been lowered in anticipation of possible mudflows.

Within 30 min, the vertical eruption column had reached a height of more than 25 km above sea level and spread laterally to form a mushroom-shaped cloud more than 50 km across. Strong, high-altitude (jet-stream) winds carried the cloud east-northeastward at about 100 km/h into eastern Washington, enveloping areas in darkness, dropping ash as much as several centimeters thick, and severely disrupting travel, communications, and many other activities. The ash cloud crossed parts of Idaho, Montana, Wyoming, and Colorado and then the central United States before continuing out to sea over New England and eventually circling the Earth. The vertical eruption column over the volcano fluctuated in height but maintained a height well above 10 km until about 5:30 P.M. The rockslides, laterally directed blast, and vertical eruption removed 2.5–3 km^3 from the volcano, leaving a horseshoe-shaped crater open to the north that measures about

Fig. 4. Large amphitheater-like crater created by the May 18 eruption. (*Photograph by C. D. Miller*)

2 km from rim to rim and 600 m from rim to floor (Fig. 4).

By midday, pyroclastic flows of hot pumice began to spill down the north flank of the volcano, and they continued intermittently through the afternoon. At the north base of the volcano, deposits of these pyroclastic flows and the debris avalanches heated lake and river water that had been engulfed, causing steam explosions that formed ash columns as much as a kilometer in height as well as numerous small craters.

Also by midday, mudflows began to develop on the debris avalanche in the North Fork Toutle valley; these poured across the surface of the avalanche deposit and downvalley. Beyond the avalanche, mudflow levels rose through the afternoon, and at the confluence of the Toutle and Cowlitz rivers reached a crest about midnight. As the mudflow reached the Columbia River, it deposited so much material that the ship channel was temporarily blocked to oceangoing vessels.

Minor, vertical-column eruptions continued through the night of May 18, and intermittently through the next week. Seismic activity lessened markedly after the cataclysmic eruption.

Eruptions after May 18. The minor eruptions of the week after May 18 became stronger on the next Saturday, and climaxed at about 2:30 A.M. on Sunday, May 25, when a vertical eruption column rose to about 14 km above sea level and pyroclastic flows again spilled down the north flank. The May 25 eruption was minor compared to that of May 18, however, and lasted only a few hours. Winds were variable but generally toward the west, and widespread ashfall west of the volcano stopped traffic and greatly hampered other activities. After May 25 the volcano again quieted, and ejected only small plumes intermittently. Seismic activity continued at a low level, but early in June gas emission increased severalfold.

On the afternoon of June 12, increasing harmonic tremor indicated the possibility of another eruption. Shortly after 7 P.M. a small vertical-column eruption occurred. Two hours later, a larger eruption column reached nearly 15 km above sea level, and pyroclastic flows again swept down the north flank. Winds blew ash over a wide arc from southeast clockwise around to west, and relatively strong winds carried ash south-southwest across the Portland–Vancouver metropolitan area. On June 15 a small lava dome—a stiff mass of new lava—was sighted in the crater; it mushroomed from the vent out over the crater floor and eventually reached a diameter of about 365 m and a height of about 45 m.

By noon on July 22 an increase in frequency of small earthquakes was recognized as a probable precursor to yet another eruption. About 5:15 P.M., a spectacular vertical eruption column climbed to about 14 km above sea level. As before, pyroclastic flows accompanied the vertical eruption. Later in the evening two more eruptive columns were also accompanied by pyroclastic flows. Winds carried the ash northeastward across Washington and northern Idaho and on into Canada.

Harmonic tremor that began about noon on August 7 preceded the next eruption, which commenced with an eruption column that rose shortly after 4 P.M. to an altitude of about 14 km. Pyroclastic flows again swept down the north side. Two more high vertical columns rose later in the evening, and the ash was carried to the east and northeast. After these eruptions, lava rose within the vent but was not voluminous enough to fill it.

During the next 2 months, several small, ash-rich plumes rose to low altitudes, but seismic activity remained low.

A series of eruptions that began on October 16 spanned nearly 2 days. Once more, an increase of small earthquakes provided a warning. About 10 P.M. a single eruption column reached an altitude of 13–14 km. About 9:30 A.M. on the 17th, another single column climbed to the same altitude, and a third similar eruption occurred about 12 h later. Two more vertical columns were erupted on October 18, but both were somewhat smaller than those on the 16th and 17th. Immediately afterward, lava filled the vent and mushroomed out over the adjacent crater floor to form another dome, somewhat smaller than that of June.

The October eruptions concluded the major explosive activity of Mount St. Helens during 1980; later ones in 1980 consisted only of small ash outbursts and further dome growth.

Aftermath. After May 18 eruptions generally decreased in explosivity and volume during 1980, and no major explosive eruptions occurred during the first half of 1981. The volcano could return to dormancy with no further significant activity, for short-lived eruptive episodes have previously occurred at Mount St. Helens. In the past, however, voluminous eruptions as well as smaller ones have occurred within extended eruptive episodes. Consequently, the possibility of both small and large eruptions in the near future must be recognized.

[DONAL R. MULLINEAUX]

Bibliography: D. R. Crandell and D. R. Mullineaux, *Potential Hazards from Future Eruptions of Mount St. Helens Volcano, Washington*, USGS Bull. no. 1383-C, 1978; D. R. Crandell, D. R. Mullineaux, and C. D. Miller, Volcanic-hazards studies in the Cascade Range of the western United States, in P. D. Sheets and D. K. Grayson (eds.), *Volcanic Activity and Human Ecology*, pp. 195–219, 1979; R. P. Hoblitt, D. R. Crandell, and D. R. Mullineaux, *Geology*, 8:555–559, 1980; P. W. Lipman and D. R. Mullineaux (eds.), *The 1980 Eruptions of Mount St. Helens, Washington*, USGS Prof. Pap. no. 1250, 1981.

E.T. Steadman

Chemistry in a Petroleumless World

Liquid products were first obtained from coal in England during the 1700s. Coal was heated in the absence of air, and the coal tar product was used by the shipbuilding industry of that day. By 1880 large German concerns were making fuel gas and metallurgical coke from coal, and the tarry by-products were refined to supply the world's demand for aromatics for pharmaceuticals, dyes, explosives, and photographic chemicals. Although lamp oil was produced from coal in the United States as early as 1850, a domestic coal chemicals industry did not develop until World War I, when Germany cut off its supply of aromatic coal tar products, particularly toluene needed for TNT manufacture.

A modern coal tar recovery system associated with a battery of coke ovens will produce 8 gal of liquid products per ton (0.03 liter/kg) of coal feed. The total volume of these products, which include benzene, toluene, xylenes, phenols, and naphthalene, is small compared with today's volume of these same chemicals produced from petroleum. There are three other coal-based processes which provide a much higher yield of chemical products than the coke ovens. In the early 1900s German scientists observed that coal differs from petroleum in that coal contains less than half the hydrogen of petroleum. Methods were developed to add hydrogen to coal at high temperature until the coal liquefied. During World War II, German industries produced 100,000 barrels (16,000 m^3) a day of trans-

J. E. Wolff is a senior research chemical engineer in Tennessee Eastman Company, with long-range planning responsibilites for feedstocks and advanced process technologies. Earlier, he specialized in engineering design and process optimization.

David M. Pond is a research associate in the Research Laboratories of Tennessee Eastman Company, where he is presently responsible for the Homogeneous Catalysis Research Laboratory. He also directed the basic catalysis research program and was involved in organic synthesis and stabilization research.

portation fuels and lubricants by using the Bergius direct liquefaction process. Although no commercial plants remain in operation today, the United States is developing processes to produce oils synthetically from coal by direct liquefaction technology. These coal liquids can be further processed by petroleum refineries into traditional fuels and petrochemical feedstocks, thereby displacing imported crude oil.

In the second process, coal is almost entirely converted to a gas by controlled combustion in an oxygen-starved environment (Fig. 1). The gaseous product contains chiefly carbon oxides, hydrogen, and methane. Coal was first gasified in the early 1800s for home lighting. A century later, coal gas became the initial feedstock for synthetic methanol and ammonia production. Eastman used coal gasifiers in the 1940s to produce a low-Btu fuel gas for ketene furnaces. The use of coal gas ended in the United States as low-priced natural gas became available in the early 1950s through pipelines from the Southwest. More than 2 dozen coal-based methanol and ammonia plants are still operating in Australia, Turkey, India, Greece, and Yugoslavia.

The final coal conversion process involves first the gasification of coal as previously described and then the conversion of the coal gas to liquid products. This process is based on the German development of Fischer-Tropsch chemistry and is called indirect liquefaction. In 1955 the South African Coal, Oil, and Gas Corporation, Ltd., or SASOL, started the only coal liquefaction plant still in operation today. A major expansion was recently completed, and another is now under construction. Coal is gasified in the well-proved Lurgi fixed-bed process, and the coal gas is then converted in a Synthol reactor to a liquid mixture of long-chain hydrocarbons in the gasoline boiling range. The indiscriminate nature of product formation in the second step makes the process unsuitable for direct chemical manufacture. The coal liquids are further refined for transportation uses and feedstocks for chemical manufacture, including ethylene. Before the coal chemicals industry fully matured, the petrochemicals industry began to flourish, and by 1920 ethylene was being made from petroleum refining by-product streams. This ethylene manufacturing technology was an extension of the thermal cracking processes developed to increase the yield of gasoline from crude oil. The first ethylene derivatives were synthetic ethanol, which displaced fermentation alcohol in industrial applications, and ethylene oxide, which provided a new family of products important to the expanding automobile industry. By the 1940s the development of new alloys permitted the construction of modern-day, high-temperature, tubular cracking furnaces. The new operating conditions increased ethylene yield and substantially lowered the cost of ethylene manufacture. Meanwhile, the oxo process had been developed in Germany to combine either ethylene or propylene with carbon monoxide and hydrogen to form aldehydes and their associated alcohols, propanol and butanol. Low-priced methane or natural gas replaced coal as the main source of synthetic methanol, acetylene, and ammonia. An extension of gasoline reforming technology led to the production of aromatics from petroleum rather than coal tar. New, more efficient processes were commercialized for ethanol and ethylene oxide. In summary, these technological breakthroughs firmly established the economic superiority of petrochemicals and led to widespread expansion of their use during the 1950s. In the 1960s another German development led to the production of acetaldehyde directly from ethylene.

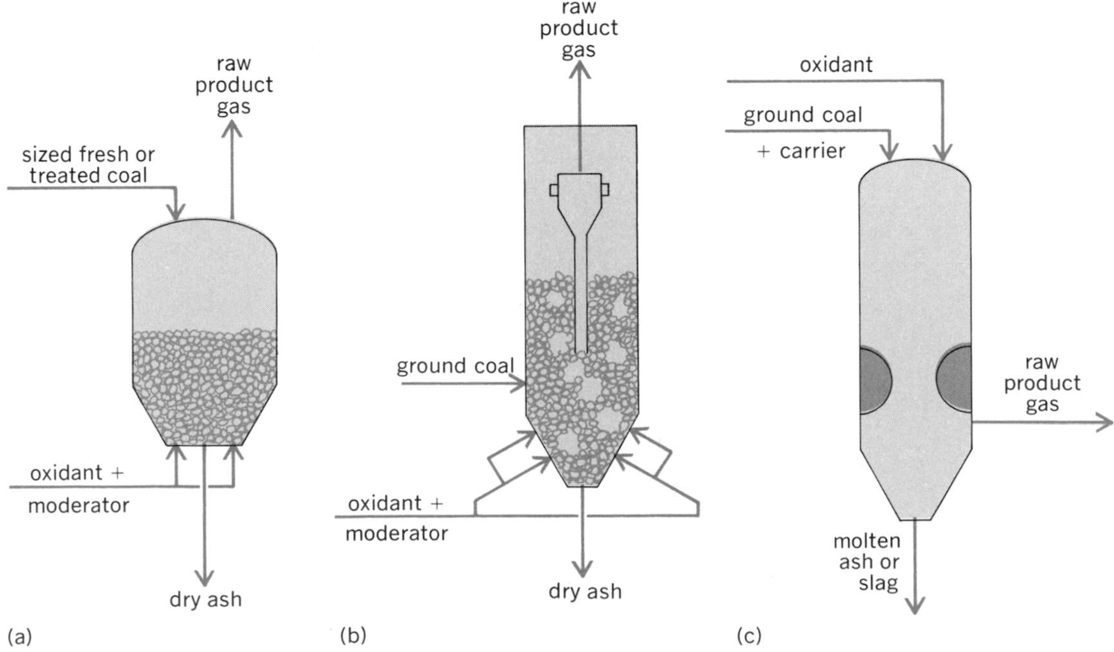

Fig. 1. Generalized configurations of coal gasification processes: (*a*) fixed bed, (*b*) fluidized bed, and (*c*) entrained bed.

SYNTHESIS GAS VERSUS PETROLEUM-DERIVED ETHYLENE

In the meantime the European chemical industry continued to rely on coal for the prodution of methane, ammonia, methanol, aromatics, and particularly acetylene. Acetylene could be produced from domestic coal resources much more cheaply than ethylene from imported petroleum. Development of the Middle East oil fields in the 1950s resulted in availability of low-priced crude oil to European markets. Refineries were expanded, and low-cost ethylene was produced by cracking naphtha, a petroleum fraction. Subsequently, ethylene displaced coal-based acetylene in most applications in Europe.

Today most of the important industrial organic chemicals are derived from petroleum. During 1980 in the United States alone, 28,000,000,000 lb (12,700,000 kg) of ethylene was consumed in the production of polyethylene, ethylene oxide and glycol, ethanol, acetaldehyde, propionaldehyde, and nonoxygenated derivatives, including ethylene dichloride for vinyl chloride manufacture and ethylbenzene for styrene production. Seventy percent of the ethylene was produced from natural gas liquids and refinery gas streams containing ethane, propane, and butane. The remainder of the ethylene was produced from naphtha and gas oil, which are primary refinery products with alternative fuel uses in transportation, industrial and utility applications, and home heating.

In the late 1960s the chemical industry stepped up its program of energy conservation and began a search for lower-cost chemical feedstocks even before the growing national concern caused by the tenfold increase in petroleum prices during the succeeding decade. An early study, conducted by Eastman in 1970, acknowledged the decline in domestic petroleum reserves and projected that coal would become a more attractive energy source and an important chemical feedstock in the longer term. Although in 1970 no one really anticipated the actual events which would take place in the Middle East, the run-up in the price of foreign oil from $3 a barrel to $12 a barrel ($19 to $75 per cubic meter) in 1973 following the Arab oil embargo, and the continued rise to over $30 a barrel in 1980 confirmed the pressing need for alternative supplies of nonpetroleum feedstocks. It was judged that the development of modern conversion methods to transform coal into more usable fuel products and chemical feedstocks would be the domain of the large energy companies. However, the chemical industry had the ability to produce the downstream chemicals now made from petroleum. Estimates were prepared which indicated that medium-Btu or synthesis gas (mixtures of carbon monoxide and hydrogen) could eventually be produced from coal gasification processes at costs substantially below those for the equivalent amount of ethylene. Coal gasification was found to be less costly, more flexible for chemical manufacture, and operable on lower-quality coals than liquefaction techniques involving the direct, high-temperature hydrogenation of coal to form complex chemical mixtures or synthetic oils suitable for processing in conventional petroleum refineries.

The composition of the gas produced depends on the design of the coal combustion reactor as well as the operating conditions. Similarly, coal can also be converted to gaseous hydrocarbons, mainly methane, and utilized as synthetic natural gas, or SNG. The methanation reaction is exothermic, so it is best to force as much as possible of that reaction to take place in the gasifier to supply heat for the endothermic reduction reactions of H_2O and CO_2 with coal. High methane yields are promoted by high pressure and low temperature, so most of the funded research for fuel gas from coal is for processes which operate below 800°C (1500°F) and above 67 atm (1000 psig or 6.8 MPa).

For synthesis gas production, the major goal is to produce a high-purity mixture of CO and H_2 by the controlled combustion of coal in an oxygen-starved environment. Therefore, in contrast to the situation with fuel gas, it is desirable to minimize methane production in the gasifier. Any methane in the product gas must be removed or converted to another form before the synthesis gas can be used for chemicals manufacture. Low methane formation is favored by high temperature and low pressure, with the temperature effect being greater than the pressure effect. Many synthesis gas units will operate above 1100°C (2000°F) and at moderate pressure. Super-atmospheric pressure is desirable to minimize the size of the gas purification equipment. Also, since moderate or high pressure is typical for the chemical reactions involving synthesis gas, some benefit is obtained from compressing the gasifier feed streams instead of the synthesis gas product. Actually, the production volume of synthesis gas already exceeds ethylene in producing downstream derivatives. With few exceptions, a synthesis gas plant is located adjacent to the consuming chemical unit, and therefore synthesis gas is not generally an item of commerce. During the last 30 years in the United States, all synthesis gas was produced from natural gas. In 1980 DuPont began operation of a partial oxidation unit in Texas that feeds residual oil, a heavy asphaltlike refinery by-product. In other areas of the world, large volumes of synthesis gas are produced from both naphtha and residual oil, particularly if low-priced natural gas is not available. Today about 3% of the world's synthesis gas is produced from coal in gasification processes described earlier.

COAL GASIFICATION

Three general configurations of coal gasification equipment are in use today: fixed or slow-moving bed, fluidized bed, and entrained bed.

Fixed-bed gasifier. Fresh coal is added to the top of the fixed-bed unit, while steam, along with air or oxygen, is injected at the bottom (Fig. 1*a*). The product gas flows upward and countercurrent to the settling bed of coal. The hot gas leaving the top of the bed strips substantial quantities of volatile compounds from the fresh coal. During subsequent cooling and scrubbing of the raw gas, by-product tars condense, thus presenting handling and disposal problems. Another drawback is finding coals which will not cake or plug the gasifier as the bed settles during depletion of the carbon content. Also, the relatively low operating temperature of this process favors the formation of methane. Most commercial gasifiers in use in the early part of the 20th century were small fixed-bed gasifiers such as the Wellman-Galusha unit. The pressurized Lurgi gasifier is another well-proved fixed-bed unit, perhaps best known for its application by SASOL.

Fluidized-bed gasifier. In a fluidized-bed gasifier, carefully sized coal is fed to the reaction vessel and suspended by the upflow of oxygen and product gas (Fig. 1*b*). As the product gas exits the top of the bed, unconverted carbon fines are entrained and swept out of the gasifier, thereby lowering the overall carbon conversion and thermal efficiency. Also, since the operating temperature must be maintained below the softening point of the coal ash, some by-product tars remain in the product gas. The fluidized-bed system was first used in the Winkler gasifier in 1926, and a number of concerns are currently working on improved operating principles.

Entrained-bed gasifier. In an entrained-bed gasifier, oxygen and either water or steam, along with finely ground coal, are introduced into the reaction vessel through a common feed nozzle and pass concurrently downward through the reactor (Fig. 1*c*). The operating temperature is maintained above the melting point of the coal ash so that the unreactive slag will drain from the unit. As a result of the relatively high operating temperature and the unique flow pattern, the gasification reaction rates are much faster, by-product tars are destroyed within the unit, and the production of methane is minimized. The product gas is mainly carbon monoxide and hydrogen, which makes this unit particularly suitable for chemical synthesis. The Koppers-Totzek process employs an entrained-bed concept and is currently used in other parts of the world to produce synthesis gas for the manufacture of methanol and ammonia.

The Texaco coal gasification process is another entrained-bed process moving rapidly toward commercialization. This coal-based process is a variation of the widely licensed, proprietary partial oxidation process that produces synthesis gas from a number of hydrocarbon feedstocks, principally low-valued, high-sulfur crude and residual oils. The Texaco gasifier can also be operated on a wide variety of low-quality coals at pressures up to 80 atm (1200 psi), which results in high throughput and single-pass conversion. Since the raw gas is generated at elevated pressures, the size of the gas clean-up equipment is reduced and the need to compress the product gas before its use as a chemical feedstock is minimized. The Texaco process, because of its inherent cleanliness, is an environmentally acceptable means of providing gaseous products from coal. Since the coal is handled as a water slurry, wet grinding eliminates the dust problems ordinarily associated with dry-grinding operations. The relatively high operating temperature within the gasifier leaves only trace amounts of hydrocarbon by-products in the coal gas. The majority of the water needed for slurrying, cooling, and scrubbing operations is reused within the process except for a small blowdown stream to prevent the accumulation of water-soluble by-products. The coal ash is drained from the reaction area as a molten slag and quenched in a water bath in the bottom of the gasifier. The coarse ash is removed as glassy pellets through a lock-hopper system. The inert ash pellets have shown very low levels of leachability and will qualify for nonhazardous waste landfill.

Feedstock selection. Choosing an economical synthesis gas feedstock is a complex process influenced by long-term cost projections, plant size and location, associated conversion processes, and the desired chemical products. There are several important considerations in selecting the most economical feedstock, other than its cost alone, for a new synthesis gas plant. The major portion of the unit cost of synthesis gas from a natural gas–based plant is the operating cost, which is chiefly for the feedstock itself. Since natural gas contains very little sulfur, the product gas does not have to be scrubbed to meet environmental requirements or to protect sensitive downstream catalysts. As the price of natural gas is decontrolled, a residual oil plant will become more attractive in overall unit cost. A more expensive plant is needed for coal operation due to the increased complexity of gasifying a solid material and the installation of proper environmental controls. As a result, the minimally economical plant size for coal operations is about 10 times larger than a natural gas reformer. Once committed, capital costs are not subject to further escalation. The relatively low annual operating cost of a coal-based plant provides an increasing economic advantage during periods of high inflation. Most experts agree that the price of petroleum and natural gas will rise much faster than coal prices, which will be closely related to mining and delivery costs. Therefore the favorable operating cost of a coal-fed plant should offset the higher capital cost and result in a lower overall cost in the next few years.

Plant location. Any synthesis gas plant should be located close to its raw material source to minimize transportation costs. Locating a synthesis gas plant on the Gulf Coast, on a natural gas pipeline, or near an inland refinery will generally favor natural gas or residual oil. The Gulf Coast chemicals industry has been developed mainly because of its proximity to these petroleum and natural gas supplies. However, the eastern coal fields are closer to the large eastern and midwestern chemicals markets and today hold promise for future utilization. Interestingly, some of the high-sulfur eastern coal, least desirable for boiler fuel, is quite attractive for gasification processes because of the lower price. During the gasification process the sulfur is converted to H_2S, which is easily removed by an acid-gas removal process. The H_2S can be removed more economically than flue gas scrubbing of SO_2, and then converted either to elemental sulfur or to sulfuric acid. The large reserves of western coal and lignite are also receiving considerable attention for synthetic fuels development where large reserve size is essential and markets for fuels are nearby.

CHEMICALS FROM SYNTHESIS GAS

The maximum incorporation of the total weight of synthesis gas into the product chemical is important to the cost efficiency of the overall process.

H_2:CO molar ratio. Synthesis gas reaches its maximum advantage over ethylene in synthesis of an oxygenate when the H_2:CO molar ratio obtained from a feedstock matches the H_2:CO required for a particular oxygenate. For example, acetic acid derives 47% of its weight from ethylene but 100% of its weight from coal-based synthesis gas. The CO and H_2 stoichiometries required to produce the oxygenates of interest results in the need for a wide range of H_2:CO ratios in the product synthesis gas (Fig. 2). Aliphatic alcohols (methanol, ethanol, and so on) require H_2:CO = 2:1, whereas aliphatic aldehydes, acids, anhydrides, and glycols require ratios between 1:1 and 1.5:1.

The precise H_2:CO molar ratio required for a specific chemical complex can be obtained by adjusting the output composition of the gasifier with a downstream water gas shift reactor. Three cases showing utilization of CO are provided in Fig. 3. Case 1 shows the approximate carbon monoxide, carbon dioxide, and hydrogen poundages obtained from 1 ton (0.9 metric ton) of coal (assume 79% carbon) in an entrained-bed gasifier. Such gasifiers typically convert over 95% of the carbon in the coal to carbon oxides with a $CO:CO_2$ ratio of 4.5:1 to 5:1. In case 2 the carbon monoxide portion of the stream has been mixed with steam and sent through a water gas shift reactor [see reaction (1)] to maximize hydrogen production. While the hydrogen content goes to 412 lb per ton of coal (0.2 kg/kg), CO_2 production rises to almost 5800 lb/ton (2.9 kg/kg), which may be utilized or vented. Maximization of CO is seen in case 3. The water gas shift reaction (1) is normally operated from left to right by utilizing

$$CO + H_2O \underset{\Delta\Delta}{\overset{\Delta}{\rightleftharpoons}} CO_2 + H_2 \quad (1)$$

either a copper metal–zinc oxide or a cobalt-molybdenum catalyst system below 570°F (300°C). The carbon dioxide is separated from the hydrogen in the acid gas removal step. Operation of the water gas shift reaction (1) in the opposite direction (right to left) is possible but requires enormous amounts of energy and normally will be economically unattractive.

Oxygenate production. Substantial research and development efforts are under way in Japan, western Europe, and the United States to discover processes that better utilize the synthesis gas that will soon be available from coal. Today there are commercial processes that utilize CO or mixtures of CO and H_2 to produce a number of oxygenates. Many of these processes also involve the use of petroleum-derived olefins, which are in fact responsible for a considerable fraction of the weight of the product. It is the long-term objective of the chemical industry to produce these kinds of compounds without the use of petroleum derivatives such as ethylene, and the number of fully developed technologies shrinks significantly. Today methanol and acetic acid and,

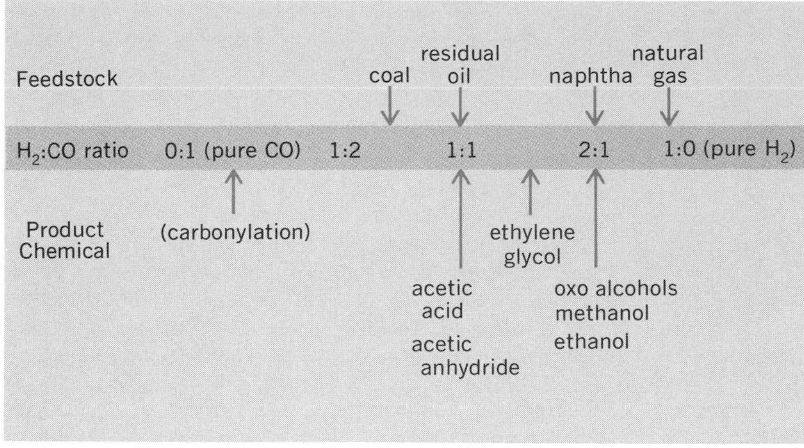

Fig. 2. The feedstocks for synthesis gas production contain widely different H:C ratios, which dictate the H_2:CO ratio that will be obtained on gasification. Natural gas produces the highest H_2:CO ratio, 4:1, whereas many coal processes produce the maximum concentration of carbon monoxide or H_2:CO = 0.8:1. Residual fuel oil and naphtha have H:C ratios that result in H_2:CO ratios of 1:1 and 2:1.

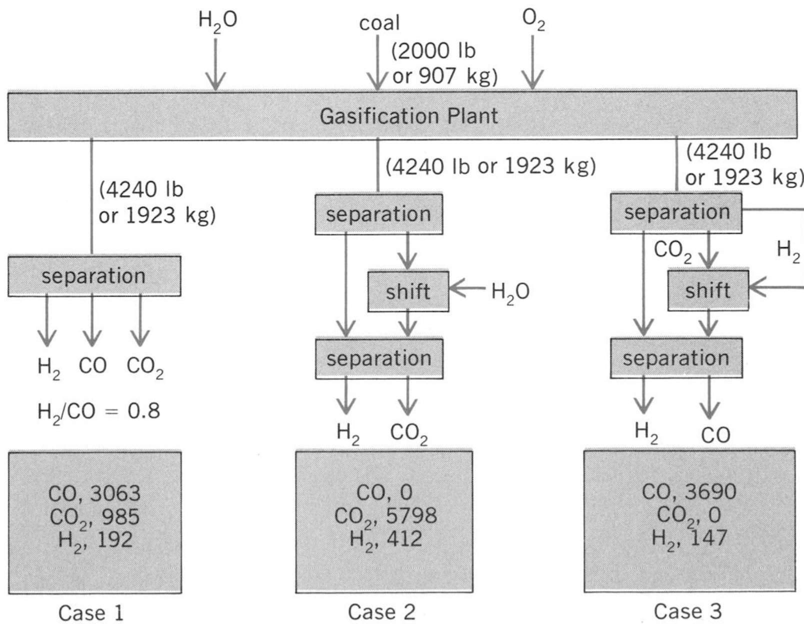

Fig. 3. Product gas mixture options from gasification of coal; the three cases are discussed in the text.

in 1983, acetic anhydride are the only oxygenates which will commercially be derived exclusively from synthesis gas.

The SASOL synthetic fuel complex produces a mixture of oxygenates as by-products in their coal-to-gasoline process. The quantities of individual oxygenates are small and are obtained only after difficult separations. This general kind of chemistry, the conversion of synthesis gas into hydrocarbons, is referred to as Fischer-Tropsch chemistry in honor of the pioneering work in the field by Franz Fischer and Hans Tropsch. The process utilizes an iron-potassium oxide catalyst system and produces a mixture of products containing 35% gasoline and diesel oil, 10% synthetic natural gas, 33% potential petrochemical feedstock, and only 8.6% oxygenates. The oxygenate fraction consists primarily of alcohols but contains small quantities of aliphatic aldehydes, ketones, and acids.

Ethylene substitutes. As the price of petroleum rises faster than that of other natural resources, naphtha-based ethylene will also become increasingly expensive. Already the production of acetic acid based on synthesis gas is reducing the demand for acetaldehyde. The Eastman acetic anhydride process via methyl acetate carbonylation will reduce the acetaldehyde demand still further. The other ethylene derivative facing near-term competitive pressure is industrial ethanol. The installation of modern fermentation processes, initially for gasohol production, may allow fermentation producers to serve a substantial portion of the industrial ethanol market by 1985. Laboratory conversions of synthesis gas directly to both ethylene glycol and oxo alcohols have been demonstrated, and separate process development programs are under way. By the late 1980s these potential processes should be commercialized, and their economic superiority will decrease the demand for ethylene oxide and oxo aldehydes from ethylene. An unconventional approach to the production of ethylene itself appears unlikely before the 1990s. The leading candidate at this time is the dehydration of sugar-derived ethanol, already being practiced in Brazil. By the 1990s economical synthesis gas-based processes for ethanol could be developed. Two additional pathways under study are the combined homologation and dehydration of methanol to olefins over a zeolite catalyst and the direct conversion of synthesis gas to olefins. Plasma-arc chemistry is being studied to produce acetylene directly from coal at costs which could compete with petroleum-based ethylene for derivative manufacture.

Methanol production. The production of methanol from CO and H_2, reaction (2), is a well-devel-

$$CO + 2H_2 \longrightarrow CH_3OH \qquad (2)$$

oped technology, dating back to the 1920s, that utilizes a heterogeneous copper metal–zinc oxide catalyst. In the 1960s both Lurgi and ICI announced modifications of these catalysts that allow for their operation in the vapor phase at temperatures of 480–520°F (250–270°C) and 50–100 atm (750–1500 psi or 5–10 MPa) pressure. With proper removal of sulfur-containing impurities in the feed gas stream, methanol is easily produced from coal-derived synthesis gas.

Long-chain alcohol production. The longer-chain alcohols (ethanol, propanols, and butanols) were also first made from synthesis gas in the late 1920s. However, severe reaction conditions, low alcohol yields, and low reaction rates discouraged commercial development of these early heterogeneous systems. In 1978 the French Petroleum Institute (IFP) discovered a catalyst system that is reported to convert synthesis gas into the linear alcohols methanol, ethanol, *n*-propanol, and *n*-butanol, reaction (3),

$$CO + H_2 \rightarrow CH_3OH + C_2H_5OH + n\text{-}C_3H_7OH + n\text{-}C_4H_9OH \qquad (3)$$

in high yield under reaction conditions similar to those employed by Lurgi and ICI in their methanol processes. The catalyst appears to be a mixed copper–cobalt oxide containing small quantities of chromium and potassium.

The synthesis of ethanol by homogeneous cobalt catalysis via reductive carbonylation (homologation) of methanol, reaction (4), was first reported in 1951

$$CH_3OH + CO + 2H_2 \rightarrow C_2H_5OH + H_2O \qquad (4)$$

by the U.S. Bureau of Mines. Since then, improvements in the catalytic activity of cobalt have been made and involve the addition of iodides, phosphines, or a second metal such as ruthenium to the catalyst system.

Ethylene glycol production. The synthesis of ethylene glycol from synthesis gas was reported in 1953. These first systems employed homogeneous cobalt catalysts and operated at 480°F (250°C) and 1400 atm (21,000 psi or 140 MPa) pressure. In 1974 Union Carbide reported a homogeneous rhodium catalyst system that converted synthesis gas into a mixture of alcohols which consisted principally of methanol, ethylene glycol, and glycerol [see reaction (5)]. This system also operates under extreme

$$CO + H_2 \rightarrow CH_3OH + HOCH_2CH_2OH + \text{trace } HOCH_2CH_2(OH)CH_2OH \qquad (5)$$

conditions—430–500°F (220–260°C) and 550–1600 atm (8000–24,000 psi or 55–160 MPa)—but apparently produces very little hydrocarbon by-product. The patent literature suggests that this catalyst system can be moderated to favor either methanol or ethylene glycol production.

The utilization of formaldehyde, available by the catalytic oxidation of methanol, to prepare an ethylene glycol intermediate has also appeared. In 1974 Ajinomoto found that synthesis gas would react with formaldehyde in the presence of a homogeneous cobalt catalyst to form glycolaldehyde. In 1980 both National Distillers and Monsanto reported similar chemistry utilizing a variety of homogeneous rhodium catalyst systems. In the rhodium-catalyzed systems, glycolaldehyde is not isolated but converted to ethylene glycol, reaction (6).

$$H_2CO + CO + H_2 \rightarrow HOCH_2CHO \xrightarrow{H_2} HOCH_2CH_2(OH)CH_2OH \quad (6)$$

Acetic acid production. The synthesis of acetic acid by methanol carbonylation, reaction (7), has been demonstrated with cobalt, nickel, rhodium, and iridium and is practiced commercially with cobalt (BASF) and rhodium (Monsanto). The process requires no hydrogen after the methanol is available. The mechanism of the rhodium-catalyzed reaction has been studied extensively by Monsanto and is shown in Figure 4. The system operates with both acid (HI) and organometallic (rhodium) cycles. The reaction is first order in rhodium and methyl iodide, and the addition of methyl iodide to the rhodium iodocarbonyl (Fig. 4*a*) is the rate-determining step. It is not known whether the rhodium acetyl (4*b*) reacts directly with water or eliminates acetyl iodide (4*c*), as shown, which in turn reacts with water to produce acetic acid (4*d*). In either case, HI is the by-product and reacts with methanol to produce methyl iodide and water, and the catalytic cycle begins again. The net reaction is methanol plus carbon monoxide combining to produce acetic acid, as shown in reaction (7).

$$CH_3OH + CO \rightarrow CH_3CO_2H \quad (7)$$

Acetic anhydride production. The carbonylation of methyl acetate, reaction (8), which is obtained

$$CH_3CO_2CH_3 + CO \rightarrow (CH_3CO)_2O \quad (8)$$

by the methanol esterification of acetic acid, to produce acetic anhydride has been reported, with palladium, cobalt, nickel, and rhodium as catalysts and reaction conditions similar to those employed in the acetic acid process. In 1980 Eastman announced plans to construct a complex that would produce acetic anhydride by methyl acetate carbonylation and utilize synthesis gas derived from entrained flow coal gasification as the feedstock.

Production of other oxygenates. Accounts of the synthesis of a number of other commercially valuable oxygenates and all obtainable ultimately from synthesis gas have appeared in recent literature without much detail. Halcon has reported that homogeneous catalysts of palladium, rhodium, or iridium can be utilized to convert methyl acetate and synthesis gas into ethylidene diacetate, a known precursor to vinyl acetate, reaction (9). Imhausen

$$2CH_3CO_2CH_3 + 2CO + H_2 \rightarrow CH_3CH(O_2CCH_3)_2 + CH_3CO_2H \quad (9)$$

Fabrik and Braca independently reported that methyl acetate can be reacted with synthesis gas in the presence of a homogeneous ruthenium catalyst system to produce ethyl acetate, reaction (10). Using a strikingly similar ruthenium catalyst system, Texaco found that acetic acid can be reacted with

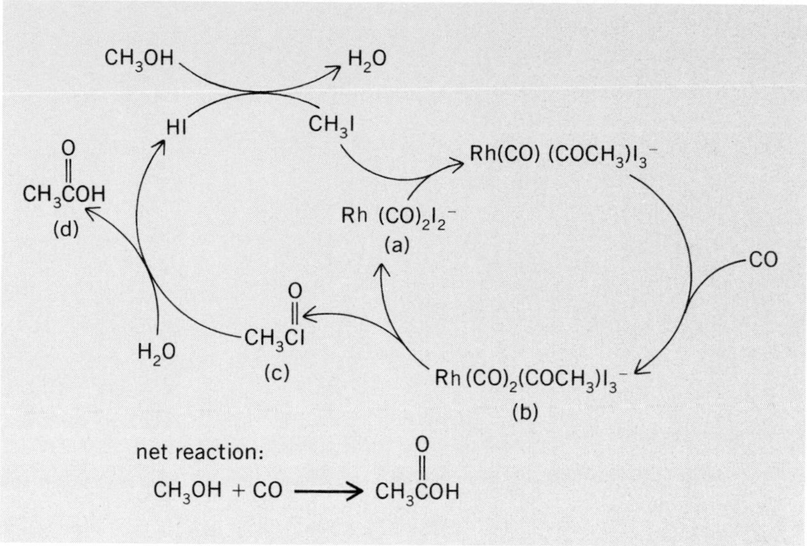

Fig. 4. Rhodium-catalyzed methanol carbonylation. The catalytic cycle for the rhodium catalyzed synthesis of acetic acid from methanol is shown. Parts *a* –*d* explained in text.

synthesis gas to produce propionic and butyric acids, reaction (11).

$$CH_3CO_2CH_3 + CO + H_2 \rightarrow C_2H_5CO_2CH_3 \quad (10)$$

$$CH_3CO_2H + CO + H_2 \rightarrow C_2H_5CO_2H + C_3H_7CO_2H + H_2O \quad (11)$$

THE FUTURE

The future looks bright for the development of chemistries and chemical processes that will utilize coal gasification as an economically attractive source of synthesis gas for direct conversion into the oxygenated compounds needed by the chemical industry. The development of small-molecule chemistry based on synthesis gas will require substantial commitments of financial and human resources to achieve the desired results.

Indeed, the most successful chemical companies of the 1980s may well be those that develop the technologies to permit the shift of their operations from petroleum-related materials toward alternative feedstocks, in particular, coal. The combination of lower-cost raw materials and new technologies will lead to new business opportunities for the chemical industry. [DONALD M. POND; J. E. WOLF]

Bibliography: J. Haggin, *Chem. Eng. News*, 59(8):39, 1981; *Handbook of Gasifiers and Gas Treatment Systems*, Dravo Corp., 1976; L. F. Hatch and S. Matar, *Hydrocarbon Processing*, parts 1–14: *From Hydrocarbons to Petrochemicals*, 1977–1979; J. E. Johnson, *Hydrocarbon Process.*, 57(7):107, 1978; J. L. Johnson, *Kinetics of Coal Gasification*, 1979; H. H. King, R. Williams, and C. A. Stokes, *Hydrocarbon Process.*, 57(6):141, 1978; New opportunities for methanol as ethylene technology advances, *Eur. Chem. News*, p. 22, Nov. 10, 1980; R. L. Pruett, *Science*, 211:11, 1981; E. Supp, *Hydrocarbon Process.*, 61(3):71, 1981; C. Y. Wen and E. S. Lee, *Coal Conversion Technology*, 1979.

DDT
E.T. Steadman

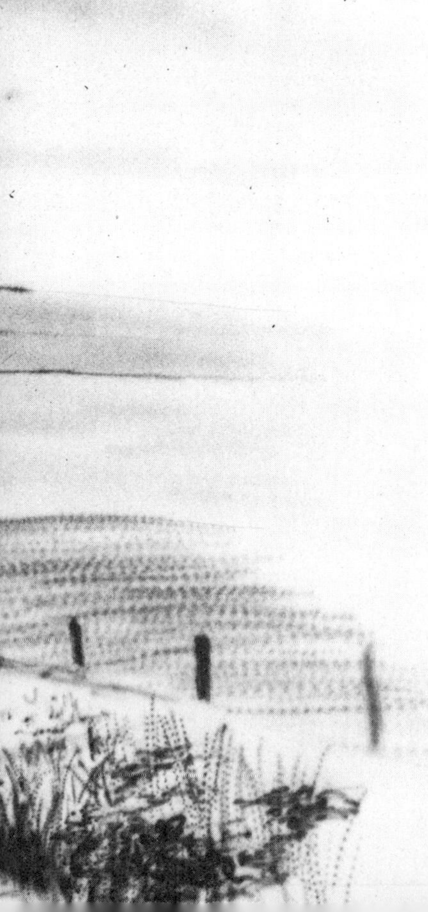

Organic Farming

Organic farming practices are based on the concept that each part of the farm operation augments the other parts to form a highly complex but efficient and sustainable food production system. Organic farming emphasizes plant nutrient cycling and the use of biological techniques for crop protection as means to achieve a greater degree of self-sufficiency in the farm operation. Certain goals, such as soil and water conservation, energy efficiency, pest control, and nutrient management, are achieved through integration of all the components of the farm management system. Thus, a legume may be grown in the rotation not only to produce forage for livestock but also to control certain pests, reduce soil erosion, improve soil structure, and conserve energy by reducing the need for nitrogen fertilizers through biological nitrogen fixation. Most organic farmers feel there is a strong relationship between the health of their animals and that of the soil and the crops. They also feel that food safety and quality and human health and nutrition depend upon that relationship.

CONCEPT AND CHARACTERIZATION

The concept of organic farming is not new. Many of its principles are practiced in various agricultural systems throughout the world. Today, in the United States and other developed countries, organic farming constitutes only a small part of the food production system compared with "conventional" agriculture, which is

Robert I. Papendick is research leader of the Land Management and Water Conservation Unit, U.S. Department of Agriculture, Agriculture Research Service, Pullman, WA.

James F. Parr is chief and supervisory microbiologist of the Biological Waste Management and Organic Resources Laboratory at the U.S. Department of Agriculture's Research Center, Beltsville, MD.

Both served on the USDA study teams which compiled the reports "Improving Soils with Organic Wastes" (1978) and "Report and Recommendations on Organic Farming" (1980).

characterized by intensive use of chemical fertilizers and pesticides with major emphasis on the production of cash grain crops. Some view contemporary organic agriculture as part of a larger social movement called alternative agriculture in which production practices and long-term goals sometimes differ from conventional agriculture with regard to the use of chemical technology, farm size, environmental concerns, and resource use.

A recent U.S. Department of Agriculture study of organic farming, which included case studies of 69 organic farms in 23 states, found that most organic farmers are guided in their practices by certain basic values and beliefs, sometimes referred to as the organic ethic. There have been a number of organic farming advocates over the years, each expressing a particular philosophical and ideological viewpoint. The principal tenets of the organic ethic are:

1. *Nature is capital*. Energy-intensive modes of conventional agriculture place people on a collision course with nature. Present trends and practices signal difficult times ahead. More concern over finite nutrient resources is needed. Organic farming emphasizes recycling of nutrients.

2. *Soil is the source of life*. Soil quality and balance are essential to the long-term future of agriculture. Human and animal health is directly related to the health of the soil.

3. *Feed the soil, not the plant*. Healthy plants, animals, and humans result from balanced, biologically active soil.

4. *Diversify production systems*. Overspecialization (monoculture) is biologically and environmentally unstable.

5. *Independence*. Organic farming contributes to personal and community independence by reducing dependence on energy-intensive agricultural production and distribution systems.

6. *Antimaterialism*. Finite resources and nature's limitations must be recognized.

Despite these principles, organic farmers nevertheless employ a wide range of technological and cultural practices. Therefore, it is difficult if not impossible to provide a precise and universally acceptable definition of organic farming. According to the USDA report, the principles and practices of organic farmers can be envisioned as a spectrum ranging from the "purist" philosophy on one extreme, to a more liberal interpretation of the organic ethic on the other to where it finally begins to merge with conventional agriculture. The most apparent difference between organic farming and the widely practiced conventional agriculture is that organic farmers avoid or restrict the use of chemical fertilizers and pesticides in their farming operations in contrast to the intensive use by conventional farmers. Except for this and the greater use of meadow crops in rotation, organic farming has much in common with conventional agriculture. Most of today's organic farmers use modern machinery, recommended crop varieties, certified seed, sound livestock management, recommended soil and water conservation practices, and innovative methods of organic waste recycling and residue management.

The definition of organic farming given in the USDA report, which encompasses the entire spectrum of organic agriculture, seems appropriate for the discussion which follows. Thus, organic farming is defined as ". . . a production system which avoids or largely excludes the use of synthetically compounded fertilizers, pesticides, growth regulators, and livestock feed additives. To the maximum extent feasible, organic farming systems rely upon crop rotations, crop residues, animal manures, legumes, green manures, off-farm organic wastes, mechanical cultivation, mineral-bearing rocks, and aspects of biological pest control to maintain soil productivity and tilth, to supply plant nutrients, and to control insects, weeds, and other pests."

Recent upsurge of interest. During the past 5 years the interest in organic farming in the United States has increased markedly. In 1979 the USDA conducted a comprehensive study to learn more about the potential of organic farming as a system for the production of food and fiber. Results of this study were published in July 1980 and, following the recommendations of the report, the Department established a permanent position for an organic resources coordinator. This person is responsible for developing liaison between organic farmers, producer associations, and the USDA. Almost concurrently with the USDA study, a 24-member team of scientists, under the auspices of the Council for Agricultural Science and Technology (CAST), was preparing another report, which was issued in October 1980.

The two reports differ widely in their findings and conclusions, and final judgment is left to the reader. However, the CAST report concludes that the organic farming movement has expanded to encompass an alternative agricultural system based on less technology, more self-reliance, and opposition to the trend toward larger farm enterprises, displacement of small farmers, and deterioration of some rural institutions. Furthermore, the CAST report concludes that it is unrealistic to consider organic farming as a significant agricultural production system but that organic gardening is considerably more attractive and feasible than is organic farming, and that most of the agricultural research done in "earlier years" is currently applicable to organic farming.

On the other hand, the USDA report concludes that more attention should be given to organic farming through research and education programs designed specifically to meet the needs and problems of organic growers. It also concludes that organic farmers have developed some unique, productive, and energy-conserving systems of farming which emphasize organic recycling, and the restricted use of chemical fertilizers and pesticides; and that much of the research relating to present-day organic farming is fragmentary. The report recommends a holis-

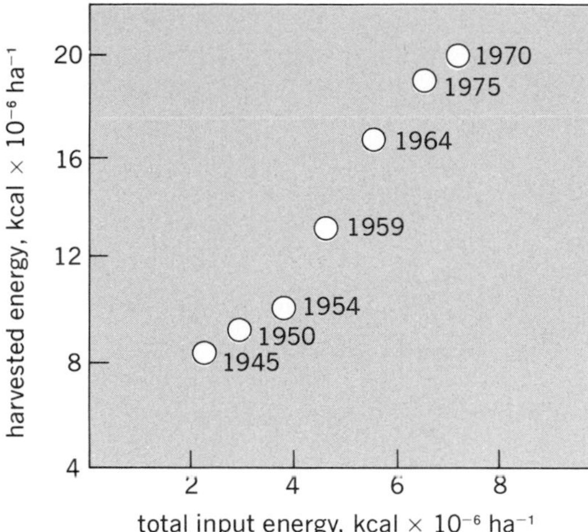

Fig. 1. Energy harvested as grain plotted against the total energy input for maize production in the United States using conventional farming methods. The energy inputs include those used for fertilizers, fuel, irrigation, machinery, seeds, drying, electricity, pesticides, transportation, and labor. The trend has been for increasing energy intensiveness, with the largest increases in energy inputs coming from increased use of nitrogen fertilizer.

tic investigation of the organic system of farming, its mechanisms, interactions, principles, and potential benefits to agriculture both at home and abroad.

Reasons for increased interest. While the large-scale, chemical-intensive practices now utilized in the United States food system have greatly increased production and labor efficiency, there is increasing concern about the heavy dependence of conventional agriculture on fossil fuel energy (Fig.1), and the adverse effects on soil productivity and environmental quality from increased soil erosion and excessive use of agricultural chemicals (Figs. 2 and 3). Neither fossil fuel nor soil is a renewable resource and, with present farming practices, both are being depleted at alarming rates which threaten the entire food system. In the Corn Belt states where much of the United States' row crops are grown, there is an annual soil loss of 8.1 tons per acre (18 metric tons per hectare) which, according to expert opinion, is twice the maximum tolerable rate for many soils that will permit sustained crop productivity. According to the USDA report, the most frequently expressed concerns about present agricultural production practices are: increased cost and uncertain availability of energy and chemicals used intensively in agriculture today; increased resistance of weeds and insects to pesticides; decline in soil productivity from erosion and accompanying loss of organic matter and plant nutrients; pollution of surface waters with agricultural chemicals and sediment; destruction of wildlife, bees, and beneficial insects by pesticides; hazards to human and animal health from pesticides and feed additives; detrimental effects of agricultural chemicals on food quality; depletion of finite reserves of concentrated plant nutrients, for example, phosphate rock; decrease in numbers of farms, particularly family-type farms, and disappearance of localized and direct marketing systems.

The relationship between many of these concerns and some of the principal tenets of organic agriculture is apparent, and is the main reason for the increased interest in organic farming.

Organic farming methods are considered to be less vulnerable to the effects of fossil fuel shortages and costs because these methods are less energy-intensive than current conventional farming systems. They reverse the trend toward farming systems that continue to depend on increased amounts of energy inputs from external sources. Moreover, organic practices would potentially have a smaller adverse environmental impact because most are far less chemical-intensive than those used in conventional farming. Organic farmers appear to be seeking the kind of changes that many scientists, economists, and environmentalists have recommended for improving the United States agricultural production system. For example, they place great emphasis on recycling of plant nutrients, protection of the environment, preservation of the soil, diversification of production systems, and achievement of self-sufficiency. Because soil resources are essentially nonrenewable, the control and prevention of soil erosion and losses of organic matter from land, including plant nutrients, are of paramount importance in farming operations.

Current status. There is only limited information on the number of organic farms, geographical distribution, farm size, and other statistics. Based on the USDA report, there are probably more than 20,000 organic farms in the United States, and these are distributed in most of the important agricultural areas with the exception of the South. Pest, soil, and climatic factors may limit organic farming more in the South than in other regions. Organic farming does not appear to be limited by scale. The

Fig. 2. Soil erosion in a field (background) in eastern Washington that lost 100 tons of topsoil per acre (224 metric tons per hectare) from erosion in one winter season. The area in the foreground which was protected by sod cover had no measurable soil loss.

Fig. 3. Sediment-laden water from runoff is the main source of pollution of surface waters draining agricultural lands, and each year it carries millions of tons of plant nutrients from United States croplands.

USDA case studies included farms ranging from 25 to 1500 acres (10 to 600 hectares) and, for the most part, the size of the organic farms studied was about the same as their conventional farm neighbors. A high percentage of the organic farmers owned either all or most of the land they farmed.

Organic farmers appear to be evenly distributed in all age groups, and many of those in the USDA case studies had gained chemical farming experience before converting to organic methods. As a group, these farmers were well educated; over 50% had attended college and several had achieved advanced degrees. The farmers were motivated toward the use of organic methods because of concern for protecting soil, human, and animal health from adverse effects of pesticides; lower production inputs; environmental protection; and protection of soil resources from erosion. Successful organic farmers were found to be good managers and dedicated to responsible husbandry of their soil, crops, and livestock.

PRODUCTION PRACTICES

Since organic farmers avoid or restrict the use of chemical fertilizers, pesticides, and additives in animal feeds, they must in some situations compensate for the nonuse of these inputs by altering crop and livestock production and protection practices.

Cropping systems and crop-livestock mix. A legume-based rotation with green manure or cover crops is an integral part of many organic farming systems. Hay crops often comprise 30–50% of the total cropland. Legumes may be used less on vegetable farms that have access to animal manures and in the lower-rainfall areas. The usual practice is to follow a heavy green manure crop with a high-nitrogen-demanding crop (corn, wheat, sorghum). For example, in the Corn Belt, a typical rotation on organic farms would be 3 years of alfalfa, 1 year of corn (or wheat), 1 year of soybeans, 1 year of corn, 1 year of soybeans, and then back to alfalfa. Monoculture cropping such as continuous corn or rotating corn and soybeans for extended periods is generally avoided. W. Lockeretz found that in the Corn Belt, where the leading crops, in order of descending importance, were corn, soybeans, hay, oats, and wheat on conventional farms, the order of hay and soybeans was reversed on organic farms. The forage produced on organic farms is usually fed to animals and thus encourages a mixed crop and livestock operation. Large-scale cash grain enterprises are relatively rare in organic farming operations.

Tillage and planting. Most organic farmers avoid the moldboard plow for primary tillage and favor disk or chisel-type implements which tend to mix rather than invert the soil. They also practice shallow tillage, often no deeper than 6–10 cm, to incorporate and concentrate organic materials such as crop residues and manures in the surface soil layers. These farmers reason that deep moldboard plowing disrupts the established and active microflora near the surface and places the organic materials at greater depths, where conditions are less favorable for decomposition and release of plant nutrients. With shallow tillage, there is greater accumulation of organic matter near the soil surface, which promotes water infiltration and erosion control. Successful farmers emphasize the importance of timely tillage and planting operations for weed control and maintenance of good soil tilth. Delayed planting is another technique sometimes used to achieve adequate weed control and to increase mineralization of organic matter and release of plant nutrients.

Nutrient and organic matter management. A major concern of many organic farmers is to obtain adequate nitrogen for crops and to maintain soil organic matter at a level that will maximize soil productivity. The principal nitrogen input in organic farming systems is from the fixation of atmospheric nitrogen by bacteria in symbiotic association with roots of legumes. In some cases, off-farm sources of manures or other organic wastes are imported. Occasionally, but very sparingly and selectively, the farmer may apply inorganic nitrogen fertilizers as a supplement for high-nitrogen-use crops. Any nitrogen deficit to crops can be decreased further by residual inorganic soil nitrogen, recycling animal manures and crop residues, and mineralization of soil organic matter. Phosphorus and potassium are usually supplied by materials of low water solubility such as rock phosphate or greensand (glauconite), respectively. Acidulated phosphate (processed forms) sources are used in some situations where either rock phosphate is not available or there is no response to rock phosphate. However, the USDA study showed that only a relatively small percentage of organic farmers were applying these unprocessed phosphate and potassium minerals to the soil. It was apparent that most of the crop requirement for these nutrients was being obtained from the breakdown and dissolution of soil minerals.

Organic farmers generally avoid repeated use of high-analysis fertilizers such as anhydrous ammonia and concentrated forms of phosphate and potassium. In some cases, less concentrated synthetic fertilizers are used (for example, ammonium sulfate), but at low application rates. Many organic farmers believe that repeated applications of concentrated chemical fertilizers can adversely affect soil microorganisms and their activities, which may lead to nutrient imbalances, reduced earthworm activity, increased soil compaction, and poor tilth.

According to the USDA study, a considerable number of organic farmers apply seaweed and fish emulsion products foliarly to a wide variety of field and vegetable crops. They believe that these products provide essential elements for plant growth and protection, and that they benefit crop yields and quality. Also, many farmers were found to have experimented with, or used, various commercial products marketed as soil and plant additives; however, this was also true of many conventional farmers.

Most organic farmers feel that the soil organic matter content is highly correlated with soil productivity. Thus, in addition to the incorporation of crop residues, they make frequent applications of animal manures, and use green manures and cover crops to maintain soil organic matter at maximum levels. Manure is sometimes composted either in windrows and aerated periodically with a turning machine, or simply in static unaerated piles. While the return of crop residues to the soil is a common practice on most organic farms, some farmers actually move residues from one part of the farm to another to increase the organic matter level where needed.

Weed and insect control. Adequate weed and insect control is often a major problem on organic farms. Growers usually must rely on a combination of control methods to reduce pest infestations to tolerable levels. Nonchemical weed control methods most commonly used by organic growers include mechanical cultivation, crop rotation, and adjusting the planting date. Biological control methods such as weed pathogens, crop competition, insects, nematodes, and animal grazing are used to a far lesser extent. Organic farmers in the Corn Belt use more mechanical cultivation to control weeds than did conventional farmers who used herbicides. Organic farmers also emphasize preventive methods (such as preventing weeds from becoming established and producing seeds) and weed sanitation (such as keeping roadsides and fence lines clean). Some organic farmers use herbicides selectively and sparingly, for example, to control localized weed patches and to occasionally support cultural and mechanical practices.

Insect control is achieved primarily through crop rotations and natural insect predators. Farmers generally experience greater difficulty in controlling insects in vegetable crops and orchards than in field crops. Organic insecticides are used to a limited extent, but many farmers find them costly and ineffective for controlling epidemic infestations.

Marketing. Based on the USDA study, most organic produce is sold through conventional marketing channels. Less than 30% of the organic farmers in the USDA case studies marketed most of their farm products as organic produce. The produce sold as "organic" is marketed in several ways. For example, small farms often sell produce directly to consumers from roadside stands, through pick-your-own or prepick operations, or through farmers' markets. Larger operations frequently market more produce to local organic food cooperatives, organic wholesalers, and organic retailers such as natural or health food stores. The primary factors that determine whether or not an organic farmer sells products as "organic" are distance to market, the likelihood of a premium price, and the difference between the proposed premium price and the conventional market price. Organic marketing channels are often not well established, and if the distance to market is great the cost of transportation may well offset any advantages of a premium price for the product. A relatively small percentage of the organic farmers in the USDA case studies received a premium price for their products, which was found to range 10–50% above the conventional market price. Organically grown meat and vegetables were most likely to bring a premium price and a higher one than organic fruit, grain, cereal, or dairy products.

FARM PRODUCTIVITY

Well-managed organic farming systems can be highly productive and efficient in the use of labor and energy. However, as is true of conventional farms, the productivity of the organic farms can vary considerably due to differences in individual management techniques and managerial skill of the farm operator.

Crop yields. It is difficult to compare crop yields of organic farms with those of conventional farms because of the differences in individual management practices. The level of production inputs and subsequent crop yields on organic farms can be equal to or greater than those for conventional farms, even though the exact nature of the inputs may differ considerably. For example, an organic farming system which utilizes off-farm sources of manure or other organic wastes (in addition to on-farm sources) to satisfy crop nutrient requirements, and which successfully controls weeds and pests, should obtain crop yields equal to those of comparable well-managed conventional farms. On the other hand, where production inputs are limited to the recycling of only on-farm sources of organic wastes, plant nutrient deficiencies could occur, resulting in lower yields of certain crops than on some conventional farms. Differences in managerial ability of individual operators can also contribute to large crop yield differences between organic and conventional farms and, for that matter, within either farming system.

Some studies have compared crop yields from organic farms with either selected conventional farms or average county yields. Most of these have been short-term studies, 3 to 5 years at most, and are based on survey-type information or harvested samples from representative fields. In a study of Corn Belt crop yields, it was concluded that yields for organic farms were much lower for wheat, moderately lower for corn, slightly lower for soybeans, and about equal for oats and hay, as compared with paired conventional farms. The margin by which crop yields were lower on organic farms was directly related to the extent of chemical fertilizer and pesticide use in the production of each crop—for example, highest with corn, less with soybeans, and negligible for hay and oats. Corn and soybean yields were about 10 and 5% less on organic farms than on the paired conventional farms. With adequate moisture conditions during the growing season, differences in corn yields were considerably greater in favor of the conventional farms, but under drier conditions the organic farmers did as well or even better than the conventional farmers.

The USDA study concluded that differences in crop yields from comparison of organic and conventional farms are somewhat inconsistent, and there is a lack of reliable research information from well-designed, replicated, field plot experiments designed specifically for comparing crop yields from organic farming and conventional systems.

Labor intensiveness. Many organic farms are highly mechanized and use only slightly more labor than conventional farms. One study in the Corn Belt showed that labor requirements averaged 3.3 hours per acre (8.2 hours per hectare) on organic farms and 3.2 hours per acre (7.9 hours per hectare) on conventional farms. However, when based on the value of the crop produced, labor use was 11% higher on the organic farms because the crop output value was lower. The labor requirements of Corn Belt organic farmers were similar to conventional farmers for corn and small grains, but higher for soybeans because of more hand weeding. The study concluded that differences in labor inputs between the two farming systems were mainly due to differences in crop mix and cultivation rather than to fundamental differences in production methods and machinery.

Other studies cited in the USDA report indicate that organic farms generally require more labor for their operations than do conventional farms. The labor requirements on organic farms depend largely on how well weeds and crop pests are controlled with mechanical or nonmechanical methods. Where hand labor is employed for this purpose, the labor inputs may rise dramatically and can be a major deterrent to the adoption of organic methods. Many organic farmers strive to avoid increased use of labor for the same reasons as conventional farmers, that is, because of high labor costs and the uncertainty of labor supply.

Energy production. The key parameters for evaluating energy production in agricultural systems are total energy output expressed as the energy output per unit of energy input. In other words, a production system with high energy efficiency but with low output, or one with high output but with low efficiency, would not be considered as desirable or, in some cases, sustainable. The agricultural production goal in terms of energy is to achieve both high output and high efficiency simultaneously.

Conventional farms consume considerably more energy than organic farms largely because they use greater amounts of petrochemicals. These products, chiefly fertilizers and pesticides, account for about 35% of the energy used in United States agricultural production. Avoidance of pesticides would not result in a significant reduction in energy inputs because relatively little energy is used in their production and field application rates are usually very low, often in the range 0.3–0.5 kg/ha. Nitrogen fertilizer accounts for the bulk of energy inputs from petrochemicals. Thus, the organic farmers' practice of using biologically fixed nitrogen and recycling organic wastes to reduce the need for chemical fertilizers offers an opportunity for significantly reducing energy inputs in agricultural production. Part of the increased savings from reduced use of fertilizers on organic farms may, however, be offset by their increased use of fuel and machinery for cultivation and the application of manure or other organic wastes which serve as plant nutrient sources.

Data collected in studying energy input-output relationships on paired organic and conventional farms in the Corn Belt showed that the organic farms were considerably more energy-efficient than conventional farms, but that their energy production might be lower. They found that the energy efficiency for corn production on the organic farms for 1974 and 1975 was 13 and 20, respectively, while for conventional farms it was 5 and 7. The total energy output based on grain yield was approximately the same for both types of farms (slightly in favor of the organic farms) in 1974, a relatively dry year, but was about 27% higher for the conventional farms in 1975, a wetter year and more favorable for corn production. Similarly, a study of wheat production in the Northeast showed that organic farmers used less energy per unit of grain yield than did conventional farmers, but their yields and hence energy production were about 22% less than that of the conventional farmers.

Comparisons of energy relationships between organic and conventional farms cannot always be properly assessed by comparing individual crops because the crop mix on the two types of farms may differ considerably. Differences in crop yield, the percentage of a crop grown per unit of cropland area, and energy value of the harvested crop must be considered in evaluating the energy input-output relationships of the two farming systems. Such an analysis was conducted, taking into account differences in the crop rotation, yields, and feed energy values of the harvested crops for the two farming systems. Again, the analysis revealed that the organic farms were about 2.5 times more energy-efficient than the conventional farms; that is, they produced about 9.4 units of feed energy per unit of energy input compared with 3.8 units of feed energy per unit of energy input for the conventional farms. However, the farm energy output per unit cropland area for conventional farms was approximately 7% higher than for the organic farms.

Organic farmers may also achieve some energy savings by applying animal manures instead of chemical fertilizers. The net reduction from using manures to replace chemical fertilizers ranges from 15 to 25% when the manure source is reasonably close (about 5 km or less) to where it will be applied.

Thus, it appears that organic farms use less total energy, achieve a greater energy efficiency, but may produce less total energy output per unit area of cropland than do conventional farms. Energy savings are achieved mainly through the limited use or nonuse of fertilizers, mainly nitrogen, but this is partly offset by increased use of fuel and machinery for extra cultivations, and for application of manures or other organic wastes.

CONSERVATION OF RESOURCES AND ENVIRONMENTAL PROTECTION

Organic farmers place much emphasis on the use of tillage and cropping practices that conserve soil and improve environmental quality. Pollution from sediment or nutrient losses is minimized by use of organic nutrient sources, inorganic nutrient sources of limited solubility, and practices that control runoff, leaching, and erosion. The plant nutrients of greatest concern with respect to water pollution are nitrogen and phosphorus.

Soil erosion control. Many practices commonly used by organic farmers are highly effective for controlling soil erosion and maintaining soil productivity. These include the use of meadow and small grain crops which decrease the percentage of row crops in the rotation (Table 1), use of cover and green manure crops, and tillage methods that conserve and maintain crop residues near the soil surface. Organic farming also emphasizes the application of animal manures and other organic wastes to maintain or increase the soil organic matter content, which in turn increases water infiltration and storage in soils, and stabilizes the soil against water runoff. Some assessment of long-term benefits of organic farming practices for erosion control can be made with use of the universal soil loss equation.

Crop rotations. Most organic farmers maintain 30–50% of their cropland acreage in meadow (sod-based crops), whereas conventional farmers on large areas grow mainly row crops, often alternating cash grain crops such as corn and soybeans. The soil erosion potential is greatly increased with increased in-

Table 1. Effect of crop rotation on soil erosion

Rotation*	Relative soil loss†
C	0.35
C–B	0.43
C–C–C–W–M–M	0.14
C–W–M	0.055
M	0.044

*C = maize; B = soybeans; W = wheat; M = meadow.

†Average soil loss relative to the loss from clean fallow which has a value of 1. The use of meadow crops (alfalfa, hay, and so on) common in organic farming systems reduces average annual erosion to a third or less, depending on the frequency of meadow in the rotation compared with continuous row cropping of maize or maize-bean rotations.

tensity of row cropping. However, with sod crops such as grasses and legumes the average annual soil loss is only one-third to one-eighth of that occurring with conventional tillage and continuous row cropping. Cover crops may also decrease erosion by as much as 50%, especially when they follow row crops that leave little residue on the surface.

Tillage. Organic farmers generally favor the use of tillage methods that conserve crop residues and other organic materials upon or near the soil surface. This is best achieved with disk- or chisel-type implements and can reduce soil erosion losses 20–75%, compared with intensive moldboard plow–based tillage. However, many conventional farmers now employ minimum tillage and no-till planting systems which are highly effective in controlling soil erosion. These conservation tillage systems are generally not a viable option for organic farming because they often require extensive use of pesticides to control weeds and insects. No-till continuous corn will reduce erosion to levels below that of meadow rotations with conventional tillage.

Soil organic matter. A high level of soil organic matter is often well correlated with a high level of soil fertility and improved physical conditions for water infiltration and tillage. Long-term experiments in both the United States and England have shown that soil organic matter losses are greatest from intensive row cropping and cultivation, and are maintained at higher levels with grass and legumes in the rotation as practiced by many organic farmers. Extensive loss of soil organic matter from intensive cropping and tillage practices generally leads to a concomitant deterioration of soil physical properties, decreased productivity, and accelerated erosion.

Organic farmers place great importance on maintaining soil organic matter at highest possible levels. They employ practices such as heavy application of animal manures, composted organic wastes, and use of grass and legumes in rotation, which can actually increase the soil organic matter content by as much as several percentage points over previous levels. For some soils a net increase of 1% in the soil organic matter content can reduce soil erosion loss by 10%. The increased soil organic matter content associated with some organic farming practices can greatly enhance soil and water conservation while maintaining or improving the soil's productivity, fertility, and tilth.

Chemical pollution abatement. Conventional farming practices such as intensive row cropping can greatly increase the potential hazard of environmental pollution from plant nutrients in water runoff or in leachate going to groundwater. Such losses can be substantially minimized when a portion of the crop's nutrient requirement is supplied by animal manures, composted organic wastes, and inorganic nutrient sources of limited solubility, and from implementation of best management practices that control runoff and soil erosion. The organic farmers' conservative use of chemical fertilizers and liberal use of organic nutrient sources minimize the pollution hazard that often exists under conditions of excessive fertilization. Moreover, practices such as the use of animal manures and other organic sources which release nutrients slowly, crop rotations, and green manure and cover crops which reduce the concentration of nitrates in the soil profile can significantly reduce nitrate-leaching losses.

Rotating crops that require low inputs of nitrogen fertilizer (soybeans or alfalfa) with crops requiring high nitrogen levels (corn or wheat), as practiced by many organic farmers, will reduce the amount of nitrate nitrogen available for leaching. Moreover, use of nonleguminous crops such as oats, timothy, and rye as cover crops can reduce nitrate leaching through plant uptake of nitrogen and by extracting soil water so that less is available for leaching nitrate from the root zone.

Soil erosion not only facilitates the transport of soil sediment in water, but is a process which selectively removes plant nutrients from agricultural land through runoff, and contributes greatly to the pollution of surface waters. It has been estimated that more than 50,000,000 tons (45,000,000 metric tons) of plant nutrients are lost annually from United States cropland through soil erosion. The cost of replacing these nutrients by chemical fertilizers today would probably range $20–25 billion. Thus, organic farming practices can effectively control soil erosion from cropland and drastically reduce plant nutrient losses as well.

Organic farming also reduces or eliminates the hazard of pesticide pollution because organic farmers discourage the use of these chemicals in their farming operations. Increased pesticide use during the past 2 decades has contributed to the pollution of surface water and groundwater from both runoff and aerial losses which may range from undetectable amounts up to 20% of the amount applied.

ECONOMIC EVALUATION

It is difficult to make meaningful comparisons of the economic performance of organic and conventional farming systems. The differences are often site-specific; that is, they may depend on such things as the type of operation, whether it is a cash grain, vegetable, or a livestock enterprise, managerial skills, soil and climatic conditions, and the nature and intensity of production inputs. However, a few studies (mainly of a survey type) have compared income, costs, and profitability of organic and conventional farms. These have shown that organic farming can be cost-efficient, and in some cases there may be little difference in production costs per unit output between the two farming systems. Where yields of certain crops are lower on organic compared with conventional farms, the net returns may be similar because of lower production costs for the organic system. In a study of selected Corn Belt farms, it was found that in years when organic farms produced between 6 and 13% less market value per unit of cropland than conventional farms, their operating costs were lower by about the same amount. Interestingly, the net income for the two farming systems differed by only 4%. However, in a year when the organic farms' gross income was 17% below that of the conventional system, the difference in operating costs was not sufficient to offset the difference in gross income.

The USDA report analyzed farm income above variable costs for several legume-based crop rotations used on organic farms in the Midwest and compared these with conventional farms rotating corn and soybeans. Three hypothetical cases were analyzed, including a 7-, 5- (each with 2 years of alfalfa), and 4- (1 year of alfalfa) year rotation on 320 acres (130 hectares) of cropland, compared with the same farm growing only corn and soybeans in rotation. Income above variable costs for conventional corn and soybean production, based on 1977 prices, was as much as 43% greater than on the organic rotation (5-year rotation, in this case). The analysis showed that the greater the substitution by other grain crops and alfalfa for corn and soybeans in the rotation, the lower the income from crop production. The analysis did not consider a combined crop-livestock enterprise. This analysis suggests that, based on crop production alone, organic farming does have an opportunity cost. However, this analysis does not take into account differences in social costs (such as soil erosion, environmental impact, and resource depletion) for the two types of farming systems.

The USDA report also assessed possible economic impacts of increased organic farming in the future, based on 1977 costs and prices. It was concluded that a large number of small farms could shift to organic farming methods with little impact on United States crop production levels and the economy. Furthermore, it concluded that a total shift of all farms to organic methods would have a major impact on crop production and food prices. For example, corn and soybean production would decrease, and prices for these crops would increase by 28 and 53%, respectively. Conversely, hay and oat production would increase, prices for these feeds would decline, animal numbers would increase, and livestock prices would decline.

An "either-or" analysis, such as this, probably has little merit because it is hardly realistic or practical. A total shift from conventional to organic methods would take years to accomplish even if there were strong incentives to do so. A significant observation in the USDA report was the number of organic farms in the United States that appeared to be operating profitably with net returns comparable to those of their conventional neighbors.

FUTURE OUTLOOK

The USDA report suggests that the future role and scope of organic farming in United States agriculture are uncertain. Much depends on the national goals and public policy pertaining to soil conservation, environmental protection, and the cost and availability of energy. Since organic farming is generally more conserving of fossil energy and petrochemicals than conventional farming, cost comparisons for crop production may improve in favor of organic methods as the cost of energy and other synthetic inputs rise or become increasingly unavailable. The time is not far away when many farmers will consider the high cost of fertilizers as prohibitive and consequently will reduce their inputs. They will have no choice but to adopt modified crop rotations with legume crops or organic recycling programs to make up the nitrogen deficit. Table 2 shows, for example, that the retail prices of the most commonly applied nitrogen fertilizers increased by over 200% during the 1970s. There are strong indications that prices will continue to skyrocket during the 1980s.

Organic farming in many ways is more conserving of natural resources and more protecting of the environment than is conventional farming. Increased public pressure over these concerns will undoubtedly generate increased interest in organic farming practices in United States agriculture.

Obstacles to organic farming. The USDA report listed a number of obstacles that may limit the development of successful and profitable organic farming systems and the adoption of organic farming practices. A discussion of some of the more important limitations follows.

Limitations of organic nutrient sources and biological nitrogen fixation. Organic nutrient sources are often in short supply, and large-scale substitution of nitrogen, phosphorus, and potassium fertilizers for these sources would be limited by both cost and

Table 2. Changes in prices of several nitrogen (N) fertilizers during the past decade

		Price*			
		1970		1981	
N fertilizer	N content, %	\$/ton	¢/lb N	\$/ton	¢/lb N
Anhydrous ammonia	82	75	4.5	247	15
Urea	46	82	9	245	27
Ammonium nitrate	34	60	9	192	28
Ammonium sulfate	21	52	12	155	37

*Average retail price of bulk fertilizer; does not include application costs.

availability. Biologically fixed nitrogen from legumes requires time and thus detracts from intensive production of higher-value crops. Moreover, legumes, which are often deep-rooted, may deplete subsoil water and in turn may reduce the yield of the following crop in low-rainfall areas.

Economic limitations. Greater diversification of crops and substitution of higher-income crops (corn and soybeans) with lower-income crops (oats and hay) would surely reduce the income of some organic farmers compared to conventional farmers who produce a greater proportion of high-value crops. Many organic farmers have reported difficulties in obtaining credit and financing because some loan officers question the viability of organic methods, especially for maximum production of cash grain crops. Organic farming in some situations may require more labor than conventional farming. Labor for large-scale operations is expensive, and often not available when needed. United States agriculture continues to strive toward reduced labor usage.

Problems of communication, understanding, and information. Lack of communication and understanding among organic farmers, agricultural scientists, and extension agents has hindered development of organic technology and transfer of research information applicable to organic farming systems. Some scientists, agricultural leaders, and extension workers for various reasons have developed negative attitudes toward organic farming or strongly believe that organic methods are neither practical nor feasible. These views are largely the result of a misunderstanding of the true character of contemporary organic farming and a lack of awareness of potential environmental- and conservation-related benefits associated with organic methods. On the other hand, organic philosophy and concepts are often not well articulated and sometimes ambiguous, which can lead to misunderstanding and rejection. For example, expressions such as "health of the soil" or the concept of "feeding the soil rather than the plant" are difficult to interpret in physically meaningful terms, and especially difficult for scientists to relate to soil physical, chemical, and biological processes.

Research and development. Many organic farmers feel strongly that there has been little research and development, past or present, to address their specific needs and problems. This consensus was reinforced by the USDA study which found that land-grant universities and state agricultural experiment stations have very few research and education programs directed exclusively to the needs and problems of organic farmers. However, a considerable number of their programs do relate to various aspects of organic farming. For example, nonchemical aspects of "integrated pest management" research, which strives for the best combination of practices to control crop pests, is of interest to most organic farmers.

The organic farming technology of today has been largely developed by the farmers themselves, and many continually experiment to improve their techniques and practices. Information exchange occurs largely through grower association–sponsored meetings and field trips, and in some cases newsletters or other published information. There are no known collections of published information on organic farming technology available for distribution to either organic or conventional farmers.

Research and education needs. Organic farmers interviewed in the USDA case studies expressed the need for specific research and education programs and for effective transfer of technology to users. Research on organic farming systems should be of benefit not only to organic farmers but to all of agriculture, and most certainly including conventional farmers. The most frequently requested areas of research include:

1. Investigate organic farming systems, using a holistic research approach to investigate the interrelationships of organic waste recycling, nutrient availability, crop protection, energy conservation, and environmental quality.

2. Determine factors responsible for low crop yields during the transition from conventional to organic farming.

3. Investigate the availability of phosphorus and potassium from low-solubility sources and their release from soil minerals in organic farming systems.

4. Expand research on biological nitrogen fixation.

5. Develop new and improved techniques for control of weeds, insects, and plant diseases using biological-nonchemical methods.

Since the USDA report was issued, questions have been raised by some scientists concerning the feasibility of minimum tillage and no-till methods in organic farming systems. Recommendations on educational programs include: Establish courses at land-grant universities on studies of self-sustaining unit systems of farming. Develop informational materials for county extension agents to assist them in providing services needed by organic farmers.

The USDA report recommended the appointment of a departmental organic resources coordinator and a multidisciplinary advisory committee on organic agriculture. It would be the duty of the coordinator to develop liaison between organic farmers, producer organizations, and the USDA. The coordinator would also collect, assemble, and disseminate information on organic technology to all farmers or interested people who might request it. The advisory committee composed of scientists and extension specialists would provide technical support to the organic resources coordinator and others requesting technical input. In July 1980 a USDA coordinator for organic resources was appointed, and a technical advisory committee was later selected. Thus, for the first time ever, a USDA program on organic farming has been initiated.

[ROBERT I. PAPENDICK; JAMES F. PARR]

Bibliography: R. Boeringa (ed.), *Alternative Methods of Agriculture: Description, Evaluation, and Recommendations for Research*, vol. 10 in *Developments in Agricultural and Managed-Forest Ecology*, 1980; Council for Agricultural Science and Technology, *Organic and Conventional Farming Compared*, CAST Rep. no. 84, 1980; W. Lockeretz, G. Shearer, and D. H. Kohl, *Science*, 211:540-547, 1981; R. C. Oelhaf, *Organic Agriculture: Economic and Ecological Comparisons with Conventional Methods*, 1978; U.S. Department of Agriculture, *Report and Recommendations on Organic Farming*, prepared by the USDA Study Team on Organic Farming, 1980; G. Youngberg, Alternative agriculturists: Ideology, politics, and prospects, in D. F. Hadwiger and W. P. Browne (eds.), *The New Politics of Food*, pp. 227-246, 1978.

Oceanic Hot Springs

Superheated thermal springs have been discovered at depths exceeding 2500 m on the crest of the East Pacific Rise (EPR). Hydrothermal fluids jet out of the sea floor at speeds of 1–5 m/s and at temperatures up to 350°C. The hottest waters issue from mineralized chimneys and are blackened by sulfide precipitates. These hydrothermal springs are the sites of actively forming massive sulfide mineral deposits. This discovery has contributed a great deal to the understanding of heat loss of the Earth, the chemical balance of the oceans, the formation of valuable mineral deposits, and the processes by which new oceanic crust is created at spreading centers. In addition, the hot springs provide the energy source for an exotic community of bottom-dwelling creatures. These unusual animals are the only known forms of terrestrial life which survive independently of sunlight and photosynthesis. First, this article will review the basic principles of plate tectonics, and then describe experiments performed by K. C. Macdonald and coworkers using the research submersible *Alvin* in the deep-sea hydrothermal field.

SEA FLOOR SPREADING

The East Pacific Rise is part of the world's largest mountain chain—the 75,000-km-long world rift system which winds around the globe much like the seams on a tennis ball. Most of the rift system lies under water, but portions are

Kenneth C. Macdonald holds a joint appointment at the University of California as associate professor of marine geophysics at Santa Barbara and associate research geophysicist at the Scripps Institution of Oceanography. He is interested in the geophysics and tectonics of spreading centers and transform faults, and is serving on the Ocean Science Board of the National Research Council.

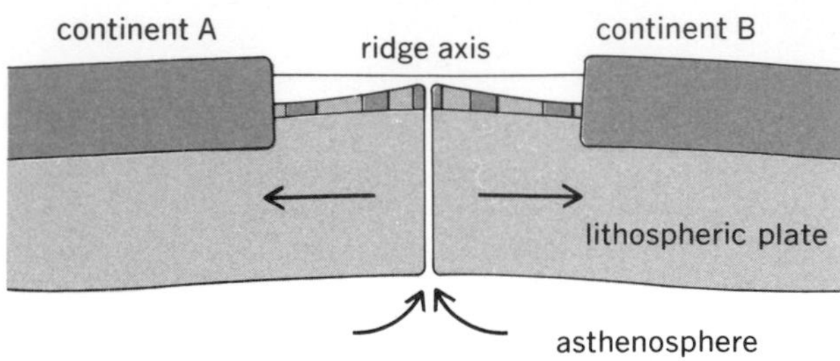

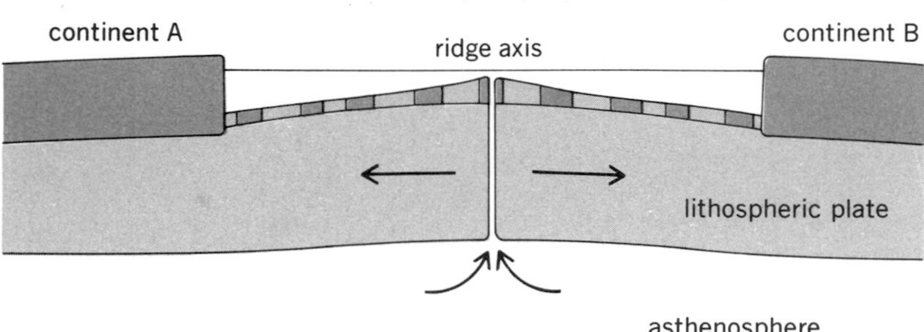

Fig. 1. Sea floor spreading involves the upwelling of magma from the asthenophere as the plates are pulled apart. As the molten rock solidifies at the spreading center, it becomes magnetized in the direction of the Earth's field. Intermittent reversals in the field direction create magnetic "stripes" on the sea floor. Based on magnetic studies, the East Pacific Rise at 21°N and the Galapagos spreading center are opening at approximately 6 cm/yr. (*After D. P. McKenzie and J. G. Sclater, The evolution of the Indian Ocean, Sci. Amer., 228:62–72, 1973*)

above sea level, for example, in the East African Rift Valley and on Iceland. According to the sea floor spreading hypothesis, the rift system marks a narrow cracked zone in the Earth's crust where the sea floor is being continuously pulled apart. These rifted zones are called spreading centers (Fig. 1). Volcanic material wells up, healing the cracks and generating new oceanic lithosphere in a conveyor belt fashion. The zone where new crust is formed at a spreading center is quite narrow, on the order of 10 km, and is marked by shallow-depth earthquakes, extensional faulting, and volcanism. Since the Earth is neither expanding nor contracting measurably, an equal amount of lithosphere must be consumed in "subduction" zones. The lithosphere descends back into the Earth in places often marked by deep-sea trenches such as the Japan Trench or Tonga Trench. These subduction zones are the locus of deep earthquakes (up to 700 km).

The theory of sea floor spreading has been strongly substantiated by the discovery of past reversals of the Earth's magnetic field recorded in the oceanic crust. One of Earth's greatest mysteries is that its magnetic field reverses direction or polarity on a time scale of every 10^4 to 10^6 years. According to the Vine-Matthews hypothesis, as volcanic material rises along the midocean rift, magnetic minerals in the magma become magnetized parallel to the Earth's magnetic field. As the magma cools and solidifies, the polarity of the magnetic field at the time is recorded as a stripe of magnetized volcanic rock along the rift system. The oceanic crust thus serves as a crude tape recorder of the Earth's field. The location of the magnetic stripes, which are due to reversals of known age, can be used to determine the rate of spreading of the ocean floor (Fig. 1). Considered a radical concept in the early 1960s, sea floor spreading is now accepted as a viable model for global geologic phenomena by all but the most quixotic objectors. It has successfully explained narrow belts of earthquakes and volcanoes

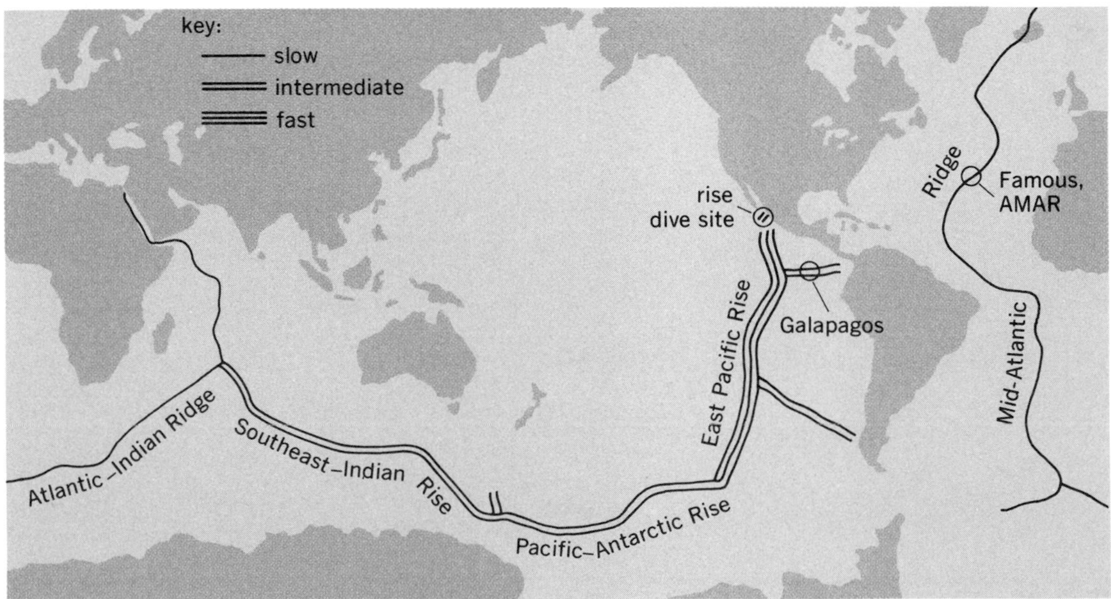

Fig. 2. Major sea floor spreading centers of the world (transform faults and subduction zones omitted). Slow (1–5 cm/yr), intermediate (5–9 cm/yr), and fast (>9 cm/yr) rates are included. Major diving expeditions are shown. (*From K. C. Macdonald, Annual Reviews of Earth and Planetary Sciences, in press*)

which encircle the globe, the creation and evolution of the sea floor, and the motions of rigid lithospheric plates which carry the continents and ocean basins across the Earth's surface.

This first-order model of spreading centers as idealized linear boundaries of crustal and lithospheric generation provides only a gross understanding of global-scale plate kinematics. As scientists attempt to understand the complexity of two-thirds of the Earth's surface, it becomes necessary to investigate processes within the spreading center rift zones. The entire ocean basin bears the imprint of these processes. To pursue this investigation, spreading centers have been studied with high-resolution instruments including deeply towed instrument packages, multi-narrow-beam bathymetric mapping, ocean floor instrument stations, and deep-diving research submersibles. Detailed studies have been completed on the Mid-Atlantic Ridge (Projects FAMOUS and AMAR), the Cayman Trough, the Galapagos spreading center, and the East Pacific Rise (Project RISE; Fig. 2). This article will consider primarily the East Pacific Rise, where the submersible *Alvin* was used in a series of sea floor geophysical experiments. In the course of these experiments and with the aid of deeply towed camera vehicles, the hottest and most pristine hydrothermal vents ever found on the sea floor were discovered.

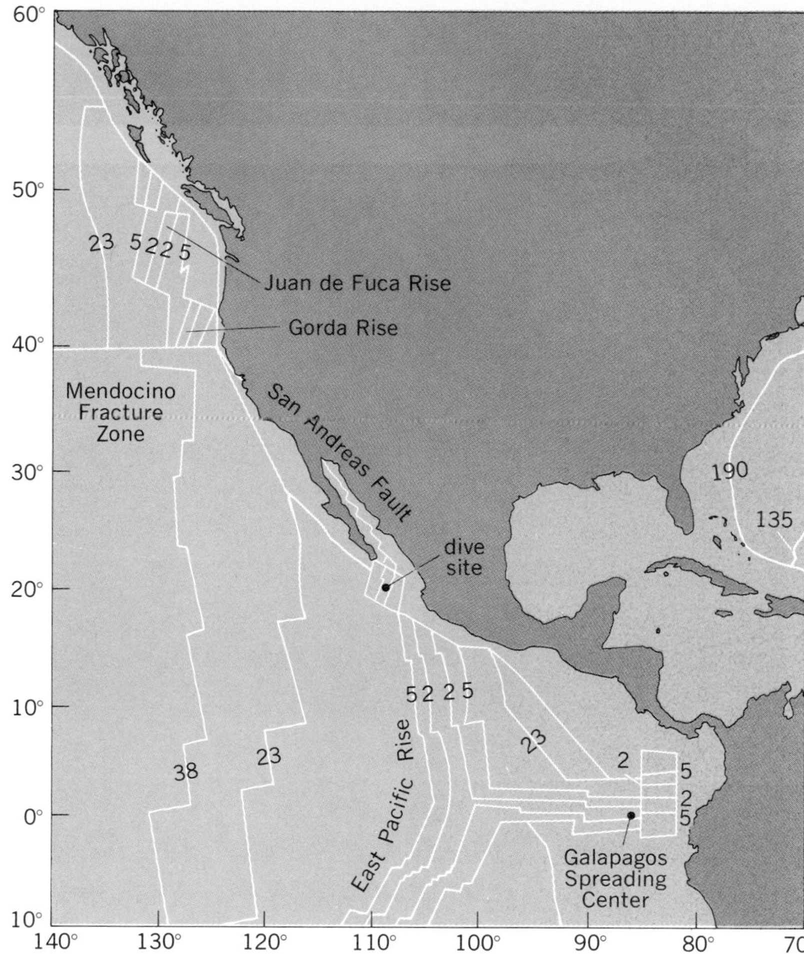

Fig. 3. The East Pacific Rise is the spreading center separating the Pacific and North American plates off the west coast of Mexico. Numbered contours indicate the age of the crust in millions of years. The dive site where the hot springs were discovered, and the earlier Galapagos site, is shown. (*After K. C. Macdonald and B. P. Luyendyk, The crest of the East Pacific Rise, Sci. Amer., 244:100–116, 1981*)

DIVING ON THE EAST PACIFIC RISE

A dive site was chosen at the northern end of the East Pacific Rise, just before it enters the Gulf of California and links with the San Andreas fault system in California (Fig. 3). This segment of the rise is spreading at 6.0 cm/yr, approximately the rate of fingernail growth. This rate is three times faster than at the Mid-Atlantic Ridge, where the FAMOUS project focused, but considerably slower than the fastest known spreading rate of 18 cm/yr near Easter Island on the East Pacific Rise. This site was chosen because it provided a classic moderate-rate spreading center for which considerable high-resolution data were in hand.

Earlier cruises had provided a clear picture of the geologic setting of the spreading center and its dimensions. Magnetic, sonar, and photographic studies from an uncrewed vehicle, the deep-tow instrument package of the Scripps Institution of Oceanography, had indicated that the actual spreading center or boundary between the Pacific and North American crustal plates may be only 1–2 km wide. A bathymetric and geologic map of the spreading center was assembled, and key dive targets were identified.

Diving program. The first phase of the diving program employed the versatile French submersible *Cyana* and focused on geologic problems along the spreading center which required visual "ground-truth" inspection and the ability to maneuver and reconnoiter outcrops and structures. The follow-up geophysical work was scheduled for 1979 using the more stable, larger submersible *Alvin*.

In 1978 a joint French-American-Mexican dive team found that the spreading center actually consists of four distinct geologic zones (Fig. 4). Zone 1 on the spreading axis is a very young volcanic zone approximately 1 km wide. Nearly all of the new volcanic material is extruded onto the sea floor within this remarkably narrow band. Pillow basalt flows and thin lava sheets here have essentially no sediment cover, exhibit a vitreous luster, and are relatively unaltered by sea-water interactions. Outside the "neovolcanic" zone, on both sides of the spreading axis, the newly formed crust begins acceleration to the half-spreading rate velocity of 3.0 cm/yr in zone 2. In so doing, the crust is stretched and cracked by fissures which line up parallel to the spreading center trend (northeast), that is, perpendicular to the spreading direction. These paired zones of crustal fissuring are each 0.5 to 2.0 km wide. Beyond zone 2 the crust is probably still undergoing some acceleration, but in zone 3 major "normal" faults develop. These faults are nearly ver-

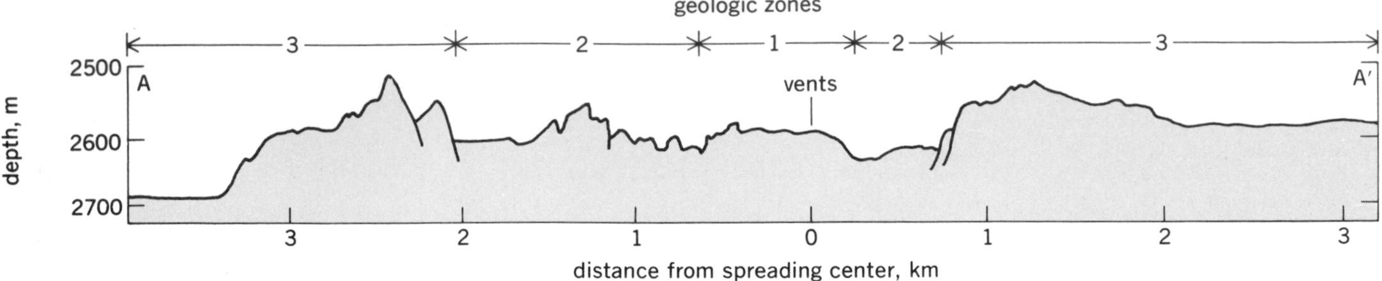

Fig. 4. Cross section of the spreading center shows the geologic zones (numbers) and location of the vents. (*After K. C. Macdonald and B. P. Luyendyk, The crest of the East Pacific Rise, Sci. Amer., 244:100–116, 1981*)

Fig. 5. The deep-diving submersible *Alvin* can carry three people to depths of 4000 m. The hot spring discovery was at a depth of 2650 m. (*Photograph by K. C. Macdonald*)

tical, or stair-stepped in cross section, and are caused by some vertical movement within a tensional stress field. Slippage along these faults gives rise to frequent earthquakes up to Richter magnitude 5.5. The fault scarps generally face toward the spreading center and are up to 70 m high. At a distance of approximately 10 km from the spreading center, active faulting (and presumably crustal acceleration) diminishes in zone 4.

The 1978 *Cyana* dives also revealed unusual lava formations and mineral deposits. "Fossil" lava lakes were discovered which were probably formed by the rapid outpouring of lava. The lakes are 5 m or more deep and hundreds of meters across. In some areas the roof of the lava lake has caved in, forming a collapse pit. Lava pillars and walls rimming the edges of the lava lake show bands of chilled basaltic glass which may record changes in lava levels within the lakes during either drainback to the axial magma chamber below, or lateral outflow.

Near the zone 1–2 boundary a chain of mounds several meters high was discovered. Geochemists found that they consist of sulfide minerals bearing zinc, iron, and copper with trace amounts of silver (400 ppm). It was thought that these mounds of sulfides were produced by sea floor hydrothermal venting. There were three other indicators that hydrothermal activity might be important in this area. In 1974 and 1977 temperature anomalies of tens of millidegrees (Celsius) were detected, and unusually high concentrations of helium-3 were measured over the spreading center (a sensitive indicator of hydrothermal activity). In addition, on one *Cyana* dive, large (20-cm) clam shells, like those seen at the Galapagos hydrothermal vents, were sighted, but none contained live animals. Ironically, this *Cyana* dive came within only a few hundred meters of the active vents.

In 1979, just before *Alvin*'s arrival and launching into the geophysical program, a brief mapping and photographic survey was conducted to the southwest of the previous *Cyana* operation and on the spreading center axis. This was done to investigate the change of geologic structure along strike, and to follow up on the tantalizing indicators of hydrothermal activity. The deep-tow instrument package was used

Fig. 6. The exotic community of animals which thrive near the hot springs include clusters of giant tube worms, clams, and white crabs. At the base of the food chain are chemosynthetic bacteria. The community is independent of sunlight. (*Photograph taken from Alvin by F. N. Spiess*)

to lay in topographic and side-scan sonar profiles, extending the chart to the southwest and pinpointing the axis of spreading. *Angus*, a heavily built camera and temperature sensor sled, was then lowered to conduct long photographic traverses a few meters above the rugged volcanic topography of the spreading axis.

Temperature elevations at several locations were detected and telemetered by *Angus* on the second lowering. The camera was raised and the film developed. A dozen frames revealed an exotic bottom-dwelling community of animals similar to those clustered around the Galapagos hydrothermal vents on the Galapagos spreading center. It appeared that the Galapagos hydrothermal vents and their unusual animals were not unique. The evidence was enough to warrant shifting the geophysics diving program to the southwest. A triangular array of ocean bottom seismometers was precisely deployed by a relay acoustics transponder in zone 1. Several preliminary dives were completed, and both seismic and gravity measurements were conducted successfully.

Hot springs. The deep research submersible *Alvin* (Fig. 5) was launched over the site of the unusual animals on the spreading center axis and landed on top of glistening volcanic rock on the axis of the East Pacific Rise spreading center. *Alvin* edged ahead slowly, at about $^1/_2$ km/h, continuing across the rough, youthful volcanic terrain, toward the vent area located earlier by towed cameras. White brachyuran crabs are seen in this area, an indicator of active hydrothermal vents. Nearby, shimmering water rises between the basalt pillows along the center of the neovolcanic zone. Large white clams up to 30 cm in length are nestled between the basalt pillows, while white brachyuran crabs scamper blindly across the volcanic terrain; there are also clusters of "giant tube worms" (vestimentiferan worms) up to 4 m long (Fig. 6). These live in dense colonies within the 2–20°C vent fluids. They wave eerily in the hydrothermal currents, their bright red plumes extending 20–30 cm beyond white protective tubes. Both the plumes of the tube worms and the tissue of the clams are bright red due to oxygenated hemoglobin in their blood. Occasionally a crab climbs the stalk of a tube worm apparently to attack its exposed plume.

The biological community is very similar to that studied on the Galapagos spreading center in 1977. The brown mussels of the Galapagos are absent, but the anemones, serpulid worms, galatheid and brachyuran crabs, clams, and giant tube worms all appear to be the same. Each community occupies an area roughly 30 m wide by 100 m long. The dense colony of animals is not attracted by the warmth of the hydrothermal fluids but by the concentrated food supply which is 300–500 times higher than the surrounding nutrient-poor waters. Bacteria form the base of the food chain by oxidizing hydrogen sulfide emitted from the vents to elemental sulfur and sulfate. These chemosynthetic bacteria then use the oxidation energy to fix carbon dioxide into organic matter. Most of the larger animals then filter-feed these bacteria, while others act as detritivores and predators. These vent communities are the first found which are independent of photosynthesis and solar energy, relying instead on the Earth's internal energy through chemosynthesis. Finding these communities at both the Galapagos spreading center 3000 km away and on the East Pacific Rise suggests that they may be globally distributed along intermediate-to-fast spreading segments of much of the world rift system.

Hydrothermal fluids and geochemistry. A subsequent *Alvin* dive was vectored directly to a hydrothermal area identified by towed cameras southwest of the first vent. Here, hydrothermal fluids blackened by sulfide precipitates blast through chimneys with 5–40-cm-wide openings (Fig. 7). The chim-

Fig. 7. The hottest of the hydrothermal vents jet out fluids through mineralized chimneys. These vents account for much of the heat loss of the Earth. (*Photograph taken from Alvin by D. Foster*)

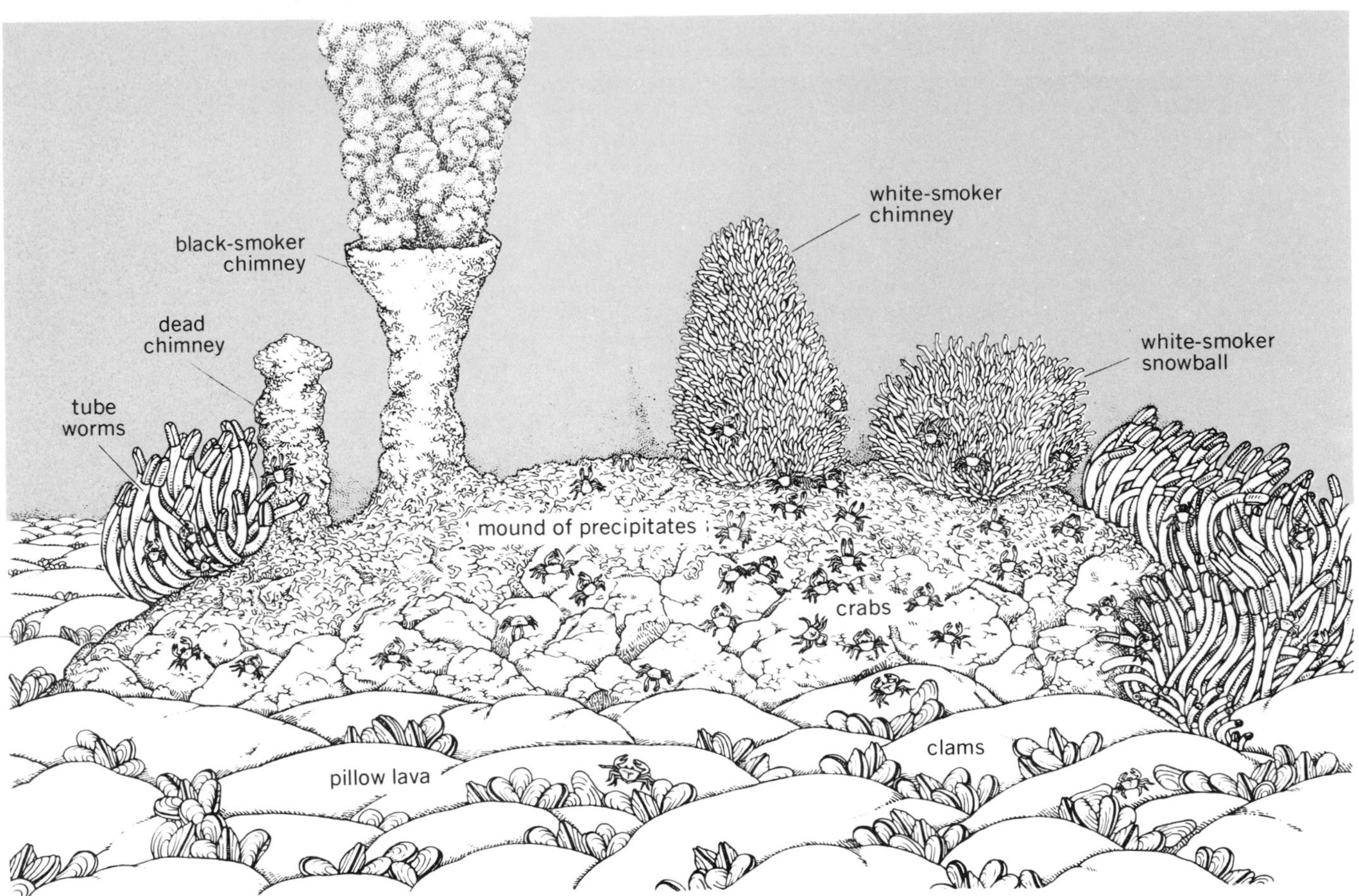

Fig. 8. Schematic illustration of the hot spring vicinity. The mound is built from mineral precipitates and organic debris on top of volcanic pillow lavas. The white-smoker chimney is built up of burrows made by the little-understood organisms called Pompeii worms. The white, cloudy fluids emitted by the white smoker have temperatures up to 300°C. The hottest water, having temperatures up to 350°C, comes from the black smokers whose chimneys are built from sulfide precipitates. (*From K. C. Macdonald and B. P. Luyendyk, The crest of the East Pacific Rise, Sci. Amer., 244:100–116, 1981*)

neys occur in clusters of up to six atop a basal mound of sulfide precipitates and are up to 10 m tall (Fig. 8).

Early attempts to measure the temperature of the black fluids were thwarted. Until then, the highest temperature ever recorded in the open deep ocean floor was 21°C, measured only 2 months earlier on the Galapagos spreading center. The temperature probe, calibrated to 32°C, should have been adequate, but when it was inserted into the first chimney, the reading sailed off-scale. The probe was pulled out and the polyvinyl chloride rod on which the probe was mounted showed signs of melting. The probe was modified and calibrated at sea to the triple point of water. Measurements on several subsequent dives indicated temperatures of at least 350°C. A subsequent series of dives equipped with a modified temperature probe documented that the fluid exit temperatures were 350 ± 6°C.

The hydrothermal vents have revolutionized theories about the oceanic chemical budget. It was formerly held that ocean waters maintained an equilibrium between processes of elemental input (chiefly river flux) and processes of uptake. The two balancing processes of chemical uptake were presumed to be deposition of detrital and chemical sediments and reactions at low temperatures between sea water and solid phases in the marine environment (that is, sediments and volcanic rocks). As more became known about mineral abundances and low-temperature reactions on the sea floor, problems arose in the bookkeeping for certain elements: for example, more Mg^{2+} and SO_4^{2-} are supplied by rivers than can be removed by sedimentation, clay formation, and basalt weathering; and the ocean floor appears to accumulate far more manganese than the rivers supply. Large-scale hydrothermal circulation of sea water along submarine rift systems introduces new possibilities for high-temperature chemical exchange between fluids and solids in the oceans. Reactions between hot sea water and basaltic rock convert aqueous SO_4^{2-} into solid sulfate and sulfide minerals and remove Mg^{2+} and OH^- from sea water into hydrothermal clays. The hot sea water is changed by these reactions into a reduced acidic solution that leaches Ca^{2+}, Si^{4+}, Mn^{2+}, Fe^{2+}, and other cations (including lithium-3) from the rocks and carries these elements back to the ocean. In this way, hydrothermal systems can supply and remove elements from the ocean in quantities sufficient to balance the budget for the major constituents of sea water and to produce the observed concentrations and distributions of numerous minor and trace constituents.

The hydrothermal effluents of the Galapagos system mix during their ascent with normal sea water within the volcanic rock pile. This mixing process lowers the temperatures of the fluids and causes mineral deposition within the rocks that alters the chemistry of the hot springs discharging onto the sea floor. The high temperature and the chemistry of the springs at EPR 21°N indicate that here the hydrothermal fluids have not mixed significantly with cool sea water on their way to the sea floor, and hence these fluids represent the pure hydrothermal contribution to the marine chemical cycle.

The undiluted hydrothermal fluids at 21°N become blackened with fine-grained Fe- and Zn-sulfide precipitates as contact is made with cold, alkaline sea water on the ocean floor. Preliminary analyses indicate that basal mounds and chimneys precipitated around the vent orifices are predominantly composed of Zn-, Fe-, and Cu-sulfides, and sulfates of Ca and Mg (Fig. 9). The precise mechanisms for the formation of the vent minerals, the rate of mineral deposition, the fluid temperatures in the subsurface, and the ratios of water to rock in the plumbing system are subjects of hot geochemical debate, but there is no question that the vents are a major geochemical discovery which will henceforth play a central role in understanding the chemical balance in the oceans.

Geophysics. Most of the geophysics program was directed toward the nature of the axial magma chamber and hydrothermal activity; fortunately the experiments were conducted in an area where hydrothermal activity is presently active. Seismic sounding, earthquake monitoring, gravity, electromagnetic, and magnetic experiments were conducted in the vicinity of the hydrothermal field to probe the subsurface structure.

Hydrothermal activity. The most spectacular accomplishment of the expedition, for geophysicists as well as the geologists, chemists, and biologists, was the discovery of the active hydrothermal vents. Although first proposed in theory 15 years ago, hydrothermal circulation at midocean ridges is poorly understood and has proved difficult to measure. The large discrepancy between measured conductive heat flow and theoretical cooling plate models suggests that at least one-third of the heat loss at midocean ridges is accomplished by nonconductive means, presumably hydrothermal circulation. On the Galapagos spreading center, hydrothermal activity was documented for the first time, but estimates of the convective heat fluxes were hampered by the diffuse nature of the circulation.

Based on film and video tapes taken of the vents, flow rates were estimated from the physical characteristics of the water and the vent dimensions. Given plume velocities of 2–3 m/s, a single vent chimney accounts for a heat flux of $(6 \pm 2) \times 10^7$ cal/s

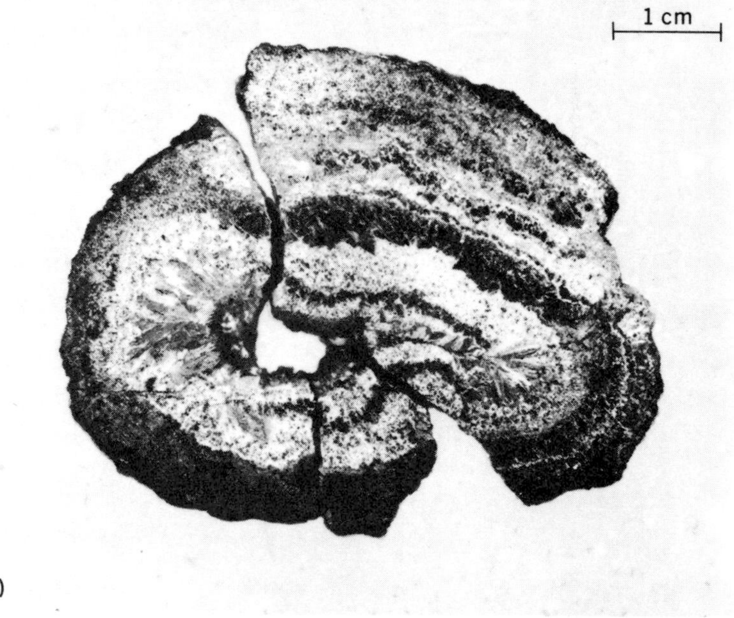

Fig. 9. The hydrothermal vents are called black smokers because of the black sulfide precipitates which cloud the hydrothermal plume. (*a*) Scanning electron micrograph of filtered precipitate from the black-smoker fluids. The hexagonal platelets are crystals of pyrrhotite, an iron sulfide. (*b*) Cross section of a small chimney vent. The concentric bands of minerals are various zinc, iron, and copper sulfides. Transitions from one mineral to another reflect changes in the chemistry of the hydrothermal fluids. (*From F. N. Spiess et al., East Pacific Rise: Hot springs and geophysical experiments, Science, 207:1421–1433, copyright © 1980 by the American Association for the Advancement of Science*)

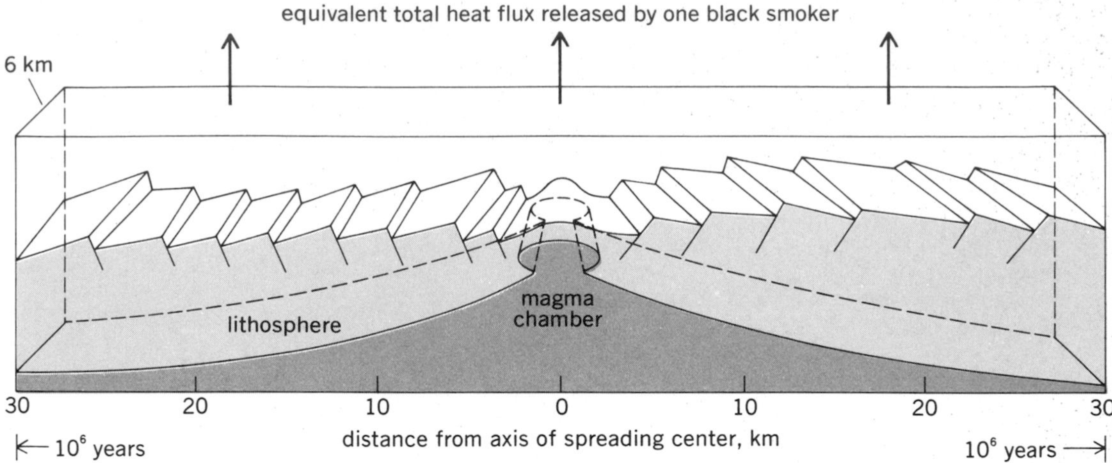

Fig. 10. The heat output of a single black-smoker vent is equivalent to the total theoretical heat flux of a segment of midocean ridge 6 km long and 60 km wide (for opening rates of 6 cm/yr). Such vents make a significant contribution to global heat flux. (*After K. C. Macdonald and B. P. Luyendyk, The crest of the East Pacific Rise, Sci. Amer., 244:100–116, 1981*)

($2.5 \pm 0.8 \times 10^8$ J/s). This is three to six times the total theoretical heat loss for a 1-km-long segment of midocean ridge out to 1 m.y. (million years), or 30 km, to either side (Fig. 10). At least 12 major distinct chimneys exist in the southwestern part of the study area, so the total heat flux is quite large. Thus it is likely that individual vents are short-lived.

The vents appear to be restricted to a narrow linear zone only 200–500 m wide and 6 km long within the neovolcanic zone (Fig. 11). Twenty-five distinct temperature anomalies were detected and photographically verified in this narrow band, and eight vents were visited and studied with *Alvin*. In general, vents toward the northeast are 2–20°C cooler, and exhibit the Galapagos-style benthic communities. The hotter, 350°C vents are near the southwestern end. They too have the exotic benthic communities, but the animals tend to reside at a safer distance (several meters) from these vents.

Seismic experiments. A high-resolution seismic refraction experiment was designed to determine the seismic velocity of the upper crust as a function of depth in order to infer the depth of crustal cracking and fissuring—a critical unknown relating to the depth of hydrothermal circulation systems in oceanic crust. Unfortunately, the spherical spreading of energy from explosive sources at the sea surface and reverberation from nearby topography smear out seismic resolution in the top 1000 m of crust. Thus it was necessary to devise a means of placing both the sensors and receivers on the sea floor. To make the experiment work, millisecond timing precision was also needed.

Alvin provided a means of solving these two difficult problems. Since explosive sources are virtually useless at the high pressures (265 bars or 26.5 MPa) on the sea floor here, a hydraulic impact hammer was attached to *Alvin* to work as a benthic "thumper" seismic source. To establish a precise time base, *Alvin* drove up to within 2 m of the ocean bottom seismometers (OBSs) for a calibration "thump" recorded by both *Alvin* and the OBS, and returned to recalibrate at the end of a dive. During four dives seismic refraction lines 1000 m long parallel to the spreading center and 800 m long across the spreading center with thumper stations every 50–100 m were completed. Following the timing calibration, stations were made by landing on the

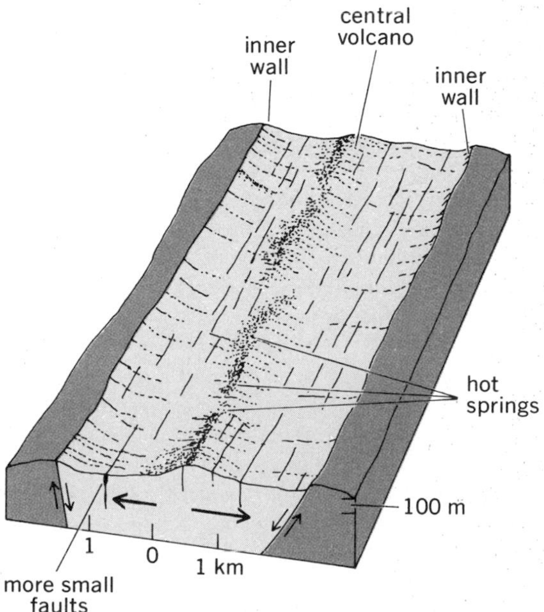

Fig. 11. Schematic of the volcanic zone and location of the hot springs on the East Pacific Rise. The zone of recent volcanism is very narrow, approximately 1 km. The hot springs occur in a narrow band within the recent volcanic zone. (*From K. C. Macdonald, Mid-oceanic ridges: Fine scale tectonic, volcanic and hydrothermal processes, Annu. Rev. Earth Planet. Sci., 1982*)

sea floor, acoustically commanding the OBSs to turn on, and then thumping 5–10 times so that arrivals could later be added together to increase the signal-to-noise ratio. The stations and the array were within zone 1 and within several hundred meters of the active vents. Detailed geologic observations were made along the track to be related to the seismic results, and rock samples were collected for later laboratory work on elastic properties.

Although complete analyses are not yet available, simple travel-time analysis was applied to the station 1 arrivals to make a preliminary determination of the velocity at the surface of the crust parallel to the strike of the spreading center. While no major fissures or faults were traversed in this line, numerous hairline cracks crossed the pillows. A velocity of 3.3 km/s was obtained. Laboratory-determined velocities for basalt at this pressure are approximately 5.5 km/s, so 3.3 km/s is quite slow. Pervasive cracking and porosity is presumably the cause. Detailed inferences related to the degree of cracking and porosity necessary to produce the observations will have to await future measurements of physical properties on rock samples and completion of seismic analyses at greater ranges. Of particular interest will be the depth at which the velocity increases above 5 km/s, indicating closure of most of the fissures, and the depth of the hydrothermal plumbing systems.

Results from an earlier seismic refraction experiment indicate the presence of a magma chamber at shallow depth beneath the hydrothermal field. The axial magma chamber provides the energy source for the hot springs as well as being the source of new oceanic crust. In this less precise but larger-scale experiment, explosive shots from a surface ship were fired at ranges of up to 60 km to a triangular array of ocean bottom seismometers. At a depth of only 2 km, beneath the sea floor, a low-velocity zone for compressional seismic waves was found, indicating the presence of partially molten rock. Ten kilometers off-axis the velocities were found to be higher or normal for oceanic basalt, so the magma chamber appears to be restricted to a 20-km-wide (or less) zone under the spreading center.

A second piece of evidence for the axial magma chamber comes from the propagation of seismic waves from earthquakes (Fig. 12). Oceanic earthquakes provide a means of studying propagation of seismic shear waves in the crust. In places where rock is partially molten as in a magma chamber, shear waves are strongly damped. For paths along the spreading center, earthquake shear waves are severely attenuated, while only 10 km off-axis, shear waves are transmitted efficiently. Again, this is clear, although indirect, evidence for a shallow and narrow magma chamber beneath the spreading center and the hydrothermal field.

Petrologic studies of basalt samples collected from *Alvin* also suggest the presence of a shallow axial magma chamber which drives the hydrothermal system. The limited composition range for basalt samples collected along a 6-km-long transect across the neovolcanic zone suggests that the samples are derived from a common parent magma by fractional crystallization of olivine and plagioclase at low pressure (<2–3 kilobars or 200–300 MPa). This would imply a magma chamber less than 6 km deep. Thus, the magma chamber has a roof which may be only 2 and no more than 6 km thick. This lid is intensely cracked and fissured, allowing percolation of sea water deep enough into the crust to be heated to at least 350°C before jetting out through the hydrothermal vents.

How deep does this circulation extend? A companion experiment to the ocean floor seismic measurements was conducted in summer 1980 to measure microearthquakes in the hydrothermal field. Seven ocean bottom seismometers were navigated into position using precise bottom-moored acoustic transponders which had been left to mark the vents. If the hydrothermal flow had some seismic signature, perhaps it would be possible to determine how deep the hydrothermal activity extends. So far the results are promising. The maximum depth for the earthquakes is very shallow, only 2–3 km. This is in excellent agreement with earlier results for depth to the roof of the magma chamber, and is a reasonable upper bound for the depth of cracking of the crust and resultant hydrothermal circulation here.

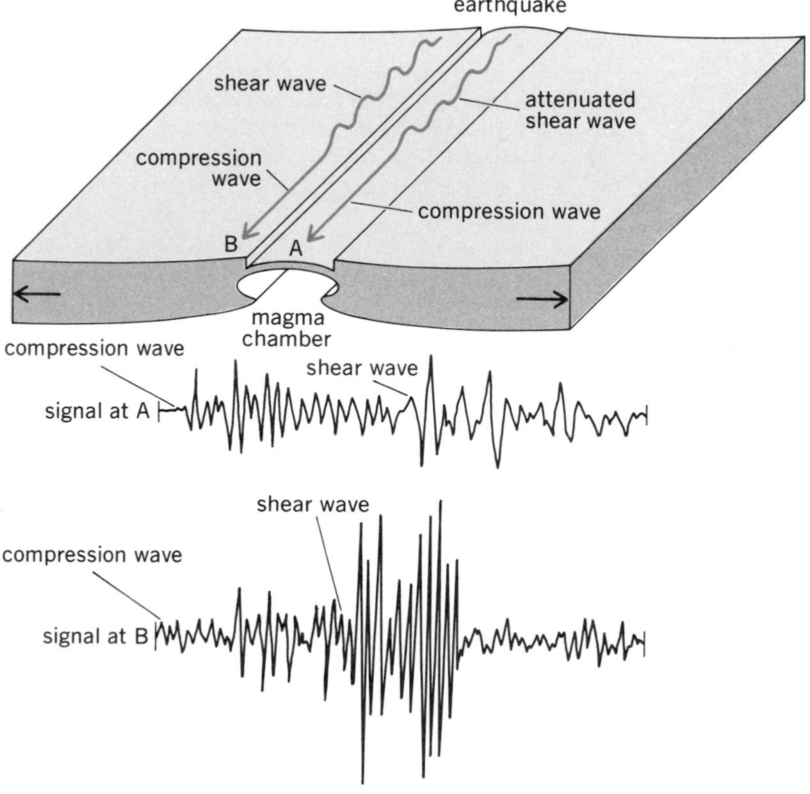

Fig. 12. Seismic shear waves from an earthquake near the spreading axis are strongly attenuated for paths along the axis, but not for paths only 10 km off-axis. This suggests that a narrow axial magma chamber exists beneath the spreading center. (*After K. C. Macdonald and B. P. Luyendyk, The crest of the East Pacific Rise, Sci. Amer., 244:100–116, 1981*)

In addition, "harmonic tremors" were recorded. These events, which are very similar to the tremors recorded at Mount St. Helens before and during its catastrophic May 1980 eruption, are a strong diagnostic for impending volcanic eruption. Just before and during the St. Helens eruption, tremors increased in occurrence until they became almost continuous. The records are very similar. It is possible that this portion of the East Pacific Rise is entering an active volcanic eruptive phase which could radically alter the hydrothermal vents, terrain, and biological communities.

Gravity measurements. Closely associated with the seismic experiments were gravity measurements conducted from *Alvin*. Combined with the seismic measurements, gravity should be sensitive to decreases in crustal density caused by fissuring of the crust or the presence of a shallow magma chamber. However, the noise level of gravity measurements made at the sea surface is 1–5 milligals (average Earth's field = 10^6 milligals) due to spurious accelerations experienced by the shipboard sensors. The expected signal from a hypothesized magma chamber is in the 1–2-milligal range. Once again *Alvin* provided a solution. Measuring gravity from *Alvin* while it sat quietly on the sea floor reduced spurious accelerations. In addition, closer proximity to the source of the expected gravity anomaly enhanced the signal measured by the gravimeter.

On-bottom gravity measurements show a pronounced negative Bouguer gravity anomaly of approximately −1.5 milligals over the neovolcanic zone. This anomaly is defined by 20 successful (out of 23) measurements on a 7-km profile crossing the neovolcanic zone well into zone 3. The negative anomaly is centered about the central volcanic ridge and occupies zones 1 and 2. The mass deficiency implied by the negative anomaly could be caused by either crustal fissuring or by a shallow low-density magma chamber. Geologic observations suggest a maximum of fissuring in zone 2 while the gravity anomaly is centered over zone 1, which is relatively unfissured. This suggests but does not require that the anomaly is caused by a shallow magma chamber.

It might be assumed that the shape of this magma chamber can be approximated by a horizontal cylinder with its axis parallel to the strike of the ridge and lying directly below the anomaly minimum. The shape of the anomaly curve then indicates that the center of this cylinder is about 1000 m below the sea floor. If it is further assumed that the chamber is filled with molten basalt, then its density would be about 0.21 g/cm^3 less than the surrounding rock. This in turn implies that the top edge of the cylinder is about 600 m below the sea floor. Assuming less of a density contrast between the chamber contents and surrounding rock requires that the magma body be larger and extend closer to the crustal surface.

Applying some geologic reasoning to this model, it is found that the cylinder is centered within the sheeted dike unit of oceanic layer 2 and that its top is probably near the base of the extrusive lava flow unit of oceanic layer 2. In fact, as suggested by the seismic results, the oceanic crust magma chamber here is probably far larger than this. Further, it must occupy the bulk of oceanic layer 3 to account for the plutonic rocks such as gabbro and cumulate rocks associated with this layer. Thus, the chamber is probably found from 2 to 6 km below the seabed, as was found from the seismic results, and is about two or three times as wide. What has been measured in the gravity data may be due to a small bump or "cupola" at the apex of the chamber. The size of this cupola suggests it occupies most of zone 1 and therefore provides the magma to feed the lava flows found here on the seabed.

The magnitude of the negative anomaly suggests a mass deficiency beneath the spreading center axis of around 9×10^7 kg per meter of ridge strike. For isostatic (buoyancy) balance to exist at the spreading center, a mass excess in the bottom topography should exist to balance the inferred mass deficiency at depth. Zone 1 displays an uplifted topographic block about 20 to 30 m higher than the surrounding region and around 1 km wide. A simple calculation shows that this block is only about one-half as high as it should be to produce mass balance at the spreading center axis. Either the mass deficiency is balanced by additional topographic features farther away from the axis, or friction along fault planes is holding down the central block out of isostatic equilibrium.

The observed gravity reading uncorrected for depth is a function of measurement depth and the average terrain density between the deepest and shallowest stations. The observed gravity-measurement depth variation indicates that the average density of the topography is 2.60 g/cm^3 ± 0.21. This value is somewhat lower than that for massive basalt rock. Measurements were made of the saturated density of 90 rock samples obtained by *Alvin* from the spreading center and magnetic reversal area. These rocks have a density of 2.89 ± 0.11, which is substantially higher than the topographic density and also is the usual value for massive basalt. The difference in the densities can be attributed to 15% porosity of the sea floor in the form of cracks, fissures, and faults. This infers high porosity in the bottom topography, as expected, given that hot waters were observed emanating from cracks in the ocean crust.

An interesting question then is: how thick is this cracked low-density layer? This is not known, but presumably some cracking extends down to the vicinity of the magma chamber because the hot springs are evidence that sea water has penetrated to substantial depths.

Electrical conductivity. A new electrical sounding experiment was developed to study the structure of electrical conductivity in the oceanic crust. The objective was to detect the presence of a conducting fluid such as sea water in the crust. This conductivity experiment could give another constraint on percolation of sea water into the crust and the depth of crustal fissuring. Electrical conductivity might also be used to detect the axial magma chamber since magma has a much higher conductivity than solid basalt. A transmitter and horizontal electric dipole antenna 800 m long were towed near the sea floor behind the R.V. *Melville* of the Scripps Institution of Oceanography. The antenna transmitted electromagnetic signals into the ocean and crust at frequencies of 0.25–2.5 Hz. The frequencies were high enough so that the signal in the ocean was rapidly absorbed and long-range signals were confined to the crust. Three receivers were placed on the sea floor near the spreading center. Since this was a prototype experiment involving towing a long antenna near the sea floor, the precipitous topography of zone 1 was avoided. The soundings apply primarily to the area 10–15 km west of the spreading axis in crust 0.3 to 0.4 m.y. old.

The conductivity pattern indicates that only 10 to 15 km off-axis the magma chamber is absent. This is strong confirmation for a narrow axial magma chamber suggested by the seismic measurements. Observations from *Cyana* in 1978 indicate that active faulting of the crust diminishes at 10–12 km off-axis also. Perhaps the magma chamber width controls the width of the zone of active tectonic faulting activity on oceanic spreading centers. A low electrical conductivity at shallow levels suggests penetration of sea water into the crust, but no deeper than 2 to 4 km.

SIGNIFICANCE OF RESEARCH

Since the general acceptance of the hypothesis of sea floor spreading nearly a decade ago, it has been difficult to unravel the important physical and chemical processes associated with the separation of the major plates. It has become clear that carefully planned, detailed investigations and experiments on the axis of spreading can help delineate important boundary conditions. In this way, considerable progress has been made in understanding the processes of creation of 70% of the Earth's surface in the form of oceanic crust and lithosphere at spreading centers. The role of an axial magma chamber and high-temperature hydrothermal activity is now recognized as important at faster-rate spreading centers. Global geochemical and heat flow budgets are coming into balance with the direct study of hydrothermal activity at midocean ridges. New measuring and sampling techniques have been developed and used successfully for the first time to approach some of these difficult problems. The research submersible *Alvin* has proved central to many of these new approaches.

[KEN C. MACDONALD]

Bibliography: R. Haymon and M. Kastner, *Earth Planet Sci. Lett.*, 53:363–381, 1981; K. C. Macdonald et al., *Earth Planet Sci. Lett.*, 48:1–7, 1980; K. C. Macdonald and B. P. Luyendyk, *Sci. Amer.*, 228:62–72, 1981; F. N. Spiess et al., *Science*, 207:1421–1433, 1980.

The Planetary Greenhouse

Life on Earth is possible because there is liquid water, a blanket of life-supporting gases, and a climate neither too warm nor too cold. Humans may now be altering this equable climate by adding carbon dioxide to the atmosphere. This could make the global climate warmer than at any other time in human history. The change would be effectively irreversible.

The atmosphere contains 75% nitrogen, 23% oxygen, with argon, water vapor, and trace gases—including carbon dioxide (CO_2)—making up the remaining 2%. Carbon dioxide is an odorless, colorless gas which constitutes about 0.035% by volume of the atmosphere but whose significance far exceeds its relative scarcity. It is essential for photosynthesis, the process by which the Sun's energy is converted into forms usable by plants and animals, and it also helps regulate the critical heat balance of the planet, thus maintaining liquid water.

Carbon dioxide affects the heat balance by acting as a one-way screen. It is transparent to incoming visible sunlight, and allows the Sun's heat to warm the oceans and land. But CO_2 molecules block some of the infrared heat radiated back into space. This reflected heat is absorbed in the lower atmosphere. This is the so-called greenhouse effect (Fig. 1), by which a portion of the Sun's heat is trapped, making the Earth's surface warmer than it otherwise would be.

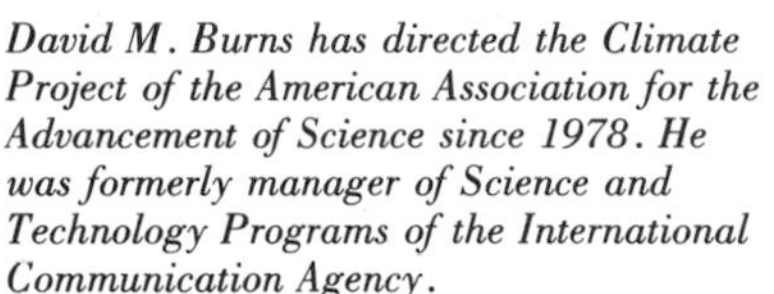

David M. Burns has directed the Climate Project of the American Association for the Advancement of Science since 1978. He was formerly manager of Science and Technology Programs of the International Communication Agency.

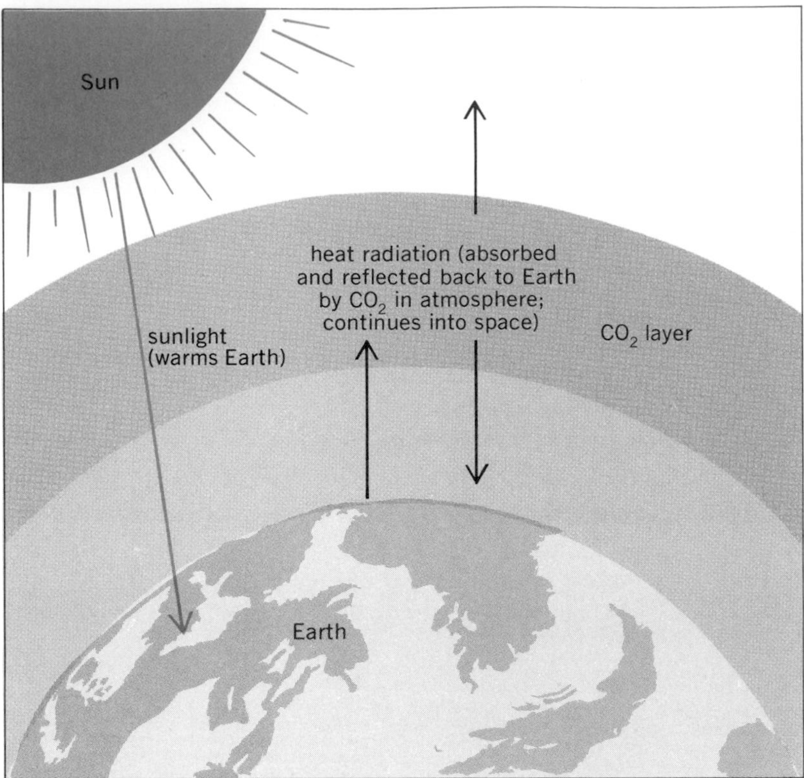

Fig. 1. How carbon dioxide buildup warms the Earth. (*From D. Sleeper, Fuel Combustion Adds to Anxiety over CO_2 Buildup, Conservation Foundation, August 1979*)

Earth's neighboring planets demonstrate the critical role of the radiative characteristics of carbon dioxide and water vapor in moderating the Sun's effect. On Mars, water is present only as ice, and its thin atmosphere allows most of the Sun's energy to escape back into space. Mars is consequently a frozen desert, with an average temperature of −23°C. On Venus, however, the atmosphere is 96% carbon dioxide, and all its water is vaporized by the Sun's heat. The carbon dioxide and thick clouds of water and sulfur dioxide cause an extreme greenhouse effect, thus pushing the surface temperatures to 482°C.

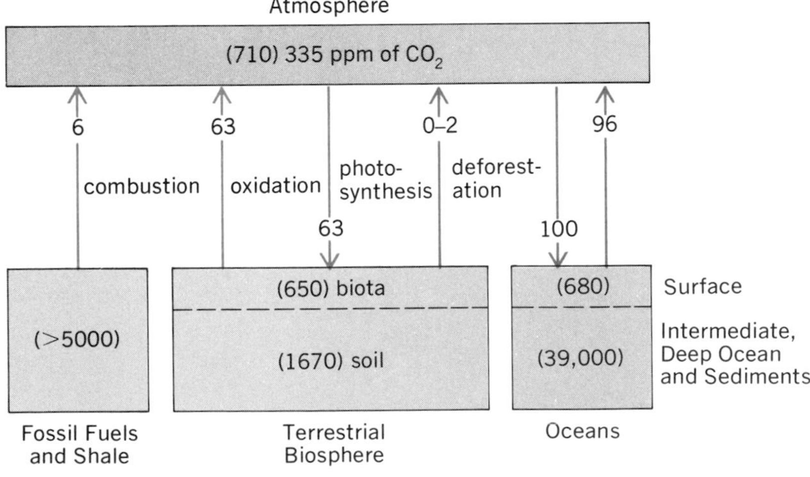

Fig. 2. Exchangeable carbon reservoirs and fluxes. (*From Carbon Dioxide and Climate Research Program, Office of Health and Environmental Research, U.S. Department of Energy, August 1979*)

Carbon dioxide is exhaled by all animals, and is used by green plants to convert sunlight and nutrients into organic matter and oxygen. As plants die and decay, their organic matter either is oxidized and returned to the atmosphere as carbon dioxide, or is stored in other natural reservoirs. Atmospheric carbon dioxide is constantly scavenged and dissolved in the oceans, which contains about 70% of the Earth's total. But it takes about a thousand years for surface and deep water to mix completely. It will thus be many centuries before the excess carbon dioxide added by humans is absorbed and stored in the oceans. The other major reservoir is geological, formed by conversion of decomposing plants and animals into hydrocarbons. The flow of carbon between these reservoirs—the oceans, the atmosphere, the biosphere, and geological deposits—is the carbon cycle (Fig. 2). Human activity is now altering this natural cycle, and an essential, familiar gas may come to be viewed as a potentially harmful pollutant.

Humans are speeding up the cycle by taking the carbon stored in plants over millions of years and burning it in just a century or two. The use of fossil fuels—coal, oil, natural gas, peat, oil shale, tar sands—has released billions of tons of carbon dioxide into the atmosphere. Deforestation and cement production (which releases carbon dioxide when limestone, calcium carbonate rock, is heated) also play a role in the buildup (Fig. 3). This rapid human injection into the atmosphere of a portion of the naturally stored carbon may alter the Earth's heat balance with effects that are difficult to foresee.

Modeling the climate with high CO_2. The possibility that slight changes in the composition of Earth's atmosphere might cause climatic variation was first put forward by J. Tyndall in 1861. The Swedish chemist Svante Arrhenius argued in 1896 that increased carbon dioxide would cause higher surface temperatures, and calculated that a trebling of atmospheric carbon dioxide would result in a rise of 9°C—a result remarkably close to modern projections. The geologist T. Chamberlin speculated in 1899 that glacial epochs could be caused by changes in atmospheric carbon dioxide resulting from geologic processes. In 1938 G. Callendar first recognized that humans, by burning carbon-based fuels, were changing the composition of the atmosphere. Roger Revelle and Hans Suess in 1957 called this "a large-scale geophysical experiment"—an experiment which Revelle has subsequently called uncontrolled.

Modern concern is based mainly on mathematical

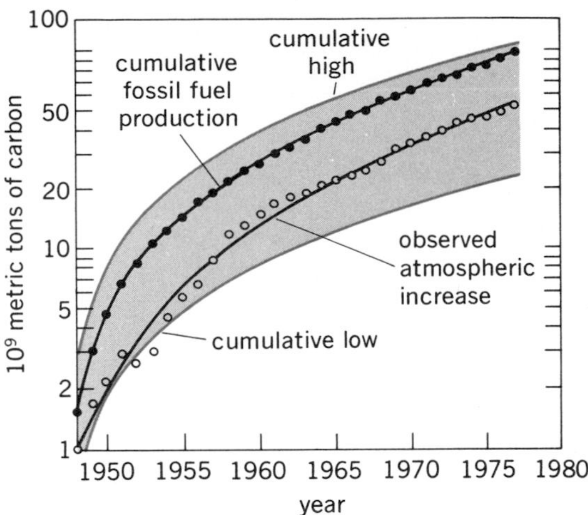

Fig. 3. Fossil-fuel carbon dioxide production and atmospheric increase, and possible carbon dioxide release (shaded area) from forest clearing. Of the total fossil-fuel carbon dioxide from 1948 to the present, approximately 53.5% remains airborne.

models. Modeling is essential because experimentation in nature is impossible: the Earth cannot be put into a test tube, and the vast array of geophysical forces which cause weather or climate cannot be marshaled. Climate modelers try to portray in numerical terms the major physical processes that operate within and between the atmosphere, oceans, land and vegetation, and ice and snow. Together with external conditions, these processes ultimately control climate on time scales ranging from seasons to eons. The modeler represents these processes by equations of motion, heat exchange, and the hydrologic cycle. Different models consider surface processes and boundary conditions in varying degrees of physical and spatial detail. The feedbacks are complex but critical (Fig. 4). Some models reproduce a three-dimensional atmosphere that appears to simulate the varying weather and average climate quite well, including the changing seasons. Such models are called general circulation models (GCMs).

Modelers recognize that these simulations are deficient in a number of important ways, and most are being constantly revised. For example, models do not yet take sufficient account of how the oceanic heat sink might delay the climatic response and alter its regional characteristics, or how altered cloud distribution might feed back to affect the response. However, even "simplified" GCMs push the most powerful computers to their limits. With increasing refinement and ingenuity of the models, and with greater speed and power in computers, researchers are sufficiently close to understanding the complex climate system to be able to predict how the atmosphere will respond to a given increase in carbon dioxide. That was the principal conclusion of "Carbon Dioxide and Climate: A Scientific Assessment" (1979), prepared by a committee of the National Academy of Sciences under Jule Charney.

The committee concluded that a doubling of carbon dioxide would result in a global average warming of "near 3 degrees C. with a probable error of

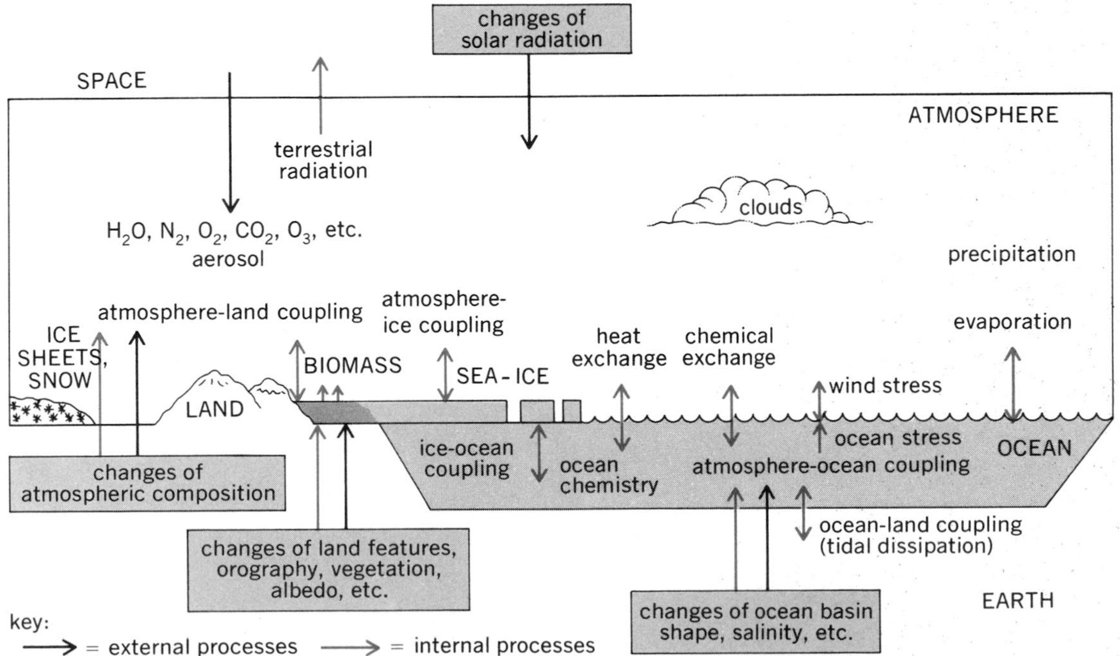

Fig. 4. Components of the coupled atmosphere–ocean–ice–land surface–biomass climatic system. (*From The Physical Basis of Climate and Climate Modeling, Report of the International Study Conference in Stockholm, 20 July–10 August 1974, World Meteorological Organization, International Council of Scientific Unions, Geneva, GARP Publ. Ser. no. 16, 1975*)

plus or minus 1.5 degrees C.," with much greater warming toward the poles.

Climate changes due to CO_2 greenhouse. There is growing confidence in the ability of the models to predict large-scale effects averaged over wide areas and considerable periods of time. At present, however, the models provide only vague indications as to what a high–carbon dioxide greenhouse climate might be like at any particular place or moment. Some of the limitations are indicated above. A few climatologists consider Earth's climate to be a system which is simply too complex for complete understanding. They argue that nature's equations will always be more subtle than those of humans, who must wait for nature to reveal the actual outcome.

Despite such reservations, modelers are attempting to provide increasingly detailed predictions. Virtually all models show that a carbon dioxide level of 600 parts per million (ppm) would result in significant shifts in climatic patterns. A warmer Earth would mean a more active hydrologic cycle. More water would be lost by evaporation and by transpiration from plants, but more water would also reach Earth through increased precipitation. Some areas which are now deserts might receive enough precipitation to become agriculturally productive. Other areas which are now productive might become deserts, unless natural precipitation was supplemented by irrigation. This alteration of patterns of precipitation, with its important consequences for farms, pastures, forests, and water supply, may well be the potential climate change of most significance to humans (Fig. 5).

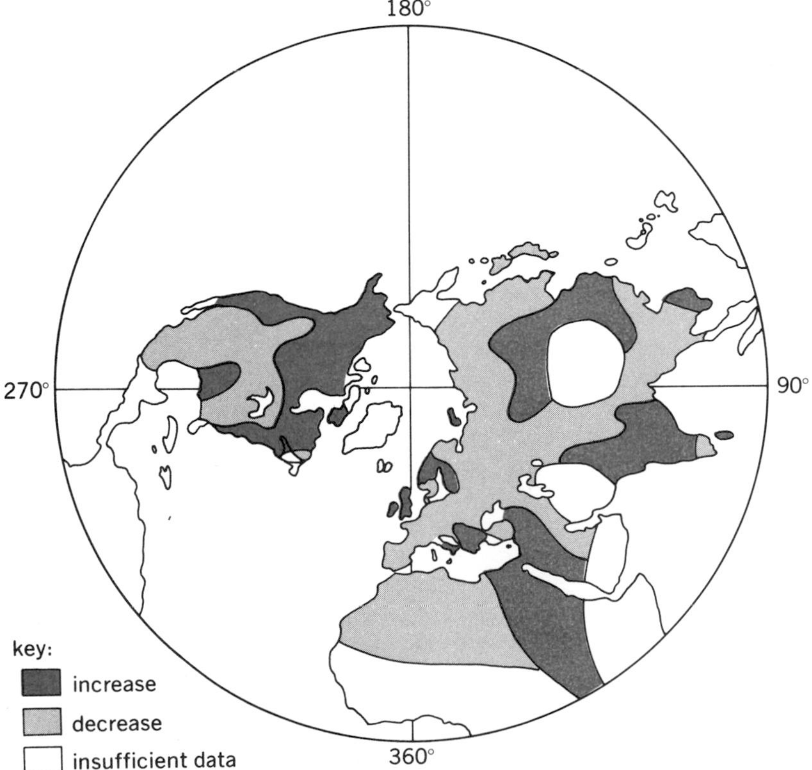

Fig. 5. Analysis of rainfall patterns in warm years since 1925 suggests that in a warmer global climate, the United States and Soviet Union grain belts get less rain, while India, western Asia (the Middle East), and most of China get more.

The warming would be greatest in the polar region, increasing there as much as 9°C, and least at the Equator. The global "heat engine" is driven by the temperature difference between the warm Equator and the frozen poles, with winds and ocean currents moving heat from warmer to cooler regions. Since there would be less temperature difference between the regions, it is assumed there would be a general lessening in the speed of the engine. Winds and currents might be moderated. This could reduce the amount of moisture-laden air carried from the oceans into the interior of the large continents. It might, for example, reduce the amount of precipitation in the Great Plains and other important agricultural areas of the United States, and the Soviet Union. [Hermann Flohn, using both numerical models and paleoclimatic evidence from times in the past when the Earth was significantly warmer than it is now, concludes that there would be a belt of decreasing rainfall around latitude 38°N accompanied by increased evaporation and temperature (Fig. 6). He finds this prediction from these two sources to be "convincingly converging" with the results of the computer models.] The slackening of wind-driven currents might also reduce the upwelling of nutrient-rich water from the deep ocean to the surface, resulting in a decline in the productivity of ocean fisheries.

The most dramatic changes could occur in the polar regions. Some models show that the Arctic Ocean pack ice might melt in the summer months. This could have important implications for ocean transport. There is also speculation that a global warming might melt the Antarctic and Greenland ice caps, causing a catastrophic rise in ocean levels. Glaciologists discount much of this, since these ice caps would respond very slowly to any warming.

The West Antarctic, however, is a "marine" ice sheet, grounded in large part below sea level. Some glaciologists believe it is unstable, and that a reduction in the extent of the ice shelves and pack ice surrounding it would remove a buttress currently impeding its "surge." Glaciologists acknowledge that a climatic warming might affect this portion of Antarctica, but they disagree as to the timing. Deglaciation might require several hundred years following a 9°C rise in temperature in Antarctica. Should deglaciation of the West Antarctic occur, it would raise ocean levels 5–6 meters, resulting in flooding of low-lying coastal areas and ports. The global economic loss would be extensive. Like many climate-induced changes, however, it is arguable that humans could adapt to a rise in sea level were it to occur slowly enough: 5 years might be a "disaster"; 500 years might be tolerable.

There are many other questions about the climate of a high–carbon dioxide world, many of them with no answers at present: Will the climate change be steady and gradual, or will it occur in stepwise "jumps" to intermediate equilibrium states (with accompanying anomalies, such as long periods of drought or flooding)? Will the change be accompanied by, or induce, greater climatic variability, or will future patterns be steadier and more predictable? This question is of great importance to agriculture, since farmers around the world tend to assume that next year's weather will be pretty much like this year's. Many agricultural experts believe farming practices, given enough time, can eventually adapt to any future climate. But greater year-to-year fluctuations could cause serious losses.

Carbon cycle and atmospheric CO_2 buildup. The carbon dioxide content of the atmosphere may have varied throughout Earth's history. Bubbles of air trapped thousands of years ago in Greenland and Antarctic ice show that during the coldest period of the most recent ice age, atmospheric carbon dioxide was about half its present level. The causes of such natural fluctuations in carbon dioxide are not known, but possibilities include variations in release of gas from volcanoes (often containing about 12% carbon dioxide), increased storage in the "standing carbon" of forests at different epochs, and fluctuations in uptake or release of carbon dioxide by warmer or cooler oceans. Ocean temperature may in turn reflect changes in the solar radiation reaching Earth, possibly caused by cyclical shifts in the geometry of the planet's orbit around the Sun. (The "Milankovitch theory" suggests that shifts in the Earth's orbit and axis of rotation may have triggered the ice ages. The change in solar radiation may be amplified by continental drift if newly formed continents and mountain uplift block the flow of ocean currents and winds and thus alter the distribution of planetary heat.)

The carbon dioxide content of the atmosphere has almost certainly increased since Europe's Industrial Revolution, when clearing of forests and the combustion of fossil fuels began on a large scale. Unfortunately, during the 19th century only scattered instrumental records of atmospheric carbon dioxide were made. There was also no standardization of measurement technique or calibration of instruments. Despite the sketchy record, there is general agreement that in 1860 the carbon dioxide level was as little as 260 to as much as 290 ppm. Atmospheric carbon dioxide is now about 16.5–30% higher than the preindustrial level. (There is currently no economically practical "technological fix." Carbon dioxide can be removed from the atmosphere by various chemical means, but the energy required to do this exceeds the energy produced initially. The most practical means of removing excess carbon dioxide is widespread global reforestation, but it would take vast areas to reduce carbon dioxide levels appreciably.)

Systematic measurement of atmospheric carbon dioxide began in 1957, when as part of the International Geophysical Year, Charles D. Keeling began monitoring levels at Mauna Loa in Hawaii and at the South Pole. Additional measurements have been made at Point Barrow in Alaska, American Samoa, and other widely scattered locations far from industrial sources. Keeling's careful work has inspired great confidence in his data. The Mauna Loa record shows a steady rise in carbon dioxide concentrations from 311 ppm in 1957 to 338 ppm in early 1981, an increase of 8.7% in 24 years. The increase continues, currently at an annual rate of between 1 and 2 ppm, and tends to parallel fossil-fuel consumption.

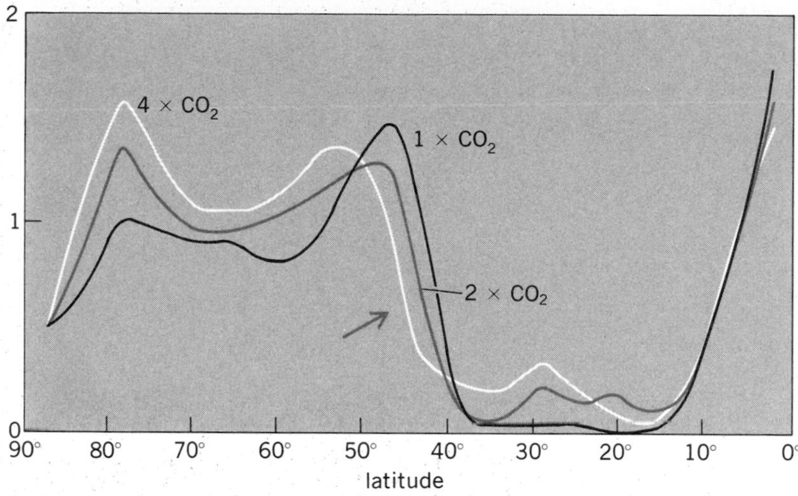

Fig. 6. Zonally averaged values of the difference between precipitation and evaporation (P-E) on land with different carbon dioxide levels. Note the strong decrease on P-E around latitude 38–49°N (arrow).

The Mauna Loa record shows a seasonal fluctuation as well as a long-term trend. Carbon dioxide levels reach a maximum in spring and a minimum in early autumn. The measurements include comparison of the ratio of two isotopes of carbon, ^{12}C and ^{13}C. Growing plants take up carbon-12 in preference to carbon-13. The carbon ratio variations suggest that the seasonal fluctuations reflect the uptake of carbon dioxide during the spring and summer growth of Northern Hemisphere deciduous plants, and the release of carbon dioxide during their fall and winter decay. However, the seasonal fluctuation may also be due in part to increased uptake of carbon dioxide by colder ocean water, and increased release of carbon dioxide as the oceans warm in spring and summer.

Controversy regarding preindustrial levels is heightened by inadequate understanding of the flow of carbon between the biosphere, the oceans, and the atmosphere. There is also uncertainty regarding the amount of carbon in the living and dead biomass and the amount in the oceans. Currently, about 5×10^9 metric tons of carbon are extracted as fossil fuel each year (releasing, when burned, roughly 2×10^{10} tons of carbon dioxide). Measurements at Mauna Loa and elsewhere reflect an annual increase in the atmospheric reservoir of only about 2.3×10^9 tons of carbon. Therefore, about half the carbon

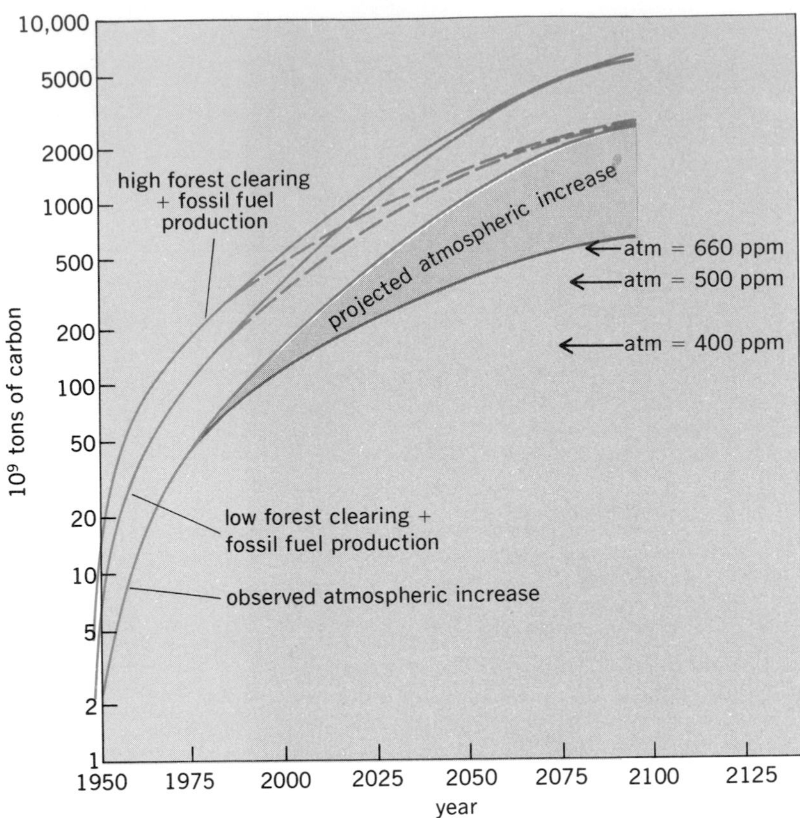

Fig. 7. Projections of future carbon dioxide emissions and atmospheric concentration. 1 ton = 0.9 metric ton. (*From S. Manabe and R. Wetherald, The effects of doubling the CO_2 concentration on the climate of a general circulation, J. Atmos. Sci., 32:3, 1975*)

does not stay in the atmosphere, and must be absorbed by the biosphere or the oceans (or possibly deposited as river-borne sediments). Calculations by ocean chemists indicate that the oceans cannot absorb as much as 2.7×10^9 tons of carbon a year.

Until the mid-1970s it was believed that limits to absorption by the oceans meant that the surplus carbon was being taken up by the biosphere. Several biologists questioned this, arguing that the global biomass appeared to be shrinking, and was possibly even a significant source rather than a sink for carbon dioxide. (Widespread land clearing, particularly of tropical forests, lent additional support to this view, as did calculations which showed that from 1850 to 1950 forest clearing could have contributed twice as much to the buildup of atmospheric carbon dioxide as did fossil fuels.) Currently, some estimates account for the missing carbon by assuming that a portion is preserved as charcoal or by regrowth of temperate forests cleared in the 19th century.

A firmer understanding of the carbon cycle and of the amount of carbon dioxide taken up or released by the various reservoirs will be required for reliable estimates of the amount of carbon dioxide that will reach the atmosphere and stay there as a result of human activity.

In a recent study, Roger Revelle wrote about the need for better data on rates and amounts of afforestation; decreases or increases of soil humus in lands that have been deforested and cultivated; the nature of ocean circulation, especially advection, convection, and turbulent mixing; the amount of organic carbon that settles out of the photosynthetic zone into deeper sea water; and the rates of accumulation or oxidation of organic-rich sediments, particularly peat, in marshes, wetlands, and estuaries. He indicates that these factors can be expected to fluctuate with changing concentrations of atmospheric CO_2.

Estimating future levels of CO_2. There are inherent uncertainties in attempting to predict how much carbon dioxide will be released during the next century and beyond. Estimates of future levels of carbon dioxide are based on assumptions about future energy demand as related to economic conditions and human behavior (Fig. 7). These projections usually begin with past trends, which are then extended into the future, with modifications of varying complexity or sophistication. However, it has proved extremely difficult to forecast future eco-

Table 1. Energy and carbon contained in fossil-fuel reservoirs

	Energy content of best estimates, terawatt-years*	Best current† estimates, 10^{15} g C	Upper limit speculations, 10^{15} g C	Historic consumption up to 1978, 10^{15} g C
Ultimately recoverable conventional petroleum resources	420	230	380	35
Ultimately recoverable conventional natural gas	331	143	230	15
Ultimately recoverable conventional coal	4,690	3,510	6,315	100
Shale oil	313	173‡	9,530‡	
Tar sands and heavy oil	136	75	200	
Total	5,890	4,131	16,655	150

*Assuming both conventional and unconventional crude oils have a density of 0.9 g/cm^3 and an energy content of 4.4 $\times 10^{10}$ joules per metric ton. 1 terrawatt-year = 29.89 quads = 1×10^{15} Btu.

†Not including that produced prior to 1978.

‡Not including carbon dioxide driven off from carbonate rock in retorting.

nomic conditions or human behavior. For example, gasoline consumption in the United States declined more sharply in response to increased prices than forecasters were able to predict only a decade ago.

According to a study by the International Institute for Applied Systems Analysis (IIASA), global energy consumption will continue to increase between now and the year 2030 at a rate varying from 1.6 to 2.4% per year depending upon economic growth. [The IIASA study, for example, projects that United States energy use will range from 98 to 123 quads (1 quad = 10^{15} Btu). This is a narrower range than estimated by the Committee on Nuclear and Alternative Energy Sources (CONAES) of the National Academy of Sciences, which said that United States use might vary from 58 to 134 quads.] The IIASA study predicts that energy demand 50 years from now will be three to six times the present world consumption of about 9 terawatts per year—90% of it from fossil fuels (Table 1).

The IIASA study notes that "only three primary sources have the potential for meeting a large fraction of the global primary energy demand over the long run—namely, fossil fuels (mainly coal), nuclear power, and solar energy. Future energy supply systems will rely heavily on the use of one or more of these sources." Nuclear and solar energy (and other renewable sources, and conservation) may thus play a critical role in future levels of carbon dioxide. Shifts in patterns of energy use are constrained, however, by inertia in energy systems. Mines, refineries, and power plants are expensive, require from 5 to 10 years for development and construction, and have working lives of from 15 to 50 years. There would thus be considerable lag time, even if carbon dioxide–induced climate changes eventually prompted a shift in energy policy.

The world's resources of oil, gas, and coal are currently estimated to total about 8×10^{12} metric tons. If this were all burned and half the carbon remained airborne, atmospheric carbon dioxide would increase about four times. A. M. Perry and colleagues believe such an increase is too much. They indicate that "acceptable" carbon dioxide levels may be no more than 1.5 to 2.5 times present concentration.

The implication is that the use of fossil fuels should somehow be restricted. When and how to do this are not clear. In addition to the uncertainties already cited, there are questions regarding the beneficial effects of carbon dioxide as a "free fertilizer" for plants and of improved climates in certain areas, the ability of farms and forests to adapt to the altered carbon dioxide–climate regime, the cost and rate of implementation of measures to limit carbon dioxide, and differing perceptions of who "wins" and who "loses" if the climate changes. Perry and colleagues conclude that it will be extremely difficult to keep the carbon dioxide concentration below 500 ppm, but it might be possible to keep it below 600 ppm if early corrective action is taken. The important climate effects will mainly arise after the year 2030. Those effects will be determined by energy use patterns and decisions taken in the near future.

There are two other significant unknowns regarding future levels of carbon dioxide: the amount and kind of energy used by the developing countries; and the amount of coal that is mined and burned. Projections of future world energy use made in the 1970s assumed that the developing countries would greatly increase their use of fossil fuel over the next 50–70 years. Population growth alone seemed sufficient justification for this conclusion. More importantly, if these countries were to make significant

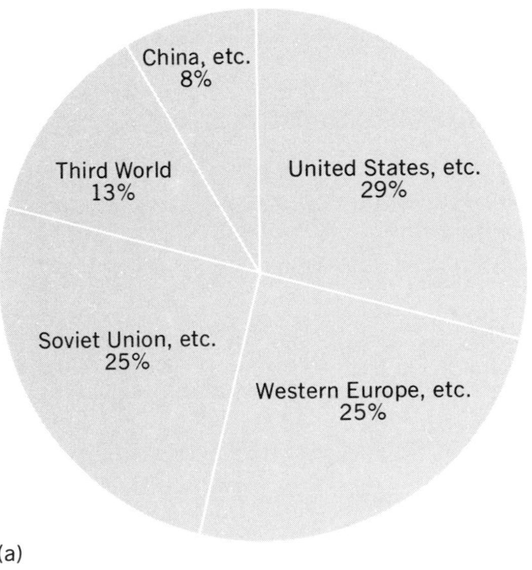

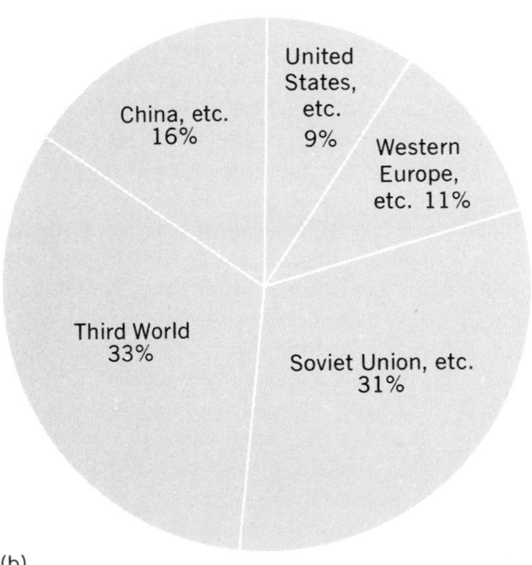

Fig. 8. Global carbon dioxide production between (*a*) 1974 and (*b*) 2025, assuming zero energy growth in the United States, 2% growth in Europe, 4% in the Soviet Union and eastern Europe, 4.5% in China, and 1.5% in the Third World. Making likely assumptions about the fossil-fuel use of each region, carbon dioxide emissions by 2025 will be six times those of 1974 (United States, etc., includes Canada; western Europe, etc., includes Japan, Australia, New Zealand; Soviet Union, etc., includes eastern Europe; China, etc., includes North Korea and Vietnam).

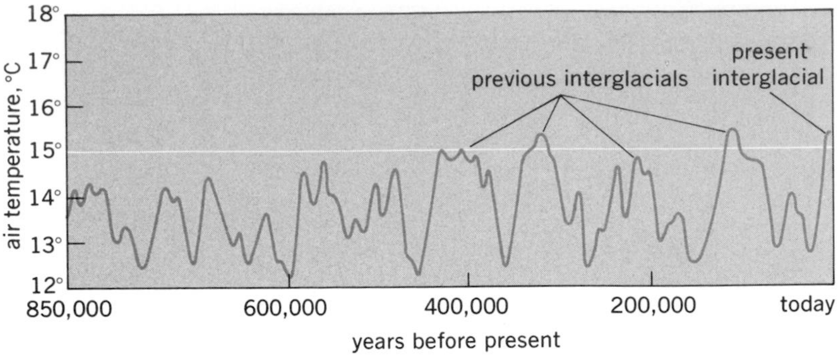

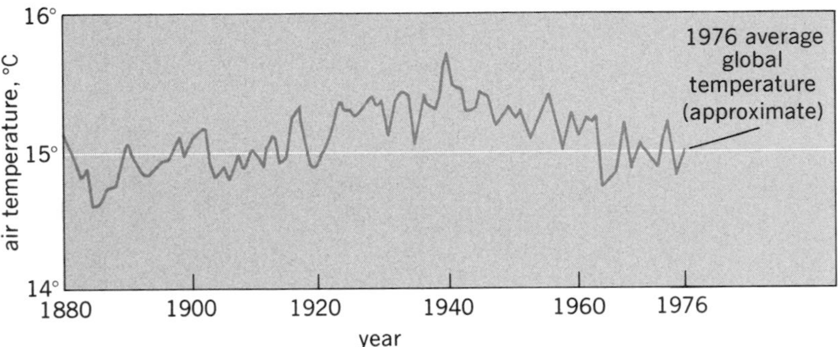

Fig. 9. Reconstruction of global temperature fluctuations showing that the Earth's climate is variable on both long and comparatively short time scales. Values over the long term are for global fluctuations; short-term values are given for the Northern Hemisphere. (*From W. O. Roberts and H. Lansford, The Climate Mandate, W. H. Freeman and Co., 1979*)

economic progress, a much greater use of energy was essential (Fig. 8). Some analysts now question these assumptions. The steep increase in oil prices has forced many developing countries to restrict their imports of oil, and the gap between their energy needs and their ability to pay seems likely to increase rather than narrow. The developing countries may add to the buildup of carbon dioxide in other ways, however, by continued logging of tropical forests, and land clearing due to population growth.

Coal is the critical factor in estimating future levels of carbon dioxide. Even if all the currently known resources of oil and natural gas were burned, atmospheric carbon dioxide would probably not exceed 500 ppm. Coal supplies 25% of the world's energy, but its share seems likely to increase. The 1980 World Coal Study projects growth in coal use of 4–5% a year, a doubling every 14 years. Three countries—the United States, the Soviet Union, and China—possess nearly 90% of the world's coal resources. There is more than enough coal to push carbon dioxide to levels sufficient to induce a climate reaction (Table 2). But coal is dangerous to mine, awkward to transport, and dirty to burn. It is not clear that a world coal trade, similar to international exports of petroleum, will develop. Engineering and technological improvements in coal liquefaction and gasification might affect this picture, however.

CO_2 warming and climates of the past. Climate fluctuates in irregular cycles, most of which are much longer than the human life-span, and this element of the environment is generally perceived as unchanging. But climatic fluctuations have profoundly shaped the evolution of humans and their societies. During the warm climates of the Middle Ages, between 900 and 1350, Arctic sea ice diminished, and the Norse cultivated Greenland and were able to grow grains above the Arctic Circle. During the Little Ice Age, from about 1450 to 1750, the climate became colder, at least in northern Europe. Glaciers expanded, Arctic sea ice returned, and human settlements in Greenland collapsed.

On the geologic time scale, the climate has been unusually warm for the past 10,000 years (Fig. 9), and it may not be accidental that great human civilizations developed during this period. If the cycle of ice ages repeats itself, the present warm "Interglacial" will presumably give way to a gradual cooling as the Earth enters another ice age. Carbon dioxide–induced warming, from this perspective, could be viewed as a sharp upward "blip" on a 10,000-year trend line of cooling.

Detection of CO_2-induced change. An unequivocal "confirmation of theory" is made difficult by the natural variability of climate. This is sometimes compared to "statistical noise." The problem is to detect a clear "signal" through this noise. T. M. L. Wigley and P. D. Jones note that the highest ratio of signal to noise occurs in summer and annual mean surface temperatures averaged over the Northern Hemisphere. However, the characteristics of warming during the early 20th century were similar to characteristics expected from increasing carbon dioxide, and the similarity may hinder early detection. Wigley and Jones conclude that decisions will possibly have to be made some time before "proof" of the effects of carbon dioxide on climate is available. Roland Madden and V. Ramanathan speculated that the surface warming predicted by the GCM models is not yet detectable "possibly because the warming is being delayed more than a decade by ocean thermal inertia, or because there is a compensating cooling due to other factors."

Table 2. Global distribution of coal resources

Area	10^9 metric tons coal equivalent
Soviet Union	4,860
United States and Canada	2,685
China	1,438
Western Europe	420
Latin America	32
Africa	172
Japan, Australia, New Zealand	272
Eastern Europe	171
Other parts of Asia	76
Total	10,126

Table 3. Experimental carbon dioxide increase effects on the biota

	Components of the biota	Primary effects	Secondary effects	Potential benefits/costs	Potential research requirement
CO_2 alone	Crops Forests Grasslands	Growth stimulation	Reduced water requirements Increased productivity	Greater yields per unit of area (but shading out of less CO_2 responsive cultivars or species)	Response of many varieties to a range of CO_2 levels Role of biota in regulation of global CO_2
	Aquatic plants	?			
	Animals Humans	None None		More forage available	
CO_2 with other atmospheric contaminants	Crops Forests Grasslands Aquatic plants Animals Humans	Closed stomata ?	Reduced injury	Reduced damage	Synergism

Information published in late 1981 suggested that effects of carbon dioxide–induced warming can perhaps already be seen. J. Hansen and coworkers found that there was an increase in global temperature of 0.2°C between the middle 1960s and 1980, yielding a warming of 0.4°C during the past hundred years. This is consistent with the theoretical values for the greenhouse effect. Cooling caused by volcanic dust in the stratosphere and possibly variation in solar luminosity seem to be the principal causes of observed fluctuations around the trend of increasing temperature. Possible effects on the climate during the next century include the creation of drought-prone regions in North America and central Asia, erosion of the West Antarctic ice sheet causing a rise in sea level around the world, and opening of the Northwest Passage due to melting of Arctic Ocean ice. Hansen and the others project a global warming for the 21st century "of almost unprecedented magnitude." On the basis of their model calculations, they estimate it to be about 2.5°C if there is slow energy growth and mixtures of nonfossil and fossil fuels are used. This temperature "would approach the warmth" of the Mesozoic.

G. Kukla and J. Gavin note that the greatest carbon dioxide–induced warming is expected to be along the marginal belt of snow and ice fields during spring and summer. They found that the summer Antarctic pack ice decreased by 2.5×10^6 km^2 (35% of average) between 1973 and 1980. They also compared the extent of summer ice as well as surface air temperatures between 1974–1978 and 1934–1938. The observations were qualitatively in agreement with the effect of the expected carbon dioxide increase. However, since it is not yet possible to establish a cause-and-effect relation, it is not known how extensively the changes can be explained by natural variability or other processes.

Impacts of a CO_2-induced climate change. Persistent uncertainties regarding estimates of future levels of carbon dioxide, and the climate's response to a doubling or other increase, make it difficult to predict the effects of a possible climate change or to estimate the costs and benefits for different regions of the world or various sectors of society. Over time, however, it may be possible to narrow the range of probable outcomes. Researchers might achieve sufficient understanding to be able to exploit favorable changes and mitigate unfavorable changes. Improved understanding of the impact of climate change might increase the likelihood that benefits will exceed costs.

A major uncertainty, for example, is the combined effects of added carbon dioxide, warmer temperatures, and changes in precipitation patterns on agricultural crop plants, domestic animals, agricultural pests, forests, other "less cultivated" biomes on land, populations of fish and other marine organisms, and the cryosphere, including permafrost, sea ice, and the West Antarctic ice sheet. The most important economic effect might be on food, forage, fiber, and forest plants. Will it be possible to select plant strains that show higher yield or other favorable responses to incremental changes in carbon dioxide? At present, many commercial greenhouses add carbon dioxide to improve production. Can plants be bred which will be take advantage of the expected ambient carbon dioxide enrichment? Similarly, can strains be developed which are more drought-resistant or otherwise tolerant of water and temperature stress (Table 3)?

Finally, can the impacts of climate changes on different societies be estimated? Can understanding of the ways that groups and institutions are likely to respond to the varied stresses be improved? The Dust Bowl of the 1930s in the United States, the

Sahelian drought in Africa in the early 1970s, and similar climatic events in the modern era and in the past reveal the kinds of social and other consequences that should be considered.

Climate change, if and when it occurs, will be superimposed on other existing future conditions—world population growth, a shrinking natural resource base, poverty and demands for economic equity, and political tension and conflict. People respond only to conditions that are perceived, and generally pay more attention to big, sudden events than to slow, incremental change.

Humans may now be bringing about an irreversible and fundamental shift in a major component of their environment. All countries of the world contribute carbon dioxide to the atmosphere. The climate changes that will almost certainly result will be felt, though in different ways, by all countries. There is no reason to believe that governments will take action—such as a global limitation on the use of fossil fuels. Additional research will clarify many questions, but major scientific uncertainties will probably remain. It may simply be necessary to await the outcome of this great geophysical experiment. [DAVID M. BURNS]

Bibliography: H. W. Bernard, Jr., *The Greenhouse Effect*, 1980; B. Bolin et al. (eds.), *The Global Carbon Cycle*, SCOPE, 1979; W. W. Kellogg and R. Schware, *Climate Change and Society: Consequences of Increasing Atmospheric Carbon Dioxide*, 1981; National Academy of Sciences, *Energy and Climate*, 1977; W. O. Roberts and H. Lansford, *The Climate Mandate*, 1979; U.S. Department of Energy–American Association for the Advancement of Science, *Environmental and Societal Consequences of a Possible* CO_2*-Induced Climate Change: A Research Agenda*, DOE/EV/10019-01, December 1980.

A–Z

Accelerometer

Electronic instruments used for measuring acceleration, called accelerometers, fall into two classes: open-loop, in which the sensing mass undergoes displacement during acceleration; and closed-loop, in which the sensing mass is held in place during acceleration by a restoring force. Closed-loop instruments, because of their complexity, are costly and are thus used primarily for precision measurements. A recently developed digital force transducer promises lower-cost, precision open-loop acceleration measurements. Called a quartz resonator force transducer, it is similar in nature to the mass-produced quartz crystals used in the electronic watch industry. Accelerometers based on the device offer an inherently digital output signal, low power consumption, and small size along with high-precision measurements.

Design. Acceleration is measured by using Newton's law $F = ma$. If the force F is measured electronically and the sensing mass m is known, acceleration a can be calculated. Single-axis sensing direction in accelerometers is accomplished by choosing one of two methods for suspension of the sensing mass, pendulous or nonpendulous. If the sensing mass is suspended from one side by a hinge so that it tries to swing in an arc, as a pendulum, the sensing direction is in the direction tangent to the arc. If the sensing mass is constrained to a linear motion by diaphragms, webs, or springs, the sensing direction is the direction of easiest motion. In both designs, the suspension system opposes the forces from acceleration in the two cross directions, leaving only the desired sensing direction available for measurement of F.

An example of a closed-loop accelerometer is the magnetic force return instrument, in which an electromagnet in a feedback loop is used to maintain the relative position of a sensing mass during acceleration. Sensing-mass suspension can be pendulous or nonpendulous. Magnitude of F is determined by measuring the magnet current necessary to hold the sensing mass in position.

In contrast, open-loop accelerometers generally measure displacement of the sensing mass electronically as it works against a known spring rate during acceleration. Again, suspension of the sensing mass can be pendulous or nonpendulous. This displacement introduces two undesirable effects: hysteresis from the mechanical suspension, and coupling between acceleration measured along the sensing axis and acceleration along the two cross axes as the geometry changes with displacement. Improved performance of accelerometers can be achieved with reduced displacement of the sensing mass during acceleration; thus, one main advantage of closed-loop designs is the effectively infinite stiffness of the sensing-mass constraint; that is, the suspension does not permit motion.

Vibrating quartz force transducer. An inherently digital force transducer with high stiffness which would lead to reduced displacement of the sensing mass during acceleration is shown in the illustration. The device is in the form of a closed-end, or double-ended, tuning fork. A flexural, mechanical resonant mode exists in the configuration in the illustration whereby the two bars bend alternately outward away from one another and then inward toward one another in a manner similar to the tines of an ordinary tuning fork. Since there are two inactive ends to the device, it is possible to apply a longitudinal force as shown in the illustration. With no force present, the resonant frequency is determined by the mass and flexural rigidity of the parallel bars, or tines. When a longitudinal force is present, the flexural vibration frequency changes much in the manner of the tuning of a guitar string: tension increases the resonant frequency, compression decreases the resonant frequency.

If a piezoelectric material such as quartz is used to fabricate the device, electrodes can be added to the surface as shown in the illustration. The piezoelectric effect allows sensing of the mechanical resonant frequency by incorporating the device into the feedback loop of a high-gain amplifier to form a crystal-controlled oscillator. Crystal-controlled oscillators are used for precise timing references in

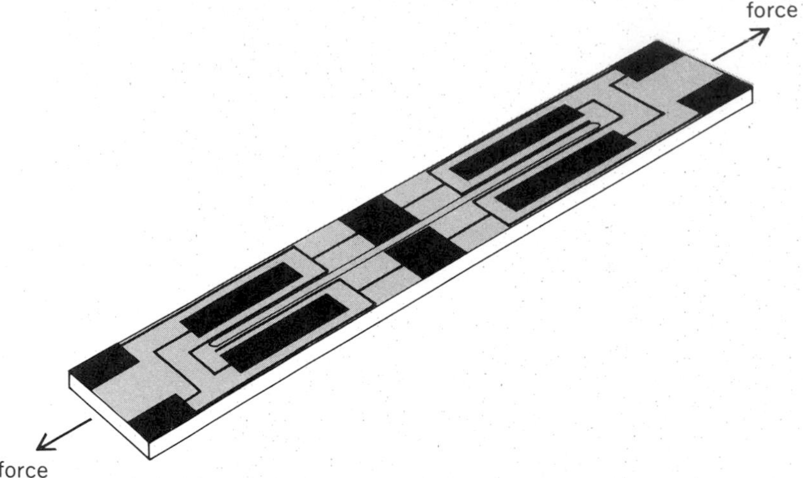

Geometry of the double-ended tuning fork force transducer, a plate with a longitudinal slot. Electrode pattern is included for incorporation into a crystal-controlled oscillator circuit when a piezoelectric material such as quartz is used. (*From E. P. EerNisse, U.S. Patent 4,215,570, 1980*)

computers, electronic counters, watches, and clocks. In the present case, the output frequency of the crystal-controlled oscillator varies proportionally with the longitudinal force applied to the double-ended tuning fork. Since frequency is readily measured precisely, and since frequency is inherently a digital measurement, a crystal-controlled oscillator incorporating the device of the illustration is a precise digital force transducer.

Performance capabilities. The inherent stability of the mechanical resonant frequency with no force applied can be as high as the stability seen in other applications of crystal-controlled oscillators employing flexural-mode quartz resonators; namely, $1/10^8$ over several minutes and $1/10^7$ over months if cleanliness is maintained in the crystal enclosure. Since the force-induced frequency changes can be larger than $1/10^2$ even with a safety factor allowed for avoiding breaking of the device, the resolution capability in terms of fraction of full scale can be better than $1/10^6$ over several minutes and $1/10^5$ over several months.

The geometry in the illustration, which is a plate with a longitudinal slot, has been chosen so that mass production is possible with photolithographic and chemical etching techniques similar to those used in the semiconductor integrated circuit and watch crystal industries. Dimensions for the device are nominally 1 cm long by 0.09 cm wide by 0.01 cm thick. The resonant frequency for the flexural resonant mode is typically 50 kHz. A representative full-scale force is 10^5 dynes (1 N). With the use of integrated circuits for the oscillator circuit, power consumption is 10^{-4} W. Stiffness of the device to the longitudinal force is typically 4×10^8 dynes/cm (4×10^5 N/m), so use in an open-loop accelerometer is attractive because actual displacement of the sensing mass will be small. Resolution for a 10-*g* accelerometer would be 10^{-4} to 10^{-5} *g*, usable for even demanding applications such as inertial guidance.

Applications. The features of an accelerometer using a vibrating quartz force transducer are high precision and resolution, low power consumption, small size, inherently digital output, and potentially low cost. Although not all of these features may be advantages in a given application, some features apply in most applications. For instance, the low power consumption and digital output are attractive in battery-powered packages for sea bed stability and geological monitoring stations. The digital output means all-digital circuitry without analog-to-digital conversion, thus preserving intrinsic measurement accuracy and allowing operation in severe temperature and radiation environments such as might be encountered in oil, gas, and geothermal exploration and in the military. Finally, the capability for mass production and low cost opens up applications where present accelerometer technologies cannot be applied.

For background information *see* ACCELERATION MEASUREMENT; ACCELEROMETER; PIEZOELECTRIC CRYSTAL; TUNING FORK in the McGraw-Hill Encyclopedia of Science and Technology.

[ERROL P. EERNISSE]

Bibliography: E. P. EerNisse, U.S. Patent 4,215,570, 1980.

Acid rain

Acid rain is defined as precipitation with a pH below 5.6–5.7; such hydrogen ion concentrations are obtained when pure water is in solution equilibrium with aerial carbon dioxide at 25°C. Most researchers consider acid rain to be of significance when pH values are at least tenfold more acid than 5.7, that is, 4.7 and below. Acid rains were common during the 18th and 19th centuries, when coal was the primary industrial source of heat and power and when the smelting and refining industries used coal. The first accurate report of acid rain appeared in 1911 in Leeds, England, where pH values approached 4.0. In North America, acid rains were noted at several locations in the northeast by 1939, and recalculations of data on earlier storms suggest that some of these precipitations may have been acidic. Alkaline rains are also found, particularly where limestone or alkaline dusts enter the atmosphere. Reports of acid rain are now common throughout the Western world and in Japan, with pH values of 3.8–4.5 becoming almost the norm. Snow, too, is now acid, as are fogs and rime ice formations.

Sources of acidity. Several natural sources can be considered. Among the products of volcanic eruptions are sulfur compounds. The hot springs of Arkansas and the fumarols in Yellowstone National Park contribute acidity to the air, and considerable amounts of oxides of nitrogen and sulfur are end products of the metabolism of several groups of bacteria. In spite of these natural air pollutants, the pH of glacial ice is close to 5.0, indicating that natural emissions of acid-forming compounds are not major sources of acid rain; anthropogenic sources are the primary contributors.

Burning of fossil fuels for heat and power is the major factor in the generation of oxides of nitrogen and sulfur which are convertible into nitric and sulfuric acids washed down in rain. Heating oils contain 0.5% nitrogen and variable amounts of sulfur; 60% of all coal has 1–3% sulfur and 1.5% nitrogen. Wood smoke, too, has significant acid-forming potential. Although gasoline contains only small amounts of nitrogen and sulfur, the modern high-compression engine can produce nitrogen oxides directly from the gaseous nitrogen that makes up about 75% of ordinary air. The combination of air compression in the cylinders, the igniting spark, and the catalytic action of cylinder metal converts N_2 into NO_x, which, in turn, can be converted into HNO_3. Pulp and paper mills use acid bisulfites, sulfurous acid, and sodium sulfide to remove lignin from wood pulp, and the oxidized sulfur compounds are vented into the air. Mining and smelting operations involve the conversion of sulfides and other complexes of heavy metals into pure copper, cadmium, zinc, or iron, with the sulfur oxidized to SO_2. Even automobile tires contain sulfur; the smell caused by a jackrabbiting hot rodder is due partly to the sulfur oxides formed when smoking rubber meets the road. Small but significant quantities of sulfur and nitrogen compounds reach the air from many industries.

Since all these sources have been active for more years than acid rain has been found, the question arises as to why it is observed now. In part, the answer is technology. Starting about 1950 the industrial world decided that "the answer to pollution is dilution." In order to reduce local deposition of acid-forming substances that devastated the land immediately downwind from Sudbury, Ontario, in many steel towns, and communities with coal- and oil-burning power-generating plants, smokestacks were built to great heights; the stack of the International Nickel Company in Sudbury is almost 400 m tall, and others are close to this height. Thus, instead of SO_2 and NO_x or heavy metals dropping within 10 km of the source with only limited air-borne time for conversion of oxides into acids, the acid-forming compounds are now carried into the upper atmosphere to be transported for distances known to exceed 1000 km. The development of electronic precipitators and filters to remove particulate matter as mandated by various national and local ordinances was also a factor, since fly ash is alkaline and potentially could neutralize acid-forming materials in the stacks. Catalytic convertors in automobiles can oxidize sulfur and nitrogen compounds directly into acids.

Extent. The most thorough studies are from Scandinavia, where the acid rain phenomenon was recognized in the early 1950s. Virtually all of northern Europe routinely experiences acid rain. Similar patterns of acid rain are now commonplace throughout the industrial world. Since airflow patterns in the Northern Hemisphere show trajectories in a west-to-east direction, acidic precipitation is experienced not only in areas of sources but downwind and across regional and even national borders. As new areas become industrialized or heavily populated, for example, the United States Sunbelt, previously unaffected areas are included in the patterns. Canadian sources contribute about 15% of the acid pollution in the northeastern United States, and United States sources are responsible for about 65% of Canadian acid rain. Virtually identical patterns are found in Europe and Japan.

The composition of the polluted rain is not completely known. Early studies indicated that the acid was primarily sulfuric, with small amounts of hydrochloric acid due to sea sprays and even smaller amounts of ammonium ion. More sophisticated analyses show that nitric acid is also present. In recent years the percentage of acidity attributable to nitric acid has been increasing both absolutely and relatively; in 1981–1982 about 60% of the total acidity in rain was determinable as sulfuric acid, and 40% as nitric acid. The physical and chemical conversion of oxides of sulfur and nitrogen into acids in the atmosphere is still under investigation, with current results suggesting that time in the air, available ozone, ultraviolet radiation, temperature, and other factors are involved in complex mechanisms and processes. Some reports indicate that acid concentrations are increasing over time, while other studies dispute this conclusion. In any case, there certainly is no decrease in the acidity of rain over the past 5–10 years. Sometimes lost sight of in the furor over acid rain is the fact—known for at least 30 years—that sulfur dioxide is itself highly toxic to plants and possibly to animals, independently of any conversion into sulfuric acid.

Consequences. The effects of acid rain on structures have been amply documented. Athens's Parthenon, Rome's Coliseum, Venice's San Marco horses, grave markers, and metal structures such as bridges have been eroded to the point at which extensive and expensive maintenance and repair work are necessary. Exfoliation of marble and limestone monuments and buildings and pitting of granitic stonework are common all over the industrial world.

Biological consequences are also evident. One phenomenon was first noted in Scandinavia in the early 1960s: many remote, pristine lakes in northern Europe are virtually sterile, devoid of fish (particularly trout and salmon), and depauperate in invertebrates and most plants; one can see directly to the bottom of such lakes through crystal-clear water. By the early 1970s, sterile lakes were reported from remote areas of Ontario, Quebec, and the Adirondack Mountains of New York State. With few exceptions, the pH of these lakes was below 4.8–5.0; prior to acidification, it had been closer to 5.5–6.0. Almost invariably, such bodies of water are on granitic bedrock, frequently Precambrian shield gneisses with low calcium and magnesium concentrations and almost nonexistent ability to neutralize acids. Lakes overlying limestones or dolomites are rarely affected, nor are lakes whose watersheds are high in acid-neutralizing compounds (high cation-exchange capacities).

Fish eggs do not hatch well in acidic water; those fry that emerge are frequently abnormal, and zooplanktonic, bottom-living animals are also affected. The aquatic plants that are the primary producers are also reduced in number. To further compound the problem, the concentration of aluminum ion (Al^{3+}) in damaged lakes is abnormally high. Aluminum ion alters cell membranes, preventing the normal cellular exchange of ions; and even slightly elevated Al^{3+} (0.1–1.0 μg/liter) blocks oxygen uptake and CO_2 release from fish gills. As soils are acidified by precipitation to pH 5 and below, hitherto insoluble aluminum complexes are converted into Al^{3+}, which moves through a watershed and is deposited in lakes. This flush of Al^{3+} is particularly noticeable in watershed waters following the melting of acidic snowpacks in the spring, a process that synchronizes with the spawning and hatching processes. Attempts to reverse the damage by liming lakes have been only partly successful and may make a bad situation worse.

Fish and other aquatic life are not the only environmental system at potential risk; a major question is whether terrestrial plants and animals are being affected. The limited studies on the effects on human and animal lungs of inhalation of air containing acidic water vapor have been poorly designed and are equivocal. Both laboratory and field studies suggest that major crops are unlikely to be adversely affected. Only when repeated additions of acidulated water at pH values in excess of those found in acid-rain monitoring studies are made is there any structural or physiological damage capable of reducing yields. Since good agricultural practices include liming, fertilizing, and control of organic matter—all of which neutralize acids and immobilize Al^{3+}—croplands and the crops growing on them are not in danger. In spite of suggestions that the nitrates and sulfates in acid rain may be useful fertilizers, simple calculations show that the fertilizing ability of acid rain is negligible and the acid load is clearly contraindicated.

A much more serious potential problem may exist in natural terrestrial ecosystems where, particularly at higher elevations, soils are poor, shallow to bedrock, and climatic conditions may be harsh and nutrients scarce in already acidic soils. Plants and the animals dependent upon the plants tend to be perpetually on the edge of disaster, with survival balances capable of being easily upset. Since people are dependent upon such natural ecosystems for timber, recreation, watersheds, and esthetic values, a few laboratories in Europe and even fewer in North America have begun to turn their attention to forested lands. No comprehensive picture has yet emerged, although research suggests that the interaction of acid rains, fogs, and snows with anthromorphically caused additions of heavy metals (Pb, Cu, Zn) and aluminum can interfere with growth of many important microflora components of forested ecosystems.

For background information *see* AIR POLLUTION; ATMOSPHERIC GENERAL CIRCULATION; FRESH-WATER ECOSYSTEM in the McGraw-Hill Encyclopedia of Science and Technology.

[RICHARD M. KLEIN]

Bibliography: F. H. Braekke (ed.), *Impact of Acid Precipitation on Forest and Freshwater Ecosystems*, Reclamo, Oslo, 1976; C. V. Cogbill and G. E. Likens, *Water Resources Res.*, 10:1133–1137, 1974; Environmental Protection Agency, *Research Summary: Acid Rain*, EPA-600/8-79-028, Washington, DC, 1979.

Activation analysis

The monitoring of wear, corrosion, and the like in both laboratory and plant environments is necessary on the one hand to understand the processes involved and on the other to establish the state of particular components in a system, in order to meet safety and operational standards. A basic parameter is the direct measurement of the loss of material from the surface exposed; yet there are few methods that offer such measurement. Thin-layer activation is virtually unique in providing an accurate and sensitive measurement of material loss which can be used in place under operating conditions. It does so by incorporating trace amounts of radioactive material into the surface. The method is now widely used

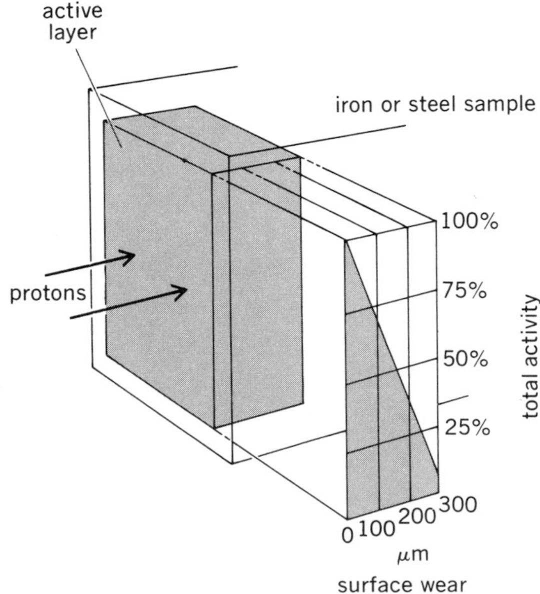

Fig. 1. Thin-layer activation of iron or steel by a proton beam. (*From T. W. Conlon, Ion beam activation for materials analysis: Methods and applications, IEEE Trans. Nucl. Sci., n.s., vol. 28, copyright 1981 by the Institute of Electrical and Electronic Engineers, Inc.*)

in Europe and the Soviet Union for measuring mechanical wear. In particular it has been extensively developed at the Harwell Laboratory (United Kingdom) for this as well as for corrosion and condition monitoring.

Technique. The principle of the technique is illustrated in Fig. 1. An ion beam (for example, protons) from an accelerator is used to label a well-defined thin layer in the surface of interest by generating trace quantities of a radioactive isotope (the label consists typically of 1 part in 10^{10} of surface material). If material thus marked is subsequently removed from the surface by any process such as wear, erosion, or corrosion, the reduction in activity as measured by the γ-rays emitted can be monitored in place, even through intervening material, and interpreted to quantify the loss. The layer depth can be precisely controlled to suit the application by varying the energy of the incident ion beam, typically in the range 25–300 μm, although layers as thin as a few micrometers up to several millimeters can be generated, as required.

The total amount of radioactivity generated is low, typically 1–10 microcuries, and comparable to that found in a luminous wristwatch or naturally occurring in a concrete block. This is possible because the activity is concentrated in that thin surface layer over a limited area. (This contrasts with the neutron activation technique, which inevitably results in the whole component being activated.) Thus only elementary handling precautions are required.

The activity is monitored, using a simple γ-ray monitor and pulse-counting electronics, in one of two ways (Fig. 2). The reduction in activity of the activated component, as material is removed from the surface, can be monitored directly or through a considerable thickness of intervening material [for example, an inch (2.5 cm) of steel], since the radiation is penetrating. The sensitivity in this case is about 1% of the layer depth, for example, 1 μm for a 100-μm layer. Alternatively, the accumulation of active debris, transported by a fluid circuit and, if possible, collected in a filter, can be monitored to achieve a much higher level of sensitivity, 0.01% of the layer depth, or a fraction of a microgram of material. In both cases, it is essential that debris be removed from the spot. The sensitivity is ultimately limited by accuracy in maintaining monitor position and the levels of natural background in the local environment.

Activation of components and plant. The ion beams are generated in the vacuum environment of an accelerator. They are transmitted through a thin metallic window into air for activation applications.

Ion beams can be accurately focused and collimated to dimensions down to at least 1 mm^2. (Indeed so-called micro beams providing spots of a few square micrometers can be useful in some applications.) By using deflectors, beams can be scanned to cover more extensive areas uniformly. Thus a wide variety of activation conditions, suitable for

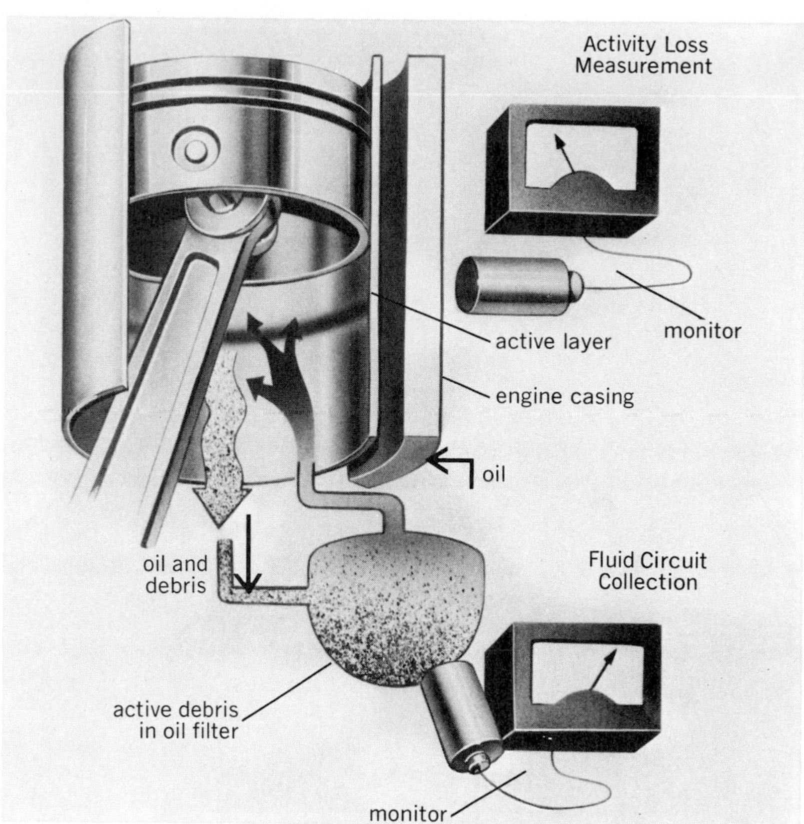

Fig. 2. Two methods of measuring activity and surface loss; dismantling is not required. (*From T. W. Conlon, Ion beam activation for materials analysis: Methods and applications, IEEE Trans. Nucl. Sci., n.s., vol 28, copyright 1981 by the Institute of Electrical and Electronic Engineers, Inc.*)

various applications in wear, erosion, or corrosion, can be achieved. Figure 3 shows some of the currently used activation geometries, in which combinations of focusing, collimation, beam scanning, and component scanning can be found. One requirement is common to all: the beam must follow a line of sight from the entrance of the component to the activation point.

Materials. A wide range of materials can be activated, including most technological metals and alloys of metals, such as iron, steel, copper, titanium, aluminum, stainless steels, and bronzes. In addition, ceramic materials and compounds such as tungsten carbide can be activated. In these materials, the ion-beam exposure required to generate suitable low levels of radioisotope activity for surface loss measurements produces no significant change in mechanical properties. For some materials, however, such as plastics, diamond, or other insulators, direct beam activation can result in significant beam-induced damage. A method devised at Harwell overcomes this problem. Radioactive atoms are generated by an ion beam in a so-called sacrificial target and become implanted in the substrate while the direct beam is caught on a beam stop. The implanted layer of radioisotope label includes only a very low level of scattered direct beam, and the direct beam damage is avoided.

Nonuniform surface loss. The standard activation technique has been widely applied for several years to surface loss measurements under wear, for example, in automobile components, for erosion by fluids and electrical breakdown, and in corrosion studies. In many areas of application, surface loss occurs fairly uniformly; under these conditions thin-layer activation provides an accurate measurement of the loss.

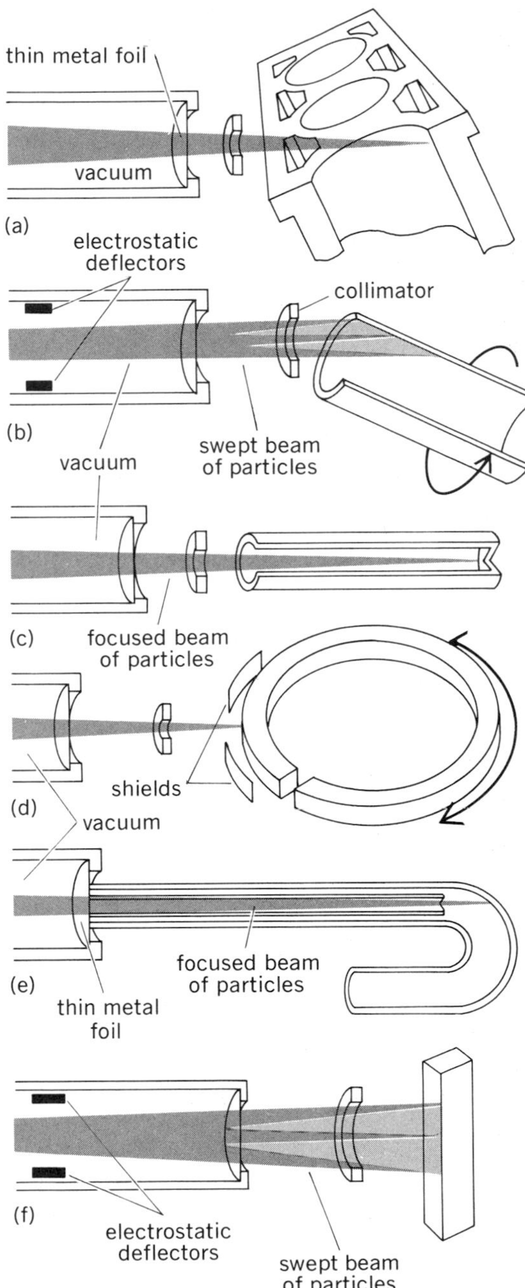

Fig. 3. Mounting and activation of different components. (*a*) Spot irradiation on each of several cylinders in an engine block. (*b*) Extended band of activity inside a rotating cylinder liner. (*c*) Finely collimated spot at the end of a narrow tube. (*d*) Activation of a segment of a large piston ring. (*e*) Activation of a spot on the bend of a tube. (*f*) Large-area activation of a plate by a swept beam. (*From T. W. Conlon, Ion beam activation for materials analysis: Methods and applications, IEEE Trans. Nucl. Sci., n.s., vol. 28, copyright 1981 by the Institute of Electrical and Electronic Engineers, Inc.*)

There are well-recognized conditions under which material loss is not uniform over the surface. In certain corrosion regimes, for example, local attack may occur in preference to uniform corrosion, resulting in pitting or undercutting. Straightforward application of thin-layer activation will underestimate the actual loss, especially when pitting penetrates beyond the labeled layer.

To overcome this problem, a double layer of radioisotope labeling, each layer having a different depth and distinctly characteristic γ-radiation, has been developed. In essence, the deep layer provides, to a first approximation, a measure of the volume loss of material, while the shallow layer indicates the fractional area of surface under attack. In this way the difference in pit morphology resulting from uniform corrosion, conical pitting, and bottle-shaped (undercutting) pitting is clearly differentiated. Thus, double-layer activation provides a sensitive indicator of the occurrence of local attack and, where the corrosion is shown to be approximately uniform, gives confidence in the quantitative determination of surface loss.

More detailed information on pit characterization can, in principle, be obtained by incorporating more than two layers in the surface. For example, a sequence of active layers, each with a distinct isotope and minimum overlap between layers, would enable the determination of the actual depth below the surface from which material had been removed. A detailed depth distribution of material loss would then be possible. This is, however, much more complex to achieve in practice.

In parallel to this approach, work is in progress to develop and apply buried layers of activity to signal the occurrence of wear or corrosion to a critical depth. This provides a similar function to the so-called sentinel holes used in plant monitoring.

Summary. Thin-layer activation provides an accurate and sensitive method of measuring surface loss in place for a wide range of materials, in many applications, and is particularly useful in monitoring and fundamental studies of wear, erosion, and corrosion. The use of double-layer activation or buried layers adds to the effectiveness of the method in applications where nonuniform surface loss presents measurement problems.

For background information *see* CORROSION; NONDESTRUCTIVE TESTING; WEAR in the McGraw-Hill Encyclopedia of Science and Technology.

[T. W. CONLON]

Bibliography: T. W. Conlon, *IEEE Trans. Nucl. Sci.*, n.s., vol. 28, April 1981; T. W. Conlon, *Wear*, 29:69–80, 1974; T. W. Conlon, *Tribology Int.*, 12:60–64, 1979; G. Essig and P. Fehsenfeld, *Proceedings of the EPS Conference on Nuclear Physics Methods in Materials Research, Darmstadt*, pp. 70–81, September 1980; V. I. Postnikov (ed.), *The Surface Activation Method in Industry* (in Russian), 1975.

Agricultural systems

Ecological concepts have recently revealed new ways of looking at subsistence agriculture. Subsistence agriculture is often used as a pejorative term even though it is not a measure of the productivity of a farming system. Yields under animal-powered substance agriculture often match or exceed those of mechanized agriculture, but if consumption equals production because of high population densities, the agriculture is at the subsistence level despite efficient production techniques. When consumption equals production, population pressures often force farmers to use poorer lands for agriculture so that averages for a particular country may give the appearance of inefficient agricultural practices. The high yields of subsistence agriculture that are often overlooked by scientists are now receiving more study.

If the operation of agriculture in developing countries with a long tradition of farming is to be understood, it is best to direct attention to the functional units of production, which are generally villages. Each village tends to be a self-supporting population of plants and animals with a fairly closed system of nutrient and energy exchange that is best called a village ecosystem.

Village ecosystems. Human energy is directed toward controlling plants in order to produce crops supplying the food, fuel, and fodder needed in a village. Villagers often come close to maintaining a self-sustained cycle of nutrients that is controlled by humans who use the solar energy stored in previous crops. Perhaps the interest in the management of village ecosystems developed in response to two concerns. First, the rejection of outside advice was perceived as a central problem in developing countries. Second, the operation of a solar-powered human ecosystem is intrinsically interesting. It is now believed that if outside advice is based on an understanding of how the village ecosystem works, it will usually be accepted, while suggestions that disrupt the pattern of labor demand, fuel supply, or fodder will be rejected. Acceptance and rejection are now seen to be rational, intelligent decisions based on the villagers' insights or traditions, and the decisions generally embody an intuitive understanding of how their ecosystem functions (Fig. 1). The regulation of village agriculture may well be a high level of ecosystem engineering based on thousands of years of cultural and biological evolution. Traditions are a guide for evaluating probabilities, such as the association of future weather with present conditions, which is often the basis for deciding upon a mix of crops that is most likely to meet the needs of the village for food, fuel, and fodder. Some combination of tradition and insight is also a guide that villagers use in deciding how a new crop or process may affect the village ecosystem (Fig. 1).

Biological components. Crops fix the solar energy that enters the ecosystem and is used to drive the system. The primary trap for solar energy is a tuber, such as yams, manioc, or potatoes, a set of grasses, or both. Three grasses—rice, wheat, or millets—support most village ecosystems. Secondary crops enrich the diet with proteins and other minor nutrients, or supply fodder or fuel. Consistently high yields can be achieved only if some components of the previous year's crop, manure, and seeds, and the work of humans and animals (Fig. 1), are invested in the next crop. Plants are the first trophic level (producers); humans and animals are the second trophic level (consumers in Fig. 1). Generally, the labor needed at the time of ground preparation and planting sets the requirements for a working population of humans and draft animals relative to crop area. The working population must reproduce to replace itself over time; hence, the minimum stable population is greater than the working population.

The well-known metabolic requirements of humans and cattle define their total energy demand. The two components of the second trophic level have different resource bases because humans feed largely on seeds and cattle feed on leaves and stems (Fig. 1).

Crop yields. Wild populations of the ancestors of wheat and barley produce as much as 800 kg/ha. Under cultivation, all grains evolve stronger stems, seeds that are strongly attached in firm heads, and seeds that separate from chaff because seeds from such plants are more likely to be replanted by humans. Besides these plant responses to the regimen of cultivation, there are the increases in yield that result from farmers' picking the largest seed heads for planting. There were only modest increases in yields during the first few thousand years of agriculture, with historical records suggesting that wheat yields were stabilized at 800–1000 kg/ha until the 20th century. Barley and oats were at the same level, and the various millets had yields 100–200 kg/ha smaller. Corn yields stabilized at twice those of wheat. Rice yields in 19th-century Asia ranged from 1500 to 4000 kg/ha, depending on the reliability and management of water. Rice yields are high because rice usually grows in floodplains where annual floods enrich the soil. Until recently the yields under mechanized agriculture were not greater, and sometimes were lower, than the yields from village ecosystems that were close to the subsistence ratio. The differences between mechanized agriculture and village ecosystems that developed during the last 50 years are largely attributable to fertilizers and plant breeding, not the techniques of farming. New varieties and fertilizer subsidies to developing countries have brought about important but less dramatic increases in yields than in developed

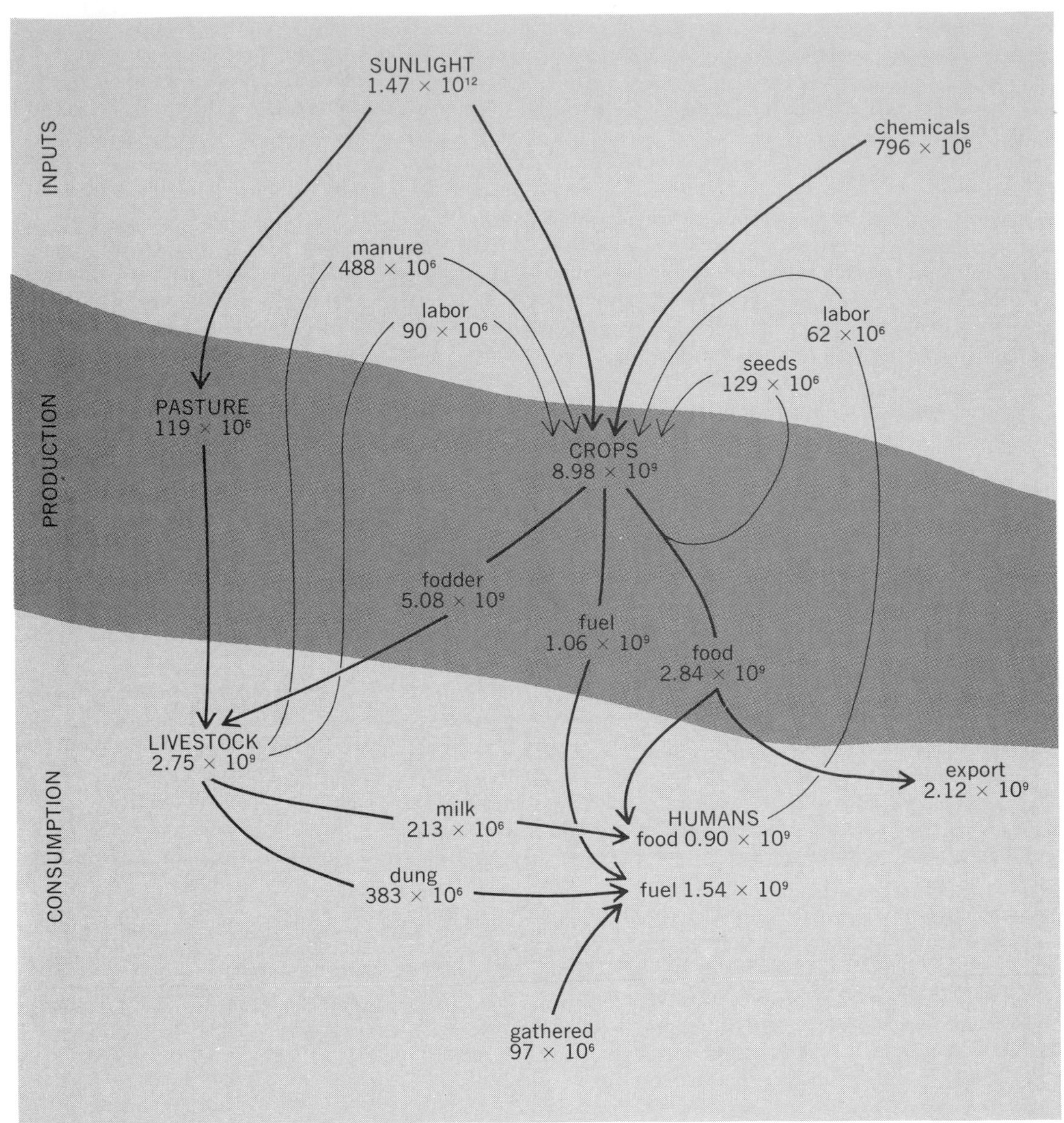

Fig. 1. Major interactions of an animal-powered village ecosystem model for Panayakurichi in southern India during 1979. The heavy arrows trace the patterns of energy exchange in kilocalories per year (1 kcal = 4.18 kJ) for a village of 994 persons with 158 ha of cropland. The light arrows show the feedbacks that maintain the ecosystem. Yields of 16,000,000 kcal/ha (67,000,000 kJ/ha) of grain are equivalent to about 4600 kg/ha. That is the average for yields of long-grain rice in the United States.

countries, in part because of the cost of chemical inputs and the need for local cultivars that are compatible with village ecosystems.

Folk technology. When labor is abundant and other resources are expensive and scarce, farmers evolve a biological technology based on the ecology of plant communities. This folk technology has not been studied so as to reveal its operation in scientific terms. Folk technology includes complex cropping systems, biological pest control, and multiple crop usage.

Complex cropping systems. When two different crops are planted together (intercropping), the combined yield is almost always greater than the yield of either crop alone. From 5 to 10 crops are usually grown together by dryland farmers in India because the absolute yields of mixtures are greater and more reliable than is the case for single crops, insect damage is lowered, and the soil resources are less strained. Intercropping usually involves a succession from low, early-bearing, shallow-rooted plants to late-bearing tall plants with deep roots. Intercropping appears to be an efficient technological exploitation of plant community ecology.

Pest management. Weeds can be exploited to benefit crops. Early in the crop cycle, weeds shade the soil surface and reduce heat stress on the crop. They reduce erosion and sequester nutrients that might leach away. When weeds begin to compete with the crop, they are removed for use as either green manure or cattle fodder. Some weeds are used as trap plants for insect pests, and others repel pests; weeds sown in the fallow season may retain or supplement soil nutrients. Weeds that indicate the quality of a fallow field are used to decide when and what to plant.

Multiple crop use. When land is abundant, cropping is done rather casually, and food for cattle, fuel for cooking and heating, and materials for construction can be collected nearby. As adjacent lands become overexploited or are committed to agriculture, the crops must supply all the needs of a village. Most grains produce equal quantities of grain, leaves, and stems; thus, the yield of fodder for draft animals is at least equal to the yield of human food. Cooking requires about 5000 kcal/day (21,000 kJ/day) of fuel, roughly twice the energy of a minimal diet, 2500–300 kcal/day (10,000–13,000 kJ/day). The small oxen in typical villages require 8000–10,000 kcal/day (31,000–42,000 kJ/day). When virtually all the land of a village is cultivated, an increase in food production will have little value unless the needs for fodder and fuel are met. The harvest of about 16 metric tons/ha of biomass in Panayakurichi (Fig. 1) is not unusual in intensive cropping.

Crops and labor inputs. The labor from humans and draft animals controls the quality and quantity of production. There is usually a limited period in which to plant a crop. It takes 20–40 days/ha to prepare the ground and plant a crop. There must be a density of at least one laborer per hectare to plant a crop within a month. Ground preparation for rice, which takes 50–70 days/ha, requires two or three laborers per hectare. Another 20–100 days are needed to care for and harvest the crop, but the time for this labor is often quite flexible.

Until the 20th century all farmers produced a variety of crops, and farmers had considerable skill in combining crops so as to stagger the labor demands over the season. Villagers still combine crops as they did in China (Fig. 2). If periods of peak labor demand are not overlapping, one laborer can farm a larger area.

A further increase in farming intensity can be achieved by planting a second crop in an already-growing crop (relay cropping). Two crops can be grown in a much shorter time, and there is often a considerable increment in the yields of the second crop.

These examples of the complexities in the farming systems of villages in developing countries have been revealed through an examination of the folk technologies for ecosystem management. Such technologies substantially increase crop yields to meet increases in the population. This evolution of self-regulated stable ecosystems is threatened by human overpopulation. A long-term view of these problems and the development of possible solutions must rest on an appreciation of and respect for the folk technology of ecosystem management that is still evolving in the villages of developing countries.

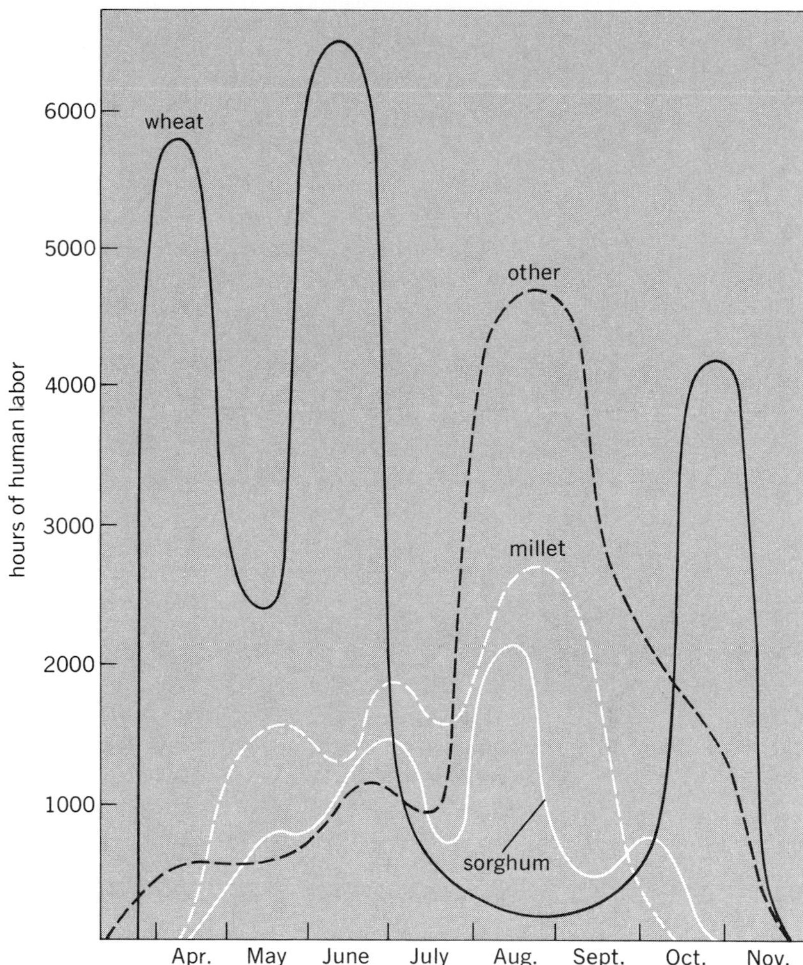

Fig. 2. Pattern of human labor inputs for the village of Pingsiang (pop. 675) in northern China, showing how the combination of crops spreads the demand for labor. A total of 710 h/ha results in yields of about 5,000,000 kcal/ha (20,900,000 kJ/ha). The yield of wheat, 1600 kg/ha, for Pingsiang in 1923 is at the lower range for United States wheat production in 1979. The United States average is 2300 kg/ha.

For background information *see* AGRICULTURAL SCIENCE (PLANT); AGRICULTURAL SOIL AND CROP PRACTICES; AGRICULTURE in the McGraw-Hill Encyclopedia of Science and Technology.

[RODGER MITCHELL]

Bibliography: L. T. Evans, *Amer. Sci.*, 68:388–397, 1980; R. Mitchell, *Analysis of Indian Agroecosystems*, 1979; D. N. Normal, in J. M. Kowal and A. H. Kassim (eds.), *Agricultural Ecology of Savannah*, 1978; D. Pimentel and M. Pimentel, *Food, Energy, and Society*, 1979.

Agriculture

Supplying agriculture with the necessary energy will be an essential part of providing the world with a secure and sustainable food supply for the future. As important as the quantity of energy required is the timeliness of its availability in relation to demand. Energy is used in agriculture to reduce the use of labor, to increase production per unit of land, and to reduce the risk of crop failure or spoilage during the growing, harvesting, or primary processing operations. Energy conservation efforts can reduce demand, but must not do so by either decreasing productivity or increasing risk.

The United States food production system is progressive and highly productive, but it is also highly energy-dependent. The entire food system uses about 17% of the nation's yearly energy. On-farm operations consume a little less than 20% of the food system's share. Preparation and consumption of food both in and out of the home account for more than 40%, while processing uses another 30% of the energy in the food system. Transportation and distribution roughly account for the remaining 10%. Because most of the energy in the food system is consumed beyond the farm gate, conservation efforts that significantly reduce the food system's share of the nation's energy use will have to be focused mainly in commercial and public sectors.

The 19-fold increase in cost of petroleum from OPEC countries since 1973 has had an impact on food prices. But far more serious than increased price would be an interruption of the energy supply to farmers. Prompt delivery and timely use of fuel are critical for most farming operations. Once set in motion, the agricultural process is virtually irreversible. A partially produced crop is not salvageable. If harvesting and initial processing cannot be accomplished when needed, all the energy inputs up to harvest will be wasted. The on-farm food system requires a modest but certain energy supply in order to function effectively.

Budget. Scarcity of energy resources in agriculture could intensify a steadily decreasing margin between production capacity and world demand for food. The world population, now 4,400,000,000 people, is projected to increase for several decades, requiring a minimum of approximately twice the food now produced. Stated another way, during the next 20 to 30 years the world must learn how to produce as much additional food each year as has been produced on a yearly basis from the dawn of history to the present time. Because most available land is in production, this increase in production must be accomplished without materially increasing the area of land farmed. This means that production from each acre farmed, on the average, must be nearly doubled.

Most inputs required to increase productivity per unit land area farmed are energy-intensive. Nitrogen fertilizer, a product derived primarily from natural gas, accounts for about one-third of the on-farm energy consumption in the United States. Energy for lifting, conveying, and pressurizing water for irrigation accounts for another one-fourth of this budget. Because of a diminishing return from each additional increment of fertilizer applied, more will be required per unit yield increase now than was required in bringing yields to today's level. Also, because the most easily developed water resources and the most easily irrigated land have already been exploited, future expansion of irrigation into nonirrigated lands will be even more energy-intensive.

Energy used for irrigation, for powering farm machinery to carry out tillage, cultural, and harvest operations in a timely manner, and for drying harvested crops tends to insulate food production from the vagaries of weather. Because only a small portion of each year's harvest is carried over to the next, increasing the stability of food production is an important strategy in meeting the world's yearly food requirements. The carrying capacity of the food production system is determined more by production during the worst years than by average yearly production.

Consideration of potential sources of energy supply for agriculture, therefore, must be concerned with energy use efficiency (production per unit of energy consumed) as well as cost efficiency (production per unit of cost), but not at the expense of long-term effectiveness of the food system. Short-term gains at the expense of long-term sustainable food production capacity should be guarded against.

Conservation. The immediate energy problem stems from a rapidly depleting supply of petroleum-based liquid and gaseous fuels. The solution to this problem involves both demand reduction and supply enhancement. The objective of the conservation strategy is to reduce demand by increasing energy use efficiency.

Energy conservation through improved efficiency has the highest near-term potential for decreasing agriculture's consumption of oil and natural gas. The most immediate savings will result simply from reducing the consumption of fuels or energy-intensive inputs in response to escalating oil and gas prices. Longer-term sustained increases in energy use efficiency will require modifying structures, machines, operations, and practices.

Substitute resources. The substitution strategy aims to enhance the energy supply to agriculture through use of renewable or noncritical energy sources such as wind, solar, hydropower, geothermal, and biomass, or fossil fuels such as coal, tar sands, or oil shale. Increasing the price of petroleum-based fuels will allow substitute energy resources to compete more effectively in the marketplace. The renewable energy resources, such as solar, wind, small-scale hydropower, and biomass, are of greatest interest as substitute fuels in agriculture for two reasons. First, in the very long term they may be the only energy resource available; and, second, some can be produced on the farm, giving agriculture a greater degree of control over its energy supply.

Numerous recent publications have raised hopes that production of fuel from farm-grown crops can solve a large part of the nation's energy requirements. Although their potential is significant, farm-produced fuels will have important limitations. Among them are: (1) The present value of food and fiber crops produced per unit of cropped area far exceeds that of any fuel crops currently being grown. This difference in value will probably continue for a considerable time. The large reserves of coal and oil shale will compete with biofuels, putting a cap on their price. There is no substitute for food produced by agriculture. (2) At present in the United States there is no unused land base of any consequence. That which is available could produce biofuels in the near term, but as food and fiber demands increase, even if higher-value fuel crops are developed, they will face stiff competition from food crops. (3) It will rarely be cost- or energy-effective to irrigate energy crops. Irrigated land is a valuable national resource that produces food when it does not rain. Existing irrigated lands that have sufficiently low energy intensity to produce net energy from biomass will probably be too valuable for food production to produce biomass on a sustained basis. While reserves of coal and oil shale last, scarce water resources can certainly produce more fuel through their conversion than by irrigating biofuel crops. (4) It is important to realize that entry of ethyl alcohol into the commercial market, which has captured the imagination of many, has done so with sizable Federal and, in some cases, state subsidy.

The Office of Technology Assessment recently evaluated (1980) the potential of biomass to contribute toward the nation's energy demands. The conclusion was that wood, grasses, agricultural crops and their residues, animal wastes, and other sources of biomass currently supply almost 2% of United States consumption, primarily from the use of wood in the forest products industry and in home heating. Depending on a variety of factors, including the availability of cropland, improving crop yields, and development of efficient conversion processes, proper resource management, and the level of policy support, bioenergy could supply up to 15–20% of current United States consumption by the year 2000. In this high-energy projection, forest products would supply 59% of this bioenergy, grasses and legume herbage (depending on the cropland needs for food production) would supply up to 29%, crop residues would supply 7%, ethanol from grain and sugar crops and from animal manures would supply 2% each, and agricultural processing wastes would supply 1%.

In the low-energy projection, energy from biological sources would supply 7% of today's United States consumption by the year 2000. No energy production from grasses and legume herbage or from grain or sugar crops was predicted, leaving 83% of the bioenergy to be produced from forest biomass. The energy produced from biomass in the year 2000 in the high-energy projection would approximately supply that required today by the entire food system. That produced on the farm (excluding forestry) would supply less than half the requirements of the entire food system, but more than double the energy requirements for on-farm use. In the low-energy projection, all biomass energy produced in the year 2000 would be about equal to that produced on the farm alone in the high-energy projection. The energy produced on the farm in this projection is less than half that required for on-farm uses today.

Assuming that the actual energy produced in the future will lie somewhere between the high and low projections, it appears reasonable to assume that the energy produced on farms from biological materials in the year 2000 will supply no more than today's requirements for the on-farm segment of the food system itself. As far as possible, technologies should be developed to convert these materials to useful fuels on the farm. Gasification of grasses, legume herbs, and crop residues for use on the farm to produce heat for crop drying, building and water heating, or for operation of stationary internal combustion engines will reduce transportation costs of such bulky materials and give the farmer a measure of control over energy supply. But these fuels alone will not meet the farmer's needs for mobile liquid fuels. The technology for direct conversion of more plentiful lignocellulosic materials, such as wood, grass, and crop residues, to liquid fuels by methanol synthesis or to ethanol through conversion to sugars followed by fermentation is not yet fully developed. Methanol synthesis appears to be the more efficient of the two processes, but it is feasible only at a scale too large to be practical in a farm setting.

Small-scale alcohol production from grain and sugar crops could permit farmers to supply their own needs for liquid fuel, but alcohol is not a direct replacement for the diesel fuel now used in most farm engines. Also, alcohol production at the farm scale is costly and time-consuming.

On the other hand, production of vegetable oil as a direct replacement for diesel fuel shows considerable promise as an emergency source of liquid fuel for farms. Initial research indicates that vegetable oil can be used in unmodified diesel engines with only minor purification. Stored oilseeds could then be marketed directly if diesel fuel were available, but crushed and extracted as a diesel fuel replacement if it were not.

Summary. The energy needs of agriculture must be met in the future if the world is to be fed. Considerable energy can be saved by practicing energy conservation on the farm, but more can be saved in the commercial and public sectors of the food system because considerably more is used there. Increasing world demand for food will probably require more energy-intensive inputs such as nitrogen fertilizer and irrigation water than have been used in the past. Meeting these energy demands to the extent possible by on-farm production and use of energy gives farmers control over their energy supply, thus decreasing the risk of supply interruption. Assuring the timely availability of energy to the agricultural system is equally as important as assuring that its quantity is sufficient. *See* the feature article ORGANIC FARMING. [STEPHEN L. RAWLINS]

Bibliography: Office of Technology Assessment, *Energy from Biological Processes*, OTA-E-124, U.S. Congress, 1980.

Aircraft manufacturing productivity

It is widely recognized that the productivity of United States manufacturing is at an alarming low. Japanese and German manufacturers are significantly increasing their productivity each year; the United States is not. Yet, the productivity of the nation's largest manufacturer of commercial jet aircraft, for example, is increasing at a rate well above the national average—in 1979 the increase in productivity from the previous year was 8.3%. During the same period United States manufacturers averaged less than a 1% increase; United States business in total experienced a productivity decline. This article discusses some of the ways that a commercial airplane company makes major productivity improvements.

Productivity is much more than a simple measurement of quantities produced versus labor consumption. Productivity is an attitude that is applied throughout a business. The gross measurement of productivity reflects the total of all the individual productivity gains and losses. Productivity is the net improvement in efficiency of a given resource and is not restricted to labor alone.

Productivity applies to all resources and may be measured in differing ways. For example, the productivity of a factory worker's labor is the quantity of products or parts produced per hour of work. The productivity of an engineer's labor is the quantity of product designs, calculations, design decisions, performance analysis, and so forth, that result per hour of work. The productivity of capital equipment is the quantity of operations performed per hour. The productivity of information is the amount, clarity, and timeliness of information produced. The productivity of money is the return on investment, or financial growth, of the money. These items, and more, contribute to the total productivity growth of a company.

Batch versus volume production. A characteristic of aircraft manufacturing is that a wide variety of parts are required to make a final product, and the number of final products made per month is low. A production rate of seven airplanes of a given model per month is considered an average. Each airplane is made up of thousands of unique parts. The result is that thousands of individual parts must be made, but the quantity to be made of any one of the parts is often 25 or less per month. This wide-variety, low-volume production is termed batch manufacturing.

In contrast, products made in large volumes, like automobiles and household appliances, constitute volume production manufacturing. Because of the very high volume of production—thousands or tens of thousands per month—specially built automated equipment is often applied. The high cost of this equipment is amortized by the high volume of products.

The economics of such amortization rarely exists in aircraft production. Because of this, the opportunities for and methods of achieving productivity improvements differ between batch and volume manufacturing.

Planned productivity improvement. Competition in the commercial aircraft field is ever increasing. United States and foreign manufacturers are seeking to increase their share of the market. Efficient manufacturing is essential if a company is to maintain its position in relation to its competitors, and efficiency and productivity improvement must be planned for and conscientiously sought. Several examples of individual productivity improvements in the commercial aircraft industry are given below.

Productivity of factory workers. Factory labor builds the individual parts and assembles them into a finished airplane. Factory labor builds tooling, stores and handles material, inspects and tests parts and assemblies, maintains equipment, and carries out numerous other tasks.

A major thrust in productivity improvement has been the improvement in efficiency of the worker's task. The development of more efficient equipment and the improvement in manufacturing tasks (methods) have been primary objectives. The application of minicomputer and microcomputer control systems to the operation of equipment and the control of processes is one method to improve productivity. An example is the simple stamping of flat aluminum parts from a strip of material with a manually operated punch press. Formerly, an operator would bolt a set of stamping dies in the punchpress, insert the material to be punched, move the material to the stamping position, activate the press to punch the part from the strip, reposition the material, and repeat the process until the desired quantity of parts was produced.

A modification to the existing press resulted in a productivity improvement of 14:1. The method of loading the stamping dies in the press was simplified. An automatic material-feeding mechanism was added to the press to eliminate the need for manual handling of the material strip. A minicomputer control system was added to control the punching process. The operation now consists of loading, but not bolting, the dies in the press, loading of several strips of material, inserting the desired number of parts in the control system, and activating the punching operation. The rest is automatic. The nonproductive elements of the worker's task have been reduced; the efficiency of the machine has been increased; the worker has time available to operate more than one machine; productivity has increased 14-fold.

Not all labor improvements occur in this manner. Systematically, each element of the manufacturing process is examined and improved where possible. Often the workers themselves suggest improvement. A formal system for employee suggestions enables the workers to offer improvement ideas and receive monetary rewards for adopted ideas.

Productivity of engineers. The productivity improvement of engineers, in product design and manufacturing, has been influenced largely by the introduction of computing systems. Large computer programs, running on giant computers in a company's data-processing network, are used by engineers to design the configuration of aircraft, determine engine requirements, simulate flight characteristics, predict structural soundness, analyze fuel consumption, correlate and analyze flight test results, and much more.

In addition to the large computing systems, minicomputer-driven interactive graphics systems have been introduced to the engineering design and drafting organizations. These systems offer the designer and drafter a more efficient means of producing part drawings and geometric descriptions than by manual drafting.

The power of the computer has greatly reduced the mathematical and data reduction tasks of the engineer. Flow time in design is reduced, errors are minimized, and a more thorough engineering job results.

The engineers in manufacturing also use large-scale systems and minicomputer systems. Process plans, which provide the step-by-step instructions to the workers to make or assemble parts, can be created and can automatically generate production orders by using a large data base system. A vast array of interactive terminals can interface to this system; aerospace manufacturers include the largest such systems of this kind in the world.

Minicomputer systems are used to design tooling for production. Programs to operate complex numerically controlled machine tools are prepared with minicomputers. Cutting tool inventories are controlled with minicomputers, and computer-driven warehousing systems are used to store and retrieve the hundreds of thousands of cutting tools that machine the parts and assemblies.

Productivity of capital equipment. The productivity of manufacturing capital equipment is increased by modification of existing equipment or development of improved equipment. As cited in an example above, productivity is greatly improved when provisions are made to automatically handle and move material and tools. Additional improvements often result through the application of minicomputer and microcomputer controls to equipment and processes.

Capital investment in new machine tools have resulted in productivity gains. Giant numerically controlled machines that make wing skins and spars have been installed that are 30% more productive than older machines.

Productivity of information. Timeliness, accuracy, and completeness of information are the keys to successful management. Decisions are based upon information. Important manufacturing decisions significantly impact productivity. For example, decisions as to how many parts to produce, when to produce, when to order material, the current status of work in process, and when to order cutting tools and shop supplies can cause production delays in one case or cause excess spending in another.

Information gathering and reporting have been greatly enhanced with computer systems. Momentary information is available upon which to make decisions. This results in more efficient flow of work, fewer delays, and reduced cost.

Productivity of money. Funding productivity improvement is an investment. Although productivity may be improved in terms of production rates per hour, the cost of productivity may or may not be favorable. Careful analysis of the investment part of productivity is an important issue not to be overlooked.

The test for financial productivity is return on investment (ROI). ROI is the amount of money to be saved over a period of time versus the cost to achieve a productivity capability. Inflation, tax laws, interest rates, and time are important factors in determining ROI. When considering possible productivity improvements, only those items are pursued which indicate both a favorable productivity gain and ROI.

For background information *see* COMPUTER-AIDED DESIGN AND MANUFACTURING; PRODUCTION ENGINEERING; PRODUCTION PLANNING in the McGraw-Hill Encyclopedia of Science and Technology.

[H. E. BUFFUM]

Airplane

In the more than 2 decades since the first 707, there have been commercial jet airliners ranging in size from the twin-engine F28 to the four-engine 747, the world's largest commercial airplane. Of course, air transportation continues to grow. And with changing needs, there are demands for airliners incorporating advanced design and technology. These demands must be met by fuel-efficient airliners, quiet enough to satisfy new standards for community noise levels. Determining the requirements for the new designs meant, as was the case with the earlier jetliners, working closely with leading airlines. This collaboration, started in the 1970s, led to the 757 and 767 designs (Fig. 1). These designs fill the gap between the wide-body trijets and the 727 airplanes.

Although the 757 and the 767 aircraft are designed to fill differing requirements, certain objectives were set relative to the commonality between the two designs. Therefore, the systems, flight decks, and handling qualities of the two airplanes are as similar as possible. About 50% of the line replaceable units on the 757 are 767-designed components and are interchangeable. This provides an advantage for airlines operating both types.

Boeing 767. The 767 is designed for medium-range routes and as such will be competing with other wide-body airplanes. The body cross-sectional width of 198 in. (5.03 m) provides seven-abreast seating in tourist class with two aisles, a choice that recognized aerodynamic efficiency.

The 300,000-lb (135,000-kg) 767 in its basic

Fig. 1. Scale models of the Boeing 757 and 767 airliners. (*Boeing*)

configuration carries 211 passengers and baggage approximately 2800 nautical miles (5200 km), with growth available for extended range. It will have a cruising altitude of 39,000 ft (11.9 km), with Mach numbers up to 0.83 (83% of the speed of sound). For increased capacity, up to 290 passengers, eight-abreast seating is also offered. The 767 flight deck arrangement allows operation with either two or three crew members. The airplane is 159 ft (48.5 m) long, and has a wingspan of 156 ft (47.5 m) and a wing area of 3050 ft^2 (283 m^2). With a total passenger and cargo payload capability of approximately 70,000 lb (32,000 kg), the 767 will carry 22 cargo containers with a total volume of 2640 ft^3 (74.8 m^3) and has an additional 430 ft^3 (12.2 m^3) of bulk storage. Two wing-mounted 48,000-lb-thrust (214-kN) engines power the 767.

Boeing 757. The 757 design retains the standard body cross section with six-abreast seating and one aisle. It grosses 220,000 lb (100,000 kg) and carries 178 passengers and their baggage almost 2200 n-mi (4100 km), with growth available. The 757, at 155 ft (47.2 m), although almost as long as the 767, has a shorter wingspan, 124 ft (37.8 m), and a wing area of 1950 ft^2 (181 m^2). This shorter wingspan allows operations at more existing airport parking gates and is thus consistent with the intended use of the airplane as a 727 replacement. The airplane has a total passenger and cargo payload capability of over 50,000 lb (23,000 kg) and bulk cargo holds with a capacity of 1700 ft^3 (48.1 m^3). The 757 flight deck is set up for operation with two crew members. Two 38,000-lb-thrust (169.0-kN) wing-mounted engines power the 757.

Technology. Both the 767 and the 757 incorporate the latest technology in aerodynamics, engines, materials, and digital equipment.

Wing design. The airfoil used for the 757 and the 767 (Fig. 2) allows the wing to have a higher thickness-to-chord ratio than current wings with the same sweepback angle and the same design speed. This

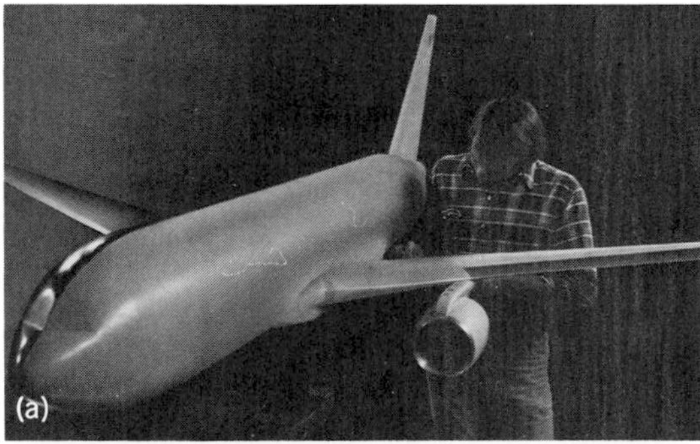

Fig. 2. Wind tunnel models of the (*a*) Boeing 767 and (*b*) Boeing 757 airliners, showing aerodynamic design. (*Boeing*)

means that structural advantages of a thicker wing section can be traded for a wingspan increase without a significant weight increase. The direct result of the increased span is aerodynamic efficiency. The increased span also results in higher cruise altitudes and a reduced time to climb to altitude. The high cruise altitude capability conserves fuel and eases air traffic congestion. The wing, with its leading-edge slats and trailing-edge flaps, is similar in takeoff and landing performance to the current 727 airplane. This allows the 757 and 767 to operate in and out of the same airports as the 727.

Engine design. Since the first commercial use of high-bypass-ratio fanjet engines on the 747, engine manufacturers have made significant advances in engine aerodynamic performance, fuel controls, and material applications. This progress, coupled with the newly designed nacelles with advanced acoustical treatments, produced quiet, fuel-efficient engines for both the 757 and the 767.

Structural materials. The advanced composites used in the secondary structure, and improved aluminum alloys used in primary structure such as the wing, also contribute to improved performance by reducing the aircraft's weight. The composite structure is graphite or Kevlar or a combination of the two (Fig. 3). All primary control surfaces, fairings, some engine nacelle components, and landing gear doors are typical applications. The new aluminums are essentially the same as the current alloys used in airplane design, but they have a higher strength resulting from a closer control of the alloy composition. All other properties, such as toughness and corrosion resistance, are equal or better.

Digital equipment. Advances in digital electronics and computer technology have created new opportunities in flight deck design and avionics. The advantages of digital systems over mechanical and analog devices grow as the number and complexity of tasks increase. The most notable application of digital avionics is in the flight deck. The flight management system is a fully integrated digital system that provides previously unavailable performance optimization and flight management capabilities to the crew while reducing airline costs. The flight management system provides fully automatic flight control from takeoff to landing while matching the flight plan to the performance of the airplane. The flight management system includes multicolor, multifeature cathode-ray-tube display of flight and airplane parameters, allowing a reduced quantity of instruments and leading to reduced work load for the flight crew.

Maintainability, reliability, and durability. The incorporation of thoroughly tested and verified new technology into the design, combined with the application of in-service experience, results in designs with a high degree of maintainability, reliability,

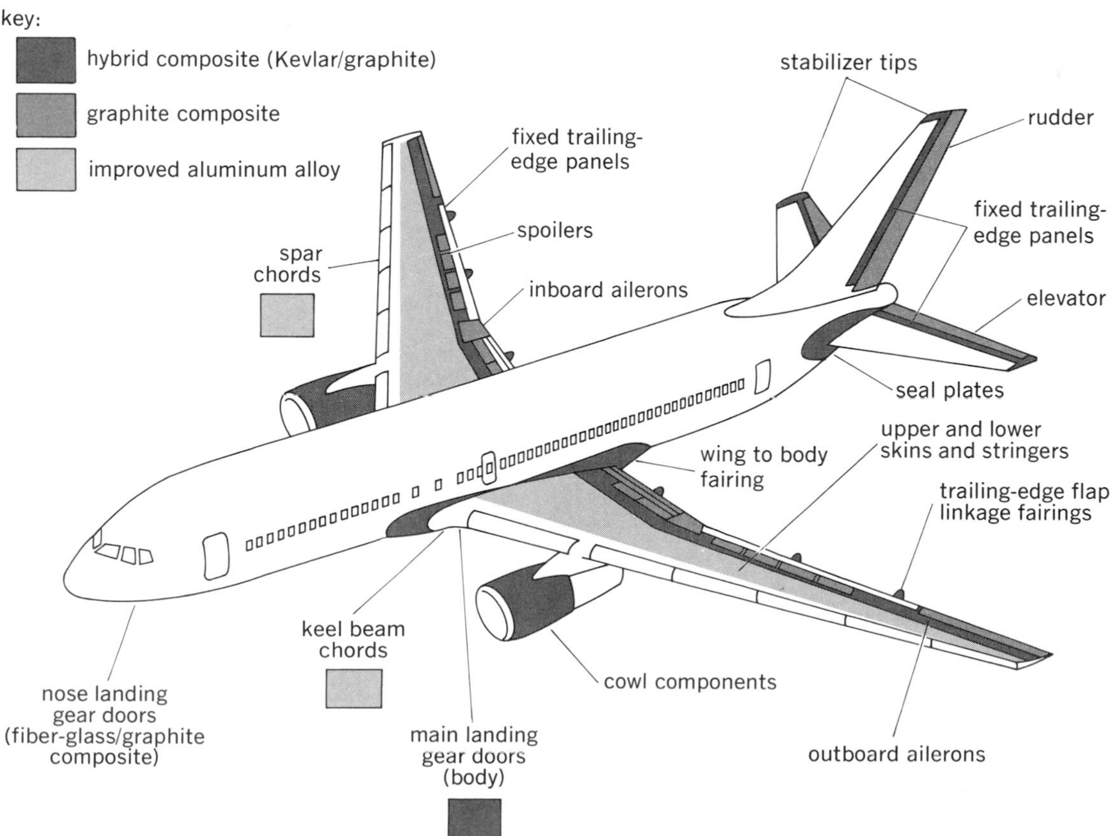

Fig. 3. Application of advanced composite materials to the Boeing 767 airliner. (*Boeing*)

and durability. In the case of durability, this in-service experience, has been reflected throughout the basic structural design, which is expected to reduce maintenance over the economic life of the airplane (approximately 20 years). The 757 and 767 designs have used combinations of advances such as digital equipment along with system simplifications to yield high reliability and maintainability.

Computer-aided design. In addition to the technology that went into the airplane design, the tools used by the designers have been greatly enhanced with the application of computer-aided design. The inherent features of computer-aided design are reduced engineering errors, easy exploration of design options, and a wide distribution of design data. The availability of design data on the computer pays large dividends in the design of manufacturing tooling and ultimately the fabrication of the airplanes. On the 757 and 767 airplanes, over one-third of the parts have been designed by using computer-aided design techniques.

Advantages. In summary, the key to successful aircraft programs is the timely application of advanced technology. The technology incorporated in the 757 and 767 provides the fuel efficiency necessary to combat the increasing costs of fuel. For instance, on a 1000-n-mi (1850-km) trip, the 757 and 767 will burn approximately 30% less fuel per seat than the 727 airliners. In fact, the 757 airliner, with 178 passengers, actually burns over 10% less total fuel than a 143-passenger 727 on the same trip. Other major advantages include reduced crew work load, reduced maintenance, higher reliability, improved airplane-handling qualities, and noise quality compatible with the new, more stringent community noise requirements.

For background information *see* AIRCRAFT INSTRUMENTATION; AIRPLANE; COMPOSITE MATERIAL; COMPUTER-AIDED DESIGN AND MANUFACTURING in the McGraw-Hill Encyclopedia of Science and Technology. [EVERETTE L. WEBB]

Alloy

The rapid increase in fuel costs since 1973 has made the need for more fuel-efficient aircraft pressing. One way of achieving this goal is to reduce weight, and recent studies have shown that this is most effectively accomplished by reducing the density and increasing the specific modulus and strength of the materials used for airframe components. Although many different materials have been considered for airframes since the Wright brothers' historic flight, none has challenged the preeminent position of aluminum. Today, on an industrywide basis, aluminum alloys make up 80% of the weight of the airframe. Consequently, metallurgists are actively pursuing ways of reducing the density of aluminum alloys.

Lithium is the lightest metallic element and, with the exception of beryllium, which has manufacturing and health-related problems, is the only metal that improves both the modulus and the density when alloyed with alluminum. Each 1% lithium added to an aluminum alloy reduces the density approximately 3% and increases the elastic modulus approximately 6% for lithium additions up to 4%.

Historical perspective of Al–Li alloys. The development of Al-base alloys containing lithium began in Germany in the 1920s and was primarily concerned with additions of small amounts of lithium to age-hardening alloys in order to increase their strength. In the 1950s metallurgists in the United States recognized that lithium increased the elastic modulus of aluminum, and they developed the high-strength Al–Cu–Li alloy designated 2020 by the Aluminum Association. Unfortunately, the alloy had low ductility and low fracture toughness in the maximum strength temper, and these limitations, along with production problems, led to its withdrawal as a commercial alloy in 1969. Interest in lithium-containing alloys then declined in the United States. At the same time research in the Soviet Union culminated in the development of an Al–Mg–Li alloy that was patented in France in 1968 and in the United Kingdom in 1969. However, the strength-ductility relationship of this alloy is less than desirable for high-performance aircraft.

Ductility problem. Insight into the low ductility of Al–Li alloys was obtained in the late 1970s by T. H. Sanders, Jr. Sanders studied the deformation and fracture behavior of Al–Li alloys at very high magnifications using transmission and scanning electron microscopy, and noted that the low ductility was due to the localization of deformation in very intense slip bands. Transmission electron microscopy revealed that the strain localization was associated with the metastable Al_3Li precipitates which were destroyed during deformation, thus producing soft zones in the aluminum matrix. In some cases, the strain localization was associated with grain boundary precipitation which resulted in soft precipitate-free zones adjacent to the grain boundaries. The intense slip bands thus developed in the soft zones caused stress concentrations which produced intergranular fracture and low ductility (Fig. 1).

Certain impurity elements such as sodium, potassium, sulfur, and hydrogen also appear to affect ductility adversely at very low levels of contamination. Large concentrations of sodium, potassium, and sulfur, introduced into the Al–Li alloy as impurities in the lithium, have been observed on fracture surfaces. These elements are virtually insoluble in aluminum, and they segregate to grain boundaries, where they enhance intergranular fracture. Hydrogen pickup in aluminum alloys occurs as a result of reaction of the melt with moisture in the melting environment. Al–Li alloys appear to pick up hydrogen more readily and are more difficult to degas than conventional aluminum alloys. The major problems resulting from hydrogen pickup are porosity in cast ingots and hydrogen embrittlement at high concentration levels. Assuming that one is able to use starting materials and casting procedures that eliminate impurity effects, improvements in ductil-

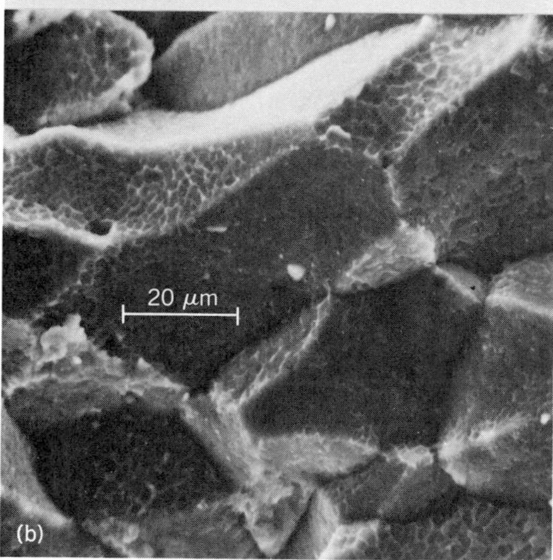

Fig. 1. Scanning electron micrographs of the tensile fractures of Al–2.8 wt % Li–0.3 wt % Mn sheet. (*a*) Intergranular fracture due to strain localization in the matrix. (*b*) Intergranular fracture due to strain localization in the precipitate-free zone.

ity of Al–Li alloys should be achieved by eliminating strain localization.

Recent developments. Current research and development of Al–Li alloys are directed toward preventing the formation of the soft zones that result in strain localization or to minimize the influence of soft zones. This can be accomplished by reducing the grain size through thermal mechanical treatments or by powder metallurgy consolidation and by adding additional precipitates called dispersoids. The Defense Advanced Research Projects Agency is supporting an investigation into the use of rapid solidification and powder metallurgy consolidation techniques for improving the mechanical properties of Al–Li–X alloys. The rapid solidification approach aims to reduce the grain size and refine the dispersoid particle size. The dispersoids form during solidification and high-temperature homogenization treatments and are primarily used for controlling grain size and shape. However, dispersoids also homogenize slip and inhibit the formation of intense slip bands. The dispersoids, acting in conjunction with a small grain size, lead to more homogeneous deformation and avert early crack nucleation due to strain localization.

The small grain size approach has also led to improvements in the ductility of the earlier Al–Cu–Li commercial alloy 2020. The original product was partially recrystallized with very large recrystallized grains which enhanced strain localization effects. F. S. Lin recently developed special thermal mechanical processing treatments for the 2020 alloy which result in a very fine recrystallized or unrecrystallized grain structure. Both microstructures give the strength-ductility relationship desired for airframe materials.

Since strain localization depends on the presence of soft zones associated with deformed metastable precipitates or precipitate-free zones, localization may be minimized by strengthening the matrix and the precipitate-free zone by alloying additions which remain in solid solution or coprecipitate with lithium. Magnesium is a potent solid-solution strengthener in aluminum alloys, and Al–Li–Mg alloys show commercial promise. In addition, magnesium suppresses the precipitation of Al–Li at grain boundaries and thus reduces the adverse effect of precipitate-free zones which accompany grain boundary precipitation. Elements like Cu, which have limited solid solubility, can have a great strengthening effect and can improve the homogenization of deformation when coprecipitated with lithium.

A schematic showing the improvements in strength-ductility relationships that have been recently obtained in Al–Li alloys is given in Fig. 2. The new alloys represented by the line to the right of the point marked 7075–T7651 show high strength, high elastic modulus, good ductility, and low density. The 7075 alloy is currently the most

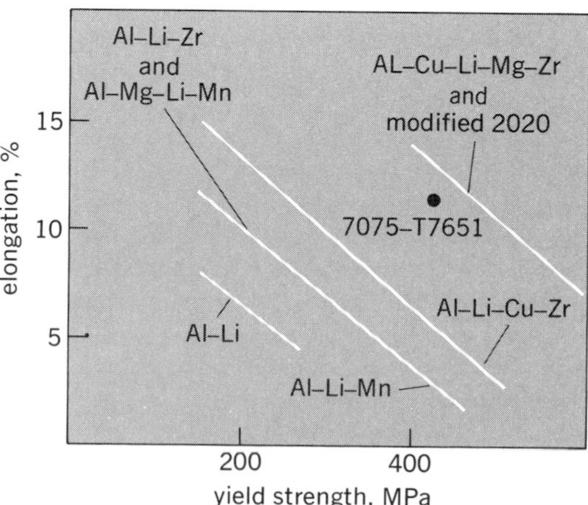

Fig. 2. Schematic of strength-ductility relationships for various Al–Li–X alloys.

widely used aluminum alloy in aircraft construction, but it has a higher density and lower elastic modulus than the new Al–Li alloys. The recent improvements make the Al–Li alloys very attractive for aircraft structures, and it appears that the promise offered by these materials is close to being realized.

For background information *see* ALUMINUM; HEAT TREATMENT (METALLURGY); METAL, MECHANICAL PROPERTIES OF; POWDER METALLURGY in the McGraw-Hill Encyclopedia of Science and Technology.

[EDGAR A. STARKE, JR.]

Bibliography: T. H. Sanders, Jr., and E. A. Starke, Jr. (eds.), *Aluminum–Lithium Alloys: Proceedings of the First International Aluminum Lithium Conference, Stone Mountain, GA, May 19–21, 1980*, Metallurgical Society of AIME, 1981.

Amphipoda

The Amphipoda are small to medium-sized peracaridan crustaceans. They are characterized principally by a laterally compressed body, total lack of carapace, first (and occasionally second) thoracic segment fused with the head. There are seven pairs of uniramous thoracic legs of which the first two pairs are basically subchelate and nonambulatory, and a six-segmented abdomen bearing three anterior pairs of biramous pleopods and three posterior pairs of rigid biramous uropods. Respiration is by means of thoracic (coxal) gills. The 5500 described species are classified in four suborders: the Gammaridea (4800 species in 19 superfamilies) is the most diverse morphologically, ecologically, and biogeographically; the Hyperiidea (about 350 species in six superfamilies) is marine, pelagic, and parasitoid; the Caprellidea (about 270 species in seven families) is marine, epifaunal, or epiparasitic on whales; and the primitive relict Ingolfiellidea (about 30 species in two families) is hypogean in fresh-water and marine sediments.

The paleohistory of the Amphipoda is largely unknown. However, an incomplete and tentative, but plausible, picture can be drawn from clues provided by (1) the fossil record of the animals themselves, or of other animals and plants with which some amphipod groups are intimately associated, (2) the Recent continental distributions of amphipod subgroups of low eurytopicity and low mobility, in relation to established chronology of plate tectonics and formation of oceanic and lacustrine basins, and (3) comparative morphological relationships of the higher amphipod taxa.

Fossil record. The fossil record of the Amphipoda encompasses 18 valid species or subspecies in seven genera, herewith relegated to three superfamilies, none earlier than Tertiary. The modern tube-constructing Corophioidea, family Corophiidae, is represented by trace fossils of *Corophium volutator* in the Quaternary of England. U-shaped fossil markings attributed to *Corophium* species (for example, *C. nathorsti*), *Corophioides*, and *Corophites*, variously from the Jurassic, Liassic, Lower Miocene, and Holocene of Europe, are probably worm burrows and have been judged inadmissible taxonomic entities. Superfamily Gammaroidea, family Gammaridae, is represented in Neogene coastal fresh-water, brackish-water, and shallow seas of western Europe and the Pontocaspian region, namely: the Pleistocene of Scotland, by *Gammarus fluviatilis*; the Miocene of Germany, by *G. oeningensis*; the Lower Miocene of the Soviet Union, by *G. o. minimus*; and the Lower Oligocene of France (Alsace), by *G. alsaticus*, *G. retzi*, and two other *Gammarus* species. However, photographic illustrations of *"Melita palmata"* show characters of the antennae, gnathopods, and peraeopods that are typical of superfamily Gammaroidea (compare the modern species of *Eulimnogammarus*) rather than of superfamily Hadzioidea, and *"Gammarus" retzi* of those same deposits has elongate antennae and peraeopods typical of the Acanthogammaridae. Family Pontogammaridae (?) is represented in Middle Miocene shallow marine deposits of the Aralocaspian region of the Soviet Union by *Praegmelina andrussovia* and *P. archangelski*. Probable members of the family Acanthogammaridae, also in Middle and Lower Miocene deposits of the Aralocaspian Basin, are *Andrussovia bogacevi* and *A. sokolovi*, *Hellenis saltatorius*, and, in the Lower Oligocene of France, *"Gammarus" retzi*. The superfamily Crangonyctoidea, family Crangonyctidae, is represented in beautifully preserved Baltic amber deposits of the Lower Oligocene–Upper Eocene by *Palaeogammarus sambiensis*, *P. balticus*, and *P. danicus*, all closely similar to Recent epigean species of *Crangonyx*. However, epigean Crangonyctidae is today absent from the Baltic region, where *Gammarus* and other (fresh-water) Gammaridae have become dominant. Possible ecological differences aside, the apparent lack of Gammaridae in Paleogene Baltic amber deposits tends to substantiate the probable greater antiquity of the Crangonyctoidea, and the relatively recent (Tertiary) origin and evolution of the fresh-water Gammaroidea, a view derived from continental drift considerations also (see below).

Geochronology. Estimated probable geological ages of subordinal, superfamily, and selected family groups within the Amphipoda are summarized in Tables 1 and 2. Ages are derived mainly from present-day group distributions in relation to continental geochronology. Assessments do not provide a true geological time span, but suggest merely the probable minimum age of the common ancestor to generic or family groups now widely separated on different continents or tectonic plates. In some instances (for example, for most Tertiary groups), the probable maximum age of the presumed common ancestor is suggested. Thus groups of more advanced morphology that are confined entirely or mainly to former Tethyan Sea margins or epicontinental areas (for example, Hadziidae, Bogidiellidae) are probably not older than Middle Cretaceous. Highly apomorphic subgroups such as Bateidae and Haustoriidae that occur within narrowly limited Tethyan marine areas (western Atlantic and Carib-

Table 1. Probable minimum geological age of amphipod subordinal, superfamily, and family groups in continental fresh waters

Geological time,* million years before present (log scale): Paleozoic (Carboniferous, Permian) — Mesozoic (Triassic, Jurassic, Cretaceous) — Tertiary (Paleocene, Neogene) — Quat; scale marks 300, 200, 150, 100, 50

Geological time	Amphipoda (higher taxa)	Distribution on continents, no.‡	Geobathic type
?	CRANGONYCTOIDEA (4)†		Epigean-Hypogean
	Crangonyctidae	1, 2	
?	Paramelitidae	3, 4, 8	
?	Neoniphargidae	5, 6, 7, 8	
	Niphargidae	2	
?	TALITROIDEA (10)		Epig.
	Hyalellidae	1, 3, 4, 8	
?	Dogielinotidae	North Pacific	
?	Najnidae	North Pacific	
	Talitridae	4, 5, 6, 8, 9, 10	
?	EUSIROIDEA (7)	1, 2, 4, 6	Epig.-Hypog.
?	Pontogeneiidae *(Paramoera, Pseudomoera)*	8	
	Calliopiidae *(Paraleptamphopus)*	10	
?	PONTOPOREIOIDEA (2)		Epig.
	Pontoporeiidae	1, 2	
?	GAMMAROIDEA (10)		Epig.
	Gammaridae	1, 2	
	Anisogammaridae	1, 2	
	Pontogammaridae	2	
	Acanthogammaridae	2	
	Typhlogammaridae	2	Hypog.
?	MELPHIDIPPOIDEA (4)		Epig.-Hypog.
	Phreatogammaridae	10	
?	HADZIOIDEA (3)		Epig.-Hypog.
	Hadziidae	1, 2	Hypog.
	Pseudoniphargus (3)	2	
?	Melitidae	1, 2, 3	Epig.
?	BOGIDIELLOIDEA (2)	1, 2, 3, 4, 7	Hypog.
	Artesiidae	1, 3	
	Bogidiellidae	1, 2, 3, 4, 7	
?	LILJEBORGIOIDEA (4)		Epig.-Hypog.
	Sebidae *(Seborgia)*	1, 8	Epig.-Hypog.
	Salentinellidae	1	Hypog.
?	*Paracrangonyx* gp.	3, 5, 10	Hypog.
	COROPHIOIDEA (9)		Epig.
?	*(Corophium, Paracorophium)*	1, 6, 8, 10	
?	INGOLFIELLIDEA (2)		Hypog. (also marine)
	Ingolfiellidae	1, 2, 3, 4	
?	Metaingolfiellidae	2	
	ANGIOSPERMS	All	Epig.
?	DECAPODA	Marine	
	ASTACURA	1, 2, 3, 5, 8, 10	
	EPHEMEROPTERA	All	Epig.
?	ISOPODA	All	
?	PHREATOICIDEA	4, 5, 6, 8, 10	Epig.-Hypog.
?	ONISCOIDEA	All	Epig.-Hypog.
?	TANAIDACEA	Marine	Epig.-Hypog.
	CUMACEA	Marine	Epigean
?	MYSIDACEA	Marine & FW	Epig.-Hypog.
?	SYNCARIDA	1, 2, 3, 4, 5, 8, 10	Epig.-Hypog.

*——— Fossil record   – – – – – – Presumed geochronology correlation   ? Probably older

‡Continents: Pangea (Laurasia-Gondwana), 180-150 m.y. before present minimum age.
Laurasia: (1) North America (including Caribbean Plate); (2) Eurasia (including Ponto-Caspian Region).
Gondwana: (3) South America (including Falklands); (4) Africa; (5) Madagascar; (6) India; (7) Antarctica (subantarctic islands); (8) Australia–New Guinea (including Tasmania); (9) New Caledonia; (10) New Zealand.

†Parenthetical numbers indicate the number of family units presently recognized in the superfamily.

Table 2. Estimated minimum geological ages of higher amphipod taxonomic groups in marine, fresh-water, and terrestrial environments

Superfamily salinity range	Geological age		
	Jurassic and Lower Cretaceous	Middle and Upper Cretaceous	Tertiary
Fresh water	Crangonyctoidea		
Fresh water, brackish, and marine	Eusiroidea (M) Talitroidea (M) (Hyalidae) Ingolfiellidea	Eusiroidea (F.W.) (*Paraleptamphopus group*) Talitroidea (F.W. & T.) (Hyalellidae, Talitridae) Oedicerotoidea Pontoporeioidea (Pontoporeiidae) Liljeborgioidea Hadzioidea (Hadziidae) Melphidippoidea (M) Bogidielloidea	Eusiroidea Gammaroidea (most groups) Pontoporeioidea (Haustoriidae) Hadzioidea (Melitidae, Carangoliopsidae) Melphidippoidea (Phreatogammaridae)
Marine	Lysianassoidea Phoxocephaloidea Leucothoidea (Pleustidae) Synopioidea	Leucothoidea (commensal groups) Stegocephaloidea Pardaliscoidea Dexaminoidea Corophioidea (Isaeidae, Ampithoidae, Podoceridae) Caprellidea (Caprellidae)	Hyperiidea Ampeliscoidea Corophioidea (apomorphic families) Caprellidea (Cyamida)

bean regions) are probably not older than early Tertiary. Species of Acanthogammaridae endemic to Lake Baikal (about 90% of the world total), and of Hyalellidae endemic to Lake Titicaca, are not older than the lake basins [maximum 25 and 10 million years (m.y.) before present, respectively]. Specialized amphipods living in close association with other organisms are unlikely to be older than their hosts. Thus Cyamidae (whale lice) must be more recently evolved than the earliest whales (fossil record to Lower Eocene); yet the high degree of host specificity and the generic diversity within the Cyamidae suggest an antiquity nearly as great as their hosts. Furthermore, terrestrial amphipods (Talitridae) of Australia and South Africa are unlikely to be older than their obligatory food and habitat plants (angiosperm leaf litter), which did not evolve until the Middle and Upper Cretaceous (Table 1).

Plesiomorphic and apomorphic groups. The geochronological data of Tables 1 and 2, despite their problematic nature, indicate that most Recent superfamilies and families of Amphipoda had evolved by Cretaceous times, and that both marine and fresh-water groups were present in the Jurassic, about 150 m.y. ago. Primitive fresh-water superfamilies and families tend to occur on high numbers of continents (Table 1); plesiomorphic marine groups are dominant in the Pacific Basin but relatively scarce in Caribbean-Mediterranean (Tethyan) seas. The so-called Jurassic groups are characterized variously by plesiomorphic features, for example, eye lobes, linear antennal calceoli, 5-dentate left mandibular lacinia mobilis, subsimilar nonsexually dimorphic gnathopods, and homopodous peraeopods. Superfamily and family groups originating in the Late Cretaceous and Tertiary periods tend to have more apomorphic features, for example, highly modified sessile eyes, 4- or 6+-dentate left lacinia mobilis, ornate "cup" calceoli (when not lost entirely), gnathopods of the dissimilar or sexually dimorphic, amplexing type, and heteropodous peraeopods (including anterolobate coxae). Members of the plesiomorphic suborders Ingolfiellidea and Gammaridea occur in both marine and fresh-water habitats; some are hypogean; and some parasitize or commensalize mainly benthic or epibenthic organisms. By contrast, members of the apomorphic suborders Hyperiidea and Caprellidea are strictly marine and epigean, and parasitize mainly pelagic animals. These ecological differences are consistent with proposals, based on morphological considerations, for an infraordinal or superfamily classification of these two units within the Gammaridea.

Adaptations. Ancestral amphipods were presumably shallow-water, mainly benthic or epibenthic, detritus feeders with a pelagic reproductive life cycle. Primordial amphipods probably evolved in high-energy coastal marine and estuarine environments, particularly along the margins of colder seas, and were suited to exploitation of food resources in confined spaces of stony, gravelly, and rubble bottoms. Such physical restrictions, and concomitant rapid and stressful fluctuations of physicochemical and nutrient regimes, would place a premium on slender, flexible, lightly chitinized bodies, on strongly ambulatory peraeopods and a forward-pushing tail fan; in these regimes the loss of stalked eyes, antennal squama, and bulky carapace would also be advantageous. Respiration and osmoregulation in high turbidities are presumably facilitated by protected coxal gills and antennal glands. Exploitation of vascular plant food resources being washed from the land into late Paleozoic seas presumably required the efficient chewing action of a compact buccal mass. The ability to nourish the young to self-maintaining size, in place, by means of lecithotrophic development of eggs in the protective thoracic brood pouch, was preadaptive in the peracaridan ancestral type. Lightly chitinized animals living in high-energy environments are unlikely to be fossilized, a probability that agrees with the absence of fossil amphipods, or their intermediate-stage progenitors, from Mesozoic deposits (especially Jurassic) that contain fossil mysids, cumaceans, tanaids, and isopods, and from Triassic and Permian deposits in which fossils of the more heavily chitinized amphipodlike phreatoicid isopods are known.

Descendants of the archetype amphipods, aided by relatively slight specializations of body form and appendages, penetrated the food resources of upper estuaries, fresh waters, and terrestrial environments, both epigean and hypogean, infaunal and neritic, from the shorelines into the deep sea, and geographically from the tropics to polar waters. Hypogean and tube-building modes of life were also facilitated by small size, slender body, and lack of carapace. More extreme modifications and sclerotization of the body and appendages and frequent loss of pelagic reproductive life cycle typify groups that have become nestlers or intimate associates with sponges, tunicates, coelenterates, wood borers, and epiparasites, or developed as armored epizoans, active predators, or free burrowers in sand.

Evolutionary relationships. Early macroevolution of the Amphipoda was almost certainly punctualistic. Gradualistic extrapolation of a precise period of origin from presumed Jurassic morphotypes is therefore not realistic (Table 1). For example, fresh-water Astacura (Decapoda), whose modern obligate continental distribution remarkably parallels that of the Crangonyctoidea (Amphipoda), occur as fossils in the Lower Triassic; yet fossil decapods of any type, including extinct groups, extend back only to the Permian, just 30 m.y. previously. The briefly flying Ephemeroptera (Insecta), whose obligate continental (nonoceanic) fresh-water nymphs co-occur with fresh-water amphipods, have changed relatively little since their fossil ancestors "suddenly" appeared in the Permian. Within the Peracarida, some Recent suborders, superfamilies, and even families of Mysidacea, Cumacea, Tanaidacea, and Isopoda occur as fossils in the Triassic and Permian; and the partly carapaced Spelaeogriphacea (now relict in hypogean fresh waters of South Africa) is closely similar to the long extinct Anthracocaridacea of the Carboniferous. The fossil record also supports the generally held view that the carapaceless condition of the Amphipoda is apomorphic or modern. Thus within the Eumalacostraca, Recent higher groups lacking a carapace (including Syncarida) occur as fossils only to the Upper Carboniferous; by contrast, corresponding taxa with a carapace (whole or in part) are found in Lower Carboniferous and Devonian deposits, and the fully bivalved phyllocaridan malacostracans are known from Cambian and early Paleozoic strata.

The Amphipoda possess several primitive malacostracan features, especially of the heart, gut, and mouthparts, and the large abdomen, that relate them to the Mysidacea and (possibly) the Euphausiacea. However, the Amphipoda are generally considered to be a highly advanced or apomorphic ordinal group within the Peracarida, particularly in the total absence of carapace, thoracic exopods, and antennal squama, the unique 3:3 subdivision of paired abdominal limbs, the internal loss of the maxillary gland, and fusion of the urosomal ventral nerve ganglia. Although amphipods are approximately similar to isopods in numbers of described species, their similarity in body form and function is considered superficial and convergent. Supporting the view of the relatively recent evolution of the Amphipoda are: (1) their lower diversity overall (four suborders of Amphipoda versus nine of Isopoda); (2) their lower diversity as parasites (about 10 gammaridean families and most Hyperiidea, as epiparasites, versus three suborders of Isopoda, one of which is internally parasitic); (3) the modern dominance of amphipods in shallow seas and in fresh waters, both epigean and hypogean, from which the amphipods were lacking (in the early fossil record); and (4) the relative scarcity of amphipods in the deep sea and on land, physiologically stressful environments that presumably require longer periods of adaptive radiation from the primitive shallow-water marine ancestors.

In summary, the most modern and specialized amphipod subordinal and superfamily groups evolved during and since Cretaceous times; the more primitive (modern) groups probably existed in the Jurassic; but the ancestral or prototype amphipods may have originated in the Triassic or possibly very late Paleozoic.

For background information *see* AMPHIPODA; ANIMAL EVOLUTION; CRUSTACEA in the McGraw-Hill Encyclopedia of Science and Technology.

[EDWARD L. BOUSFIELD]

Bibliography: C. G. Adams, An outline of Tertiary palaeogeography, in P. H. Greenwood (ed.), *The Evolving Earth*, 1981; E. L. Bousfield *Trans.*

Roy. Soc. Can. 4th ser. 16:343–390, 1979; R. R. Hessler, Peracarida, in R. C. Moore (ed.), *Treatise on Invertebrate Paleontology*, pt. R.: *Arthropoda 4*, vol. 1, 1969; M. K. Howarth, Palaeogeography of the Mesozoic, in P. H. Greenwood (ed.), *The Evolving Earth*, 1981; J. Just, *Steenstrupia*, 3:93–99, 1974; J. Kukalova-Peck, A phylogenetic tree of the animal kingdom (including orders and higher categories), *Nat. Mus. Natur. Sci. (Ottawa) Publ. Zool.*, vol. 8, 1973; G. L. Stebbins and F. J. Ayala, *Science*, 213:967–971, 1981.

Analytic hierarchies

The analytic hierarchy process (AHP) is a problem-solving framework. It is a systematic procedure for representing the elements of any problem. It organizes the basic rationality by breaking down a problem into its smaller constituents and then calls for only simple pairwise comparison judgments, to develop priorities in each level.

The analytic hierarchy process provides a comprehensive framework to cope with intuitive, rational, and irrational factors in making judgments at the same time. It is a method of integrating perceptions and purposes into an overall synthesis. The analytic hierarchy process does not require that judgments be consistent or even transitive. The degree of consistency (or inconsistency) of the judgment is revealed at the end of the analytic hierarchy process.

Human reasoning. People generally provide subjective judgments based on feelings and intuition rather than on well-worked-out logical reasoning. Also, when they reason together, people tend to influence each other's thinking. Individual judgments are altered slightly to accommodate the group's logic and the group's interests. However, people have very short memories, and if asked afterward to support the group judgments, they instinctively go back to their individual judgments. Repetition is needed to effect deep-rooted changes.

People also find it difficult to justify their judgments logically and to explicate how strong these judgments are. As a result, people make great compromises in their thinking to accommodate ideas and judgments. In groups, there is a willingness to compromise. If truth is to be an objective reality, reality must be very fuzzy because the search for truth often ends in compromise. What is regarded as truth may often be essentially a social product obtained through interaction rather than by pure deduction.

Logical understanding does not seem to permeate judgment instantaneously. It apparently needs time to be assimilated. However, even when logical understanding has been assimilated, people still offer judgment in a spontaneous emotional way without elaborate explanation. Even when there is time, explanations tend to be fragmentary, disconnected, and mostly without an underlying clear logical foundation.

Outline of the process. People making comparisons use their feelings and judgment. Both vary in intensity. To distinguish among different intensities, the scale of absolute numbers in Table 1 is useful.

The analytic hierarchy process can be decomposed into the following steps. Particular steps may be emphasized more in some situations than in others. Also as noted, interaction is generally useful for stimulation and for representing different points of view.

1. Define the problem and determine what knowledge is sought.

2. Structure the hierarchy from the top (the objectives from a broad perspective) through the intermediate levels (criteria on which subsequent levels depend) to the lowest level (which usually is a list of the alternatives).

3. Construct a set of pairwise comparison matrices for each of the lower levels, one matrix for each element in the level immediately above. An element

Table 1. Scale of relative importance

Intensity of relative importance	Definition	Explanation
1	Equal importance	Two activities contribute equally to the objective
3	Slight importance of one over another	Experience and judgment slightly favor one activity over another
5	Essential or strong importance	Experience and judgment strongly favor one activity over another
7	Demonstrated importance	An activity is strongly favored and its dominance is demonstrated in practice
9	Absolute importance	The evidence favoring one activity over another is of the highest possible order of affirmation
2, 4, 6, 8	Intermediate values between the two adjacent judgments	When compromise is needed
Reciprocals of above nonzero numbers	If an activity has one of the above numbers assigned to it when compared with a second activity, the second activity has the reciprocal value when compared to the first	

in the higher level is said to be a governing element for those in the lower level since it contributes to it or affects it. In a complete simple hierarchy, every element in the lower level affects every element in the upper level. The elements in the lower level are then compared to each other, based on their effect on the governing element above. This yields a square matrix of judgments. The pairwise comparisons are done in terms of which element dominates the other. These judgments are then expressed as integers according to the judgment values in Table 1. If element A dominates element B, then the whole number integer is entered in row A, column B, and the reciprocal (fraction) is entered in row B, column A. Of course, if element B dominates element A, the reverse occurs. The whole number is then placed in the B,A position with the reciprocal automatically being assigned to the A,B position. If the elements being compared are equal, a one is assigned to both positions. The numbers used express an absolute rather than an ordinal relation.

4. There are $n(n-1)/2$ judgments required to develop the set of matrices in step 3 (taking into account the fact that reciprocals are automatically assigned in each pairwise comparison), where n is the number of elements in the lower level.

5. Having collected all the pairwise comparison data and entered the reciprocals together with n unit entries down the main diagonal (an element is equal to itself, so a "one" is assigned to the diagonal positions), the eigenvalue problem $Aw = \lambda_{\max} w$ is solved and consistency is tested, using the departure of $\lambda_{\max}$ from n (see below).

6. Steps 3, 4, and 5 are performed for all levels and clusters in the hierarchy.

7. Hierarchical compositon is now used to weight the eigenvectors by the weights of the criteria, and the sum is taken over all weighted eigenvector entries corresponding to those in the next lower level of the hierarchy.

8. The consistency ratio of the entire hierarchy is found by multiplying each consistency index by the priority of the corresponding criterion and adding them together. The result is then divided by the same type of expression, using the random consistency index corresponding to the dimensions of each matrix weighted by the priorities as before. The consistency ratio should be about 10% or less to be acceptable. If not, the quality of the judgments should be improved, perhaps by revising the manner in which questions are asked in the making the pairwise comparisons. If this should fail to improve consistency, it is likely that the problem should be more accurately structured; that is, similar elements should be grouped under more meaningful criteria. A return to step 2 would be required, although only the problematic parts of the hierarchy may need revision.

If the exact answer in the form of hard numbers was actually available, it would be possible to normalize these numbers, form their ratios as described above, and solve the problem. This would result in getting the same numbers back, as should be expected. On the other hand, if firm numbers were not available, their ratios could be estimated to solve the problem.

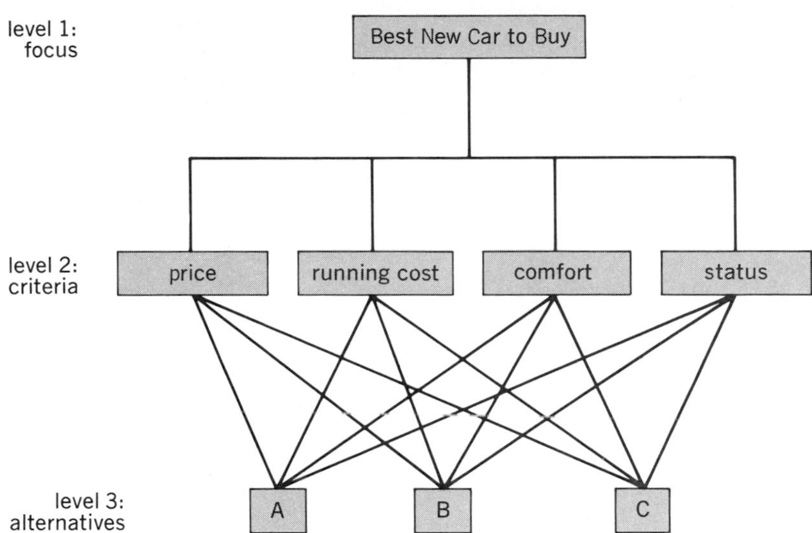

Analytic hierarchy used to assist a family to buy a new car.

Example of the process. In the following example the analytic hierarchy process is used to assist a young family (a father, a mother, and a child) of specified income to buy a new car, say, either model A, B, or C. The choice will be determined through four important criteria.

The hierarchy of such a decision often takes the form shown in the illustration. In this hierarchy, level 1 is the single overall objective: Best New Car to Buy. On level 2 are the criteria which are perceived to compose what is meant by Best New Car, such as Price and Running Cost (operating and maintenance). On level 3 are the various alternative cars from which the family will choose.

This downward decomposition format can easily be used on a wide class of problems. In addition, a slight further modification to incorporate feedback loops will cover an even wider range.

The questions asked when comparing each criterion are of the following kind: Of the two alternatives being compared, which is considered more important by the family buying a car and how much more important is it? The comparison matrix of Table 2 is then formed. Since Price and Running Cost have the highest priorities, the other factors are discarded and only these two are used in continuing the process. Care must be taken in doing this. The original priorities are then normalized by dividing each by their sum to obtain the new relative priorities in Eqs. (1).

$$0.67 = \frac{.58}{.58 + .28} \qquad 0.33 = \frac{.28}{.58 + .28} \tag{1}$$

The process is then repeated for the third level, where each car is compared with respect to the two

Table 2. Comparison matrix comparing criteria for buying a car

Decision to Buy a Car	Price	Running cost	Comfort	Status	Priorities
Price	1	3	7	8	.582
Running Cost	1/3	1	5	5	.279
Comfort	1/7	1/5	1	3	.090
Status	1/8	1/5	1/3	1	.050

λ_{max} = 4.198
C.I. = .066
C.R. = .073

Table 3. Comparison matrices comparing alternative cars

Price	Car A	Car B	Car C	Priorities
Car A	1	2	3	.540
Car B	1/2	1	2	.297
Car C	1/3	1/2	1	.163

λ_{max} = 3.009
C.I. = .005
C.R. = .008

Running cost	Car A	Car B	Car C	Priorities
Car A	1	1/5	1/2	.106
Car B	5	1	7	.745
Car C	2	1/7	1	.150

λ_{max} = 3.119
C.I. = .059
C.R. = .103

high-priority factors from the second level, as shown in Table 3.

The two priority columns are then recorded as in Table 4. All entries of the first column are then multiplied by .67, the priority of Price, and those of the second column by .33, the priority of Running Cost, and added. This gives the third column. Thus Car B was selected for its efficient operation even though its initial price is considerably higher than Car A. A car dealer in the income neighborhood of this family may find it profitable to stock up on the cars in the respective proportions.

Table 4. Composition of priorities

	Price (priority .67)	Running Cost (priority .33)	Composite priority of cars
Car A	.540	.106	.396
Car B	.297	.745	.445
Car C	.163	.150	.159

One of the most powerful contributions that the analytic hierarchy process makes is to test out the degree of inconsistency or incompatibility of new ideas or new policies adopted with older, more familiar, better tried successful methods. For example, in doing the above problem the participants were not sure whether the judgments for Price over Running Cost should be 7, 5, or 3. Each one was tried separately, and it was found that 3 yielded the highest consistency. Those who voted for 3 won that argument.

Priorities and consistency. An easy way to get a good approximation of the priorities is to use the geometric mean. This is done by multiplying the elements in each row and taking their nth root, where n is the number of elements. Then, normalize the column of numbers thus obtained by dividing each entry by the sum of all entries. Alternatively, normalize the elements in each column of the matrix and then average each row.

The consistency index can also be determined by hand calculations. Add the numbers in each column of the judgment matrix, multiply the first sum by the first priority, the second by the second, and so on, and add. For the first matrix the column sums (1.60, 4.40, 13.33, 17) are obtained, and multiplying by (.582, .279, .090, .050) gives 4.20. This number is denoted by λ_{max}. The consistency index is given by Eq. (2).

$$\text{C.I.} = \frac{\lambda_{max} - n}{n-1} \qquad (2)$$

The consistency is now checked by taking the ratio (C.R.) of C.I. with the appropriate one of the set of numbers in Table 5 to see if it is about 10% or less (20% may be tolerated in some cases but not more). Otherwise the problem must be studied again and judgments revised. The consistency of the hierarchy in the above example, as given by Eq. (3), is .06, which is good.

$$\frac{.066 \times 1 + .005 \times .67 + .059 \times .33}{.900 \times 1 + .580 \times .67 + .580 \times .33} = \frac{.089}{1.48} = .06 \qquad (3)$$

Judgment formation. When several people participate, judgments are often debated. Sometimes the group accepts a geometric average of their combined judgments. If there is strong disagreement, the different opinions can each be taken and used to ob-

Table 5. Random consistencies

n:	1	2	3	4	5	6	7	8	9	10
Random consistency:	0	0	.58	.90	1.12	1.24	1.32	1.41	1.45	1.49

Table 6. Comparison matrix comparing perceived brightnesses of chairs

	Chair 1	Chair 2	Chair 3	Chair 4	Brightness ratios
Chair 1	1	5	6	7	0.61
Chair 2	1/5	1	4	6	0.24
Chair 3	1/6	1/4	1	4	0.10
Chair 4	1/7	1/6	1/4	1	0.05

λ_{max} = 4.39
C.I. = 0.13
C.R. = 0.14

tain answers. Those which subsequently display the highest consistency within the group are the ones usually retained.

The analytic hierarchy process incorporates equally both tangible factors, which require hard measurements, and such intangible factors as comfort, which require judgment. Eventually one finds that so-called hard numbers have no meaning in themselves apart from their utilitarian interpretations. In the above example, buying a $10,000 car is more than twice as "painful" as buying a $5000 car.

The interdependence of criteria, such as Comfort and Price, have to be considered carefully since there may be some perceived overlap. For example, higher price buys more comfort, but it also buys other desirable attributes. Judging the relative importance of such things as price and comfort, therefore, must be done as independently as possible with avoidance of overlaps.

Validation by physical laws. Using the scale 1–9 has been justified and demonstrated by many examples. However, the following simple optics illustration, carried out with small children, shows that perceptions, judgments, and these numbers lead to results which can be validated by laws of physics. In this example, four identical chairs were placed at distances of 9, 15, 21, and 28 yards (1 yd = 0.9144 m) from a floodlight. The children stood by the light, looked at the line of chairs, and compared the first with the second, the first with the third and then with the fourth, and so on for the second, third, and fourth chairs. Each time, the children said how much brighter one chair was, compared to the other.

Their judgments were entered in the matrix of Table 6 to record the relative brightness of the chairs. The reciprocals were used in the transpose position.

The inverse-square law of optics is now used to test these judgments. Since the distances are 9, 15, 21, and 28 yd, these numbers are squared and their reciprocals calculated. This gives .0123, .0044, .0023, and .0013 respectively. Normalization of these values gives .61, .22, .11, and .06, which are very close to the brightness ratios obtained in the test using the analytic hierarchy process.

Structuring a hierarchy. There are no rules for structuring a hierarchy. However, a typical analytic hierarchy for allocating resources—either by measuring costs or by measuring benefits—will often be stratified roughly as follows. The top level will include the overall objectives of the organization or system. Benefit-cost criteria may appear in the next level. A subordinate level may further clarify these criteria in the context of the particular problem by itemizing specific tasks which are to be accomplished at some level of performance. This is followed by the alternatives being evaluated.

Capabilities. Designing an analytic hierarchy—like the structuring of a problem by any other method —is more art than science. It necessitates substantial knowledge of the system in question. A very strong aspect of the analytic hierarchy process is that the knowledgable individuals who supply judgments for the pairwise comparisons usually also play a prominent role in specifying the hierarchy.

Although a hierarchy to be used in resource allocation will tend to have the vertical stratification indicated above, it can also be much more general. The only restriction is that any element on a higher level must serve as a governing element for at least one element (which can be the element itself) on the immediately lower level. The hierarchy need not be complete; that is, an element at an upper level need not function as a criterion for all the elements in the lower level. It can be partitioned into nearly disjoint subhierarchies sharing only a common topmost element. Thus, for instance, the activities of separate divisions of an organization can be structured separately. As suggested above, the analyst can insert and delete levels and elements as necessary to clarify the task or to sharpen a focus on one or more areas of the system.

The analytic hierarchy process has already been successfully applied in a variety of fields, including: planning the allocation of energy to industries; designing a transport system for the Sudan; planning the future of a corporation and measuring the impact of environmental factors on its development; designing future scenarios for higher education in the United States; selecting candidates and winners in elections; setting priorities for the top scientific institute in a developing country; solving a faculty promotion and tenure problem; and predicting oil prices.

For background information *see* DECISION THEORY; SYSTEMS ENGINEERING in the McGraw-Hill Encyclopedia of Science and Technology.

[THOMAS L. SAATY]

Bibliography: T. L. Saaty, *The Analytic Hierarchy Process*, 1980; T. L. Saaty, *Decision Making for Leaders*, 1981; T. L. Saaty, *J. Math. Psychol.*, 15:234–281, 1977; T. L. Saaty and L. Vargas, *The Logic of Priorities*, 1981.

Analytical chemistry

At some stage of almost all scientific and technological activity, measurements must be made. In the chemical sciences the measurements generally refer to the elucidation of molecular structure, the identification of materials, or the determination of the concentrations of various species, and often involve costly spectroscopic instruments—for example, nuclear magnetic resonance spectrometers, mass spectrometers, and ultraviolet, visible, and infrared spectrophotometers. Because of the constantly increasing complexity of new materials and processes, the nature of the measurements and the instruments to make them become more and more sophisticated and costly. In addition, the rapid advance in instrumentation technology not only has accelerated the rate of obsolescence of research-grade instrumentation but also has produced instruments of a power vice. The regional instrumentation facilities sup- mentation is vital to the creation of new knowledge, but is so expensive and complex that few research groups can afford to acquire and maintain it.

Centralized instrumentation. Centralized chemical analytical instrumentation laboratories were established in research universities to effect the economies resulting from sharing the equipment and the professional expertise required to operate and maintain it. Although the instruments in such centers are primarily for the use of the students and faculty of the chemistry department, researchers in other areas of the university also have access to them. For example, biochemists, molecular biologists, geologists, and environmental scientists often need to make use of instruments such as those available in the center, and can ill afford to purchase or to maintain them in their own laboratories.

Centralized instrumentation laboratories in chemistry departments have existed for at least 25 years and probably stem from efforts initiated at Purdue in the early 1950s. Several other facilities were started in the 1960s, but in recent years there has been a surge of interest in establishing measurements centers for the reasons stated above. The general objective of these centers is to provide the chemistry department and other parts of the university with the instrumentation and training to support the research faculty. The nature of support covers the spectrum from primarily service with a modest amount of analytical research to primarily instrumentation research with some analytical service. The staffs of the centers vary similarly in capability from skilled technicians to versatile professionals.

University Laboratory Managers Association. In late 1980 the managers from a number of university analytical service laboratories met at Northwestern University to hold what became the first annual conference of the University Laboratory Managers Association (ULMA). Also at the conference were scientists from industry and government and from universities interested in starting a centralized analytical instrumentation laboratory. On the basis of information from the conference and with the criteria that a centralized laboratory must contain more than one type of instrumentation and be headed by a manager or director, it appears that about 20 to 25 centralized chemical instrumentation facilities were in existence in 1980 in the chemistry departments of as many universities. The criteria suggested above exclude a number of facilities which consist only of nuclear magnetic resonance spectrometry or mass spectrometry and which have a research faculty member nominally in charge of service. The regional instrumentation facilities supported by the National Science Foundation also are not included.

Northwestern University laboratory. The analytical services laboratory of the chemistry department of Northwestern University are an example of a centralized chemical instrumentation facility.

Staff. The staff of the facility consists of a laboratory director, a nuclear magnetic resonance specialist, a gas chromatography–mass spectrometry specialist, a Fourier transform–infrared spectroscopy specialist, a generalist analytical chemist responsible for the optical and separation instruments and for special problems, and a skilled technician for microchemical carbon, hydrogen, and nitrogen determinations. In addition to instruction, the staff is charged with maintaining equipment in operating condition and is available for consultation. The laboratory director is responsible for everyday administration, for keeping abreast of the latest developments in analytical instrumentation, and for recommending equipment acquisitions.

Instruments. The instruments available in the Northwestern University Analytical Instrumentation Laboratory can be classified as nuclear magnetic resonance units, mass spectrometers, separations, optical, and "other," which roughly corresponds to the responsibility of the staff. After suitable instruction by the laboratory staff members, researchers (graduate and undergraduate students, postdoctoral fellows, and faculty) may use most of the equipment on 24-h-per-day basis. Because of its complexity and "unforgiving" nature, the mass spectrometer is used only by the staff; and the C, H, and N determinations are done only by the skilled technician.

Users. The instruments are arranged by the type

of user. One area is primarily for the research students, another for the undergraduates from a special integrated laboratory course, and still another for the staff. The arrangement is also for the convenience of the user. For example, there is considerable clearance around the instruments so that carts with special experimental set-ups can be moved up to an instrument for complex experiments. Also to facilitate the use of movable special equipment, the utilities (power, water, nitrogen, compressed air) are dropped from the ceiling. The absence of water lines and cords on the floor naturally enhances safety and aids in maintaining a clean, bright, spacious appearance.

Each incoming class of graduate students learns about the use of the analytical instrumentation laboratory. They are given written information about all of the instruments in the laboratory with suggested sample sizes and other information needed to help them decide which instruments to use for a given problem. They learn about the scope and services of the laboratory, the procedures to be followed, and the charges for operating each instrument. With the advice of a staff member or a research director, the student carefully plans an experiment to obtain the maximum amount of useful information for the minimum cost of instrument usage. The user fees associated with each instrument provide an educational value in that the student is taught to use the least expensive piece of equipment capable of yielding the needed data.

National Science Foundation regional instrumentation centers. In 1978 the National Science Foundation extended the concept of sharing costly instruments to include geographical regions of the country by establishment of the NSF Regional Instrumentation Facilities program. In fiscal 1978 the foundation funded six centers, and in the following year eight additional centers. The location of the centers and the measurement areas are shown in the table. These facilities are intended to provide a service for qualified scientists in the particular specialized area of the facility.

National Science Foundation regional instrumentation centers

Measurement area	Location
NMR spectroscopy	University of South Carolina
	Colorado State University
	California Institute of Technology
	Yale University
	University of Illinois
Mass spectroscopy	University of Nebraska
	Johns Hopkins University
Laser	University of Pennsylvania
	Massachusetts Institute of Technology
	University of California, Berkeley, jointly with Stanford University
Carbon-14 dating	University of Arizona
Electron microscopy	Arizona State University
Surface science	Montana State University
	University of Minnesota

Future developments. The existing centers have demonstrated their value to the research and teaching functions of university chemistry departments. There is little doubt that the trend toward the centralization of instrumentation will continue for the foreseeable future because continued availability of sophisticated instruments must involve sharing. Sharing is dictated not only by the costly instruments themselves, but also by the financial support required for professional specialists to operate, maintain, instruct in the use of, and supervise the instrumentation. A new class of experts is developing for career positions in analytical chemical instrument centers.

For background information *see* ANALYTICAL CHEMISTRY; INSTRUMENTATION in the McGraw-Hill Encyclopedia of Science and Technology.

[CLAUDE LUCCHESI]

Bibliography: L. Berlowitz and R. Zdanis, *The Scientific Instrumentation Needs of Research Universities: A Report to the National Science Foundation by the Association of American Universities with Support Provided by NSF under Contract Number C-PRM-7225829*, 1980; C. Walling et al., *A Study to Improve the Management of Costly Instrumental Centers: A Report to the National Science Foundation with Support Provided by NSF under Grant Number GP 42932*, 1976.

Antibiotic

The main scope of antibacterial agents is to eradicate bacteria from the site of infection. Consequently, most investigators in the field of antibiotics directed their efforts toward the effects of antibiotics which interfere with bacterial survival. Since such effects are expected only at high concentrations, prior to the early 1970s only a small number of incidental observations on the effects of low concentrations of antibiotics were reported. More recently, however, it has become evident that while high concentrations of antibiotics do produce dramatic effects on bacteria, antibiotics at lower concentrations, contrary to logical expectations, do not produce milder effects, but effects different in kind. These effects appear to be significant in the prevention and treatment of infections.

The systematic investigation of the effects of antibiotics at low concentration, below the minimum inhibitory concentration (MIC), was initiated in the early 1970s. A few advances in technology, such as growth of bacteria on filter membranes, interference phase-contrast microscopy, and readily available transmission electron microscopy, contributed to an accelerated documentation in this area.

MBC and MAC concepts. Since the MIC expresses the effect of the antibiotic so long as the bacteria remain in contact with the antibiotic, the effect on the viability of the organisms remained undefined. As a consequence of the need to define and express such viability, the concept of minimum bactericidal concentration (MBC) was developed. It also became clear that concentrations of antibiotics below the MIC do have definite effects on bacteria, and the concept of the minimum antibiotic concentration (MAC) was developed. The MAC, like the MBC, expresses the viability of the organisms after exposure to an antibiotic. The MBC is defined as the concentration which kills 99.9% of the exposed bacteria. The MAC is defined as the concentration which reduces the number of exposed bacteria by 90% (one log) relative to bacteria that were not exposed. The MAC also produces changes in the morphology or ultrastructure of bacteria. Obviously, the MAC is related to the MBC in that it expresses the end point of activity at the lowest concentration, whereas the MBC expresses the end point of activity at the highest concentration. The MIC is therefore pivotal to both of these expressions of viability.

Range of activity. The series of concentrations from the MAC through the MBC is the "range of activity." Each group of antibiotics has a characteristic range of activity. There are, however, differences from one genus of bacteria to another. Penicillins show a wider range of activity than do cephalosporins. Colistin and rifampin have very narrow ranges of activity; below one-third the MIC they show practically no antibacterial effect. Some beta-lactam antibiotics such as mecillinam or cefsulodin have a very large range of activity. Aminoglycosides show a wide range of activity with gram-negative bacilli, but a narrow range with gram-positive cocci. The determination of the range of antibacterial activity of a drug could be important for the characterization of new drugs as well as for the treatment of patients with relatively toxic drugs.

Mechanisms of action. The beta-lactam antibiotics at concentrations equal to or higher than the MIC bind in certain proportions to proteins of the cell membrane (penicillin-bound proteins, or PBP). According to their binding ratio to PBP, the beta-lactam antibiotics bind to the PBP in different proportions and inhibit the autolytic systems. This results in bacterial cells that are divided but cannot separate. In this manner, sub-MICs of penicillin allow staphylococci to divide, but the resultant group of divided cells remain bound together by thick cross walls as a large cluster which consists of 8–16 staphylococci. Similarly, streptococci grow into chains consisting of divided but unseparated streptococci. In both of these instances, in the absence of an operable lytic system, the septa become thick and fibrous, and the clusters of unseparated cells assume a bizarre morphology. In some instances the wall breaks, producing a single point of rupture with an extruding cytoplasm (Fig. 1). By the same basic mechanism, gram-negative bacilli divide but cannot separate, and the result is a filament that can sometimes attain lengths up to 150 μm or giant spherical cells (Fig. 2). These abnormal forms of bacteria are encountered in specimens from patients who have been subminimally treated with antibiotics as a result either of a low dose or of intermittent therapy. Most drugs produce different concentrations in different organs and tissues. Such concentrations may be far below the level obtained in the serum, and therefore could not reach the MIC. Low concentrations of antibiotics at the site of infection can produce abnormal forms of bacteria. Filaments of gram-negative bacilli have been observed in blood cultures, urine sediments, and pleural effusions. Bizarre forms of staphylococci or pneumococci have been observed in sputa and spinal fluids. The abnormal morphology of these bacteria can be confused with other genera of bacteria, and therefore the clinical microbiologist should be aware of the possible existence of these abnormal organisms in specimens received from patients treated with drugs. Moreover, these organisms could indicate antibiotic intake which might not be suspected otherwise.

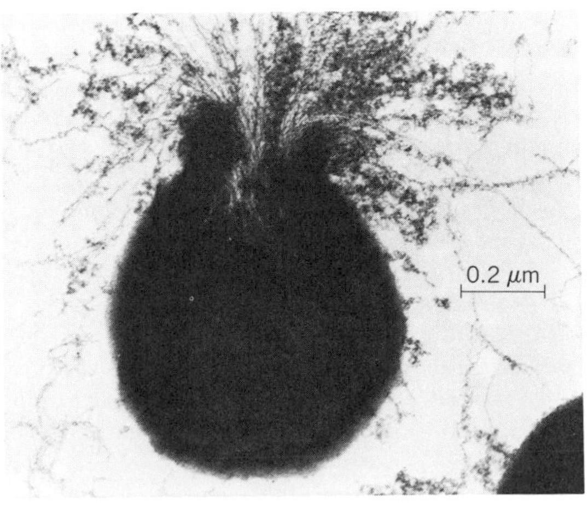

Fig. 1. *Staphylococcus aureus* exploding in the presence of sub-MIC of a cell-wall-active antibiotic.

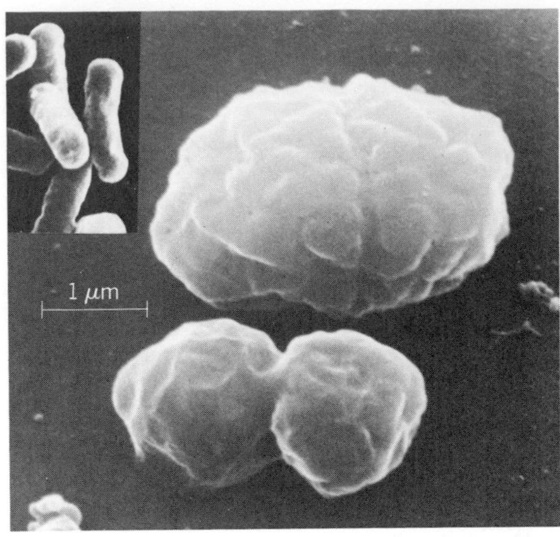

Fig. 2. *Proteus mirabilis* grown in the presence of sub-MIC of mecillinam. Inset is the normal organism. *(From V. Lorian, in V. Lorian, ed., Antibiotics in Laboratory Medicine, Williams and Wilkins, 1980)*

Virulence factors. The expression of virulence factors of bacteria can be influenced by sub-MICs of antibiotics. One of the most significant changes in the production of toxins generated by bacteria exposed to sub-MICs of antibiotics is an increase in beta-hemolysin produced by staphylococci and pneumococci grown in the presence of beta-lactam antibiotics at concentrations equal to one-third of the MIC. In the presence of sub-MIC of clindamycin, staphylococci have been shown to produce less coagulase and DNase. Another factor also related to the virulence of bacteria is the capacity to adhere to epithelial surfaces. The adherence of gram-negative bacilli to cells is due mostly to their fimbriae. Fimbriae of *Escherichia coli* bind to receptors of epithelial cells; mannose and some antibiotics at sub-MIC can inhibit bacterial binding to these cells. Ampicillin at one-fourth of the MIC inhibits the binding of *E. coli* to epithelial cells of the urinary tract. Streptomycin and penicillin have a similar effect on streptococci in the pharynx. It is therefore possible that sub-MICs of certain antibiotics, in addition to the effect on the viability of bacteria by interfering with their binding to cells or tissues, reduce their virulence. The effect of the complement and phagocytosis on the killing of bacteria which are exposed to sub-MICs of antibiotics is expected to be different from that of normal bacteria. Staphylococci exposed to any beta-lactam antibiotic at one-third to one-fifth the MIC change into "clusters" which consist of staphylococci held together by thick cross walls. These clusters are phagocytized by polymorphonuclear cells at about the same rate as normal staphylococci. During the first hour, clusters of staphylococci are 30% less affected by bactericidal enzymes than the normal staphylococci. In contrast, streptococci exposed to sub-MICs of clindamycin lose their M substance, which has an antiphagocytic effect, thereby enhancing the phagocytosis of streptococci which have been exposed to sub-MICs of clindamycin. Gram-negative bacilli which grow into filaments in the presence of sub-MIC of certain beta-lactam antibiotics are slightly more resistant to the effect of either serum of phagocytosis than the normal bacilli. While the mechanism of this higher resistance is not clear, it appears that the long filaments are not subject to total destruction by complement or phagocytosis. If one small segment of those filaments is spared, it results in the growth of at least one colony. While both gram-negative bacilli and gram-positive cocci grown in the presence of sub-MICs of antibiotics appear to be more resistant to the immunodefense mechanisms than normal bacteria, the differences are small. Consequently, for therapeutic purposes it is expected that the inhibitory effect of antibiotics at the MAC largely outweighs the possible negative effects, which may be related to the bactericidal activities of serum complement and phagocytosis. Animal experiments have confirmed this view.

Therapeutic activity. Of utmost importance is that sub-MICs of antibiotics have shown therapeutic activity. Patients with serious gram-negative bacilli infections, mostly bacteremias, have been treated with gentamicin and tobramycin with currently used dosages. In a small but significant number of these patients, the antibiotic concentration in the serum was below the MIC. The cure rate, however, for these patients was similar to those patients who had antibiotic concentrations equal to or higher than the MIC. Patients with urinary infections were treated with ampicillin at a total dose of 10 mg per day (this is about 200 times less than the currently used minimal dose). This dose produced urinary concentrations about one-third the MIC of the infecting organisms (most were *E. coli* with an MIC of 1.3 μg/ml). These patients also received 2 liters of liquid per day. A control group received only the 2 liters of liquid. Seven-day posttreatment evaluation showed that 80% of patients receiving the subminimal dose of ampicillin were cured, while none of the controls receiving only liquids were cured. It is speculated that sub-MICs of ampicillin in the urine interfere with the adherence of the *E. coli* on the bladder epithelial cells, and the large amounts of excreted water produced a large urinary flow which practically "washed out" the bacteria from the urinary tract.

For background information *see* ANTIBIOTIC in the McGraw-Hill Encyclopedia of Science and Technology. [VICTOR LORIAN]

Bibliography: V. Lorian, *Bull. N.Y. Acad. Med.*, 51:1046–1055, 1975; V. Lorian, in V. Lorian (ed.), *Antibiotics in Laboratory Medicine*, pp. 342–408, 1980; V. Lorian and B. A. Atkinson, *Antimicrob. Agents Chemother.*, 9:1043–1055, 1976; W. Stille and E. G. Helm, The activity of subinhibitorial concentrations of antibiotics, *8th International Congress on Chemotherapy*, Abstr. vol. A, 1973.

Antifungal agents

Chemotherapy of the human mycoses has been severely restricted in terms of the number of available drugs which are both clinically effective and safe. This can be attributed to three major factors: the diverse nature of mycotic infections; the limited market represented by certain mycotic infections such as those occurring primarily in third-world nations; and the unique host-parasite and biochemical relationships involved in treatment of infections in eukaryotic hosts caused by eukaryotic pathogens. Until recently, development of effective antifungal agents has been the result of serendipitous rather than deliberate discoveries. However, the last decade has seen a reversal of this process as new developments in antifungal chemotherapy have emerged through four distinct avenues of research: synthesis or development of new agents based upon previously described substances with known antifungal properties; chemical modifications of previously described substances; use of previously described substances in synergistic combinations; and, finally, discovery of new and novel substances.

Synthesis or development of new agents. Thiabendazole, an antihelmintic, was the first in a se-

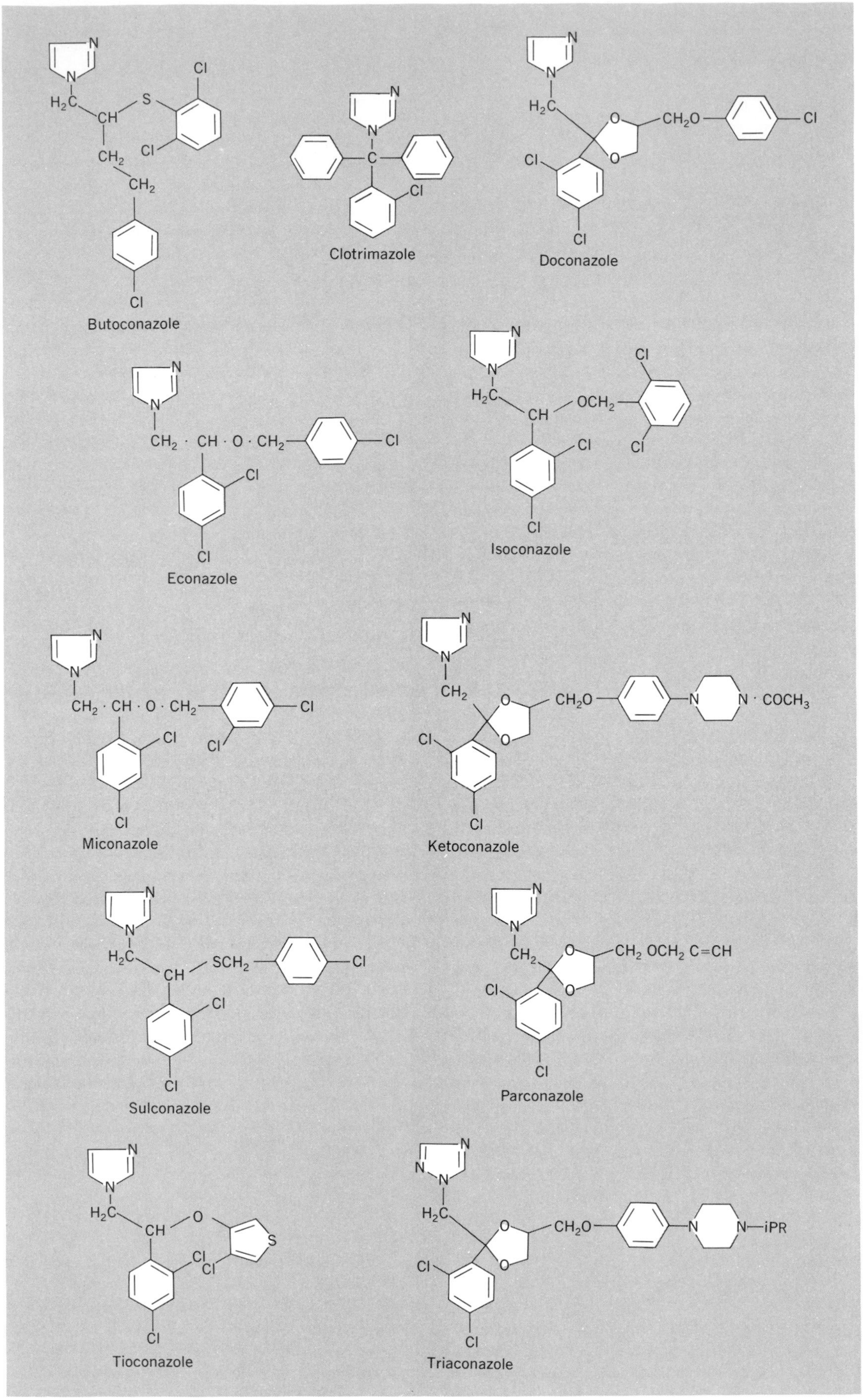

Butoconazole

Clotrimazole

Doconazole

Econazole

Isoconazole

Miconazole

Ketoconazole

Sulconazole

Parconazole

Tioconazole

Triaconazole

Fig. 1. Clinically useful and experimental imidazoles.

ries of substances, known collectively as the imidazoles, to have clinically useful antifungal activity. Thiabendazole has been used topically in the treatment of both ringworm infections and chromomycosis (chromoblastomycosis, verrucous dermatitis). Exploitation of the antifungal properties of the imidazoles led to the development of two related derivatives: clotrimazole [bisphenyl-(2-chlorphenyl)-1-imidazolyl methane] and miconazole [1-{2,4-dichloro-β-[(2,4-dichlorobenzyl)oxy]phenethyl} imidazole]. Both substances have wide spectra of activity against a variety of fungi in culture, including yeasts, molds, and dimorphic species and dermatophytes.

Clotrimazole and miconazole. Clotrimazole has been found to be highly effective as a topical agent for treatment of dermatophytic and yeast infections. Problems with both reduced bioavailability due to enzymatic degradation of the drug and toxicity associated with oral administration preclude other routes of administration. Miconazole, in contrast, has proved to be successful both topically and parenterally. Topical preparations are available for treatment of yeast and dermatophytic infections, and an intravenous preparation is available for treatment of opportunistic and systemic infections. While a high degree of clinical success has been associated with the topical preparations, results obtained with intravenous miconazole have been less than fully satisfactory. Limitations of the intravenous form include poor penetration into cerebrospinal and synovial fluids, frequent clinical relapses, and the requirement for prolonged periods of treatment often exceeding 6 months.

The successes associated with miconazole and clotrimazole stimulated the development of several second- and third-generation imidazole substances (Fig. 1). Many of these now either are in clinical use or are undergoing clinical trial. Included are econazole, triaconazole, sulcanazole, butoconazole, tioconazole, and isoconazole. All of these latter substances represent only minor variations on the "miconazole theme."

Ketoconazole. The most significant development within the imidazoles appears to be ketoconazole. One of a series of dioxolane derivatives including parconazole and doconazole, ketoconazole combines the broad spectrum of activity characteristic of the imidazoles with improved water solubility, and is orally active against a number of diverse mycotic infections. Favorable results have been obtained with ketoconazole in treatment of dermatophytic infections, *Candida* infections including vaginal as well as systemic candidiasis, histoplasmosis, coccidioidomycosis, and blastomycosis. Less favorable results have been reported in treatment of sporotrichosis and cryptococcosis. Bioavailability studies with oral ketoconazole show that the drug is well absorbed and that serum levels in excess of 5 μg/ml are attainable following a single oral 200-mg dose. All these results suggest that ketoconazole may represent an effective oral form of therapy for both dermatophytic and systemic mycotic infections.

Polyenes. The polyenes are a group of related macrolide compounds with certain common characteristics: conjugated olefinic chromophores; large lactone rings consisting of rigid lipophilic, conjugated double bonds; a flexible hydrophilic chain of hydroxylated carbon bonds; and a linked mycosamine amino sugar (Fig. 2). The polyenes are antifungal because of their ability to bind with sterols in cell membranes and thus produce permeability dysfunctions. This binding is the basis of both the antifungal and toxic properties of the polyenes, as sterols are present in membranes of both fungal and mammalian cells.

The polyenes of historical importance include nystatin, amphotericin B, and candicidin. Nystatin, the first clinically useful polyene, is used topically in the treatment of yeast infections such as vaginal candidiasis. The clinical spectrum of nystatin does not include dermatophytic infections. Amphotericin B is used intravenously in the treatment of opportunistic and systemic fungal infections. The toxicity of the drug restricts its use in all but acute or life-threatening infections. Candicidin, like nystatin, has a limited spectrum of clinical usefulness; this is restricted to vaginal infections caused by *Candida* species.

More recently, the list of clinically useful polyenes has been expanded by the addition of natamycin (pimaricin). The mode of action of natamycin is similar to those of amphotericin B and nystatin, except that natamycin causes permanent lytic alterations in fungal membranes as opposed to the reversible changes caused by low concentrations of amphotericin B and nystatin. The spectum of natamycin in culture includes a variety of fungal pathogens; the clinical spectrum is restricted to topical treatment of ocular fungal infections caused by susceptible yeasts and fungi. These include, in particular, infections caused by *Fusarium* and *Cephalosporium* species.

Modification of previously described agents. Two major disadvantages shared by the polyene antifungal agents are their toxicity and lack of water solubility. Attempts to bypass these have been made through the development of a variety of water-soluble derivatives. A methyl ester of amphotericin B has been shown to have less nephrotoxicity in humans than the parent component. D-Ornithyl salts of the methyl ester not only are highly water-soluble, but also possess higher therapeutic ratios in mice than either amphotericin B or the methyl ester.

Use of synergistic combinations. A once universally fatal disease, cryptococcal meningitis today can be successfully treated with several antifungal agents. These include amphotericin B and flucytosine (5-fluorocytosine). Earlier studies revealed that these two agents can act synergistically against pathogenic yeasts such as *Candida* and *Cryptococcus neoformans.* This synergism provided the basis for combination therapy with the two drugs in treatment of human cryptococcal disease which now has become the standard form of therapy.

The success obtained with combination therapy

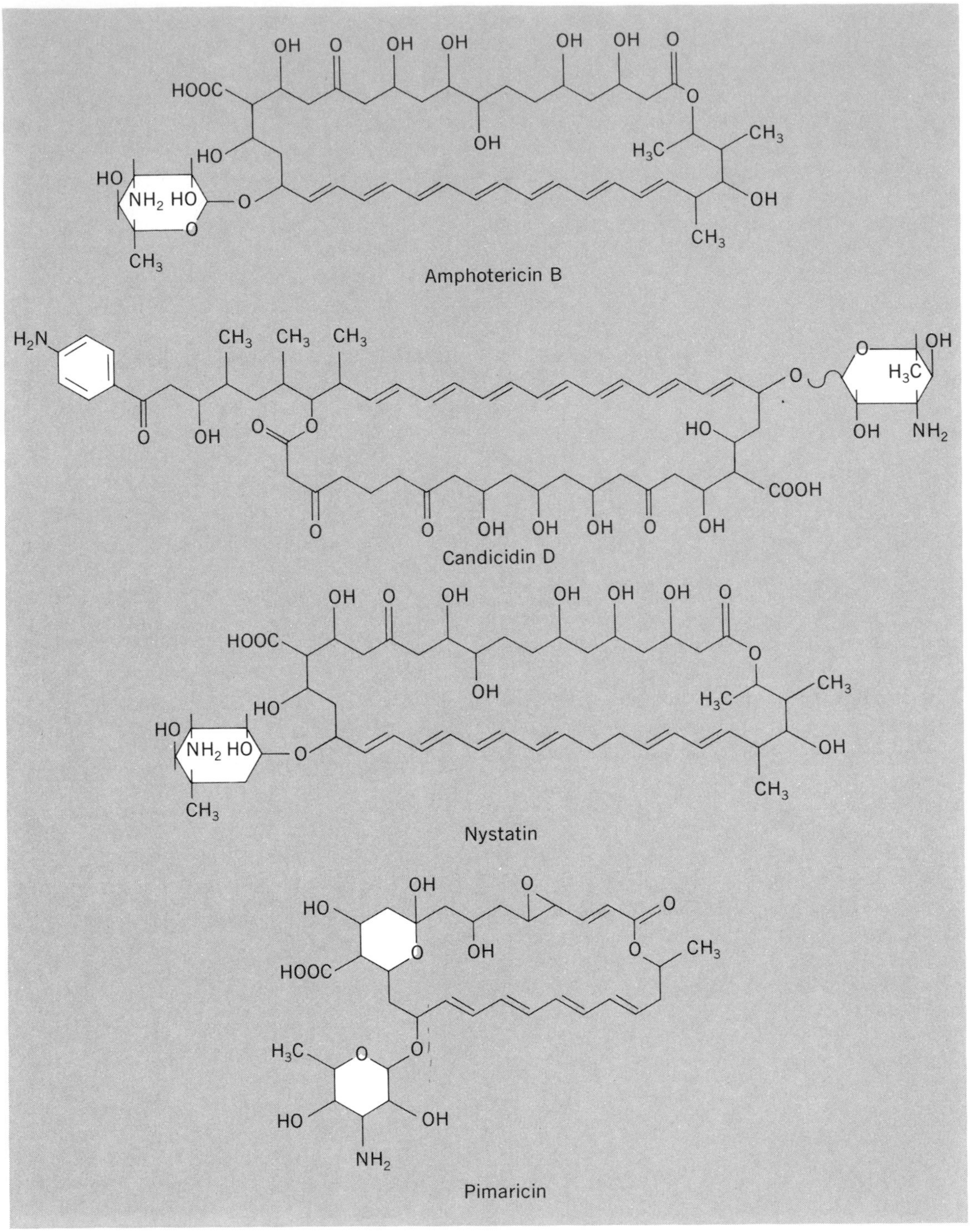

Amphotericin B

Candicidin D

Nystatin

Pimaricin

Fig. 2. Polyene antifungal antibiotics used in clinical medicine.

for cryptococcal meningitis stimulated studies of other potential synergistic antifungal combinations. Several such combinations have been identified. Minocycline acts synergistically with amphotericin B against pathogenic yeasts in culture; amphotericin B is synergistic with the antitubercular rifamycins, and this combination has been used in the treatment of fungal eye infections.

The opposite of synergism, antagonism, also exists between antifungal agents. Most importantly, the imidazoles act antagonistically against amphotericin B. This is believed to involve inhibition by the imidazoles of a demethylation step involved in biosynthesis of ergosterol which results in a lowered cell membrane sterol content.

Discovery of new agents. A variety of new antifungal agents have been described (Fig. 3). These include both synthetic compounds and naturally occurring substances. None, however, have been developed clinically as new therapeutic agents.

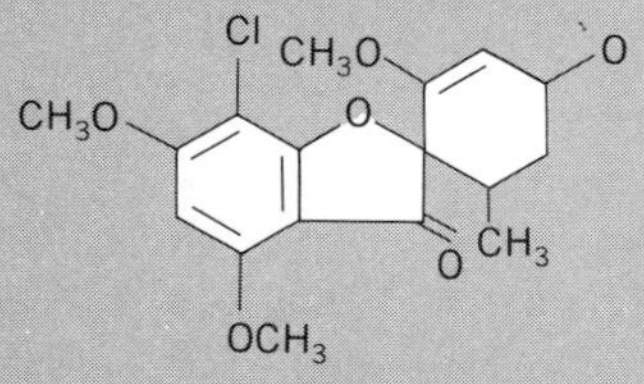

Griseofulvin

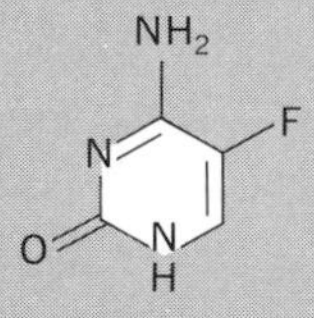

Flucytosine

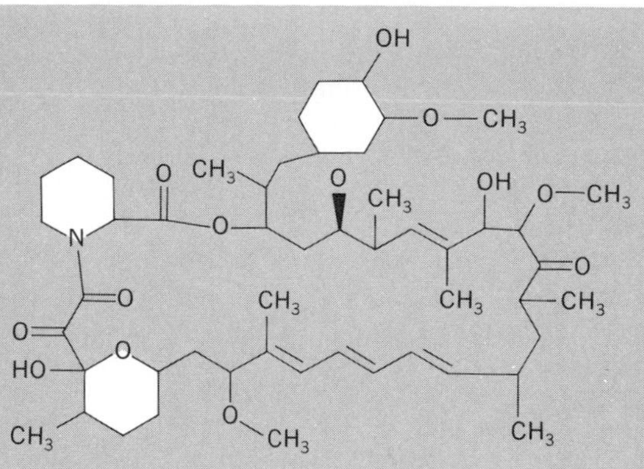

Rapamycin

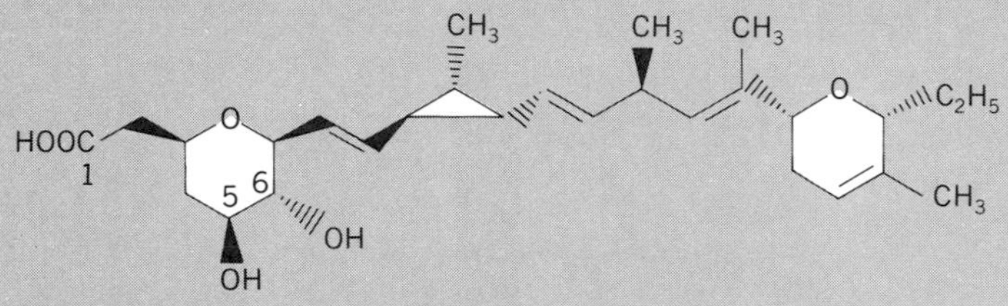

Ambruticin

	R	R[1]	R[2]	R[3]	R[4]	R[5]
a (−) Phaseollinisoflavin	H	OH	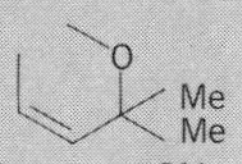 (ring)	Me, Me	H	H
b Sativan	H	OMe	H	OMe	H	H
c Vestitol	H	OH	H	OME	H	H
d Demethylvestitol	H	OH	H	OH	H	H
e 5-Methoxyvestitol	OMe	OH	H	OMe	H	H
f Astraciceran	H	OMe	H	O-CH$_2$—O		H
g Mucronulatol	H	OMe	OH	OMe	H	H
h "3-Hydroxymaackiainsoflavin"	H	OH	H	O-CH$_2$—O		OH

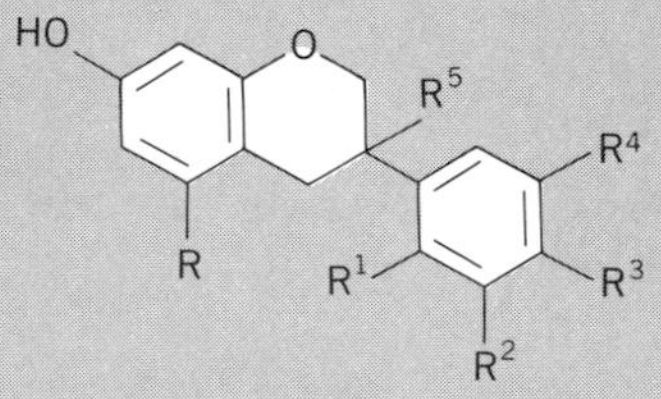

Isoflavonoids

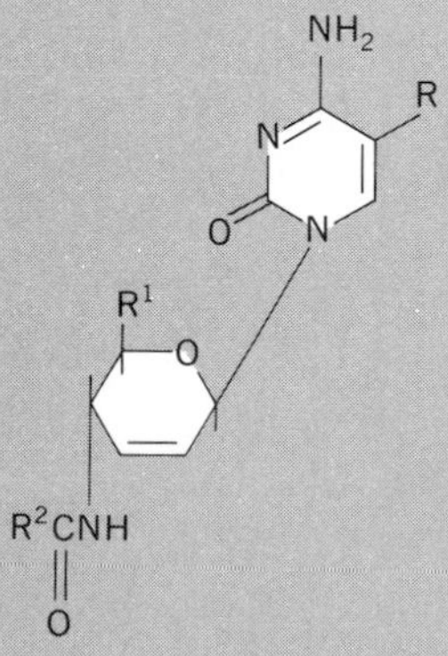

	R	R[1]	R[2]
a Blasticidin-S	H	—COOH	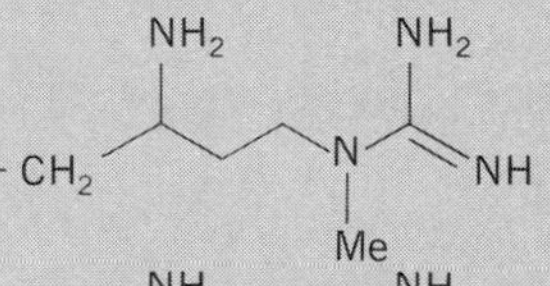 —CH$_2$ (side chain with NH$_2$; N(Me)C(NH$_2$)=NH)
b Demethyl blasticidin-S	H	—COOH	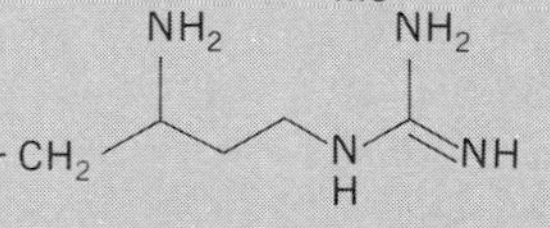 —CH$_2$ (side chain with NH$_2$; NHC(NH$_2$)=NH)
c Mildiomycin	—CH$_2$OH	HO, COO$^-$ —C (side chain with HO; HN, NH$_3^+$, NH)	—CH(NH$_2$)CH$_2$OH

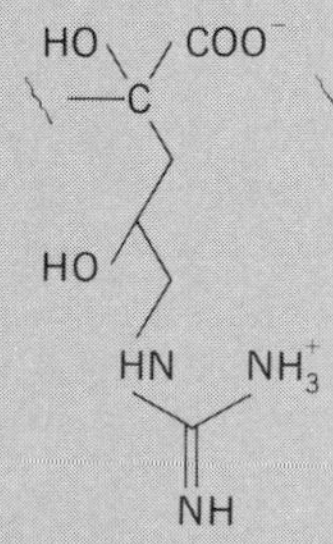

Blasticidins (polyoxins)

Fig. 3. Miscellaneous antifungal agents. Only flucytosine (5-fluorocytosine) and griseofulvin are used in clinical medicine.

Rapamycin. Among the polyenes, one of the more significant discoveries is rapamycin. Rapamycin differs from previously described polyenes in that its mode of action involves inhibition of nucleic acid synthesis rather than affecting cell permeability. Rapamycin also has significant antitumor activity. It is active orally against *Candida*.

Ambruticin. Ambruticin is a novel antifungal agent in two ways. First, it was isolated from a species of *Myxobacteriales*, or slime bacteria, a microbial source not previously known to produce antimicrobial agents. Second, ambruticin represents a new chemical series of biologically active agents: the cyclopropyl-pyranic acids. Ambruticin is active both in cultures and in animals against systemic as well as dermatophytic fungi.

Phytoalexins. The phytoalexins represent an interesting group of antifungal compounds with some potential clinical value. Phytoalexins are formed by plants in response to some form of external insult—physical, chemical, or biological. Biological insults include infections caused by plant fungal pathogens (phytopathogens) against which the phytoalexins act as natural antifungal agents in limiting or eliminating the infections. The phytoalexins are composed of four major chemical groups of compounds: polyacetylenes, isoflavonoids, sesquiterpenes, and dehydrophenanthrenes. While a great wealth of information has been accumulated regarding the activity of phytoalexins against phytopathogenic fungi, little exists regarding their activity against the zoopathogenic species. One series of compounds has been so studied. These are isoflavonoids produced by legumes. Of these, phaseollin isoflavan from *Phaseolus vulgaris* was the most active against a broad selection of human pathogenic fungi in culture.

Polyoxins. Polyoxins are pyrimidine nucleoside peptide antibiotics which act by inhibition of the enzyme chitin synthetase. Synthesis of chitin represents one of the few biochemical processes which are unique to fungal cells. Thus, this represents a unique chemotherapeutic target specific for such eukaryotic cells. Several of the polyoxins are employed commercially as fungicides in the Orient in prevention and treatment of plant diseases such as rice blast caused by *Piricularia oryzae*. One polyoxin, blasticidin S, has been shown to have activity in mice experimentally infected with *Candida*. However, toxicity precludes its use in human medicine.

For background information *see* CHEMOTHERAPY; FUNGISTAT AND FUNGICIDE in the McGraw-Hill Encyclopedia of Science and Technology.

[SMITH SHADOMY]

Bibliography: G. S. Kobayashi, and G. Medoff, *Ann. Rev. Microbiol*. 31:291–308, 1977; J. F. Ryley, et al., Experimental approaches to antifungal chemotherapy, in *Advances in Pharmacology and Chemotherapy*, vol. 18, 1981; S. Shadomy, H. J. Shadomy, and G. E. Wagner, Fungicides in medicine, in M. R. Siegel and H. D. Sisler (eds.), *Antifungal Compounds*, vol. 1, pp. 437–461, 1977; D. C. E. Speller, *Antifungal Chemotherapy*, 1980.

Antimicrobial agents

Diarrhea is one of the more frequent side effects of antimicrobial chemotherapy. It ranges in severity from a mild self-limited diarrheal illness to—albeit rare—a fulminant colitis. In the latter form, focal necrotic lesions can be found on the mucosa of the colon. These lesions gave the disease the name pseudomembranous colitis. They consist of fibrin and polymorphonuclear leukocytes exuding from a focal area of mucosal necrosis. The single plaques have a diameter of approximately 5 mm. Adjacent mucosa is normal. The whole colon may be covered with such mucosal changes. The patients have abdominal pain and tenderness which may overshadow the concomitant diarrhea. Fever and leukocytosis are also common findings. Pseudomembranous colitis may be self-limiting; in some cases, however, it may lead to the death of the patient. Mortality rates for pseudomembranous colitis of up to 27% have been reported, but this figure is probably far too high owing to the fact that many milder cases are not diagnosed.

Nearly all antimicrobial agents have been found to cause diarrhea or colitis or both. These include penicillins, especially the broad-spectrum aminopenicillins, the cephalosporins, tetracycline, chloramphenicol, cotrimoxazole (sulfamethoxazole-trimethoprim), lincomycin, and clindamycin. Pseudomembranous colitis is relatively rare. The disease has been reported in all age groups, but it is distinctly unusual under the age of 16–20 years despite the extensive use of antimicrobial drugs in pediatric practice.

Etiology. Until 1977 the etiology of antibiotic-associated diarrhea and pseudomembranous colitis was unknown. The possible suspects were, among others, *Staphylococcus aureus*, viruses, and vascular insufficiency. In 1977 a toxic substance was found in the feces of patients with pseudomembranous colitis. This toxin caused cytopathogenic changes in tissue cultures of live cells such as those used in growing viruses. Toxin-producing strains of *Clostridium difficile*, an anaerobic spore-forming bacterium, were identified as the causative agent of colitis in hamsters by the following means: (1) consistent development of colitis after clindamycin therapy; (2) transfer of the disease to healthy animals by inoculation of cecal contents from affected animals or of cecal contents incubated with clindamycin; (3) induction of the disease in healthy animals by inoculation of cell-free filtrates of cecal contents from hamsters with colitis; and (4) production of colitis by inoculation of the previously isolated *Clostridium* in pure culture and by inoculation of cell-free filtrates of broth cultures of the *Clostridium*.

Clostridium difficile and its toxin. *Clostridium difficile* was first isolated in 1935 as part of the normal fecal flora of many infants. It was rarely isolated from other sources such as wound infections, and it was considered as being of low pathogenicity. *Clostridium difficile* is not thought of as being a permanent resident of the human large intestine except in infants. It is suggested that *C. difficile* increases in number during antimicrobial therapy because of the resistance of many strains to these agents. This approach is probably too simplistic, and there may be other factors leading to multiplication of *C. difficile* in the intestine and production of its toxin.

The detection of the toxin produced by *C. difficile* in 1977 provided the impetus for extensive studies. This work showed that cell-free filtrates of *C. difficile* and of stools from patients with pseudomembranous colitis produce identical cytopathogenic changes in tissue culture. The toxic effect of both cell-free filtrates of cultures of *C. difficile* and of stool filtrates from patients can be neutralized by combining the broth or stool filtrate with *C. difficile* antitoxin. This antitoxin is produced in animals by injecting them repeatedly with formalin-inactivated toxin, that is, a toxoid. The toxin of *C. difficile* is heat-labile and has a molecular weight of approximately 550,000.

The diagnosis of antibiotic-associated diarrhea and colitis caused by *C. difficile* is made by the demonstration in stool filtrates of the toxin which can be neutralized by specific antibodies directed against it. Extensive studies have shown that almost all cases of pseudomembranous colitis and approximately 20% of cases of antibiotic-induced diarrhea are caused by *C. difficile*.

Treatment. The first recommendation is judicious use of antimicrobial agents. If feasible, the offending drug should be discontinued in patients who have pseudomembranous colitis or incapacitating diarrhea. Symptoms generally subside rapidly when, after the diagnosis has been established, the drug is discontinued immediately. Some patients lose large amounts of liquid, requiring replacement of fluids and electrolytes.

Vancomycin, an antibiotic never reported as having caused colitis, has proved to be a therapeutic success, and the clinical benefits are occasionally dramatic. Vancomycin is not reabsorbed by the gastrointestinal tract. Only oral administration gives adequate fecal levels. In treated cases *C. difficile* and its toxin disappear in 2–5 days, paralleling the diminution of diarrhea and pain. However, relapses after cessation of vancomycin therapy have been reported. These were probably due to spores of *C. difficile* surviving the vancomycin course and reappearing after vancomycin had been stopped. These *C. difficile* spores were still susceptible to vancomycin, and such relapses responded well to repeated courses of vancomycin. To date, no vancomycin-resistant strains of *C. difficile* have been reported.

[JÜRG WÜST]

Bibliography: J. G. Bartlett, *Rev. Infect. Dis.*, 1:530–539, 1979; W. L. George, R. D. Rolfe, and S. M. Finegold, *Gastroenterology*, 79:366–372, 1980; W. L. George, V. L. Sutter, and S. M. Finegold, *J. Infect. Dis.*, 136:822–828, 1977; H. E. Larson, *J. Infect.*, 1:221–226, 1979.

Ape language

Apes have the capacity to use symbols representationally. This function is basic to semantics, a major requisite to language. Whether apes are capable of syntax remains an open question.

Syntax research. The possibility that the ape might have language skills was strongly suggested by the initial results of Project Washoe, conducted by Beatrice Gardner and Allen Gardner in 1971, in which American Sign Language (ASL) was used for communication. It was claimed that the ape Washoe mastered a vocabulary of 132 signs by the age of 5 years. Roger Fouts extended the Gardners' research and demonstrated that the molding of apes' hands into the proper physical configurations is superior to other methods of teaching animals to sign. Francine Patterson also used ASL in her project with a female gorilla named Koko, and a computer terminal so that Koko could produce spoken words. *See* SIGN LANGUAGE.

Herbert Terrace's signing project with the chimpanzee Nim failed to support the conclusions of other ape-signing projects and, consequently, stirred much controversy. Nim was raised as a human child and was encouraged to communicate concerning all aspects of daily life. On the basis of analyses made of video tapes, Terrace and his coworkers ultimately concluded that Nim, primarily through imitation, was only simulating language. These same methods of analysis resulted in similar conclusions when applied to the results of others who had used ASL as a medium of communication with apes.

Nim was very young (4 years) even when Terrace's study ended. Whether the chimpanzee used signs representationally was not tested. Consequently, that Nim did not use syntax productively is no basis for concluding that apes are incapable of syntax, for competent use of syntax requires semantics.

By contrast to the ASL training approach, David Premack and Duane Rumbaugh independently employed methods in which the physical attributes of visual symbols served as the functional equivalents of words. Premack used a variety of plastic chips; Rumbaugh and associates used a computer-monitored keyboard upon which the keys had a variety of symbols. Premack claimed that his chimpanzee, Sarah, learned a variety of linguistic functions and that she had evidenced sensitivity to syntax in at least a very rudimentary way; that is, her ordering of the plastic chips or tokens was said to be constrained by observable world events.

Rumbaugh, H. Warner, and E. C. von Glasersfeld incorporated computer technology into the research effort in order to record objectively each linguistic event (a very real problem with the signing approach) so that it might be subject to subsequent computerized and statistical analysis. An artificial visual language of geometric figures, called lexigrams, was used. The system included keyboards for the chimpanzee and for the experimenter. Both keyboards were equipped with rows of projectors immediately above them. As each key on the board was activated, a facsimile of the lexigram on its surface was produced on a projector, the result being that sentences were formed for reference and for transmission between human and ape.

The first chimpanzee of this project was Lana. Her training emphasized stock sentences that had unique functions so far as the computer's operation of vending devices was concerned (see illustration). Through modifications of the stock sentences, that is, by substitution of words in various slots of the various stock sentences, it was intended that Lana would come to learn meanings for each and every word used. The locations of the keys were changed frequently so that Lana eventually had to attend to the distinctive lexigrams that embossed the surfaces of the keys rather than just to the locations of the keys.

Results with Lana indicated that she had become sensitive to the rules of Yerkish grammar (developed at the Yerkes Regional Primate Research Center), which had a correlational base. By 1978 Lana had come to formulate lexigram word strings, some containing 10 or 11 lexigrams. Many of the strings given within the context of experimenter-chimpanzee conversation would, at face value, suggest comprehension on Lana's part regarding what had been said by the experimenter and what had to be said by her so as to achieve some desired incentive. That impression notwithstanding, she lacked retention of specific meanings for words and phrases which she used rather adroitly in those strings. Thus, while it was clear that Lana's performance indicated a competence for syntax in at least an ordering sense, her performance failed to convince the researchers that she had competence for syntax in its fullest sense; that is, that her productions were discourse-dependent. The question of her competence for productive use of syntax was never resolved with a clear, positive answer. Consequently, it was evident that the inquiry into language acquisition needed to be directed away from syntax and toward semantics.

Lana chimpanzee observes a slide of a ball. She activates the projector by stating at the keyboard, "please machine make slide." The lexigrams for this statement can be seen both in the row of projectors above the keyboard and on the keys which she uses on the board.

Semantics research. Sherman and Austin were introduced to the Yerkes project in 1975. By contrast to Lana, they were not taught to produce strings of symbols. Rather, their training emphasized the meaning of each lexigram which they encountered. They were reared and trained in a social preschoollike setting. From the very beginning of their work, communication was emphasized, first with persons and then with each other. Both animals were taught far more than the production skills that had characterized previous ape-language subjects. In addition to the production of words in response to contexts and exemplars, the apes were taught receptive and indicative skills. Sherman and Austin have been the first chimpanzees to communicate their needs for specific tools and desires for specific foods to each other through use of learned lexigrams. They also have been the first chimpanzees to accurately categorize lexigrams for foods and tools through use of two other lexigrams, one for "food" and one for "tool." In short, they have demonstrated that the chimpanzee can use learned symbols referentially, that is, to represent things not present; and reference is critical to semantics, a requisite of language. A review of certain steps of this study with Sherman and Austin will contribute to an appreciation of the significance of its results.

Initial sorting. The categorizing of foods and tools was initiated by having the animals sort three foods (orange, bread, and beancake) into one bin and three tools (key, money, and stick) into a second bin. As none of the foods or tools resembled each other physically, a match-to-sample strategy for the sorting was not possible.

Categorizing objects with labels. The generic lexigrams for "food" and "tool" were introduced. The chimpanzees' task was now to sort a given food or a tool into the appropriate bin and then to select the lexigram representing either "food" or "tool" as the case might be. Next, the bins were dispensed with, and the task then was to categorize, through use of just the generic lexigrams, each of the six training objects.

At this point it would have been possible to account for the animals' performance in either of two ways: the animals have formed a specific association between each item and its appropriate lexigram; or they have formed a rule, such as "this is the lexigram for things I eat and this other lexigram is for the items I do not eat." Only if some concept or rule of the latter type was underlying the animals' performances might they use the generic symbols for "food" and "tool" to categorize still other items with no additional training. The generalizability of their categorization skills was tested by presenting them with five additional foods and five additional tools for which they already knew specific lexigrams or names. Austin was without error, and Sherman made only one. The item for which he was incorrect was sponge. (One possible explanation for this error is that, as he sucks and chews on the sponge, which is used to obtain liquids from inaccessible places, he frequently ingests portions of it. That is not the case with Austin.) Thus, it appears that Sherman and Austin had acquired in the initial training the concepts of food and of tool that were functionally based, generalizable, and symbolically encoded.

To determine whether or not learned or second-order food-food associations or tool-tool associations could account for Sherman's and Austin's test performances, they were presented next with 28 items (14 of which were foods, the others, tools) with which they were familiar but for which they had not learned specific "names." They were, therefore, not associated with lexigrams. Sherman categorized 24 of the 28 items correctly; Austin categorized 25 of the 28 items correctly. All but one of the errors resulted from classifying tools used in the preparation of food (knife, cutting board, and so forth) as foods.

Labeling photographs. In the next experiment of the series, the initial training items of three foods and three tools were used once again. Initially in this experiment, photos of these objects were taped to the objects, and the chimpanzee subject was asked to label them. Once accuracy was attained, the objects were deleted and only the photographs were used. Once they had achieved the criterion of performance in working with this set of six training photographs, they were then presented with novel photographs under blind test conditions.

Categorizing lexigrams. The final experiment of this series is the most critical of all for it is the one in which an unequivocal test of referential use of the symbols was made. The test included the following constraints: (1) it required a completely novel response, that of categorizing (labeling) one lexigram, which was the specific name of a food or tool, with another lexigram which in the generic sense represented either "food" or "tool"; (2) it was administered under tightly controlled conditions so as to preclude unintentional cuing; and (3) it required cognitive reference to the items represented by lexigrams. Cognitive referencing would have to be functioning, for the chimpanzee would have to classify a series of lexigrams, each standing for either a specific food or tool, with either the lexigram "food" or "tool" as the case might be on a given trial. Such labeling would have to be based on the referent's characteristics as recalled upon sight of each lexigram.

Initially the original training group of three foods and three tools was employed. Again, the lexigrams for these items were at first taped to the photographs of them in order to provide a bridge. The photographs were then removed as the chimpanzees achieved accuracy. The chimpanzees were then shown only the printed lexigrams.

Test trials were administered with the experimenter out of the room and with the novel word-lexigrams being interspersed randomly with the lexigrams for the training items. It is important to note that prior to this test, these chimpanzees had never been asked to make categorical assessments of these specific lexigrams through the use of generic lexigrams, nor had they ever observed any human teacher do so. For all of the lexigrams used in this study, red was the common background color. Consequently, it would have been impossible for the animals to categorize on the basis of color background differences.

Sherman categorized the novel lexigrams correctly on 15 out of 16 first-trial presentations; Austin's performance was 100% correct. The possibility that the animals had in some manner matched the lexigrams for specific foods and tools on the basis of physical similarity with the generic lexigrams for "food" and "tool" was controlled for, so that if Sherman and Austin made their choices on that basis they would have been correct only half of the time. Between the two animals, only one error was made on the entire test, and that error was again "sponge"—which was called a food by Sherman. This error would not have been predicted on the basis of lexigram physical similarity.

Conclusion. Sherman and Austin were able to use "food" and "tool" as representational labels and to use these labels to categorize novel lexigram exemplars. The authors believe these chimpanzees were able to do so because of the specific type of training over the 4 years prior to the conduct of this study. That training had encouraged functional symbolic communication between the chimpanzees.

That Sherman and Austin were able to achieve such a level of symbolic functioning says a great deal about what the chimpanzee can do. It is not to be concluded, however, that because of this the behaviors of all other chimpanzees—in or out of language projects—reflect representational symbolism.

The results of this study not only serve to reinforce the claim that in the past researchers have failed to appreciate fully the chimpanzee's high level of intelligence relative to that of other animals; they also emphasize the potential value of the ape as an animal model for the study of the processes germane to the cognitive processes of *Homo sapiens.*

Sherman's and Austin's data support the conclusion that the chimpanzee can master referential use of learned symbols, the meanings of which have been arbitrarily assigned to lexigrams, and this reinforces the conclusion that apes are capable of mastering semantics. It is contended that Sherman and Austin, in fact, were using their learned symbols as words with semantic definitions. Their use of learned symbols, at least in a restricted sense, was the same as humans' use of words; that is, they used their symbols representationally. Can they use syntax? Can they use syntax productively in a linguistic sense? The answers to these questions are still unknown. And those answers are not to be presumed to be negative on the basis of Terrace's report with Nim chimpanzee.

Although it cannot now be argued that the ape is capable of language in the human sense, it is clear that through the study of apes during language training a great deal has been learned about what human language is and is not. Neither human language itself nor words, which are basic to human language, are considered to be all-or-none phenomena. Syntax and semantics are no longer seen as being divorced from behaviors and the organizations of behaviors in nonlinguistic contexts. It is also clear that there is a very close relationship between the acquisition of language skills and the acquisition of nonverbal communication skills.

For background information *see* COGNITION; INTELLIGENCE; LEARNING THEORIES in the McGraw-Hill Encyclopedia of Science and Technology.

[DUANE M. RUMBAUGH; E. SUE SAVAGE-RUMBAUGH]

Bibliography: B. T. Gardner and R. A. Gardner, in A. M. Schrier and F. Stollnitz (eds.), *Behavior of Nonhuman Primates*, vol. 4, 1971; D. Premack in A. M. Schrier and F. Stollnitz (eds.), *Behavior of Nonhuman Primates*, vol. 4, 1971; D. M. Rumbaugh, H. Warner, and E. C. von Glasersfeld, in D. M. Rumbaugh, *Language Learning by a Chimpanzee: The LANA Project*, 1977; E. S. Savage-Rumbaugh et al., *Science*, 210:922–925, 1980; E. S. Savage-Rumbaugh and D. M. Rumbaugh, *Brain Lang.*, 6:265–300, 1978; E. S. Savage-Rumbaugh, D. M. Rumbaugh, and S. Boysen, *Science*, 201:641–644, 1978; H. S. Terrace et al., *Science*, 206:891–900, 1979.

Atmospheric sampling

The development of the elemental analysis technique of particle-induced x-ray emission (PIXE) during the 1970s has revolutionized methods for atmospheric sampling of aerosol particles and certain trace gases. The high sensitivity of the PIXE method and the small dimensions of the sample area analyzed during proton bombardment (a few millimeters or less) have permitted miniaturized samplers and vacuum pumps to be used. The modest electric power requirements, low sampler unit cost, and simplicity of operation of PIXE-compatible equipment enable sampling to be carried out in locations inaccessible to large field equipment, such as on aircraft, on remote mountain peaks, and on ships at sea, and help in carrying out multisite network sampling inexpensively. Cascade impactors for particle-size fraction sampling, time-sequence filter samplers, and treated filters for trace-gas sampling are used in these investigations.

Particle-size fraction sampling. The small dimensions of the particle beam from a Van de Graaff accelerator used to generate characteristic x-rays in a sample during PIXE analysis are readily compatible with aerosol samples sorted according to their aerodynamic diameters by cascade impactors of the single-orifice type (Fig. 1). The airflow drawn through the impactor increases in velocity from orifice to orifice of progressively decreasing diameter, enabling progressively smaller size ranges of particles to be collected on the greased impaction surface behind each of the orifices. Thus stage 7 collects the largest particles, greater than 16-μm diameter, stage 6 the next largest, 8–16 μm, and so on, until the airstream reaches the after-filter, which collects particles smaller than 0.06-μm diameter. Figure 1 shows the inclusion of two optional low-pressure stages, L2 and L1, placed in the low-pressure region after stage 1, which fractionate particles smaller than 0.25-μm diameter.

This cascade impactor, which is compatible with the PIXE method, operates at a flow rate of 1 liter/min of incoming air, controlled by the small orifice of stage 1 which functions as a limiting flow orifice. Small-capacity vacuum pumps which can maintain a high degree of vacuum can be used. Unless stages L2 and L1 are used, the pumps may be miniature in size and operated by dc or ac power, in remote locations if necessary. The small size and weight (1

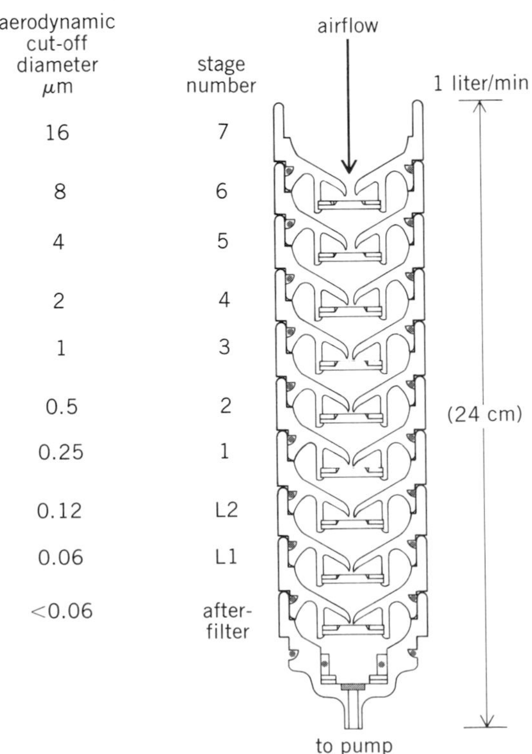

Fig. 1. Schematic diagram of cascade impactor of single orifice design. (*PIXE International Corp.*)

kg) and low cost of the impactor make this an attractive sampler for field use. Moreover, particle-size distributions within the fine (<1-μm diameter) range, which are important in air-pollution research, can be investigated.

Time-sequence air sampling. The small beam dimensions for PIXE analysis also permit the analysis of small filter areas. This has made possible the design of a time-sequence filter sampler consisting of a smooth Nuclepore polycarbonate filter and tightly adhering, sucking orifice. A continuous sliding motion by means of a clock motor as air is drawn through the filter produces a sample streak along the filter. This so-called streaker sampler exists in both linear and circular form so that the filters, usually representing 1-week periods, may be analyzed stepwise, for example, in increments corresponding to 2-h time averages.

Figure 2 illustrates a commercially available circular streaker with an exposed filter. The banded structure of the streak sample represents natural short-term variability in the density of black particulate material in the atmosphere. The elemental composition of the aerosol as determined by PIXE also generally exhibits marked variability in time, although not necessarily in correlation with the visible blackness. The 2-h time resolution given by the streaker sampler corresponds to variability in characteristics of the atmosphere, making the streaker sampler convenient for routine measurements in sampling networks. Moreover, the 84 separately analyzed time steps of a week-long sample streak on a single filter constitute a data set sufficiently large for statistical analysis of aerosol composition trends. This analysis may lead to resolution of the ambient aerosol into distinct components characteristic of the most important pollution and natural sources in the area investigated.

Trace-gas sampling. Chemically treated filters may be used for the sampling of gaseous sulfur dioxide and inorganic halogen species. For example, Nuclepore filters coated with aqueous potassium hydroxide efficiently collect trace concentrations of SO_2 in moderately polluted atmospheres. These filters may be used to distinguish gaseous from particulate atmospheric sulfur by placing them as afterfilters following particle collection in cascade impactors or sequential filter samplers. The analysis of the gas-absorbing filter may be carried out by PIXE or other methods. Such filters are well suited to measurement of SO_2 at the low concentration levels which may be found at long distances downwind of pollution source areas. For example, concentrations of less than 1 part per billion by volume have been reported over the Gulf of Mexico using air sampling volumes of 0.5–1 m^3 in 2–4 h sampling time. The filter treatment procedure may also be applied to the Nuclepore filter used in the streaker sampler for acquiring a measure of the variability of gaseous SO_2 concentrations as a function of time.

Alkali-treated filters have also been used in sampling trace quantities of inorganic chlorine-containing compounds, for example, HCl. Particulate chloride is removed from the airstream by means of an impactor or filter sampler, and the gaseous chlorine is then absorbed by the treated filter. This technique has been applied to the investigation of gaseous chlorine released from marine aerosol particles above the Gulf of Mexico as a result of reaction with acidic air-pollution substances.

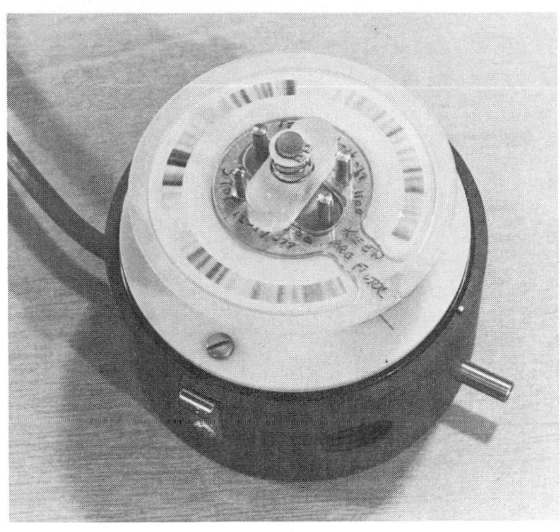

Fig. 2. A time-sequence streaker filter sampler of circular design. (*PIXE International Corp.*)

Pollution aerosol investigations. Aerosol sampling by cascade impactors and streaker filter samplers generally shows good agreement in the concentrations of the elemental constituents measured by PIXE. In March 1980 the concentrations of six elements were measured by streaker in 4-h time steps together with the sums of concentration over eight particle-size fractions sampled by cascade impactor in air from a rural area downwind of the city of Beijing, China. Agreement was best for the elements potassium, sulfur, and zinc, elements which are characteristically found mainly in submicrometer particle sizes. For calcium and iron, elements characteristic of coarse particle dust, the impactor measurements were slightly higher than those of the streaker, reflecting a greater efficiency of coarse-particle sampling by the impactor under field conditions.

The streaker measurements of particulate chlorine were considerably lower than those made by the impactor, probably reflecting chemical reactivity and volatilization of gaseous HCl from the filter because of acidic air pollutants; this effect was not observed in the impactor samples because of the different airflow pathway through the sampler and the possible coating of impactor particles by the retaining grease. These results illustrate how the simultaneous use of different atmospheric sampling devices can increase understanding of the aerodynamic and chemical characteristics of different aerosol constituents.

The data included detailed particle-size distributions of sulfur, chlorine, potassium, and calcium measured at the same sampling location. During the study period, coal combustion aerosol was prevalent and was the principal source of fine-particle sulfur, chlorine, and potassium. Calcium, on the other hand, was most abundant in coarser particles and was derived mainly from the dispersion of soil dust. In this study both S and Cl occurred with similar size distributions, having maximum concentrations in the 0.5–1-μm-diameter range. They were associated with the same coal combustion aerosol source, and their fine particle size indicates a gas-to-particle aerosol formation process in the atmosphere. On the other hand, the maximum concentration of Ca was in the 4–8-μm-diameter range, due to the dispersion characteristics of soils. The distribution of K over particle size exhibited intermediate characteristics. By assuming K to represent the sum of two modes, one of which is coarse and similar to the distribution of Ca, it was possible to resolve a fine-particle mode, K_x. This fine K_x component was found to be similar to the distributions of S and Cl, and is considered to be associated with the coal combustion aerosol, whereas the coarse K component is a soil dust constituent. These results illustrate that cascade impactor sampling can give a detailed measure of particle-size distribution in the submicrometer range, where combustion and other pollution aerosol constituents are frequently found. Natural dusts occur characteristically in coarser-size fractions, and aerosol sampling by impactors permits the resolution of coarse from fine components based on particle-size distributions. Combined with streaker sampling, the total information can lead to the identification of both natural and pollution source types and eventually their locations.

For background information *see* AIR-POLLUTION CONTROL; GAS AND ATMOSPHERIC ANALYSIS; SAMPLING TECHNIQUES in the McGraw-Hill Encyclopedia of Science and Technology.

[JOHN W. WINCHESTER]

Bibliography: S. Bauman, P. Houmere, and J. W. Nelson, *Nucl. Instr. Meth.*, 181:499–502, 1981; W. W. Berg and J. W. Winchester, *J. Geophys. Res.*, 82:5945–5953, 1977; J. W. Nelson, Proton-induced aerosol analyses: Methods and samplers, in T. G. Dzubay (ed.), *X-ray Fluorescence Analysis of Environmental Samples*, pp. 19–34, 1977; S. Tanaka et al., *Nucl. Instr. Meth.*, 181:509–515, 1981; J. W. Winchester et al., *Nucl. Instr. Meth.*, 181:391–398, 1981.

Automotive vehicle

At the turn of the century, the first horseless carriages were primitive battery-powered electric automobiles. Within a few years, these vehicles were made obsolete by the internal combustion engine and the mass production technology developed by Henry Ford. It is ironic that today's gasoline-powered automobiles may disappear during the next few decades, with high-performance electric automobiles becoming the predominant mode of personal transportation in the 21st century. The major reasons for this revolutionary change are expected to be increased cost and scarcity of liquid fuels, and vast technical improvements in batteries and electric cars.

Liquid fuel shortages and electric cars. One of the most serious energy problems faced by the United States during the remainder of the 20th century will be shortages of liquid fuels for transportation. The magnitude of this problem will be reduced through the development of down-sized, more efficient vehicles, by reducing nonessential travel, by producing synthetic fuels from coal or biomass, or by exploiting oil shales and tar sands. However, each of these is only a partial solution, and serious liquid fuel shortages are possible even in the face of dramatic success in each partial solution.

An attractive solution to the transportation problem would be the widespread use of electric vehicles. Because electric power–generating plants can use a wide variety of energy sources, many of them abundant for the foreseeable future, replacing conventional vehicles with electric ones reduces the demand for high-grade liquid fuels. Intensive efforts by government and industry to develop acceptable electric vehicles have produced several technical advances over the past 10 years, including efficient motor-controller systems, reductions in aerodynamic drag and rolling resistance, more efficient transmissions, and a variety of lightweight components. These advances pave the way for high-performance electric cars of the future, capable of impressive speed and acceleration. Only one obstacle to the commercialization of electric cars remains in the 1980s—the lack of an advanced battery of suitable performance, ruggedness, and cost.

Advances in battery technology. The very real possibility of an immense transportation market has created a technological race to develop a suitable battery for electric vehicles. Dozens of laboratories in western Europe, Japan, and the United States have entered this competition, the result of which appears to be a worldwide renaissance in battery technology. For example, lead-acid battery technology, which was dormant for 50 years, has made exciting progress during the past five years—about 50% improvement in specific energy (measured in watthours per kilogram) and specific power (measured in watts per kilogram) and 100% in cycle life. Similarly, the performance capability of nickel-iron and nickel-zinc batteries has improved by about 50% in the past few years. The zinc-chloride battery, which was invented only 12 years ago, has advanced to the point at which full-size batteries are being tested in electric cars. The lithium–iron sulfide battery, invented in 1973, has advanced to the engineering stage with specific energy approaching 100 Wh/kg. In addition, sodium-sulfur, aluminum-air, iron-air, and zinc-bromine batteries have experienced major technical advances during the past 5 years. Each of the above-mentioned battery systems is a candidate for electric vehicle markets.

However, the technical barriers facing battery developers are numerous and difficult, and the rate of progress is uncertain in every case. It is expected that one of these batteries will eventually emerge as the winner and will dominate vehicle markets.

Lead-acid batteries. The lead-acid battery is the most widely used electrochemical system. Lead-acid cells consist of positive and negative electrodes that are immersed in an electrolyte solution of sulfuric acid. As the cell is discharged, lead dioxide in the positive electrodes and elemental lead in the negative electrodes are converted to lead sulfate.

Recent technical improvements have produced lead-acid batteries capable of 40 Wh/kg and lifetimes of 500–800 cycles. Even with these improvements, lead-acid batteries are not well suited for use in electric cars because of bulkiness, marginal specific energy, and poor power late in the discharge. At present, the lead-acid battery is the only battery system available for any kind of electric vehicle. In general, it appears to be most attractive for applications in which a limited range (less than 100 mi or 160 km) is acceptable. One can therefore expect that the principal markets for lead-acid vehicles will be local commercial fleets, especially vans, trucks, and buses.

Nickel-zinc batteries. Nickel-zinc batteries have been likened to a rocket that shot into the air with great dazzle and noise, but soon fell back to Earth and was forgotten. Just 2 years ago, nickel-zinc batteries represented the leading candidate for electric cars, and General Motors was preparing to mass-produce them for public use. Unfortunately, nagging problems with battery lifetime have diminished the enthusiasm of most developers, and nickel-zinc research is now considered to be a high-risk, exploratory proposition.

In a nickel-zinc battery, the overall cell reaction is given by reaction (1). An inherent problem with

$$2NiOOH + H_2O + Zn \rightleftarrows 2Ni(OH)_2 + ZnO \qquad (1)$$

this system is that ZnO is partly soluble in the KOH electrolyte; this causes shape changes, dendrite formation, and densification of the electrode during cycling, so that the battery life is shortened. A wide variety of potential solutions to this lifetime problem are under investigation, but recent progress has been imperceptible.

The specific energy presently attainable in nickel-zinc batteries is about 65 Wh-kg, with an additional 25% improvement expected by the year 2000. Peak power and volumetric energy density are both excellent and steadily improving with time. The nickel-zinc battery is well suited for the transportation application and easily adaptable into a wide variety of vehicle designs. However, a major advance in battery life must be achieved before the nickel-zinc electric car will have a significant effect upon society.

Nickel-iron batteries. Recent advances in the Soviet Union, the United States, Japan, Sweden, and Bulgaria have revived interest in the nickel-iron battery developed by Thomas Edison in 1901. The active materials in a nickel-iron battery consist of finely divided hydrated nickel peroxide for the positive plate and finely divided iron for the negative plate.

At present, the nickel-iron battery system appears to be greatly underrated in the United States. It is, in fact, the only system among the major contenders that has demonstrated ruggedness and long life. Currently, nickel-iron batteries can store 50–55 Wh/kg, and there appears to be no major technical barrier to increasing the specific energy to 60 Wh/kg.

The major problems associated with nickel-iron batteries are high initial cost, nickel import requirements, and hydrogen evolution. Their cost, however, may well be competitive, because of their excellent service life. They are also quite bulky, and thus less attractive for use in private cars; for commercial fleets (buses, light trucks, and so on) their bulk should not be a problem.

Lithium–iron sulfide batteries. In 1973 researchers at Argonne National Laboratory invented a battery with iron sulfide positive electrodes and lithium-aluminum negatives. The overall cell reaction for the case of FeS_2 is given by reaction (2).

$$4Li + FeS_2 \rightleftarrows 2Li_2S + Fe \qquad (2)$$

This battery technology has made impressive progress during the past few years, but many difficult problems must be overcome before the Li–Al/FeS_x battery is a practical reality. The most difficult challenges for the developers are to develop a low-cost separator and to achieve satisfactory ruggedness and service life in high-performance batteries. Prospects for overcoming these technical barriers appear to be fairly good, but successful development of the system is not expected until the 1990s.

The Li–Al/FeS_x battery is very well suited for use in electric vehicles. This battery will probably be the most compact battery by 1990, with volumetric energy densities expected to exceed 200 Wh/liter. The specific energy is expected to be about 20–40% higher than that of nickel-zinc systems, and peak power should be satisfactory. Although successful development is far from certain, the Li–Al/FeS_x battery has emerged as a leading candidate for the electric vehicle application.

Zinc-chlorine batteries. Until recently, the zinc-chlorine battery, long recognized as offering high energy and low cost, was not exploited because chlorine is hard to manage. In 1969 Energy Development Associates (EDA) introduced a zinc-chlorine concept in which chlorine is stored as chlorine hydrate ice. This Zn/Cl_2 battery concept was rapidly brought to the engineering stage of development, with electric vehicle testing initiated in the 1980s.

The Zn/Cl_2 battery developed by EDA represents a unique and very complex system involving a circulating (pumped) electrolyte and a refrigeration system for production and storage of frozen chlorine hydrate. The electrolyte is an aqueous solution of zinc chloride, and the operating temperature is about 40–50°C. For discharge, warm electrolyte circulates through the hydrate store, picking up chlorine, which is then circulated to the battery stack for reaction in the cells.

Chlorine safety is a nagging problem since certain types of accidents could result in serious releases of toxic chlorine to the environment. On the positive side, there is little doubt that Zn/Cl_2 batteries with specific energies of about 100 Wh/kg and service lifetimes of several years will be developed. A strong point in its favor is that the battery raw materials are inexpensive and readily available. The Zn/Cl_2 battery is a clear contender for future electric vehicle markets; its use, however, may be restricted to vans, trucks, and buses because of problems in scaling down to small sizes.

Sodium-sulfur batteries. Major programs have been under way for more than 10 years in the United States, the United Kingdom, France, and Japan to develop a sodium-sulfur battery. The essential feature of this system is the ceramic electrolyte beta alumina ($Na_2O \cdot 11Al_2O_3$), which is highly conductive to sodium ions at 300–350°C. The active materials, sodium and sulfur, are both liquid at the operating temperature, and the solid electrolyte serves as the separator.

Typically, sodium-sulfur cells involve a central tube of beta alumina containing liquid sodium surrounded by a layer of graphite felt loaded with liquid sulfur and polysulfide. This cell assembly is encased in a metal housing that also acts as the positive current collector. Corrosion of the positive current collector is a serious problem, especially for batteries to be used in vehicles. Other technical barriers include marginal durability of the ceramic electrolyte tubes and sodium safety problems. A crash-worthy battery casing may be necessary to minimize the likelihood of sodium fires or explosions. On the bright side, all the raw materials are intrinsically inexpensive and are available domestically. The sodium-sulfur battery has been considered a candidate for electric vehicles for over a decade; however, worldwide interest in this battery system seems to be declining.

Other battery systems. During the past 5 years, zinc-bromine and aluminum-air batteries have become viable candidates for future electric vehicle markets. Zinc-bromine batteries are similar to zinc-chlorine systems in that a circulating electrolyte transports the halogen reactant to and from graphite battery stacks where the cell reactions occur. Important advantages of this Zn/Br_2 battery include efficient bromine storage in a liquid medium, and the ability to achieve wide power variations at constant electrolyte pumping rates. As in the case of Zn/Cl_2, the Zn/Br_2 battery is a highly complex chemical plant which must be compactly packaged for use in vehicles. The principal features of Zn/Br_2 batteries include low-cost, available raw materials, moderate electrical performance (60–80 Wh/kg), and a safety problem related to potential bromine releases. Worldwide interest in Zn/Br_2 batteries appears to be accelerating as a result of recent technical advances.

The aluminum-air battery is the only electrochemical system with realistic prospects for achieving performance equivalent to that of gasoline-fueled vehicles. The concept involves mechanical replacement of the aluminum electrodes about every 1000 mi (1600 km) and electrolyte additions every few hundred miles. Presumably, these operations could be accomplished in less than 15 min at a service station, in effect providing the electric vehicle with unlimited range. Good progress has been made recently, including demonstration of the cell chemistry and development of full-scale cells. The principal problems relate to system complexity, service life, a very poor energy efficiency of less than 40%, and the need for an infrastructure of specialized service stations and aluminum recycling plants. The technology is in the exploratory stage, but continued progress could catapult aluminum-air batteries into contention for future electric vehicle markets.

Among other exploratory battery contenders are iron-air batteries and sodium-sulfur batteries with glass fiber electrolytes. Both of these systems offer the promise of low cost and superior performance but have severe problems related to abbreviated service life. A major breakthrough in either of these exploratory batteries could make most of the other electric vehicle battery candidates obsolete.

Prospects for electric cars. It is impossible to identify the eventual winners in the battery competition, since each system has a less than 50% probability of successful development and commercialization. However, the present rate of technical advance in battery technology is historically unprecedented, and it is highly probable that at least one of the advanced batteries will be successfully developed. By the year 2000 there are likely to be a large number of electric vehicles on the road, possibly millions. The actual number will depend on the severity of future shortages of liquid fuels as well as on the progress in battery research.

For background information *see* BATTERY; ENERGY SOURCES; STORAGE BATTERY in the McGraw-Hill Encyclopedia of Science and Technology.

[WILLIAM J. WALSH]

Bacteria

Many bacteria produce compounds or structural components that increase their pathogenicity. Proteins that may be released into the surrounding milieu are one example. A variety of destructive enzymes, toxins, and substances which inhibit growth of other bacteria (bacteriocins) have been identified. Outer bacterial layers that impede normal host defensive mechanisms are another example. Bacterial capsules have been implicated in resistance to engulfment (phagocytosis) by specific host cells which are perhaps related to the capsules' negative charge repelling the negative charge of mammalian cells.

As a final example, certain bacterial components facilitate attachment of bacteria to receptor surfaces, allowing bacteria to colonize and subsequently infect the host.

Much research has been conducted in this area for many years, and production of such compounds or structural components can be altered by viral infection of the bacteria, nonchromosomal pieces of genetic material, the surroundings, and so on. Some recently studied examples are described in greater detail below.

Proteases. Microorganisms frequently have initial mammalian host contact at mucous membranes. Secretory immunoglobulins (antibodies) of the IgA type are synthesized by surrounding plasma cells, and function as a host defensive mechanism. Several independent groups of researchers have described the production of bacterial extracellular enzymes (proteases) which cleave IgA into Fab and Fc fragments. Bacteria as diverse as *Neisseria*, *Haemophilus influenzae*, and streptococci can produce these enzymes. The more virulent *Neisseria* (the species that usually causes gonorrhea and meningitis) produce IgA proteases, but generally the *Neisseria* species that are usually nonpathogenic do not produce IgA proteases. Similar findings have been reported for the pathogenic species of *Haemophilus* when compared with the less-often-pathogenic species.

Production of IgA proteases by bacteria could well be a virulence factor which enables pathogenic bacteria to survive in the environment of hosts' mucous membranes. Bacteria which do not attach to mucous membranes may be removed by propulsive mechanisms and the normal flow of secretions. Antibodies to bacteria can prevent attachment; secretion of IgA proteases might enable pathogenic bacteria to remain and subsequently cause infection.

Toxins. Diseases such as botulism, tetanus, cholera, and diphtheria have been known for many years to be associated with bacterial toxin production. Recently two other disease processes have been associated with production of toxins by bacteria.

Diarrhea. Enterotoxigenic strains of *Escherichia coli* are now considered to be the cause of many cases of childhood diarrhea, diarrhea in travelers, and other cases of diarrhea. Much elegant work has identified two toxins produced by these bacteria, heat-labile and heat-stable. Production of these toxins is under genetic control by DNA in transferable pieces of chromosomal material known as plasmids. These toxins promote secretions by the intestines of animals. Intestinal ganglioside mucosal receptors have been identified for heat-labile toxin. The biochemical basis of activity for heat-labile and heat-stable toxins appears to be through activation of the intestinal cell enzymes adenylate cyclase and guanylate cyclase respectively. Production of other toxins has been suspected, but not yet proved.

Toxic-shock syndrome. This syndrome was first described in 1978 and has more recently been associated most often with tampon usage during or shortly after menstruation. Persons with this syndrome develop a combination of fever, rash, hypotension, involvement of multiple organ systems, and subsequent sloughing of the skin of the palms and soles. Colonization of the vagina or cervix or local infection with *Staphylococcus aureus* is associated with this syndrome. Treatment with antibiotics active against this organism seems to hasten recovery and lower the chance of recurrence. Certain of the bacteria isolated from patients initially described with this syndrome produce an exotoxin that causes dermatological changes in mice. Although research is not conclusive, absorption of toxin produced by bacteria in the vagina or elsewhere is likely to be an integral part of the pathogenesis of this syndrome.

Bacterial attachment. The attachment of bacteria to a host surface can allow utilization of nutrients and growth at that site without removal by the flow of secretions or propulsive forces. Infections of the genitourinary tract, respiratory tract, gastrointestinal tract, and heart valves in mammals have been best correlated with attachment of virulent bacteria prior to establishment of infection. Attachment or adsorption of bacteria to surfaces is a very general phenomenon, however, and specific bacteria (as well as other microorganisms) can stick to mammalian surfaces, inert substances, plants, and other surfaces.

In certain mammalian infections, selectivity of attachment by bacteria appears to act as a virulence factor. For example, bacteria which commonly cause heart valve infections (endocarditis) attach to pieces of canine heart valve in greater numbers than other bacteria, and bacteria which commonly cause urinary tract infection attach to urinary tract epithelial cells better than other bacteria. Mechanisms of attachment are not being defined. Host and bacterial factors are involved in this process.

Host factors. Host factors include antibodies to specific bacterial antigens which can decrease or prevent attachment, and layers of mucus surrounding host membranes which can decrease attachment of bacteria to the underlying epithelial cells. Also, engulfment (phagocytosis) and killing of bacteria by particular host cells such as white blood cells and sloughing of external layers of epithelial cells by the host remove bacteria.

Putative receptors on mammalian host cells include glycoproteins and glycolipids. *Mycoplasma pneumoniae* binds to erythrocytes and isolated glycophorin (a glycoprotein) from the membranes of erythrocytes. Treatment of receptor cells with proteolytic enzymes, lectins which bind to sugars on glycoproteins, and use of other techniques suggest that sugar containing proteins or lipids is involved in the adherence of *Mycoplasma*, *E. coli*, and other bacteria to receptor cells. It is not clear yet whether different receptors exist on mammalian cells for a variety of organisms or what other structural or functional purposes these receptors serve.

Bacterial factors. Piliated bacteria use their hair-like external structures in attachment. Morphological type I pili (7-nm diameter) of *E. coli* and other members of the Enterobacteriaceae bacterial family consist almost exclusively of amino acids and appear to be the ligand which associates these bacteria with red blood cells, epithelial surfaces, and inert substances. Other pili antigens, sometimes also under genetic control, are involved in attachment to mammalian cells. These include colonization factor antigens of certain diarrhea-causing strains of *E. coli* and pili on the gram-negative bacterium *Pseudomonas aeruginosa* (which frequently causes infections in humans).

Common disease-causing gram-positive organisms such as staphylococci and streptococci are not piliated. Teichoic acids make up a substantial part of the outer cell wall of these bacteria. They are phosphodiester-linked polymers of sugar alcohols such as ribitol and glycerol, and seem to be the ligands which mediate adherence to host cells.

For background information *see* BACTERIAL APPENDAGES; BACTERIAL ENZYME; DIARRHEA; IMMUNOGLOBULIN; INFANT DIARRHEA; TOXIN in the McGraw-Hill Encyclopedia of Science and Technology.

[BARRETT SUGARMAN]

Bibliography: G. Bitton and K. C. Marshall (eds.), *Adsorption of Microorganisms to Surfaces*, 1980; R. F. Fisher et al., *Ann. Intern. Med.*, 94:156–163, 1981; M. H. Mulks, S. J. Kornfeld, and A. G. Plaut, *J. Infect. Dis.*, 141:450–456, 1980; R. B. Sack, *J. Infect. Dis.*, 142:279–286, 1980.

Bacterial genetics

In bacteria a variety of natural mechanisms lead to gene exchange between different members of a species, or even between different species. Conjugation is a mating involving cell fusion and so resembles somewhat the sexual processes of higher life-forms. In other genetic processes, fragments of the chromosome of one cell (the donor) either may be carried into a recipient cell by a benign virus, which previously grew on the donor (a process called transduction), or may enter the recipient as naked DNA, after being liberated by bursting of the donor cell (the phenomenon of transformation). In the last few years, artificial procedures for transferring genes between bacteria, and of inserting nonbacterial genes into them, have enormously broadened the scope of bacterial genetics.

The streptomycetes hold a special position in bacterial genetics because, although a few other bacterial types have been studied in greater depth, there is now available in *Streptomyces* the widest range of different techniques for genetic recombination. The actinomycetes, to which the streptomycetes belong, differ from other bacteria in two important features. One is morphological: Instead of existing as individual cells which separate from each other soon after cell division, the actinomycetes form a mycelium of branched tubular cells. Cross walls spaced at intervals divide the mycelium into compartments, but the cells remain connected until eventually some of them differentiate into chains of grape-shaped spores which are readily dispersed in air or water, so disseminating the organisms in their natural habitats of soils and composts. The other special feature of the actinomycetes is their ability to produce an amazing variety of antibiotics and other unusual compounds, over 500 being already known. One reason for interest in the genetics of streptomycetes is to try to understand the mechanisms of differentiation and antibiotic production, as well as to produce new strains better suited to making valuable antibiotics.

Conjugation. Although there are examples of natural transduction and transformation among the actinomycetes, conjugation is by far the most common method by which the organisms exchange genes. Conjugation in streptomycetes—as in the well-known *Escherichia coli*—is mediated by sex plasmids, small circular DNA molecules which the bacteria possess in addition to the much larger circle of DNA that constitutes the main chromosome. However, conjugation in *Streptomyces* is now known not to be the same process as in *E. coli*. Recently sex plasmids have been found in *Streptomyces* that are less than half the size of the smallest *E. coli* sex plasmids. These *Streptomyces* plasmids contain

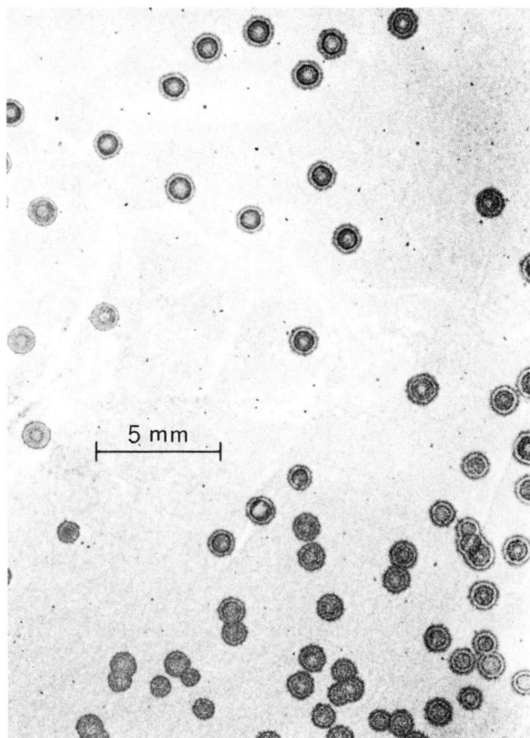

Fig. 1. Pocks produced by the growth of spores of a strain of *Streptomyces coelicolor* carrying the SCP2 plasmid on an agar plate spread with a confluent culture of a strain lacking the plasmid. Each pock is produced when the plasmid is transferred by mating from SCP2^{+} to SCP2^{-} cells, whose development is then slowed down relative to the rest of the culture, thus giving rise to circular areas of retarded growth. (*Photograph courtesy of M. J. Bibb*)

fewer than 10,000 base pairs, and so enough genetic information for only about 10 normal-sized genes. Moreover, only a few, perhaps no more than two, of these genes seem to play a direct role in conjugation, whereas in *E. coli* the complex mating machinery is assembled by the products of more than 20 genes. Perhaps plasmid-mediated conjugation in streptomycetes is a rather simple fusion between cells.

A surprising recent finding is that many *Streptomyces* strains carry two, three, or even more different sex plasmids. They can be detected in several ways. For example, they are seen by the examination of DNA extracted from cells, a method that can be used for any bacteria. However, many *Streptomyces* plasmids are also revealed by a special reaction when a colony that contains a plasmid grows on an agar plate confluently inoculated with a plasmid-free culture. The plasmid causes the cells to mate, and plasmid copies then enter the previously plasmid-free cells, retarding their growth for some unknown reason. Thus small circular inhibition zones, called pocks, are seen (Fig. 1).

In *Streptomyces coelicolor* A3(2), genetically the most-studied streptomycete, four sex plasmids are now known. The first two, SCP1 and SCP2, can be detected by their ability to promote new gene combinations in matings between strains of the organism carrying genetic differences. SCP2, which contains about 30,000 base pairs of DNA, can also be studied easily by physical examination outside the cells, while SCP1 is larger and much more difficult to handle without breakage. The other two plasmids, SLP1 and SLP4, were not detected in *S. coelicolor* itself, but only when matings were made between *S. coelicolor* and a related species, *S. lividans*. The two plasmids were transferred to individuals of the new host, where they revealed themselves by giving rise to pocks, and also by promoting genetic recombination in *S. lividans*. SLP1 exists in *S. coelicolor* integrated into the main chromosome. It "loops out" from the chromosome during interspecific mating, bringing variable segments of the chromosome with it, to yield circles of DNA of a range of sizes which can readily be isolated from the *S. lividans* recipient. There is no certainty that these four plasmids represent the complete set of *S. coelicolor* sex plasmids; still others may await discovery.

Protoplast manipulations. *Streptomyces* belongs to the gram-positive bacterial group, in which the outer layers of the cells have a simple structure, with just one layer—the rigid cell wall—outside the delicate cytoplasmic membrane, which is the main selective permeability barrier of the cell. This fact is important for genetic manipulation, since gram-positive cells can readily be converted to protoplasts by removing the cell wall, whereas gram-negative bacteria (such as *E. coli*) have an extra "outer membrane" outside the wall, and this is hard to remove completely. Protoplasts are crucial for the artificial genetic manipulation of *Streptomyces*.

Streptomyces protoplasts are made by treating the mycelium with the enzyme lysozyme, which digests the cell walls (Fig. 2). This is done in a sucrose solution that balances the osmotic pressure inside the cells and so prevents the protoplasts from bursting when the supporting wall is removed. In order to be useful for genetic experiments, protoplasts must be capable of being regenerated back to a cellular culture; fortunately, this occurs when the protoplasts are placed on a specially prepared agar me-

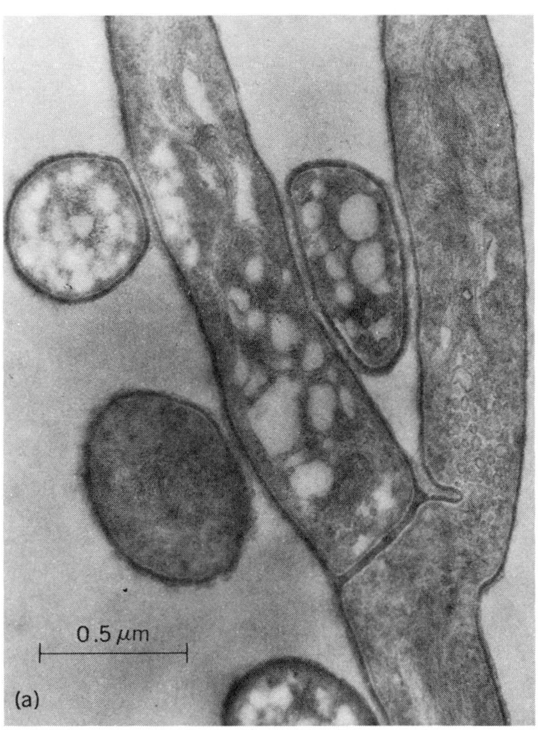

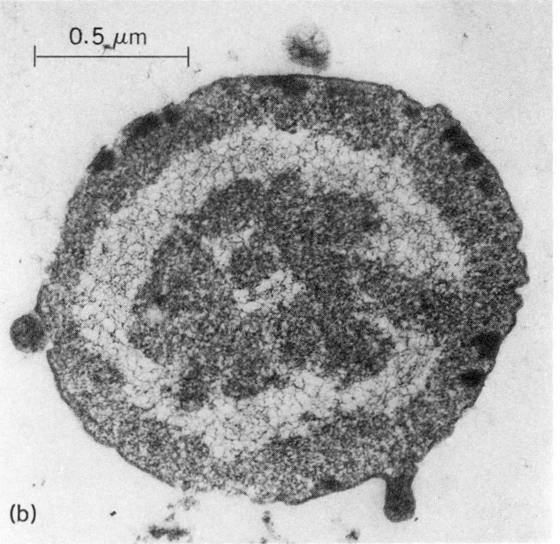

Fig. 2. Electron micrographs of thin sections of *Streptomyces coelicolor*. (*a*) Normal mycelial cells (*photograph courtesy of H. Wildermuth*). (*b*) A protoplast (*courtesy of I. G. Stevenson*). Note that the thick cell wall seen in *a*, which gives the cells their tubular, branching slope, has been removed in *b*; the protoplast is surrounded only by the delicate cytoplasmic membrane.

dium. Three different artificial genetic procedures depend on protoplasts: protoplast fusion; the transformation of naked plasmid or virus DNA into protoplasts; and the introduction of fragments of chromosomal DNA into protoplasts by encapsulating the DNA in artificial lipid vesicles, called liposomes, that can be fused with natural protoplasts.

Protoplast fusion. In its simplest form, protoplast fusion is very straightforward. Protoplasts of two strains are mixed, then briefly exposed to a solution of polyethylene glycol, and spread on regeneration medium. During incubation for a few days, the fused protoplasts regenerate to yield a sporulating culture. When these spores are allowed to grow into separate colonies, up to 20% have new combinations of the genes of the parents. This occurs if the parents are closely related, for example, by being substrains of one original culture. If they are different species, the proportion of recombinants is usually much lower.

There are several variations of the simple, two-parent fusion. One is to use up to four genetically different strains as parents in a multiple fusion. When this is done, many of the progeny acquire genes from three or four parents which have all fused together. Another trick is to kill the protoplasts of one strain with ultraviolet light, before fusing them with untreated protoplasts of the second strain. Recombinants still arise at a high frequency.

Transformation of protoplasts. Streptomyces cells usually are unable to take up DNA, but if protoplasts are mixed with plasmid or virus (bacteriophage) DNA and exposed to a polyethylene glycol solution, many of the protoplasts take up the DNA: they are said to be transformed (or transfected, if the DNA is that of a virus). Transformation or transfection can easily be detected and measured because every regenerating colony that contains plasmid DNA shows itself as a pock in the surrounding mass of regenerating untransformed growth, while transfection by bacteriophage DNA yields a burst of virus particles which attack surrounding cells and leave a clear area—a phage plaque.

This technique is fundamental in current genetic work with *Streptomyces* because it allows genetic engineering to be done. The DNA of a plasmid or phage, outside the host, is cut at defined positions by treating it with special restriction enzymes which cleave DNA only at precise base sequences, and segments of other DNA—from another streptomycete, an unreleated bacterium, or even a higher organism—are spliced into the plasmid or phage DNA to generate new hybrid molecules. These can then be introduced into a *Streptomyces,* where propagation of the plasmid or phage allows the foreign genes to be inherited indefinitely, to give a clone of genetically altered cells. These techniques, pioneered in *E. coli* over the last few years, are now well developed in *Streptomyces*.

Natural plasmids and phages of *S. coelicolor* and *S. lividans* have been genetically engineered to make them suitable vectors for foreign genes. This has involved identifying segments of plasmid DNA which are inessential, and which can therefore have foreign genes inserted into them without destroying the ability of the plasmid to replicate; SLP1 is useful here because the segments of chromosomal DNA naturally attached to the plasmid can be used as sites for DNA insertion. Genes conferring antibiotic resistance, taken from other antibiotic-producing *Streptomyces* species, were attached to these regions of the plasmid to provide ideal markers for the section of transformed individuals. A current vector consists of SLP1 carrying thiostrepton resistance from *S. azureus* (which makes thiostrepton) and neomycin resistance from *S. fradiae* (which makes neomycin). Suitable sites for inserting foreign DNA lie within the neomycin-resistance gene. Therefore, when new artificial recombinants are to be made, they are found among transformed progeny which are resistant to thiostrepton because they contain the plasmid vector, but sensitive to neomycin because the neomycin-resistance gene has been disrupted by inserting foreign DNA. To develop a phage as a cloning vector, inessential phage genes have to be recognized, cut out and discarded, to make space for inserting foreign DNA (a phage DNA has to be a certain size to be packaged by coat proteins to make an infectious particle). This has been done for a phage of *S. coelicolor* and *S. lividans* called ϕC31. What is more, a cloning vector plasmid from *E. coli,* pBR322, has been artificially inserted into ϕC31 to make a shuttle vector able to exist as a plasmid in *E. coli* and as a phage in *Streptomyces*. Shuttle vectors, which have also been made by combining pBR322 with the SLP1 plasmid, are very useful for studying the effects of the same genes in different kinds of bacteria.

Liposomes. Purified chromosomal DNA from a donor *Streptomyces* strain has been entrapped in liposomes and introduced into a recipient strain carrying different genetic markers by fusing the liposomes with protoplasts of the recipient in the presence of polyethylene glycol. Very high frequencies of progeny were obtained that had inherited donor genes. Since up to 10% of the progeny had any one of several donor genes, nearly all the protoplasts of the recipient strain must have been transformed for at least one donor gene. Without the use of liposomes, transformation by chromosomal DNA (as distinct from that of plasmids or phages) is at least a million times less efficient. Thus this technique opens up new possibilities for the analysis and manipulation of important chromosomal genes.

Applications of the techniques. These four methods of manipulating the genetic makeup of *Streptomyces* have varied applications, both in analyzing the genetic control of interesting genes involved in differentiation and antibiotic biosynthesis, and also in the practical objective of improving antibiotic production.

Conjugation is the best way to map genes on the chromosome, or to show that they lie on plasmids. This has revealed the fact that while the pathways of biosynthesis of some antibiotics are controlled by groups of genes on the main chromosomes, the synthesis of other antibiotics is coded by plasmid genes. For example, the SCP1 plasmid of *S. coelicolor* codes for production of methylenomycin—with resistance to the antibiotic also being plasmid-coded, an obvious protective mechanism for any strain carrying the plasmid.

The artificial techniques involving protoplasts promise to be valuable in strain improvement. Protoplast fusion gives a high frequency of new combinations of the genes of two (or more) parents. Thus fusion between strains which have been laboriously selected over a long period, by successive rounds of mutation and screening, to contain combinations of genes enhancing the yield of an antibiotic can generate new strains combining many of the desirable genes from separate selection lines. One application of liposome-mediated transformation is its potential for focusing mutagenesis on particular regions of the chromosome. Donor DNA could be intensely mutagenized and then introduced into recipient protoplasts, with selection for a donor gene in a particular region of the chromosome. Inheritance of this gene by transformed progeny would bring with it a segment of adjacent DNA carrying many new mutations focused in this region of the chromosome. The same thing may happen when natural protoplasts that have been heavily irradiated are fused with viable protoplasts. Genetic engineering, through the transformation or transfection of protoplasts with artificially constructed plasmid and phage vectors, is potentially a most powerful tool. Increasing the number of copies of a cloned gene should allow one to ease a biosynthetic bottleneck that limits the yield of an antibiotic. More exciting, the addition of foreign genes coding for enzymes able to modify antibiotic structures, or to add new side chains to them, should aid the search for new antibiotics and other pharmacologically active compounds.

Thus, with the availability of this series of new techniques for genetic manipulation of *Streptomyces*, strain improvement has changed from a largely empirical, or hit-and-miss, operation into one in which rational genetic approaches can increasingly be taken.

For background information *see* BACTERIAL GENETICS; GENETIC ENGINEERING; TRANSDUCTION (BACTERIA); TRANSFORMATION (BACTERIA) in the McGraw-Hill Encyclopedia of Science and Technology.

[D. A. HOPWOOD]

Bibliography: D. A. Hopwood, *Curr. Adv. Plant Sci.*, 10:467, 1978; D. A. Hopwood and K. F. Chater, *Phil. Trans. Roy. Soc.*, B290:313, 1980; D. A. Hopwood and H. M. Wright, *Mol. Gen. Genet.*, 162:307, 1978; C. J. Thompson et al., *Nature*, 286:525, 1980.

Biomagnetics

The body produces its magnetic fields in two main ways: by electric currents and by ferromagnetic particles. The electric currents are the ion currents generated by muscles, nerves, and other organs. For example, the same ion current generated by heart muscle, which produces the electrocardiogram, also produces a magnetic field over the chest; or the same ion current generated by the brain, which produces the electroencephalogram, also produces a magnetic field over the head. The ferromagnetic particles are insoluble contaminants of the body; the most important of these are the ferromagnetic dust particles in the lungs, which are primarily Fe_3O_4 (magnetite) particles. Because these fields, due either to currents or particles, can give information about the internal organs not otherwise available, they have potential application in research and in clinical diagnosis. Thus there is increasing interest in their use, and they are now being measured by various groups around the world.

These magnetic fields are very weak, usually in the range of 10^{-10} to 10^{-5} gauss; for comparison, the Earth's field is about 1 gauss. The fields at the upper end of this range (stronger than 10^{-6} gauss) can be measured with a simple but sensitive magnetometer called the fluxgate; the weaker fields are measured with the extremely sensitive cryogenic magnetometer called the SQUID (superconducting quantum interference device). The levels of the body's fields are shown in Fig. 1. These fields, whether they are fluctuating or steady, are seen to be orders of magnitude weaker than the fluctuating or steady background fields. They can, however, be measured by using either a magnetically shielded room or two detectors connected in opposition so that much of the background is canceled, or a combination of both methods.

Some recent developments concerning magnetic fields of the body are described below. These involve the lungs, the brain, the liver, and steady currents in the body.

Lungs. Inhaled dust in the lung is deposited either in the airways (bronchi and so on) or in its ter-

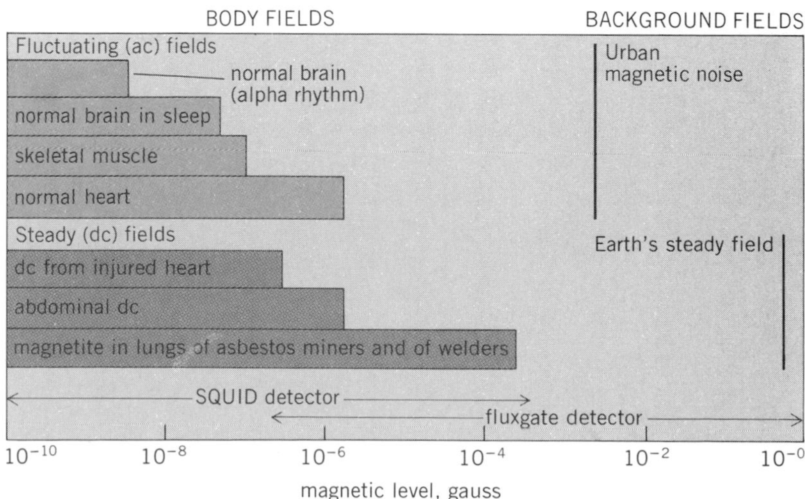

Fig. 1. Levels of the body's magnetic fields and background magnetic fields, as well as the range encompassed by the detectors.

minal chambers (alveoli). The dust deposited in the airways is known to be cleared out by the mucociliary carpet in a matter of some hours; this rate, called short-term clearance, has been well measured in the human lung by using short-lived radioactive dust. The dust deposited in the alveoli is cleared out much more slowly, in a matter of weeks or months; this rate, called long-term clearance, had not previously been well measured, because of the hazard of using long-lived radioactive dust. However, it was recently measured by using the ferromagnetic dust Fe_3O_4, which is nonhazardous; it is classified only as a nuisance dust. The rate in smokers was compared with that in nonsmokers.

In a pilot study, three subjects who were heavy smokers and nine nonsmoking subjects performed a one-time inhalation of about a milligram of this dust. A clearance curve was then determined for each subject, from periodic measurements of the amount of this dust remaining in the lung. For each measurement, the dust in the subject's lung was first magnetized by an externally applied magnetic field. Then, after the external field was removed, the remanent field of the magnetized dust was measured around the torso with a SQUID in a well-shielded room; this yielded the amount of dust in the lung. As shown in Fig. 2, the clearance in the

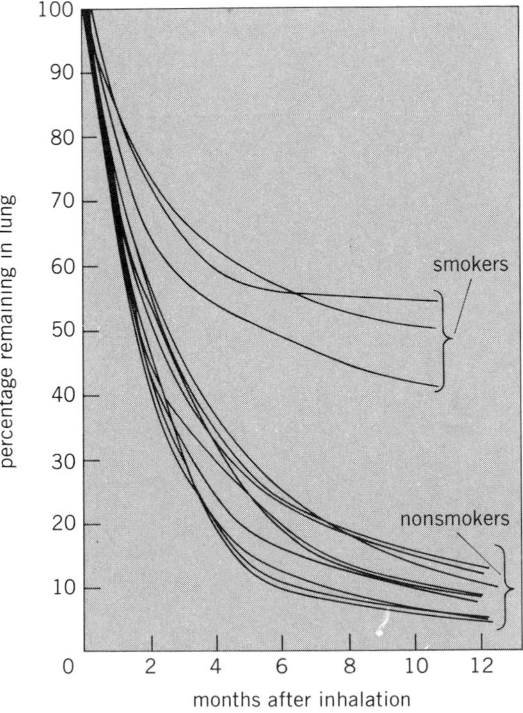

Fig. 2. Long-term clearance curves of Fe_3O_4 dust in smokers versus nonsmokers. The 100% point of each curve is begun after the completion of any short-term clearance, some days after the one-time inhalation. (*From D. Cohen, S. F. Arai, and J. D. Brain, Smoking impairs long-term dust clearance from the lung, Science, 204:514–517, copyright 1979 by the American Association for the Advancement of Science*)

three smokers was found to be considerably slower than in the nonsmokers. After 11 months, about 50% of the dust remained in the smokers, while only 10% remained in the nonsmokers; the smokers therefore retained five times more dust. This impaired clearance in smokers, seen here for the first time, suggests that smoking impairs the clearance of dust other than Fe_3O_4. In that case, the impaired clearance of carcinogenic dust may contribute to the high incidence of lung cancer in smokers.

Brain. Previously, theory suggested that a measurement of the brain's magnetic field, called the magnetoencephalogram (MEG), could reveal some extra, new information of the brain not available on the electroencephalogram (EEG); this new information involved spatial arrangements of the electrical sources in the brain. For example, the MEG would be able to reveal side-by-side opposing sources (called current dipoles), which produce an enhanced magnetic field but a suppressed surface potential, in comparison to a single-dipole source. In a sense, therefore, the MEG would be able to reveal more than the EEG, for special arrangements of sources. Recently, however, extensive computer calculations have suggested that for many source arrangements the MEG would show less than the EEG. In particular, the MEG would "see" a smaller area of sources on the cortex, and would respond only to a narrow depth of sources nearest the skull. This suggested ability of the MEG to detect less than the EEG may be most useful. This is because the EEG, although it is an important tool, sees too much, in that it detects sources from too large an area and too thick a layer of brain; therefore, the EEG usually cannot localize the source. The MEG, although it sees less, should therefore be able to better localize the particular sources it can indeed detect. This would have applications in both brain research and clinical diagnosis. For example, the MEG may clinically be able to localize an epileptic focus, and thereby be an aid to medical intervention.

The suggested smaller area and depth of response of the MEG is due to the relatively high resistivity of the skull. This smears out the EEG signals on the scalp, while it has no effect on the MEG signals, which are due to the magnetic field produced by only deep currents; the skull has no magnetic effect and is transparent to the magnetic field. However, these computer calculations are based on a simplified model of the head, in which the brain, fluids, skull, and scalp are assumed to be perfect spheres, and the calculations are only as valid as this assumption. If the assumption is not valid, for example, if the fissures of the brain are important (they are ignored in these calculations), then the conclusions of the calculations are also not valid. No reliable experimental comparison of the MEG versus the EEG has yet been made, and efforts are under way to do a detailed mapping of the EEG and MEG over the scalp; a precise, known electrical source in the human cortex is used. This type of mapping will show if indeed the MEG can better localize the source than can the EEG.

Liver. Unlike the lung or the brain, the liver is measured by measuring its magnetic susceptibility. This presents a new aspect of magnetic fields of the body, in which the body can be considered to produce a magnetic field in a third way. An external magnetic field is applied to the body, and each element of the body produces its own weak magnetic field in response to the applied field. The susceptibility measurement consists in determining the amount and polarity of this weak field in the presence of the applied field; the weak field can be in the same or opposite direction as the applied field, and the element producing the response is called paramagnetic or diamagnetic, respectively. The liver is normally a diamagnetic organ; however, it is the main localized storage site of iron in the body, and its susceptibility depends on the amount of iron it contains within its organic molecules. In the normal liver, a few parts in 10^4 consist of iron, but this concentration may rise a hundredfold in iron overload disease, or drop by a similarly large factor in iron-deficiency disease; the iron concentration changes the liver susceptibility so that the liver becomes paramagnetic for iron concentrations greater than 6 parts in 10^3, which is within the overload range. While departures from normal are associated with widespread medical problems, no reliable noninvasive method has previously existed to assess this iron store. An estimation of the iron concentration from a magnetic susceptibility measurement would therefore be of some value.

In the first use of a SQUID in a hospital, initial steps to evaluate this method were made by measuring the susceptibility of the liver of seven subjects. The iron concentration, estimated from this susceptibility, was compared with the concentration determined from a direct chemical analysis of their liver tissue obtained by biopsy. The two estimates were in substantial agreement over a range extending from normal to some 30 times normal. This result suggests that liver susceptibility measurements may become the first clinical application of magnetic fields produced by the body. Efforts are under way to extend the method to the case of iron deficiency in the liver, and to iron abnormalities in general of the heart.

Steady currents in the body. The internal organs of the body can generate both steady (dc) and fluctuating (ac) current. While the potentials of the fluctuating current can readily be measured on the body's surface, as with the EEG, the potentials of the steady current cannot reliably be measured because of large dc potentials generated by the skin; these mask the dc from the underlying organs. This is unfortunate because knowledge of the steady current generated by internal organs would be valuable, not only in the investigation of normal physiological processes in the body but also in medical diagnosis of abnormality; for example, the damaged heart can generate a steady "current of injury."

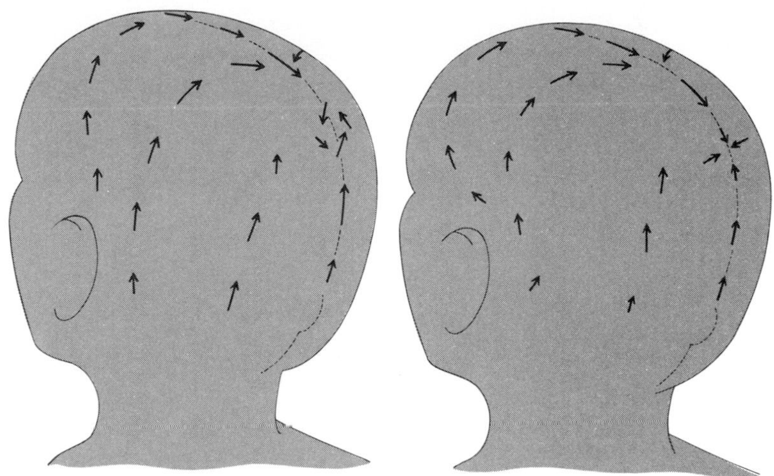

Fig. 3. Patterns of steady magnetic fields measured over the scalp of two full-haired young men; these are typical of all full-haired subjects. The length of each arrow is proportional to the local current which produced the field. The direction of each arrow coincides with the local tilt pattern of the hair follicles of the scalp; this was a clue that follicles generated the steady currents which produced these fields. The currents (hence the fields) are induced by applying light pressure to the scalp, and decay in several seconds after pressure is removed. (*From D. Cohen et al., Magnetic fields produced by steady currents in the body, Proc. Nat. Acad. Sci. U.S.A., 77(3):1447–1451, March 1980*)

However, in contrast to measurements of the surface potential, measurements of the steady magnetic field produced by steady internal currents usually suffer no interference by steady electrical events of the skin. This is because the steady potentials of the skin have no significant currents (hence no magnetic fields) due to both a special geometry of their sources and the high resistance of the skin. Steady magnetic measurements can therefore be reliably used to measure the steady current from internal organs. The steady magnetic fields over the normal human torso were recently mapped as a first step in learning to use this technique, and to explore the amount of steady current generated by the various organs.

For this mapping, a new SQUID arrangement was used, called the 2-D system, which produced a direct display of the currents producing the magnetic field; this facilitated the location of the sources of the current. The mapping showed reproducible steady fields over the head and other extremities of the body; over the torso proper, the field was weaker, except over the abdomen where it was strong and variable. The field over the head, shown in Fig. 3, was due to an unexpected source: steady current generated by the hair follicles in the scalp. Although the skin over the remainder of the body produces no magnetic field, the scalp is special in that its resistivity is lower (hence currents are greater) and the hair follicle sources are oriented in directions favorable for producing a magnetic field. These sources have not been seen in steady potential measurements of the scalp, presumably because they are masked by the larger potentials generated by the sweat glands (which produce no magnetic field). The steady fields over the arms and legs are due to currents generated by muscles; these currents are presumably caused by variations, in the interstitial fluid, of the K^+ concentration along the length of the fibers, which produce variations in the internal resting potential. Steady fields of comparable magnitude are not seen over the skeletal muscles of the torso proper, probably because of the different architecture of these muscles; the fibers are shorter and may be unable to attain large resting potential differences. The relatively strong and variable steady field over the abdomen is caused by large ion currents in the various organs of the gastrointestinal tract; these currents are associated with the phases of digestion.

[DAVID COHEN]

Bibliography: D. Cohen, *Phys. Today*, pp. 34–43, August 1975; D. Cohen et al., *Proc. Nat. Acad. Sci., U.S.A.*, 77(3):1447–1451, 1980; D. Cohen, S. Arai, and J. Brain, *Science*, 204:514–517, 1979; D. Farrell et al., *IEEE Trans. Mag.*, 16(5):818–823, 1980.

Bivalvia

Shell sections of several species of bivalve mollusks reveal the presence of fine growth lines. To study these lines, shell valves are embedded in plastic and then cut radially from the umbo to the growing edge. When acetate peels, prepared from the polished and etched cut surface of the shell of the intertidal cockle *Cerastoderma* (= *Cardium*) *edule*, are viewed in the optical microscope, they reveal a series of closely spaced bands which vary in thickness along the length of the shell. Under semidiurnal tidal conditions, it can be proved that one band is laid down at every emersion. In continuously submerged conditions a weak pattern of bands with an approximate semidiurnal periodicity is still observed. It is suggested that an endogenous rhythm of shell deposition exists which is entrained in intertidal animals by periods of emersion but which persists when the animals are kept continuously immersed. Following this interpretation of the pattern, the bands have been used to obtain growth rates from samples of cockle shells.

In 1963 John Wells proposed that fossil shells might be used as geochronometers. He suggested that the growth lines on fossil coral skeletons preserved a useful record of long-term changes in astronomical periodicities. Recently Wells's concept has been applied to living bivalve mollusks. The peel from a cockle shell section is shown in Fig. 1. The peel replicates three layers, the organic periostracum covering the outer surface of the shell including a spine, the peripheral or outer crossed-lamellar layer, and an inner crossed-lamellar layer. Growth bands appear as dark lines with thicker, more transparent growth increments between. It is evident that the increments and bands are not constant but vary with the position along the shell.

Tidal bands. The periodicity of the growth lines has been established directly from experiments using living cockles. Cockles marked by cold shock produce a cleft in the shell which acts as a dated point to which all subsequent growth can be related. Experiments either involved holding marked animals for varying periods of time at different levels on the shore or maintaining them in the laboratory in a tidal apparatus. Others involved continuously immersed conditions on a raft and in the laboratory. The results from several experiments are shown in the table. The total number of bands laid down during each experiment is given. In the shells of the intertidal cockles there was an almost exact coincidence between the number of tidal emersions and the number of bands. A peel from a cockle shell grown at low water is shown in Fig. 1. The cold shock cleft marks the beginning of the experiment period from which 54 tidal bands can be counted up

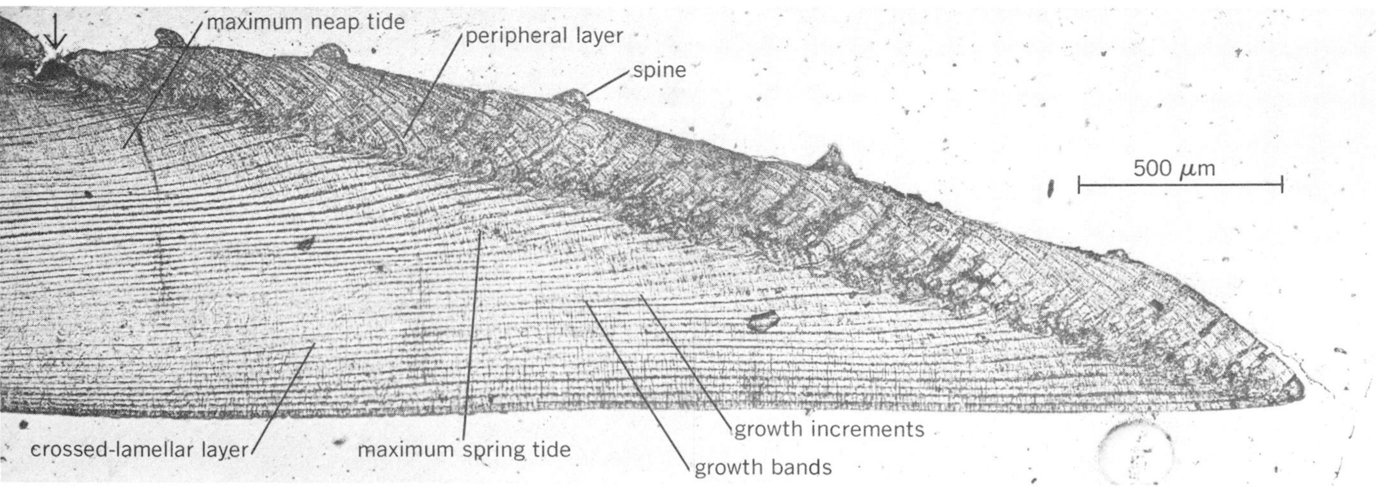

Fig. 1. Acetate peel from *Cerastoderma edule* grown at low-water neap tide level. Fifty-four bands can be counted from the cleft (shown by arrow). (*From C. A. Richardson, D. J. Crisp, and N. W. Runham, Tidally deposited growth bands in the shell of the common cockle, Cerastoderma edule (L.), Malacologia, 18:277–290, 1979*)

Growth bands in shells of *Cerastoderma edule* kept under various conditions of exposure

Conditions of exposure	Number of shells examined in peripheral (P) and crossed-lamellar (C) layers	Duration of experiment, days	Number of emersions	Total number of bands, mean ± standard error
Intertidal, low-water neap tides, 1976	P 10 C 10	29.0	54	54.1 ± 0.8 52.9 ± 0.5
Raft, continuously immersed submarine, daylight, 1977	P 12 C 16	45.0 (= 87 tidal cycles)	0	99.3 ± 3.0 71.1 ± 3.8
Laboratory, continuously immersed, 12 h light, 12 h dark, 1978	P 11 C 11	28.0 (= 54 tidal cycles)	0	62.3 ± 3.3 52.9 ± 1.9
Laboratory, tidal emersion, twice a day, 12 h light, 12 h dark, 1976	P 6 C 10	30.5	61	61.2 ± 1.3 59.0 ± 0.7
Laboratory, tidal emersion, once a day, 12 h light, 12 h dark, 1978	P 6 C 6	23.0	23	33.5 ± 1.9 32.7 ± 1.2

to the growing edge of the shell on the right.

When the peels of cockles which had been transferred from an intertidal habitat to the raft were examined, a sudden change in the appearance of the pattern of banding could be seen (Fig. 2). In place of the clearly marked darker and lighter stripes of the intertidal pattern were fainter and less regular bands. Immediately after the shells were placed under continuous submersion, there was in many shells a zone of almost continuous deposition in which few bands were laid down. After a few days bands reappeared but were weaker in appearance. The number of bands formed in the shells was closer to the number of tidal periods than to the number of light and dark periods. Peels from laboratory experiments in essence confirmed those from the field and the raft.

The experiments on the raft and under constant immersion in the laboratory ought not to have produced any growth bands if emersion were the sole factor responsible for the formation of bands. Nevertheless, bands with an approximate tidal periodicity were formed in these shells. No periodic influences could be suggested, except gravitational force, which seems an unlikely cue, that might cause the animals to form bands in their shells. Thus there emerged the possibility that an endogenous rhythm of shell deposition exists which is entrained in intertidal animals by emersion but which persists when the animals are continuously immersed. Further experiments were undertaken to determine whether the process of band formation could be entrained by emersion of a different period. Animals subjected to a 24 h period between emersions produced regular bands with some faint ones and more than the number of expected tidal emersions (see table). Some of the shells displayed a pattern of alternately strong and weak bands, whereas those from the laboratory semidiurnal tidal regime had regular strong bands. The stronger bands represented times of emersion, and the occasional weaker ones were formed when there was no emersion; that is, the latter were due to the endogenous component.

Alternating bands. Shells collected from the lower shore show an alternation in the thickness of the growth bands (Fig. 1). In North Wales the most marked alternations coincided with or slightly preceded the spring tide maxima. The peels of the shells offered strong evidence that the thinner, weaker bands were formed on the early morning ebb, and the alternating stronger and darker bands on the afternoon ebb. It is likely that during and after the heat of a summer day the stresses of high temperature and potential desiccation will cause the shells to close more firmly and for a longer time than in the cool of the morning. Tight closing of the shell valves causes a decrease in the pH of the pallial

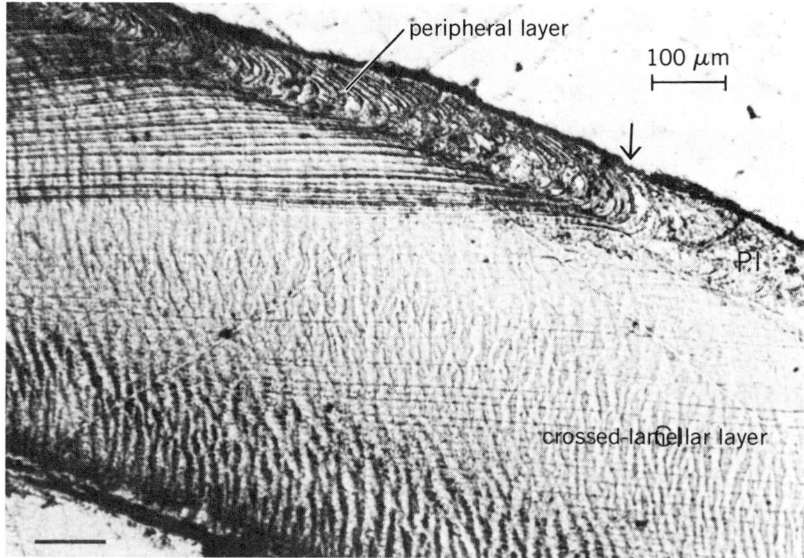

Fig. 2. Acetate peel from an experiment where *Cerastoderma edule* was transferred from the intertidal zone to a raft with continuously submerged conditions. Arrow marks transition. (*From C. A. Richardson, D. J. Crisp, and N. W. Runham, Tidally deposited growth bands in the shell of the common cockle Cerastoderma edule (L.), Malacologia, 18:277–290, 1979*)

and extrapallial fluids and decalcification of the newly formed shell. Though during neap tides the cockles will be exposed to air and high temperatures only briefly, in North Wales they will receive a much greater difference in illumination than at spring tides. However, the banding with fainter, less evident alternations corresponded with the neap tide period.

Banding patterns. Following the interpretation of the internal pattern, the bands have been used to obtain growth rates from the shells. The width of the growth increments varies along the length of the shell (Fig. 1). Since each band can be dated, the time of formation of the band during the spring-neap lunar cycle is known. Measurements of the width of groups of increments over periods of several months showed a small but significant difference in shell growth (5–10%) between the spring and neap lunar cycle. Animals at high water of neap tides and mean tide level showed an increased growth rate during spring tides and reduced growth at neaps. But shells collected at low water of spring tides were anomalous in that they showed faster growth at neaps than at springs.

During the winter months shell growth slows down, resulting in the formation of an annual ring or check on the shell surface. These rings have been used as a means of estimating the age of the animal. Generally the first-formed ring is weak, while disturbance rings may form as a result of an environmental disturbance such as a storm. Although the disturbance rings are generally weaker in definition than the truly annual ones, they can cause some confusion when estimating the age of shells. Sections prepared in the region of the annual rings revealed the reason why the first ring was less conspicuous than the other annual checks. In first-year individuals the bands are laid down close together, but shell growth does not cease altogether, with the result that a distinct cleft (or first annual ring) does not develop on the shell surface. Thus in some shells the first annual ring might be overlooked unless shell sections are examined. Figure 3 shows a section from a cockle shell taken in the region of the cleft formed during the second winter. The bands are so close together as to be almost unrecognizable. In the third winter, shell growth is so slow that the bands are laid down close together to form a zone of almost continuous deposition. Inevitably only an approximation can be made of the band numbers in these areas.

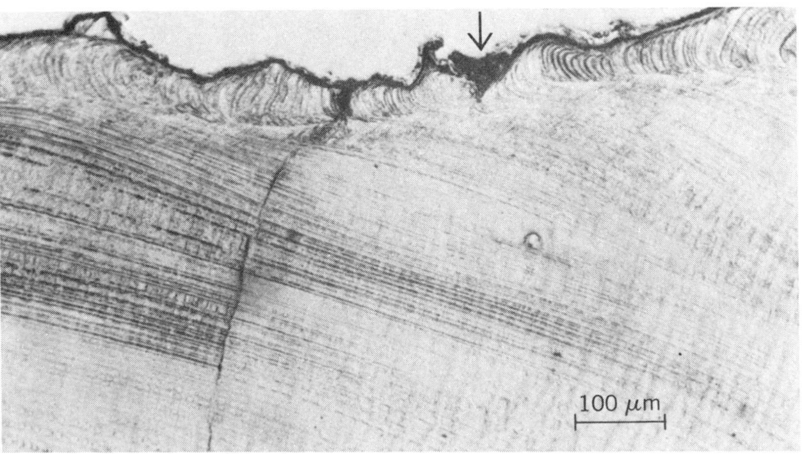

Fig. 3. Acetate peel from *Cerastoderma edule* collected about mean tide level and sectioned through the second annual check. Growth is from right to left. The arrow shows how the narrow increments appear as an almost continuous zone of deposition. The peripheral layer is reduced, forming a cleft at the winter check mark. (*From C. A. Richardson et al., The use of tidal growth bands in the shell of Cerastoderma edule to measure seasonal growth rates under cool temperate and sub-artic conditions, J. Mar. Biol. Ass. U.K., 60:977–989, 1980*)

Shells of cockles subjected to attack by predators also leave evidence of this event in the form of a check on the shell surface. When a shell section is examined, the disturbance ring can clearly be distinguished from an annual ring by the fact that the peripheral layer banding is broad on both sides, whereas in an annual check the bands become progressively more closely spaced on both sides of the cleft corresponding to the gradual changes in growth during autumn and spring.

The bands seen in fossil shells from different geological strata have been used to deduce the length of the year, and provide evidence that the rotational velocity of the Earth has changed through time. Work on living bivalves has shown that under certain conditions, notably in winter, the production of bands in the shell may temporarily cease, so that the number of tidal (daily) bands counted does not necessarily equal the number of tides (days in a year). The existence of bands resulting not only from external stimuli but also from endogenous rhythms must throw some doubt on the use of fossil shells for estimating the primeval calendar of past eras.

For background information *see* BIVALVIA in the McGraw-Hill Encyclopedia of Science and Technology. [C. A. RICHARDSON]

Bibliography: C. A. Richardson, D. J. Crisp, and N. W. Runham, *J. Mar. Biol. Ass. U.K.*, 60:991–1004, 1980; C. A. Richardson, D. J. Crisp, and N. W. Runham, *Malacologia*, 18:227–290, 1979; C. A. Richardson et al., *J. Mar. Biol. Ass. U.K.*, 60:977–989, 1980; J. W. Wells, *Nature*, 197:948–950, 1964.

Bonsai

The construction of a mature, very dwarfed tree in a relatively small container requires the interaction of art and science. Although bonsai is formally an esoteric branch of horticulture, it is firmly based in the social, historical, and cultural ethic of Oriental peoples and its introduction to Western cultures is a development of the 20th century; indeed, Westerners had not even seen bonsai plants prior to the early decades of this century.

At the fundamental level, the word bonsai means merely "plant in a pot," and its origin can be traced back to China's Chou dynasty (900–250 B.C.), when emperors built miniaturized gardens in which soil, rocks, and plants from each province were installed as dwarfed representations of the lands held by the central government. These gardens included

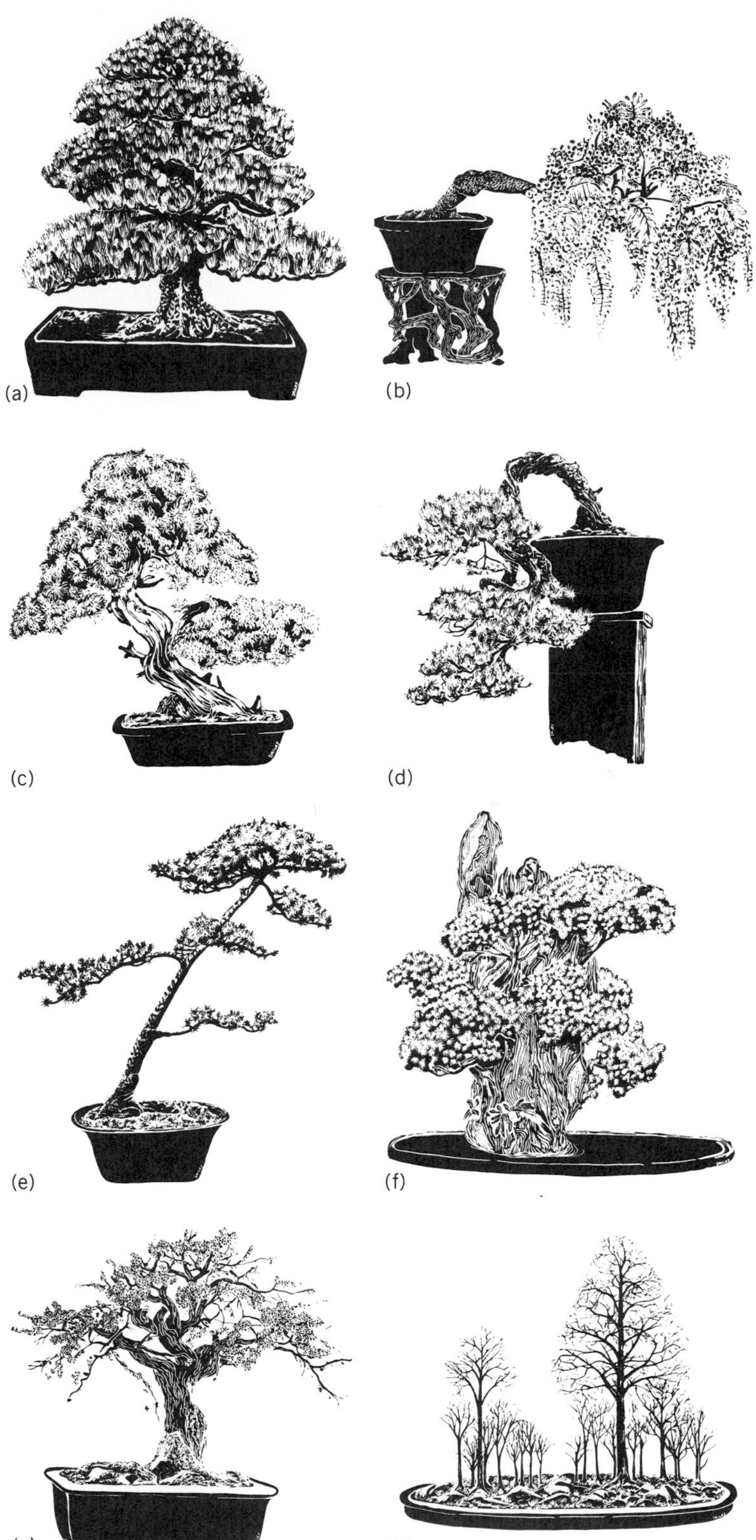

Major bonsai styles. (*a*) Formal upright (*Chokkan*); black pine. (*b*) Semicascade (*Han-Kengai*); wisteria. (*c*) Informal upright (*Moyogi*); juniper. (*d*) Cascade (*Kengai*); white pine. (*e*) Slanting (*Shakan*); spruce. (*f*) Rock clinging (*Ishezuke*); azalea. (*g*) Domed (*Gaito-Kengai*); plum. (*h*) Forest (*Yose-uye*); beech.

pun-ching (potted plants within landscapes) and *shea tse ching* (plants in tubs) or, more commonly, *pun-sai* (plants in small containers). Selected species were usually those with mythic, symbolic, and esthetic significance—plum, cypress, and pines plus bamboo—or plants representing seasons of the year—peony, cherry, and maple. The introduction of Buddhism into China, with the religious stress on self-awareness and a concept of the eternal and agelessness as exemplified by trees—concepts in keeping with tenets of Confucianism—shifted the emphasis of dwarfed potted trees from its political and esthetic aspects toward that of the construction as a representation and visual image of the eternal. The Buddhist cult of Chan (the Japanese Zen) utilized *pun-sai* as objects of contemplation to be viewed during the meditations that constituted a basic part of Chan ritual.

Up to the 11th century A.D., China was the cultural and governmental model for most Oriental cultures, and other nations sent officials, priests, and artists to China to learn the intellectual and practical skills which could be introduced into their own countries. Early Japanese visitors commented on the *pun-sai*, and some were carried back to Japan by the 12th century. During the Kamakura period (1192–1333), a significant commerce in dwarfed plants was noted in official Japanese documents, and bonsai (a direct transliteration of *pun-sai*) were included in paintings. Books and plays of the 14th century included bonsai as main or subsidiary themes, facilitating the diffusion of knowledge and appreciation of the plants and their significance. Bonsai in Japan received a major impetus when, in the 17th century, a Chinese official became disaffected with the Ming court and decamped to Japan, taking an extensive collection of plants and manuals on their construction and maintenance. A fairly complete historical record of bonsai styles, species used, and types of containers can be followed in Japanese woodblock prints dating from the 16th century.

In spite of the long-standing interest in bonsai in Japan and growing interest in Western nations, there have been surprisingly few controlled scientific studies related to the plants. As has long been true of most ornamental horticulture throughout the world, techniques and concepts have been passed by word of mouth or through apprenticeships, rather than having been codified and tested under controlled scientific conditions.

Esthetic aspects. There are two major aspects involved in the construction and maintenance of a bonsai plant, only one of which is susceptible to direct study. Esthetic aspects, which partake of the cultures from which the form is derived plus individual tastes, are not easily analyzed, although they undoubtedly play a more important role than do technological aspects.

Over several centuries the Japanese codified not only the plants to be used as bonsai subjects, but also the styles and sizes of the plants in their final form. In general, styles are based on observable natural growth patterns in various ecological situations. A solitary tree growing in an open field will develop a much different form than a genetically identical plant surrounded by other trees in a forest, and a tree exposed to constant winds coming from one direction will grow differently from one not so exposed. Major styles are shown in the illustration, and others have also been recognized. The rules for styles are flexible and variations are common. In Japan, some individual bonsai plants have been known as that particular plant for several centuries and are emulated and honored. Sizes, too, have been codified. Plants less that 15 cm (pot included) are called *mame-bonsai* (hold-in-the-palm-of-one-hand), those from 15 to 30 cm tall are *ko-bonsai* (hold-in-one-hand), plants 30 to 60 cm tall are *chie-bonsai* (hold-with-two-hands), and those taller than 60 cm are *dai-bonsai* (two-people-needed).

The pot or container is also a matter of esthetic concern. Form, color, size, and decoration are selected to complement, enhance, or contrast with the plant. Until the 18th century, containers were usually Chinese antiques and correspondingly rare and expensive, but the growing enthusiasm for bonsai as both a religious form and a secular hobby caused the development of indigenous ceramic industries devoted to the manufacture of containers. These are usually clay or porcelain, high-fired and appropriately glazed, usually in earth tones of brown, blue, green, or rust. Sharp reds, purples, or other bright colors are usually avoided, although there are important exceptions.

Since, in Japan, bonsai was at first restricted to the priests and to associated nobility, monks were usually trained to make and maintain the plants. As interest rose, individuals adept at the procedures began to develop their own nurseries and to conduct classes on the technique. Some families have continued this tradition and, with formal examinations, can be designated as bonsai masters. Many have settled in Omiya, a suburb of Tokyo, where they not only keep large bonsai nurseries and teach, but also may keep precious individual plants for families who will have a plant delivered for viewing on a special holiday or for a few days when the plant is in bloom. Such viewings, particularly among the wealthy, are occasions for a tea ceremony. A national viewing and competition is held yearly under the patronage of the Emperor.

Technological aspects. To a large extent, the technology of bonsai is derived from classical horticultural procedures which are the product of long tradition firmed up with considerations of many areas of plant physiology and ecology. For optimum growth of any plant, adequate levels of water, air, nutrients, light, and temperature must be presented. Good horticultural practice is as necessary in bonsai as it is in any other aspect of gardening, although the restrictions of small containers and carefully controlled growth patterns necessitate additional care.

There are, however, some techniques which are unique to bonsai. To control and restrict top growth, nutrient levels, particularly those of nitrogen and phosphorus, require careful regulation. Most potted plants are viewed favorably when they are large and lush, while bonsai plants are admired when small, gnarled, and restrained. High levels of nitrogenous fertilizers, which facilitate rapid growth, are contraindicated, while higher P/N ratios promote woody growth with smaller leaves and better flowering. Because of small container volumes and restricted root systems, bonsai plants require a freely draining soil kept close to field capacity.

To maintain shape, not only must growth be controlled, but the branching patterns must be planned. This is done by selective pruning and by pinching out of terminal buds to release axillary buds from apical dominance. Root prunings, done at intervals as short as 2 or 3 months in some species, promote the development of a tight root ball with large numbers of young lateral roots whose ability to take up water and minerals is great.

Branch and stem position is controlled by selective pruning and by the wiring techniques developed by the Japanese in the 18th century. Copper wire is wound around the branch or stem, allowing the part to be moved into the desired position without cracking or breaking. Not only does the wire permit bending, but the stresses imposed by the wire and the new position induces the formation of reaction wood, xylem laid down by the cambium that fixes the branch in the new position, so that when the wire is removed, the branch remains in position. Weighting or tying is used for the same reasons.

For background information *see* PLANT GROWTH; PLANT MINERAL NUTRITION in the McGraw-Hill Encyclopedia of Science and Technology.

[RICHARD M. KLEIN]

Bibliography: T. Kawamota, *Saikei: Living Landscapes in Miniature*, Kodansha, Tokyo, 1967; J. Y. Naka, *Bonsai Techniques*, Bonsai Institute of California, Santa Monica, 1975; Y. Yoshimura and G. M. Halford, *The Japanese Art of Miniature Trees and Landscapes*, 1957.

Brachiopoda

The brachiopods are a wholly marine group of macro-invertebrates that are rare to uncommon members of the shallow to abyssal epifauna. Interest in brachiopods derives chiefly from the fact that during the middle and upper Paleozoic (Middle Ordovician through Permian), a time interval of about 470 to 230 million years (m.y.) ago, they were very abundant as individuals, and were represented by an unusually large number of genera and species as well as families, superfamilies, and orders, as contrasted with the post-Permian to present situation of greatly reduced diversity at all levels.

Taxonomy. A great deal is known about the taxonomy of the brachiopods, particularly members of the class Articulata (the other class of this phylum is Inarticulata). There are currently about 600 workers studying fossil brachiopods, but only a handful are concerned with living brachiopods, with almost all attention being devoted to skeletal morphology and taxonomy. R. E. Grant has provided some idea of the pace of recent taxonomic advances in pointing out that the number of described genera has almost doubled since 1965. The fast pace of modern taxonomic work being carried out in previously little-known areas such as the People's Republic of China suggests that this rate of increase in numbers of genera will certainly keep up for the next few decades. Researchers are beginning to have a fairly good idea of the taxonomy of the articulate brachiopods. Unfortunately, very little attention has been devoted to the inarticulate brachiopods, particularly those of the post-Ordovician, although this is partly because of the relative paucity of the material as contrasted with the articulates.

The scientific potential of the brachiopods is twofold. (1) For the geologist they provide a means of dating ancient strata, for learning more about ancient marine environments by means of community ecology studies, and for learning more about ancient, global, shallow marine current circulation patterns from brachiopod-based biogeographic data. (2) For the biologist the brachiopods are a group whose evolution may be studied morphologically from the base of the Cambrian to the present—there is no other fossil group having this potential. The biologist also has the possibility with the brachiopods of studying the community ecological constraints on evolutionary processes through this same time interval—this again is a unique opportunity. In addition, the brachiopods of the middle and late Paleozoic have great evolutionary potential from a biogeographic point of view, that is, in learning more about the biogeographic constraints on evolutionary processes.

Nontaxonomic aspects. A brief review of pathologic conditions reported by A. J. Boucot in both living and fossil brachiopods revealed that not a single disease has ever been recognized within the phylum. This dubious distinction provides an index of how little is known concerning the basic biology of the brachiopods. The most recent summary of brachiopod biology, chiefly gross anatomy, is by L. M. Hyman. In recent years there has been a certain amount of attention devoted to the ecology of the living brachiopods, chiefly in order to better comprehend the environmental significance of the fossils. Little attention has been given to brachiopod physiology, but H. M. McCammon has devoted considerable effort to elucidating some aspects of their digestive physiology. Virtually nothing is known about the genetics and molecular biology of the brachiopods.

Paleobiogeography. In addition to the taxonomic work cited earlier, there is beginning to be a real effort devoted to the global paleobiogeography of the articulate brachiopods. Researchers are also seeing a real beginning in the systematic documentation and description of brachiopod-rich and brachiopod-dominated communities of the past, particularly those of the Paleozoic.

Some useful generalizations to be drawn from the paleobiogeographic and paleoecologic studies are as follows: (1) Cold- and cool-water, brachiopod-dominated communities tend to include fewer taxa—from the ordinal to the specific level—than do warm-water communities. (2) Shelf-margin brachiopods belonging to the same taxic group as nearer-shore representatives tend to be significantly smaller than do the nearer-shore shells. (3) Rough-water shells tend to be far more massive, and generally larger, than quiet-water shells, with much of the additional shell material deposited in the beak regions. (4) Brachiopods of the past, as of the present, tend to behave in a stenohaline manner—this is particularly true for the articulates. There is little evidence that articulate brachiopods commonly were able to tolerate either hypo- or hypersaline conditions, as contrasted with other groups such as some bivalves and ostracods. (5) Some articulate brachiopods are coevolved to exist in varied organic complexes, including those of differing reef types from the Ordovician to the present, as well as varied-level bottom complexes, including crinoidal thickets, and ahermatypic coral thickets. (6) The spacing of different taxa, including sheetlike blankets, patches, and nests, as well as scattered individuals, is a taxonomically stable character at about the family-to-subfamily level in many instances.

Cambrian-Ordovician. Brachiopods are rare both as individuals and as taxa at every level in the Early and Middle Cambrian. This fact alone helps to account for the ignorance of Early and Middle Cambrian brachiopod biogeography, but there is no reason to suspect that the trilobite-based biogeography of the time interval would be incompatible with brachiopod distributions. Brachiopods become more abundant as individuals in the Late Cambrian, but the number of taxa is still relatively low, although much higher than for equivalent Early and Middle Cambrian time intervals. For the Late Cambrian too, nothing has been done to analyze the biogeography of the brachiopods, although there is probably enough material to at least find out if the available data are permissively in agreement with the well-known, trilobite-based biogeography. The Early Ordovician situation involves different brachiopod genera for the most part, but the overall situation is about the same as that for the Late Cambrian. Most of the Cambrian and Early Ordovician brachiopods occur in what are estimated to be inner- to mid-shelf positions, with few being in an outer-shelf to shelf-margin equivalent depth position.

The Middle and Late Ordovician see a great taxonomic diversification of the articulate brachiopods, and also a great increase in their abundance relative to other macrofossils. During the Middle and Late Ordovician it is easy to recognize a cool- to cold-water biogeographic unit for which the included brachiopods are largely distinctive at the generic level, as are many other fossil groups. This biogeographic unit is commonly termed the Atlantic Realm. The warmer-water faunas of the time interval, commonly associated with limestone and other calcareous rocks, may be divided into a number of biogeographic units—possibly as many as four based on present knowledge—and the brachiopods appear to fit comfortably into these units. In general, the overall level of provincialism declines from the beginning of the Ordovician to the end of this period. Possibly the most widespread Ordovician fauna is the terminal Ordovician, Ashgillian fauna, the so-called *Hirnantia* fauna, which probably reflects a cool-water expansion of the Atlantic Realm in largest part.

Silurian. The Silurian is a time interval, similar to the Ordovician, in which there is a major division into a cool- or cold-water realm, the Malvinokaffric Realm, on the south, and a warmer-water realm, the North Silurian Realm. The North Silurian Realm is divided into two regions—the North Atlantic Region and the Uralian-Cordilleran Region. The major terminal extinction event at the end of the Ordovician produced a very much reduced earlier Silurian brachiopod fauna at the generic and specific level. Near the end of the Early Silurian there was a major dispersal event involving the migration of many taxa from the Uralian-Cordilleran Region into the North Atlantic Region. This dispersal event resulted in important restructuring of many community groups. Similar major extinction events affecting the brachiopods occurred at the end of the Cambrian and at the end of the Early Ordovician.

Devonian. Brachiopod biogeography during the Devonian is far more complex, and better known, than that of earlier time intervals. The latter part of the Early Devonian sees a major provincial high—similar highs are probably reached only in the later Carboniferous–Early Permian, as well as in the Late Cretaceous and Cenozoic. The Late Devonian is a time interval of almost highest cosmopolitanism, with the Early Triassic being the only more cosmopolitan time interval, and is also marked by one of the limited number of truly major terminal extinction events which occurs in the middle of the series. The Middle Devonian is intermediate biogeographically between the Early and Late conditions of provincialism and cosmopolitanism. During the Early and Middle Devonian there is a well-marked cool- to cold-water, Southern Hemisphere Malvinokaffric Realm, just as in the Silurian, but there is no evidence for this realm in the Late Devonian, when global climatic conditions may have been much lower gradient. The highly provincial Early Devonian is divided into two realms additional to the Malvinokaffric, as well as a number of their regional and provincial subdivisions. The number of such extra–Malvinokaffric Realm subdivisions decreases progressively through the Middle Devonian.

Carboniferous-Permian. The Carboniferous and Permian are a time of increasing shallow marine provincialism, which affects the brachiopods as well as most other groups, that culminates in the Permian after the fairly cosmopolitan conditions of the early Carboniferous, with the late Carboniferous being intermediate. The late Carboniferous and early Permian see the development of a highly distinctive cool- to cold-water, largely Southern Hemisphere, Gondwana Realm. The Gondwana Realm is associated with tillites and glacial-marine deposits in many places. The extra-Gondwana regions of the late Carboniferous and early Permian are warm-water in type, and include important reef community complexes in some areas. The extra-Gondwana of this time interval is further subdivided into a number of biogeographically important units.

Mesozoic-Paleozoic. The biogeography of the Mesozoic brachiopods has not been done on a truly global scale. But available information suggests that the brachiopod genera occur in a manner permissively in agreement with biogeographic schemes worked out for more abundant groups such as the ammonites and belemnites, as well as the bivalves. The same is true for the Cenozoic. One item of interest today is that there are more articulate brachiopod genera from the temperate zones than from the low latitudes. This state of affairs is very unlike that characteristic of most of the geologic record. Any consideration of post-Paleozoic brachiopod biogeography must also contend with the fact that the bulk of the Paleozoic superfamilies and orders vanished at the end of the Permian, and the post-Paleozoic orders and superfamilies are very reduced in numbers. This ordinal and superfamily anomaly is similar in some regards to that characteristic of the pre-Middle Ordovician.

Phanerozoic articulate communities. The past decade has seen a tremendous surge in the study of brachiopod-rich and brachiopod-dominated benthic communities—particularly those of the Paleozoic. However, community-oriented information is available on only a very small fraction of the potentially knowable community units. With this situation in mind, the following summarizes what appears to be the Cambrian-to-present picture.

For the Early and Middle Cambrian there is very little information except that brachiopod-rich and brachiopod-dominated communities are very uncommon. Most of the data seem to indicate that most of these communities contain a single species only. There is no evidence for communities containing a large number of articulate brachiopod genera.

In the Late Cambrian overall abundance of articulate brachiopods increases drastically, but brachiopods are still relatively infrequent members of the benthos. Brachiopod-rich and brachiopod-dominated communities still continue to consist of only one or possibly two species. It is important to point out that in the Cambrian there are only a few orders and superfamilies present among the articulate brachiopods.

In the Early Ordovician the articulate brachiopod community situation is essentially similar to that for the Late Cambrian, except that the genera and species are chiefly new and that more genera are present per time interval.

The Middle and Late Ordovician see a major surge in number of articulate orders and superfamilies, as well as in numbers of genera and species. The Middle and Late Ordovician also see the articulate brachiopods occupying a number of new niches, as suggested by their far more varied gross morphology, than was the case during the pre-Middle Ordovician. Brachiopod-dominated and brachiopod-rich communities become relatively widespread and abundant for the first time. It is not uncommon to find 5–10 species belonging to as many genera of brachiopods in any one community during this time interval, although there are during this time interval, as well as during all others, many single-species communities which probably reflect restrictive conditions.

During the Silurian there is an earlier Llandovery interval (A_1 through about C_2) in which brachiopod-rich and brachiopod-dominated units have about 5–10 genera as a high, followed by a later time interval (C_3 part of the Late Llandovery through Pridoli) during which 10–15 genera are not uncommon per community unit.

It is also notable, beginning in the Middle Ordovician, that shelf-margin depth equivalent region communities tend to be dominated by relatively small species of genera which in nearer-shore conditions reach a significantly larger size. The reasons for this size differential are not understood.

It is also worth commenting on that more data are available concerning Silurian and Devonian brachiopod-dominated and brachiopod-rich communities than for any other periods. This situation reflects the interests of students, rather than any intrinsically necessary situation.

The Devonian brachiopod communities have a high level of continuity with those of the Silurian. In fact, there is no truly major change in community character from about C_3 time through the lower half of the Late Devonian (Frasnian).

At the end of the Frasnian the major terminal extinction event markedly decreased the number of brachiopod species per community, with a half dozen species being a normal high for the latter half of the Late Devonian (Famennian).

In the Carboniferous-Permian there are some major changes regarding the brachiopod communities. Many of them tend to consist of communities rich in molluscan taxa, as contrasted with the Silurian-Devonian when molluscan taxa were far less abundant as individuals in most situations. There are still many high-diversity brachiopod communities in the Carboniferous-Permian, in which 15–20 species is a normal number. The Carboniferous-Permian, however, sees additional niche diversification of the brachiopods, despite their somewhat overall diminished numerical abundance (consider the somewhat oysterlike oldhaminid brachiopods and the somewhat corallike richthofenid brachiopods).

The end of the Permian sees the total extinction of most brachiopod taxa from the ordinal down to the specific levels. The Early Triassic is a time of few brachiopod taxa and very few brachiopod communities. The number of brachiopod-rich communities is low, and few species per community is the norm.

The Middle Triassic sees an important upsurge in numbers of brachiopod taxa at the generic and specific levels, as well as in overall abundance. But the brachiopods of the Middle and Late Triassic never reach the abundance present in the Middle Ordovician through Permian. Communities with three to five or so species are present.

The Jurassic and Cretaceous situation is about the same as is that for the Middle and Late Triassic, although the genera and species are mostly different.

In the Cenozoic it is very common to find communities with only a single species, and it should be noted that brachiopods are overall very subordinate members of the fauna. The Cenozoic situation is reminiscent of that present in the Late Cambrian and Early Ordovician. The only exceptions are a few regions like New Zealand and possibly Antarctica, where Mesozoic-like brachiopod community situations persist.

For background information *see* BRACHIOPODA in the McGraw-Hill Encyclopedia of Science and Technology. [ARTHUR J. BOUCOT]

Bibliography: F. J. Ayala, *J. Paleontol.*, 49:1–9, 1975; A. J. Boucot, *Principles of Benthic Marine Paleoecology*, 1981; R. E. Grant, *J. Paleontol.*, 54:499–507, 1980; J. Gray and A. J. Boucot (eds.), *Historical Biogeography, Plate Tectonics, and the Changing Environment*, 1979; J. Gray, A. J. Boucot, and W. B. N. Berry (eds.), *Communities of the Past*, 1981; L. M. Hyman, *The Invertebrates: Smaller Coelomate Groups*, vol. 5, pp. 516–609, 1959; H. M. McCammon and W. A. Reynolds, *Mar. Biol.*, 34:41–51, 1976; J. R. Richardson, *J. Roy. Soc. N.Z.*, 9:415–436, 1976; A. Williams et al., Brachiopoda, in R. Moore (ed.), *Treatise on Invertebrate Paleontology*, pt. H, vols. 1 and 2, 1965.

Breeding (animal)

With artificial insemination it is possible for bulls to father more than 10,000 calves each year instead of a maximum of several hundred with natural breeding. Embryo transfer technology has a similar role in amplifying reproductive rates on the female side. Normally, a cow has one calf each year (a few percent have twins) following 9 months' gestation. With embryo transfer technology, the embryos are removed from the reproductive tract of genetically valuable cows (donors) early in pregnancy before they become attached to the uterus, and are placed into the uterus of less valuable cows (recipients). Thus the donor can become pregnant temporarily during each reproductive cycle, or about every 3 weeks. In practice, the reproductive cycle is modified by giving the donors fertility drugs (usually follicle-stimulating hormone) which result in an average of six to eight embryos per recovery attempt, and sometimes up to 30 or more instead of the usual 1. This technique is called superovulation.

Procedures. Superovulation treatment is followed by breeding 5 days later, almost always artificially instead of by natural mating. About a week later, the fertilized ova (embryos) are recovered with a device called a Foley catheter. This instrument, made of latex rubber, is placed into the uterus through the vagina and cervix (opening between uterus and vagina), much as artificial insemination is done. The Foley catheter has three channels. One leads to a balloon which is inflated within the uterus to hold the instrument in place; the other two channels are for inflow and outflow of the fluid used to wash the embryos from the uterus. The reproductive tract is flushed six to eight times with 30 to 200 ml of this fluid to recover the embryos, which are about 0.005 in. in diameter. Next, the embryos are located with the aid of a microscope and examined for normality and stage of development. Generally, embryos are kept in an incubator in small covered dishes until they are transferred to recipients. The embryos can be kept for about 24 h in this manner with little loss in viability. Alternatively, they can be frozen in liquid nitrogen and stored this way for years and probably for centuries, but the freeze-thaw process kills about one-third of them.

The next step is to transfer the embryos to the reproductive tracts of recipients, usually one embryo per recipient. For high rates of success, recipients must be at the same stage of the reproductive cycle as the donor. The transfer may be accomplished nonsurgically by methods nearly identical to artificial insemination, or surgically with local anesthesia through a small incision in the flank of the cow. With surgery, the embryo is placed into the uterus with a fine pipette through a small puncture wound in the uterine wall.

Pregnancy is diagnosed 2–3 months after embryo transfer by feeling the fetus through the rectal wall. Pregnancy rates are about 60% with surgical transfer, nearly the same as those with artificial insemination. They are slightly lower with nonsurgical transfer methods. The recipient accepts the calf as her own; there are no more abnormalities than with conventional reproduction.

Applications. The most important application of this technology is amplification of reproduction of valuable cows. With three courses of superovulation and embryo transfer per year, a cow can have an average of about 10 calves instead of the normal one. Some cows have had more than 50 calves in a year with these methods, but others have only a few calves because individual cows respond very differently to superovulatory treatment.

The second most important application of embryo transfer in cattle is circumventing infertility. Frequently, after a cow has had many calves, the uterus wears out, and pregnancies do not go to term even though normal embryos are produced. The solution is to remove the embryos from these old cows and transfer them to a younger uterus for gestation.

There are several other applications such as exporting embryos to foreign countries. This is much less expensive than shipping animals, and there is much less risk of transmitting disease. Embryo transfer is also used to test animals for genetic defects.

This industry is only a decade old but is growing very rapidly. More than 25,000 calves will be born from these procedures in North America in 1981. Although well established, embryo transfer is not likely to be applied on a scale similar to artificial insemination for at least another decade, primarily because costs are high and procedures are complex. Nevertheless, it is a valuable tool for the animal breeder.

For background information *see* BREEDING (ANIMAL) in the McGraw-Hill Encyclopedia of Science and Technology.

[GEORGE E. SEIDEL, JR.]

Bibliography: G. E. Seidel, Jr., *Science*, 211:351–358, 1981; G. E. Seidel, Jr., and S. M. Seidel, *Adv. Anim. Breeder*, 26:6–10, 1978.

Breeding (plant)

Plant breeders, concerned with improving crop production, have developed methods for producing new crop varieties by manipulating the genetic components as well as the structures of certain plants. This article describes the uses of genetic traits from wild populations and the cultivation of leafless peas.

USE OF GENETIC TRAITS FROM WILD POPULATIONS

Trends toward food production as contrasted with hunting and food gathering from the wild began at least 12,000 years ago. The human species has become progressively dependent on domesticated plants and animals ever since. Many people today assume that humankind depends entirely on the produce of the farm for food and fiber and that wild plants contribute very little to the support of society. This is far from the case: wild populations continue to be essential for survival and well-being. Wild plant populations contribute germ plasm directly to: forestry (managed stands for lumber, veneer, pulp, biomass, and so on); forage and range plants (mostly grasses and legumes for livestock); ornamentals (such as tulips, irises, roses, orchids, and flowering shrubs); technical crops (such as rubber, pyrethrin, and derris); drug and medicinal plants (sources of steroids, alkaloids, and so on); and some specific crops such as cacao, pecan, lupine, the American octoploid strawberry, and rootstocks for grapes, citrus, and others.

Equally important is the use of wild genetic resources to support major food and commercial crops. Such crops as sugarcane, tomato, tobacco, and potato could not maintain commercial status without the genetic contributions of their wild relatives. For example, tomatoes could not be grown commercially in the United States without resistance to both Fusarium and Verticillium wilts. Resistance to these diseases has been obtained from wild tomato species, as well as resistance to bacterial canker, bacterial wilt, gray leaf spot, leaf mold, Septoria leaf spot, curly top virus, mosaic virus, three kinds of spotted wilt virus, root knot nematode, and probably more. Resistance to major diseases has also been isolated from wild relatives and inserted into tobacco, sugarcane, and potato, permitting their survival as major crops.

When useful traits are simply inherited, that is, conditioned by one or two mendelian genes, they are usually incorporated one at a time and usually by the backcrossing method. The most common and simple model is the transfer of a gene for disease resistance from a wild source to a specific cultivar. The hybrid is made and crossed back to the cultivar. Each succeeding generation is crossed back to the cultivar with concurrent selection for the desired trait. After about six backcross generations, the desired product is obtained. If more than one gene is involved the model becomes more complex, and if the desired trait is quantitatively inherited incorporation may be long and difficult or compromises must be accepted. For example, wild cereals and grain legumes are often higher in protein than domesticated cultivars, but protein content is often negatively correlated with yield. One may need to compromise on maximum yield of protein rather than maximum gross yield.

While the use of wild relatives as sources of disease resistance is the most simple and common, wild populations have also contributed many other traits. Insect and nematode resistances may be incorporated much as with disease resistance, although clear-cut mendelian genes are less common. Some other uses follow.

Improved quality. Wild sources have increased protein content in cassava, and efforts are under way to improve rice, rye, oats, wheat, maize, sunflower, and soybean in a similar way. Increased vitamin C and soluble solids (mostly sugars) have been introduced into tomato from wild species. Greater unsaturation of oil has been transferred to African oil palm by crossing with an American relative. Greater fiber strength has been transferred to cotton from wild sources. Better quality, aroma, and lower nicotine has been transferred from wild species to cultivated tobacco. The use of wild species and races of several fruit and nut species have improved flavor or keeping qualities.

Wider adaptation. Increased cold tolerance has been introduced into wheat, pineapple, tea, cassava, sugarcane, rye, potato, grape, lupine, and other crops. Increased heat or drought tolerance has similarly been incorporated into a number of crops, including wheat, pea, rice, tomato, and grape. Among the more spectacular recent developments is the use of a wild tomato that grows along the shores of the Galapagos Islands and is highly tolerant of salt. By crossing it to cultivated tomato, lines were selected that could grow and bear fruit when irrigated with sea water only somewhat diluted. Wild sources have been used to produce earlier, shorter-season varieties that could be extended into the Arctic or into desert fringes with short rainy seasons. Examples are soybean and rye in the Soviet Union and lupine in Australia.

New cytoplasms. An array of cytoplasms is especially important where a cytoplasmic male sterility system is used to produce hybrid seed for commercial plantings. The uniform susceptibility of a cytoplasm to race T of southern corn leaf blight led to a severe epidemic of the United States corn crop in 1970. Wild relatives are the main sources of additional cytoplasms. They have been incorporated as female parents in species hybrids in sunflower, cotton, barley, wheat, rye, and other crops.

Alternative modes of reproduction. Rye is a cross-fertilizing crop, but a wild relative, *Secale vavilovii*, is strongly self-pollinating. The character is inherited and has been transferred to cultivated rye. In some kinds of apomixis, seeds are produced without a sexual process and inheritance is strictly maternal. Such a system is especially useful where hybrid vigor can be exploited. The system has been successfully exploited in some forage grasses, and attempts have been made to introduce apomixis into corn from a wild relative, *Tripsacum dactyloides*. So far, an operational system has not yet been transferred.

Chromosome manipulation. The use of doubled haploids to produce completely homozygous lines has become widely popular in recent years. One method of producing haploids is to culture immature anthers, followed by induction of plantlets from the callus so produced. Another is to use alien but related pollen. Barley (*Hordeum vulgare*) and even wheat (*Triticum aestivum*) pollinated by *Hordeum bulbosum* will produce haploids at a low but useful frequency. Haploid (dihaploid) potatoes can be obtained by pollinating common tetraploid potatoes with pollen from any of several wild diploid species of potato. In these cases, fertilization does not occur, but the female plant is induced to produce an embryo parthenogenetically. Chromosome numbers may also be raised by use of alien pollen. In some interspecific hybrids, the only viable plants produced are derived from an unreduced female gamete. In such cases, the hybridization of two diploids will produce a triploid, and the hybridization of two tetraploids will produce a hexaploid and so forth. Sugarcane is an especially good example of the use of wild species conferring disease resistance or other useful traits and increasing the chromosome number by addition at the same time. Crossing species with different chromosome numbers may, of course, result in changes in number and structure of chromosomes.

Greater yield. In general, plant breeders expect a reduction in yield when high-performance cultivars are crossed with wild populations. This is not always the case. Modern, complex hybrids of sugarcane that incorporate genomes from wild grasses are far more productive and dependable than the old standard "noble canes." Increases of 25–30% in oats have been obtained through the use of wild *Avena sterilis* germ plasm. The modern, high-yielding octoploid strawberry is a classic example of the manipulation of wild sources resulting in striking yield increases. Much of the germ plasm that goes into modern potato cultivars derives from wild sources because that is where most of the diversity is to be found. Wild sunflower sources have increased yield of cultivated sunflower. There are enough additional examples to suggest that wild sources have been underestimated and underexploited as to their yield potentials.

Serendipity or the unexpected bonus. Thomas Kerr once showed that lint strength in upland cotton could be increased by use of *Gossypium thurberi*, which has no lint, and Charles Rick found that the B gene in *Lycopersicon hirsutum* directs carotenoid synthesis entirely to B-carotene when inserted into cultivated tomato. This was not predictable because *L. hirsutum* does not produce colored carotenoid pigments.

Conservation of genetic resources of wild species has become a major global issue in recent years. Frequently, they do not lend themselves well to the same techniques of collection and storage that are applied to domesticated plants. They are much more difficult and expensive to maintain, and rejuvenation of seed collections poses special problems. In many cases some form of in-place conservation in parks and reserves would be most desirable.

[JACK R. HARLAN]

NEW PLANT MODELS FOR PEAS

Particular forms of peas (*Pisum sativum*) exist in which the leaves and associated structures have been drastically altered; they have recently become known as leafless peas. The changes arise as a result of the expression of two recessive nuclear genes *af* and *st*. The former converts the leaflets of the normal leaf into tendrils, and the latter reduces the large leafy stipule at the base of the leaf to a small vestigial structure (Fig. 1). Another and less drastically altered form is the semileafless pea; in this the *af* gene is expressed, so that the plant has tendrils instead of leaves, but has normal stipules (Fig. 2). These two mutations have long been known as interesting curiosities to geneticists and pea breeders, but more recently their potential value as crop plants has been recognized.

Fig. 1. Leafless pea (genetic constitution *afafstst*).

Fig. 2. Semileafless pea (genetic constitution *afafStSt*).

The two recessive genes were found as spontaneous mutations. As far as is known, they do not directly affect any other aspect of the plant's structure or development, although the changes may have other consequential effects as will be seen later. The retention of such mutant forms in germ plasm collections highlights the value of conserving a wide range of genetic forms, forms which at any given time may appear to have no value but which subsequently prove to be of considerable importance in the production of new crop varieties.

Reduced foliage and standing capacity. The potential agronomic significance of leafless peas for the damp conditions of northern Europe was first recognized with the realization that a primary defect of the pea, and particularly of the crop grown for dry seed, was its tendency to fall or lodge. Within these temperate regions the pea and field bean (*Vicia faba*) are the two most important seed legume crops, and the contribution of peas to the total production of protein within the area is very important. Nevertheless, the traditional pea crop for dry seed remains a difficult one for the farmer. First, although plant breeders have bred new varieties which are much shorter than their immediate ancestors—the garden peas, the lack of stem strength combined with the weight of leaves causes most crops to lodge prior to maturity. This has several consequences. The acute bend in the stem reduces translocation of material from roots to shoots. Second, if pods are immersed in a mass of foliage or make contact with the ground, product quality falls. Third, harvesting is difficult, and crops often have to be cut and partly dried prior to being threshed in a combine harvester. The leafless model was expected to have better standing capacity because of the reduction in foliage, and because the tendrils which replaced the leaves intertwined with each other to form a mutually supportive crop.

A very large acreage of peas is grown for harvesting other than as mature dry peas; this involves the use of the immature seed for freezing or canning. A different and less severe set of limitations prevails with this crop. Its harvesting involves separating the seed from the remainder of the plant, in the field, by using mobile viners. Because of the reduction in the amount of foliage, as well as the slight increase in standing capacity, the semileafless pea could be an attractive alternative to conventional forms of this crop. At present there does not seem to be any convincing advantage for growing leafless forms for this use.

Yields of leafless plants. Early evaluations of the leafless model, using single plants grown in glasshouses and experimental fields or in small-scale plots, showed that the yields of leafless plants were usually lower than yields of leafed plants. As more seed became available and a range of strains of leafless peas was generated, larger-scale field trials became possible, and a different picture emerged. Most of this work has involved the leafless dry pea variety Filby, which became commercially available in the United Kingdom from late 1980 and was essentially a by-product of a genetic program. The results of 48 trials, extending over a 4-year period in the United Kingdom, Canada, and Finland, showed a mean yield of 3.34 metric tons per hectare, which compares very well with the average yield of conventional dry peas in the United Kingdom in recent years. A direct comparison could be made in 11 trials between the variety Vedette, a conventional form, and Filby, and in these the mean yield of the latter was 3.63 ± 0.54 and of Vedette 3.41 ± 0.87 tons per hectare. Clearly the yield potential of leafless forms grown on a field scale is equivalent to that of present commercial varieties. But it is felt that while breeding can lead to increases in the yield of both types, the clear advantage of the improved standing ability of the leafless model should give more scope for improvement of the dry pea crop.

Physiology of leafless plants. As the leafless peas differ so drastically in morphology from their leafed counterparts, it is probably as well to consider them as an entirely separate crop. Physiological studies undertaken on single plants and on crops confirm this. When the photosynthetic performances of tendrils and leaflets are compared, there is surprisingly no obvious disadvantage, and indeed there may be a slight benefit for the former in terms of net fixation of carbon dioxide per unit area of photosynthetic surface. The anatomy of the tendrils is markedly different from that of leaves in lacking the typical large intercellular spaces of the latter; the stomatal frequencies and stomatal characteristics of the two structures remain to be studied in depth. However, photoassimilation in crowded plants within a crop will be expected to differ markedly. The lower leaves of conventional plants are shaded due to the efficient interception of light by the upper leaves. In contrast, tendrils are less effective in this respect, and more light is available for the lowermost photosynthetic structures. Since a subtending leaf or tendril is a primary source of photoassimilate for a given pod, an adequate level of light for pods developing at lower nodes is important; therefore the leafless forms are at an advantage in this respect. The available evidence indicates that leafless peas are not likely to be restricted in their yield capacity as a direct result of a limitation in photosynthetic capacity. Neither do these forms suffer any disadvantage in the translocation of photoassimilate.

A comparison of the water requirements of single plants of leafless and conventional plants shows no significant difference in their relative response per unit of dry seed produced. The absolute amount of water utilized from seedling stage to maturity, however, was lower in leafless types, this being related to their lower yield and biomass in the particular experimental regime used.

Planting densities. Recent studies at a range of planting densities shows that the response of leafed and leafless peas to crowding is markedly different. It is found that at the lower end of the range the leafed plants can compensate for the increased space available by increasing their size, but at the upper end where the plants are densely crowded the plants compete intensely at later stages of growth. The result is that the total biological yield per unit area decreases slightly with increasing planting density. The leafless genotypes at a low density cannot compensate to the same extent to make use of the increase in available space. As a result, the total biological yield per unit area is much lower than that of conventional ones at this planting density. At high densities the leafless plants in general compete less with each other, and as a result the biological yield per unit area can exceed that achieved with leafed forms. The proportion of the total biological mass directed to the production of seed remains somewhat comparable in the two types of plants at these high densities. While these conclusions are valid as generalizations for the two classes of plants, there are a number of other factors which have been shown to influence the competitive behavior and performance of the plants, including the size of seed, flowering time, and extent of branching. Leafless plants grown at high densities can give higher yields per unit area, and to maximize this capacity plant breeders should select forms which are weak competitors. To be uncompetitive the genotypes must have a low growth rate, and this seems to be an intrinsic feature of the leafless forms.

The selection of poor competitors may be difficult for the plant breeder when dealing with a range of segregating genotypes, and new selection procedures may need to be adopted for identifying them. The picture that now emerges is that future leafless varieties will be tolerant of high densities to maximize yields, will flower early and over an extended period to spread the demand for photoassimilate, will waste little resources in the production of branches, and will have nonshattering pods.

Semileafless forms. While the emphasis in this article has been on the production of leafless peas for dry seed production, where the overriding advantage of the standing ability of this model has been emphasized, further research could show that the semileafless model is also attractive. A possible role in the production of varieties for freezing has already been mentioned, but further studies are under way to examine aspects of the behavior of the semileafless forms when grown under different conditions and at different densities. If the cost of seed for planting the leafless crop at high densities is excessive, the semileafless crop may have to be used; almost certainly it can be planted at slightly lower densities. Adoption of a semileafless form would involve some compromise on the standing capacity of the crop.

Summary. The leafless and semileafless peas provide an important illustration of the way in which it is possible to restructure a crop plant. The fact that such forms are radically different from those that have evolved in nature need not cause concern; the conditions under which plants are currently grown are vastly different from those in which their ancestors evolved. Increasingly, plant breeders may be forced to consider equally drastic alternatives in other crops to those being considered for peas, as they strive to maximize the return from agricultural and horticultural resources.

For background information *see* BREEDING (PLANT) in the McGraw-Hill Encyclopedia of Science and Technology.

[D. R. DAVIES]

Bibliography: D. R. Davies, *Sci. Prog.*, 64:205-218, 1977; O. H. Frankel and M. E. Soule, *Conservation and Evolution*, 1981; J. R. Harlan, *Crop Sci.*, 61:328–333, 1976; B. Snoad, *ADAS Quart. Rev.*, 37:69–86, 1980; H. T. Stalker, *Adv. Agron.*, 33:111–147, 1980; A. C. Zeven and A. M. van Harten (eds.), *Proceedings of the Conference Broadening the Genetic Base of Crops*, Pudoc, Wageningen, 1979.

Carboniferous

Progress in coal-related sciences has reached a threshold in determining climatic trends of rainfall, runoff, and evapotranspiration and their control of the origin of coal in the Pennsylvanian Period (3.15 to 2.8×10^8 years ago). This "Age of Coal" in the eastern half of the United States accounts for more than 80% of the total identified bituminous coal resources of the country and virtually all the anthracite; it has a similar significance in Europe, where it is called the Late Carboniferous. The eastern United States, maritime Canada, and western Europe were juxtaposed with the same kinds of vegetation (Euramerican floral province) extending into the Donets Basin of the Soviet Union. The paleoequatorial belt arched across these areas during the Pennsylvanian. The amount of fresh water available for wetlands was generally abundant but differed in regions and changed through time.

Paleoecology. Coal swamp ecology utilizes plants as environmental indicators. Estimates of identified coal resources in the United States also provide stratigraphic and regional indicators of wetness. Although the distribution of swamp plants and coal deposits depends on climate, there are numerous differences between coal basins. These variables alter climatic imprint on the vegetation of a region. Where peat-accumulating environments existed, there are close relationships between the abundances of certain kinds of plants and the derived quantities of coal. Plant habitats and peat accumulation reflect the same kinds of swamp environments. The wettest peat-forming environments, such as standing water with a rising water table (balanced deltaic subsidence), yield the most peat and form the most aquatically adapted forests. Habitats subjected to lower water tables, wet-dry seasonality, or ephemeral wetlands (nonfloating herbaceous vegetation) yield less peat generally because of the rates of degradation or of productivity.

Vegetation. Vegetational patterns of Euramerican swamps were summarized in 1974 based on anatomically preserved peat stages in coal ball concretions and on spore floras from coals primarily in the Illinois Basin coalfield. Five kinds of trees account for more than 95% of the swamp peats and spores. Lycopods (*Lepidodendron* and *Lepidophloios*) dominated most of the swamps during Early and Middle Pennsylvanian time except in the physiologically drier environments, which were either quite brackish or contained lower water tables. These kinds of Middle Pennsylvanian environments were habitats of seed plants (cordaites and seed ferns) and some tree ferns. In the transition to the Late Pennsylvanian the highly adapted aquatic trees (*Lepidodendron* and *Lepidophloios*) disappeared abruptly across all the Euramerican swamps. *Psaronius* tree ferns generally became dominant, along with abundant *Medullosa* seed ferns and some *Calamites* (sphenopsids) and *Sigillaria* (lycopods). There were also ephemeral marshlands dominated by a robust herb, *Poly-*

sporia (lycopod). The most striking change in swamp vegetation at the Middle to Late Pennsylvanian transition coincides with changes in terrestrial floras and is attributed to a major climatic shift on a multicontinental scale. That shift is diminution in rainfall.

Coal ball peats. An unusual aspect of paleobotanical studies of Pennsylvanian swamps is the preservation of the plants. Coal ball concretions were formed during peat accumulation or shortly after burial of the swamp. Entombed within them is the beautifully preserved anatomy of successive communities, essentially where they grew. The precipitation of calcite or dolomite within peat layers of more than 65 coal seams in the Euramerican Province provides most of the anatomically preserved vascular plants of the entire geologic record. From these assemblages paleobotanists have reconstructed plant assemblages and their habits, growth, and reproductive structures.

Coal palynology. As plant sources of dispersed spores in coal became known, the biostratigraphic use of spores for coal correlations shifted to paleoecology. Coal palynology is now the principal means of determining vegetational changes in Pennsylvanian swamps. In the only interval of the early Middle Pennsylvanian of the Illinois Basin in which *Lepidophloios* and *Lepidodendron* waned, marshlands developed with successive introductions of tree ferns, seed plants, and other plants adapted to drier environments. Later these plants characterize Late Pennsylvanian wetlands.

Plants as environmental indicators. The presence of growth rings in the wood of certain seed plants and their progymnosperm ancestors are indicative of seasonal climates as in warm-cold temperate regions or in the wet-dry tropics. This phenomenon extends back to the Devonian Period (more than 3.5×10^8 years), but it is not known to occur in lower vascular plants such as the tropical lycopod and calamites trees. Frequent occurrences of unusual growth rings in cordaitean roots (but not evident in the stem) in the Early and early Middle Pennsylvanian of Europe and the Appalachians suggest a wet-dry seasonality (Fig. 1). This finding is consistent with the exceptional mixture of lycopod trees adapted to the wettest swamps, such as *Lepidophloios* and *Lepidodendron*, and those adapted to the driest ones, *Sigillaria*. The known coal ball peats from this interval consist mostly of roots.

Reproductive biology. Euramerican wetlands were the habitats in which lycopods attained their maximum abundance, and such environments literally became their refuges as seed plants and ferns became the dominant vegetation outside the swamps. Lycopods exhibit a spectrum of aquatic to wet-dry adaptations in their sexual reproduction. The seedlike structures of *Lepidophloios* and *Lepidodendron* are boatlike adaptations for fertilization in and dispersal by water. In contrast, the small granular-walled megaspores of *Paralycopodites*, *Sigillaria*, and the herbaceous *Polysporia* are adapted to drier swamp habitats or wet-dry fluctuations with repeated reproduction during the life cycle. The abundances and succession of such genera are considered indicators of fresh-water availability.

Fig. 1. Transverse section of cordaitean root (*Amyelon radicans*). (*From the Felix Collection, Museum für Palaontologie, Humboldt-Universität zu Berlin*)

Community analyses. In coal swamp communities in the Herrin Coal Member, located in the upper Middle Pennsylvanian of Illinois, *Lepidophloios* accounted for more than half the peat, and such communities showed the least diversity. *Lepidodendron*, along with *Medullosa* and *Psaronius*, formed multistoried forests, usually with abundant ground cover. At the driest extreme, *Sigillaria* and *Paralycopodites* communities exhibited exceptional diversity but accounted for only a small percentage of the peat.

Coal resources. Regional patterns in the United States for coal seams containing more than 10^8 short tons (1 short ton = 0.9 metric ton) are compared using data based on stratigraphic plots of identified bituminous resources for the Pennsylvanian (Fig. 2). In the Midcontinent coalfields (Interior Coal Province), 97% of the coal resources are in the Middle Pennsylvanian, as are about two-thirds of those in the Appalachians. The next stratigraphic interval, the lower Upper Pennsylvanian, contains the least, or about 2% of some 6×10^{11} short tons. The wettest interval was followed by the driest interval. Information on climate from the upper Middle Pennsylvanian (Desmoinesian) of the Interior Province involves about half of the bituminous coal resources and most of the occurrences of coal ball peats of the United States. At the beginning of the Desmoinesian, cordaites dominated swamps in Iowa and Kansas in the Western Coal Region and were subdominant to *Lepidodendron* in the Illinois Basin (Eastern Coal Region). There was a subsequent westward

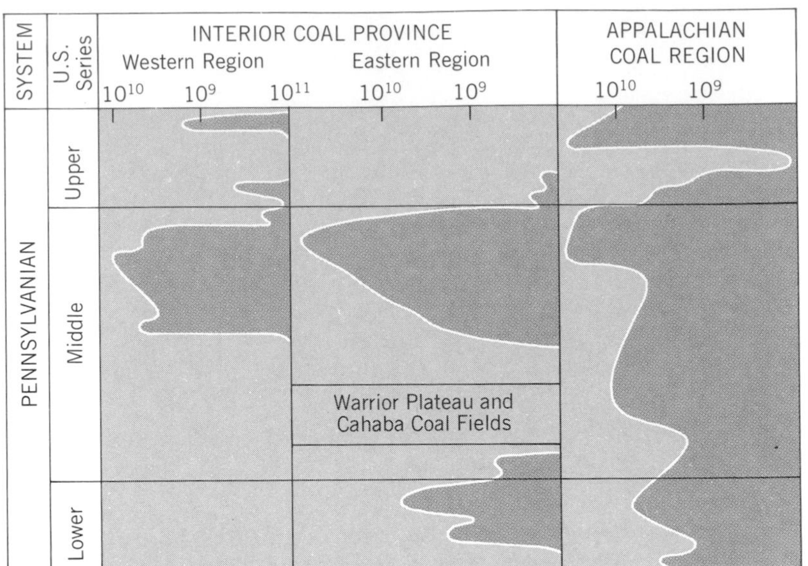

Fig. 2. Stratigraphic patterns of identified bituminous coal resources in the Pennsylvanian System of the eastern half of the continental United States. Plots are on a log scale for regional seams with more than 10^8 short tons (1 short ton = 0.9 metric ton). The Warrior, Plateau, and Cahaba coalfields are plotted separately from the rest of the Appalachian Coal Region because of limitations in stratigraphic correlations.

shift of *Lepidodendron* and then *Lepidophloios* as the dominant trees in the Western Region, while *Lepidophloios* dominated in the Illinois Basin. There was also a synchronous increase in coal resources of these two areas. The peak of *Lepidophloios* dominance occurs in the major seams of the Midcontinent, the Springfield and Herrin Coals and their equivalents. Subsequent diminutions in coal resources occur toward the end of the Desmoinesian, first in the Western Region, then in the Illinois Basin, and last and less abruptly in the Appalachians. Both *Lepidophloios* and *Lepidodendron* disappear. It is generally conceded that the Appalachians received more rainfall during the Pennsylvanian than did the Interior Coal Province. This is consistent with the 79 major coals seams scattered across every interval of the Pennsylvanian in the Appalachians. However, the Illinois Basin contains the largest coal seams, and factors such as large basinal size, slower subsidence rates, and drainage patterns are pertinent. In particular, the Michigan River System supplied additional water to the Illinois Basin from portions of the eastern Canadian Shield and northern Appalachians.

Climatic changes. Fresh-water availability in the eastern half of the United States exhibits short- and long-term changes. Seasonal rainfall patterns are suggested for portions of the Appalachians during the Early and early Middle Pennsylvanian; this observation is in agreement with similar evidence from Europe. The first drier interval in the early Middle Pennsylvanian shows a major reduction in *Lepidodendron* and *Lepidophloios* and significant introductions of seed plants and ferns into the wetlands. Onset of marine transgressions across the Interior Province with abundant cordaites apparently coincides with increased fresh-water supply from the east, flushing out coastal swamps and gradually shifting *Lepidophloios* and *Lepidodendron* westward as dominant trees. The diminution of fresh-water availability in the transition to Late Pennsylvanian was much more severe, resulting in extinctions of dominant lycopods. On the basis of coal resources, it can be inferred that the easternmost basins in the Appalachians received more rainfall than elsewhere during the driest interval, and that there was a resurgence of wet climate toward the end of the Pennsylvanian. This began with the famous Pittsburgh Coal and then diminished toward the Permian. Causes of the major climatic shift in the Middle to Late Pennsylvanian transition in Euramerica were rising orographic barriers such as the Appalachians to the east of the major coal regions of the United States; similar barriers developed for western Europe and the Donets Basin. These resulted from plate collisions as continents approached a Pangean (united from pole to pole) arrangement in the Permian Period and brought the Euramerican "Age of Coal" to an end.

For background information *see* CALAMITALES; CARBONIFEROUS; COAL; COAL BALLS; COAL PALEOBOTANY; LYCOPODIOPHYTA; PALEOECOLOGY; TREE-RING HYDROLOGY in the McGraw-Hill Encyclopedia of Science and Technology.

[TOM L. PHILLIPS]

Bibliography: R. A. Peppers, *Depositional and Structural History of the Pennsylvanian System of the Illinois Basin*, pt. 2, Illinois State Geological Survey, 1979; T. L. Phillips, *Biostratigraphy of Fossil Plants*, 1980; T. L. Phillips, *Rev. Palaeobot. Palynol.*, 27:239–289, 1979; T. L. Phillips and W. A. DiMichele, *Paleobotany, Paleoecology and Evolution*, vol. 1, 1981; T. L. Phillips et al., *Science*, 184(4144):1367–1369, 1974; T. L. Phillips, J. L. Shepard, and P. J. DeMaris, *Geol. Soc. Amer. Abs.*, 12(7):498–499, 1980.

Carbynes

Carbynes are carbon polymorphs containing triple bonds. A number of forms have been recognized since they were first identified by Soviet and American workers in 1968.

The term carbyne was coined by A. M. Sladkov and Y. P. Kudryavtsev, who prepared one form by dehydrogenation of acetylene using a copper chloride catalyst. As the reaction proceeded, they were able to confirm the continuing presence of the triple bonds until they obtained the final product of essentially pure carbon. Sladkov and Kudryavtsev proposed a carbyne structure consisting of $(-C{\equiv}C-)_n$ chains stacked in a hexagonal array.

About the same time, A. El Goresy and G. Donnay discovered a new carbon form (chaoite) in a sample taken from the Reis meteor crater in Bavaria. Their x-ray diffraction data indicated that it was a hexagonal form with unit cell parameter a_0 = 8.948 A (0.8948 nm) and c_0 = 14.078 A

(1.4078 nm). However, the Soviet scientists claimed that the El Goresy and Donnay data came from a mixture of two carbon forms which they called α- and β-carbyne. This controversy probably will not be resolved until a crystal is found that is large enough for structure determination by x-ray diffraction.

Subsequently, investigators in the Soviet Union and the United States began to realize that there were several carbyne forms, but because of their acicular or tabular crystal habit, it was not possible to determine all the unit cell parameters by ordinary electron diffraction methods. If the carbyne forms consisted of $(\text{—C}\equiv\text{C—})_n$ chains stacked in hexagonal array, as had been proposed, the crystals would be held together by van der Waals forces in two dimensions. Such crystals would be weak and softer than graphite. Often crystal fragments observed in the electron microscope are decomposed rapidly by the electron beam even at 60 kV before the diffraction pattern can be obtained. These could be the forms that produce acicular crystals.

Some carbynes are quite hard (they easily scratch sapphire, and at least one can scratch cubic boron nitride), indicating cross-linking between the chains, a decrease in the number of triple bonds, and a more complex crystal structure. Laser Raman studies have confirmed that some of the triple bonds remain in the hard forms. Shortly after the discovery of carbynes, it was found that they could be produced synthetically by heating graphitic carbon above ~2600 K, followed by quick cooling to trap the high-temperature form. Generally, carbynes produced in this manner are hard tabular hexagonal crystals showing trigonal symmetry.

Terrestrial forms. Natural carbon deposits reveal that carbynes are widely distributed and occur in small amounts in all carbon sources studied. Carbon crystals occurring in marble deposits were examined in some detail, with several interesting results obtained by ion probe analysis. The carbon negative-ion spectra of carbynes differs considerably from that of graphitic carbon. In addition, the carbyne ion etch patterns are all different from the etch pattern formed in graphite crystal. The carbyne crystals found in marble consist of layers of different carbynes, each epitaxially attached to one another, sometimes with a hard carbyne form attached to a soft form.

Examination of these ion-etched areas by the scanning electron microscope showed that the carbyne layers had varying degrees of secondary electron emission; hence, some layers appeared brighter than adjacent layers. Further study by the electron microscope (both transmission and scanning) showed that carbynes were more transparent to electrons than graphitic carbon, and it was not possible to find small carbyne crystals by using the scanning electron microscope.

Space forms. A recently proposed carbon phase diagram (illustration) includes a carbyne region. According to this diagram, the carbyne forms are stable in the temperature range from 2600 to 3800 K. The low end of this range corresponds approximately to the temperatures that exist in the outer atmosphere of carbon stars. Hence, it is possible that these stars produce carbynes and release them into the surrounding space. It has been known for many years that the regions near carbon stars emit several infrared lines or bands that do not correspond to any known terrestrial material. One of these lines occurs at 3.1 μm. A. Webster of the Cavendish Laboratory has shown theoretically that if the carbynes consist of $(\text{—C}\equiv\text{C—})_n$ chains, they would have an asymmetric emission line at 3.1 μm. The line shape would be steep on the short wavelength side and less steep on the long wavelength side. This line shape has been observed from the interstellar dust. Carbyne films prepared by quenching carbon gas also show a line at ~3.1 μm with the proper shape. Other interstellar unidentified infrared lines occur at 6.2, 7.7, 8.6, and 11.3 μm; and the carbyne films show bands at 5.8, 6.2, 7.3, 8.5, and sometimes at 11.3 μm. The coincidence of lines further supports the notion that carbynes exist in the space around carbon stars, but there are some differences to be resolved. It is possible that the carbyne forms in interstellar space are not the same as those in the carbyne films. Only further research can reveal whether there are actual or only apparent discrepancies.

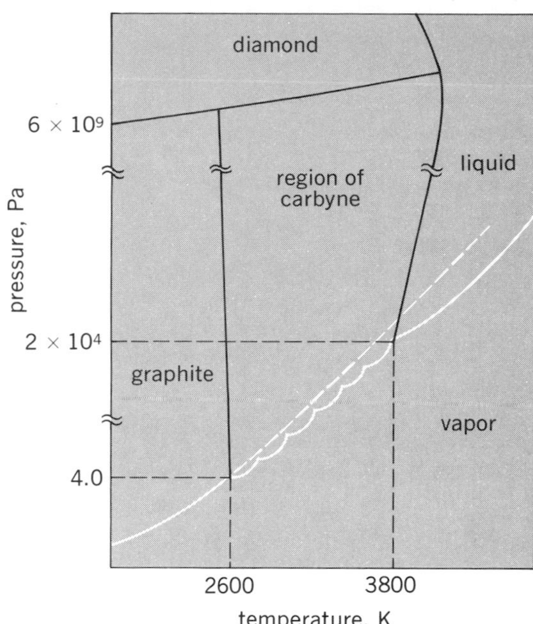

Proposed carbon phase diagram. The broken curve indicates the vapor pressure that graphite would exhibit if it were the stable form above 2600 K. (*From A. Greenville Whittaker, Carbon: A new view of its high-temperature behavior, Science, 200:763–764, copyright 1978 by the American Association for the Advancement of Science*)

Additional information on the existence of carbynes in space has come from the study of meteorites, which are considered to be solar nebula condensates carrying information on the processes which led to the formation of planets. The carbonaceous chondrites are believed to be the oldest of the meteorites and therefore the most primitive in structure. This class of meteorite contains about 0.3% of a carbon phase, part of which can be separated by solvent extraction. Some of the residue consists of graphitic carbon, with a nongraphitic fraction that appears to be an unusual polymer. Although the carbon phase constitutes a very small part of the meteorite, it is very important because it is the carrier of the primordial noble gases, which give significant information on presolar nucleosynthesis processes. Consequently, it is important to get a better understanding of the nature of the nongraphitic phase.

The nongraphitic phase from the Allende and Murchison meteorites was studied by electron diffraction and by ion probe analysis. Sharp electron diffraction patterns were obtained, showing the presence of carbynes in both meteorites. Also, several euhedral crystals of carbyne were found in the carbon phase from Allende. Ion probe analysis gave a strong carbon negative-ion spectrum characteristic of carbyne, in support of the electron diffraction results. The negative-ion spectrum also showed the fragments CN, C_2N, C_3N, and C_5N, indicating that the carbon phase contains polycyanoacetylenes, substances closely related to carbynes chemically. This suggests that they occur together in interstellar space. An Auger study of carbynes by Soviet scientists showing that they strongly absorb oxygen molecules indicates that carbynes have active surfaces and supports the notion that carbynes can be the host phase for the noble gases. Another feature of the nongraphitic phase is its distinctive ion etch pattern, a vermicular pattern that has not been observed in any terrestrial carbon form so far. This is further evidence that the Allende carbon phase has a unique structure.

During the study of the Allende and Murchison meteorites, it was found that carbynes can be produced by a low-temperature synthesis. R. Hayatsu and coworkers at the University of Chicago found that carbynes can be produced by the reaction $2CO \rightarrow CO_2 + C$ on chromite, a Fischer-Tropsch catalyst. Under these conditions, carbynes can be produced in the temperature range 300–400°C. This is about the maximum temperature reached by meteorites; hence, the carbynes could be formed in place. Also, carbynes could be produced in interstellar space by this process because CO is the most abundant gas in space, and much of the dust consists of Fischer-Tropsch catalysts. This could also account for the existence of polycyanoacetylenes in interstellar space and could provide a solution to one of the current astrophysical problems.

Spherules of possible extraterrestrial origin were found in old (~ 400 years) high-altitude glacial ice. They are ~100 μm in diameter, and many of them contain carbon. Because of the location of the glacier and the age of the ice, these spherules could not be some sort of anthropomorphic debris. Many were glassy in appearance and contained many minerals and only a small fraction of carbon. Others were essentially pure carbon. These spherules were analyzed by many techniques, including electron diffraction study, ion probe analysis, and laser micromass analysis. The last technique is much like ion probe analysis, except that a laser is used to remove material from a sample rather than an ion beam. Again, electron diffraction gave sharp single-crystal patterns of the carbyne forms, and the ion probe showed the characteristic carbon negative-ion spectra. The spectra obtained by the laser micromass analyser were in complete agreement with the ion probe results, and both showed the CN, C_3N, and C_5N lines, probably from polycyanoacetylenes. Finally, the ion etch pattern on the carbon spherules was very much like the unusual pattern found in Allende carbon phase. These results show that the spherules are of extraterrestrial origin and could be either cometary matter or ablation material from carbon-containing meteorites that entered the Earth's atmosphere.

For background information *see* INTERSTELLAR MATTER; METEORITE; SECONDARY ION MASS SPECTROMETRY (SIMS) in the McGraw-Hill Encyclopedia of Science and Technology.

[A. G. WHITTAKER]

Bibliography: R. Hayatsu et al., *Science*, 209:1515, 1980; A. Webster, *Mon. Not. Roy. Astron. Soc.*, 192:7P, 1980; A. G. Whittaker, *Science*, 200:763, 1978; A. G. Whittaker et al., *Science*, 209:1512, 1980.

Catalysis

An important factor in the slowness of many organic reactions is the lack of homogeneity of the reaction mixture. This is particularly the case with nucleophilic substitution reactions, depicted in (1), where

$$RX + Nu^- \rightarrow RNu + X^- \quad (1)$$

RX is the organic reagent. The nucleophilic reagent (Nu^-) is frequently an inorganic anion which is soluble in water (in which the organic substrate is insoluble) but is insoluble in the organic phase. The encounter rate between Nu^- and RX is consequently low, as they can meet only at the interface of the heterogeneous system. The water-soluble anion is also frequently highly solvated by water molecules, which stabilize the anion and thus reduce its nucleophilic reactivity. These problems have been overcome in the past by the use of polar aprotic solvents, which dissolve both the organic and the inorganic reagents, or by the use of homogeneous mixed-solvent systems, for example, water-ethanol or water-dioxane. Homogeneous reaction systems can also be established by the use of surfactants, which disperse the organic reagent in the aqueous medium through micellar formation. Rapid agitation of the heterogeneous reaction system increases the interfacial surface area, as does mechanical emulsification. Although all these procedures can increase the rate of reaction, they have disadvantages such as difficulties in the isolation and purification of the products, and on an industrial scale, they can be costly in terms of solvents and energy.

Phase-transfer catalysis. Phase-transfer catalysis involves the transportation of the inorganic anion (Nu^-) from the aqueous phase into the organic phase by the formation of a nonsolvated ion pair with a cationic phase-transfer catalyst (Q^+). C. M. Starks originally proposed that this process involved the formation of the ion pairs in the aqueous phase, followed by their partition between the aqueous and organic phases (Fig. 1). The rate of the reaction is enhanced, as the encounter rate of the nucleophile (Nu^-) with the organic reagent (RX) in the single phase will be significantly higher than at the interface. Moreover, as the anion is transferred without water of solvation, its nucleophilic reactivity can be considerably higher in the organic phase than in the aqueous phase. Rate enhancements of greater than 10^7 have thus been observed.

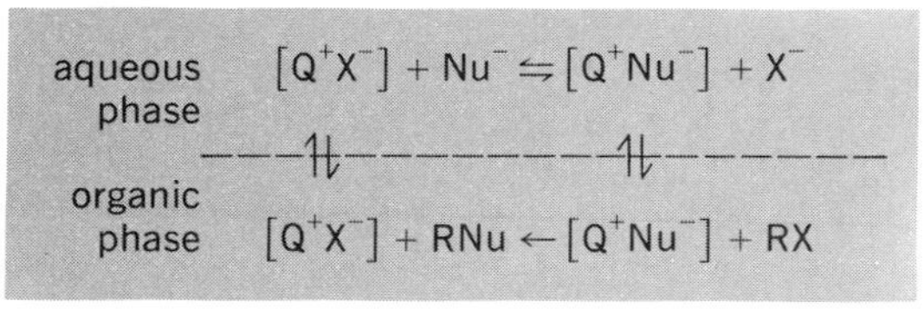

aqueous phase $[Q^+X^-] + Nu^- \rightleftharpoons [Q^+Nu^-] + X^-$

organic phase $[Q^+X^-] + RNu \leftarrow [Q^+Nu^-] + RX$

Fig. 1. Reaction sequence for the formation and partition of the ion pairs in a phase-transfer catalysis process.

The efficiency of the catalyst depends upon the ability of the cation (Q^+) to transfer the anion across the interface of the heterogeneous system; that is, the rate should be proportional to the partition coefficient of the ion pair $[Q^+Nu^-]$. It has been found that bulky quaternary ammonium cations (R_4N^+), for example, tetra-*n*-butylammonium and methyltrioctylammonium ions, and phosphonium cations (R_4P^+), for example, tetra-*n*-butylphosphonium cations, have a high propensity to form strong ion pairs which have a high solubility in organic solvents and are excellent phase-transfer catalysts.

In the reaction in Fig. 1, if the water solubility of $[Q^+X^-]$ is higher than that of $[Q^+Nu^-]$, the cationic catalyst is returned to the aqueous phase after the nucleophilic reaction has occurred in the organic phase. The reaction will continue, therefore, until all of either the nucleophile (Nu^-) or the organic reagent (RX) has been consumed.

The original concept of the ion-pair formation occurring in the bulk of the aqueous phase, followed by transportation to and across the interface, requires modification for highly lipophilic quaternary ammonium catalysts such as methyltrioctylammonium chloride, which has a low solubility in water. It has been suggested that the reactive ion pair $[Q^+Nu^-]$ is formed at the interface between the aqueous and organic phases, followed by rapid transportation into the bulk of the organic phase (Fig. 2). The efficiency of the catalyst still depends, however, upon the relative values of the partition coefficients of $[Q^+Nu^-]$ and $[Q^+X^-]$.

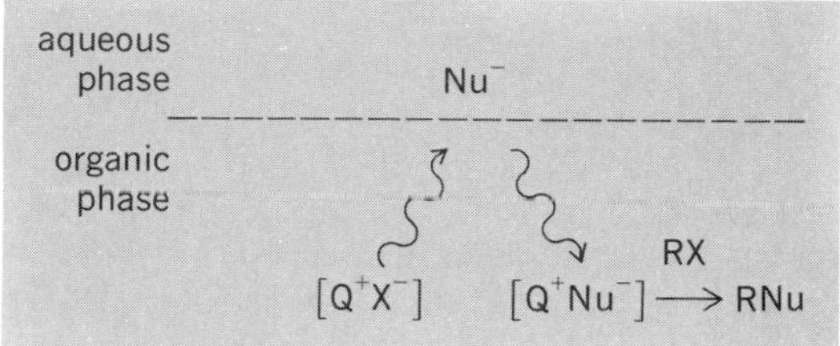

Fig. 2. Formation and transportation of the reactive ion pair in phase-transfer catalysis.

As the partition coefficient of the ion pair is a significant factor in the catalytic effect, the choice of the organic phase is important. The most commonly used organic phases are the semipolar solvents dichloromethane and 1,2-dichlorobenzene and the less-polar solvent toluene. Ethyl ethanoate is also suitable, but chloroform should not be used in the presence of hydroxide ions.

An alternative phase-transfer catalytic procedure utilizes polyethers, which are capable of complexing alkali metal cations. Ion pairs of the complexed cation and the nucleophilic anion are formed and are transported across the water–organic phase interface in the same manner as the quaternary ammonium and phosphonium salts. The polyethers (Fig. 3*a*) may be acyclic (glymes, I), monocyclic (crown ethers, II), or polycyclic (cryptands, III). The most commonly used polyether catalysts are the crown ethers, for example, dicyclohexano-18-crown-6, which complex readily with potassium and sodium ions to form complexes of type (IV), shown in Fig. 3*b*.

Applications. Phase-transfer catalysis procedures can be employed in a wide variety of applications such as nucleophilic substitution reactions, oxidation-reduction reactions, generation of reactive carbanions, generation of carbenes, asymmetric induction, solid-state–liquid phase heterogeneous reactions, and triphase catalysis.

Nucleophilic substitution reactions. Using either onium salts or polyether catalysts, it is possible to enhance the rate of most nucleophilic substitution reactions (1) through the transfer into the organic phase of a wide range of nucleophilic anions, for example, F^-, Cl^-, Br^-, I^-, CN^-, NCO^-, HO^-, RO^-, HS^-, RS^-, RCO_2^-, NO_2^-, and NO_3^-.

In contrast to the nucleophilic activity of the hydrated halide anions in aqueous solution, kinetic studies have shown that the nucleophilicity of the

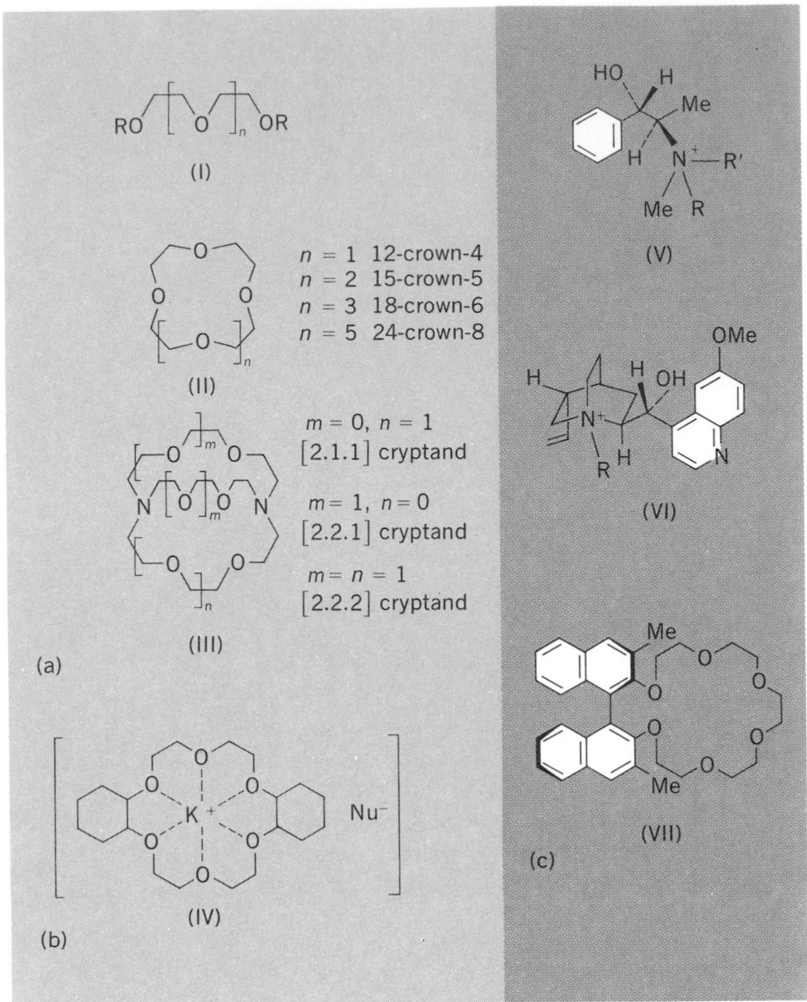

Fig. 3. Structures of some species involved in phase-transfer catalysis. (*a*) Typical polyether catalysts used in phase-transfer catalytic procedures. (*b*) A typical complex formed with a polyether catalyst. (*c*) Chiral catalysts used in asymmetric induction.

nonsolvated anions, produced upon dissolution of the quaternary onium halides in organic solvents, follows the order F > Cl > Br > I. Under these conditions the fluoride ion is also a strong base and can be used in preference to the hydroxide ion in base-catalyzed reactions.

Oxidation-reduction reactions. Numerous examples have been reported of oxidations in organic solvents through the catalyzed phase transfer of permanganate, dichromate, and hypochlorite anions. "Purple benzene or dichloromethane" and "yellow benzene or dichloromethane" are produced by the dissolution of permanganate and dichromate ions, respectively, in the otherwise colorless organic solvents by partition from aqueous potassium permanganate or dichromate using, for example, tetra-*n*-butylammonium bromide or 18-crown-6 as phase-transfer catalysts. Under these conditions, oxidation of a wide range of organic systems can be effected in high yield with minimal side reactions. Although isolation of crystalline tetraalkylammonium permanganates and dichromates has been reported, it is important to note that such compounds are potentially explosive.

Hydride reduction of aldehydes and ketones can be conducted in organic solvents by using tetraalkylammonium borohydrides.

Treatment of the tetraalkylammonium borohydride solutions in dichloromethane with simple alkyl halides, for example, iodomethane or bromoethane, liberates diborane, B_2H_6, as shown in reaction (2).

$$2[Q^+BH_4^-] + 2CH_3I \rightarrow 2CH_4 + 2[Q^+I^-] + B_2H_6 \qquad (2)$$

The dichloromethane solution of diborane has a similar activity to that of ethereal solutions produced by traditional methods, and can be used for all the usual reduction and hydroboration reactions.

Generation of reactive carbanions. Reactive carbanions, which require strictly anhydrous reaction conditions when prepared by traditional methods, can be generated by using the phase-transfer catalytic technique in the presence of the aqueous phase, as shown in reaction (3).

$$\underset{(org)}{RCHZ_2} + \underset{(aq)}{HO^-} \xrightarrow{[R'_4N^+X^-]} \underset{(org)}{[RCZ_2^-R'_4N^+]} \xrightarrow{R''X} \underset{(org)}{\overset{R}{\underset{R''}{>}}CZ_2} \qquad (3)$$

Z = acyl, alkoxycarbonyl, cyano, nitro
R = H, alkyl, aryl

Similarly, other reactive systems, for example, RCOCN, which are labile in the presence of water, can be synthesized by the phase-transfer catalytic procedure.

Carbenes. Tetraalkylammonium hydroxides are soluble in and react with chloroform to yield, initially, the trichloromethyl carbanion (CCl_3^-), which loses a chloride ion to generate the highly reactive dichlorocarbene ($\ddot{C}Cl_2$) [reaction (4)]. The catalyst of

$$\underset{(org)}{CHCl_3} \xrightarrow{HO^-(aq)} \underset{(interface)}{CCl_3^- + H_2O} \xrightarrow{[Q^+Cl^-]} \underset{(org)}{[Q^+CCl_3^-]} \qquad (4)$$

$$[Q^+CCl_3^-] \rightarrow \underset{(org)}{[Q^+Cl^-]} + \underset{(org)}{\ddot{C}Cl_2}$$

choice for this reaction is benzyltriethylammonium chloride, which is extremely inefficient in its ability to transfer the hydroxide anion across the water-chloroform interface. Consequently, it has been proposed that the abstraction of the proton from the chloroform occurs at the interface and that the role of the catalyst is the transfer of the trichloromethyl carbanion into the bulk of the organic solvent.

In the absence of the hydroxylic solvent, which is present in the classical procedure for the synthesis of carbenes, the dichlorocarbene reacts rapidly and in high product yield with a wide range of organic substrates.

Other dihalogenocarbenes have been obtained by similar procedures from the appropriate trihalogenoalkanes.

Phase-transfer catalytic conditions have also been

used successfully in the generation of the less readily accessible vinylidene and vinyl carbenes.

Asymmetric induction. Chiral ammonium catalysts having both a chiral quaternary nitrogen atom and chirality within the carbon skeleton, and some degree of rigidity in their structure, for example, (V) and (VI) in Fig. 3*c*, have been used successfully in the stereochemical control of reactions involving carbonyl-containing compounds. A critical feature of the structure of the chiral catalyst is the presence of a β-hydroxyethyl substituent on the chiral nitrogen atom. It is apparent that hydrogen bonding between the carbonyl group of the organic substrate and the β-hydroxy group is important in holding the organic substrate so that one face of the system is preferentially presented for reaction. Reactions have been observed in which the enantiomeric excess is >60% for Michael additions to α,β-unsaturated ketones, the reduction of ketones, and the alkylation of the α-methylene position of ketones.

Recently, stereochemical control of organic reactions has also been accomplished using chiral crown ethers, such as (VII) in Fig. 3*c*.

Solid-state–liquid phase heterogeneous reactions. The solubilization of solid inorganic salts by crown ethers in organic solvents is a simple extension of liquid-liquid phase-transfer catalysis. Similarly, quaternary ammonium catalysts aid the transfer of inorganic anions from the solid state into organic solvents. This technique is frequently successful when the conventional liquid-liquid phase-transfer process fails as, for example, the catalyzed dissolution by 18-crown-6 of potassium carbonate in a wide range of polar and nonpolar organic solvents.

Triphase catalysis. In a recent elaboration of liquid-liquid phase-transfer catalysis, the quaternary ammonium group or the crown ether has been attached to a polymeric support. Although the efficiency of the water–polymeric catalyst–organic solvent triphase system is generally lower than that of the conventional phase-transfer process, the procedure has potential advantages over the two-phase system in that the insoluble polymeric catalyst is more readily separated from the liquid phases, compared with the soluble monomeric catalysts. Also, with further development, triphase catalysis has the important possibility of becoming a continuous instead of a batch process on an industrial scale.

For background information *see* CATALYSIS; CROWN ETHERS AND CRYPTANDS; HETEROGENEOUS CATALYSIS; HOMOGENEOUS CATALYSIS; REACTIVE INTERMEDIATES in the McGraw-Hill Encyclopedia of Science and Technology.

[R. ALAN JONES]

Bibliography: E. V. Dehmlow and S. S. Dehmlow, *Phase-Transfer Catalysis*, 1980; C. M. Starks, *J. Amer. Chem. Soc.*, 93:195–199, 1971; C. M. Starks and C. Liotta, *Phase-Transfer Catalysis: Principles and Techniques*, 1978; W. P. Weber and G. W. Gokel, *Phase-Transfer Catalysis in Organic Synthesis*, 1977.

Cell membranes

Enclosing all cells is a thin covering, the membrane. The life of a cell is absolutely dependent upon the integrity and function of this membrane. Three key properties of biological membranes are the high degree of plasticity; the wide range of enzymic and energy-coupling capabilities; and the capacity to receive, transmit, and respond to external stimuli. Before dealing with these three parameters, it will be useful to define biological membranes more explicitly and specify their structural components.

General properties. The membranc separates each cell from other cells and from the aqueous phase which envelops each cell. It is continuous and so no chemical species can enter the cell without penetrating the membrane. Membrane thickness usually varies from 6 to 10 nanometers. Each membrane has two faces—one exposed to the fluid phase bathing the exterior of the cell, and the other exposed to the fluid phase bathing the interior. The faces exposed to the external or internal aqueous phase are hydrophilic or polar in character; the sector of the membrane between these two faces is highly nonpolar.

Bacterial cells have one membrane which exhibits the full gamut of membrane capabilities. Plant and animal cells have, in addition to the external plasma membrane, a set of internal membrane systems (organelles), and there is specialization of membrane function. The plasma membrane of plant and animal cells has some but not all the properties of bacterial membranes. The "missing" properties are found in the membranes of the organelles.

Structural components. Membranes are mixes of two kinds of structural components—lipids and proteins. Both the lipids and the proteins from which the membrane continuum is built have one invariant characteristic—bimodality. One sector is hydrophilic and another sector is hydrophobic. Figure 1 shows how this bimodality allows lipids and intrinsic protein to nest with one another and thus form a membrane continuum. Strictly speaking, lipids in bilayer are trimodal. The two outer faces are hydrophilic and the sectors between faces are hydrophobic. The same is true for the proteins in membranes. The lipids and proteins can line up only as shown in Fig. 1. This is the state of least energy. Any other arrangement would be energy-requiring and therefore prohibited. Each transmembrane unit can be either a single protein or a complex of several proteins. When the unit is a complex, it is the complex that is trimodal and not necessarily any of the component proteins. The ratio of lipid to protein varies greatly from membrane to membrane. For example, the lipid content of myelin is 90% by weight, and that of the mitochondrial inner membrane is only 33% by weight.

Lipid and protein can spontaneously form membranes under appropriate conditions. When a mem-

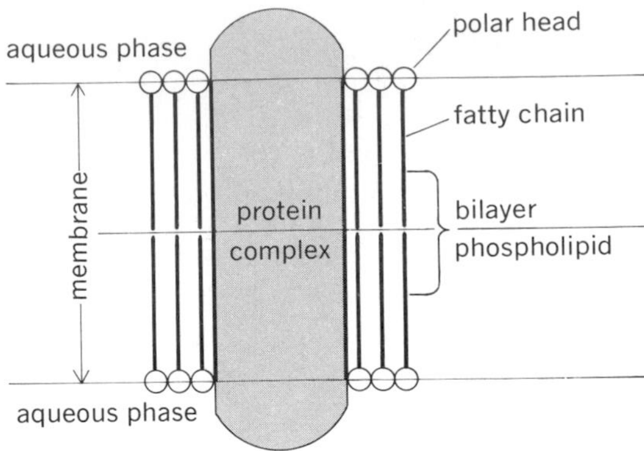

Fig. 1. Nesting of bilayer phospholipid and a protein complex in the membrane continuum. The hydrophobic sectors of both lipid and protein are emphasized by thickened lines.

brane is dispersed by detergents into its solubilized components, whether lipid or protein, and then the detergent is removed, these components can spontaneously reform a membrane continuum. This capability for spontaneous reformation of membranes accounts for the extraordinary resistance of biological membranes to rupture and perforation.

The polar sector of bimodal lipids usually contains glycerophosphate linked by an ester group to choline, ethanolamine, or inositol (Fig. 2). There are also bimodal lipids such as the gangliosides and cerebrosides that are not phospholipids. Depending upon the cell, the lipid composition of biological membranes is highly variable, not only in the nature of the polar sector but also in the number of carbon atoms in the fatty chain, and in the number and positions of the double bonds. Each cell typed has a distinctive lipid mix, and the properties of a

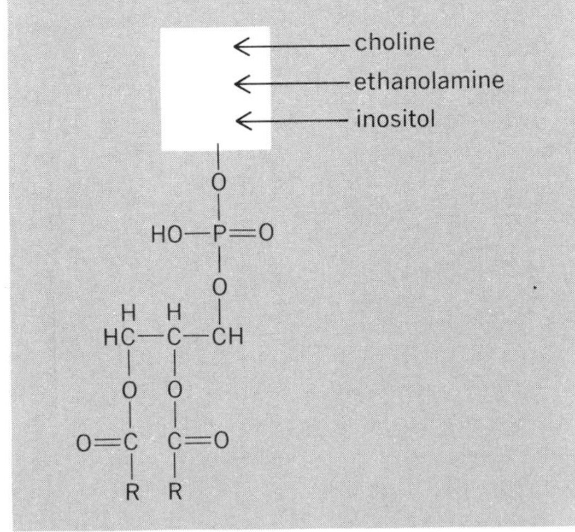

Fig. 2. Common structure of three common phospholipids (phosphatidyl choline, phosphatidyl ethanolamine, and phosphatidyl inositol). RCOO is a fatty acyl residue; R is a linear chain of 17 carbon atoms.

membrane are profoundly affected by its lipid composition.

Plasticity. To appreciate the remarkable plasticity of biological membranes, it is instructive to examine with a light microscope the gyrations of the membrane of free-swimming fibroblasts. The membrane throws out long streamers and then these streamers are retracted. The cells undergo compression to a fraction of their original diameter and then extension to many times their original length. All these changes take place with great rapidity and many repetitions, and without compromise of the integrity of the membranes.

In the presence of Ca^{2+}, two cells can fuse to form a single hybrid cell without any separating membrane. The membranes of the original two component cells coalesce to form a single continuous membrane. During this maneuver, there is no loss of internal contents from either cell. Another characteristic maneuver is the extension of two separated folds of the membrane which next coalesce to form a vesicle that can be brought interiorly through the membrane and then pinched off and left as a vesicle inside the cell. This maneuver makes it possible to encapsulate droplets of the external aqueous phase and bring these droplets into the interior of the cell. When viruses invade cells, they utilize this encapsulating property of the plasma membrane to inject their nucleic acid into the cell interior.

Enzymic and energy-coupling capabilities. Cell activities require a considerable number of molecules and ions to be brought into the interior from the outside, and a comparable number of the products of chemical change to be released into the aqueous phase on the exterior side of the membrane. This inward and outward flux of dissolved substances is mediated by transport devices that are protein in nature. In the absence of such devices the membrane acts as an impenetrable barrier to all but a few relatively small molecules. These transport devices are usually specific for a particular molecule or ion. In facilitated transport an equilibrium is established between the concentration inside and outside the cell of a particular molecule. In energized transport the transported species is pumped inward or outward and, in principle, can be concentrated exclusively on one side of the membrane. The energization of transport requires a set of specialized enzymes and a molecular strategy (energy coupling) by which the transport process is driven.

Classical enzyme catalysis takes place exclusively in an aqueous medium and involves the rupture of two bonds and the formation of two new bonds. Apart from the chemical change accomplished by this cycle of paired bond rupture and bond formation, no other process takes place. Transport enzymes by contrast carry out their catalysis in the precincts of a membrane continuum, and not only do they catalyze paired bond rupture and formation but also they mediate the transmembrane movement of the transported species. Long-range movement of

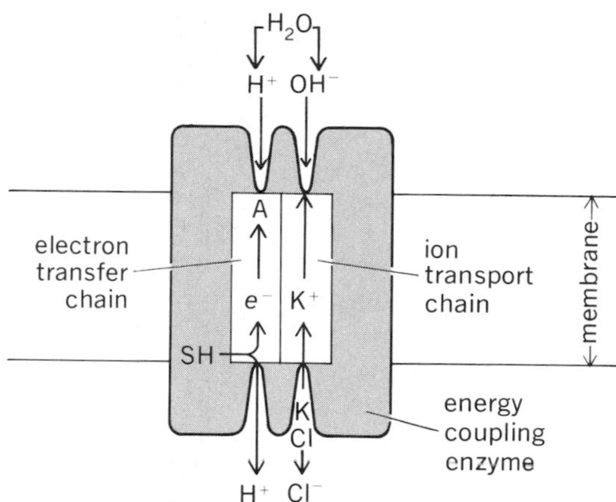

Fig. 3. Paired charge flow of the electron (e^-) and the cation (K^+) during coupling of the oxidation of hydrogen donor (SH) by hydrogen acceptor (A) to the transport of K^+ in an energy-coupling complex in the membrane. A is reduced to AH; SH is oxidized to S; K^+ recombines with OH^- to form KOH. The indentations leading to the electron transfer chain and ion transport chain on the two sides of the membrane represent aqueous channels.

the reacting species is the new property of membrane-centered enzymes, and this property depends upon a unique molecular strategy for executing enzymic catalysis.

Everyone is familiar with macromachines found in the home, driven by electricity, that execute work performances such as radio-wave reception, homogenization, refrigeration, and illumination. Membrane enzymes are micromachines, driven by the biological equivalent of electricity, that execute work performances such as transport, heat production, and synthesis of ATP from ADP and P_i. The technique by which micromachines in the membrane continuum drive any of these work performances is referred to as energy coupling. The unit of energy coupling is a transmembrane protein complex or a close association of two such complexes. In each unit two coupled sequences take place very close to each other—both beginning at one side of the membrane and terminating at the other side. Sequences which span the membrane are referred to as vectorial sequences. There is always a driving and a driven sequence in each energy-coupling system. The free-energy change in the driving sequence pays the energy bill for the driven sequence. When a molecule or ion is pumped across a membrane against a concentration gradient, energy is required to form and expand such concentration gradients.

In electrically driven machines the motive force is provided by the electron. The electrons enter the machine at a high energy level and leave the machine at a low energy level. The work performance of the machine is at the expense of the energy level of the electron. In transport energy–coupling enzymes the situation is analogous. In the driving sequence a moving electron or a moving negatively charged species is the motive force. The driven sequence involves a moving positively charged species. Coulombic electrostatic force compels the coupling of the movement of the positively charged species to the movement of the negatively charged species (Fig. 3). By this technique of coulombic coupling, it becomes possible to force the transport of positively charged species either into or out of the cell.

Any one of three sequences can drive transport. These are, respectively, an oxidative sequence such as the oxidation of *D*-lactate by molecular oxygen, a sequence involving pyrophosphorolysis of ATP to ADP and P_i, and finally a photochemical oxidative sequence in which light-activated chlorophyll or rhodopsin is the reductant. The oxidative and photooxidative sequence generates an electron as the driving charged species; ATP pyrophosphorolysis generates a negatively charged species (inorganic phosphate stripped of a proton) as the driving ion.

Movement of charged species across membranes is mediated by molecular devices such as an electron transfer chain for electrons and an ionophoric pore for cations. The essence of such devices is that they provide a mechanism whereby a charged species can move in discrete steps from one reaction site to another in an array of sites that spans the membrane (Fig. 4).

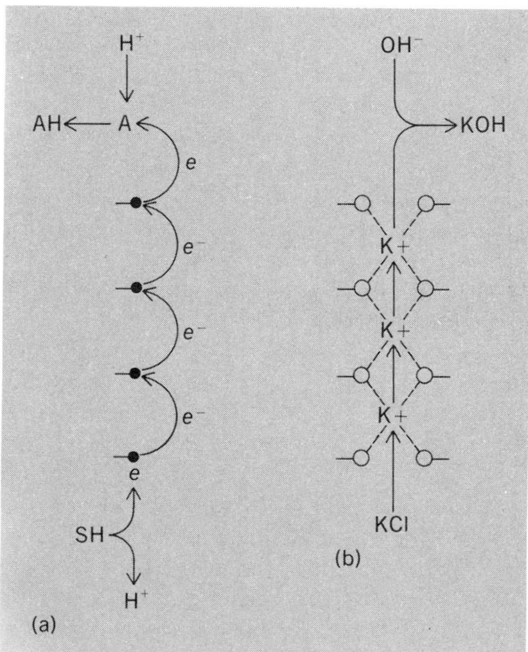

Fig.4. Diagrammatic representation of (*a*) an electron transfer chain and (*b*) an ion transport channel. The residues in the electron transfer chain that react with the electron are metal-containing residues. The electron can hop or tunnel from residue to residue. The residues in the ion transport channel are usually carbonyl oxygens that can form a coordination bond with the cation.

Reception, transmission, and response to external stimuli. Energized transport of cations can generate large ion gradients. For example, the Na^+/K^+ ATPase can build up a high internal K^+ gradient and a high external Na^+ gradient. These gradients are a source of osmotic energy which can be transduced into coulombic energy. Given an ionophoric channel allowing only passage of Na^+ across the membrane, a Na^+ gradient can drive the separation of Na^+ from its cation, with Na^+ moving across the membrane and the anion remaining behind on the other side. This charge separation generates a membrane potential which at equilibrium is equivalent in free energy to the osmotic energy of the Na^+ gradient. In the nervous system of vertebrates a membrane potential developed in one nerve cell can propagate an impulse that will move from one nerve cell to the next and finally to the brain.

Neurotransmitters such as acetylcholine and various amines can actuate gradient-driven channels to generate a membrane potential that can propagate a nerve impulse. Similarly stimuli such as infrared radiation or visible light or even pressure can trigger the action of sensory receptors that can generate a gradient-dependent membrane potential and finally a nerve impulse. These receptors are turned on specifically by a unique kind of stimulus.

Hormones in many cases exert their effects by actuating transport enzymes in a latent form. For example, insulin turns on glucose transport in susceptible cells, and in this fashion hormones can initiate a major metabolic sequence that requires the availability of glucose in the interior of the cell.

Membrane enzymes. Nature has selected the membrane as the medium in which to carry out some of the most fundamental enzymic processes of living cells—processes such as oxidative and photosynthetic phosphorylation, energized transport, and the formation of ion gradients. The membrane enzymes that execute these processes are variations of classical enzymes adapted for energy coupling. The molecular strategy of paired vectorial sequences requires specialized structural devices in the membrane such as electron transfer chains, ionophoric channels, and transmembrane multiprotein complexes. The membrane thus has to be conceived of not only as a protective envelope for the cell but also as the powerhouse of the cell.

For background information *see* CELL (BIOLOGY); CELL MEMBRANES; ENZYME; ION TRANSPORT in the McGraw-Hill Encyclopedia of Science and Technology. [DAVID E. GREEN]

Bibliography: D. E. Green, *J. Theor. Biol.*, 62:271–285, 1976; D. E. Green and H. Vande Zande, *Proc. Nat. Acad. Sci. USA*, 78:5344–5347, 1981; G. Vanderkooi, *Ann. N.Y. Acad. Sci.*, 195:6–15, 1972.

Cement

The characteristic property of a hydraulic cement is that it sets and hardens by chemical reaction with water. Based on readily available mineral raw materials, and relatively cheap to manufacture, cement is widely used in the construction industry. For the most part, it is incorporated in mortars and concretes to provide the adhesive filler between sand and crushed rock or graded pebble aggregates. In these terms, cement is recognized as a very convenient and versatile bulk structural material. Interest is growing, however, in the possibility of exploiting the advantages of cement in more diverse and unconventional ways. In particular, efforts are being made to improve the mechanical properties so that products based on cement can be developed as viable alternatives to other common materials which are most costly in economic or ecological terms.

Relation to other materials. Current world production of cement approaches 10^9 metric tons per year, and of this more than 95% is portland cement, made by calcining limestone or chalk and clay. This figure very substantially exceeds that of any other structural material, including steel. High-alumina cement, made from limestone and bauxite, is more expensive than portland cement and has more restricted and specialized applications.

As a basis for comparison, the table lists some physical properties of a number of common materials together with their energy cost to manufacture. With current emphasis on energy conservation, the cost advantage of portland cement over the materials is clear. When cement is used, there are additional advantages in that the paste can be easily cast or shaped prior to setting and that it hardens (albeit rather slowly) at ordinary temperatures.

Some cements, in particular high-alumina cement, can be used for refractory purposes since they form a ceramic bond when fired at high temperatures. A hardened portland cement, however, cannot sustain high temperatures without dehydration and consequent degradation in strength. Thus, although nonflammable and nontoxic, portland cement is restricted to fairly low-temperature uses.

The main limitation of portland cement, and indeed of hydraulic cements in general as structural materials, lies in their poor mechanical properties. The potential advantages of a relatively low density and reasonably high value for Young's modulus (a measure of the rigidity) are never properly realized because cement as normally made is very weak and brittle. Although a hardened cement can support quite appreciable loads under compression (typically about 70 meganewtons per square meter), it has a very low tensile strength coupled with a very low work of fracture (a measure of the fracture toughness). It is recognized by engineers that allowances have to be made for these mechanical deficiencies. The reinforcing and prestressing of structural concrete members with steel bars is normal practice, and the engineering design of buildings has to be such that cement or concrete sections are not subjected to any appreciable tensile stress.

Since strength and toughness are two parameters that often determine the range of use of a material, the development of improved cements has focused

Comparison of the physical properties of common materials, with their energy cost to manufacture

Materials	Density, g/cm^3	Young's modulus GN·m^{-2}	Tensile strength, MN·m^{-2}	Work of fracture, J·m^{-2}	Energy cost to manufacture*
Portland cement	2.4	20	5–10	10	1
Ordinary glass	2.5	70	70	10	1.3
Aluminium + alloys	2.8	70	150–600	10^4	32
Cast iron	7.3	150	150–350	10^3–10^4	17
Steel (plain carbon)	7.8	200	400–1000	10^4–10^5	20
Polystyrene	1.1	3	50–120	10^3	6
Wood	0.7	10	10–100	10^3	—

*Figures given per unit volume, relative to cost of cement.

in this area. The problem is that strength and toughness do not necessarily go hand in hand. As experience shows, a high-strength material can also be very brittle, and vice versa. It is not surprising to find, therefore, that two basic methods have been followed to improve the mechanical properties. The first, more fundamental approach involves an understanding of the chemical processes during hydration that are responsible for hardening, and a recognition of the physical sources of weakness in the microstructure. This is an effort to improve the inherent strength of the material. The second approach which is essentially remedial, involves making cement composites with a reinforcing phase. In principle, mortars and concretes are cement composites, but in this context one refers to composites made mainly with fiber reinforcement.

Hardening and sources of weakness. When cement and water are mixed, a complex series of chemical reactions occur which result in the precipitation of insoluble hydration products (Fig. 1). These initially coat the cement grains to provide a loose coagulational network bridging the cement grains at their points of contact—thus leading to a loss of plasticity of the cement paste (setting). Secondary precipitation reactions then occur which progressively intrude into the interstitial water-filled spaces between the cement grains to give a cohesive matrix binding the paste into a rigid solid (hardening). In the case of portland cement, which consists mainly of reactive calcium silicates, about 70% by volume of the hydration products consist of a colloidal calcium silicate hydrate gel and about 20% crystalline calcium hydroxide, the residue being mainly aluminate hydrates. In high-alumina cement, which hardens more rapidly, the main products are crystalline calcium aluminate hydrates, although appreciable quantities of gelatinous hydrated alumina are also present.

In both cases, the cohesive strength of the ce-

(a)

(b)

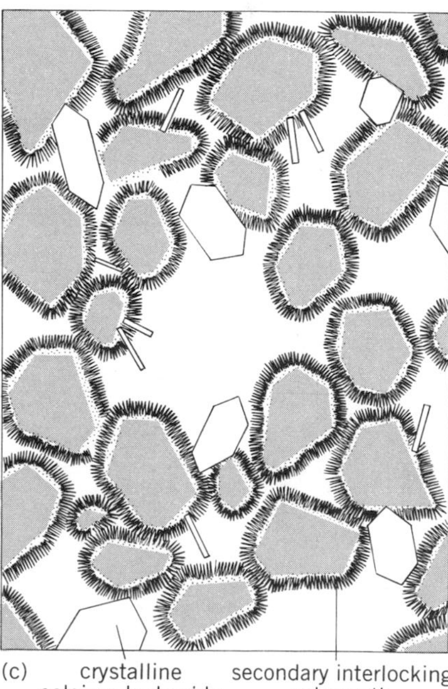

(c)

Fig. 1. Schematic representation of hydration of portland cement. (*a*) Cement grains in water. (*b*) Setting; surface coatings of gel/microcrystalline hydration products bridge the cement grains at their points of contact. (*c*) Hardening; interlocking microstructure formed by secondary growth of calcium silicate hydrate gel around surfaces of cement grains and crystallization of calcium hydroxide.

ment derives from surface and chemical bonds formed between the residual cement grains and the hydrate particles (colloidal and crystalline) and also by mechanical bonds formed simply by physical interlocking of the microstructural constituents. Over longer periods during aging, the gelatinous products of hydration tend to polymerize or crystallize to provide further consolidation of the structure.

The main source of weakness in a hydrated cement is the residual porosity in the microstructure. Cement grains are irregular in shape and size (typically ranging from 10 to 70 μm) and do not pack efficiently when mixed into a paste with water. Although the hydration reactions produce a net increase in volume of solids (by a factor of about 2 in the case of portland cement), the hydration products do not effectively fill the interstitial spaces between the cement grains. A hardened cement paste typically has a residual porosity of about 20%.

It has long been recognized that there is a direct correlation between the total porosity and the final strength of a hydrated cement. For this reason, in mixing a cement paste, it is desirable to keep the water/cement ratio as low as possible. The limitation of loss of workability can be overcome to some extent by use of rheological aids. Commercially, these are organic-based water reducing or plasticizing additives (lignosulfonates and hydroxy-carboxylic acids, for example) which function by reducing the surface tension or viscosity of the paste.

In detail, however, the situation is more complex than this. Recent developments have shown that the main limitation in strength is not so much the total porosity but the pore-size distribution. This conclusion follows from classical theory relating to tensile failure in homogeneous brittle materials such as glass (Griffith theory). This predicts an inverse relationship between the tensile failure stress (σ) and the size (c) of the largest flaw in the microstructure

$$\sigma \propto \sqrt{\frac{\gamma . E}{c}}$$

where γ is fracture surface energy and E is Young's modulus of elasticity. Although there is some uncertainty about the extent to which this idealized relationship applies in the case of such an inhomogeneous material as cement, the general principles are clear. In cements the principal sources of weakness and the flaws providing sites for the easiest propagation of cracks are the largest pores in the microstructure. As normally made, a hydrated cement has pore sizes ranging from about 10^{-3} to 10^{-8} m. Practice has shown that if the largest pore sizes can be eliminated, very substantial increases in strength can be achieved—as the Griffith equation implies. The microstructural modification thus need not necessarily involve a reduction in total porosity, but may involve a shift in the porosity distribution to the smaller size range.

High-strength cements. Successful methods of improving the strength of cement generally involve a reduction in the microstructural porosity. Vibrocompaction of the paste prior to setting has proved advantageous. Compaction under high pressure of cements made with a minimal amount of water (so-called cement compacts) have produced increases in strength by a factor of about 5. Even higher strengths have been achieved by autoclaving (steam curing under pressure), a process which not only serves to compact the cement paste but also, by virtue of the high temperatures involved, increases the rate of chemical reaction and produces more minerallike (and possibly intrinsically stronger) hydration products. From a commercial point of view, however, autoclaving is an elaborate and relatively costly process. Ideally, a lower temperature route for improving the strength is required. In this context, the most successful example to date is macrodefect free (MDF) cement. By using a patented method involving physical processing of the cement paste and chemical modification with rheological aids, the pore-size distribution has been reduced to less than 100 μm, and resulting flexural strengths in the hardened product have been in excess of 60 MN · m^{-2}. These strengths, which are generally higher than those achieved by autoclaving methods, represent an increase by about an order of magnitude over those of normally made cements.

Cement composites. Most commercial interest in improved cement materials has concentrated on the method of reinforcement with fiber admixtures. Asbestos cement, now generally regarded as a potential health hazard, is one of the earliest examples. More recently the use of glass-fiber reinforcement,

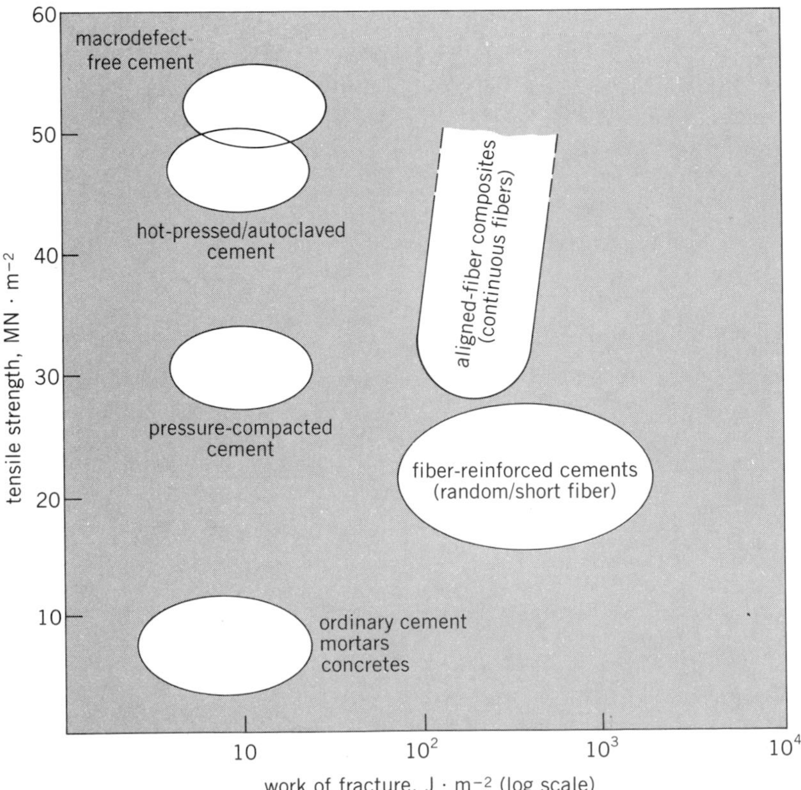

Fig. 2. Tensile strength and fracture toughness of various improved cement-based materials.

originally developed in the Soviet Union, has been tested in the United Kingdom and commercially exploited as Cem-FIL. There has also been interest in the United States and the United Kingdom in polypropylene fibers. In general, many countries have investigated the use of various fibers, including steel, alumina, nylon, carbon, and cellulose-based materials. Methods of making the composites have included pressing, injection molding, and estrusion of the premixed constituents as well as spray-coating and impregnation of the fiber mat with cement. Considerations of rheology and flow characteristics of the mixed material, as well as the longer-term possibility of corrosion of the fibers in the highly alkaline cement matrix (particularly in the case of steel and glass fibers), have introduced practical problems. The question of cost has also been a very relevant factor.

The mechanical properties developed in these composites depend on the type and volume fraction of the fibers and on the way in which they are incorporated (random short lengths, aligned fiber bundles, interwoven mesh, and so on). The main advantage of fiber reinforcement has been in the very substantial improvements that have been achieved in the fracture toughness, in some cases by about two orders of magnitude (Fig. 2). An addition of as little as 2% steel fibers can increase the work of fracture to 500 joules$\cdot$m^{-2}. In glass- and carbon-fiber composites with about 3% by volume of fibers, works of fracture of about 10^3 J$\cdot$m^{-2} are possible.

Some very high tensile strengths (in excess of 200 MN$\cdot$m^{-2}) have been achieved in aligned composites using high-volume fractions and continuous lengths of fiber tested in tension parallel to the fiber axis. In general, however, the tensile and flexural strengths of random- or short-length fiber composites have not shown substantial increases. In commercial glass-fiber cements, for example, ultimate tensile strengths are of the order 20 MN$\cdot$m^{-2}. The basic problem with these composites lies in the mechanical deficiency of the cement matrix. An effective fiber composite can be achieved only if the matrix is capable of transferring appreciable load to the fibers and, from premature tensile cracking and debonding, cement has not proved adequate. The problem of premature cracking is one that relates to the inherent strength of cement. The problem of debonding can be mitigated by surface roughening or twisting of the fibers or by use of meshlike arrangements. Some limited chemical reaction between the cement and the fiber is desirable for adhesion, but plastics for the most part are chemically inert. Even after catastrophic failure of the cement matrix, it can still be an advantage that the fibers provide some cohesion and postcracking ductility to the composite.

For background information *see* CEMENT; COMPOSITE MATERIAL; PORTLAND CEMENT in the McGraw-Hill Encyclopedia of Science and Technology.

[D. D. DOUBLE; N. McN. ALFORD]

Bibliography: J. D. Birchall, A. J. Howard, and K. Kendall, *Nature* 289:388, 1981; D. J. Hannant, *Fibre Cements and Fibre Composites*, 1978.

Cephalopoda

The cephalopod orders of squids, octopuses, and cuttlefishes are well known in coastal waters, but the great diversity of the squids, in particular, can be seen only when deep-water varieties are captured. Within the great mass of water exceeding 200 m in depth which constitutes 92% of the total area of the oceans and 72% of the Earth's surface, there exists species that are unknown and many more which have only recently been recognized as important in the deep-sea world. Because of the inaccessibility of that world and the particular abilities of squids, somewhat bizarre methods must be employed to study them. While bottom trawls, seines, and line fishing with special hooks or jigs are adequate to supply commerce with increasingly lucrative cephalopod products from the continental shelves, such devices are poor tools for catching the sharp-eyed, fast-moving squids in the crystal-clear water of the deep ocean, where their movement is unrestricted by either bottom or sea surface as in shallow seas. Technology has not advanced sufficiently to catch more than the young, the small, and, occasionally, the infirm squids in the deep ocean. However, many marine mammals, seabirds, and oceanic fishes live entirely on cephalopods and many more include them in their diet. There is much to be learned from predators' capture techniques but, in the meantime, a great deal can be discovered about biology of squids from the contents of predators' stomachs. This article will outline some conclusions which are possible from a study of cephalopod, mainly squid, remains from the stomachs of sperm whales.

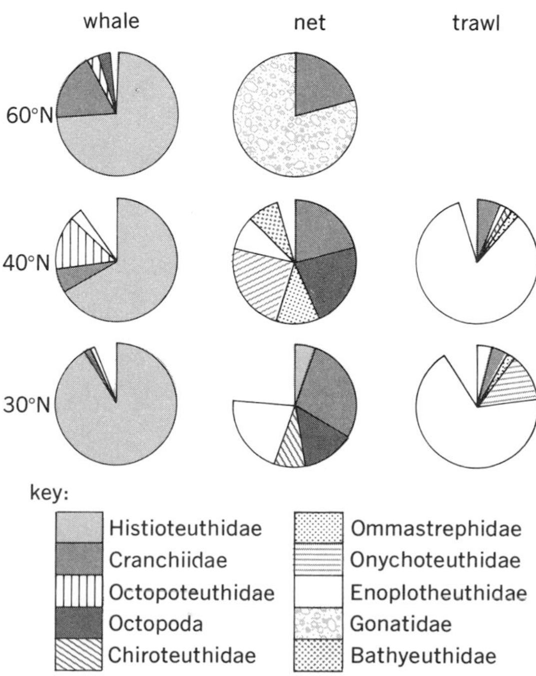

Fig. 1. Families of oceanic cephalopods (mainly squid) caught by sperm whales, research nets (9-m^2 mouth), and commercial trawls in three areas of the North Atlantic.

Distribution of families and species. Sperm whale stomachs usually contain numerous horny squid jaws or beaks and some flesh which may be anything from small pieces attached to beaks to complete heads or bodies; only rarely are completely intact squids collected. The lower beaks are often characteristic in shape, structure, and size for a species, genus, and family, and patient collection of beaks from intact or partially intact squids over many years has enabled identification of the majority of the genera and many of the species of squids represented by beaks in stomachs of sperm whales. Providing they can be identified, lower beaks are specially useful for providing biological information about squids since up to 8000 (an average of 2000) can accumulate in a whale's stomach before they are got rid of by vomiting.

From the lower beaks it has been found that the majority of species eaten by sperm whales are never, or only rarely, caught by underwater samplers such as research nets or jigs. If a comparison is made between the cephalopod families caught by sperm whales and by nets in the same region, it is found that the same families are present but in very different proportions. For example, in the North Atlantic over 70% of the squids eaten by whales are histioteuthids (Figs. 1 and 2), research nets up to 9 m^2 in mouth area catch mainly gonatids in high latitudes, cranchiids in all latitudes, and several other families including octopods in lower latitudes, while the large commercial trawls, used in deep water, catch mainly small enoploteuthids.

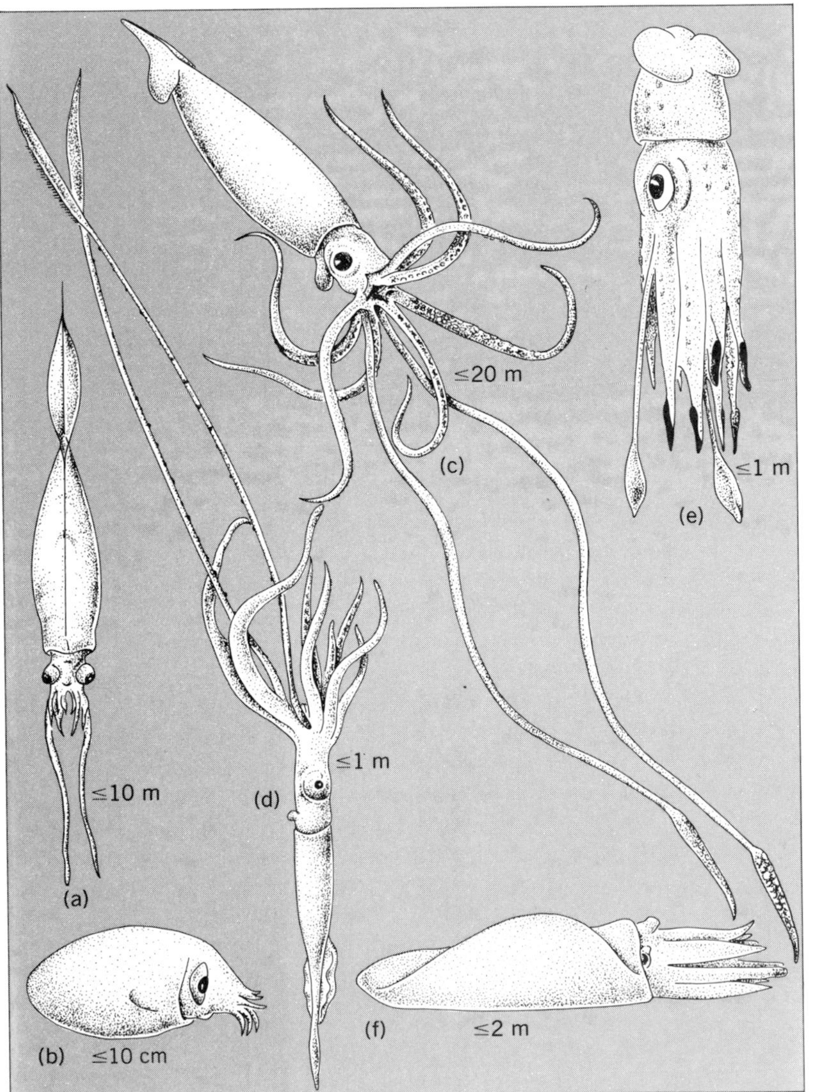

Fig. 2. Selection of deep-water cephalopods. (*a*) A cranchiid. (*b*) An octopod. (*c*) *Architeuthis.* (*d*) *Chiroteuthis.* (*e*) *Histioteuthis.* (*f*) An octopoteuthid, *Taningia danae.* Maximum sizes are indicated but are approximate.

By noting when flesh is attached to beaks, the broad distribution of squid species can be established because the available evidence shows that flesh is digested rapidly and the whales cannot move many miles before the flesh passes from their stomachs. By counting the lower beaks of species known from flesh remains, the relative abundance of local species in the whale's diet can be found for the regions investigated (Fig. 3*a*). In northern temperate regions and Iceland, histioteuthids are very dominant (comprising 30–91% of the squids eaten) except for the Vancouver Island region, where they form a negligible part (5%) of the diet. In most temperate regions of the Southern Hemisphere the octopoteuthids (Fig. 2) are also well represented (10–33%; Fig. 3). The Vancouver region of the North Pacific is unusual in having large proportions of gonatids (32%), also common in high-latitude North Atlantic research nets (Fig. 1); onychoteuthids (24%); and cranchiids (26%). The last two families are the main groups in the Antarctic Ocean samples, where they constitute 53% and 23% of the diet respectively. West of South America samples differ from those of all other regions by having appreciable numbers (16%) of the chiroteuthids (Fig. 2). While these broad family differences indicate large differences between the squid fauna of some oceans, smaller species differences indicate changes over smaller geographical distances. While some species (such as *Histioteuthis bonnellii*) span over 10,000 mi (16,000 km), others seem to be much more restricted and, at present, there is little idea of the environmental barriers limiting them.

Local distribution relative to the continental slope has been studied off South Africa by plotting the abundance of flesh remains according to bathymetric depth, and this shows that some species prefer the upper slope to more stable oceanic waters. Collections spanning the Tasman Sea show that many squids occur far from continental slopes and are not limited to slope areas where upwelling often occurs and sperm whales tend to congregate.

The numbers of squids represented by flesh in stomachs distinguish species which the whale usually catches singly (such as *Taningia danae*) from those which are caught in groups (such as *H. bonnellii*).

Size and age of species. Size distribution of the beaks can be found for each species, and the state

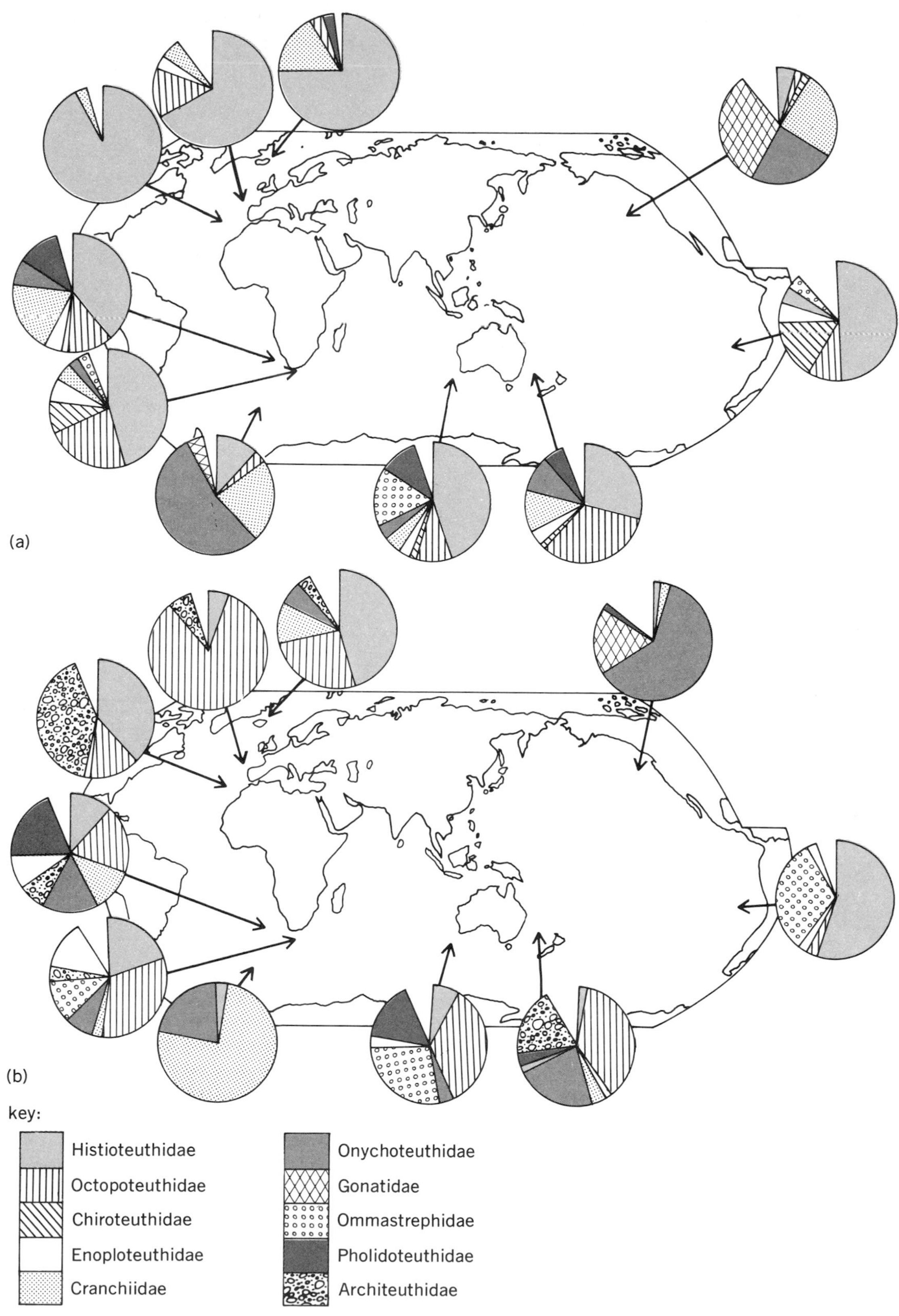

Fig. 3. Squid families in the diet of sperm whales caught in various regions of the world. (*a*) Composition from the number of lower beaks. (*b*) Composition by weight estimated from the numbers and sizes of beaks.

of maturity can be recognized by the extent of darkening of the horn or chitin of the beaks. It is remarkable that for many species the size distributions eaten by whales are closely similar over considerable geographic ranges. For example, 17 species from whales caught from western South Africa to north of New Zealand, a distance of over 6250 mi (10,000 km), show very little variation in size. While the whales eat small species of squid, they often eat rather few small individuals of species growing to a large size. Perhaps this is because the young of large species are not present at the whales' sampling depth or they exhibit some structural feature which makes them unnoticed or unwanted by the whales.

From seasonal changes in the size distributions of beaks the growth of several fast-growing squids has been found to be either 1 year (such as *Cycloteuthis akimushkini*) or 2 years (such as *Kondakovia longimana*). The season of spawning is shown by the disappearance of the largest, old beaks from the samples since, as far as is known, in nearly all species females die at spawning. Maturation processes have been studied for males and females in the commoner species of which fairly intact specimens have been collected. Such specimens have also provided much information on specializations such as buoyancy mechanisms and light organs. Some 53–78% of cephalopods taken by whales in the Southern Hemisphere attain neutral buoyancy by retaining ammonium ions in their tissues or coelom, and 41–97% have light-producing organs. These percentages vary from region to region.

In the most commonly occurring families sufficient complete squids have been collected to relate beak size with body length and total wet weight. Such estimates provide a rough means to find the relative biomass in the sperm whales' diet of each family represented by beaks in each region (Fig. 3*b*). Because histioteuthids are much smaller than the common octopoteuthid *T. danae* (Fig. 2), they are less important by weight than the octopoteuthids off Spain, South Africa, and Australia. In the Tasman Sea and the Atlantic Ocean the architeuthids, the kraken or sea monsters of legend (Fig. 2), are large enough to be important in the diet (up to 40% according to region) even though they are few in number. Other families such as the ommastrephids, enoploteuthids, and pholidoteuthids are also important by weight but not by number in southern temperate seas.

From the size of lower beaks the mean weight of all the squid eaten by whales is about 8 kg in the Antarctic Ocean, 0.6 kg off South Africa, 2.3 kg off western Australia, 3.6 kg off eastern Australia, and 1.3 kg off western South America. Thus, sperm whales do not live exclusively on giant squid as popular belief and whalers' tales would suggest; to the contrary they eat many squids which to a 50-ton sperm whale must be like a herring eating a copepod.

Value as food resource. Marine biologists hope ultimately to assess the available sea food resources and the effects depletion of these resources will have. These researchers move toward these goals by estimating biomass of all the important groups in the sea and by understanding their interactions. To estimate biomass, biologists need to know the numbers and weights of the different types of animals in some volume of water. As nets take only very small cephalopods (from research nets deep-sea squid probably average less than 0.002 kg, and from commercial trawls used in deep water less than 0.1 kg) and sperm whales catch much larger ones (average 0.6–8 kg), it is clear that biologists will have an exceedingly low estimate of biomass of cephalopods if nets and volume of water filtered are used.

Can samples from sperm whales help in this dilemma? From estimates of the sperm whale population, food consumption, and average body weight, the weight of squid consumed per year can be estimated at well over 100,000,000 metric tons. This figure compares with 70,000,000 tons for the total world annual catch of fish and probably approaches the total biomass of humans. This is the weight actually *consumed* by sperm whales and can only be a fraction of the weight of the population supporting the predation. There are also all the other predators of cephalopods to be considered. The sperm whale is the largest which almost exclusively eats squid, and its very large weight and quite large numbers make it a particularly important predator and good example for showing the inadequacies of sampling data. These comparisons raise serious doubts concerning the value of midwater nets to assess relative or absolute numbers and biomass of any large, active, deep-sea midwater animals, including fish. Until squid can be caught which predators show to be ubiquitous, little confidence should be placed in sampling methods.

For background information *see* BIOMASS; CEPHALOPODA; DEEP-SEA FAUNA; MARINE FISHERIES in the McGraw-Hill Encyclopedia of Science and Technology. [MALCOLM R. CLARKE]

Bibliography: M. R. Clarke, *Discovery Report*, vol. 37, pp. 1–324, 1980; M. R. Clarke, *Symp. Zool. Soc. Lond.*, 38:89–126, 1977; M. R. Clarke, E. J. Denton, and J. B. Gilpin-Brown, *J. Mar. Biol. Ass. U.K.*, 59:259–276, 1979.

Cephalosporins

Cephalosporins as antibiotics have been referred to as agents in search of diseases to treat. Many infectious disease experts, particularly those brought up with penicillins, say that there are no diseases for which these agents are the drug of first choice. These drugs are always placed in the second column of any compendium which lists antibacterial agents. This is unfortunate, for the cephalosporins have many of the best attributes of the penicillins and often have fewer of the adverse effects that are associated with the use of the β-lactam antibiotics.

The cephalosporins have been effective in eradicating streptococcal, pneumococcal, staphylococcal, *Klebsiella*, *Neisseria*, and enteric gram-negative rod bacteria that produce pulmonary, skin and soft tissue, bone and joint, endocardial, surgical, urinary, and bacteremic infections. They have been used most often in a preventive or prophylactic fashion at the time of various surgical procedures.

Table 1. Classification of cephalosporins

First generation	Second generation	Third generation
Cefaclor	Cefamandole	Cefmenoxime
Cefadroxil	Cefmetazole	Cefotaxime
Cefatrizine	Cefotiam	Cefoperazone
Cefonicid	Cefoxitin	Cefsulodin
Ceforanide	Cefuroxime	Ceftazidime
Cephalexin		Ceftizoxime
Cephaloridine		Ceftriaxone
Cephalothin		Moxalactam
Cephapirin		
Cephradine		

Activity. Cephalosporins, like the penicillins, have a bactericidal action against both gram-positive and gram-negative bacteria. The cephalosporins of the first generation (cephalothin, cefazolin, cephalexin, cephradine, cephapirin) and the second-generation agents (cefamandole, cefaclor, and cefoxitin) all inhibit gram-positive cocci such as *Streptococcus pyogenes* (group A), *S. pneumoniae* (pneumococci), the *viridans* group of streptococci, and *Staphylococcus aureus* and *S. epidermidis.* These agents do not inhibit the *Streptococcus pneumoniae* resistant to penicillin or the *Staphylococcus aureus* resistant to the semisynthetic penicillins (methicillin-resistant). Cephalosporins do not inhibit the anaerobic gram-positive cocci such as peptostreptococci or the peptococci better than penicillin does.

In the 1960s and early 1970s Enterobacteriaceae were readily inhibited by all the cephalosporins, but some institutions have found in the 1980s that 20% of the *Escherichia coli* and *Klebsiella* are resistant to cephalothin and cefazolin. The older cephalosporins did not inhibit *Enterobacter*, *Serratia*, indole-positive *Proteus*, and *Bacteroides fragilis*. As medicine has made progress, the latter group of bacteria have been rediscovered. Although *Haemophilus influenzae* was inhibited by the older drugs of this class, the activity in laboratory cultures did not correlate well with clinical results. None of the agents had any activity against *Pseudomonas aeruginosa*, which has become an important pathogen.

In the past few years a large number of new agents became available (Table 1). Drugs such as cefoxitin and cefamandole used parenterally have an activity in culture which overcomes some of the aforementioned defects, and some new oral cephalosporins have good anti-*Haemophilus* activity. Cefoxitin has excellent β-lactamase stability and so inhibits many β-lactamase-producing *Escherichia coli*, *Klebsiella*, *Proteus*, and *Bacteroides* organisms which would be resistant to penicillins such as ampicillin, carbenicillin, and cefazolin. Cefamandole inhibits some β-lactamase-positive strains, but it is hydrolyzed by inducible β-lactamases which act primarily as cephalosporinases. Cefotaxime was the first of these new agents to be approved for clinical use by the Food and Drug Administration. These new agents can be classified as aminothiazolyl-oxyimino-acetamido cephalosporins (such as cefotaxime, ceftizoxime, cefmenoxime, ceftriaxone, ceftazidime), or as oxa-cephems (such as moxalactam). Agents such as cefsulodin and cefoperazone fall into neither category.

The culture activity of the new agents is similar in most instances, but exceptions do occur. All the agents are relatively stable to hydrolysis by β-lactamases of both gram-positive and gram-negative bacteria. Moxalactam possessing both a methoxy side chain at position 7 of the ring structure and a carboxy group at position 10 is the most β-lactamase–stable of the new agents. Cefoperazone is hydrolyzed by high concentrations of the most common plasmid β-lactamase, TEM-1 (Table 2).

The overall activity of the new agents is shown in Table 3. It is apparent that most of the Enterobacteriaceae are inhibited by concentrations below 1 μg/ml for cefotaxime, ceftizoxime, cefmenoxime, ceftazidime, cefoperazone, and moxalactam. Cefsulodin in contrast has minimal activity against the Enterobacteriaceae, but it does inhibit *Pseudomonas aeruginosa* and *Staphylococcus aureus*. Al-

Table 2. Stability of new cephalosporins to β-lactamases; relative hydrolysis*

β-Lactamase	Cefamandole	Cefoxitin	Cefotaxime Ceftizoxime	Ceftazidime	Moxalactam	Cefoperazone
TEM-1	75	<1	<1	<1	<1	50
TEM-2	50	<1	<1	1	<1	60
Oxa 1	490	<1	10	5	<1	20
Oxa 2	150	<1	<1	5	<1	80
Oxa 3	140	<1	<1	5	<1	50
P99-*Enterobacter*	118	<1	15	5	<1	1
Klebsiella	130	<1	10	5	<1	5
SHV-1	40	<1	0	<1	<1	70
Pseudo-SA	4	<1	1	<1	<1	3
Staphylococcus aureus	50	<1	0	0	<1	30
PSE-1	100	25	0	0	0	15
PSE-2	160	30	20	30	10	160
PSE-3	250	<1	15	10	10	230
Bacteroides fragilis	160	<1	10	1	<1	0

*Based on rate of 100% for cephaloridine. Comparisons can be made only across the table.

Table 3. Activity of new cephalosporin agents against various bacteria, range of MIC 90%, μg/ml

Organism	Cefotaxime	Ceftizoxime	Cefmenoxine	Ceftazidime	Ceftriaxone	Cefoperazone	Moxalactam	Cefsulodin
Staphylococcus aureus	1.6–4	1–4	1–4	4–16	3.1–6	1.6–4	6–8	1.6–4
Haemophilus influenzae	0.008–0.20	0.008–0.1	0.1–0.03	0.02–0.1	0.004–0.1	0–0.8	0.03–0.3	
Neisseria gonorrhoeae	0.005–0.1	0.007–0.4	0.01–0.03	0.02–0.1	0.01–0.03	0.05–0.8	0.05–0.3	
Escherichia coli	0.02–2	0.04–0.25	0.1–0.5	0.4–8	0.1–1.6	1–50	0.1–0.8	
Klebsiella pneumoniae	0.1–0.4	0.05–0.1	0.2–1	0.25–2	0.1	1–128	0.4–0.8	
Enterobacter	0.5–25	0.12–0.8	0.5–6.3	0.25–50	1.6–50	0.5–>100	0.5–25	
*Proteus, Providencia**	0.06–8	0.1–6.3	0.2–2	0.1–2	0.1–3.1	0.1–50	0.06–12.5	
Serratia marcescens	0.4–25	0.1–12	1–8	0.5–8	0.5–50	8–>100	8–50	
Pseudomonas aeruginosa	16–>100	16–>100	16–>100	4–25	32–>100	16–100	16–>100	16–>100
Bacteroides fragilis	100–>100	16–>100	16–>100	>100	>100	50–>100	6–100	>100

**Providencia vulgaris, Morganella, P. rettgeri, P. stuarti, P. inconstans.*

though many of these agents inhibit *P. aeruginosa* isolates, only cefoperazone, cefsulodin, moxalactam, and ceftazidime could be considered consistently active agents against *Pseudomonas*.

Activity of the newest agents against the anaerobic bacteria is variable. Only moxalactam is stable to all of the *Bacteroides* β-lactamases, but its activity is only minimally better than that of cefoxitin.

Although the new agents have increased gram-negative activity, they all have less activity against staphylococci than does cephalothin. However, the activity of some of the agents against the streptococci and *Streptococcus pneumoniae* remains excellent and surpasses on a weight basis that of cephalothin or even penicillin. None of the new agents is an effective antienterococcal agent; *S. faecalis* organisms are not inhibited.

In analyzing the activity of the new agents individually (Table 3), it is apparent that in each group of bacteria some organisms are resistant to one or another of the new agents. Local differences in susceptibility are due to the presence of certain plasmids and to local use of antimicrobial agents. Table 4 gives an overview of what the new agents offer against clinically relevant bacteria.

Pharmacology. The pharmacology of the early cephalosporins was such that the agents had to be administered by vein in frequent intervals in serious infections. This problem has been overcome since most of the new agents have much longer half-lives (Table 5). Cefotaxime which has an acetoxy side chain, undergoes metabolism in the body to a less active desacetyl derivative. This derivative is also converted in urine to totally inactive lactones and other by-products. Some metabolism of other agents occurs but is of no major significance.

Toxicity. All the new cephalosporins penetrate well into tissues, and antibacterially active levels in various body fluids (pleural, peritoneal, synovial, biliary) and tissues such as bone are excellent. the toxic potential of the agents, considering their broad antibacterial spectrum, has been minor, with little hematologic, hepatic, gastrointestinal, renal, or nervous system damage.

Toxicities which are seen with the new cephalosporins are those of bleeding due to vitamin K depletion, which has been encountered most often with moxalactam, cefoperazone, and cefmetazole. These agents, which possess a similar nucleus at position 3 of the dihydrothiazine ring, also cause antabuse reaction.

Treatment of upper respiratory and oral infections. Streptococcal pharyngitis and tonsillitis are still best treated with penicillin. Whether the new oral cephalosporin, cefaclor, is a superior agent to treat sinusitis and otitis media will have to be established. Serious life-threatening infections, such as epiglottitis due to *Haemophilus influenzae*, would still be best treated with parenteral chloramphenicol. The newer agents such as cefotaxime, moxalactam, ceftizoxime, cefsulodin, cefoperazone, ceftriaxone, cefmenoxime, and ceftazidime will probably not be a major help in these infections. They will be superb agents, but they will still be second choice, unless more resistance to chloramphenicol developes.

Treatment of lower respiratory infections. Cephalosporins have always been used as a form of treatment of pneumonitis due to *Streptococcus pneumonia*. All the cephalosporins have good track records

Table 4. Activity of new agents against major classes of bacteria compared to all agents

Bacteria	Comparison
Staphylococcus aureus	No advance
S. epidermidis	No advance
Streptococci	No advance
S. faecalis	No activity
S. pneumoniae	No advance
Haemophilus influenzae	Improved
Neisseria meningitidis	No advance
N. gonorrhoeae	Improved
Escherichia coli	Improved
Klebsiella	Improved
Enterobacter	Improved
Proteus mirabilis	Improved
Proteus, indole-positive	Improved
Citrobacter	Improved
Serratia	Improved
Providencia	Improved
Acinetobacter	No advance
Pseudomonas aeruginosa	Improved
Clostridia	No advance
Bacteroides fragilis	No advance

Table 5. Pharmacology of the new cephalosporins

	Half-life, hours	Excreted via	Metabolized*	Serum level,† μg/ml
Cefotaxime	1	Renal-tubular	Yes	100
Cefotaxime (desacetyl derivative)	1.6	Renal-tubular	Yes	—
Ceftizoxime	1.6	Renal-tubular	No	100
Cefmenoxine	1.6	Renal-tubular	No	100
Ceftazidine	1.4–1.8	Renal-glomerular filtration	No	120
Ceftriaxone	6–8	Liver, renal	No	150
Cefoperazone	2	Liver, renal	No	125
Moxalactam	2.2	Renal-glomerular filtration	No	100
Cefsulodin	1	Renal-tubular	No	70

*Metabolism considered only if > 15% of original compound.
†After 1 gram intravenously.

with uncomplicated pneumonia, but they are not any better than procaine penicillin. The role of cephalosporins in treatment of acute infectious exacerbations of bronchitis which are due to *S. pneumoniae* and *H. influenzae* also is not very impressive for either the oral or parenteral agents. They are effective, but they are not better than ampicillin or tetracycline and probably not as useful as trimethoprimsulfamethoxazole in difficult cases.

Cephalosporins are useful to treat *Klebsiella* pulmonary infections. The third-generation β-lactams (cefotaxime, ceftizoxime and cefoperazone, and moxalactam) inhibit *Klebsiella* in the test tube at concentrations below 1 μ/ml, and it is easy to achieve blood levels 100 times that. These agents will cure serious gram-negative pneumonitis due to *Klebsiella*, *Escherichia coli*, *Proteus*, or *Providencia* that was caused by strains resistant to the older cephalosporins, but mortality will still be high because the people who get these infections are poor risks. The effectiveness of agents such as cefoperazone, cefsulodin, and ceftaxidime in cases of *Pseudomonas* pneumonitis is difficult to ascertain since the patient often has so much structural lung disease that he or she does poorly regardless of the antimicrobial regimen.

Treatment of abdominal infections. Although the original cephalosporins have been widely used to treat peritoneal infection which follows perforation of abdominal viscus by trauma or at surgery, they really should not be used alone. The results with cefoxitin therapy as a single agent in such infections are good, and results equal to those achieved with two agents—one an aminoglycoside and the other an antianaerobic agent—have been reported. But even here one must be cautious in treating the late postoperative infection since *Pseudomonas* may be present. Ironically the best new anti-*Pseudomonas* cephalosporins have some antibacterial weaknesses against some of the *Bacteroides fragilis*, which are high producers of β-lactamase; so thus it may be necessary to use them with another antibiotic. It certainly will be of interest to see whether the agents still in the test tube, such as thienamycin-type agents, which readily inhibit *Staphylococcus aureus*, enterococci, *Escherichia coli*, *Klebsiella*, and *Pseudomonas*, will be effective as combined therapy.

Treatment of urinary tract infections. Cephalosporins have been used to treat urinary tract infections since they inhibit the common infecting organisms and provide high urinary concentrations of the antibiotic in the bladder. But almost any agent will cure a lower urinary tract infection if the organism is susceptible. Pyelonephritis is the more complex problem. Treatment for 7 to 10 days with these agents may cure only 50% of patients; urine cultures taken 1 to 4 weeks after therapy reveal the same organism. Cephalosporins can be used to treat a urinary infection because these new agents, particularly those with broad activity in culture, are an alternative to use when hospital-acquired upper tract disease is encountered and aminoglycoside would otherwise have been used. Therapy can be completed in 4 to 6 weeks with an oral agent or one of the long-acting agents for outpatients.

Treatment of gynecological infections. Cephalosporins did not play a major role in gynecological infections prior to the introduction of cefoxitin. Considering the bacterial flora involved in infection in the pelvis, the agents under investigation in the United States probably will not be any more useful than cefoxitin in this type of infection. The new cephalosporins should probably be utilized rather than the high-dose penicillin plus another drug—tetracycline or chloramphenicol or an aminoglycoside—which is common today.

Treatment of septicemia. Many cases of septicemia due to gram-positive cocci have responded to therapy with cephalosporins. Septicemia due to susceptible *Escherichia coli*, *Klebsiella*, and *Proteus mirabilis* has also been effectively managed with cephalosporins. However, an aminoglycoside is commonly used in combination with a cephalosporin when therapy has been initiated for sepsis of unknown origin. The extensive antibacterial spectrum of the newer agents under investigation should permit the use of a single agent, if the patient is not neutropenic.

Cephalosporins have proved useful in the treatment of endocarditis, but the drugs are not superior to the semisynthetic penicillins in the treatment of *Staphylococcus aureus* endocarditis. Vancomycin would seem to be a more appropriate alternative agent to use than a cephalosporin in treatment of endocarditis in view of its excellent antistaphylococcal and streptococcal activity. Clearly the new β-lactam compounds do not offer anything in this area, since none of the drugs is more effective against *S. veridans* or *S. aureus*, and they do not inhibit *S. faecalis*.

Treatment of bone and joint infections. Agents such as cephalothin and cefazolin have proved to be extremely useful in treating osteomyelitis and septic arthritis due to *S. aureus*—the most common infecting organism in young adults and older patients as well. It is not certain whether these agents should be used in preference to a semisynthetic penicillin.

Treatment of neurological infections. Until now cephalosporins have proved to be a major disappointment as a form of therapy for meningitis or brain abscess. Unsatisfactory results have followed use of cephalosporins for meningococcal, pneumococcal, and *Haemophilus* meningitis. The initial encouraging results with cefamandole have not been borne out, and it also should not be considered as an agent to treat meningitis.

Both cefotaxime and moxalactam have been successful in the treatment of meningitis in children and have cured meningitis due to chloramphenicol- and ampicillin-resistant *Klebsiella* and *Escherichia coli* in adults. Thus these agents may be an important form of therapy for meningitis of the newborn due to *E. coli*, ampicillin-resistant *H. influenzae* of childhood, and the hospital-acquired meningitis due to antibiotic-resistant gram-negative bacteria. The last is a real problem since chloramphenicol often fails in such patients.

Summary. There are many other areas in which cephalosporins have been used. New roles in skin and soft tissue infections have been advocated for the drugs, but the agents still are a second choice and the investigational agents offer most for the uncommon multiresistant bacterial infection. The role of agents such as cefazolin or cephalothin when administered at the time of operative procedures is well established. Whether the new agents will be more effective in the one area in which older cephalosporins are inadequate, colonic surgery, is to be established.

Cephalosporins are now available which can treat not only streptococcal, staphylococcal, *E. coli*, and *Klebsiella* infections but also *Haemophilus*, cephalothin-cefazolin–resistant *E. coli*, *Klebsiella*, and other organisms such as *Enterobacter*, indole-positive *Proteus*, *Serratia*, and *Bacteroides fragilis*. Perhaps it is time to reevaluate some of the concepts and consider that in some situations the new cephalosporins with their increased antibacterial activity, while retaining the excellent pharmacokinetic properties and low toxic potential of the older cephalosporins, may be drugs of first choice.

For background information *see* ANTIBACTERIAL AGENTS; ANTIBIOTIC in the McGraw-Hill Encyclopedia of Science and Technology.

[HAROLD NEU]

Bibliography: H. C. Neu, *J. Antimicrob. Chemother.*, 6(Suppl.A):1–11, 1980.

Climatic change

Small particles, or aerosols, composed of substances such as sulfuric acid, ammonium sulfate, soot, soil, and hundreds of organic compounds are found everywhere in the Earth's atmosphere with concentrations so large that an urban dweller may inhale 10^8 particles with a single breath. These aerosols scatter and absorb sunlight, and they scatter, absorb, and emit infrared energy radiated by the Earth (Fig. 1). Aerosols in the stratosphere are concentrated about 20 km above the Earth's surface. They absorb light from the Sun and scatter sunlight back toward space. They also absorb terrestrial infrared energy and radiate it back toward the surface. The absorbed sunlight and infrared light warm the stratosphere. The absorption of sunlight and the backscattering of sunlight to space prevent light from reaching the surface and lead to a cooling there. Tropospheric aerosols, concentrated in the atmosphere just above the Earth's surface, cool the surface by scattering sunlight, but warm it by absorbing sunlight. Whether the net effect is warming or cooling depends upon the ratio of absorption to scattering. This ratio depends upon the highly variable and poorly known composition of the aerosols. The radiation balance between solar energy entering the atmosphere and terrestrial infrared energy leaving the atmosphere controls the climate. Aerosols modify the radiation balance, and they may thereby affect the climate.

Studies of the climatic effects of tropospheric

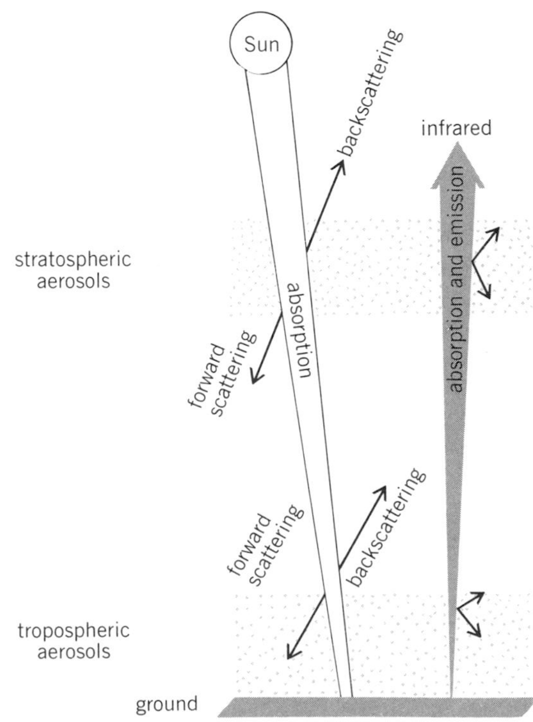

Fig. 1. Schematic drawing of the interactions between aerosols and sunlight and between aerosols and infrared light. (*From O. B. Toon and J. B. Pollack, Atmospheric aerosols and climate, Amer. Sci., 68:268–278, 1980*)

aerosols, which are found in the lower atmosphere, differ from studies of the climatic effects of stratospheric aerosols, which are found in the cloud-free region of the atmosphere at altitudes above about 12 km. Explosive volcanic eruptions produce vast numbers of stratospheric aerosols over large areas of the globe. The climatic changes caused by these particles can be relatively easily observed and fairly simply calculated. Following the May 18, 1980, eruption of Mount St. Helens in Washington State, numerous studies of stratospheric aerosols were made that greatly improved understanding of their properties. Since tropospheric aerosols are highly variable in space and time, their effects on climate are difficult to observe or to predict. Different types of tropospheric aerosols have different effects on climate. The complicated nature of tropospheric aerosols has only recently been fully recognized.

Stratospheric aerosols. Very explosive volcanic eruptions inject vast quantities of rock and gas into the atmosphere. When Mount St. Helens erupted, it threw several cubic kilometers of rock into the lower atmosphere, and it injected a million tons of sulfur dioxide gas into the stratosphere. A small fraction of volcanic rock hurled into the atmosphere enters the stratosphere, but the rock fragments are generally so large that most of them fall out of the stratosphere within a few weeks and so they have little opportunity to affect climate. The sulfur dioxide gas is photochemically converted into sulfuric acid vapor, which condenses to form small sulfuric acid particles. These particles remain suspended for several years in the stratosphere, where they are quickly dispersed over the hemisphere of the Earth in which the volcano is located, or for low-latitude eruptions, over the entire Earth. The particles in the stratosphere interact with visible and infrared light. Their effects on sunlight are easily observed as brilliant twilight colors, blue or green suns and moons, whitish rather than blue skies, and large rings around the Sun and Moon.

Benjamin Franklin was the first person to observe that the direct sunlight reaching the surface after passing through a stratospheric volcanic "dust veil" was less intense than normal. Following a large eruption in Iceland, he found, during the summer of 1783, that it was more difficult than usual to ignite a piece of paper with a magnifying glass. Franklin reasoned that the smaller amount of sunlight reaching the surface was responsible for cooler temperatures and for the harsh winter of 1783–1784.

An abnormally small number of large eruptions has occurred during the 20th century. The only large one since 1912 was that of the Indonesian volcano Mount Agung in 1963. Following this event, surface temperatures declined by several tenths of a degree Celsius, while stratospheric temperatures increased by several degrees for a few years. The stratosphere warmed because the particles in the stratosphere absorbed both sunlight and infrared radiation. The surface cooled because the absorbed solar energy and the energy scattered back to space no longer reached the surface.

Prior to 1912, explosive volcanic eruptions were more common. Although meteorological data are sparse for this era, it has been found that the Earth's surface cooled by several tenths of a degree for a year or two after most of the large eruptions for which temperature data are available. The most dramatic cooling followed the largest eruption of recorded history, that of the Indonesian volcano Tambora in 1815. The year 1816 became known as the "year without summer" because New England and western Europe suffered remarkably cool summers. In New England, snow fell in June and frost occurred every few weeks during the summer, killing crops and creating a famine.

Single volcanic eruptions affect climate for only a few years because that is as long as the volcanic particles remain in the stratosphere. During the period from 1500 to 1900, eruptions occurred so frequently that they may have had a continuous impact on climate. This period is often called the Little Ice Age because mountain glaciers, arctic sea ice, and wintertime river freezing were more extensive than now. Volcanic activity reached a maximum about 1800, when sea ice and mountain glaciers also reached their peak. When volcanic activity ceased at the beginning of the 20th century, hemispheric average temperatures rose to their highest levels in over 500 years (Fig. 2).

Numerous difficulties remain in understanding the impact of volcanoes on climate. Although most records of climate variations during the previous few hundred years correlate with the record of volcanic activity, some indicators of climate compare better with the record of variations in sunspot numbers. Perhaps both volcanic activity and solar activity affect climate. Also, the record of volcanic activity needs to be improved. The critical factor in linking an eruption to the resulting climate change is the amount of sulfur dioxide entering the stratosphere.

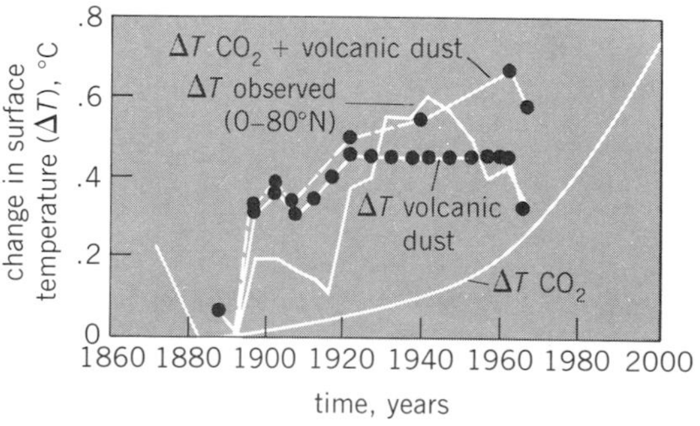

Fig. 2. Calculations of the temperature changes over the last century due to the observed increase in CO_2, the observed decrease in dust from volcanic activity, and the combined changes in CO_2 and dust compared with the observed changes. The comparison suggests that the decline in volcanic activity made a major contribution to the observed climate warming in the Northern Hemisphere between 1880 and 1940. Estimates show that CO_2 warming may become significant during the next several decades. (*From J. B. Pollack et al., Stratospheric aerosols and climatic change, Nature, 263(5578):551–555, Oct. 14, 1976*)

However, the geologic evidence for an eruption is mainly related to the rock component of the eruption, and it is difficult to infer the amount of sulfur dioxide emitted. Many large eruptions such as that of Mount St. Helens have injected surprisingly small quantities of sulfur dioxide into the stratosphere, apparently because the erupted magma contained little sulfur. Such eruptions are not significant ones as far as changing the climate is concerned. A promising supplement to geologic studies of volcanic activity is the measurement of sulfate levels in polar ice cores. These cores record the amount of sulfate actually in the atmosphere at the time that the ice was deposited on the polar cap and may thereby give clues to the amplitude of volcanic activity over the past 10^5 years.

Calculations of the effects of volcanic eruptions on climate have been hampered by lack of knowledge of the physical properties of the stratospheric particles. In order to calculate how much the surface cools and how much the stratosphere warms after an eruption, it is necessary to know the size of the particles and how transparent the compounds composing the particles are to visible light. After the Mount St. Helens eruption, NASA U-2 aircraft captured volcanic stratospheric particles. Laboratory study revealed these particles to be composed of sulfuric acid and volcanic glass which were quite transparent to visible light. Theoretical models using these measured particle properties, and the number of particles appropriate to a larger eruption than that of Mount St. Helens, do yield global temperature changes of the size and magnitude of those observed.

Current climate models deal with the change of a few tenths of a degree in the global or hemispheric mean temperature following an eruption. The mean temperature change is an average of many local changes. It is observed that when some regions undergo large temperature or precipitation decreases, other regions undergo large temperature or precipitation increases. Very tiny changes in global mean climate may represent larger variations in local weather. Weather anomalies such as the summertime snow and frost of 1816 are of great potential significance and need to be studied theoretically in the future. Unfortunately, calculating local weather changes is much more difficult than calculating changes in global climate.

Tropospheric aerosols. The properties of the particles found in the lower atmosphere vary greatly from place to place. Fortunately, various classes of aerosols whose properties are relatively well defined have been recognized. Each class is defined on the basis of its composition. The properties of the aerosols at any one location are specified by the relative abundances of the various classes.

One prevalent aerosol is composed of soil. These particles are typically larger than 10^{-4} cm and are moderately transparent to visible light. During droughts in semiarid regions and constantly in desert regions, winds lift large quantities of dust from the surface and create great dust storms. During the summer over 2×10^8 tons (1.8×10^8 metric tons) of dust may be blown off the Sahara Desert and out over the Atlantic Ocean, where optically dense dust clouds are commonly observed. Some of this dust actually reaches the Caribbean before being removed from the atmosphere. Humans contribute substantial soil debris to the atmosphere by using agricultural practices that expose the soil to wind erosion.

Another common type of aerosol is composed of sea salt. These transparent particles are generally larger than 10^{-4} cm. Since cloud droplets preferentially form on sea salt particles, these particles generally do not extend far above the ocean surface or far inland before they are washed from the sky.

Sea salt and soil particles are injected directly into the atmosphere. Another class of aerosols forms within the atmosphere from chemical reactions of minor gases. The most important compounds are the sulfates such as sulfuric acid and ammonium sulfate. These transparent particles are usually about 10^{-5} cm in radius. The efficiency with which particles scatter light depends upon their surface area. A given mass of small sulfate particles is more efficient at scattering light than the same mass of large soil or salt particles. Sulfates dominate the light scattering in several American cities, throughout the midwestern and southern United States, and in large areas of Europe. Human activity is thought to be responsible for about half the sulfates in the atmosphere.

Although there are many types of aerosols present in small quantities, the climatically most significant minor component is composed of small (10^{-6} to 10^{-5} cm), opaque, graphitic carbon or soot particles. Recently it has been found that aerosols in many urban areas are very opaque and that aerosols in rural areas and even in some remote areas such as the arctic are moderately opaque. Surprisingly large quantities of soot, corresponding to several percent of the mass of rural aerosols and several tens of percent of the mass of some urban aerosols, have been discovered. The precise source of this soot is not known, although it is clearly generated by combustion. Some soot may come from forest fires, but humans may also be directly responsible for soot production through the use of diesel engines and inefficient heating processes.

One debate that has gone on for many years is centered on whether tropospheric aerosols warm or cool the Earth. Observations after volcanic eruptions show that stratospheric aerosols tend to cool the Earth. However, tropospheric aerosols vary so rapidly in time and space that their effects on the climate are not easy to observe and to separate from normal meteorological variations. Theoretically, it has been found that transparent materials such as sulfates will generally create a radiational cooling in the region where they are located by scattering sunlight back to space. Even moderately absorbing particles, however, will cause a radiational heating because they absorb energy that otherwise would have been reflected to space by the Earth's surface. Whether an aerosol tends to warm or cool the Earth depends upon how opaque the material composing the aerosol is and upon the reflectivity of the surface below the aerosols. There is no single answer to the question of whether tropospheric aerosols warm or cool the Earth. Some types of particles, such as sulfates, tend to cool; some types, such as soot, tend to warm; and some, such as soil debris, may have either effect.

Another debate about tropospheric aerosols concerns whether or not human activities have increased the amount of aerosols. Pollution-control devices may recently have stabilized the input of some chemicals to the atmosphere. However, many materials such as soil are not controlled. Aerosol concentrations measured at remote mountaintop sites for several decades have not revealed any long-term trends. Measurement at remote sites may underestimate the level of pollution, however, because most aerosols are removed from the atmosphere before they have the opportunity to spread over much of the globe. There is evidence of large upward trends in aerosol concentrations during the past several decades at many rural sites which are near cities, industrialized areas, or agricultural areas, in locations including parts of Japan, the eastern and southwestern United States, and the Soviet Union.

Much further work remains to be done on the relation between tropospheric aerosols and climate. Rather than affecting the temperature locally, the radiation changes caused by aerosols may instead alter the moisture and the wind fields in the troposphere. Furthermore, aerosols are critical to the formation of clouds, so that changes in aerosol properties could lead to changes in cloud structure, which in turn might modify climate. Studies of the interactions between aerosols, dynamics, and clouds, as well as more complete studies of the properties and sources of aerosols in the atmosphere, will probably be the main focus of future studies of tropospheric aerosols and climate.

For background information *see* AIR POLLUTION; CLIMATIC CHANGE in the McGraw-Hill Encyclopedia of Science and Technology.

[OWEN B. TOON]

Bibliography: J. E. Hansen et al., *Ann. N.Y. Acad. Sci.*, 338:575–587, 1980; H. Rosen et al., *Appl. Opt.*, 17:3859–3861, 1978; O. B. Toon and J. B. Pollack, *Amer. Sci.*, 68:268–278, 1980; S. Twomey, *Atmospheric Aerosol*, 1977.

Clinical bacteriology

Hospital-acquired (nosocomial) infections are a major public health problem. About 5% of the 35,000,000 patients admitted to hospitals in the United States become infected. Nosocomial infections increase both mortality and the cost of hospital care. One approach to controlling the problem is by surveillance to detect outbreaks or changes in both infection rates and the bacteria isolated from these infections. Unraveling the complex relationships of these bacteria with the environment and each other is frequently aided by laboratory methods which can "fingerprint," or type, individual strains of a species. While characteristics used to define a species need to be consistent among strains, those used for typing must show some variability. Characteristics used include bacteriophage susceptibility, serotype, biotype, bacteriocin production, bacteriocin susceptibility, and antimicrobial susceptibility patterns. The purpose of a typing system is to ascertain whether one bacterial isolate is the same as another.

Uses of typing systems. Typing bacteria has been used as a tool in many areas of microbiology, particularly in medical and clinical microbiology. For some bacterial species there is a relationship between virulence and subtype, one of the more familiar examples being strains of capsular type b *Haemophilus influenzae*. Sometimes certain types of a species are associated with a particular site of infection. Again, *H. influenzae* is a good example: the majority of blood and cerebrospinal fluid isolates are biotype 1.

In the hospital environment, typing techniques may be used to compare several isolates of the same species from a single patient. This may be useful if there is a question of reinfection versus recrudescence or even to determine the significance of several isolates from a single source.

Typing systems are extensively used by public health authorities, especially in investigation of *Salmonella* and *Shigella* outbreaks. Subtype determination provides a means of relating cases of food poisoning to each other and to a possible source.

If a species has a group characteristic such as the group antigen of group A beta hemolytic streptococci or the susceptibility of *Salmonella* to bacteriophage 01, a typing technique can be used in the routine clinical laboratory for species determination. Typing methods even have a place in the research laboratory, since it is important to confirm periodically that one is indeed working with a particular strain.

However, the most common and familiar use of typing techniques is in the investigation of nosocomial infection problems, both to detect and confirm a common source outbreak and to establish relationships of patient isolates to each other and the environment.

Methods of typing bacteria. Ideally, a typing method should be easy to perform, inexpensive, and reproducible, and it should allow subdivision of the species into a large enough number of fairly evenly distributed types to distinguish outbreak strains. It is also important that the majority of isolates be typable by a selected procedure. To permit interlaboratory use, methods must be carefully standardized.

Serotyping. For many years serologic typing systems have been devised for many species of bacteria and are widely used to examine isolates of both public health and nosocomial significance. Antigens examined include O (cell wall polysaccharide), H (flagellar), and K (capsular). Procedures used depend somewhat upon the antigen and include agglutination, fluorescent antibody, quellung, and H immobilization tests. Generally, serotypes are very reproducible, and many species contain a large number of different types (for example, members of the Enterobacteriaceae). The major disadvantages of serotyping are the high cost of producing or purchasing antisera, time-consuming procedures, and frequently the necessity of titrations to resolve cross-reactions among closely related strains.

Bacteriophage typing. A successful bacteriophage typing system depends upon the isolation and selection of a manageable number of phages that produce a variety of lytic patterns among the majority of isolates tested. Phage typing sets have been devised for many bacteria, including most of the Enterobacteriaceae and *Staphylococcus aureus.* Phage typing is technically often easier than complete serotyping. It does require maintenance of indicator and propagating strains as well as preparation and titration of phage lysates. It is less expensive than serotyping but not as reproducible. Some variation of phage patterns, particularly of weak reactions, frequently occurs among related cultures. Two isolates are considered to be different if one is lysed strongly by two phages which do not lyse the other isolate at all.

Bacteriocin typing. Bacteriocins are proteins produced by one strain of a species which are bactericidal for other strains of the same species. Bacteriocin typing can be done in two ways: (1) bacteriocins produced by the strain to be typed are tested against a set of indicator strains, giving a production pattern; or (2) bacteriocins from a standard set of known bacteriocin producers are tested against the strain to be typed, resulting in a susceptibility pattern. Like bacteriophage typing, a successful, sensitive bacteriocin scheme depends on the selection of a workable number of appropriate indicator strains. Frequently, this requires screening a large number of isolates and then attempting to choose the best strains. This analysis has been simplified by the use of a computer program.

Bacteriocin typing by both production and susceptibility has been used most successfully with *Serratia marcescens* and *Pseudomonas aeruginosa.* Once standardized, bacteriocin typing is easy to perform and requires only the indicator strains, media, and reagents commonly found in the laboratory. Use of an ACCU-DROP applicator allows the inoculation of 24 standardized drops at once and can be used for both bacteriophage and bacteriocin typing. Day-to-day variation in patterns does occur with both methods, and all related isolates should be tested at the same time. In fact, some investigators feel that reaction-difference rules should also be used for bacteriocin typing.

Biotyping. Variability of biochemical characteristics of strains within a species has a special appeal as a typing method. In the last few years, attempts have been made to use the same system for both species identification and biotype determination so that most hospital laboratories could routinely type most isolates, thus allowing for day-to-day surveillance of many bacterial species. Use of the commercially available microsystems designed for identification of Enterobacteriaceae for biotyping has not been successful because the resulting biotypes, for the most part, are not reproducible. However, some biotyping methods may be quite reproducible and very useful. A biotyping scheme has been developed for *Serratia marcescens,* based on variations in utilization of eight carbon sources, reduction of tetrathionate, pigment production, and hemolysis of horse blood, that is highly reproducible. Reproducibility of biotyping methods depends on a number of variables including inoculum size, incubation time, the use of growth versus a color change as an end point, and the biochemical reactions included in the scheme.

Antibiograms. Another rather tempting approach is to use antimicrobial susceptibility patterns (antibiograms) for typing since they are readily available, but such data must be interpreted carefully. Generally, resistance patterns are consistent within an individual hospital, so that many strains with the same antibiogram often represent a number of different types, and conversely, isolates of the same strain based on other typing methods may have varying susceptibility patterns. To add to the confusion, outbreaks due to the presence of a particular plasmid carrying antibiotic resistance genes can occur. This resistance can involve plasmid transfer among strains of a single species or dissemination of a plasmid into several or even multiple species of Enterobacteriaceae. On the other hand, a distinctive change in the antibiogram of a particular species may indicate a nosocomial problem. Outbreaks of methicillin-resistant *Staphylococcus aureus* and aminoglycoside-resistant Enterobacteriaceae and *Pseudomonas aeruginosa* have been detected by surveillance of antibiograms.

Combined methods. If random isolates of a species are typed by more than one method, they generally can be even further subdivided. Thus, a number of isolates of a single serotype can contain a number of different bacteriocin types, bacteriophage types, and biotypes. In contrast, isolates from an outbreak are usually found to be the same strain even if several typing methods are used. That is, the isolates are the same bacteriophage type, bacteriocin type, serotype, and biotype.

Significance of typing bacteria. Although detection of transmission of nosocomial infection might be enhanced by typing all isolates from these infections, the techniques are too time-consuming and costly for the routine laboratory. Most outbreaks are, in fact, detected initially by routine surveillance of bacteria isolated from patients. However, substantiation of an outbreak and its eventual control often require typing the implicated species. The ability to differentiate strains of a species is necessary for further studies of their ecology and epidemiology.

For background information *see* CLINICAL MICROBIOLOGY in the McGraw-Hill Encyclopedia of Science and Technology. [SALLY JO RUBIN]

Bibliography: J. J. Farmer, III, in J. E. Prier and H. Friedman (eds.), *Opportunistic Pathogens*, pp. 49–63, 1974; A. P. Grimont and F. Grimont, *J. Clin. Microbiol.*, 8:73–83, 1978; T. R. Oberhofer and A. E. Black, *J. Clin. Microbiol.*, 10:168–174, 1979; T. F. O'Brien et al., *Antimicrob. Agents Chemo.*, 17:537–543, 1980.

Cogeneration

Cogeneration is potentially an important form of industrial energy conservation, achieved by increased thermal efficiency during power generation. Its potential depends to an extent on overcoming political and institutional barriers rather than technical problems, since it uses mainly conventional power-producing technology. Cogeneration was given significant importance in the National Energy Plan announced by President Jimmy Carter in 1977. The goal of the plan was to reduce the United States' dependence on imported oil and achieve economic growth in an era of limited energy resources.

This article describes the current status of cogeneration in producing the electric power requirements of industry in the United States, and covers the main forms of cogeneration, namely, industrial cogeneration and district heating. The application of the basic cogeneration concept in nuclear power plants and newer technologies such as solar and fuel cells is also considered, as are the changing utility attitudes and recent regulatory actions to encourage cogeneration.

Cogeneration is the simultaneous production of electric or mechanical energy and useful thermal energy from the same fuel source. Cogenerating electricity and steam presents an attractive opportunity to conserve energy because about 40% of industrial energy is used to produce low-pressure process steam. The additional fuel required to do both jobs by cogeneration is about half that required by the most efficient single-purpose utility plant. Cogeneration also produces lower thermal discharges than conventional single-purpose systems.

Cogeneration, like other types of conservation, does not require a major technological breakthrough but, rather, represents a regaining of a path that was abandoned. At the beginning of the 20th century, nearly all industries cogenerated; in 1920, 30% of the electricity in the United States came from cogeneration. This figure had declined by 1950 to 15%, and the decline continued so that by 1973 only 5% was being generated by industry, a figure that differed markedly from that in other countries. West German industries, for example, supply 12% of their electricity needs through cogeneration. British industry produces 20% of its own electricity needs.

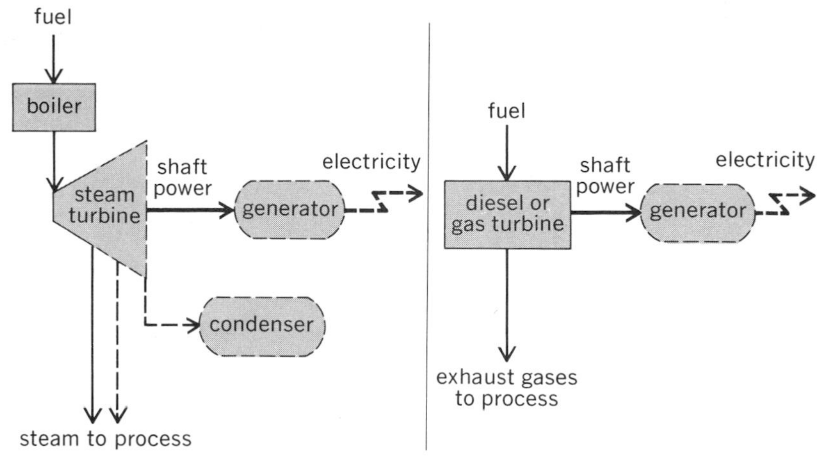

Fig. 1. Configuration of a direct-process topping cogeneration system without heat pumps. (*From Industrial Cogeneration Optimization Program, Final Report, September 1979, published January 1980*)

Industrial process cogeneration. The type of cogeneration with by far the most significant potential is the combined production of electricity and steam for an industrial site. The system may be operated either as a topping system or as a bottoming system. Topping uses thermal energy conventionally treated as reject heat from an electricity-generating process (Fig. 1). In bottoming, thermal energy is generated for an industrial process and any energy which would normally be rejected is used to generate electricity.

The cogeneration system for an industrial plant must satisfy two different types of energy requirements—electrical and thermal. The diversity of process energy needs in different industries, and the significant differences in the mix of thermal and electrical energy produced by different cogeneration technologies, necessitate evaluation of industrial cogeneration applications on a site-specific basis (a particular site can have a significant effect on economics, energy savings, and environmental effects, and so cogeneration is "site-specific").

The U.S. Department of Energy conducted optimization studies in 1979 on industries that offered the best potential for cogeneration. They concentrated on five major industries that require a lot of steam in manufacturing, namely, pulp and paper, food, textiles, chemical, and petroleum refining. Two criteria were regarded as most significant in this optimization study: (1) the return on investment (ROI) of the system required to meet a minimum specified value (ROI hurdle rate) for each industry; (2) the maximum energy savings, especially of oil and gas, must be consistent with the ROI hurdle rate.

The first criterion is given higher importance since economics is usually the primary factor in most business decisions. The ROI in various industries is usually 20–25%, much higher than in the regulated electric utility industry, in which it is usually around 12–15%.

Options are available for matching or decoupling electric and thermal loads, and these also affect the economics of cogeneration. The first option is isolated operation. In this mode, the cogeneration system is not linked with any other electrical or thermal energy source and is designed to be large enough to satisfy all plant needs. Considerable redundancy of equipment is required to achieve the desired reliability.

A second option is to provide only a portion of plant electrical needs by cogeneration and to purchase supplemental power from the utility or, alternatively, to produce more power than the plant needs and export the excess to the utility. These options can be combined to decouple the electrical loads by allowing both import and export of electric power as needed. The cogeneration system can then satisfy thermal loads and interchange electric power with the grid, and have the needed reliability without investment in redundant equipment.

District heating. In the second type of cogeneration, steam (or hot water) from a power station is delivered by pipes to homes and offices to provide heat and hot water. Such systems are called district heating. District heating and cooling systems may be small-scale, such as those in hospital complexes, educational facilities, industrial complexes, shopping centers, and residential or commercial developments, or they may be large-scale, as an entire downtown or metropolitan area.

District heating was started in the United States in the late 1870s, and a steam company began serving customers in New York City from 1879 onward. Early district heating systems used boilers that functioned just to supply steam, and a major advance came with the first dual-use systems, supplying both steam and electricity. Steam from boilers drove generators to produce electricity, and exhaust steam that would have gone to waste was used for district heating purposes. By 1909 an estimated 150 companies sold heat in this fashion across the United States.

After several decades of success, district heating declined in the United States. One reason was technological progress in the production and transmission of electricity; another was that large electrical plants were being located at great distances from urban centers rendering heat transmission inefficient.

In many European countries district heating is more widespread than in the United States. The principal reasons for this difference include: (1) fewer domestic energy sources, necessitating more efficient fuel utilization, (2) higher fuel prices, (3) scarcity of land for sanitary landfill operations, (4) fewer governmental and institutional barriers, and (5) more governmental sponsorship for implementation of district heating. In Sweden, Denmark, and other Scandinavian countries, cogeneration is the rule for new electricity-generating stations.

Nuclear power cogeneration. Cogeneration has also been employed in nuclear power plants. Midland Nuclear Power Station in Michigan is the first large nuclear plant built for commercial cogeneration in the United States. Generally, close-in siting is required for any power plant with a potential for cogeneration, and this causes some special regulations and social intervention for nuclear units. Several design features have been added to this Michigan plant, in view of its being sited within the city, for added assurance of public health and safety. Of particular interest is the tertiary heat-exchange system for the process steam to assure that any possible primary-to-secondary radioactive leakage cannot contaminate the process steam supply.

The two units at Midland illustrate the advantage in thermal efficiency offered by cogeneration. Unit 2 generates only electricity and carries a rating of 810 MWe. Its thermal efficiency is in the 30–32% range, which is normal for these units. In contrast, Unit 1 can supply up to 4,000,000 lb/h (1,800,-000 kg/h) of process steam to the adjacent Dow Chemical Company Midland Industrial Complex, in addition to generating electricity for the grid, as shown in Fig. 2. This unit is rated at 523 MWe, but its thermal efficiency is in the 40–42% range. This increase in thermal efficiency by one-third is a significant contribution to energy conservation, leading to reduction of both waste heat disposal and cost of electricity.

Other examples of nuclear plant cogeneration include Ontario Hydro in Canada, which has been using 3,000,000 lb/h of steam from its Bruce Nuclear Power Units to operate its heavy-water plant, also located on the Bruce Nuclear Power Development.

High-temperature gas-cooled reactors (HTGRs) offer another possible cogeneration option. Work is proceeding on this type in several countries, such as the United States, England, West Germany, and Japan, with a view to using the high-temperature

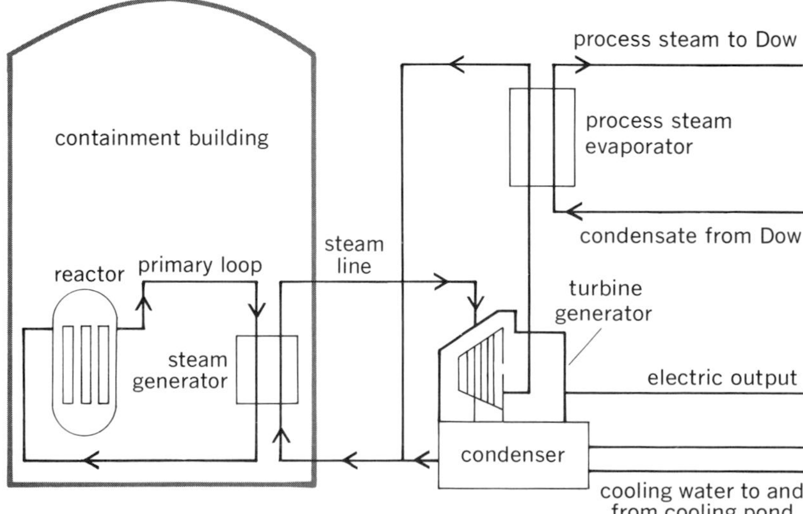

Fig. 2. Schematic flow diagram of the cogeneration cycle at the Midland nuclear power plant in Michigan. Extraction steam from the main turbine goes to process steam evaporators, and then back to the secondary system condensate return line. (*From F. C. Olds, Power plant construction: State of the art, Power Eng., 85:50–58, March 1981*)

gases and steam from the reactor for emerging synthetic fuel technologies and steelmaking operations, in addition to producing electrical power. Fort St. Vrain, CO, is presently the only commercial HTGR plant in operation in the United States; light water reactors (LWRs) are receiving more attention in the United States.

New technologies. New technologies such as combined-cycle cogeneration or fuel cells with heat recovery are likely to be attractive technical options because of the possibility of decoupling the electrical and thermal outputs. Research is continuing into other new technologies, including solar and geothermal, which can be used to generate electricity and thermal energy.

Utility attitudes and concerns. Until recently, few utilities have actively encouraged or participated in industrial cogeneration projects, viewing cogeneration as competition and being concerned about the loss of their base load. Utilities also expressed concern regarding the reliability and maintenance of cogeneration equipment, interconnection costs, the availability of cogenerated power when needed by the utility, and the need for standby capacity.

A number of significant changes in the economic and institutional aspects of power generation occurred in the late 1970s that resulted in a trend toward increased interest in, and acceptance of, industrial cogeneration by utilities. Some of these changes include: constraints on siting large central station plants; increasing capital costs of new generating capacity and difficulties in raising capital; heightened customer resistance to electricity rate increases; slower load growth, which makes it more attractive to increase capacity in small increments rather than large increments; Federal regulatory changes (described below), which have made cogeneration more attractive to industry.

These changes not only have led utilities to consider industrial cogeneration in their planning for future capacity needs but have also resulted in the growing recognition of cogeneration systems as a business opportunity for utilities. Figure 3 shows the change in price of central generating capacity between 1967 and 1978. The average cost per kilowatt for an on-site cogeneration plant was about \$275 in 1967, or 2.2 times the cost per kilowatt of new central capacity. By 1978 the cost per kilowatt—\$600—of an on-site plant was less than 50% of the cost of new central capacity. The difference is even greater today; thus utilities have a considerable economic incentive to proceed with cogeneration in certain regions of the United States. *See* ELECTRICAL UTILITY INDUSTRY.

Relevant Federal legislation. The National Energy Act (NEA) of 1978 contains a number of provisions attempting to remove institutional barriers to cogeneration. The most important provisions are in the Public Utility Regulatory Policies Act (PURPA), which provides the following requirements for cogeneration facilities that qualify by meeting certain operating and efficiency requirements. (For a cogeneration facility to qualify, no more than 50% of the firm may be owned by an electric utility.) (1) Utilities must purchase any and all power that the qualifying cogenerator (QC) wants to sell. (2) The rate offered by the utility for such purchase of power must be equal to the avoided cost of the utility. (3) The rate charged by a utility to a QC for standby or backup power must be nondiscriminatory. (4) The QC is exempted from utility regulation under the Federal Power Act and the Public Utility Holding Company Act and state regulations related to rates and financial reporting.

For background information *see* DISTRICT HEATING; ELECTRIC POWER GENERATION; NUCLEAR POWER in the McGraw-Hill Encyclopedia of Science and Technology.

[VIJAY K. DATTA]

Bibliography: *Energy in Transition 1985–2010: Final Report of the Committee on Nuclear and Alternative Energy Systems*, 1980; R. Stobaugh and D. Yergin (eds.), *Energy Future: Report of the Energy Project at the Harvard Business School*, 1979; *Power Engineering*, published by Technical Publishing (of Dun and Bradstreet Corporation), a monthly magazine.

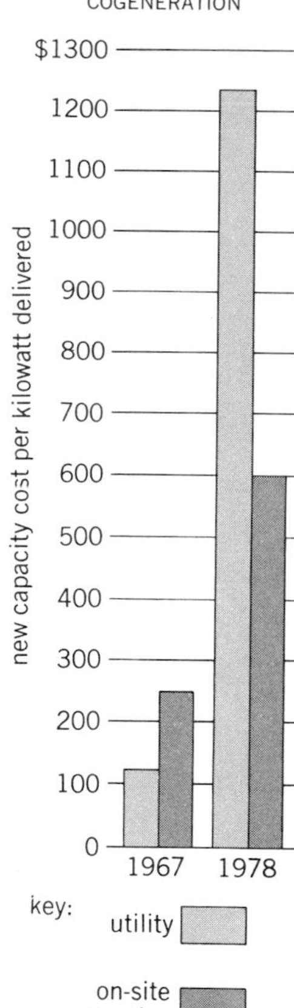

Fig. 3. Relative costs for electricity-generating capacity in 1967 and 1978.

Computational chemistry

Computational chemistry is the calculation of molecular structure, properties, and reactions. Computers assist chemists in many ways (including collecting, analyzing, storing, and retrieving laboratory data, often via sophisticated graphic displays); this article focuses on the computer as the central source of data. The equations of physics are combined with the art and science of applied mathematics to produce directly from the computer the chemical data of interest. Rather than assisting in the chemists' laboratory, the computer *is* the laboratory. In common with all laboratory data, results obtained computationally vary in accuracy, reliability, and the ease of obtaining them. Nevertheless, there are problems of considerable practical importance for which the computer provides the best way to obtain the desired information. Computational methods of proved reliability already exist to solve important classes of chemical problems. The costs of computation continue to drop as other laboratory costs increase. It seems inevitable that computational chemistry will become an increasingly attractive research technology.

Mathematical tools. Virtually all industrial chemistry can be described with nonrelativistic quantum mechanics, adding electron spin. The Schrödinger equation is the formulation most familiar to chemists, having the form of an operator-eigenvalue equation: $H\Psi = E\Psi$. The hamiltonian operator H for the system to be described (for example, the electrons and nuclei making up a molecule or collection of interacting molecules) is a partial differential operator depending on the coordinates of all the particles in the system. The

Schrödinger equation must be solved for the wave function Ψ, a state vector completely describing the observable properties of the system, and the set of allowed energies E. For bound systems such as molecules, only discrete values of E will permit solutions of the Schrödinger equation to exist. These energy eigenvalues are the steady-state energies of the system. There are slightly more general formulations of the Schrödinger equation that are appropriate for the interaction of molecules with radiation and the interaction of unbound systems (that is, molecular collisions and chemical reactions). The net result of this highly formal theory is a quantitative theory of chemistry.

Once the wave function Ψ and the spectrum of allowed energies E are known, all the information about the system has been determined. The barrier to practical application and quantum engineering is the fact that the Schrödinger equation is very hard to solve for molecules. Even rather small molecules (for example, C_6H_6) have many particles to be considered (12 nuclei, 42 electrons). Hence, the magnitude of the partial differential equations rapidly becomes unmanageable. This was the primary barrier that prevented computational chemistry from emerging as a recognized scientific discipline. It was widely considered to be a purely formal exercise leading to results of dubious reliability, at best, and for rather simple chemicals (that is, H_2). Several things happened in the 1970s to change this. Most important, the digital computer made impressive progress, permitting the extremely laborious arithmetic to be carried out by small, inexpensive minicomputers.

As important as the advance in computing (and decrease in cost) has been the development of better mathematical procedures. Even the most advanced computer today could not solve very many problems if the most direct, brute-force methods were employed to solve the Schrödinger equation. Also, it has become clear that there is not yet in existence a single general-purpose computer code for all chemical problems. It is true that a complete knowledge of the wave function would yield complete knowledge of all observables in the system. However, no computational method is completely exact. There are always errors inherent in the use of finite precision arithmetic, finite grids, finite series, and so forth. Thus, in practice, a computational method must be selected that is most appropriate to the desired answers. This is a large hurdle.

Methods of computational chemistry are inherently rather esoteric. A working chemist rarely has the time or inclination to become an expert. If a collaborator skilled in the computational arts and interested in the problems can be located, progress may be made. But failing that, it is difficult to make a reasoned choice of method and approach. The next major advance will be in education about computational chemistry. Interactive helping programs need to be constructed that will take the place of a skilled computational chemist as collaborator.

Table 1. Comparison of predicted and experimental dissociation energies (kcal/mole)*

Molecule	Predicted $H°_R$, 300	Experimental $H°_R$, 300
CH_3OH	99.8	100.9
		104.0
CH_3O	22.2	—
CH_2O	86.5	88.2 ± 1.6
		87.5 ± 1.0
CHO	12.9	15.5 ± 1.5
		17.0 ± 2.0

*1 kcal/mole = 4.184 kJ/mole.

Rather than dealing with a professional colleague, the chemist can have a dialog with the program itself to decide what method to use, how to use it, and what accuracy to expect. Such developments are coming from many places, but have yet to make a major impact on the practice of computational chemistry. Even though the underlying mathematical technology is quite complex, the interface with the user can be made much more accessible.

Case study: computational thermochemistry. In computational chemistry, as in any experimental science, the results are the key. An example in thermochemistry can serve to illustrate the ability of modern computational chemistry to obtain data inaccessible to experiment.

Combustion frequently proceeds through unstable transient species which are difficult to study experimentally. One example is the methoxy radical CH_3O, the thermochemistry and geometry of which are not known from experiment. A series of compu-

Table 2. Optimized structural parameters for molecules (nanometers, degrees)

	Computed	Experiment
CH_3OH		
r_1	0.1093	0.10937
r_2	0.1093	0.10937
α	107°9′	108°32′
methyl tilt	2°12′	3°12′
S_1	0.1432	0.14214
t_1	0.0963	0.0963
β	107°2′	108°2′
CH_3O		
r_1	0.1081	—
r_2	0.1085	—
α_1	108°32′	—
α_2	112°6′	—
β_1	111°30′	—
β_2	103°54′	—
S_1	0.1405	—
CH_2O		
r_1	0.1102	0.1099
S_1	0.1211	0.1203
	116°11′	116°30′
CHO		
r_1	0.1111	0.1125
S_1	0.1188	0.1175
	124°	124°57′

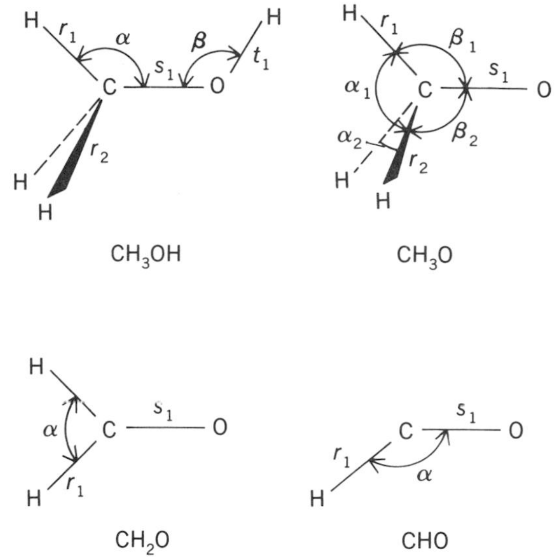

Definition of molecular parameters for the species listed in Table 2.

tations were carried out by G. F. Adams and colleagues on $CH_3O H$, CH_3O, CH_2O, and CHO to determine the geometries and heats of formation. The results of these calculations are shown in Tables 1 and 2 and are compared with experimental data where available. The molecular parameters are shown in the illustration. The calculations for all systems are of comparable accuracy. Therefore, since the computations are demonstrably accurate for all but CH_3O, the results can be accepted with confidence, even without laboratory data.

Many rocket fuels and other types of propellants contain borane. A great deal of thermochemical information for some of the more exotic fuels is not available. Also, detailed kinetics of the fundamental reactions involved in these combustion processes is very hard to acquire, frequently because the basic reaction steps involve transient species such as free radicals. A good example is the species BH_3, which does not exist in the usual sense, since it cannot be trapped in an inert gas matrix. This fact has caused a great uncertainty in a problem as superficially simple as determining the binding energy of diborane (B_2H_6). Since information on BH_3 is not available, the reactions $BH_3CO \rightarrow Bh_3 + CO$ and $B_2H_6 \rightarrow 2BH_3$ are studied together experimentally to eliminate a need for information on BH_3. With this procedure, two widely differing sets of experimental values were obtained for heats of formation of diborane (57 and 36 kcal/mole or 238 and 151 kJ/mole) and for borane carbonyl (34 and 22 kcal/mole or 142 and 92 kJ/mole). Each so-called experimental value had been obtained at least three times. Calculations have predicted 35.5 ± 1.5 kcal/mole (148.5 ± 6.3 kJ/mole) and 22.8 ± 1.5 (95.4 ± 6.3 kJ/mole) respectively, which lends support for one set of experiments. There is no experimental value for the binding energy of BH_3NH_3, but the computational value of 30 kcal/mole (126 kJ/mole) is a prediction that experiment will very likely verify.

For background information *see* QUANTUM CHEMISTRY; NONRELATIVISTIC QUANTUM THEORY; THERMOCHEMISTRY in the McGraw-Hill Encyclopedia of Science and Technology.

[GEORGE T. S. WOLKEN, JR.]

Bibliography: G. F. Adams et al., *Chem. Phys. Lett.*, 81(3):461–466, August 1981; C. Evans, *The Micro Millennium*, 1979; F. L. Pilar, *Elementary Quantum Chemistry*, 1968.

Computer-aided engineering

Computer-aided engineering involves the combined use of design standards; computer software, hardware, and graphics; automated static and dynamic analysis; and simulated operations and physical testing for the overall purposes of improving engineering accuracy, effectiveness, and productivity.

Many steps in the product design and development process are time-consuming and error-prone when manual methods are used. For example, millions of hours are spent annually by designers and drafters laboriously putting their ideas on paper. And thousands of computations are made manually and by hand calculators.

The traditional design process is very dependent on the use of hardware models and prototypes. Typically, the first step is to draw and then construct a physical model of the new product. The model or prototype is then tested under operating conditions until weaknesses are discovered. A series of alterations are made until the product performs satisfactorily.

Because it is difficult to get the product completely correct during the design engineering phase, additional changes are often made during the manufacturing engineering phase after tooling is built, during production, and even after the product is in the marketplace. This is an expensive, inefficient process.

During the late 1970s a number of engineering methods were developed which removed some of the tedium and added more insight to product development: Finite-element analysis permitted analysts to predict stresses, dynamic properties, and lifetimes of components prior to construction of physical models. Modal testing and analysis enabled test engineers to characterize the dynamics of existing components and systems. Mechanism design and analysis programs were used to synthesize and analyze simple and complex mechanisms without building prototypes. Life estimation software predicted survival expectancy via fatigue theory. Turnkey design-drafting systems took designers off away from the drafting board and put them in front of a cathode-ray tube (CRT) to improve their productivity by factors of 2, 5, or more. Systems analysis software predicted the performance of systems of interconnected components and subsystems. Computer-aided manufacturing (CAM) software was the basis for automated machinery control, bills of materials, and materials handling requirements.

These engineering methods yielded substantial gains in engineering productivity and product performance in those few cases when they were used. However, these methods were used mainly to streamline the old procedures. The engineering environment of the 1980s needs more. A new method can be adopted: computer-aided engineering :CAE). CAE makes use of the methods developed in the 1970s but goes on to generate the improvements needed in the environment of the 1980s. In the CAE process, initial prototypes are replaced by computer simulation.

Simulating product performance. The purpose of early system modeling is to assemble a computer-based model of the entire product which meets design standards established by market needs and competitive forces. A computer-based model is constructed from three data sources: test data from existing components, analysis data from proposed components, and a data base of existing component performance data. In a sequence of logical steps, this model is revised and updated until product performance meets design standards.

Modal analysis software is used to collect, analyze, and display data from operating and artificial excitation tests. This software also determines mode shapes, natural frequencies, modal masses, and damping values of components and subsystems. Modal analysis software typically runs on minicomputer-based testing systems especially designed for this purpose. A good test-analysis system has two or more input channels for rapid acquisition of data, a graphic display terminal, a hardcopy printer/plotter, and a 16-bit central processing unit.

Finite-element analysis begins with the creation of two- and three-dimensional models at the drafting board or by graphics, geometric modeling, and mesh generation software. The interactive versions of this software are most convenient to use, and a menu capability makes it easy to learn to use the software. The output of these programs can be formatted to match the input of the finite-element analysis program.

The finite-element analysis program processes the inputs of nodes, elements, physical properties, material properties, and boundary conditions to generate natural frequencies and modal data for use in the system model. Finite-element analysis programs are usually designed to run on large mainframe computers. However, several of the programs have been converted to run on 32-bit superminicomputers.

New testing and analysis may not be necessary for system components used in previous designs. Data banks may contain dynamic data for existing components and subsystems. The combination of modal testing, finite-element analysis, and data banks provides the data necessary for construction of the system model.

Evaluating product performance. After characteristics of the components are determined, they are mathematically assembled in the computer to form a system model—a mathematical representation which replaces the first prototype. This system model can then be used to simulate product performance. Required design changes can then be made simply and quickly on the mathematical model without going through time-consuming and costly hardware changes on a prototype.

The advantage of using system models is that design changes and performance tests can be performed in the computer and the response of the structure can be animated on the CRT. Thus, the designer can minimize the costly, time-consuming, trial-and-error testing of prototypes. The computer simulation method not only saves time and money in product development but also permits the user to develop a more satisfactory design configuration.

System modeling has been used in many situations to predict product performance and to aid the designer in developing better configurations. Computer-simulated automobiles, for example, can be put through their paces while still in the conceptual stage. The entire automobile—existing only as a computer model—can be "driven" over computerized highways so that ride quality, handling, and body rigidity can be assessed before a costly prototype is built. Or a designer can watch a bulldozer blade deflect and vibrate as it bites into the ground, with all the action provided by a computer-generated model.

If the product model does not perform as it should, design changes are made in the computer models of the components. Only when satisfactory performance is achieved will the design move on to the next stage. It is much easier to optimize the computer model than to build and rebuild prototypes. Further, the system model gives the designer a bonus: as a result of the system analysis, the designer knows the loads on each component. These loads can be applied to finite-element models of the components to determine their ability to perform in the final product.

Finite-element analysis allows a designer to predict structural responses (stress, deflection, heat transfer) to loadings on components. It works even for complex structures that hand calculations cannot handle. In finite-element analysis, the structure is divided into finite elements connected together at nodes. Loads, of known magnitude, direction, and position, are then applied. Through the solution of a group of equations, prediction of structural deformation, stresses, natural frequencies, and mode shapes are obtained.

Modal testing, finite-element analysis, and system modeling are good predictors of current performance, but the thoughtful designer also requires estimates of fatigue life. Product life estimation is considered by many to be a mysterious art. But recent progress in theory, hardware, and software has made the task more reasonable. Current minicomputer-based systems help the designer acquire, reduce, store, display, and interpret the great quantities of data needed for life estimation. Stress and strain information from physical testing, modal analysis, and finite-element analysis combine with material behavior properties to predict crack initiation and to display cycle and damage histograms.

Preparing component designs and computer models. Computer-aided design systems make it easier than ever to create and store component shapes in an intelligent data base. These shapes can be easily revised if necessary; they also provide the basis for drafting and documentation. Graphics systems are also used to build finite-element system models.

Geometric modeling software provides the capability to create, present, edit, and view models on a CRT display terminal. Programs usually handle both two- and three-dimensional models, and several offer a digitizing capability which saves even more time for the designer. For digitizing, a sketch or drawing of the component of interest is positioned on the tablet. Nodes are entered by digitizing locations on the surface of the drawing. Elements are then defined by selecting nodes on the CRT screen. To make it easier to add finishing touches, graphics software designers have programmed a generous supply of display and editing features for checking and modifying the model. Advanced features such as mesh generation reduce modeling time even further.

Documenting results. When a designer is satisfied with the computer prototype, the computer-aided design system is used for drafting and documentation. Drawings created and stored during the component design phase are recalled and detailed for construction of the initial prototype and preliminary tool design.

Evaluating prototype performance. Computer models speed the design process, reduce design costs, and give the designer a better understanding of the product. But in spite of the most confident computer-based predictions, the prototype will probably exhibit some structural difficulties. With hundreds of natural frequencies and dozens of forcing functions in a complex mechanical system, several may possibly cause vibration or failure problems. A "discover, analyze, fix, and verify" process is used to debug the prototype. The sophisticated modal analysis and system modeling tools used in the earlier stages of the design process are useful again in prototype refinement.

The first step is to conduct operating tests. These tests pinpoint the forcing functions and natural frequencies which are causing the problem. Artificial excitation tests are performed to yield additional data: modal mass, damping, natural frequencies, and mode shapes. These fundamental dynamic properties are extracted for each significant natural frequency of the system. Animated mode shapes help the troubleshooter visualize the effects of dynamic weaknesses. Then approaches to solving the problem are considered. Modal parameters can be changed (for example, modal damping, stiffness, or mass). This procedure determines the sensitivity of the mechanical system to each type of modification but does not necessarily reflect the effects of physical modifications to the structure. Another option is to physically alter the prototype. This is expensive and time-consuming and usually does not yield the desired result, however.

For most problems, system modeling is a better approach. A mathematical model of the prototype is made, using the modal data base developed during artificial excitation. Then, instead of cutting the prototype, the engineer alters the mathematical model. It is easier and less expensive to alter the software prototypes than the physical prototype. When a near-optimal redesign is developed, operating data (road profile data in the case of vehicle design) are applied to the model. If the operating data are satisfactory, the next step is to look at the whole spectrum again. If the redesign has not created any new problems, the physical prototype can be altered. Finally, new operating and artificial excitation tests are performed on the altered prototype as verification of satisfactory dynamic performance. After the prototype is refined sufficiently, the final configuration is released to manufacturing.

Designing a CAE facility. The optimum CAE facility is a distributed hardware system. This facility consists of one or more large mainframes for batch computing, large-system analyses, and corporate (or division) archival engineering data management systems. Superminicomputers are used as product or project computers and inferfaced with mainframes via network communications systems. Intelligent graphics work stations are dedicated to particular engineering or manufacturing functions. Testing work stations are specially designed to collect, catalog, and analyze test data. These interface with the midsize product computers via high-speed cable communications. For multidivision firms, a worldwide communications network provides access to common capabilities and applications software for all divisions or departments around the world.

All these components are integrated into systems which generate, process, format, communicate, and store data for each function and application in the total CAE product development process.

Better products. Computer-aided engineering is a product design and development philosophy which integrates design, analysis, testing, drafting-documentation, and related manufacturing considerations into each phase of the product development process. At the heart of CAE is the system model, which reduces the time and cost of product development and encourages innovation. The system model provides the ability to evaluate alternative concepts in the computer before hardware is built, allowing the engineer to apply personal creativity more effectively. The resulting products are more innovative, more competitive, and of greater value to the end user.

For background information *see* COMPUTER-AIDED DESIGN AND MANUFACTURING; FINITE ELEMENT

METHOD in the McGraw-Hill Encyclopedia of Science and Technology.

[JOHN Q. PARMATER]

Computer storage technology

The digital optical disk recorder developed as an extension of the technology of the consumer video disk player first described in 1973. The recorders use laser light to machine micrometer-sized holes (Fig 1*a*) on a thin tellurium alloy film, providing a permanent spatial representation of the data on a rotating disk. Gas lasers are used when maximum data rate and disk data capacity are desired, and semiconductor diode lasers are used when low cost and compactness are more important. These recorders can store between 10^{10} and 10^{11} bits of user data on a 30-cm-diameter disk (about the size of a typical phonograph record), and data can be retrieved from any location in less than 1 s. Data can be recorded or played back at rates as high as 10^7 bits per second with an error rate of 1 in 10^9 bits. This article describes a gas laser recorder and optical disk which exist in prototype form.

Disk. The sensitive material on the disk, a tellurium alloy, is deposited as a 30-nm-thick microcrystalline film on a plastic substrate. The sensitivity of the material for hole burning is between 100 and 300 millijoules/cm^2 for all visible laser light wavelengths. The film is considered suitable for use in archives requiring a record lifetime in excess of 10 years.

The disk is constructed to protect the sensitive material from dust without degrading sensitivity and to allow normal handling by the operator. The disk protective mechanism, called the Philips air sandwich (Fig. 1*b*), consists of two disks separated by ring spacers at the inner and outer radii of the information band. The annular cavity between the two disks is essentially a miniature clean room protecting the tellurium alloy from dust. Because the information is recorded and played back through the plastic substrate, approximately 1 mm thick, dust and fingerprints on the outer surfaces are out of fo-

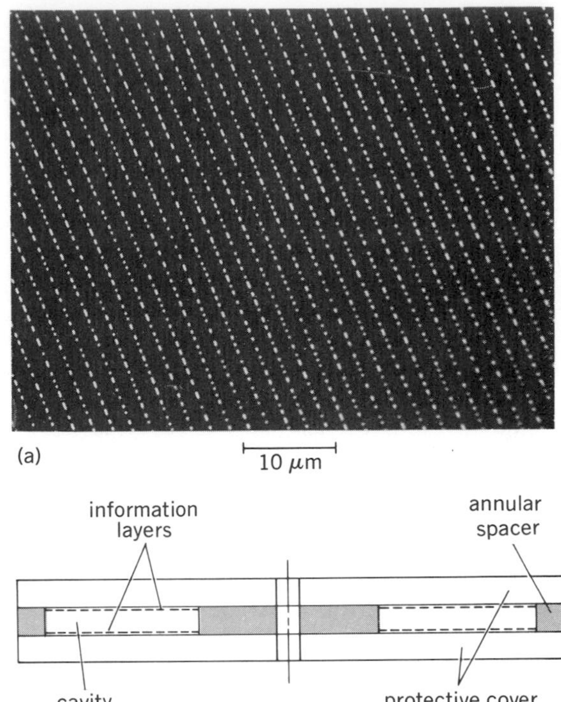

Fig. 1. Disk used for data storage in optical recorder. (*a*) Detail of pits. (*b*) Air sandwich structure.

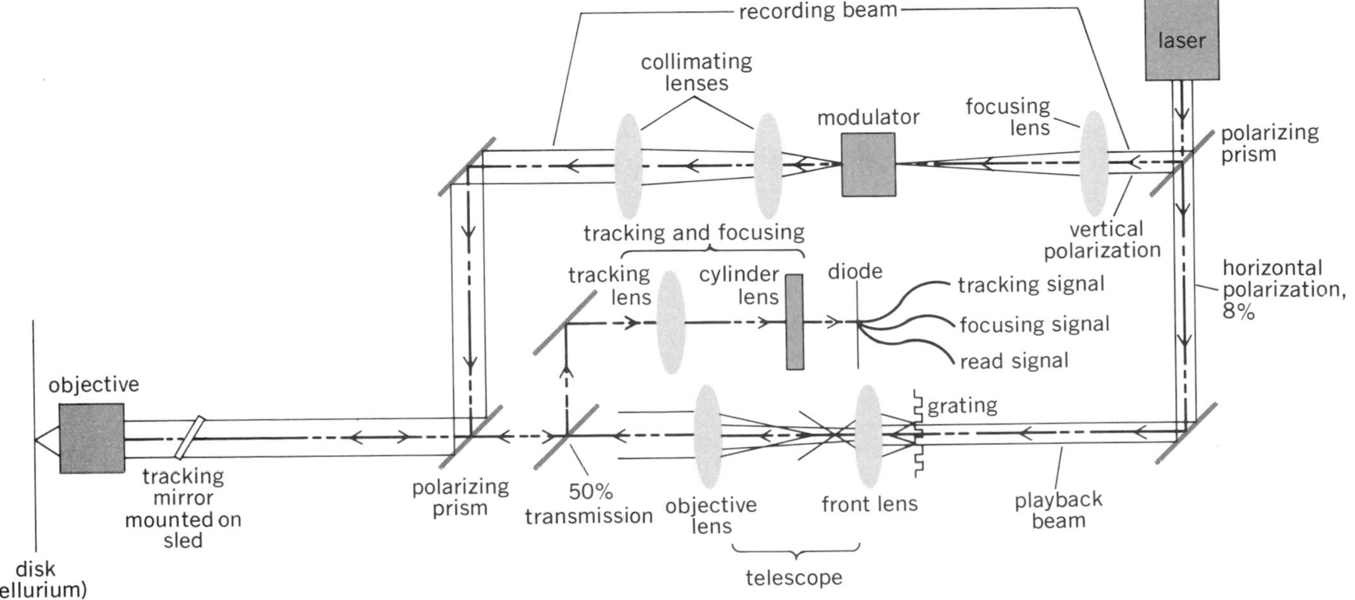

Fig. 2. Optical layout of recorder.

cus and are essentially transparent to the system.

Drive. The recorder drive converts electrical recording signals to a spiral of holes on the active material of the disk, and retrieves the data from the disk, converting it to electrical playback signals. The driver consists primarily of a turntable and linear sled or translator to generate the spiral and optics for recording and playback.

Optical system. In the optical system, two light beams are generated, one for recording and one for playback. These beams share a common objective lens for focusing on the disk.

In the optical layout of Fig. 2, light from the laser is elliptically polarized by a half-wave plate and divided by a polarizing beam-splitting prism into a recording beam and a playback (read) beam. Usually 8 to 10% of the total laser light is in the playback beam. The intensity of the playback beam is limited by the need for nondestructive retrieval of data; the intensity of the recording beam is maximized for high-data-rate recording.

The recording beam is focused by a lens to a narrow beam waist at the modulator. The modulator is of the acoustooptic type, which is suited for hole burning because of its high extinction ratio (contrast). For digital recording, the modulator is excited by a series of high-frequency bursts in synchronism with the digital data.

The modulated recording beam is collimated by a pair of lenses and is reflected off a mirror and a second polarizing beam-splitting prism to the tracking mirror and objective lens. The tracking mirror and objective lens are mounted on the linear sled and, therefore, move with respect to the other optical elements as the read head travels from inner to outer radius.

The read beam passes through a phase grating to produce a three-beam array. The zero-order (undiffracted) beam is used for reading and focusing. The two first-order diffracted beams are used for radial tracking. The three-beam array is expanded and collimated by a pair of lenses forming a telescope, and is focused on the disk by the objective.

Electronic subsystems. The rest of the recorder consists of electronic subsystems for modulation, signal processing, error detection and correction, and computer control. An intermediate modulation of the recording signal (Miller modulation) is required before optical recording to provide for self-clocking of retrieved data and for a spectral match of the recording channel. Signal processing is required to compensate for the nonlinear recording process and to reshape the playback waveform for clock recovery.

Error correction. Error correction is accomplished by three means. First, the data are encoded for double error detection in a limited string of bits; then the encoded data are interleaved to provide immunity to long-burst errors caused by scratches. Finally, the recorded data of each block or sector are verified bit by bit by examining the playback signal which is obtained after a small delay (direct read after write, or DRAW). If the recorded data quality falls below the correcting capacity of the system, the block is rewritten.

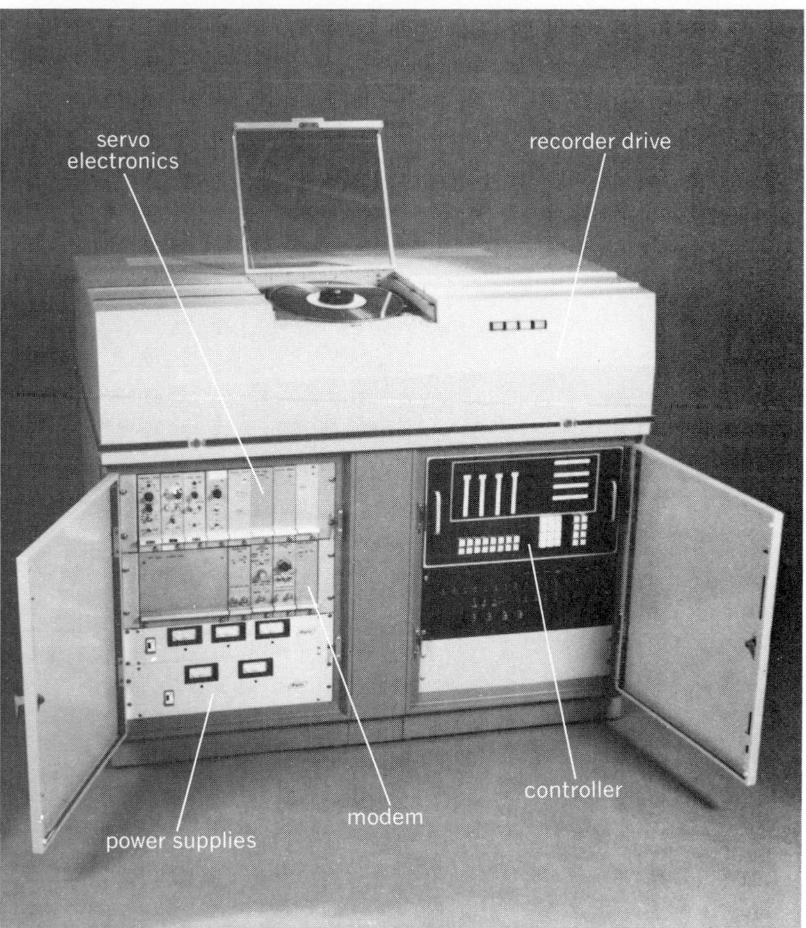

Fig. 3. Optical recorder showing major components.

Complete recorder. The completed recorder is shown in Fig. 3. In addition to the subsystems already described, there is a controller which allows the recorder to communicate with the PDP 11 minicomputer with random access on playback. The system records 10^{10} user bits of data per disk with record and playback rates of 5 Mbit/s. Random access to any data block (1 kilobyte) is accomplished on an average of 1 s.

For background information *see* COMPUTER STORAGE TECHNOLOGY; VIDEO DISK RECORDING in the McGraw-Hill Encyclopedia of Science and Technology. [ROBERT MC FARLANE]

Bibliography: R. Bartolini et al., *IEEE Spectrum*, 15(8):20–28, 1978; K. Broadbent, *J. SMPTE*, 83(7):553–559, July 1974; K. Bulthius et al., *IEEE Spectrum*, 16(8):26–33, 1979; K. Compaan and P. Kramer, *Philips Tech. Rev.*, 33(7):178–180, 1973; G. Kenney et al., *IEEE Spectrum*, 16(2):33–38, 1979.

Conducting polymers

The first covalent polymer exhibiting the electronic properties of a metal but containing no metal atoms, polymeric sulfur nitride, $(SN)_x$, was actively studied in the latter 1970s. In 1977 the first covalent organic polymer, polyacetylene, $(CH)_x$, which could be doped through the semiconducting to the metallic regime was reported. This has served as a prototype for the synthesis and study of other conjugated dopable organic polymers, several of which have been discovered during the last 3 years. Polyacetylene has been investigated much more extensively than any other conducting polymer.

Polyacetylene. Polyacetylene is the simplest possible conjugated organic polymer. It can be prepared in the form of lustrous, silvery, polycrystalline films by the catalytic polymerization of acetylene gas, HC≡CH, at less than 1 atm (101.3 kPa) pressure. It exists in two isomeric forms or as a mixture of the two forms depending on the temperature at which it is synthesized. The structures of the two forms, *cis*- and *trans*-$(CH)_x$, are shown below. The carbon-carbon bond lengths in the trans isomer are almost equal and are intermediate between those expected for a single and double bond. The cis isomer is converted to the more thermodynamically stable trans isomer on heating to 150–200°C or by chemical or electrochemical doping and "undoping" (compensation) at room temperature.

```
 H          H  H          H
  \        /    \        /
   C=C          C=C
  /    \      /    \      /
        C=C          C=C
       /    \      /    \
      H      H  H        H
                Cis

    H     H     H     H
    |     |     |     |
    C     C     C     C
  /  \\ /  \\ /  \\ /  \\ /
      C     C     C     C
      |     |     |     |
      H     H     H     H
               Trans
```

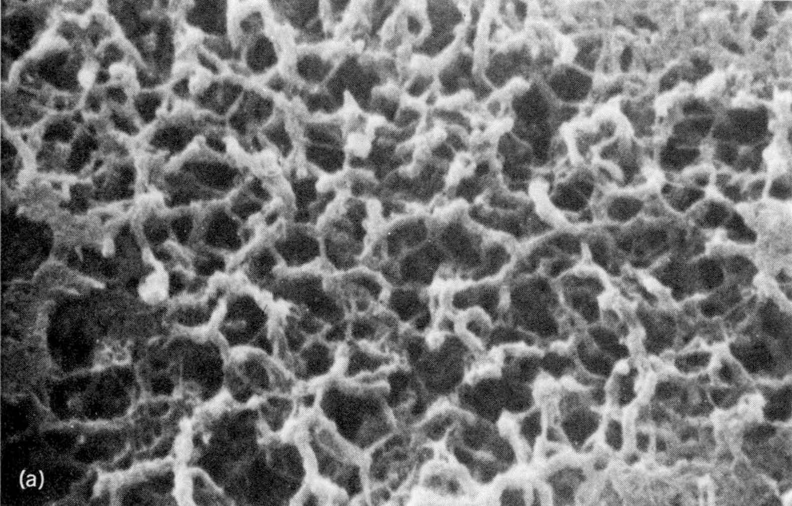

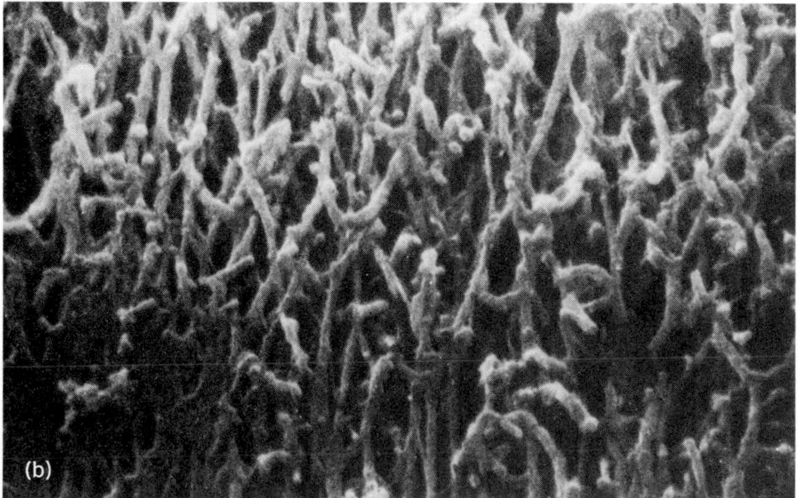

Fig. 1. Scanning electron micrographs of polyacetylene $[(CH)_x]$ films. (*a*) As-formed film. (*b*) Stretched film showing partial alignment of individual fibrils. Average fibril diameter is approximately 20 nm.

Electron microscopy studies show that the as-formed $(CH)_x$ films consist of randomly oriented fibrils (typical fibril diameter about 20 nm). The bulk density is approximately 0.4 g/cm^3 compared with 1.2 g/cm^3 as obtained by flotation techniques. This shows that the polymer fibrils fill only about one-third of the total volume. X-ray studies show that the films are polycrystalline with principal interchain spacing (in the cis polymer) of 0.439 nm. The surface area of the films is approximately 60 m^2/g.

The room-temperature conductivity of films of polyacetylene depends on the cis-trans content, varying from 10^{-5} ohm^{-1} cm^{-1} at room temperature for the trans material to 10^{-9} ohm^{-1} cm^{-1} for the cis isomer. In view of the sensitivity of the material to impurities or defects as demonstrated by doping studies, it appears likely that the intrinsic conductivity of pure polyacetylene is even lower.

Doping of films. *p*-Type doping, which involves oxidation of $(CH)_x$, occurs rapidly when either the cis or trans film is exposed to a variety of chemical reagents such as Br_2, I_2, AsF_5, $HClO_4$, H_2SO_4, and $(NO_2)(PF_6)$. It may also be accomplished conveniently by electrochemical methods by which anions such as $(ClO_4)^-$, $(PF_6)^-$, and so forth are introduced. The film becomes doped with the species and its electrical, optical, and magnetic properties change greatly. With many dopants, the electrical conductivity increases rapidly through the semiconducting to the metallic regime to reach an ultimate maximum value in the range of 10^2 to 10^3 ohm^{-1} cm^{-1} with the best dopants. Typical maximum extents of doping, with values of conductivity at 298 K given in parentheses, are: $[CHI_{0.20}]_x$ (1.6×10^2 ohm^{-1} cm^{-1}), $[CH(AsF_5)_{0.10}]_x$ (1.2×10^3 ohm^{-1} cm^{-1}), $[CH(H_2SO_4)_{0.1}]$ (1×10^3 ohm^{-1} cm^{-1}), and $[CH(ClO_4)_{0.065}]_x$ (9.7×10^2 ohm^{-1} cm^{-1}). Doping can be terminated conveniently at any degree of lower doping level desired, with corresponding lower conductivity.

n-Type doping, which involves reduction of $(CH)_x$, occurs rapidly when the film is immersed in a solution of lithium, sodium, or potassium naphthalide in tetrahydrofuran solution. Cations such as Li^+, Na^+, and K^+, respectively, are thereby introduced to give compositions such as $[Li_{0.3}(CH)]_x$ (conductivity at 298 K = 20 ohm^{-1} cm^{-1}). n-Doping may also be accomplished conveniently by electrochemical methods with the incorporation of cations such as Li^+, $[(n\text{-}C_4H_9)_4N]^+$, and $[(C_6H_5)_4P]^+$.

Electronic properties of doped $(CH)_x$. When *cis-* or *trans-*$(CH)_x$ films are p-doped (by acceptor species) or n-doped (by donor species), their conductivity increases sharply over many orders of magnitude at low concentration, then saturates at higher dopant levels, above approximately 1%, at which level the material undergoes a semiconductor-metal transition. This transition has been confirmed by a sharp decrease in the activation energy for the conduction process, and by thermoelectric power, electron paramagnetic resonance, reflectance, far infrared transmission, and visible/near-infrared absorption studies.

Cis-rich films of $(CH)_x$ can be stretched readily at room temperature up to three times their original length with partial alignment of the $(CH)_x$ fibrils. This effect is illustrated in Fig. 1. The conductivity in the direction of alignment increases sharply to values well in excess of 2000 ohm^{-1} cm^{-1} after doping with AsF_5, while the conductivity in directions perpendicular to alignment decreases, consistent with the conduction process occurring primarily along the fibrils.

The electronic properties of $(CH)_x$ in both the semiconducting and metallic regimes have been utilized in the fabrication of a variety of *pn* and Schottky rectifying and photovoltaic junctions. The electrochemical doping and undoping of $(CH)_x$ has resulted in the use of $(CH)_x$ in lightweight, high-energy-density, high-power-density rechargeable storage batteries.

Bonding and conduction mechanism. Most studies to date have concentrated on elucidating the nature of *trans-*$(CH)_x$. The currently most widely accepted theory for explaining the electronic properties and bonding in *trans-*$(CH)_x$, and its doped forms, involves the concept of neutral, positive, and negative solitons. A positive soliton, present in p-doped $(CH)_x$, is shown in Fig. 2. It consists of approximately 15 (CH) units over which the positive charge of unity is spread. The charge diminishes in magnitude from the center to the ends of the soliton, and is designated by δ^+ symbols of decreasing size. The greatest charge is located nearest the dopant anion, A^-. The carbon-carbon bond lengths are believed to be identical at the center of the soliton and to gradually become more and more unequal as the ends of the soliton are approached. In *trans-*$(CH)_x$, three main electronic regimes are obtained:

1. In the semiconducting regime, where the dopant concentration is small (less than about 0.1%), the properties are similar to those of a doped semiconductor. The conduction process proceeds by a hopping mechanism involving either a movement of a positive soliton [in p-doped $(CH)_x$] or negative soliton [in n-doped $(CH)_x$] along a chain, or the capture of a mobile electron from a neutral soliton in one chain by a relatively nonmobile positive (or negative) soliton in an adjacent chain.

2. In the metallic regime, where the dopant concentration is greater than about 6%, the electronic properties are characteristic of a true metal, displaying high conductivity, a temperature-independent Pauli paramagnetic susceptibility, a linear dependence on temperature of the thermoelectric power and low-temperature heat capacity, and metallic infrared and visible/near-infrared absorption spectral properties. There is strong evidence that in the metallic regime the carbon-carbon bond lengths have become equal since, both for p- and n-doped $(CH)_x$, the bandgap transition disappears. At this degree of doping, the solitons may be regarded as having overlapped with each other to form a conduction band which fills in the semiconductor bandgap.

3. In the transitional region the properties are intermediate between those of the semiconducting and metallic regimes. This intermediate regime is in many ways the most interesting; for example, although the electrical conductivity is high, the temperature-independent Pauli susceptibility is very

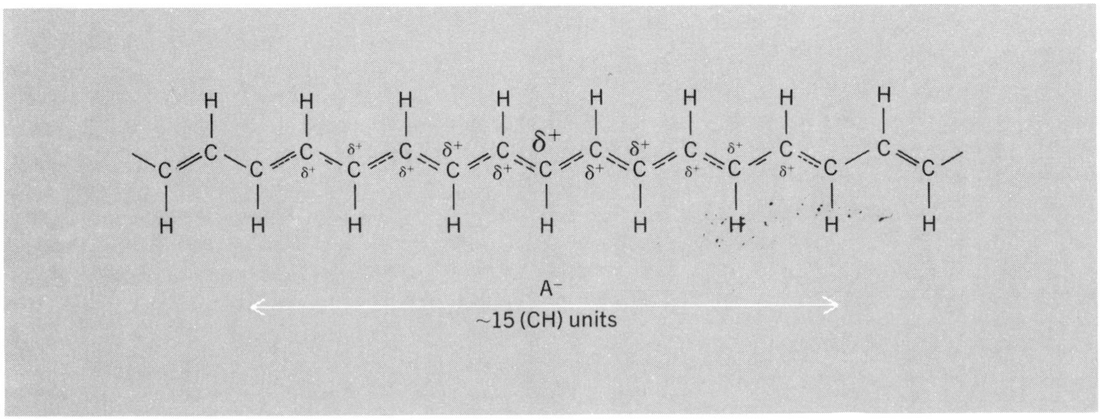

Fig. 2. Positive soliton in p-doped polyacetylene $[(CH)_x]$.

small, suggesting the possibility of a nontraditional transport mechanism.

Other covalent organic polymers. The second dopable covalent organic polymer, polyparaphenylene,

$[-C_6H_4-]_x$

(conductivity at 298 K = 10^{-15} ohm^{-1} cm^{-1}), was reported in 1979. It can be synthesized in a variety of ways as a brown powder having a high surface area. It can be *p*-doped chemically to the metallic regime by AsF_5, HSO_3F, and so forth, and electrochemically with the incorporation of anions such as $(BF_4)^-$ and $(PF_6)^-$. The highest conductivity (500 ohm^{-1} cm^{-1} at 298 K) is obtained with AsF_5 dopant. It may also be *n*-doped with lithium or potassium naphthalide in tetrahydrofuran to give materials having conductivities in the range 5–20 ohm^{-1} cm^{-1}. Analogously to $(CH)_x$, it is believed that the *p*- and *n*-doped materials contain organic cations or anions of the type $[(C_6H_5)^{+y}A_y^-]_x$ and $[M_y^+ \ (C_6H_5)^{-y}]_x$, respectively. The doped material has been used in the fabrication of high-energy-density and high-power-density rechargeable batteries.

Polyphenylene sulfide,

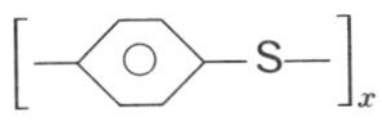

$[-C_6H_4-S-]_x$

can also be *p*-doped with AsF_5 to give material having a conductivity of approximately 10 ohm^{-1} cm^{-1}. The doping is accompanied by a complex chemical reaction, and recent results show that the chief compound formed,

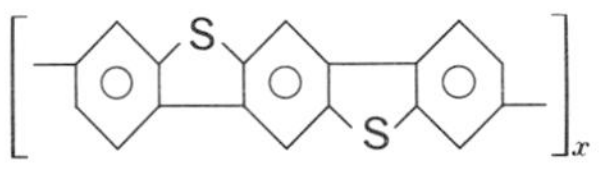

$[\ldots]_x$

is actually responsible for the high conductivity observed.

Other polymers, which have been less extensively investigated, but which can be doped chemically or electrochemically to moderately high conductivities, include the following:

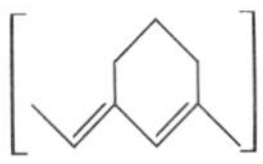

Poly(1,6-heptadiyne)

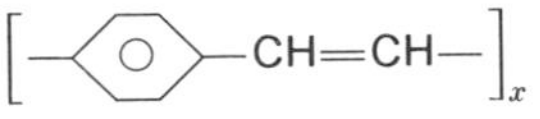

$[-C_6H_4-CH{=}CH-]_x$

Polyphenylenevinylene

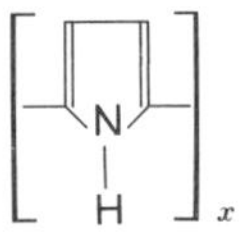

Polypyrrole

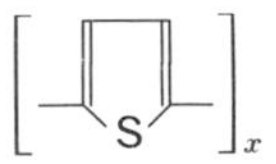

Polythiophene

The enormous increase in research in the field of dopable conducting organic polymers during the past few years and the wealth of new fundamental scientific concepts resulting from these studies indicate that this class of material is now well established as an important area of condensed matter science.

For background information *see* SOLITON in the McGraw-Hill Encyclopedia of Science and Technology. [A. G. MAC DIARMID; A. J. HEEGER]

Bibliography: International Conference of Low Dimensional Synthetic Metals, Helsingør, August 10–15, 1980, *Chem. Scripta*, 17:115–196, 1981; A. G. MacDiarmid and A. J. Heeger, *Synthetic Met.*, 1:101–118, 1979/1980; L. W. Shacklett et al., *J. Chem. Phys.*, 75:1919–1927, 1981.

Control systems

The advent of the NASA space shuttle heralds a new era in the exploration and utilization of space. This space transportation system makes it feasible to consider construction and operation of very large spacecraft and satellites in the relatively near future. Applications would include antennas and other structures for communications, remote sensing, astronomy, and energy generation, as well as space platforms for experimentation and the development of commercial processes not possible on Earth.

The size of these large space structures (LSS) may range from 20 m to 20 km in their largest dimension. The large size of the structure is the result in part of its purpose, for example, a high-resolution multibeam radar antenna, and in part of the readily available electricity which can be generated by large arrays of solar cells which can operate the various support systems on board. In fact, the "solar power satellite" concept for the generation of electric power for Earth use would be a rectangular array of solar cells nearly the size of Manhattan Island (and possibly as hard to handle); the electricity would be sent to Earth via lasers or microwaves.

No structure of this magnitude has ever been attempted in space before. A great deal of new technology must be developed, and a very high level of experience (gained from theory, computer simulation, and experiments in space) must be attained in order to guarantee the success of such a venture. Nevertheless, the next decade should witness the deployment of LSS in the 20–100-m range; these would be collapsed (for example, like an umbrella) to fit into the space shuttle payload bay and then erected in orbit. Other, even larger, concepts like the solar power satellite would require the capability to manufacture and construct structures in space from raw materials carried into orbit.

Because of their construction from lightweight materials (such as graphite-epoxy composites) and their size, it would be impossible to build or operate LSS on Earth due to their severe lack of rigidity. However, in the very-low-gravity environment of space, such highly flexible structures become reasonable and, indeed, even economical. To get an idea of how flexible these structures would be and how difficult to control, try to balance a piece of standard bond typing paper on the fingertips of both hands and, without pulling on the edges of the paper, try to keep the surface reasonably flat.

It is likely that LSS will exhibit the following characteristics: high total mass but low mass density (mass distributed throughout the structure over a large area), a very large number of low-frequency resonances (many below 1 Hz), and very low natural damping (hence, easily excited resonances).

Because of their high level of flexibility, LSS will tend to react to forces such as thermal and gravity gradients, solar pressure, and small impulse loadings in ways which were heretofore ignorable in smaller space structures. In particular, the solar power satellite will be especially vulnerable to these disturbance forces and yet, simultaneously, be expected to meet stringent performance requirements for shape fidelity, orientation, alignment, vibration suppression, and pointing accuracy; however, all LSS will face similar difficulties.

Active control of structures. Active control must be used to augment LSS in order to meet performance specifications. The basic element of active control is feedback. Sensors at various critical locations on the structure produce electronic signals proportional to local disturbance-induced errors in position, velocity, or acceleration. These signals are processed by control algorithms implemented in an on-line digital computer. The computer generates new electronic signals which are control commands applied to control actuators such as thrust engines or torquing devices. The control actuators, placed at various locations on the structure, exert forces to bring the structure back into line and reduce the effects of any disturbances. Although the placement of appropriate actuators and sensors is an important item, the control algorithm, implemented in real-time by the on-line computer, is the principal component in this feedback link from sensors to actuators. It must be carefully designed to carry out its complex tasks within the physical limitations of the actuators and sensors and the computational limitations of the computer. Control issues have always been present in aerospace applications, but they are substantially more critical for LSS.

Structural dynamics and control. The essential element of the control algorithm is a model of the structural dynamics. This model is used to predict the behavior of the structure and compare it with the actual sensor measurements. It uses any errors it detects to improve its perception of the instantaneous structural behavior and to generate control commands to cause this behavior to meet the system specifications (for example, to keep the microwave system of the solar power satellite accurately pointed at the Earth). For the rather rigid structures which make up current spacecraft and satellites, such models are relatively simple; however, LSS require very complicated sets of partial differential equations and corresponding boundary conditions to predict the behavior at every location on the structure. An LSS is an example of what is called a distributed parameter system, which in theory must be described by an infinite number of modes of vibration; but even in standard engineering practice a very large number of these modes is required to adequately approximate the dynamical behavior of the structure.

An additional complication is that the equations which describe the structural dynamics are sufficiently complicated that the true vibration modes can be approximated only by some numerical technique. The most popular such technique is the finite element method, which can be implemented by various computer codes, including NASTRAN, developed for structural analysis at NASA.

Model reduction and reduced-order control. The number of modes included in the structure model is related to the computer memory capacity and its access time, that is, how fast the computer can calculate. If only a few modes are included in the controller's model, the computer can do the calculations very rapidly and there is imperceptible delay in the feedback system. However, if a large number of modes is included in order to accurately model the structural behavior, the computer responds more slowly and the actuator feedback commands are delayed and arrive too late to correctly reduce the disturbance effects. In some cases, they arrive so late that they increase the disturbance effects and cause the system to be unstable.

This dilemma of how many modes to keep (and which ones to throw out) and still be able to compute control commands rapidly is called model reduction; it is the crux of the LSS active control problem. Even if all the low-frequency modes of a structure were retained in the controller model, they would more than exhaust the available computer capacity. Consequently, a model reduction must be performed and a trade-off made: some modes are retained and others (thought to be less critical for performance) are left out of the controller's model.

When a control algorithm is generated based on a model reduction of the actual structure, it is called a reduced-order controller. In theory, it is impossible to obtain a controller for a structure without model reduction. This is because an infinite number of vibration modes would be needed in the model; consequently, because of finite word length, no digital computer could achieve this. In practice, even though only a finite number of modes would be retained to approximate the structural dynamics, this number would be so large that there would be no practical way to implement the controller.

Controller-structure interaction problem. One of the most basic LSS control problems is the controller-structure interaction. By necessity, a controller must be designed which actively controls the structure using a reduced model. The controller computer is aware only of the modes included in its model of the structure. The residual (that is, unmodeled) modes are an intrinsic part of the actual structure, but the controller has no internal knowledge of them. When a particular sensor detects a local displacement error, this error is made up of contributions from all modes—modeled and residual; the part of the sensor output caused by residual modes is called observation spillover. The control computer accepts the sensor signal and compares it with its reduced internal model to produce control commands. Such a command causes a force to be applied at an actuator; this force excites all modes—modeled or residual. The excitation of the residual modes is called control spillover (control energy spills over into the residual modes).

Consequently, through the structure controller, feedback takes place both around the modeled modes of the structure (which is desirable) and around the unmodeled modes through the control and observation spillover terms (which is undesirable). The latter is the controller-structure interaction, and it must not be ignored in the design and evaluation of LSS control. An actively controlled structure is quite different from a passive structure; by electronic feedback, the active controller changes the structure's characteristics, and its response to disturbances. It is intended that this change will improve the performance of the structure; however, if the controller-structure interaction assessment is not considered as a fundamental part of any active control design for LSS, the sought-after performance improvement will be lost. In fact, in some cases, the controller interaction with the unmodeled modes can cause them to be unstable, which would mean an eventual loss of control of the structure.

Parameter estimation and adaptive control. An equally important basic problem in LSS control is that the control is only as good as the model. The effects of the unmodeled modes have been discussed above; however, even the modeled modes may be rather poorly described due to lack of knowledge of the zero gravity behavior of large structures and the possible changes in the configuration of the structure because of construction in space. Consequently, some of the parameters of the structure may be very poorly modeled; hence, there is a need for on-board estimation of structural parameters. Furthermore, it may be necessary to use an adaptive approach to control, that is, one which incorporates the ability of the control algorithm to make use of these parameter estimates to adapt (tune) itself to the structure it is controlling.

Prospects. Fundamental ideas in the control of LSS are related to a variety of technical areas, including partial differential equations, structural dynamics, and distributed parameter control theory. These (and some related issues) are topics of current research; their resolution is essential to the successful operation of large structures in space. A major outgrowth of this research will be a better understanding of highly flexible structures (not only aerospace structures but civil engineering structures such as suspension bridges and tall buildings as well) and their interaction with active control—an understanding which will be greatly aided by laboratory and space shuttle–based experiments in the future.

For background information *see* ADAPTIVE CONTROL; CONTROL SYSTEMS; FINITE ELEMENT METHOD; SOLAR ENERGY; SPACECRAFT STRUCTURE in the McGraw-Hill Encyclopedia of Science and Technology.

[MARK J. BALAS]

Bibliography: *AIAA Astronaut. Aeronaut. Mag.*, October 1978; M. Balas, *IEEE Trans. Automat. Control*, vol. AC-27, June 1982; M. Balas and C. R. Johnson, Jr., Toward adaptive control of large structures in space, in K. S. Narendra and R. Monopoli (eds.), *Applications of Adaptive Control*, 1980; S. Brand (ed.), *Space Colonies*, 1977; T. Heppenheimer, *Colonies in Space*, 1977; G. O'Neill, *The High Frontier*, 1978; *Outlook for Space: Report to the NASA Administrator*, NASA Rep. SP-386, January 1976; *Satellite Power System: Concept Development and Evaluation*, DOE/NASA Rep. DOE/ER-0023, October 1978.

Crab

Animals living in the intertidal zone, the interface between land and ocean, are dramatically affected by the changes in water level as the tidal height varies in a highly predictable fashion. It is therefore not surprising that many of these animals have internal rhythms in behavior, physiology, and reproduction which are timed in relation to the occurrence of a particular phase of the tide. In contrast, planktonic animals which live in the ocean are usually unaffected by tides because the water column does not change markedly over a tidal cycle. These animals lack tidal rhythms.

An exception to this generalization about zooplankton, however, is observed in an estuary. During a tidal cycle in an estuary there can be changes in water quality factors such as salinity and temperature due to the mixing processes driven by tidal currents. Recent detailed studies of the planktonic larval stages of the estuarine crab *Rhithropanopeus harrisii* have shown that the behavior of this species varies with the stage of the tide.

This crab lives on the bottom of the estuary as an adult but has larval stages that swim freely. The larvae are not uniformly distributed vertically in the water column, nor are they continuously clumped at one particular depth. Instead, they show changes in vertical position in relation to environmental cycles of light and dark, salinity, and current speed. The tidal vertical migration is quite extensive. The pattern, observed at the time in the lunar month when the greatest tidal amplitudes occur, consists of an ascent toward the water surface during rising tides and descent during falling tides.

This pattern could result from behavioral responses to tidal changes in environmental factors (for example, salinity), and, in fact, many studies have shown that crab larvae are very responsive to changes in environmental factors. However, when an animal displays overt rhythmic changes in nature, one suspects that these cycles are controlled by both responses to environmental change and an internal clock. Consequently, endogenous tidal rhythms might be expected to be found in crab larvae.

Laboratory studies indicate that the tidal migration pattern of these larvae is indeed controlled by an internal clock. This was demonstrated by collect-

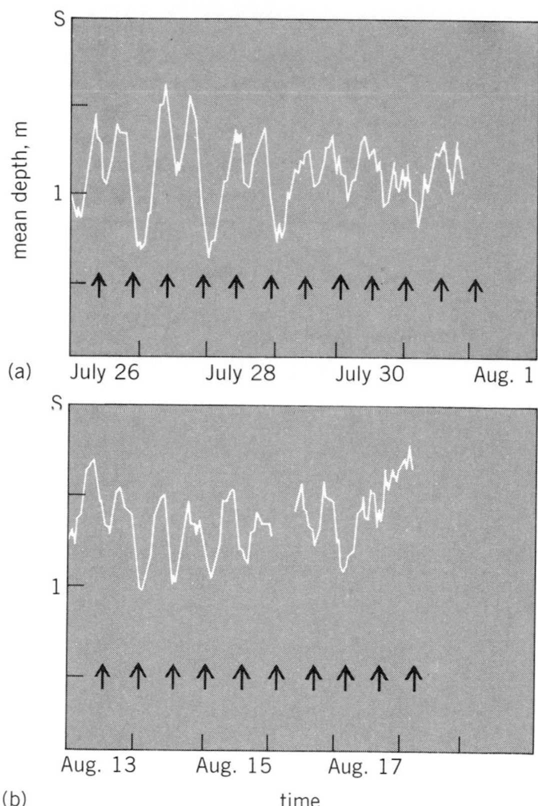

Fig. 1. Movements of the center of the larval distribution for animals caught at time of (*a*) spring and (*b*) neap tides. S is the surface, and the arrows show the time of low tide at the collection site. (*From T. W. Cronin and R. B. Forward, Jr., Tidal vertical migration: An endogenous rhythm in crab larvae, Science, 205:1020–1021; copyright 1979 by the American Association for the Advancement of Science*)

ing larvae in the field at times in the lunar month when the greatest and lowest tidal amplitudes occur (spring and neap tides, respectively). The larvae were placed in a 1.9-m-tall vertical column under conditions of constant darkness, temperature, and salinity. Their depth in the column was monitored over time by far-red light using a closed-circuit television system.

Under these conditions the larvae showed a regular vertical migration, the timing of which was related to actual tide times in the area where the larvae were captured (Fig. 1). They ascended at the time of rising tide and descended at the time of falling tide. This pattern is similar to that observed in the field near the times of spring tide. Since the animals are not exposed to natural tides or other tidal influences in the laboratory and a tidal migration pattern persists, an internal clock must be controlling this migration pattern.

Cycles in activity and predator avoidance. Further studies have shown that these crab larvae also have a tidal rhythm in activity. If larvae are captured in the field and maintained under constant conditions in the laboratory, they show a cyclic variation in swimming speed (Fig. 2). The pattern shows highest activity during rising tides, with a maximum usually several hours after the time of low tide at the capture site. The lowest activity occurs during falling tides, with the minimum recorded several hours after high tide. It is likely that the observed tidal vertical migration pattern results from this cycle in swimming.

During vertical migration the larvae actually move over a short vertical distance (1 to 1.5 m in the estuary). Considering their relatively rapid swimming speeds (about 7–11 m/h) and sinking rates (about 30 m/h), it is evident that the larvae are not swimming directly up during the ascent or simply sinking during the descent. Instead, depth changes correspond with alterations in activity. Assuming that at an average swimming speed the larvae tend to maintain a constant depth in the water, when the swimming speeds are above the average, the larvae would rise. Conversely, when swimming speeds are below the average, a descent would occur. During the activity rhythm, average swimming speeds occur at times very near those of high and low tide (Fig. 2), which correspond to the times when migration is at its highest and lowest limits (Fig. 1). During rising tides swimming speeds are above average, and an ascent is predicted. This is the situation observed in the column experiments. The opposite occurs during falling tides. Thus the changes in swimming speed probably lead to the laboratory-observed tidal vertical migration pattern.

Upon ascending to the surface area at high tide, the larvae are vulnerable to predation by ctenophores and jellyfish, which float at the surface and

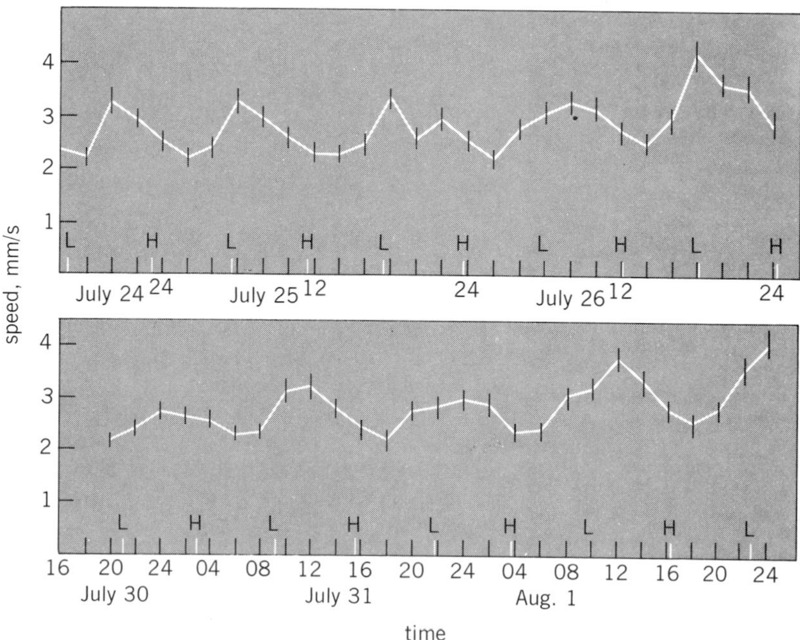

Fig. 2. Rhythms in swimming speed over time. Mean speeds and standard error are shown. L and H are the time of low and high tides, respectively. (*From R. B. Forward, Jr., and T. W. Cronin, Tidal rhythms in activity and phototaxis of an estuarine crab larvae, Biol. Bull., 158:295–303; copyright 1980 by the Marine Biological Laboratory*)

engulf zooplankton that become entrapped in their tentacles or feeding currents. Detailed observations of larval behavior using closed-circuit television indicate that the larvae have a predator-avoidance response in which the predator's shadow evokes sinking and, in certain situations, swimming away from the direction of the shadowed light source. This response is especially effective for avoiding ctenophores, which are one major predator on the zooplankton of estuaries. Freshly collected ctenophores usually have large numbers of zooplankton in their stomachs, but they rarely contain crab larvae. Recent studies have shown that this shadow response also varies over a tidal cycle. It is strongest on rising tides, when the larvae are likely to be near the predators at the surface, and decreases on falling tides, when larvae would be moving downward away from predators.

Functional advantages. An estuary is characterized by an overall seaward flow of water due to the entry of water from its fresh-water drainage basin. Animals living in the water column encounter the continual problem of potentially being exported into adjacent coastal areas. This problem is especially severe for species like *R. harrisii*, which live on the bottom as adults but which have larval stages in the water column. The limited swimming capabilities of the larvae prevent them from moving against seaward-flowing currents; yet their survival depends upon their being in the estuary near the adult habitat at the time when they settle out of the water column onto the bottom in preparation for metamorphosis into an adult. It has long been known that the larvae of various estuarine species, including *R. harrisii*, remain in the estuary through all stages of development. However, the method by which retention is effected is frequently unclear. The observed tidal rhythm of vertical migration could play an important part in retaining these larvae in the estuary.

Stratified, partially mixed estuaries have a layer of high-salinity water near the bottom. The net movement of this water over a complete tidal cycle is landward. In contrast, the upper layer of water has a lower salinity and shows net seaward movement. If an animal partitions its time between these layers, it is possible to reduce horizontal movement in the estuary. The observed tidal migration would serve this purpose. On rising tides, activity would increase, and larvae would move up in the water column and be transported up the estuary. As the tide falls, activity is reduced and larvae descend. Seaward transport would be reduced because the slowest seaward-flowing currents occur at greater depths. In this way the larvae would be retained near the parent population.

Tidal rhythms in behavior are thus an important adaptation in these estuarine larvae, since they permit the establishment of a permanent population of adult crabs in the estuary. The presence of tidally timed behaviors in other estuarine zooplankton or larvae is not yet established, nor are the synchronizing events which set the larval crab's clock or, indeed the location of the clock itself. These will be fruitful areas for future research, since tidal rhythms are potentially highly adaptive for the zooplankton of estuaries.

For background information *see* PERIODICITY IN ORGANISMS in the McGraw-Hill Encyclopedia of Science and Technology.

[RICHARD B. FORWARD, JR.; THOMAS W. CRONIN]

Bibliography: T. W. Cronin and R. B. Forward, Jr., *Science*, 205:1020–1022, 1979; R. B. Forward, Jr., and T. W. Cronin, *Biol. Bull.*, 158:295–303, 1980; J. D. Palmer, *Biol. Rev.*, 48:377–418, 1973.

Diabetes

Insulin is the body's principal hormone regulating glucose metabolism and the primary determinant of the concentration of glucose in the blood. Acting in concert with other hormones such as glucagon, the catecholamines, cortisol, and growth hormone, insulin maintains the concentration of glucose in the blood within a relatively narrow normal range at all times. It accomplishes this by balancing the rate of glucose removal from and entry into the blood, and through its actions to stimulate glucose uptake and metabolism in the cells of its peripheral target tissues (skeletal muscle and adipose tissue) and to inhibit glucose production in and output from the liver.

In type I diabetes mellitus (insulin-dependent diabetes or juvenile-onset-type diabetes), the pancreas fails to produce and secrete insulin. As a result of an absolute deficiency of this hormone, the concentration of glucose in the blood becomes abnormally high, and hyperglycemia and ketoacidosis ensue. This type of diabetes accounts for perhaps 5–10% of the total diabetic population.

The overwhelming majority of diabetics have type II diabetes (non-insulin-dependent diabetes or adult-onset-type diabetes). These individuals have normal or even excessive levels of insulin and appear to suffer a more mild abnormality of glucose metabolism than do those with type I diabetes. Their diabetes is characterized by either glucose intolerance or mild degrees of frank hyperglycemia and the absence of ketosis. Since the absolute amount of insulin in many of these individuals is not reduced, their abnormal glucose metabolism appears to be the result of an impaired ability of insulin to produce its normal metabolic effect on the cells of its target tissues—in other words, to a decreased insulin action on skeletal muscle, adipose tissue, and the liver. This condition is referred to as insulin resistance, a condition which appears to characterize many individuals with type II diabetes and which contributes significantly to their abnormal glucose metabolism. Insulin resistance also occurs frequently in obesity and rarely in other unusual metabolic diseases of humans. This article briefly summarizes some current knowledge about the mechanism of insulin action in the normal state, in diabetes, and in some other disorders in which insulin resistance is present.

Insulin receptor. The initial interaction of insulin with its target cells occurs at the cell surface, through specific receptors in the cell's plasma membrane. Insulin receptors have been identified on almost all cells of vertebrate species. Although the receptor has not yet been chemically characterized, considerable information about it and its reaction with insulin has been obtained through measurements of the binding of radioactively labeled insulin to cells or cell membranes using standard tissue culture techniques.

Insulin binding to its receptor is rapid, saturable, and reversible. There is a high degree of specificity of the receptor for insulin since it has a stronger attraction (affinity) for insulin than other hormones and substances, it binds insulin in preference to other hormones and substances, and its interaction with insulin results in the known specific biological effects of that hormone. The number of specific cellular insulin receptors appears to be regulated, at least in part, by insulin itself. Studies suggest that the plasma insulin concentration inversely regulates the number of insulin receptors, a phenomenon referred to as down regulation.

The insulin receptor enables a cell to recognize insulin in the blood, to bind it, and to form an insulin-receptor complex. The insulin-receptor complex then somehow activates or triggers an orderly sequence of events throughout the cell which culminates in a multiplicity of biological responses among which are glucose uptake and metabolism. The final biological effect of insulin on the cell, then, is a function, at least in part, of the ability of insulin to bind to its specific plasma membrane receptor and to form the insulin-receptor complex. The amount of insulin-receptor complex depends, in turn, upon the concentration of insulin, the concentration of receptors, and the affinity of the receptor for the hormone. Alterations in any one of these can influence the biological effect of insulin on the cell.

Thus, in insulin-deficient states such as type I diabetes, insulin binding and formation of the insulin-receptor complex do not occur because insulin is not available; the biological effect of insulin is absent because of the absolute lack of insulin. On the other hand, in type II diabetes and in obesity, more than enough insulin is usually present, yet the biological effect of insulin on glucose metabolism is blunted.

In an effort to determine whether abnormalities in the insulin receptor are responsible for the insulin resistance of type II diabetes and obesity, extensive biochemical and physicochemical investigations of the insulin-receptor interaction have been undertaken. These studies have provided new and critical insights into the importance of the insulin receptor in the normal mechanism of insulin action and the nature of alterations in insulin-receptor interactions in diabetes and other disorders of insulin action. At the same time a number of uncertainties have arisen concerning the nature of the insulin receptor and the role of the insulin-receptor interaction in mediating the biological effects of insulin in health and disease.

For example, it is uncertain whether a single class of specific receptors exists on the plasma membrane with which insulin interacts to control all of the subsequent metabolic steps, or whether more than one class of specific receptors resides in the plasma membrane with different affinities for insulin, each controlling separate metabolic functions of the cell. Moreover, although it is well established that the initial interaction of insulin with its target cells occurs at the cell surface receptor, it is now apparent that the insulin-receptor complex is internalized within the cell, resulting in the association of the hormone with intracellular particles at several different loci. Furthermore, a portion of the insulin bound to plasma membrane receptors is degraded and therefore unavailable for insulin action on the cell. Whether internalization or degradation plays an important role in mediating insulin action in the normal and diabetic state is not known.

There are many more insulin receptors on the plasma membrane than are necessary for the full action of insulin. Only a relatively small fraction (10%) of the total number of receptors are occupied when maximal biological activity occurs, after which additional receptor occupancy does not induce a further response to insulin. A marked reduction in either the number of insulin receptors or the amount of insulin would be required if insulin resistance were the result of decreased insulin binding. Such abnormalities do appear to exist in certain diabetics and in some insulin-resistant states. On the other hand, in other diabetics and other insulin-resistant states, decreased insulin action exists in the face of hyperinsulinemia and normal insulin binding. It is apparent that the underlying defect in these individuals resides at some site other than the insulin receptor.

Insulin's interaction with its receptor is an important, indeed critical, component of the process by which insulin influences the cell, but it is by no means the only event which determines the final biologic effects of the hormone. A series of postreceptor reactions combine to produce the ultimate biological response of the cell to its environment. Consequently, considerable attention has focused on factors beyond the receptor, that is, on postreceptor events.

Postreceptor events. Once insulin interacts with its cell surface receptor, the insulin-receptor complex somehow triggers an orderly sequence of postreceptor events throughout the cell involving transport, enzyme activation and deactivation, and other metabolic events which combine to produce the ultimate biological response of the cell to insulin.

Each of these events, as well as others, is a critical factor in determining the total biological effect of insulin on the cell. Alterations in any of them could, independent of the insulin receptor, influence the cell's response to insulin, and could be responsible for an impairment in insulin action and the development of insulin resistance.

Although the mechanisms through which insulin binding to its receptor accomplishes this highly integrated process are poorly understood, several events probably are involved, including transformation of the insulin-receptor interaction into a transmembrane signal (coupling of insulin binding to effector systems), generation of an interacellular messenger, and subsequent alteration of transport systems and enzymes within the cell, which lead to the final biological effects of the hormone. Recent studies have given rise to the hypothesis that the insulin-receptor complex may generate a peptide mediator in the plasma membrane which enters the cell and directs transport and enzyme systems to carry out the actions of insulin. Other investigators have presented evidence that the insulin-receptor complex may comprise several individual regulatory components or subunits, each with a specific function in transforming and transmitting the binding signal.

Whatever the transformation and signaling process, it seems apparent that alterations in any of these postreceptor events could impair insulin's ability to affect the cell even in the presence of normal insulin-receptor interaction. These alterations could be the result of a faulty coupling of insulin binding to effector systems in the face of normal insulin binding or of an abnormality at one or more of these postreceptor events, independent of the insulin-receptor interaction.

Type II diabetes mellitus. It has become increasingly clear that individuals classified as having type II diabetes mellitus are a metabolically heterogeneous group. Many (approximately 85%), but not all, of these individuals are obese. Abnormalities of glucose metabolism in type II diabetes may range from mild degrees of glucose intolerance to much more serious derangements, including severe fasting hyperglycemia. Plasma insulin concentrations may vary considerably from low to normal to excessive: the more mild the abnormality in glucose metabolism, the higher the plasma insulin levels. Most individuals with mild glucose intolerance are hyperinsulinemic, while those with severe fasting hyperglycemia tend to have lower than normal plasma insulin concentrations. Insulin resistance is present in some, but not all, type II diabetics, and it is believed that this tissue resistance to insulin is at least partially responsible for the abnormal glucose tolerance and hyperglycemia in these patients. Most investigators believe that in many of these individuals the obesity is responsible for and contributes importantly to the development of the insulin resistance, since it is well established that obesity causes insulin resistance and that weight loss restores to normal the ability of insulin to influence tissue glucose metabolism. Obesity is, in fact, probably the most common form of insulin resistance in humans. Yet, insulin resistance also exists in type II diabetic patients who are not obese. Thus, insulin resistance appears to be characteristic of most patients with type II diabetes, regardless of whether or not they are obese, and it appears that both obesity and diabetes per se can contribute to the insulin-resistant state.

Insulin binding to various types of cells from type II diabetics has been shown to be decreased due to a reduction in the number of insulin receptors per cell; receptor affinity for insulin is normal. Similar observations have been made in obese persons without diabetes. The decrease in receptor number has been demonstrated in both nonobese and obese individuals with type II diabetes and in individuals with mild to more severe forms of the disease with or without fasting hyperglycemia. Thus, it has been postulated that, at least in some individuals with type II diabetes, insulin resistance is due to a decreased number of receptors. Indeed, in many type II diabetics with mild abnormalities in glucose metabolism there is a highly significant inverse correlation between the number of insulin receptors and the degree of insulin resistance.

One hypothesis advanced to explain the decreased receptor number postulates that the hyperinsulinemia of type II diabetes causes a reduction in the number of insulin receptors on the cell—that decreased insulin binding is due to down regulation of receptor number by insulin itself. While this may provide an explanation in some individuals, clearly this is an oversimplification and cannot explain the insulin resistance in all type II diabetics, since, for example, insulin resistance can be demonstrated in type II diabetics with hypoinsulinemia. Other explanations seem to be equally plausible, including a primary reduction in receptor number leading to insulin resistance, hyperglycemia, and, secondarily, to hyperinsulinemia.

While a decreased number of cellular insulin receptors may characterize many individuals with type II diabetes, the existence of decreased receptor number does not necessarily establish it as the cause of the insulin resistance in these patients. Indeed, there is increasing evidence from studies in both laboratory animals and in humans that insulin resistance in many diabetics may be related to abnormalities beyond the insulin receptor, that is, to postreceptor events. Thus, in some diabetic and obese persons and animals, receptor number is not decreased or there is a lack of correlation between the degree of insulin resistance and insulin binding. Moreover, it has been demonstrated that insulin's ability to stimulate the transport of glucose into the cell may be impaired in this form of diabetes, independent of insulin's interaction with its receptor. Studies of tissues from obese animals and inviduals with abnormal glucose metabolism also demonstrate alterations in specific intracellular pathways of glucose metabolism independent of the character of insulin-receptor interaction.

Thus, defects at several levels of insulin action, including the insulin receptor, glucose transport, and intracellular glucose metabolism, have been described in cells from type II diabetics and the obese. It is possible that different defects in insulin action may exist in different individuals, in keeping with the apparent metabolic heterogeneity of this disease. On the other hand, multiple defects in insulin action may exist in type II diabetics either in the same cells (for example, combined receptor and postreceptor defect in muscle or adipose cells) or in different cells (for example, a receptor defect in hepatic cells and a postreceptor defect in the peripheral tissues).

The insulin resistance in the tissues of type II diabetes can be improved in a variety of ways. Weight reduction in obese diabetics is accompanied by a fall in plasma insulin concentration, improvement in glucose tolerance, and increased ability of insulin to stimulate glucose metabolism in the tissues. These changes may be paralleled by an increase in insulin receptor number, an enhanced rate of glucose transport, or a change in the intracellular pathways of glucose metabolism in the cells of insulin's target tissues. Similarly, treatment of type II diabetics with oral hyperglycemic agents has been shown to result in improved glucose tolerance, sensitivity to insulin, and increased receptor number, glucose transport, and intercellular glucose metabolism.

Type I diabetes. These individuals have an absolute deficiency of insulin due to some as yet unknown primary pancreatic abnormality. Insulin-deficient animals, and untreated insulin-dependent diabetics, have either an increased or normal number of insulin receptors. Postreceptor changes, such as decreased glucose transport, have been observed in these animals.

Rare unusual types of insulin resistance. A syndrome of severe insulin resistance due to circulating antibodies (IgG) to the insulin receptor has been described in humans. The antireceptor antibodies lead to decreased insulin binding, hyperinsulinemia, and hyperglycemia. Extreme degrees of insulin resistance due to a primary reduction in receptor number have also been observed in some patients. Finally, some forms of insulin resistance in which insulin binding is normal and a postreceptor defect is observed have been reported.

For background information *see* DIABETES; INSULIN in the McGraw-Hill Encyclopedia of Science and Technology.

[LESTER SALANS]

Bibliography: J. M. Olefsky, The insulin receptor: Physiology and pathophysiology, in *Current Concepts*, 1980; G. M. Reaven, *Metabolism*, 19:445–454, 1980; J. Roth et al., *Recent Prog. Horm. Res.*, 31:95–126, 1976; L. B. Salans and S. W. Cushman, Relationships of adiposity and diet to the abnormalities of carbohydrate metabolism in obesity, in H. M. Katzen and R. J. Mahler (eds.), *Diabetes, Obesity and Vascular Disease: Metabolic and Molecular Relationships*, pp. 267–302, 1978; L. J. Wardzala, S. W. Cushman, and L. B. Salans, *J. Biol. Chem.* 253:8002–8005, 1978.

Diamond

Diamond, the king of minerals, consists of the same element as graphite, namely carbon. The difference between diamond and graphite is the crystal structure, that is, the manner in which the carbon atoms are arranged in space and bonded together. In graphite, the atoms are arranged in a hexagonal way, whereas in diamond the atoms are in a close-packed cubic arrangement. It is this relatively simple crystal structure that imparts to diamond its unique hardness and dispersion, as well as other important properties. These two attributes make diamond the ideal gem, but several of the other properties, such as high thermal conductivity and semiconducting character, make it useful in numerous industrial ways.

Interestingly, under the ambient conditions of temperature and pressure at the Earth's surface, diamond is metastable and should invert to graphite. Luckily this transformation is so slow as to be nonexistent at room temperature, and it is known that diamonds have existed at the Earth's surface for over 1.7×10^9 years without change. Yet at only 800°C, diamonds will oxidize in a few hours.

The reverse transformation, graphite to diamond, is a reaction that historically many alchemists wished to perform. Although it was known that high temperatures and pressures were necessary for the reaction to occur, it was not until the early 1950s that attempts to synthesize diamond were successful. The first documented laboratory production of diamond was by Allman Svenska Elektriska Aktiebolaget (ASEA) in Sweden, and this was closely followed by General Electric (GE) in the United States. Temperatures and pressures of the order of 1700°C and 55 kilobars (5.5 MPa) are necessary, as well as the presence of a transition-metal catalyst. Today factories in the United States, Ireland, South Africa, and the Soviet Union produce large quantities of synthetic diamonds for industrial uses.

In spite of this large output of synthetic material as well as the experimental, but not commercial, production of small gem-quality diamonds, nature has retained the secret for the conception and gestation of natural diamond. Mineralogists, geochemists, and physicists have undertaken considerable effort to unravel the enigma of the genesis of natural diamond, but to little avail. One problem is that the depths in which diamonds form are beyond the capability of humans to sample directly. Thus it is necessary to rely on the whim of nature to provide samples. Luckily this has been done by the host rock in which diamonds occur—kimberlite.

Kimberlite. Kimberlite is named after the town of Kimberley, South Africa, where this type of rock was discovered. The exact relationship between diamond and kimberlite is still a matter of speculation, although evidence suggests that kimberlite is the vehicle by which diamond is transported to the surface from depths of about 200 km. Otherwise, the two are not genetically related; that is, kimberlite in a molten state is not the mother liquor from which diamond crystallized.

The name kimberlite actually refers to a special type of rock that in an earlier molten and fluidized state transported accidental fragments of mantle and crustal rocks to the Earth's surface. In many respects kimberlite is an unusual rock and, though it is possible to define the now consolidated rock within general terms, the mineralogical and chemical nature of kimberlite at depth is unknown. Kimberlite is now considered to be an intrusive inequigranular rock which contains large crystals of olivine, phlogopite mica, pyrope-almandine garnet, pyroxene (diopside), and magnesian ilmenite set in a finer-grained matrix of serpentine, phlogopite, calcite, chlorite, perovskite, spinels, and monticellite. One or more of the above minerals may be dominant or absent in any given kimberlite. The presence of diamond is not necessary for a given rock to be referred to as a kimberlite, and more nondiamondiferous kimberlites are known than diamondiferous ones.

Kimberlite usually occurs in small (a few acres; 1 acre = 0.4 ha) pipelike bodies, or as thin dikes, but rarely as sills. The common occurrence is in pipes which are in effect the eroded sections of old volcanoes. However, these volcanoes did not produce lava as in the case of Hawaii or Vesuvius. Instead they erupted with great velocity vast amounts of ash and other debris, including, presumably at times, diamonds. These volcanoes were generally confined to very old stable continental, or shield, areas, and thus kimberlite is normally found only in these stable cratons. Nevertheless, kimberlites have intruded these cratons several times throughout the geologic record. For example, in southern Africa, kimberlites erupted in the Precambrian (1.2×10^9 years), Devonian (3×10^8 years), and Cretaceous (9×10^7 years). Generally a kimberlite pipe is conical in shape, tapering at depth to a narrow dikelike outlet. The structural control of kimberlites has not yet been elucidated, and this is possibly due to the fact that kimberlite pipes are the surface manifestation of deeper upper mantle structures that are not otherwise reflected in the crustal rocks.

The speed of ascent of the kimberlitic material was sufficiently fast that blocks of rock were torn from the sides of the vent or conduit. These blocks represent fragments of the crust and upper mantle through which the kimberlite has passed during its ascent to the surface. Detailed mineralogical and geochemical study of these exotic blocks, or xenoliths as they are correctly termed, has shown that several of them were incorporated in the kimberlite at depths of the order of 150 to 200 km.

Xenoliths. The xenoliths in kimberlite range from upper mantle rocks to crustal sedimentary rocks equivalent to those at the present surface in which the kimberlite outcrops. For the purposes of this discussion, only those xenoliths of upper mantle origin will be considered.

The mineralogy of these upper mantle xenoliths is relatively simple; olivine, clinopyroxene, orthopyroxene, and garnet are the major phases, with lesser amounts of spinel, ilmenite, or mica. However, depending upon the percentage of the various minerals, there is quite an abundance of petrographic and mineralogic types. For example, garnet lherzolite, harzburgite, and websterite (Table 1) may be present in a single kimberlite. Complex sil-

Table 1. Rocks and minerals associated with kimberlite

MINERALS		
Family name	Variety	Composition
Olivine		$(Mg,Fe)_2SiO_4$
Pyroxene	Enstatite (orthopyroxene)	$(Mg,Fe)_2Si_2O_6$
	Diopside (clinopyroxene)	$CaMgSi_2O_6$
	Jadeite (clinopyroxene)	$NaAlSi_2O_6$
(Omphacite = combination of diopside and jadeite molecules)		
Garnet	Pyrope	$Mg_3Al_2Si_3O_{12}$
	Almandine	$Fe_3Al_2Si_3O_{12}$
	Grossularite	$Ca_3Al_2Si_3O_{12}$
Mica	Phlogopite	$K_2(Mg,Fe)_6Al_2Si_6O_{20}(OH,F)_4$
Spinel	Chromite	$(Mg,Fe)(Al,Cr,Ti)_2O_4$
Ilmenite	Picroilmenite	$(Mg,Fe)TiO_3$

ROCKS		
Group	Rock name	Major minerals
Ultramafic	(Garnet) Lherzolite	(Garnet: pyrope-almandine), olivine, enstatite, diopside
	Websterite	Enstatite, diopside
	Harzburgite	Olivine, enstatite
	Dunite	Olivine
	Wehrlite	Olivine, diopside
Eclogitic	Eclogite	Garnet (almandine-pyrope), omphacite

icate-oxide intergrowths, or large single crystals of pyroxene or garnet up to 20 cm (xenocrysts or megacrysts), may also occur, together with xenoliths of eclogite (or griquaite)—a garnet and omphacitic pyroxene rock.

Examination of the mineral chemistry of these various xenoliths has shown that all three types (ultramafic, eclogitic, and megacryst) are chemically distinct and different. Furthermore, they all differ chemically from similar mineral types that are present in kimberlite. Xenoliths are not now considered to be genetically associated with kimberlite, but the possibility cannot be completely ruled out that there is an association between some of the xenoliths and a very early "proto-kimberlitic" magma.

In spite of the above problem, a considerable amount of information regarding the chemistry and mineralogy of the upper mantle has been gained from studying these xenoliths. It is now obvious that the mantle is very heterogeneous, both chemically and physically. Also, the ability to place constraints on the temperature and pressure of equilibration of the constituent minerals of the xenoliths has helped in providing petrologic models of the mantle.

Inclusions in diamond. A relatively common feature of natural diamonds is that they contain small (about 100 μm), mostly monomineralic inclusions (see illustration). These inclusions may have either formed prior to the diamond and subsequently been incorporated, or crystallized contemporaneously with the host diamond. Other inclusions, often alteration products, are later than the diamond and are not discussed further. It is significant that the pre- or syngenetic mineral inclusions have been unable to react chemically or reequilibrate with external phases because the host diamond has acted as an armor. These inclusions are thus pristine samples of the upper mantle. Typical mineral inclusions are listed in Table 2.

Inclusions in a diamond. The largest inclusion is about 200 μm.

Table 2. Mineral inclusions in diamond*

Pre- or syngenetic	Epigenetic
Olivine	Serpentine
Garnet	Graphite
Enstatite	Goethite
Diopside	Hematite
Diamond	Amphibole
Mica	Xenotime
Kyanite	Sanidine
Coesite	
Zircon	
Rutile	
Ruby	
Ilmenite	
Spinel	
Pyrrhotite	
Pentlandite	
Quartz	

*Identification by x-ray diffraction or electron microprobe analysis.

Overall the chemistry of the inclusions is somewhat similar to that of the comparable minerals in kimberlite, and the ultramafic and eclogitic xenoliths. However, in detail there are significant and characteristic differences that set these inclusions apart. In summary, here are two important characteristics of the inclusions. First, two distinct and different mineral suites exist; one suite consists of minerals similar to those in ultramafic rocks, and the other resembles the mineral constitutents of eclogite. Both suites are mutually exclusive.

Second, although inclusions may be from different geographic localities and geologic times, the majority of inclusions of a specific mineral type have relatively limited compositional ranges.

The origin of those diamonds that contain eclogitic suite inclusions must be related to that particular suite, since a large number of eclogite xenoliths in kimberlite have been found to contain diamond. In contrast, only a few samples of ultramafic xenoliths are known that contain diamond, in spite of the abundance of diamonds with ultramafic suite inclusions. Furthermore, the difference in chemistry between these latter inclusions and the comparable minerals in the kimberlite and ultramafic xenoliths suggests that diamond (with ultramafic suite inclusions) has formed in an environment distinct to kimberlite. Evidence is also provided by the fact that isotopic studies have shown that diamonds are considerably older than the kimberlite in which they occur.

A problem of both scientific and economic interest is why the majority of kimberlites are not diamond-bearing. Although kimberlites often occur in groups and are spatially closely related, not uncommonly only one of the group will have a sufficiently high diamond content to be economically important. Possibly the answer to this question lies in the mechanism of eruption of kimberlite, particularly involving such parameters as temperature and oxygen fugacity. Undoubtedly, the amount of carbon in the source area and depth of origin play important roles.

Uses of diamond. The word diamond seems to convey to most people the image of polished gemstones. However, the majority of diamonds are used for industrial purposes—approximately 10^8 carats per year (1 carat = 0.2 g). Interestingly, over 50% of this amount is supplied by synthetic diamond manufacture. Synthetic diamond has an advantage over natural diamond for industrial uses in that the former can be produced in specific sizes, shapes, and quality. Thus synthetic diamond powder has none of the inherent cracks that can be present in a natural diamond powder which has been crushed to size.

The industrial uses of diamond are diverse. Probably it is best known for its role in the machining to high tolerances of ferrous and nonferrous metals, including both lapping and polishing. In mining, it has an almost unparalleled use in drill bits, particularly in the oil industry. At the other end of the scale, miniature diamond drills are used in dental work. The grinding, cutting, and polishing of glass, ceramics, natural and artificial stone, and concrete are areas in which industrial diamonds have significant applications. In fact, the grooved runways of most airports have been produced by diamond-impregnated cutting wheels. Finally, the use of diamonds in the electronics industry is important, not only for cutting and grinding but also as heat sinks upon which the tiny electronic parts are mounted. Diamond stylus cartridges in record players are a common feature.

Nevertheless, in spite of the large number of industrial uses, the most familiar use of diamond is that of a gem. In cooperation with nature, the cutter and polisher patiently unlock the fire and brilliance of these fragments from the Earth's hearth to shine in true magnificence.

For background information *see* DIAMOND; GRAPHITE; XENOLITH in the McGraw-Hill Encyclopedia of Science and Technology. [HENRY O. A. MEYER]

Bibliography: J. B. Dawson, *Kimberlites and Their Xenoliths*, 1980; H. O. A. Meyer, *Rev. Geophys. Space Phys.*, 17:776–778, 1979.

Dinosaur

New evidence, new hypotheses, and discoveries of new kinds of dinosaurs have changed scientists' view of these animals in recent years. Long classified as reptiles, dinosaurs have traditionally been interpreted as cold-blooded like all living reptiles. Recent speculations, however, suggest that they probably were warm-blooded like modern mammals and birds. Dinosaur extinction has always been a mystery. A new theory attributes their demise to an explosion that resulted from the collision of a giant meteor or asteroid with the Earth. Dinosaurs are supposed to have disappeared completely, leaving no descendants. But evidence has been presented recently which indicates they did leave descendants that survive today—birds! Dinosaurs are noted for their large sizes, with estimated weights of several metric tons, even tens of tons. Therefore, discovery in Argentina of a dinosaur skeleton the size of a robin is noteworthy. But discoveries of giant bones in Colorado reaffirm dinosaur gigantism. These fragmentary remains represent animals that might have approached 100 tons in live weight. The rarity of young dinosaurs has always puzzled paleontologists, leading them to conclude either that dinosaur young were greatly outnumbered by adults or that dinosaur hatchlings lived in regions different and far removed from the adult habitats where most fossil remains have been found. Discovery in Montana of several clutches of eggs and nests of dinosaur hatchlings now seems to confirm the latter hypothesis.

Ectothermic or endothermic? Modern reptiles are ectothermic, which means their body temperature is largely dependent on the environmental temperature. They bask in the sun to warm themselves to optimum body temperatures at which they can be active, or move into the shade to prevent overheating. Birds and mammals are endothermic. They produce and maintain elevated body temperatures by internal metabolic processes and release excess heat by perspiring or panting. Although both methods are successful strategies of thermoregulation, a common misconception holds that endothermy is superior to ectothermy. This has led some to theorize that the 150,000,000-year success of the dinosaurs was due to their being endothermic. The reasoning is that dinosaurs could not possibly have been so successful if they had been "mere" ectotherms. Evidence cited in support of dinosaurian endothermy are: upright posture (today only endotherms are able to stand and walk upright); elevated head position well above the heart level—indicative of a four-chambered heart (no living ectotherm has an elevated head or a fully divided four-chambered, double-pump heart); microstructure of dinosaur bone consisting of dense compact bone rich in blood vessel channels (Haversian canals), much like that of modern endotherms; and a census of dinosaur kinds from fossiliferous strata revealing a very low ratio of carnivorous dinosaurs relative to plant-eating kinds, a predator/prey ratio that is similar to modern mammalian community ratios of carnivores/herbivores (endothermic animals with their high metabolic rates require more food than similar-sized ectotherms with their lower metabolic rates). These different lines of evidence for dinosaurian endothermy are very suggestive, but far from conclusive. Yet, another piece of evidence—birds—suggests that at least one group of dinosaurs may have been endothermic, or very close to it.

Birds as descendants of dinosaurs may seem absurd, but recent studies of the famed specimens of *Archaeopteryx*, the oldest known bird, show a remarkable number of anatomical features that are present in only one other kind of animal, the small carnivorous dinosaurs known as coelurosaurs. Although there are some critics of this hypothesis, it is now generally accepted that birds are descended from the carnivorous dinosaurs. If correct, this has interesting implications for the possibility of dinosaurian endothermy. The specimens of *Archaeopteryx* clearly show that this animal was feathered. This means that it almost certainly was not ectothermic. It seems reasonable that the ancestors of *Archaeopteryx*, the coelurosaurs, were also endothermic, or at least approaching that condition.

Extinction of the dinosaurs has been attributed to disease, a sudden heat wave, a sudden cold snap, egg-eating mammals, a nearby supernova, and many other imaginative causes. Discovery in 1979 of a 65,000,000-year-old thin clay layer at Gubbio, Italy, containing an unusually high concentration of the metal iridium, has generated an exciting theory. Iridium is a rare substance on Earth, but occurs in higher concentrations in some meteorites. This led Luis Alvarez and Walter Alvarez to conclude that this iridium-rich layer (which has now been found at several other places) is the result of an explosive impact of a giant asteroid—a rocky object 6–10 mi (10–16 km) or more in diameter—colliding with the Earth at a speed of at least 60,000 mph (27 km/s). The resulting explosion, according to this hypothesis, shrouded the Earth in a huge dust cloud that blocked out sunlight for several years or more, causing plants to die, followed by the plant-eating dinosaurs, and then their flesh-eating kin. It is known that large meteorites have collided with the Earth (Meteor Crater in Arizona is conspicuous proof), and it is highly probable that huge objects like asteroids have crashed into the Earth in the past with explosive consequences. Perhaps the Gubbio clay layer does include the debris from such an explosive impact. But the fossil evidence does not seem to support the asteroid impact–extinction theory. There is no evidence of massive changes in the flora at the end of the Cretaceous. On the other hand, there is evidence of a progressive general decline in the number of dinosaur kinds that lived during the several million years before the dinosaurs died out. *See* EXTINCTION (BIOLOGY).

Gigantism. The sauropod *Brachiosaurus*, known from several incomplete specimens from the American West and from the nearly complete skeletons from Tanzania, has been labeled as the largest land animal of all time at an estimated live weight of 70–80 tons. In 1974 James Jensen unearthed several bones of even larger sauropods that may have been 10% or more larger than *Brachiosaurus*. Jensen dubbed these creatures "Supersaurus" and "Ultrasaurus." On the other end of the spectrum, until 1976 the smallest reported dinosaur specimens were the classic chicken-sized specimen of *Compsognathus* from Bavaria and a few juvenile specimens of *Protoceratops* and *Psittacosaurus*. Discoveries by J. Bonaparte in Argentina have now added several tiny specimens, one of which appears to be a prosauropod. This is the group which is judged to have given rise to the largest of all land animals, the sauropods. Most remarkable of Bonaparte's discoveries is a nearly complete 6-in.-long (15-cm) skeleton of a prosauropod which he named *Mussaurus* (see illustration), the skull of which measures about 3 cm, or less than $1^1/_4$ in. (31.8 mm). The discoveries of *Mussaurus*, *Protoceratops*, and *Psittacosaurus* represent some of the most important dinosaur finds because specimens of young dinosaurs are very rare.

Mussaurus patagonicus, one of the smallest dinosaur specimens known, from Late Triassic rocks of Argentina. It is probably the skeleton of a very young prosauropod dinosaur—perhaps a hatchling. (*Courtesy of José Bonaparte of the Museo Argentina de Ciencias Naturales*)

In 1978 a team led by John Horner found the first of several nests of dinosaur eggs. Even more important, they discovered nests of very young hatchlings of the duck-billed variety (hadrosaurs). These young duckbills were about the size of a chicken (compared to the adult size ranging up to 2 or 3 tons). Of particular interest is the fact that the teeth of these young dinosaurs were well worn, indicating that they are not fresh hatchlings. Some specialists have interpreted the worn teeth as evidence of parental care. But, as Walter Coombs has pointed out, the great size disparity between tiny hatchlings and adults makes it highly unlikely that these dinosaur young were fed by their parents. Much more important than these speculations, though, is the location of Horner's discovery. He appears to have found that upland nesting ground of dinosaurs mentioned earlier. In the 3 years since the original find, remnants of more than a dozen clutches of eggs or nests of young dinosaurs have been found. They represent at least three different kinds. The surprising abundance of eggshell fragments, clutches of eggs, and nests of young hatchlings confirms earlier suspicions that dinosaurs deposited their eggs and raised their young in a special breeding ground.

For background information *see* DINOSAUR; EXTINCTION (BIOLOGY); THERMOREGULATION in the McGraw-Hill Encyclopedia of Science and Technology.

[JOHN H. OSTROM]

Bibliography: J. Bonaparte and M. Vince, *Ameghiniana*, 16:173–182, 1979; W. Coombs, *Nature*, 283:380–381, 1980; J. H. Ostrom, *Biol. J. Linnean Soc.*, 8:91–182, 1976; R. D. K. Thomas and E. C. Olson, *A Cold Look at the Warm Blooded Dinosaurs*, 1980.

Doppler navigators

Many types of navigation and guidance equipment have been developed for aircraft, including helicopters. One type, incorporating a Doppler radar velocity sensor, has been in use in various forms since the mid-1950s. A Doppler navigator determines aircraft velocity from measurements of the shift in frequency in the radio energy that it transmits toward and receives back from the Earth's surface. The shift in frequency is proportional to the speed of the transmitting and receiving antenna in the aircraft relative to the ground. Modern Doppler navigators transmit and receive three or four non-coplanar beams and use their frequency shifts to reconstruct the three perpendicular components of aircraft velocity. These three components can be used directly by some avionic systems on the aircraft, but they are usually converted into the horizontal, north and east components when used for dead-reckoning navigation. A computer in the aircraft or in the Doppler navigator performs this conversion by using heading, pitch, and roll information from other aircraft sensors. The change in aircraft present position is calculated by using velocity and time information; this change is then added to the initial present position (at takeoff or at a fix or update point) to arrive at actual present position.

Helicopter navigation requirements. Doppler radars for use in helicopters should be able to operate properly during hover and backward flight, as well as in forward flight. Helicopters usually fly more slowly than fixed-wing aircraft and often fly at very low altitudes, particularly when used for military operations. Very-low-speed, hover, and backward flight results in a frequency shift that may be zero or negative, and thus the Doppler navigator requires careful design to achieve accurate and reliable operation throughout the helicopter flight regime.

Helicopters and low-speed, fixed-wing aircraft are generally small, lightweight, and inexpensive, and are usually purchased in large quantities by military forces. Navigation equipment for such aircraft should have comparable qualities. Doppler navigator velocity sensors for helicopters achieve low weight and cost by the use of several design techniques. A fixed antenna avoids the expensive and heavy gimbals used in previous Dopplers to physically stabilize the antenna. The radome, whose function is to protect the antenna from environmental effects, is made an integral part of the antenna, thereby reducing installation costs. The low altitude and speed of helicopter flight reduce the signal strength required for normal operation, permitting the use of solid-state devices, such as Gunn or Impatt diodes as sources of transmitter power. The four beams of electromagnetic energy are usually transmitted and received sequentially, and thus only one set of transmit, receive, and Doppler shift–measuring hardware is needed. This time-sharing or multiplexing of hardware significantly reduces the cost and complexity of a Doppler radar.

Transmitter modulation. Helicopters can fly at low speeds and within a few feet of the Earth's surface. A Doppler navigator for helicopters should therefore be able to operate reliably at virtually zero altitude. One approach to solving this problem is to use continuous-wave (CW) transmission and separate transmitting and receiving antennas. The cost, weight, and installation penalties of separate transmitting and receiving antennas can be avoided by using pulse modulation, although this technique is not generally employed for helicopter Dopplers since operation down to 1- or 2-ft (0.3- or 0.6-m) altitudes would require transmitter turn-on and turn-off times of less than 2 nanoseconds. The most frequently used transmitter modulation technique of modern helicopter Doppler navigators is frequency modulation (FM) of a continuous-wave signal in which one or more of the Doppler-shifted sidebands of the CW signal are tracked. This approach also permits the use of a single antenna for transmission and reception and avoids the problem of extremely short transmission rise and fall times of pulse systems.

At certain altitudes FM-CW Doppler navigators experience a loss of signal and hence can no longer operate. This occurs because at these altitudes (called altitude holes) the phase of the frequency modulation in the energy returned from the ground is the same as the phase of the FM in the transmitted signal, resulting in a cancellation of the received signal. Several techniques have been used to circumvent this problem, including: (1) modulating the CW transmission at two or more frequencies so that only one frequency at a time will lose signal; (2) using a single low modulation frequency the first altitude hole of which is above the operational altitude range of the helicopter; and (3) using a high modulation frequency with many narrow holes.

In an FM-CW system the returned energy consists of the carrier and sidebands located at multiples of the modulation frequency above and below the carrier. The sidebands are usually referred to as Bessel sidebands since they can be described in terms of the Bessel mathematical function. The carrier, J0, and the Bessel sidebands, J1, J2, J3, . . . , experience the Doppler shift, and thus any one or more of the sidebands can be selected and used for velocity measurements. The use of J1, the first-order sideband, has the advantage of enabling operation down to zero altitude. Higher sidebands can experience difficulty at very low altitudes, but are sometimes desirable because of their reduced sensitivity to unwanted leakage of the nonshifted sidebands into the transmitter signal. An example of a Doppler navigator for helicopters in which these various system design elements have been successfully combined is the AN/ASN-128, shown in Fig. 1. Large-volume production of this system began in 1978 for several types of helicopters. The system utilizes a single fixed transmit-receive antenna, FM-CW transmission, and a single low-modulation frequency (30 KHz), and it processes the resultant J1 sideband for velocity measurement. The system op-

Fig. 1. AN/ASN-128 Doppler navigation system. From left to right, the elements comprise a receiver-transmitter antenna, control and display unit, and signal data converter. (*Singer Kearfott Division*)

erates reliably from 0 to over 10,000 ft (3000 m) above the Earth's surface. Thus it operates over the full range of most helicopters. It also operates while the craft hovers or operates backward or sideways. The velocity-sensor portion of the total system weighs only 22 lb (10 kg), including a built-in radome.

Signal processing. The Doppler frequency–shifted signals processed by a typical Doppler navigator are proportional to the components of helicopter velocity along the directions of the three or four antenna beams. These components are first transformed into three perpendicular components along the helicopter's roll, pitch, and yaw axes. Several types of avionics equipment can utilize these components of Doppler radar velocity directly, but for navigation and other functions the velocities are usually further transformed through pitch, roll, and heading coordinates into earth or geographic coordinates. Navigation accuracy therefore depends not only upon the Doppler velocity sensor but also upon the accuracy of the pitch, roll, and heading sensors. Modern Doppler navigators have velocity errors on the order of 0.25% (1 sigma) of the helicopter's velocity. Typical low-cost heading sensors such as magnetic compasses have errors of about 1° (1 sigma) which correspond to a velocity error of 1.74% in a direction normal to the flight direction or track. The heading reference is thus the major contributor to error in a Doppler navigation system for helicopters. Pitch and roll cause very small navigation errors compared with the Doppler navigator's contribution, and they can therefore be neglected. A low-cost Doppler navigation system provides position data with sufficient accuracy for most helicopter functions. A typical modern Doppler radar velocity sensor, when used to navigate with a heading sensor which has an error of 1° (1.74%), will produce a position error of less than 1 nautical mile (1.8 km) after 50 n-mi (90 km) of flight. By careful calibration after installation, the helicopter's heading reference error may be reduced to 0.5° and the position error to only 3000 ft (0.9 km) after 50 n-mi (90 km) of flight. Position accuracy can also be improved by overflying known checkpoints and updating the computed present position coordinates to correspond with those of the checkpoint. The low speeds and altitudes of most helicopter missions enable this updating process to be performed accurately.

Fig. 2. AN/ASN-128 control and display unit. (*Singer Kearfott Division*)

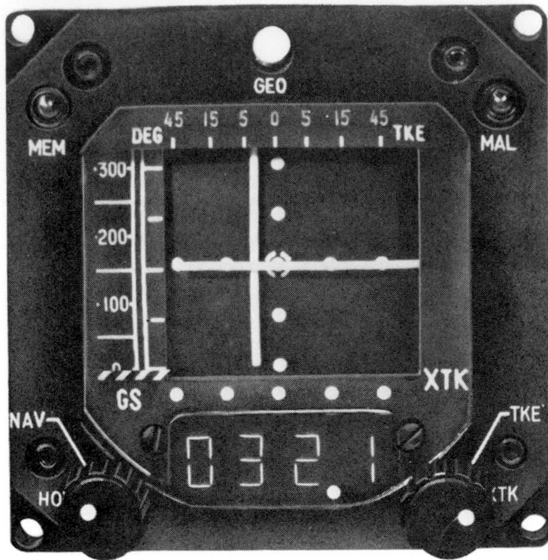

Fig. 3. Steering/hover display unit for AN/ASN-128. (*Singer Kearfott Division*)

Data computation and display. Flight crews of helicopters and fixed-wing aircraft require essentially the same navigation and guidance data; present position, speed, and direction of flight; coordinates of destinations or waypoints; left/right steering signals; and distance-to-go to the selected destination. These data should be displayed in a form that can be assimilated nearly instantly by the flight crew. Several helicopters employ two displays, one for entering and reading out data and another for displaying steering signals to the desired destination. Figure 2 shows the ASN-128 data entry (control) and display unit for helicopters. This unit also contains an LSI-chip computer that performs all computations for the ASN-128. One requirement unique to helicopters is the ability to display velocity during very-low-speed flight or hover. This enables the flight crew to maintain the helicopter virtually stationary over a desired point with only minimal reference to that point, a capability which is particularly important at night, under conditions of limited visibility, or over water. Figure 3 shows a unit that displays the three perpendicular components of hover velocities. This unit also displays left/right steering signals, and distance-to-go for use during flight between destinations.

Computations for Doppler navigator velocity processing and for navigation and guidance are now being performed by microprocessors using solid-state memories which may be contained within the Doppler system. These computers also perform test routines on the various major subassemblies of the Doppler navigator, the interface electronics, and the computers themselves. Results of these tests are analyzed for their impact on system operation, and suitable messages are displayed to the flight crew. These messages may include system status and identification of any failed subassemblies. Use of computer-based testing has reduced the need for flight-line test equipment for Doppler navigation systems. The elimination of this equipment is very desirable considering the frequency with which helicopters operate in remote areas and airfields.

For background information *see* DOPPLER RADAR in the McGraw-Hill Encyclopedia of Science and Technology.

[HEINZ BUELL]

Bibliography: F. B. Berger, *IRE Trans.*, ANE-4:176–196, 1957; H. Buell, *J. Inst. Navig.*, 27(2):124–131, 1980; M. Kayton and W. R. Fried, *Avionics Navigation Systems*, pp. 207–280, 1969.

Earthquake

In recent years there have been many advances in earthquake prediction research, although the goal of being able to predict with a high degree of certainty the magnitude, location, and time of occurrence of impending major seismic events is still far from realization. While theoretical developments, including the dilatancy-diffusion model, have provided a qualitative understanding of some of the changes in geodetic, geophysical, and geochemical phenomena that have been reported prior to large earthquakes, in the past few years much also has been learned from the examination of a rapidly growing, empirical data base. The growth of this data base largely has been the result of improvements in instrumentation which have allowed the continuous or near-continuous monitoring of many different natural phenomena that may show significant changes before a large earthquake. An example of such a system is the recently developed automated radon-thoron monitor. A network of these devices now routinely records radon and thoron concentrations in several boreholes in southern California, and attempts are under way to correlate these data with the seismicity observed in the region and with other more direct measures of strain.

Precursory phenomena. A wide variety of instrumentally measurable precursors have been reported before major earthquakes. Certainly one of the most useful factors in earthquake prediction is the availability of complete, up-to-date seismic information for regions with high risk. Changes in the pattern of background seismicity often have been noted prior to large earthquakes. Sometimes these changes have taken the form of increased activity in a formerly quiet seismic gap, that is, an area of minimal seismic activity surrounded by a region of higher activity. The onset of such foreshocks played an important role in the successful prediction by Chinese scientists of the Haichen earthquake in 1975. At other times an area with a normally high background level of small earthquakes has experienced a sudden quiescence or drop in activity prior to a large event. The computer automation of the earthquake detection and location process is a recent major advance. With these computer-operated systems, seismologists can study subtle changes in the patterns of background seismicity almost as rapidly as they occur.

Relatively sudden deformations in the crust of the Earth are a direct measure of changes in strain patterns, and they often have been reported prior to large earthquakes. They have taken the form of uplifts or downwarpings of the surface, or compressions or extensions (dilatations) of the crust. During the past decade improved laser rangefinding equipment has been used to track the temporal development of some of these changes. Since crustal deformation is a direct measure of strain accumulation, this information is a particularly valuable component in an earthquake prediction program.

Numerous geophysical precursors also have been reported before large earthquakes. These include changes in the ratio of seismic compressional wave (P-wave) velocity, changes in the local gravity field, changes in local magnetic field, and changes in the electrical resistivity of the crust. Anomalous changes in the electrical resistivity of the crust were noted before a number of large earthquakes in China. Changes in water level in wells can be an indication of compression or dilatation, and such changes have been reported shortly before many large earthquakes.

In addition to changes in geodetic and geophysical parameters, many geochemical changes have been reported before large earthquakes. These have included changes in the chemical composition, in the ratios of stable isotopes, and in the concentrations of helium and other gases in groundwater. A trace gas present in groundwater which frequently has been used for earthquake prediction studies is radon.

Radon monitoring. Radon is a radioactive gas which is found, usually in very minute concentrations, in all groundwater and soil gas. It arises from the decay of radium, which itself is a decay product of the uranium that is found in very small quantities in all rocks and soils. The uranium and radium usually are found in inclusions of accessory minerals which are not well integrated into the crystalline structure of the host rock. Radon has a short half-life, only 3.8 days, so that an individual radon atom cannot move very far from the location at which it is produced. The concentration of radon in groundwater depends on several factors. These include the concentration of the radium and uranium progenitors, the amount of pore space in the rock matrix, the degree to which the pore volume is filled with water, and the rate at which groundwater moves through the pores, seams, cracks, and joints in the rock.

A change in one or more of these factors can lead to rather sudden variations in the concentration of radon in the groundwater. Since changes in the strain environment can affect the amount of pore volume, the degree of saturation of this volume with water, and the bulk flow of fluids through the rock, rather rapid changes in the radon concentration can result. If the changes in the strain environment are caused by tectonic forces, the radon variations can be useful in earthquake prediction.

During the past decade almost 100 radon anomalies preceding earthquakes have been reported at locations in various parts of the world. Some of the most spectacular have been recorded in China. A tabulation of data was recently presented from the Kutzan station, which is located near the junction of a Y-shaped fracture zone in western Szechuan Province. In a 4-year period, radon spikes of short duration were observed several days before eight earthquakes ranging in magnitude from 5.2 to 7.9. All of these earthquakes occurred within the fracture zone. The epicenter of the closest event for which an anomaly was observed was 54 km from the Kutzan station, while the furthest occurred 345 km from the station. During the same period one radon spike was observed with no subsequent earthquake, and a magnitude-6.5 earthquake occurred on the same fracture zone with no precursory anomaly. Because of its short half-life, it is not possible for a radon atom to migrate more than several tens of meters before it decays. Thus the observation of radon anomalies at large distances from the epicenters of subsequent earthquakes can be explained only by changes in the strain regime at these distances.

Monitoring in southern California. More than 125 years have passed since the last great earthquake on the "locked" segment of the San Andreas Fault, which lies to the northeast of the heavily populated Los Angeles metropolitan area. Recent geological evidence suggests that the average recurrence interval for such great earthquakes is about 150 years; however, the deviations from this average are large. Nevertheless, the probability for the occurrence of the next great earthquake in southern California is growing with each passing year. This has spurred interest in earthquake prediction in the region, and the U.S. Geological Survey is supporting several projects related to prediction, including radon measurements.

Discrete monitoring of radon in soil gas and groundwater has been carried out in California for several years by a number of groups. Recently scientists developed an automated instrument controlled by an inexpensive microcomputer for monitoring radon and thoron. The first version of the instrument was deployed at the Kresge Seismological Laboratory in northwest Pasadena in early 1977, and additional units have been added at several southern California sites since then.

Each monitor is attached to a borehole which is drilled in to rock to a depth of 30–60 m. The borehole is cased through the overburden so that only radon from the groundwater in contact with rock is measured. Three times each day, on command from the microcomputer within the instrument, a pump bubbles air through the water standing in the borehole. The bubbling action removes radon gas from the water and also produces a copious quantity of aerosol particulates in the air space above the water in the borehole. As the radon undergoes radioactive decay, it produces several short-lived isotopes. These decay products are electrically charged, and they attach themselves quickly to the larger aerosol particles. The aerosol particles are trapped on a continuous filter paper strip where their radioactivity is counted with a small Geiger counter. This provides a signal related to the concentration of radon in the borehole which is stored in the memory of the microcomputer. On a daily basis, a central laboratory computer calls each of the field monitors over ordinary telephone lines. In response to a call, each field monitor transmits its data to the central computer, where it is stored on magnetic disk, printed, and analyzed.

From 1975 through the first half of 1978 the level of seismicity in southern California was unusually low. No earthquakes of magnitude greater than 5 were recorded. Radon data from the Kresge monitor which were recorded during this period showed a yearly cycle that was attributed to thermal and hydrological effects (Fig. 1).

Early in July 1979 large increases in radon levels were recorded at two monitors of the network. One of these was the Kresge unit, and the other was located at Big Dalton Canyon about 30 km to the east in the foothills of the San Gabriel Mountains. At the Kresge site, where the longest baseline existed, radon levels quickly rose some 400% above their previous values and exhibited large and rapid fluctuations. At the Big Dalton Canyon location, frequent sharp spikes were observed in the radon level. Data from this period are shown in Fig. 2. The unusual radon levels continued for several months.

On October 15, 1979, a major earthquake (6.6 magnitude) occurred in the Imperial Valley some 290 km to the southeast of the Caltech radon monitoring network. At the time it was not clear that any connection existed between the radon anomalies observed near Pasadena and the Imperial Valley earthquake. However, as more data became available, it appeared that several different anomalies were occurring at the same time at scattered locations over a wide area of the Transverse Ranges (which run in an east-west line north of the Los Angeles Basin). These included a relatively rapid strain reversal (from compressional to tensional) measured by laser geodimeter techniques for baselines crossing the San Andreas Fault near Palmdale, a small change in gravitational acceleration recorded in the vicinity of Pasadena, a relatively large change in the electrical resistivity of the crust measured just north of the "big bend" in the San Andreas Fault, changes in the magnetic field over a wide segment of the Transverse Ranges, and anomalies in the levels of radon and helium seen in discrete samples from Arrowhead Hot Springs just north of San Bernardino.

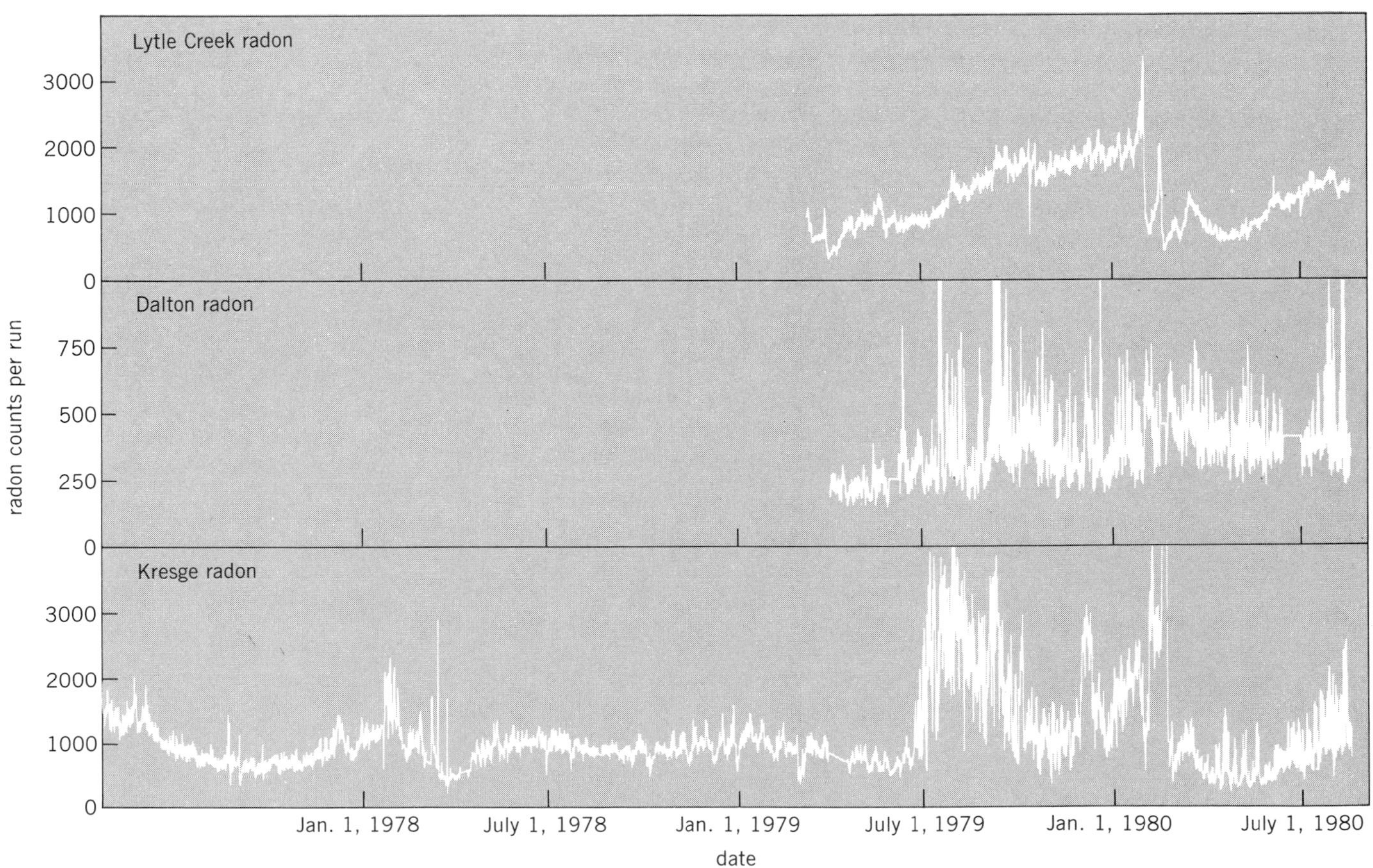

Fig. 1. Radon levels recorded at three stations of the Caltech automated radon-thoron monitoring network from the inception of monitoring through July 1980. (*From M. H. Shapiro et al., Relationship of the 1979 southern California radon anomaly to a possible regional strain event, J. Geophys. Res., 86:1725–1730, 1981*)

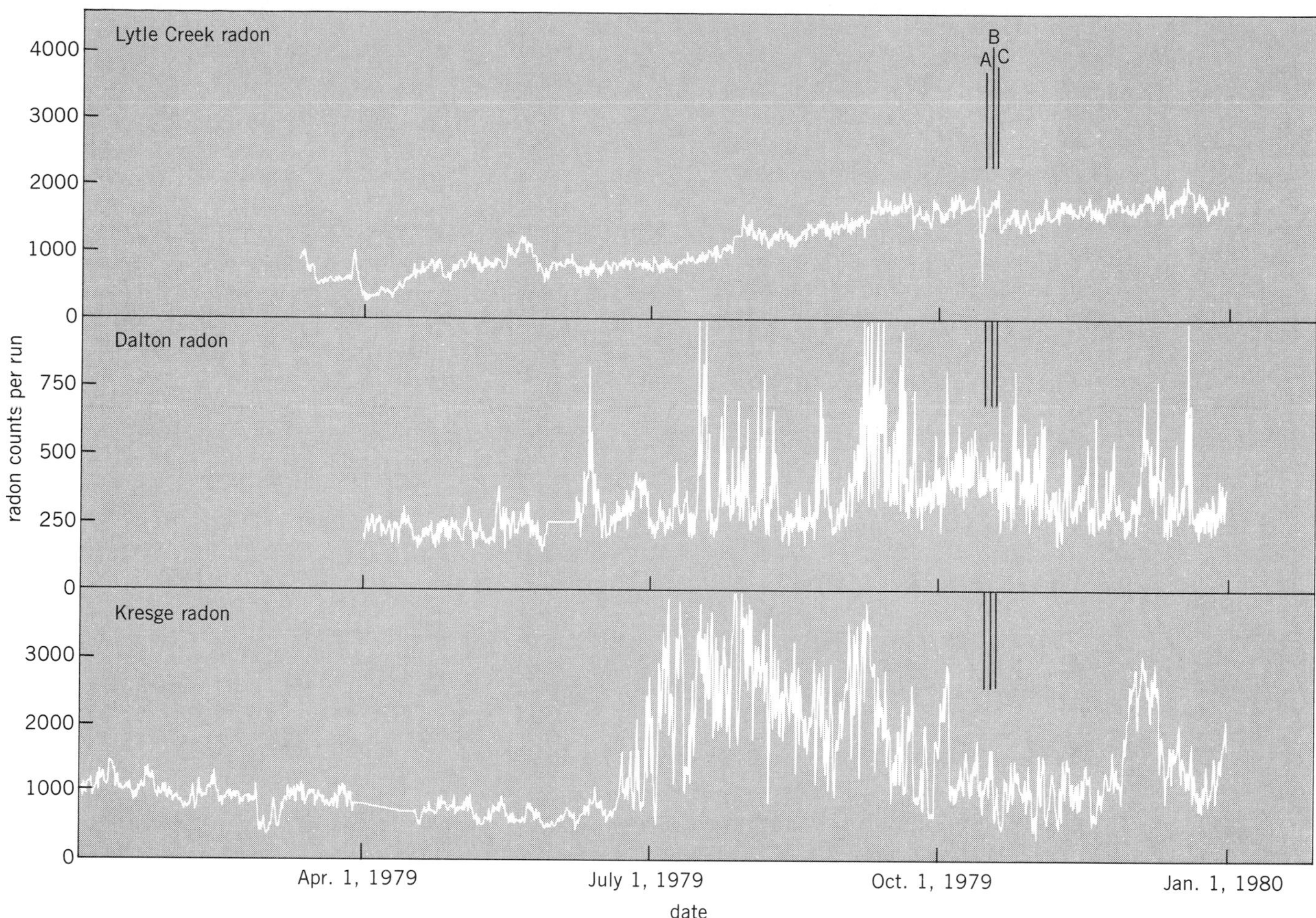

Fig. 2. Radon levels recorded at three stations of the Caltech automated radon-thoron monitoring network during 1979. A, B, and C denote the times at which the Imperial Valley earthquake (magnitude 6.6) and two moderate local earthquakes occurred. (*From M. H. Shapiro et al., Relationship of the 1979 southern California radon anomaly to a possible regional strain event, J. Geophys. Res., 86:1725–1730, 1981*)

While geodetic, geophysical, and geochemical anomalies were being recorded at several locations in the Transverse Ranges, the only relatively local anomaly that appeared before the Imperial Valley earthquake was a drop in background seismicity of about 40% which coincided in time with the radon anomaly seen in the Transverse Ranges. Uncertainty still surrounds the interpretation of these data; however, there have been suggestions that changes in the rate of motion for the Pacific Plate relative to the North American Plate cause episodic strain events extending over wide areas of California, and that these strain events induce many of the observed anomalies and initiate the larger earthquakes in the region.

There is growing evidence that radon anomalies often precede moderate or large earthquakes, and that they can be of help in forecasting the time of occurrence and, perhaps, the magnitude of the coming event. However, they appear to give only a regional estimate of the location. Reliable earthquake predictions will require the integration of several different kinds of information as well as advances in understanding the mechanisms that cause the precursory phenomena.

For background information *see* EARTHQUAKE in the McGraw-Hill Encyclopedia of Science and Technology. [MARK H. SHAPIRO]

Bibliography: M. H. Shapiro et al., *J. Geophys. Res.*, 86:1725–1730, 1981; M. H. Shapiro, J. D. Melvin, and T. A. Tombrello, *J. Geophys. Res.*, 85:3058–3064, 1980; K. Sieh, *J. Geophys. Res.*, 83:3907–3939, 1978; T. Teng, *J. Geophys. Res.*, 85:3089–3099, 1980.

Ecological interactions

Although there are many and diverse interactions between organisms, one of the more interesting to come to light in recent years is the role of viruses in certain parasitoid-host relationships.

Parasitoid viruses. The largest group of parasitic insects occur in the order Hymenoptera and are generally referred to as parasitoids. Most parasitoids belong to the suborder Ichneumonoidea, which is

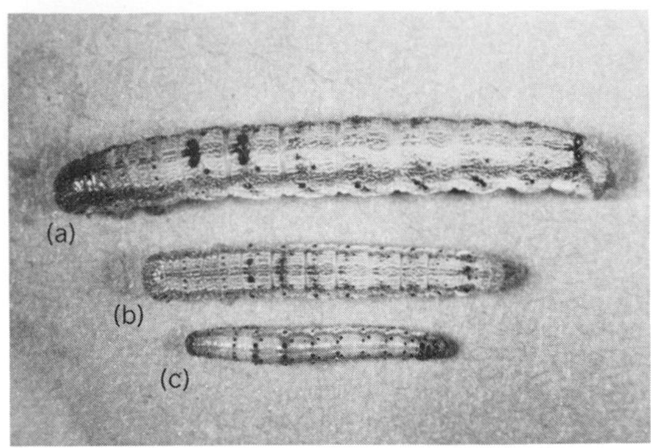

Fig. 1. Differences in the growth of hosts due to parasitoidism. (*a*) Healthy tobacco budworm larvae. (*b*, *c*) Larvae of similar age to *a* but parasitoidized 5 days earlier by *Cardiochiles nigriceps* and *Campoletis sonorensis*, respectively.

divided into two major families, Braconidae and Ichneumonidae, containing over 60,000 species. Although parasitoids are free-living as adults, females deposit their eggs on or in other insects, and the eggs and resulting larvae develop parasitically on or in these hosts. The parasitoid larvae grow and develop at the host's expense and ultimately cause the death of the host. Just prior to or at the time of host death the fully developed parasitoid larva emerges and pupates, an event which marks the beginning of the parasitoid's free-living existence.

The growth, physiology, and behavior of host insects are altered soon after egg deposition (parasitoidism). Changes such as reduced host growth (Fig. 1) have been attributed to the growth and development of the parasitoid within the host, to toxins released by the egg or developing parasitoid larvae, and to venoms injected by the ovipositing female. Parasitoid-host relationships have received increased attention since the recent finding that changes in growth, physiology, and behavior of the host often involve a symbiotic relationship between the parasitoid and a virus. The virus, which occurs in the ovary of the parasitoid, is injected into the host together with the parasitoid egg at oviposition. When the virus alone is manually injected into the host, the physiological changes in the host are identical to those resulting from parasitoidism. This finding indicates that changes in the host once attributed to the developing parasitoid are due to the viral symbiont.

The virus. The virus is produced in the nucleus of the epithelial cells of the calyx, a region of the reproductive system between the egg tubes and the lateral oviduct. The mature virus migrates from the nucleus into the cytoplasm of the epithelial cell and is released into the lumina of the calyx. The virions are so numerous in the calyx that they constitute a particulate fluid (calyx fluid) which can be expressed as a thick paste upon dissection of the calyx. The virus surrounds the mature eggs (Fig. 2) and is injected together with the egg into a host during oviposition. There is presently no evidence which indicates that the virus is produced in tissues of the parasitoid other than the calyx epithelial cells.

Viruses have been found associated with the reproductive system of 22 of 36 genera of braconids and ichneumonids thus far examined. These genera include endoparasitoids of Lepidoptera, Coleoptera, and Hymenoptera. Viruses have not been found in the few ectoparasitoids or parasitoids of Diptera or Hemiptera that have been examined. The morphology and viral biology of the viruses are invariant in females of all affected parasitoid species, and yet are different in each different species.

Nucleocapsid. In Braconidae the viruses resemble baculoviruses in that they consist of cylindrical nucleocapsids that are generally uniform in width ($\simeq$ 40 nm) but variable in length ($\simeq$ 25–250 nm) and enclosed by a unit membrane envelope. The nucleocapsids of viruses from different braconids vary in morphology; for example, the virus in *Cardiochiles nigriceps* consists of singly enveloped nucleocapsids ranging from 25 to 50 nm in length. The viral nucleocapsids of *Chelonus insularis* exhibit greater variations in their length, which ranges from 50 to 100 nm. The virus of *Apanteles melanoscelus* consists of several nucleocapsids, 30–100 nm in length, enclosed within a single viral envelope. The

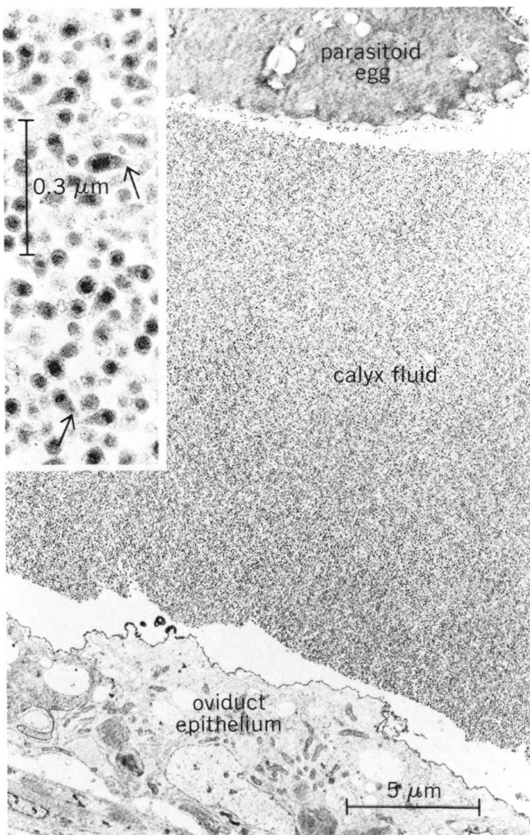

Fig. 2. Section through the calyx of *Chelonus insularis* showing the calyx fluid consisting of virions, the parasitoid egg, and the oviduct epithelium. The inset shows the virus with a single dark core or nucleocapsid; arrows show membrane protrusions.

nucleocapsids of the *Apanteles* virus have a tail assembly similar to that of the *Oryctes* baculovirus (a viral disease of a beetle) and some bacteriophage particles. The unit membrane envelopes that enclose the nucleocapsids of braconid viruses possess a definite protrusion that may be involved in infecting host cells.

The viruses in Ichneumonidae are larger and consist of a quasicylindrical nucleopcapsid (85 × 330 nm) surrounded by two unit membranes. As in the braconids, the viruses vary in different species of Ichneumonidae. The virus in some species such as *Campoletis sonorensis* have one nucleocapsid within the unit membranes, while others such as *Glypta* sp. have several nucleocapsids within the inner membrane.

DNAs. One of the most significant characteristics of parasitoid viruses is the presence of covalently closed circular DNAs of highly variable molecular weight (polydisperse). In *Apanteles* sp. the viral DNAs range in size from 2 to 25 megadaltons, and the virus contains approximately 18 to 20 structural polypeptides which range in molecular weight from 16,000 to 105,000. Synthesis of the structural polypeptides would require a minimum DNA coding capacity of 20 megadaltons, which could be coded for by a single molecule of the higher-molecular-weight circular DNA species of their virus. In comparison, the circular DNAs isolated from the virus of *Cardiochiles nigriceps* are less than 10 megadaltons in size, while the composition of the structural polypeptides of the *C. nigriceps* virus is similar to that of the *Apanteles* virus. Because of the small size of the circular DNAs from the *C. nigriceps* virus the synthesis of the viral polypeptides would appear to require the coding information of several different size classes of *C. nigriceps* circular viral DNAs.

The DNAs of the ichneumonids are also polydisperse in size. For the virus from *Hyposotor exiguae* most of the DNA molecules are less than 4 megadaltons in size. Further, each size class of circular DNA from the virus of *Campoletis sonorensis* and *Hyposotor exiguae* have been shown to consist largely of unique DNA sequences. It is reasonable to assume that several of the different size classes of DNA would be required for virus replication and synthesis of structural polypeptides. The data suggest that the extent of molecular heterogeneity in size and the unique sequence of encapsulated parasitoid viral DNA are unprecedented when compared to any other DNA virus.

Fate of virus in host. The virus enters various host cells, particularly those of the fat body, after injection into the host's hemocoel. The virus appears to gain entrance to host cells by a membrane fusion event resulting in the nucleocapsids' entering the cytoplasm. The nucleocapsids of the braconid viruses interact with nuclear pores of the host cell apparently by inserting the viral tail and releasing the DNA into the nucleus. In contrast, the entire nucleocapsid of the ichneumonid viruses appears to enter the nucleus, where it is then uncoated.

At present, there is no evidence to suggest that such viruses replicate within the host. However, it can be speculated that the viral DNAs code for messenger RNAs, which redirect the host cells' metabolic machinery to form specific viral gene products that are important to the development of the parasitoid within the host.

How the virus is transmitted from one generation of parasitoid to the next is also unknown. Preliminary evidence suggests that the viral genome is part of the genome of the parasitoid, but confirmation of this hypothesis is yet to be provided.

Effects of parasitoid virus on host. There are many changes in a host after parasitoidism that have been attributed to the developing parasitoid egg or larva, or to venoms injected by the female during oviposition. These changes include alterations in behavior, respiration, growth, development, blood proteins, blood carbohydrates, and immune response. Only a few of these host changes have been examined to determine whether the viruses are responsible, and only a few of the parasitoid viruses have been examined with regard to these changes. Studies have revealed that virus from *Campoletis sonorensis*, *Hyposotor exiguae* (Ichneumonidae), *Cardiochiles nigriceps*, and *Chelonus insularis* (Braconidae) injected into their respective host insects causes reduced growth and development of the host that mimic the effects observed after parasitoidism.

The trehalose (an insect blood sugar) level in the hemolymph of the host *Heliothis virescens* (Lepidoptera: Noctuidae) is elevated three times above normal when larvae are parasitoidized by the braconid *Microplitis croceipes*. The virus from the calyx of *M. croceipes* is responsible for the trehalose elevation. The elevated trehalose is important to the developing parasitoid larva, which absorbs trehalose needed for growth from the host's hemolymph through a special organ, the anal vesicle.

One of the most intriguing questions in invertebrate zoology is how parasitoids evade the immune response of their host. Insects respond to invading microorganisms by hemocytic phagocytosis. However, foreign material such as parasitoid eggs or larvae which are too large to phagocytize are encapsulated. Hemocyte capsules are formed by the aggregation of several layers of hemocytes around the foreign object. Although the mechanisms that result in the immune response in insects is as yet unknown, as are the means by which the parasitoid evades the host's immune response, it appears that the parasitoid viruses are involved in this evasion. Parasitoid eggs implanted without the virus into host insects are encapsulated; however, the addition of the virus prevents the encapsulation of the parasitoid progeny and an adult parasitoid eventually emerges. The virus appears to act in a specific way, since the host retains the ability to encapsulate or phagocytize other foreign material. Whether the effect of the virus is direct or indirect is as yet unknown. However, it is clear that the virus and parasitoid have evolved a symbiotic relationship in which the virus needs the parasitoid to replicate and the parasitoid needs the virus to alter the biochemical machinery of the host to its advantage.

For background information *see* ECOLOGICAL INTERACTIONS; INSECT PATHOLOGY; PARASITOLOGY in the McGraw-Hill Encyclopedia of Science and Technology.

[S. B. VINSON]

Bibliography: K. M. Edson et al., *Science*, 211:582–583, 1981; P. J. Krell and D. B. Stoltz, *Virology*, 101:408–418, 1981; D. B. Stoltz and S. B. Vinson, *Adv. Virus Res.*, 24:125–171, 1979; S. B. Vinson et al., *J. Invert. Pathol.*, 34:133–137, 1979.

Electric charge conservation

Like mass and angular momentum, electric charge is one of the fundamental properties of the elementary particles. Electric charge is remarkable in several respects. It is quantized in units of the electron charge. The neutrality of atoms implies that the magnitude of the electric charge of the proton and that of the electron are the same to within 1 part in 10^{19}. Electric charge also seems to be conserved; that is, the net charge of the universe does not change with time. This conservation law has been tested recently to extremely high levels of sensitivity.

The law of conservation of electric charge follows from Maxwell's equations of electricity and magnetism. Another way of saying this is that charge conservation is a consequence of the invariance of the electromagnetic field under a gauge transformation of the first kind. Nevertheless, the validity of any physical law must be tested experimentally.

Searches for electron instability. The electron is the lightest known charged particle. Thus a test of the stability of the electron against decay into lighter particles is a test of electric charge conservation. This idea has formed the basis of a number of experimental tests of charge conservation. The first such test was performed by E. der Mateosian and M. Goldhaber in the 1950s. The basis of their experiment is shown in Fig. 1. Suppose that an electron in the innermost shell (*K* shell) of a heavy atom were to decay. A hole or vacancy would thus be created in this shell. The vacancy would be quickly filled by an electron from an outer shell (*L* shell). The transition of the electron from the *L* shell to the *K* shell would produce an x-ray, which could be detected as a signal for charge nonconservation.

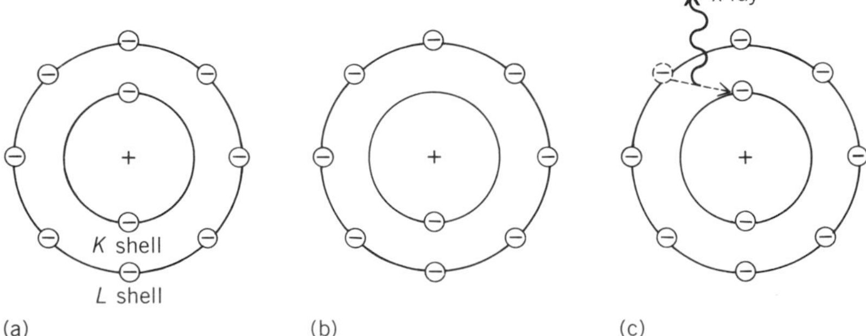

Fig. 1. Production of an x-ray following the decay of an electron in the *K* shell of a heavy atom. (*a*) *K* shell and *L* shell are filled in a normal heavy atom; (*b*) a vacancy in the *K* shell is created by the decay of an electron; (c) the *K*-shell vacancy is filled by an *L*-shell electron and an x-ray is produced. (*After G. Feinberg and M. Goldhaber, The conservation laws of physics, Sci. Amer., 209(4):36–45, 1963*)

There are a number of less exotic mechanisms that can also produce x-rays. Most materials contain small amounts of long-lived radioactive atoms such as ^{40}K, ^{232}Th, and ^{238}U. The alpha, beta, and gamma radiations produced by the decays of these atoms can all produce x-ray emission. Also, the Earth is continually bombarded by cosmic-ray particles whose interactions in matter can also produce x-rays. Thus the apparatus used in experimental tests of charge conservation must be well shielded. A clever technique which has been used in all the experimental tests of electron stability helps to minimize the background produced by ordinary charge-conserving processes. In these experiments the sample in which the search for electron decays is made serves as both the source and detector.

In the experiment of der Mateosian and Goldhaber, and in subsequent experiments by M. Moe and F. Reines and by E. Kovalchuk and his collaborators, electron stability is tested with electrons in the *K* shells of iodine atoms which are located in a sodium iodide (NaI) scintillator detector. In the case of iodine, the energy released in filling a *K*-shell vacancy by an outer-shell electron and the subsequent rearrangement of the other atomic electrons is 33.2 keV. Thus if an electron decayed into particles that escaped from the detector without interacting, the signal from the NaI detector would correspond to an energy of 33.2 keV.

To reduce the effects of external radiation sources, der Mateosian and Goldhaber used a shielded 10 × 12.7 cm detector. Counting was done for 6½ h. No excess of counts above background was observed in the vicinity of 33 keV. From these measurements the lifetime of the electron, τ_e, was established to be greater than 10^{17} years.

In the 1965 experiment of Moe and Reines, the background counting rate was further reduced by placing a 7.6 × 7.6 cm detector at the bottom of a 585-m deep salt mine. All of the materials used in this experiment were selected for their low levels of radioactivity. The detector was surrounded with a layer of mercury and a 23-cm-thick iron shield. Counting in this mode was done for a total of 110 h. Again, no excess of counts over background was observed. The limit on the electron lifetime deduced from this experiment was $\tau_e > 5 \times 10^{19}$ years.

A slightly different technique was used in the 1975 experiment of R. Steinberg and his collaborators. Instead of a NaI scintillator, a solid-state lithium–drifted germanium [Ge(Li)] detector was used, and a search was made for the 11.1-keV radiation that would be produced by the decay of a germanium *K*-shell electron. The energy resolution of this type of detector is far superior to that of a NaI detector. This means that the possible signals of charge nonconservation would be contained in a much narrower energy window than was the case in the previous experiments. The 66.1-cm^3 Ge(Li) detector used in this experiment was located aboveground, but was shielded from external radiation by an active NaI anticoincidence shield. Data were taken for 1185.3 h, and a limit of $\tau_e > 5.3 \times 10^{21}$ years was established for the electron lifetime.

In 1979 E. Kovalchuk, A. Pomansky, and A. Smolnikov reported the results of an experiment performed with a 7 × 40 cm NaI detector located underground at a depth equivalent to 660 m of water. The detector was further shielded with tungsten, Plexiglass, copper, rock, and concrete. Counting was done for 515 h. No evidence of electron decay was observed, and a limit of $\tau_e > 2 \times 10^{22}$ years was established.

The above limits on the electron lifetime refer to decay modes that produce nonionizing particles, that is, particles that would escape from the detector without interacting (for example, neutrinos). However, one can imagine that an electron might decay into a neutrino and a photon. If this process occurred inside a detector of the type used in the above experiments, it would be likely for the photon to be absorbed by the detector. Because the photon and neutrino are both massless, the energy each would have as a result of this decay is one-half the rest mass energy of the electron, 255 keV. Thus the signature for this decay would be a signal from the detector corresponding to an energy of 255 keV. The experiments of der Mateosian and Goldhaber, Moe and Reines, and Kovalchuk, Pomansky, and Smolnikov were also sensitive to this decay mode. However, none of these experiments showed any evidence for such a decay, and the lower limits deduced for the lifetime of the electron against $e^- \rightarrow \nu + \gamma$ were 10^{19}, 4×10^{22}, and 3.5×10^{23} years, respectively.

Searches for charge-nonconserving neutron decay. Searches for electron instability are not the only tests of charge conservation that have been performed. Another charge-nonconserving mechanism that has been sought is the decay of a neutron into a proton plus neutral particles. If such a process did occur, certain nuclear transitions which cannot proceed via a conventional charge-conserving beta decay would be possible. For example, as shown in Fig. 2, ^{87}Rb normally undergoes a beta-minus decay to ^{87}Sr in which a neutron decays into a proton plus an electron and an antineutrino with a decay energy of $Q_{\beta-} = 273$ keV and a half-life of $t_{1/2} = 4.8\ 10^{10}$ years. Thus it is not energetically possible for ^{87}Rb to decay to the 2.8-h isomeric level, ^{87}Sr^m, via a normal beta decay. However, if a charge-nonconserving neutron decay did occur in which the neutral particles were massless, such as photons or neutrinos, then ^{87}Rb could decay to ^{87}Sr^m with a decay energy of 396 keV.

The charge-nonconserving decay of ^{87}Rb to ^{87}Sr^m has been searched for by radiochemical techniques. In 1960 A. Sunyar and M. Goldhaber searched for such decays in a 30-g sample of RbF. They chemically separated the Sr fraction and used a shielded 7.6 × 7.6 cm NaI detector to search for the 388.4-keV gamma ray that is emitted in the decay of ^{87}Sr^m. No evidence of ^{87}Sr^m decays was observed, and a lower limit of 1.8×10^{16} years was established for the lifetime of neutrons against such charge-nonconserving decays.

In 1979 this experiment was repeated by E. Norman and A. Seamster. A 400-g sample of Rb_2CO_3 was used in this experiment. The Sr fraction was chemically separated, and the search for ^{87}Sr^m decay was conducted with a well-shielded 79-cm^3 solid-state Ge(Li) detector. This procedure was repeated five times during a 5-day period. Again, no such decays were observed, and a lower limit of 1.9 × 10^{18} years was established for the lifetime of neutrons against charge-nonconserving decays.

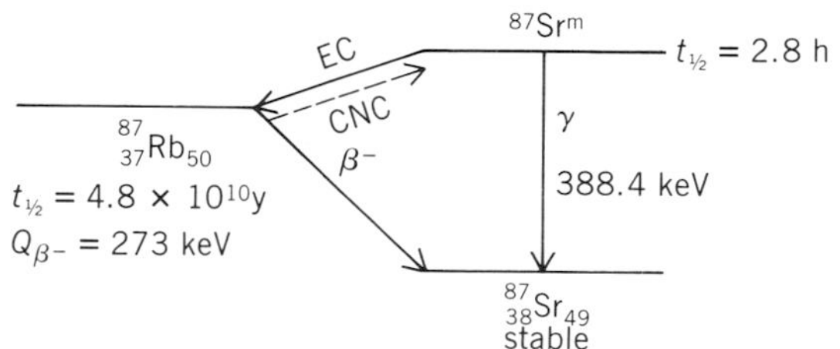

Fig. 2. Decay schemes of ^{87}Rb and ^{87}Sr^m. The solid lines indicate the normal β^--decay of ^{87}Rb and the γ-decay and electron-capture (EC) decay of ^{87}Sr^m. The dashed line shows the possible charge-nonconserving (CNC) decay of ^{87}Rb. (*From E. B. Norman and A. G. Seamster, Improved test of nucleon charge conservation, Phys. Rev. Lett., 43:1226–1229, Oct. 22, 1979*)

Indirect observations. Direct experimental searches have thus placed very strong constraints on charge-nonconserving decays of elementary particles. Even tighter limits can be deduced from indirect observations. Pomansky has shown that if the total electric current in the atmosphere is assumed to be due to the decay of electrons in the Earth, then the electron lifetime is greater than 5×10^{22} years. As pointed out by R. Lyttleton and H. Bondi, the observed expansion of the universe could be accounted for by a charge imbalance of 2 parts in 10^{18}. Moe and Reines have thus argued that if electrons disappear without an accompanying disappearance of positive charge, then the ~ 10^{10}-year age of the universe implies the electron lifetime is greater than 10^{28} years. Further improvements in direct experimental tests of charge conservation will come with the next generation of solar neutrino detectors. In addition to helping solve the present solar neutrino problem, the proposed ^{71}Ga neutrino detector would have a sensitivity equivalent to neutron charge-nonconserving decay lifetimes on the order of 10^{26} years.

For background information *see* ELECTRIC CHARGE; ELECTROSTATICS; MAGNETIC MONOPOLES; SYMMETRY LAWS (PHYSICS) in the McGraw-Hill Encyclopedia of Science and Technology.

[ERIC B. NORMAN]

Bibliography: J. N. Bahcall, *Rev. Mod. Phys.*, 50(4):881–903, October 1978; G. Feinberg and M. Goldhaber, *Sci. Amer.*, 209(4):36–45, October 1963; M. Goldhaber, *Proc. Amer. Phil. Soc.*, 119(1):24–28, February 1975; F. Reines and H. W. Sobel, *Trans. N.Y. Acad. Sci.*, ser. II, 40:154–165, September 1980.

Electric power systems

The security of an electric power system at any time is a measure of its ability to undergo a major disturbance without breaking down. A system can be made more secure in two ways: by designing (and operating) it with greater spare capacity in both its energy conversion equipment (generating plants) and its energy delivery equipment (transmission lines), or by increasing the ability of its control systems to sense (even to anticipate) disturbances and to make control adjustments to counteract those disturbances before their impact on the system develops. The former approach has been followed generally since electric energy systems were first established; it has been effective and successful, but it has been getting increasingly expensive to the point at which questions of the worth of reliability have begun to surface at the same time that utilities have been finding themselves unable to afford all the investment in generation and transmission equipment for which their historic standards would have called. The second approach has been slow to develop because of the amounts of data involved in keeping track of the status of the systems in necessary detail, and, even given that data, the complexity of analyzing it and prescribing proper control countermeasures in the few seconds available for such action to be effective. Fortunately in this period when rising costs, resource shortages, and environmental constraints are making the traditional approach to reliable system performance less practicable, improvements in engineering systems theory (including information theory, communication theory, and control theory) are now making the second approach more feasible, and revolutionary changes in power system operation are taking place quietly but effectively.

States. In order to help operators respond effectively to changing circumstances while being responsible for a wide variety of objectives, power system engineers have categorized those circumstances into five major divisions, called states: normal, alert, emergency, in extremis, and restorative. The awareness that the system is in a given state gives the system operator guidance as to how much risk exists that a major system blackout might occur and hence how drastically he or she should depart from adherence to normal control objectives in order to ensure or restore the continued physical integrity of the system.

Normal control is concerned with generating and delivering electric energy to satisfy customers' continually changing requirements, at the lowest cost, without violating constraints (such as environmental ones) and keeping prepared for unexpected occurrences. In the alert state, while conditions are still ostensibly normal, some change in circumstances has occurred, whether internal or external to the system or both, whereby the possibility of some credible disturbance resulting in a major disruption of system performance has become significant. Assurance of uninterrupted system performance depends on keeping such possibilities at a very low level of probability.

An emergency is indicated by a change in system conditions such that one or more major system components, such as a generator, a high-voltage transformer, or a transmission line, is caused to operate beyond safe limits. In such a case, considerations of economic operation are set aside completely until the limit violation is removed.

In all these circumstances, whether the system is in the normal, the alert, or even the emergency state, the customer is likely to be completely unaffected; the only likely exception to this would be if a serious shortage of system capacity relative to customer demand resulted in a need for voluntary or mandatory load reduction to avoid the system's being dragged down by its inability to keep pace. Transition of the system to the in-extremis state, however, will be only too obvious, for this consists of a loss of system integrity and the onset of disintegration of the system as an operating entity. Under these circumstances, automatic devices will operate to remove from service any major components which are threatened with physical damage owing to operation outside safe limits, and often to interrupt service to large blocks of load in order to forestall unacceptable overloading of remaining equipment.

Once any such disruption has run its course, the system is in the restorative state, wherein operating personnel must repair damage and restore normal operation.

Security control. The task of security control, that is, of forestalling incipient blackouts, is given in terms of a few elements: (1) security monitoring—collecting valid data on system conditions at frequent intervals (every few seconds); (2) security assessment—determining, on the basis of that data, the state of the system; (3) security enhancement—if the system is not in the normal state, determining what is the best control action, or sequence of control actions, that can be taken to return the system to the normal state; and (4) active control—implementing the chosen control actions and monitoring their effectiveness (that is, returning to the first element above).

Security monitoring. Modern control centers are increasingly well equipped to carry out the first of these functions, security monitoring. Advanced control center telecommunication and computer facilities enable system operators to keep aware of conditions on their system and to take quick effective action to counter serious disturbances. Every few seconds, equipment located at scores of scattered locations will provide to the control center computer the exact values of perhaps thousands of critical variables, such as voltages, currents, power flows, and switch positions. These data are processed for a variety of purposes, including automatic normal control, estimating and predicting load requirements, and taking control action on dozens of generating units to maintain the necessary continuing balance between load and generation in the most economical manner consistent with security and environmental constraints.

Security assessment. Although security assessment might seem straightforward—its purpose being to determine at any time what state a system is in, and to detect transitions from one state to another—this is in general not so, if only because of the size and complexity of the system. It is particularly difficult to recognize when a system is in the alert state. Those systems which attempt to make that determination do so by first postulating a list of possible major disturbances and then, using extensive mathematical models of their networks, calculating periodically (for example, every half hour) whether or not the actual occurrence of any of those disturbances would indeed result in an emergency. This is an intricate, time-consuming calculation, even for large modern high-speed computers, yet there is little reason to expect that any list of postulated disturbances will include what actually will happen. The procedure does serve to encourage continual operator attention to keeping the system as robust as possible vis-à-vis potential upsets. Some research attention has been directed toward development of techniques for assessing the changing probabilities of disturbances and the related probability of the system's being able to ride through them, but realization of such capability is still in the future.

Security enhancement and active control. The third and fourth elements of security control are still the responsibility of the human operator, who must make the best possible use of information on the state of the system provided by the computer. A further stage of automation is sometimes projected, wherein the computer will be enabled to make recommendations to the operator as to what preventive action should be taken. In the light of this evolutionary process, complete automatic security control, without operator intervention, may be envisioned.

At present, emergency control, that is, coordinated automatic control of the system under emergency circumstances, is also not feasible. Too much happens too quickly over too broad an area, and invariant underlying patterns, if there are any, which could serve as a unifying basis for automatic control are not known. With most systems, automatic control in emergencies is limited to local relay actions which isolate faulted or endangered equipment for protection and to contain the disturbance.

It would seem impossible, now and for the foreseeable future, for data transmission and analysis technologies to achieve a level which would enable a computer to maintain continuous instantaneous awareness of the status of every device on the system, and continuous precognition of every demand about to be made on the system, and of every disturbance about to impinge on the system, and to calculate and carry out effective control strategies. A way out of this impasse might be to enable computers to do explicitly what operators have always done (or seemed to do) intuitively: to notice only relevant information and to assess it probabilistically. An ability to design computers for such a function will require significant advances not only in fields such as information theory and pattern recognition but also in analytical (as contrasted with intuitive) understanding of power system characteristics and behavior.

For background information *see* ELECTRIC POWER SYSTEMS; ELECTRIC POWER SYSTEMS ENGINEERING in the McGraw-Hill Encyclopedia of Science and Technology.

[LESTER H. FINK]

Electrical utility industry

The year 1981 witnessed a major change in the business philosophy of the investor-owned sector of the utility industry in the United States. With interest rates for their bonds exceeding 18% in many cases and new equity selling at an average of 75% of book value, investor-owned utilities have increasingly announced their intention to withdraw from the capital markets until regulatory relief permits a rate of return that will again make expansion profitable. The sustained high interest rates and inflation have inflicted considerable damage on an industry that is the most capital-intensive of all major industries and that requires extremely long construction time for major projects. The decision to withdraw amounts to a decision to curtail construction of generating capacity that will be needed 10 years hence.

The inability to earn a rate of return on equity that is acceptable to stockholders has also sparked intense interest among utilities in diversifying into businesses where profits are not controlled by regulation. The Edison Electric Institute, the utility industry's trade association, has published a listing of 247 business activities in which utilities are now engaged, outside their primary business of generating and distributing electricity. Many of these businesses involve the exploration for or the mining and transportation of fuel for the parent company or for sales to others. Other businesses, however, are as varied as operating a shopping mall or selling financial services.

Ownership. Ownership of electrical utilities in the United States is pluralistic, being shared by investor-owned corporations, customer-owned cooperatives, and public bodies on the city, district, state, and Federal levels. The industry is dominated in all essential measures by the investor-owned sector. This sector serves 77.6% of the 93.5 million electric customers in the United States. Municipal, state, and district publicly owned entities serve only

11.9% of the total customers. This is almost the same as the cooperatives, which serve 10.4%. Federal utilities are basically wholesalers and do not serve retail customers directly.

Investor-owned utilities also own and operate 77.7% of the nation's installed generating capacity. Municipal, state, and other publicly owned systems own 10.4% of all installed capacity, while cooperatively owned companies own only 2.3%. Federal capacity is comparable to the other public sector portion at 9.7%.

The seeming discrepancy between the percentage of customers served by the cooperatives and the much smaller percentage of capacity owned by them reflects the fact that most such organizations are distribution companies which buy their power from others at wholesale rates and redistribute it to their customers.

The extraordinary financial pressure on investor-owned utilities has, in recent years, led to a strong trend of selling part ownership in their large, new generating units to cooperatives and municipal utilities. The investor-owned utilities are anxious to sell because they can, in effect, gain access to the capital needed to complete the projects; the others are anxious to buy the lower-cost energy flowing from the economy of scale of the newest, very large and efficient units. A typical example of such an arrangement is that between Duke Power Company of Charlotte, NC, and groups of 10 North Carolina and 5 South Carolina cooperatives. Duke has sold 75% of its Catawba nuclear unit No. 1 to these two groups. Duke will construct the plant, then operate it under contract to the coowners.

Because privately owned utilities cannot profitably finance new capacity in today's financial situation, there is also a trend toward construction of large, new units by cooperatives. Because they are unregulated and have access to either government loans or government guarantees for loans, they can borrow at market rates well below those of the private utilities. These large units generally are beyond the needed capability of the cooperatives that build them, but the excess output can be profitably sold to the private utilities.

Cajun Electric Cooperative in Louisiana, for example, is building three 550-MW units with a fourth planned, though its own system load is only about 1000 MW. Similarly, the Seminole Electric Cooperative in Florida, which has only 11 wholesale customers with a system demand of 1100 MW, is building two 500-MW units, whose excess output will displace purchases from neighboring investor-owned utilities in Florida.

Capacity additions. Utilities had a total generating installed capability at the end of 1980 of 613,582 MW, having added 18,882 MW during that year. Utilities of all types added 15,561 MW during 1981, bringing total industry installed capability to 629,143 MW (see table).

United States electric power industry statistics for 1981*

Parameter	Amount	Increase compared with 1980, %
Generating capability, $\times 10^3$ kW		
Conventional hydro	64,971	2.6
Pumped-storage hydro	13,625	4.4
Fossil-fueled steam	434,072	2.2
Nuclear steam	59,651	5.5
Combustion turbine and internal combustion	56,824	1.1
TOTAL	629,143	2.5
Energy production, $\times 10^6$ kWh	2,348.8	1.2
Energy sales, $\times 10^6$ kWh		
Residential	729,900	1.2
Commercial	523,400	2.7
Industrial	803,300	1.5
Miscellaneous	76,000	3.0
TOTAL	2,132,600	1.8
Revenues, total, $\times 10^6$ dollars	109,500	19.5
Capital expenditures, total, $\times 10^6$ dollars	39,280	9.7
Customers, $\times 10^3$		
Residential	83,300	2.4
TOTAL	93,485	1.9
Residential usage, kWh	8,912	0.7
Residential bill, ¢/kWh (average)	5.6	4.3

*From 32d annual electrical industry forecast, *Elec. World*, 195(9)73–88, September 1981; and extrapolations from monthly data of the Edison Electric Institute–The Association of Electric Companies.

The composition of the capacity additions during 1981 was 1675 MW of conventional hydroelectric, 570 MW of pumped-storage hydroelectric, 9548 MW of fossil-fueled steam, 3125 MW of nuclear steam, and 643 MW of diesel and combustion turbine.

Of this added capacity, 9342 MW was installed by investor-owned utilities, 861 MW by municipal, state, and public power districts, 3145 MW by cooperatives, and 2213 MW by Federal agencies.

The composition of total plant by type of generation at the end of 1981 was 434,072 MW, or 69.0%, of fossil-fueled capacity; 64,971 MW (10.2%) of conventional hydroelectric; 13,625 MW (2.1%) of pumped-storage hydroelectric; 59,651 MW (9.4%) of nuclear; 51,263 MW (8.2%) of combustion turbines; and 5561 MW (0.9%) of internal combustion engines, essentially all diesels (see illustration).

Fossil-fueled capacity. The Fuels Use Act of 1974 dictated that all new fossil-fueled capacity should fire coal, though there were specific exceptions for certain units then in the final stages of construction. The results are apparent in the capacity coming into service during 1981. All new steam units actually entering service were coal-fired. In addition to the fossil steam additions, 19 MW of internal combustion engines and 624 MW of combustion turbines were brought into service. Although geothermal units are not rigorously classed as fossil-fired units, they are so classified here for the sake of convenience. No new units entered service in 1981, but installed capacity now is 2.8 MW.

Utilities spent $11.6 billion on construction of fossil units during 1981. Investor-owned utilities spent $8.5 billion; cooperatives, $1.8 billion; municipals, state, and other public bodies, $0.7 billion; and Federal agencies, $0.6 billion.

Nuclear power. Utilities have not ordered a new nuclear unit since 1978, and no planned orders have been announced for the foreseeable future. However, a nuclear unit in the United States takes 10 years to build if no unusual delays are encountered, so many units will be entering service between now and the end of the decade as a result of the orders placed before the hiatus began. Three nuclear units entered service in 1981, with an aggregate rating of 3125 MW. This raises the total installed capacity of nuclear units in the United States to 59,651 MW, residing in 76 units in 51 individual stations.

All of the units added in 1981 were pressurized water reactors (PWR); none were boiling water reactors (BWR). The total for PWRs and BWRs now in service is 48 units and 28 units, respectively.

Current utility plans call for 84,050 MW of new nuclear capacity to enter service between now and 1990. Present projects of growth indicate that by that date more than 20% of total installed capacity in the United States will be nuclear.

The Three Mile Island nuclear reactor on the Metropolitan Edison system in Harrisburg, PA, that was severely damaged in an accident in March 1979 resulting in a partial core meltdown, is still out of service. Removal of the radioactive water in the reactor containment building has begun. The Federal government has agreed to take the high-level waste generated by the cleanup, and the utility industry has agreed to contribute more than $200 million toward the cleanup, estimates of which now run as high as $1 billion. Nuclear utilities have also set up a self-insurance pool to protect against some of the exorbitant costs of such accidents in the future.

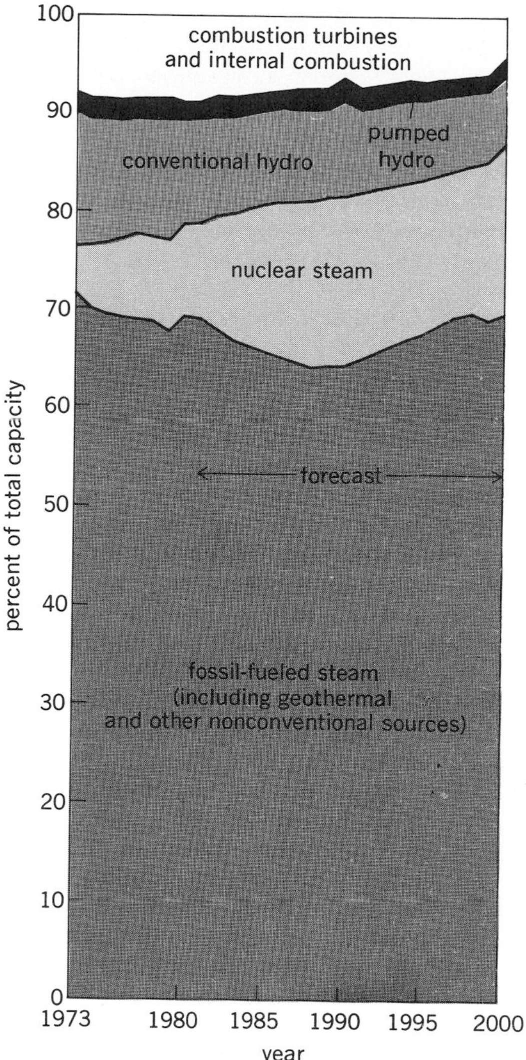

Probable mix of net generating capacity. (*From 32d annual electric industry forecast, Elec. World, 195(9):73–88, September 1981, used with permission of the Edison Electric Institute–The Association of Electric Companies*)

Combustion turbines. Combustion turbines have historically been installed by utilities for use at times of peak demand to supply up to 10–15% of that peak. The low capital cost of such machines, currently about $275 compared to $600/kW for coal-fired units, more than offsets their high fuel consumption when used for approximately 200 h/yr in this way. Combustion turbines, which use either natural gas or oil as fuel, have heat rates of about 13,000 Btu/kWh (3.8 joules of heat per joule of electric energy) compared to 9000 Btu/kWh (2.6 joules of heat per joule of electric energy) for a modern fossil unit. Further, they have quick-start capability which permits them to be brought up to full load in 2–3 min and provides flexible capacity for emergency conditions. They are used also to provide startup power for generating stations that have experienced a complete shutdown, as during a blackout of an entire area or system.

Although individual units as large as 125 MW have been installed, units mostly favored are in the 25–50MW range. In 1981 utilities added 625 MW of combustion turbine capacity to raise the total industry capacity to 51,263 MW, which is 8.2% of total capacity installed.

Combustion turbines are also used in combination with steam turbines in highly efficient combined-cycle operation. In this mode, the 900–1000°F (482–538°C) gas-turbine exhaust produces steam in a heat-recovery boiler which supplies a steam turbine. Efficiencies of this cycle may run as high as 60%, compared to 34–35% for today's most efficient steam cycles.

Utilities spent $157 million on combustion turbines in 1981.

Hydroelectric installations. Utilities brought 1675 MW of conventional hydroelectric capacity into service in 1981, raising the total now installed to 64,971 MW. Hydro units of this type, that is, hydroelectric turbines driven by water impounded behind a dam or by the natural flow of a river, constitute 12.4% of total installed capacity of all types of utility systems in the United States.

Future plans call for an additional 5700 MW of capacity to be built over the next 10 years, though locations for the required dams are becoming increasingly difficult to find and to license. The few major sites suitable for high dams are, in general, in areas where the environmental effects of the resulting lake are unacceptable. Utilities spent $866.8 million on this type of installation during 1981.

The future of hydroelectricity may rest with small-scale projects that either can be incorporated in existing dams or can be powered by the normal flow of the river. These units carry ratings of 1.5–50 MW. Such units have become especially popular in the West, where the extensive irrigation canal systems often have locks that create heads of 15–25 ft (4.5–7.5 m), adequate to power the small tube- or bulb-type hydroelectric units.

Pumped storage. Pumped storage represents one of the few possible methods for storing large amounts of energy from electrical generators. Water is pumped from a body of water on which the generating unit is located into a reservoir elevated some distance above. This is normally done during off-peak periods when large, efficient base-load units that would ordinarily be shut down for lack of demand are available. During the subsequent peak demand period, this water is released through the pumps, which can be reversed to act as turbines, recovering only 65% of the energy originally expended but reducing the need for the equivalent capacity to be provided at peak.

Utilities installed an additional 570 MW of pumped-storage capacity during 1981, raising total installed capacity in the United States to 13,625 MW, or about 2.2% of the nation's total capacity.

Utilities spent $324.6 million for pumped-storage facilities in 1981. The investor-owned companies spent $299 million, the major portion of that total. Federal agencies spent only $23.7 million, and the cooperatives the remaining $1.8 million.

Rate of growth. After a substantial growth of 6.6% in peak demand in the summer of 1980, growth sagged to only 0.4% in 1981. It is normal for a year of lower growth to follow one of high growth, but the flat response this year was primarily attributable to the persistent recession. It also reflects to some extent the negative price elasticity generated by the continuing rise in the cost of electricity.

The loss of another full year's growth will again reduce long-range forecasts, with some industry analysts now projecting long-term growth of peak demand on the order of 2.5–3.0%.

A long-term decreasing trend naturally arises from the mix of demographic factors that characterize a maturing society such as the United States. Population is growing at a decreasing rate, and passing beyond the years of peak consumption. This effect will be compounded by price-induced conservation. Because of the high technological content of the utility industry's plant, the preponderance of skilled labor employed, the high cost of capital engendered by unresponsive regulation, and escalating fuel costs, electricity prices should rise slightly more than inflation over the coming decade.

Industry and commerce have also embraced the concept of demand control. In this technique, electrical equipment is computer-controlled to minimize peak demand on the entire manufacturing plant or building within the constraints of required production. This reduces the demand charge from the utility which is a substantial part of energy cost to industry. This past year has seen the first load-shedding agreements between utilities and groups of industrial customers acting as a load bloc which, in return for a reduced rate for electricity, agree to reduce demand at peak upon request.

The pattern of load growth for utilities mimics to no small degree the economic growth patterns of the country. The strongest regional growth occurred in the Southwest, with individual utilities there experiencing rises in demand of about 10%. Conversely, the utilities supplying the older, heavy industrial areas of the Midwest saw their peaks drop 4%, and New England and the North Atlantic states experienced negative growth of around 2% an average.

The overall national declining pattern of growth has a major effect on reserve margins—that is, the excess of installed capacity over demand— on a national basis. A rule of thumb is that average national reserve margin should be about 25%. It is now about 33%. Though utilities are delaying the construction of or canceling many major generating units, some started years ago will continue to come into service, supporting this elevated level of reserve. Margin will not decline to the 25% level until 1990.

Usage. Sales of electricity rose at a much more substantial rate than did peak demand. Overall sales to all classes of customers gained 1.8%. Total national usage was 2133 × 10^9 kWh. The total was depressed by the continuing weakness in the economy, and especially by the extraordinarily low number of new residences coming onto the lines.

Commercial sales held up best in this past year, rising 2.5% to a total of 523.4 × 10^9 kWh. This figure was consistent with the performance over the last few years in this sales category.

Industrial sales did recover somewhat from the dismal drop of last year, but still managed only 1.5% increase due to the economic stagnation. Sales in this category were 803.3 × 10^9 kWh.

Residential sales dropped sharply from last year, primarily because of the absence of the extraordinarily hot and prolonged heat waves that blanketed much of the South and East in the summer of 1980. Summer 1981 was normally warm, pushing sales up just 1.2%. This equates to a usage of 729.9 ×10^9 kWh.

Despite the continual increases in the cost of electricity, electric heating for residences continues to make gains. Slightly over one-half of all new homes constructed in the United States in each of the last 10 years have been electrically heated, and in 1981 heating energy sales topped 151 ×10^9 kWh.

Residential use per customer declined slightly from 1980's use of 8976 kWh/yr to 8912 kWh. This gave utilities 5.6 cents/kWh and resulted in an annual average bill per residential customer of $499.

Total revenue for the entire industry was $109.5 billion.

Fuels. The effect of escalating prices for gas and oil and the long-range threat of the Fuels Use Act of 1974 have moved utilities strongly toward coal as a fuel. Consumption in 1980, the last year for which good figures are available, rose 18%, to 569.4 × 10^6 tons (516.6 × 10^6 metric tons). Oil concurrently dropped 19.5% to 421.3 × 10^6 bbl (66.9 × 10^3 m^3). Not all the shift from oil was absorbed by coal, but some was absorbed by natural gas, consumption of which rose 5.4% to 3679.7 × 10^{12} ft^3 (104.1 × 10^{12} m^3).

The same type of shift was seen in actual energy generated. Coal generated 1161 × 10^{12} kWh or 50% of the total. Oil accounted for 245.8 × 10^{12} kWh or 10.6%. Gas was used to generate 345.9 × 10^{12} kWh or 14.9%. Nuclear power put out 251.1 × 10^{12} kWh, representing 10.8% of all generation. The rest was primarily hydroelectric, but with a very small contribution from other sources such as geothermal.

Distribution. Distribution capital expenditures for 1981 amounted to $5.8 billion, and an additional $2.0 billion was spent maintaining existing plants. During the year, 18,500 miles (29,800 km) of three-phase equivalent overhead lines and 9550 three-phase equivalent miles (15,400 km) of underground lines came into service at voltages ranging from 4.16 to 35 kV. The majority of this mileage was at 15 kV, which accounted for 14,200 three-phase equivalent miles (22,850 km) of overhead and 7350 three-phase equivalent miles (11,800 km) of underground circuitry. The percentages for overhead construction held by other voltage classes were 8.5, 10.5, and 4.5% for 35, 25, and 4 kV, respectively. For underground construction, the equivalent percentages were 8.5, 11.0, and 3.5%. During 1981, utilities energized 19,000 MVA of distribution substation capacity and expended $700 million in capital for substation construction.

Transmission. Utilities spent $3.7 billion in capital accounts in 1981. During the year, they spent $772 million for overhead lines at 345 kV and above, and $884 million for overhead circuits at 220 kV and below. For underground transmission construction, which can cost on an average eight times more than equivalent overhead construction, capital expenditures amounted to $35.3 million at voltages of 220 kV and higher, and $15.0 million for circuits at 161 kV and below. Utilities installed 2870 mi (4619 km) of overhead lines at 345 kV and above, but 5883 mi (9468 km) at 220 kV and below. Looking at the capacity of those lines rather than the mileage gives a different picture of the place of the different voltage classes. Of the overhead lines installed in 1981, lines at 345 kV and higher totaled 2850 GW-mi (4600 GW-km), while lower voltage lines contributed only 800 GW-mi (1300 GW-km). The picture is completely different in the underground sector, because current cable technology costs favor the lower voltages. In 1981, only 17 mi (27 km) of cables operating at or above 230 kV came into commercial service, and 852 mi (1371 km) at or below 161 kV.

Utilities brought 52.6 GVA of transmission substation capacity into service in 1981 and spent $978.6 million for substation construction. Maintaining existing transmission plant cost $539.5 million.

Capital expenditures. Total capital expenditures in 1981 rose to $39.3 billion despite utility efforts to curtail construction programs. Of this total, $27.8 billion went for generating facilities, $3.7 billion for transmission, $5.8 billion for distribution, and $2.0 billion for miscellaneous facilities such as headquarters buildings and vehicles which cannot properly be associated directly with other categories. Total assets held by the investor-owned segment of the industry were $263 billion at the end of 1980.

For background information, *see* ELECTRIC POWER GENERATION; ELECTRIC POWER SYSTEMS; ENERGY SOURCES; TRANSMISSION LINES in the McGraw-Hill Encyclopedia of Science and Technology.

[WILLIAM C. HAYES]

Bibliography: Edison Electric Institute, *Statistical Yearbook of the Electric Utility Industry*, 1980,

1981 annual statistical report, *Elec. World*, 195(4): 73–104, April 1981; 32d annual electrical industry forecast, *Elec. World*, 195(9):73–88, September 1981; 21st annual steam station cost survey, *Elec. World*, 195(11):69–84, 1981.

Electron

The electron, discovered in 1895, is the oldest known elementary particle. Nevertheless, until recently an individual electron was never permanently isolated, confined, and brought to rest unperturbed in essentially free space, so that it could be subjected to detailed study. Finally, in 1973 an electron was continuously observed while contained in a large magnetic field plus a superimposed weak parabolic electric potential, realizing the "monoelectron oscillator." This technique, after being adapted to make use of a novel form of axial Stern-Gerlach effect and important improvements, has yielded a rich harvest of experimental results. Most important is the measurement of the spin magnetic moment of a free electron via spin-dependent changes in classical trajectories, a task once declared impossible in principle. The 1976 experiment in which this measurement was first carried out also marked the advent of high-resolution monoparticle spectroscopy. New measurements in 1979 and 1981 on the electron/positron mass ratio and the positron magnetic moment, both with unprecedented accuracy, were also of great importance.

Orbits. The simplest orbit of an electron in a homogeneous magnetic field $\vec{B}$ is the constant-velocity linear motion parallel to the field. This orbit is modified when a weak electric field created by a positive charge $+Q$ on a ring electrode and two negative charges, $-Q/2$ each, on two cap electrodes (Fig. 1) are superimposed on the magnetic field. The orbit remains a straight line if it coincides with the symmetry axis (here chosen to be the z axis) of the so-called Penning trap thus formed. However, repulsion by the negative charges on the caps turns a slow electron around each time it comes too close to one of them, and a harmonic oscillatory motion of frequency ν_z results.

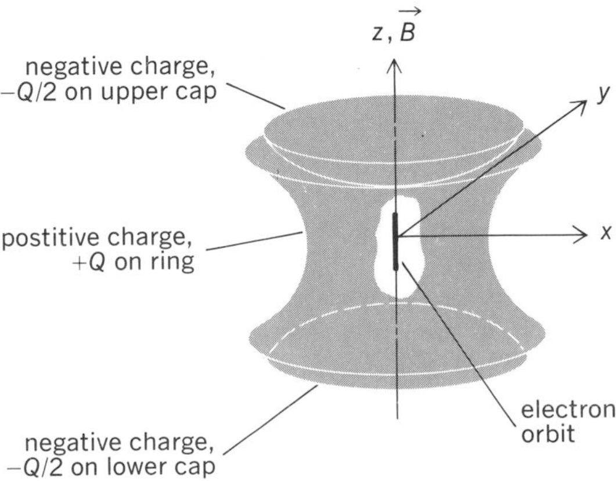

Fig. 1. Monoelectron oscillator mode of the so-called geonium atom. The electron moves only parallel to the magnetic field $\vec{B}$ and along the symmetry axis of the electrode structure.

Another simple orbit in the zero-electric-field case is the fast circular cyclotron motion of frequency ν_c with fixed center in the xy plane, given by Eq. (1). Here e and m are the charge and mass

$$2\pi\nu_c = eB/mc \qquad \nu_c >> \nu_z \tag{1}$$

of the electron and c is the speed of light. For the special case in which the center of the cyclotron orbit coincides with the center of the trap structure (the origin $x = 0$, $y = 0$, $z = 0$), turning on the electric field hardly changes the orbit, except for a slight change in the frequency from ν_c to ν_c'. This electric shift is negative, because the attraction of the electron by the positive charges on the ring electrode slightly weakens the centripetal force exerted by the magnetic field which forces the electron to move in a circle. Taking the axial frequency ν_z as a measure of the strength of the electric (quadrupole) field, Eq. (2) can be derived, which makes it possible to correct for the undesirable electric shift.

$$\nu_c = \nu_c' + \nu_z^2/2\nu_c' \tag{2}$$

A third simple orbit in the trap (Fig. 1) is a very slow circular motion at ν_m, the "magnetron" frequency (important in microwave ovens and other microwave devices) in the xy plane centered on the origin. For this motion the $\vec{B}$ field and the perpendicular, radial $\vec{E}$ field form a velocity filter so that ν_m does not depend on charge or mass of the trapped particle, and is given by Eq. (3).

$$\nu_m = \nu_z^2/2\nu_c' \qquad \nu_m << \nu_z \tag{3}$$

The most general orbit may be obtained by superimposing all three simple orbits described above. The most remarkable feature of the Penning trap is that all three frequencies ν_c', ν_z, ν_m are constants of the trap; that is, they do not depend on the location of the electron in the trap. For the parameter values $\nu_c' \cong 51$ GHz, $\nu_z \cong 60$ MHz, $\nu_m \cong 35$ kHz, and for thermal (or Brownian motion) excitation of cyclotron and axial motions of the single electron at liquid helium temperature, the general orbit consists of a very fast cyclotron motion of about 60-nm diameter around a more slowly moving guiding center. This guiding center in turn executes a fast axial oscillation of about 0.1-mm peak-to-peak amplitude and simultaneously carries out a very slow circular motion around the origin of about 30-μm diameter.

The small magnetron orbit diameter could be achieved only by the specially developed "sideband" cooling technique discussed below. Finally, the spin angular momentum of the electron, which is only capable of pointing either parallel or antiparallel to the magnetic field $\vec{B}$, must be considered. The energy difference between these two states allowed by quantum mechanics is $h\nu_s$, where h is Planck's constant, with the spin-resonance frequency $\nu_s \cong \nu_c$. The microscopic system (the electron in a Penning trap) can be regarded as a kind of synthetic atom, geonium, with quantum-mechanical energy levels given by Eq. (4), where E is the en-

$$E/h = m\nu_s + (n + 1/2)\nu_c' + (k + 1/2)\nu_z - (q + 1/2)\nu_m \quad (4)$$

ergy. Here the quantum numbers m, n, k, and q for spin, cyclotron, axial, and magnetron orbits can take only the values $m = \pm 1/2$; $n, k, q = 0, 1, 2, 3$, etc. For $k = q = 0$ the levels are shown in Fig. 2. Typically only the four lower cyclotron levels are populated appreciably at 4 K.

Apparatus. Current technology makes communication with the trapped electron via its axial frequency $\nu_z \cong 60$ MHz most convenient. Figure 3 shows a plot of the axial resonance signal versus time for a bunch of seven electrons initially injected

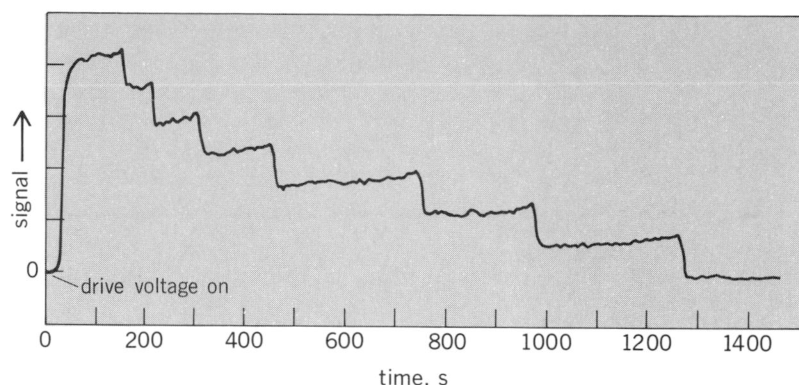

Fig. 3. Recorder trace of forced-oscillation signal versus time. The signal near 60 MHz for an initially injected bunch of electrons decreases discontinuously as the electrons are successively boiled out of the trap by the rf drive. The last plateau corresponds to a single electron. (*From D. Wineland, P. Ekstrom, and H. Dehmelt, Monoelectron oscillator, Phys. Rev. Lett., 31:1279–1281, 1973*)

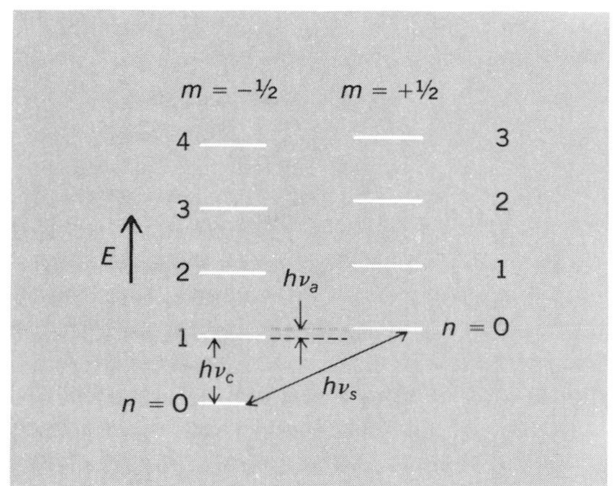

Fig. 2. Lowest energy levels of geonium. For simplification $k = q = 0$ has been assumed. Only the few cyclotron levels shown here are appreciably occupied under thermal excitation at 4K.

into the Penning trap. By judiciously adjusting the amplitude of the approximately 60-MHz radio-frequency drive amplitude, the electrons could be made to strike trap electrodes at random intervals. This resulted in the roughly exponential stepwise decay of the electron number until, for the last plateau reached after 970 s, only an individual electron was left. This particular electron survived in the trap only about 5 min and gave a signal about 20 times the noise. By halving the drive amplitude, the lifetime could be extended indefinitely to months without serious reduction of the signal.

A later version of the apparatus is shown in Fig. 4. The homogeneous confining magnetic field (up to 5 teslas) is produced by a superconducting solenoid. The axial parabolic electric field is created by two hyperbolic cap electrodes at dc ground potential and a hyperbolic ring electrode at about +10 volts. The axial oscillation of the electron is excited by applying an rf drive voltage near 60 MHz to the lower cap electrode. The motion of the electron in turn causes an rf displacement current to flow from the upper cap electrode to ground. This current is then picked up and amplified by a superheterodyne (radio) receiver. The trap structure is housed in a sealed-off Pyrex envelope submerged in liquid helium and exhausted to about 10^{-15} torr (10^{-13} Pa).

A most important part of the apparatus is the magnetic bottle used for producing the "nondestructive" axial Stern-Gerlach effect, that is, the macroscopic spatial separation in different orbitals for the two spin states. This bottle is created by a ring of magnetic dipoles realized by a nickel wire wound around the ring electrode and magnetized to saturation by the magnetic field. It causes a slight spin-dependent shift in the axial frequency whose value in the 1976 experiment, for example, is given by

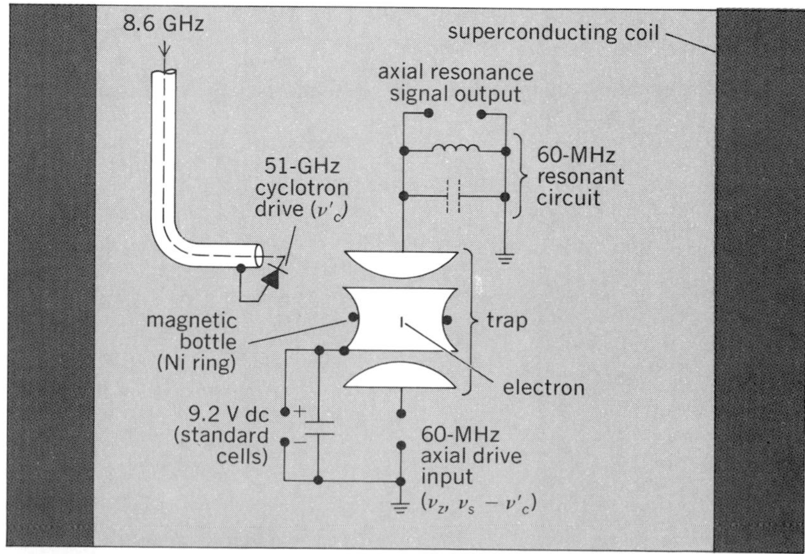

Fig. 4. Apparatus for geonium spectroscopy experiment. (*From R. S. Van Dyck, Jr., P. B. Schwinberg, and H. G. Dehmelt, Precise measurements of axial, magnetron, cyclotron, and spin-cyclotron-beat frequencies on an isolated 1-meV electron, Phys. Rev. Lett. 38:310–314, 1977*)

Eq. (5). This frequency shift becomes an important observable.

$$\delta\nu_z = m \cdot 2.5 \text{ Hz} \qquad (5)$$

Spectroscopy. The first spin flip seen with the geonium apparatus is shown in Fig. 5, depicting the time variation of the shift $\delta\nu_z$. The pen-recorder was allowed to sweep forward and then backward for successive three-minute intervals. Because of the random fluctuations in the thermally excited cyclotron motion, the axial frequency shift $\delta\nu_z$ associated with the magnetic bottle shows a corresponding unsymmetric fluctuation, always staying above a fixed floor for a given spin direction. This floor level, however, suddenly changes by 2.5 Hz during the backward (second) sweep, indicating a spin flip from $m = +{}^1/_2$ to $m = -{}^1/_2$. These data alone make possible a crude determination of the magnetic moment, since the parameters of the bottle field are known, and also confirm the spin value of $s = {}^1/_2$ for the free electron. For a precision determination of the magnetic moment, resonant spin flips are induced not by means of an applied magnetic rf field at $\nu_s = 51$ GHz but by shaking the electron axially at the spin-cyclotron difference frequency $\nu_s - \nu_c' \equiv \nu_a$ through the magnetic bottle field by means of an auxiliary electric rf drive. This causes the electron to see effective rotating magnetic fields at the sidebands of the cyclotron frequency $\nu_c' \pm (\nu_s - \nu_c')$, the upper one falling on ν_s and inducing spin flips. Cyclotron and magnetron resonances at ν_c' and ν_m are also detected via the axial frequency shift $\delta\nu_z$, for which the full expression is given by Eq. (6).

$$\delta\nu_z = [m + n + {}^1/_2 + (\nu_m/\nu_c)q] \cdot 2.5 \text{ Hz} \qquad (6)$$

The presence of the magnetic bottle requires a careful centering of the electron in the trap to guarantee reproducible magnetic field values seen by the electron. Centering is achieved by the technique of sideband cooling, first demonstrated in the 1976 geonium experiments, in the following fashion: an inhomogeneous rf field at $\nu_z + \nu_m$ excites the damped ν_z motion. This is possible because the ν_m motion through the rf field causes the electron to see a sideband of its frequency at $(\nu_z + \nu_m) - \nu_m = \nu_z$. For each quantum of energy $h(\nu_z + \nu_m)$ absorbed from the field, $h\nu_z$ goes into the axial motion and $h\nu_m$ into the magnetron motion, shrinking the radius of the latter. This cooling mechanism has also found application in the laser sideband cooling of stored atomic ions. The mechanism is also essential in a continuous scheme to catch and thermalize positrons in a Penning trap, developed in 1980.

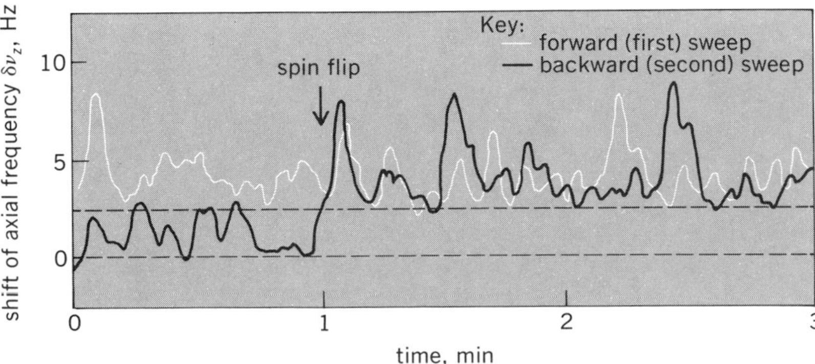

Fig. 5. First spin flip seen in the monoelectron oscillator (geonium). (*From R. Van Dyck, P. Ekstrom, and H. Dehmelt, Axial, magnetron, and spin-cyclotron beat frequencies measured on single electron almost at rest in free space (geonium), Nature, 262:776–777, 1976*)

Results. According to Eq. (2) the value of the measured cyclotron frequency ν_c' must be corrected for the electric field shift to give ν_c. Analogously the measured anomaly or spin-cyclotron beat frequency ν_a' must be corrected to yield Eqs. (7) and (8).

$$\nu_a = \nu_a' - \nu_z^2/2\nu_c' = \nu_s - \nu_c \qquad (7)$$

$$\nu_s = \nu_a + \nu_c \qquad (8)$$

Now, the dimensionless structural parameter ν_s/ν_c may be formed which may be identified with the value of the spin magnetic moment in units of the Bohr magnetron, and also one-half the g factor. The latest result is given by Eq. (9). The quoted error

$$\tfrac{1}{2}g(e^-) = 1.001\,159\,652\,200 \pm 0.000\,000\,000\,040 \qquad (9)$$

here is about 100 times smaller than in the best data obtained by other techniques. The g value is the most precisely determined parameter of an elementary particle. In 1981 analogous data were obtained for an individual positron, given by Eq. (10), which

$$\tfrac{1}{2}g(e^+) = 1.001\,159\,652\,222 \pm 0.000\,000\,000\,050 \qquad (10)$$

has error limits about 20,000 times smaller than the best previous data. The close agreement of the measured g factors for electron and positron, given by Eq. (11), constitutes the most severe test of the

$$g(e^+)/g(e^-) = 1.000\,000\,000\,022 \pm 0.000\,000\,000\,064 \qquad (11)$$

CPT theorem for a charged particle–antiparticle pair to date.

For background information *see* ATOMIC STRUCTURE AND SPECTRA; ELECTRON; ELECTRON SPIN; QUANTUM ELECTRODYNAMICS; QUANTUM MECHANICS in the McGraw-Hill Encyclopedia of Science and Technology.

[HANS DEHMELT]

Bibliography: H. Dehmelt, *Adv. Atom. Mol. Phys*., 3:53–72 and 5:109–154, 1967 and 1969; H. Dehmelt, *Atomic Physics 7*, pp. 337–372, 1981; P. Schwinberg, R. Van Dyck, Jr., and H. Dehmelt, *Phys. Rev. Lett*., 47:1679–1682, 1981; D. Wineland, P. Ekstrom and H. Dehmelt, *Phys. Rev. Lett*., 31:1279–1281, 1973.

Electronic camera

Commercial television is quite low in resolution. The high resolution of deep-space and ocean-bottom images is made possible by solid-state integrated circuit advances that affect both television camera design and, more importantly, the computer memories being used to manipulate or store the images. As solid-state chips acquire more capacity and their production costs drop, high-resolution electronic still cameras for the consumer will be developed.

Operation. The heart of the camera will be a charge-coupled device (CCD), also known as a silicon imaging device (SID), the most complex solid-state circuit ever built (Fig. 1). One CCD behind the lens of the camera will function both as an electronically reusable "piece of film" and as the camera's shutter. CCDs consist of horizontal rows of photodiodes known as pixels. The smallest separations are 2 micrometers wide. Resolution in the first generation of CCDs is limited by the photolithography process. Using electron-beam or x-ray lithography, it will be possible to size pixels on the atomic level, thereby matching or surpassing the finest-grain 35-mm film. While an alternating current is passing through the CCD, the pixels register the light falling upon them. At the end of the "exposure," the readings of each pixel are passed vertically down through the picture taking area of the CCD into a short-term storage area. Existing CCDs are being used only in television cameras, so the signals exiting these CCDs are processed as conventional video signals to be transmitted, displayed on a monitor, or recorded on video tape. The picture information from a high-resolution CCD still camera will probably continue into a solid-state memory unit consisting of stacked CCDs, random-access memories (RAMS), or bubble memory chips. If solid-state high-density memory development does not keep pace with the imaging chips, a small video recorder can supply a permanent record.

Fig. 2. Oceanographic submersible *ALVIN*, equipped with CCD color video camera. (*National Geographic Society*)

Fig. 1. Silicon imaging device (SID). (*RCA*)

Production of still images. Even though a high-resolution electronic still camera is several years away from production, currently available video technology is being used in new ways to make still images. A prototype RCA CCD color video camera, provided for deep dives into Galapagos hot water vents by the submersible *ALVIN* (Fig. 2), produced macro stills of newly discovered animals that would have been impossible to acquire otherwise. The stills (Fig. 3) are obtained from the 1-in. (2.5-cm) video recordings. They are transcribed directly and electronically from the video tape to the film, and are not photographed from a monitor. During computer processing, the data lines of the video signal are swelled so that they will touch each other and close over that area of non-information known

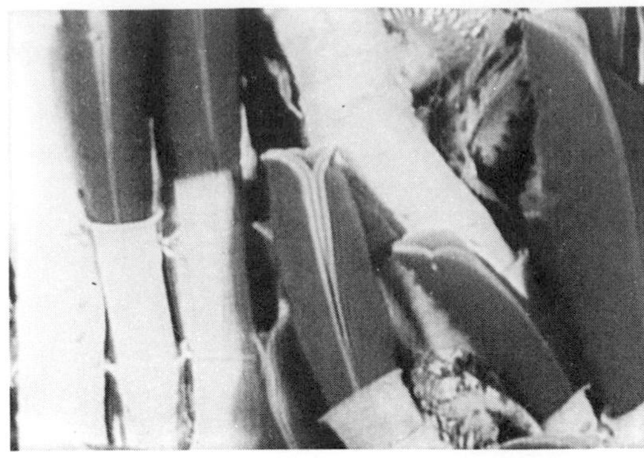

Fig. 3. Electronic still taken from video tape made by CCD camera on the *ALVIN*, showing deep-sea animal (giant tubeworm) found in hot-water vents of the Galapagos rift at a depth of approximately 2600 m. (*National Geographic Society*)

as scan lines. Then the edges of all the objects in the picture are electronically hardened, giving the illusion of more sharpness than actually exists within the camera/recording system. Finally, the image is written by an electron beam onto a piece of film in a vacuum chamber.

Another electronic camera, the silicon intensified tube (SIT) camera, has an effective ASA of 200,000—far greater than any film. This camera has been used to study bird migration and horseshoe crab mating at night. This camera has been integrated with a digital memory, and, working again with *ALVIN*, has made the largest-area bottom photographs ever seen in the ocean. With strobe illumination, the camera's sensitivity increased the amount of coverage by a factor of 10 over that done by the highest-energy film camera system.

Recording technology. Electronic cameras have been used chiefly to do these kinds of projects that cannot be done with conventional cameras. Electronic cameras have an unassailable advantage in sensitivity, some imaging tubes approach film for resolution, and the cameras allow the user to observe the field of view while it is being photographed. The biggest drawbacks of electronic cameras have centered on the recording technology. Commercial high-band video recorders and single-frame storage devices have been used to hold the images. Besides being expensive, these are barely portable systems. The mass-produced, home video recorder with its rapidly shrinking design, is being studied as a solution. Weighing less than 10 lb (5 kg), portable home units process millions of bits of information a second while producing the 30 low-resolution pictures necessary for television. If this capacity were concentrated in, say, three pictures a second, a new high-resolution video still standard would be established. An engineering project has been undertaken to demonstrate a 1000 × 1000 still recording standard. Initially in black and white, a small portable high-resolution system would be very useful to the newspapers and wire services. One wire service has been routing its pictures through an "electronic darkroom" in New York for several years. By using laser print scanners, computer disk storage, and editing on cathode-ray tubes, only the photographer's roll of film has to be transported, developed, and printed. An electronic still could move immediately to New York by satellite, microwave, or wire. In fact, the picture would not need to exist as hard copy until it came off the press.

Household electronic system. The electronic still camera will probably become part of the household computer/information/entertainment/television center. The electronic camera is an amalgamation of computer and television technology, and only as a part of a strong central system will it reach its full potential as a successor to the photochemical camera. The home computer will be the most expensive part of the camera, but since it will be used to handle so many other household functions, it will already be present. The computer will store and arrange pictures for showing on the television, by then a large high-resolution flat screen. Pictures will be transmitted directly to the computers of other selected households. Prints, if needed, will be available from a home printer or through a computer link to a commercial outlet. The ease of storing, sorting, viewing, and sharing electronic images should increase photography's popularity.

Commercial camera. Polaroid has taken out the initial patents on an electronic still camera that will record, display, and make all the prints wanted of any single exposure. The video cassette can then be stored or erased. Sony has just demonstrated such a system.

For background information *see* ASTRONOMICAL PHOTOGRAPHY; TELEVISION CAMERA; TELEVISION CAMERA TUBE; UNDERWATER PHOTOGRAPHY in the McGraw-Hill Encyclopedia of Science and Technology. [EMORY KRISTOF]

Energy sources

Energy shortages have sparked a great interest in biomass as a source of energy to replace fossil fuels. Ideally such products could be used as stand-alone fuels or converted to alcohol fuels to be used in a blend with gasoline to produce gasohol. An attractive feature of biomass resources is that they are renewable. The principal biomass resources available are wood and a variety of agricultural commodities, including grain and sugar crops, and crop residues, grass, and legume herbage, collectively called lignocellulosics. While interest has developed in several specialty crops, their potential as a fuel source is now considered minor. Only agricultural commodities are considered here.

In spite of recent interest in biomass fuels, their present contribution to total energy use remains small. Exploitation of the estimated potential of renewable resources for energy will be dependent on a great many economic, social, and environmental factors. Agricultural commodities now are almost totally relegated to other important uses such as food, animal feed, and export, and diversion of large quantities to fuel uses would have serious implications.

Sources. As energy sources, grain and sugar crops are used primarily for the production of ethanol by fermentation. Since production technology is well understood and is currently in use, considerable capacity could be developed in a few years. The goal for 1981 was 5×10^8 gal (1.9×10^6 m^3) or about ½% of gasoline used. However, production costs make ethanol only marginally economical in the absence of considerable investment and tax subsidy. Continuing research might improve technology and make ethanol more competitive with gasoline.

Sugar crops have the highest ethanol yields per acre, but costs per gallon are substantially higher than for grain. Sugarcane, a crop that supports a large ethanol industry in Brazil, can be grown only in limited areas of the United States. Therefore sweet sorghum has more potential in the United States. Among the grain crops, corn seems to be first choice on the bases of wide availability and ethanol price and yield per acre. Most ethanol now produced from grain in the United States comes from corn.

Lignocellulosics have a number of potential uses in the energy scheme according to a recent study by the Office of Technology Assessment (OTA). Among these are direct combustion, gasification, liquid fuels such as ethanol and methanol, and anaerobic digestion to produce methane. In contrast to grain and sugar crops, the technology for utilizing lignocellulosics in energy production is not yet commercially available. Additional research and development are required, particularly for conversion to liquid fuels.

Potentials. The OTA study indicates that 1 ton of grain could produce 93 gal (388 liters per metric ton) of ethanol and that 1 ton of dry grass, legume herbage, or crop residue could produce an equal amount of ethanol or methanol. Thus, a 105-bu-per-acre (9.1 m^3/ha) yield of corn plus the 1 ton (2.24 metric tons per hectare) of crop residue that could be safely removed could together produce about 375 gal (3.5 m^3/ha) of ethanol. A 4-ton-per-acre (9 metric ton-per-hectare) yield of grass or legume hay would yield an equal amount. Theoretically, each acre of corn, grass, or legume could yield the equivalent of about 8 barrels of crude oil (3.1 m^3/ha).

Corn acreage in the United States is about 8×10^7 acres (3.2×10^7 ha). If all corn grain were converted to ethanol, about 2.2×10^{10} gal (8.3×10^7 m^3) could be produced. This represents about 22% of the total gasoline use. Blending ethanol with gasoline on a 1:9 ratio to produce gasohol for all current uses of gasoline would require 1.1×10^{10} gal (4.2×10^7 m^3) of ethanol or about half the present corn production. However, all corn now produced has an alternative end use with which ethanol production would compete.

The OTA report indicates that about 7.8×10^7 tons (7.1×10^7 metric tons) of crop residues could be utilized in energy production without serious consequences to soil erosion, soil structure, or soil fertility. These crop residues are concentrated in the Corn Belt, with more than two-thirds in the north-central region. The study also indicates that 1.25×10^8 acres (5.0×10^7 hectares) of pasture and hay land in the humid regions of the United States could produce an additional 2 tons per acre (4.5 metric tons per hectare) if known technology was applied. Of this, 10^8 acres (4.0×10^7 ha) could be harvested, providing a potential availability of about 2×10^8 tons (1.8×10^8 metric tons) of lignocellulosic material for use in energy production. Nearly 70% of this material is in the humid, cool temperate part of the country, while the balance is in humid subtropical areas where warm-season grasses predominate. Thus, 2.78×10^8 tons (2.52×10^8 metric tons) of lignocellulosic residues could be available for energy use by means of direct combustion, gasification, or conversion to alcohol fuels. Theoretically, this could produce about 2.6×10^{10} gal (9.8×10^7 m^3) of ethanol or methanol.

Crop residues or grasses and legumes could be combusted directly in power plants. This possibility was studied in some detail by B. C. English and coworkers in Iowa and found to be relatively uneconomical; however, price changes in competing fuels or benefits from low sulfur content could make lignocellulosics desirable. They could be used as fuels for boilers or home heating or gasified to produce intermediate-Btu gas that could fuel retrofitted boilers now using fuel oil or natural gas. Methanol production probably would be the most economic liquid fuel from lignocellulosics, but it would be more costly than methanol from coal according to the OTA. Methanol probably could be produced with a process similar to current production from wood. Economic ethanol production from lignocellulosics requires further technological developments. Bulkiness is a major problem, and transportation costs would be high unless economic densification technology is developed. Lignocellulosics may have to be utilized in close proximity to production areas.

Implications. Large-scale production of ethanol from grain implies the creation of a new market which would compete vigorously with present feed, food, and export uses. Grain prices probably would be set by petroleum prices, thus creating a direct fuel-food or fuel-feed competition. The OTA reports indicate that beyond the 2×10^9 gal (7.6×10^6 m^3) level, production of ethanol from grain would increase food prices by \$3–4 per gallon (\$0.80–\$1.05 per liter). It is very probable that these impacts could occur at even lower levels of ethanol production. This is a seemingly high price to pay for 1 or 2% of gasoline consumption.

The effect on food prices would result mainly from the direct competition between ethanol production and feed usage. Almost 60% of current corn production is fed to livestock, which in turn provides about 42% of the food nutrients for the United States population. Nearly 40% of the corn feed is given to ruminants, which supply 32% of the food. Sixty percent of the corn is fed to hogs and poultry. Increases in feed prices resulting from competition with ethanol production and increasing exports would greatly affect production of meat and dairy products.

Lignocellulosics provide about 75% of the feed consumed by ruminants. Increasing the digestibility of this component by 5–10% could double ruminant productivity from the forage. Thus, research to increase digestibility of forage crops could release sufficient corn from ruminant feed usage to provide the base for 2×10^9 gal (7.6×10^6 m^3) of ethanol and greatly reduce impact on food costs to consumers.

A second serious result of the development of a large fuel market for grain would be the threat of increased soil erosion. A recent U.S. Department of Agriculture report estimates that 1.35×10^8 acres (5.5×10^7 ha) of cropland could be developed from land now in other uses, mostly grazing and forest land, although productivity would probably be lower and erosion susceptibly greater on most of that land. Soil erosion is above acceptable levels on nearly 1.5×10^8 acres (6.1×10^7 ha) of present cropland. Proper land use probably means that fewer acres of present cropland should be planted with crops such as corn, soybeans, and cotton which enhance soil losses unless proper soil and water management is practiced. A new fuel market for grain would increase the use of erosion-susceptible land and thus would be expected to intensify soil-erosion and water-quality problems. If soil erosion continues at the present rate, the ability to produce renewable biomass resources for any end use will be threatened.

If technology for using lignocellulosics were developed, such biomass could produce as much or probably more fuel per acre as grain. New cash-crop markets would develop and forages would compete successfully for land with soil-depleting crops. Further, energy inputs in lignocellulosic production using perennial species are lower than in grain production, particularly if legumes are used. Most importantly, soil erosion problems would be largely solved if the most erosion-prone land produced lignocellulosic biomass for energy production.

Biomass could contribute to energy needs in the short run, especially if the necessary technology were developed. However, the rapid increase in world food needs seems to indicate that such use could not be justified in the long term. Other solutions to world long-term nonfood energy needs must be found.

For background information *see* BIOMASS; ENERGY SOURCES in the McGraw-Hill Encyclopedia of Science and Technology. [HARLOW J. HODGSON]

Bibliography: B. C. English et al., *CARD Report no. 88*, Iowa State University, 1980; H. Hodgson, *Agronomy Abstracts*, American Society of Agronomy, Madison, 1980; Office of Technology Assessment, *Energy from Biological Processes*, Library of Congress Cat. no. 80-600118, 1980; U.S. Department of Agriculture, *Soil and Water Resources Conservation Act*, Appraisal, parts I and II, and Program Report, review draft, 1980.

Environmental chemistry

Pesticides, which are chemicals that are employed to control a variety of organisms usually referred to as pests, may be classified according to their use that is, insecticides, herbicides, fungicides, and so on. They can also be classified according to a chemical family, such as the chlorinated hydrocarbons, the organophosphates, the carbamates, and the chlorophenoxyacids. The era of synthetic organic pesticides began in the early 1940s with the introduction of DDT (dichlorodiphenyltrichloroethane), a chlorinated hydrocarbon. Because of their persistence and eventual accumulation along the food chain, this group of pesticides has been progressively replaced by less persistent chemicals such as organophosphates.

Recently several important advances have been made in the field of pesticide analysis. In the early years, extensive use was made of spectrophotometry for the quantitative determination of pesticides. As the number of chemicals increased, so did the need for more selective and sensitive techniques. This has involved a tremendous amount of research, particularly in the field of separation techniques, such as thin-layer chromatography and gas chromatography. Now the use of high-performance liquid chromatography (HPLC) has become more popular. There also have been major changes in the extraction procedures involved in residue analysis.

High-performance thin-layer chromatography (HPTLC). When selectivity became essential in pesticide residue analysis, separation techniques were introduced. Because of its greater adaptability, thin-layer chromatography became popular at the expense of paper chromatography. An advantage of thin-layer chromatography was that quantitative measurements could be done directly on the chromatogram with proper equipment. Recent developments have been directed toward perfecting the thin-layer chromatographic film itself rather than improving the measuring apparatus.

Resolution in HPTLC has been much improved by increasing the number of theoretical plates fivefold over conventional thin-layer chromatography. This has been accomplished by using special precoated plates with a surface of small particles and narrow particle size distribution. In practice, control is also exercised on solvent delivery and composition of the atmosphere within the unit through repeated developments. HPTLC plates are fast, and excellent resolution is often obtained in a little more than 1 min. Some substances can even be detected in the picogram (10^{-12} g) range with fluorescence as the detection method.

However, HPTLC has not yet had widespread acceptance in the field of pesticide residue analysis, although there have been some applications. It is anticipated that in the near future, methods using HPTLC will be developed, particularly for those compounds difficult to determine by gas-liquid chromatography and also as a screening technique that is still relatively simple and inexpensive.

Capillary gas chromatography. Until now the technique of gas-liquid chromatography has provided the analyst with a magnificent tool for pesticide residue analysis. Its principal advantages are high resolution, selectivity, sensitivity, capacity, and speed. In practice, traces of pesticides (even 10^{-14} g) may be detected using selective detectors. For instance, it is possible to analyze for pesticides in water in the parts per trillion (ppt) range. The provision for automatic injection permits very rapid processing of samples. The most serious disadvantage is that the sample has to be vaporized; thus nonvolatile or heat-labile compounds cannot be determined directly. This problem is partially circumvented by proper derivatization, a technique requiring one additional step.

An important advance has been the recent development of capillary columns (small-diameter packed columns), wall-coated open tubular columns, and support-coated open tubular columns. The wall-coated and support-coated columns are also small in diameter, but they can generate over 100,000 theoretical plates as compared with conventional columns (Table 1). Capillary gas chromatographs have been developed that incorporate these new designs as well as corresponding changes in injection systems and detectors. This new instrumentation is characterized by greater resolution over conventional models. A great deal of recent research has involved adapting capillary gas chromatography to pesticide residue analysis.

High-performance (pressure) liquid chromatography (HPLC). While the potential of this technique had been known for a long time, it was only in the late 1960s that pumps became available which were capable of forcing a liquid through an analytical column to achieve chromatogram peaks that could be resolved satisfactorily. The lack of suitable detectors also impeded development. While most of the 1970s were spent improving the instrumentation, it is expected that the 1980s will witness major developments in applications, particularly in the field of pesticide residue analysis. The major advantage of HPLC is that the sample is not heated, eliminating the problems associated with lack of volatility or thermal instability so inherent in gas-liquid chromatography. Columns are now available for almost any chemical, whether it has a high or low molecular weight, is polar or nonpolar, is ionic or nonionic, or is a combination of these factors. Thus separations may be based on molecular weight, absorption, adsorption, and ion exchange. The compounds may be proteins (high molecular weight), ions (such as amino acids and metal ions), and low-molecular-weight chemicals (including pesticides and herbicides).

In fact, most recent research in pesticide residue analysis involves HPLC. Most of the modern instruments are equipped with ultraviolet–visible or fluorescence detectors which lack selectivity in comparison with the highly selective detectors available with gas chromatographs (for example, the flame photometric detector). Nonetheless, progress is being made in this field.

Extraction. Traditionally, pesticides in water were often determined directly by spectrophotometry through measurement of the natural absorbance in the ultraviolet–visible or by production of highly absorptive species (through complexation). The establishment of lower tolerance levels for pesticides in environmental substrates prompted the development of extraction and concentration procedures. Organic solvents are used extensively for retrieving pesticides from water. With more difficult substrates the extraction solvent is usually miscible with water, but the extract has to be cleansed by column chromatography or liquid-liquid partition, or both, prior to the quantitation step.

Table 1. Relative comparison of various gas-liquid chromatography columns*

Type	Wall-coated open-tubular column	Support-coated open-tubular column	Micropacked capillary	Conventional packed
Graphic description				
Coil diameter	130 mm	130 mm	130 mm	Variable
Outside diameter	0.75 mm	1.0 mm	1.5 mm	3.18–6.35 mm
Inside diameter	0.25 mm 0.50 mm	0.50 mm	0.6–10.8 mm	2 mm and 4 mm
Typical length	20–100 m	15–30 m	3–6 m	1–3 m
Relative efficiency per column	100,000	50,000	28,000	4,000
Type of sample injection system	Splitting	Splitting or splitless	Splitting or splitless	Conventional
Sample size	10^{-4}–10^{-3} μl	10^{-3}–10^{-2} μl	10^{-2}–10^{-1} μl	Microliter and larger

*From *Chromatography Supplies*, Catalog 26, Alltech Associates, Inc., Arlington Heights, IL.

Table 2. Some properties of amberlite XAD resins*

Resin	Chemical nature	Porosity volume, %	Surface area, m^2/g
XAD-1	Polystyrene	37	100
XAD-2	Polystyrene	42	330
XAD-4	Polystyrene	51	750
XAD-7	Acrylic ester	51	450
XAD-8	Acrylic ester	52	140

*From *Laboratory Reagents and Biochemicals*, BDH Chemicals, Montreal, Canada.

One of the most recent developments in this field has been the introduction of polymeric adsorbents. In particular, special (XAD) resins such as polystyrenes (nonpolar) and acrylic esters (semipolar) having various degrees of porosity and greater contact surface areas (Table 2) have proved very effective. Pesticides in ultratrace quantities can now be recovered from water simply by passing the required volume of sample through a column containing the resin. Sometimes the pesticide is stable on the column, thus ensuring sample preservation. There is also some saving in operation cost and time. This approach is finding wide acceptance for field monitoring of air and water samples.

Summary. The field of pesticide residue analysis has become very sophisticated. Most modern instruments incorporate data systems that have memory capabilities, permitting, for instance, automatic analysis of large quantities of samples with excellent reliability. Whenever necessary, identification of unknown components is greatly simplified through the use of separation techniques and mass spectrometry.

In the future it is probable that techniques for direct processing of field samples will be developed. This would save a great deal of time by eliminating the long and tedious extraction and preconcentration processes still required with most substrates.

For background information *see* CHROMATOGRAPHY; EXTRACTION; GAS CHROMATOGRAPHY; PESTICIDE in the McGraw-Hill Encyclopedia of Science and Technology.

[G. G. GUILBAUT; V. N. MALLET]

Bibliography: R. D. Davies, *J. Chromatogr.*, 170:453–458, 1979; V. N. Mallet, Quantitative thin-layer chromatography of pesticides by *in situ* fluorometry, in J. Harvey, Jr., and G. Zweig (eds.), *Pesticide Analytical Methodology*, no. 136, 1980; V. N. Mallet et al., *J. Chromatogr.*, 160:81–88, 1978; L. G. M. Th. Tuinstra, W. A. Traag, and A. J. Van Munsteen, *J. Chromatogr.*, 204:413–419, 1981.

Extinction (biology)

During the early 1980s new sources of data on biological extinction have been contributed by geochemists, paleontologists, and nuclear physicists with the help of micropaleontologists. Studying the chemical compositions of sedimentary deposits laid down during the time of mass extinction at the end of the Cretaceous, 6.5×10^7 years ago, these scientists have come up with remarkably firm evidence that the great dying was caused by the collision of the Earth with an extraterrestrial body, probably a 10^{12}-ton comet. The new ideas postulating an extraterrestrial cause for biological extinction have found little acceptance among paleontologists, who adhere to the theory of steady-state evolution.

Prior to the 19th century, living organisms on the Earth were believed to have been periodically wiped out by catastrophes. This premise is the basic tenet of catastrophism. The development of natural sciences during the 19th century brought an end to this school of thought. Charles Lyell's principle of uniformitarianism, has been so deeply entrenched as to induce most modern geologists to shy away from what has been considered to be the heresy of catastrophism.

Uniformitarianism. The Lyellian dogma of uniformitarianism has two basic postulates: physical and chemical laws have not changed with time, and geologic processes on Earth have been operating at a uniform rate. While the first is still a basic premise of science, the second, as many geologists have begun to realize, may be an inadequate assumption. While the logic of science is incompatible with the assumption of catastrophes caused by divine intervention, there is no reason to assume that natural catastrophes of extraordinary proportions could not have happened during the billions of years of the Earth's existence. Meteor falls are natural phenomena. They are mostly small, but a statistical study of craters caused by metorites impact indicated that some very large ones must have fallen during the last few hundred million years.

Darwin was a uniformitarianist. In his theory of evolution he advocated slow changes of the evolving organisms through natural selection. He was bothered, however, by the apparently sudden extinction of swimming marine shells, called ammonites, at the end of the Cretaceous Period, 6.5×10^7 years ago. Subsequent studies of fossils did not help to resolve Darwin's dilemma. Instead, there has been more and more evidence that there was a catastrophe of mass mortality at that time.

Evidence in deep-sea sediments. The catastrophe did not exterminate all living organisms. Studies of land plants, fresh-water mollusks, small land animals, and numerous kinds of marine invertebrate animals all showed that their evolutionary changes were slow and gradual at the end of the Cretaceous Period. On the other hand, the record of ancient deep marine sediments is impressive. Those sediments consist almost entirely of skeletons of small one-called animals called foraminifera and calcified remains of even smaller one-celled plants called nannoplankton. When evolutionary changes were slow, the microfaunal and nannofloral composition of the sediments changed very little from one geological epoch to another. One might notice a gradual decrease in the number of specimens of some species in a sample because of an evolutionary decline of the species; and extinction is signified by the total absence of the species in the sediments above a

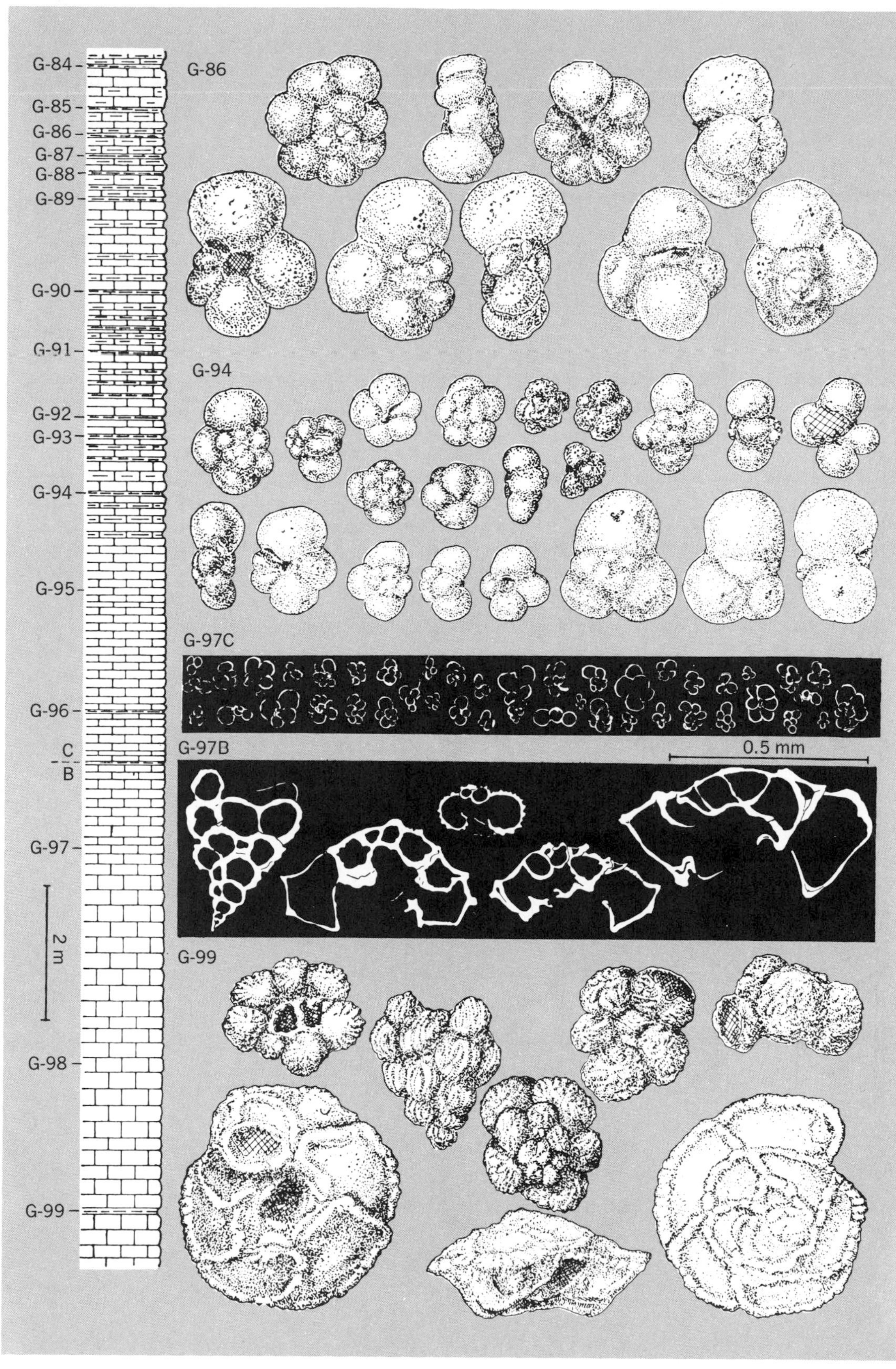

Change in oceanic faunas at the end of the Cretaceous. The foraminiferal faunas of the Cretaceous ages are mostly large, robust forms. All the large and robust species died out during a catastrophe 6.5×10^7 years ago which marked the end of the Cretaceous Period (and of the Mesozoic Era). Only scattered tests of the very small foraminifera *Globigerina eugubina* survived the catastrophe, and they became the ancestors of the Tertiary foraminiferal faunas. Note the remarkable change in size of the foraminifera; those in sample G97-B were the last Cretaceous forms, and they are about 10 times larger than the first Tertiary forms, in samples G97-C. (*From H. P. Luterbacher and I. Premoli-Silva, Biostratigrafia del limite Cretaceo-Terziario nell'Appennino Centrale, Riv. Ital. Paleontol., 70:67–128, 1964)*

last-appearance horizon. However, the change of the paleontological record at the end of the Mesozoic, from the Cretaceous to the Tertiary Period, is quite unusual, as Hanspeter Luterbacher and Isabella Premoli-Silva learned during the 1960s through their study of the rock formations near the community of Gubbio, in Tuscany, Italy (see illustration). They noted that the last Cretaceous sediment is a limestone containing a rich and diversified fauna of foraminifera, but the first Tertiary sediment is a red clay, which is almost devoid of any kind of fossils. In the sediment about 1 cm above the Cretaceous-Tertiary boundary, skeletons of very small foraminifera called *Globigerina eugubina* are present. This species is probably the ancestor of all the formaniferal faunas which live in the ocean today. The record is clear: there was a mass extinction with rare survivors that evolved into many new forms by means of explosive evolution.

The other major components of a deep-sea sediment are the nannoplanktons. In the late 1950s William Bramlette discovered that the Cretaceous nannoplanktons are almost entirely different from those of the Tertiary. Katharina Perch-Nielsen studied a sedimentary sequence in Denmark which yielded detailed evidence of the catastrophic change in the world of nannoplanktons across the Cretaceous-Tertiary boundary. More recently, Jan Smit made a detailed analysis of the Caravaca sequence in Spain; he found that only a few of about 60 foraminifera species survived the catastrophe at the end of the Cretaceous.

Geochemical evidence. One of the most significant discoveries on the extinction problem was made in 1980 by Luis Alvarez and Walter Alvarez. They and their coworkers discovered an abnormally high concentration of the heavy metal iridium in the sediments at Gubbio. Subsequently they found anomalous concentrations of iridium in sediments of the same age elsewhere. The Alvarezes considered the iridium anomaly proof that a large meteorite collided with the Earth 6.5×10^7 years ago, and they believed that this collision was responsible for the terminal Cretaceous mass extinction.

Kenneth Hsü had come to a similar conclusion from a completely different approach. He was inspired by a talk given by Nicholas Shackleton on the isotopic composition of sediments deposited during the time of the so-called great dying 6.5×10^7 years ago. Geochemists have been using the method of measuring the relative abundance of two oxygen isotopes, ^{18}O and ^{16}O, in fossil skeletons to determine the temperature of ocean water in which the fossil organisms once lived. Shackleton found that the temperature increase across the Cretaceous-Tertiary boundary was 5°C. Such a sudden increase could be triggered only by an extraordinary event, probably extraterrestrial. In 1980, recalling suggestions by D. J. McLean and others that sudden increases of atmospheric temperatures should have been lethal to dinosaurs, Hsü revived the earlier speculations of D. M. W. de Laubenfels and Harold Urey that the extinction of those giant reptiles may have been caused by atmospheric heating during the fall of a giant comet. However, floating or swimming organisms in the oceans should not be as sensitive to temperature changes as the dinosaurs on land. Invoking a recent discovery of cyanides in comets, Hsü postulated that those marine organisms may have been killed off by the poison of a fallen comet that had polluted ocean currents.

Extraterrestrial collision. Two critical questions are involved. Did a meteor (or cometary) impact indeed take place? Could such an impact have caused the mass extinction? The occurrence of an iridium anomaly in terminal Cretaceous sediments worldwide is very good evidence for the extraterrestrial event. Denis Kent suggested that such an anomaly could possibly have resulted from some special chemical reactions between sediments and seawater. However, the Alvarezes ruled out this alternative explanation when they found anomalously high iridium concentrations also in sediments deposited on land, at a horizon just above the last fossil skeletons of dinosaurs. Furthermore, R. Ganapathy found that the iridium-rich sediments also contain many other trace-element metals, and they are present in proportions similar to those found in meteorites. Finally, two craters, 6.5×10^7 years old, one 25 km and the other 3 km in diameter, have been found at Kamensk and Gusev in the southern part of the Soviet Union; they could be considered the concrete evidence of an extraterrestrial impact at the time of Cretaceous extinction.

There is less agreement on the question of why a collision with an extraterrestrial object should have caused the mass extinction. The Alvarezes thought that the fallen object was a 10^{11}-ton (10^{17}-g) asteroid. The dust ejected from the crater caused a blackout (or brownout) that lasted for 3–5 years. Plants in the oceans died out because their photosynthesis was suppressed; terrestrial plants survived because their seeds germinated after this period of limited light, enabling the continuation of the species. The dinosaurs, according to the Alvarezes, died out because of starvation. This asteroid scenario cannot satisfactorily explain the observed pattern of selective extinction, as the paleobotanist Leo Hickey pointed out. In fact, Frank Kyte and associates questioned whether the fallen object could be an asteroid, because the dust ejected from a crater by its impact should consist mostly of iridium-poor terrestrial materials; the extraterrestrial body itself should be buried in the crater. Since an iridium-rich dust was distributed worldwide as fallout, it seems that the fallen body had been largely disintegrated during its entry; such a body is most probably a comet, not a solid asteroid.

Sediment core studies. An opportunity to study the detailed environmental and evolutionary changes consequent upon a terminal Cretaceous catastrophe was provided in 1980, when an oceanographic cruise by the drilling ship *Glomar Challenger* obtained sediment cores from the South Atlantic Ocean. Those sediments were deposited at a very rapid rate and thus provided and amplified signals on the fertility and temperature changes of the oceans. A team of scientists consisting of Hsü and colleagues studied those samples. They could repeat the observations that had been made elsewhere. A rich and diversified microfauna and nannoflora were practically wiped out at the end of Cretaceous. The first sediment laid down after the catastrophe is a very thin lamina of red clay less than 1 cm thick. The clay is sterile and devoid of fossils. Scattered foramaniferal tests belonging to some dwarf species of *Globigerina*, identical to those described by Luterbacher and Premoli-Silva at Gubbio, are present in the sediments above the sterile horizon. The succession of nannofloras is also identical to that described by Perch-Nielsen in Denmark; the rare survivors belonged mostly to a species which left calcified cysts.

Analyses of the chemistry of the sediments across the Cretaceous-Tertiary boundary provided even more revealing tales. An anomalously high iridium concentration was found in the last Cretaceous sediments, as predicted. A sterile sediment, devoid of fossil skeletons, was deposited when the biomass of the ocean was almost eliminated by a catastrophe. The fertility of nannoplanktons and of foraminifera was then much reduced. Dissolved carbon dioxide in seawater was not utilized by nannoplankton for photosynthesis, so that the seawater became more acidic than ever. Calcified remains of the meager fauna and floras of the oceans were mostly dissolved in this acidic ocean, so that only red clays were sedimented. The isotopic analyses indicated a steady rise of the relative abundance of the light carbon-12 atoms in ocean waters. Their rise can be explained by the fact that those light carbon atoms should be taken up by the photosynthesis of marine plants, but they would become excessive if the ocean were a desert almost devoid of nannoplanktons. The isotopic analyses of the oxygen isotopes confirmed the rise of temperature after the terminal Cretaceous catastrophe. However, it was a slow rise which took about 50,000 years to reach the maximum, and such a slow increase cannot be related to atmospheric heating during the fall of a meteor or comet. More probably, CO_2 in the oceans, not utilized by photosynthesis, escaped into the atmosphere. The excess CO_2 worked like a greenhouse to trap solar radiation reflected from the Earth's surface, thereby causing an increase in the temperatures of the atmosphere and of the oceans. The investigations of Hsü's team have thus given credence to his earlier postulates that the mass mortality of ocean life was caused by poisoning, and the dinosaurs became extinct when those heavy-bodied animals could no longer regulate their body temperature and died of heart failure.

Implications. Many further investigations are needed to confirm the hypothesis of catastrophic extinction consequent upon cometary impact. However, if the idea is correct, there may be a reevaluation of the basic idea underlying some aspects of Darwin's theory of evolution. The catastrophe hypothesis interprets evolutionary changes in terms of chance survival rather than as a struggle for existence and natural selection. The terminal Cretaceous event caused the extinction of large terrestrial reptiles and marine planktons. Another catastrophe could have caused another set of catastrophic changes and resulted in another round of chance survivals. Furthermore, mass extinction could have liberated many of the previously occupied ecologic niches and thus provided the impetus for explosive evolution, as is witnessed by the paleontological record for the beginning of the Tertiary.

For background information *see* EXTINCTION (BIOLOGY); FORAMINIFERIDA; MICROPALEONTOLOGY; PALEONTOLOGY in the McGraw-Hill Encyclopedia of Science and Technology. [K. J. HSÜ]

*Bibliography:*L. W. Alvarez, *Science*, 208:1095–1108, 1980; K. J. Hsü, *Nature*, 285:203–205, 1980; H. P. Luterbacher and I. Premoli-Silva, *Riv. Ital. Paleontol. Strat.*, 70(1):67–128, 1964; J. Smit and J. Hertogen, *Nature*, 285:198–200, 1980.

Fern

There has been great interest in the scientific community concerning the tiny aquatic fern azolla as a source of biologically fixed nitrogen, especially since its use in the agricultural systems of China and Vietnam has been described. This interest was heightened by the energy crisis in 1973, which caused a rapid increase in the price of nitrogen fertilizer.

Azolla is a genus of small floating aquatic ferns that have become naturalized in many places in the tropics and subtropics, and in the temperate zone. A member of the Salviniales, *Azolla* comprises some seven recognized living species, *caroliniana*, *filiculoides*, *mexicana*, *microphylla*, *nilotica*, *pinnata* (syn. *imbricata*), and *rubra* (Fig. 1). There is

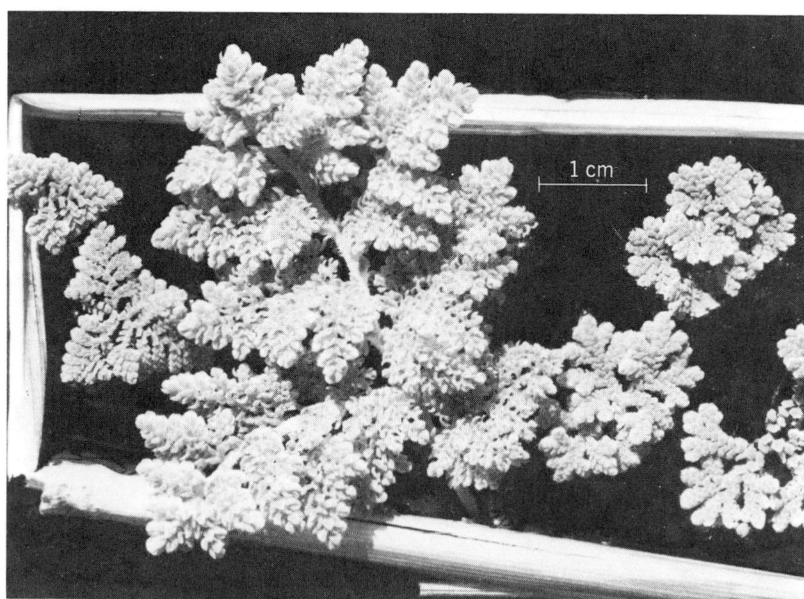

Fig. 1. Plants of *Azolla* species. For a size comparison, the plants are enclosed within a border made up of a rice plant stem.

disagreement among botanists as to the taxonomic status of some of the species. The genus is divided into two subgenera, *Azolla* (three floats) and *Rhizosperma* (nine floats). *Azolla* fronds are triangular or polygonal in shape, and float on the water individually or in mats, often giving the appearance of a dark green to reddish carpet.

In the *Azolla–Anabaena* symbiosis the delicate azolla provides nutrients and a protective cavity in each leaf for algal colonies (Fig. 2) in exchange for fixed atmospheric nitrogen and possibly other growth-promoting substances. The cavity in each leaf is formed by an epidermal cell growth covering a depression on the basal half of the lobe's ventral surface. Over the center of the depresson, epidermal cells meet and form a large pore which may allow gaseous exchange between the cavity and the atmosphere. *Anabaena azollae* cells sheltered in the shoot apex are entrapped by the enclosing epidermal cells, and begin colonizing the cavity. The inner surface of a mature leaf cavity is lined with transfer hairs which may be organs of metabolic exchange between the fern and the alga.

Azolla has been used as a green manure for rice and other crops, as a fodder or feed for animals, and as a weed suppressor in flooded rice. In some situations azolla itself may be considered as a weed, for example in lakes, ponds, waterways, and aquatic crops.

Azolla has come into prominence mostly because of its use as a green manure. Hundreds of years ago, Chinese peasants discovered that the fern was an effective green manure for increasing rice yields. Today azolla is used for more than 1,000,000 hectares of rice in southeastern China and some 500,000 hectares of rice in Vietnam.

Green manure production systems. There are three major green manure production systems for azolla in rice. In the most common system, azolla is grown as a monocrop on flooded rice fields before the rice crop is planted. Depending upon the time available before rice planting, one or more crops of azolla may be grown on the water surface and plowed or harrowed into the soil as soon as a sufficient azolla mat (15–25 metric tons per hectare) forms. The more mats of azolla that are incorporated into the soil, the more nitrogen that is available for the rice. This system is referred to as basal azolla.

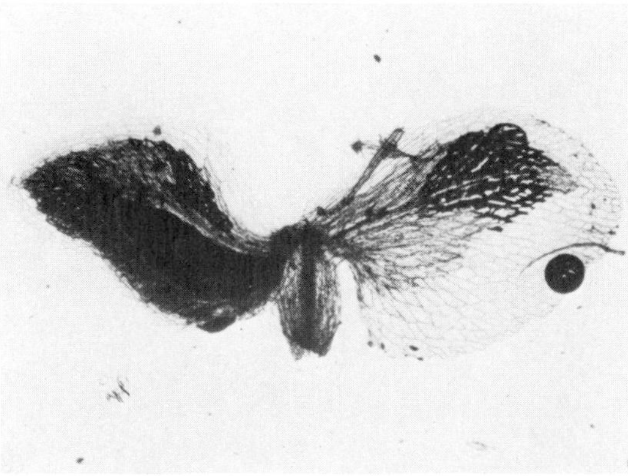

Fig. 2. An azolla leaf. On the left is the thick aerial dorsal lobe with cells containing chlorophyll. The very dark central portion is the central cavity which contains the blue-green alga, *Anabaena*. To the right is the slightly larger but thin, floating ventral lobe of the leaf. The nearly achlorophyllous ventral lobe floats on the water surface and supports the frond.

In another system, azolla is grown as an intercrop with rice. It either is allowed to die naturally or is incorporated into the soil at the time of weeding, or by special azolla incorporation measures. Often the azolla is pressed or pushed into the soil by hand. Incorporating azolla two or three times before the maximum tillering stage of rice can raise rice yields markedly. When it is allowed to die naturally through disease or insect attack or by shading of the rice, the nitrogen release is not very effective. These systems are known as topdressing azolla.

The third system is a dual basal/topdressing system in which azolla is grown alone, is incorporated (one or more times) into the soil before rice planting, and is also grown under the rice as an intercrop.

The benefits of azolla green manuring in rice can be significant. According to the results of more than 1500 experiments in seven Chinese provinces, the cultivation of *Azolla imbricata* increased rice yields by an average 600–700 kg/ha above the yields of control (zero nitrogen) plots.

An azolla crop may produce 3 or more kg N/ha/day under good growing conditions. Thus a 20-day-old crop may produce about 60 kg N/ha, and two basal crops of 20 days each could produce almost 120 kg N/ha for the following rice crop.

Azolla cultivation requires careful management year-round. Because it is vegetatively propagated, the fern must be kept alive during the harsh weather of both winter and summer. Special nurseries are often required to maintain sufficient starter stocks. Also, large-scale field multiplication is required to ensure sufficient planting material for the main rice production seasons.

N_2 fixation. Fixation of atmospheric nitrogen is a function of the blue-green alga (cyanobacterium) partner in the *Azolla–Anabaena* symbiosis. *Anabaena azollae* colonies, living inside the azolla leaf cavity, have the appearance of long strings, made up of two sizes of beadlike cells. The smaller and most common (70–80%) of the cells are vegetative cells, whose main function is photosynthesis. The larger ovoid cells, called heterocysts, make up 20–30% of the total and are evenly spaced individually in the strings. Heterocysts have thick walls and are known to be the main repository of the enzyme for nitrogen fixation, nitrogenase. In a low-oxygen environment, nitrogenase has the ability to convert inert atmospheric N_2 molecules into ammonia molecules, which can be used for the synthesis of amino acids. The ammonia produced by the *Anabaena* either is used or stored by the alga or is exported to the azolla partner. This internal supply of ammonia gives azolla a competitive advantage in environments where low levels of available nitrogen are a constraint to the growth of other aquatic plants.

Although azolla usually meets its nitrogen requirement through a combination of absorption from the environment and nitrogen fixation, the *Anabaena* leaf colonies are capable of meeting its entire nitrogen requirement, even at growth rate doubling times of 2.5 days or less. The rate of nitrogen fixation is affected by the concentration of available nitrogen in the water, particularly ammonia, and has been measured indirectly by the use of acetylene reduction–gas chromatography and directly by the use of $^{15}N_2$. However, these techniques cannot be used to estimate annual or even daily values of fixed nitrogen, since they require disruption of the plant's natural growth environment. Estimates of nitrogen fixation are best made by determining total nitrogen accumulated over time. Through this method, estimates of nitrogen fixation range from 3 to more than 15 kg N/ha/day, under optimum conditions.

H_2 evolution. Many nitrogen-fixing organisms have demonstrated nitrogenase-catalyzed H_2 evolution. Scientists are interested in this phenomenon because of the potential use of biologically evolved hydrogen as a source of fuel. Hydrogen evolution by the *Azolla–Anabaena* symbiosis is constantly occurring with the heterocysts of the *Anabaena*, but does not reach significant levels unless the plant is grown in an artificial N_2-free environment. Within the heterocyst, the evolution of H^+ may help in maintaining the low partial pressure of O_2, necessary for the functioning of nitrogenase. Hydrogen evolution is strongly inhibited by N_2, completely suppressed by C_2H_2, but unaffected by CO. In an argon atmosphere, light-dependent hydrogen evolution by azolla may reach levels approaching 800 nanomoles H_2 per gram fresh weight per hour. However, under these conditions, azolla will eventually die from nitrogen starvation unless nitrogen fertilizer is supplied in the growth solution or the argon atmosphere is alternated with one containing N_2.

For background information *see* NITROGEN FIXATION in the McGraw-Hill Encyclopedia of Science and Technology.

[DONALD L. PLUCKNETT; THOMAS LUMPKIN]

Bibliography: T. A. Lumpkin and D. L. Plucknett, *Econ. Bot.*, 34:111–153, 1980; A. W. Moore, *Bot. Rev.*, 35:17–35, 1969; G. A. Peters, and B. C. Mayne, *Plant Physiol.*, 53:820–824, 1974.

Flower

Flowers are the sexual reproductive shoots of angiosperms. They serve as the best example of the great diversity of form in this dominant group of land plants. The study of floral form can be facilitated by a knowledge of floral function in the breeding system. In many cases the intricacies of floral structure appear to be adaptations to attract insects or other animal vectors in order to effect cross-pollination. Since changes in form are due to alterations in development, comparative ontogenetic study can be used to detect the development processes responsible for the great diversity of floral forms in the angiosperms. Floral morphogenesis is characterized by sequential organ production from a meristem which gets used up in the process of producing the flower. The usual sequence of organ production is calyx, corolla, androecium, and gynoecium. Variation in the patterns of cell division and cell expansion during organ initiation, growth, and differentiation result in the different shapes and sizes of mature floral organs.

Species which produce two floral forms with different breeding systems are useful subjects for comparative study. One such example is that of cleistogamous species which produce two hermaphroditic floral types on a single plant. One, referred to as cleistogamous, remains closed and self-pollinates in the bud. The other, called chasmogamous, opens at anthesis and may be visited by an insect vector, providing a means of cross-pollination for the species. Thus, two floral forms with different functions are produced by a single plant, providing a comparative system for the study of developmental processes responsible for alterations in form that may result in changes in the breeding system.

Cleistogamy. There are at least 287 species in 56 families of angiosperms that produce both cleistogamous and chasmogamous floral forms. The cleistogamous flower is a source of inbred, genetically invariable seed; the chasmogamous flower is a means for providing genetic variability via the production of outbred seed. Seasonal environmental cues such as temperature or photoperiod may trigger production of either floral form. The perennial species *Viola odorata* (sweet violet) produces chasmogamous flowers during the short, cool days of spring, and then switches to cleistogamous flowers (after a progression of intermediate forms) during the long, warm days of summer. Annual cleistogamous species have a shorter period for flowering, usually extending over only one season. These annuals may produce cleistogamous and chasmogamous flowers in succession in their inflorescences, the percentage of each flower type depending on both genetic and environmental factors. Usually, poor growth conditions like drought, low nutrient supply, and low light intensity will induce predominantly cleistogamous flower production. When plants are grown under controlled growth conditions, two populations of flowers, one cleistogamous and the other chasmogamous, are produced for comparative developmental study.

The advantages of using such a system to study floral morphogenesis and function are many. First, the two forms can be produced by a single genotype. Knowledge of the developmental processes responsible for such a switch can allow predictions about how analogous form changes occurred in the past evolution of new species.

Second, the two floral forms usually have different functions in the breeding system of the plant. The chasmogamous flower has a form typical of the genus; the cleistogamous flower is modified to ensure self-pollination or inbreeding. Study of the developmental and physiological processes responsible for such a change in breeding systems may enable researchers to understand the evolution of inbreeding in the angiosperms, a strong evolutionary trend in recent times.

Third, the morphological differences between the cleistogamous and chasmogamous floral forms are not qualitatively great. Typically, the cleistogamous flower has a reduced corolla that fails to expand at anthesis and a reduced androecium (fewer or smaller anthers). So the modifications in development are minor, yet the mature forms appear very different and may have opposite functions in the breeding system. This is an example of how slight changes early in development can lead to diverse mature floral forms and functions.

Study of such a system can lay a foundation from which experiments can be devised to determine the mechanisms which trigger such form changes in floral development. Knowledge of these mechanisms may allow predictions as to how such changes must have occurred in the evolution of the reduced floral forms that characterize autogamous (predominantly inbred) species.

Flower development in Lamium amplexicaule. *Lamium amplexicaule*, a cleistogamous mint, was shown by E. M. Lord to produce a succession of flowers on one plant, progressing from cleistogamous through intermediate forms, then to chasmogamous forms, and finally back to cleistogamous forms before senescence. This ontogenetic shift in floral form occurs on most plants growing in the long, warm days of spring in California. During the short, cold days of winter, plants may produce only cleistogamous and intermediate closed floral forms (Fig. 1*a* and *b*). Apparently, long days, warm temperatures, and high light intensities induce chasmogamous floral production.

The cleistogamous flower is a reduced form of the chasmogamous flower; neither the calyx nor the ovary is altered in form or size, but the stamens, style, and corolla are smaller (Fig. 1*c–f*). The closed cleistogamous flower, which is always the predominant form on a plant, self-pollinates and produces abundant seed. The larger, open chasmogamous flower may be visited by honeybees (*Apis mellifera*) which collect nectar produced by a nectary at the base of the gynoecium. In the process of flower visitation, pollen may be transferred from another plant, effecting cross-pollination. Most of the chasmogamous flowers will self-pollinate if no vectors are available.

The first produced flower is always cleistogamous, with a progression of intermediate closed floral forms separating it from the chasmogamous flowers produced at the later nodes. This fixed pattern of floral expression can be referred to as heteroblasty, a term usually used to describe successive changes in leaf form which characterize the developing shoot systems of all seed plants. This positional localization has allowed the study of floral development from inception of the two extreme forms, cleistogamous and chasmogamous. Results from these studies indicate that only slight modifications in developmental rates in the corolla and stamens of the two flower types are responsible for their distinctively divergent mature forms.

At initiation, the two floral types are indistinguishable in their form and size. Three whorls of primordia arise successively from the floral meristem, beginning with the calyx and followed by the corolla and finally the four stamens. The rest of the floral meristem is used up in the production of the gynoecium, which appears initially as a collar at the apex of the meristem (Fig. 1*g*). The gynoecium is composed of two carpels which will fold together on themselves, each becoming two-lobed to form a total of four nutlets, each containing one ovule, a condition typical of the mint family. Histological study of these growth stages demonstrated that floral organ initiation and growth were similar in both flower types. The first detectable divergence in form in the two floral buds occurs when they are about 0.5 mm long. At this time sporogenous tissue in the anthers is entering prophase of meiosis, a process that will lead to the formation of four haploid microspores from each sporogenous cell. Each microspore will develop into a male gametophyte, the pollen grain. By the time meiosis is about to occur, counts of these original sporogenous cells reveal that the cleistogamous flowers have fewer; this results in their anthers' producing fewer pollen grains at maturity than are formed in the chasmogamous flower. This reduction in pollen number can be due to either a slowing down of the mitotic rate or a shortening of its duration, resulting in fewer cells in a smaller anther at maturity.

A feature that characterizes other cleistogamous species as well as *Lamium* is the germination of pollen grains inside the cleistogamous anthers (Fig. 2*a–c*). The pollen remains inside the anther, sending the pollen tube out to the stigmatic surface nearby through a slightly opened anther sac. In this manner, pollination occurs in the closed cleistogamous floral buds. In contrast, the anthers of the chasmogamous flower do not open until anthesis, and pollen does not germinate until it is placed on the stigmatic surface (Fig. 2*d* and *e*).

In 1904, Karl Goebel suggested that cleistogamy in *L. amplexicaule* was due merely to an arrest of corolla expansion in the cleistogamous flower. This implied that a lack of cell expansion alone could convert a chasmogamous to a cleistogamous floral form. Developmental studies by Lord have demonstrated, however, that a more fundamental difference exists between these flowers: a reduction in cell number occurs early in development of the cleistogamous stamen.

If corolla development is followed in the same manner, differences in shape can also be detected

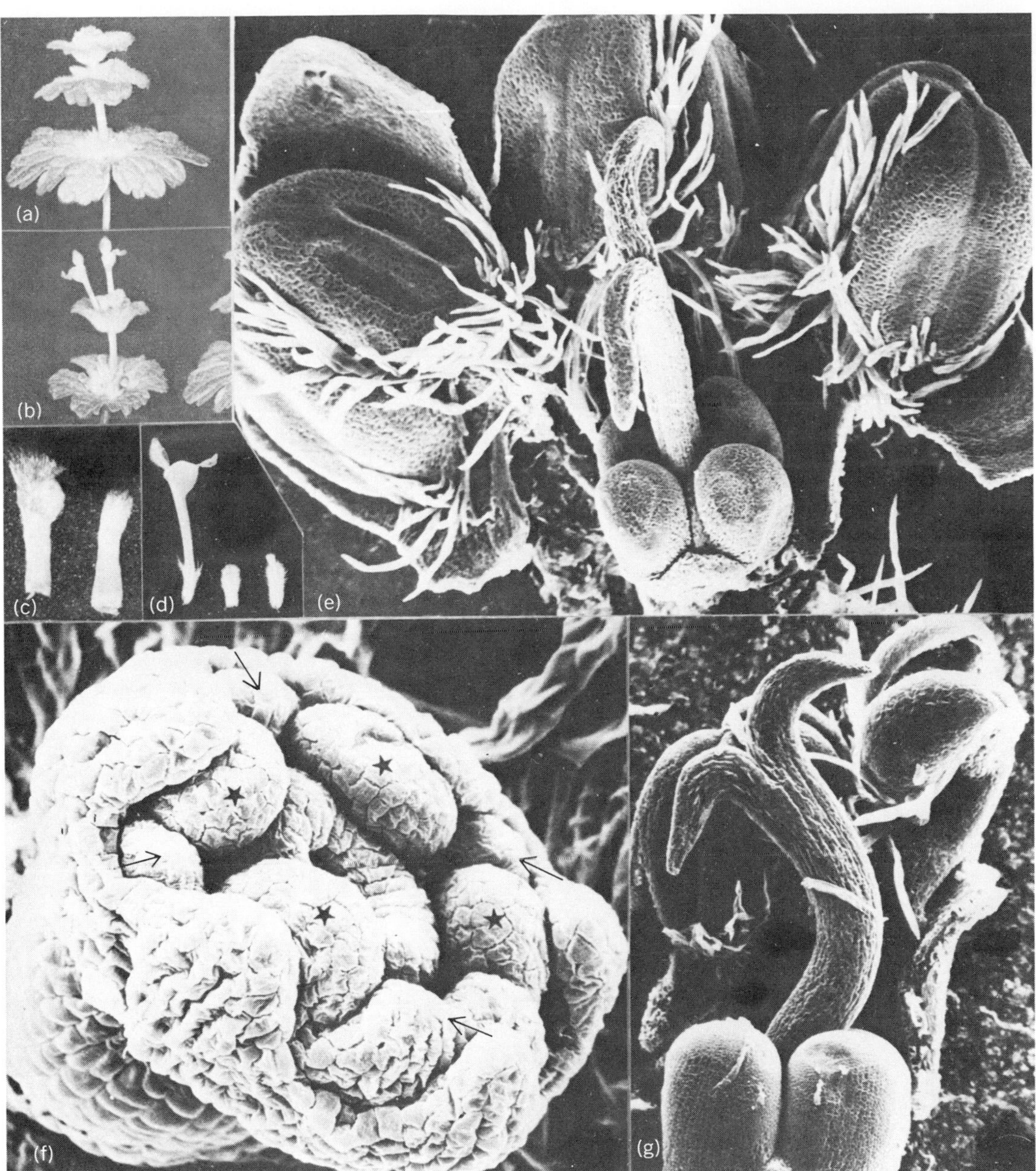

Fig. 1. Flowers of *Lamium amplexicaule*. (*a*) Shoot bearing cleistogamous flowers. (*b*) Shoot bearing mostly chasmogamous flowers. (*c*) Corollas from chasmogamous (left) and cleistogamous (right) floral buds of the same age. (*d*) Chasmogamous flower at anthesis (left); chasmogamous bud prior to anthesis (center); and cleistogamous bud of same age (right). (*e*) Chasmogamous floral bud dissected open; note larger size of anther sacs in relation to those of cleistogamous flower in *g*. (*f*) Floral primordium with all organs visible (stamens starred; arrows on corolla primordia). (*g*) Cleistogamous floral bud dissected open. (*From E. M. Lord, The development of cleistogamous and chasmogamous flowers in Lamium amplexicaule* (*Labiatae*)*: an example of heteroblastic inflorescence development, Bot. Gaz., 140*:*39–50, 1979*)

following stamen divergence in buds as small as 1-mm. The upper half of the cleistogamous corolla, which encloses the anthers, slows down in its increase in width relative to corolla length (Fig. 1*c*). At maturity the cleistogamous corolla is 3 mm long as compared to the 18-mm chasmogamous corolla (Figs. 1*d* and 3). If cell counts are taken in the two regions indicated by the arrows in Fig. 3, it becomes apparent that lack of cell expansion alone cannot fully describe the mature cleistogamous corolla form. All the cells in the cleistogamous corolla are smaller than those in equivalent regions of the chasmogamous corolla, but there are also fewer cells in the circumference of the cleistogamous upper corolla region. The cell numbers are the same in the corolla base region of both floral forms. Thus the reduced cleistogamous corolla form is due to both a lack of cell expansion at anthesis and a re-

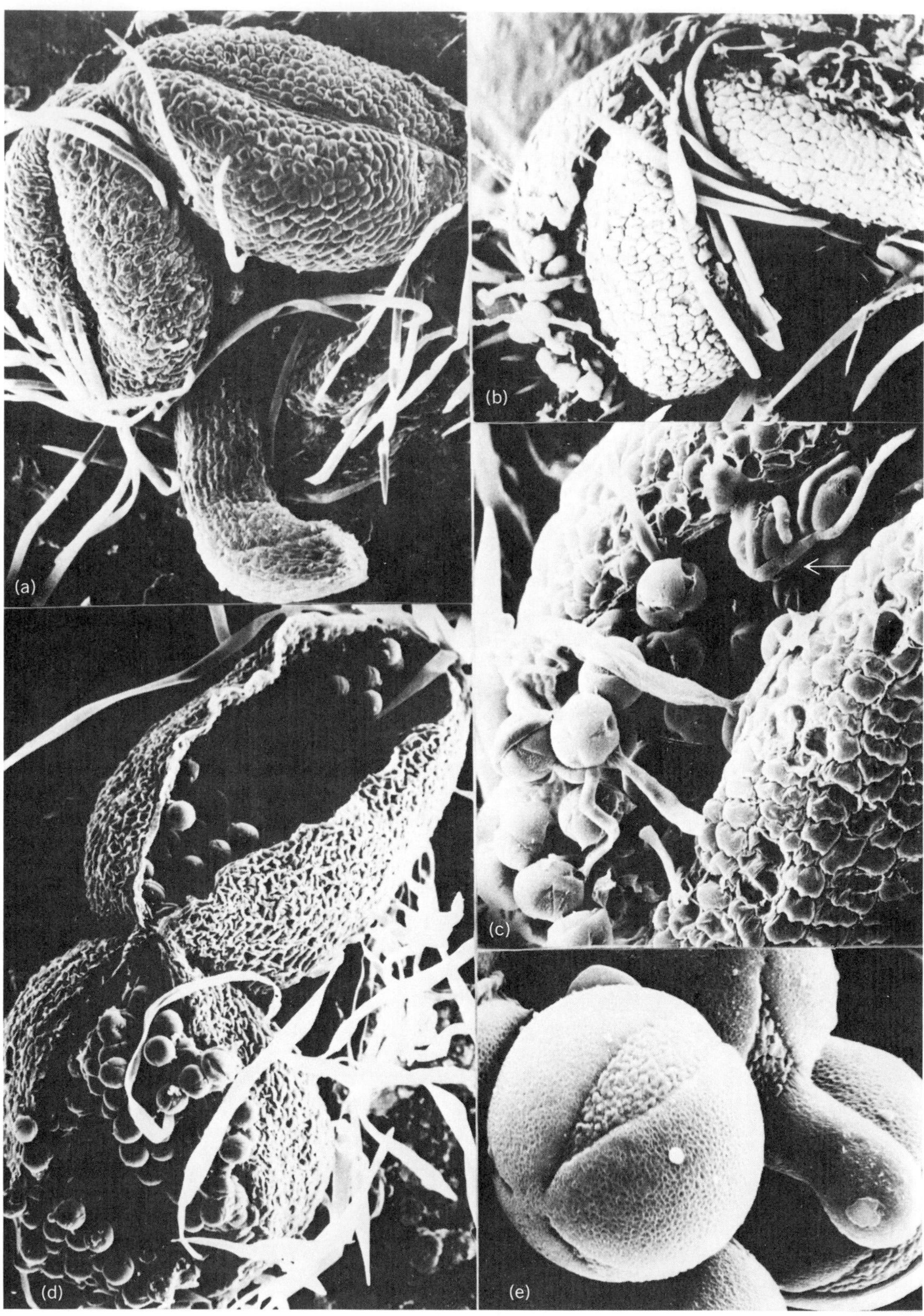

Fig. 2. Anthers and pollen of *Lamium amplexicaule*. (*a*) Unopened cleistogamous anther. (*b*) Open cleistogamous anther sacs with germinating pollen grains inside. (c) Higher magnification of *b*, arrow pointing to pollen tubes. (*d*) Chasmogamous anther at anthesis. (*e*) Pollen tube emerging from germination pore of pollen grain. (*From E. M. Lord, The development of cleistogamous and chasmogamous flowers in Lamium amplexicaule* (*Labiatae*)*: An example of heteroblastic inflorescence development, Bot. Gaz., 140*:*39–50, 1979*)

duction in cell number in the upper half of the corolla.

Detectable differences in rates of growth in the various dimensions of the organs of cleistogamous and chasmogamous flowers can direct researchers to examine, histologically, specific stages in development when divergent growth patterns are initiated. Once the stage in growth when divergence actually occurs is known, attempts can be made to manipulate the fate of floral buds in order to determine the mechanisms of control on their form.

Hormonal mechanisms. Other investigators have shown that gibberellin is present in the anthers of a number of species and can be implicated in the elongation of floral parts at anthesis. In *L. amplexicaule* the smaller anther size in the cleistogamous flowers may result in gibberellin levels that are too low to stimulate expansion of floral organs at anthesis.

To test this hypothesis, gibberellin was applied to the shoot apex of a plant producing only cleistogamous flowers. Every subsequently produced flower was chasmogamous. This suggests that gibberellin is the switch mechanism responsible for the shift in floral form from cleistogamous to chasmogamous in the developing inflorescence. However, if these gibberellin-induced chasmogamous flowers are examined closely, it becomes apparent that the hormone has caused cell expansion to occur but not the increase in cell number necessary to switch a cleistogamous into a chasmogamous flower. The gibberellin-induced chasmogamous corolla and anthers have cell numbers typical of a cleistogamous flower. Gibberellin is probably a late developmental trigger for floral expansion at anthesis in chasmogamous flowers. The switch mechanism responsible for early divergence in anther and corolla development in the cleistogamous flower has yet to be elucidated.

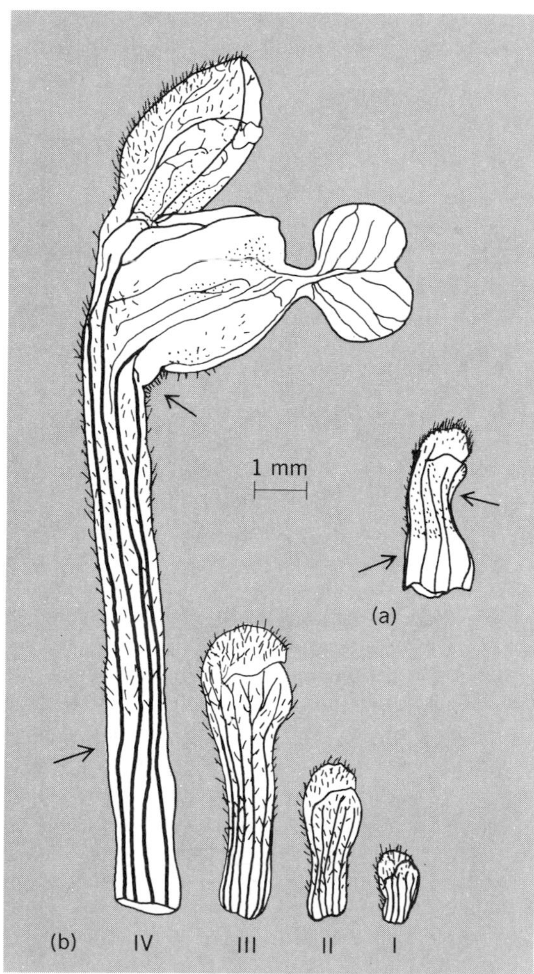

Fig. 3. Corollas of *Lamium amplexicaule*. (*a*) Mature cleistogamous corolla which never achieves anthesis. (*b*) Stages in development of the chasmogamous corolla (I–IV) which expands and opens at anthesis. The arrows point to morphologically equivalent regions of the two corolla types where cell size and number measurements were taken. (*From E. M. Lord, An anatomical basis for the divergent floral forms in the cleistogamous species, Lamium amplexicaule L. (Labiatae), Amer. J. Bot., 67:1430–1441, 1980*)

For background information *see* FLOWER; GIBBERELLIN; PLANT MORPHOGENESIS in the McGraw-Hill Encyclopedia of Science and Technology.

[ELIZABETH M. LORD]

Bibliography: K. Goebel, *Biol. Centralbl.*, 24:673–697, 737–753, 769–787, 1904; E. M. Lord, *Amer. J. Bot.*, 67:1430–1441, 1980; E. M. Lord, *Bot. Gaz.*, 140:39–50, 1979; Uphof, *J.C.T. Bot. Rev.*, 4:21–49, 1938.

Food manufacturing

An important feature of modern agriculture is the link provided by the food processing industry between production and consumption of food. Processing operations such as sterilization, drying, and freezing help to assure uniform supply of foods with minimal product loss. These operations require energy to accomplish processes that involve heat and mass transfer and mechanical handling and processing. In recent years introduction of mechanization to achieve high processing capacity at reduced cost of labor has added to the requirement of energy. Since the mid-1970s the food industry, like other manufacturing industries, has been seeking ways to reduce energy consumption and energy costs of food processing. In this article, an energy accounting method developed specifically for the food processing industry is presented.

Studies have shown that 12–17% of the total United States energy consumption is used for the food system. The food system encompasses such sectors as farm production, food processing, in-home food preparation, out-of-home food preparation, and wholesale and retail trade. Within the United States food system, the food processing sector is the most energy-consumptive sector. Approximately 18% of energy within the food system is expended for production, whereas 30% is spent on processing. These facts emphasize the importance of the food processing industry in terms of its energy requirement. Accounting of energy in the food processing industry is a useful approach to obtain relevant information that may be used for such purposes as energy conservation and improving energy conversion efficiencies.

Energy accounting. A method has been developed specifically for accounting of energy in the food processing industry. There are seven procedural steps: determination of the objective, selection of a system boundary, charting a process flow diagram, identification of all mass and energy inputs, quantification of the mass and energy inputs, identification of all mass and energy outputs, and quantification of all mass and energy outputs. Once the objective has been determined, energy accounting will involve either all or some of the other steps.

Determination of the objective. An energy accounting study is often designed to accomplish specific objectives. For example, information may be sought to develop energy use profiles for a given food processing plant. On the other hand, an energy accounting objective may involve investigating the feasibility of energy conservation modifications on a specific processing equipment.

Selection of a system boundary. After an objective is selected, the system to be considered must be clearly defined. A system boundary allows a choice of the items that will be considered or neglected in the accounting study. The importance of this step lies in correct interpretation of the results by the users. In addition, the system boundary assists in determining the total cost of the energy accounting study.

Process flow diagram. This assists in identifying various units that will be studied for energy accounting. The diagram is also useful in presenting the results of the accounting study. Symbols useful in drawing an energy accounting flow diagram are shown in Fig. 1.

Identifying mass and energy inputs. In a food processing plant, energy may be used in various forms, such as electricity, natural gas, steam, fuel oil, or coal. Similarly, mass flow, in addition to the primary product being processed, may involve such items as sugar, salt, or water. Any mass and energy input crossing the system boundary must be correctly identified.

Quantifying mass and energy inputs. Actual measurements of mass and energy inputs to the system are required. Flow rates of energy inputs such as steam, fuel oil, or natural gas must be measured along with product flow over a reasonable period of time, selected as a basis, to account for any unique variations associated with the process. The total time of measurement should be sufficient to allow observation of variations. This step requires installation of flow-measuring equipment. Depending on the system being studied, the cost of monitoring instruments can be substantial if flow measurements are required at a large number of locations.

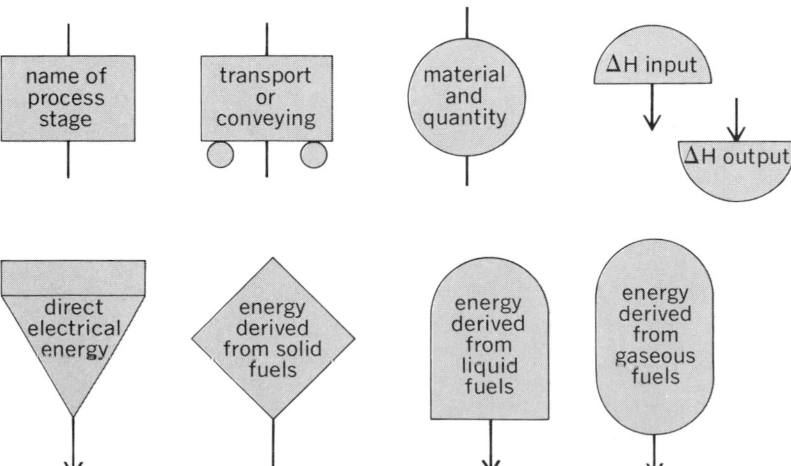

Fig. 1. Symbols for use in drawing energy accounting diagrams of food processing operations. ΔH = enthalpy. (*From R. P. Singh, Energy accounting in food process operations, Food Tech., 32(4):40–45, copyright 1978 by the Institute of Food Technologists*)

Identifying mass and energy outputs. This step is important for objectives that involve energy conservation or energy recovery. As energy is used (or converted) in a system, the form in which it exits across the boundary is identified in this step. Similarly, any waste product discarded or product losses from the system should be identified.

Quantifying mass and energy outputs. Energy flow out of the system must be either measured or estimated by using standard engineering calculations. Measurements of waste products exiting the system are necessary to account for energy losses with such streams.

Energy and mass flow measurements. This represents the major portion of an accounting study. A large number of energy monitoring instruments useful for energy accounting are currently available in the market. For example, steam flow can be measured by installing an orifice plate or a pitot tube–type sensor in the steam pipe. The differential pressure output from the sensor can be converted to either a pneumatic or an electrical signal by the use of a differential pressure transmitter. An electrical signal is preferred if computerized data collections are being made from a large number of measurement stations. Consumption of electricity can be measured continuously by using wattmeters or units that convert consumption of electricity to a signal which may be fed into a recording device. Often the amount of data that are collected in an accounting study becomes too large. It is important therefore, in the beginning of the study, to consider computer-assisted data collection and analysis. The procedure can be modified to assure that the collected data are compatible and amenable to the available computer. Desk-top computers are quite useful and often contain sufficient memory for use in accounting studies.

Case studies. The usefulness of energy accounting in food processing is illustrated with two case studies. In the first case study, energy accounting was conducted on a processing line operated to can whole-peeled tomatoes. The accounting study was conducted to identify energy-intensive operations. The second case study illustrates the use of energy accounting in seeking ways to modify an energy-intensive operation, an atmospheric retort, to make it more energy-efficient.

Energy accounting of a canning plant. The objective of an energy accounting study conducted in a tomato canning plant was to determine the relative importance of energy input to the various units used in processing raw tomatoes to obtain canned peeled tomatoes. The system boundary was drawn across the canning line. Energy input to the canning line was in the form of electricity and steam. The energy accounting diagram is shown in Fig. 2. Two operations, namely lye bath peelers and retorts, consumed steam in addition to electricity, which was used for all other operations. The accounting diagram contains the energy use on an 8-h shift basis. The mass input to the canning line was 200 tons (180 metric tons) of whole tomatoes and 84 tons (76 tons) of tomato juice. The mass output was 272 tons (245 metric tons) of canned tomatoes with juice, and 12 tons (11 metric tons) of tomatoes was discarded as waste product. The relative energy input is evident from the diagram. Analysis of the data indicated that retorting and lye bath peeling consumed 99% of total energy input to the canning line. Only 1% of the energy was consumed by all other operations. Thus, the accounting study revealed that the lye bath peelers and retorts are highly energy-intensive and should be examined further for possible modifications to reduce energy consumption.

The data presented in the accounting diagram (Fig. 2) can be used to determine the direct energy consumption per unit mass of the raw or processed product. Assuming a boiler efficiency of 70% and using a fossil fuel conversion factor of 10,799 kJ/kWh to account for the inefficiency of the generation of electricity at power plants, it can be calculated that for the canning line studied the total energy consumption for peeled tomato processing is 1300 kJ/kg of product received. An important use of this analysis is to compare energy requirement per unit quantity of product processed by different canning lines. Such comparison can help identify inefficiencies between different processing lines within the same plant or between different plants processing similar products. Continuous monitoring of energy consumption and analysis of data, preferably by a computerized system, can provide timely information useful in better energy management.

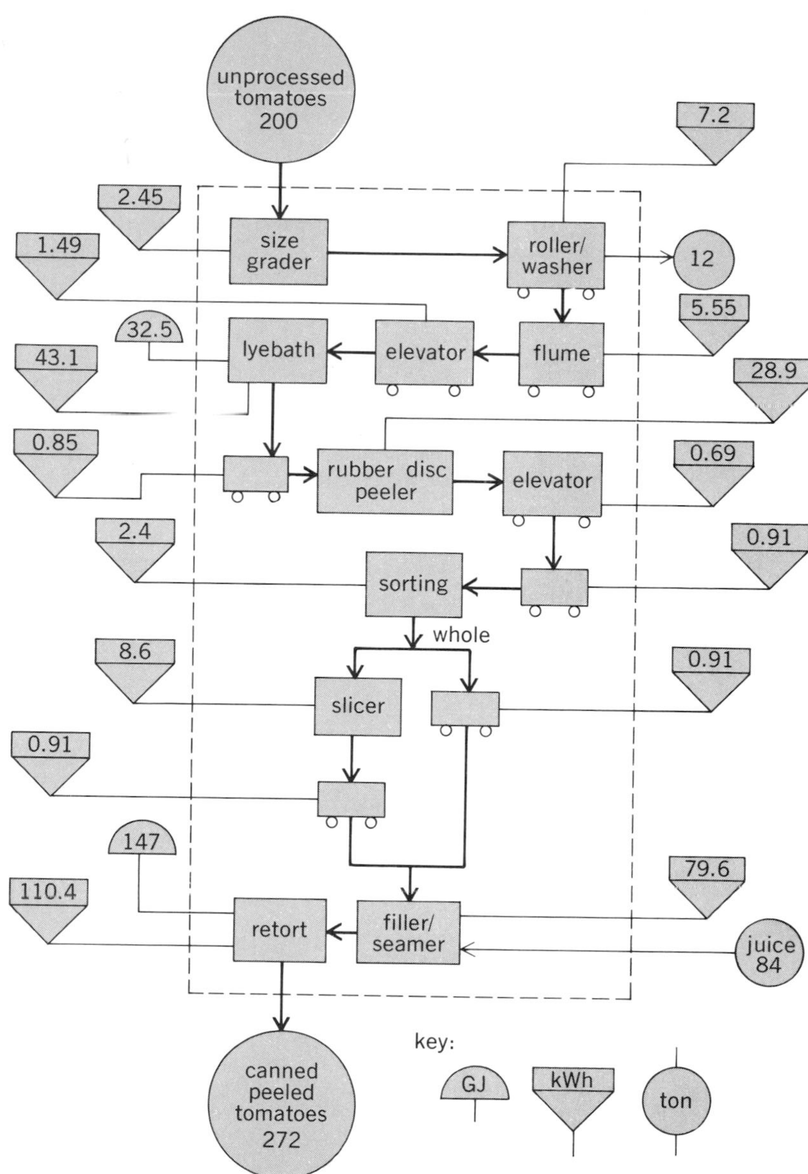

Fig. 2. Energy accounting diagram of canning whole peeled tomatoes; basis is an 8-h shift. 1 ton = 0.9 metric ton; 1kWh = 3.6×10^6 J. (*From R. P. Singh et al., Energy accounting in canning tomato products, J. Food Sci., 45(3):735–739, copyright 1980 by the Institute of Food Technologists*)

Energy accounting of an atmospheric retort. Atmospheric retorts are commonly used to sterilize canned food of low pH that require sterilization temperatures of less than 100°C. A typical atmospheric retort is a large cylindrical vessel (2–3 m in diameter and 10–15 m in length) filled to two-thirds of its volume with water. A steam spreader at the bottom of the vessel is used to heat the water. Cans are conveyed continuously through the retort on a reel conveyer. The temperature and residence time of the cans inside the retort are controlled to achieve the desired sterilization process.

An energy accounting study was conducted to identify possible means of energy conservation. Figure 3 is an energy accounting diagram of the retort for a 1-h operating basis. As is evident, thermal energy enters the retort with steam and with the canned food, since the canned food entering the retort is at a temperature of around 40°C. Electrical energy is used to turn the reel conveyer inside the retort. Thermal energy exits the retort with heated canned food, condensate, and both radiative and convective losses from the shell surface. The energy measurements shown in Fig. 3 were obtained when the retort was operated at 109 cans per minute or 87% of rated capacity. The analysis shows that a major amount of energy is lost due to incomplete steam condensation. Equipment modifications to reduce this loss should assist in reducing energy requirements. In addition, energy can be conserved by insulating the retort shell to reduce radiative and convective losses.

Summary. The energy accounting analysis yields quantitative data that can be used for computing en-

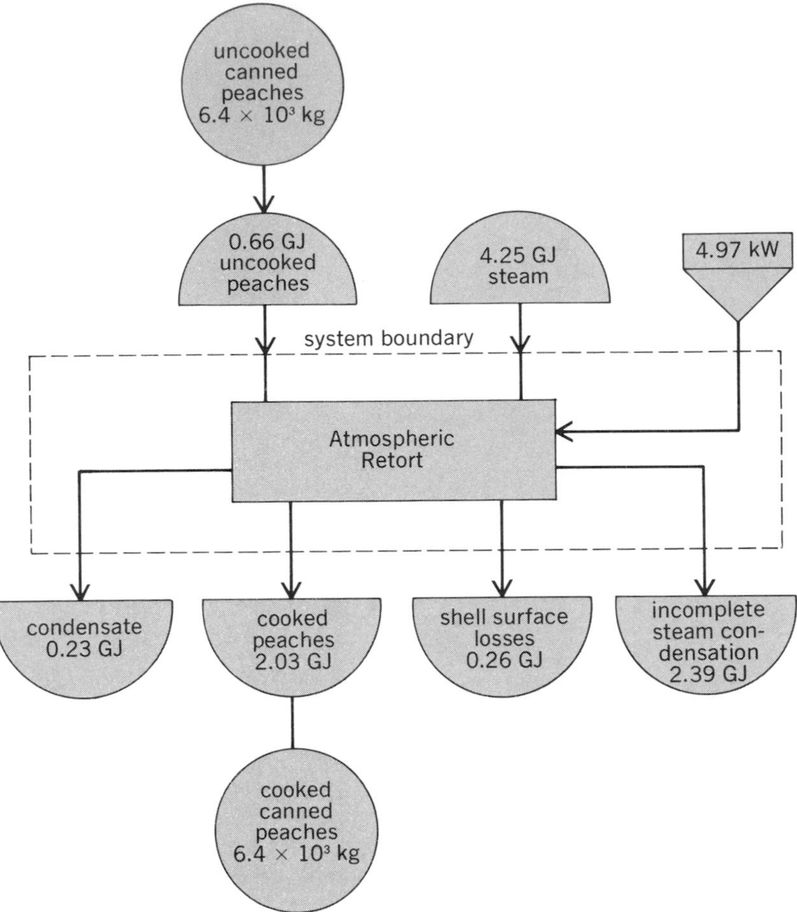

Fig. 3. Energy accounting diagram of an atmospheric retort; basis is 1 h. (*From R. P. Singh, Energy accounting in food process operations, Food Tech., 32(4):40–45, copyright 1978 by the Institute of Food Technologists*)

ergy cost savings possible with different modifications. These calculations allow estimation of economic payback periods that are useful to the plant management in deciding which of the suggested modifications should receive funding priority.

Efficient use of energy in food processing is a topic of major interest to the food industry. The energy accounting method, as illustrated in the case studies of the canning plant and the atmospheric retort, provides a quantitative approach to collecting and analyzing energy-related data. The information obtained through this method is useful to both the operators of equipment and the plant management in selecting process and equipment alternatives that should assist in achieving high energy-use efficiencies.

For background information *See* FOOD ENGINEERING; FOOD MANUFACTURING in the McGraw-Hill Encyclopedia of Science and Technology.

[R. PAUL SINGH]

Bibliography: G. H. Brusewitz and R. P. Singh, *Trans. Amer. Soc. Agr. Eng.*, 24(2):533–536, 1980; M. S. Chhinnan and R. P. Singh, *Lebensm. Wiss. Technol.*, 14:122–126, 1981; R. P. Singh, *Food Technol.*, 32(4):40–46, 1978; R. P. Singh et al., *J. Food Sci.*, 45(3):735–739, 1980.

Gamma-ray astronomy

The most intense gamma-ray burst ever observed was detected by nine spacecraft on March 5, 1979. The burst rose to its peak strength in less than a millisecond, which allowed the source location to be determined from the various detection times for the widely separated spacecraft to an accuracy of less than a minute of arc. The direction coincided with that of N49, a supernova remnant (Fig. 1) in the Large Magellanic Cloud, which is at a distance of 55 kiloparsecs (180,000 light-years, or 1.7×10^{18} km) from Earth. This was the first time that a gamma-ray burst location error box of small size had included any obvious candidate for the source. If this gamma-ray burst occurred at such a distance, its peak intensity amounted to more than 10^{44} erg · s^{-1} (10^{37} W), much larger than had previously been considered reasonable. Theorists are in disagreement as to whether this event represents a fortunate breakthrough in understanding gamma-ray bursts, an untypical gamma-ray burst of a new breed, or an unfortunate coincidence in direction between an otherwise inconspicuous and nearby source and a conspicuous galaxy (Large Magellanic Cloud) in the Local Group.

Location of March 5, 1979, burst. The discovery of gamma-ray bursts by means of the Vela spacecraft was first announced in 1973. Several hundred bursts have now been detected, but there is still no clear understanding of their origin. It is fortunate that at the time of this very intense burst (GB790305) there were so many operational spacecraft able to detect it. In addition to three Vela spacecraft and *Prognoz 7* near the Earth, *Helios 2*, *ISEE 3*, and *Venera 11* and *12* were all at considerable distances from Earth, and *Pioneer Venus Orbiter* was around Venus. By precise timing of the gamma-ray peak as it arrived at the various spacecraft, with differences up to 25 s, the location of the source was determined to be, in galactic coordinates, $l^{\mathrm{II}} = 276.09°$, $b^{\mathrm{II}} = -33.24°$. The burst source may also be designated, in the standard method based on equatorial coordinates, as GBS0525-661.

Oscillating x-ray emission. To the astonishment of the various teams of observers, the enormous peak of the outburst, which had a total effective pulse length of 130 ms (Fig. 2*a*), was followed by an oscillating, softer, and several-hundred-times weaker x-ray yield with a period of 8.0 s, lasting for at least 200 s (Fig. 2*b*). Thus, this gamma-ray burst source behaved like a temporary x-ray pulsar, with alternating high and low pulses in the x-ray yield spaced about 4 s apart. The sharpness of the periodicity is characteristic of x-ray pulsars, believed to be rotating neutron stars, although the gradual decay of intensity with a characteristic time of approximately 50 s, with the alternating pulses becoming more equal in height, is not. There is also some evidence (see Fig. 2, left) for a 23-ms periodicity during the decaying portion of the initial outburst, the origin and certainty of which are not clear.

Energy spectrum. This gamma-ray burst was somewhat softer in photon energy than most, with an initial spectrum similar to that of a 30-keV (3.5×10^8 K) blackbody, except for a high-energy tail, with a broad peak at approximately 430 keV. This is lower in energy than the 511-keV line produced by the annihilation of positrons and electrons, in agreement with the calculated gravitational red shift of light produced at the surface of a neutron star of 1 solar mass and a radius of 10 km. This spectral line thus furnishes additional evidence that the gamma-ray burst was produced by a neutron star, collapsed to a size only a few times larger than its Schwarzschild radius. Such gravitationally red-shifted lines have also been observed in several other gamma-ray bursts.

Recurrences. Surprisingly, GB790305 was followed by three recurrences from the same direction, on March 6, April 4, and April 24, 1979, as detected by the *Venera 11* and *Venera 12* spacecraft. These were considerably weaker than the original burst, which released approximately 10^{43} ergs (10^{36} J) if it was at the distance of the Large Magellanic Cloud and was isotropic in radiation. This is as much energy as would be produced by the complete annihilation of 10^{22} g of matter, such as would be contained in a 100-km-radius asteroid.

Source. The 8.0 s period of the x-ray oscillations, if due to rotation of the source, could be produced only by an astronomical body of density at least 2.2×10^6 g/cm^3, which is only marginally consistent with a white dwarf as source. Such a period is typical, however, of x-ray pulsars, believed to be relatively slowly rotating neutron stars in binary systems, closely related to the more rapidly rotating radio-emitting pulsars. Presumably the source in this case had two oppositely placed hot source regions at the magnetic poles, coming alternately into view as it rotated. The very rapid rise and short duration of the initial burst are quite consistent with an object of such size (10-km radius, corresponding to a light travel time of 0.033 ms, and an average density of 5×10^{14}).

The supernova remnant N49 in the Large Magellanic Cloud, the apparent source of this gamma-ray burst, is one of the stronger x-ray sources in that galaxy, with a temperature of 0.4 keV (4×10^6 K) and a strength of approximately 10^{37} erg · s^{-1} (10^{30} W). It is also a weak radio source. Its age (the time since the supernova explosion which formed it, and presumably created a neutron star) is estimated, on the basis of its observed diameter of 16 parsecs (50 light-years) and the density of gas in its vicinity, to be approximately 5500 years.

Its x-ray strength is an order of magnitude less than the Eddington limit for an accretion-powered neutron star x-ray source, approximately 10^{38} erg · s^{-1} (10^{31} W). The Eddington limit is that level of emitted radiation which would repel infalling matter by radiation pressure with a force balancing that of gravity. Thus an intensity greater than the Eddington limit cannot be powered by accretion of matter, at least not in a steady-state situation. An intensity many orders of magnitude greater than this limit, as was apparently the case for GB790305, would very quickly give relativistic ejection velocities to any infalling matter as well as to the radiating surface, if not prevented from doing so by, for example, a very strong magnetic field. The x-ray pulsarlike sequel to the gamma-ray burst was also apparently well above the Eddington limit in intensity, which may be associated with its relatively rapid decay.

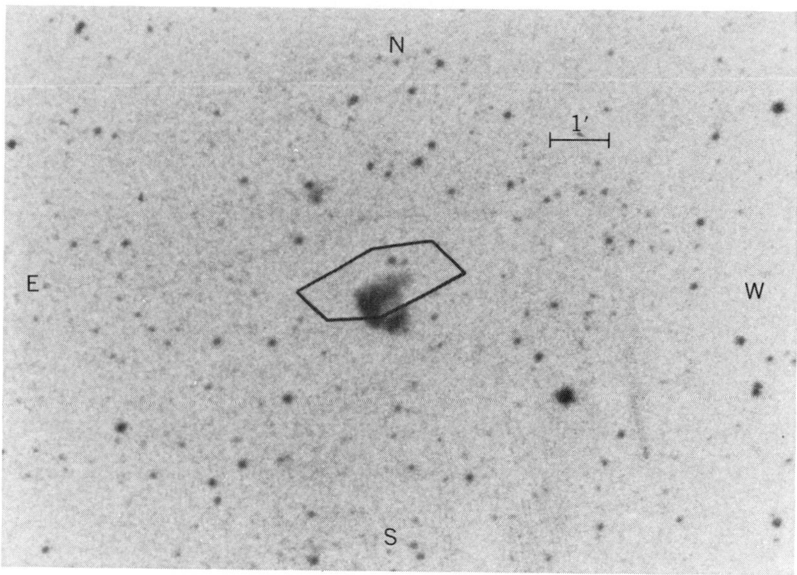

Fig. 1. Original error box (later reduced in size) for GB790305, superimposed on a photo of the supernova remnant N49 in the Large Magellanic Cloud. Photo is a Cerro Tololo Inter-american Observatory Curtis-Schmidt plate taken by H. E. Epps of the University of California, Los Angeles. (*From W. D. Evans et al., Location of the gamma-ray transient event of 1979 March 5, Astrophys. J., 237:L7-L9, published by the University of Chicago Press; copyright © 1980 the American Astronomical Society*)

The 8-s period of this gamma-ray burst source, if it is the rotation period of a neutron star in the Large Magellanic Cloud, is not what might be expected from the 5500-year age of this supernova remnant. It is of course possible that this particular neutron star was formed with unusually small angular momentum, and did not need a long period to slow down in rotation.

Models for origin. A great many models have been proposed for the origin of gamma-ray busts, from pulsar glitches to thermonuclear explosions or the infall of an asteroid on the surface of a neutron star. Although there is no agreement yet on the nature of these bursts, it seems to be generally accepted that neutron stars are the most likely source. The detonation of accumulated carbon in a thermonuclear explosion on the surface of a neutron star was proposed as a model in 1976, with yield calculated to be approximately 10^{40} ergs. The infall of a comet to the surface of a neutron star was suggested as an explanation in 1973. An energy yield of approximately 10^{37} ergs (10^{30} J) would be produced by the fall of a comet or asteroid of 10^{17}-g mass and a radius of approximately 1 km onto the surface of a neutron star, at the relativistic speed

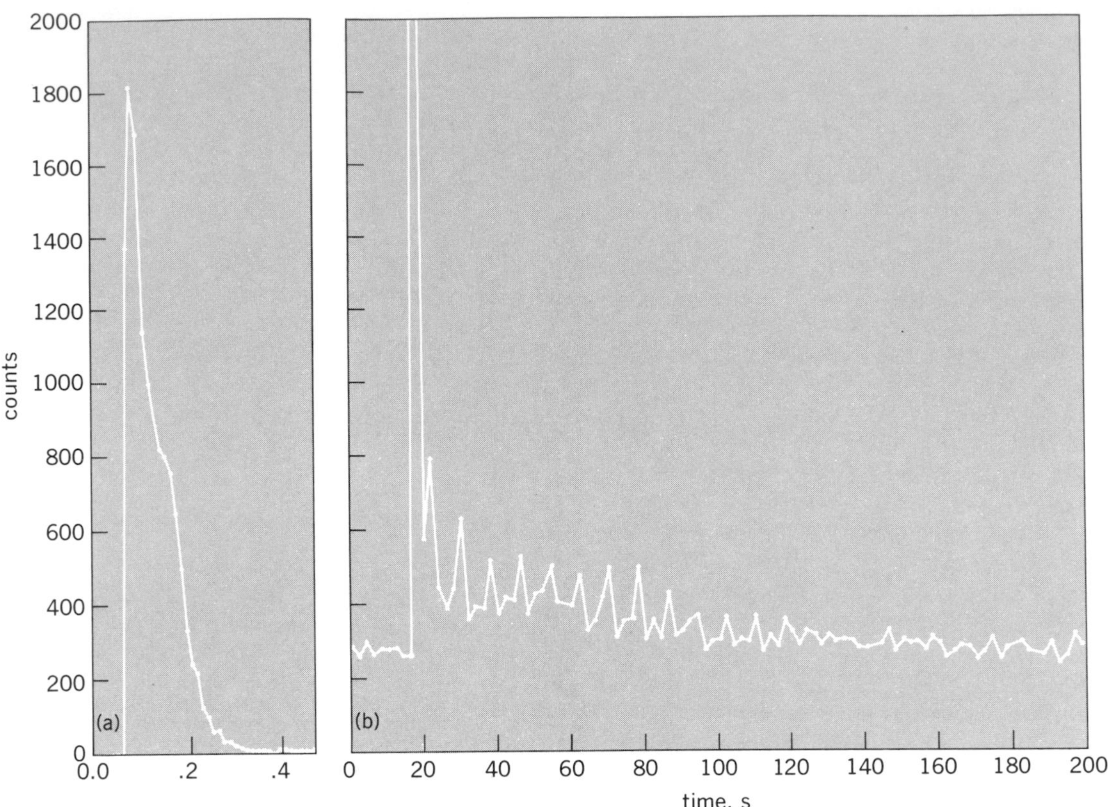

Figure 2. Time history of the March 5, 1979, gamma-ray burst as seen by *Pioneer Venus Orbiter*. The energy range covered is 60–1200 keV. (*a*) Initial time history, with counts grouped into uniform 11.72-ms time intervals. (*b*) Longer time history with counts grouped into 2-s bins. For clarity, the actual peak count of 13,000 is not fully shown. (*From J. Terrell et al., Periodicity of the gamma-ray transient event of 5 March 1979, Nature, 285:383–385, 1980*)

imparted by the enormously strong gravitational field of such a collapsed object.

Asteroid infall. Detailed explanations of the March 5, 1979, gamma-ray burst have recently been presented in terms of an infalling nickel-iron asteroid of high tensile strength (to shorten the outburst). Although many features of the event can be accounted for in these terms, the observation of three recurrences from the same source is a somewhat awkward feature for such a model. For these models, and for almost all the others, it seems necessary to assume that the source of GB790305 is not in the Large Magellanic Cloud, but is a local galactic object perhaps 1000 times closer, in order to keep the required x-ray and gamma-ray energy yield from becoming embarrassingly large. Thus for such models the apparent association with the Large Magellanic Cloud, and with the supernova remnant N49, is merely an unfortunate coincidence.

Internal restructuring. Perhaps the only reasonable source for the amount of energy emitted by GB790305, if it did indeed occur in the Large Magellanic Cloud, is gravitational energy released by a phase transition or other internal restructuring within a neutron star. In 1974, it was proposed that gamma-ray bursts could be produced by neutron starquakes or "glitches"—sudden changes in the rotational speed of the star. Such glitches have been observed on several occasions to occur in radio pulsars. It was proposed that the ensuing readjustments in the immense magnetic field of the neutron star would accelerate charged particles to high energies and thus produce gamma rays. It has been proposed that up to 10^{39} ergs (10^{32} J) of elastic energy might be released more or less directly by crust-quake surface heating of a neutron star. These models may be capable of accounting for larger energy releases. A proposed model of GB790305 is a very energetic phase transition in the interior of a neutron star, releasing gravitational energy in the form of vibrations which could shake the magnetosphere and lead to the emission of approximately 10^{44} ergs of gamma rays. An attractive feature of this model is that the 0.13-s effective pulse length of this burst is in good agreement with the characteristic time required for the damping of quadrupole and higher-mode vibrations by gravitational radiation. The later recurrences of this gamma-ray burst might be accounted for as aftershocks of the original, stronger outburst.

Objection to extragalactic origin. The principal objection of some theorists to such models, or any models for GB790305 based on the assumption of the distance of the Large Magellanic Cloud, is the great intensity of gamma- and x-ray emission which is required—approximately 10^{44} erg · s^{-1} (10^{37} W). This is a million times larger than the Eddington limit for a neutron star, and is an order of magnitude larger than what could be radiated by such a small surface at the indicated temperature of approximately 30 keV (3×10^8 K). Such an intense flux of radiation could eject the radiating surface with enormous acceleration. Two factors which might prevent this are a sufficiently strong magnetic field, such as the 10^{12}-G (10^8-T) strength believed to be common for neutron stars, and rapid cooling of the surface due to adiabatic expansion, balanced by the rate at which energy could be transferred into the surface from the interior.

Cyclotron radiation. The observed x radiation may be cyclotron radiation emitted by relativistic electrons spiraling around magnetic field lines. The required field would then be approximately 10^{11} G (10^7 T), perhaps barely enough to hold the surface in. In such a model a very hot photon-electron-positron plasma undergoes rapid synchrotron cooling in the magnetic field and finally emits a relatively small amount of pair annihilation radiation, observed as the 430-keV red-shifted peak in the spectrum. The calculated spectrum is in good agreement with observation.

Relativistic expansion. Another possibility is that the magnetic field is not quite strong enough to restrain the extremely hot surface from expanding at relativistic speed. Even a modest and brief expansion during which the surface cooled rapidly would allow the emission of the peak x-ray flux, which (Fig. 2) rose to its maximum and fell to a lower level in about 30 ms. Relativistic expansion would greatly speed up the initial rise of the pulse; the surface would appear optically thin toward the observer in less than a millisecond, permitting red-shifted annihilation radiation to come through from the residual neutron star surface.

Burst in direction of Small Magellanic Cloud. The arguments against the possibility that this gamma-ray burst actually came from the supernova remnant N49 in the Large Magellanic Cloud may be weakened somewhat by the recent discovery that another gamma-ray burst, on July 23, 1974, came from the direction of the Small Magellanic Cloud, near the x-ray pulsar SMC X-1. This may indicate that neutron stars are capable of emitting very large bursts of gamma rays, since the Small Magellanic Cloud and the Large Magellanic Cloud are at similar distances. It is not clear whether such gamma-ray bursts as GB790305 are a special class of events, perhaps more powerful than other gamma-ray bursts which come from a still undetermined distance, or are produced by much more local neutron stars close to the Earth in the Galaxy, and merely coincidentally in the same direction as external galaxies.

For background information *see* GAMMA-RAY ASTRONOMY; X-RAY ASTRONOMY in the McGraw-Hill Encyclopedia of Science and Technology.

[JAMES TERRELL]

Bibliography: T. L. Cline et al., *Astrophys. J.*, 237:L1-L5, 1980; W. D. Evans et al., *Astrophys. J.*, 237:L7-L9 1980; E. P. Mazets et al., *Nature*, 282:587–589, 1979; Symposium on Cosmic Gamma-Ray Bursts, Toulouse, France, November 26–29, 1979, *Astrophys. Space Sci.*, 75:1–224, 1981.

Gas chromatography

Derivatization is the application of microchemical reactions to purposefully change the properties of a compound which could not be chromatographed, separated, or determined with sufficient selectivity or sensitivity without such treatment. Derivatives are used in gas chromatography for one or more of the following reasons: to improve the thermal stability of a compound, to improve the volatility of a compound, to correct peak distortion on gas chromatographic analysis (that is, reduce adsorption and chemical interactions with the chromatographic system), to improve the separation of compounds by changing their elution order on the column to minimize peak overlap, and to aid in the detection of compounds by introducing specific groupings (these can be considered to be detector-oriented chemical tags or labels) which enhance the selectivity or sensitivity of the detection of the compound when used in conjunction with selective gas chromatographic detectors.

Selection of derivatizing reagents. Figure 1 summarizes the criteria used to select a derivatizing reagent for a particular application. The reagent typically has two distinct properties, which are not completely independent of each other. First, the reagent must contain a reactive group which will perform the chemical transformation by which the organic chain is attached to the derivatized compound. Such attacking groups are chosen on the basis of their ability to react rapidly, and preferably to completion, with a functional group or on the basis of their ability to react selectively with a certain type or arrangement of functional groups to the exclusion of any others which may be present in the same compound or in a mixture of compounds. Secondly attached to the reactive group is an organic chain chosen to aid the separation of the derivatized compound by changing its volatility or polarity, ef-

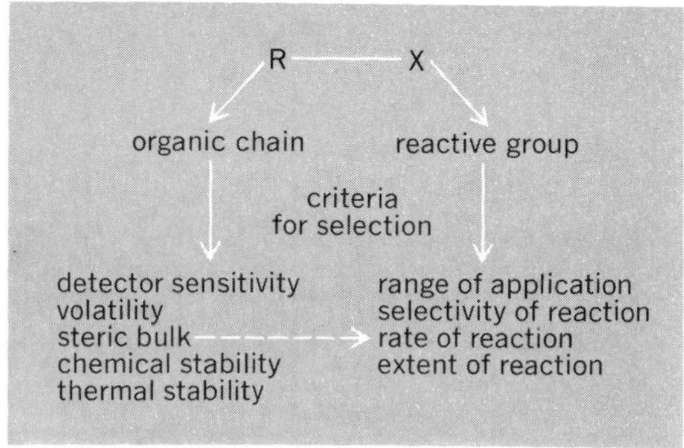

Fig. 1. Anatomy of a derivatizing reagent.

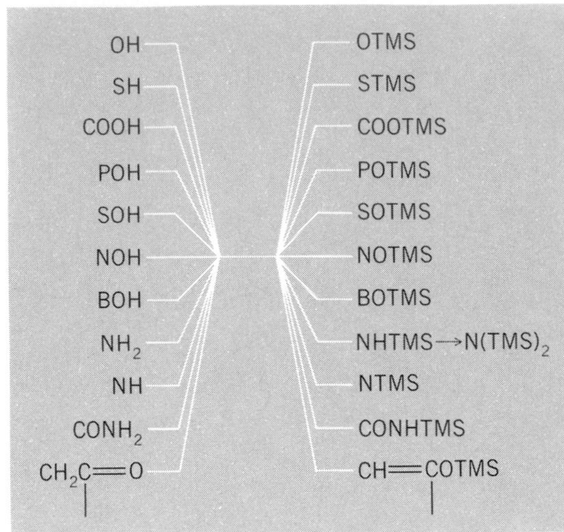

Fig. 2. Functional groups forming trimethylsilyl derivatives.

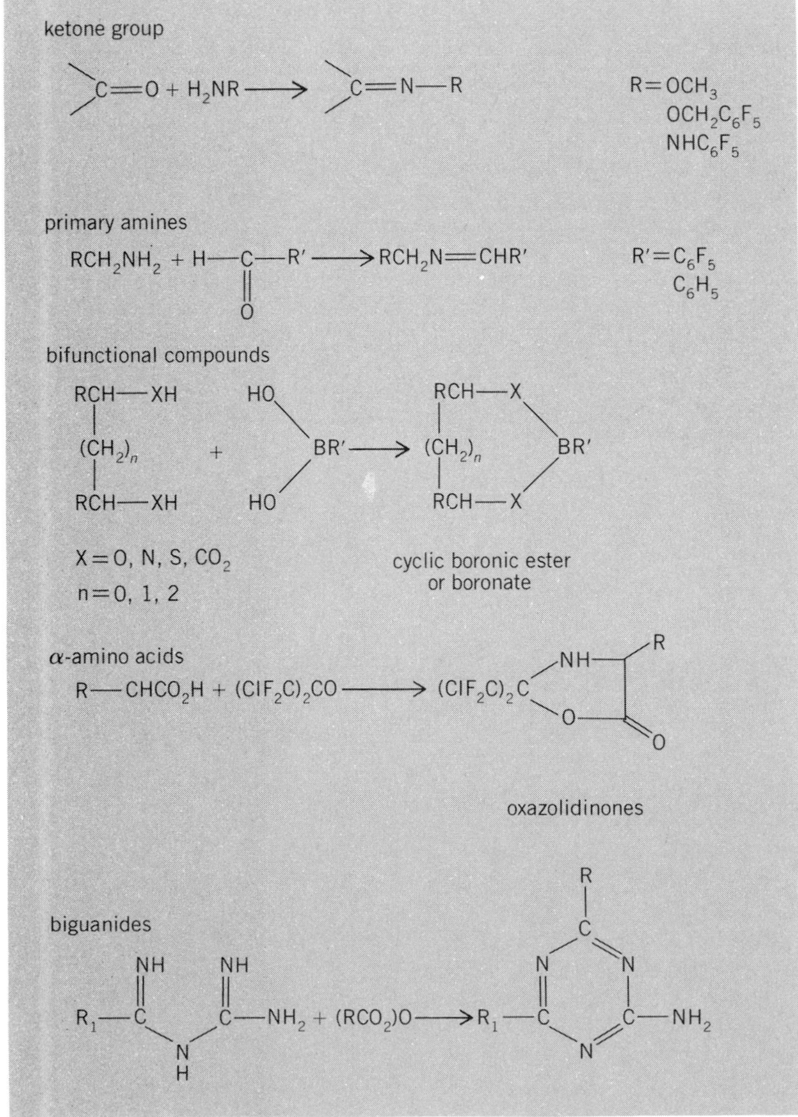

Fig. 3. Examples of reactions useful in derivatization.

fects which alter the elution order of a compound on gas chromatography, or to introduce into the derivative a group which is oriented to respond to one of the range of selective and sensitive detectors available for gas chromatographic detection. Such detector-oriented tags would include: electronegative groups responding selectively to the electron capture detector; nitrogen- or phosphorus-containing groups responding to the alkali flame ionization detector, an instrument selective to nitrogen and phosphorus (also known as the NP detector or the thermionic detector); and sulfur- or phosphorus-containing groups responding to the flame photometric detector (selectivity is based on the nature of the chemiluminescent reaction of sulfur- and phosphorus-containing compounds in a cool hydrogen flame). By far the most important of these detectors is the electron capture detector because of its great sensitivity, the commercial availability of a wide range of reagents for use with it, and the fact that it is inexpensive and in routine use in many laboratories. A mass spectrometer can also be considered to be a selective and sensitive gas chromatographic detector. When used in this mode, the mass spectrometer is tuned to determine a single ion during the course of a chromatographic run (single-ion monitoring or SIM) or by rapidly switching between several ions (multiple-ion detection or MID). Choosing ions which are specific or descriptive of the derivative being separated makes it possible to obtain both specificity and sensitivity for the compound of interest. Derivatization reagents substituted with stable isotopes (for example, ^{2}H, ^{18}O) can be used to help unravel mass spectral fragmentation mechanisms. The rather special nature and high purchase and operating costs of mass spectrometers tend to limit mass spectrometric techniques to research laboratories or to centers with multiple users. The majority of derivatives used in gas chromatography are determined with either the universal flame ionization detector or the selective and sensitive electron capture detector.

Classes of derivatizing reagents. It is convenient to divide the reagents used for forming derivatives in gas chromatography into two classes: general reagents and selective reagents.

General reagents. These are characterized by the lack of discrimination by which they react with polar functional groups. Examples are the trialkylsilyl ethers and esters and the acyl derivatives formed with acid anhydrides. The most often encountered derivatives in gas chromatography are the trimethylsilyl (TMS) ethers. Functional groups which can be converted to trimethylsilyl derivatives are summarized in Fig. 2. If the trimethylsilyl derivatives have a weakness, it is their poor hydrolytic stability, which can be a problem if the derivative is to be subjected to sample clean-up techniques prior to separation by gas chromatography. Reagents which are higher alkyl homologs have been developed to minimize this problem, the most important being the *tert*-butyldimethylsilyl ethers. Also, reagents designed for use with the electron capture detector are available commercially. The most important of these are the folphemesyl (pentafluorophenyldimethylsilyl) derivatives. Reagents in the class discussed above are of most use for exhaustive derivatization of compounds containing multiple functional groups or for the analysis of mixtures in which the individual constituents contain a variety of different functional groups.

Selective reagents. These provide chemical specificity for either a limited number of functional group types (for example, alcohols for carboxylic acids) or the spatial arrangement of more than one functional group in a compound (for example, boronic acids for bifunctional compounds). Some examples of well-known reactions are shown in Fig. 3.

These kinds of reactions have two general areas of usefulness. They can be used to identify the presence of a certain functional group or spatial arrangement of functional groups in an unknown compound by noting either the change in retention increment after derivatization or the characteristic changes in the mass spectra of the compound on derivative formation. Selective derivatization reactions also provide a method for the determination of certain compounds in a complex mixture without the need for extensive sample clean up. This can be achieved by a combination of the chemical specificity of the derivatization reaction and by the selection of reagents which provide a selective response to a discriminating detector such as the electron capture detector.

Volatile organofluorine derivatives. For the analysis of high-molecular-weight compounds or thermally labile volatile compounds, it is important that the derivative used for their analysis contribute to the overall increase in volatility of the compound. Here, advantage can be taken of the unique properties of organofluorine compounds for derivative formation. In spite of their high molecular weight, organofluorine compounds containing closely bound fluorine atoms show remarkable volatility (accredited to the small intermolecular bonding forces observed in fluorocarbons) and chemical and thermal stability. Typical reagents for the preparation of perfluoroacyl derivatives are trifluoroacetyl, pentafluoropropionyl and heptafluorobutyryl anhydrides, and imidazole reagents. These reagents are fairly nonspecific (although not as reactive as the trimethylsilyl reagents) in their range of reaction with protonic functional groups (the exception being carboxylic acids, which do not form derivatives with these reagents). The type of functional groups which can be derivatized and the products formed are shown in Fig. 4. To increase their response to the electron capture detector, substitution of fluorine by another halogen atom is possible, but these derivatives lose the volatility advantages enjoyed by the perfluorocarbon derivatives. Special mention should be made of the imidazole reagents, which because of the weakly amphoteric nature of the leaving group, imidazole, can often be used to derivatize sensitive molecules which are degraded or dehydrated by the use of the acid anhydrides.

As well as forming volatile derivatives, the halocarbonacyl derivatives impart useful electron-capturing properties to the compound derivatized. The halocarbonacyl derivatives may be detected at the low picogram level ($\times\ 10^{-12}$ g) with an electron capture detector, making them of value for problems in trace analysis. As it is the reagent which transfers its electron-capturing properties to the derivative, this technique is commonly employed for the trace-level analysis of compounds which do not have native electron-capturing properties and would be difficult to detect at low levels without the aid of detector-oriented derivative formation.

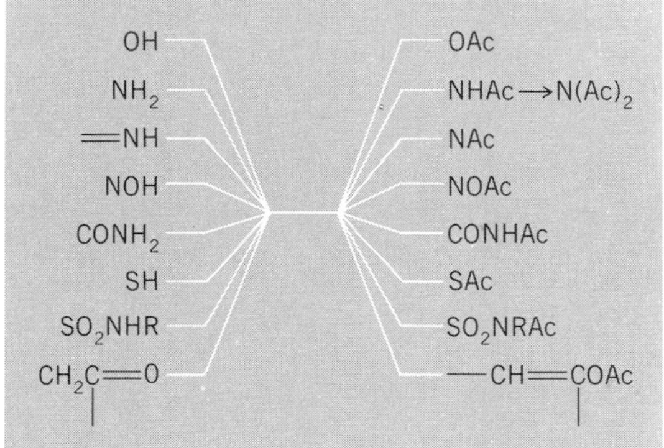

Fig. 4. Functional groups forming halocarbonacyl derivatives.

Fluorinated aromatic compounds have much higher electron capture detector responses than their alkyl analogs with a similar number of fluorine atoms but retain the excellent volatility advantages enjoyed by the latter. In making a comparison of the electron capture detector response of the halocarbonacyl derivatives with those of the pentafluorophenyl-containing derivatives, it is necessary to keep in mind that the response of the former is due to the presence of both the perfluorocarbon chain and the ketone portion of the derivative. Reagents for introducing the pentafluorophenyl group into organic compounds are summarized in the table. Changing the reactive group leads to development of reagents which are general or specific in their area of use. The pentafluorophenyl-containing derivatives are characterized by good volatility, chemical and thermal stability, and ease of preparation, and they have excellent electron-capturing properties. They are becoming more widely used and are replacing conventionally used reagents (for example, halocarbonacyl reagents) for many applications.

Separations. Derivatization also provides a means by which inorganic cations and anions can be separated by gas chromatography. In their original ionic and non-volatile form these substances are unsuitable for this technique, but conversion to neutral derivatives permits achievement of separations. Chelation of metal ions is the most common technique,

Reagents for the introduction of the pentafluorophenyl group into organic molecules

Reagent	Functional group type
Pentafluorobenzoyl chloride	Amines, phenols, alcohols
Pentafluorobenzyl bromide	Carboxylic acids, phenols, mercaptans, sulfonamides
Pentafluorobenzyl alcohol	Carboxylic acids
Pentafluorobenzaldehyde	Primary amines
Pentafluorobenzyl chloroformate	Tertiary amines
Pentafluorophenacetyl chloride	Alcohols, phenols, amines
Pentafluorophenoxyacetyl chloride	Alcohols, phenols, amines
Pentafluorophenylhydrazine	Ketones
Pentafluorobenzylhydroxylamine	Ketones

of which the determination of Be(II), Cr(III), Al(III), and Cu(II) as their β-diketone derivatives are the most important from a practical point of view. The alkylation of inorganic mercury using a variety of reagents is an important method for detecting Hg(II) in environmental samples. Reagents for derivatizing anions tend in the main to make use of specific chemical reactions in order to form neutral products. Among the anions which can be derivatized and analyzed through gas chromatography are the halogens, oxyanions (for example, CO_2^{2-}, PO_3^{3-}, SO_4^{2-}, NO_3^{-}), sulfide, cyanide, and thiocyanate.

Asymmetric (optically active) compounds can be separated as their diastereoisomers by gas chromatography. For this purpose, the derivatization reagent must itself by asymmetric) otherwise no separation will be achieved. This is a very specialized form of derivatization, and the reagents in use tend to be synthesized or selected with a particular application in mind. In very general terms, acid-base reactions, condensations, or complex-forming reactions are employed with a chiral reagent to form a diastereoisomeric mixture which can then be separated by partition differences in the chromatographic system and the original optical purity of the sample discerned.

For background information *see* GAS CHROMATOGRAPHY; TRACE ANALYSIS in the McGraw-Hill Encyclopedia of Science and Technology.

[COLIN F. POOLE]

Bibliography: K. Blau and G. S. King (eds.), *Handbook of Derivatives for Chromatography*, 1978; D. R. Knapp, *Handbook of Analytical Derivatization Reactions*, 1979; C. F. Poole and A. Zlatkis, *Anal. Chem.*, 52:1002A–1016A, 1980; C. F. Poole and A. Zlatkis, *J. Chromatogr.*, 184:99–193, 1980.

Gel

A gel is a form of matter intermediate between a solid and a liquid. It consists of polymers cross-linked to create a tangled network and immersed in a liquid medium. The properties of the gel depend strongly on the interaction of these two components. The liquid prevents the polymer network from collapsing into a compact mass; the network prevents the liquid from flowing away. It has recently been discovered that drastic changes in the state of the gel can be brought about by small changes in the external conditions. For example, when the temperature is lowered, the polymer network loses its elasticity and therefore becomes increasingly compressible. When a certain critical temperature is reached, the elasticity falls to zero and the compressibility becomes infinite. The gel can also swell or shrink by a factor of as much as several hundred when the temperature is varied. Under some conditions the swelling or shrinking is discontinuous, so that an infinitesimal change in temperature can cause a large change in volume. Temperature is not the only factor that can give rise to such transformations; they can also be brought about by altering the composition, the pH, or the ionic strength of the solvent in which the gel is immersed or by imposing an electric field across the gel.

A familiar gel is the dessert Jello, where the network is made up of polymers derived from animal protein. The network constitutes only about 3% of the volume of Jello; the rest is colored, flavored, and sweetened water. There are a great many other natural and artificial gels. The vitreous humor that fills the interior of the eye is a gel, and so is the material of the cornea. A large portion of the blood vessel walls, the connective tissues, and the synovial fluid that lubricates the skeleton joints are gels. In such biological gels the liquid component allows for the free diffusion of oxygen, nutrients, and other molecules, whereas the polymer network provides a structural framework that holds the liquid in place. Gels are also important intermediates in the manufacture of polymers such as rubber, plastics, glues, and membranes. In chemistry and biochemistry gels are employed in the analytic methods of chromatography and electrophoresis, where molecules are separated according to the speed with which they percolate through the pores of the gel.

The gel in which the phase transition was observed is made of cross-linked polyacrylamide. It consists of two monomers: acrylamide, which terminates in an aminocarbonyl ($-CONH_2$) group and constitutes the linear polymer chains; and bisacrylamide, which consists of two acrylamide monomers that are linked through their aminocarbonyl groups, and serves as a cross-linking molecule. Polymerization of these monomers is permanent. In order that the volume transition of the gel be discrete, it is necessary for a portion of the aminocarbonyl groups

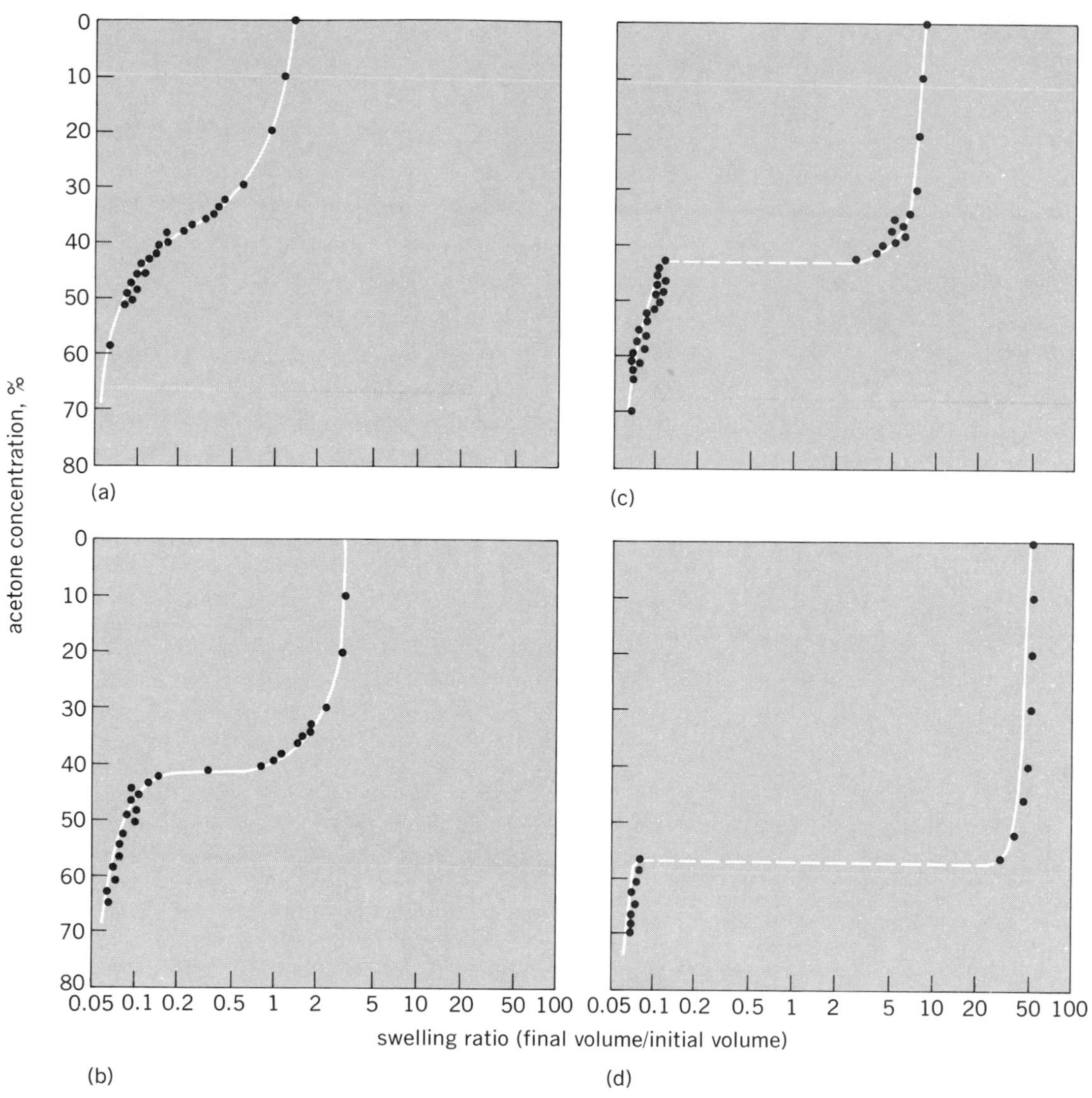

Fig. 1. Swelling curves recording the change in the volume of a gel placed in solvents with varying concentrations of acetone. The extent of the change in volume and the shape of the swelling curve depend strongly on the degree of hydrolysis. (*a*) No hydrolysis. (*b*) Hydrolysis for 2 days. (*c*) Hydrolysis for 6 days. (*d*) Hydrolysis for 60 days. (*After T. Tanaka, Gels, Sci. Amer., 244(1):124–138, January 1981*)

to be converted by hydrolysis into a carboxyl group (–COOH). In solution some of the carboxyl groups are spontaneously ionized to yield H^+ and $-COO^-$ ions. The polymer network is negatively charged, but as a whole the gel is neutralized by the presence of the counterions H^+.

Phase transition. The phase transition is observed when the partially hydrolyzed gels are immersed in mixtures of acetone and water having different proportions (Fig. 1). For lower acetone compositions the gel is quite swollen and has a large volume. As the acetone composition increases, the volume becomes gradually smaller until at a certain composition the gel collapses discretely into a small compact mass. Further increase of acetone composition induces only a slight and continuous decrease in the gel volume. The phenomenon is reversible, and the shape of the gel remains the same before and after the transition. The volume change at the transition increases with the degree of hydrolysis of the acrylamide groups and can be as large as 500-fold. The gel without hydrolysis shows only a continuous volume change. Such discrete volume transitions are also observed upon changing the temperature or pH of the solution or by adding a salt.

The discrete volume transition is a manifestation of the competition among three pressures acting on the polymer network: positive pressure of counterions, negative pressure due to the affinity among polymers, and the so-called rubber elasticity which acts to keep the network in a moderate expansion. The balance of these three pressures determines the equilibrium volume of the gel. The temperature, pH, and added salt affect only the former two pressures, whereas the solvent composition influences only the negative pressure due to polymer-polymer affinity. Altering one of these factors creates an excess nonzero pressure, positive or negative, which brings the gel to a new equilibrium volume. For example, at higher temperatures the positive pressure of counterions dominates and the network is quite swollen when the total pressure balances to zero, whereas at lower temperatures the negative pressure due to polymer-polymer affinity wins and the network is shrunken. At a particular temperature there exist two different gel volumes giving zero total pressure. This is the temperature at which the discrete volume transition takes place.

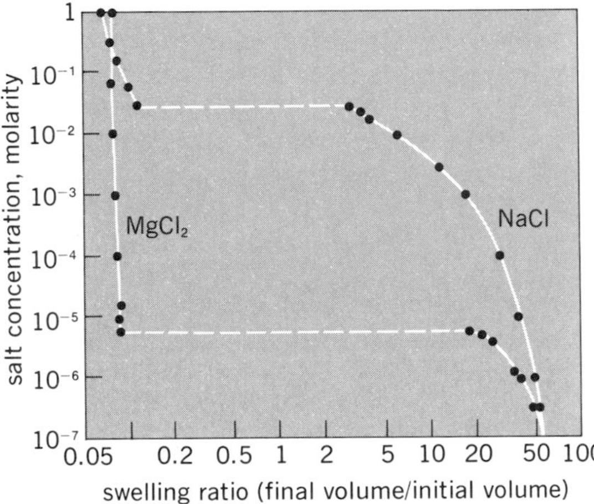

Fig. 2. Dependence of gel volume on concentration of dissolved salts, showing phase transition involving collapse of the gel. The salt concentration needed to induce the phase transition is 4000 times smaller with divalent magnesium ions than it is with monovalent sodium ions. (*After T. Tanaka, Gels, Sci. Amer., 244(1):124–138, January 1981*)

The effect of a salt on the phase transition depends strongly on the valency of the salt (Fig. 2). The salt concentration needed to induce collapse is 4000 times less for the divalent magnesium ion than it is for the monovalent sodium ion. Surprisingly, the large difference can be explained by the fact that only half as many divalent ions are needed to neutralize the network as monovalent ions. Divalent ions have an essential role in the contraction of muscle, a process that in some respects seems to resemble the collapse of a gel.

Electric field–induced collapse. Still another way to induce a phase transition is by applying an electric field across a gel. The polymers, which are negatively charged, are drawn toward the positive electrode. The force gives rise to a negative pressure in the gel. A pressure gradient is set up, ranging from strongly negative pressure adjacent to the positive electrode to zero pressure at the other end of the gel. The pressure varies smoothly along the length of the gel, but the response of the gel is nonetheless discontinuous. Where the negative pressure is sufficient to bring on a phase transition, the gel collapses; where the pressure is below the transition threshold, the volume of the gel is scarcely changed. Hence, the gel is shaped like a wine bottle, with a thick body, a shoulder, and a thin neck. Changing the magnitude of the field adjusts the proportions of the body and the neck. When the field is removed entirely, the gel resumes its original form.

Critical phenomena. Critical phenomena are observed in the density fluctuations of the polymer network as the gel approaches a particular condition called the critical point where the discontinuous volume transition disappears. At the critical point not only the total pressure balances but also its derivative with respect to gel volume is zero. This means an infinite compressibility; the network volume can change without any change in the pressure. Near the critical point the polymer-density fluctuations, like those of other critical systems, grow larger in amplitude and scale, and their rate of relaxation slows. At the critical point these measures diverge toward infinity.

The technique of laser-light scattering offers direct experimental access to the critical fluctuations of gels. As the temperature is lowered toward the critical point, the intensity of the scattered light increases. At the critical temperature the gel becomes almost opaque. Random fluctuations in the intensity of the scattered light can also be measured; they indicate the relaxation rate of the corresponding fluctuations in polymer density. The rate becomes indefinitely slow at the critical point.

Possible applications. A drastic change in volume that can be elicited in response to a considerable variety of small stimuli suggests certain possible applications for ionized gels. For example, a chemical engine might be designed in which the prime mover is a gel. Cycles of expansion and contraction could be powered by changes in temperature, solvent composition, pH, ionic strength, or electric field. Such a device is particularly appealing for use as an artificial muscle. Because the change in volume is discrete and predictable, a gel might also serve as a memory element or a switch.

An important factor in most such applications is the speed of the phase transition. A gel with dimensions of about a centimeter requires several days to reach a new equilibrium volume. The time required, however, is proportional to the square of the linear dimensions of the gel, and so it can be greatly reduced by making the gel small or finely divided. A cylindrical gel 1 μm in diameter, the size of a muscle fibril, would swell or shrink in a few thousandths of a second.

In the eye retinal detachment is caused by a collapse of the vitreous gel; the collapse allows the retina to pull away from the back of the eye. Corneal edema, another common cause of human blindness, is an opacification of the gel of the cornea; it is thought to result from a separation of fluid domains. Further investigation of the physics of phase transitions in gels may improve understanding of these conditions. Apart from these potential applications, a common but curious state of matter has come to be better understood, and the mathematical techniques for describing critical phenomena have been shown to apply to yet another physical system.

For background information *see* CRITICAL PHENOMENA; GEL; PHASE TRANSITIONS in the McGraw-Hill Encyclopedia of Science and Technology.

[TOYOICHI TANAKA]

Bibliography: T. Tanaka, *Sci. Amer.*, 244(1): 124–138, January 1981; T. Tanaka et al., *Phys. Rev. Lett.*, 45(20):1636–1639, 1980.

Geotextiles

Geotextile fabrics are a new and valuable addition to construction technology. They are defined by the American Society for Testing and Materials as any permeable textile used with foundations, soils, rock, earth, or other geotechnical material as an integral part of a manufactured project, structure, or system. These textile products are, in fact, fabrics made of synthetic fibers or yarns, constructed into woven or nonwoven fabrics that weigh anywhere from 3 to 30 oz/yd^2 (100 to 1000 g/m^2). Geotextiles are more commonly known by other names, for example, filter fabrics, civil engineering fabrics, support membranes, and erosion control cloth—and their applications are as varied as the names. These materials are used in constructing roadways, embankments, drains, erosion control systems, and a variety of other earthwork structures. The function of a geotextile in these structures is separation, reinforcement, or filtration; the benefits of geotextile

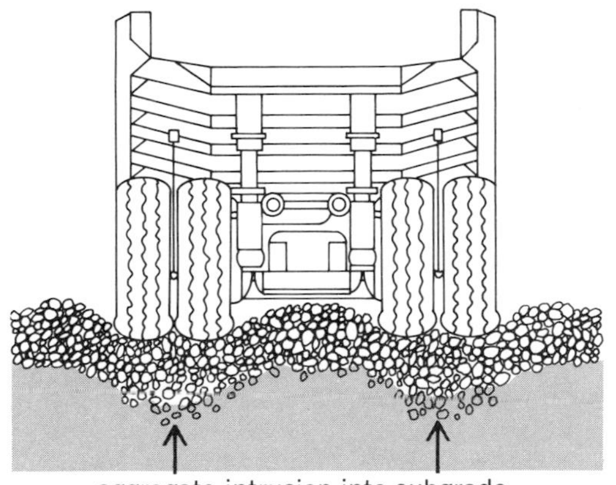

(a)

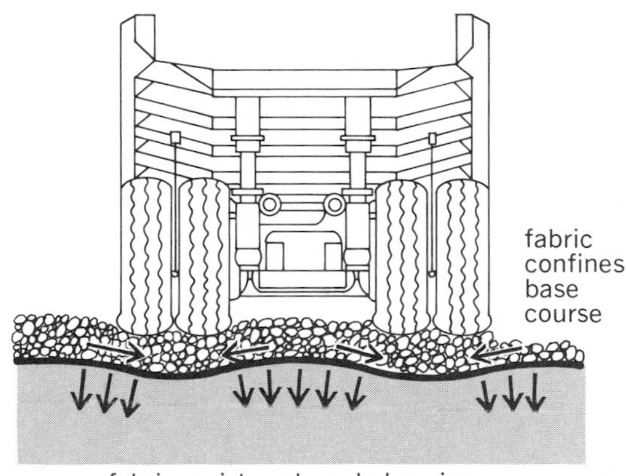

(b)

Fig. 1. Geotextiles are used to stabilize soils. (*a*) A rutted and unstable surface results from subgrade deformation and intermixing between subgrade soil and aggregate base. (*b*) Geotextiles resist rutting through separation and reinforcement (*from Mirafi Fabrics for the Mining Industry, Celanese Fibers Marketing Co.*). (c) A highway base course constructed with geotextile between subgrade and aggregate base (*from Mirafi Family of Construction Fabrics/MPB8, Celanese Fibers Marketing Co.*).

use are improved performance at little or no extra cost.

In earth structures, geotextiles perform three basic functions: separation (the fabric provides a boundary that segregates materials); reinforcement (the fabric imparts tensile strength to the system, thereby increasing its structural stability); and filtration (the fabric retains soil particles while allowing water to pass through). Geotextiles can thus be adapted to numerous applications in earthwork construction. The major end-use categories are: stabilization (for haul roads, highways, parking lots, and other structures built over soft ground); drainage (of subgrades, foundations, embankments, dams, or any earth structure requiring seepage control); erosion control (for shoreline, riverbanks, steep embankments, or other earth slopes to protect against the erosive force of moving water); and sedimentation control (for containment of sediment runoff from unvegetated earth slopes).

No one geotextile is suited for all these applications. Each use dictates a specific fabric requirement to resist installation stresses and to perform its function once installed. Geotextile manufacturers are sensitive to these fabric requirements, and most producers offer several fabric styles, each intended for specific end uses. In the United States there are more than 50 geotextiles from which the design engineer can choose.

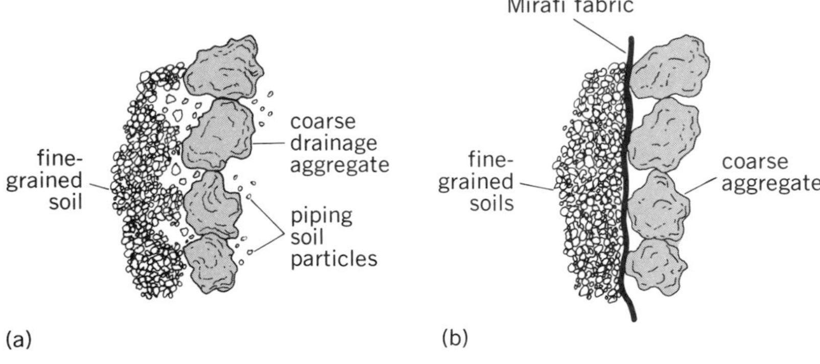

Fig. 2. Geotextiles find wide use for drainage. (*a*) Without a filter, soil particles may wash into and clog the drain. (*b*) Drainage fabric provides particle retention at the soil-drain interface, while allowing water to pass through (*from Mirafi Fabrics for the Mining Industry, Celanese Fibers Marketing Co.*). (*c*) Drain trench lined with a geotextile and backfilled with coarse aggregate.

Stabilization applications. Soft or low-strength soils on a project site present costly problems in the construction and maintenance of haul roads, storage yards, railroads and other areas which must support vehicular traffic. Poor soil conditions also cause rapid deterioration of paved and unpaved roads, city streets, and parking lots. The problems are caused by subgrade deformation and intermixing between subgrade soil and aggregate base. The result is a rutted and unstable surface that impedes or even prohibits traffic flow (Fig. 1*a*).

Geotextiles can eliminate or reduce the effect of these soft soil problems through separation and reinforcement. When placed over a soft soil, the geotextile provides a support membrane for placement and compaction of aggregate base. The fabric barrier prevents aggregate particles from intruding into the soft soil and prevents soil particles from pumping up into the aggregate layer. As a continuous membrane between soil and aggregate base, the geotextile helps confine the aggregate against lateral and vertical movement. This confining action maintains the density and hence the load-distributing characteristics of the aggregate. The fabric also resists the upward heaving of subgrade between wheel paths. If the subgrade is extremely soft and will not support vehicle loads, the fabric will act as a reinforcing membrane to assist the subgrade in supporting loads (Fig. 1*b* and *c*). Reinforcement from a geotextile can be achieved only if the fabric's deformation modulus, that is, the ability to resist deformation or strain, is greater than that of the subgrade and the aggregate it separates. Therefore, a stabilization fabric should have high modulus to assure optimum performance.

The geotextile functions of separation and reinforcement result in significant benefits to the structure both during and after construction, for example, 25–50% less aggregate thickness required to achieve a stable base, less maintenance required after construction, and longer performance life.

Drainage applications. Soil moisture control through drainage is essential to achieve and maintain stability in pavement foundations, cut slopes, and earth dams. Drainage is accomplished by providing a trench or blanket of porous rock for soil moisture to seep into. A perforated pipe is often installed within the porous rock to collect the moisture and transport it to an outlet. To ensure effective performance and long life, drain structures need a filter to retain soil while allowing water to pass into the drain. Without a filter, soil particles may wash into and clog the drain.

Graded aggregate filters are conventionally used to prevent soil particles from washing into a drain. A properly designed graded aggregate will confine

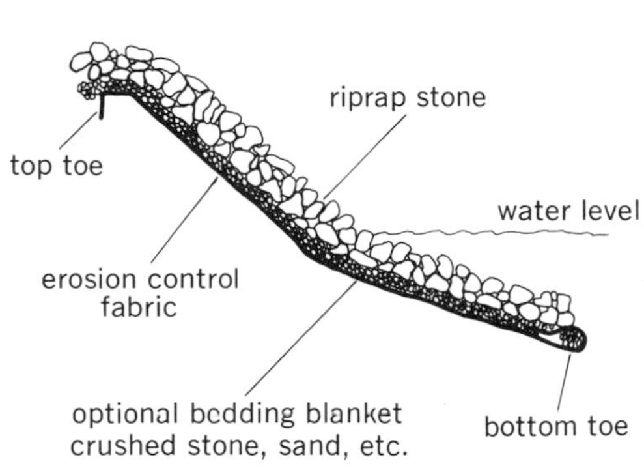

Fig. 3. Slopes along waterways can be protected by geotextiles and armor stone. (*Mirafi Family of Construction Fabrics/ MPB8, Celanese Fibers Marketing Company*)

or retain the soil, thus preventing significant particle movement. If, however, the drainage aggregate is too coarse, the voids between rocks at the soil-aggregate interface will be too large for soil particles to bridge across. The resulting lack of soil particle confinement will result in erosion when water seeps out of the soil, that is, soil piping (Fig. 2*a*).

The pore structure and permeability of some geotextiles are similar to those of graded aggregate filters. These fabrics can provide the same particle retention at the soil-drain interface while permitting unrestricted flow of water from the adjacent soil. Geotextiles can thus eliminate the need for graded aggregate filters in drains. When drainage fabric is used, no special aggregate gradation is required, because the fabric prevents soil from washing into the drain (Fig. 2*b*).

Performance of geotextiles for drainage has been proved through research and thousands of drain installations during the 1970s. These years of drainage fabric experience have shown several advantages for fabrics over more conventional graded aggregate filters. Drain fabrics are more readily available than selected aggregate gradations. The installed cost of drain fabric is typically less than that of graded aggregate filter, at least 20% less in most cases. In addition, drain size can be reduced because the thick, graded aggregate layer is eliminated. Consequently, less excavation and aggregate are required to construct the drain (Fig. 2*c*).

Erosion control applications. Embankments along coastal shore lines and inland waterways are subjected to wave and current action that can cause severe erosion, instability, and even destruction of the earth slope. To protect against erosion, earth slopes have heretofore often been covered with armor (rip rap, concrete blocks, concrete slabs) to resist the force of water on the soil's surface. Despite the armor covering, water can still come in contact with the soil and cause erosion, thereby undermining the armor's foundation. To assure long-term stability and performance, the erosion control structure must include a barrier that shields the soil surface from scouring. In addition, this barrier should be permeable so that any moisture seeping from the soil slope can escape through the barrier without buildup of hydrostatic pressure. Traditionally, granular filters of specially graded sand, gravel, or stone have been used to prevent slope erosion beneath the armor. But granular filters are expensive, particularly when not locally available, and even when properly installed are subject to erosive forces that can wash them away, leaving the soil slope unprotected.

Some geotextiles are ideal erosion control barriers. Erosion control fabrics will shield a soil slope from the erosive force of moving water, and they are permeable so that seepage from the earth slope can pass through freely. These fabrics will remain intact, covering the soil slopes as long as the armor stone remains in place above it. In contrast to granular filters, erosion control fabrics are readily available everywhere. In addition, the installed cost of erosion control fabric is less than for a granular filter, especially when the proper aggregate gradation is not locally available (Fig. 3).

Sedimentation control. Severe erosion can occur during earthwork construction or mining operations when protective vegetation is removed and soil slopes are left temporarily unprotected. Sediment runoff that results from this erosion can create serious downstream damage, for example, contaminated waterways, clogged storm drains, or sediment-covered forests or pastures. Government agencies at the Federal and state level have recognized the problems associated with sedimentation and have passed legislation requiring the control of sediment runoff from any disturbed land area. As a result, earthwork contractors and mine operators are faced with the responsibility and cost of controlling sediment runoff from their work sites.

Silt fences constructed with geotextiles offer a cost-effective solution to the problem of sedimentation control. A silt fence is a fabric-lined fence structure installed on or at the base of an unvegetated slope. The silt fence acts as a water-permeable barrier that retains sediment runoff from the slope, allowing water to seep through the fabric while eroded soil particles settle out on the upstream side of the fence. The silt fence can be thought of as a temporary impoundment structure that forms a sediment pond aboveground (Fig. 4). When installed along the perimeter of a construction site, a silt fence can prevent sediment from leaving the disturbed area. By installing silt fences along stream banks, sediment can be kept from reaching the waterway.

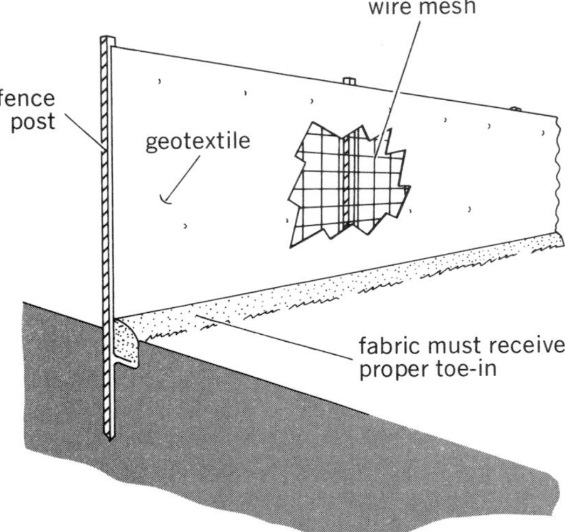

Fig. 4. A silt fence is a fabric-lined fence structure. When lined with a geotextile, it retains sediment runoff, preventing contamination of adjacent waterways. (*Mirafi Fabrics for the Mining Industry, Celanese Fibers Marketing Co.*)

Stabilization, drainage, erosion control, and sedimentation control are the major applications for geotextiles, but their functions of separation, reinforcement, and filtration mean that these fabrics can be adapted to an endless array of uses. Geotextile researchers have developed test methods to evaluate fabric performance, design guidelines for various applications, and specification criteria for fabric selection. And the state of the art in this new geotextile technology continues to expand as the civil engineering community and the textile industry sponsor new research programs every year.

The popularity of this new construction material is growing rapidly as more and more design engineers and contractors experience the cost savings and improved performance that result from geotextile use. Before 1970 the construction industry had never heard of geotextiles; by 1990 geotextiles will be as common on construction sites as bulldozers and hard hats.

For background information *see* ENGINEERING GEOLOGY; EROSION; RIVER ENGINEERING; SOIL MECHANICS; SYNTHETIC FIBER in the McGraw-Hill Encyclopedia of Science and Technology.

[ROBERT G. CARROLL, JR.]

Grain boundaries

The rapid growth of the electronics industry during the past quarter century is due in large part to the development of processes for growing large, high-quality, single crystals of semiconductors such as germanium, silicon, and gallium arsenide. However, several recent applications of semiconductors—low-cost flat-plate solar cells and varistors (voltage-controlled resistors)—are committed, either for economic reasons or by the physics of the device operation, to the use of polycrystalline semiconductors. Since the regions between grains—the grain boundaries—largely control the electronic properties of polycrystalline materials, a considerable amount of basic research has been carried out during the past several years in order to understand the properties of these boundaries. As a result, sufficient progress has been achieved to enable the major electronic properties of these intergrain regions to be predicted with some degree of accuracy.

Atomic arrangements. By definition, a grain boundary is the region of intersection of two perfect but orientationally mismatched crystallites. In this region the atomic arrangements deviate from a perfect periodic lattice. Recent studies on single grain boundaries of germanium and silicon have shown that the most commonly observed grain boundary structure consists of a narrow (less than 5 nm) band of regularly spaced edge dislocations. Associated with this regular array are alternating regions of compression and dilation which account for the considerable elastic energy stored in the boundary structure. The boundary surface between two crystallites is sometimes curved rather than planar; this happens because of differences in the atomic arrangements at different points along the interface. This nonuniformity is thought to be caused by variations in the growth conditions as the boundary is formed.

As has been observed in metals, grain boundaries in semiconductors can act as precipitation sites for any impurities present in the material. For example, oxygen and carbon have been detected at silicon grain boundaries in amounts in excess of bulk concentrations.

Electronic structure. From the preceding discussion it is clear that rather severely perturbed electronic energy levels might be expected in the vicinity of grain boundaries. For instance, when edge dislocations are present, a significant number of grain boundary atoms find themselves in environments with less than the normal number of nearest neighbors. This leads to severe disruption of the normal crystal bonding, or sharing of valence electrons, and usually gives rise to states for electrons in the forbidden energy gap of the semiconductor. The consequences of this are shown in Fig. 1*a*, which illustrates a thin, electrically neutral, grain boundary region as it would exist if it could be placed between two semi-infinite, nearly adjacent crystals without any electrons being allowed to transfer from one to any other of the parts, so that the grain boundary region remains in a neutral state. The adjacent crystals are shown as being separated physically from the grain boundary region in Fig. 1*a* to emphasize that no charge flow to the boundary has yet occurred. Here E_V and E_C are the energies of the top of the valence band and the bottom of the conduction band, ζ is the energy separation between the highest filled (Fermi) level and the bottom of the conduction band. E_{FG} is the energy of the highest filled electron state in the grains, and E_{FB} is the corresponding quantity for the boundary region. Because of the defective structure of the grain boundary region, it contains numerous empty electron states (drawn as open circles) lying below the filled level of the crystalline regions. An *n*-type semiconductor which has electrons as the majority carriers is shown here, but similar phenomena occur in *p*-type materials.

If the hypothesis is continued, and the electrons in all three parts are allowed to find a common energy level (by joining), the structure of Fig. 1*b* results. Electrons (shown as minus signs) flow from the crystals to fill some of the empty states of lower energy in the grain boundary. This flow results in a negative charging of the center of the grain boundary and the formation of an electrostatic repulsive barrier which stops further electron flow by raising the energy levels of all the electron states at the center of the boundary until the energy of the highest filled state is the same everywhere. Here ϕ_B is the amount of electrostatic bending of the energy bands. If some of the defect states at the center of the boundary remain unfilled, they will play an important role in the electronic behavior of the boundary.

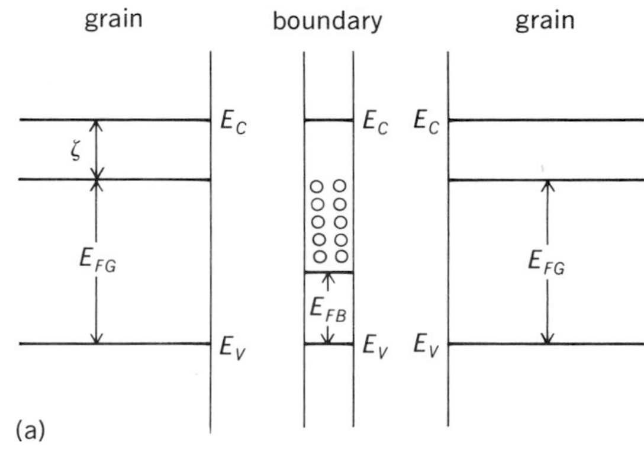

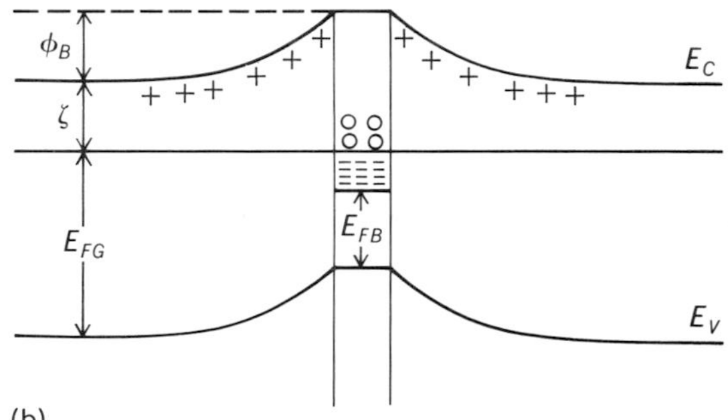

Fig. 1. Process of grain boundary charging in semiconductors. (*a*) Electronic energy levels of two adjacent crystallites and their intervening grain boundary region, as they would exist if electron transfer between the three regions could not take place. (*b*) Electronic energy levels as they actually exist, after the three regions are joined.

Transport of charge. The accumulation of majority carriers at semiconductor grain boundaries has been verified in a variety of materials, including silicon (Si), germanium (Ge), gallium arsenide (GaAs), cadmium sulfide (CdS), lead sulfide (PbS), and zinc oxide (ZnO). This charging effect has a profound influence on the electrical properties of polycrystalline semiconductors. In order for current to flow across such charged boundaries, the carriers must have an energy considerably in excess of their normal thermal energy (0.025 eV at room temperature). The number of electrons having this extra energy at any time is small, and hence the conductance such a potential barrier has when current flows across it is low and strongly temperature-dependent. The temperature dependence frequently seen is an exponential function of inverse absolute temperature—much like the leakage current across a *pn* junction device. Because of the small magnitude of this grain boundary conductance, polycrystalline aggregates frequently offer conductance to current flow which is many decades lower than that of single crystals of similar purity.

The exponential temperature dependence of the

electrical conductance G measured across a single silicon grain boundary is shown in Fig. 2. The grains in this case are doped n-type to a level of 1.3×10^{16} cm^{-3}. The data points fit a formula of the type given below (indicated by the straight line

$$G = G_0 \exp [-(\phi_B + \zeta)/kT]$$

in Fig. 2), where k is Boltzmann's constant, G_0. is a constant, T is the absolute temperature, and $(\phi_B + \zeta)$ is 0.62 eV in this case. This result shows that electrons crossing the grain boundary are thermally emitted over a potential energy barrier of total height $(\phi_B + \zeta)$. This is the energy difference between the highest filled energy state and the top of the barrier (Fig. 1*b*).

Because of the narrow width of the grain boundary space charge regions, the capacitance of such a structure can also be large. This effect is put to use in boundary-layer capacitors, which are polycrystalline devices made of materials like barium titanate. These are useful for high-capacitance-value applications where circuit element size is an important consideration.

Another useful property of grain boundaries—employed in devices called varistors—is their voltage-dependent resistance. As voltage is applied across a grain boundary, the presence of unfilled trapping states causes the potential barrier for current flow to remain large. However, continued application of larger and larger voltages eventually fills these states, and further bias causes a rapid reduction in this barrier. Drastic increases in current occur above this filled trap threshold, and the grain boundary turns on. Varistors made of materials such as polycrystalline zinc oxide exhibit this effect. They are used as voltage regulators and surge protectors in a variety of applications.

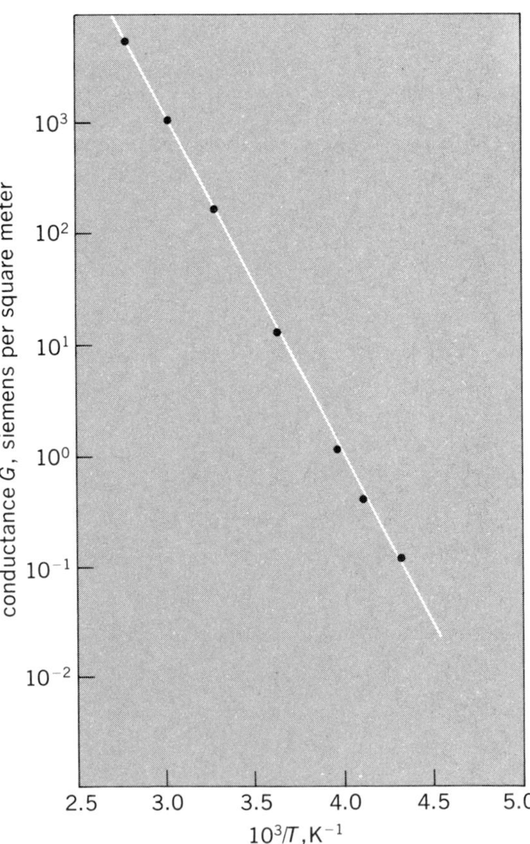

Fig. 2. Temperature dependence of the electrical conductance measured across a single silicon grain boundary.

Minority carriers are also strongly influenced by the presence of grain boundaries. An example is the photovoltaic cell, in which light-generated minority carriers must diffuse substantial distances to a collecting junction for current to be generated; here the presence of grain boundaries is quite harmful. The same grain boundary potential barrier which repels majority carriers attracts minority carriers and traps them until they recombine. This results in a short minority carrier lifetime, a property which has traditionally made polycrystalline semiconductors less than ideal for a variety of semiconductor devices such as transistors and photovoltaic cells.

Recent research has shown that hydrogen diffused into silicon removes grain boundary defect states, thereby considerably improving the electrical properties of polycrystalline silicon; chemical methods to alter grain boundary properties in other semiconductors are also under investigation. It is believed that improved polycrystalline devices will result from these efforts.

For background information *see* GRAIN BOUNDARIES; SEMICONDUCTOR; SOLAR CELL; VARISTOR in the McGraw-Hill Encyclopedia of Science and Technology.

[C. H. SEAGER]

Bibliography: H. K. Charles and A. P. Ariotedjo, *Solar Energy*, 24:329–334, 1980; L. L. Kazmerski and P. J. Ireland, *Solar Cells*, 1:178–182, 1980; G. D. Mahan, L. M. Levinson, and H. R. Phillip, *J. Appl. Phys.*, 50:2799, 1979; C. H. Seager and G. E. Pike, *Appl. Phys. Lett.*, 35:709–711, 1979.

Grasslands

Natural ecosystems—whether grassland, forest, or desert—are determined primarily by the interactions of climate, soils, and vegetation. Climate determines the nature of both soils and vegetation, but vegetation interacts with the climate by modifying the environment immediately above the soil and thus the soil itself. Temperature and water are controlled primarily by climate, although topographic factors (such as slope and exposure) and soil factors (such as soil water-holding capacity and infiltration rate) can greatly influence air and soil temperature and available water for plants. The location of any vegetation zone is primarily related to climatic variables such as air temperature and rainfall, whereas the differences in vegetation type and amount within a vegetation zone are caused primarily by local differences in slope and exposure and type of soil. The principal grasslands of North America are shown in the illustration.

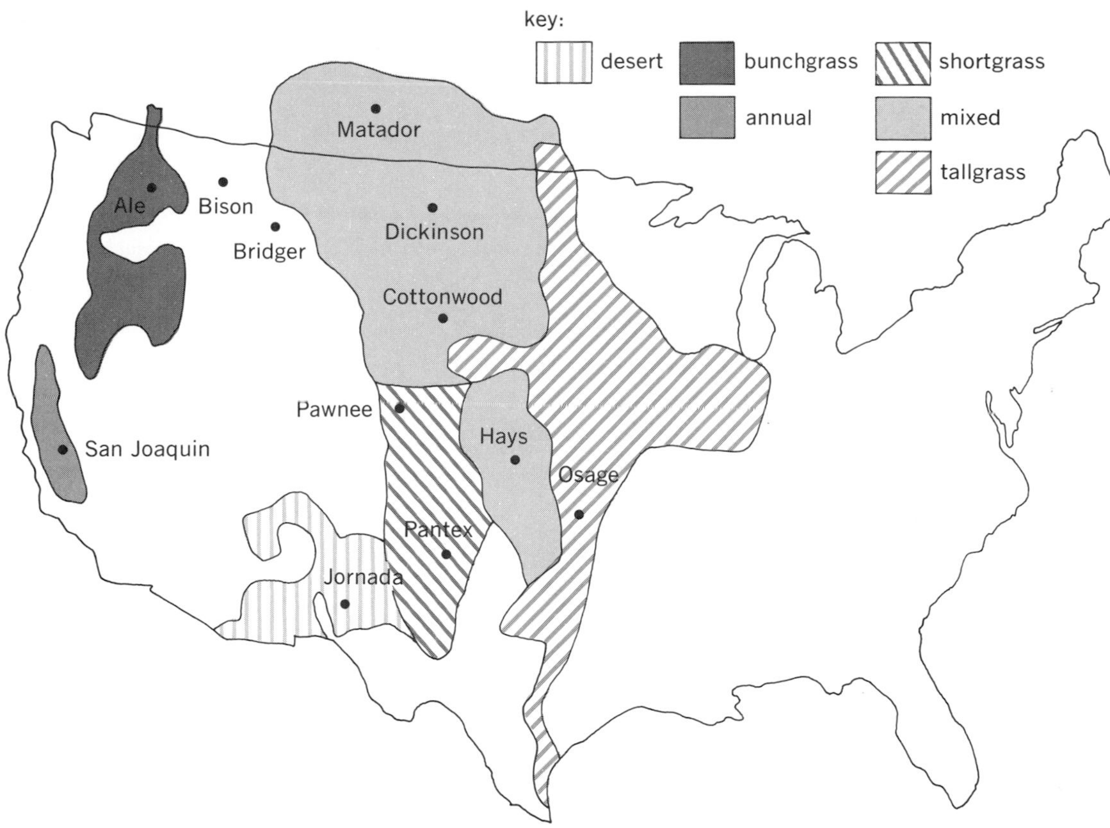

Major grassland types in North America.

Grasslands differ from other types of ecosystems because grasses and forbs are the dominant life forms. In perennial grasses and forbs, the crowns, rhizomes, or other types of stem modifications that function as storage organs for carbohydrates and nutrients are at or beneath the ground surface and are thus well protected from drying during dormant periods. Annual grasses and forbs are adapted to survive in deserts and the drier grasslands because they avoid drying by completing their life cycles in the growing intervals between drought- or cold-induced dormant periods. Although the "plant" phase of the cycle is not resistant to drying or freezing, the seeds are. Annuals constitute the major form of vegetation in many deserts and are the dominant life form in one North American grassland, the annual grasslands of California.

Grasslands grade into deserts at the dry extreme of their range and into forests at the wet extreme. Deserts differ from grasslands in that the dominant perennial plants are xerophytic (drought-resistant) shrubs and succulents with specialized mechanisms for drought resistance or avoidance. Forests differ from grasslands in that they are dominated by life forms (tree and mesophytic shrubs) whose perennial organs are elevated substantially above the ground surface and thus are subject to drying in a dry atmosphere. In general, the more mesic grasslands are dominated by perennial grasses and forbs because trees and mesic shrubs that might otherwise be present cannot survive the periods of aerial drying characteristic of grassland climates. Soils under grassland also are greatly affected by climate. For example, in the Great Plains the soils range from shallow (30 cm) with relatively small amounts of organic matter in the semiarid west to deep (greater than 2 m) with large amounts of organic matter in the subhumid east.

Shortgrass prairie. The shortgrass prairie is characterized by relatively low amounts of precipitation (25–40 cm/year) and very high potential water loss as evaporation and transpiration (60–110 cm/year). However, a substantial portion (two-thirds) of the water falls during the growing season. The period of greatest precipitation is spring.

The drought-resistant native vegetation is dominated by warm-season species (principal growth during summer) and cactus, which occur in response to the low precipitation. The vegetation is usually less than 20 cm high, and typical annual herbage production is 1000–2000 kg/ha of dry matter. On isolated sites, soils or topographic position may modify the prevailing water balance by reducing evaporation so that the more mesic species (mixed-grass prairie species) can persist.

In general most precipitation is stored near the surface of the soil, and much of it is lost by evaporation. The soils are typically shallow and light-colored because the organic matter content is low (less than 2%), the consequence of relatively low plant production, frequent drought, and large temperature extremes.

Tallgrass prairie. Tallgrass prairie occurs in areas with relatively high rainfall (more than 60 cm/year) and high summer temperatures. The precipitation is relatively evenly distributed throughout the growing season when compared with other grassland zones of North America. Potential evaporation and transpiration range between 60 and 80 cm/year.

Native vegetation is dominated by tall (more than 1 m) warm-season grasses that use rainfall during the growing season and stored soil water during the nongrowing season. Annual herbage production generally exceeds 4000–5000 kg/ha dry matter.

During the growing season the soils store nearly all of the water that falls, making most of it available for plant growth. The deep, dark-colored soils result from extensive weathering of geologic materials caused by precipitation combined with large amounts of plant material, abundant water, and favorable temperatures. Organic matter concentrations are often 4–6%.

Northern mixed-grass prairie. The distinctive features of the northern mixed-grass prairie are cold winter temperatures and the broad annual precipitation range (from 28 cm in the west to about 64 cm in the east). Most precipitation occurs in late winter and spring. However, potential evaporation and transpiration are relatively low (45–60 cm/year) compared with those of southern sites because northern latitudes receive less radiation.

The vegetation reflects the temperature-precipitation-evaporation-transpiration interactions that favor cool-season (maximum growth during spring), midheight (about 0.5 m) grasses and forbs. Annual dry-matter production of herbage is often between 2000 and 4000 kg/ha. Water-holding capacities of many of the soils are high, and the water from melted snow and spring rain is available for growth of plants as soon as temperatures allow. Because summer drought is common, most soil water is utilized by the early-growing, cool-season vegetation. Relatively small amounts of soil water are available during the summer to support warm-season species.

Soil development is intermediate between that in shortgrass prairie and that in tallgrass prairie. Sufficient water is present to cause soils that are relatively deep, though not as deep as those in tallgrass prairies. Organic matter accumulates to 4–7% because production of plant material is relatively high and decomposition rates are relatively low owing to low temperatures during the principal periods of ample soil water storage.

Southern mixed-grass prairie. Southern mixed-grass prairie occurs in areas of moderate rainfall (40–80 cm/year), high temperatures, and a long growing season. Precipitation occurs throughout the year. Potential evaporation and transpiration are high, ranging between 60 and 100 cm/year.

Vegetation consists principally of warm-season, midheight grasses and forbs. In most areas the water-holding capacities of the soils are sufficiently large to store all of the available water. Summer precipitation, combined with sufficient water-holding capacities, favors warm-season species over cool-season species.

Soils are well developed because relatively large amounts of rainfall, combined with abundant soil water and favorable temperatures, speed weathering. Because of the high temperatures and greater decomposition rates, however, the organic matter concentration (3–4%) is somewhat less than that in the northern mixed-grass prairie.

Desert grasslands. Low precipitation (15–25 cm/year) and hot summer temperatures characterize desert grasslands. Precipitation often comes as intense rainstorms during the summer. Potential evaporation and transpiration are very high, ranging between 100 and 110 cm/year.

Vegetation is dominated by short, drought-resistant, warm-season grasses and short and tall shrubs. Several soil characteristics, however, reduce the amount of water available to plants: low water-holding capacity, the frequent occurrence of an impermeable layer (caleche) near the surface, and frequently high sodium content that causes clays to disperse, thereby impeding the infiltration of water. Annual herbage production is often about 1000 kg/ha or less.

Because of low precipitation, frequent drought, and low plant productivity, soils are poorly developed except for the well-developed caleche layer. Organic matter concentrations are often less than 1%.

Annual grasslands. The California annual grasslands are in a Mediterranean climate characterized by cool, wet winters and hot, dry summers. Precipitation ranges from 20 to 100 cm/year, and potential evaporation and transpiration can exceed 110 cm/year.

The interactions of climate, grazing by domestic herbivores, and soils result in herbaceous vegetation dominated by exotic annual cool-season plants that originated near the Mediterranean Sea. These species were introduced in the 18th and 19th centuries, but many scientists think that perennial grasses dominated the original herbaceous vegetation. At present, the climate has a profound influence on the herbaceous vegetation. Seed germination, or the breaking of dormancy, occurs after autumn rains commence. The relatively warm winter temperatures (mostly above freezing) and winter and early spring precipitation stimulate the plants to rapid growth. Generally, precipitation ceases by late spring, and transpiration and evaporation processes dry the soil. Without precipitation during the summer, the annual plants die and perennial herbaceous vegetation, where it does occur, enters dormancy. Annual herbage production is highly variable but may be 4000–6000 kg/ha during favorable years.

Soils are characteristically shallow, coarse, poorly developed, and highly permeable, and have a low water-holding capacity. Organic matter concentrations are highly variable but are often <2%. In most years soils saturated with water in winter or spring discharge excess water and soluble nutrients into groundwater and streams. Thus, not all the water that falls as precipitation is available for biological activity.

Bunchgrass steppe. The bunchgrass (cluster growth form) steppes of North America are characterized by cool, wet (20–40 cm/year) winters and springs and hot, very dry summers. Potential evaporation and transpiration are 80–100 cm/year. Precipitation is usually lower than that in annual grasslands.

Vegetation is dominated by cool-season perennial bunchgrasses and forbs and cool-season shrubs. Adequate precipitation falls during the winter and spring and is stored in the soils. This stored water supports perennial species that grow early and are drought-resistant. Annual herbage production often ranges between 1000 and 2000 kg/ha.

Soils are poorly developed because of low rainfall and low soil organic matter content (less than 1%). Organic matter content is low because much of the plant material is woody and decomposes slowly in the arid climate.

Mountain grasslands. Mountain grasslands develop in climates with cold, snowy winters and springs and cool summers. Precipitation and potential evaporation and transpiration are extremely variable from site to site, often reflecting the ranges of all other grassland types.

The perennial cool-season vegetation of mountain grasslands results from the interactions between climate and soils and thus is also highly variable. Vegetation growth begins early because the soil is charged with water from melting snow accumulated during the winter. Most rain occurs in spring and early summer, but, though only sporadic, late summer rains can significantly recharge the surface layers of the soil. Also, summer temperatures are low, favoring cool-season plants both directly and by reducing evaporation and transpiration.

The dominant characteristic of the soils is depth. During snowmelt periods of some years, sufficient water is available to saturate the soil; the excess water is discharged to groundwater. Soil organic matter accumulates to quite high levels (3–7%) because of high production of plant tissues, and decomposition rates are low due to low soil temperature.

Summary. The grasslands of North America show a wide range in precipitation, evaporation and transpiration, temperatures, vegetation types and amounts, and soil characteristics. In general, the more mesic grasslands are very fertile and hence productive. Due to their high fertility, most of the mesic grasslands of North American have been cultivated and now constitute an area of enormous importance in world food production. For example, the tallgrass prairie now is largely known as the Corn Belt. Many other crops are also grown in that zone. The northern mixed-grass prairie produces much of the world's wheat, and the southern mixed-grass prairie produces a variety of crops, including sorghum.

The more arid or xeric regions of North American grasslands are mostly untilled and are principally used for livestock production. Where grazing has not been excessive, native plant species predominate. As world food demand increases, much of the grain grown in the mesic grasslands that now is used to feed livestock will be directly converted to human food. As that conversion takes place, the arid and semiarid ranges of western North America will become increasingly important in meat production.

For background information *see* DESERT ECOSYSTEM; ECOSYSTEM; GRASSLAND ECOSYSTEM; PLANTS, LIFE FORMS OF; WORLD VEGETATION ZONES in the McGraw-Hill Encyclopedia of Science and Technology.

[R. G. WOODMANSEE; W. J. PARTON; J. L. DODD]

Bibliography: H. Jenny, *The Soil Resource: Origin and Behavior*, 1980; H. Walter, *Vegetation of the Earth*, 2d ed., 1979.

Gravitropism

Plants and animals have evolved in the presence of a strong gravitational force at the surface of the Earth. The most obvious response of living things to this force of 1 *g* is orientation. Thus, shoots of higher plants tend to grow vertically away from the center of the gravitational field, while roots tend to grow toward it. Botanists have been interested in this phenomenon for over a centruy, and there is now renewed interest and intensive research due to advances in space science. Space vehicles orbit at altitudes where the force of Earth's gravity is negligible. Although the force is never totally absent, the low values and the virtual weightless conditions created result in a near-zero-gravity situation. The response of plants to the near-zero-gravity condition is not always predictable on the basis of knowledge gained from plant physiology experiments on the surface of the Earth. In all Earth-bound experiments the 1 *g* force is always present. True weightlessness can be achieved only in outer space. It is for this reason that a number of United States and Soviet biosatellite programs have carried plants aboard their space vehicles. Knowledge of plant growth in outer space is essential for short- or long-term colonization in space. Successful completion of a plant life cycle, from seed germination to seed formation, is essential for production of food for the crew, and possibly raw materials for some products.

Plant responses to gravity. Effects of gravity on plant growth, development, and morphology are generally termed gravimorphic effects. Gravimorphism is a field that is little understood at this time. The special case where plant organs exhibit oriented growth in relation to the center of a gravitational field is termed gravitropism. This term is preferable to the more commonly used term geotropism, which specifically implies an Earth-centered response to gravity. A force of acceleration, such as produced in a centrifuge, can substitute for, and can have an effect equivalent to, the Earth's gravity. Gravitropic reactions are readily observed in most plants. The vertical growth of the main stem, as well as the positioning of the branches at various angles, is a response to gravity. A stem growing away from the center of the gravitational field is said to be negatively gravitropic. A primary root, on the other hand, is positively gravitropic and grows toward the center of the gravitational field. Branches of shoots and roots that grow at different angles from the vertical are said to be plagiogravitropic, while the special case of organs growing horizontally is termed diagravitropic.

Gravitropism. The gravitropic responses are found in the more than 220,000 species of flowering plants as well as in other vascular plants such as the confiers and ferns. A number of bryophytes (mosses and liverworts), some fungi, and algae also show gravitropic responses. As plant life moved from water to land, some 600,000,000 years ago, it became necessary for these terrestrial plants to properly orient their shoots to intercept incident solar energy and their roots to absorb water and minerals from the soil. Thus, plants have evolved by making efficient use of the presence of gravity as a "guiding system" for their functions.

Light has an independent effect on the oriented growth of plants. Often, this phototropic effect (one of growing toward or away from a source of light) is combined with the gravitropic effect to produce the final growth form of a plant. That gravitropism can be independent of light is easily demonstrated by orienting plants in darkness and observing their gravitropic responses.

Bending response. Experimentally, the response of a plant to gravity is studied by placing a shoot or root in a horizontal position and following its bending reactions. The organ most studied in this manner is the leaf sheath, called a coleoptile, that covers developing regions of seedling grass shoots. Starting with Charles Darwin and his son, Francis Darwin, the coleoptiles in seedlings of canary grass, oats, wheat, corn and other grasses have been well investigated. The regions above and below the cotyledons of seedlings of other plants, such as peas, cucumber, sunflower, and others, are also favorite material for the study of gravitropic responses. The positive gravitropic responses of roots of corn, peas, and bean seedlings have also been studied in detail. All these materials are easily obtained by simply germinating the seeds. But gravitropism is not confined to seedlings only. Specialized organs called pulvini, located at the base of the leaf sheaths of grass shoots, also respond to gravity. There is much interest in this organ because of its importance in erecting cereal grasses that may fall down due to the action of wind or water (called lodging).

Fig. 1. Shoot segment of barley photographed once every 3 h after placing horizontally. Bending occurs at the leaf-sheath pulvinus near the node.

Instruments. The bending response of shoots or roots is often studied by photographing at regular time intervals the position of the bending organ (Fig. 1). Recently, electronic devices have become available that can record the continuous gravitropic bending reaction with the help of angular position transducers. The study of gravitropism also involves the use of several other instruments. Clinostats are devices that rotate the plants in a horizontal axis. Such horizontal rotation exposes all sides of the plant organ to gravitational stimulus, thus preventing a unilateral exposure which is necessary for any bending reaction to occur. This omnilateral stimulation has the effect of compensating for the unilateral action of the Earth's gravity. This method of study is the only means of approaching the question of the effect of weightlessness on plants in Earth-bound laboratories, where 1 g gravitational force is ever present. The centrifuge is a device similar to the clinostat but rotates at considerably higher speeds to generate significant g forces. This instrument permits study of the effects of more than 1 g on plants. Centrifuges will be common instruments on board future spaceships. In outer space, the centrifuges will help create precise amounts of g forces, as desired, to study their effects on plant and animal development. Thus, in the absence of a centrifuge the space vehicle itself will offer a weightless environment that the best of clinostats can never achieve on Earth.

Because the response of plants to gravity is manifested in terms of biochemical and morphological changes, biochemical analyses, particularly for plant hormones, and light and electron microscopy have also become essential in unraveling these changes that underlie gravitropic responses.

Perception and reaction to gravity. When a plant is reoriented from its normally vertical position to a horizontal position, it is being exposed to a 1-g force, which is equivalent to a constant acceleration of 982 cm · s^{-2}. A minimum time of exposure is necessary before a reaction is noticed in the shoot or root. This is the presentation time. Presentation times as low as 12 or 18 s have been recorded for roots, and 30 s for coleoptiles. Being a threshold phenomenon, an inverse correlation exists between force (g) and time (t) to produce a constant (k) minimum threshold stimulus before the reaction can occur ($gt = k$). Although plants are usually stimulated at 90° to the vertical, the optimum angle of stimulation seems to be about 130°.

The response of the organ, positive or negative, results because of asymmetric cell elongation across the organ (Fig. 2). This asymmetry of growth is a precisely controlled response, where, in an organ such as the grass pulvinus, no growth occurs on the upper region but a gradation of growth occurs below this region, ending in maximum elongation at the lower surface (Fig. 3). Every portion of a cylindrical organ seems to recognize its unique position in relation to the gravity vector. The major problem in gravitational plant physiology is trying to understand how the gravitational stimulus is perceived and translated into this precise asymmetric cell growth. *See* PLANT GROWTH.

The presence of heavy bodies (statoliths) inside the cells in the regions of gravity response has long been recognized. These statoliths are specialized plastids (amyloplasts) that contain several starch grains in each plastid and occur in shoots and root caps. Their sedimentation times correlate well with the gravitropic reaction times, and their removal (as by the removal of the root cap) results in the loss of the gravitropic response. In spite of these correlations, scientists are still not sure how these statoliths can bring about a differential growth response. In the unicellular rhizoids of an alga, *Chara*, heavy bodies that act as statoliths are made up of barium sulfate crystals.

The growth response of the cells are now known to be due to an unequal distribution of plant hormones. Indole-3-acetic acid (IAA) has a major role in shoots. However, recent investigations have implicated gibberellic acid and ethylene as other hormones which play a role in gravitropism. It is believed that IAA is present in an inactive form, as conjugates of amino acids (peptidyl-IAA) or sugar alcohols (inositol ester of IAA) in the unstimulated organ. Gravity stimulation seems to release the active, free IAA. The same occurs with gibberellins and their conjugates. Root gravitropism was once thought to be controlled by IAA alone. But recent research has not firmly confirmed this hypothesis. Instead, another plant hormone, abscisic acid, is known to be involved in root gravitropism. Thus, response to gravity stimulation is a complex phenomenon involving more than one plant hormone.

Space botany. Crewed and uncrewed space flights have already carried different plants to outer space for different experiments. The United States Biosatellite and Soviet Kosmos, Soyuz, and Salyut satellite experiments have confirmed the necessity of a gravity vector for oriented growth and normal development of plants. In its absence, plants grow in various directions. Among the plants observed so far in outer space are: pine, wheat, onion, green pepper, arabidopsis, radish, cucumber, lettuce, peas, tomato, fennel, parsley, dill, garlic, mush-

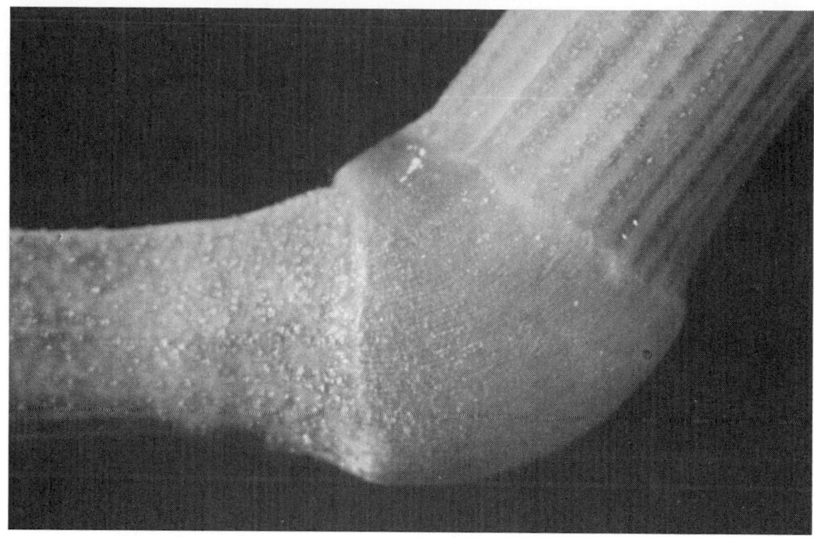

Fig. 2. Closer view of the nature of asymmetric growth response that occurs in the leaf-sheath pulvinus, here in oats, in response to gravity stimulation. The bending reaction causes the apical portion of the shoot eventually to grow upright.

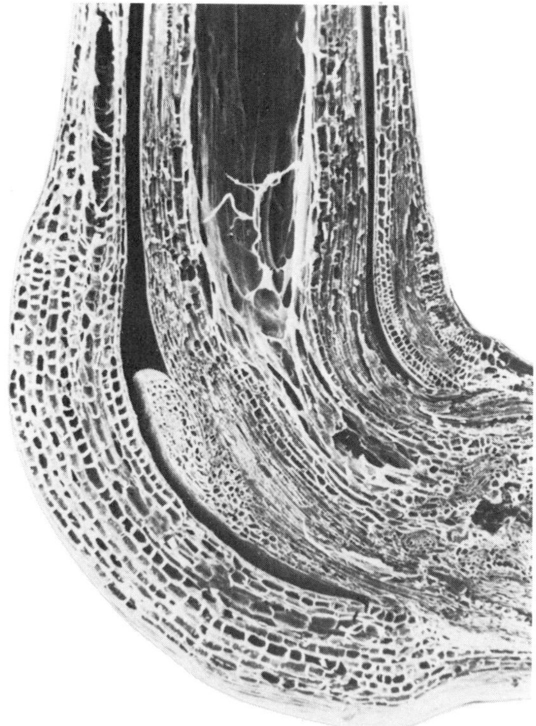

Fig. 3. Scanning electron micrograph of a longitudinal section of a grass pulvinus that has bent upright in response to gravity stimulation. The central stem and axillary bud are surrounded by the leaf-sheath pulvinus, which shows the differential cell elongation on either side.

rooms, and orchids. Weightlessness seems to interfere with normal fertilization and seed setting, probably by interfering with the normal polarity of the fertilized egg. In several cases flowering is also inhibited. But weightlessness does not affect plantlet formation in isolated somatic carrot cells, thus offering the potential for plant regeneration from tissue cultures. The successful launching of the space shuttle, and the future experiments planned with it and the European Spacelab, will offer further insight into the way in which gravity controls plant growth and development.

For background information *see* PLANT GROWTH; PLANT HORMONES; PLANT MOVEMENTS; SPACE BIOLOGY; WEIGHTLESSNESS in the McGraw-Hill Encyclopedia of Science and Technology.

[PETER B. KAUFMAN; P. DAYANANDAN]

Bibliography: *AIBS/NASA Space Biology Symposium*, Life Sciences Division, NASA, October 1980; B. E. Juniper, *Proc. Roy. Soc. Lond. B.*, 199:537–550, 1977; L. Reinhold, Phytohormones and the orientation of growth, in Letham et al. (eds.), *Phytohormones and Related Compounds: A Comprehensive Treatise*, vol. 2, 1978; M. B. Wilkins, *Bot. Mag. Tokyo*, Special Issue, 1:255–277, 1978.

Grinding

Creep-feed grinding is an abrasive machining process capable of heavy stock removal of "difficult-to-machine" materials. The process has the ability to hold very close tolerances of both size and form. Thermal damage to the workpiece surface is a serious limitation of the creep-feed grinding process. However, it has been shown that by combining the creep-feed principle with a continuous dressing action, not only is the risk of thermal damage virtually eliminated but the stock removal rate can be increased by a factor of 20. This recent development is revolutionizing the abrasive machining industry.

Development. Creep-feed grinding was adopted as a production process mainly due to the needs of the aerospace industry. In order to improve the performance and reliability of gas turbine aircraft engines, the engine manufacturers developed turbine blade alloys (nickel-based superalloys) with high strength, high temperature creep resistance, and corrosion and abrasion resistance. These material properties directly affect machinability. The refractory nature of these superalloys causes high temperature gradients across the cutting tool/workpiece surface interface which may cause surface cracking. Many of the superalloys work-harden, and precipitated carbides in the grain boundaries act as an abrasive on the cutting tool edge, causing rapid wear and tool failure. It is these factors which classify materials as "difficult to machine."

In order to achieve the accuracy of size and form in superalloy materials, grinding was the only process available to the manufacturing engineer. The traditional reciprocating grinding process can be costly and time-consuming due to the need for frequent redressing of the form on the grinding wheel periphery and the ineffective air-cutting time at the end of each stroke of the machine table. Creep-feed grinding was therefore developed as an economic alternative to the traditional processes. Much of the research and development of the creep-feed grinding process has been carried out in Europe. Over the past decade the University of Bristol, in England, has emerged as the center for research into the creep-feed grinding process which is now widely used in Europe, even outside the aerospace industry, for machining parts such as typewriter spacing racks, thread chasing dies, deep-groove gear tooth forms, and slotting hydraulic pump rotors. Creep-feed grinding is generally reserved for surface grinding. Research is being carried out using creep-feed in the cylindrical mode; however, the economics appear to be marginal.

Process. Creep-feed grinding is analogous to the milling process where a grinding wheel is used in the place of the milling cutter (Fig. 1). The cutting forces are high when creep-feed grinding, hence the machine tool must be designed for rigidity and must be capable of closely controlled slow table speeds, typically in the range 10 to 500 mm/min. One of the most successful table drive systems utilizes a preloaded ballscrew drive powered by a dc motor. There are a number of hydraulically driven creep-feed machine tables, but they appear satisfactory only in the up-grinding mode where the grinding wheel rotates in the direction which opposes the machine table motion. Vibration is not a serious problem with creep-feed grinding as the long arc of cut tends to dampen any regenerative chatter which might occur.

Creep-feed grinding wheels are very special. The most common type of grinding wheel used for creep-feed grinding is composed of vitrified aluminum oxide grit which has porosity induced into the mix by the addition of a material (such as naphthalene granules) which can be released from the wheel matrix in the green state, prior to firing. These grinding wheels are very "soft," and the large pores in

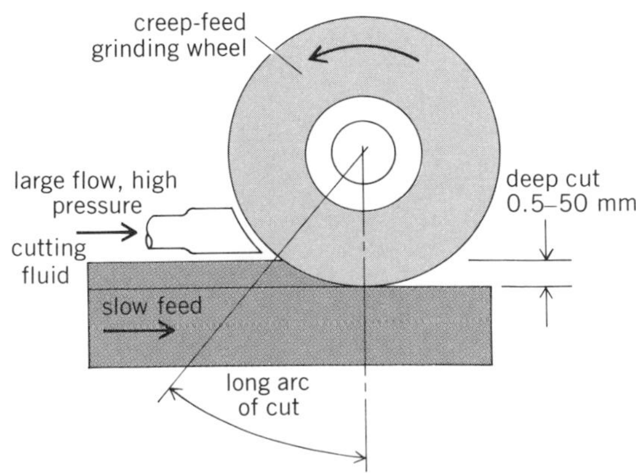

Fig. 1. Schematic diagram of the creep-feed grinding process.

the wheel allow the cutting fluid to be taken up and into the arc of cut. Vitreous bonded wheels are not the only wheels used for creep-feed grinding. Cubic boron nitride resin-bonded grinding wheels are also used when their cost is not prohibitive.

Dressing of the grinding wheel periphery is most important as it is this operation which conditions the grinding wheel for cutting. Accurate dressing is required in order to achieve a precise form and an effective cutting action that will not create adverse thermal conditions. A diamond roll dresser is most often used to dress the vitrified grinding wheels. This is a metal-bonded diamond-impregnated wheel which rotates on the grinding wheel periphery, machining the form onto the grinding wheel. Diamond rolls are most commonly used where the most accurate forms are required. It is in this area of dressing that the recent advances have been made.

Heat transfer. The creep-feed grinding process generates a great deal of heat which has to be dissipated, hence the application of the cutting fluid is more important. Typically a water-based cutting fluid is injected into the porosity of the grinding wheel at pressures approaching 5 bars (500 kPa) and at flow rates in the order of 6 liters/s. Research has shown that the cutting fluid increases in temperature as it traverses the long arc of cut. The heat transfer mechanism is such that the cutting fluid accepts heat from the process along the arc of cut in the regime of nucleate boiling. The heat generated by the process increases as the grinding wheel dulls; correspondingly the heat flux increases until the threshold of transference from nucleate boiling to film boiling. Once film boiling occurs, there is virtually no heat transfer from the process into the cutting fluid, hence all the heat energy is conducted into the workpiece surface resulting in thermal damage. At the onset of thermal damage the input of the process heat to the workpiece will cause the material to expand, increasing the normal force on the grinding wheel. The increase in force at this stage will usually dress the grinding wheel periphery sufficiently to affect a return to satisfactory grinding; however, as the grinding wheel wears further, the frequency of thermal damage increases until burning occurs and the workpiece discolors. The onset of thermal damage is recognized by a series of irregularly spaced "surge" marks on the ground surface of the workpiece, which should not be confused with the very regular marks of chatter, caused by vibration.

Energy partition. Three sources of heat energy are generated by the grinding process (Fig. 2). The cutting energy is that energy created by the shearing action of the active grinding wheel grits on the workpiece material. The plowing energy is that energy created by the upsetting of the workpiece surface as the grit moves through the material. No material is removed by plowing. The rubbing energy is that energy created by friction between wear flats on the grinding wheel periphery and the workpiece arc of cut. Wear flats are areas on the grinding wheel periphery which have evolved due to the attritious wear of the grit or the loading of workpiece material around the grits.

Research has shown that of the total creep-feed grinding energy 97% is due to rubbing and 3% due to cutting and plowing. More importantly, the three energies may be further partitioned into the amount of each energy which is conducted into the workpiece surface as heat. There has been much study on the heat balance in grinding, experimental values conflicting with analytical values; however, it may be accepted that 90% of the rubbing energy is conducted into the workpiece, virtually 100% of the plowing energy remains in the workpiece surface, and 97% of the cutting energy goes off in the grinding chip.

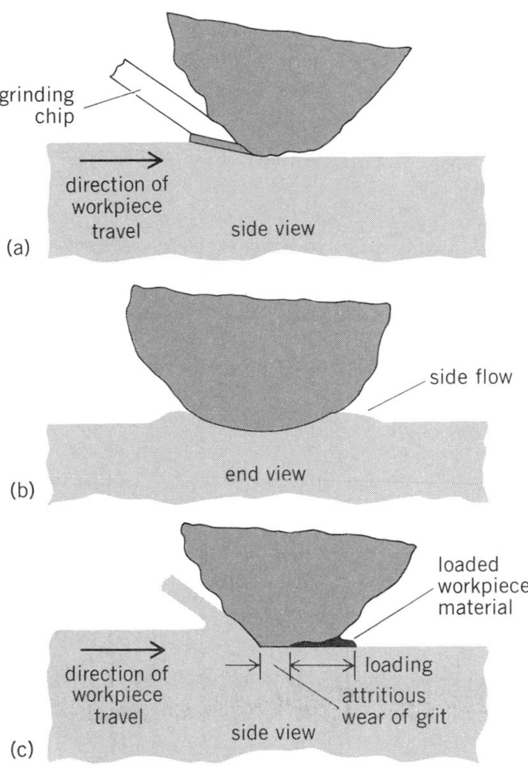

Fig. 2. The three sources of grinding energy: (*a*) Cutting. (*b*) Plowing. (*c*) Rubbing.

The partition of the total creep-feed grinding energy shows that by far the most detrimental energy is that due to rubbing. The rubbing energy component could virtually be eliminated if the grinding wheel were kept sharp throughout the process. Continuous dressing of the grinding wheel with a diamond roll dresser has been shown not only to keep the grinding wheel sharp, but also to maintain the exact form on the grinding wheel periphery.

Continuous dressing. The continuous dressing process results in very low process specific energy, which enables the stock removal rate to be signifi-

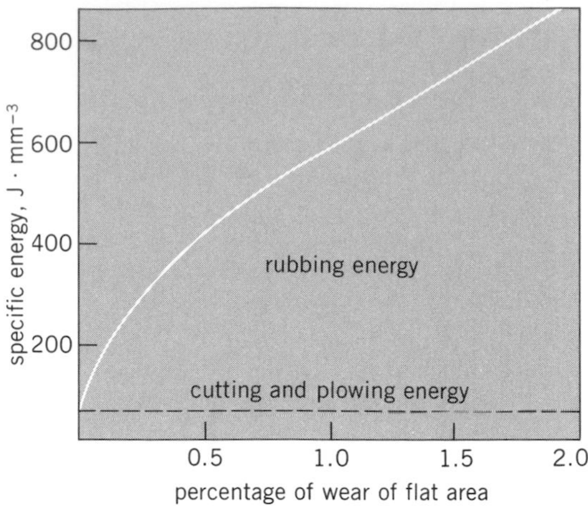

Fig. 3. Relationship between the specific energy of the creep-feed grinding process and the percentage of wear-flat area.

cantly increased, in the order of 20 times over conventional creep-feed grinding (Fig. 3). The limitation of creep-feed grinding, as mentioned earlier, is thermal damage to the workpiece surface, which in the aerospace industry may have grave consequences. The continuous dressing process, however, has the limitation of grinding wheel breakdown which protects the workpiece from thermal damage.

Future. The two factors of continuous dressing combined with the creep-feed grinding principle, namely, vastly increased productivity and a safe operating limit, are changing the philosophy of the abrasive machining industry. New and revolutionary machine tools are beginning to appear on the market to launch a new era in abrasive machining.

For background information *see* ALLOY; MACHINABILITY OF METALS; GRINDING in the McGraw-Hill Encyclopedia of Science and Technology.

[STUART SALMON]

Bibliography: T. D. Howes, Creep-feed grinding, *Eng. Dig.*, 40(12):9–13, December 1979; *Proceedings of the International Creep-Feed Grinding Conference*, University of Bristol, England, September 1979.

Gyroscope

The nuclear magnetic resonance gyroscope (NMRG) is an inertial angle sensor that obtains its information from the dynamic angular motion of atomic nuclei. Interest in the development of an NMRG comes from its potential to be a small, extremely accurate, and highly reliable strap-down gyro capable of accommodating large input rates with no acceleration sensitivity. The NMRG is expected to be smaller in size than a ring laser gyro (RLG), with comparable performance and with no fundamental dependency between angular resolution and size. It is anticipated that the NMRG will have completed development and be available for military applications in navigation systems during the 1990s.

Rotation sensitivity using NMR was first successfully demonstrated in 1958. This demonstration was a free precession experiment utilizing a resonance cell containing water to sense rotational motion. The experimental apparatus demonstrated a sensitivity of 180°/min. Experimental models currently in operation (Fig. 1) utilize gaseous resonance cells in a closed-loop operation (spin generator) and have demonstrated sensitivities in the region of 0.1–0.2°/h. It has been projected that sensitivities in the region of 0.001°/h can be achieved if research efforts continue into the three fundamental parameters of the gyro: signal to noise, relaxation times, and bias instabilities.

Theory of operation. The gyroscope utilizes nuclear species which possess angular momentum, and hence a magnetic moment to sense rotation. The magnetic moment is forced to precess about a preferred direction (the sensitive axis; see Fig. 2). Rotational information is observed as a phase shift in the precessional frequency of the magnetic moment. The implementation of a practical gyro requires additional phenomena: the establishment, sustained precession, and optical detection of a macroscopic nuclear magnetic moment.

Initially a weak dc magnetic field $\vec{H}_0$ is applied to a gaseous ensemble of randomly oriented nuclei. This establishes an extremely small static and thermal equilibrium magnetic moment along the direction of $\vec{H}_0$, defined as the sensitive axis. The establishment of a net macroscopic nuclear magnetic moment is accomplished by a technique known as optical pumping. A beam of circularly polarized light is directed at the ensemble along the direction of $\vec{H}_0$. The absorption and reemission of the light result in an exchange of angular momentum between the light and the ensemble of atoms, establishing a net macroscopic nuclear magnetic moment.

Sustained precessional motion is achieved by applying an ac magnetic field perpendicular to the $\vec{H}_0$ field. This produces a torque which tilts the nuclear magnetic moment $\vec{M}$ away from the direction of $\vec{H}_0$. The nuclear magnetic moment will then precess about the direction of $\vec{H}_0$ owing to a torque exerted by $\vec{H}_0$. The precessional (resonance) frequency is called the Larmor precessional frequency and is given by Eq. (1), where γ is the gyromagnetic ratio.

$$\omega_L = \gamma H_0 \qquad (1)$$

The ac magnetic field then maintains the precessional motion of the nuclear magnetic moment in an orbital plane about the $\vec{H}_0$ field direction.

There are several techniques that can be implemented for sensing the precessional motion of the nuclear magnetic moment. The two techniques presently being used are the Faraday and Dehmelt techniques. The Faraday technique utilizes a linearly polarized readout beam whose plane of polarization rotates back and forth with a modulation proportional to the position of the nuclear magnetic moment as it precesses about the cell. By means of a suitably oriented analyzer, the plane of polarization of the readout beam is transformed into amplitude modulation, which is a measure of the rate of rotation of the nuclear magnetic moment. With the Dehmelt technique, the precessional motion of the nuclear magnetic moment causes variation in the absorption of the circularly polarized light in the resonance cell. The light becomes intensity-modulated as it traverses the cell because of the relative position of the nuclear magnetic moment with respect to the incident light. This modulated signal is proportional to the rate of rotation of the resonance cell.

If vehicle motion is directed about the sensitive axis, a phase shift in the Larmor frequency occurs. This phase shift is equal to the rate of rotation of the vehicle. The shift ω in the Larmor frequency is given by Eq. (2), where ω_R is the rotational rate of

$$\omega = \gamma H_0 + \omega_R \tag{2}$$

the vehicle about the sensitive axis. In a practical gyroscope, the dependency of H_0 must be eliminated because of its stability requirement, 10^{-9} parts per million. This requirement can be eliminated by placing two isotopes into a single sample cell. Two frequencies are then sensed, given by Eqs. (3) and (4). The mechanization is such that a

$$\text{Isotope 1} \qquad \omega_1 = \gamma_1 H_0 \pm \omega_R \tag{3}$$

$$\text{Isotope 2} \qquad \omega_2 = \gamma_2 H_0 \pm \omega_R \tag{4}$$

cancellation occurs and eliminates the H_0 sensitivity, as shown by Eq. (5). The dependency of ω_R on

$$\omega_R = \frac{\omega_1 - \omega_2 \left[\dfrac{\gamma_1}{\gamma_2}\right]}{1 - \left[\dfrac{\gamma_1}{\gamma_2}\right]} \tag{5}$$

the ratio of gyromagnetic ratios is eliminated by employing a dual-cell mechanization. In this mechanization the direction of the $\vec{H}_0$ field in the second cell is opposite to that of the first cell. The nuclear magnetic moments in the second cell will then precess in the opposite direction to those in the first cell. ω_R will then subtract from ω_L as indicated by the negative sign in Eqs. (3) and (4). The rotational rate for this mechanization is then given by Eq. (6), where

$$\omega_R = 0.25\left[(\omega_1 + \omega_2) - (\omega_3 + \omega_4)\right] \tag{6}$$

ω_3 and ω_4 are the rotational rates from the second cell. In principle, the phases ϕ of the precessing nuclei, rather than their frequencies, are compared and used for gyro control and signal processing. The phases are the time integrals of Eq. (2), given by Eq. (7).

$$\phi = \int \gamma H_0 \, dt + \phi_R \tag{7}$$

Fig. 1. A nuclear magnetic resonance gyroscope, the experimental model A-1 (EMA-1). (*Singer Kearfott Division*)

Mechanization. A typical design incorporates one or two resonance cells, pump and readout lamps,

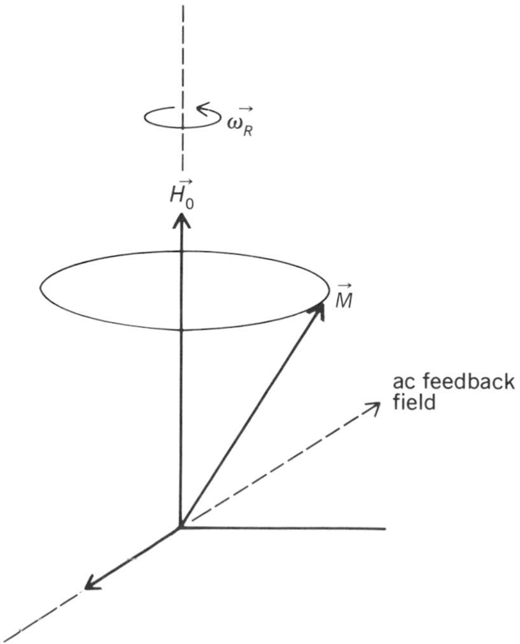

Fig. 2. Forced precession of magnetic moment about $\vec{H}_0$.

optics (Fig. 3), and associated electronics for control and signal processing. Two odd isotopes are encapsulated in the resonance cell, which is fabricated from quartz or Pyrex. For certain isotopes, buffer gases are required in the resonance cell to improve translation time across the cell. These gases also help to reduce sticking at and permeation through the cell wall. The resonance cell is centrally positioned in the magnetic field coil structure. This structure is small because the required magnetic fields are weak. In the earlier liquid resonance cells, large field coil structures were required to produce large-magnitude magnetic fields. Field uniformity problems, electrostatic coupling, and the coil size made this approach impractical for gyro application.

The $\vec{H}_0$ field is generated by a stable current source which is adjusted by signal control electronics to maintain the field constant. The ac or feedback field is achieved by feeding back a portion of the detected signal. This feedback produces a closed-loop oscillator which sustains the precession of the nuclear magnetic moment.

The pump lamp encapsulates a single isotope whose spectral characteristic matches those of the isotopes to be optically pumped in the resonance cell. A high-frequency oscillator ignites the pump lamp and produces unpolarized light. The light is directed by a lens through a linear polarizer and $\lambda/4$ plate to produce circularly polarized light for the optical pumping process. The Dehmelt technique utilizes this circularly polarized light incident at 45° with respect to the $\vec{H}_0$ field. In this manner, a portion of the light is used for optical pumping. The remaining portion becomes intensity-modulated for readout at a photodetector. If the Faraday technique is used, a separate lamp for readout is incorporated into the mechanization. The isotope selected for the readout lamp has spectral characteristics different from those of the isotopes in the resonance cell. This ensures that no absorption of the readout light will occur. The lamp is ignited in the same manner as the pump lamp. The unpolarized light is again directed by a lens through a linear polarizer and then to the resonance cell, where it interacts with the precessing nuclei, resulting in a rotation of its plane of polarization. After translation through the resonance cell, the linearly polarized light is converted into intensity-modulated light by an analyzer. The intensity-modulated light is detected by the photodetector. The signal is amplified, conditioned, and demodulated to produce the correct signal for control and information processing.

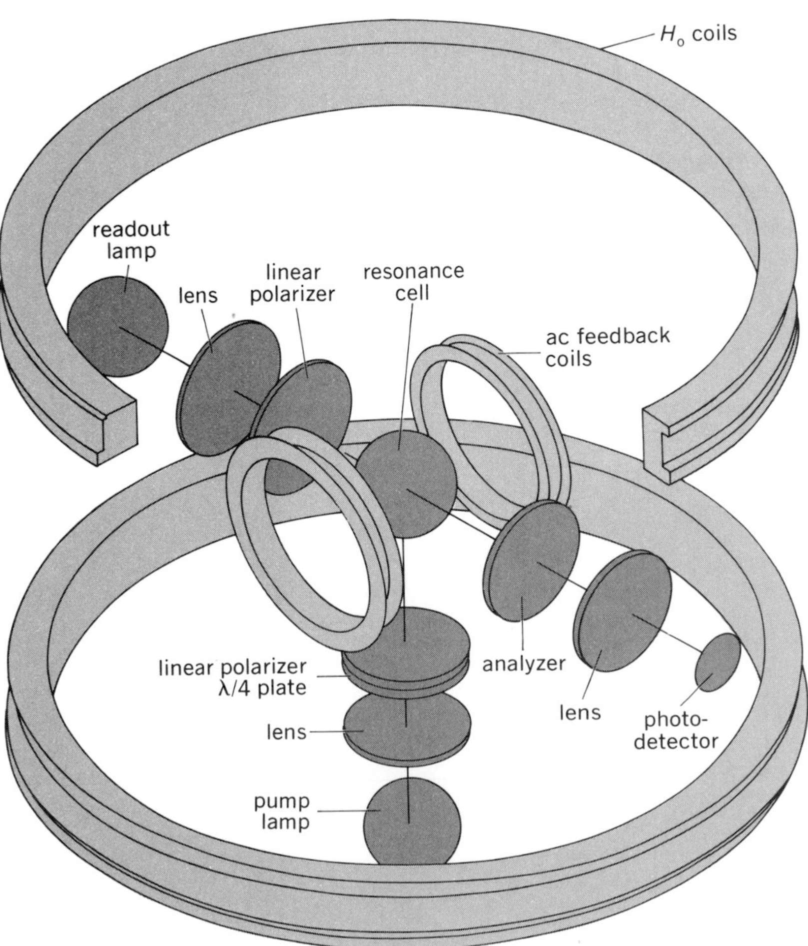

Fig. 3. Nuclear magnetic resonance gyroscope configuration.

Signal-to-noise ratio. The signal-to-noise ratio is a fundamental parameter which affects the noise-induced random drift and the ability to determine drift angle in the NMRG. Improvements in signal to noise are achieved by investigating lamps, optics, and the photodetection and feedback electronics to determine a means for improving signal strength and reducing noise. To this end, efforts are being directed at developing laser diodes for pumping and readout, along with the use of coating to minimize reflective losses from optical components.

Relaxation time. Relaxation time is another fundamental parameter which describes the effective usefulness of the resonance cell. Relaxation time investigations examine the behavioral characteristics of the isotopes in the resonance cell for the purpose of increasing the relaxation time. Two relaxation times are associated with each isotope, longitudinal (τ_1) and transverse (τ_2). The longitudinal relaxation time describes the decay of the macroscopic magnetic moment along the sensitive axis. The transverse relaxation time describes the loss of coherence between the rotating moment and the feedback field. To improve relaxation time efforts are being directed at investigating the interaction of the quadrupole moment with the electric field gradient at the inner wall of the resonance cells.

Rate bias instabilities. Rate bias instabilities are caused by shifts of the precessional frequency of the macroscopic magnetic moment. The investigation into rate bias instabilities is separated into two areas: the shifts associated with the resonance cell and the shifts attributed to instrumentation (electronics and mechanization). The resonance cell investigation is concerned with the interaction of the isotopes with their environment and the subsequent effect of this interaction on the precessional frequency. The instrumentation investigation examines circuit phase shifts, optical, magnetic field, and geometric stability, and their effect on the precessional frequency. Both the resonance cell and instrumentation investigations are directed at determining the causes of phase shifts of the precessional macroscopic magnetic moment and developing a means to control or eliminate the causes of the phase shifts. Specifically, investigations are being conducted into light-induced frequency shifts, beam instability, and resonance cell temperature effects.

For background information *see* GYROSCOPE; MAGNETIC RESONANCE in the McGraw-Hill Encyclopedia of Science and Technology.

[FRANCIS A. KARWACKI]

Bibliography: A. Abragram, *The Principles of Nuclear Magnetism*, 1961; E. Kanegsberg, Proceedings of the SPIE Conference on Laser Inertial Rotation Sensors, vol. 157, pp. 73–80, August 1978; J. Simpson, *Proceeding on Unconventional Inertial Sensors*, p. 202, December 1962; C. P. Slichter, *Principles of Magnetic Resonance*, 2d ed., 1978.

Hydroelectric power

Many new developments in the small hydropower field have appeared during the past 5 years. Some have been brought about by the increasing costs of alternative energy forms (notably fossil fuels); others have been related to continuing advances in technology, and certainly the high cost of money has influenced today's design thinking about lower-capital-cost installations. Parallel to these developments is the significant effect of recent legislation requiring public utilities to purchase power offered to them at least at avoided-cost levels. Tax credits available to homeowners have spurred efforts to conserve energy by improving insulation in both new and older homes. Electric heating is now feasible, and excess power from small hydroelectric plants now has a ready market. The combined effect is a renewed interest in small hydroelectric plants.

Comparative heating costs. Prior to about 1976 homeowners employed electric heating only as a luxury item since resistance heating yields only 3413 Btu/kWh (1 joule of heat energy per joule of electric energy). In the 5-year period ending in 1981 the price of electric fuel for home heating increased dramatically, so that homeowners along the Atlantic coast of the United States may now pay above 6 cents per kilowatt-hour. Thus a modest home requiring 50,000 Btu (52.8 MJ) each hour would consume $21 per day using purchased electric fuel at 6 cents per kilowatt-hour. But new heat-pump technology mated with heat availability from hydraulic turbine exhaust water opens a new area of low-cost heat for rural homeowners.

Fuel oil costs also increased during the 5-year period at an even more rapid rate, from about 36 cents per gallon to $1.36 per gallon (9.5 to 36 cents per liter). The gallon contains about 140,000 Btu (148 MJ), of which only 70%, or 98,000 Btu (103 MJ) on the average, are actually released in the home, the remainder being lost as the result of heating combustion air to room temperature and the large amount of relatively hot gases flowing out the chimney. Thus at 50,000 Btu/h (14.65 kW), oil heat would cost this homeowner about $16.60 per day.

The use of heat pumps has increased the utilization efficiency of electric power for heating. When exhaust of a water turbine is at the homesite, the temperature lift is from 32°F or 0°C (turbine exhaust water temperature) to 72°F (22°C) inside-house temperature. This 40°F (22°C) temperature lift persists even when the outside temperature is zero or lower, since the heat source is turbine exhaust water temperature (not air). Performance coefficients as high as 5 or better can be achieved with such a system.

What this really means is that 5 × 3413 Btu/h (5 kW) may be released to the home for each kilowatt of power consumed. Neglecting the increased capital and maintenance costs, the actual fuel costs would be $4.21 per day versus $21.00 per day for resitance heating and $16.60 for oil heating. This is predicated on having a homesite at (or very near) a site where hydropower may be developed at a reasonable cost (Fig. 1).

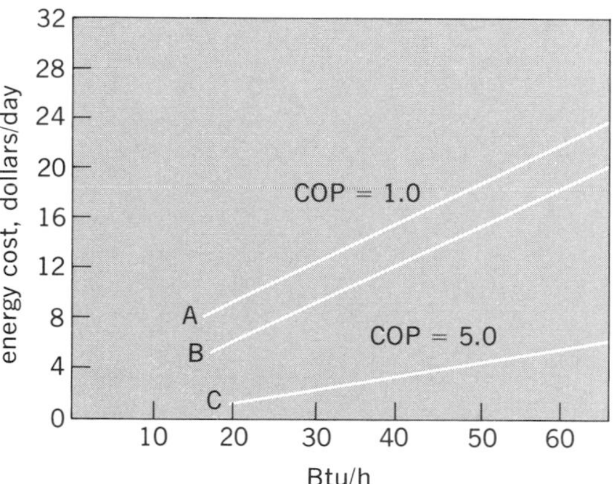

Fig. 1. Coefficient of performance (COP) for electric heating (curve A) versus oil heating with and without heat pumps (B). Curve C shows heat pumps from hydrosites.

Plant design. Figure 2*a* shows a very new design of hydroelectric plant that is able to recover usable power from low head locations. Heads may be as low as 4 ft (1.2 m). This turbine is usually built with horizontal inlet and vertical discharge for ease of installation. The location of the diffuser section is upstream of the discharge elbow in order to lower losses in the change from horizontal flow to vertical discharge.

This turbine has a hydraulically operated inlet butterfly control valve. When used with an electronic load diverter, this control valve is placed in full open position and load is placed on the generator to hold the speed at 60 cycles. Thereafter the electronic control maintains that same load and speed through substitution of parasitic (usually resistance) load for any decrease in load demand at the system busbar.

After the water passes through the control valve, its pressure energy is converted to velocity energy by passage through the fixed-inlet vane section. This high-velocity water impinges on the propeller blades of the runner, forcing it to turn even against the resistance of the generator being driven from the turbine propeller shaft. The work done on the propeller results in a decrease in water velocity, with the output power of the shaft being proportional to the change in velocity of the water.

The unique thing about this design is its size. The next-closest tube turbine is about 100 times larger. Now the advantages of the tube turbine design are available to the homeowner and small power plant operator.

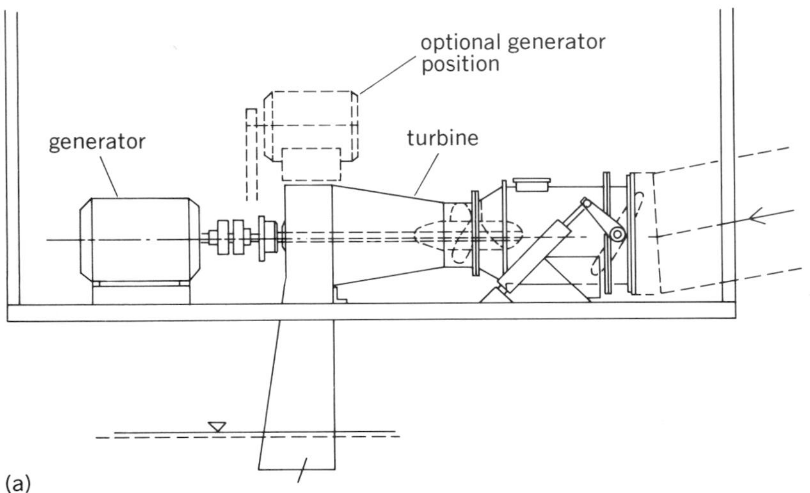

Fig. 2. Small hydroelectric plants. (*a*) Propeller turbine uses fixed-inlet vane tube-type turbine in sizes from 5 to 500 kWh. (*b*) Runner and nozzle design impulse turbine for heads above 20 m and power outputs from 10 to 500 kWh. (*Peltech*)

For many years the impulse or Pelton-type turbine (Fig. 2*b*) has been used for high head (above 50 ft or 15 m) applications. Many have been installed for heads up to 1500 ft (450 m) or more. Again, the head (pressure) is converted to velocity in a jet not unlike a garden hose nozzle or nozzles used on fire hoses. The velocity produced is equal to $\sqrt{2gh}$, where g = acceleration due to gravity (32.2) and h is the head in feet available at the site. Velocity calculated from this equation is in feet per second (or meters per second).

Originally this turbine was produced by bolting semicircular buckets to the rim of a wheel and directing the flow from the nozzle into these half-moon–shaped buckets. It can readily be seen that with a given nozzle velocity, revolutions per minute will increase as the rim diameter is reduced. Bucket velocity will be half of nozzle velocity.

The unique aspect of the new design is a change in the shape of the bucket from semicircular to semioval. The wider bucket can receive much more water from a larger nozzle discharging into a reduced diameter. Many of these turbines are belt-driven for safety reasons and not for speed increase.

Several other technical advances appear in this new design. Again, electronic load diversion may be used for constant speed control with varying bus bar load demand. Wheel stresses are reduced in case of runaway due to reduced runner diameter. Electrically operated spill valves are placed in the line to each nozzle to waste the water supply to the nozzles whenever any unsafe condition of speed, voltage, oil pressure, or load exists. These units are difficult to damage, even in unattended operation. New technology has thus made turbines smaller, safer, easier to handle, and less costly.

Hydropower developments. There are several new developments in hydropower generally. Using or needing a dam to create head and storage volume is no longer a requirement for a feasible hydrosite. A simple diversion ditch or power canal may be dug along the side of a hill at lesser gradient than that of the flowing stream. When sufficient difference in elevation is developed between the water in the canal and the water in the stream bed below, the canal water may be discharged via a penstock through the hydraulic turbine back to the stream bed. It is safe to say that had today's construction machinery (notably the backhoe) been available to constructors in the early 1900s, many of the 49,000 dams in the United States would not have been built. Even today, fewer than 2500 of these dams are being used to generate power. Though many unused dams do exist, greater opportunity exists for many landowners along small rural streams in America and elsewhere. Over 200 such systems have been designed by K. Grover for Third World nations throughout the world. China reports 80,000 small units in operation, with only a few using dams (Fig. 3).

The concept of needing a speed governor costing several thousands of dollars has also been outdated by technology. When head and flow to the turbine remain unchanged, the turbine speed will remain unchanged if the power demand is held constant. Thus the turbine is always operated at full load and full speed by electronically diverting unwanted power to water heating or hydrogen producing (fuel), or simply wasting power to resistance elements submerged in the stream bed. Even full-output diversion will warm the water only a very few degrees, given the relatively large flows discharged by the turbine.

Run-of-stream hydro stations. Stream flows vary widely from hour to hour and over longer time cycles. Run-of-stream hydro stations whose output is to be used for seasonal heating should be sized for minimum expected seasonal flows, even though it means that there will nearly always be excess water wasted even at well-chosen sites. There is little reason to pay for expensive high-efficiency turbines only to have heat energy going to waste at a site. Yet durability need not be sacrificed when efficiency is disregarded—quite the contrary, since the elimination of governors, movable gates and linkage, and so on, makes the turbine now as durable and maintenance-free as large water pumps, many of which have operated a quarter century and longer in municipal water plants.

It is easily seen that small hydro stations do not need governors or top efficiency for absolute minimum water consumption, and can in fact use converted water pumps running in reverse direction as excellent drivers for power generators. Pump manufacturers are becoming increasingly interested in converting pumps for turbine duty. Any single-stage centrifugal pump will produce power when operated in reverse as a turbine, provided any mechanical problems associated with reverse rotation are recognized and handled accordingly. When properly applied; it will yield more power as a turbine than it consumed as a pump.

Perhaps most significant is the change in attitudes toward energy during the past 5 years. Both government regulatory agencies and utilities have changed their stance from resistance to assistance. A public power utility whose return is based on investment has little incentive to purchase power at avoided cost if no increased investment appears in its rate base.

Under the Public Utilities Regulatory Policies Act (PURPA), utilities do purchase power from small hydro producers. In part this is due to public pressure, and in part due to the forthcoming shortage of conventional energy forms, but certainly it is a significant, and welcome, change in attitude.

Many energy reports exist which do not even discuss recoverable hydropower from stations in the 10- to 100-kW range. Yet, from experience in 20 third-world countries, it is estimated that well over 100,000 locations in the United States could develop 50 kW or more and several times that number in very small sizes. While this may still be less than 5% of the total energy requirement, it is a step in the right direction since the costs are minimal (about $1000 per kilowatt, installed); the dollars are spent within the economy, and hydropower availability is everlasting.

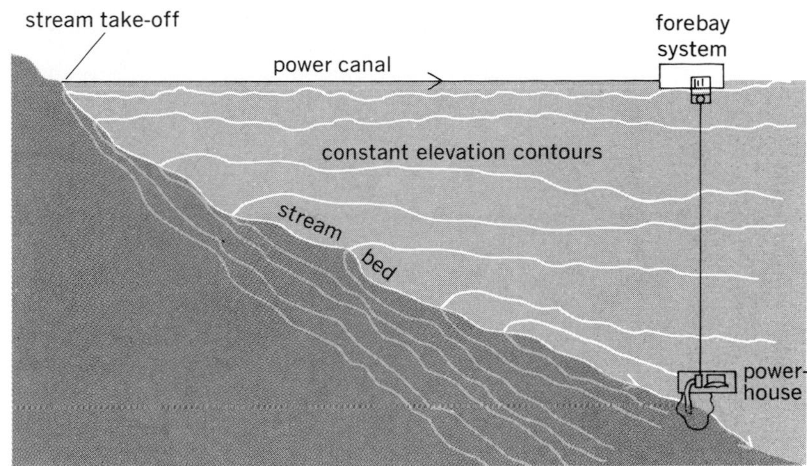

Fig. 3. Power canal for hydropower generation. At stream takeoff there is an offtake canal at 90° to stream flow, a stop log frame to stop flow in canal, and a settling basin with 1/2 ft/s (0.15 m/s) velocity at canal inlet. Velocities in the power canal are up to 2 ft/s (0.6 m/s). The forebay system, with velocities of 1/4 ft/s (0.075 m/s), includes a trash rack assembly, a head gate assembly, and a penstock. The powerhouse contains a water-to-wire turbine-generator package.

Single-stage centrifugal pumps make excellent drivers for electric power generators at a fraction of the cost of hydraulic turbines. The secret is knowing how to choose an inexpensive pump turbine for the location. Heating of living and working spaces is an ideal application for energy utilization since the heating season and best water flow season coincide in a major part of the United States. Substitution of low-cost hydroenergy for high-cost fossil fuels is the most economical approach to high heating costs in many rural areas where small hydro systems can be economically constructed. Once the hydro heat system is installed, its low output cost is noninflating since the energy is free.

New technology offers hydroelectric systems capable of recovering energy from water differentials as low as 4 ft (1.2 m). This is the type of substitute energy source that may be developed by the stream site owner with a minimum of outside assistance. It was done in the early 1900s, and it must be done again.

For background information *see* HYDROELECTRIC GENERATOR in the McGraw-Hill Encyclopedia of Science and Technology. [KEN GROVER]

Hydrogen, spin-polarized atomic

Under ordinary circumstances pure hydrogen exists only as a diatomic molecule (H_2). In this form its properties are quite similar to those of the heavy rare gases (neon, argon, krypton, and xenon). Recently it has been shown experimentally that a system which consists of hydrogen atoms, not molecules, can be prepared by cooling hydrogen to a very low temperature in a very high magnetic field. Because this magnetic field is so high, the spins of the electrons in almost all of the hydrogen atoms will be antiparallel to the magnetic field. For this reason, the system is called spin-polarized atomic hydrogen (H↑). In contradistinction to H_2, H↑ is expected to exhibit quantum properties on a macroscopic scale. In particular, it is predicted to be gaseous at all temperatures, even at absolute zero, except at pressures sufficiently high to cause it to solidify; it is predicted to have no liquid phase. Further, it is predicted that this gaseous phase will become superfluid at a sufficiently low temperature and will therefore exhibit many properties that have heretofore been observed only in liquid helium-4. It is thus appropriate to call H↑ a quantum gas.

Interaction between hydrogen atoms. The interaction between two hydrogen atoms depends strongly on the electronic spin state of the two atoms; that is, whether they are in the singlet or triplet spin state. If the two atoms are in the singlet spin state, the interaction is strongly attractive so that the two atoms can form a bound state with a binding energy of approximately 4.5 electronvolts; this bound state is the hydrogen molecule. If, however, the two hydrogen atoms are in the triplet spin state, their interaction is strongly repulsive for internuclear distances less than 0.37 nanometer and is weakly attractive for larger distances. Since a hydrogen atom has such a small mass, this triplet-state interaction is too weak to produce a bound state between two hydrogen atoms. Thus, if a system of hydrogen atoms can be maintained in a state such that no pair of hydrogen atoms is in a singlet electronic state, there will be no hydrogen molecules formed. Thus, H↑ is a collection of hydrogen atoms constrained (for example, by a large magnetic field at a low temperature) so that each pair of hydrogen atoms interacts only via the triplet interaction.

Preparation of H↑. It is difficult to prepare H↑ because two hydrogen atoms react so readily to form a hydrogen molecule. It turns out that it is possible to slow down this reaction by coating the walls of the container with a film of liquid helium-4. This works because a hydrogen atom is only weakly bound to a helium-4 film (the binding energy is approximately 1 K). Thus, for temperatures above a few tenths of a kelvin, the hydrogen atoms do not spend enough time on the helium-4 surface to recombine.

A typical scheme for preparing H↑ involves several steps. First, a nonpolarized flow of hydrogen atoms is generated by dissociating the molecules in a flow of hydrogen gas. Next, this flow is cooled to approximately 4 K by passing it through a Teflon-coated tube in contact with a 4 K temperature bath. Finally, the flow is passed into a sample chamber whose walls are coated with a helium-4 film, all of which is in a large magnetic field (approximately 10 teslas). The effect of the field is to attract into the sample chamber those atoms whose spins are antiparallel to the field, and to repel those atoms whose spins are parallel to the field. This is the point at which the atoms are polarized. The first workers to succeed in preparing H↑ were I. F. Silvera and J. Walraven at the University of Amsterdam.

Hyperfine states and rate processes. For hydrogen, as opposed to deuterium, there are four hyperfine states. In a strong magnetic field these states are the so-called mixed and pure states. In the pure state, both the electron and proton spins are antiparallel to the field. The mixed state is a superposition of two states, one in which the electron spin is antiparallel to the field with the proton spin parallel to the field, and one in which the electron spin is parallel and the proton spin antiparallel; this latter state is admixed with a small amplitude given approximately by $0.025/B$, where B is the external magnetic field in teslas.

It turns out that recombination between two hydrogen atoms in a large magnetic field occurs only if one of them is in the mixed hyperfine state. Hence, the presence of the magnetic field reduces the recombination rate below that experienced in zero-field by a factor of $[0.025/B]^{-2}$, which is approximately 10^5 if $B = 10$ T. This is another reason why it is possible to prepare H↑.

There are three rate processes shown below which

$$\mathrm{H + H + X \rightarrow H_2 + X} \quad (1)$$

$$\mathrm{H + H + surface \rightarrow H_2 + surface} \quad (2)$$

$$\mathrm{H\ (pure) \rightarrow H\ (mixed)} \quad (3)$$

are important for spin-polarized atomic hydrogen (where X is any atom or molecule). The table shows characteristic times for these processes for various densities at 0.3 K in a magnetic field of 10 T. For densities less than 10^{16} atoms/cm^3, the system will last for more than a quarter of an hour. At densities about 10^{18} atoms/cm^3, process (3) becomes the rate-determining process, so that at this and higher densities all atoms in the mixed state will recombine and only atoms in the pure state will remain. They will decay slowly into atoms in the mixed state, which will rapidly recombine to form H_2.

Characteristic times for the three important rate processes in spin-polarized atomic hydrogen for three densities at a temperature of 0.3 K in an external magnetic field of 10 T

Density, atoms/cm^3	Characteristic time, s		
	Process (1)	Process (2)	Process (3)
10^{16}	5×10^5	10^3	2×10^4
10^{18}	50	10	2×10^2
10^{20}	5×10^{-3}	10^{-1}	2

Quantum statistics. It is predicted that H ↑ will obey Bose-Einstein statistics, whereas spin-polarized atomic deuterium (D ↑) will obey Fermi-Dirac statistics. A hydrogen atom is a tightly bound system of two fermions (one electron and one proton), whereas a deuterium atom is a tightly bound system of three fermions (one electron, one proton, and one neutron). The reasoning is the same as that for the isotopes of helium. Helium-4 consists of an even number of fermions and has been shown experimentally to obey Bose-Einstein statistics, whereas helium-3 consists of an odd number of fermions and obeys Fermi-Dirac statistics.

Effect of the surface. A hydrogen atom is bound to the surface of a helium-4 film with a binding energy of approximately 1 K. There is only one bound state, so that the hydrogen atoms will form a monatomic layer on the surface of the helium-4 film. At a sufficiently low temperature, it is predicted that the hydrogen atoms will preferentially fill up this monatomic layer before forming the bulk gas. Thus, the balance between the fraction of the atoms on the surface and those in the bulk depends sensitively on temperature. Surface recombination [process (2)] is faster than bulk recombination [process (1)] at densities of less than 10^{18} atoms/cm^3, so that the lifetime of H ↑ also depends sensitively on the temperature. It is thought that the decay of the pure state into the mixed state will not be severely affected by the surface.

Properties of H ↑. Since it appears that it will be possible to prepare H ↑ only in the pure hyperfine state, only the properties of bulk or a monatomic layer of atoms in this state will be considered. Both the bulk and the monatomic layer are predicted to be gaseous unless placed under sufficiently high pressures to cause them to solidify. Neither system is predicted to have a liquid phase; that is, there will be no liquid-gas coexistence curve and no critical point in either system.

Both the bulk gas and the monatomic-layer gas are predicted to exhibit superfluid transitions. In the bulk this transition is expected to be closely related to the Bose-Einstein condensation. The low-temperature phase is expected to be a superfluid with properties like those exhibited by superfluid liquid helium-4. This transition is predicted to take place at a density of approximately 10^{19} atoms/cm^3 at 0.3 K. Given the lifetimes shown in the table, it will be extremely difficult, if not impossible, to observe the Bose-Einstein condensation in gaseous H ↑.

The monatomic-layer gas is also predicted to exhibit a transition to a superfluid phase, an example of a Kosterlitz-Thouless transition. This transition is predicted to occur at a density of 10^{13} to 10^{14} atoms/cm^2 at a temperature of 0.1 or 0.2 K. Experimental observation of this transition is thought to be much more likely than that in the bulk gas.

If either the Bose-Einstein condensation in the gas or the Kosterlitz-Thouless transition in the monatomic layer can be observed, it will be an important advance toward developing an understanding of the properties of quantum systems which obey Bose-Einstein statistics. Further, the study of these transitions may provide clues to long-standing open questions regarding these transitions.

For background information *see* LIQUID HELIUM; QUANTUM STATISTICS; SUPERFLUIDITY in the McGraw-Hill Encyclopedia of Science and Technology.

[LEWIS H. NOSANOW]

Bibliography: I. F. Silvera and J. Walraven, *Sci. Amer.*, 246(1):66–74, 1982.

Immunology

In the early 1940s K. Landsteiner and M. W. Chase first demonstrated the transfer of immune functions in guinea pigs by injection of live lymphocytes from an immune donor to a nonimmune recipient. In 1949, H. S. Lawrence demonstrated that, in humans, the skin reactivity to PPD (protein purified derivative of tuberculin, an antigenic component of *Mycobacterium tuberculosis*) could be transferred by using viable leukocytes from a sensitive normal donor to a nonsensitive normal recipient. Since PPD reactivity is mediated by T lymphocytes, involved in cell-mediated hypersensitivity processes, it was obvious that the transfer of viable leukocytes was somehow able to influence the recipient's T lymphocytes, so that they would become hypersensitive to PPD, an antigen to which the recipient had not been previously exposed.

Dialyzable leukocyte extracts. In 1955 Lawrence showed that dialyzable extracts of leukocytes from sensitive donors were as effective in transferring skin sensitivity to PPD as viable leukocytes. This was an extremely significant discovery, since the transfer of dialyzable extracts, containing substances that by their ability to cross a conventional dialysis membrane must be of very small size (molecular weight below 10,000 daltons), is practically innocuous, without the risks of graft-versus-host reactivity that are implied in the transfer of viable leukocytes, or of antigenic stimulation, since very small molecules are not able to induce antibody formation. Besides low molecular weight, other important physiochemical characteristics of the activity of dialyzable leukocyte extracts are their heat lability and resistance to enzymes such as trypsin (that destroys protein molecules), DNase (that destroys nucleic acids), and some (but not all) RNases.

Transfer factor. Lawrence and coworkers proposed the term dialyzable transfer factor for the factor contained in dialyzable leukocyte extracts that was responsible for the transfer of specific delayed-type (cell-mediated) hypersensitivity processes. This type of activity was the only one known until culture assay methods were developed for these extracts. Studies with culture assays developed in the late 1970s showed, not surprisingly, that dialyzable leukocyte extracts are very heterogeneous and contain, at least, a nonspecific adjuvant fraction, able to stimulate the cell-mediated responses against a diversity of antigens, and an antigen-specific moiety that would be responsible for the transfer, for example, of PPD hypersensitivity from a sensitive donor to a nonsensitive recipient. The term transfer factor is reserved for this last moiety.

Transfer factor is believed to be a "derepressor" of the immune function of normal lymphocytes; however, it is not known whether this involves the transfer of an informational molecule from donor cell to recipient or simply the presentation of a "trigger" or "derepressor" which allows the recipient cell to synthesize and liberate molecules of similar character. Many questions remain, such as the precise structure of transfer factor; whether it can induce RNA production in the recipient cells; whether it triggers a metabolic cascade leading to immune reactivity; and the intracellular target of transfer factor.

In spite of all these questions, countless studies attest to the fact that transfer factor transfers specific cell-mediated immunity. This is usually verified by skin testing with several different types of antigens, such as PPD, coccidiodin (obtained from a pathogenic mold, *Coccicioides immitis*), streptokinase-streptodornase (obtained from *Streptococcus* species), inactivated mumps virus, and keyhole limpet hemocyacin. The transfer factor is derived form leukocytes obtained from donors showing reactivity to these antigens and administered to individuals showing no reactivity; a few days later, a repeat skin test of the recipients will show positivity. More important is the fact that dialyzable leukocyte extracts containing transfer factor activity have been used efficaciously in the treatment of immuno-deficiency states, disseminated infections, and malignancies.

Clinical use of transfer factor. Transfer factor has been used in so-called broad-spectrum immunodeficiency states—situations in which the patients suffer from repeated infections caused by a wide variety of microorganisms, particularly when due to defects of cell-mediated immunity. The best results have been obtained in the Wiskott-Aldrich syndrome, a disease characterized by eczema, bleeding due to low platelet counts in the blood, and recurrent infections with polysaccharide-containing bacteria, such as pneumococcus, meningococcus, and *Haemophilus influenzae*, against which these patients are unable to form antibodies. Clinical improvement is observed in 50% or better of patients treated with dialyzable leukocyte extracts.

The administration of dialyzable leukocyte extract has also proved beneficial in patients with antigen-specific immunodeficiency, that is, patients with recurrent or persistent infections with one given antigen, particularly in coccidiodomycosis (a systemic infection due to the mold *Coccidioides immitis*), progressive primary tuberculosis, leishmaniasis, and infections with atypical mycobacteria. Of particular interest are the studies in patients with acute leishmaniasis and *Mycobacterium fortuitum* infection.

Leishmaniasis. In acute cutaneous leishmaniasis (a parasitic infection of the skin) a controlled therapeutic trial was undertaken to determine whether antigen-specific transfer factor or nonspecific adjuvant factors contained in dialyzable leukocyte extracts were primarily responsible for the patient's improvement. One group of patients was treated with a placebo (physiological saline), a second group with nonspecific dialyzable leukocyte extract obtained from donors with no history of exposure to the parasite, and a third group with specific transfer factor obtained from patients who had previously recovered from leishmaniasis infections. None of the patients who received placebo showed improvement. Of 13 patients treated with nonspecific dialyzable leukocyte extracts, only 3 showed healing of the lesions, and all of them had positive skin tests to the parasite (indicating the existence of cell-mediated immunity) prior to therapy. Strikingly, 18 patients treated with specific transfer factor showed marked improvement with complete healing by 12 weeks.

Mycobacterial lung infection. In a case of lung infection by *Mycobacterium fortuitum* (a bacteria closely related to the agent of tuberculosis) the use of dialyzable leukocyte extract obtained from donors sensitive to *M. tuberculosis* PPD but not to *M. fortuitum* PPD failed to induce clinical improvement, while administration of dialyzable leukocyte extract from a donor highly sensitive to *M. fortuitum* PPD and not to *M. tuberculosis* PPD resulted in a dramatic clinical improvement, with disappearance of all evidence of infection. Similarly, in cases of chronic infection of the skin and mucosal membranes by the mold *Candida albicans*, clinical improvement is only evident when dialyzable leukocyte extracts from donors immune to this agent are used.

Coccidioidomycoses. The importance of using specific transfer factor from donors showing strong immunity against the causative agent of the disease has also been exemplified in animal experiments, using cows and mice. The animals were infected with 50% of a lethal dose of the mold *Coccidioides immitis*, and protection was attempted both with nonspecific dialyzable leukocyte extracts (obtained from animals not previously exposed to this microbe) and with specific dialyzable leukocyte extract (obtained from animals previously exposed to *C. immitis*). The mortality was reduced from 90 to 10% only in those animals treated with specific leukocyte dialyzable extracts that can be assumed to contain specific transfer factor.

Conclusions. The above results suggest that the antigen-specific transfer factor moiety of dialyzable leukocyte extracts is clinically useful. Efforts to isolate and characterize structurally this moiety, having in sight the possibility of synthesis in culture, and to clarify its mechanism of action are under way. The skepticism that has surrounded the action of transfer factor has slowly been replaced with the belief that this may be one of the most potentially useful weapons in immunotherapy, given its almost total innocuity and the unquestionable beneficial effects that have been obtained with it. But the full clinical potential of transfer factor will be realized only when the basic questions concerning its structure and mechanism of action are answered and methods to obtain large amounts of well-standardized product become available.

For background information *see* ANTIBODY; ANTIGEN; CELLULAR IMMUNOLOGY; IMMUNOLOGY; SKIN TEST in the McGraw-Hill Encyclopedia of Science and Technology. [H. H. FUDENBERG; G. VIRELLA]

Bibliography: K. Landsteiner and M. W. Chase, *Proc. Soc. Exp. Biol. Med*, 49:688, 1942; H. S. Lawrence, *J. Clin. Invest.*, 34:219, 1955.; H. S. Lawrence, *Proc. Soc. Exp. Biol. Med.*, 71:516, 1949; M. Sharma et al., in S. Khan, C. H. Kirkpatrick, and N. O. Hill (eds.), *Immune Regulators in Transfer Factor*, p. 563, 1979.

Infant botulism

The botulinal nature of what is now called infant botulism was first recognized in 1976. The illness differs from the classically known food poisoning botulism in being restricted to infants and in the source of the causative toxin. Infant botulism is a toxicoinfection in which the toxin is formed by *Clostridium botulinum* multiplying in the intestinal tract, whereas food poisoning is a toxemia in which already formed toxin is ingested with foods. Since infant mice and rats have an age-dependent susceptibility to enteric *C. botulinum* colonization analogous to that of humans, they are used to study infant botulism. The results of these studies indicate the microbial ecological basis of this disease. The toxicoinfection is limited to infants because their intestinal microflora lack the bacteria which make the gut of older individuals unsuitable for growth of the pathogen.

Clinical aspects. Botulinum toxin inhibits the release of acetylcholine at neuromuscular junctions and thereby blocks the conduction of nerve stimuli from the nerve ending to the muscle. Since food poisoning and infant botulism are the effects of the toxin, the illnesses are similar in each—a hypotonia developing as a descending, symmetrical flaccid paralysis.

The illness occurs in infants of normal birth and development. It starts with constipation which lasts through the acute phase and beyond. The infant fails to thrive and is lethargic. Loss of appetite and pooling of oral secretions accompany diminished suck and gag reflexes. When paralysis is more general, marked loss of head control results in a "floppy" baby. The illness can progress to respiratory difficulty and arrest, requiring immediate mechanical ventilation.

Hospital care has been limited to supportive measures, the extent depending on severity of the illness and complications. Botulinum antitoxin is not used. Unless sepsis is proved, antibiotics are not recommended; transitory aggravation of existing paralysis can follow soon after aminoglycosides are administered. Convalescing patients can be released from the hospital while excreting the organism and toxin.

Infant botulism cases range from essentially asymptomatic to those which become life-threatening within a few days. Except for a very few, almost all cases are infants between 1 and 6 months of age, with the peak incidence occurring in 2- to 4-month-old infants. This peak coincides with the characteristic age incidence of the sudden infant death syndrome. Additionally, *C. botulinum* has been found in autopsy specimens from 10 (toxin also present in 2) of 280 sudden infant deaths, but not in similar specimens from 68 control infants whose deaths could be attributed to specific causes. The observations suggest that a rapidly lethal form of infant botulism is responsible for some of the infant deaths that are now classified as sudden infant deaths.

Toxicoinfection. The laboratory confirmation of infant botulism is the presence in feces of *C. botulinum* and its toxin, which are both excreted during the acute illness and in some instances for weeks during convalescence. Up to 100,000,000 *C. botulinum* organisms have been quantitated in 1 gram of feces, and 366,000 mouse minimal lethal doses of toxin have been found in the stool obtained after a laxative was given. Only multiplication of the organism in the intestinal tract would offer an explanation of the quantities and persistence of organism and toxin.

Epidemiology. Except for one type F, all recorded cases have been caused by *C. botulinum* types A or B. The preponderance of type A cases west and type B east of the Mississippi River corresponds to the known geographical prevalence of the *C. botulinum* types.

Individual infant botulism cases are epidemiologically unrelated, involving only the infant of a family. The failure to find a toxin-containing food as a cause is circumstantial evidence for the toxicoinfectious nature of infant botulism. Since the infection is limited to the gut, the organism (most probably spores) is being swallowed. However, not all the means by which infants acquire the organism are known.

One known source of *C. botulinum* is honey. The feeding of honey and infant botulism cases are statistically correlated, and *C. botulinum* spores are present in this food. Since honey is not a necessary food, feeding it to infants less than 1 year old involves a risk that could be avoided.

Since the illness affects infants fed only mother's milk or formula, honey is not the only vehicle by which infants acquire the organism. These other vehicles are not known, but their *C. botulinum* spores probably originate in the soil, which is a natural habitat of the organism. *Clostridium botulinum* spores of the type that cause infant botulism have been found in the homes of patients in samples such as dust in a vacuum cleaner, soil of a potted plant, and yard dirt.

For background information *see* ACETYLCHOLINE; BOTULISM; FOOD POISONING; TOXIN in the McGraw-Hill Encyclopedia of Science and Technology.

[HIROSHI SUGIYAMA]

Bibliography: S. S. Arnon, *Ann. Rev. Med.*, 31:541–560, 1980; R. A. Feldman (ed.), *Rev. Infect. Dis.*, 1:611–700, 1979; L. J. Moberg and H. Sugiyama, *Infect. Immunity*, 25:653–657, 1979; L. J. Moberg and H. Sugiyama, *Infect. Immunity*, 29:819–821, 1979.

Insect physiology

Hormones are blood-borne chemical messengers which are secreted into the bloodstream by endocrine tissues; they regulate many aspects of development and physiology in both vertebrates and invertebrates. The endocrine system often complements the action of the nervous system, the other major communication system in multicellular organisms. Hormones, however, frequently exert a slower, more sustained action on a target tissue than does neuronal regulation. The neural-endocrine interaction extends to the fact that large, highly specialized cells in the nervous system called neurosecretory cells are sites of hormone production and under neural control which is exerted in response to environmental or physiological cues. Hormones may also modify the activities of the nervous system directly, and thereby have a profound effect on behavior.

A hormone may alter or induce a behavior in more than one way. When it has a direct action on the nervous system, it is said to act as a releaser. Rather than having a rapid, direct effect, a hormone may induce a behavioral change indirectly by affecting the activity or development of a nonneural target or possibly the development of the nervous system. This type of endocrine influence, often called a primer or modifier effect, is usually not immediate, requiring the intervening response of the nonneural target or a relatively long-term developmental change in the nervous system.

The insect nervous system is simple in comparison with that of the vertebrate. In particular, there are considerably fewer cells per gram body weight in the insect central nervous system. It is also somewhat decentralized, having a number of ventral ganglia, which are sufficient to control some behaviors, along the ventral nerve cord in addition to the main cerebral ganglia that make up the brain. Most insect behavior is highly stereotyped and programmed into the neural circuitry of the brain or the ventral ganglia. Many behavioral patterns seem to be genetically determined and merely released by appropriate environmental or physiological cues. Yet insects are capable of amazingly complex behavior. Foraging worker honeybees, for example, can learn some kinds of mazes more quickly than the average laboratory rat. Monarch butterflies accomplish a round-trip migratory flight each year from Canada to their overwintering sites in the mountains north of Mexico City, a feat equivalent to the flights of many migratory birds. One theory which has been advanced to partially explain the rather remarkable behavioral repertoire of insects suggests that interaction between the neural and endocrine systems imparts an added dimension of complexity and flexibility to the insect nervous system which is not obvious from the neural anatomy alone. The endocrine effects on insect behavior are many and varied, often differing between even closely related species.

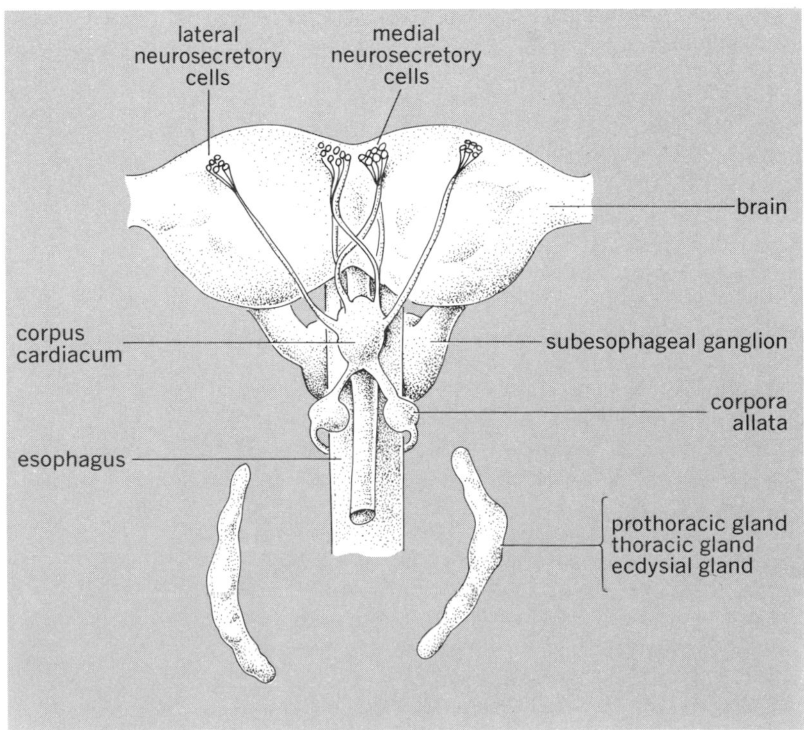

Fig. 1. Primary elements of the insect neuroendocrine system.

Insect endocrine system. Before discussing hormonal control of insect behavior, it may be useful to review the major components of the insect endocrine system (Fig. 1). The primary endocrine glands are the corpora allata (small, primitively paired glands which lie behind the brain, connected to it by neural connectives) and the prothoracic glands (usually situated in the anterior part of the thorax). The corpora allata produce juvenile hormone, a sesquiterpene derivative which determines the type of molt which will take place as the insect grows to maturity (if juvenile hormone is present in high titers at the appropriate time preceding a molt, the molt will be to the pupal stage if one exists, and if a molt is stimulated in the absence of juvenile hormone, it will result in the development of an adult). In response to the prothoracicotropic hormone produced by brain neurosecretory cells, the prothoracic glands produce the steroid hormone ecdysone. Ecdysone is a growth hormone as well as the primary initiator of the molting process. Although the prothoracic glands degenerate after the adult molt, ecdysone may be produced in the adult by the ovary or other tissues in relation to various aspects of vitellogenesis in some species.

The main neuroendocrine components of the insect endocrine system are the brain neurosecretory cells, which occur as small, heterogeneous clusters of cells in the medial and lateral areas of the forebrain; and the corpora cardiaca, a pair of retrocerebral glands which lie between the brain and the corpora allata, receive the axons of many of the brain neurosecretory cells and act as neurohemal organs. The corpora cardiaca also contain intrinsic neurosecretory cells which produce hormones. The brain–corpora cardiaca neuroendocrine complex, which has often been compared to the hypothalamic pituitary axis of vertebrates, produces hormones which regulate many aspects of physiology and behavior, including prothoracic gland activity, cuticle tanning, lipid mobilization, blood sugar levels, heartbeat rate, water balance, mating, eclosion, oviposition, and calling behavior. Neurosecretory cells are also found in many of the other ganglia of the central nervous system and release their products through paired, segmentally arranged, neurohemal organs called perivisceral or parasympathetic organs.

Hormonal control of reproductive behavior. As in the vertebrates, some of the earliest demonstrations of the hormonal control of behavior in insects involved changes in reproductive behavior. Aspects of reproductive behavior which will be discussed here are female mating behavior, mosquito biting behavior, calling behavior, oviposition behavior, and male reproductive behavior.

Female mating behavior. Females often undergo several clear behavioral shifts associated with onset of reproductive development. This is especially true in insects with a long-lived adult whose life cycle may or may not include adult diapause under some environmental conditions. In such insects the newly emerged or sexually immature female is frequently unreceptive to the male's attempts at copulation. As she becomes sexually mature, receptivity develops. After mating, it is common to observe an abrupt switch in behavior again to an unreceptive state associated with the onset of oviposition behavior. The cycle may be repeated several times. Hormonal control has been demonstrated for each of these switches in female reproductive behavior, although not necessarily for all components of the cycle in a given organism.

Although juvenile hormone is responsible for maintaining the juvenile morph in the immature insect, the corpora allata become active again after the adult molt, and are necessary for some aspect of ovarian development in most insects. It is perhaps not surprising, therefore, that juvenile hormone is involved in the control of reproductive behavior as well. Franz Engelmann discovered in 1960 that female *Leucophaea maderae* (a cockroach) would not mate if the corpora allata had been removed. Allatectomized females, though they were vigorously courted by males, refused to extrude their ovipositors so that the male could accomplish the necessary connection. Reimplanting loose corpora allata restored mating behavior to most allatectomized females. It was later shown that other species of cockroaches, in particular *Byrsotria fumigata*, display a similar lack of mating behavior after allatectomy because allatectomized females fail to produce a female pheromone and are therefore not attractive to or courted by male roaches. If such females are put in a container with pheromone-treated filter paper, they will be courted and will mate readily. It is interesting that juvenile hormone is necessary for normal sexual behavior in both species, but the control is accomplished in different ways.

A similar study was done in the housefly in which allatectomy immediately after adult emergence inhibited both ovarian development and mating. Allatectomized females were mounted by the male flies, which suggests that production of the female pheromone was not stopped by allatectomy, but females refused to extrude the ovipositor to make contact with the male claspers. Similarly in the grasshopper, *Gomphocerus rufus*, the change from primary defensive behavior to receptivity that normally occurs in females at sexual maturity can be prevented by allatectomy and restored by corpora allata implantation into allatectomized females.

Mosquito biting behavior. A blood meal is necessary for oogenesis in many species of mosquitoes. The meal initiates secretion of a neurohormone from the corpora cardiaca which stimulates the second stage of vitellogenesis. The female then discontinues biting or host-finding behavior until after oviposition. It was recently shown that juvenile hormone induces biting behavior in at least two species of mosquitoes. Rising titers of the hormone at the beginning of oogenesis apparently stimulate blood feeding as well as the first stage of ovarian development.

Recent studies suggest that juvenile hormone is not the only hormone involved in feeding behavior. In *Aedes aegypti* and *Anopheles freeborni*, ovaries with developing eggs secrete a hormone that suppresses host-seeking or biting behavior between gonotrophic cycles. Whether this ovarian hormone decreases juvenile hormone synthesis to prevent biting is not known but is a possibility.

Changes in the antennal chemosensory neurons such as the lactic acid–excited neurons (lactic acid is a normal host attractant) occur in response to transfusion of hemolymph from gravid female mosquitoes to blood-starved females, indicating that hormonal changes which alter host-seeking behavior may do so by an effect on the peripheral nervous system.

Calling behavior. Adult silk moths live only a few days. They emerge from the cocoon with fully developed eggs, mate, oviposit, and then die usually within 7–10 days. Juvenile hormone is not necessary for oogenesis and does not affect mating behavior of either sex. Yet there is hormonal involvement in reproductive behavior, for when a virgin female silk moth produces her sex pheromone to attract a mate, she assumes a calling posture, extending her last two abdominal segments and thereby exposing glands which produce and release the sex pheromone. The calling behavior is induced by a hormone produced by the intrinsic neurosecretory cells of the corpora cardiaca. The target for the calling hormone appears to be the abdominal nervous system because isolated abdomens from virgin *Hyalophora cecropia* females respond to injection of blood from calling females or corpora cardiaca extracts within 5–10 min by an extension of the terminal abdominal segments in the typical calling position. This behavioral response is turned off in mated females apparently by a hormone from the mated bursa copulatrix (sperm receptacle in the female) since implantation of bursae from mated females into virgin females eliminates the calling response to corpora cardiaca extracts. Like many of the behavioral effects of corpora cardiaca hormones, this seems to be a true releaser action, triggering a predetermined motor program which results in calling.

Oviposition behavior. The act of mating may bring about subsequent changes in female behavior such as lack of receptivity to further mating attempts and onset of oviposition activity. In some cases, as the mosquito *Aedes aegypti*, these changes are caused by substances from the male accessory glands introduced with the spermatophore at mating. In other insects such changes seem to occur by way of the female endocrine system. In many insects histological examinations have shown a release of neurosecretory material at the time of oviposition which has been taken to indicate a causal relationship. Actual physiological studies have been done in only a few species but have revealed the involvement of a variety of humoral factors.

In the desert locust, blood or the corpora cardiaca from ovipositing females cause oviposition movements in virgin females. Electrical stimulation of the central nervous system or enforced hyperactivity causes release of neurosecretory material and induces oviposition. Neurosecretory cells in the brain, glandular lobes of the corpora cardiaca, and the thoracic ganglion seem to be involved. Similarly, in the stick insect, *Carausius*, extracts of neurosecretory cells of the brain, subesophageal ganglion, and the thoracic and first four abdominal ganglia have direct myotropic effects on the ovarioles, an effect which can be mimicked by serotonin.

In the silk moth, *Hyalophora cecropia*, intrinsic cells of the corpora cardiaca are involved in oviposition behavior. It seems to be the presence of sperm in the female tract which initiates the mating effect. Apparently, interaction of sperm with the wall of the bursa copulatrix (seminal receptacle of the female) causes release of humoral factors which in turn cause release of an oviposition-stimulating hormone from the corpora cardiaca.

A similar situation seems to prevail in the bloodsucking bug, *Rhodnius prolixus*, in which sperm act on the female spermatheca (sperm reservoir sac) and cause the release of a humoral substance. The spermathecal factor stimulates the medial neurosecretory cells of the brain to release a hormone which in turn can directly stimulate contraction of the ovariole musculature and has a direct releaser action on the phallic nerves.

In the Indian plant bug, *Iphita*, a single copulation often lasts for several weeks, and during this period oogenesis occurs. When the eggs are mature, the female terminates the mating and begins oviposition behavior. A factor produced by the mature ovary apparently acts on brain neurosecretory cells, causing the release of an oviposition-stimulating substance. Implanting two sets of brain neurosecretory cells into mating females rapidly stimulates oviposition behavior or quivering movements of the genital plates, indicating an immediate and probably direct effect on the nervous system or phallic muscles.

Male reproductive behavior. Male reproductive behavior has less frequently been shown to be under hormonal control. However, control does occur, particularly in insects with an adult diapause. For example, allatectomy of male *Schistocerca gregaria* or *Nomadocris septemfasciata*, both locusts which display an adult diapause, results in the absence of sexual behavior and normal yellow coloration in the operated animals. Treatment of operated animals with naturally occurring juvenile hormone I restores normal sexual behavior and coloration. Interestingly enough, treatment with juvenile hormone III, which is thought to be the specific orthopteran juvenile hormone, has virtually no effect as replacement therapy on either sexual behavior or the development of the yellow coloration (Fig. 2). In the migratory locust, *Locusta migratoria*, both the brain neurosecretory cells and the corpora allata seem to be necessary for full expression of normal sexual behavior. Similarly, allatectomized male dung flies do not court females, nor do males of several species of Hemiptera in which the corpora allata seem to be necessary for the normal development of male sexual behavior.

In most cases the hormonal effects on mating behavior in both males and females seem to be modifier effects in which the corpora allata are necessary for the development of normal sexual behavior. In male praying mantises and cockroaches, however,

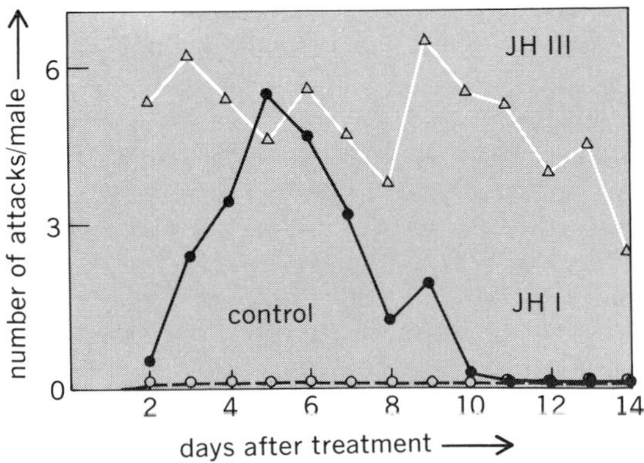

Fig. 2. Sexual activity of juvenile hormone I– and JH III–injected crowded allatectomized males. Intact males are used for control. (*From E. P. Amerasinghe, Effects of JH I and JH III on yellowing, sexual activity and pheromone production in allatectomized male Schistocerca gregaria, J. Insect Physiol., 24:603–611, copyright 1978 by Pergamon Press, Ltd.*)

a neurohormone from the corpora cardiaca is capable of releasing copulatory behavior within 15 min. Motor output to the phallic musculature is normally inhibited by centers in the head. Decapitation or application of corpora cardiaca extracts to the intact insect or to the nerve cord results in rhythmic bursts in the phallomere motor neurons. Apparently the corpora cardiaca hormone acts on the brain centers to release the inhibition of the phallic motor program. The phallic nerve–stimulating hormone is a peptide which is presumably released during courtship in response to stimuli from the female, although this has not been specifically demonstrated. It is worth noting that many hormonal effects on the phallic nervous system such as those just mentioned, as well as induction of oviposition and calling behavior, seem to be rapid, direct releaser effects of a neurosecretory product from the corpora cardiaca.

Flashing in fireflies is a very important aspect of courtship behavior. The light signals which are given are species-specific and allow males to locate receptive females. Flashing occurs only at night; it is inhibited when the insect is exposed to bright light by a unique hormonal mechanism. In response to neural stimulation due to exposure to light, the cortical granules of the testis produce a hormone, thought to be noradrenaline, which inhibits flashing by the firefly lantern. Inhibition of the flash occurs even if the nerves to the lantern are stimulated, indicating a peripheral action of the hormone. It is not clear whether a similar mechanism exists in the female.

Hormonal control of developmentally important behaviors. Included in this category are wandering behavior, eclosion behavior, caste-associated behaviors, and locomotor and migratory behaviors.

Wandering behavior. Prior to pupation, the last instar *Manduca sexta* enters a wandering stage. In the field this results in movement of the caterpillar off the food plant and its burrowing into the ground, where pupation occurs. A peak of ecdysone occurs approximately 24 h prior to the onset of wandering behavior. Both release of ecdysone and wandering behavior are gated events; that is, they will occur only during a certain period of time after lights on. Ecdysterone infused into intact larvae induces the onset and controls the duration of wandering behavior 24 h later in a dose-dependent manner. In 24-h cultures of the isolated nervous system ecdysterone can stimulate neural activity characteristic of wandering behavior in segmental motor neurons. When ecdysone is applied to the isolated ventral cord without the brain, no such effect is observed, indicating that the brain is the target for the hormone in the control of this behavior. Although the hormonal effect is apparently directly on the central nervous system, it does not result in the characteristic increase in motor output until 24 h later, an interesting intermediate between a rapid releaser and a slower developmental response to hormone stimulation.

Eclosion behavior. One of the best-studied systems of direct hormonal control of insect behavior is the hormonal release of ecdysis behavior in Lepidoptera.

The eclosion hormone was discovered some 10 years ago by J. W. Truman and coworkers. They found that at the end of adult development just prior to the shedding of the pupal exoskeleton, the pharate moth begins a stereotyped motor program which results in the emergence of the new adult from the old pupal cuticle.

The preeclosion motor program is triggered by the eclosion hormone which is released from the corpora cardiaca into the blood at this time. The hormone is produced in the medial brain neurosecretory cells, is a peptide of about 8500 molecular weight, and is heat-stable but inactivated by proteases.

The eclosion motor program in *Hyalophora cecropia* consists of three phases of stereotyped abdominal movements which last for about 1 ¼ h. The behavior or its neural correlate can be elicited even in an isolated abdomen or isolated nervous system within 10–15 min of application of corpora cardiaca extract or eclosion hormone. Only a very brief exposture (as little as 2.5 min) to the hormone is necessary to evoke the entire behavioral pattern.

Recent experiments with the isolated central nervous system have confirmed that the hormone is indeed acting directly on the nervous system to release a motor program which is already fully mature but not expressed until hormonal stimulation triggers the "read-out" of the eclosion motor "tape." It has recently been discovered that the action of eclosion hormone occurs by way of cyclic 3′5′-guanosine monophosphate (cGMP), presumably at the level of the target nerve cells. The cGMP will trigger the preeclosion and eclosion motor patterns in the isolated central nervous system. The amount of cGMP in the central nervous system doubles within 10 min after eclosion hormone treatment. This is the earliest known effect of the hormone and precedes the start of eclosion behavior by 5–10 min, as one would expect if cGMP were acting as a cellular "second messenger" in this system, as it does with certain vertebrate peptide hormones. It is also interesting to note that the action of eclosion hormone requires the presence of extracellular calcium as do neurotransmitters and some other neurochemicals.

There is now clear evidence for the involvement of eclosion hormone in all of the molts, not just the molt to the adult as was previously believed. There is also some evidence that it is not solely a lepidopteran hormone but that it may occur in other orders as well. In earlier molts, eclosion hormone does not appear to be released from the corpora cardiaca as it is at adult ecdysis. It seems that at larval and pupal ecdyses eclosion hormone is released from the ventral nerve cord rather than the brain or corpora cardiaca. The release site may depend on whether or not the release is under circadian control. The eclosion hormone system is currently the best demonstration of the mechanism of action of a hormone on the insect nervous system.

Induction of caste-associated behaviors. Injections of juvenile hormone mimic induces changes in honeybee behavior, causing them to move out of the brood nest to storage combs before controls. Injected workers begin foraging at an earlier age and to a greater extent than controls. Extremely high doses of juvenile hormone mimic result in many more flights, less pollen collection, and significantly reduced longevity in the treated animals. Juvenile hormone also increases dominance behavior

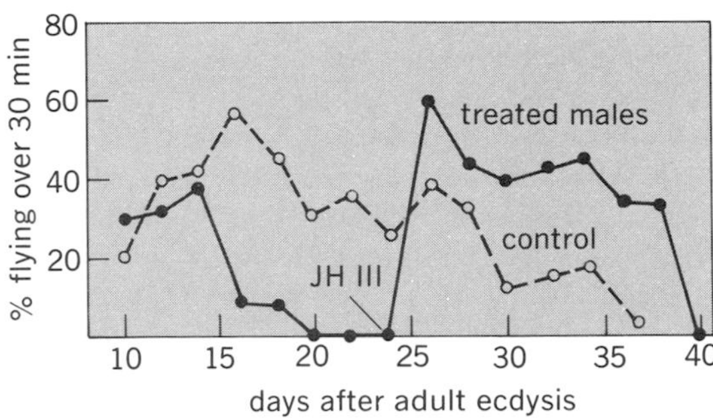

(a)

(b)

Fig. 3. Flight behavior of male insects treated with precocene. (*a*) *Oncopeltus fasciatus* treated with 100 μg of precocene I every 5 days beginning on day 10 after the adult molt, and subsequently treated with 10 μg of juvenile hormone III on days 25, 30, 35 (*from M. Rankin, Effects of precocene I and II on flight behaviour in Oncopeltus fasciatus, the migratory milkweed bug., J. Insect Physiol., 26:67–73, 1980*). (*b*) *Hippodamia convergens* treated with 10 μg of precocene II every 2 days followed by 10 μg of juvenile hormone mimic (JHM) replacement therapy in one group on day 4 (*from S. Rankin and M. Rankin, The hormonal control of migratory flight behaviour in the convergent ladybird beetle, Hippodamia convergens, Physiol. Entomol., 5:175–182, copyright 1980 by Pergamon Press, Ltd.*)

(a queen characteristic) in treated worker honeybees.

Hormonal control of locomotor and migratory behavior. Migration is, for some insects, an alternative to diapause as a mechanism to escape unfavorable environmental conditions as well as a mechanism for dispersal. Migration is most often a prereproductive phenomenon of the young adult and is physiologically similar to adult diapause. Onset of reproduction often causes an inhibition of migratory behavior.

In the milkweed bug, *Oncopeltus fasciatus*, and the ladybeetle, *Hippodamia convergens*, juvenile hormone is necessary for long-duration tethered flight (presumed migratory behavior). Treatment of these insects with precocene II, a drug which causes degeneration of the corpora allata and cessation of juvenile hormone production, causes inhibition of long-flight behavior in both of the above species (Fig. 3). Injection of juvenile hormone can induce very rapid restoration of flight activity in *O. fasciatus*, suggesting that in this instance juvenile hormone may act as a behavioral releaser. If so, this would be the only example of such an immediate effect of juvenile hormone (possibly on the nervous system). Recent unpublished evidence indicates, however, that one action of juvenile hormone in *Oncopeltus* may be to mobilize lipid flight fuel rapidly rather than releasing any neural activity program.

Hormones have been isolated from the corpora cardiaca of locusts and several other insects which mobilize fat body lipid during long flight. The primary adipokinetic hormone in locusts seems to be produced and released from the glandular lobes of the corpora cardiaca in response to stimulation by neurons from the protocerebrum. Recent evidence indicates that the transmitter at this synapse is octopamine. Adipokinetic hormone also seems to enhance transport of fuel substrates into flight muscle mitochondria. Whether it also has a direct effect on the nervous system has not been determined.

There is also some evidence that ecdysone can inhibit locomotor activity by direct action on the central nervous system both at the time of the nymphal molts and in the newly emerged solitary (nonmigrant) phase locust.

Inactivity following feeding is usual among animals which take discrete meals, and there are numerous references to such a reduction in locomotor activity of insects. Moreover, responsiveness to at least some environmental stimuli is reduced, including those associated with food. In locust nymphs the reduction in activity may be apparent within a few minutes of the end of feeding. Distention of the foregut leads to release of humoral factors from the storage lobes of the corpora cardiaca. Thus it would appear that a corpora cardiaca factor can also bring about a reduction in locust locomotor activity.

For background information *see* CHEMICAL ECOLOGY; ECDYSONE; ENDOCRINE SYSTEM (INVERTE-

BRATE); INSECT PHYSIOLOGY in the McGraw-Hill Encyclopedia of Science and Technology.

[MARY ANN RANKIN]

Bibliography: R. W. Meola and R. S. Petralia, *Science*, 209:1548–1550, 1980; S. M. Rankin and M. A. Rankin, *Physiol. Entomol.*, 5:175–182, 1980; J. W. Truman et al., *Nature*, 291:70–71, 1981; J. W. Truman, S. M. Mumby, and S. K. Welch, *J. Exp. Biol.*, 84:201–212, 1979.

Interferon

Interferons were discovered some 25 years ago by A. Isaacs and J. Lindenmann as they were investigating the mechanism whereby one virus-infected cell can protect other cells from infection by the same or other viruses. The protein factor liberated from the infected cell and able to interfere with virus replication was called interferon. There are many distinct forms of interferons produced by human cells, and these interferons can induce a number of diverse modifications in cells in addition to causing them to become refractory to viruses. The diversity of interferons themselves coupled with the multiplicity of their effects on cells makes them exciting as a family of natural biological response modifiers with wide clinical prophylactic and therapeutic potentials. Though the enormous clinical potentials of interferons were recognized immediately after their discovery, it has taken some 2 decades of research on production and purification of these proteins to obtain sufficient quantities and qualities to allow researchers to begin testing them against a few diseases. In these limited trials they have shown sufficient promise to stimulate further research and clinical evaluations.

Proteins. Each species of animal, or at least of vertebrates, is able to produce interferons, apparently in a variety of forms. Thus, human cells produce interferons alpha, beta, and gamma (and probably others as yet unidentified) which are each antigenically distinct and have a number of other important distinguishing physicochemical and biological properties (see table).

Alpha interferons, usually produced from suspensions of white blood cells exposed to a virus, are a family of antigenically closely related proteins that are efficient at entering the blood stream following intramuscular injections. To date, over a dozen human alpha interferons have been identified which have subtle differences in their primary (amino acid sequence) structures; these slight differences have significant effects on certain biological properties of the molecules.

Beta interferons, which are usually produced from fibroblast cultures of human skin cells exposed to synthetic double-stranded RNAs, are antigenically distinct from alpha interferons and are very inefficient at entering the bloodstream when injected into the muscle or under the skin, apparently because the beta interferons are locally inactivated. So far only two forms of beta interferons have been identified.

Gamma interferons, which are usually produced by T lymphocytes exposed to mitogenic stimuli (or specific antigens to which they have been sensitized), are antigenically distinct from alpha and beta interferons and, unlike these interferons, are unstable in strongly acidic solutions. There appear to be at least two different molecular forms of the human gamma interferons, but as sequence information is available for only one of these, it is possible that the gamma forms differ only in degrees of glycosylation of a common polypeptide.

Thus, for human interferons, there are at least

Distinguishing properties of human interferon forms

Property	Interferon type: Alpha	Beta	Gamma
Predominant cell source	B lymphocytes	Fibroblasts	T lymphocytes
Principal inducer	Viruses	Double-strand RNA	Mitogens
Molecular weights	15,000–25,000	20,000–25,000	20,000–50,000
Amino acid sequence heterogeneities identified	⩾15	⩾2	?
Stability at pH 2	+	+	−
Neutralization by antiserum against:			
Alpha	+	−	−
Beta	−	+	−
Gamma	−	−	+
Efficient entry into circulation from intramuscular injection	+	−	+
Predominantly stable in anionic detergents	+	+	−
Cross-reactivity on heterologous cell species	+ + + +	±	− −

three distinct antigenic classes, each likely composed of a number of subtly heterogeneous forms. Similar heterogeneities of interferons also seem to exist in other animal species. Some of these subtle distinctions have been shown to profoundly affect the level of activity of the interferons on specific cell species, but as yet the physiological significances of the multitude of interferons produced by the human body in response to various stimuli are not understood. It is conceivable that the interferons have target-tissue specificities and defined functions as yet unrecognized.

Activities. For several years interferons were investigated strictly as proteins whose sole function was to confer virus resistance in cells. However, over the last decade it has become apparent that interferons are able to induce a phenomenal variety of alterations in cells. In addition to inducing resistance against virtually all viruses, interferons can inhibit cell growth, alter surface morphology of cells, activate macrophages and cytotoxic lymphocytes, modulate a variety of immune functions, induce febrile responses, and enhance production of several constitutive and inducible cell products and inhibit production of others. That such pleotypic alterations were attributable to a single protein met with considerable skepticism for many years, but these multiple non-antiviral activities of interferons have recently been substantiated in studies using several homogeneously pure native interferons as well as interferons derived from genetically engineered microorganisms. Thus, interferons are a family of heterogeneous proteins, each able to induce a wide variety of alterations in cells; it is the immense clinical significance of certain of these alterations that makes interferons particularly promising for the treatment of viral diseases and malignancies.

Clinical experiences with interferons. It was only after several years of extensive research on the production and purification of human interferons that sufficient quantities of the materials were available to begin even meager clinical trials. Thus, in the latter 1970s it was reported that interferons could prophylactically protect against the common cold, influenza, and vaccinia viruses and could exert therapeutic effects against hepatitis, herpetic keratitis, multiple sclerosis, and warts in humans. Interferons have also been shown to exert antitumor activities in patients with osteosarcoma, juvenile larynx papilloma, multiple myeloma, breast carcinoma, melanoma, and a variety of other malignancies. To date, most of these trials are still ongoing, and it is anticipated that interferons will be tested in an increasing number of diseases as supplies of these substances increase. Increased availabilities would be largely due to the parallel advances in recombinant DNA technology, which makes it feasible to produce these rare, trace human proteins in relatively vast quantities as compared to their availabilities from cultured human cells. As such trials are expanded in more diseases at escalating doses, and with administration of each of the particular forms of interferons, either singly or in combination therapy with each other or with other agents, there are many reasons for great expectations. It is already apparent that the gamma interferons are significantly more potent than the alpha or beta interferons as cell growth inhibitory agents and as antitumor agents in animals; it is now known that alpha interferons are more potent than the beta interferons against certain tumors, and that the beta interferons have certain pharmacological limitations for systemic administrations. As work continues, and as awareness of the particular merits of each of the various forms on interferons increases, their potentials can be more appropriately exploited for particular clinical applications.

For background information *see* INTERFERON in the McGraw-Hill Encyclopedia of Science and Technology. [WILLIAM E. STEWART, II]

Bibliography: K. Cantell, Why is interferon not in clinical use today?, in I. Gresser (ed.), *Interferon-1979*, 1979; A. Isaacs, and J. Lindenmann, Virus interference, I. The interferon, *Proc. Royal Soc. B*, 147:258–267, 1957; W. E. Stewart, II, *The Interferon System*, 2d ed., 1981; W. E. Stewart, II, Varied biologic activities of interferons, in I. Gresser (ed.), *Interferon-1979*, 1979.

Ion exchange

The exceptional chemical and thermal stability of perfluorinated ion exchange polymers, such as Nafion, has opened new opportunities in the fields of industrial electrochemistry and catalysis. Principal applications are as cell dividers in chlor-alkali cells and other electrochemical devices.

Composition and properties. A melt-fabricable precursor polymer is prepared by copolymerization of tetrafluoroethylene (C_2F_4) with perfluoro-3,6-dioxa-4-methyl-7-octenesulfonyl fluoride [$CF_2{=}CF$-$OCF_2CF(CF_3)OCF_2CF_2SO_2F$]. Since it is a material that can be extruded, this copolymer can be fabricated into the desired form, such as a film, or a reinforced laminate, which then can be hydrolyzed by contact with a solution of KOH followed by acid exchange to yield the free sulfonic acid, Nafion, with the structure shown.

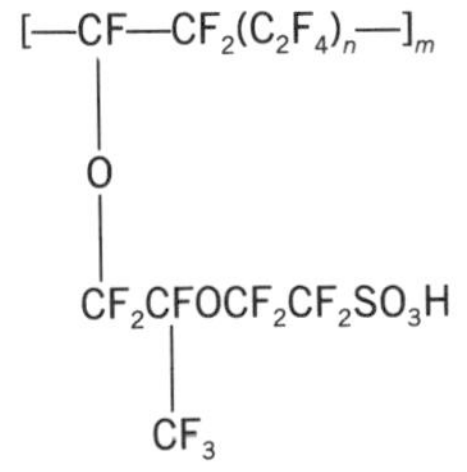

$$
\begin{array}{l}
[\text{—}CF\text{—}CF_2(C_2F_4)_n\text{—}]_m \\
\quad | \\
\quad O \\
\quad | \\
\quad CF_2CFOCF_2CF_2SO_3H \\
\qquad\quad | \\
\qquad\quad CF_3
\end{array}
$$

Nafion is stable at elevated temperatures in concentrated bases and acids such as nitric acid, chromic acid, and aqua regia. The upper operating temperature is 160 to 220°C depending on the environment.

Room temperature properties of Nafion 117 and Nafion 125 and after 30 min of boiling in water

Property	Nafion 117 50% relative humidity	Nafion 117 Wet	Nafion 125 50% relative humidity	Nafion 125 Wet
Equivalent weight	1100	1100	1200	1200
Thickness, mm	0.18	0.20	0.13	0.14
% H_2O by weight	8	35	7	28
Density, g/cm^3	1.98	1.65	1.98	1.7
Area resistance, Ω cm^2	—	1.5	—	1.6
Tensile strength, psi*	4000	2700	4400	3000
% ultimate elongation	Approx. 200		Approx. 200	

*1 psi = 6.9 kPa.

Nafion is swollen by water and polar organic solvents. In this swollen state the polymer is an ionic conductor (solid electrolyte) as a result of the mobility of the cations. This conductivity, particularly combined with a high degree of selectivity (preferential transport of cations compared to anions), is the basis for the use of Nafion in film or composite membrane form as a divider in electrochemical devices. Typical properties of Nafion films are shown in the table.

Although Nafion is transparent in both the dry and wet state, it is believed that there is a degree of phase separation into hydrophilic ionic clusters embedded in a hydrophobic fluorocarbon matrix.

Applications. Unreinforced film such as Nafion 117 or 125 has been used mainly as a solid polymer electrolyte, sandwiched between two porous electrodes, which also support the film. Examples are separators for fuel cells and cells for the electrolysis of water or aqueous hydrochloric acid.

For many other electrochemical applications the mechanical properties of Nafion 117 and 125 are improved by lamination to a reinforcing fabric made of Teflon TFE fluorocarbon fiber, resulting in composite membranes. These membranes can be heat-sealed to form cylinders or other shapes. Membranes of this type are used as cell separators in electrowinning of metals, the regeneration and purification of spent chromic acid solutions, the electrochemical synthesis of potassium gold cyanide, and many other applications.

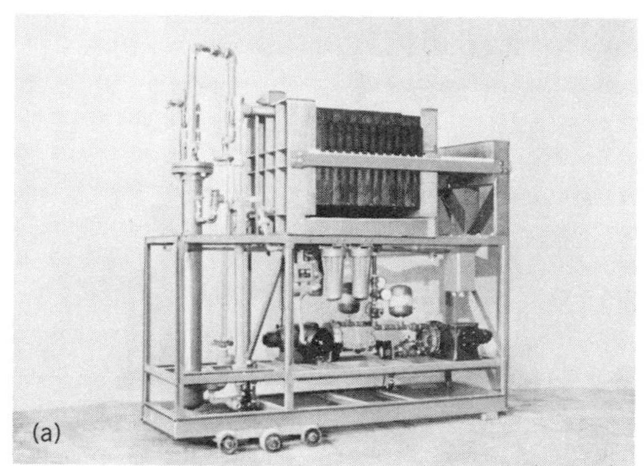

Fig. 2. Chlor-alkali production cell. (*a*) View of the complete unit. (*b*) Exploded view of the components of a type model for an individual cell. (*Ionics, Inc.*)

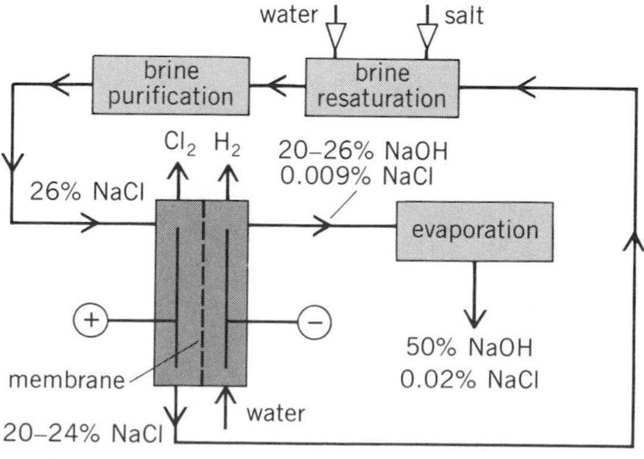

Fig. 1. Flow diagram of a chlor-alkali cell. (*From W. Grot, Nafion membrane and its application, in International Symposium on Electrochemistry in Industry—New Directions, Case Institute of Technology, 1980*)

By far the largest use for Nafion is as a cell separator in the electrolytic production of chlorine and caustic. Figure 1 shows schematically the operation of a chlor-alkali cell. Figure 2 shows a unit for chlorine and caustic production. This model contains a stack consisting of 12 individual cells electrically in series. The total membrane area is about 2.8 m^2 and has a chlorine production capacity of 140 kg. Chlorine capacities range from very small cells for supplying chlorine for swimming pools (1 kg/day) to large industrial plants (>100 metric tons/day). For this application, Nafion membranes of outstanding selectivity (that is, ability to prevent current transport by OH^- ion migration) have been developed in order to satisfy the requirement of high current efficiency at high caustic concentration. The improved current efficiency is achieved through the use of a thin barrier layer on the catholyte (caustic) side of the membrane. This barrier layer may consist of a sulfonamide derivative of Nafion, created by treating one surface of the membrane in the $—SO_2F$ precursor form with ethylene diamine prior to hydrolysis with KOH. Or, the barrier layer may also consist of Nafion of a higher equivalent weight. In both cases, the support layer is of low equivalent weight (1100 to 1150) for maximum conductivity.

A new high-performance chloro-alkali membrane has been developed which will operate at efficiencies that will average 94% or greater for the first 2 years of operation and produce caustic at a concentration of 33% and at a very high purity. The use of Nafion membranes in chlor-alkali cells not only eliminates the environmental problems associated with the use of mercury or asbestos in the older processes, but also reduces overall energy consumption.

Nafion also has applications as a catalyst. Because of the fluorine substitution, Nafion is one of the strongest acids known. This, in combination with the excellent chemical and thermal stability, has led to its use in a variety of acid-catalyzed organic reactions.

For background information *see* ELECTROCHEMISTRY; ELECTROLYSIS; ION EXCHANGE in the McGraw-Hill Encyclopedia of Science and Technology.

[WALTHER G. GROT]

Bibliography: A. Eisenberg and H. L. Yeager (eds.), *Perfluorinated Ionomer Membranes*, ACS Symposium Series, 1981; W. G. Grot, *Chem. Ing. Tech.*, 47(14):617, 1975; W. G. Grot, *Chem. Ing. Tech.*, 50(4):299-301, 1978.

Ion-selective membranes and electrodes

Ion-selective field-effect transistors (ISFET) are integrated solid-state devices for the detection and measurement of activity of ions in solution. They are a combined product of ion-selective electrode (ISE) technology and solid-state integrated microcircuits. They are a subcategory of chemically sensitive solid-state devices (CSSD) and can be regarded as second-generation ion-selective electrodes. These devices offer several advantages over conventional electrodes, particularly for biomedical measurements.

Principles. The principle of operation of a field-effect transistor is shown in Fig. 1. Drain voltage (V_D) is applied between the drain and source electrodes, which are made by doping the substrate with suitable impurities. If the substrate is *p*-type silicon (conduction is done by positive-charge carriers—holes), the drain and source are doped to be *n*-type silicon. The third metal electrode is separated from the substrate by one or more layers of insulator; this metal-insulator-semiconductor structure is called a gate, and the device is often referred to as an insulated-gate field-effect transistor (IGFET). If a positive gate voltage (V_G) is applied to the gate electrode, the electric field repels the holes in the substrate from the surface of the *p*-type semiconductor and electrons are attracted to the layer adjacent to the insulator. When a voltage (V_D) is applied between the drain and the source, these electrons will enable the drain current (I_D) to flow. This process, which is the basis of operation of several types of semiconductor devices, is called a semiconductor surface field effect. Since there is no direct electrical contact between the gate electrode and the substrate, the device presents an infinitely high impedance to a dc input signal. In addition, the intensity of the electric field in the insulator is the result of all potentials in the circuit between the substrate and the gate electrode. Thus, if the potential difference between the substrate and the metal electrode is changed by a process such as absorption of gaseous molecules into the gate electrode, the electrical field will change accordingly and so will the drain current. This is, in fact, the basis of operation of a field-effect transistor sensitive to molecular hydrogen in which the metal gate is made of palladium. The device represents a transition between conventional metal-insulated gate transistor and a chemically sensitive solid-state device.

The schematic diagram of a true ISFET is shown in Fig. 2. The main structure of this device is identical with that of the IGFET of Fig. 1, except for the gate metal, which has been replaced by an ion-selective membrane. The electrical contact in the gate circuit is completed by the measured solution and a reference electrode. The analytical information is obtained from the potential (E) developed at the solution-membrane interface which for a typical

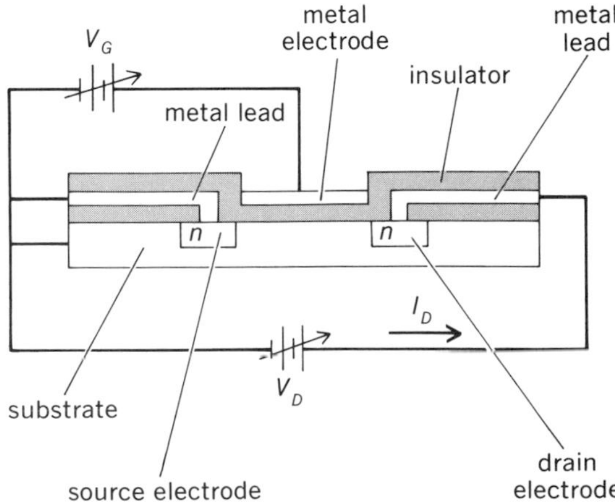

Fig. 1. Schematic diagram of an insulated-gate field effect transistor. (*From J. Janata and S. D. Moss, Chemically sensitive field-effect transistors, Biomed. Eng., 11(7):241–245, July 1979*)

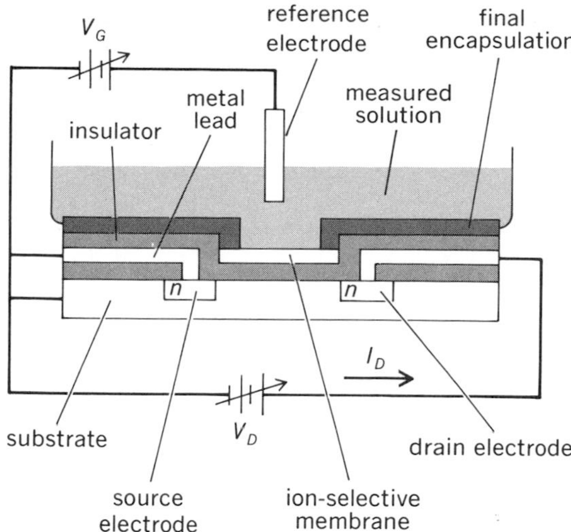

Fig. 2. Schematic diagram of an ion-selective field-effect transistor. (*From J. Janata and S. D. Moss, Chemically sensitive field-effect transistors, Biomed. Eng., 11(7):241–245, July 1979*)

ISE membrane has the form of the Nernst equation (1), where S is the slope and a_i is the activity of

$$E = \text{constant} + S \ln (a_i) \quad (1)$$

the measured ion. It is evident from the foregoing discussion that the drain current (I_D) is a function of both V_D and V_G. The semiquantitative equation expressing this relationship is normally written as shown in Eqs. (2) and (3), where α is the factor

$$I_D = \alpha V_D \left(V_G - V_T - \frac{V_D}{2} + S \ln a_i\right) \quad \text{for } V_D < V_G - V_T \quad (2)$$

$$I_D = \alpha(V_G - V_T + S \ln a_i)^2 \quad \text{for } V_D > V_G - V_T \quad (3)$$

related to the geometry of the transistor and V_T is the so-called threshold (or turn-on) voltage, which is related to the materials of the transistor. Equations (2) and (3) represent a functional relationship between the drain current (I_D) and the activity (a_i) of the measured ion. The application of different membranes then yields ISFETs sensitive to different ions. Because it is customary to relate solution activity to potential rather than to current, the preferred mode of operation of ISFETs is at constant current. In this mode the Nernst potential change (ΔE) at the membrane-solution interface is compensated for by the externally applied voltage (ΔV_G) of the same magnitude but of the opposite polarity.

Advantages. The process by which the analytical information is obtained with ISFETs is identical with the operation of a conventional ion-selective electrode. Indeed, the selectivities and sensitivities of the ISFETs described to date match those obtained with the corresponding ion-selective electrodes. The main difference is the missing internal reference electrode system of the conventional ion-selective electrode; in ISFETs the sensing membrane is placed directly at the input insulator of the field-effect-transistor preamplifier. The advantage of these devices over conventional ion-selective electrodes must, therefore, follow from this difference.

One important advantage is that ion-selective electrode membranes usually have resistance up to 100 megohms. The intimate contact of the preamplifier with this high-resistance signal source represents an in-place impedance transformation which results in a significantly improved signal-to-noise ratio. In addition, the sensing areas (gates) are very small (approximately hundreds of square micrometers); therefore, several hundred of these devices can be accommodated on a single chip with an area of a few square millimeters. This opens up the possibility for a miniature multiprobe, differential measurement, and combined data acquisition–data processing monolithic circuits. It also makes ISFETs compatible with computers. Other advantages are that their all-solid-state construction makes them rugged and readily sterilized by gases such as ethylene oxide; temperature and light sensitivity can be compensated for with relative ease by the differential mode of measurement; and a low manufacturing cost of ISFETs is anticipated because of the highly developed fabrication technology and the saving of expensive membrane materials. The field-effect transistor is basically a charge-measuring device. The gate can be regarded as a capacitor. Thus, when charge density on the solution side of the insulator changes, the condition of charge neutrality dictates that the charge on the semiconductor side of the insulator change as well. This can serve as

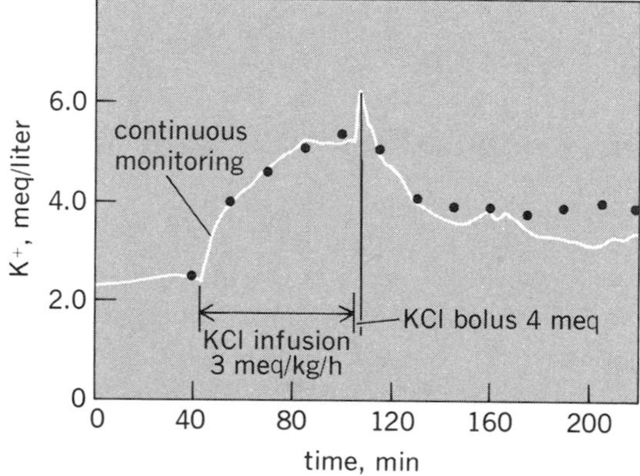

Fig. 3. Venous potassium activity monitored continuously during venous KCl solution infusion in a dog with a precalibrated ISFET probe compared with arterial serum samples analyzed for potassium concentration K^+ using an emission flame photometer. An increased infusion rate near the end of the 60-min infusion period is reflected by the ISFET probe continuous trace but not by flame photometer samples, indicated by the dots. (*From B. A. McKinley, et al., In vivo continuous monitoring of K^+ in animals using ISFET probes, Med. Instrum., 14(2):93–96, March–April 1980*)

the basis of a new class of devices based on direct charge measurement.

Outstanding problems. ISFETs are relatively new devices. The main problem in their future application is the fact that these solid-state circuits are in part exposed to ionic solutions. Reliable automatic encapsulation methods must be developed before these devices can gain commercial acceptance.

Biomedical applications. Although ISFETs have potential applications in other fields, they are particularly attractive as biomedical devices. Figure 3 shows a representative result of continuous monitoring of K^+ activity in a dog's vein. It is clear that continuous monitoring is preferable to discrete sampling. Simultaneous measurement of K^+, Ca^{2+}, and pH in various body fluid compartments have been carried out. In addition, a transistor sensitive to penicillin has been developed and tested.

Future developments. It is expected that the development of chemically sensitive solid-state devices will initially mimic the existing conventional electrodes and that the new measuring possibilities based on charge detection, pattern recognition, multiparameter sensing, and bioelectrochemical detection will follow.

For background information *see* ION-SELECTIVE MEMBRANES AND ELECTRODES; TRANSISTOR in the McGraw-Hill Encyclopedia of Science and Technology. [JIRI JANATA]

Bibliography: S. Caras and J. Janata, *Anal. Chem.*, 52:1935, 1980; J. Harrow et al., *Proc. EDTA*, 17:179, 1980; J. Janata and R. J. Huber, in H. Freiser (ed.), *Ion-Selective Electrodes in Analytical Chemistry*, vol. 2, 1981; B. A. McKinley et al., *Crit. Care Med.*, 9:333, 1981.

Iron metallurgy

Direct reduction is defined as any process which converts iron oxide into metallic iron without melting the material. The metallic iron product is called direct reduced iron, and is used as a high-quality feed material in steelmaking and ironmaking.

The traditional method of converting iron ore into metallic iron requires the material to be melted in a blast furnace. In the blast furnace process, energy costs are relatively high, pollution problems of associated equipment are severe, and the capital investment requirements are, in many cases, enormous.

In comparison with the blast furnace method, the direct reduction process yields a cold product, which can be easily stored and shipped. It permits a wider choice of more economical fuels, is environmentally clean, and is much lower in capital cost.

Although the first commercial direct reduction plants were built over 20 years ago, only within the past 10 years has direct reduction achieved successful commercial application. In 1980, approximately 7,150,000 metric tons of direct reduced iron were produced. Production of direct reduced iron is expected to reach 24,000,000 tons per year by 1985 and 50,000,000 tons per year by 1990.

Direct reduction process. Currently there are 12 commercially available direct reduction processes. The field is largely dominated by the MIDREX and HyL (HYL Iron & Steel Technology) processes, which currently account for 56% and 32% respectively of world direct reduced iron production.

The commercial direct reduction processes are often categorized according to the type of vessel used for the reduction step. A shaft furnace is used in the MIDREX, Armco, Purofer (Thyssen Purofer GmbH), NSC (Nippon Steel Corporation), HyL III, and Kinglor-Metor processes. Four batch reactors are used in the HyL I process. Four fluidized-bed reactors are used in the FIOR (Davy-McKee Corporation) process. A rotary kiln is used in the SL/RN (Lurgie Chemie und Huettentechnik GmbH), ACCAR (Allis-Chalmers Corporation), CODIR (Krupp Industrie- und Stahlbau), and DRC (Direct Reduction Corporation) processes.

Processes based on shaft furnaces, batch reactors, or fluidized-bed reactors normally use a gaseous reducing agent (hydrogen and carbon monoxide). The reducing agent can be generated by reforming natural gas, partially oxidizing fuel oil, gasifying coal, or recovering a by-product gas, such as coke-oven gas. There are exceptions, such as the MIDREX Electrohermal direct reduction process and the Kinglor-Metor process, which utilize low-cost coal directly in the shaft furnace without gasification.

Rotary kiln processes normally utilize a solid re-

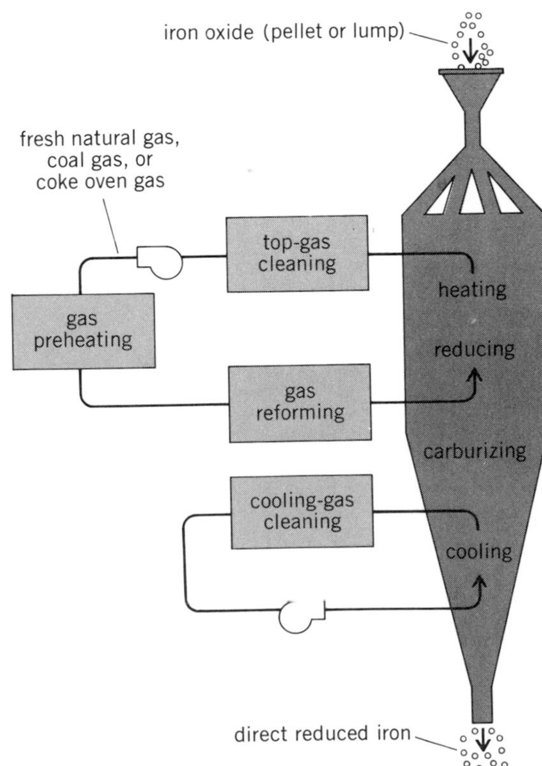

Fig. 1. Flow sheet for MIDREX direct reduction process. (*MIDREX Corp.*)

Fig. 2. Series of three MIDREX direct reduction modules at the plant constructed for C.V.G. Siderurgica del Orinoco (SIDOR) by Lurgi Chemie und Huettentechnik GmbH. (*MIDREX Corp.*)

ductant, such as charcoal, lignite, subbituminous coal, bituminous coal, or anthracite. The ACCAR process can also use natural gas or fuel oil as a reductant.

The chemical reactions in all the direct reduction processes are basically the same. Regardless of the initial reductants used, the end reducing agents are gaseous hydrogen and carbon monoxide.

With respect to energy efficiency, continuous-type processes are naturally more efficient than batch-type processes, and processes which recycle process gases and use heat recovery systems are more efficient than those that do not. Shaft furnace–type processes tend to be most efficient because the gas-solids reduction reactions are much more uniform and controllable, and result in a much more uniform quality product.

A flow sheet for a typical commercial direct reduction process is shown in Fig. 1. Iron oxide (pellet or lump ore containing about 67% iron) is fed into a shaft furnace. Inside the shaft furnace, the iron oxide is heated to 850° C and reacted with hot reducing gas, composed of hydrogen and carbon monoxide [reactions (1) and (2)]. The reducing gas

$$Fe_2O + 3H_2 \rightarrow 2Fe + 3H_2O \quad (1)$$

$$Fe_2O_3 + 3CO \rightarrow 2Fe + 3CO_2 \quad (2)$$

removes oxygen from the iron oxide, leaving behind a solid metallic product called direct-reduced iron (containing 92–94% iron). The direct-reduced iron is cooled and carburized [about 1.5% carbon is added as shown in reactions (3) and (4)] and then

$$3Fe + CO + H_2 \rightarrow Fe_3C + H_2O \quad (3)$$

$$3Fe + CH_4 \rightarrow Fe_3C + 2H_2 \quad (4)$$

discharged from the bottom of the furnace. Most of the spent reducing gas (top gas) is recycled through a reformer, where it is reacted with natural gas to form fresh reducing gas [reactions (5) and (6)]. The

$$CH_4 + H_2O \rightarrow CO + 3H_2 \quad (5)$$

$$CH_4 + CO_2 \rightarrow 2CO + 2H_2 \quad (6)$$

remaining top gas is used as a fuel for heating the reformer and process gas streams. Figure 2 shows a series of three MIDREX direct reduction modules.

Direct reduced iron. Direct reduced iron is similar in size and shape to the particles, pellets, and lumps of ore that are charged into the direct reduction furnace. When the ore charge consists of fine particles, as in the fluidized-bed processes, the metalized fines usually are agglomerated into briquettes.

Chemically, direct reduced iron consists of metallic iron, residual iron oxides, the gangue present in the original ore, and other metallic elements present in the original ore. In addition, direct reduced iron usually contains some carbon in the form of iron carbide.

It is common practice to describe the quality of direct reduced iron in terms of its content of metallic iron, as in formula (7).

$$\text{Degree of metallization (\%)} = \frac{\text{metallic iron content (\%)}}{\text{total iron content (\%)}} \times 100 \quad (7)$$

The degree of metallization depends upon an analytical determination of the metallic iron content in the product. MIDREX direct reduced iron has an average metallization of 92–94%, permitting movement in holding areas by industrial magnets (Fig. 3).

For further processing into steel, direct reduced iron must be melted, reduced further to eliminate residual iron oxides, slagged to remove the gangue, and refined.

Advantages of direct reduced iron. Most direct reduced iron is converted into steel in electric arc furnaces. Usually scrap iron is also an important part of the electric arc furnace charge. The amount of direct reduced iron in the charge may range from 0 to 100%. With the development of continuous charging for direct reduced iron, an increase in productivity is achieved compared with productivity of the same furnaces on all-scrap charges. Other desirable effects of continuously charging direct reduced iron include decreased arc flicker and a corresponding decrease in disruption of power systems, decreased arc noise, and the opportunity to eliminate heavy charging cranes. In addition, the use of direct reduced iron containing low levels of undesirable tramp elements, such as copper, produces higher-purity steel.

Direct reduced iron has been used as a coolant in the basic oxygen furnace steelmaking process. In this case, direct reduced iron offers the advantageous cooling effect of scrap while offering the purity and handling characteristics of iron ore.

Foundries have used direct reduced iron to replace scrap in cupolas or induction furnaces. The low content of undesirable tramp elements in direct reduced iron makes this material especially useful in the production of high-purity steels and ductile cast irons.

Direct reduced iron used in blast furnaces and electric smelting furnaces has resulted in increased productivity, as well as decreased energy consumption.

Future developments. The rapid growth of demand for direct reduced iron is largely due to the rapid growth of electric arc furnace steelmaking. Direct reduction can be combined with the electric arc furnace to offer an efficient, integrated steelmaking route, which has significant advantages over the conventional blast furnace/basic oxygen furnace route.

Blast furnace/basic oxygen furnace steelmaking requires coking coal, which is rapidly becoming more expensive, while pollution control laws are making it prohibitively expensive to build and operate coke ovens in some parts of the world. In comparison, fuel for direct reduction plants remains readily available in the form of natural gas or noncoking coal.

A direct reduction/electric arc furnace facility can be installed for about 60% of the cost of a comparable blast furnace/basic oxygen furnace facility and in about half the time. It is feasible to build direct reduction/electric arc furnace facilities in increments as small as 250,000 metric tons per year, compared to 1,500,000 tons per year for a blast furnace/basic oxygen furnace facility, thus placing less strain on capital, labor, infrastructure, and the environment. Also, it is not necessary to have the direct reduction plant and electric arc furnace at the same site. A large direct reduction plant can be built to supply several regionally located steel mills, each with electric arc furnaces.

The fact that the direct reduction/electric arc furnace facility can be operated more efficiently than the blast furnace/basic oxygen furnace facility during periods of fluctuating world steel demand is of great economic significance.

Based on worldwide steelmaking trends, direct reduction/electric arc furnace steelmaking likely will play a major role in expansion of the world's steel industry during the 1980s.

For background information *see* IRON METALLURGY; STEEL MANUFACTURE in the McGraw-Hill Encyclopedia of Science and Technology.

[ECKART E. GOETTE]

Fig. 3. Direct reduced iron being moved in a holding area by an industrial magnet. (*MIDREX Corp.*)

Krill

Seldom has the development of a new fishery engendered as intense an interest and as much concern as now attends the development of the Antarctic krill fishery. The attention given this 4- to 5- cm shrimplike crustacean (*Euphausia superba*) stems from estimates of the potential krill harvest at between 50,000,000 and 60,000,000 metric tons per year, an amount nearly equal to the world's total yearly catch of marine fish and shellfish. To a burgeoning and protein-hungry world population, the prospect of a krill harvest of that magnitude takes on a dramatic aspect.

Krill may in fact be one of the world's largest untapped sources of protein, but it is also the key organism in the Antarctic food web, providing food directly or indirectly for most of the whales, seals, penguins, winged birds, fish, and squid of the Antarctic region (Fig. 1). Research on krill, therefore, has gone in two distinct directions: one, scientific, attempting to understand the role krill plays in the delicate ecological balance of the Antarctic Ocean, and the other, technological, exploring methods for harvesting and utilizing a potentially vast human and animal food resource.

Historical studies. Most early studies of krill were more exploratory and scientific than technological and were part of investigations on winged birds, penguins, seals, and the great baleen whales. One of the most comprehensive early studies, including krill investigations, was undertaken in the late 1920s and 1930s by the *Discovery* committee for the British government. Almost all that is known about krill comes from the *Discovery* expedition. Although the existence of large stocks of krill in the Antarctic has thus been known for many years, interest in their commercial exploitation arose in the early 1960s with the decline of whaling in the Southern Hemisphere.

The Soviets and the Japanese started experimental krill fishing in 1961–1962. In recent years, several other countries have joined in, notably Poland, West Germany, South Korea, and Taiwan. Because of the high cost of krill fishing, however, only the Soviet Union, Poland, and Japan, all of which are substantially subsidized by government agencies, have continued large-scale technological research on krill fishing and utilization. At the present time, Antarctic krill are being harvested at the rate of about 250,000 metric tons annually.

Scientific research on krill, on the other hand, has been ongoing and cooperatively undertaken by a large scientific community including Argentina, Australia, Chile, Japan, West Germany, France, Poland, South Africa, the United Kingdom, the Soviet Union, and the United States. The most significant of these cooperative international studies is being conducted as part of the Biological Investigations of Marine Antarctic Systems and Stocks (BIOMASS) program, which is sponsored by the Scientific Committee on Antarctic Research (SCAR), the

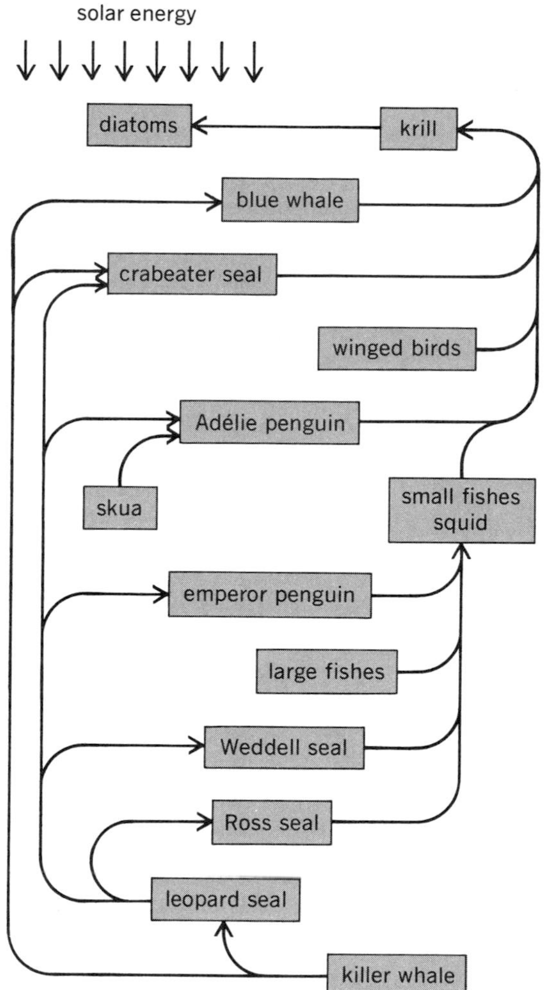

Fig. 1. Food cycle of marine animals in the Antarctic region is based on the krill, *Euphausia superba.*

Scientific Committee on Oceanic Research (SCOR), the International Association of Biological Oceanographers (IABO), and the Advisory Committee on Marine Resources Research (ACMRR) of the Food and Agriculture Organization (FAO). These organizations have recently (1980–1981) sponsored the First International BIOMASS Experiment (FIBEX) to the Antarctic, the primary purpose of which was to study the distribution and abundance of krill. Data from this expedition, which included 11 ships from 10 nations, was evaluated in a workshop conducted in Hamburg, in October 1981. The organizers of BIOMASS are planning a Second International BIOMASS Experiment (SIBEX) for 1983–1984, the exact objectives of which will be determined after the FIBEX data of 1980–1981 have been interpreted.

Technological investigation. Krill generally occur in dense concentrations (patches or swarms) which would support a high volume of commercial catch. The chief areas of krill abundance are near South Georgia, the South Sandwich Islands, the South Shetlands, and the South Orkneys,

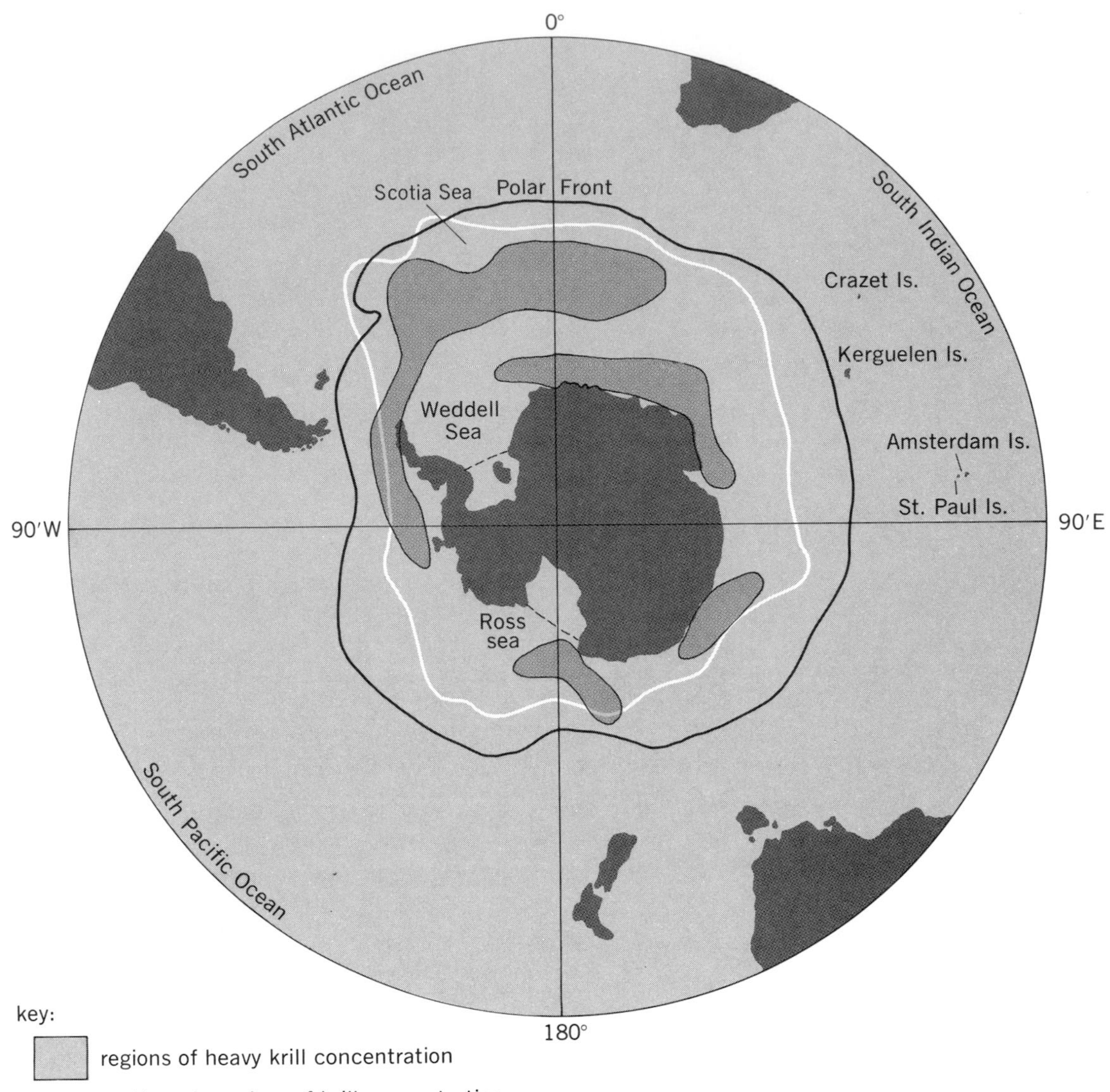

Fig. 2. Living resources of the Southern Ocean. (*From S. Z. El-Sayed, Biological Investigations of Marine Antarctic Systems and Stocks (BIOMASS), vol. 1: Research Proposals, SCAR and SCOR Scott Polar Research Institute, August 1977*)

in the Scotia Sea and in the Bransfield Strait (Fig. 2).

There are several obstacles to krill fishing, the most obvious of which is the Antarctic climate. Even in the austral summer (from November to April) ice covers 11% of the total ocean surface, and winds are constantly high. During winter, spring, and autumn, weather conditions are unpredictable, and at its greatest extremes, ice covers 60% of the total ocean area. Such conditions restrict krill fishing virtually to the summer months, although in late 1979 a Finnish shipyard designed a trawler specifically for harvesting krill, which could lower its warps under ice in fields up to 60 cm thick. This ship has not been built, but nonetheless the technology does exist.

Other major obstacles to krill harvesting come in locating krill, determining the size of the crustacean, and differentiating the krill from other marine life. These problems have largely been solved by the use of the vertical echo sounder with an acoustic and transmission frequency between 100 and 200 kHz. Such an instrument, in skilled hands, can be used efficiently to locate and differentiate krill. To date, the most efficient catches are made by using fine-mesh nets (8–24 mm) trawling at midwater and shot by single trawlers.

These methods for catching krill are efficient and justify large-scale harvesting, but the technology for processing krill has not caught up with the fishing techniques.

Krill processing, like krill harvesting, was pioneered by the Soviets and the Japanese, who in the early 1960s, produced krill meal and krill protein concentrate. There are various problems that attend krill processing, most notably that unless they are processed within a few hours, krill undergo texture, taste, and color changes which render

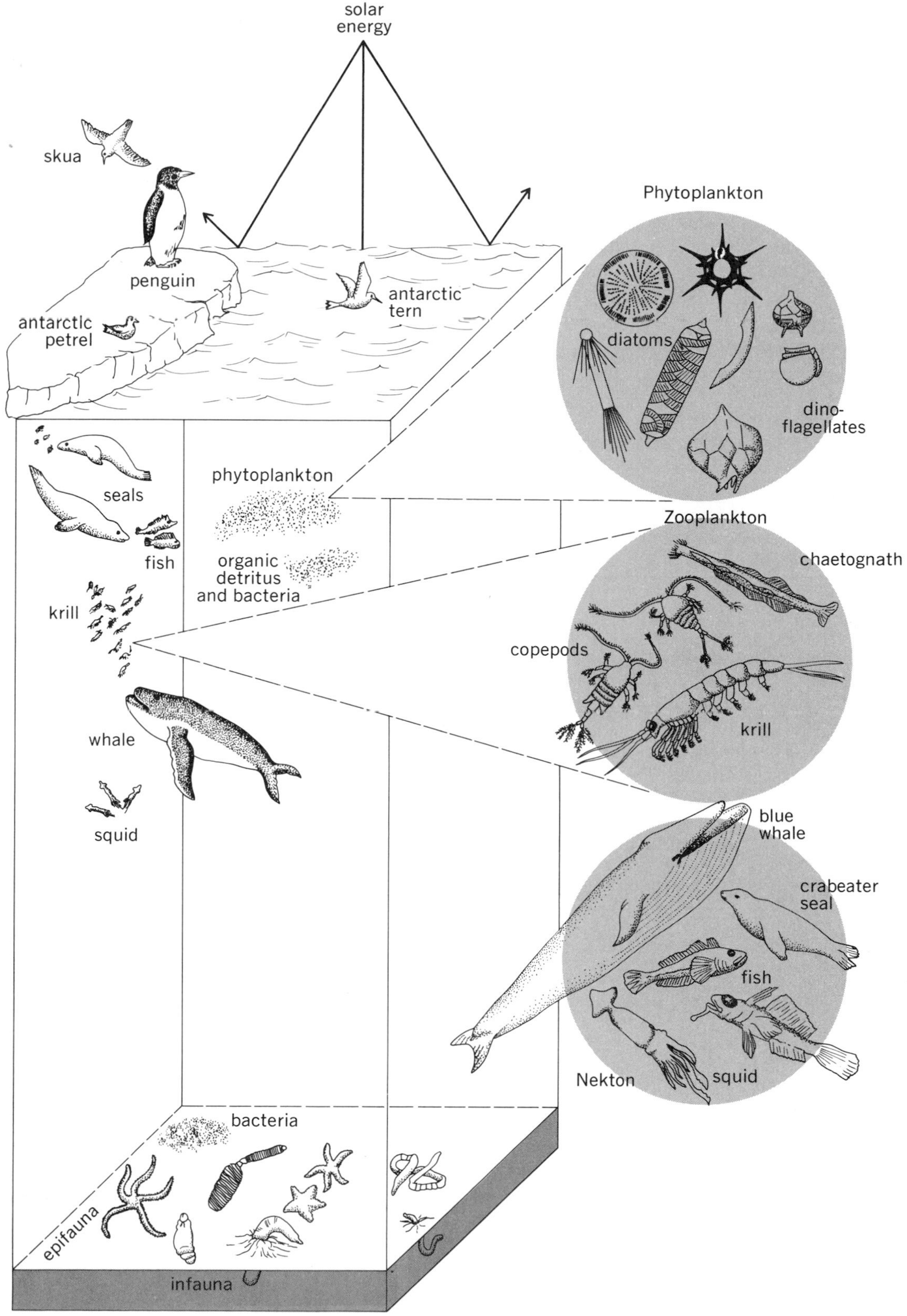

Fig. 3. Major biotic components in the Antarctic marine ecosystem. (*From S. Z. El-Sayed, Dynamics of Trophic Relations in the Southern Ocean, in L. O. Quam, ed., Research in the Antarctic, American Association for the Advancement of Science, 1971*)

them unfit for human consumption.

On-ship processing, therefore, has proved to be highly desirable. Several countries have experimented with methods of processing krill; the Soviet Union has concentrated on producing a krill paste which can be used with cheese, butter, and mayonnaise and to fortify other foods. Such products have not been produced on a large-scale basis for human consumption. The Soviets also claim medicinal values for krill paste, ranging from treating hyperacidity of the stomach to arteriosclerosis. Japan has devised methods of processing krill tail meat which involve cooking and freezing whole krill aboard ship and then drying it later on shore. It can thus be used as a substitute for the small dried shrimp currently well established as part of the Japanese diet. However, such technology has not yet been used on a large-scale consumer basis.

Chile has developed over 20 krill products including minced krill, dried krill, krill paste, and krill sticks which have been marketed with limited success. The West Germans produced a krill mixture with the consistency of a bruhwurst, containing cooked krill forcemeat and milk protein.

Most other nations have dropped experimentation with krill processing because the expense involved cannot justify the commercial viability of the products obtained.

No method devised thus far for processing krill for human consumption is very efficient, the maximum edible yields seldom being above 20% of the total weight of the catch, but uses for other parts of the krill, besides the edible tails, might still make processing economically feasible. The most promising of these by-products is chitin, a substance obtained from the krill exoskeleton. This substance and its deacetylated derivative chitosan can be used in waste water treatment and as replacement for arteries, bones, and cartilage in medical procedures. Krill meal, derived from whole krill, could provide another potentially profitable use of krill, since it could provide a high-protein food source for animals.

Although the technological problems of krill processing have not been entirely solved, enough progress has been made to predict that within the next 10 years the large-scale harvesting of krill will be not only possible but probable.

Scientific investigations. Sufficient information has now accumulated from various Antarctic expeditions to allow scientists to draw a fairly accurate picture of the food chain relationships in Antarctic waters (Fig. 3), and to show the key role that krill plays in the food chain dynamics of the Antarctic oceans.

In the interest of the most efficient management of a living resource, there is a need to know many details of the life history of the species targeted for exploitation. While the broad outlines of the life history of krill are known, many fundamental questions remain unanswered, specifically concerning physiology, distribution, population dynamics, standing stocks, and productivity.

Scientific investigations can aid in determining the size, location, abundance, and migratory habits of krill which would be invaluable information for fishing interests, and they can also provide data to help in the wise management of the exploitation of krill to avoid disturbance of the ecosystem and to ensure the renewability of krill resources.

Conclusion. It is clear that the soundest and most profitable use of the krill resources of the Southern Ocean can best be effected by international collaboration and by cooperation between scientific and technological interests in krill research. Without such cooperation, the krill population and the entire Antarctic ecosystem could be doomed to the fate of the Antarctic great whales, but with it, krill could both maintain the bountiful Antarctic ecosystem and provide a rich source of food for the human population.

For background information *see* FISHERY CONSERVATION; MARINE FISHERIES in the McGraw-Hill Encyclopedia of Science and Technology.

[SAYED Z. EL-SAYED]

Bibliography: G. O. Eddie, The Southern Ocean: The harvesting of krill, in *Southern Fisheries Programme*, Food and Agriculture Organization of the United Nations, p. 76, 1977; S. Z. El-Sayed, *Oceanus*, 18(4):40–49, 1975; S. Z. El-Sayed and M. A. McWhinnie, *Oceanus*, 22(1):13–20, 1979; G. J. Grantham, The Southern Ocean: The utilization of krill, in *Southern Fisheries Programme*, Food and Agriculture Organization of the United Nations, p. 61, 1977.

Laminar flow

The increase in cost of petroleum fuels in the past several years, greater than the general inflation level, has raised the percentage of aircraft operating expense due to fuel to the point where fuel efficiency is the key design goal for future derivative and new aircraft. NASA and the aircraft industry have identified technologies that can significantly reduce aircraft fuel consumption in coming years and are actively pursuing research and technology development programs. One of these technologies is the attainment of extensive regions of laminar airflow over aircraft surfaces. Although this offers the greatest potential for fuel conservation of any single new technology, it involves the greatest challenges in the areas of aircraft manufacture, maintenance, and operational procedures required to provide acceptable reliability at reasonable cost.

Flow phenomenon. When a fluid moves past a solid surface, a thin layer develops adjacent to the surface where frictional forces tend to retard the motion of the fluid. This layer is defined as the boundary layer. The boundary layer generally exists in one of two states: laminar, where fluid elements travel along well-ordered nonintersecting paths, and turbulent, where fluid elements from adjacent layers become totally mixed. The state of the boundary layer, in the absence of disturbing influences, is directly related to the speed and the distance the fluid has traveled along the surface. Laminar flow, however, is an inherently unstable condition that is easily upset, and transition to turbulent flow can occur prematurely as a result of amplification of disturbances emanating from any of a number of

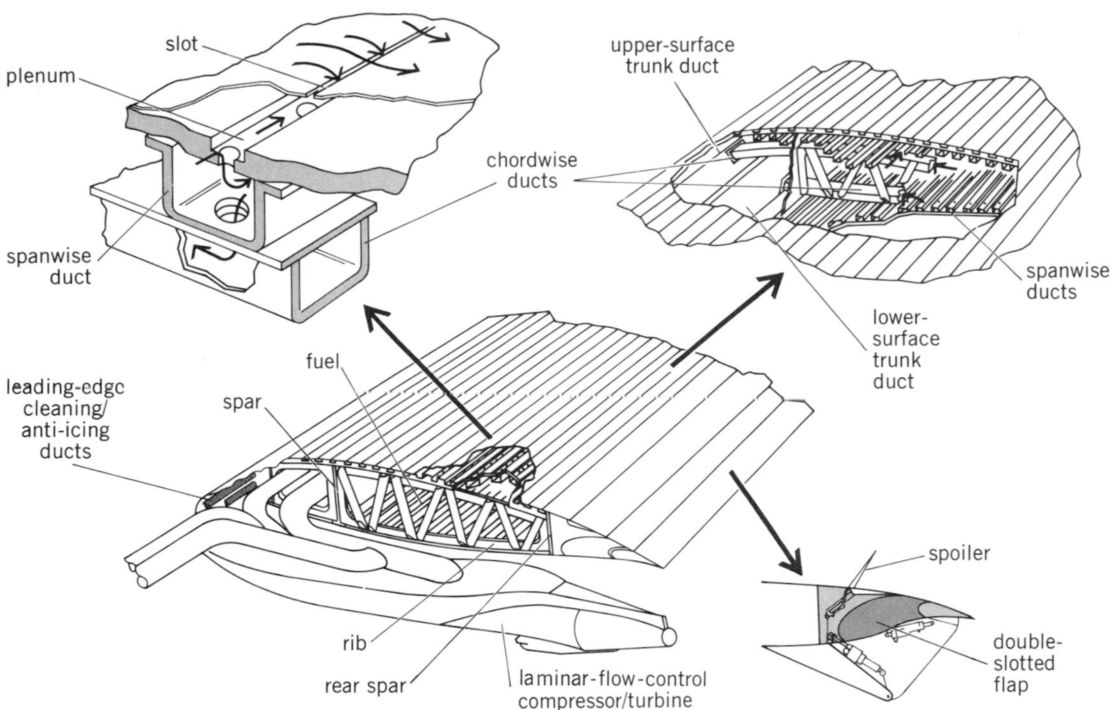

Portion of aircraft wing with laminar flow control system, showing details of suction surface, ducting system, and wing trailing edge.

sources, for example, surface irregularities and noise.

Associated with the transition of the boundary layer from laminar to turbulent is a large increase in the frictional force between the fluid and the surface. This force is referred to as friction or viscous drag and is a major contributor to fuel usage.

The attainment of larger regions of laminar flow with its associated low value of skin friction drag has long been a goal of the aerodynamicist. Much has been accomplished, both analytically and experimentally, in establishing an understanding of transition from laminar to turbulent flow and in developing passive (natural laminar flow, NLF) and active (laminar flow control, LFC) means of delaying the transition phenomenon. To obtain laminar flow passively over part of an airplane wing with little sweep, the wing contour must be designed to have decreasing local pressures over the surface in the direction of the airflow. To obtain laminar flow on swept wings, required for efficient flight at higher speeds, or in the rear portions of any wing where the local pressures must necessarily increase in the flow direction, laminar flow control must be employed. The most effective type of laminar flow control involves the removal of a small amount of the boundary-layer air by suction through narrow surface slots, small perforations, or porous materials.

The basic theories are well established, verified in wind tunnel tests, and demonstrated in various flight tests. The challenge of laminar flow has been the ability to manufacture and maintain the aircraft surfaces within permissible tolerances considerably smaller than required for turbulent aircraft and at acceptable cost.

Technology development programs. The primary goal of the current NASA/industry programs on the attainment of extensive regions of laminar flow is to develop and demonstrate the technologies required for economically feasible application to future aircraft of various types.

Laminar flow control. The largest program concerns fuel reduction through the application of laminar flow control by suction to future commercial air transports. During cruise of current transports, about half of the engine power needed to maintain level flight is used to overcome the turbulent friction drag on the external aircraft surfaces. Because about 70% of the fuel on a transcontinental trip is consumed during cruise, attainment of large regions of laminar flow on the wings and tail surfaces of long-range transports can, in combination with other advances in technology, reduce the fuel by a very significant amount (60% less than current airplanes).

Basically, the laminar flow control system includes a suction surface through which a portion of the boundary-layer air is taken into the airplane, a system for metering the level and distribution of the ingested flow, a ducting system for collecting the flow, and pumping units which provide sufficient compression to discharge the suction flow at a velocity at least as high as the airplane velocity (see illustration). The program is focused on the prime airframe and airline industry concerns, that is, reliability of operation in an airline environment, and life cycle costs of laminar flow control aircraft. Concerns about both of these aspects of laminar flow control must be allayed before the industry will be in a position to implement it on commercial transports. Realistic manufacturing costs will be determined from fabrication of full-scale representative hardware with demonstrations that modern processes can economically accommodate the required exceptional tolerances in surface shape and smoothness. Acceptable means to prevent or control contamination, corrosion, and erosion of surfaces and suction systems by insect debris, dirt, ice, and so forth, must be found, and acceptable maintenance costs, derived from a flight test program reflecting operations in an airline environment, must also be demonstrated. The program builds upon existing laminar flow control technology but utilizes new materials, methodology, and manufacturing techniques where applicable to enhance feasibility. The systems investigated will be compatible with anticipated technological advances in other areas under development in the Aircraft Energy Efficiency Program such as active controls, advanced aerodynamics, composite materials, and advanced, quieter, more efficient engines. The program is structured to maximize the participation of industry with particular emphasis on involvement of manufacturers of large commercial transports.

The program, which was initiated in 1976, involves the following elements in a phased approach:

1. System studies, technology development, and aircraft concepts.

2. System design and development, and component and system testing.

3. Research aircraft modifications, proof-of-concept flight tests, and simulated in-service validation.

The phased approach allows progress in each phase to be evaluated prior to initiation of the next phase. The development of this approach was due in part to the high risk associated with the program and the large resource commitments required to demonstrate technology readiness. It was recognized that the step from an experimental status to technology readiness for application would require solutions to many difficult problems.

The first step was to identify potential missions for future commercial laminar flow control aircraft and identify a baseline configuration which would provide a realistic basis for cost-benefit evaluations of alternative concepts for the various systems. Basic to these concepts was determination of an efficient structural arrangement that successfully integrates the smoothness and suction requirements with acceptable weight and cost penalties.

Two basic structural concepts remain under consideration. One involves the use of an outer-surface glove panel attached to a primary wing structure utilizing either conventional aluminum material and manufacturing techniques or advanced graphite/epoxy composite construction. The other involves an integration of the suction ducting into elements of the primary wing structure fabricated with graphite/epoxy where the ducting contributes to the strength and stiffness of the wing structure. Each approach, as is usually the case, has its individual advantages and disadvantages. The external-surface candidates still under consideration after numerous specimen and component tests of various types are titanium with very small perforations produced with the recently developed electron-beam manufacturing technique, and titanium with fine slots either cut or assembled as strip inserts. Development of design data and establishment of manufacturing feasibility are being accomplished through design, fabrication, and tests of subscale and full-size components.

An area of aerodynamics receiving considerable emphasis is that of advanced airfoils (sections of wings) having large regions of supercritical flow, that is, local flow velocities over the wing surface greater than the local speed of sound. These airfoils, when applied to future turbulent aircraft, will result in improved performance as measured by either weight reductions or higher aerodynamic efficiency. To ensure that a laminar flow control aircraft retains the projected performance advantage over an advanced turbulent aircraft, it must be capable of utilizing these advanced airfoils. Analytical methods, utilizing the tremendous advances in computer technology, have been used to define supercritical airfoils that are best suited for laminar flow control applications. Because suction levels required for laminarization are highly dependent upon the airfoil characteristics, substantial reductions in suction power and system complexity may be realized. A representative airfoil will be tested in a swept configuration at high subsonic speeds in a NASA Langley Research Center wind tunnel with suction through slots and perforations.

Minimization of surface roughness, especially in the leading-edge regions of airplane components, must be accomplished if laminar flow is to be expected. Measurements have indicated that the height of insect residues in these regions is large enough to cause premature transition from laminar to turbulent flow. A means of eliminating or reducing insect excrescence must, therefore, be found. Two promising approaches are under development. One of these integrates the insect-protection function and the anti-icing function into a single liquid-ejection system. A freezing-point depressant ejected in the leading-edge region has been shown by wind tunnel and flight tests to prevent accumulation of insect residue. The second utilizes a deployable shield which also acts as a leading-edge high-lift device for improved takeoff and landing characteristics. Back-up spray nozzles behind the shield are provided for additional leading-edge protection, if needed. Wind tunnel tests verify the shield effectiveness as predicted by insect trajectory analyses.

Effort is also being directed toward developing refined analysis methods required in design application. Based on the latest numerical techniques and computer technology, methods have been developed for calculating swept-wing boundary-layer development and stability with surface suction. These tools are necessary for evaluating the different types of suction surfaces (porous, perforated, slotted) and establishing optimum suction requirements. Experiments are being conducted to validate these design tools.

Pending continued success in the development program, an integration of the various laminar flow control subsystems into a demonstrator aircraft will be required to provide industry a sound basis for decisions on the possible application to future airplanes. Modification of the wings of an existing airplane to accommodate LFC structures and systems, and flight tests in a simulated operational environment will provide: data on interface problems between laminar flow control systems and the other airplane systems; operational experience and verification of maintenance, servicing, inspection, and repair requirements and techniques; and a direct measure of airplane economy and performance.

Factors that will probably determine whether the design and development of any new high-speed transport should include application of laminar flow control are: the economic advantage over a competitive turbulent transport; the existence or real possibility of government-imposed fuel allocations; and unique mission requirements. The economic advantage will depend upon two rather unpredictable trends between now and the time of consideration of laminar flow control: the relative trend of fuel price with the general inflation of the other airplane operating costs; and the magnitude of fuel reductions from other technology advances equally applicable to turbulent airplanes. Fuel allocations, of course, will magnify the importance of fuel savings relative to economic benefits. Because the benefits increase with airplane range, laminar flow control may represent the only available technology for meeting extremely long-range or extremely long-endurance missions, of possible importance to the military. These factors will determine not only whether application of laminar flow control will be attempted but also whether commercial or military airplanes will be the first beneficiary.

Natural laminar flow. More limited regions of laminar flow than that possible with laminar flow control may be attainable through design with favorable pressure gradients over part of the wing (decreasing pressures in the direction of the flow). Adverse cross-flow effects induced by sweep of the wing leading edge, necessary for cruise at high subsonic speeds, limit the extent of laminar flow attainable. Recent boundary-layer stability analyses, however, indicate that wing sweep angles up to about 17°, corresponding to a lower cruise speed than today's transports, may be permissible. Verification of the analytical predictions is being pursued in a flight test program at the NASA Dryden Flight Research Center of a natural laminar flow airfoil incorporated in partial gloves on the wings of the NASA TACT F-111 research airplane. The variable wing-sweep feature of this airplane permits investigation in flight through the range of sweep angles of interest. Results will help define the types of airplane for which natural laminar flow may be feasible. For any type, smoothness of surface finish and contour must be provided as well as leading-edge protection from accumulation of insect residue and ice.

Hybrid laminar-flow control (HLFC). A combination of the principles for laminar flow control and natural laminar flow may find application to large high-speed transports. Suction in the leading-edge region, where the cross flow due to sweep is large, may be used to control the cross-flow disturbances, and favorable pressure gradients, not large enough to induce unacceptable cross flow, may be used aft of the front spar to maintain natural laminar flow to the vicinity of midchord. Such a hybrid approach may provide more extensive laminar flow than possible with natural laminar flow at high sweeps, and has the advantage of avoiding the complexities associated with providing suction in the region of the wing torsion box where fuel is stored. Analysis of the feasibility of this hybrid concept is under way.

For background information *see* BOUNDARY-LAYER FLOW; SUPERCRITICAL WING in the McGraw-Hill Encyclopedia of Science and Technology.

[ALBERT L. BRASLOW]

Bibliography: D. M. Bushnell and M. H. Tuttle, *Survey and Bibliography on Attainment of Laminar Flow Control in Air Using Pressure Gradient and Suction*, NASA Ref. Publ. 1035, vol. 1, September 1979.

Laser imaging

The application of lasers to the production of graphic arts and printed products is increasing significantly. The laser is becoming the dominant light source in new products that digitally (by computer) form images of text, line art, and halftones. One extensive area of application is in the production of newspapers. Lasers are also playing a key role in electronic color separations, and are the major image source for the rapidly evolving intelligent copiers (high-speed nonimpact printers). As the printing and publishing industries evolve into full electronic processing of the material to be printed, the laser will become the dominant digital light source.

Advantages. During the past 10 years much has been heard about the use of lasers in graphics arts, and specifically in the production of newspapers. Several thousand units of laser-based equipment are already operational in newspapers throughout the world in a variety of applications, including optical character recognition (OCR), electronic cameras, fascimile senders and receivers, electronic color separations, news photo service, newspaper counting, and typesetting.

There are several properties of the laser that will lead to its widespread use in the evolving electronically produced printed page. Laser light is colli-

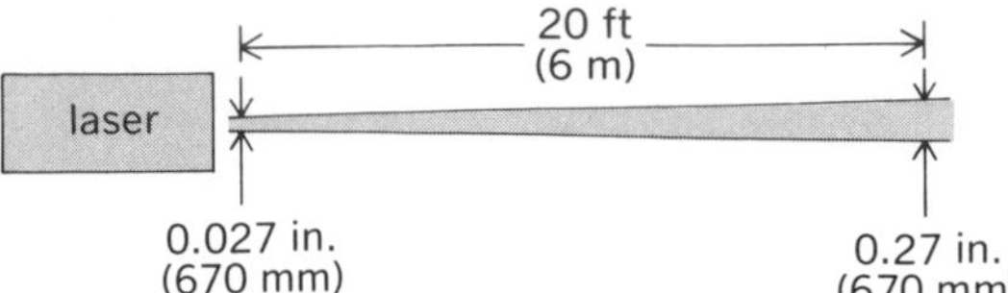

Fig. 1. Collimated laser light diverges very little and therefore allows for accurate resolution.

mated, allowing for very accurate and high-resolution focusing (Fig. 1). It is this property that will improve the quality of the printed product. Finer-quality text and pictures (both black and white and process color) can be anticipated. The availability of electronic color processing of pictures will accelerate the use of process color throughout the newspaper industry. Even the major metropolitan newspapers will have substantial process color by 1985, although only a few use significant amounts of process color at this time.

Another important feature offered by laser imaging systems is that they will allow the printing and publishing industries to move away from silver-based imaging materials (photopaper and film). Silver, a volatile commodity of limited supply, currently is the basis of the primary image transfer technology for composing the page. The laser offers the possibility of nonsilver technologies, including the computer imaging of photopolymer, diazo, thermal, and electrophotographic printing plates (thus eliminating prepress composing as it is now known).

There are many different lasers that allow a whole range of non-silver-based materials to be imaged (Fig. 2). These include ultraviolet lasers to expose diazo and photopolymer emulsions, such as Dylux and printing plates; blue and red lasers to expose photofilm, photopaper, and electrostatic media; and infrared lasers with sufficient power to thermally develop (machine) an image.

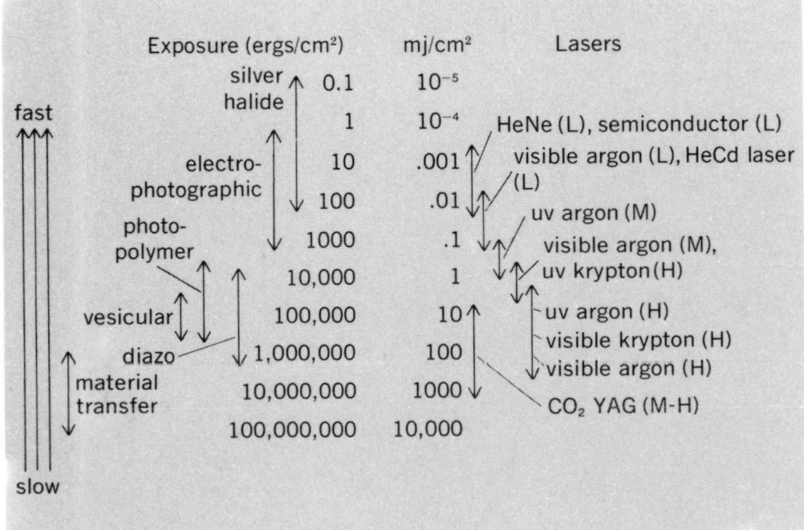

Fig. 2. Types of lasers and non-silver-based materials used for laser imaging. L = low power, M = medium power, and H = high power, relative to the specific laser indicated.

Costs vary widely from laser to laser, and in many cases must be taken into account when analyzing the operating and maintenance cost of a piece of laser-based production equipment. There is a basic trade-off in labor and materials savings brought about by laser imaging systems versus the cost of acquiring and maintaining such systems. In 1975 these trade-offs were not substantially in favor of laser imaging. By 1980 these trade-offs had shifted substantially in favor of laser imaging systems through the development of new imaging materials, primarily electrophotographic and thermal systems.

Finally, a laser is the logical digital light source for generating combined images of text, line art, and halftone pictures. Significant here is the capability of the laser to image halftones at the same speed as text. Current cathode-ray-tube (CRT) typesetting systems slow down substantially when they are used to set halftones. The output speed (for text, line art, and halftones) of current CRT typesetters is 0.08 to 0.33 in.^{2}/s (0.5 to 2.1 cm^2/s), in comparison with laser typesetters at 2.1 to 4.5 in.^{2}/s (13.5 to 29 cm^2/s). Further, the Xerox 9700 laser printer scans at 96 in.^{2}/s (619 cm^2/s), and Litton has under development a laser imaging system (for the U.S. Post Office) that images in excess of 480 in.^{2}/s (3097 cm^2/s).

System components. A laser imaging system consists of four basic components: laser, modulator, focusing optics, and scanner. The selection of the laser is basically a trade-off involving imaging speed, the sensitivity (speed) of the material to be imaged, and overall operating costs. The modulator is used to turn the laser "on and off" in response to a video or digital signal. Actually, this is a misstatement since most modulators deflect the beam to achieve the effect of on/off (Fig. 3). In addition to the modulator, the laser beam is focused to a small spot on the order of 0.8 to 2.0 mils (20 to 50 μm). This spot is then scanned by a variety of techniques to compose the image spot-by-spot within a scan line, and line-by-line until the full page is imaged. The types of basic scan mechanisms are the external drum, internal drum, and Flatfield.

Newspapers. As the newspapers begin to electronically compose full pages (with text, line art, and halftones), the implications will be dramatic. Facsimile will be accomplished with high-speed laser typesetters at the receiver site (as is currently done with national news magazines). Typeset commands as well as data-compressed line art and continuous-tone pictures will be transmitted from site to site. This will lead to several significant advantages over current facsimile networks: improved quality (every output will be an original); reduced bandwidth and thus reduced cost of transmission; and improved speed and throughput.

The availability of this type of equipment will al-

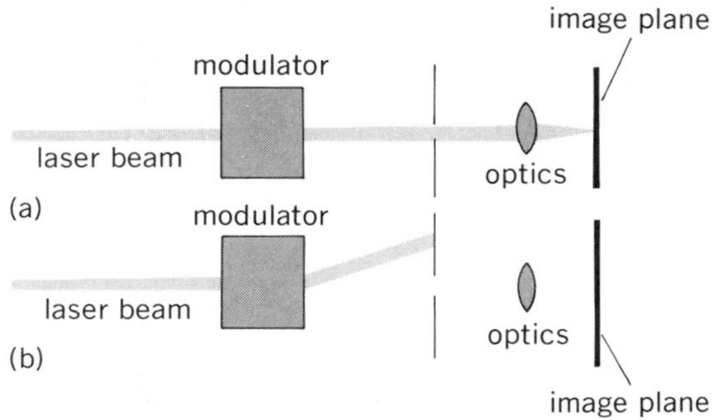

Fig. 3. Modulator operating in a laser imaging system in the (*a*) on and (*b*) off mode.

low for highly regionalized newspapers, since the production of the newspaper will be essentially a data base management task. All of this will be possible because of the speed of digital imaging that lasers make possible. It is the regionalization that will form the backbone of the newspapers' competition with home TV services (Yellow Pages, classifieds, and news).

Intelligent copiers. Electrophotographic copiers with laser imaging devices driven by computers are rapidly entering the nonimpact printing market. These devices are currently low-resolution (180–300 lines per inch) and medium-quality. However, the technical evolution of this process will lead to copier systems with capabilities such as high resolution (1250 lines per inch), use of 150-line screen, and process color.

With these developments by 1985–1986, it will be difficult to distinguish an intelligent copier from a typesetter or a color scanner. As a result, the only significant differences between the fully developed intelligent copier system and the fully developed electronic publishing system for offset printing will be run length and printing speed. By 1985 these systems will routinely begin displacing typewriters and daisy wheel printers associated with word processing. The new laser-based printers for word processing will have medium-quality typography, thus allowing two or three times as much printed information per page with an improved visual appearance.

Another impact of laser imaging copiers will be to displace short-run offset printing for run lengths of up to 1000 copies (and maybe up to 5000 copies).

Commercial printing. Here lasers are being pioneered for several critical applications, including laser gravure, laser flexography, steel rule dies, typesetting, and electronic color separation. Again the advantage of the laser system is the ability to image pictures, line art, and text at high speed and with excellent quality.

Perhaps the Crosfield Laser Gravure is one of the most exciting possibilities for the gravure industry. Here several hundred watts of CO_2 laser power is used to engrave a variable-depth spiral groove in a polymer-coated gravure cylinder. After chrome coating, these cylinders have shown run lengths in excess of 1,000,000 impressions. The reduction in cylinder preparation costs is dramatic and should stimulate growth of gravure printing.

For electronic color separation and four-color electronic page makeup, the blue line (488 nanometers) of the argon laser has become the most popular imaging source, whether for electronic or contact screening. The HeCd laser may come into more favor for these applications during the next 5 years due to lower purchase and maintenance costs. Also, the semiconductor laser will find wide-ranging applications in electrophotographic systems.

The commercial printing industry is also going to experience a wave of facsimile transmission for distributed printing, much in the same fashion as newspapers. Already most national weekly news magazines are printed in multiple plants via facsimile. The application of laser imaging facsimile networks will improve speed and quality of these systems and allow the transmission of process color display ads.

For background information *see* PRINTING in the McGraw-Hill Encyclopedia of Science and Technology. [S. THOMAS DUNN]

Bibliography: S. Thomas Dunn, Printing plates, in *McGraw-Hill Yearbook of Science and Technology*, pp. 308–310, 1979; *Lasers in Graphics*, Quarterly Reporting Service, Dunn Technology, Inc., Vista, CA, 1979–1980, 1981.

Leguminous trees

Members of the legume family commonly known to North Americans and Europeans are annuals such as soybean, peas, beans, and peanuts, and non-woody perennials such as clover or alfalfa. However, most legume genera occur in the subtropics and tropics in such diverse forms as 50-cm-tall annuals which may have large tubers, vines, shrubs, and 45-m-tall trees. Leguminous trees such as black locust (*Robinia pseudoacacia*) and honey locust (*Gleditsia triacanthos*) occur in temperate regions; mesquite (*Prosopis* species) or *Acacia* species dominate much of the semiarid regions of North and South America, Africa, the Near East, and Australia; and large trees such as *Albizzia* species, *Dalberghia* species, and *Parkia speciosa* occur in the wet tropics.

Managed orchards of selected strains of leguminous trees show potential for producing protein, carbohydrate, and fuel resources that require less water, nitrogen, and tillage than traditional crops. Tree- and shrub-based ecosystems such as forests typically lose less soil and soil nutrients than row-crop agricultural systems.

Root nodules on these leguminous trees convert atmospheric nitrogen into a form suitable for plant use, thus avoiding nitrogen fertilizers synthesized from fossil fuels. Machinery requirements for tree legume orchards could be less than for conventional agricultural systems because intensive site preparation and planting operations would occur only once in the life of the orchard, at the time of planting. Mature leguminous tree orchards require only light tillage for weed control, which does not have to be carried out at a specified time of the year. Tree legumes exist with taproots that can easily reach water at 5–10 m depths and thus avoid transient moisture stress. Some tree legumes photosynthesize and fix nitrogen at lower soil moisture levels than annual legumes by holding the water in their tissues more tenaciously than is possible by annuals (from a leaf water potential of 2800 to 3400 kilopascals). Tree legumes exist that can grow well in salinities equivalent to half sea water and show potential for use in highly saline coastal environments.

Leguminous trees can provide large quantities of livestock or human food in the form of pods, seeds, or leaf material. *Prosopis* pods range from 10 to 17% protein and 17 to 44% sucrose, are free of toxic compounds, and were the major food source for Indians in the desert of southern California and Arizona. *Leucaena leucocephala* can produce up to 15 dry metric tons/ha of highly palatable leaves containing 28% crude protein useful as a protein supplement for livestock.

A recently identified *Prosopis* selection contained 43.5% sugar that was easily fermented to alcohol. Mature *Prosopis* trees (17 years old) in southern California have produced 40 kg of oven-dry pods per tree, and an exceptional tree produced 73 kg. The greatest pod-producing selection of 3-year-old trees spaced 1.2 m (4 ft) apart in Riverside, CA, had a yield of 3500 kg/ha dry-pods. Future yields may be lower because of the tendency of trees to bear heavy fruit crops every other year (alternate bearing). Tree legume pod production shows great potential for human or livestock food or for fermentation to alcohol, as highly promising single-tree selections have been made; but commercially viable pod production systems are still in their infancy.

A brief description of the major uses of leguminous trees follows.

Fig. 1. *Acacia albida* trees in recently harvested peanut fields in Senegal, West Africa.

Increased soil fertility. Farmers have pruned and cared for naturally regenerating *Acacia albida* trees in millet, sorghum, and peanut fields in the West African countries of Senegal, Mali, Niger, and Chad (Fig. 1). The farmers have high regard for these trees because they increase yields of annual crops grown beneath the canopies. Scientists have demonstrated severalfold increases of soil nitrogen, organic matter, and associated fertility parameters under *A. albida*'s canopy, and that *A. albida* can raise the average millet grain yield from 500 to 900 kg/ha. The fertility increase seems attributable to large quantities of nitrogen-rich leaf litter supported by nitrogen fixation in *A. albida*'s root system.

A similar situation occurs in the western Rajasthan region of India, where farmers maintain *Prosopis cineraria* trees in densities as high as 150 per hectare in their pearl millet fields. Soil chemistry studies have found increased organic matter and nitrogen content under the trees. Scientists have not yet examined millet yields under the trees, but increased fertility parameters indicate they should be increased. *Prosopis cineraria* has been included in the religious customs of this area, indicating significant importance to the people.

In semiarid regions of West Africa it was observed that nitrogen fixation in annual legumes such as peanuts occurs in bursts following rainfall events. P. Felker and coworkers have demonstrated that *Prosopis* can fix nitrogen in moist soil at 2.7–3.0-m depths when covered by 2.7 m of dry soil (top 0.5 m greater than 2.2 mega pascals). It seems reasonable that deep-rooted tree legume nitrogen fixers are able to integrate moisture additions over long periods of time to avoid transient moisture stresses which limit nitrogen fixation in annual legumes. This is probably the reason *A. albida* can increase yields of annual legumes such as peanuts.

Production of livestock food. *Leucaena leucocephala* has been grown in Hawaii, the Philippines, northern Australia, and other areas in the moist tropics, where its high-protein (28%) leaves have been used for livestock production. Some strains of leucaena can grow into a tall tree, but grazing or harvesting at several months' intervals can maintain leucaena in a leafy form with a low percentage of woody stems. High leaf dry matter yields of 15–20 metric tons/ha/year have been achieved containing over 3000 kg protein per hectare. Cattle graze leucaena, and it has been harvested for sale as an alfalfa-meal substitute.

Prosopis tamarugo supports sheep-raising activities in the most arid (less than 0.7 mm rainfall/year) and highly saline regions of the world by existing on groundwater within 10 m of the surface. The Chilean-government-supported corporation CORFO has established 22,000 ha of tamarugo trees in this region by transplanting in 0.5 × 0.5 × 0.5 m holes through the salt crust (Fig. 2). Approximately 6000 kg of tamarugo pods are obtained per hectare per year, but their protein content (10%) is not sufficiently high or digestible to support good production of sheep without supplementation. Nevertheless, this level and quality of production is remarkable, given the aridity and salinity of the region. Greenhouse experiments have demonstrated that *P. tamarugo* can grow (albeit very slowly) in a nitrogen-free media with salinity levels equivalent to sea water.

In Mexico in 1965, 40,000 tons of mesquite pods were sold for incorporation into rations for cattle, sheep, goats, horses, donkeys, and mules. The sale of mesquite pods, which were all collected from native stands, often provided much-needed revenue for low-income farmers in dry summers when other crops failed.

Wildlife food resources. Tree legumes have made substantial contributions to wildlife food resources. *Acacia xanthophlea*, *A. tortilis*, and *A. hockii* may contribute 5000 kg/ha/year to the diets of giraffes. Several species of *Acacia* in South Texas brushland are high in protein and highly preferred by deer. The tree legume genera *Brachystegia* and *Julbernardia*, which dominate the Central African woodlands of major parts of Zambia, Zimbabwe, Angola, and Tanzania, have been observed to be useful to tall browsers such as elephants and giraffes. High-sugar *Prosopis* pods are so palatable to desert rodents, cattle, horses, deer, and even toy French poodles that several weeks after as much as 40 kg have fallen from a single tree a pod cannot be found.

Human food. Mesquite pods were among the most important food for Indians in the California and Arizona deserts. Mesquite pods served mainly as a carbohydrate or energy source since the hard undamaged seeds containing much of the protein were removed from the ground pods, leaving a flour containing 30–35% sugar but only 5% protein. Nevertheless, mesquite pods were important as they were the only really abundant food source in these desert regions receiving less than 150 mm (6 in.) annual rainfall. Similar uses for *Prosopis* occurred in the semiarid regions of Argentina.

In the semiarid and savannah regions of West Africa the pods of *Parkia biglobosa* and *P. clappertoniana* were important human food resources. Seeds of these 25-cm-long pods were fermented into a cheeselike product, and the yellow mesocarp surrounding the seeds was mixed with cereals, meat, or soup and pressed into cakes for storage.

In the moist tropics of Indonesia, seeds of *Parkia speciosa* and *Pithecellobium lobatum* (Djenkol bean) were sufficiently regarded as a food delicacy that they were not used for livestock food. In spite of special preparation required prior to being eaten, *P. lobatum* seeds show exceptional potential for human food because of higher levels of the nutritionally limiting amino acids cysteine, methionine, and tryptophan than are also present in soybeans and milk casein.

Prevention of desertification. The local government in Andhra Pradesh, India, established shelter belts by using the tree legumes *Prosopis juliflora*, *Pithecellobium dulce*, and *Albizzia lebek*, and the nonleguminous but nitrogen-fixing tree casuarina to prevent encroaching sand dunes from covering agricultural fields, irrigation canals, rods, and parts of villages. *Prosopis juliflora* halted this sand invasion, relieved a critical wood fuel shortage, and provided pods needed in the dry season for cattle food. P. Felker is currently working with several agencies in the Sudan to accomplish similar goals.

Fig. 2. Goats foraging fallen *Prosopis tamarugo* pods and leaves on the 22,000-ha plantings of *P. tamarugo* in northern Chilean salt deserts.

Luxury quality timber. Many of the tree legume woods are hard and durable with strikingly rich colors and prominent figure. Species with valuable wood such as desert ironwood (*Olneya tesota*), mesquite, and screwbean (*Prosopis pubescens*) are adapted to harsh desert conditions, while *Dalberghia*, *Pterocarpus*, and *Afrormosia* are adapted to the wet humid tropics. Lumber of some of the wet humid tree legumes is more costly than African mahogany (*Khaya* species). Wood carvings, parquet flooring, and gunstocks are the predominant products of desert tree legume woods such as *Olneya* and *Prosopis* because of lack of large pieces.

Fuelwood production. The United Nations Food and Agricultural Organization wood use surveys indicate that the consumption of firewood often exceeds 80% of all wood use (including timber) in developing countries and that firewood consumes 65% of the world's yearly wood production. In the northeastern United States wood fuel use has increased rapidly. Economic studies indicate that a 30–50% annual return on the investment can be achieved by conversion of existing gas- or oil-fired electrical generating facilities to burn wood. Economic studies on large-scale farming of the tree legumes *Leucaena leucocephala* (Hawaii Giant K-8) in Hawaii and of selected *Prosopis* strains in South Texas indicate that wood can be produced from the trees at an energy-equivalent price competitive with alternative fuel sources, including coal. Marginal, infertile, semiarid lands exist in sufficiently large areas to support the feedstock requirements for moderate-sized electrical generation facilities or chemical plants. Less environmental problems would arise from wood than coal or nuclear power because wood burning yields virtually no sulfur, only one-tenth the ash of coal, and no radioactive waste.

[PETER FELKER]

Bibliography: P. Felker, *Econ. Bot.*, 35:174–186, 1981; P. Felker et al., *Screening Mesquite (Prosopis spp.) for Energy Production on Semi-Arid Lands: Final Report to the U.S. Department of Energy*, National Technical Information Services, 1981; N. Vietmeyer (Coordinator), *Tropical Legumes: Resources for the Future*, National Academy of Sciences, 1979.

Life, origin of

In the search for the beginnings of life on Earth, spectacular progress has recently been made in pushing back the origin of life to the earliest period of the Earth's history. The hints of a biota in the sediments in Isua, Greenland, which are the oldest known on Earth and dated at 3.8×10^9 years, were soon followed by the observation of stromatolites in the Western Australian sequences dated at 3.5×10^9 years. If these recent findings hold up to further scrutiny and can be substantiated by the barrage of tests which these rocks are undergoing, the evidence will compel acceptance of the fact that life is as old as the oldest known sediments on the Earth. The implications of such a finding for the theory of chemical evolution are momentous. Since this newly found early life appears to be photosynthetic and thus well organized, it must be concluded that life seems to have started almost as soon as the Earth was ready for it.

Dating methods. In attempting to determine when life began, the student of chemical evolution has various avenues. First, the dating of the rocks must be incontrovertible. Today geochronology has an array of precise dating techniques, utilizing the uranium-238/lead-204, rubidium-87/strontium-86, lead-206/lead-204, uranium-238/lead-204, and samarium-147/neodymium-143 methods. In some instances, single zircons have been isolated from rocks and the uranium/lead ratios determined. These figures appear to be more reliable than those obtained in whole rock dating. Since the carbonaceous materials cannot be directly dated by these methods, the silicate intrusions of the sediments provide the evidence of the time of deposition.

Molecular fossils. In searching for evidence of the earliest life, molecular fossils are useful. These are molecules of biological importance, whether from the proteins or the nucleic acids, or from the cell walls of an organism retained by the ancient sediments. It is important in the study of these molecular fossils that all possibility of external and recent contamination be removed and that information be sufficient to establish that these molecules are not later contaminants but syngenetic with the rocks themselves.

Along with the search for molecular fossils, there is another useful criterion: the fractionation of the stable isotopes of carbon. Carbon exists in the form of several isotopes, radioactive as well as nonradioactive. The nonradioactive forms are carbon-12 and carbon-13. Biological processes tend to concentrate the lighter isotope. The existence of biological processes such as photosynthesis can be resolved by the ratio of carbon-12 to carbon-13 in the isotopes. Thus, if a ratio of carbon-12 to carbon-13 can be obtained in the study of ancient organic matter, it may be possible to determine whether a particular sample has a biological or nonbiological origin. This intrinsic standard is a powerful evidence for life.

Microfossils. It is possible to examine a thin section of a sediment by an electron microscope and see whether microorganisms are mineralized. Although in some instances there may be no evidence of molecules such as amino acids or hydrocarbons, the structure appears to be preserved. Sometimes these microstructures appear in the form of colonies, so that a whole grouping of them may provide evidence of an ancient biota—for example, the stromatolites.

Perhaps the most exciting recent information has come from a region known as Isua, 150 km northeast of Goddhab, on the western coast of Greenland. Isua, in the Eskimo language, means "the farthest one can go" geographically. This is the farthest geologists have been able to go chronologically. While conducting an economic geological survey of the formations along the Greenland coast, Vic McGregor found sediments in Isua which appeared to be most ancient, based on the tectonic formations and convolutions. When the first dates were examined, the age of the sediments was found to be around 3.7×10^9 years. Later searches and more careful analyses have dated rocks from this formation at 3.83×10^9 years.

Unfortunately, the Isua formation has been extremely metamorphosed, providing no evidence of microstructures. Microfossil evidence seems to have been wiped out completely. In spite of painstaking searches, no convincing structural evidence of ancient microbiota has been found. Searches have also been made for organic molecules. Here again, the results have been disappointing, since fragments have been extremely small, prohibiting the determination of the biological or prebiological origin of the material. However, graphite has been isolated from many of the samples. The graphite must have originated from some carbonaceous material which could have been biological or prebiological. In some instances the graphite appears to be somewhat amorphous as determined by x-ray crystallography, suggesting that some of the graphite has not been completely metamorphosed. Indeed, it was in these fragments that the rudimentary evidence of organic molecules was found.

A large number of these graphite samples have

been analyzed for their isotope fractionation, and in most cases the numbers suggest that the material may be biological in origin. However, there is a spread of numbers which goes from what appears to be truly biological to the nonbiological. There is a possibility that the nonbiological ratios might have resulted from metamorphism. Further study is required to establish this fact. So far, the examination of the Isua rocks seems to indicate that the carbonaceous material in the sediments is biological in origin and to suggest that life may have occurred at the time of deposition of these formations.

Microfossil evidence has been accumulating from some samples from Western Australia and the Archean of Zimbabwe. Examination of the formation at the North Pole called the Pilbara Block has revealed microfossils that have been dated from 3.4 to 3.5 $\times$ 10^9 years. The examination of the Western Australian sediments revealed extensive colonies of microstructures. A similar discovery of stromatolites in limestones assigned a minimum age of 3.5 billion years in the Sebakwian Group of Zimbabwe has also been reported. These formations are a result of the deposition of sediment material over layers of microorganisms and are even visible to the naked eye. This evidence is perhaps the most convincing for the existence of life 3.5 $\times$ 10^9 years ago.

The information from these studies has pushed the date of the origins of life back to the earliest sediments on Earth. New findings of rocks 3.5 $\times$ 10^9 years old reported from India seem to suggest that further avenues may be open for study. In general, geologists have not expected discoveries of material of biological origin in rocks older than those at Isua. The data from India, however, has raised hopes that perhaps some unmetamorphosed material may be found in that region.

Implications. The Earth is 4.6 $\times$ 10^9 years old. If about a half billion years is allowed for the accretion of the Earth and the formation of the oceans, only 2 $\times$ 10^8 years was available for the origins and evolution of life. While this appears small on the geological time scale, from the viewpoint of prebiological chemistry, 2 $\times$ 10^8 years is a very long period, sufficient for the accumulation and interaction of organic molecules giving rise to the first living entity. Indeed, the possibility that many of these processes took place in a much shorter period of time is warranted by laboratory simulation studies, and biological organization may perhaps be independent of geological time.

For background information *see* LIFE, ORIGIN OF; STROMATOLITE in the McGraw-Hill Encyclopedia of Science and Technology. [CYRIL PONNAMPERUMA]

Bibliography: D. R. Lowe, *Nature*, 284:441–443, 1981; S. Moorbath, *Chem. Geol.*, 20:151–187, 1977; J. L. Orpen and J. F. Wilson, *Nature*, 291: 218–220, 1981; C. Walters and C. Ponnamperuma, Characterization of graphite isolated from 3,800 m.y. old metasediments of Isua, Greenland: Implications for organic geochemical studies, *Nature*, in press.

Lupus erythematosus

Systemic lupus erythematosus (lupus) is a chronic systemic disorder classified as an autoimmune disease because patients develop antibodies against one or more components of their own tissues. The underlying cause of this disease and the mechanisms leading to its development are unknown, but recent evidence indicates that a combination of genetic, immunologic, and environmental factors interact to cause the characteristic clinical abnormalities (see illustration).

The clinical course of lupus is typically one of alternating periods of activity and quiescence. Systemic symptoms such as fever, malaise, weight loss, and weakness are common, and localized manifestations are often present in skin, joints, kidneys, and heart. In fact, signs and symptoms may be referable to any organ system or combination of organ systems. Kidney involvement (glomerulonephritis) and brain involvement (cerebritis) account for most deaths in the natural course of the disease.

Autoantibodies and immune complexes. Patients with lupus form antibodies against a wide variety of cellular components, particularly nuclear ones (see table). Although normal individuals will develop small amounts of antinuclear antibodies, particularly during acute illness, only, and evidently all, lupus patients will eventually produce large quantities of antibody against native (double-stranded) DNA.

The antibodies in lupus cause several forms of tissue injury. Some bind directly with cell membrane antigens, mediating destruction of such cells as platelets, red blood cells, and lymphocytes. Most target antigens in lupus, however, are intracellular and must be released after cell death before they can be bound by antibodies. These antigen-antibody aggregates, or immune complexes, cause injury through activation of complement and other enzyme systems that mediate the inflammatory process.

In many instances, immune complexes form in the blood and subsequently lodge in tissue, often blood vessel walls, and cause an inflammatory reaction. When immune complexes localize in the renal glomerulus, they initiate the events of glomerulonephritis. The glomerulus is particularly prone to immune complex deposition because it consists of a network of high-pressure capillaries specifically adapted for plasma protein ultrafiltration.

Several patterns of glomerular injury are described in lupus. This variability, like that of the other clinical features of the disease, reflects the diversity of the antigen-antibody complexes that may occur. The relative proportions of antigen and antibody, the predominant immunoglobulin type and class and its ability to bind complement, the nature of the antigen, and the net electrochemical charge of the complex are among the variable factors that determine the site of immune complex deposition in the glomerulus and the destructiveness of the in-

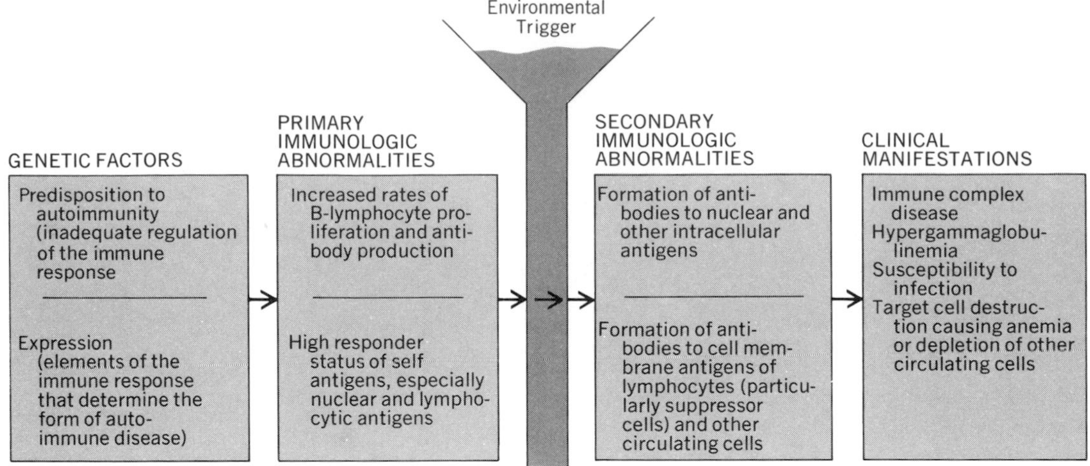

Theoretic scheme for the mechanisms involved in the development of lupus. The primary immunological abnormalities are direct manifestations of the genetic abnormalities and therefore are detectable in all lupus patients, whether the disease is active or not. The secondary immunologic abnormalities are mediated or triggered by the effects of environmental factors and are proportionate in degree to the activity of the disease. The types and activities of autoantibody systems among patients determine the individual patterns of clinical presentation.

flammatory response that ensues.

Lupus is typical of many other diseases in which circulating immune complexes cause glomerulonephritis. In some of these, as in lupus, antigens are endogenous, being released from injured or otherwise abnormal tissues. In many other instances, antigens are exogenous, originating from infecting organisms or chemical substances such as therapeutic drugs.

Some immune complexes in lupus form in the tissues instead of the circulation. Apparently the antigen becomes fixed to extracellular sites before antibody binding occurs. Such is the case of the antibodies that are detected in lupus patients as a linear band along the dermal-epidermal junction of the skin. Because both intact and partially denatured (single-stranded) DNA bind nonspecifically to basement membranes, this lupus band probably represents complexes of anti-DNA antibodies and basement membrane-fixed DNA.

Antibody deposits may be detected in small amounts in the skin and renal glomeruli in patients with no clinical evidence of disease, an indication that even in the inactive phase of the disease a small number of antigen-antibody complexes are formed. The source of antigen for this phenomenon is presumably the products of cell breakdown released from tissues with normally high turnover rates. Any process, whether toxic, infectious, or otherwise, that increases the rate of cell death will accelerate tissue antigen release and immune complex formation. For example, ultraviolet radiation injures epidermal cells of the skin and increases nuclear antigen release. It is not surprising, therefore, that sun exposure in lupus patients results in heightened local immune complex formation and an inflammatory reaction that causes a rash. Thus, the immunologic phenomena of lupus may perpetuate and amplify any tissue injury and lead to a period of heightened disease activity.

Among the many varieties of antigens and antibodies (serotypes) encountered in lupus patients, some are identical to those that are characteristic for other autoimmune diseases such as scleroderma and Sjögren's disease (see table). A lupus patient with such a shared serotype will sometimes develop a disease complex that includes features of both disorders.

Defined antigens reactive with lupus sera*

Nuclear antigens
Native DNA
Histones (drug-induced lupus)
Nucleoprotein
Nonhistone protein
RNA protein (scleroderma, mixed connective tissue disease)
HA-glycoprotein (Sjögren's syndrome)
Sm-glycoprotein
Cytoplasmic antigens
La-RNA (Sjögren's syndrome)
Ro-glycoprotein (Sjögren's syndrome)
Ribosomal RNA protein
Cell surface antigens
Erythrocytes
Granulocytes
Lymphocytes
Platelets
Miscellaneous antigens
Clotting factors
Tissue-specific antigens (thyroid, liver, muscle, stomach, adrenal)

*More than one of these antigen-antibody systems may be present in one patient. In parentheses are other autoimmune disorders that share the same immunologic system. Ha, Sm, La, and Ro are abbreviations of the surnames of the patients in whose sera these antibodies were first found.

Immunogenetic factor. Autoantibodies and formation of immune complexes are the proximate causes of disease in lupus, but they are the expressions of an as yet incompletely defined immunogenetic disorder. Invaluable insight into the pathogenesis of lupus has been gained by studying similar, if not identical, diseases in NZB and NZB/W mice and in dogs. Besides forming autoantibodies, patients (and animals) with lupus have elevated levels of circulating antibodies to many exogenous antigens, including some that form spontaneously. This hyperactivity of antibody-forming cells (B-lymphocytes) and the predilection to make autoantibodies probably reflect a genetically determined defect. The production of any specific antibody, including autoantibody, is coded for by an individual immune-response gene and is normally controlled, at least in part, by other cells such as suppressor lymphocytes and their products. Incomplete control by suppressor agents could result in the immunologic abnormalities seen in lupus, partly because the mechanisms by which autoimmunity is normally controlled would be substantially limited. Immunologic controls would be stressed further by the production of antilymphocytic antibodies that preferentially destroy suppressor lymphocytes, a phenomenon that has been described in active lupus.

Apparently, a state of tenuous regulation of the immune system exists, and any stimulus, such as infection, that further augments B-lymphocyte activity may push the system beyond the limits of control and result in a state of autonomy. Whatever other regulatory functions that remain will not significantly offset this hyperfunctional state, but may suppress new, protective immune responses. Therefore, despite excessive antibody production, the patient is susceptible to infection. The formation of antibodies against "helper" lymphocytes, cells that assist in the initial sensitization process of the immune response, may cause further impairment of the patient's resistance to infection. Drugs that suppress the immune system are often effective in the treatment of lupus, probably because they interrupt these abnormalities of immune system function. Unfortunately, these drugs also increase the already heightened predisposition to infection.

In the NZB mouse, lupus is the expression of genetic factors; the same is probably true in humans. Apparently, at least two separate factors are required: one, as described above, may predispose to autoantibody formation; the other may determine the modes of expression of the disease. Differences in the latter would account for the immunologic and clinical variations among the autoimmune diseases. Absence of either factor precludes the development of disease.

Environmental factors. The relationship between virus infection and the pathogenesis of lupus remains speculative. Autoantibody formation and the development of lupus occur with higher-than-expected frequency among household contacts of lupus patients, even including in some instances the family dog. Such observations support the idea that an environmental, possibly infectious, factor is important in triggering the disease. Virus infection is an attractive candidate for such a factor. Viruses are effective nonspecific B-lymphocyte stimulants, and some induce changes in lymphocyte membranes that increase their antigenicity and initiate antilymphocytic immune responses. Individuals with lupus develop high antiviral antibody titers; immunoglobulin extracted from the kidneys of lupus patients with glomerulonephritis often has strong antiviral activity.

A number of studies have linked lupus in animals and humans to infection by C-type viruses. These RNA viruses are typically transmitted vertically; that is, they are inherited as genes from parent to offspring hosts. C-type viruses can provoke the production of antinuclear antibodies in experimental animals, and studies have shown that animals and patients with lupus have both detectable C-type viral antigens in the membranes of circulating lymphocytes and antibodies against those viruses. It has been proposed that the viral genome affects the expression of host immune control genes, allowing the production of excessive amounts of autoantibodies and the development of lupus. Alternatively, the presence of C-type virus genome in lupus patients may be incidental and expressed because of the lupus-associated abnormalities in gene regulation. Antibodies against the viral genome might then develop as a subset of the antinuclear antibodies. As a rule, antiviral antibodies in lupus patients are reactive to multiple organisms, and it is likely that the high titers of these antibodies reflect the susceptibility that these patients have to infection coupled with an augmented immune responsiveness.

For background information *see* IMMUNE COMPLEX DISEASE; IMMUNOPATHOLOGY; LUPUS ERYTHEMATOSUS in the McGraw-Hill Encyclopedia of Science and Technology.

[DONALD C. HOUGHTON]

Bibliography: A. Chubick, *Adv. Intern. Med.*, 26:467–487, 1980; S. Cohen, P. A. Ward, and R. T. McClusky (eds.), *Mechanisms of Immunopathology*, 1979; J. L. Decker et al., *Ann. Intern. Med.*, 91:587–604, 1979; F. W. Quimby and R. S. Schwartz, *Pathobiol. Ann.*, 8:35–59, 1978.

Magnetic ferroelectrics

Recent research efforts on materials which are both magnetically ordered and spontaneously polar have enabled considerable advances to be made in understanding the interplay between magnetism and ferroelectricity. The existence of both linear and higher-order coupling terms has been confirmed, and their consequences studied. They have given rise, in particular, to a number of magnetically induced polar anomalies and have even provided an example of a ferromagnet whose magnetic moment per unit volume is totally induced by its coupling via linear terms to a spontaneous electric dipole moment.

Most known ferromagnetic materials are metals or alloys. Ferroelectric materials, on the other hand, are nonmetals by definition since they are materials which can maintain a spontaneous electric moment per unit volume (called the polarization) which can be reversed by the application of an external electric field. It therefore comes as no surprise to find that there are no known room-temperature ferromagnetic ferroelectrics. In fact, there are no well-characterized materials which are known to be both strongly ferromagnetic and ferroelectric at any temperature. This is unfortunate since not only would a study of the interactions between ferromagnetism and ferroelectricity be valuable as basic research but such interplay could well give rise to important device applications.

Most antiferromagnetic materials, however, are nonmetals, so there is no apparent reason why ferroelectricity and antiferromagnetism should not coexist. A study of antiferromagnetic ferroelectrics would also provide much information concerning the interplay of magnetic and ferroelectric characteristics even if the device potential were very much reduced. Somewhat unaccountably, antiferromagnetic ferroelectrics are also comparative rarities in nature. Nevertheless, a few are known, and among them the barium–transition-metal fluorides are virtually unique in providing a complete series of isostructural examples. First characterized in the late 1960s, they have the chemical composition $BaXF_4$ in which X is a divalent ion of one of the $3d$ transition metals, manganese, iron, cobalt, or nickel. Nonmagnetic, but still ferroelectric, isomorphs (that is, isostructural equivalents) also exist in the series in which X is magnesium or zinc.

The antiferromagnetic ferroelectrics $BaMnF_4$, $BaFeF_4$, $BaNiF_4$, and $BaCoF_4$ and their nonmagnetic magnesium and zinc counterparts are orthorhombic and all spontaneously polar (that is, pyroelectric) at room temperature. For all except the iron and manganese materials, which have a higher electrical conductivity than the others, the polarization has been reversed by the application of an electric field, so that they are correctly classified as ferroelectric, although their ferroelectric transition (or Curie) temperatures are in general higher than their melting points. The elementary magnetic moments of the manganese, iron, cobalt, and nickel structures are thermally disordered at room temperature, and long-range antiferromagnetic ordering sets in at temperatures somewhat below 100 K. Structurally the materials consist of XF_6 octahedra which share corners to form puckered xy sheets which are linked in the third dimension z by the barium atoms. The magnetic interactions are very dominantly within the xy layers, leading to two-dimensional magnetic characteristics, with the magnetic spins eventually aligning themselves antiferromagnetically within the planes. The axis of the spontaneous polarization is also contained within the planes.

The importance of these magnetic ferroelectrics is the opportunity they provide to study and to separate the effects of a variety of magnetic and nonmagnetic excitations upon the ferroelectric properties and particularly upon the spontaneous polarization. Measurements are often made via the pyroelectric effect p, which is the variation of polarization P with temperature T; that is, $p = dP/dT$. This effect is an extremely sensitive indicator of electronic and ionic charge perturbations in polar materials. Through these perturbations the effects of propagating lattice vibrations (phonons), magnetic excitations (magnons), electronic excitations (excitons), and even subtle structural transitions can all be probed with precision.

Phonons. At sufficiently low temperatures the dominant contribution to the pyroelectric effect p comes from thermal excitation of acoustic-phonons, with p varying as T^3. At higher temperatures optic-phonon-mode excitations contribute additional components of exponential form and tend to dominate the acoustic terms. In the magnetic ferroelectrics these contributions must compete with the components arising from the thermal disruption of the ordered antiferromagnetic state. These low-temperature magnetic components take different forms depending on the character of the low-temperature magnetic ordering (in particular, depending on whether the magnetic anisotropy is large, as in the iron and cobalt systems, or small as in the manganese and nickel analogs). The two types of phonon terms are present, of course, in both the magnetic and the nonmagnetic materials, with the additional magnon contributions present only in the magnetic systems. Although the basic contributions are readily detectable in the pyroelectric response, a full separation of all three contributions has not yet been convincingly obtained at low temperatures.

Magnons. The magnetic contributions to p are seen most easily near the magnetic transitions (that is, the Néel points). Since the magnetic systems undergo very drastic changes, here the perturbation of p is large and easily recognizable. The Néel temperature anomalies take on a cusplike form and arise physically because both the polarization and the magnetic interactions couple strongly to crystal strain, the former via the piezoelectric effect and the latter via magnetrostriction. The details of the magnetically produced cusps differ for the different magnetic materials depending on the details of the magnetic anisotropy, but they are now very well understood. Detailed theories have been given, and the quality of agreement between theory and experiment indicates that the magnetic to ferroelectric coupling mechanisms are now known in considerable detail. Since the magnetism is basically of a two-dimensional nature, a significant vestige of the Néel point p-anomalies persists to temperatures well in excess of the transition. This reflects the well-known persistence of short-range spin correlations in two dimensions.

Excitons. The optical absorption spectrum in the magnetic ferroelectrics $BaXF_4$, is dominated throughout the visible region by the d-state-to-d-state localized electronic excitations on the magnetic ions. The approximately cubic crystal field environment splits the magnetic d states into two subgroups which are typically separated in energy by values corresponding to the visible or near-visible range. The excited states on the individual magnetic ions tend to be long-lived at low temperatures, and the excitation can propagate through the crystal via magnetic (exchange) interactions with its magnetic neighbors. These excitations are known as Frenkel excitons, and they couple to the spontaneous polarization via the elementary electric moment perturbation associated with each exciton due to the different charge distributions in the ground and excited *d* states. The electrical response of the magnetic crystals to short-duration optical pulses clearly shows the exciton contributions to *p* and their relaxation as the excitations gradually decay back to their ground states. In $BaMnF_4$, for which the most detailed exciton work has been performed, the excited-state polarization is of opposite sign to the spontaneous polarization, enabling a particularly convincing verification of the exciton origin of the effect to be made.

Structural transitions. Of all the X ions present in the series $BaXF_4$, the largest is Mn^{2+}. As the temperature is reduced from room temperature, the fluorine cages contract and eventually the divalent manganous ion becomes too big for its cage, precipitating a structural transition at 250 K. For the rest of the X ions this overcrowding effect does not occur. The resulting transition in $BaMnF_4$ is quite a complicated one, and produces a cell doubling in the *xz* plane and a repeat distance in the *y* direction which is incommensurate with the high-temperature unit cell. The transition involves a complex series of rigid rotations of the fluorine octahedral cages. The subtle atomic rearrangements which take place perturb the local charge environment in such a way that the response is again easily picked up by the pyroelectric effect. A pronounced but rather diffuse peak is seen in *p*, although the structural transition is known to be sharp from the findings of other probes such as neutron or light scattering, and fluorine nuclear magnetic resonance. A detailed interpretation of this pyroelectric response has not yet been forthcoming.

Magnetoelectric effect. One of the more interesting effects of the incommensurate phase transition at 250 K in $BaMnF_4$ is that it produces a lower-temperature phase with a crystal symmetry low enough to support the existence of the linear magnetoelectric effect. The couplings between magnetism and ferroelectricity discussed above all involve quadratic spin terms. The magnetoelectric effect in contrast is a linear coupling between magnetization and polarization. It is forbidden by symmetry in most magnetic systems, but its presence when allowed always raises the possibility of pyroelectrically inducing a ferromagnetic moment in an antiferromagnetically ordered system. Below the antiferromagnetic transition at 26 K in $BaMnF_4$ this linear coupling does indeed produce a canting of the antiferromagnetic sublattices through a very small angle (of order 0.2 degree of arc). The result is a spontaneous, polarization-induced magnetic moment. At low temperatures $BaMnF_4$ is therefore technically a weak ferromagnet, although the resultant magnetic moment is extremely small, and it is more usually referred to as a canted antiferromagnet. Pyroelectrically driven ferromagnetism is extremely rare, and this is the only well-categorized example. The magnetoelectric effect also produces additional perturbations of the dielectric susceptibility near the magnetic Néel temperature for the manganese material which are not seen in the other magnetic ferroelectrics of the series for which the linear coupling term is absent for symmetry reasons.

For background information *see* ANTIFERROMAGNETISM; CRYSTAL STRUCTURE; EXCITON; FERROELECTRICS; MAGNON; PHONON in the McGraw-Hill Encyclopedia of Science and Technology.

[MALCOLM E. LINES]

Bibliography: D. L. Fox and J. F. Scott, *J. Phys. C: Solid State Phys.*, 10:L329–L331, 1977; A. M. Glass et al., *Commun. Phys.*, 2:103–107, 1977; A. M. Glass et al., *Ferroelectrics*, 22:701–704, 1978; J. F. Scott, *Rep. Progr. Phys.*, 12:1056–1084, 1979.

Marine geology

The continental margin along the Atlantic seaboard of North America is constructed of sedimentary rocks that overlie a basement of crystalline (igneous and metamorphic) rocks. The greatest thickness of these sedimentary rocks, over 10 km, was deposited in narrow, deep basins that parallel the margin (Fig. 1). These marginal basins began to form about 1.80 to 2 × 10^8 years ago as North America and Africa broke apart, with large blocks of crystalline rock dropping into the resulting rift. This rift zone was similar to the present East African Rift System. Initially, the sedimentary material eroded off the continent was trapped in these marginal basins, which are found on both the North American and African sides of the Atlantic Ocean. As North America and Africa continued to separate, a deep ocean basin—the Atlantic—was formed of new igneous rock emplaced seaward of these marginal basins. Eventually these marginal basins were unable to contain the continued influx of sedimentary material, and the surplus spilled onto the deep ocean floor. At this time the continental margin began to acquire the familiar morphology that remains today of a continental shelf, slope, and rise (Fig. 2). The continental slope, a narrow zone of rapidly deepening sea floor, separates the flat, shallow continental shelf from the gradually dipping, deeper continental rise that extends into the deep sea.

Sedimentary material continued to be deposited on the shelf as the margin slowly sank under the weight of these accumulating sedimentary rocks and as a result of changes in the thermal and chemical properties of the underlying crystalline rocks. The shape of the sea floor changed gradually as this new material buried the old sea floor. In response to changes in sea level, varying ocean current circulation patterns, and variations in the supply of sed-

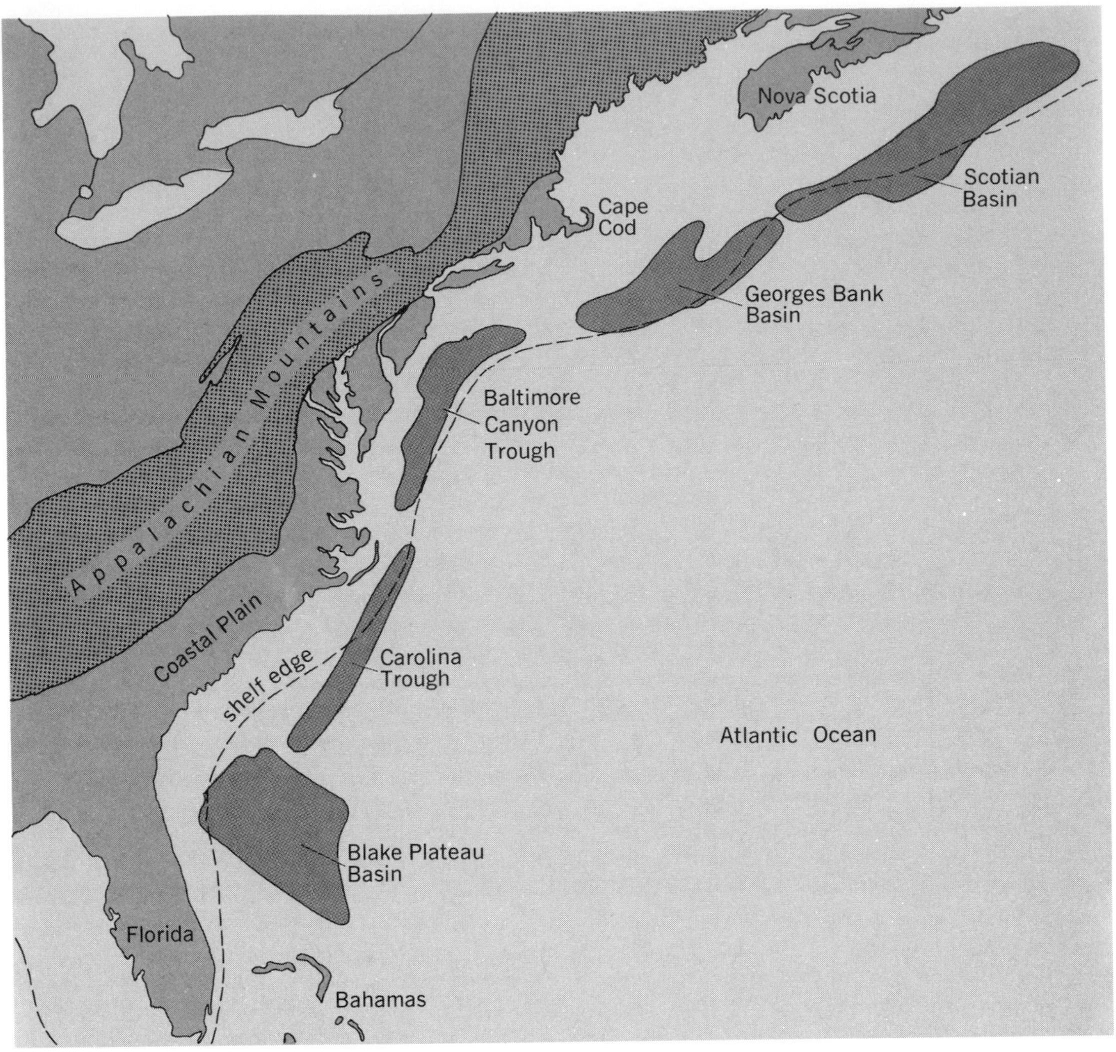

Fig. 1. North American continental margin with locations of sediment-filled marginal basins.

imentary material, the position of the continental slope moved landward or seaward while the shelf depth was kept just below sea level. Along most of the margin, the zone within which the slope position oscillated is at least 30 km wide.

Recent investigations of the Atlantic continental margin by the United States Geological Survey, the Geological Survey of Canada, and various oceanographic institutions have delineated the deep structure of these marginal basins and mapped the pattern of ancient continental shelves, slopes, and rises that overlie these marginal basins. The primary tool in these studies is the seismic reflection system, developed by the oil industry to facilitate the search for gas and oil. Acoustic energy is transmitted from a ship and a long array of acoustic receivers towed behind the ship is used to record the reflected energy. Any surface that forms the contact between materials of differing acoustic properties will reflect some of the energy while allowing the rest to pass through it. Besides the interface between water and sedimentary rocks at the sea floor, the contacts between sedimentary layers of different rock types, such as shale, limestone, and sandstone, will reflect some acoustic energy. The seismic reflection record (Fig. 3), a graphic output of the system, is a map of the depth to these acoustic reflectors along the ship's track, providing a cross-sectional picture of the traversed region. These seismic reflection records are compared with drill hole information in the area to identify the age and type of sedimentary rock between the strongest of these acoustic reflectors.

Marginal basins. There are five major sedimentary basins on the Atlantic margin between Florida and Newfoundland (Fig. 1). From north to south, these marginal basins are the Scotian Basin offshore Nova Scotia, Georges Bank Basin east of Cape Cod, the Baltimore Canyon Trough east of New Jersey and Delaware, the Carolina Trough off North and South Carolina, and the Blake Plateau Basin east of Georgia and Florida and north of the Bahamas. Basin widths vary from less than 100 km to over 300 km, and sedimentary rock thicknesses ex-

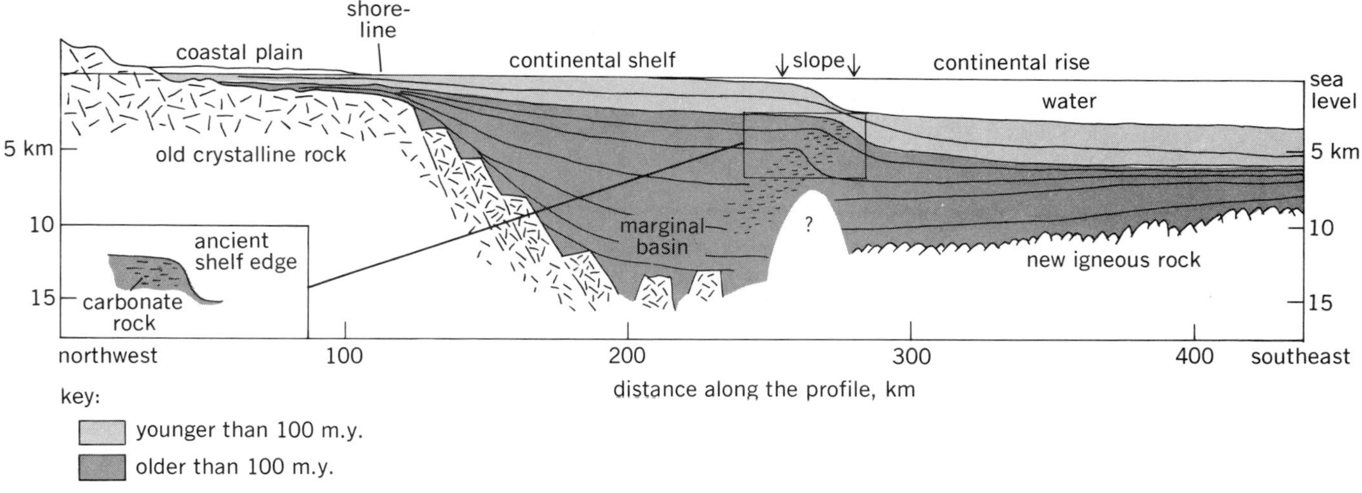

Fig. 2. Schematic cross section of the North American continental margin. The question mark indicates an area of the basement not yet characterized.

ceed 10 km in all of them. The base of these basins is thought to be formed of a mixture of igneous rock emplaced at the time of rifting and blocks of older igneous and metamorphic rocks, such as granites and slates, that broke from the edge of the continent at the time of rifting. The landward edge of these basins is a block-faulted zone where basement deepens rapidly from a few kilometers to over 10 km (Fig. 2). Landward of this point, the continental crust was relatively unaffected by the rifting. The overlying sedimentary rock thins to the west on the coastal plain until crystalline rock of the Appalachian Mountains are exposed at the surface. The initial seaward edge of the basins is in the vicinity of the ancient continental slope location, but its actual position is obscured on the seismic reflection records (Fig. 3) by the overlying complex structure of the ancient shelf edges.

The rift zone was initially heated and uplifted as the result of the rifting process and the injection of hot igneous rock. The subsequent downdropping of large blocks to fill the space that resulted from the separation of North America and Africa produced a deep rift valley that later became the base of the marginal basins. As the zone of igneous emplacement moved seaward, keeping near the center of the newly forming deep-ocean Atlantic Basin, the source of heat was removed and the rock beneath the marginal basin began to cool. The reduction in volume associated with this cooling resulted in the

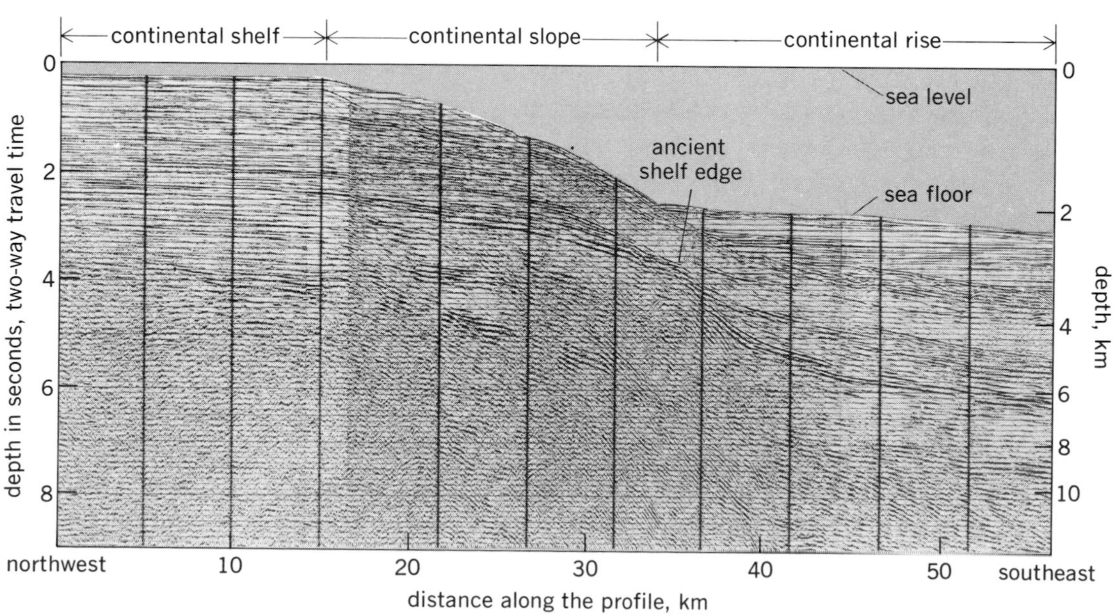

Fig. 3. Seismic reflection record across the shelf edge on the western end of Georges Bank Basin showing the buried ancient shelf edge. Vertical exaggeration is about 2:1.

sinking of the basin. This sinking allowed the continued accumulation of sedimentary material on the shelf area, keeping it near sea level. Later, as the rate of cooling, and thus the sinking, decreased, the locus of sediment accumulation shifted seaward, broadening the continental shelf, pushing the continental slope seaward, and building the continental rise.

The locus of sediment accumulation also shifted back and forth between the shelf and rise in response to changes in sea level. The sediment accumulation and erosion process on the margin tended to keep the shelf depth just below sea level. With a constant sea level and more than sufficient supply of sedimentary material to keep the slowly sinking shelf just below sea level, the shelf would build seaward. When the sea level rose, most of the sedimentary material would be deposited on the shelf, building it vertically. A drop in sea level would expose part or all of the shelf to erosion by wave action, rapidly removing the uppermost sedimentary rocks until the shelf was once more below sea level. The eroded material would be deposited on the continental rise. This sudden flood of new material, plus the influx of sedimentary material that normally spilled over the shelf edge, tended to produce a second zone of fairly thick sedimentary rocks beneath the continental rise.

Ancient shelf edges. In general, the present and the ancient shelf edges at the continental slope are the most impressive and economically the most important structures on the continental margin. During the first 10^8 years of the Atlantic margin development, limestones, dolomites, and other carbonate rock accumulated at the shelf edge. At some points along the margin, carbonate banks and reefs developed. This type of environment is associated with gas and oil in many other parts of the world. During this period the shelf built upward and seaward. The youngest of the ancient shelf edges associated with this period is clearly seen in the seismic reflection records (Fig. 3). The tendency of the massive carbonate rock to reflect a great deal of energy makes the upper surface appear prominently on the seismic record. Unfortunately this same rock absorbs a great deal of the energy that tries to pass through it, obscuring any of the structures below it. Later the changes in sea level and erosion of the shelf caused the shelf edge to move landward and then seaward again, enhancing the acoustic difference in the sedimentary layers above and below this 10^8-year-old shelf edge. With respect to the present shelf edge, the shelf edge that existed 10^8 years ago lies about 10 km landward for the Scotian Basin, at the same location for the Georges Bank Basin, up to 30 km seaward for the Baltimore Canyon Trough, and about 10 km seaward for the Carolina Trough. The Blake Plateau, which was unable to remain at sea level, lies in about 1 km of water depth, and its ancient shelf edge lies about 350 km seaward of the present shelf edge. This ancient shelf edge is the location of a significant (but not yet commercial) gas discovery in the Baltimore Canyon Trough and the site of most of the proposed future United States Atlantic offshore petroleum exploration.

For background information *see* CONTINENTAL MARGINS; GEOPHYSICAL EXPLORATION; MARINE GEOLOGY; SEISMIC EXPLORATION FOR OIL AND GAS in the McGraw-Hill Encyclopedia of Science and Technology. [KIM D. KLITGORD]

Bibliography: J. S. Schlee et al., *Oceanus*, 22:40–47, 1979; M. Talwani, W. Hay, and W. B. F. Ryan (eds.), Deep drilling results in the Atlantic Ocean: Continental margins and paleoenvironment, *American Geophysical Union Maurice Ewing Series*, vol. 3, 1979; J. S. Watkins, L. Montadert, and P. W. Dickerson (eds.), Geological and geophysical investigations of continental margins, *Amer. Ass. Petrol. Geol. Mem.*, no. 29, 1979.

Marine mining

The 1970s witnessed the mushrooming of interest in deep-sea manganese nodules as a result of detailed geochemical and oceanographic research, with more attention shown to this potential ore deposit by mining companies, mineral economists, environmentalists, and the United Nations Law of the Sea Conference. Recent measurements on the potato-shaped concretions of ferromanganese oxide have elucidated the origin of these metal-rich deposits, the complex growth histories of which are revealed by the textures of nodule interiors shown in Fig. 1. Techniques such as electron microprobe analysis, scanning electron microscopy, and high-resolution transmission electron microscopy have enabled geochemical and mineralogical measurements to be made on individual metal-rich layers inside the nodules.

Fig. 1. Reflected-light photograph of the polished surface of a sectioned manganese nodule showing the complex growth history of the concretionary deposit (diameter 4 cm).

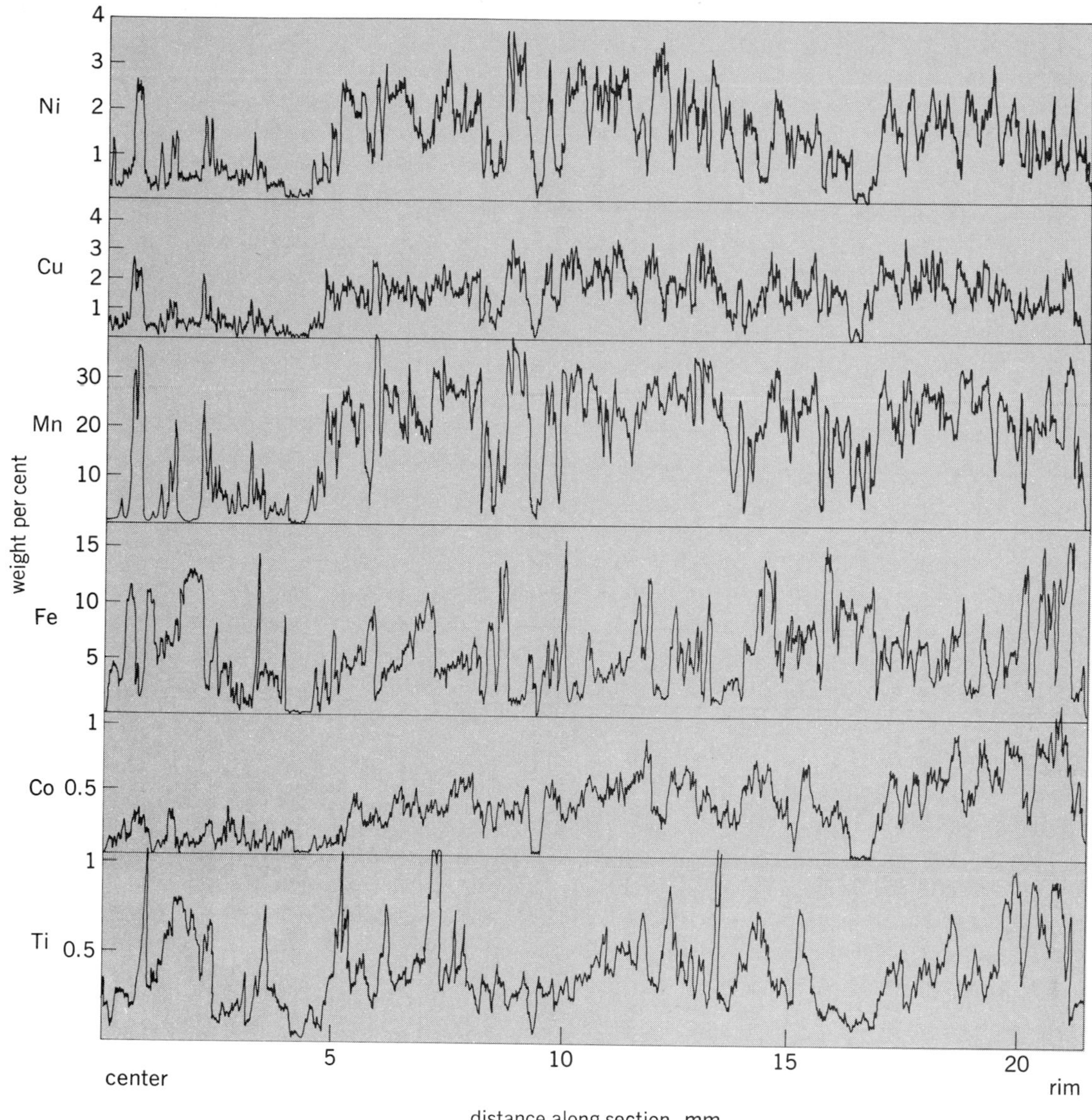

Fig. 2. Electron microprobe traverses a Ni + Cu–rich manganese nodule from the north equatorial Pacific. The chemical analyses of areas only a few micrometers in diameter reveal the enrichments of Ni + Cu in manganese oxide layers. The fluctuating metal concentrations reflect the periodic rolling-over of the manganese nodule on the sea floor.

Concretions of manganese and iron oxides on the floors of many oceans were first described and classified by J. Murray during the historic voyage of HMS *Challenger* in 1873–1876. However, it was 80 years later, during doctoral dissertation research of J. L. Mero at Berkeley, CA, that the economic potential of marine manganese nodules became apparent: nodules from certain regions are significantly enriched in nickel, copper, cobalt, zinc, molybdenum, and other elements so as to make them important reserves for these strategic metals. In recent years, oceanographic surveys have delineated areas of the world's sea floor where nodule abundances and metal concentrations are highest, while geochemical research has revealed information on the mechanism of enrichment of Ni, Cu, Co, and so on, into manganese nodules.

Regions of metal-rich nodules. Although manganiferous nodules and crusts have been sampled or observed on most sea floors, attention has focused on the nickel-plus-copper–rich nodules (2–3 wt% metals) from the north equatorial Pacific in a belt stretching from southeast Hawaii to Baja, California, as well as the high-cobalt nodules from seamounts in the Pacific Ocean. Manganese nodules from the Atlantic Ocean and from higher latitudes in the Pacific Ocean have significantly lower concentrations of the minor strategic metals. However, current surveys of the Indian Ocean are revealing metal-enrichment trends comparable to those found in the Pacific Ocean nodules; high Ni + Cu–bearing nodules are found adjacent to the Equator.

The metal-rich nodules of the north equatorial Pacific are underlain by siliceous ooze sediments composed primarily of skeletons of radiolarians and diatoms. The sea floor here is generally deeper than the so-called carbonate compensation depth, below which the $CaCO_3$ in tests of foraminifera, coccoliths, and other calcareous organisms is dissolved at the high pressures (≈500 atm or 50 mPa) and low temperatures (≈2°C) of the deep ocean. South of the Equator, calcareous ooze sediments predominate and abundances of manganese nodules are considerably lower. At higher latitudes, away from the high biological productivity surface waters which occur near the Equator, pelagic red clays underlie the manganese nodules which generally have lower Ni + Cu, but higher Fe, contents than equatorial nodules.

The growth rate of marine manganese nodules, determined by decay measurements of traces of radioactive isotopes such as ^{230}Th, ^{231}Pa, ^{10}Be, and ^{40}K contained in the nodules is very slow, averaging a few millimeters each million years. Since the sedimentation rates of accompanying siliceous ooze or pelagic red clay sediments are almost a thousand times faster than the nodule growth rates, an enigma in manganese nodule geochemistry is how they avoid being buried by sediment raining down on them. Time-lapse sea floor photography and sedimentological studies have demonstrated the presence of surprisingly high populations of benthic organisms living in the inhospitable, dark, high-hydrostatic-pressure, and low-temperature environment on deep sea floors. Profiles through box cores raised from the sea floor containing cubic-meter samples of the uppermost sediments reveal bioturbation structures in the underlying layers, indicating the burrowing habits of benthic organisms. Such movements are believed to roll manganese nodules

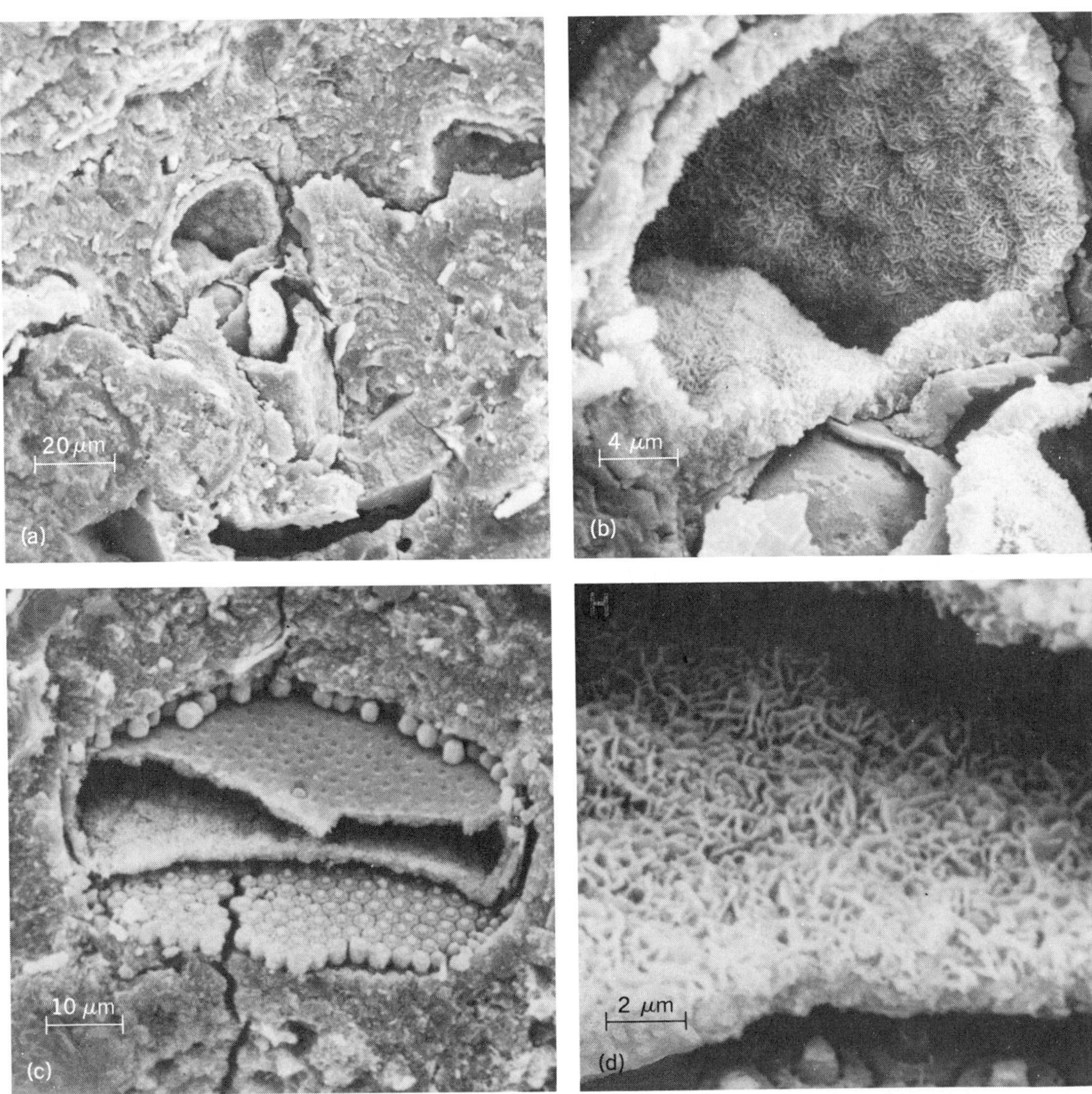

Fig. 3. Scanning electron microscopy photographs showing the growth of Ni + Cu–rich todorokite inside a manganese nodule from the north equatorial Pacific. The meshworks of tiny todorokite crystals (*a*, *b*) line a cavity and (*c*, *d*) replace biogenic debris in a fossil cast inside the nodule. (*From R. G. Burns and V. M. Burns, Manganese oxides, in R. G. Burns, ed., Marine Minerals, Mineralogical Society of America Publication Series Reviews in Mineralogy, vol. 6, pp. 1–46, 1979*)

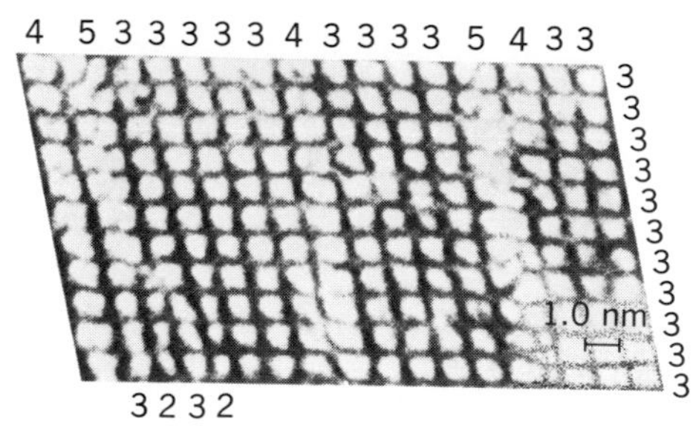

(a)

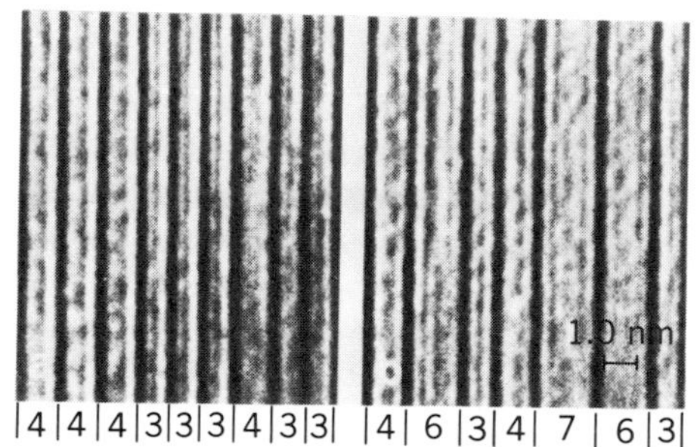

(b)

Fig. 4. High-resolution transmission electron microscope photographs of the lattice of todorokite crystals. The structure consists of tunnels whose walls are formed by linkages of $[MnO_6]$ octahedra, most commonly three octahedra wide. (*a*) A view down the tunnels. (*b*) A view parallel to the tunnels. The numbers refer to the dimensions of the tunnels in terms of the number of edge-shared $[MnO_6]$ octahedra forming the walls. (*From S. Turner and P. R. Buseck, Todorokites: A new family of naturally-occurring manganese oxides, Science, 212:1024–1027, copyright 1981 by the American Association for the Advancement of Science*)

over periodically and to keep them at the sediment–sea water interface.

Chemistry of the nodules. Microchemical analyses have revealed that chemical differences exist between the outermost top (exposed to sea water) and bottom (immersed in sediment) layers of manganese nodules. Surfaces buried in underlying sediments are generally higher in Mn, Ni, and Cu contents, compared to the more Fe+Co–rich surfaces exposed to sea water. Episodic rolling-over of a nodule accounts for fluctuating concentrations of Mn, Fe, Ni, Cu, Co, and other metals across sectioned manganese nodules, demonstrated by electron microprobe measurements such as those shown in Fig. 2. Observations by scanning electron microscopy point to a biological source of the metals and diagenetic remobilization of them in the underlying sediment (Fig. 3). Very-fine-grained, highly dispersed clay minerals and oxides of iron and manganese derived from the continents are buried with biogenic debris from deceased organisms living in surface waters of the equatorial photic zone. Traces of the strategic metals Ni, Cu, Co, and so on, either are adsorbed onto the surfaces of clays or Mn and Fe oxides or are organically complexed in biological matter settling to the sea floor. A reducing environment is set up at depths in the buried sediments, and organic debris is oxidized by the Mn(IV) and

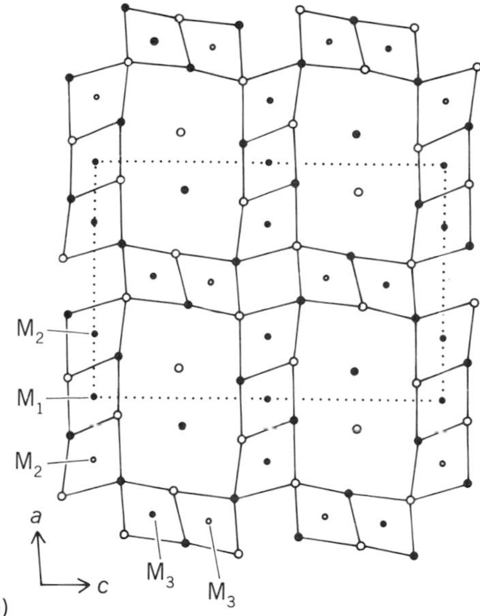

(a)

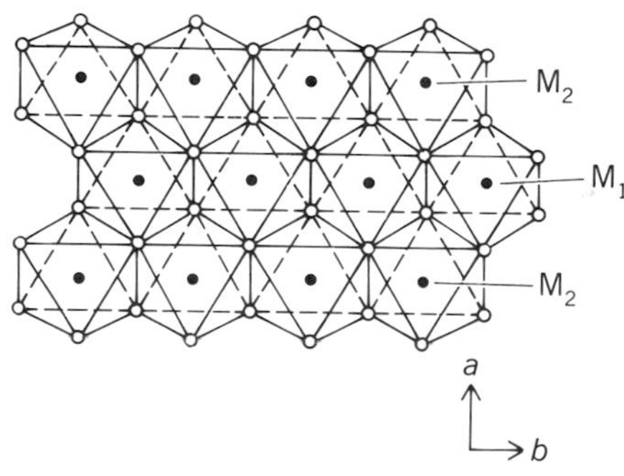

(b)

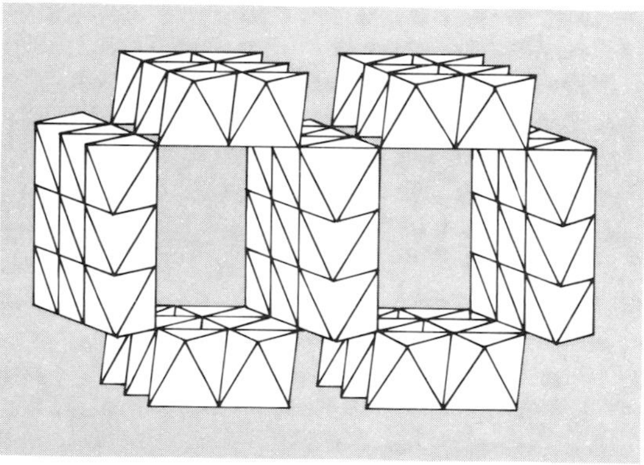

(c)

Fig. 5. Crystal structure of the manganese oxide mineral psilomelane, $(Ba,H_2O)Mn_5O_{10}$, which has a tunnel structure analogous to that of todorokite. (*a*) Projection onto the (010) plane showing four linked tunnels. Mn^{4+} ions are located in the M1 and M3 positions, and divalent cations such as Mn^{2+} and Ni^{2+} occupy the M2 octahedra. Large cations, such as Ba^{2+} and K^+, and H_2O molecules occur inside the tunnels. (*b*) A treble chain of $[MnO_6]$ octahedra parallel to the *b* axis. (*c*) Schematic representation of the 3 × 2 tunnels between chains of edge-shared $[MnO_6]$ octahedra. (*From R. G. Burns and V. M. Burns, Manganese oxides, in R. G. Burns, ed., Marine Minerals, Mineralogical Society of America Publication Series Reviews in Mineralogy, vol. 6, pp. 1–46, 1979*)

Fe(III) oxides, leading to remobilization of soluble cations such as Mn^{2+}, Fe^{2+}, Ni^{2+}, and Cu^{2+} in pore waters of the sediments. Oxidation by aerated pore water or ocean bottom water near the sediment–sea water interface, first of iron and then of manganese at shallower depths in the sediments, produces insoluble authigenic hydrated Mn(IV) oxide phases which accumulate onto underlying surfaces of manganese nodules accreting in the sediments. The Ni^{2+}, Cu^{2+}, and other ions are bound as essential constituents in the crystal structures of the nodule-forming manganese oxides, such as the todorokite crystallites $[(Ca,Na,Ba,Ag)(Mg,Mn^{2+},Zn)Mn_5{}^{4+}O_{12}\cdot H_2O]$ in Fig. 3.

Minerals in manganese nodules. A major development during the past few years has been the characterization of the host minerals in manganese nodules which are responsible for the enrichment of Ni, Cu, Co, and so on. The bane of mineralogical research is the presence of extremely small, poorly crystalline phases in the nodules, studies of which have been facilitated by electron microscopy techniques. In the cobalt-rich nodules from seamounts and relatively shallow sea floors, vernadite (delta manganese dioxide) accommodates Co^{3+} ions which have ionic radii identical to those of Mn^{4+} ions. However, todorokite is the predominant mineral concentrating Ni^{2+} and Cu^{2+} ions in the metal-rich nodules from siliceous ooze sediments. Lattice imaging by high-resolution transmission electron microscopy has confirmed earlier predictions that todorokite has a zeolitelike framework consisting of intergrowths of multidimensional tunnels formed between triple or quadruple chains of edge-shared $[MnO_6]$ octahedra (Fig. 4). The crystal structure accommodates divalent transition metals in these octahedra, while larger cations (K^+, Ba^{2+}, Ca^{2+}) and H_2O molecules occupy voids in the tunnels (Fig. 5). Ions such as Ni^{2+}, Cu^{2+}, and Zn^{2+} stabilize the todorokite structure. However, Mn^{2+}-bearing todorokites in continental ore deposits and in manganese nodules growing in hemipelagic sediments adjacent to continents are prone to oxidize to birnessite. Iron occurs as a separate mineral, feroxyhyte, an oxide hydroxide polymorph of ferric iron, and may be epitaxially intergrown with vernadite in nodules from pelagic red clays and seamounts.

For background information *see* AUTHIGENIC MINERALS; DIAGENESIS; MARINE MINING; MARINE SEDIMENTS in the McGraw-Hill Encyclopedia of Science and Technology.

[ROGER G. BURNS]

Bibliography: V. M. Burns and R. G. Burns, in *La Genèse des Nodules de Manganèse*, Colloq. Int. Centre Nat. Rech. Sci. no. 289, pp. 387–404, 1979; R. G. Burns and V. M. Burns, in C. Emiliani (ed.), *The Sea*, vol. 7, pp. 875–914, 1981; D. S. Cronan, *Underwater Minerals*, 1980; G. P. Glasby (ed.), *Marine Manganese Deposits*, 1977; W. S. Moore et al., *Earth Planet. Sci. Lett.*, 52:151–171, 1981; S. Turner and P. R. Buseck, *Science*, 212:1024–1027, 1981.

Marine sediments

Recent work involving marine sediments has yielded fundamental data concerning seasonal changes in deep-sea sedimentation. This article discusses these findings as well as research involving organic compounds in deep-sea sediments and their possible relationships to oil and gas deposits.

Seasonal changes. Seasonal changes in temperature, salinity, and photosynthetic production of organic matter by free-swimming microscopic algae in large parts of the surface ocean have been recognized for a long time. Only recently was it discovered, however, that expressions of these changes at the sea surface are propagated to the permanently dark and cold depths of the ocean through changes in the quantity and composition of the rain of particles sinking into the deep sea. This rain of particles sustains life in the deep sea and is responsible for the formation of deep-sea sediments far away from the influence of river-borne debris shed by continents and islands. The particles are mostly skeletal remains of tiny plants and animals, fecal particles, and wind-blown dust from the continents.

Sampling. The findings were made possible by a novel method of collecting samples in the deep ocean. The method involves the use of so-called sediment traps (Fig. 1), devices which intercept in

Fig. 1. Sediment trap being recovered in the Sargasso Sea after being recalled to the surface following 2-month collection period at a depth of 3200 m. The sediment sample is in the transparent cup at the bottom of the large collecting funnel.

mid-water some of the rain of particles approaching the sea floor. Sediment traps provide the best measurements thus far of the vertical flux of particles in the ocean. The measurements are expressed in units of mass per unit area and unit time, for example, milligrams per square meter per day ($mg \cdot m^{-2} \cdot d^{-1}$). Seasonal variability in this flux was detected when long-term deployment of a sediment trap with bimonthly sample recovery was first attempted in the Sargasso Sea, in the vicinity of Bermuda.

Quantitative differences. It soon became apparent that in late winter and spring the trap collected about three times as much material per unit time as it did in late summer and fall. This pattern has now been observed for 3 consecutive years and appears to be a consistent characteristic of sedimentation in the Sargasso Sea and probably in much of the world's oceans. Data on seasonal fluctuations of the primary productivity of phytoplankton (microscopic algae) in the same area indicate that those fluctuations are largely responsible for the changes in the intensity of the rain of particles in deep water (Fig. 2). Greater productivity sustains more plant and animal life in the surface water, which in turn means that more particles, both living and dead, sink into deep water. The match between the productivity fluctuations and the flux variations of the major particle constituents (calcium carbonate, silicate, and organic matter) in deep water is very close. It is so close, in fact, that it suggests very rapid sinking of the majority of the particles collected in sediment traps. Their average sinking rate appears to be of the order of 100 m per day.

Interpretation. These findings provide explanations for hitherto puzzling phenomena in the realm of marine biology. Annual breeding cycles have been reported for a number of animals living on the floor of the deep sea, and some clams, believed to be growing very slowly, exhibit what appear to be annual growth bands. Neither of these observations could be reconciled with the traditional view of a never-changing, uniformly cold and dark deep-sea environment. However, an explanation may lie in the influence on the animals of a seasonally fluctuating food supply reaching them in the form of particles sinking from above. Growth and reproduction require energy beyond that needed for mere maintenance of normal body functions. That extra energy is more easily expended at times of greater food availability. Therefore, growth or reproduction, or both, in some deep-sea animals may be triggered at times of seasonally increased food supply.

Surprisingly, even particles of strictly inorganic origin, such as clay delivered to the open ocean as air-borne dust, arrive in the deep sea in seasonally fluctuating amounts. The only way in which these tiny dust particles can reach the deep sea at a rate fast enough to preserve any trace of seasonal change occurring in the surface water is as constituents of larger, rapidly sinking particles. Sinking individually, they would require years to reach the bottom

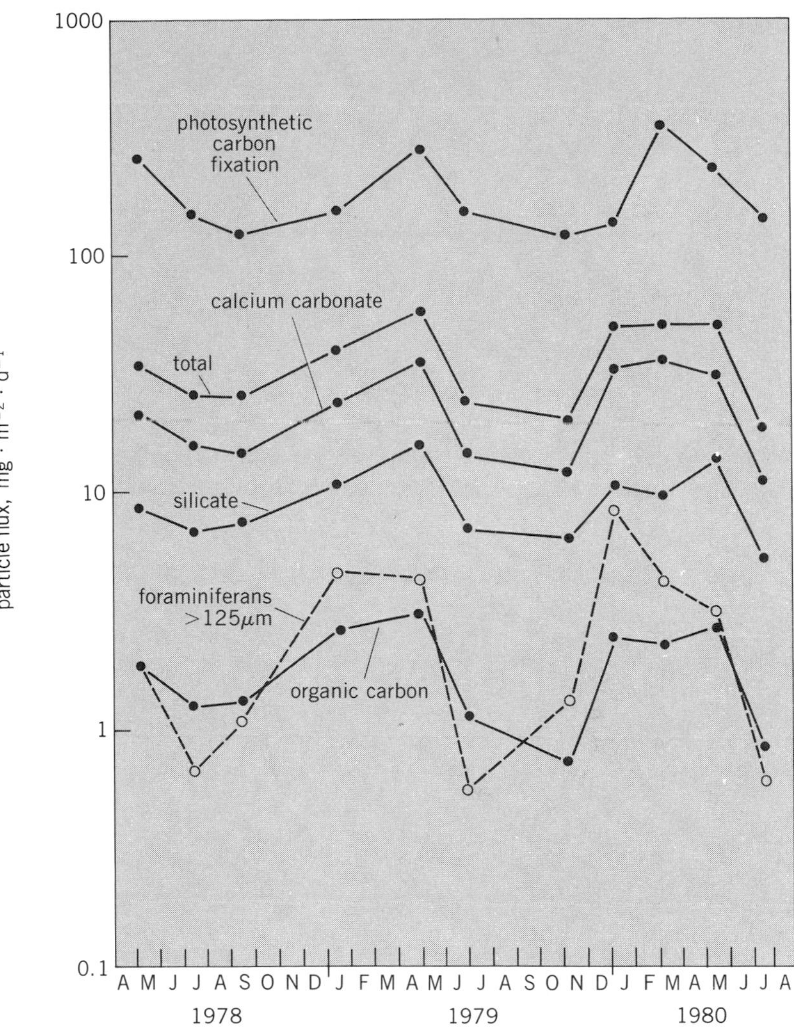

Fig. 2. Seasonal fluctuations in major components of particle flux at a depth of 3200 m in the Sargasso Sea. The average net photosynthetic carbon fixation in surface water (1957–1960) for intervals equivalent to trapping periods is shown for comparison. (*From W. G. Deuser, E. H. Ross, and R. F. Anderson, Seasonality in the supply of sediment to the deep Sargasso Sea and implications for the rapid transfer of matter to the deep ocean, Deep-Sea Res., 28A:495–505, copyright 1981 by Pergamon Press, Ltd.*)

of the deeper parts of the ocean. The prime particle makers in the sea are organisms, and a type of particle commonly observed in sediment traps is fecal pellets of small planktonic animals. The pellets consist of indigestible matter which has passed through the gut and is held together by mucus or organic membranes. Many animals in the sea constantly filter sea water in their search for food and retain both digestible and indigestible particles. The latter are eventually ejected as parts of fecal pellets. It is advantageous to the animals to expend some energy to make firmly packaged pellets: they sink rapidly out of the animals' domain and, therefore, are not likely to be reprocessed repeatedly. Reprocessing indigestible matter is useless expenditure of energy.

The sediment-trap data indicate that more clay sinks into the deep water when organic productivity in the surface water is high. They also indicate that the clay flux in the deep water, except near the bottom where it may be resuspended, accounts for the atmospheric input at the surface. It therefore appears that the organisms are extremely efficient in filtering near-surface water and in packaging its fine-particle load into fast-sinking fecal pellets. That high degree of efficiency suggests that other particles, including anthropogenic ones, and all chemicals associated with particulate matter can be removed from surface water and transmitted to the deepest parts of the ocean within a few weeks. Certain pollutants may thus reach the remote deep sea shortly after having settled on the sea surface.

The new data also show clearly that sedimentary accumulations in any but the tropical regions may be strongly biased toward representing the productive part of the year. For example, even in the subtropical Sargasso Sea daily sedimentation during the productive spring period is three times that during the fall. Even greater differences may be expected for temperate and subarctic regions.

Qualitative differences. Besides quantitative changes in the particle flux to the deep sea through the year, there are also qualitative differences. Different groups of organisms thrive in the surface water at different times of the year. The sequence of living populations is reflected in the arrival in the deep water of skeletal parts and other life remnants having different morphologies and chemical compositions. Most interesting are the hard parts of organisms which as a population persist throughout the year, but as individuals are so short-lived that the composition of successive generations reflects the seasonally changing environment in which they form their skeletons. Foraminifera are one group of such organisms. They are single-celled animals which construct intricate skeletons made of calcium carbonate (Fig. 3). The oxygen in the carbonate is of an isotopic composition dependent only on the isotopic composition of sea water and on the temperature at which the animals build their skeletons. This temperature-dependent composition has been used widely in determining climatic conditions of the past through the isotopic analysis of foraminifera separated from deep-sea sediment cores. The recent results on seasonal changes in sedimentation make it possible also to extract information on summer-winter temperature contrast in the past by measuring the isotopic composition of species which flourish at different times of the year and can now be identified as such in the sediment. Such information is helpful in reconstructing more precisely the climatic conditions and wind and current patterns of the past. Recognition of regularities in the past will aid in understanding the present position in the sequence of climatic cycles and in predicting future climate trends. [W. G. DEUSER]

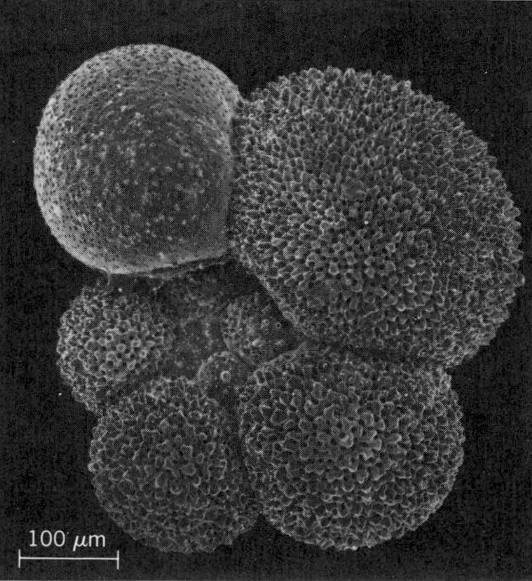

Fig. 3. Calcium carbonate skeleton of the planktonic foraminiferan *Globigerinella aequilateralis*, commonly found in the subtropics.

Organic sediments. Oceanic sediments are a repository for organic matter produced by both marine and terrestrial biologic activity. This organic matter is made up of a mixture of many organic compounds, most of which are complex in chemical structure. After deposition in oceanic sediments, much of the organic matter can undergo thermal alteration that may lead to the generation of petroleum. Oil and gas deposits are possible where thick accumulations of oceanic sediment occur, such as on continental margins and in small oceans. In these areas, as elsewhere, the basic requirement for petroleum occurrence is a rich source of organic matter connected to a reservoir in which thermally generated oil and gas can be trapped.

Sources of organic matter. Phytoplankton (plants) and zooplankton (animals) constitute most of the life in the oceans, and these microscopic organisms are the source of much of the organic matter deposited in oceanic sediments. Most phytoplankton are single-cell algae and are commonly consumed by zooplankton. Both phytoplankton and zooplankton are, in turn, eaten by invertebrates and fish. In addition to marine sources, considerable quantities of organic matter from land plants are transported to the marine environment by rivers and wind. Organic matter that reaches the sea floor from all of these sources may be used by bottom-dwelling organisms and bacteria that contribute their own products to the sediment record. Less than 1% of the organic matter that originates in or is transported to the marine environment is eventually incorporated into oceanic sediment.

Anoxic conditions. In general, waters of the oceans are well oxygenated, so much of the organic matter is oxidized and never incorporated in sediment; where there is rapid sedimentation, however, organic matter may be preserved even where oxic conditions prevail. In some marine environments, oxygen is so depleted that virtually no oxidation of organic matter occurs. These anoxic conditions occur where the chemical and biochemical demand for oxygen exceeds the supply. Important areas where such conditions can be found are in silled basins, which are local depressions on the sea floor; in areas of upwelling bottom water; and in the oxygen-minimum layer of the open ocean (Fig. 4). Anoxic environments in oceanic sediments favor the preservation of organic matter which, under appropriate circumstances, can serve as the source material for petroleum.

Organic compounds. Organic matter deposited on the sea floor is composed of the major polymeric substances that make up living systems, such as proteins, carbohydrates, lipids, waxes, and lignins. Marine sources contribute all of these polymers except lignin and waxes, which come from land plants. The polymeric substances undergo biochemical, physical, and chemical alterations in a process known as diagenesis. Diagenesis takes place at temperatures below 50°C and leads to many of the organic compounds that have been found in oceanic sediment. The compounds range from simple molecules composed only of carbon and hydrogen (hydrocarbons), including alkanes, alkenes, aromatics, and isoprenoids, to complex cyclic compounds such as steranes, sterenes, diasterenes, triterpanes, and triterpenes. In addition to hydrocarbons, they include oxygen- and nitrogen-containing organic compounds. The organic acids are alkanoic, alkenoic, dicarboxylic, hydroxy, isoprenoid, methyl-branched (iso or anteiso), monocarboxylic, naphthenic, and terpenoid acids. The alcohols include fatty alcohols and sterols. Amino compounds are present in the form of amino acids, amino sugars, or amines. There are sterones (steroid ketones) as well as isoprenoid and pentacyclic ketones. Among the pigments are carotenoids, chlorins, and porphyrins. The sediments from marine sources also include sugars and humic substances.

These organic molecules are basically molecular fossils that provide clues to: original sources of organic matter; paleoenvironmental conditions that once existed in the area of deposition; and postdepositional processes that have altered the composition and structure of the compounds. Besides these compounds, the major part of the organic matter formed in sediment during diagenesis is kerogen, a very complex polymeric substance that cannot be extracted from sediment with any common organic or aqueous solvents. Kerogen forms by the condensation and polymerization of many different organic molecules.

That organic compounds and kerogen are common in oceanic sediment has been amply demonstrated by sampling conducted by the Deep Sea Drilling Project. Since 1968, this project has collected sediment samples from all of the world's oceans except the Arctic, and at many locations (Fig. 5) the sediments have been shown to contain organic compounds. The record of organic compounds at some of these locations goes back to the oldest sampled oceanic sediments, deposited during the Jurassic Period about 150,000,000 years ago.

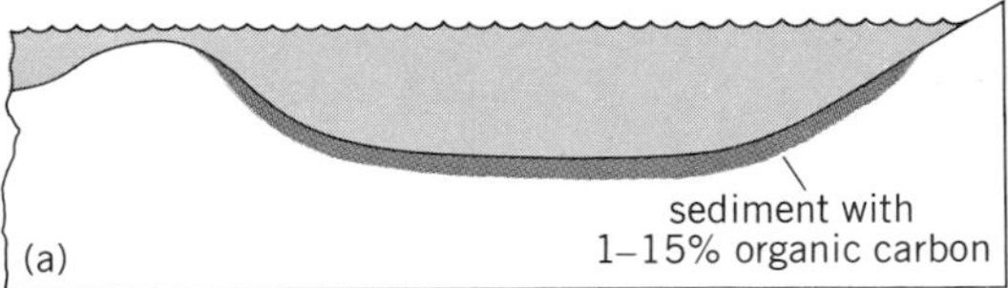

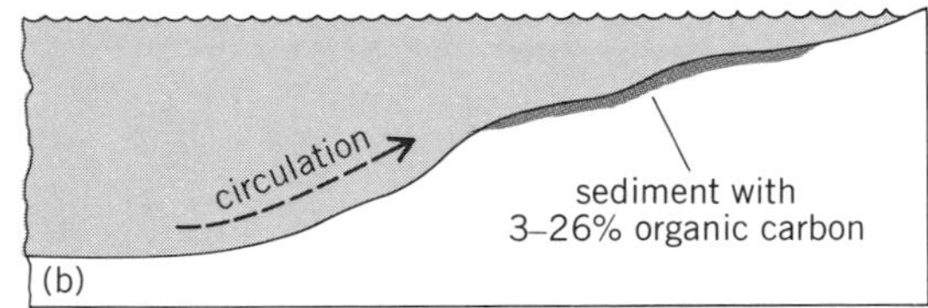

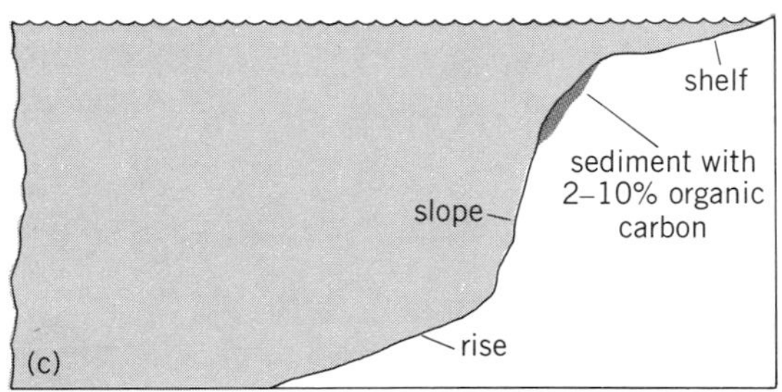

Fig. 4. Diagrams showing three marine settings in which anoxic conditions develop. (*a*) Silled basins (Black Sea, Baltic Sea, Saanich Inlet, and Lake Maracaibo). (*b*) Upwelling (Southwest African shelf and Peru shelf). (*c*) Oxygen minimum layer (Indian Ocean and northwestern Pacific Ocean). (*From G. J. Demaison and G. T. Moore, Anoxic environments and oil source bed genesis, Amer. Ass. Petrol. Geol. Bull., 64:1179–1209, 1980*)

Petroleum potential. Increasing temperatures dictated by the geothermal gradient transform organic matter as it is buried; at temperatures of more than 50°C, organic matter is altered by a process called catagenesis. Most petroleum is formed during catagenesis. If sufficient organic matter is present, oceanic sediments that undergo this process are potential petroleum sources. As a general rule, deeply buried marine organic matter yields mainly oil, whereas land-plant material yields mainly gas.

Samples from the Deep Sea Drilling Project provide an excellent basis for evaluation of the petroleum potential of oceanic sediments. While drilling the first few holes of the project, oillike hydrocarbons were recovered from sediment associated with a salt dome at the Challenger Knoll in the Gulf of Mexico. Very small amounts of liquid hydrocarbons were also detected in sediment from the Shatsky Rise in the western Pacific Ocean, from the Balearic Abyssal Plain in the western Mediterranean Sea, and from the Norwegian Sea in the North Atlantic Ocean.

Particular attention has been given to the evaluation of the petroleum potential of middle Cretaceous (about 100,000,000-year-old), organic-rich black shales that are widespread in areas of the Atlantic Ocean. Geochemical studies have shown that the origin and the potential for petroleum generation of the organic matter in these shales is highly variable.

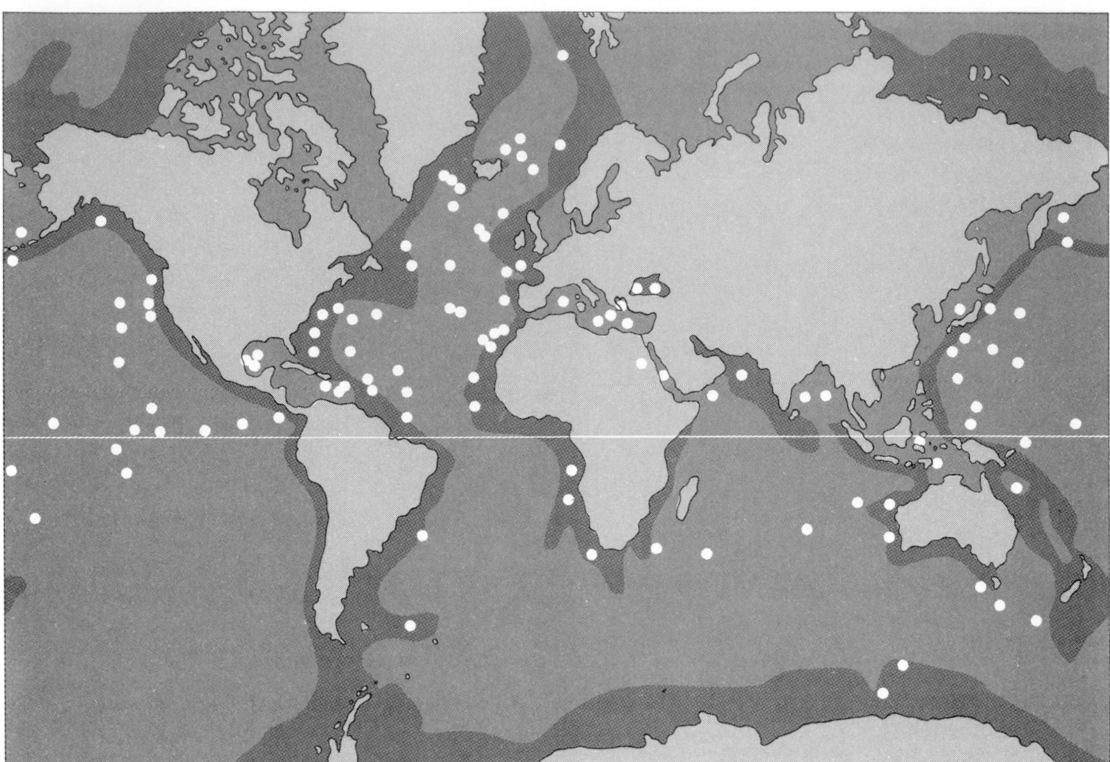

Fig. 5. Map showing locations (solid circles) where samples obtained by the Deep Sea Drilling Project contain organic compounds, and the offshore regions of the world (shaded areas) where petroleum accumulations are likely to occur. (*From H. D. Hedberg et al., Petroleum prospects of the deep offshore, Amer. Ass. Petrol. Geol. Bull., 63:286–300, 1979*)

An evaluation of the kerogen in these black shales shows that three main types of organic matter are present: marine planktonic material deposited in anoxic environments; land-plant material that has been partially degraded; and residual organic matter that has been oxidized or recycled from older sediment. Most of the organic matter that has been found in sampling is thermally immature; that is, it has not undergone sufficient catagenesis to reach levels of alteration considered essential for petroleum generation. However, where sediment is more than 1 to 1.5 km thick on continental margins (outer continental shelf, slope, and rise), organic matter is buried to sufficient depths to be heated to temperatures greater than about 50°C, and catagenesis begins. This deeply buried organic matter, which cannot yet be sampled because of the limited capabilities of current drilling technology, probably has reached a level of thermal maturity at which petroleum generation takes place. Thus, it is possible to evaluate the petroleum potential of oceanic sediment by considering the nature of the organic matter found in shallow oceanic sediment. In basins on the margin off southern Africa in the South Atlantic, the petroleum potential of black shales is good, and it is good to excellent in the basins on the margin off northern Africa in the North Atlantic. These black shales have a very low petroleum potential in the basins on the western margins of the North Atlantic Ocean because much of the organic matter there is recycled from older sediments and, therefore, is not capable of generating either oil or gas.

Petroleum prospects. Geochemical studies of many Deep Sea Drilling Project samples have shown, therefore, that sources of petroleum are possible in thick sediment sections offshore, where organic matter has been buried to depths suitable for catagenesis. The central parts of the bottom of the ocean are not likely to have petroleum deposits because the layer of sediment is too thin, the content of organic matter is too low, and the probability that reservoirs are present to trap the petroleum is very small. The continental margins (shelf, slope, and rise, including deep-sea fans), however, are quite favorable for the generation and entrapment of petroleum (Fig. 5). Sediment at some locations is thick enough that conditions appropriate for petroleum generation have developed, and geochemical studies show that source sediments for hydrocarbons are present. Sedimentation histories of some continental margins indicate that reservoirs and traps are likely to be present.

Finally, small ocean basins, such as the Gulf of Mexico, Black Sea, Norwegian Sea, and the Sea of Japan, are among the most promising areas for petroleum in oceanic sediments (Fig. 5). These basins have thick sections of sediment, and significant amounts of organic matter from both marine and terrestrial sources have accumulated under anoxic conditions. In addition, reservoirs and traps for petroleum are present. These conditions suggest that petroleum accumulations will eventually be found, especially along the borders of these basins.

For background information *see* ANOXIC ZONES; DIAGENESIS; MARINE BIOLOGICAL SAMPLING; MARINE SEDIMENTS; PETROLEUM, ORIGIN OF; PHYTOPLANKTON in the McGraw-Hill Encyclopedia of Science and Technology.

[KEITH A. KVENVOLDEN]

Bibliography: G. J. Demaison and G. T. Moore, *Amer. Ass. Petrol. Geol. Bull.*, 64:1179–1209, 1980; W. G. Deuser and E. H. Ross, *Nature*, 283:364–365, 1980; W. G. Deuser et al., *Palaeogeogr. Palaeoclimatol. Palaeoecol.*, 3:103–127, 1981; W. G. Deuser, E. H. Ross, and R. F. Anderson, *Deep-Sea Res.*, 28A:495–505, 1981; H. D. Hedberg, J. D. Moody, and R. M. Hedberg, Petroleum prospects of the deep offshore, *Amer. Ass. Petrol. Geol Bull.*, 63:286–300, 1979; S. Honjo, J. F. Connell, and P. L. Sachs, *Deep-Sea Res.*, 27:745–753, 1980; D. W. Menzel and J. H. Ryther, *Deep-Sea Res.*, 7:282–288, 1961; B. R. T. Simoneit, The organic chemistry of marine sediments, in J. P. Riley and G. Skirrow (eds.), *Chemical Oceanography*, vol. 7, pp. 233–311, 1978; B. Tissot et al., *Amer. Ass. Petrol. Geol. Bull.*, 64:2051–2063, 1980.

Mass spectrometry

Recent advances in mass spectrometry have resulted in development of the new instrumentation and techniques of ultrasensitive mass spectrometry and Fourier transfer mass spectrometry. This article describes these methods, both of which offer unprecedented analytical utility.

Ultrasensitive mass spectrometry (USMS). This is a recent addition to the many tools available for the analysis of materials. By coupling the established technology of mass spectrometry with a detection system that is centered on a nuclear particle accelerator, ultrasensitive mass spectrometry provides unprecedented sensitivity for specific atoms in the near-surface region of a material.

Just as in classical mass spectrometry, the atoms to be analyzed are produced as ions by directing high-velocity particles onto the surface of the material. As the incident particles slow down in the top atomic layers, the kinetic energy is transferred to the atoms adjacent to the point of entry. This energy transfer can be much higher than the chemical binding energy of the atoms to the material lattices, causing atoms and molecular fragments to be emitted from the surface by the process called sputtering. A substantial fraction of these sputtered secondary atoms and molecules are electrically charged so that it is possible to accelerate them and measure their mass.

Elimination of molecular interference. One basic limitation of classical secondary ion mass spectrometry arises from the background ions that are produced at the sample from molecules that have equal or nearly equal mass to that of the element of interest. For example, if the spectrum of secondary ions produced from the surface of a geological sample in the region around mass 88 is examined, where only the minor components rubidium and strontium would be expected with any frequency, it is found that the whole spectrum from mass 67 to 116 is cluttered with intense peaks which originate from molecules formed from combinations of the lighter components sodium, calcium, aluminum, silicon, and oxygen. These molecules often reduce the effective sensitivity of classical secondary ion mass spectrometry by orders of magnitude.

In 1976 it was pointed out that these molecular interferences could be eliminated by the use of higher ion energies. The method is based on the fact that if particles at energies of a few megaelectronvolts are passed through a thin gas cell or foil, both molecular and atomic ions readily lose several of their outer chemical electrons. Atoms, of course, can lose all of their electrons and still remain stable, but the same is not true of molecular ions. If a molecule loses just two of its binding electrons, the atomic constituents of the molecule frequently become dynamically unstable, repel each other, and dissociate into two or more fragments. It appears that, with the exception of a few exotic species or long-chain organic molecules, when stable molecules are stripped of three electrons the molecule has a very short lifetime (on the order of 10^{-9} s) against fragmentation into lower-mass components.

Procedure. The above principle of molecular dissociation is used for an ultrasensitive mass spectrometer having no molecular backgrounds. The procedure involves the following steps: (1) Secondary ions are sputtered from the sample. (2) Particles having the wanted mass are selected. (3) The selected particles are accelerated to megaelectronvolt energies, where they are dissociated by stripping some electrons from the outside of each ion, causing atoms to assume a high-positive-charge state and all molecular components to dissociate into fragments of lower mass. (4) The resultant particles are reanalyzed for mass. Atoms will have the same mass as that originally selected; molecular fragments will have a lower mass. Thus, molecules can be completely rejected from the detector.

Low-mass-resolution requirement. In such a system, because of the elimination of molecular ions, the resolution needed at each of the two mass selection stages is only one mass unit. This requirement for only modest resolution, compared to the high mass resolution of conventional mass spectrometers of between 3000 and 10,000, allows the construction of a mass analyzer with very high transmission and large solid angles for ion acceptance. Large solid angles lead to high acceptance fluxes from the ion source, making possible analysis at very high sensitivities because the wanted particles arrive at a statistically acceptable rate.

Primary beam diameter and intensity. In a typical

ultrasensitive mass spectrometry system, an ion source produces a well-directed stream of cesium ions, each of which has a kinetic energy of about 30 keV. This ion beam is focused at the surface of the sample to a diameter as small as 100 μm. With some reduction in sensitivity, this diameter could be easily reduced to 1 μm. This small area of beam impingement permits the probing ion beam to explore elemental distributions at very low levels by rastering the beam across the sample. When high spatial resolution is not essential, the investigator can use primary beam currents of cesium greater than 200 μA or 10^{15} particles per second. Each of these primary particles sputters several secondary ions from the surface so that atoms are eroded from the surface at a rate close to 10^{16} particles per second. Even at concentrations of $1{:}10^{16}$, wanted particles are emitted at the rate of one per second.

Applications. Since 1977, a number of demonstrations of the technique have been made, with most of the experiments focusing on the detection of rare radioactive nuclei, such as ^{14}C. Radioactive nuclei have often been selected for these experiments rather than stable atoms because it is easy to prepare uncontaminated low-concentration samples ($1{:}10^{15}$). ^{14}C is an excellent candidate because in recent biological samples the isotopic ratio of ^{14}C to ^{12}C is equal to 1.2×10^{-12}. Thus, each milligram of recent material has distributed throughout its volume 6×10^7 ^{14}C atoms. For each 5730 years that have elapsed since the biological material was formed, this concentration of ^{14}C atoms reduces by one-half so that samples of any concentration can be easily found.

For dating purposes, the advantage of counting ^{14}C atoms directly by ultrasensitive mass spectrometry rather than detecting their presence from radioactivity is that a statistically significant result can be obtained from a very small sample. In practice, the sample size needed for direct counting measurements of ^{14}C is less than one-thousandth that needed for radioactivity measurements.

Several groups have detected other long-lived radionuclides, including ^{10}Be, ^{26}Al, ^{36}Cl, and ^{129}I, each at mass-abundance concentrations of approximately $1{:}10^{15}$. In addition, stable isotope measurements for a few elements have been carried out at concentrations below 1 part in 10^9. For platinum and iridium, these measurements indicated sensitivities close to 1 part in 10^{11}. For most atoms, sensitivities up to a million times greater than possible with conventional mass spectrometry are possible by using ultrasensitive mass spectrometry instrumentation that is becoming commercially available.

[KENNETH PURSER]

Fourier transform mass spectrometry. Recent dramatic advances in digital electronics and computer technology have made possible development of new analytical instrumentation to carry out measurements in ways which were previously impossible. The new technique of Fourier transform mass spectrometry, based upon the phenomenon of ion cyclotron resonance, is one such example. With this method, mass spectra of unprecedented high resolution have been obtained, as have moderate-resolution spectra (10,000–20,000), with measurement times of 1 s or less. [Spectral resolution ($m/\Delta m_{1/2}$) is defined as the ratio of the frequency of a given mass to its peak width at half height.] Further, the applicability of the new technique to gas chromatography–mass spectrometry and low-pressure chemical ionization mass spectrometry has been demonstrated.

Principles. When ions are placed in a magnetic field, they travel in circular paths perpendicular to the field lines. Their angular frequencies ω_c in rad/s are equal to qB/m, where q is the charge of the ion, B the magnetic field strength, and m the mass.

Early ion cyclotron resonance spectrometers. Commercial ion cyclotron resonance mass spectrometers used in the late 1960s detected ions in the presence of a magnetic field by applying a constant radio-frequency signal and varying the magnetic field to bring ion frequencies ($\omega_c/2\pi$), equal to the applied radio frequency, sequentially into resonance. Instantaneous power absorption of sample ions was measured by a marginal oscillator detector as the ions drifted through an analyzer region of the sample cell after being formed in a source region. In addition to slow scan speeds (dictated by the magnetic field scanning technique), these instruments possessed limited resolution and had restrictive upper mass range limits (approximately unit resolution at m/z 200). Subsequently, the original scanning mode instrument was improved by the use of a trapped ion cell design introduced in 1970. This cell prolonged ion residence times and permitted unit resolution at values of the mass-to-charge ratio (m/z) as high as 500.

Later developments. The next major development in ion cyclotron resonance mass spectrometry was the demonstration in 1974 of the feasibility of Fourier transform mass spectrometry. In this initial experiment, CH_4^+ ions from methane were measured with a signal-to-noise (S/N) ratio of 8:1 and a peak width at half height corresponding to 0.005 dalton, using a measurement period of 25.6 ms. Such rapid measurement time is possible due to the realization of the Felgett (or multiplex) advantage, which derives from the fact that when a time domain signal (obtained after suitable excitation of the ions) is sampled at N equal intervals, the S/N is enhanced by a factor of $\sqrt{N}$ for a given measurement time. Thus, to achieve a desired value of S/N, using a time domain measurement of all ion signals simultaneously, observation time can be reduced accordingly. Enhanced resolution is obtained in Fourier transform mass spectrometry by using ion trapping, achieved in this case by application of small positive potentials to two walls of a cubic cell, to prolong observation times, as had earlier been shown to be possible.

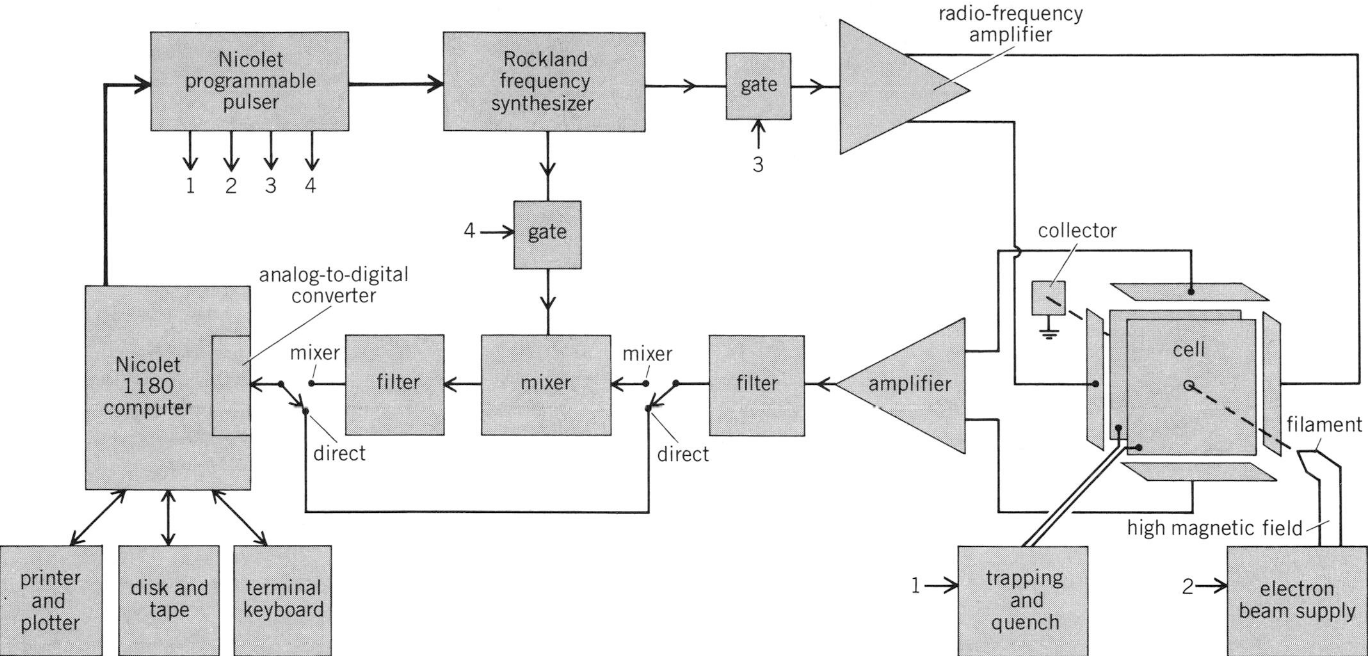

Fig. 1. Block diagram of Fourier transform mass spectrometer showing the cell and electronics. Pulses from the Nicolet programmable pulser are sent to the trapping and quench (1), the electron-beam supply (2), and the gates (3, 4). (*From S. Ghaderli et al., Chemical ionization in Fourier transform mass spectrometry, Anal. Chem., 53:428–437, copyright 1981 by the American Chemical Society*)

As shown in Fig. 1, the cubic cell of a Fourier transform mass spectrometer is positioned in a high magnetic field (generally from 1 to 6 teslas). Ionization of the sample contained in the cell at pressures between 10^{-10} and 10^{-6} torr (1 torr = 133 pascals) is accomplished by briefly gating electrons (5–10 ms) from a rhenium filament into the cell. As mentioned above, positive ions are trapped by application of potentials between 0.1 and 1.0 V to two of the cell walls as indicated. With high magnetic fields, ions may be trapped for periods ranging from seconds to hours. These same walls can be used for the application of a much more positive quench pulse, which can be used to remove ions prior to beginning a measurement. Excitation follows either immediately or after a variable delay which can be utilized to yield spectra corresponding to different times after the electron beam is stopped (Fig. 2). This excitation, which is a varying radio frequency signal applied to two of the remaining cell walls, causes ions to move from their small-radius thermal paths (about 0.1 mm or less) to the much larger excited paths required to preserve their constant angular frequencies (ω_c). If an entire spectrum is to be recorded, this excitation will sweep through the entire range of expected ion frequencies in approximately 0.5 ms. Subsequently, detection is accomplished by monitoring potentials induced on the remaining two cell walls by ion image currents. High-speed analog-to-digital converters (1–5 MHz) are often used for this purpose.

Alternatively, by means of monitoring a difference frequency between the natural ion frequency and a reference, slower analog-to-digital converters can be used. The latter technique (called a mixer mode measurement) is employed when ultrahigh-resolution measurements are made. The difference between the two measurement modes, direct and mixer, is best understood by considering a simple example. Benzene molecular ion (m/z 78) is found to have a resonance frequency of 236.129 KHz in a 1.2-tesla magnetic field when a 0.25-V trapping voltage is employed. A direct mode measurement of this frequency would require sampling at least twice per cycle with an analog-to-digital rate of 472.258 KHz. If 32,768 (32K) computer memory locations

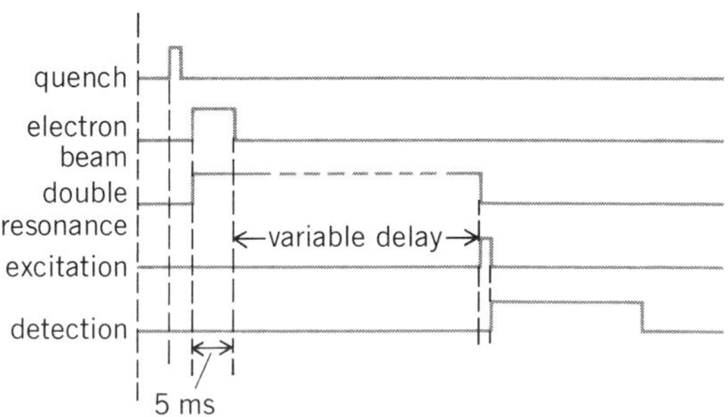

Fig. 2. Timing diagram for a Fourier transform mass spectrometry experiment. (*From S. Ghaderli et al., Chemical ionization in Fourier transform mass spectrometry, Anal. Chem., 53:428–437, copyright 1981 by the American Chemical Society*)

were available for data storage, 69.4 ms would be the maximum time that the signal resulting from excitation could be observed. Alternatively, an excitation sweep from 235.989 to 236.269 KHz could be used, with the higher frequency serving as a reference. Now the spectral bandwidth would be 280 Hz and the signal could be digitized at 560 Hz. Thus, the same 32,768 memory locations could contain data obtained during 58.5 s. Because resolution is a linear function of observation time, the latter mixer mode measurement is, for the example chosen, capable of yielding resolution over 800 times better than the direct-mode measurement. In fact, such experiments at 1.2 teslas were shown in 1980 to yield a resolution of 760,000 for benzene molecular ion and 220,000 for the isobaric doublet of a mixture of benzene-d_6 and cyclohexane molecular ions (m/z 84). These results should be compared with conventional mass spectral measurements, with which maximum resolution of about 200,000 at these masses is obtainable only with great difficulty.

As mentioned previously, it is possible to selectively eject specific ions. Furthermore, if a delay is imposed after ion formation and prior to observation, chemical reactions between ions and neutrals may occur. The time-resolved spectra thus possible make Fourier transform mass spectrometry an extraordinarily powerful tool for the study of gas-phase ion molecule reactions, as well as the photodissociation of ions.

Rapid-scan ion cyclotron resonance. A variant of the basic Fourier transform mass spectrometry scheme, called rapid-scan ion cyclotron resonance spectroscopy, was introduced in 1977. In this method, which is similar to correlation nuclear magnetic resonance spectroscopy, a continuous-wave excitation frequency is rapidly scanned across the spectrum of cyclotron frequencies and the transient response of ions is detected by the use of a capacitance bridge detector. Measurements with this method require an order of magnitude more time (10–20 s) and appear to offer lower resolution than Fourier transform mass spectrometry.

Analytical use of Fourier transform method. Analytical application of Fourier transform mass spectrometry takes advantage of its inherently high mass spectral resolution (for example, $m/\Delta m_{1/2} = 1.5 \times 10^6$ for m/z 166 from tetrachloroethylene using a 4.7-tesla superconducting magnet) and its capability for high measurement speed at reduced resolution. Of particular value in the latter context is the use of Fourier transform mass spectrometry in conjunction with gas chromatographic separations. In 1980 the successful coupling of a Fourier transform mass spectrometer with a support-coated open-tubular capillary gas chromatography column was reported, and using a jet separator to remove carrier gas, it demonstrated $m/\Delta m$ of 9000 at m/z 106 for the isobaric xylene/benzaldehyde doublet. Thus Fourier transform mass spectrometry can be used with high-resolution gas chromatography. In 1981 it was demonstrated that Fourier transform mass spectrometry was practical as a tool for low-pressure chemical ionization mass spectral analysis. A variety of chemical ionization reagent gases were used, including methane, isobutane, water, and ammonia at pressures of 10^{-6} torr for analytes with partial pressures of 10^{-8} torr or less. The possibility of rapidly switching between electron impact and chemical ionization measurements for both positive and negative ions was also demonstrated. In addition, the further analytical utility of time-resolved measurements was shown. It appears that both gas chromatography–mass spectrometry and low-pressure chemical ionization capabilities of Fourier transform mass spectrometry will make it a powerful analytical method.

For background information *see* GAS CHROMATOGRAPHY; MASS SPECTROMETRY; MASS SPECTROSCOPE; RADIOCARBON DATING; SPECTROSCOPY in the McGraw-Hill Encyclopedia of Science and Technology. [C. L. WILKINS]

Bibliography: M. B. Comisarow and A. G. Marshall, *Chem. Phys. Lett.*, 25:282–283, 1974; S. Ghaderi et al., *Anal. Chem.*, 53:428–437, 1981; E. B. Ledford et al., *Anal. Chem.*, 52:2450–2451, 1980; R. L. White et al., *Anal Chem.*, 52:1525–1527, 1980.

Mass wasting

The submarine slopes of the offshore Mississippi River delta are subject to a variety of mass movement processes which provide design constraints for oil and gas engineering. Since the late 1950s, a large number of offshore platforms and pipelines have been installed in the area, and some have been adversely affected by active submarine landsliding. Recently, a comprehensive and uniquely detailed geologic and geophysical survey was completed for the entire submarine delta and adjacent continental shelf. This assessment of the distribution of instability processes and their mechanisms provides a base line for the evaluation of future process events. Maps of the submarine geology of the delta, showing all mass movement features and near-surface sediment deformations, are now available at scales of 1:48,000 and 1:100,000 from the Bureau of Land Management.

Survey methods. Maps of mass movement features were prepared from high-resolution geophysical and side-scan sonar surveys, supplemented by precision bathymetry, bottom samples, and data from soil foundation borings. The geophysical surveys used high-frequency (about 110 kHz) acoustic sources for bathymetry, 3.5–12-kHz sources for near-surface (less than 75 m) subbottom profiling, and lower-frequency (50–100 Hz) sparkers for deeper penetration. Side-scan sonar data were acquired with range settings of 150 to 200 m, giving overall swath widths of 300 and 400 m, respectively. Track line spacing ranged from 250 to 360 m, providing continuous overlapping side-scan imagery for the whole delta, and very-high-density seismic coverage. Adjacent sonograph swaths composed of digitally acquired, scale- and speed-corrected data were referenced to navigation grids, and large-area, scale-true mosaics of the sea floor were constructed. These sonar mosaics provide the essential detail of the morphology of instability features. The mosaics are thus analogous to air photo mosaics as a basic terrain evaluation tool, except that the former is produced by acoustic reflectivity and acoustic albedo differences in sea-floor sediments and morphology.

Types of mass movement. Offshore slope angles in the delta region are extremely small, rarely exceeding 1.5° and generally averaging 0.5°. In water depths of 5–80 m, bottom slopes range from 0.7° to 1.5°, and in 80–150 m the slopes are less than 1.0°. In general, hydrographic data show extremely irregular topography, the bottom displaying a large number of gullies and chutes leading downslope to broad, flat terraces.

Recent surveys have revealed a number of different types of slope instability features, including collapse depressions, mudslide depositional lobes, and a variety of slumps and faults (Fig. 1). The last are most common in shelf edge locations, seaward of the modern deltaic sediments.

Collapse depressions and bottleneck slides occur primarily in the shallow-water areas of interdistributary bays. They are associated with slopes of 0.1–0.2° and sedimentation rates relatively small in comparison with other parts of the delta. The depressions range in diameter from 50 to 150 m and are typically subsidence features bounded by curved or near-circular escarpments up to 3 m in height. On the upslope margins of the depressions, crack systems extend into adjacent stable sediments, while the internal area of subsidence is usually composed of irregularly shaped blocks.

On slightly steeper slopes (0.2–0.4°), in relatively shallow water areas, there are bottom features that are similar to collapse depressions except that the boundary scarps are not totally closed around the perimeter of the instability. Rather, they have narrow openings at the downslope margins through which debris is discharged over surrounding intact slopes. This narrow chute leads to displaced debris arranged as distinct undulatory lobate fans, and the term bottleneck slide, analogous to similar instability features in quick clay, is applied.

Extending radially from each of the main deltaic distributaries, in water depths of 10–100 m, are major elongate systems of sediment instabilities referred to as delta-front gullies caused by mudslide processes (Fig. 2). Side-scan sonar records show that these emerge from within an extremely disturbed area of shallow rotational sliding high on the delta. Each mudslide possesses a long, narrow chute or channel which links a depressed, hum-

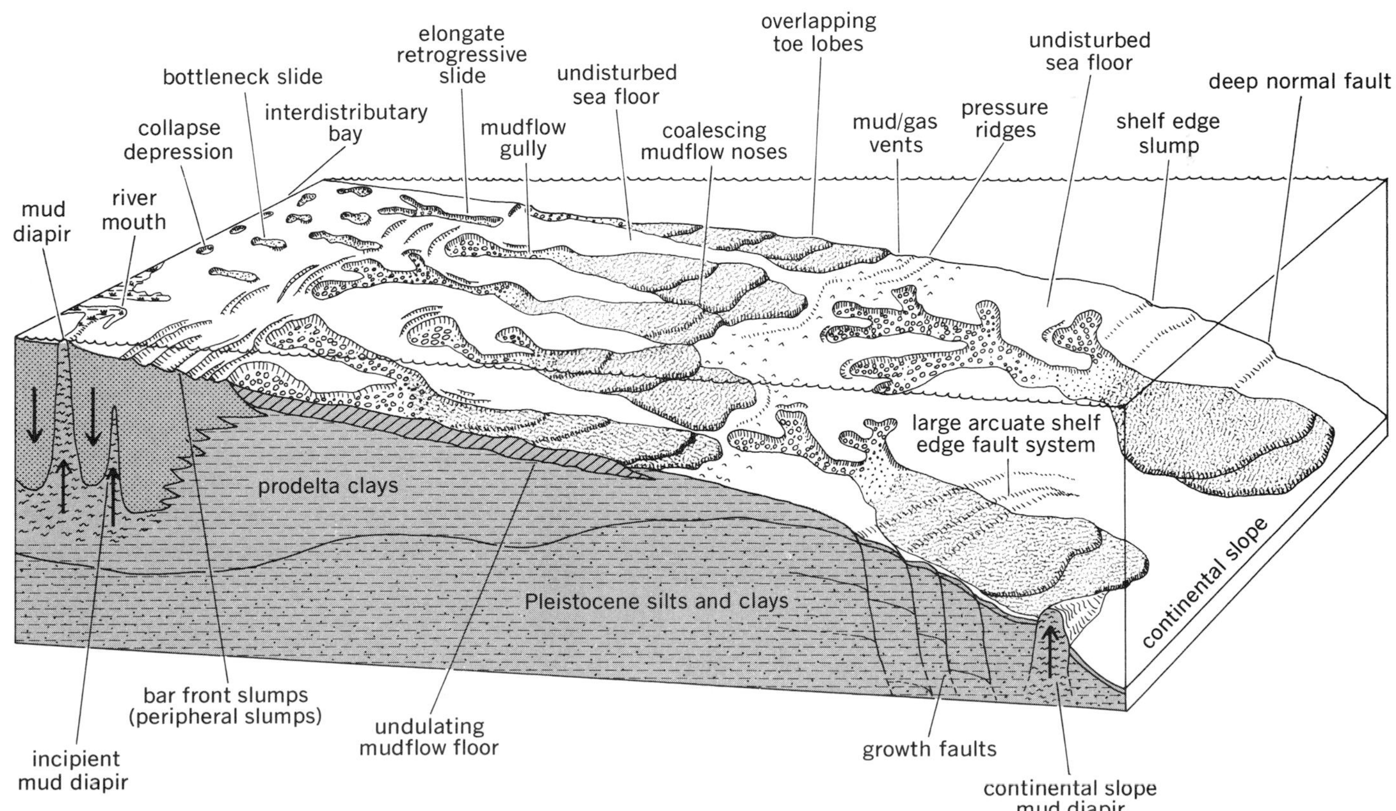

Fig. 1. Schematic representation of the distribution of mass movement processes on the Mississippi delta-front slope. (*From D. B. Prior and J. M. Coleman, Beneath the Mississippi Delta, Geogr. Mag., 52:281–285, January 1980*)

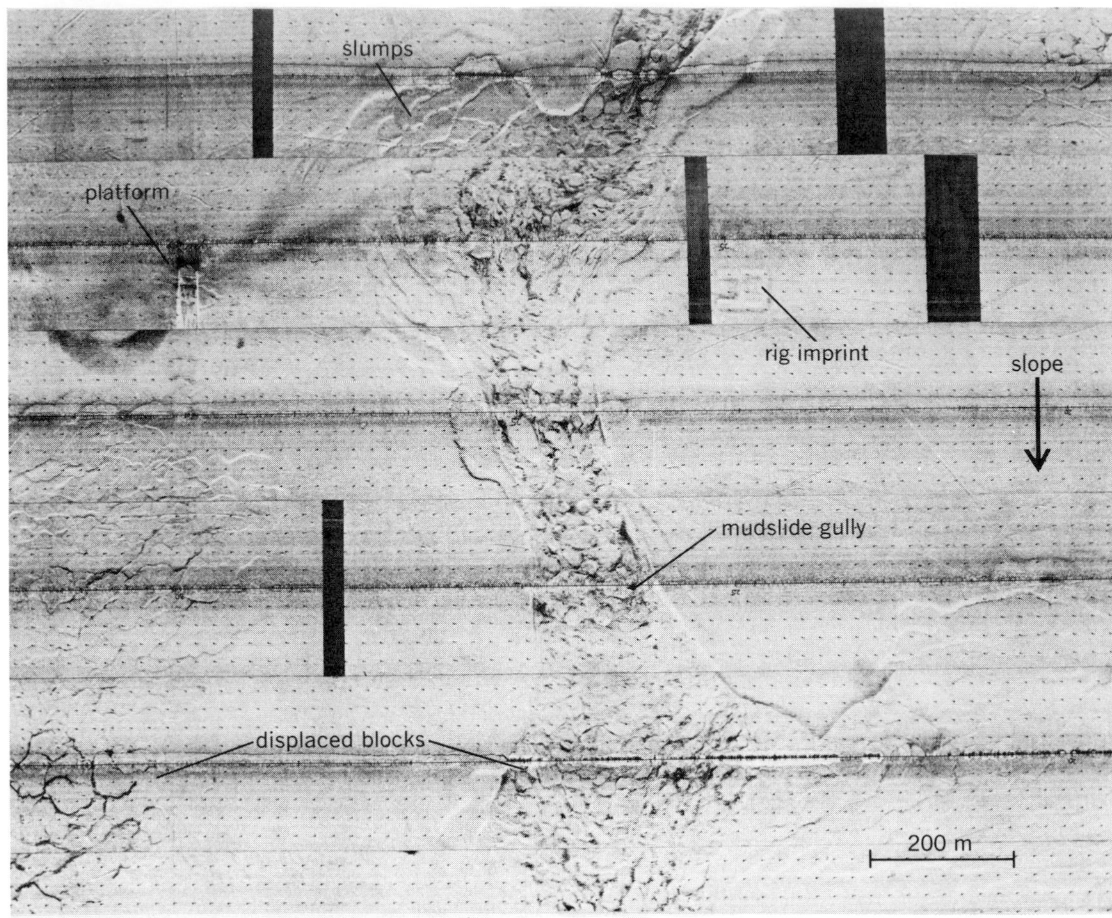

Fig. 2. Side-scan sonar mosaic composed of adjacent scale-rectified sonar swaths showing a variety of sea-floor features in the offshore delta region.

mocky source area to composite overlapping depositional lobes or fans on the seaward end. The gully floors lie from a few meters to 20 m below the adjacent intact sea floor. In plan view these features are rarely straight and commonly are markedly sinuous with alternating narrow constrictions and wider bulbous sections. The widths of individual gullies vary considerably, from 20–150 m at their narrowest to 1200–1500 m in extreme cases. Adjacent gullies often coalesce to form branching tributary systems, and their junctions may be discordant, marked by accumulations of lobes of debris discharged from a tributary into the main channel.

At the seaward or downslope ends of the mudslide gullies there are extensive areas of irregular bottom topography composed of blocky redeposited deltaic sediments. This debris, discharged from the gullies, forms widespread, overlapping lobes or fans. The thicknesses of the individual lobes are difficult to determine but may be 5–15 m; however, because of overlapping and cumulative buildup, mudflow deposits can approach 50–60 m. Peripheral to the lobes, the effects of loading on the preexisting shelf sediments can be seen as arcuate pressure ridges and numerous mud and gas vents, as a result of upward release of gas and excess pore water pressures within the sediment voids.

In the deeper waters along the outer continental shelf and upper continental slope fronting the delta, a variety of slumps and faults are found at the shelf edge slope break. Large arcuate-shaped families of slide features and deep-seated contemporaneous or depositional faults are present. These often cut the modern sediment surface, forming distinct scarps on the sea floor.

Magnitude and frequency of movement. The evidence from platform damage, pipeline breaks, and systematic resurveys of mass movement features in the delta region shows conclusively that the slope instability processes are presently active. For example, damage to offshore installations was due to substantial bottom sediment movement accompanying Hurricane Camille in 1969. Also, pipeline breaks associated with subsidence of collapse depressions and downslope movement of sediment in bottleneck and elongate slides occur quite frequently, often without the effects of major hurricanes. Resurveys of particular areas allow comparisons of bathymetric profiles with those made in earlier years, beginning in 1874, when the first comprehensive bathymetric survey of the delta was completed. Analysis of such historical data shows that new mudslide gullies frequently are formed after major flood and sediment discharges from the river. For example, 15 new chutes were formed following the 1922 flood, 7 in 1927, and 10 were observed after the 1945 flood event. Also, recent annual surveys have shown substantial advances of mudslide depositional lobes, and in the period 1975–1976 one lobe prograded a distance of 19 km. Repeated annual side-scan sonar mapping since 1974 of particular gullies has also revealed bottom changes. Feature enlargement, offsets of anchor scour marks and trawler drags, and comparisons of identifiable blocks of sediment have indicated local gully movements of up to 100 m per year. More recently, a test resurvey area of approximately 70 km^2 has been reexamined by using digitally acquired side-scan sonar on a regular basis every 3–4 months. In some areas there were substantial changes in the geometry of existing features between surveys. For example, one gully had retrogressed and lengthened in an upslope direction a distance of more than 80 m, and a surge of debris down a chute exceeded 100 m of downslope transport.

Factors contributing to mass movement. The features identified in the delta region (Fig. 3) are believed to be the result of interaction of many variables rather than the product of any single factor. The basic conditions for failure of the submarine slopes exist when the stresses exerted on the sediment are sufficient to exceed its strength. This initiation of movement can be due to temporary or sustained stress increases, short-term or progressive strength reduction, or combinations of these effects. The major characteristics of the offshore delta region that influence these stress/strength relationships include high rates of sedimentation, which result in excessive sedimentary loading and oversteepening, especially in the bar-front areas; deposition of coarse-grained sands and silts near the distributaries and finer-grained deposits farther seaward, causing differential loading; high water contents, generally low strengths, and underconsolidation of the deltaic deposits; rapid biochemical degradation of organic material in the deposits, which results in formation of large quantities of in-place sedimentary gases; and annual passage of winter storms or hurricanes, which result in cyclic wave-loading processes.

The factor interaction and complex interrelationships can be considered in terms of the principle of effective stress. For example, rapid sedimentation leads to the generation of excess pore water pressures within the sediments (underconsolidation). Such excess pore pressures may migrate within the sediments. The total effect of such large pore water pressures, which have been measured by piezometer experiments, is to reduce the effective stress, and results in weakening of the soil's resistance to applied stresses. Similarly, it is believed that generation of sedimentary methane gas can, at particular pore pressures, lead to the formation of bubble-phase gas in the sediment voids. This is believed to reduce the strength properties of the sediments, although the precise effects are as yet unknown. Surface wave activity affects the stress/strength equilibrium in two complementary ways. Surface waves during storms and hurricanes exert cyclic loading forces on the weak sediments and are responsible for pore water pressure increases. Once again, this process has actually been observed during long-term piezometric monitoring. In addition, surface water wave perturbation of the sediments on a sloping surface exerts an increased downslope component of stress. Thus, the initiation of failure can be the result of a combination of these factors and processes, operating on different time scales. Individual failure thresholds may be achieved by quite different combinations of the same basic factors over time and space. For example, storm waves may have a potentially greater effect if reduction in sediment strength is well advanced by other factors. Alternatively, generation of excess pore fluid pressures may result in apparently spontaneous failure without any external changes in stress conditions.

Significance. Research continues in the Mississippi River delta region in an effort to quantify the precise causes of mass movement. Similarly, the existence of detailed maps of the instability features is promoting continued study as mudslide activity modifies already identified feature distributions and geometries. The significance of these active delta-front processes is twofold. Engineering projects must, of necessity, consider the potential applied

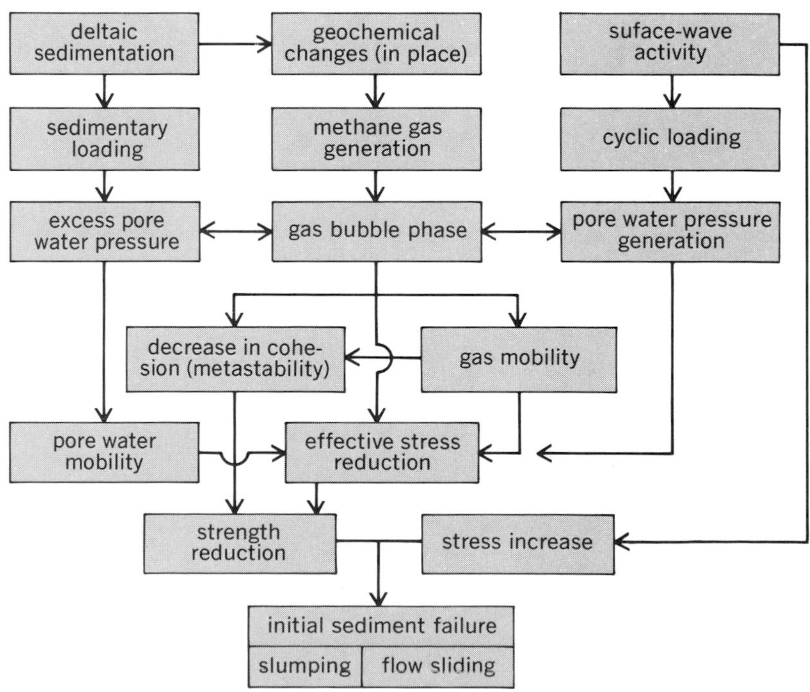

Fig. 3. Conceptual framework of factors and their interrelationships which may initiate mass movement activity on delta-front slopes. (*From D. B. Prior and J. M. Coleman, Disintegrating retrogressive landslides and very low angle subaqueous slopes, Mississippi Delta, Mar. Geotech., 3 (1):37–60, 1978*)

forces on structures due to continued mass movement activity. Also, it is abundantly clear that these processes cause significant effects on the mechanisms of deltaic progradation and sedimentation. Shallow-water sediments are transported downslope, leading to thickened depositional wedges of sediment on the lower delta-front slopes. Sediment texture, lithology, and faunal assemblages and distributions are all affected by these processes. Indeed, recent surveys and research in other deltaic regions suggest that mass movement processes, such as have been identified in the Mississippi Delta, are not uncommon. They may perform a much more important role in modern and ancient deltaic sedimentary sequences than has hitherto been recognized.

For background information *see* DEPOSITIONAL SYSTEMS AND ENVIRONMENTS; MARINE MINING; MARINE SEDIMENTS; MASS WASTING; OIL AND GAS, OFFSHORE; SONAR in the McGraw-Hill Encyclopedia of Science and Technology.

[D. B. PRIOR; J. M. COLEMAN]

Bibliography: J. M. Coleman and D. B. Prior, *Amer. Ass. Petrol. Geol. Continuing Educ. Ser. 15*, 1980; J. M. Coleman, D. B. Prior, and L. E. Garrison, *Bur. Land Manag. Open File Rep. 80-01*, 1980; D. B. Prior, J. M. Coleman, and L. E. Garrison, *Geology*, 7:423–425, 1979; D. B. Prior and J. N. Suhayda, *Proc. Offshore Tech. Conf.*, Pap. 3472, pp. 1055–1061, 1979.

Material requirements planning

Material requirements planning (MRP) is a formal computerized approach to inventory planning, manufacturing scheduling, supplier scheduling, and overall corporate planning. The MRP system provides the user with information about timing (when to order) and quantity (how much to order), generates new orders, and reschedules existing orders as necessary to meet the changing requirements of customers and manufacturing. The system is driven by change and constantly recalculates material requirements based on actual forecast orders. It makes adjustments for possible problems prior to their occurrence, as opposed to traditional control systems, which looked at more historical demand and reacted to existing problems. The ability of an MRP system to meet the various needs of manufacturing, materials, and marketing personnel within a changing business environment contributes to its growing implementation by manufacturing companies.

Functional uses. Coping wih a business environment characterized by slow growth and high inflation makes efficient management of inventories critical from thc standpoint of cash flow and profitability. However, low inventories or inventories of the incorrect items create material shortages which are costly in terms of lost sales, added expense, and increased manufacturing time. The constant updating of priorities by te MRP system helps planners maintain lower inventory balances of the correct items. In addition, updated priorities assist manufacturing supervisors with job sequencing, leading to increased productivity and efficiency. Good priority planning also helps purchasing to schedule the supplier deliveries, which improves supply relations and leads to subsequent cost savings. While up-to-date priorities help manufacturing and materials departments, several firms are finding that the information contained in the MRP data base can also aid in other areas of corporate planning. Simulation of various production plans can assist corporate planners to forecast cash flows, labor needs, facility needs, and capital expenditures.

Dependent and independent demand. The logic of the MRP system is based on the principle of dependent demand, a term describing the direct relationship between demand for one item and demand for a higher-level assembly part or component. For example, the demand for the number of wheel assemblies on a bicycle is directly related to the number of bicycles planned for production; further, the demand for tires is directly dependent on the demand for wheel assemblies. In most manufacturing businesses, the bulk of the raw material and in-process inventories are subject to dependent demand. An important characteristic of dependent demand items which affects timing is known as lumpy demand. Since most manufacturers assemble their products in batches or specific lot sizes to fulfill customer requirements, dependent demand items at lower levels do not exhibit uniform usage but are subject to extreme fluctuation. This creates situations in which no demand exists for weeks, and then a large quantity is required. Figure 1 illustrates the concept of lumpy demand on three styles of drills. While demand for the end drills is constant, the plant produces them in various lot sizes, thereby creating discontinuous and discrete demands. Lumpy demand is handled in MRP through the calculation of future requirements at all levels of the assembled product. Dependent demand quantities are calculated, while independent demand items are forecast. Independent demand is unrelated

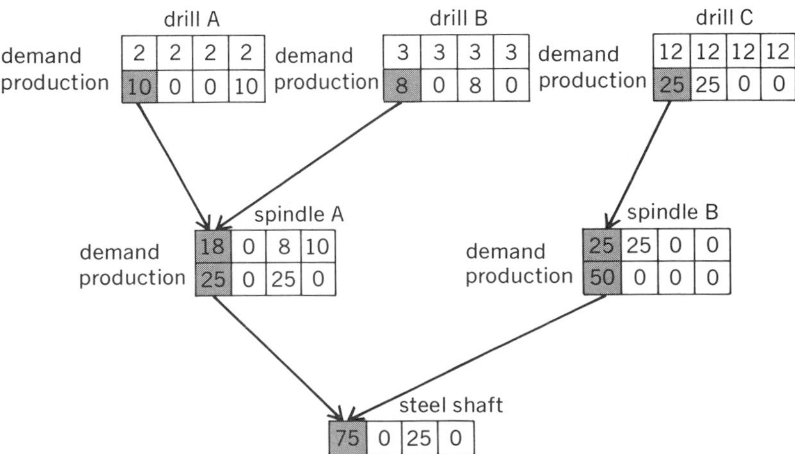

Fig. 1. Lumpy demand in manufacturing for three types of drills. The boxes represent equal time periods.

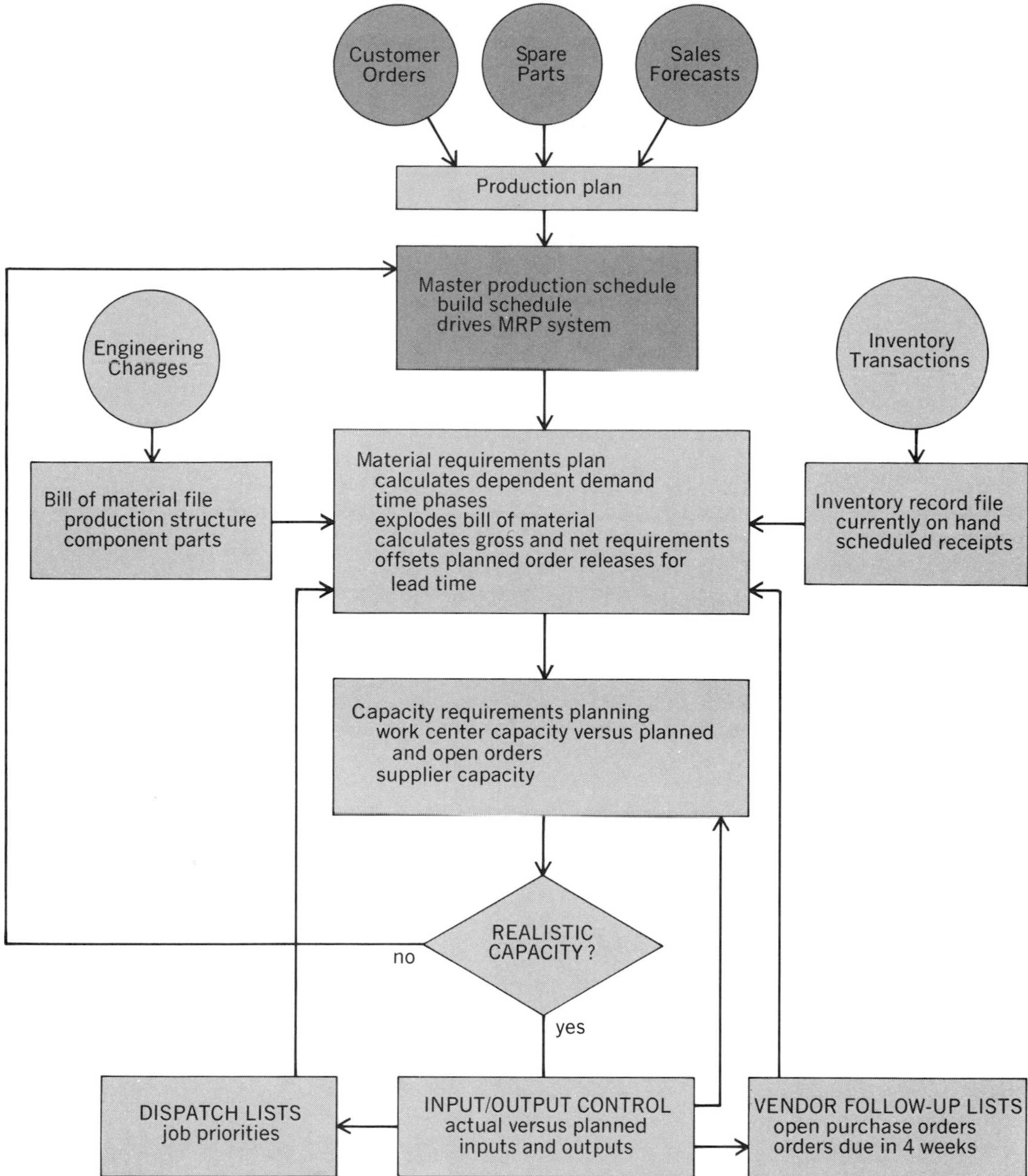

Fig. 2. Flow chart for a model MRP system.

to a higher-level item which the company manufactures or stocks. Generally, independent demand items are carried in finished goods inventory and subject to uncertain end customer demand. Spare parts or replacement requirements for a drill press are an example of an independent demand item.

MRP system. A model MRP system is shown in Fig. 2. The first step in the MRP process involves obtaining input from marketing via sales forecasts or firm orders, or a combination of both forecasts and firm orders. Demand inputs on the typical firm can originate from several sources: customers, warehouses, service parts, safety stock, and interplant orders. Except for make-to-order firms, the marketing forecast is adjusted periodically and reviewed by the plant manager as well as functional managers from marketing, manufacturing, and materials. The marketing forecast is then translated into a production plan. This plan establishes an aggregate production rate which considers planned inventory increases or decreases. In a make-to-order company, production planning adjusts the customer order backlog; in a make-to-stock firm, the inventory levels are adjusted.

Master production schedule. While the production plan is a general plan, the master production schedule is the shop-build schedule which drives the MRP system. The master production schedule is stated in bill-of-material numbers, and is used to generate the material requirements plan, and also generates the capacity requirements plan. This master schedule also lists all the end items (the items as sold by various users) by date (planned shipping period) and their quantity. Most master schedules are prepared for a certain horizon (period of future time, for example, 12 or 18 months) which must be at least as long as the cumulative lead time. The time horizons are usually split into short and long horizons. The short horizon serves as input into the material requirements plan for the production of components, planning of priorities, and planning of short-term capacity. The longer horizon is used as a basis for estimating long-term demand on company resources.

The master production schedule gives management an approximation of long-term capacity utilization, labor needs, the impact of new product plans in the shop load, and current priority status.

Bill-of-material file. This contains information about all parts and shows the dependent demand relationship of subassemblies or end items (finished product) down to components and raw material level. For example, a bill of material for a screwdriver would show the plastic handle, forged shaft, and steel from which the shaft was made. If users are to obtain valid system outputs, the bill-of-material file must be accurate and up to date, and must be suitably defined for customers, manufacturing, and purchasing. Users must ensure that the latest engineering changes are incorporated into the bill of material.

Inventory record file. This contains a record of the actual inventory level of each item and part as well as lead times. Inventory accuracy is critical in avoiding erroneous system ouputs. Most users strive for 95–99% accuracy and monitor actual inventory through cycle counting.

Material requirements plan. The function of this plan is to explode the master schedule against the bill-of-material file to determine the gross requirements necessary to make the product. These gross requirements are then compared to inventory on hand to determine net requirements. The net requirements at the higher level become gross requirements at the next lower level, and so on, down through the bill of material. These requirements are offset for lead time by subtracting the lead time from the date of the net requirements. Figure 3 shows the MRP format and how lead times are offset. In Fig. 3 the term projected requirements refers to the number of items necessary to make up the next-higher assembly (for example, 40 skateboards require 80 axles). The term scheduled receipts refers to material already on order or scheduled to come in. The inventory of the item is represented by the term on hand. Planned order release (net requirements) means the number of items manufacturing must make or purchasing must buy to meet higher-level requirements. Level 0 is the highest level and represents the end item. Lead time is the time required to acquire an item after it is ordered or manufactured.

Capacity requirements plan. This part of the system enables manufacturing to determine what capacity will be required by work centers during each time period to meet the master prduction schedule. The MRP system will indicate which items are required and the timing of the items. The actual work load is then compared to work center capacity. If the available capacity falls short in meeting the master schedule, a decision must be made either to reduce the schedule or to expand capacity by working overtime, subcontracting, or purchasing additional capacity.

Input/output control. This has the function of monitoring the levels of work into and out of each work center. Control is achieved by ensuring on-schedule job flow through various work centers. If a job slips behind schedule, its priority should be increased. When input to the factory begins to exceed

Master Schedule		week 1	2	3	4	5	6
level 0 skateboards	Planned order release	40	0	50	0	0	60
level 1 axles lead time = 2	Project requirements	80	0	100	0	0	120
	Scheduled receipts						200
	On hand 200	120	120	20	0	0	80
	Planned order release				200 ← offset		
level 2 steel rod lead time = 3	Projected requirements	0	0	0	200	0	0
	Scheduled receipts				200		
	On hand 10	10	10	10	10	10	10
	Planned order release	200 ← offset					

Fig. 3. Materials requirements plan mechanics and lead-time offsetting for skateboards.

output, this indicates that the production schedule has been overstated and should be reduced or the capacity expanded.

Dispatch lists and vendor follow-up lists. These help manufacturing and purchasing supervisors determine accurate priorities. The dispatch list will print out jobs requested to be run and jobs available to be run, and the sequence in which to run them. Dispatch lists, usually generated daily, continually revise priorities on the shop floor. Follow-up lists give purchasing a tool to ensure that vendors will deliver on time by providing information about requirements on a short- and long-term basis.

Regenerative versus net change. There are two basic types of MRP systems, regenerative and net change. The essential difference is in the frequency of replanning and what triggers the replanning process. Regenerative systems reexplode all material requirements periodically when the master schedule is revised. The most common regeneration period is weekly. Gross and net requirements for each level are recomputed, and the planned order schedule is re-created.

In net change planning systems only partial explosions of material requirements are performed. Part of the master production schedule is subject to frequent explosions, and a full explosion is done periodically over longer intervals. Net change permits frequent replanning and minimizes the extent of the replanning at any one time.

Advantages. A computerized MRP system provides many benefits to users. These benefits include: improved priority planning, professional purchasing, better customer service, and increased productivity. Improved priority planning permits items to be deexpedited as well as expedited. The process of purchasing becomes more sophisticated. Under traditional inventory systems, purchasing would place orders and then expedite. Under MRP, vendors are given much more information about when to ship and how much to ship. Purchasing spends less time expediting and more time searching for cost savings through alternate supply sources, value analysis, and negotiations. Better customer service is provided by meeting customer shipping dates much more frequently and in a shorter period with less premium freight. If a scheduled date is not met, the customer will be notified in advance. Increased productivity is achieved because dispatch lists and priority planning help supervision. The workforce will be more productive since there is less need to break into a schedule and run different parts. By use of the computer, MRP is able to manipulate massive amounts of data to keep schedules up to date and priorities in order.

The technological advances in computing and processing power, the benefits of on-line capabilities, and reduction in computing cost make computerized manufacturing planning and control systems such as MRP powerful tools in operating modern manufacturing systems productively.

For background information *see* INDUSTRIAL COST CONTROL; INDUSTRIAL ENGINEERING; INVENTORY CONTROL; SYSTEMS ENGINEERING in the McGraw-Hill Encyclopedia of Science and Technology.

[LARRY C. GIUNIPERO]

Bibliography: J. G. Miller and L. G. Sprague, *Harvard Bus. Rev.*, 53(5):83–91, September–October 1975; J. Orlicky, *Material Requirements Planning*, 1975; O. W. Wight, *Production and Inventory Control in the Computer Age*, 1974.

Metamorphic petrology

Metamorphic petrology is the branch of earth science that deals with solid-state transformations in mineralogy or chemical composition of rocks in response to changing temperature or pressure within the Earth. It has come to involve the study of rocks ranging from barely recrystallized sedimentary material to rocks deep within the Earth's crust at temperatures of 700–1000°C which are at the verge of melting to form silicate magma. The last 2 decades have seen considerable changes in the ways metamorphic rocks have been studied. These changes are due both to the development of plate tectonics, a revolution in geologists' overall view of earth processes, and, more importantly, to development of new analytical techniques and to refinement of physicochemical characterization of minerals and mineral reactions. The capability of the electron microprobe for chemical analysis of micrometer-sized spots on minerals has allowed a far deeper understanding of compositional variations in zoned minerals, and the recent application of transmission electron microscopy to metamorphic rocks has begun to illuminate specific mechanisms of metamorphic reactions. Experimental studies with precise control of pressure and temperature have greatly enlarged knowledge of stability fields of many minerals as well as understanding of the thermodynamic properties of these minerals. This article summarizes recent developments in important areas of metamorphic petrology.

Geothermometry and geobarometry. A goal of metamorphic petrologists for many years has been a method of deducing from a rock's mineral assemblage and mineral chemistry the precise physical conditions at which the rock formed. Recent experimental investigations of a number of key mineral equilibria, along with improved calibration of mineral thermodynamic parameters, have allowed much better estimates of the temperature and pressure of metamorphism than were previously possible. The two principal techniques for thermometry and barometry involve: (1) identification of mineral assemblages that have very limited ranges of stability, in either pressure or temperature, and for which experimental calibration exists; and (2) use of measured partitioning of two elements between two coexisting minerals for which a calibration exists. The first technique is typically used for determination of either temperature or pressure, while the second is much more commonly a thermometric method because most element-partitioning processes are largely or virtually pressure-independent.

A mineral assemblage that is useful for the first technique is typically one which contains a relatively large number of minerals for the number of chemical components. According to the Gibbs phase rule, if the number of minerals exceeds the number of chemical components, it is likely that several minerals will exist in a reaction relationship; the pressure and temperature dependence of many of these reactions have been determined. If a reaction has a relatively large volume change associated with it, it will likely be much more pressure-dependent than temperature-dependent. If a rock contains both reactant and product minerals, then it can be said that the rock formed in a restricted pressure range. Conversely, a reaction with a small volume change will typically be temperature-dependent, and the mineral assemblage representing that reaction will indicate a restricted temperature range.

The second technique depends upon having only two minerals rather than a large assemblage. If the minerals are solid solutions which share two elements but which have different ratios of the two elements, the minerals are said to partition the two elements. The partition coefficient which reflects the different ratios in the two minerals is dependent upon the relative sizes of the atoms and of the crystallographic sites, and also dependent upon the relative energies of the two sites. The most common pair of elements used for partitioning study are Fe and Mg because they are about the same size and occupy similar sites in different minerals, and also because they occur in a large number of important rock-forming minerals. The most commonly used partition thermometer is based on the distribution of Fe and Mg between garnet and biotite, which has been calibrated experimentally (Fig. 1).

Volatiles and devolatilization reactions. Most of the metamorphic reactions that are the basis for the classification and mapping of metamorphic grade are reactions that evolve either H_2O or CO_2 with increasing temperature. The precise temperature of a dehydration or decarbonation reaction is dependent upon the concentrations of H_2O and CO_2 in the ambient metamorphic fluid. It is therefore important to be able to determine or estimate fluid composition in order to understand the thermal implications of a particular metamorphic mineral assemblage. For example, the mineral reaction calcite + quartz = wollastonite + CO_2 (a reaction observed in marbles) has been shown by experiment to occur at 600–700°C in the presence of CO_2-rich fluid, while it may occur several hundred degrees lower in H_2O-rich fluid.

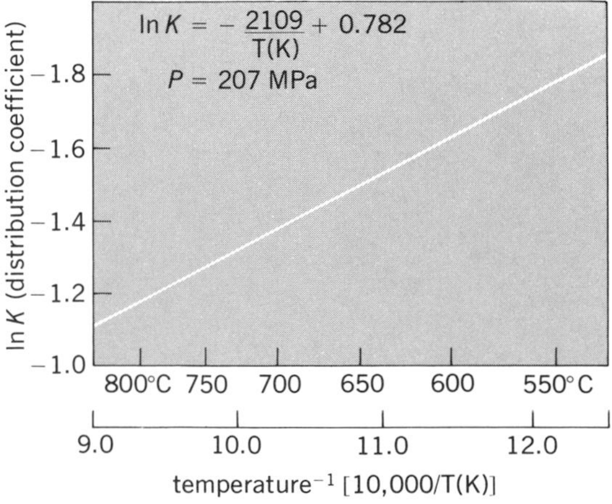

Fig. 1. Temperature dependence of Fe-Mg distribution between garnet and biotite. (*From J. M. Ferry and F. S. Spear, Partitioning of Fe and Mg between biotite and garnet, Contrib. Mineral. Petrol., 66:113–117, 1978*)

Conversely, a devolatilization reaction may have the capability of altering, or buffering, the composition of the fluid if it generates a large enough volume of fluid of different composition. For buffering to operate, the rock must generally behave as a "closed" chemical system; that is, the ambient fluid cannot be interconnected to a large external reservoir of fluid which controls the composition.

Considerable effort is now being expended by metamorphic petrologists in the estimation and measurement of compositions of metamorphic fluids. Indirect estimation of fluid compositions can be accomplished by using experimental data on a variety of mineral assemblages, while direct measurement of fluid composition can be done by observation of fluid inclusions in mineral grains. These fluid inclusions are small trapped bubbles of fluid, typically in quartz grains, which were incorporated in the recrystallized grains during metamorphism. Inclusions are commonly multiphase, with a gas bubble trapped in the liquid. A special heating and cooling stage mounted on a petrographic microscope is used to homogenize the liquid and gas phases, and the homogenization temperature may be used to determine the composition (for example, the CO_2/H_2O ratio) of the inclusion.

Diffusive mass transfer and kinetics. Many petrologic processes occur over such long times that thermodynamic equilibrium is presumed to have been achieved. Strict heterogeneous equilibrium, however, requires homogeneity of each mineral in the rock, and it is commonly the case that minerals are compositionally zoned. This zoning is called growth zoning and is due to changing chemistry of the immediate environment as a crystal grows larger. The preservation of these disequilibrium concentration profiles is possible because very slow diffusion rates in many silicate minerals require impossibly long times for homogenization. At highest metamorphic temperatures, however, diffusion rates may be rapid enough for partial equilibration to occur. Recent work in metamorphic petrology has attempted derivation of mathematical expressions for diffusive processes for the purpose of calculating the actual times for certain processes to occur.

Attempts have been made to characterize not only the kinetics of intracrystalline diffusion, as in the case of zoned minerals, but also intercrystalline or

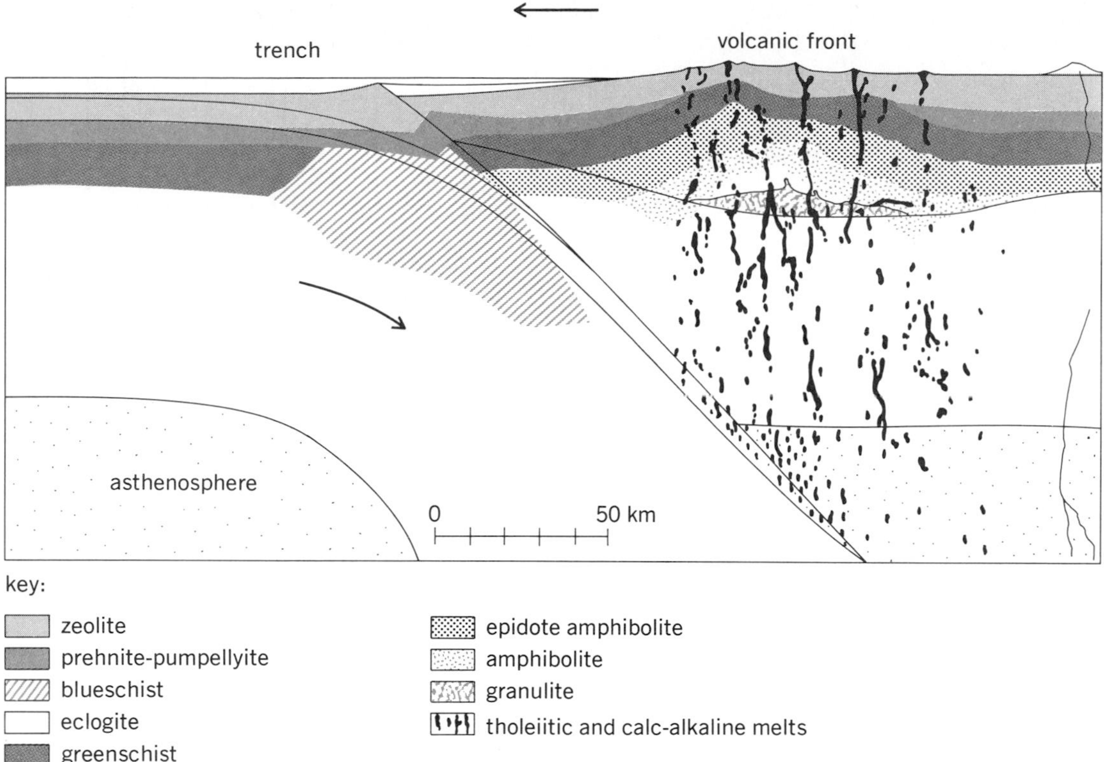

Fig. 2 Distribution of metamorphic facies types near a convergent plate margin. (*After W. G. Ernst, Petrologic Phase Equilibria, W. H. Freeman, 1976*)

intergranular diffusion by which mass transfer of chemical components can occur over distances considerably greater than the average grain size of the rock. Intergranular diffusion is the basis for much of what has traditionally been called metasomatism, that is, the process of large-scale change in bulk composition of rocks. The likeliest mechanism of this type of diffusive mass transfer is by movement of ions or complexes through the intergranular fluid in the rock. In order to better understand this process, metamorphic petrologists are currently studying the physical, chemical, and thermodynamic properties of metamorphic fluids. Precise calculation of rates in the case of both intracrystalline and intercrystalline diffusion has been hampered by imprecise knowledge of coefficients of diffusivity. Much current research is focused on refinement of diffusion coefficients and specific mechanisms of diffusive mass transfer.

Metamorphic and plate tectonic regimes. Increasing acceptance of plate tectonics as the governing mechanism of petrologic processes in the upper several hundred kilometers of the Earth has made it imperative that metamorphic petrologists fit their observations into this new framework. Regional metamorphism is almost always associated with moderate to intense deformation as part of a mountain-building episode. The plate tectonic regime appropriate to this type of metamorphism is a convergent plate margin regime, where two lithospheric plates come together at a combined rate of 10 or more centimeters per year. In a typical convergent margin the plate composed of lithosphere capped by dense, basaltic oceanic crust descends at a shallow angle beneath the less dense continental plate. Over a period of time, there are numerous minienvironments created in which rocks are subjected to widely different pressure-temperature paths. Figure 2 shows in schematic form the distribution of metamorphic facies types near a convergent plate margin. An apparently early collisional effect is the generation of high-pressure, low-temperature environments in which blue schists or glaucophane schists are produced from sedimentary or igneous rocks that were carried deep into the collision zone. These unusual rocks may be preserved (as in California) if they are then rapidly uplifted before they are heated up. More typically, blue schists are destroyed by heating as the basaltic slab descends to a deeper level and starts to melt. Ascending magma bodies heat up material carried into the subduction zone and create intermediate pressure-temperature metamorphic rocks of amphibolite facies. The potential end product of regional metamorphism at a convergent plate margin is the generation of virtually anhydrous rocks of granulite or eclogite facies through long exposure to very high temperatures and pressures. Careful characterization of areal and temporal distribution of the above metamorphic facies types is essential information for reconstruction of the exact type of plate tectonic environment from which a mountain belt was produced.

For background information *see* GEOLOGIC THERMOMETRY; HIGH-PRESSURE MINERAL SYNTHESIS; METAMORPHISM; PETROLOGY; PLATE TECTONICS in the McGraw-Hill Encyclopedia of Science and Technology.

[ROBERT J. TRACY]

Bibliography: W. G. Ernst, *Metamorphism and Plate Tectonic Regimes*, 1975; J. M. Ferry and F. S. Spear, *Contrib. Mineral. Petrol.*, 66:113–117, 1978; D. M. Kerrick, *Amer. Mineral.*, 59:729–762, 1974; S. K. Saxena and S. Bhattacharji, *Energetics of Geologic Processes*, 1977.

Metastable solid phases

A major recent trend in solid-state research has been the development and exploitation of techniques of synthesizing materials in metastable solid forms, for example, amorphous solids, supersaturated solutions, compositionally modulated films, and metastable intermetallic compounds. Much of the stimulus for this activity, especially in metallurgy, came from the demonstration in 1960 that certain molten metal alloys, when quenched at extremely high rates (greater than about 10^{6}°C/s), solidify to amorphous solid forms not previously recognized in metals. Besides rapid quenching, various condensation (for example, vapor, electro-, sputter, and chemical deposition) and irradiation (for example, ion implantation and ion beam mixing) methods have been developed and effectively applied to the synthesis of a wide variety of metastable metallic and semiconductor structures.

Principles of synthesis. The general technique in metastable structure synthesis is to energize a material (for example, by melting, dissolution, or irradiation) and then quench it to some configurationally frozen state (that is, a state of fixed atomic short-range order) in which the atomic transport rates are imperceptible over the period of use or observation of the material. Such freezing can occur whenever the atomic motions needed to reach equilibrium are thermally activated. However, there are certain interfaces which move without thermal activation in a diffusionless way and thereby, as in martensite-type transformations, prevent the configurational freezing of some structures.

A truly metastable material, though its free energy is higher than that of the same mass in stable form, is in internal equilibrium. For a mass to remain in internal equilibrium, the frequency of local atomic rearrangements within it must be high enough for it to pass freely through the many configurations constituting its macrostate. In such a high mobility regime a metastable body is highly susceptible to transformation to a more stable state. Thus, to be of technical use, metastable materials must be quenched to configurationally frozen states in which they are, in the strict sense, thermodynamically unstable with respect to changes in atomic short-range order. However, it is customary to apply to a body the label of the state it was in when its local order froze.

When an energized mass is quenched, there may be several paths of structural evolution which are permissible thermodynamically, including, of course, that to the most stable state. Metastable structures form only when they lie on the kinetically preferred paths of structural change. Actually, in evolving toward equilibrium, systems often do pass through a succession of metastable states of decreasing free energy. It has been proposed that this stepwise progression of states reflects a general principle, called the step rule. However, in the situations where the rule seems to apply, the entropy and internal energy, as well as the free energy, decrease in steps. There is no fundamental justification for such a rule, but the behavior on which it is based perhaps can be interpreted best on the basis of a qualitative principle that the paths of structural evolution most favored kinetically will be those requiring the lesser correlations of atomic positions and motions. Thus, structures which are disordered form more rapidly than highly ordered ones. For example, at temperatures below their configurational freezing points, materials generally condense as amorphous solids, in which the positional correlations are short range, rather than in their more stable crystalline forms.

Supersaturation of semiconductors. One of the areas of metastable structure synthesis in which there is great current interest is that of the formation of supersaturated overlays on semiconductors. Impurities are injected into the semiconductor by ion implantation or by ion beam mixing of thin overlays of the impurity material with the underlying semiconductor. These processes severely damage or even amorphize the crystal, and so the perfection often essential for semiconducting application has to be restored by epitaxial regrowth of the undamaged region of the crystal beneath the overlay. If regrowth occurs without impurity segregation, the overlay can retain concentration levels of electrically active impurities which may be orders of magnitude higher than the near-equilibrium levels achieved in normal crystal growth processing.

In the implantation technique, ions of the desired dopant are energized and driven into the crystalline semiconductor, which is held at a temperature far below that of configurational freezing. It is presumed that the ion disorders a radial zone, a few atoms across, along its path before it is deenergized and brought to rest within the body. In ion beam mixing, the energetic ion, for example Kr^{+}, upon crossing an interface between two mutually insoluble materials, creates so much disorder that the materials become atomically interdispersed along the narrow path of the ion. The deenergization, occurring within picoseconds, is so rapid that the intermingled atoms cannot resegregate before configuration freezing. The extent of intersolution achieved by this means is much greater than that from mixing the molten materials, indicating that the degree of atomic disordering in the wake of the ion is far more than that in the liquid state.

Epitaxial regrowth of semiconductors. In germanium and silicon epitaxial regrowth of the undamaged crystal into amorphized overlays can be effected by furnace annealing at relatively low temperatures, in the range 0.4 to 0.6 of the absolute melting temperature T_m. The growth is thermally activated and proceeds at rates of the order of nm-per second. There is negligible redistribution of those impurities which enter the crystal substitutionally in this process, reflecting the extremely low solid-state solubility of such impurities in the temperature regime of regrowth as well as the large decrease in free energy attending the crystallization. Thus the crystal retains the highly nonuniform impurity distribution which it acquired during implantation.

More uniform distribution of the supersaturated impurity may be achieved by high-energy pulsed annealing of the damaged overlay. The energy can be deposited by irradiating the semiconductor with high-energy laser pulses for short periods, on the order of a few nanoseconds. Amounts of energy are deposited which are sufficient to convert the overlay to the molten metallic phase of germanium or silicon if thermal equilibrium is reached in the period of the pulse. There is some controversy on the time τ required for thermal equilibration of the electron-hole plasma, produced by the irradiation, with the silicon or germanium ions. Much of the experimental and theoretical evidence seems to point to τ values no larger than a few picoseconds, but it has been argued that τ for silicon exceeds 100 ns. However, from the rapid redistribution of impurity there can be no doubt that the pulsing produces a state in the overlay, whether consisting of cold or hot ions, having the atom transport properties of a liquid metal. In what follows, it is assumed that the heavy ions are in thermal equilibrium with the charge carriers.

In the high-energy pulsing, epitaxial regrowth occurs at rates in the range of meters-per-second. These rates contrast sharply with the nanometers-per-second rates of the lower temperature regime and are of the order of those for regrowth of metal crystals in their undercooled pure melts. This growth occurs in a sequence of at least two steps: an atomic rearrangement at the interface between the crystal and melt, followed by transport of the heat of crystallization, and possibly impurities as well, away from the interface. The heat transport into the bulk of the semiconductor in epitaxial regrowth is driven by the thermal gradient (grad $T)_i$ at the interface, while the interfacial rearrangement is driven by the free energy of crystallization at the actual interface temperature T_i. In magnitude this free energy will generally be proportional to the departure $T_m - T_i$ of T_i from the equilibrium melting temperature T_m, provided $T_m - T_i << T_m$.

When crystals grow in undercooled liquid metals, the interfacial rearrangement process is so rapid that, with thermal gradients of the usual magnitude, its net rate will match that of heat flow when T_i is displaced only slightly from T_m. In this case $T_m - T_i << T_i - T_a$, where T_a is the ambient temperature, and the growth is said to be diffusion-limited. However, in laser pulsing a large quantity of energy is deposited in a thin section coupled to a massive heat sink so that (grad $T)_i$ will be orders of magnitude greater than in ordinary crystal growth. Thus the heat flow rate will be greater by a corresponding amount, and much larger displacements of T_i from T_m will be needed in order for the interfacial rearrangement rate to match it. It is this larger $T_m - T_i$ which provides the free energy required for the generation of metastable structures, specifically supersaturated solid solutions, during regrowth.

Impurity trapping in crystal growth. Thermodynamic and kinetic analyses indicate that the degree of supersaturation reached in regrowth should be determined by the ratio of the interface velocity u to the maximum diffusive speed D_i/λ (where D_i is the diffusivity of the impurity in the interfacial region and λ is an interatomic spacing) of impurity away from the interface. The larger this ratio is, the nearer will the supersaturation approach the limit set by the implantation.

In liquid metals, including molten germanium and silicon, D_i/λ may approach levels near 10 m/s. When ordinary thermal gradients are imposed, $T_i - T_m$ and u will be relatively small so that there would be little nonequilibrium retention of impurity in the crystal. However, as (grad $T)_i$ is increased, $T_m - T_i$ increases so that the interface is driven faster, and with sufficiently high (grad $T)_i$ it may reach velocities in the meters-per-second range. In this regime the retention of impurity beyond equilibrium increases sharply with $u\lambda/D_i$. When (grad $T)_i$ is increased beyond a certain point, the interface temperature falls below that T_o at which "diffusionless" crystallization becomes thermodynamically possible; that is, the free energy of crystallization to the melt composition becomes negative. In this regime, $T_i < T_o$, the whole of the implanted impurity may be trapped within the crystal by the fast running interface even though its chemical potential at T_i is thereby increased. It is this regime which is approached or entered in regrowth after high-energy pulsing, with the result that good crystalline overlays are formed which are quite uniformly doped to levels much higher than achievable by the usual processing.

Amorphous silicon from melt quenching. Recently it was observed that, after energization by appropriately designed picosecond laser pulses, sections on (111) surfaces of pure silicon crystals quenched to the amorphous semiconducting state of silicon. It was presumed that the energization produced a thin layer of molten silicon on the (111) face and that the extreme thermal gradient at the crystal-melt interface during the quench was sufficient to reduce T_i to a temperature well below that T_{al} at which the undercooled metal liquid phase would be in equilibrium with the metastable amorphous semiconducting silicon phase. T_{al} was estimated to be about 0.8 of T_m of crystalline silicon. When T_i falls below T_{al}, silicon may regrow in either its stable crystalline or metastable amorphous semiconducting form. Observations have indicated that regrowth to the less stable amorphous state actually occurs, in accord with the principle suggested above that the paths most preferred kinetically are likely to be those leading to the more disordered states.

For background information *see* AMORPHOUS SOLID; ION IMPLANTATION in the McGraw-Hill Encyclopedia of Science and Technology.

[DAVID TURNBULL]

Bibliography: M. J. Aziz, *J. Appl. Phys.*, 53: 1158–1168, 1982; J. C. Baker and J. W. Cahn, Thermodynamics of solidification, in *Solidification*, pp. 23–58, American Society of Metals, 1970; J. M. Liu et al., *Appl. Phys. Lett.*, 38:617–619, 1981; D. Turnbull, *Met. Trans.*, 12A:695, May 1981.

Microkelvin temperatures

The ambition of physicists to investigate matter under extreme conditions and to look for new phenomena under these conditions has led to considerable progress in refrigeration. The third law of thermodynamics demonstrates the impossibility of reaching absolute zero (0 K = −273.15°C), but scientists have by now far surpassed nature. Whereas the lowest temperature in nature is the temperature of the cosmic background radiation (3 K), recently samples have been refrigerated and investigated down to 0.00004 K (40 μK). The applied refrigeration method is the nuclear magnetocaloric effect, the adiabatic demagnetization of nuclear magnetic moments.

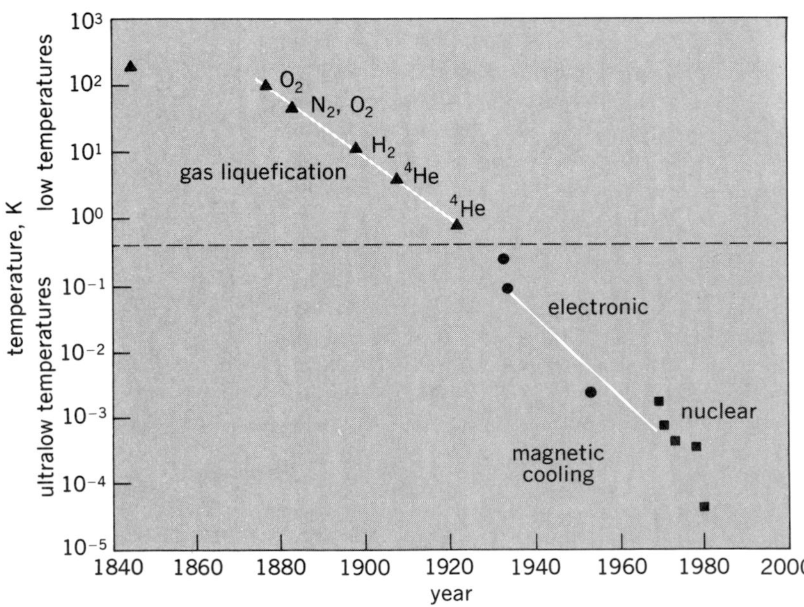

Fig. 1. Historical development of the minimum temperature to which samples have been refrigerated by liquefication of the indicated gases or by adiabatic demagnetization of electronic magnetic moments and of nuclear magnetic moments, respectively.

Development of refrigeration. The historical development of refrigeration to successively lower temperatures is shown in Fig. 1. In 1908 H. Kamerlingh Onnes succeeded in liquefying ^4He at 4.2 K, enabling him to discover the superconductivity of metals in 1911. By 1926 Kamerlingh Onnes was able to reduce the minimum temperature T of his experiments to 0.7 K by pumping on a liquid helium bath with an enormous battery of pumps; for the first time a temperature below the minimum temperature in nature was reached in the laboratory.

In the same year P. Debye and W. F. Giauque suggested a new cooling method, the adiabatic demagnetization of paramagnetic salts. The first step into ultralow temperatures, a temperature of about 0.25 K, was reached by this magnetic cooling method in 1933. The next step, a temperature of 0.085 K, followed in 1934, and eventually 2 mK was reached in 1953. Similar low limits exist for other cooling techniques developed since then, namely $^3He-^4He$ dilution refrigeration ($T \gtrsim 2$ mK) and cooling by solidification of liquid ^3He (Pomeranchuk cooling; $T \gtrsim 1$ mK). The only known method for refrigeration to temperatures appreciably below 1 mK is an advanced version of magnetic cooling; adiabatic demagnetization of *nuclear* magnetic moments, suggested in 1934 by C. J. Gorter and by N. Kurti and F. Simon.

The limit for cooling by adiabatic demagnetization of paramagnetic salts is set by the spontaneous magnetic ordering of the electronic magnetic moments when their interaction overcomes the thermal disorder. Because nuclear moments are about 3 orders of magnitude weaker, their ordering usually occurs at substantially lower temperatures. Unfortunately, the price of cooling to these lower temperatures is a large number of substantial experimental problems. The first experiment, performed in 1956 by Kurti and colleagues, made these problems obvious. Starting from a magnetic field of 3 T and a temperature of 12 mK, they cooled copper *nuclei* to about 1 μK, but the copper electrons and the lattice stayed at 12 mK and the nuclei warmed up to the starting temperature within a few minutes. The problem in nuclear refrigeration is not the cooling of nuclei but the transfer of the ultralow temperature from the nuclei to the electrons and the lattice of the refrigerant, and eventually from there to an experimental sample and to thermometers. To solve this one must, above all, reduce heat leaks so that they are balanced by the cooling capacity of the refrigerant, and so that the ultralow temperature can be sustained for a sufficiently long time.

The magnitude of acceptable heat leaks for experiments at ultralow temperatures is extremely small by normal standards. In the refrigerator to be described below, as little as 10^{-12} W heat input into the thermometer was enough to cause a 2% error in the thermometry at 50 μK. In order to cool the sample and thermometers to 50 μK, the refrigerant had to be at 5 μK, a factor of 10 colder, because of a total heat leak of merely 10^{-9} W. For comparison, the tiny anount of power required by a quartz crystal wristwatch to run for a year on a small battery amounts to about 5000 × 10^{-9} W.

Nuclear adiabatic demagnetization. Figure 2*a* is a schematic representation of a modern single-stage nuclear refrigerator as developed particularly by O. Lounasmaa's group. The nuclear stage consists of 10–40 moles of copper in the form of insulated wires of a few hundred millimeters in length. The wires are long enough to cool a sample located outside the high field required for the magnetization and demagnetization process. The bottom ends of the wires are located in a superconducting solenoid. After the solenoid has been energized to about 8 T, the copper is precooled by a dilution refrigerator located above it to, typically, 15 mK. These rather demanding starting fields and temperatures have been made accessible only by the development of superconducting magnets and ^{3}He – 4He dilution refrigerators in the last 2 decades. The limits set by the technology of these two tools determine the initial entropy reduction in the nuclear refrigerator, which usually is only about 3–5%. This entropy reduction determines the cooling capacity of the refrigerator. The thermal contact from the nuclear refrigerator to the dilution refrigerator is via a heat switch which consists of a piece of superconducting metal. In the superconducting state, the thermal conductivity of this switch is very low; but if the metal is driven into the normal state by a magnetic field, its thermal conductivity is as high as that of any normal metal.

When the precooling by the dilution refrigerator is complete, the heat switch is made superconducting in order to thermally isolate the copper; the demagnetization is then begun. It continues slowly, to reduce eddy current heating and to keep thermal equilibrium between nuclei and electrons, until the desired temperature is reached. If the demagnetization is performed adiabatically, without losses, the entropy of the copper nuclei and the ratio B/T remain constant so that the temperature T decreases proportionally to the field B.

About 20 such nuclear refrigerators were built between 1970 and 1980. Some can cool samples to about 0.5 mK, and the best one refrigerated to 0.2 mK. A great number of exciting experiments in the newly accessible temperature range have been carried out with the aid of these nuclear magnetic refrigerators—particularly to investigate the superfluidity of liquid ^{3}He below 2.7 mK (detected in 1972) and the nuclear antiferromagnetism of solid ^{3}He below 1 mK (detected in 1974). Heat leaks and the small values of the initial entropy reductions limit the abilities of these refrigerators. But there are important experiments which require lower temperatures, demanding new concepts in refrigerator design. Some new ideas have been realized in the Jülich double-stage refrigerator, which demonstrates the feasibility of performing experiments in solid-state physics down to 40 μK. In addition, a nuclear refrigerator has been built in Helsinki, particularly for polarization of nuclear spins; a record nuclear spin temperature of about 50 nK was reached in 1979.

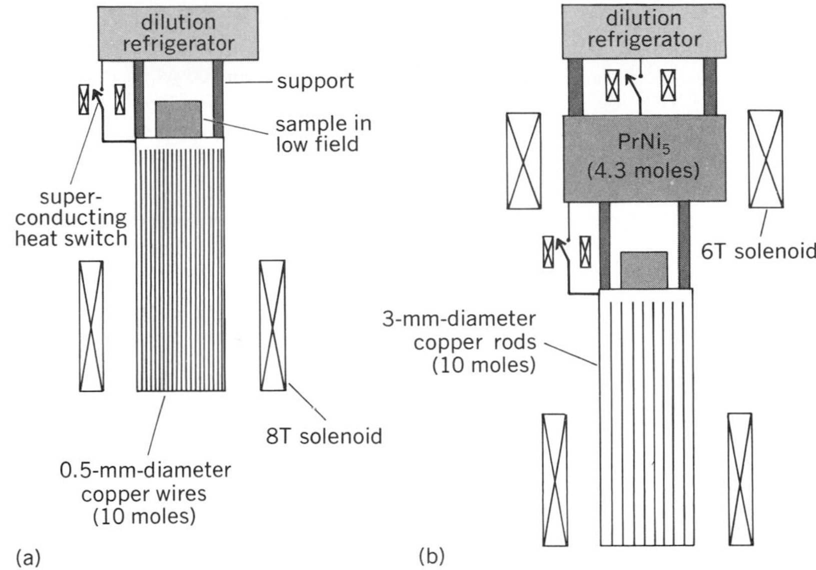

Fig. 2. Configuration of nuclear demagnetization refrigerators. (*a*) Single-stage refrigerator. (*b*) Jülich double-stage refrigerator.

Double-stage nuclear refrigerator. The cooling capacity of a nuclear refrigerator can be increased only by increasing the initial entropy reduction. This requires lowering the temperature of the copper or increasing the magnetic field at the start of the demagnetization. Unfortunately, the cooling power of dilution refrigerators deteriorates rapidly with decreasing temperature. Also, a substantial increase of the magnetic field above the usual 8 T with a superconducting magnet is difficult and expensive. Therefore other approaches for improvement had to be taken.

The improvements largely depend on having two nuclear demagnetization stages which operate in cascade; this approach was first realized in 1977. The purpose of the first stage is to precool the second stage, which is ultimately demagnetized to ultralow temperatures. The first stage is distinguished by doing all of its work at rather "high" temperature, above 5 mK, but here it must absorb extremely large amounts of heat. Copper could meet these requirements, but the large quantity required would fit only in an expensive, massive solenoid which also might seriously interfere with sensitive experiments. There are much better materials for this requirement, namely the Van Vleck paramagnets. In these materials a large hyperfine field is induced at the nuclei when an external field is applied. Together, the applied and induced fields are more effective in aligning the nuclei than is the applied field alone; therefore the reduction of entropy and the cooling power are greater. One of the best Van Vleck paramagnets for nuclear demagnetization is the compound $PrNi_5$.

It is tempting to consider using $PrNi_5$ alone for nuclear refrigeration. It is, in fact, excellent for the purpose, but temperatures below about 0.2 mK cannot be produced because spontaneous nuclear ferromagnetic ordering of the praseodymium nuclei occurs at around 0.4 mK. Spontaneous nuclear ordering is also expected in copper, but only at a temperature near 10^{-7} K because no large hyperfine field acts on the copper nuclei.

The manner in which a $PrNi_5$ stage is added to the system is depicted schematically in Fig. 2*b*. Besides the extra refrigerant, another superconducting magnet and heat switch are required. During precooling, both heat switches are in the high-thermal-conductivity, normal state and both solenoids are generating full field. About 70% of the entropy of the $PrNi_5$ is already removed at 25 mK in a field of 6 T. The stages are then thermally isolated from the dilution refrigerator, and the $PrNi_5$ is demagnetized. At the end of the demagnetization, the copper temperature is about 5 mK. At this temperature, 28% of the nuclear entropy of copper is removed, in comparison to 4% if the copper is cooled only to 15 mK. This greatly increases the amount of heat the copper can absorb when it is demagnetized.

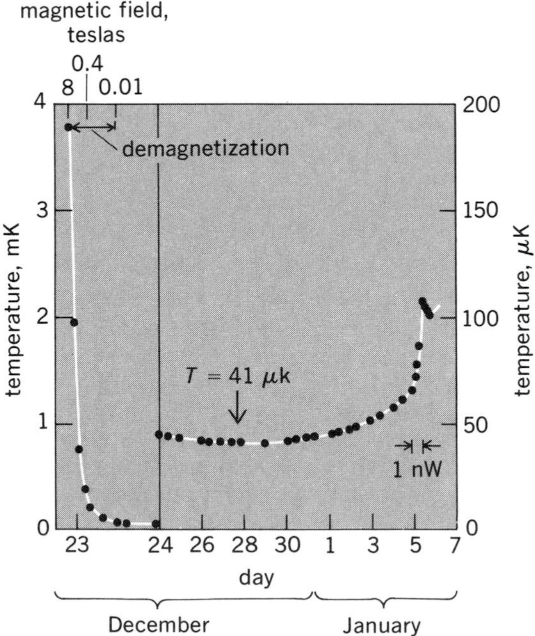

Fig. 3. Temperature measured at the location of samples in the Jülich double-stage nuclear refrigerator as a function of time. The left part of the graph shows the decrease of temperature (in mK) of the sample when the magnetic field on the copper refrigerant is decreased exponentially in time from 8 to 0.01 T. The right part shows, in an expanded temperature scale (in μK) and a compressed time scale, the development of the sample temperature after demagnetization.

Besides decreasing the minimum achievable temperature, lengthening the time the refrigerator can remain cold, and decreasing the amount of heat transported from outside, the large entropy reductions possible in the double-stage refrigerator permit valuable changes in the design of the copper stage. The performance of the refrigerator is shown in Fig. 3, which presents the temperature measured at the location of the experimental samples as a function of time. In particular, it shows the very long relaxation times typical for ultralow temperatures. Only four days after the demagnetization was completed, the thermometer indicated that the experimental samples had been relaxed to 41 μK, which is the lowest conduction electron temperature ever measured. The sample stayed below 50 μK for 10 days. The nuclei of the demagnetized copper were at 5 μK. The sample is so much hotter than the nuclei because of the heat leak, small as it is, between 0.1 and 1 nW, the source of which is unknown. On January 5, an additional heat input of 1 nW was supplied by the heater to the sample to accelerate its warmup.

If a new temperature regime is entered, thermometry is as important as the refrigeration—and often as difficult. The applied thermometry method in the microkelvin range is nuclear magnetic resonance. At present, there exists no established temperature scale in the milli- and microkelvin range. The given temperatures are believed to be accurate to about 5%.

The microkelvin range now accessible opens the possibility for detection of new phenomena and decisions about unsolved problems. Among these are the questions whether all metals either must order magnetically or become superconducting if they are sufficiently cold and clean, and whether ^{3}He diluted in liquid ^{4}He may become superfluid (so that two different, interpenetrating superfluid systems would exist); the various aspects of magnetic ordering of nuclear magnetic moments are also significant. All these phenomena are under investigation.

For background information *see* ADIABATIC DEMAGNETIZATION; CRYOGENICS; LOW-TEMPERATURE PHYSICS; LOW-TEMPERATURE THERMOMETRY in the McGraw-Hill Encyclopedia of Science and Technology. [FRANK POBELL]

Bibliography: D. S. Betts, *Refrigeration and Thermometry below 1 K*, 1976; O. V. Lounasmaa, *Experimental Principles and Methods below 1 K*, 1974; K. Mendelssohn, *The Quest for Absolute Zero*, 1966; R. M. Mueller et al., *Cryogenics*, 20:395–407, July 1980.

Microscopic solid structures

During the last 2 decades the electronics industry has been involved in an ongoing revolution in the large-scale integration (LSI) of both analog and digital systems. The digital revolution is leaving a permanent imprint on all aspects of life, especially as the implementation of microelectronics has spread to the consumer industry. As a result of the drive to less expensive, but more complex and sophisticated, integrated systems, the complexity of digital LSI circuits, in terms of the number of individual devices or individual functions on a chip, has approximately doubled each year since the late 1960s. There are, of course, several factors which contribute to this increase in complexity. However, the three major effects arise from increased die size, increased skill in circuit design, and reduced device size. The last factor, reduction of the individual feature size in a device, is of paramount importance, and dimensions of laboratory systems are currently down to the submicrometer range. Preproduction systems have been developed that have typical source-to-drain, or gate-length, dimensions of 1.0 μm. However, researchers have produced devices in both silicon and gallium arsenide that have gate lengths in the 0.3–0.4-μm range, and metal oxide semiconductor field-effect transistors (MOSFETs) have been built with gate lengths of only 0.1 μm.

Progress in the microelectronics area is tied inevitably to the ability to continue to put an ever-increasing number of smaller devices on a chip; that is, movement toward very-large-scale integration (VLSI) is of greatest importance to this continued progress. Consequently, it is reasonable to assume that extrapolation of today's technology will produce individual devices whose dimensions are of the order of 0.1–0.2 μm or less. High-resolution electron, x-ray, molecular, and ion-beam lithography may well lead to the fabrication of individual features on the molecular scale of 10–20 nm. It becomes feasible then to conceive of device structures so small that the basic properties of the host material may be significantly less important than effects related to size and environment. In particular, once these small dimensions are reached, quantization due to the small device dimensions can alter the properties of carriers in the host material, thus leading to new effects.

Not only is there a considerable amount of work to be done to understand the physics and technology of semiconductor devices whose dimensions are below 0.2 μm, but these devices also become experimental systems for the investigation of numerous current problems in many-body physics. The advance of technology toward this dimensional scale is thus of great importance for solid-state physicists as well as for electrical engineers.

Size quantization. One can visualize the onset of size quantization by simply realizing that the classical theory of inversion layers leads to extremely thin layers at the surface. It is generally assumed in the classical model that $n(z)$, the surface concentration of charge carriers (electrons or holes) in the inversion layer, decays with distance from the surface z according to Eq. (1), where q is the electron

$$n(z) = n_s \exp\left[-qV(z)/k_BT\right] \tag{1}$$

charge, $V(z)$ is the electric potential, k_B is Boltzmann's constant, and T is the thermodynamic temperature. Thus, $n(z)$ falls to $1/e$ of its surface value when $qV(z)$ changes by only k_BT. Thus $<z>$, the value of z averaged over all charge carriers, is given approximately by Eq. (2), where E_s, given by Eq. (3), is the surface field just inside the semiconduc-

$$<z> \cong k_BT/qE_s \tag{2}$$

$$E_s = \frac{q(N_{depl} + n_s)}{\varepsilon_{sc}} \tag{3}$$

tor, ε_{sc} is the permittivity of the semiconductor, and $N_{depl} = N_A z_{depl}$ is the sheet concentration of the depletion charge. Here N_A is the density of acceptor atoms and $z_{depl} \cong (\varepsilon_{sc}V_s/qN_A)^{1/2}$ is the depletion depth, for $V_s = V(0)$. For N_A in the range 10^{16}–10^{17} cm^{-3}, $<z>$ is of the order of 0.7–1.0 nm, an exceedingly short distance.

To ascertain whether or not quantization effects are present, this distance is compared with the electron wavelength in the inversion layer. The electron's de Broglie wavelength λ_e is given by Eq. (4),

$$\lambda_e = h/p \tag{4}$$

where h is Planck's constant and p is the mean momentum. For a thermal energy of k_BT, the electron wavelength in silicon at 300 K is of the order of 10 nm. Thus the classical theory breaks down and the inversion layer is a quantum system in the direction perpendicular to the surface. The thickness of the inversion layer and the properties of the electrons within this layer must be calculated by self-consistent quantum-mechanical methods. In Fig. 1, one such calculation is shown and is compared with the classical value given by Eqs. (2) and (3).

Because of the quantization of the surface, the inversion layer is now a quasi-two-dimensional electron system. Moreover, the Fermi energy E_F (the electrochemical potential in equilibrium) lies within the conduction band, so that the inversion layer is almost metallic in many of its properties. This is useful in another sense. Recent theories of localization and correlation at low temperatures have pri-

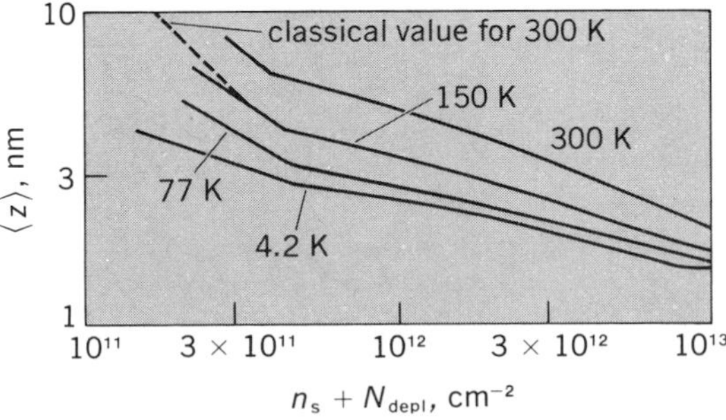

Fig. 1. Average electron penetration $<z>$ into silicon for a (100) surface with $N_A = 1.6 \times 10^{16}$ net acceptors per cubic centimeter.

marily focused on one- and two-dimensional systems. The MOSFET is vital to experimental studies since both $<z>$ and E_F (through n_s) can be controlled separately by gate and substrate biasing. Thus, this two-dimensional system, complete with random potentials due to oxide charge and interface roughness, is a classical test system to study the properties of metal-insulator transitions. Understanding of these quantum-mechanical properties is necessary to the understanding of submicrometer devices when the gate length drops below 0.1 μm.

Transport on short time scales. In the very small semiconductor devices envisaged here, the active region length is only a fraction of a micrometer. For this situation, the average electron velocity cannot easily be represented as an explicit function of the electric field through a mobility. Rather, the electric fields are very high (1 V across 0.1 μm gives a field of 10^5 V/cm), and the important time processes such as transit time, energy and momentum relaxation times, and screening time are all comparable in magnitude. In this generally nonlinear, nonequilibrium, and nonstationary many-body problem, the transport parameters must be calculated from quite complicated retarded transport equations, which generally are non-markovian in nature.

If a system of electrons is injected at the source and subjected to the combined influence of an electric field and scattering centers, then the drift velocity v_d of the particles rises as given by Eq. (5),

$$v_d = \frac{eE}{m^*}\int_0^t \phi'_{\Delta v}(0, t')\, dt' \qquad (5)$$

where $\phi'_{\Delta v}(0,t) = \phi_{\Delta v}(0,t)/\phi_{\Delta v}(0,0)$ is the normalized velocity autocorrelation function of the nonstationary system. Here, $\phi_{\Delta v}(t,t') = \langle[v(t) - <v(t)>][v(t') - <v(t')>]\rangle$, and the averages are ensemble averages. This function describes the velocity correlation of the ensemble due to collisions. The system is a nonequilibrium and nonstationary one during the transient dynamic response phase, and the response is nonlinear. The decay of $\phi_{\Delta v}(t_0, t_0 + t)$ is in general not a simple exponential, as in the linear case. Rather, it is a nonpositive definite function, as shown in Fig. 2. The behavior of $\phi_{\Delta v}(t, t_0)$ is quite important (the two-time behavior here is a significant departure from equilibrium stationary statistical mechanics). When the ensemble sees the electric field at $t = 0$, it begins to respond instantaneously by a shift in momentum space corresponding to the field and time. Thus the velocity begins to rise in a ballistic manner immediately, but this linear rise lasts for only a very short time. However, spreading of the ensemble, characteristic of the thermal motion of the individual carriers and their mean energy, does not begin until collisions begin to break up the correlations with the initial equilibrium state. As shown in Fig. 2, that $\phi_{\Delta v}(t, t_0)$ passes through zero and is negative for an appreciable portion of the total duration. This leads to an overshoot in $v_d(t)$. Thus the velocity rises past its steady-state value and then settles back toward it. The same behavior is seen in the transient diffusion coefficient given by Eq. (6), where the integral on

$$D(t) = \frac{1}{2}\frac{d}{dt}[<(x - <x>)^2>] = \int_0^t \phi_{\Delta v}(t, t_0)\, dt_0 \qquad (6)$$

the right-hand side is over the initial times of the autocorrelation function (again, this is a departure from equilibrium stationary statistical mechanics).

Most field-dependent velocities and values for the saturated velocity assume that steady-state conditions are reached. Clearly, this is not the case in very small devices. The variation in $\phi_{\Delta v}$ lasts for as much as 0.25 picosecond in silicon. An electron traveling at 2×10^7 cm/s can cover 0.1 μm in 0.5 ps. It is evident that over a great portion of the device the velocity and diffusion currents never reach their steady-state values. This effect is more pronounced in gallium arsenide and other III–V materials in which the time variation of $\phi_{\Delta v}$ can last for as long as 1–2 ps. From these considerations, it follows that the velocity in very small devices is not likely to be the same as in large devices, and indeed can be much larger due to overshoot effects. This increase in the effective velocity allows higher currents to flow through the device, which in turn allows a logic gate to switch faster through a faster charging of the node capacitance. Figure 3 shows the enhancement expected in the effective saturated velocity for some typical semiconductors. *See* TRANSISTOR.

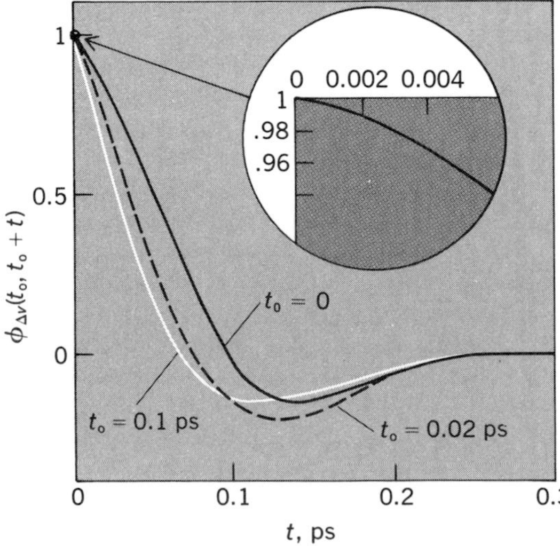

Fig. 2. Normalized velocity autocorrelation function for electrons in silicon at 300 K. Three different initial times t_0 are shown for an electric field of 50 kV/cm; $t_0 = 0$ is used in Eq. (5).

Oxides. As device dimensions shrink further, materials other than silicon are expected to have advantages for applications to high-speed logic. This exemplifies the results shown in Fig. 3. The high values of velocity will allow larger drive currents, and thus faster circuit switching. However, present very-large-scale integration (VLSI) technology is based predominantly on silicon metal-oxide-semiconductor (MOS) technology, and if MOS technology is to carry over the new materials, a good oxide must be present. However, the oxides on the III–V materials of interest show strange physical behavior themselves.

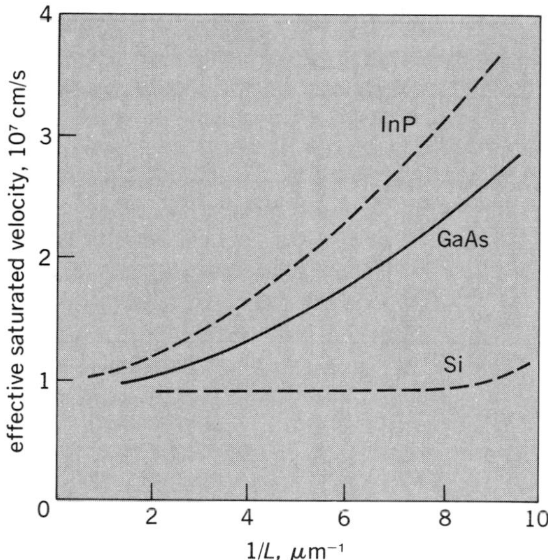

Fig. 3. Effective, or time-of-flight, saturated velocity as a function of inverse channel length, $1/L$.

The rather good qualities of the silicon/silicon dioxide interface and of silicon dioxide have been a fortuitous advantage to the semiconductor industry. Such is not the case with the III–V materials such as gallium arsenide and indium phosphide. While oxides on the latter compound have been made sufficiently good to allow inversion layer transport, the quality of the oxide is not well characterized. While probably not a technologically adequate oxide, the anodic oxide indium phosphide remains a standard and can provide a stable oxide. It represents interesting physics that also characterizes some of the problems of these compound semiconductor oxides.

Anodic oxidation is in many ways a highly desirable method of forming the insulator of a metal-insulator-semiconductor (MIS) insulator on indium phosphide, as on other indium-based compounds, since it is formed near room temperature in a few minutes. This method avoids evaporation or damage of the surface often caused by insulator deposition. In the past, however, the anodic oxides have been poor insulators. This is somewhat surprising since P_2O_5 is thought to have a band gap near that of SiO_2. For previous oxides this could be explained by the low concentrations of P_2O_5 in the grown films. More recently, anodic oxides with good insulator properties have been obtained when the P_2O_5/In_2O_3 ratio approached 1.6. This ratio is controlled by the pH of the anodization solution, and the sharp increase of oxide resistivity at a ratio of 1.6 is suggestive of a phase change or the occurrence of some type of conductivity transition. X-ray analysis of these and other III–V oxides indicates that the oxides are amorphous, and an amorphous to crystalline transition is unlikely. Even if such a transition could occur, it would probably not cause any large change in the resistivity. A more plausible model envisions islands of In_2O_3 becoming surrounded by P_2O_5 as the P_2O_5/In_2O_3 ratio increases. For the moment, the In_2O_3 islands are allowed to be of arbitrary size, ranging from a single molecule to an island containing thousands of molecules. It is known that In_2O_3 has a band gap of approximately 3.7 eV and tends to be a conductive n-type semiconductor; the donor doping arises from oxygen vacancies. Unfortunately, little is known about the electrical properties of P_2O_5. However, from the positions of phosphorous and oxygen in the periodic table, it is surmised that the band gap is between 6 and 10 eV. The nature of the P_2O_5 bonding suggests that there should be few free carriers, and thus it would appear that P_2O_5 should be a wide-band-gap insulator.

From the above data, a conceptual atomic model may be constructed for the oxide with a wide-band-gap host material (P_2O_5) containing potential wells from the In_2O_3. This model is similar to one proposed for amorphous materials. Conduction through this type of medium would result in hopping or tunneling of the free electrons from one In_2O_3 island to another. This type of transport is not necessary until the In_2O_3 islands are completely surrounded by P_2O_5. It is postulated that this occurs at a P_2O_5/In_2O_3 ratio of approximately 1.6, and that it is the cause of the dramatic increase in resistivity.

Simple analysis tends to verify this hypothesis and leads to the expectation that the large change in resistivity is due to a percolation threshold. In the alloy percolation theory, the conductance above threshold is proportional to $(p - p_c)^n$, where p corresponds to the percentage of In_2O_3, and p_c is the critical concentration. Evaluation of these parameters gives $p_c \approx 0.38$, that is, a P_2O_5/In_2O_3 ratio of approximately 1.6, and $n \approx 4.5$. Unfortunately, percolation calculations have been carried out only for the simpler crystal structures. Crystalline In_2O_3, however, has T_h^7 symmetry, a much-modified fluorite structure. Each indium atom sits in an incomplete octahedral site and has six oxygen neighbors, while each oxygen has only four indium neighbors. On the other hand, P_2O_5 has space group C_{2v}^{19}. Here, it is thought that each phosphorous atom has four neighbors, with one being double-bonded. Clearly, the mixed structure will be even more complicated in the amorphous phase. For the threshold observed, the concentration $p_c \approx 0.38$ is not incompatible with the basic structures of these oxides.

The value of n, however, is considerably larger than the values of 2.23–2.36 expected from three-dimensional renormalization group and percolation theory. These latter calculations were performed on simple cubic structures, and it is not clear whether the theoretical results will carry over to the complicated structures of the oxides. However, it appears that such considerations lend credence to the model proposed above. In any case, it is clear that the mixed oxide provides another controllable system for the study of metal-insulator transitions, this time for a three-dimensional system.

For background information *see* BAND THEORY OF SOLIDS; INTEGRATED CIRCUITS; MICROWAVE SOLID-STATE DEVICES; QUANTUM MECHANICS; SEMICONDUCTOR; STOCHASTIC PROCESS in the McGraw-Hill Encyclopedia of Science and Technology.

[DAVID K. FERRY]

Bibliography: D. A. Baglee et al., *Inst. Phys. Conf. Ser.*, 56:259, 1981; D. K. Ferry and J. R. Barker, *J. Phys. Chem. Solids*, 41:1083, 1980; B. G. Levi, *Phys. Today*, 34(5):19–21, May 1981; F. Stern, *Crit. Rev. Solid State Sci.*, 4:499, 1974.

Mineralogy

Historically, quantitative chemical bonding studies in mineralogy have been limited to classical electrostatic potential methods with which Madelung energies, site potentials, lattice energies, and crystal field effects are evaluated. Over the past decade there has been an increased utilization of modern bonding formalisms, based on quantum-mechanical principles, in mineralogical research. Investigations have included studies of equilibrium molecular conformations, modeling of electrical and optical properties, calculations of x-ray photoelectron and x-ray emission spectra, ESCA (electron spectroscopy for chemical analysis) spectral shifts, lattice and site energies, equilibrium bond length calculations, force constant and bulk moduli determinations, and thermochemical properties.

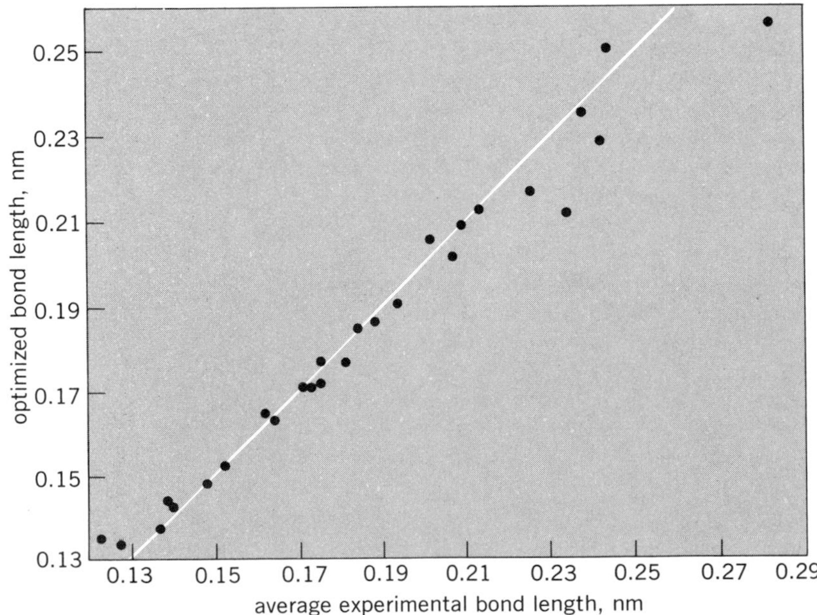

Comparison of the optimized bond lengths for various first- and second-row-atom oxide and sulfide molecules versus the average experimental bond lengths taken from oxide and sulfide solids. (*From G. V. Gibbs et al., in M. O'Keeffe and A. Navrotsky, eds., Structure and Bonding in Crystals, pp. 195–225, Academic Press, 1981*)

There has been a general feeling in mineralogy that because minerals are bulk solids, a successful bonding formalism must include both short- and long-range atomic interactions. Unfortunaely, in a quantum-mechanical context, large atomic groups require large numbers of basis (wave) functions to adequately describe the molecular electronic environment. The large number of basis functions required by a molecule containing only a hundred light atoms creates an enormous computational task even for modern computers. Because of this practical limitation, most modern bonding theories are developed for bulk (crystal) energy calculations or for local atomic cluster calculations. Recent local atomic cluster calculations have revealed that the latter method can be successfully applied to studies of molecular configurations as well as chemical and physical properties of minerals. These investigations indicate that studies of local bonding forces can provide insights about bulk solids that were previously thought untenable.

Crystal energies. Madelung energies and closed-shell repulsive energies, calculated within the context of classical ionic electrostatics, are summed to yield lattice energies. However, accurate lattice energies have been calculated for only the simplest of minerals by these methods. For most minerals the method fails because of series termination effects or because the repulsive-energy parameters are poorly defined. Recent investigations have attempted to determine the repulsion parameters by assigning physically more realistic partial charges to the ions and by fitting the parameters to minimize errors between calculated and observed lattice energies. While such efforts show promise, these methods must be improved further if accurate lattice energies of complex mineral structures are to be determined.

An alternative method of calculating lattice energies has developed recently and is referred to as the modified-electron-gas ionic model. In this method the electron density at each point is the sum of the free ion densities, and each free ion is described by a highly accurate Hartree-Fock wave function. In addition, each ion pair interaction is calculated using Coulomb's law, with a free-electron-gas approximation employed to evaluate exchange and correlation energies. In effect, this method is a combination of classical electrostatics and quantum-mechanical formalisms and has been demonstrated to yield accurate lattice energies, heats of formation, and site energies.

Atomic cluster calculations. In the early 1970s mineralogists began applying semiempirical methods such as extended Hückel molecular orbital theory and the "complete neglect of differential overlap" to studies dealing with equilibrium molecular configurations in silicates, phosphates, sulfates, arsenates, and other oxyanion groups. Although these are approximate methods, they have provided considerable insight in numerous crystal chemical investigations over the past decade. Of the two, the "complete neglect of differential overlap" method is somewhat more sophisticated, although various approximations are called upon to simplify the large computational effort in evaluating certain electron repulsion integrals, and is semiempirical, since some terms are taken from experimental atomic data or from fitting to accurate Hartree-Fock data. The two electron integrals, which depend on overlap of charge densities of different atomic basis orbitals, are neglected. Nevertheless, the computation approximates important electron repulsion integrals, evaluates the repulsion energy terms between cores, and refines the molecular orbital expansion coefficients to self-consistency.

Although the "complete neglect of differential overlap" is an approximate method, it has been successfully applied to studies of the origins of shared polyhedral-edge distortions in close-packed solids, energy-optimized molecular geometries in isolated and corner-shared atomic polyhedra, compressibilities of the silica polymorphs, and network-modifying effects of common ions in the polymerization of silicate and aluminate tetrahedra in glasses, melts, and aqueous solutions.

Perhaps the most serious shortcoming of the "complete neglect of differential overlap" method is its inability to accurately predict equilibrium bond lengths. More sophisticated methods, which accurately reproduce bond lengths and bond angles for molecules involving first- and second-row atoms, have recently been coded and are available through the Quantum Chemistry Program Exchange. One such method is an ab initio self-consistent field–molecular orbital calculation which utilizes gaussian expansions of Slater-type orbitals. Although the calculations are considerably more expensive than those of the "complete neglect of differential overlap" method, this type of calculation has been used in numerous mineralogical investigations devoted to determinations of equilibrium stereochemistries for first- and second-row-atom oxide, sulfide, and silicate clusters. The illustration shows the agreement between equilibrium bond lengths computed by this method for a variety of molecular groups and those observed in solids. In addition, equilibrium bond lengths and force-constant calculations have recently been undertaken at simulated high pressures.

Another self-consistent field–molecular orbital method which has been successful in constructing one-electron energy diagrams and in the interpretation of emission, absorption, and photoelectron spectra is the χ_α scattered-wave method. The χ_α approach avoids using large atomic basis sets required by other methods by subdividing the molecular group into areas within which Schrödinger's equation is solved. These solutions are then combined to yield self-consistent sets of molecular orbitals and energies. Mineralogical applications have dealt mainly with oxides, silicates, and sulfides.

For background information *see* CHEMICAL BONDING; CRYSTAL FIELD THEORY; MOLECULAR ORBITAL THEORY in the McGraw-Hill Encyclopedia of Science and Technology.

[E. P. MEAGHER]

Bibliography: M. Catti, *Phys. Chem. Minerals*, 7:20–25, 1981; G. V. Gibbs et al., in M. O'Keeffe and A. Navrotsky (eds.), *Structure and Bonding in Crystals*, pp. 195–225, 1981; J. A. Tossell, *Amer. Mineral.*, 65:163–173, 1980; D. J. Vaughan and J. A. Tossell, *Can. Mineral.*, 18:157–163, 1980.

Missile guidance

The United States has begun to deploy a family of tactical and strategic weapons known as cruise missiles. These missiles fly at high subsonic speeds and at near-treetop levels to avoid enemy detection and effective defensive measures. The air-launched cruise missiles (ALCM) are launched from carrier aircraft, the sea-launched cruise missiles (SLCM) from surface ships or submarines, and the ground-launched cruise missiles (GLCM) from a ground-mobile transporter system. As opposed to intercontinental ballistic missiles with their large nuclear warheads, cruise missiles carry either a small nuclear warhead or conventional explosives. Thus the ultimate effectiveness of the weapon depends on precise missile navigation and guidance to the target.

All cruise missiles use inertial navigation systems for postlaunch guidance to the target. However, inertial navigation systems develop navigation errors that increase with time from launch due to instrumentation errors within the navigation system and to initial condition errors, such as position and velocity errors, inherited from the launch vehicle navigation system. A scheme referred to as terrain contour matching (TERCOM) is used to partially remove these navigation errors at certain locations along the cruise missile trajectory.

Principles of TERCOM. TERCOM uses barometric and radar altimeters in conjunction with terrain elevation profile data to provide periodic measurements of the missile's position on its trajectory. In this technique, the radar altimeter measures the height of the missile above the terrain for a specified distance d. During the radar altimeter's period of operation, the inertial navigator with the barometric measurement calculates the altitude of the missile above an altitude reference such as sea level. The difference between the two altitudes is the terrain profile below the missile. Prior to the missile flight, predicted terrain profile data are stored in the missile computer, and by properly comparing and processing the real-time information collected during the sensing distance d, it is possible to determine the missile position very accurately. This position information is used to update the inertial navigation system, which then guides the missile to the next TERCOM update area along the missile's flight path.

Implementation. In one method of implementing TERCOM, the terrain profile data stored in the missile computer cover a ground track f which is the sum of the actual sensing distance d and an allowance for the inertial navigation along-track error e, such that it is almost certain that the sensed profile will match part of the stored profile in the missile computer. Typically, the inertial system errors are assumed gaussian, and e is chosen as six times the standard deviation of the predicted inertial system along-track error at the time of sensing, thus effectively guaranteeing with 0.99+ probability that the sensed profile will correspond to some portion of the stored profile. In a similar manner, the cross-track dimension of the stored profile data may be chosen as six times the standard deviation of the predicted inertial system cross-track error at the time of sensing. This method of implementation is called the long-matrix, short-sample technique because the radar altimeter is used only during d.

An alternate short-matrix, long-sample implementation of TERCOM minimizes the amount of along-track terrain profile data that must be stored in the missile computer. In this case, the along-track sensing distance is set at f and the stored profile data has an along-track dimension d. The cross-track dimension of the stored profile data is chosen as before, thus making the reference profile data larger in the cross-track direction than in the along-track direction.

Trajectory planning. Because of the need for stored terrain profile data, cruise missile mission trajectory planning is an extensive process which begins by determination of acceptable areas on the Earth's surface that can be used for a TERCOM update. Neither flat terrain (no signal) nor terrain with too many altitude variations (too much noise) is acceptable. Furthermore, of the areas that do appear acceptable, it must be determined that every possible cruise missile trajectory (considering the navigation errors) over the area will result in an unambiguous (unique) terrain profile measurement to have a satisfactory navigation update. Thus, acceptable TERCOM reference areas tend to be somewhat limited in length and width. Based on the reference area dimensions, the mission planning procedure can then determine the allowable cross- and along-track missile navigation errors. From the allowable errors and the knowledge of the statistical growth rate of the missile navigation errors between updates, the distances between update areas can be specified.

Initialization. The function of the launch vehicle navigation system is to provide the missile navigation system with the proper initial conditions in position, velocity, and attitude such that the growth of missile navigation errors after launch will be less than the allowed cross- and along-track errors at the first TERCOM update area. Because the ALCM flys a considerable distance between launch (typically over the ocean for a strategic missile) and the first update area, the initial condition errors must be kept quite low and a high-quality navigation system must be used on the carrier aircraft.

During 1981, initial development tests were successfully conducted from B-52 carrier aircraft using a new navigation system for cruise missile initialization. The system consists of a precise inertial navigation system, a position-fixing radar used for updating the inertial system, and a Doppler velocity radar also used for updating. Aircraft positional inaccuracies have been reported to grow only at a rate of 600 to 1200 ft (180 to 370 m) per hour of carrier aircraft flight subsequent to a radar position update, thereby allowing many hours of aircraft flight prior to launch of a cruise missile.

Immediately before missile launch, navigation information must be transferred from the carrier system to the missile navigation system. This initialization is accomplished by comparing outputs from the two navigation systems during aircraft maneuvers and level flight. Any difference in the outputs is attributed to missile navigation system errors, which are then corrected in the missile computer prior to launch. After launch, the missile navigates to the first TERCOM update area.

Additional sensors and systems. The success of the many flight tests for the ALCM has allowed the planned deployment of the weapon system to proceed on schedule. Tests have also been conducted where an additional sensor is used in the target area for a high-precision update. This sensor performs an optical match with a reference photograph of the terrain below the missile, enabling a very precise update of the missile navigation system. In the future, sensors may be developed that look from the cruise missile at the target itself and thereby enable almost a direct hit. Other foward-looking sensors may allow detection and avoidance of unknown obstacles (such as power lines) during the missile flight. Sensors may be developed that detect mobile enemy threats (ground radars, aircraft and so forth), which the cruise missiles would automatically avoid. It is possible that other missile sensors would be developed to determine if the target has already been destroyed, thus allowing the missile to attack an alternate target. Better navigation systems may be developed for the carrier aircraft and for the missile navigator to minimize the need for TERCOM updates. All of these possibilities indicate that the full versatility and potential of the cruise missile weapon system has yet to be realized.

For background information *see* GUIDANCE SYSTEMS; GUIDED MISSILE; INERTIAL GUIDANCE SYSTEM in the McGraw-Hill Encyclopedia of Science and Technology. [GEORGE T. SCHMIDT]

Bibliography: G. T. Schmidt and R. H. Setterlund, *AIAA J. Guidance Control*, 3(6):487–493, November–December 1980.

Mitochondria

Mitochondria are found in all eukaryotic cells capable of using oxygen. Often referred to as the powerhouse of the cell, they are the site of cellular res-

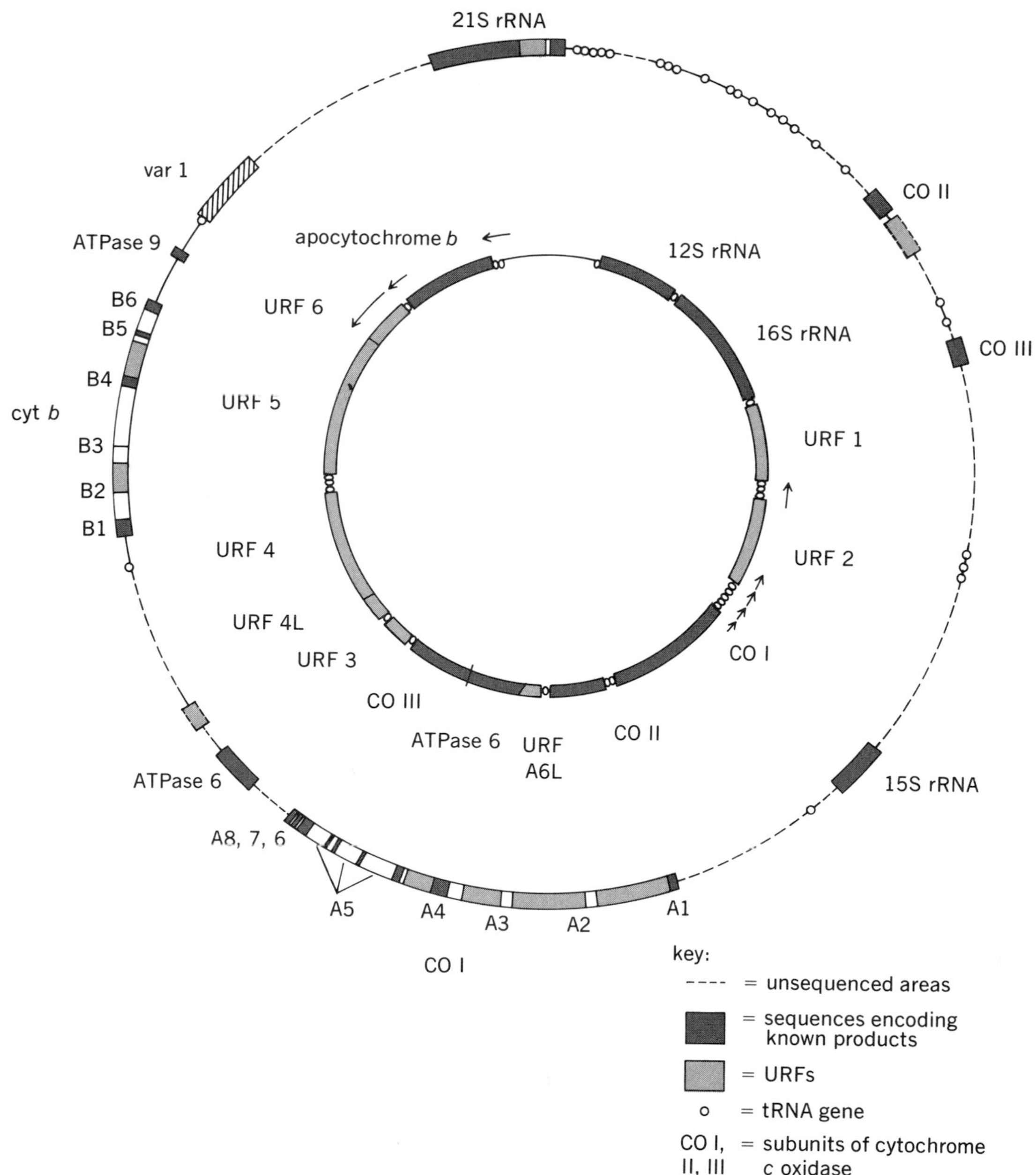

Fig. 1. Comparison of the organization of mitochondrial genes in the mtDNAs of humans (16569 bp) and in the yeast *Saccharomyces cerevisiae* strain KL 14-4A (78000 bp). In the split genes for apocytochrome *b* and subunit I of cytochrome *c* oxidase, introns not containing a URF are shown as open bars. Transcription is in a clockwise direction, unless indicated otherwise by arrows, which denote that the relevant gene is located on the DNA strand opposite to the one encoding the remaining genes. For reasons of clarity, the human genome has been drawn approximately twice its real size relative to yeast.

piration and are responsible for capture of the energy liberated during the oxidation of the simple organic compounds generated by the breakdown of foodstuffs. Synthesis and assembly of a functional mitochondrion—unlike other organelles in the cell, such as lysosomes and microbodies—is dependent on a contribution from a genetic system located in the organelle itself. Thus, while the bulk of mitochondrial proteins are specified by nuclear genes, synthesized on cell sap ribosomes, and imported into the mitochondrion, a small minority (about 10%) are encoded by mitochondrial DNA (mtDNA) and synthesized in the mitochondrion by a genetic apparatus which differs in many respects from that in the nucleus and cytoplasm.

The distinguishing feature of mtDNA in all organisms is its simplicity compared with its nuclear counterpart; most mitochondriologists have hoped that study of mitochondrial genes would be a relatively simple affair, so that mtDNA might form a

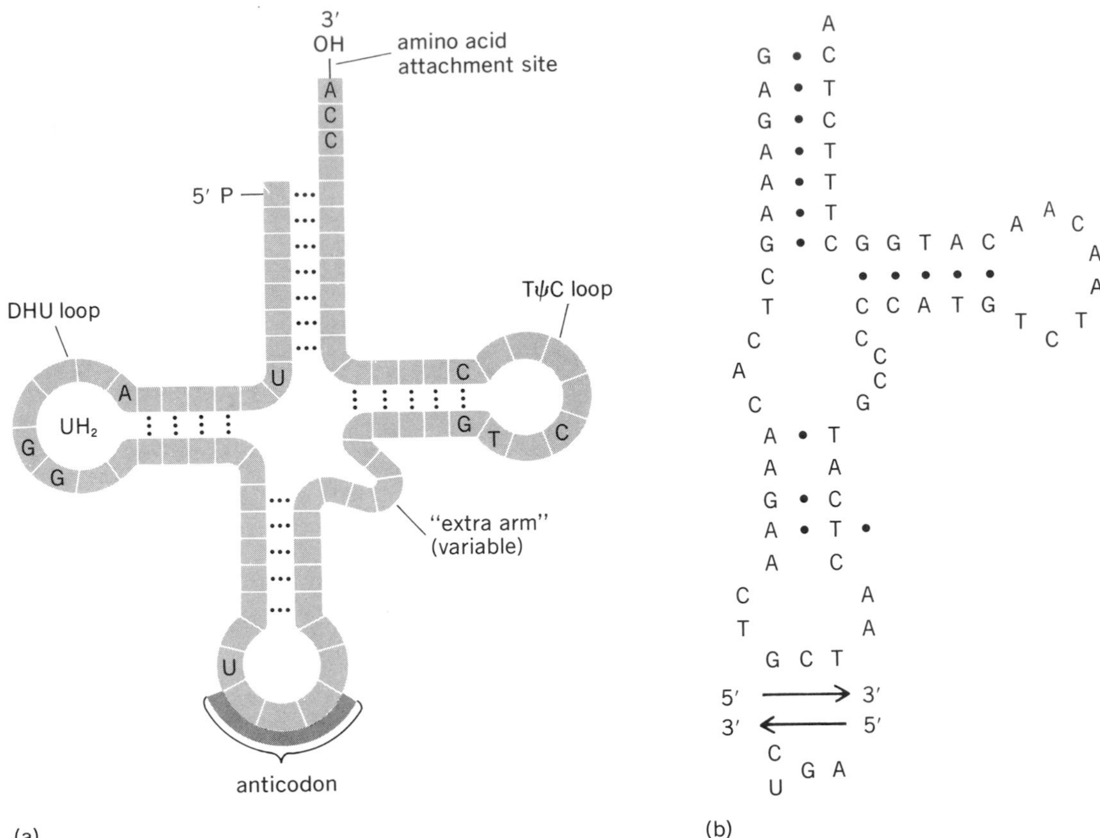

Fig. 2. Unusual tRNA structure in human mitochondria. (*a*) Structure of tRNAs involved in protein synthesis generally fit into the cloverleaf pattern. About half the nucleotides are base-paired and a number of invariant, or strongly conserved, bases are sited at set locations in the DHU and TψC arms. (*b*) The DNA sequence of the tRNAser is an extreme example of the way in which mammalian mitochondrial tRNAs fail to conform to the common pattern. The tertiary structures of this and other unusual mitochondrial tRNAs have yet to be determined.

back door to the study of gene expression in an otherwise extremely complex eukaryotic cell. In the past 2 years, advances have been made in the understanding of this simple genome. These have been brought about mainly by the application of molecular genetics and rapid DNA-sequencing techniques. The resulting picture gives remarkable insight into the way in which mtDNA contributes to the synthesis of mitochondrial membrane proteins, and at the same time teaches a number of lessons relevant for molecular biology in general. A broad outline of some of the major developments is given below.

Human mtDNA. Figure 1 shows the gene map of the human mitochondrial genome. The most striking feature of this 16569-bp (base pairs) molecule is the extreme economy of its organization. Study of other DNAs had led to the expectation that in order to be functional, genes must be flanked by DNA sequences carrying signals for the initiation and termination of RNA and protein synthesis, with the signals for the latter being present in the messenger RNA (mRNA) transcribed from the gene. This is clearly not the case in the human mitochondrial genome. Genes for 13 different proteins, 2 ribosomal RNAs, and 22 transfer RNAs are packed so tightly that there are few or no noncoding bases between them. In addition, the mRNAs lack nontranslated 5′-leader and 3′-trailing regions, and over half of them lack a complete stop codon to signal termination of protein synthesis. In these cases, the U or UA at the 3′ terminus has to be converted to the UAA stop codon by addition of a poly (A) tail to the mRNA after completion of its synthesis.

Transcription-translation machinery. The transcription-translation machinery necessary for expression of these genes also seems to have been reduced to a bare minimum. From a detailed analysis of transcripts it is known that a single site for initiation of RNA synthesis is present on each DNA strand, and that both are completely and symmetrically transcribed. The transcripts of individual genes are later generated by extremely precise cleavage of the resulting giant RNAs at the 5′ and 3′ termini of the transfer RNA (tRNA) sequences that flank virtually every gene. The basis of this tRNA-punctuation processing is probably that the tRNA sequence present in the primary transcript can fold up and assume a characteristic structure recognizable by the enzyme responsible for cleavage. Curiously, antisense tRNA sequences (that is, transcripts of tRNA genes located in the opposite DNA strand) also seem to be recognized, but in these cases cleavage is restricted to the 3′ side. One intriguing problem is how the differential expression of mitochondrial genes is achieved when opportunities for transcriptional controls are so limited. At least three mechanisms seem to be operative. First, a stop signal for RNA synthesis (transcription attenuator) situated immediately downstream of the ribosomal RNA (rRNA) genes provides almost a hundredfold excess of rRNAs over other transcripts by ensuring that correspondingly fewer RNA polymerase molecules proceed beyond this point. Second, controls probably exist on the cleavages required to generate individual transcripts, and finally, the various mature transcripts may differ in stability.

Protein synthetic machinery. Economy is also evident in the mitochondrial protein synthetic machinery. Thus the rRNAs have been reduced to a size (1559 and 954 nucleotides) not seen outside mitochondria; 5-S rRNA is absent, and only 22 tRNAs read all codons. This is possible because some tRNAs read four codons, and in these the first position of the anticodon is occupied by a U, permitting an interaction with any of the four bases in the 3′ position of the codon. Besides these unusual decoding abilities, nearly all human mitochondrial tRNAs have unconventional structures. All but one differ from the more than 200 previously sequenced tRNAs of other organisms in that they deviate from the standard clover leaf pattern and its associated common features in the DHU and TψC arms (Fig. 2*a*). Perhaps the most extreme example of such deviation is the tRNAser shown in Fig. 2*b*, in which the DHU arm has been amputated. How this and the other unusual tRNAs function is still a mystery, but it is clear that the mitochondrial ribosome and the decoding mechanism must have undergone compensatory changes to accommodate them. These changes are probably still going on: in contrast to the strong conservation of ribosomal RNAs seen in other genetic systems, the mitochondrial rRNAs of cow, mouse, and human show only about 70% homology, and similar low homologies are found when individual mitochondrial tRNAs in the three organisms are compared. Evidently, the function of the mitochondrial protein synthetic machinery is compatible with a high fixation rate of mutations, and as a consequence it is evolving rapidly.

Genetic code and URFs. Closer examination of the nucleotide sequence has yielded two additional discoveries. The first is the finding that the genetic code in human mitochondria differs from that used by other genetic systems (see table): UGA is read as tryptophan rather than stop; AGA and AGG are read as stop rather than arginine, and AUA as methionine rather than isoleucine. In addition, AUA or AUU can be used as initiation codons for protein synthesis instead of AUG. Thus, contrary to the prediction made by some authorities, the genetic code is not universal.

The second discovery is that not only are the genes subunits of major mitochondrial membrane enzyme complexes, but the genome contains information for eight unidentified proteins in the form of long open-reading frames (unassigned reading frames, or URFs). Although these URFs have not yet been definitely identified as products of mitochondrial protein synthesis, there seems little doubt that they represent protein-coding genes. Not only is it unlikely that such a compact genome would carry nonfunctional sequences, but comparative DNA sequence studies on cow, rat, and mouse mtDNAs show strong evolutionary conservation of the URFs, and furthermore discrete RNA transcripts of each exist. It has been suggested that the URF-encoded proteins might be RNA-processing enzymes or mitochondrial ribosomal proteins, but of course many other roles are possible, and defining these will be just one of the tasks for the future.

Yeast mtDNA. The findings made with human mtDNA can probably be extrapolated to other animal mtDNAs. The sizes of these DNAs are similar, and in those cases studied so far, the relative positions of genes for rRNAs, tRNAs, and many of the minor transcripts are comparable. A very different

Genetic code in human mitochondria*

First position (5′ end)	Second position: U	C	A	G	Third position (3′ end)
U	Phe	Ser	Tyr	Cys	U
	Phe	Ser	Tyr	Cys	C
	Leu	Ser	stop	*Trp (stop)*	A
	Leu	Ser	-(stop)	Trp	G
C	Leu	Pro	His	Arg	U
	Leu	Pro	His	Arg	C
	Leu	Pro	GlN	Arg	A
	Leu	Pro	GlN	Arg	G
A	Ile	Thr	AsN	Ser	U
	Ile	Thr	AsN	Ser	C
	Met Ile	Thr	Lys	*Stop (Arg)*	A
	Met	Thr	Lys	*Stop (Arg)*	G
G	Val	Ala	Asp	Gly	U
	Val	Ala	Asp	Gly	C
	Val	Ala	Glu	Gly	A
	Val	Ala	Glu	Gly	G

*Deviations from the universal code are shown in italics, with the normal assignments given in parentheses. Boxes of four codons are each read by a single tRNA with U in the first position of the anticodon; boxes of two codons are read by tRNAs with G:U wobble anticodons. Some uncertainty still exists as to the number of methionine tRNAs and their codon response. These have therefore been left unboxed.

situation holds for yeast, however, and in order to bring out some of the contrasts, the two genomes are juxtaposed in Fig. 1.

Distinctions. The nucleotide sequence of a large part of yeast mtDNA is known. This shows that the genome carries more or less the same set of genes for major mitochondrial membrane proteins as human mtDNA and that the amino acid sequence of these proteins has been strongly conserved. The similarity ends there, however. Whereas the human genes are crammed into a bare 16569 bp, the same information in yeast is spread over some five times this amount of DNA, with no obvious logic to the order and with vast expanses of virtually pure AT interspersed between the coding sequences.

Next, three of the yeast genes are interrupted by intervening sequences, whereas all the human genes are continuous. [Many genes in the nuclear genomes of higher organisms have recently been found to be split. In these, the DNA sequences coding for mRNA (termed exons, for expressed regions) are interrupted by noncoding segments, referred to as intervening sequences, or introns. During expression of such genes, both coding and intervening sequences are transcribed into RNA, which then undergoes a form of processing, called splicing, in which the intervening sequences are excised, and the coding sequences are linked together to form a translatable mRNA.] Further, nearly all the yeast mRNAs have long, untranslated 5′-leader or 3′-trailing sequences, while the human mRNAs contain at most a short poly (A) tail added posttranscriptionally to the 3′ end. Yeast also uses UGA to specify tryptophan, rather than stop, but the other unusual assignments of the AGA, AGG, and AUA codons are not used (see table). Instead, the CUN family of leucine codons is recognized as threonine, so that there is not even a common mitochondrial genetic code. Finally, yeast mtDNA also contains a number of URFs, located in the spaces between genes and in some of the introns of the split genes, but no homology with the human URFs is apparent.

Cut-and-splice mechanism. As in human mitochondria, RNA processing probably plays an important regulatory role in gene expression in yeast mitochondria, but for different reasons. Yeast is unlikely to make the same elegant use of tRNA punctuation processing, first because the genome contains several promotors for the initiation of RNA synthesis, and second, because most tRNA genes are clustered in one quadrant of the DNA. On the other hand, the presence of introns in some genes requires the existence of a cut-and-splice mechanism to remove these sequences from the initial RNA transcript, with the linkage of the appropriate coding sequences. In itself, such a mechanism is not unusual; similar events are involved in the maturation of transcripts of split genes in the nucleus. What is unusual, however, is that in one gene at least—that for apocytochrome *b*—splicing of an intron is catalyzed by a protein specified by the intron itself.

This intricate situation was first predicted from an elegant genetic analysis, but has since been confirmed directly by DNA sequence analysis, and the currently accepted picture of this mosaic gene is shown in Fig. 3. There are two versions of the gene in laboratory yeast strains. The long gene contains six exons (E_{1-6}), five introns, and two long open-reading frames in introns I_2 and I_4 (DNA sequence analysis of I_3 is not yet complete). The "short" gene lacks the first three introns. An early step in the processing of the primary transcript of the long gene is the removal of intron I_1, to link exon E_1 with E_2 (only 14 bp long) and the long open reading frame fused to it. The resulting RNA acts as messenger for the synthesis of a protein resembling in some ways the chimaera of Greek mythology. Its amino-termi-

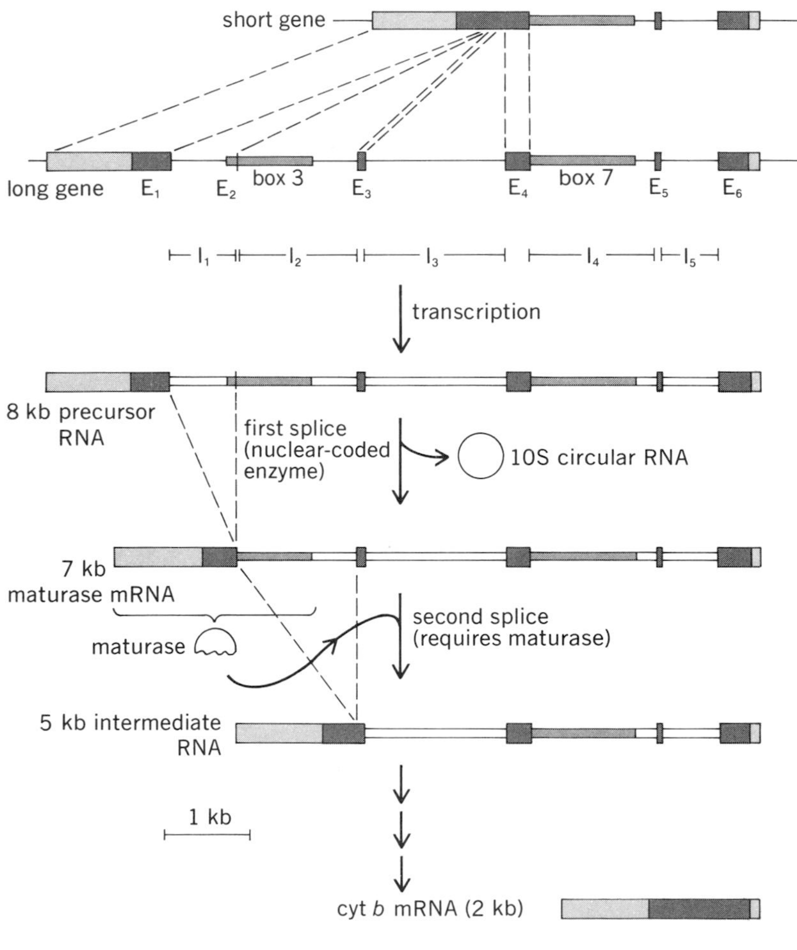

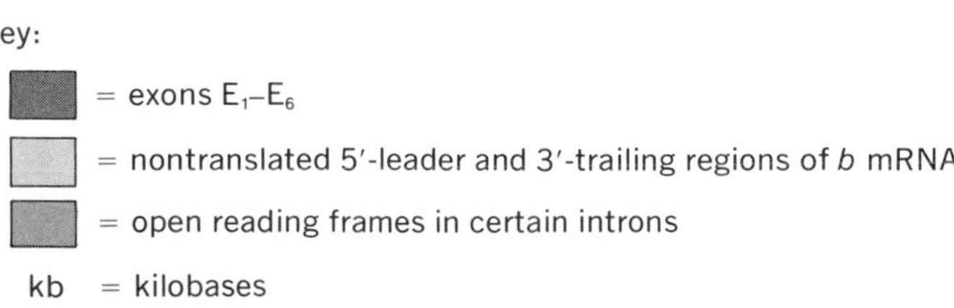

Fig. 3. "Long" and "short" versions of the gene for apocytochrome *b* in yeast mitochondria. Two versions exist as a result of the presence or absence of "optional" introns. The figure also shows the mRNAs transcribed from the "long" gene.

nal "head" belongs to apocytochrome *b*, and its carboxy-terminal "tail" is encoded by the intron. This protein, the *box*3 maturase, helps to catalyze the excision of intron I_2. In so doing, it destroys the mRNA which directed its synthesis and thus controls its own level within the cell—a phenomenon called splicing homeostasis. In wild-type yeast, the maturase is present at concentrations too low to be detectable by pulse-labeling experiments, but from the DNA sequence it would appear to be a highly unusual protein. Of its 423 amino acid residues, 143 belong to the extremely hydrophobic amino-terminal part of apocytochrome *b*. The remainder, encoded by the intron, are either hydrophobic or basic, and almost 60% of them are specified by codons consisting only of A and U. Nevertheless, it is this part of the protein that contributes most to maturase function, since both missense and nonsense mutations in the intron reading frame result in an inability to excise I_2, and lead to an overproduction of defective maturase. In contrast, missense mutations in exon E_1 generally have little effect.

Splicing homeostasis may not be restricted to the *box*3 maturase. The existence of URFs in other introns, such as I_4 of the cytochrome *b* gene and the first four introns of the gene for subunit I of cytochrome *c* oxidase, could mean that these code for other maturases, each specific for the excision of a particular intron. Suggestive evidence for this in the case of I_4 (*box*7) is given by the finding of mutants in which protein synthesis in the URF is prematurely terminated. These mutants cannot excise I_4, and like corresponding mutants in I_2, they accumulate novel proteins which might represent defective maturases. Major differences are that these new proteins are not related to cytochrome *b*, and they are shorter than expected for a fusion protein consisting of the exons E_{1-4} plus I_4, so that there could be yet further twists in an already complicated story, in terms of either posttranslational cleavage of such a fusion protein or a reinitiation of protein synthesis inside the intron itself. Both possibilities are at present being studied.

One curious aspect of the *box*3 maturase system is that it is dispensable. In yeast strains with the short version of the cytochrome *b* gene, the first four exons are fused together, and the information in introns I_{1-3} is lacking, so that in long strains it would almost seem as if the *box*3 maturase exists only to control its own synthesis. In fact, introns I_{1-3} are just three of a class of "optional" introns discovered in yeast mtDNA. Others are located in the genes for the large ribosomal RNA and for subunit I of cytochrome *c* oxidase. Of course, classification of an intron as optional is dependent on finding a yeast strain that lacks it, and since only a limited number have been looked at, all introns may eventually turn out to be optional. This variation raises intriguing questions: Why are some genes split, while others are not? Why should the same gene exist in split and continuous forms? It is clear that many of the ideas put forward to explain the existence of split genes in the nucleus do not apply in the mitochondrion. It is possible either that the presence of introns confers some kind of selective advantage, perhaps in terms of flexibility of metabolic response, or that the introns, by increasing the length of the genes they occupy, speed up evolution by permitting the construction of new genetic combinations through more effective recombination. A radically different alternative is that the introns are invaders from outside, perhaps originally in the form of a piece of mobile DNA, such as an insertion element or transposon. In this scheme, the intron would not confer any particular advantage, but the cell probably lacks an effective mechanism for ridding itself of such elements. This hypothesis has two attractive elements. First, it accounts neatly for the high sequence homologies displayed by some introns: in the gene for subunit I of cytochrome *c* oxidase, introns 1 and 2 are homologous, and intron 4 is homologous with I_4 in the cytochrome *b* gene. Second, the theory suggests an origin for maturase activity: bacterial transposons usually contain information for the synthesis of a transposase, an enzyme which catalyzes movement of the transposon to a new site. Transposition involves cleavage exactly at the ends of the transposon, just like the precise cuts made during excision of an intron from RNA. The maturases may therefore have evolved from transposases.

Mitochondrial DNA and evolution. From this brief comparison, it will be clear that while the major products of mitochondrial protein synthesis have been largely conserved in evolution, the organization and mode of expression of the genes specifying them have not. This lack of restraint on sequence evolution may be related to the simplicity of the mitochondrial genetic system. Its only task is to synthesize a limited number of proteins in relatively constant amounts, and as such it may be better able to accommodate potentially deleterious mutations than a system geared to the differential regulation of very many genes. Obviously, selective pressures have pushed yeast and the other fungi on the one hand and animal mtDNAs on the other into different, extreme solutions to the problem of gene expression, and discovery of what these pressures are is a task for the future. Hopefully, the further study of these and other extremes will define the minimal requirements of a functional genetic system and provide information on the evolutionary pathways that might have been followed in the development of the sophisticated mechanisms of gene expression used by present-day bacteria and eukaryotes.

For background information *see* CYTOPLASMIC INHERITANCE; DEOXYRIBONUCLEIC ACID (DNA); MITOCHONDRIA in the McGraw-Hill Encyclopedia of Science and Technology.

[L. A. GRIVELL]

Bibliography: S. Anderson et al., *Nature*, 290:457–465, 1981; P. Borst, in Th. Bücher, W. Sebald, and H. Weiss (eds.), *Proceedings of Mos-*

bach Colloquia, pp. 27–41, 1980; P. Borst and L. A. Grivell, *Nature*, 289:439–440, 1981; J. Lazowska, C. Jacq, and P. P. Slonimski, *Cell*, 22:333–348, 1980; D. Ojala, J. Montoya, and G. Attardi, *Nature*, 290:470–474, 1981.

Neutrino

Neutrinos have provided a steady stream of surprises in elementary particle physics ever since Wolfgang Pauli proposed their existence in 1930. These particles are electrically neutral and highly penetrating—a low-energy neutrino from nuclear β decay can pass through several light-years of lead before it interacts. Lately, there has been great interest in the possibility that neutrinos have mass. Until recently, most physicists believed that neutrinos had zero mass and consequently moved with the speed of light. However, there is no compelling theoretical reason for neutrinos to be exactly massless. In fact, in recently developed grand unified theories (GUTS), which unite the weak, electromagnetic, and strong interactions, it is natural for neutrinos to acquire nonzero masses which are considerably smaller than the masses of the quarks and the charged leptons. (Another consequence of grand unified theories is the prediction that the proton is unstable.) There are also new experimental results indicating that neutrinos may have masses on the scale of tens of electronvolts. If confirmed by further measurements, this has extremely interesting implications for astrophysics and cosmology, concerning dark matter in clusters of galaxies and the expansion of the universe. For nondegenerate neutrino masses, neutrino identities may interchange with time. Such neutrino oscillations may explain a longstanding puzzle regarding a deficiency of solar neutrinos.

Beta-decay experiments. One source of information about the neutrino mass is the shape of the energy spectrum of electrons emitted in nuclear beta decay. The basic process is the change of a neutron (n) into a proton (p) with the emission of an electron (e^-) and an electron antineutrino ($\bar{\nu}_e$): $n \rightarrow p + e^- + \bar{\nu}_e$. Pauli's suggestion that the $\bar{\nu}_e$ carries the apparent missing energy and momentum in beta decay was put on a quantitative theoretical footing by Enrico Fermi in 1933. The shape of the energy spectrum of emitted electrons is sensitive to the neutrino mass only near its end point, where the electron carries the maximum available energy. In the end-point region the decay rate is small, so high precision measurements are necessary to detect neutrino mass effects. Until 1980, all measurements of the spectrum agreed with theoretical curves for zero electron-neutrino mass within the limits of experimental error, and it could be concluded that the rest mass of that neutrino was very small compared to the rest mass of the electron (511,003 eV). The upper limit of $m(\nu_e) < 60$ eV had been obtained by K. Bergkvist at Stockholm. The latest result from an experiment at the Institute of Theoretical and Experimental Physics (ITEP) in Moscow which analyzes the beta decay of tritium, ${}^3H^0 \rightarrow {}^3He^+ + e^- + \bar{\nu}_e$, indicates an electron-neutrino mass somewhere in the range of 14 to 46 eV. The tritium in the ITEP experiment is contained in a valine molecule ($C_5H_{11}NO_2$). To eliminate uncertainties in the theoretical calculation for a molecular source, a similar experiment is planned at Los Alamos using a monoatomic tritium beam.

Neutrino oscillations. In addition to the electron neutrino (ν_e) associated with electrons (e), a muon neutrino (ν_μ) is produced in association with muons (μ), and a tau neutrino (ν_τ) presumably also exists as a companion to the tau lepton, although direct evidence for ν_τ has not yet been seen. The best upper limit on the mass of the muon neutrino from the charged pion decay, $\pi^+ \rightarrow \mu^+ + \nu_\mu$, is 570 keV. The upper limit on the ν_τ mass obtained from tau decay into leptons is even less stringent, 250 MeV.

Charged leptons (e^-, μ^-, τ^-) are always accompanied in reactions by their antiparticles (e^+, μ^+, τ^+) or by their associated antineutrinos ($\bar{\nu}_e$, $\bar{\nu}_\mu$, $\bar{\nu}_\tau$), which suggested that three conserved additive quantum numbers exist—electronic, muonic, and tauonic numbers. It then seems plausible that ν_e, ν_μ, and ν_τ are distinct. However, this hitherto immutability of different neutrinos may break down if neutrinos have mass. The wave function of each of the neutrino flavors (ν_e, ν_μ, ν_τ) may be a linear superposition of three neutrino states (ν_1, ν_2, ν_3) of distinct masses. Then, according to quantum mechanics, an initially pure beam of, say, electron neutrinos could be observed at a later time as a superposition of electron, muon, and tau neutrinos. The interchange of neutrino identities in flight is known as neutrino oscillations, a phenomenon first considered in 1958.

In the simple example of two neutrinos, the weak eigenstates ν_e and ν_μ at time $t = 0$ are related to the mass eigenstates ν_1 and ν_2 of masses m_1 and m_2 by Eqs. (1), where α is called the mixing angle.

$$\begin{aligned} \nu_e &= (\cos\alpha)\nu_1 + (\sin\alpha)\nu_2 \\ \nu_\mu &= (-\sin\alpha)\nu_1 + (\cos\alpha)\nu_2 \end{aligned} \tag{1}$$

Starting from an initial ν_e source, the probability for finding a ν_μ at time t is given by Eq. (2), where

$$P(\nu_e \rightarrow \nu_\mu) = \sin^2 2\alpha \sin^2 [(E_1 - E_2)t/2\hbar] \tag{2}$$

E_1 and E_2 are the energies of the neutrinos, and $\hbar$ is Planck's constant divided by 2π; the probability of still finding a ν_e is given by Eq. (3). To an

$$P(\nu_e \rightarrow \nu_e) = 1 - P(\nu_e \rightarrow \nu_\mu) \tag{3}$$

excellent approximation, the oscillation argument $(E_1 - E_2)\, t/2\hbar$ in radians is equal to 1.27 δd L/E, where $\delta = m_1^2 - m_2^2$ in units of eV^2, E is the average neutrino energy in MeV, and L is the distance of the detector from the source in meters. The behavior of $P(\nu_e \rightarrow \nu_e)$ versus L/E is shown in the illustration, which has a logarithmic L/E scale. If L/E is small compared to $1/\delta$, oscillations have little effect. The amplitude and phase of oscillations can be most readily mapped out for L/E of the order of $1/\delta$. At L/E much greater than $1/\delta$, only average probabilities can be measured. The approximate lower limits on δ which could be probed by different experiments are 10^{-12} for solar neutrino experiments, 10^{-5} for deep mine experiments, and 0.1 for experiments at accelerators, reactors, and meson factories. The difficulty in detecting oscillations for a given δ depends on the size of the mixing angle α. In one class of grand unified theories, the minimal 0(10) model, neutrino mixing is expected to be the same as quark mixing, in which case α would be identified with the Cabibbo angle, $\sin^2 2\alpha \cong 0.2$.

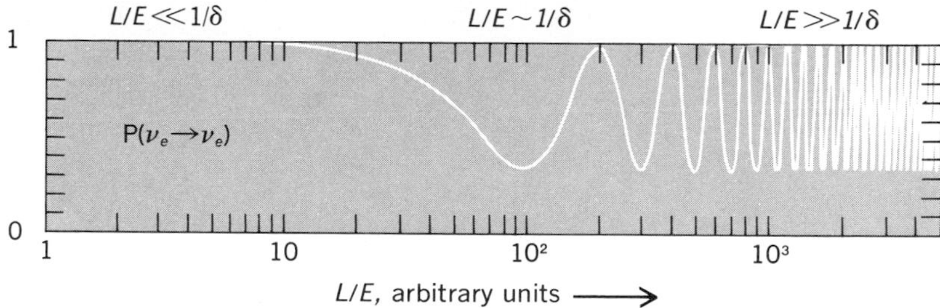

Oscillation probability of $\nu_e \to \nu_e$ versus L/E, where L is the distance from the neutrino source to the detector and E is the neutrino energy.

Solar neutrinos. Neutrino oscillations are a possible explanation of a long-standing puzzle concerning an apparent deficiency of electron neutrinos from the Sun. In a Brookhaven National Laboratory experiment by Ray Davis and collaborators, which spans the last decade, the flux of solar neutrinos has been measured deep underground in the Homestake Gold Mine in South Dakota. The capture of electron-type neutrinos converts chlorine atoms in the detector to argon atoms. The observed capture rate is about three times smaller than the expected rate based on conventional wisdom about the nuclear fusion chain that powers the Sun. If neutrinos oscillate, the neutrino flux which was initially electron-type neutrinos could arrive at the Earth as a mixture of the three neutrino species. The muon or tau neutrinos would escape detection because the neutrino energy is below the minimum required for the production of muons or taus. Hence, if the neutrino mixing is large, oscillations could account for the missing neutrino flux. The apparent deficiency of solar neutrinos in the chlorine experiment depends critically on calculations by John Bahcall and collaborators of the flux from a rare nuclear reaction. Future experiments are planned with gallium or indium detectors to reduce the uncertainty in the calculated flux by measuring neutrinos from the dominant lower energy components of the fusion processes. Solar neutrino experiments have a sensitivity to neutrino masses as small as 10^{-6} eV, provided that there is considerable neutrino mixing. If neutrino masses are indeed this tiny, neutrinos could undergo just one oscillation between the Sun and the Earth, and a significant time variation in the solar flux could result from the change in the Sun-Earth distance. The existing measurements are not sufficiently precise to resolve annual variations.

Cosmic neutrinos. Other experiments located in deep mines are designed to detect muons produced by ν_μ interactions in the surrounding rock. The source of the neutrino flux is the interaction of cosmic radiation with the Earth's atmosphere, giving a calculable composition of muon- and electron-type neutrinos. An experiment in a deep mine in Johannesburg, South Africa, reported evidence of a deficiency in the observed flux compared to that expected. However, preliminary results from an experiment at Baksan in the Soviet Union, which detects neutrinos that pass through the Earth's diameter, are consistent with no oscillation effects. New deep mine experiments, with large water detectors designed to search for proton instability, will detect the cosmic flux of both muon and electron neutrinos, allowing another test of oscillations over long distances.

Reactor antineutrinos. Conflicting results have been reported from reactor experiments on the presence of neutrino oscillations. A reactor provides an intense source of electron antineutrinos whose flux can be calculated from knowledge of nuclear processes in the core. The antineutrino flux can be measured at some distance from the reactor by detection of the inverse beta-decay process $\bar{\nu}_e + p \to e^+ + n$. If oscillations occur over such distances, then fewer electron antineutrinos would be observed than expected. The ratio of observed to calculated flux gives the oscillation probability versus the $\bar{\nu}_e$ energy. With a detector located 11.2 m from the reactor core center of the 2000-MW Savannah River reactor in South Carolina, a University of California at Irvine (UCI) experiment reported evidence for an oscillation, with a difference in neutrino mass-squares $\delta \cong 1$ eV^2. However, a similar experiment with a detector 8.75 m from the 57-MW ILL Grenoble reactor in France found no oscillation effect. The results of the two experiments could possibly be compatible if $\sin^2 2\alpha \cong 0.3$, but the evidence is regarded as inconclusive at present. Future reactor experiments at larger distances are in progress.

The UCI experiment also measured deuteron breakup by antineutrinos in both charged current (CC) $\bar{\nu}_e + d \to e^+ + n + n$ and neutral current (NC) $\bar{\nu} + d \to \bar{\nu} + p + n$ channels. The number of NC events is immune to flavor oscillations. The CC/NC event ratio divided by the theoretical ratio for no oscillations determines $P(\bar{\nu}_e \to \bar{\nu}_e)$ averaged over $\bar{\nu}_e$ energies. Evidence for an oscillation effect was found, at the 2 to 3 standard deviation level.

Accelerator limits. The solar and reactor experiments can only detect oscillations through the disappearance of neutrino flux. With neutrino beams at accelerators, a search can also be made for the appearance of different neutrino types due to oscillations. For example, a beam of ν_μ neutrinos interacting with a nucleon target could produce electrons if ν_μ to ν_e oscillations occur over the distance between source and detector, typically 1 km. To date, no convincing evidence for oscillations has been found in accelerator experiments. Upper limits on oscillations parameters are summarized in the table for maximum mixing or for large δ.

Upper limits on neutrino oscillation parameters determined from accelerator experiments

Oscillation channel	δ, for $\sin^2 2\alpha = 1$	$\sin^2 2\alpha$, for $\delta > 100$ eV^2
$\nu_\mu \rightarrow \nu_e$	< 0.6 eV^2	< 0.007
$\nu_\mu \rightarrow \nu_\tau$	< 3.0 eV^2	< 0.05
$\nu_e \rightarrow \nu_x$ $(x \neq e)$	< 0.14 eV^2	< 0.07

By correspondence with quark masses, a neutrino mass hierarchy is commonly expected, with each mass well separated from the next. Consequently, for a neutrino mass scale of order 20 eV, as indicated by the ITEP beta-decay experiment, mass-squared differences much greater than 1 eV^2 are anticipated. Since accelerator experiments are on the verge of ruling out substantial oscillations in this δ range, neutrino masses have unexpected near-degeneracies, or neutrino mixing is very small, or the mass scale is in fact considerably lower than 20 eV.

Majorana neutrino. The neutrino mass may be of two types—a Dirac mass term conserving lepton number and a Majorana mass term violating lepton number by two units. If a Majorana mass term exists, the neutrino mass eigenstates are self-charge-conjugate and neutrino-less double beta decay can occur, in which two nucleons exchange a neutrino in the process of emitting two electrons. Laboratory limits on the no-neutrino modes for ^{82}Se and ^{76}Ge yield an upper limit of 15 eV on a Majorana neutrino mass. However, there is a discrepancy with geochemical results on double beta decay.

When both Dirac and Majorana neutrino mass terms exist, a second class of oscillations may arise in which the ordinary neutrinos (ν_e, ν_μ, ν_τ) oscillate into sterile neutrinos that have essentially no interactions with ordinary matter. These oscillations are characterized by a disappearance of initial flux and can be distinguished through studies of neutrino interactions in which charged leptons are not produced.

Cosmology and astrophysics. A neutrino mass larger than a few electronvolts has great influence on the structure and expansion of the universe. The number of neutrinos left over from the primordial big bang should be comparable to the number of background photons, about 450 per cubic centimeter. Since the number of these relic neutrinos is about 10^8 times the number of nucleons (10^{80}), the neutrino mass density exceeds the nucleon density for a neutrino mass more than a few electronvolts. The universe is then neutrino-dominated. A neutrino with mass greater than 20 eV would give a gravitational mass density sufficient to slow down the current expansion of the universe to a halt and to cause its eventual collapse.

A neutrino mass of order 30 eV could account for the dark mass found in clusters of galaxies. Masses of galaxies are determined by observation of orbital velocities around the galaxies or by counting stars and using the known relation of their mass to their luminosity. The mass inferred from the orbits comes out much greater than that deduced from the emitted light. The missing mass cannot be rocks, planets, black holes, or neutron stars since the correspondingly higher nucleon density at the time of primordial nucleosynthesis would imply the wrong abundances of deuterium and helium. A plausible hypothesis is that the dark mass is associated with massive neutrinos that cluster gravitationally with the galaxies.

Prospects. Numerous new experiments are under way which will either confirm the present indications for neutrino mass and oscillations or push the limits to lower levels. Considering the importance of neutrino mass in particle physics, astrophysics, and cosmology, a major experimental effort is warranted. The pattern of neutrino masses and mixing may provide vital clues about grand unified theories and the Higgs bosons which give masses to the quarks and leptons. Neutrino oscillations would also provide another opportunity to search for *CP* (charge conjugation–space inversion) violating effects, whose origin in the quark sector is still not understood.

For background information *see* COSMOLOGY; FUNDAMENTAL INTERACTIONS; NEUTRINO; NONRELATIVISTIC QUANTUM THEORY; RADIOACTIVITY; SOLAR NEUTRINOS; WEAK NUCLEAR INTERACTIONS in the McGraw-Hill Encyclopedia of Science and Technology. [VERNON BARGER]

Bibliography: V. Barger and D. Cline (eds.), *Proceedings of 1980 Neutrino Mass Miniconference at Telemark, WI*, University of Wisconsin–Madison Rep. 186, 1980; L. Durand and L. Pondrom (eds.), *High Energy Physics: 1980 Proceedings of the 20th International Conference, Madison*, AIP, 1981 (reports by V. Barger, pp. 334–340; P. Langacker, pp. 483–487; S. Pakvasa, pp. 1165–1194; H. Sobel, pp. 326–333); B. Pontecorvo, Fifty years of neutrino physics, in E. Fiorini (ed.), *Neutrino Physics and Astrophysics*, 1982.

Nitrogen cycle

Nitrous oxide (N_2O), one of the trace constituents of the atmosphere, is present in the troposphere at a level of approximately 300 parts per billion. This level may be rising as a result of anthropogenic activity. An increase in the atmospheric N_2O concentration could lead to a partial destruction of the stratospheric ozone layer, which protects the Earth's surface from ultraviolet radiation, and a small but significant increase in the temperature at the Earth's surface. Models predicting the environmental consequences of increased atmospheric N_2O levels require inputs of N_2O source and sink strengths and knowledge of the behavior of N_2O in the atmosphere. Recent studies have also examined the role of soil as a major source of N_2O.

Sources. Atmospheric N_2O is largely of biological origin, being formed during microbial transformations of inorganic forms of nitrogen in terrestrial and aquatic environments. The generation processes include reduction of nitrate ion (NO_3^-) and oxidation of ammonium ion (NH_4^+). Reduction of NO_3^- to N_2O is accomplished by the denitrifiers, aerobic microorganisms which utilize NO_3^- as electron acceptor in the absence of oxygen (O_2). Most denitrifiers can also reduce N_2O to nitrogen (N_2) and so have the potential to be both a source and a sink for atmospheric N_2O. The reduction sequence (1) is regu-

$$NO_3^- \rightarrow NO_2^- \rightarrow NO \rightarrow N_2O \rightarrow N_2 \quad (1)$$

lated biochemically so that reduction of N_2O occurs following depletion of NO_3^-. In heterogeneous systems, such as soil, N_2O may be generated and reduced at the same time. The proportion of N_2O released to the atmosphere can vary considerably, but overall less than 10% of the nitrogen denitrified in soils is released as N_2O. Only small amounts of N_2O may be released following denitrification in aquatic systems where the long residence time of N_2O in the dentrifying zone favors reduction to N_2.

Generation of N_2O during nitrification of NH_4^+ to NO_3^- occurs during, or as a result of, conversion of NH_4^+ to NO_2^-. The yield of N_2O varies with oxygen level and is highest at low oxygen concentrations. T. J. Goreau and coworkers found that pure cultures of marine *Nitrosomonas* sp. generated N_2O at levels ranging from 0.25 to 10% of the NO_2^-—N formed as O_2 partial pressures were reduced from 0.2 to 0.005 atmosphere (20 to 0.5 kPa), respectively. Studies with soils suggest that 0.1 to 1% of NH_4^+—N oxidized is usually liberated as N_2O, with formation increasing as soil moisture content increases, that is, as O_2 declines. Higher levels of N_2O, perhaps as high as 10% of NH_4^+—N oxidized, can be liberated when crops are fertilized by injection of anhydrous ammonia into soil. In the localized zone of fertilizer placement, oxidation of NO_2^- to NO_3^- can be retarded, allowing NO_2^- to accumulate. Nitrite can react in soil in various ways and can form N_2O in addition to other substances, although the precise mechanism is unclear.

Measurements of N_2O emissions from terrestrial and aquatic environments have been made only in the last few years, and the data base is almost too limited to project source strengths on a global scale. Nevertheless, such projections are necessary because of the potential environmental consequences of increased emissions. The best estimates presently available are given in the table. It is likely that emissions of N_2O from the soils and oceans of the world are comparable. The value of 10^{10} kg N_2O—N yr^{-1} represents an upper limit for both soil and ma-

Estimated source and sink strengths for atmospheric N_2O

Location or process	Annual N_2O flux 10^9 kg N
Sources	
Soils	4–10
Marine	<10
Biomass burning	13
Fossil fuel combustion	3.5
Industrial emissions	0.5
Human waste treatment	1
Total	<31–38
Sinks	
Stratosphere	7–20
Troposphere	3
Total	10–20

rine environments. Earlier estimates that marine sources generated up to 10^{11} kg N_2O—N yr^{-1} are no longer considered possible. The scant literature on fresh-water systems suggests that N_2O emissions from large bodies of water, such as the Great Lakes in North America, are similar on a unit area basis to the ocean and so would contribute small amounts of N_2O relative to the marine environment.

The Potomac River estuary in the United States, which receives large inputs of NH_4^+—N from a sewage treatment plant, is estimated to contribute about 4×10^5 kg N_2O—N annually to the atmosphere. Extrapolation of this data suggests that human waste could result in a global source as large as 10^9 kg N_2O—N yr^{-1}.

Biomass burning has been identified as a potentially large source of atmospheric N_2O, but the estimate of 1.3×10^{10} kg N_2O—N yr^{-1} must be considered tentative because it is based on measured N_2O concentrations in the plume of only two forest fires. The contribution of 4×10^9 kg N_2O—N from fossil fuel combustion is much less than the biomass burning estimate, and industrial emissions are even smaller.

Sinks. Based on present knowledge, total annual additions of N_2O to the atmosphere are probably in the range of 2.5 to 4×10^{10} kg N yr^{-1}, which may be up to three times the estimated annual destruction rate. Only one substantial sink for N_2O, photolysis in the stratosphere, has been documented. Various estimates place this sink strength in the range of 0.7 to 2×10^{10} kg N yr^{-1}. Tropospheric chemical reactions may remove a further 3×10^9 kg N yr^{-1}.

Biological reduction of N_2O to N_2 is commonly observed during denitrification in soils and aquatic environments but, in general, this process only acts as a sink for N_2O generated in place and not for atmospheric N_2O. The reason is that transport of atmospheric N_2O to sites of N_2O reduction does not occur when the systems are net sources of N_2O to the atmosphere, which is almost always the case.

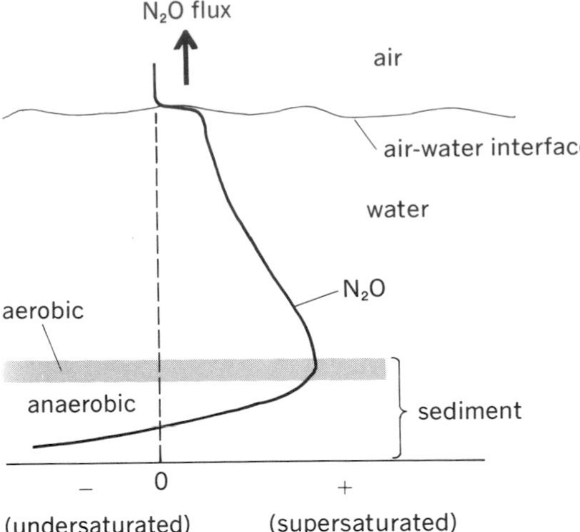

Fig. 1. Model nitrous oxide profile in a shallow-water system. The broken line represents the saturation point of N_2O in the shallow-water system.

To illustrate, Fig. 1 shows a typical N_2O distribution profile in a shallow-water system which is acting as both a source and a sink for N_2O. Generation of N_2O by nitrification of NH_4^+ occurs in and above the thin aerobic sediment layer, and has a maximum concentration within this zone. Nitrous oxide diffuses, and its concentration declines, in all directions away from the zone of generation. Nitrous oxide diffusing into the anaerobic sediment is reduced to N_2 by denitrification, and the concentration falls below the value for water in equilibrium with air; that is, the water is undersaturated with respect to the atmosphere. No atmospheric N_2O diffuses into the anaerobic sediment, however, because the water above the zone of N_2O generation is supersaturated with respect to atmospheric N_2O, and overall the system is a net source of N_2O to the atmosphere.

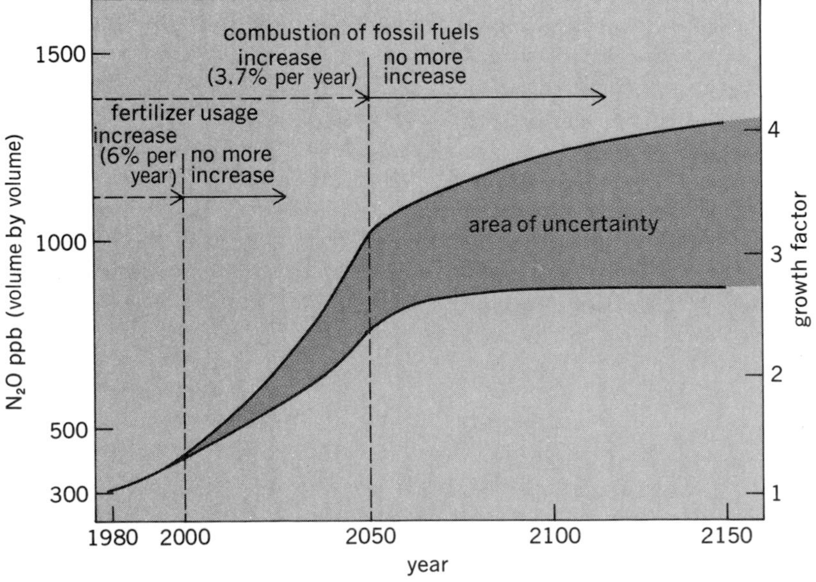

Fig. 2. Projected increase in tropospheric N_2O due to increased use of fertilizer and fossil fuels. Upper curve utilizes a N_2O residence time of 50 years and assumes that 20% of the fertilizer nitrogen returned to the atmosphere is N_2O. The corresponding values for the lower curve are 18 years and 30%. (*From W. Bach et al., eds., Man's Impact on Climate, Elsevier, 1979*)

Although biological reduction of N_2O to N_2 may only rarely be a sink for atmospheric N_2O, the importance of this process in reducing the amount of N_2O released to the atmosphere from denitrifying environments cannot be overemphasized. There is a growing body of evidence that denitrification is a minor contributor to aquatic sources of atmospheric N_2O, because the reduction sequence (1) is substantially completed to the terminal product, N_2. In soils, less than 10% of the N_2O generated by denitrification escapes reduction to N_2. Nitrification is now considered as the major source of N_2O released from aquatic systems, and is recognized as an important contributor to soil sources of atmospheric N_2O.

Recent interest in N_2O has been stimulated by the potential effects of increased inputs of N_2O into the atmosphere on stratospheric ozone levels and on the Earth's climate. The stratospheric ozone layer acts as a protective shield by limiting the amount of ultraviolet radiation reaching the Earth's surface. Nitrous oxide plays an important role in regulating stratospheric ozone levels through a series of photochemically catalyzed reactions which result in reduction of ozone (O_3) to oxygen (O_2). Nitrous oxide is relatively unreactive in the troposphere but reacts with atomic oxygen $O(^1D)$ at altitudes above about 15 km. Reaction (2) is the major pathway of N_2O destruction, but the nitric oxide (NO) formed by reaction (3) initiates further reactions which reduce

$$N_2O + O(^1D) \rightarrow N_2 + O_2 \quad (2)$$

$$N_2O + O(^1D) \rightarrow 2\,NO \quad (3)$$

ozone to oxygen. The nitrogen dioxide (NO_2) formed in (4) reacts with atomic oxygen to form more NO, in reaction (5).

$$NO + O_3 \rightarrow NO_2 + O_2 \quad (4)$$

$$NO_2 + O(^1D) \rightarrow NO + O_2 \quad (5)$$

The NO_2 formed in reaction (4) may also form NO_3^-, which is eventually transported back to the Earth's surface. One estimate suggests that 10% of the N_2O—N reaching the stratosphere is returned to the Earth as HNO_3. It should be noted that NO_2 and NO arising from fossil fuel combustion and industrial emissions at the Earth's surface react within the troposphere and do not enter into the stratospheric ozone question.

Significance. Early mathematical models suggested that a doubling of N_2O inputs into the atmosphere would reduce stratospheric ozone levels by 20%. Present thinking is that the reduction would be about 5%. Much uncertainty exists in the models presented to date, and further work is needed to clarify this issue.

Atmospheric nitrous oxide has a small but significant effect on the temperature at the Earth's surface due to absorption of emergent radiation in the infrared region. Using a model which included the uncertainties associated with the thermal structure of the Earth's surface, J. Hahn calculated that a doubling of the atmospheric N_2O concentration would cause a temperature increase between 0.2 and 0.8 K.

How much and on what time scale the concentration of N_2O in the atmosphere will increase is unclear. Attempts have been made to assess the impact of increased fixation of nitrogen for agricultural purposes on N_2O emissions from soil sources. The assessment includes increased use of nitrogen-fixing plants as well as commercial nitrogen fertilizer. From the measured emissions of N_2O from soils, it can be tentatively concluded that present-day agriculture has already increased N_2O emissions from soils by a maximum of 50% above the global baseline level. Further study may, however, scale this estimate downward. No estimates have been made of the effect of anthropogenic activity on N_2O emissions from marine sources and biomass burning sources.

Predictions of future increases in atmospheric N_2O levels are essential but highly speculative. The most important variables in such models are the projected increases in N_2O inputs into the atmosphere and the residence time of N_2O in the atmosphere, which is dependent on sink strengths for N_2O. Figure 2 shows the projection of Hahn, which is based on annual increases in fertilizer use of 6% until a constant consumption rate is reached in the year 2000 and on an annual increase in fossil fuel consumption of 3.7% until consumption levels off in the year 2050. The upper and lower limits incorporate a range in tropospheric turnover times of 50 and 18 years, respectively, and a variation in the proportion of fertilizer nitrogen returned to the atmosphere as N_2O from 20 to 30%, respectively. Within these limits, substantial changes in atmospheric N_2O levels are predicted over the next hundred years even though the range is great. Fortunately, the analytical capability to detect small changes (0.5 to 1 ppb) in atmospheric N_2O concentrations already exists, and measurements will quickly establish the actual rate of increase, at least on a short-term basis. This information and a better understanding of the environmental effects of increased levels of atmospheric N_2O are urgently needed, for changes once observed may not be quickly reversed.

[JOHN DUXBURY]

Nitrous oxide from soil. Atmospheric N_2O directly affects heat exchange processes of the Earth and participates in reactions governing the quantity and distribution of ozone (O_3) in the atmosphere, and so any change in its quantity affects both heat exchange and the penetration of ultraviolet radiation to the Earth surface. The soil is a major source of N_2O, particularly from the denitrification process in which nitrate ion (NO_3^-) is converted to gaseous nitrogen (N_2) and N_2O and, under some conditions, from the nitrification reaction in which ammonium ion (NH_4^+) is oxidized to nitrite ion (NO_2^-) and NO_3^- by soil microorganisms. Both of these reactions yield energy to the organism under the conditions in which they are carried out, and so they are important reactions of the nitrogen cycle and involve large quantities of nitrogen annually. Equivalent reactions take place in the ocean, and although these have not been studied as much as the terrestrial ones, there are indications that they are comparable in intensity, particularly where N_2O production is concerned. There are theoretical reasons for believing that N_2O production by nitrifiers in the ocean may involve a large proportion of the NH_4^+ oxidized than it does on land.

Earlier studies of denitrification were motivated both by interest in it as a biological reaction and the realization that it constituted a significant loss of valuable soil nitrogen which had to be replaced either by the application of industrially fixed nitrogen or by the use of plant-microbial associations (such as leguminous crops). More recently, as some of the implications of N_2O production have been realized, these studies have been intensified. The increased cost of fossil fuels and problems attending their use have provided additional incentive. Changing patterns of population distribution have greatly changed methods of food production and utilization, further affecting the nature and extent of N_2O production. Sewage- and other waste-processing procedures can be particularly strong sources of N_2O.

As a consequence of these developments, studies have concentrated on several fronts: the influence of nitrogen fertilizer on N_2O production in denitrification, factors affecting N_2O production in nitrification and denitrification, conditions affecting the ratio of N_2O to N_2 in the denitrification reaction, the extent to which soil (or other systems) can serve as a sink for N_2O, and means for limiting denitrification (and N_2O production).

Soil measurements. The measurement of processes of the nitrogen cycle, including N_2O production and utilization, has been aided by the use of several tools in combination. Rates of nitrogen fixation can be estimated by means of the acetylene reduction technique, in which acetylene at low concentrations is put in contact with a volume of soil (or other system) and the rate at which ethylene is produced is followed. The nitrogen-fixing enzyme system preferentially reduces acetylene to ethylene, and so the rate of ethylene production is a measure of nitrogen-fixing ability. The technique is highly sensitive, yet simple enough so that it can be taken to the field. Gas chromatography provides a sensitive means for determining small quantities of acetylene, ethylene, N_2O, nitric oxide (NO), and other gases of interest in the nitrogen cycle.

Because there are simultaneous or successive reactions bringing nitrogen into the system (rainfall, nitrogen fixation, fertilizer addition) and removing nitrogen from the system (leaching, denitrification, crop harvest), it is sometimes difficult to determine their comparative significance. Two techniques have aided in this determination. The lysimeter, a tank holding soil and plants, makes it possible to measure the quantity of water leaving the system by leaching and the solutes contained in that water. The use of isotopically labeled nitrogen (usually either enriched or depleted in ^{15}N) makes estimation of nitrogen turnover possible. These two techniques require more elaborate instrumentation, and so their use is limited by the cost of instruments, isotopically labeled nitrogen, and skilled technical labor. They are the only means of obtaining some information, however, and are used in conjunction with the more flexible techniques of gas chromatography, acetylene reduction, and conventional chemical analysis.

Ratio of N_2 to N_2O in denitrification. The denitrification reaction, in which microorganisms under conditions of limited oxygen supply obtain their energy by the oxidation of an organic (or sometimes inorganic) material using NO_3^- as the oxidant or electron acceptor, can be expressed by a statement such as (6). However, reaction (7) also takes place,

$$NO_3^- + 1.25\,\underset{\text{Carbo-hydrate}}{[HCHO]} \rightarrow 0.5N_2 + 0.75H_2O + 1.25CO_2 + OH^- \quad (6)$$

$$NO_3^- + [HCHO] \rightarrow 0.5N_2O + 0.5H_2O + OH^- + CO_2 \quad (7)$$

resulting in N_2O production. The ratio of N_2O to N_2 in the reaction is a function of a number of factors, including the type of microorganisms present, their past history, concentration of NO_3^- and organic substrate, and others. Under field conditions, where most of the variables are not subject to control, probably the most important determinant is the quantity of available NO_3^-. Field measurements have shown that of the nitrogen lost in denitrification, usually not more than 10–15% appears as N_2O. If NO_3^- concentrations are high, however, the proportion of N_2O can be higher. Thus, although denitrification is to be avoided wherever possible as a nitrogen-conserving measure, the quantity of N_2O produced in the reaction can be lessened, if the reaction does take place, by factors which limit the concentration of NO_3^-. These include methods of fertilizer placement, frequency of fertilizer application, quantity of fertilizer applied, use of so-called slow-release fertilizers, and other practices.

N_2O produced in denitrification can serve also as an oxidant (electron acceptor) for most of the denitrifying microorganisms, being further reduced to N_2 according to reaction (8). Reaction (8) yields energy and supports abundant growth. Thus an anoxic environment conducive to denitrification is a potential sink for N_2O. Soils and waters frequently are found to contain less N_2O than equilibrium with the atmosphere would dictate.

$$2N_2O + [HCHO] \rightarrow 2N_2 + CO_2 + H_2O \quad (8)$$

N_2O from nitrification. The nitrification reaction in which microorganisms obtain their energy from the oxidation of NH_4^+ is commonly written in two steps [reactions (9*a*) and (9*b*)], because two different groups of organisms have been identified with the separate steps.

$$NH_4^+ + 1.5O_2 \rightarrow NO_2^- + H_2O + 2H^+ \quad (9a)$$

$$NO_2^- + 0.5O_2 \rightarrow NO_3^- \quad (9b)$$

The organisms carrying out this reaction sequence are classified as autotrophic because they subsist on inorganic materials as their only requirement for growth. Among the heterotrophic organisms of soil, those which commonly subsist on organic compounds, some are also capable of nitrification. It is not known whether they realize any energy in doing so, but probably they do.

At least one organism carrying out the first step in the above nitrification sequence, *Nitrosomonas*, is known to produce some N_2O in the process. This can be expressed as reaction (10). Although the nitrification reaction has been studied for years, and scientists frequently were unable to account for all of the nitrogen-supplied laboratory cultures, only recently has this reaction been recognized as a possibly significant source of atmospheric N_2O. Field studies indicate that the quantity of N_2O produced in nitrification is also concentration-dependent. Therefore, while the quantitative significance of this source is uncertain, it also is influenced by the management practices that affect N_2O production in denitrification.

$$2NH_4^+ + 2.5O_2 \rightarrow N_2O + 4H_2O \quad (10)$$

Trends in atmospheric N_2O concentration. Only recently has it been possible to make measurements of atmospheric concentrations of N_2O with the accuracy required to determine concentration trends. Some data indicate that the present concentration is higher than that 2 decades ago, but the methods are quite different from those used earlier and it is difficult to interpret the differences seen. At the low concentrations involved, small influences such as the adsorption of gases on the walls of sampling containers, for example, can introduce systematic errors which yield deceptively good reproducibility. This generates confidence in the result but makes it difficult to relate values to others obtained at different times by different techniques. Nevertheless, it appears that the concentration of N_2O in the atmosphere is rising, and soon the trend will be discernible in a more quantitative way. This will make possible a more accurate evaluation of both the extent of human influence on the process of N_2O formation and the role of N_2O in the nitrogen cycle.

For background information *see* ATMOSPHERIC OZONE; NITROGEN CYCLE in the McGraw-Hill Encyclopedia of Science and Technology

[C. C. DELWICHE]

Bibliography: J. M. Bremner and A. M. Blackmer, *Science*, 199:295–296, 1978; J. R. Burford and R. C. Stefanson, *Soil Biol. Biochem.*, 5:133–141, 1973; C. C. Delwiche (ed.), *Denitrification, Nitrification, and Atmospheric Nitrous Oxide*, 1981; C. C. Delwiche and B. A. Bryan, *Ann. Rev. Microbiol.*, 30:241–262, 1976; T. J. Goreau et al., *Appl. Env. Microbiol.*, 40:526–532, 1980; J. Hahn, in W. Bach, J. Pankrath, and W. Kellogg (eds.), *Man's Impact on Climate*, 1979; W. J. Payne, *Trends Biochem. Sci.*, 1:220–222, 1976; P. F. Pratt et al., *Climatic Change*, 1:109–135, 1977.

Nitrogen fixation

Much of the world's food production is dependent on heavy inputs of fertilizers, especially nitrogen. Most fertilizer nitrogen is made by the energy-intensive Haber-Bosch method of ammonia production, which consumes large quantities of fossil fuels. During the energy crisis of the mid-1970s, the cost of nitrogenous fertilizers escalated sharply as the price of oil and natural gas was increased and available supplies in the United States were reduced. This dramatic increase in fertilizer costs stimulated a keener interest in developing and better exploiting "new" and existing biological alternatives for providing nitrogen for crop production.

Rhizobium-legume symbiosis. The primary alternative is biological nitrogen fixation in which dinitrogen gas (N_2) from the atmosphere is converted into ammonia, by an enzyme complex called nitrogenase, and incorporated into cells of microorganisms. A variety of N_2-fixing systems have developed in the biological realm. The most prominent is the symbiosis between legumes (such as peas, beans, and soybeans) and root nodule bacteria (*Rhizobium* species). The *Rhizobium* bacteria invade legume roots and cause the production of small tumorlike growths called nodules. It is in the nodule that the bacteria and the plant "cooperate" to bring about substantial rates of N_2 fixation. Up to 100–300 kg per hectare of nitrogen may be fixed in one growing season.

Associative symbioses. In addition to the *Rhizobium*-legume symbiosis, there is a wide variety of other soil bacteria which enter into loose associations with an equally wide variety of nonleguminous plants such as grasses. Many of these grass-type plants are important cereal crops or forage crops such as wheat, corn, rice, and grain sorghum. The loose associations between the roots of these plants and nitrogen-fixing soil bacteria are often referred to as associative symbioses. In the associative symbioses nonsymbiotic nitrogen-fixing soil bacteria live on or sometimes in the plant root or in the soil immediately surrounding the root (an area called the rhizosphere) and fix nitrogen at these sites. The bacteria are provided with food (carbohydrates and such) and energy sources in the form of excretions from the plant root, and in turn, some of the nitrogen fixed by the bacterial cells is eventually transferred to the plant. This transfer probably occurs when the bacterial cell dies and its cellular constituents are made available through the actions of other microbes. Thus, these associations are much more casual than the well-developed association between legumes and their root nodule bacteria. In fact, some free-living nitrogen-fixing bacteria such as species of *Azotobacter* and *Azospirillum* are associated with roots of many different plants. These associative symbioses are not nearly as specific as the legume system.

Development of grass-bacteria systems. Because of the enormous benefits that could be derived if these associative symbioses could be improved and manipulated, research efforts toward this goal were intensified in the 1970s. Still, these efforts were but a fraction of the effort which has gone into improving and utilizing the legume systems since the clear-cut proof of N_2 fixation by legumes was provided nearly a century ago. The development of grass-bacteria systems capable of deriving nitrogen for crop growth through biological N_2 fixation is an important goal. Much of the world population and their domestic animals rely on these crops for sustenance, and the principal factor limiting productivity is, in many cases, lack of sufficient quantities of nitrogen. Many of the farmers of developing nations do not have access to inexpensive fertilizers and must rely on the natural systems such as nitrogen fixation for maintaining soil fertility. For these reasons, the development or improvement of associative nitrogen fixation in nonleguminous crops warrants continued investigation.

Limits of associative systems. Much has been learned in the past decade regarding actual rates or amounts of nitrogen fixed by these associative systems. While reports in the late 1960s and early 1970s indicated that up to 90 kg of nitrogen per hectare (approximately 90 lb/acre) was fixed by some tropical grasses, subsequent investigations indicated that substantially lower rates of N_2 fixation were occurring in both tropical and temperate regions of the world. Many studies indicated that these associative grass-bacteria systems fixed 1–10 kg of nitrogen per hectare in one growing season. An exception to these low rates can be found in lowland rice. Rice grown in paddies may derive more nitrogen from associated microorganisms because of the unique features of the environment. The waterlogged paddy soils provide more favorable environments for the nitrogen-fixing bacteria than do the well-drained upland farming systems.

Energy supply. Why are these associative systems limited to the low rates of biological nitrogen fixation now so frequently observed? Clearly, more work is required to answer this question satisfactorily. Nevertheless, a number of factors have been shown to place constraints on associative nitrogen fixation. Some of these constraints may be difficult to overcome, for they stem from the inherent limitations of the enzyme system (nitrogenase) itself. Just as the chemical fixation of nitrogen is energy-intensive, so too is the biological fixation in terms of energy (as adenosinetriphosphate, or ATP) demands on the bacteria to carry out the process. The nitrogen-fixing bacteria require large amounts of carbonaceous substrates to grow and fix nitrogen. As an approximation, 1 ton of glucose would be required to fix 100 lb of nitrogen (1 metric ton to fix 50 kg). Thus, lack of a sufficient energy supply in most soils is a major barrier to higher rates of nitrogen fixation. Superimposed on this limitation is the fact that a myriad of nonnitrogen-fixing microorganisms is competing for the same supply of carbon and energy sources, and this competition reduces the supply for the nitrogen fixers which may be less competitive.

Combined nitrogen levels. Another factor which may limit associative nitrogen fixation is the level of combined forms of nitrogen such as ammonium (NH_4^+) and nitrate (NO_3^-) in the root zone of the plant. This limitation is due to the regulation of the enzyme, nitrogenase, by the level of ammonium. In situations where sufficient levels of ammonium are present, the nitrogen-fixing organisms do not produce the enzyme and hence nitrogen fixation is precluded. This is a difficult problem to overcome since even minimal levels of fertilizer nitrogen applied to crops may be sufficient to prevent or limit nitrogen fixation. It appears, however, that associative N_2 fixation may become important late in the growing season when the supply of soil nitrogen in the root zone becomes depleted and the plant must rely on other sources.

Oxygen concentrations. The nitrogen-fixing enzyme complex is also very sensitive to oxygen (O_2). The enzyme is "poisoned" by too much oxygen and rendered inoperative. This poses a problem for bacteria which require oxygen for respiration (aerobic bacteria) but still must protect the enzyme against too much oxygen. Aerobic bacteria thus fix nitrogen more efficiently at reduced partial pressures of oxygen, and are said to be microaerophilic (preferring reduced O_2 concentrations). For example, organisms such as *Azotobacter* and *Azospirillum* will fix N_2 in cultures exposed to air (0.2 atm or 20 kPa O_2), but they will do so with much greater efficiency when the O_2 tension is reduced to much lower levels (0.002–0.007 atm or 0.2–0.7 kPa). Thus, in well-aerated agricultural soils, exposure to oxygen may inhibit or limit rates of N_2 fixation. However, it is also possible that oxygen consumption by other microorganisms in the rhizosphere (root zone) serves to lower the partial pressure of oxygen to levels suitable for efficient N_2 fixation. The enhanced level of N_2 fixation associated with roots of rice grown in paddies is no doubt due in part to the waterlogged environment with low oxygen content.

A variety of other environmental factors can limit nitrogen fixation in associative symbioses. These include the environmental variables of soil moisture, temperature, and pH (acidity or alkalinity).

Enhancing associative systems. Of prime importance from the standpoint of using nitrogen-fixing bacteria as soil inoculants is whether or not they can become established in the soil population and remain competitive. Unfortunately, the record for successful inoculation trials is not a strong one. Most attempts at increasing crop yields through the application of associative nitrogen-fixing bacteria have met with little success. Generally, the yield increases are not substantial enough to justify the efforts required to produce and deliver the bacteria to the field. There is also little evidence to indicate that any observed yield increases are indeed due to the applied inoculum. Since most soils harbor populations of nitrogen-fixing bacteria, the necessity for inoculation remains questionable. A more plausible approach to enhancing N_2 fixation might be to manipulate those populations already present in a given soil. More research on the behavior of these populations in soil will be required before researchers are able to bring about changes which will lead to enhanced nitrogen fixation.

Grass crops. The prospects for enhancing associative nitrogen fixation in grasses are not clear at this time. Some reports indicate that certain cultivars (breeding lines or genotypes) of crop plants such as rice, corn, and pearl millet exhibit higher rates of associative N_2 fixation than others. Implicit in these reports is that there may be some potential for breeding plant varieties with enhanced capabilities for associative nitrogen fixation. Plant breeding programs have already led to the development of enhanced N_2 fixation by legumes such as alfalfa. Such a program might involve attempts to alter the quantity or quality (specific substrates) of energy sources translocated to the roots. It might also involve other ways of making the root environment more suitable for colonization by nitrogen-fixing bacteria.

Genetic engineering. Another approach to enhancing associative nitrogen fixation lies in the development of bacteria with altered properties, making them more suitable for the task. For example, it is possible to eliminate the regulation of the nitrogen-fixing enzyme complex by ammonium. Such a deregulation has been accomplished through genetic manipulation. These modified bacteria continue to fix nitrogen in the presence of levels of ammonium which would ordinarily inhibit the process. It is also possible to move the genes responsible for N_2 fixation into bacteria which did not previously possess them, and such bacteria gain the ability to fix nitrogen. It might be possible to add the nitrogen-fixing system to bacteria which are normal competitive inhabitants of plant roots, and have them become established more easily. The powerful tools of genetic engineering are now available to assist in making modifications to both host plants and associative bacteria in an effort to increase the efficiency of these systems. Only time will tell what fruits these lines of research will bear.

Prospects. The one certain thing is that complete understanding of all the factors involved in controlling the establishment and functioning of the associative symbioses is well in the future. On the one hand, this is frustrating because there is an acute need to extend the ability for nitrogen fixation to nonleguminous crops for food production. This need will only intensify as the world's energy supplies become scarcer and the population continues to grow. On the other hand, current limitations in the understanding of this process provide a source of encouragement to continue toward a complete understanding. As this goal is approached, it should be easier to make more accurate assessments of the potential for developing these systems for use in agricultural production. There is, clearly, much to be learned and perhaps even more to be discovered. The search for new and better nitrogen-fixing associations must be continued, and agriculturists must learn to exploit those systems already available to a greater degree. Nitrogen fixation, in its many variations, will continue to play an important part in the production of food for a growing world population.

For background information *see* FERTILIZER; NITROGEN FIXATION in the McGraw-Hill Encyclopedia of Science and Technology.

[DAVID A. ZUBERER]

Bibliography: W. J. Brill, *Sci. Amer.*, 245(3):198–215, 1981; R. H. Burris, Overview of biological N_2 fixation, in *Report of the Public Meeting on Genetic Engineering for Nitrogen Fixation*, National Academy of Sciences, 1977; R. W. F. Hardy and W. S. Silver (eds.), *A Treatise on Dinitrogen Fixation*, sec. III: *Biology*, 1977; W. E. Newton and W. H. Orme-Johnson (eds.), *Nitrogen Fixation*, vol. II: *Symbiotic Associations and Cyanobacteria*, 1980; C. A. Neyra and J. Doberiner, *Advanc. Agron.*, 29:1–38, 1977; P. van Berkum and B. B. Bohlool, *Microbiol. Rev.*, 44(3):491–517, 1980.

Nobel prizes

The Swedish Royal Academy announced 11 recipients of the Nobel prizes for 1981.

Medicine or physiology. Half of this prize was presented to Roger W. Sperry of the California Institute of Technology for his pioneering research in brain hemisphere specialization. From work with split-brain patients Sperry concluded that the right hemisphere of the brain, originally believed to be less important than the left, is actually dominant for various spatial and perceptual tasks. The other half of the award was presented to the American David H. Hubel and the Swede Torsten W. Weisel, colleagues at Harvard Medical School, for their research on the brain's function in vision. By measuring electrical impulses in nerve cells, the two neurobiologists discovered a complex hierarchy of cortical neurons, each of which responds to a different combination of visual variables.

Physics. Half of this award was shared by Nicholaas Bloembergen of Harvard University and Arthur Schawlow of Stanford University for their work in the field of laser spectroscopy. The other part of the prize was awarded to Kai Siegbahn of Uppsala University for his contribution to the development of high-resolution spectroscopy. Both electron and laser spectroscopy are methods of examining the behavior of excited atoms to detect specific properties of matter that are otherwise unobservable.

Chemistry. Kenichi Fukui of Kyoto University and Roald Hoffmann of Cornell University shared this award for their independent work using quantum mechanics to predict chemical reactions. Fukui is best known for his frontier orbital theory, which proposes that the chemist can predict the combinative properties of a molecule from the shape and density of its loosely bound outer electrons. Hoffmann's work extends that of Fukui to predict whether a reaction is possible, thereby allowing the chemist to select proper pathways for developing new drugs.

Economics. James Tobin, professor at Yale University, was cited for his analysis of how fiscal policy interacts with real-world investment. A presidential economic adviser during the Kennedy administration, Tobin is best known for his portfolio selection theory, which states that both individuals and financial institutions seek to balance high-yielding investments with risk factors.

Literature. Bulgarian-born Elias Canetti was honored for "writings marked by a broad outlook, a wealth of ideas, and artistic power." Raised in Western Europe, Canetti fled the Nazis in 1939 and settled in London. Cannetti's writings, including novels, plays, and a major study of mass movements, were originally published in German.

Peace. The peace prize was awarded to the Office of the United Nations High Commissioner for Refugees for its humanitarian efforts in aiding 25,000,000 homeless people during the 30 years of its existence. The agency has been directed by former Danish prime minister Poul Hartling since 1977.

Nonlinear acoustics

Nonlinear acoustics is that portion of physics devoted to study of the behavior of nonlinear sonic and ultrasonic disturbances and the various phenomena associated with their propagation. The word nonlinear implies that nonlinear differential equations are necessary for an adequate mathematical description of the phenomena. Oftentimes nonlinear phenomena are observed only at high acoustical amplitudes, and thus are referred to as finite-amplitude effects. The word macrosonics also is used in this connection. Subjects of investigation in nonlinear acoustics include waveform distortion of finite-amplitude sonic and ultrasonic waves, weak shock-wave propagation, collinear and noncollinear wave interactions of sound with sound, parametric phenomena in sound waves, acoustic radiation pressure, acoustic streaming, and cavitation.

Historical development. Although nonlinear acoustics was established as a separate branch of physics only in the 1970s, a number of studies fundamental to its basic structure have been carried out over the last 2 centuries. The fundamental modern experimental work was carried out by A. L. Thuras, R. T. Jenkins, and H. T. O'Neil (1935), who demonstrated the waveform distortion of sound waves propagating in air, and by V. A. Krasilnikov, V. V. Shklovskaya-Kordi, and L. K. Zarembo (1956) and M. A. Breazeale and E. A. Hiedemann (1958), who obtained direct evidence of waveform distortion in liquids. In 1955 the discrepancy between theoretical and measured values of ultrasonic wave absorption in liquids led F. E. Fox and W. A. Wallace to develop a theory of sound propagation in liquids which predicted phenomena similar to those previously predicted for gases by R. D. Fay (1931).

The definitive modern work relating theory and experiment for finite-amplitude sound waves in fluids (gases and liquids) was carried out in 1960 by W. Keck and R. T. Beyer. The same approach was used in 1965 by Breazeale and J. Ford to describe the propagation of a finite-amplitude longitudinal wave in a solid. For an isotropic solid or along the principal crystallographic directions in a cubic crystal, they showed that the propagation can be described by a differential equation whose solution takes the same form as that describing finite-amplitude wave propagation in liquids or gases. By identifying the nonlinearity parameter, it thus is possible to describe gases, liquids, and solids by differential equations having the same mathematical form. Much of the recent progress in nonlinear acoustics has been in measuring the nonlinearity parameters of gases, liquids, and solids.

Derivation of basic equation. As an illustration of the approach of nonlinear acoustics, the nonlinear distortion of a finite-amplitude wave will be considered. Since nonlinear distortion has been observed to increase with frequency, it is common to consider ultrasonic frequencies (greater than 20 kHz); however, there is no fundamental limitation in this respect.

The basic equation of nonlinear acoustics can be derived either from hydrodynamics or from nonlinear elasticity, depending upon the medium under consideration. It is possible to start with the equations of hydrodynamics and derive the basic equation of nonlinear acoustics. Alternatively, it is possible to consider nonlinear elasticity and show that the basic equation of nonlinear acoustics also comes from this derivation if the appropriate restrictions are made. The end result is a nonlinear differential equation appropriate to the description of finite-amplitude distortion of waves in gases, liquids, and solids.

Hydrodynamical equations. The medium under consideration, of course, is made up of molecules with empty space between them. However, it is convenient to assume that the propagating medium is a continuum—except at boundaries or at interfaces.

Equations of continuity. Consider a volume V of fluid through which flow is taking place. If V is considered relative to a cartesian coordinate system, the flow can be defined in terms of the volume V, or in terms of the surface S of the same volume. The conservation of mass states that the net rate of mass flow out of the surface S is equal to the rate of decrease of mass inside V. A mathematical statement of the conservation of mass results in the equation of continuity, which can be expressed in differential form.

Equations of motion. The conservation of linear momentum can be expressed by applying Newton's second law of motion to a volume element of a fluid medium. In addition, deformation of the volume element can be accounted for by a rate-of-deformation tensor consisting of a symmetrical part (the strain rate tensor) and an antisymmetrical part (the rotation tensor).

The resulting equations of motion are differential equations involving the shear viscosity coefficient η and the dilatational viscosity η'. If it is assumed that η and η' are constants and are related through the Stokes relation, $\eta' = -2\eta/3$, then these differential equations reduce to the so-called Navier-Stokes equations. If the viscosity is small or the derivatives of velocities with respect to directions other than the velocity direction are small, the Navier-Stokes equations in turn reduce to the so-called Euler equations, which sometimes are referred to simply as the equations of motion. Abuse of these equations has occurred in the past, and considerable care must be exercised in applying the appropriate form of the equations of motion in nonlinear acoustics to avoid paradoxical results.

Energy equation. In considering the motion of a volume V, the total energy per unit mass associated with this volume can be divided into (1) internal energy associated with molecular and atomic behavior, (2) kinetic energy, and (3) potential energy associated with the location of V in the Earth's gravitational field. The law of conservation of energy states that the net rate of increase of total energy in the control volume is equal to the rate at which heat flows into the volume V, plus the rate of internal heat production minus the rate at which the fluid in the control volume does work on its surroundings. A mathematical statement of the conservation of energy, plus the definition of some thermodynamic quantities, leads to an energy equation.

Equations of state. For a specific fluid, a number of thermodynamic equations of state may be written. Since the propagation of sound is an isentropic process, it is desirable to write the equation of state in the form of Eq. (1), where P is the pressure, ρ is

$$P = P(\rho, s) \tag{1}$$

the density and s is the entropy. The equation of state can then be written as a function of the single variable ρ, since the entropy s remains constant. The most useful form is Eq. (2), where P_0 and ρ_0

$$P - P_0 = A\left[\frac{\rho - \rho_0}{\rho_0}\right] + \frac{B}{2!}\left[\frac{\rho - \rho_0}{\rho_0}\right]^2 + \ldots \quad (2)$$

are the equilibrium values of pressure and density, and the coefficients A and B are defined by Eqs. (3), with c_0 being the velocity of sound evaluated for infinitesimal amplitude disturbances.

$$A = \rho_0\left[\left(\frac{\partial P}{\partial \rho}\right)_S\right]_{\rho=\rho_0} = \rho_0 c_0^2$$
$$B = \rho_0^2\left[\frac{\partial^2 P}{\partial \rho^2}\right]_{\rho=\rho_0} \quad (3)$$

Since the propagation of sound involves perturbations of the thermodynamic variables about a static value, the form of Eq. (2) is especially useful for describing sound-wave propagation in general, and nonlinear acoustical phenomena in particular. The ratio B/A is an essential parameter in nonlinear acoustics, and it can be written as Eq. (4), where

$$\frac{B}{A} = 2\rho_0 c_0\left[\left(\frac{\partial c}{\partial P}\right)_T\right]_{\rho=\rho_0} + \left(\frac{2c_0 T\beta}{c_P}\right)\left[\left(\frac{\partial c}{\partial T}\right)_P\right]_{\rho=\rho_0} = \left(\frac{B}{A}\right)' + \left(\frac{B}{A}\right)'' \quad (4)$$

c is the sound velocity, c_P is the specific heat at constant pressure, T is the temperature, and β is the coefficient of thermal expansion.

This equation suggests that B/A can be evaluated from the isothermal rate of change of sound velocity with pressure and the isobaric rate of change of sound velocity with temperature.

The relationship between Eq. (2) and more common forms of the equation of state may be evaluated directly. For example, the equation of state for an isentropic process in an ideal gas may be written as Eq. (5). Substituting this into the first two terms of

$$P = P_0\left(\frac{\rho}{\rho_0}\right)^\gamma \quad (5)$$

Eq. (2) leads to the result that Eq. (6) holds for an ideal gas, from which the ideal gas value of B/A

$$\gamma = \frac{B}{A} + 1 \quad (6)$$

can be found for a monatomic, a diatomic, and a polyatomic gas. For a liquid, B/A can be evaluated, for example, from experimental determination of the quantities in Eq. (4). Likewise, the equivalent quantity can be evaluated for solids, but the derivation must be made from the fundamentals of nonlinear elasticity rather than hydrodynamics.

Perturbation analysis. In principle, it should be possible to combine the equation of continuity, the equations of motion, the energy equation, and the equation of state to predict the way in which specific thermodynamic variables change as a sound wave passes through a fluid medium. In practice, the general solution cannot be written. Rather, one usually obtains a specialized approximate solution under assumptions which may be quite restrictive. Thus, it is worthwhile to consider a perturbation analysis, which proves to be an efficient tool for the evaluation of the significance of the terms of the fundamental equations, and which can lead to equations in which a consistent approximation is maintained. In order to apply the perturbation analysis, the dependent variables in the equations of continuity, the equations of motion, and the energy equation are replaced by the sum of their equilibrium values (zero-order values) and their first-order and second-order variations components, as already has been done in the equation of state (2). Then, terms of equal order are collected to form the equations to be satisfied by the variables of the order in question.

In effect, the procedure often used in acoustics heretofore has been to keep only the terms in the perturbation analysis which lead to the linear wave equation in the particle displacement and to ignore higher-order terms. In nonlinear acoustics, second-order acoustical phenomena not covered by the first-order acoustical equations are investigated. This means that higher-order terms in the perturbation analysis must be evaluated and compared with experimental results. At present the complete solution to the general system of equations is not available, although many approximate solutions are. The more successful of these approximate solutions have tended to be either exact solutions of approximate differential equations or approximate solutions of almost exact differential equations.

Finite-amplitude waves. The recent increase in interest in nonlinear acoustics has been brought about by the increase in the number of practical occurrences and applications of waves of finite amplitude and the increasing possibilities for the evaluation of numerical solutions to complicated nonlinear equations.

One can derive a nonlinear differential equation capable of describing a number of nonlinear acoustical phenomena by applying perturbation analysis to the equation of continuity, the equations of motion, the energy equation, and the equation of state (2) and keeping second-order terms in the dependent variables.

In applying the perturbation analysis it is important to realize that the dependent variables are determined by time and their cartesian coordinates in a fixed reference frame. This is the eulerian (or spatial) coordinate description. In nonlinear acoustics it oftentimes is more convenient to express the dependent variables in terms of a coordinate system in which each particle has its own set of identification coordinates describing its position at time $t = 0$.

Subsequent motion of the particle can then be described in terms of the deviation from its equilibrium position. This is the lagrangian (or material) coordinate description. The difference between eulerian and lagrangian coordinates is of fundamental significance in nonlinear acoustics, even though the difference vanishes in the linear approximation.

One can use lagrangian coordinates to derive the wave equation for adiabatic propagation of a plane wave in a nonviscous fluid. In this one-dimensional fluid motion, let a be the equilibrium position of a fluid "particle" (position at $t = 0$) and let x be its instantaneous position. The difference $x - a = \xi$ describes the displacement of the particle from its equilibrium position. This dependent variable ξ constitutes a convenient quantity for the description of displacement, velocity, and acceleration of the fluid particle in lagrangian coordinates. With specific designation of the independent variables, one can write Eqs. (7), where u is the particle velocity,

$$\begin{aligned} \xi(a,t) &= x(a,t) - a \\ \frac{\partial \xi}{\partial t} &= \frac{\partial x}{\partial t} = u^{(L)}(a,t) \\ \frac{\partial^2 \xi}{\partial t^2} &= \frac{\partial u^{(L)}}{\partial t} \end{aligned} \tag{7}$$

and the superscript (L) is used to emphasize the fact that lagrangian coordinates are used.

A fluid element originally (at time t_0) bounded by the planes a and $(a + \delta a)$ at a later time t will be bounded by the planes $(a + \xi)$ and $(a + \delta a + \xi + \delta\xi)$. The "length" of the fluid element changes from the value δa at time t_0 to the value $(\delta a + \delta\xi)$ at time t. According to the equation of continuity (or the conservation of mass), the mass of a fluid element of unit cross section is given by Eq. (8), from which Eq. (9) follows.

$$\rho_0 \delta a = \rho^L(\delta a + \delta\xi) = \rho^L \delta a \left(1 + \frac{\partial \xi}{\partial a}\right) \tag{8}$$

$$\rho_0 = \rho^L\left(1 + \frac{\partial \xi}{\partial a}\right) \tag{9}$$

The equation of motion for a fluid element of unit cross section and originally bounded by the planes a and $(a + \delta a)$ is given by Eq. (10) or (11).

$$\rho_0 \delta a \left(\frac{\delta^2 \xi}{\partial t^2}\right) = -\delta P^L \tag{10}$$

$$\rho_0 \frac{\partial^2 \xi}{\partial t^2} = -\frac{\partial P^L}{\partial a} \tag{11}$$

If the process is assumed to be isentropic, then the equation of state can be used in the general form, Eq. (2), to relate pressure variations with the corresponding density variations. (Since the medium is assumed to be nonviscous, it is not necessary to include the energy equation.) Thus, keeping only the first two terms in Eq. (2), and combining with Eq. (11), results in Eq. (12).

$$\rho_0 \frac{\partial^2 \xi}{\partial t^2} = -\frac{1}{\rho_0}\left[A + \frac{B}{\rho_0}(\rho^L - \rho_0)\right]\frac{\partial \rho^L}{\partial a} \tag{12}$$

Using Eq. (9), ρ^L can be eliminated to obtain Eq. (13), where $c_0^2 = A/\rho_0$. This is the nonlinear equation describing the propagation of a plane wave in a nondissipative fluid described by the general form of the equation of state (2).

$$\frac{\partial^2 \xi}{\partial t^2} = c_0^2 \left[1 - B/A + \frac{B/A}{1 + \dfrac{\partial \xi}{\partial a}}\right] \frac{\dfrac{\partial^2 \xi}{\partial a^2}}{\left(1 + \dfrac{\partial \xi}{\partial a}\right)^2} \tag{13}$$

There are two additional forms of Eq. (13) which emphasize the similarity of finite-amplitude wave propagation in gases, liquids, and solids. First, the function $(1 + \partial\xi/\partial a)^{-1}$ can be expanded into a series with the result that the term in the square brackets in Eq. (13) is given by Eq. (14). But the first two terms on the right side of Eq. (14) are the same as the series representation of the function in Eq. (15).

$$\left[1 - B/A + \frac{B/A}{1 + \dfrac{\partial \xi}{\partial a}}\right] = \left[1 - B/A \frac{\partial \xi}{\partial a} + \cdots\right] \tag{14}$$

$$\left[1 + \frac{\partial \xi}{\partial a}\right]^{-B/A} = [1 - \frac{B}{A}\frac{\partial \xi}{\partial a} + \cdots] \tag{15}$$

Since the series converges rapidly, the bracketed term in Eq. (13) can be replaced by $(1 + \partial\xi/\partial a)^{-B/A}$, to give Eq. (16).

$$\frac{\partial^2 \xi}{\partial t^2} = c_0^2 \frac{\dfrac{\partial^2 \xi}{\partial a^2}}{\left(1 + \dfrac{\partial \xi}{\partial a}\right)^{B/A+2}} \tag{16}$$

This is a generally recognized form of the nonlinear wave equation. If the equation of state for an ideal gas (5) had been used, the nonlinear wave equation would be Eq. (17), which is the identical mathematical form as Eq. (16) with $B/A + 2$ replaced by $\gamma + 1$, and the velocity of a wave of infinitesimal amplitude given by $c_0^2 = \gamma P_0/\rho_0$.

$$\frac{\partial^2 \xi}{\partial t^2} = c_0^2 \frac{\dfrac{\partial^2 \xi}{\partial a^2}}{\left(1 + \dfrac{\partial \xi}{\partial a}\right)^{\gamma+1}} \tag{17}$$

The second form of Eq. (13) is obtained by expanding both $1 + \partial\xi/\partial a$ and $(1 + \partial\xi/\partial a)^2$ into a power series. The result is Eq. (18). This form is

$$\frac{\partial^2 \xi}{\partial t^2} = c_0^2\left[\frac{\partial^2 \xi}{\partial a^2} - (B/A + 2)\frac{\partial \xi}{\partial a}\frac{\partial^2 \xi}{\partial a^2} + \cdots\right] \tag{18}$$

convenient for comparing with the nonlinear wave equation describing the propagation of a finite-amplitude wave in a nonlinear solid.

Finite-amplitude waves on solids. The derivation of the nonlinear wave equation to describe finite-amplitude wave propagation in solids proceeds along different directions from that used for fluids, although for special circumstances the resulting differential equation has the same form, with a set of slightly more complicated coefficients. For many situations, then, the same mathematical description of the propagation of finite-amplitude waves is appropriate for gases, liquids, and solids.

To derive the nonlinear differential equation for describing finite-amplitude waves in a solid, Lagrange's equations for continuous media are used. This derivation requires only the definition of the strain and the assumption that the elastic energy is a function of strain alone. The latter assumption does, however, require that the medium be a lossless continuum in which attenuation and dispersion are negligible.

Consider an elastic continuum and let a_i, $i = 1, 2, 3$, be the cartesian coordinates at time $t = 0$ of a point in the solid in the unstrained state. Let x_i, $i = 1, 2, 3$, be the cartesian coordinates at time t of the same point in the strained solid. Then, a_i and t are the independent (lagrangian) variables upon which the variables x_i depend.

It is convenient to choose to orient the a_1 axis along the propagation direction. This choice simplifies the equations derived by the lagrangian procedure. The price paid for this simplification lies in the fact that choosing the a_1 axis along the propagation direction requires the use of some special elastic constants. The usual elastic constants are defined as coefficients in the expansion of the elastic energy in powers of strain when the strain is calculated with the a_k axes parallel to the symmetry axes of the crystal. The special constants must be derived from these by transformation of a strain matrix under a rotation matrix which rotates the $\bar{a}_1$ axis (parallel to the crystal symmetry axis) to the a_1 axis (parallel to the propagation direction).

If the x_k axes are chosen parallel to the a_k axes, the particle displacement can be written in a manner similar to that used with fluids, Eq. (19).

$$\xi_k = x_k - a_k \tag{19}$$

Use of the lagrangian procedure, followed by rotation operations appropriate to the crystal symmetry, then gives Eq. (20), where the A_{ij} are known

$$\rho \frac{\partial^2 \xi_i}{\partial t^2} = \sum_{j=1}^{3} A_{ij} \frac{\partial \xi_j}{\partial {a_1}^2} + \sum_{j,l=1}^{3} B_{ijl} \frac{\partial \xi_j}{\partial a_1} \frac{\partial^2 \xi_l}{\partial {a_1}^2} \tag{20}$$

linear combinations of second-order elastic constants, and the B_{ijl} are known linear combinations of second-order and third-order elastic constants. Equation (20) is the general form taken by the nonlinear wave equation in a crystal; however, it can be further simplified and brought into a form that can be compared with the corresponding equation for fluids by specializing to cubic crystals.

If cubic crystals are considered and the nonlinear terms in Eq. (20) are ignored, it is found that pure longitudinal waves may propagate in the [100], [110], and [111] directions (neglecting directions equivalent by symmetry). Pure transverse waves may propagate not only in these three directions, but also in the directions: (1) $k_1 = k_2$; k_3 arbitrary; and (2) $k_3 = 0$; k_1, k_2 arbitrary; where k_1, k_2, and k_3 are the three components of the propagation vector. If nonlinear terms are included, it is found that pure transverse modes do not exist. The transverse wave is always accompanied by a longitudinal wave. However, pure longitudinal modes continue to exist for all three directions [100], [110], and [111]. Therefore, the equation most appropriate for comparison with the behavior of fluids is the one describing the propagation of longitudinal waves. For this comparison, it is convenient to separate out those terms in B_{ijl} which contain second-order elastic constants. Let K_2 represent a linear combination of the appropriate second-order elastic constants, and K_3 represent a linear combination of third-order elastic constants. Then the nonlinear equation for plane longitudinal waves propagating in either the [100], the [110], or the [111] direction in a cubic crystal takes the form of Eq. (21).

$$\rho_0 \frac{\partial^2 \xi}{\partial t^2} = K_2 \left(\frac{\partial^2 \xi}{\partial a^2} + 3 \frac{\partial \xi}{\partial a} \frac{\partial^2 \xi}{\partial a^2} \right) + K_3 \frac{\partial \xi}{\partial a} \frac{\partial^2 \xi}{\partial a^2} \tag{21}$$

For the three principal directions in a cubic crystal, the linear combinations of second-order elastic constants K_2 and the linear combinations of third-order elastic constants K_3 are given in Table 1. Although there is a slight complication resulting from the number of elastic constants entering K_2 and K_3, it nevertheless is clear that the general form of Eq. (21) is the same as Eq. (18). Therefore, the general behavior of finite-amplitude waves in solids should be the same as that in fluids. In fact, Eqs. (16),

Table 1. K_2 and K_3 for the principal directions [110], [110], and [111] in a cubic crystal

Direction	K_2	K_3
[100]	C_{11}	C_{111}
[110]	$\dfrac{C_{11} + C_{12} + 2C_{44}}{2}$	$\dfrac{C_{111} + 3C_{112} + 12C_{166}}{4}$
[111]	$\dfrac{C_{11} + 2C_{12} + 4C_{44}}{3}$	$\dfrac{C_{111} + 6C_{112} + 12C_{144} + 24C_{166} + 2C_{123} + 16C_{456}}{9}$

Table 2. Parameters entering into the description of waves of finite amplitude in gases, liquids, and solids

Parameter	Ideal gas	Liquid	Solid
c_0^2	$\frac{\gamma P_0}{\rho_0}$	A/ρ_0	K_2/ρ_0
Nonlinearity parameter	$\gamma + 1$	$B/A + 2$	$-(K_3/K_2 + 3)$
B/A	$\gamma - 1$	B/A	$-(K_3/K_2 + 5)$

(17), (18), and (21) demonstrate that the general features of the propagation of longitudinal waves of finite amplitude in ideal gases, liquids, and solids (pure-mode directions only) are identical. Therefore, the discussion below is devoted to the propagation of finite-amplitude waves in a solid; however, the discussion also is appropriate to the propagation of a finite-amplitude wave in an ideal gas or a liquid if the equations are modified by making the appropriate substitutions given in Table 2. These parameters enter into the differential wave equation, its solution, and the definition of the discontinuity distances.

Solution of nonlinear equation. The solution of Eq. (21) describing the situation resulting from a sinusoidal driver at $a = 0$ will be discussed. An exact solution can be written as Eq. (22), where ω

$$u = 2(\delta) \sum_{n=1}^{\infty} \frac{J_n(na/L)}{na/L} \sin n(\omega t - ka) \quad (22)$$

and k are the frequency and wave vector of the wave, δ is an arbitrary constant, and J_n is the nth Bessel function. The discontinuity distance L is a parameter of considerable significance to nonlinear acoustics. In a crystalline solid the magnitude of L depends on the direction of propagation of the wave with respect to the crystalline axes, as it will be shown to be a function of K_2 and K_3. In fluids (either liquids or gases) it depends on the equation of state.

By making a power series expansion of this solution, the particle displacement ξ as a function of the distance a from a sinusoidal driver is found to be given by Eq. (23), where A is the amplitude of the fundamental wave.

$$\xi = A \sin(ka - \omega t) - \left[3 + \frac{K_3}{K_2}\right] \frac{A^2k^2a}{8} \cos 2(ka - \omega t) + \cdots \quad (23)$$

This solution is of fundamental importance in the study of finite-amplitude waves in solids, as it demonstrates the fact that (for small distances a) the second harmonic of an initially sinusoidal disturbance grows linearly with distance and quadratically with initial amplitude A and quadratically with frequency (since $k = \omega/c$). Measurement of the amplitude of the second harmonic with propagation distance in three different directions in a cubic crystal allows one to determine three distinct combinations of third-order elastic constants. This fact is very important to the study of solid-state physics, since this is the most accurate way to determine this set of third-order elastic constants.

The solution predicting growth of harmonics in liquids and gases [obtained from Eq. (23) by substituting either $(B/A + 2)$ or $(\gamma + 1)$ for $-[3 + K_3/K_2]$] has been of fundamental significance to the verification of the correctness of the theory; however, because of the effect of attenuation, there are more accurate ways to evaluate γ or B/A than measurement of harmonic generation.

Discontinuity distance. Implicit in the expansions resulting in the different forms of Eq. (21) for gases, liquids, and solids has been the assumption that $-(K_3/K_2 + 3)\,\partial\xi/\partial a << 1$, or equivalently, $(\gamma + 1)\,\partial\xi/\partial a << 1$ for gases, or $(B/A + 2)\,\partial\xi/\partial a << 1$ for liquids. This assumption is satisfied quite well in most situations, and there then exists an implicit solution for the particle velocity u appropriate to the boundary condition $u = u_0 \sin \omega t$. Calculation of the space derivative of this solution, $\partial u/\partial a$, reveals that $\partial u/\partial a$ has an infinity at a distance $a = L$, where L is given by Eq. (24).

$$L = \frac{2c_0^2}{-[3 + K_3/K_2]\omega u_0} \quad (24)$$

Since Eq. (25) is valid, where ξ_0 is the amplitude evaluated at $a = 0$, L is also given by Eq. (26).

$$\left[\frac{\partial \xi}{\partial t}\right]_{a=0} = u_0 = \omega\xi_0 \quad (25)$$

$$L = \frac{2c_0^2}{-[3 + K_3/K_2]\omega^2\xi_0} \quad (26)$$

The discontinuity distance L is an extremely important measure of the relative nonlinearity of different media (even though its meaning becomes less definite when one takes attenuation into consideration). In fluids it often is useful to express the discontinuity distance in terms of the Mach number $M = u_0/c_0$. Thus, L is also given by Eq. (27)

$$L = \frac{2}{(B/A + 2)Mk} \quad (27)$$

Waveform distortion. The implicit equation for u mentioned above can be used to derive Eq. (28) for

$$c_{ph} = c_0\left\{1 - \frac{3(3 + K_3/K_2)}{2}\frac{u}{c_0}\right\}^{1/3} \quad (28)$$

the phase velocity, which shows that the phase velocity depends upon the particle velocity u. Since the quantity K_3 is inherently negative in most solids and in general has a magnitude greater than $3K_2$, it follows that an initially sinusoidal wave will progress as indicated in the illustration, and will distort as it propagates. At the discontinuity distance $a = L$, the wave front is predicted to have a vertical tangent. Beyond $a = L$, the theoretical solution begins to predict a multivalued function, a physically unrealistic situation. The actual physical situation is that if the dissipation is small enough, the waveform becomes a repeated shock wave.

Up to this point the description of the progressive distortion of a finite-amplitude wave is the same for the case of liquids or gases as for solids, and the substitutions indicated in Table 2 can be made to transform the equations into those appropriate to liquids or gases. Now, caution is necessary. Careful attention to thermodynamics would be necessary to progress further in the description of fluids, and if the equations were actually used to verify experimental results, it would be necessary to pay closer attention to the approximations made in the hydrodynamical derivations. A single exceptional example will suffice to illustrate the point.

It happens that fused silica, an amorphous substance, can be described by K_2 and K_3 appropriate to the [100] direction. In this case, however, K_3 is positive. This means that a sinusoidal wave in fused silica will distort, but the minus sign in Eq. (28) means that the regions of maximum particle velocity will travel slower than c_0. The result is a progressive waveform distortion in the opposite sense from that in the illustration. Although this unique situation in fused silica has been experimentally verified, its analog in liquids or gases is not to be expected since such a situation would violate fundamental principles of thermodynamics.

With this reservation it is possible to understand the general features of waveform distortion in solids, liquids, and gases on the basis of the theory, and to show that the experimental situation nominally is predicted by the above analysis.

Theories including dissipation. Although the theory developed by neglecting dissipation is useful as a general guide, and actually is quantitatively correct for distances very small compared with the discontinuity distance, it obviously will not agree with experiment in situations in which viscous and heat-conducting terms are nonnegligible in fluids, or situations in which other attenuation mechanisms such as dislocation interactions are important in solids. Therefore, the attempt to make a general approach even to the understanding of nonlinear distortion of an initially sinusoidal wave must now be abandoned in favor of specific approaches to solution of specific problems.

In general, however, in most media, attenuation increases with frequency (in many instances the attenuation coefficient is proportional to the square of the frequency). Therefore, the higher harmonics generated according to Eq. (22) or (23) will be attenuated more rapidly than the fundamental, and a point can be reached at which the harmonic is attenuated as rapidly as it is produced. This situation

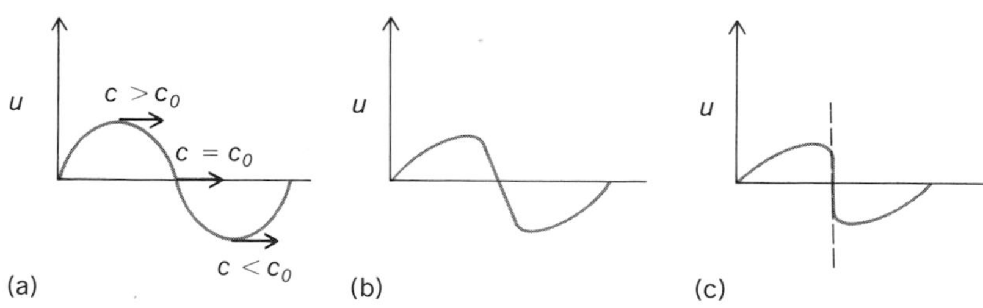

Progressive distortion of an initially sinusoidal finite-amplitude wave. (*a*) $a = 0$. (*b*) $a > 0$. (*c*) $a = L$.

often is referred to as the most stable waveform.

In general, then, an initially sinusoidal finite-amplitude wave produces harmonics and becomes distorted until the most stable waveform changes only slightly, the higher harmonics being attenuated more rapidly than the fundamental frequency. This situation remains for some distance until the higher harmonics are attenuated. Finally, there remains a small-amplitude fundamental wave which continues to propagate until it too is completely attenuated.

For background information *see* ACOUSTIC RADIATION PRESSURE; ELASTICITY; HYDRODYNAMICS; PARAMETRIC ARRAYS; SOUND; SOUND ABSORPTION in the McGraw-Hill Encyclopedia of Science and Technology.

[M. A. BREAZEALE]

Bibliography: R. T. Beyer, Nonlinear acoustics, in W. P. Mason (ed.), *Physical Acoustics*, vol. 2, pt. B, 1965; R. T. Beyer, Nonlinear acoustics (experimental), in D. E. Gray (ed.), *American Institute of Physics Handbook*, 3d ed., pp. 3–206 through 3–210, 1972; D. T. Blackstock, Nonlinear acoustics (theoretical), in D. E. Gray (ed.), *American Institute of Physics Handbook*, 3d ed., pp. 3-183 through 3-205, 1972; L. Bjørnø, Nonlinear acoustics, in R. W. B. Stephens and H. G. Leventhall (eds.), *Acoustics and Vibration Progress*, vol. 2, pp. 101–198, 1976; W. Keck and R. T. Beyer, Frequency spectrum of finite amplitude ultrasonic waves in liquids, *Phys. Fluids*, vol. 3, no. 3, May–June 1960; H. Lamb, *The Dynamical Theory of Sound*, 2d ed., 1960.

Nonlinear integrated optics

Light can be guided for centimeter distances in thin films deposited on low-refractive-index substrates; such systems are called integrated optics waveguides. For typical films of thickness 0.05 to 1 μm, high optical power densities are produced in the film and adjacent media with only moderate total optical powers. Both the long interaction length and high power density have proved to be ideal for nonlinear optical interactions. For two incident codirectional waves, second harmonic generation, sum and difference frequency parametric mixing, and parametric amplification have been obtained with efficiencies previously unachievable in bulk (versus thin-film) media. A new effect, second harmonic generation by two counterpropagating guided waves, has been demonstrated and applied to the nonlinear processing of optical signals on the scale of 10^{-10} to 10^{-12}s. Furthermore, new directions of research are expected which essentially involve the mixing of three rather than two optical beams.

Nonlinear planar optical waveguides. The physics on which integrated optics is based is simply classical electromagnetic theory, and the dispersion relations which govern the guided modes are easily derivable from Maxwell's equations and the usual electromagnetic boundary conditions. When light is incident from a high- onto a low-refractive-index medium, there is a range of incident angles for which the light is totally reflected back into the high-index medium. If total internal reflection occurs at both film interfaces, and a constructive interference (resonance) condition is satisfied across the film, light is guided by the film in the form of discrete modes, one for each allowed solution to the dispersion relations. In fact, films are not necessary for waveguiding, but rather a region of high refractive index (formed by diffusion, doping, and so forth) surrounded by regions of lower refractive index is the crucial requirement; one of the most important nonlinear waveguides is formed by diffusing titanium ions into lithium niobate (Ti:$LiNbO_3$).

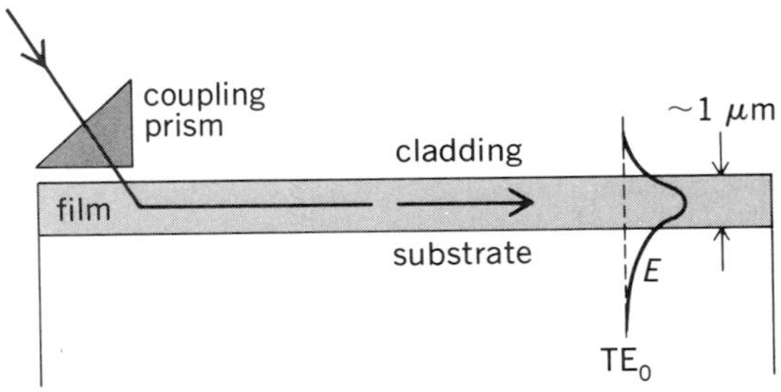

Thin-film optical waveguide. The electric field distribution E corresponds to the lowest-order TE mode, TE_0. The prism is used to couple light into and out of the waveguide with up to 90% efficiency.

A typical electric field distribution associated with the lowest-order mode is shown in the illustration, along with the coupling prism used to inject the light into the waveguide with efficiencies of up to 90%. Losses are caused by scattering of the guided waves from film imperfections typically limiting propagation distances to a few centimeters. Guiding can also be achieved in rectangularlike channels inside which the refractive index is higher than the surrounding media, and light is confined by total internal reflection at all four channel boundaries; typical dimensions are $10 \times 2-5$ μm in the lateral and depth dimensions respectively. A characteristic of guided waves important to nonlinear interactions is that the propagation velocity depends on the refractive indices of all of the media in which the wave exists, and on the product of the film thickness and wave frequency. That is, the guided wave velocity can be adjusted by varying the film thickness. Another important point is that the existence of a number of discrete modes with different propagation velocities can be used to achieve phase matching of nonlinear interactions.

It is relatively simple to produce the high power densities required for nonlinear interactions with moderate total powers. For example, 1 W of laser power in a 1-mm-wide beam when injected into a 0.1-μm-thick waveguide produces a power density of 10^6 W/cm^2. In the 10-μm-wide channel, 10^9 W/cm^2 results. The power densities are sufficiently large that nonlinear interactions occur in any of the media in which a guided wave field exists, that is, as in the illustration, the cladding, the film, or the substrate regions. Fortunately waveguide fabrication processes have been refined to the point that power densities of tens of gigawatts per square centimeter can be transmitted in a number of waveguides without damage occurring.

Nonlinear phenomena are inherently very small over the distance of an optical wavelength, and guided wave versions are no exception. Therefore nonlinear effects usually have to be accumulated over long distances, and in the waveguide case, parallel to the surface. When two waves mix, there is a nonlinear polarization source field created which travels along the waveguide at a velocity intermediate between the two incident wave velocities. If the amplitude of the nonlinearly produced wave is to grow linearly (and the energy quadratically) with propagation distance, its velocity must match that of the polarization field; this is commonly called phase matching. In the waveguide case, only the velocities parallel to the surface must be the same, and there is increased flexibility in achieving this goal because of velocity dispersion with film thickness as well as mode number.

Second-order nonlinear interactions. There are essentially two types of nonlinearities of current interest in nonlinear optics, second and third order in the incident fields. The first, which has been experimentally pursued in waveguides, is the mixing of two waves in a medium which is noncentrosymmetric, that is, a material with no center of symmetry, a requirement for the appropriate nonlinear coefficient to be nonzero. This restricts the phenomenon to certain types of crystals and eliminates waveguides consisting totally of amorphous media, the most common case. To date, nonlinear waveguides have consisted of amorphous films deposited on appropriate crystalline substrates, as well as waveguides produced by diffusing ions and so forth into crystalline media. All of the recent work has been carried out in waveguides formed by diffusing titanium ions (or a similar metal) into lithium niobate, which is a well-known nonlinear material used for bulk media applications.

The most vigorously pursued problem has been

the efficient frequency doubling of light in thin-film and channel waveguides, that is $\omega_1 + \omega_1 \rightarrow 2\omega_1$. Although most of the early work was based on nonlinear substrates, by far the most impressive results have been obtained with $LiNbO_3$. In planar thin film waveguides, 45-W peak power pulses of 1.08-μm radiation have been harmonically doubled into the visible with an efficiency of approximately 25%. These results were achieved with approximately 0.1-mm-wide beams and interaction distances of 17 mm, and by using mode conversion to optimize the phase-matching condition. For channel waveguides 10 μm wide and 4 μm deep, 0.7% conversion efficiency has been demonstrated with 60-mW input power and 1.5×10^{-2}% with 2-mW input power. These efficiencies are not possible with such low incident powers unless waveguides are used to achieve and maintain high power densities over centimeter distances. Furthermore, since there are other nonlinear materials with figures of merit orders of magnitude larger than $LiNbO_3$, large conversion efficiencies (50%) can be expected for milliwatt input power levels. Once these levels have been achieved, they will be used to double radiation from near-infrared semiconductor lasers efficiently into the visible region of the spectrum and to produce inexpensive ultraviolet sources.

Sum ($\omega_1 + \omega_2$) and difference ($\omega_1 - \omega_2$) frequency generation with efficiencies comparable to the second harmonic case have also been achieved. Again the most impressive results have been obtained in channel $LiNbO_3$ waveguides. Furthermore, the state of the art has advanced to the point that parametric amplification has been reported. Although the gain was relatively modest (1.75), it is clear that a highly efficient parametric oscillator can be constructed.

The most unusual nonlinear waveguide interaction reported to date has been second harmonic generation by the mixing of oppositely propagating guided waves. This phenomenon does not occur in bulk nonlinear optics since it requires a combination of the confinement of the interaction region to an optical wavelength in a dimension orthogonal to the propagation direction of the incident waves and/or the presence of a material boundary. When two waves of the form $E_{\pm} \exp{[i(\omega t \pm \beta x)]}$, where t is time and x is distance along the waveguide, mix in an optical waveguide, a polarization field proportional to$E_+ E_- \exp{(2i\omega t)}$ is produced throughout the volume of the waveguide. Since this source field has no periodicity parallel to the surface and is very thin in the depth dimension, it can radiate only normal to the surface. If the two mixing fields are in the form of optical pulses, second-harmonic radiation occurs only during their overlap, and the total output signal is instantaneously proportional to the degree of overlap. Hence the harmonic signal $S(t)$ is given by the notation below, which corresponds

$$S(t) \propto \int |E_+(t-\tau)E_-(\tau)|^2 d\tau$$

to the mathematical convolution of the two pulses. The ability to carry out such an operation has enormous ramifications on the processing of very fast optical signals in an all optical communications system.

This phenomenon has been demonstrated experimentally in $LiNbO_3$ waveguides with optical pulse waveforms of approximately 10^{-11} s duration. In the most sophisticated version, the convolution signal has been captured on an array of detectors (CCD array) placed the waveguide surface. Such a device corresponds to a picosecond transient digitizer with subpicosecond resolution.

Third-order nonlinear phenomena. This class of nonlinear interactions has not been studied experimentally yet in the context of optical waveguides. (This is in contrast to the optical fiber case, where all of the nonlinear phenomena investigated fall into this category.) Progress is expected in the near future in a number of areas, probably phase conjugation, optical bistability, and surface-coherent anti-Stokes-Raman scattering. *See* OPTICAL BISTABILITY.

For background information *see* INTEGRATED OPTICS; NONLINEAR OPTICS in the McGraw-Hill Encyclopedia of Science and Technology.

[GEORGE I. STEGEMAN]

Bibliography: R. Normandin and G. I. Stegeman, *Appl. Phys. Lett.*, 36(4):253–255, 1980; W. Sohler and H. Suche, *Appl. Phys. Lett.*, 37(3):255–257, 1980; N. Uesugi, *Appl. Phys. Lett.*, 36(3):178–180, 1980; A Yariv, *Sci. Amer.*, 241(1):64–72, 1979; F. Zernike and J. E. Midwinter, *Applied Nonlinear Optics*, 1973.

Nuclear orientation

Optical spectroscopic techniques have traditionally played an important role in the development of nuclear physics by probing the small but well-defined nuclear effects in atomic spectra. The nucleus–atomic electron interactions give rise to the optical hyperfine structure and to the optical isotope-isomer shifts; thus optical measurements may yield information on nuclear spins, nuclear magnetic and quadrupole moments, and the radii and shapes of nuclei. The use of tunable dye lasers in nuclear physics, a relatively recent development, has greatly extended the range, sensitivity, precision, and applicability of these optical techniques. Continuous-wave (cw) tunable dye lasers, when tuned to resonance with a suitable optical spectral line, may be used to repeatedly excite the same atom or atoms, a technique called optical pumping. Nuclear orientation or polarization, a preferred alignment in space of the nuclear spin vector, may be achieved in the optical pumping process when polarized light, either circularly or linearly polarized, is used. This technique, called laser-induced nuclear polarization (LINUP), has provided important nuclear structure information for a class of nuclides called spontaneous-fission isomers. This technique has near single-atom sensitivity and may be used to great advantage in nuclear physics research.

Nuclear effects in atomic spectra. It has been known since the early days of atomic spectroscopy that many fine-structure lines can be resolved into a number of closely spaced components. Since this splitting is typically much smaller than the fine structure, the new structure became known as the hyperfine structure. In 1924 W. Pauli suggested that the hyperfine structure arose through interactions of the atomic electrons with the magnetic dipole moment of the nucleus. Pauli's suggestion was soon shown to account for the experimental data, and for the first time systematic studies of nuclear spins and magnetic moments became possible.

In the modern theory of the hyperfine interaction, the atomic and nuclear angular momentum vectors $\vec{J}$ and $\vec{I}$ are coupled to form total angular momenta $\vec{F}$. According to the rules of quantum mechanics, a number (either $2I + 1$ or $2J + 1$, whichever is smaller) of distinct total angular momentum states F is formed. The energy splittings between these various states depend on the properties of the electron state and on the electric and magnetic moments of the nucleus. The largest part of the splitting is generally due, as Pauli suggested, to the nuclear magnetic dipole moment μ, but the effect of the electric quadrupole moment Q is often large enough to provide a good measure of Q. Higher nuclear electromagnetic moments contribute in principle, but these small effects are difficult to observe.

The nucleus also has other effects on atomic spectra; these effects include the isotope and isomer shifts. If the same electronic transition is studied in atoms of the same species (with the same number of protons but with different numbers of neutrons, that is, different isotopes), the frequency of the transition is sometimes found to change. Part of this frequency shift is due simply to the change in mass of the nucleus (mass shift), but an additional part can arise because of a change in nuclear size (field shift). As neutrons are added to a nucleus, the radius of the nucleus generally increases. This increase changes the binding energy of some atomic states and hence contributes to a frequency shift in the atomic transitions. It turns out that the field shift is almost proportional to the root-mean-square (rms) nuclear radius. This means that the field shift occurs, not only as the volume of the nucleus changes, but also as the shape of the nucleus is changed, keeping the volume constant. Observations of field shifts, which were larger or smaller than expected because of the change of nuclear volume with neutron number, have been used for over 30 years as a measure of the change in nuclear shape from isotope to isotope.

The shape of a nucleus can also change between the nuclear ground state and an excited state of the same nucleus. This also results in a shift in transition frequency, the isomer shift, but with no mass shift component. Systematic studies of isomer shifts give information about changes in nuclear shape (via the rms radius) from state to state within the same nucleus.

Orienting nuclei with lasers. According to quantum mechanics, atoms can exist only in quantized states of total angular momentum labeled by integral or half-integral values of the quantum number F (corresponding to angular momentum $F\hbar$, where $\hbar$ is Planck's constant divided by 2π.). The orientation of the angular momentum vector in space is also quantized in such a way that the projection of the angular momentum vector on some arbitrary axis (the quantization axis) must range from $-F$ to F in integral steps. The projection of $\vec{F}$ on the quantization axis is labeled by M_F. In the absence of external electromagnetic fields, these projection substates of an atom all have the same energy. Under normal conditions, in a large group of atoms, an atom is equally likely to be found in each of the projection substates, so that the group of atoms is said to be randomly oriented or isotropic. If an imbalance in the population of M_F substates is achieved, the group of atoms is said to be oriented or polarized. One technique for achieving this imbalance is to induce transitions in the atom, that is, optically pump the atom, with polarized light. Tunable lasers can provide very intense beams of polarized light which can be matched (tuned) to the exact frequency of an atomic transition. If, for example, right circularly polarized light is used and the quantization axis is chosen along the laser beam, a transition to the excited state of the atom is induced which changes the value of M_F by $+1$. The excited state will subsequently decay back to the ground state by a transition that can change M_F by $+1$, 0, or -1. With the high intensities available with laser beams, this cycle can be repeated many times in a short period and will produce a predominance of atoms in projection substates near $M_F = F$. If the angular momenta of the excited and ground states are favorable, all the atoms exposed to the laser beam can be pumped into the single substate $M_F = F$, corresponding to orientation of the atoms such that their angular momenta are parallel to the laser beam. Linear-polarized laser light can also lead to atomic orientation, but the nature of the resulting orientation depends on the angular momenta of the atomic states involved.

The optical pumping process thus results in a net orientation of the total angular momentum vector $\vec{F}$. Since this vector is formed from the nuclear and electronic angular momentum, coupled by the hyperfine interaction, a net orientation of $\vec{F}$ leads to a related orientation of $\vec{I}$, the nuclear spin. Optical pumping with lasers is, with appropriate atomic species, a very efficient method of achieving nuclear spin orientation. With commercially available tunable dye lasers tuned to excite a typical allowed atomic transition with frequency in the visible region, a complete polarization of all nuclei within the laser-beam volume can be achieved within a few microseconds under favorable experimental conditions.

Study of fission isomers with LINUP. Laser optical methods, like the LINUP technique, have a sensitivity making them suitable for investigating nuclei far from nuclear stability. The use of conventional optical methods is usually precluded for these nuclides because of the extremely low production rates, often measured in atoms per hour, and the short nuclear half-lives that are characteristic for nuclides far from stability. In this class are the spontaneous-fission isomers, which are very short-lived nuclei in excited states that decay by the spontaneous-fission process, that is, by splitting into two approximately equal fragments. About 35 such isomers are known among the nuclides of the heaviest elements, uranium ($Z = 92$) through berkelium ($Z = 97$), with nuclear half-lives that range from 10^{-9} to 10^{-2} s. At Oak Ridge National Laboratory the LINUP method has been applied to the study of fissioning isomers by measuring the optical isomer shift for the 0.94 ms isomer ^{240m}Am. Although all nuclei in the actinide region are permanently deformed in the nuclear ground state, that is, prolate ellipsoids, spontaneous-fission isomers were thought to be abnormally deformed, and thus it was thought that a determination of the isomer shift would yield information on the nuclear shape. The experimental arrangement for this experiment is shown in the illustration.

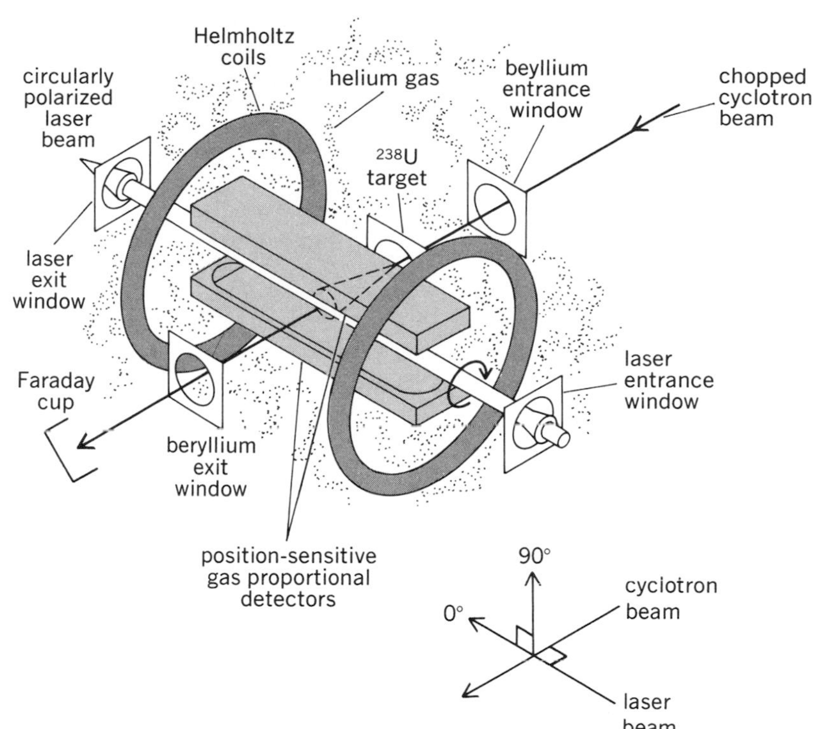

Experimental arrangement for the study of spontaneous-fission isomers using the LINUP method. (*From C. E. Bemis, Jr., et al., Optical isomer shift for the spontaneous-fission isomer* ^{240m}*Am, Phys. Rev. Lett., 43:1854–1858, 1979*)

The fission isomers ^{240m}Am are produced in nuclear reactions of a ^{238}U target with 48-MeV ^{7}Li ions, and the recoiling product nuclei are thermalized in a buffer gas of helium so that they come to rest within the volume of an intersecting polarized laser beam. When the laser is tuned to exact optical resonance with the isomer-shifted atomic transitions characteristic of the abnormally deformed fission isomer, nuclear orientation is achieved via optical pumping. The alignment or quantization axis is the laser-beam propagation axis, and when the isomer subsequently fissions, the fission fragments are preferentially emitted along this laser-beam direction. This anisotropic fission decay from the oriented isomers is detected by using the position-sensitive counters located above and below the laser-beam propagation axis. The exact laser wavelength, at which this occurred for the fission isomer ^{240m}Am, was noted and compared with the wavelength for the normal isotopes of americium yielding an optical isomeric shift. The wavelength for the particular atomic transition, the $^{8}S_{7/2} \rightarrow {}^{10}P_{7/2}$ transition normally at ~640.5 nm, was found to be red-shifted by 0.26 nm, which is the largest optical isotope or isomer shift ever observed. This large shift is indicative of the abnormally large deformations expected for fission isomers, and provided a direct measure of the deformation. Using this nonoptical detection of optical resonance conditions with LINUP, the sensitivity of optical techniques may be extended, as illustrated by the fission isomers, to studies of nuclear properties of other very-short-lived nuclides not amenable to investigation by conventional methods.

Future applications. In addition to its use in the study of short-lived nuclides and those nuclides far from stability, nuclear orientation with lasers (LINUP) is applicable in other areas of nuclear physics. Nuclear reactions with charged-particle beams, particularly with beams of heavy ions like ^{12}C, ^{16}O, . . . ^{238}U, can be performed only by using unpolarized targets and beams and polarization-independent measuring techniques. Since microscopic nuclear reaction theories describe reactions in terms of states of specified nuclear spin orientation and thus must be averaged over all possible spin directions for comparison with experiment, potentially important details of the nuclear reaction may be lost. For deformed nuclei, even macroscopic theories predict large differences for different orientations of the nuclear spin orientation, which is usually identical to the nuclear symmetry axis of the deformation. To probe these interesting and predicted spectacular deformation effects in nuclear reactions, polarized targets or beams are required, and the LINUP method can be used. Several groups in the United States and in Europe are using lasers to produce polarized or oriented beams and targets for investigations of this type in nuclear physics.

For background information *see* HYPERFINE STRUCTURE; ISOTOPE SHIFT; NUCLEAR MOMENTS; NUCLEAR ORIENTATION; OPTICAL PUMPING in the McGraw-Hill Encyclopedia of Science and Technology.

[JAMES R. BEENE; CURTIS E. BEMIS, JR.]

Bibliography: C. E. Bemis, Jr., et al., *Phys. Rev. Lett.*, 43:1854–1858, 1979; P. Jacquinot and R. Klapish, *Rep. Progr. Phys.*, 42:773–832, 1979; D. E. Murnick and M. S. Feld, *Ann. Rev. Nucl. Particle Sci.*, 29:411–459, 1979; E. W. Otten, *Nucl. Phys.* A354:471c–496c, 1981.

Optical bistability

Optical bistability is an expanding field of research because of its potential application to all-optical logic and because of the interesting physical phenomena it encompasses. A bistable optical device can function as a variety of logic devices. It is called bistable because it can have two stable output states labeled 0 and 1 for the same input, as shown in Fig. 1*a*. Thus it can serve as an optical memory element. Under slightly modified operating conditions, the same device can exhibit the optical transistor characteristic of Fig. 1*b*. For input intensities close to I_{gain}, small variations in the input light are amplified, in much the same way that a vacuum tube triode or transistor amplifies electrical signals. The characteristic of Fig. 1*b* can also be used as a discriminator; inputs above I_{gain} are transmitted with far less attenuation than those below. Finally, there is limiting action above I_{gain}: large changes in the input hardly change the output. There is hope that bistable devices will revolutionize optical processing, switching, and computing. Since the first observation of optical bistability in a passive, unexcited medium in 1974, bistability has been observed in many different materials, including tiny semiconductor etalons. Current research is focused on optimizing these devices and developing smaller devices of better materials which operate faster, at higher temperatures, and with less power.

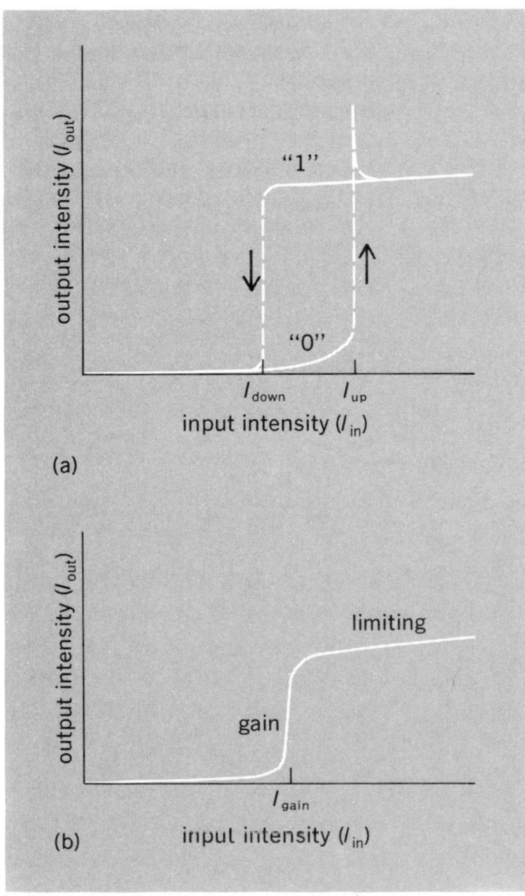

Fig. 1. Transmission of a typical bistable optical device, under conditions of (*a*) bistability (memory) and (*b*) high ac gain (optical transistor, discriminator, or limiter).

Optical bistability has also attracted the attention of physicists interested in fundamental phenomena. In fact, a bistable device often consists of a nonlinear medium within an optical resonator; it is thus similar to a laser except that the medium is unexcited (except by the coherent light incident on the resonator). Such a device constitutes a simple example of a strongly coupled system of matter and radiation. The counterparts of many of the phenomena studied in lasers such as fluctuations, regenerative pulsations, and optical turbulence can be observed under better-controlled conditions in passive bistable systems.

All-optical systems. The transmission of information as signals impressed on light beams traveling through optical fibers is replacing electrical transmission over wires. The low cost and inertness of the basic materials of fibers and the small size and low loss of the finished fibers are important factors in this evolution. Furthermore, for very fast transmission systems, for example, for transmitting a multiplexed composite of many slow signals, optical pulses are best. This is because it is far easier to generate and transmit picosecond optical pulses than electrical pulses. With optical pulses and optical transmission, the missing component of an all-optical signal processing system is an optical logic element in which one light beam or pulse controls another. Because of the high frequencies of optical electromagnetic radiation, such all-optical systems have the potential for subpicosecond switching and room-temperature operation. Although any information processing and transmitting system is likely to have electrical parts, especially for powering the lasers and interfacing to humans, the capability of subpicosecond switching appears unique to the optical part of such a system.

Bistable optical devices. Recently, bistable optical devices have been constructed which have many of the desirable properties of an all-optical logic element. A device is said to be bistable if it has two output states for the same value of the input over some range of input values. Thus a device having the transmission curve of Fig. 1*a* is said to be bistable between I_{down} and I_{up}. This device is clearly nonlinear; that is, I_{out} is not just a multiplicative constant times I_{in}. In fact, if I_{in} is between I_{down} and I_{up}, knowing I_{in} does not reveal I_{out}. To accomplish this behavior, an all-optical bistable device requires feedback. Even though a nonlinear medium is essential, the nonlinearity alone only means that I_{out} versus I_{in} is not a straight line. The feedback is what permits the nonlinear transmission to be multivalued, that is, bistable.

Fabry-Perot interferometer. The most nearly practical devices so far are tiny semiconductor etalons, that is, tiny Fabry-Perot interferometers consisting of a gallium arsenide (GaAs) or indium antimonide (InSb) crystal with flat parallel faces sometimes coated with dielectrics to increase the reflectivity to about 90% (Fig. 2). In these etalons the Fabry-Perot cavity provides the optical feedback, and the nonlinear index of refraction n_2 is the intensity-dependent parameter. The bistable operation of such a Fabry-Perot cavity containing a medium with a nonlinear refraction can be pictured as follows. In the off or low transmission state, the laser is detuned from one of the approximately equally spaced transmission peaks of the etalon and most of I_{in} is reflected (Fig. 3*a*). The refractive index is approximately n_0, its value for weak light intensity. In the on state the index is approximately $n_0 + n_2I_c$, where the intensity inside the cavity is I_c. This change in index shifts the etalon peak to near coincidence with the laser frequency, permitting a large transmission and a large I_c. Clearly there must be a consistency between the index and the laser frequency; each affects the other through the feedback. As the input is increased from low values, the frequency ν_{FP} of the Fabry-Perot peak begins to shift when n_2I_c becomes significant. But this shift increases I_c, which further increases n_2I_c, and so on. This positive feedback continues until, at I_{up}, the effect runs away, the device turns on, and the transmission reaches a value on the negative feedback side of the etalon peak consistent with $n = n_0 + n_2I_c$ (Fig. 3*b*). Once the device is turned on, I_c is larger than I_{in} because of the storage property of the cavity. Therefore I_{in} can now be lowered to a value below I_{up}, and the large I_c will keep the device on. Thus the hysteresis of the bistability loop arises from the fact that, for the same input intensity, the intracavity intensity and index contribution are small in the off or detuned state and large in the on or in-tune state.

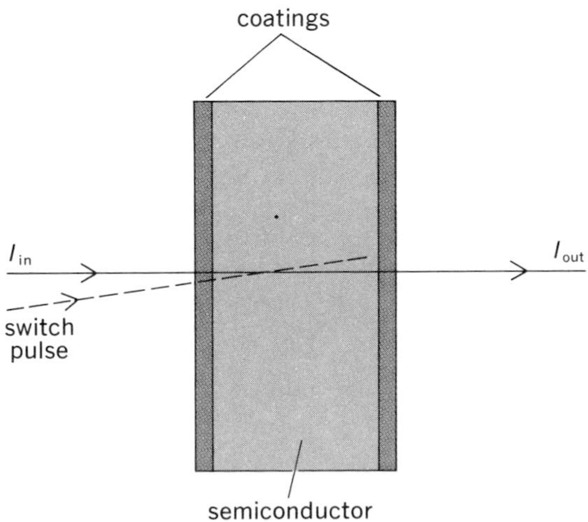

Fig. 2. Bistable optical etalon.

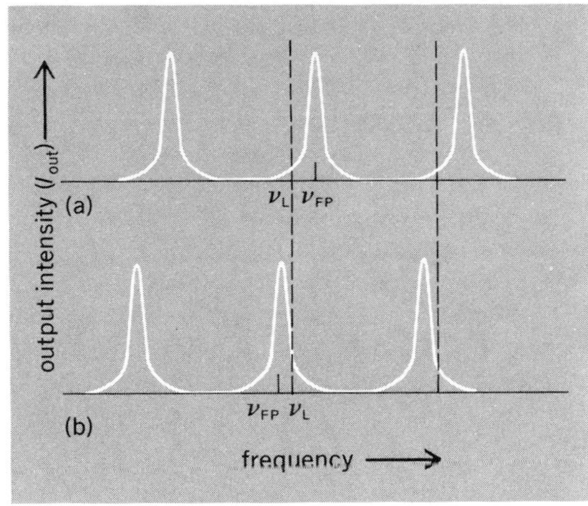

Fig. 3. Transmission function of a nonlinear Fabry-Perot interferometer: (*a*) device off, index $n \cong n_0$, and (*b*) device on, $n \cong n_0 + n_2I_c$. The laser frequency ν_L is fixed, but the cavity peak frequency n_{FP} changes with light intensity via n_2.

Clearly, a device with the transmission characteristic of Fig. 1 can serve as an optical memory. If I_{in} is maintained between I_{down} and I_{up}, the value of I_{out} reveals the state of the device. Light pulses can be used to switch the device on and off, just as for an electrical flip-flop. A simple analysis of a nonlinear Fabry-Perot cavity and experiments show that by changing the initial detuning this device can perform a whole host of optical operations. Bistable optical devices have been operated as an all-optical differential or ac amplifier (also called an optical transistor and a transphasor), limiter, discriminator, gate, oscillator, and pulse shaper.

Properties. The GaAs device was only 5 μm thick, and the laser beam diameter on the etalon was only 10 μm—about one-tenth the diameter of a human hair. The device was switched on by a 10-ps pulse in a time shorter than the 200-ps detector response time, and a time of a few picoseconds is believed possible. The GaAs device operates at a convenient wavelength (0.83 μm) and is constructed of a material used for electronics and diode lasers, facilitating interfacing and integration. But all the properties of the demonstrated device are not ideal. The switch-down time is about 40 nanoseconds, presumably arising from the lifetime of the free carriers and excitons created by the intense light in the on state. The minimum input power is about 100 mW, and the minimum switch-on energy about 1 nanojoule, far more than a practical device can have. And the highest operating temperature is 120 K, because the free exciton resonance used to produce n_2 decreases, and the undesirable band-tail absorption increases, with increased temperature. However, measurements of the intrinsic properties of pure GaAs reveal that the switching times and energy should be greatly reduced in optimized etalons.

The InSb etalons are longer (a few hundred micrometers), operate at longer wavelengths ($\approx$ 5 μm), and hence have larger transverse dimension limitations than GaAs, have demonstrated switching times of a few hundred nanoseconds, operate at least up to 77 K, and have functioned with as little as 8-mW input power. And, just as for GaAs, it is unlikely that the InSb devices are optimized.

The results of these first experiments seeking practical devices point to obvious areas currently under intense research. The physics of semiconductors is being challenged to identify giant nonlinearities at convenient wavelengths and at higher temperatures. In addition to the GaAs free exciton and InSb below-edge band-to-band saturation mechanisms already observed, electron-hole plasma and biexciton mechanisms have been proposed. Other bistable configurations are being sought to better utilize the nonlinearities and minimize the power required: thin etalons, nonlinear interfaces, self-focusing devices, and guided-wave structures.

Fundamental studies. In addition to work on the application of optical bistability to practical devices, a comparable effort is being directed toward the properties of these devices, whose operation depends on the nonlinear coupling between electronic material and light. Figure 1 shows a discontinuity in the transmission, reminiscent of a first-order phase transition. The input light maintains the system in equilibrium which is far from thermal. Fluctuations and transient behavior of such systems are of great interest. A sluggishness in response, called critical slowing down, has already been observed for I_{in} close to I_{up}. Regenerative pulsations have also been seen in which competing mechanisms cause the device to switch on and off repeatedly, forming an all-optical oscillator.

Hybrid devices. Many of these fundamental studies have been conducted with hybrid (mixed optical and electronic) bistable devices in which the intensity dependence of the intracavity index results from applying a voltage proportional to the transmitted intensity across an intracavity modulator. In fact, the cavity is not necessary, since electrical feedback is present. Placing the modulator between crossed polarizers provides the required nonlinearity. Integrated hybrids are being considered for practical devices, but the ultimate in shortening the detector-to-modulator wire length is to place the detector inside the cavity. The distinction between hybrid and intrinsic then fades. The best device may have a voltage across the semiconductor to increase its speed and sensitivity. Arrays of hybrid devices have been used to study image processing using bistable elements; eventually bistable arrays could be used for parallel computing, for example, for propagation problems with transverse effects.

Optical turbulence. Hybrid devices have been considered to be completely analogous to intrinsic devices. Recently it has been predicted that an intrinsic ring-cavity device subjected to a steady input will exhibit periodic oscillations and turbulence or chaos if the medium response time is short compared with the cavity round-trip time. The hybrid analogy is to delay the feedback by a time longer than the detector-feedback-modulator response time. Optical chaos, observed in such a hybrid, has been used to study the evolution, from a stable output, through periodic oscillations, to chaos as the input is increased. The optical turbulence studies interface to phenomena in many other disciplines, such as mathematics, genetics, and hydrodynamics, emphasizing the basic similarity of these seemingly diverse phenomena, in agreement with some recent mathematical findings.

With such important potential applications, with such significant possibilities for improvements, and with so many opportunities for experiments of fundamental interest, the field of optical bistability is likely to continue its explosive growth.

For background information *See* INTERFEROMETRY; NONLINEAR OPTICS; OPTICAL COMMUNICATIONS in the McGraw-Hill Encyclopedia of Science and Technology. [HYATT M. GIBBS]

Bibliography: H. M. Gibbs, S. L. McCall, and T. N. C. Venkatesan, *Opt. Eng.*, 19:463–468, 1980; H. M. Gibbs et al., *Phys. Rev. Lett.*, 46:474–477, 1981; S. D. Smith and D. A. B. Miller, *New Sci.*, 85:554–556, Feb. 21, 1980; P. W. Smith and W. J. Tomlinson, *IEEE Spectrum*, 18(6):26–33, 1981.

Osteoporosis

Postmenopausal osteoporosis, a condition of deficient bone mass (skeletal osteopenia), is associated with fractures of the spine, forearm, and hip. Although not conclusively proved, an increase in bone mass would be expected to lessen the incidence of fractures; current therapeutic regimens for postmenopausal osteoporosis are consequently directed at maintaining or increasing total bone mass.

Although numerous factors contribute to the development of postmenopausal osteoporosis (age, estrogen deficiency, diet, racial factors, and so on), the basic pathophysiologic abnormality appears to be some disturbance of the normal bone remodeling sequence. To understand the pathogenesis of this condition (and also to understand its treatment), some knowledge of bone remodeling is therefore necessary. As noted in the illustration, bone is constantly being resorbed, then replaced with new bone. The initial event in osteoporosis is an increase in bone resorption mediated by the osteoclast and (to a lesser extent) the osteocyte. This event is generally followed by an increase in bone formation mediated by the osteoblast. The processes of bone resorption and formation are normally (perhaps homeostatically) coupled: an increase or decrease in resorption produces a corresponding increase or decrease in formation, such that the net change in bone mass is zero.

In postmenopausal osteoporosis (illustration *c* and *d*), however, normal remodeling sequences appear disrupted either by an increase in bone resorption over normal bone remodeling levels without a corresponding increase in bone formation, or by a decrease in bone formation without a corresponding decrease in resorption. Both disruptions may lead to a net loss in bone mass; bone remodeling is negatively uncoupled.

In terms of bone remodeling, maintaining, or increasing, total bone mass may obviously be accomplished by two therapeutic maneuvers: a decrease in bone resorption or an increase in bone formation. To prevent osteopenia (maintaining bone mass), a decrease in resorption may suffice if bone formation is maintained at a normal level. However, the restoration of bone mass previously lost (increasing bone mass) requires ideally a decrease in resorption as well as an increase in formation; an increase in resorption with a greater increase in formation, or a decrease in resorption with a lesser decrease in formation, also results in the desired net positive bone mass change. In other words, these therapeutic maneuvers attempt to uncouple (positive uncoupling) the normal coupling mechanism of bone remodeling.

A number of therapeutic agents are available to slow the loss of bone mass, principally by decreasing bone resorption. In addition, a few agents appear capable of restoring previously lost bone mass, and possibly of preventing further fractures. Currently available therapeutic agents are calcium, vitamin D_2, anabolic steroids, estrogens, sodium fluoride, synthetic salmon calcitonin, diphosphonates, and exercise.

Calcium. It appears that the average diet of the postmenopausal female is deficient in calcium; the pre-, peri-, and postmenopausal female should probably consume at least 1000 mg of oral elemental calcium daily. Oral calcium is apparently of short-term benefit (probably by decreasing bone mass loss by decreasing bone resorption through a depression of parathyroid hormone), but its potential long-term value is unproved. It is unlikely that oral calcium alone can restore previously lost bone mass, as following a decrease in bone resorption, normal coupling mechanisms may eventually result in a decrease in bone formation; adequate calcium intake should, however, be mandatory to any postmenopausal osteoporosis combination treatment program.

Vitamin D. A mild vitamin D deficiency (mild osteomalacia) may exist in many elderly postmenopausal osteoporitic females, and a multivitamin with 400 International Units is a reasonable treatment program inclusion; this amount of vitamin D has little effect on postmenopausal osteoporosis per se, however. High-dose (greater than 1000 IU daily) vitamin D may indeed be deleterious in the individual with postmenopausal osteoporosis, perhaps due to a primary stimulation of bone resorption. Studies of the efficacy of various vitamin D congeners (25-hydroxycholecalciferol, 1,25-dihydroxycholecalciferol, and so on) in postmenopausal osteoporosis are ongoing; these agents act primarily by increasing calcium absorption in the gut and subsequently decreasing bone resorption, although a primary increase in bone formation also remains a possibility.

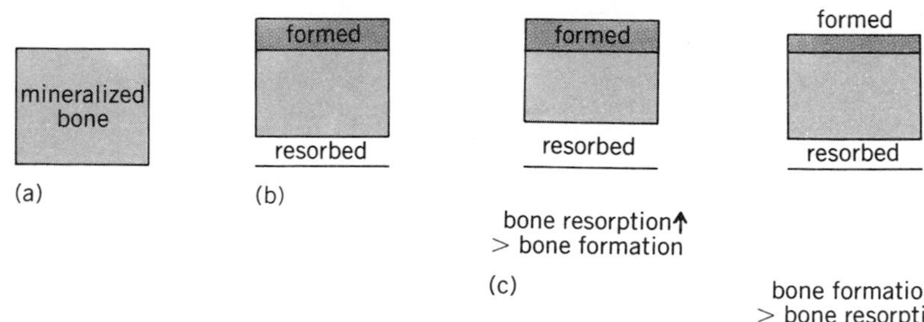

Osteoporosis due to disruption of normal bone remodeling sequences. (*a*) Bone trabecula at zero time and (*b*) normal bone turnover. (*c, d*) Two possible mechanisms of abnormal bone remodeling.

Anabolic steroids. These modified male hormones appear to offer primary efficacy when combined with oral calcium in postmenopausal osteoporosis, with a significant increase in bone mass above pretreatment levels following 26 months of therapy with the anabolic steroid stanozolol. This beneficial effect could be explained by a primary decrease in urinary calcium, or by a combined primary decrease in urinary calcium plus an increase in the rate of bone formation. Substantiation of an increase in bone formation is documented from iliac crest bone biopsy data, and from other data noting a significant increase in the bone isoenzyme of alkaline phosphatase following stanozolol therapy. In addition, the increase in bone mass noted in the stanozolol-treated individuals was associated with lack of further spinal compression fractures; the positive uncoupling apparently persisted throughout the period of medication usage. While these agents do appear to be quite efficacious in postmenopausal osteoporosis, potential side effects may occur with their usage, and the need for an apparently effective agent must be balanced against the possibility of less desirous drug actions.

Estrogens. Estrogen preparations most likely act by decreasing bone resorption (perhaps by decreasing bone responsiveness to parathyroid hormone), and consequently slow bone mass loss; one study demonstrates efficacy through 8 years of therapy (in normal oophorectomized females), but also notes accelerated bone mass loss following estrogen cessation. There is little evidence that estrogens can significantly replace bone mass previously lost (persistently maintain positive uncoupling), and it is unclear whether fracture rate is reduced with this therapy. In addition, significant side effects (endometrial carcinoma, for example) are associated with estrogen usage.

Fluoride. Sodium fluoride, an experimental medication for postmenopausal osteoporosis, appears to directly increase bone formation and theoretically not only would slow bone mass loss but would also restore bone mass previously lost. However, concern has arisen regarding the structural integrity of bone produced while on fluoride therapy; recent data suggest an increased fracture incidence following long-term fluoride administration, possibly due to the formation of fluorapatite (rather than hydroxyapatite) bone crystal. Whether a fluoride-induced increase in bone mass results ultimately in a reduced fracture incidence requires further study.

Calcitonin. Synthetic salmon calcitonin, an experimental modality, acts primarily by decreasing bone resorption by osteoclastic inhibition with a subsequent increase in total bone mass. Whether the beneficial drug effect persists beyond 12–18 months is, however, unclear; restoration of normal coupling mechanisms may have occasioned an observed reduction in bone mass at 24 months. In addition, it is unclear whether fracture rate is diminished with calcitonin therapy. Calcitonin does exhibit proven efficacy in Paget's disease of bone (through 18 months) and is a safe medication without side effects.

Diphosphonates. The diphosphonate dichloro-MDP (clodronate disodium) is currently undergoing clinical trials. Although not well tested to date in postmenopausal osteoporosis, diphosphonates also have a proved value in Paget's disease, and act also by decreasing bone resorption (probably by coating the hydroxyapatite crystal and neutralizing osteoclast resorption) and thereby slowing bone mass loss. Diphosphonates may prove to be an effective postmenopausal osteoporosis therapy.

Exercise. An exercise program appears of definite value in postmenopausal osteoporosis and seems to decrease bone mass loss by increasing bone formation to a greater extent than it increases resorption.

Parathyroid hormone. Studies note a favorable response (increased trabecular bone mass, as assessed by iliac crest bone biopsy) to daily subcutaneous injections of a synthetic subfragment of human parathyroid hormone; this subfragment of the hormone molecule appears to stimulate bone formation in contradistinction to the bone resorption action of the entire molecule. Trabecular bone is the type most frequently affected by postmenopausal osteoporosis, as opposed to the more structurally secure cortical bone. Further work with this hormone fragment is indicated prior to its acceptance as a therapeutic modality, especially in light of the suspicion that trabecular bone mass increase may be at the expense of cortical bone mass.

Coupling factor. An exciting new development in the postmenopausal osteoporosis field, with potential ramifications for therapy, is the discovery of a putative bone coupling factor. A protein with a molecular weight of 83,000 has been isolated and purified to homogenetic purity from human bone matrix; this substance is apparently produced in a laboratory-culture system of embryonic chick tibiae in response to bone resorbing stimuli, and appears to stimulate bone formation as evidenced by an increase in osteoblast number. Antibodies prepared against the purified factor show cross-reactive material in human serum, and a substance in serum with similar heat and acid stability to the purified factor does increase bone cell proliferation.

Summary. A number of therapeutic modalities are currently available for the treatment of postmenopausal osteoporosis. While the major beneficial effect of most of the currently available therapeutic agents is in decreasing bone mass loss, two (fluoride and stanozolol) appear to increase bone mass by an increase in bone formation. Even with these latter agents, however, the observed increase may be insufficient to substantially improve bone strength or, in the elderly osteoporotic female, to fully replete skeletal bone mass within the expected life span. Nevertheless, osteoporosis is an extremely morbid and incapacitating disease; any decrease in bone mass loss, or increase in total bone mass, will be of benefit. In addition, it should be noted that the prophylactic potential of these medications (either singly or in combination) has not been defined but may be substantial. Overall, osteoporosis would now appear to be a treatable disease; as further research proceeds, it may become a preventable one as well.

For background information *see* BONE; CALCIUM METABOLISM; ENDOCRINE MECHANISMS; PARATHYROID HORMONE; SKELETAL SYSTEM; SPECIALIZED TISSUE in the McGraw-Hill Encyclopedia of Science and Technology. [CHARLES H. CHESNUT, III]

Bibliography: C. H. Chesnut, D. J. Baylink, and W. B. Nelp, *Clin. Res.*, 27:85A, 1979; C. H. Chesnut, D. J. Baylink, and W. B. Nelp, *Clin. Res.*, 27:363A, 1979; G. A. Howard et al., *Proc. Nat. Acad. Sci. USA*, 78:3204–3208, 1981; B. L. Riggs et al., *JAMA*, 243:446, 1980.

Particle accelerator

The microtron is a particle accelerator that can provide high-energy electrons by recirculating them repeatedly through an accelerating section in which there is a time-varying electric field. In the simplest microtrons, the accelerating section is small enough to fit into the uniform magnetic field that guides the electrons in circular orbits. Microtrons of this type provide electrons in the energy range from about 5 to 50 million electronvolts (MeV). Higher energies can be achieved by splitting the magnetic field in order to make space for larger accelerating sections, which increase the energy gain of the electrons during each traversal. Such split-field microtrons are now being planned to increase electron energies to about 1000 MeV.

Microtrons were used only rarely before 1976 because they had not been developed as fully as other electron accelerators. The medical, industrial, and physics research uses of high-energy electrons were mainly satisfied by betatrons, electron linear accelerators (linacs), and electron synchrotrons. However, the higher energies and other improvements in microtrons have increased their popularity.

Acceleration conditions. The electrons in a microtron gain energy each time they pass through an accelerating section in which the electric field is varying with time. The curve in Fig. 1 shows how the energy gain of an electron depends upon its arrival time at the accelerating section. The vertical distance between the curve and the horizontal axis represents the energy change that an electron would experience if it arrived at different times. Electrons which arrive when the curve is above the axis gain energy; if electrons arrived when the curve was below the axis, they would lose energy. The curve repeats itself with a time interval $T_{\text{r.f.}}$, which is the reciprocal of the frequency of the electromagnetic field. For a frequency of 1 gigahertz, the repetition time is 1 nanosecond.

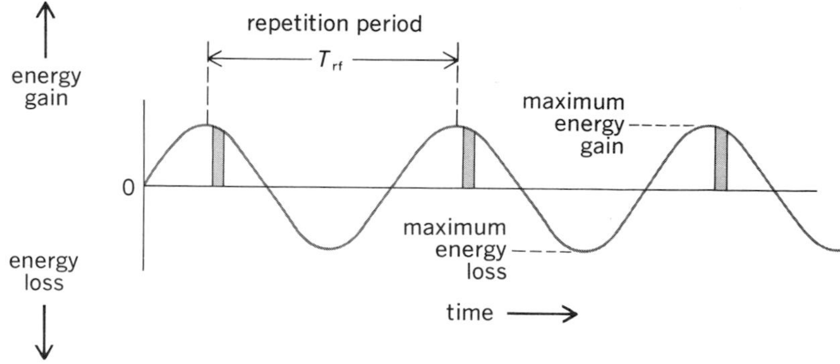

Fig. 1. Dependence of an electron's energy gain in a microtron on its arrival time at the accelerating section.

The recirculation system of a microtron returns the electrons to the accelerating section only at those special times when the electrons will gain the correct energy. The acceptable time intervals are shown by the shaded regions of Fig. 1. These time intervals are preferred because they correspond to very high energy gain while minimizing the variation in energy between different electrons. Electrons with energy slightly above average will travel on a larger recirculation path, will arrive a bit late, and will gain a bit less energy so that they will tend to have the average energy. An electron with slightly less than average energy will travel on a smaller recirculation path, will arrive a bit early, and will gain a bit more energy, thereby moving its total energy closer to the average. This ability of a microtron to produce a high-energy electron beam with a particularly small energy spread is one of the causes of its increasing popularity.

The microtron and the cyclotron have some similarities as particle accelerators, but they also have an essential difference. In both accelerators, the particles gain energy from time-varying electric fields, and the particles are recirculated at the appropriate times with the aid of a constant, uniform magnetic field. The particles in both accelerators move in larger circles in uniform magnetic fields when their energies increase. The main differences between microtrons and cyclotrons originate from the large differences in the speeds of the particles being accelerated. The particles in a cyclotron are moving slowly with respect to the speed of light, and their speeds increase at the same rate as does the size of their orbits. Therefore, even though the cyclotron orbits increase in size with increasing energy, the time required for the particles to traverse the orbits remains the same; the particles move more quickly over the larger orbits.

The electrons in a microtron require successively longer time to travel over successive orbits. Because these electrons are traveling with speeds which approach the speed of light, their speeds increase much more slowly than does the size of the orbits as the energy changes. However, it is still possible to bring the electrons back to the accelerating section at the correct time by arranging for the energy increase to add to the orbit traversal time exactly either one electrical field repetition period or an integral number of periods. Consider, for example, a microtron that operates at a frequency of 1 GHz, which corresponds to a repetition period of 1 ns. The electrons gaining energy could take precisely one extra nanosecond for each successive orbit; they might take 2 ns to complete their first orbit, 3 ns for the second, 4 ns for the third, and exactly $(n + 1)$ ns to complete the nth orbit. Under these conditions the electrons would always return to the accelerating section only during the allowable time intervals such as those indicated by the shaded areas in Fig. 1.

Uniform field microtrons. The microtron was first proposed in 1944 by the Soviet physicist V. I. Veksler, who referred to it as an electron cyclotron. A group of Canadian physicists produced the first working model in 1948. It was called a microtron because the high-frequency electromagnetic fields that were used are generally referred to as microwaves.

The simplest microtrons were developed during the 1950s and 1960s mainly in the Soviet Union at the Physics Laboratory of the Academy of Sciences (Institute for Physical Problems) in Moscow. In these microtrons, a small accelerating section is placed in a uniform magnetic field. The orbits are circles with a common tangent at the accelerating section as illustrated in Fig. 2. The time required for an electron to travel over the uniform circle in

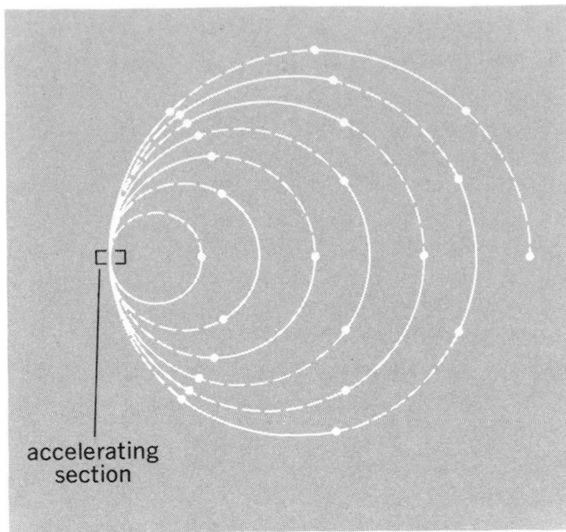

Fig. 2. Electron orbits in a uniform field microtron.

which it moves in a uniform magnetic field is proportional to the total relativistic energy, which is the sum of the kinetic energy and the rest energy of the electron. The electrons gain an equal energy each time they pass through the accelerating section, and this causes their traversal time to increase by exactly the same amount after each traversal. The rest energy of an electron is m_0c^2, where m_0 is the rest mass of an electron and c is the speed of light in a vacuum. The orbits shown in Fig. 2 would be followed by an electron which had a kinetic energy equal to m_0c^2 after the first passage through the accelerating section. This would correspond to a total energy of $2m_0c^2$ during the first orbit, a total energy of $3m_0c^2$ during the second orbit, and so forth. The traversal time of the electrons in Fig. 2 increases by exactly one repetition period after each traversal, so that the electrons travel over the first circular orbit in two periods, over the second orbit in three periods, and so forth; the dots in Fig. 2 indicate the electron positions at the end of each period. The return of electrons to the accelerating section at the correct time after each orbit can be accomplished with other values of the energy gain for each traversal of the electron through the accelerating section. The magnetic field and energy gain can be adjusted so that the additional traversal time for each orbit beyond the first is an integral number of repetition periods. An independent control is available to assure that the electrons will arrive at the proper time after the first orbit.

Such microtrons have been built with a varying number of orbits and energy gains per traversal of the accelerating section. The energy gains per traversal have ranged between about 0.5 and 1.5 MeV, the number of orbits between 10 and 56, and the maximum energies between 6 and 44 MeV.

Split-field microtrons. Improved microtrons were produced recently after the uniform magnetic field was separated to make room for more versatile accelerating sections. Split-field microtrons had been suggested in the 1950s, and small working models had been reported in 1967. Interest in such microtrons grew rapidly following the success of a six-traversal accelerator that has been used at the University of Illinois since 1973. This Illinois microtron is shown schematically in Fig. 3. Electrons are injected into a linac in which they gain an energy E_0. These electrons move in a semicircle when they enter the uniform magnetic field in one of the two matched end magnets. The electrons then travel past the linac on a line parallel to the linac axis, traverse a semicircle of the same size in the second end magnet, and return toward the linac along its axis. The electrons gain the same amount of energy E_0 when they pass through the linac, move through larger semicircles in the end magnets, and return to the linac at exactly the correct time to receive additional energy. In the Illinois microtron, this process continues until the electrons have passed through the linac six times, thereby gaining an energy equal to 6 E_0. This recirculation system was used first in 1973 with an electron linac which added 3 MeV of energy for each traversal; the final energy gain was 6 × 3 MeV = 18 MeV. Since 1978 this microtron has been operating at Illinois with a linac that increases the energy by 11 MeV per traversal, resulting in a final electron energy of 66 MeV.

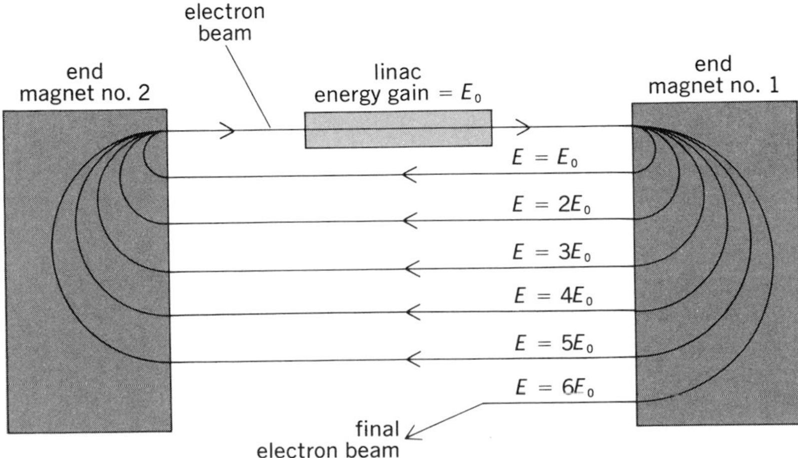

Fig. 3. Six-traversal split-field microtron. The electrons travel in the clockwise direction shown by the arrows on the recirculation paths. The energies that electrons have after each of the traversals through the linac is indicated by the value of the energy E given above each orbit.

Continuous electron beams. The Illinois microtron is particularly useful for nuclear research because it provides a steady stream of electrons in contrast to the intermittent bursts of electrons that are produced by almost all other electron accelerators. In most linacs the accelerating electric fields are sustained for short time intervals separated by much longer time intervals during which the heat can be removed from the linac walls. Linacs in which the accelerating fields can be maintained continuously are expensive to build and to operate. Microtrons are well suited to obtain higher energy from such linacs because the recirculation of the electrons through a linac adds energy without increasing the operating cost.

Two methods have been used to produce linacs that can be operated continuously. A group of physicists at Stanford University in California developed a superconducting linac made with the metal niobium. When it is cooled to within a few degrees of absolute zero, this linac can sustain accelerating electric fields continuously while generating only a few watts of heat in the walls. The 11-MeV linac in use at Illinois is a 6-m-long niobium structure designed and built at Stanford. Several of these superconducting linacs together with a recirculation system called a recyclotron are available at Stanford and may provide continuous electron beams.

The second method of producing continuous electron beams uses specially designed linacs which have less energy gain per unit length to reduce the heat generated in the linac walls. These linacs are being incorporated into a microtron project at the University of Mainz in Germany. The Mainz plan envisioned a series of three microtrons that would produce a high-energy continuous electron beam using 250 kW of power. The lowest-energy microtron at Mainz, which uses 20 recirculations to provide 14-MeV electrons, has been operating since 1979. The energy of these 14-MeV electrons will be increased to 180 MeV by a 40-MeV traversal system now under construction. The original Mainz plan for the third microtron suggested 80 traversals through a 9-MeV linac to add an energy of 720 MeV to the electrons, but the design parameters might change before the system is built.

Several other noteworthy split-field microtrons are being constructed or planned. A 100-MeV, 20 traversal, split-field pulsed microtron is being constructed at the University of Lund in Sweden. The Lund plan is to convert these pulsed electrons into a continuous source by injecting them into a storage ring from which they can be extracted slowly. A similar microtron is being built to serve as an injector for the synchrotron radiation storage ring being constructed at the Synchrotron Radiation Center at the University of Wisconsin. The National Bureau of Standards and the Los Alamos Scientific Laboratory are collaborating on an accelerator development project designed to test the amount of current that can be recirculated by a microtron through a linac operating at high temperature. This development project might help decide the feasibility of using a microtron to provide continuous electron beams with an energy near about 1000 MeV for nuclear research. Meanwhile, the Illinois group has plans to add a second microtron stage to provide continuous electron beams with an energy of about 400 MeV. The impressive advances in split-field microtron developments make it clear that these electron accelerators will be used much more widely in the future.

For background information *see* MICROWAVE; PARTICLE ACCELERATOR in the McGraw-Hill Encyclopedia of Science and Technology.

[PETER AXEL]

Bibliography: S. P. Kapitza and V. N. Melekhin, *The Microtron*, 1969, English transl., 1978.

Petrology

Early interpretation of seismic refraction data led to the traditional geophysical view of the continental crust as a simple structure composed of one or two thick continuous layers. These notions produced theories of crustal formation and evolution which were thought to apply on a global scale but probably were oversimplified. Recent analysis of data from several disciplines is beginning to reveal petrologic and structural heterogeneity of the middle and lower crust. Such developments require new approaches to concepts of crustal formation and evolution. These modern interpretations stem from the investigation of sites in mountain belts which can be geophysically documented as lower and middle crust thrust to the surface; examination of xenolith suites (samples of deep crust entrained by magmas during ascent to the surface); and interpretation of deep seismic reflection profiles in light of known physical properties of rocks. Indirect methods such as these are required because midcrustal levels are presently inaccessible to the drill bit.

Cross sections of crust in mountain belts. There are many high-grade metamorphic terranes throughout the world which equilibrated under the pressure and temperature conditions of the middle and lower crust some time in the course of the Earth's history. Now exposed at the surface as a consequence of uplift and erosion, these terranes provide clues about the petrology, geochemistry, and structure of deep crustal levels at the time the rocks were in an undisturbed position. More importantly, a few metamorphic terranes can be geophysically traced di-

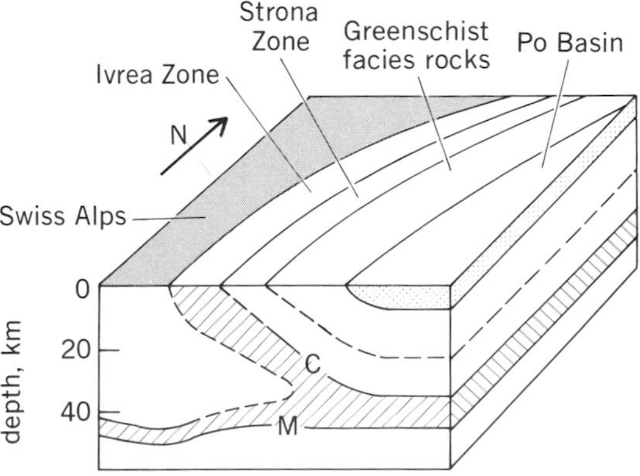

Fig. 1. Schematic diagram showing setting of the Ivrea and Strona zones in the southern Alps. Deep crustal structure deduced through analysis of seismic and gravity data shows the zones to be directly connected with the lower and middle crust under the Po Basin. The line at M is the Moho (base of the crust), and C marks the top of the lower crust as identified in that region.

rectly into the present lower crust of a given region. Thus these terranes represent cross sections of the crust apparently brought to the surface along great thrust faults in mountain belts. The best known of these is the combined Ivrea-Verbano and Strona-Ceneri zones in the southern Alps of Italy. Seismic refraction and gravity data demonstrate that the rocks exposed in these zones can be traced into the subsurface and are indeed continuous with the present-day lower and middle crust under the Po Basin (Fig. 1).

Excellent geologic mapping and petrologic investigations in the Ivrea and Strona zones permit construction of a schematic cross section (Fig. 2*a*) which illustrates several important petrologic characteristics of the continental crust. The cross section shows a zonation in metamorphic grade, with the highest-grade rocks (granulite facies) comprising the lower crust. Two major rock types within this level are garnet-bearing mafic granulites and intermediate-composition, aluminous gneisses. The latter represent aluminous sedimentary rocks metamorphosed under high-grade conditions. Enclosed within these are mafic and ultramafic complexes which were probably intruded into the deep crust of this area early in its history. Thus the lower crustal levels are composed of both metamorphosed igneous and sedimentary rocks which range widely in chemical composition. The presence of metasedimentary rocks in the deep crust implies that crustal formation involved transport of surficial materials to great depths. Other rocks at this level may be residual phases (restites) remaining in the lower crust following a partial melting event. Rock units in the Ivrea Zone are interlayered and folded into a complex geometry. Also evident are both vertical and lateral changes of rock composition over relatively short distances.

Midcrustal levels range from mid- to upper-amphibolite-facies metamorphic rocks and host unme-

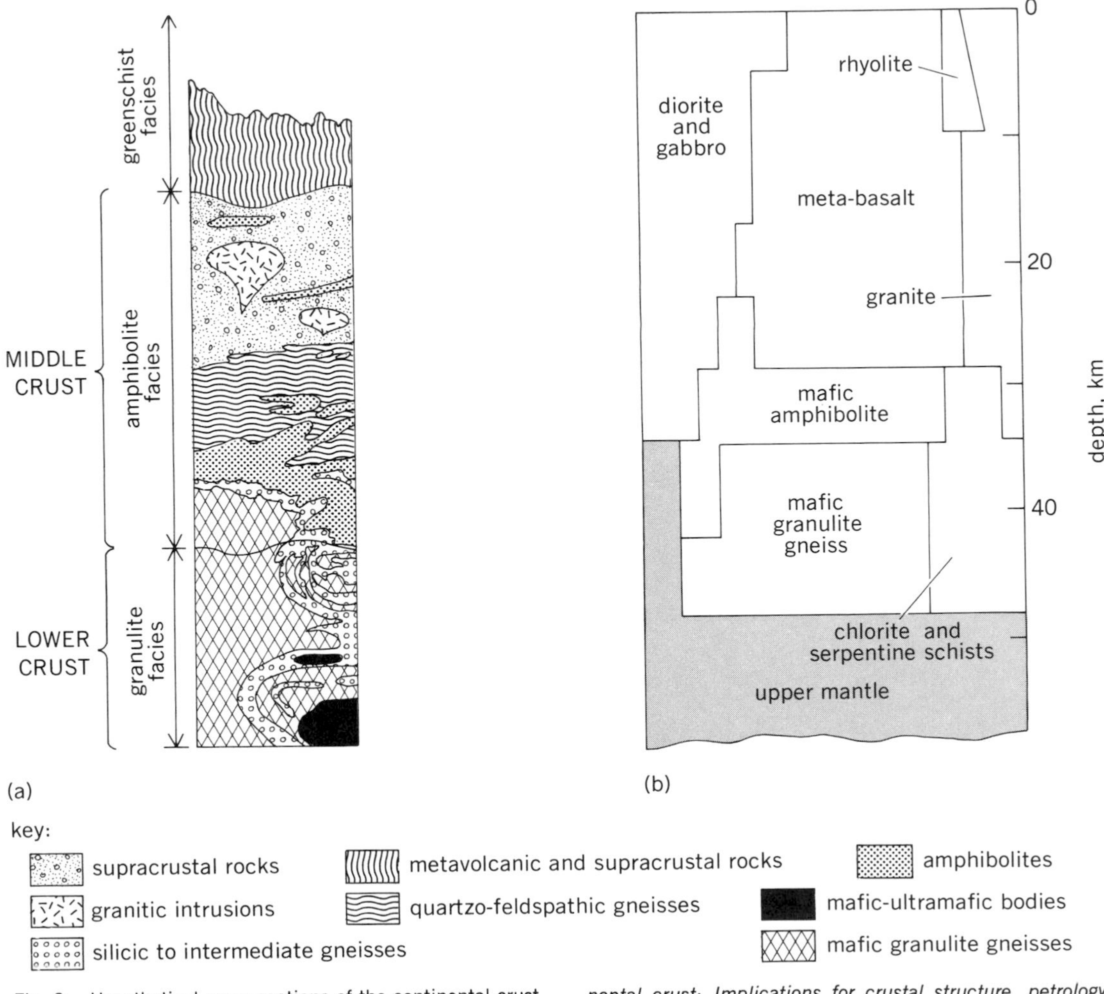

Fig. 2. Hypothetical cross sections of the continental crust (*a*) based on exposures of rock in the Ivrea and Strona zones of the Alps and (*b*) under the Colorado Plateau based on xenoliths collected in the region. The area of rock types in each cross section is proportional to their assumed volumetric abundance in the crust. (*Part a from D. M. Fountain and M. H. Salisbury, Exposed cross-sections through the continental crust: Implications for crustal structure, petrology, and evolution, Earth Planet. Sci. Lett., 56:263–277, 1981; part b from T. R. McGetchin and L. T. Silver, A crustal-upper-mantle model for the Colorado Plateau based on observations of the crystalline rock fragments in the Moses Rock Dike, J. Geophys. Res., 77(35):7022–7037, 1972*)

tamorphosed granitic plutons. Hornblende- and pyroxene-bearing gneisses of apparent igneous origin are common within the granulite-amphibolite-facies transition zone. Marbles and aluminous schists, once a sequence of limestones and shales, and prevalent above that transition, are complexly folded and migmatic in many places. The rocks are interlayered, and individual units are laterally continuous for only a few tens of kilometers.

The general picture of the deep crust that emerges from analysis of the Ivrea and Strona zones is one of great variations in rock composition on all scales in both horizontal and vertical directions. Superimposed on these compositional variations are mineralogic changes induced by increasing metamorphic grade with depth, thus producing a petrologically complex crust. Unmetamorphosed igneous bodies rest within this matrix of metamorphic rocks. These general themes are also evident in other high-grade terranes which constituted the deep crust at some time in the past.

The Ivrea Zone is only one example of deep continental crust and therefore may not be representative of the crust in general. However, its existence suggests that there may be other such terranes exposed in orogenic belts, and identification of those will provide the basis for more refined crustal models in the future.

Xenolith studies. Xenoliths are rocks which were mechanically dislodged from the walls of magma conduits and transported upward when molten material rose through the crust and upper mantle. They are regarded as direct samples of the levels through which the magmas passed. Although xenoliths represent a scrambled sample, they are very useful in defining characteristics of the deep crust. Notably, lower crustal xenoliths are usually high-grade metamorphic rocks (generally granulite facies) and tend to be basaltic in composition, although a wide range in rock composition is found. Igneous and metaigneous rocks are preponderant in xenolith suites, but metasedimentary rocks are usually present. The diversity of rock types in xenolith suites suggests that the crust beneath any one area is likely to be vertically and laterally heterogeneous. Metasedimentary rocks in xenolith collections imply that surficial rocks were carried down to deep crustal levels as an integral process of continental crustal formation. Although many investigators regard the lower crust as anhydrous, the occurrence of hydrous minerals such as hornblende in high-grade xenoliths suggests water may be an important constituent of the lower crust in some regions.

Specific tectonic environments appear to be characterized by different suites of xenoliths. Samples collected in intraplate areas are compositionally diverse with metasedimentary rock types subordinate to metaigneous types. Basaltic composition rocks are common within xenolith suites from magmatic arcs along converging plate boundaries. Rift valley suites consist of granulite facies rocks, basaltic composition intrusive rocks, and some restites.

Fig. 3. Petrologic and structural model of the continental crust derived by S. B. Smithson and coworkers. Shaded areas represent large bodies of igneous rocks. (*From S. B. Smithson, P. N. Shive, and S. K. Brown, Seismic velocity reflections and structure of the crystalline crust, in J. G. Heacock, ed., The Earth's Crust, Amer. Geophys. Union Monogr. no. 20, pp. 254–270, 1977*)

Such variations support the idea that there are many mechanisms of crustal formation and evolution which leave a distinct imprint on crustal petrology.

Because xenoliths are scrambled out of their original sequence in the crust during transport to the surface, it is difficult to reconstruct a crustal cross section from the samples. However, T. R. McGetchin and L. T. Silver constructed a hypothetical crustal model for the Colorado Plateau (Fig. 2*b*) by stacking the xenoliths on the basis of size and degree of abrasion. In general, this cross section reinforces many of the points presented above.

Interpretation of seismic reflection data. Recent application of seismic reflection methods to the study of deep continental crust by the Consortium for Continental Reflection Profiling (COCORP), a multiuniversity project headquartered at Cornell and funded by the National Science Foundation, provides data which support the concept of crustal heterogeneity and complexity discussed above. Profiles of several regions show numerous pronounced seismic reflections in the deeper portions of the crust. There are probably many causes of these reflectors. Some possibilities include magma bodies, large low-angle fault zones, large folds, and considerable variation in rock composition over small vertical distances. S. Schilt and coworkers conclude that the nature of the seismic reflections indicates that the crust is petrologically heterogeneous at scales of a few to tens of kilometers. The type of petrologic and structural complexity found is comparable to that displayed in high-grade terranes such as the Ivrea Zone, although it is probably premature to assign lithologies to specific reflecting horizons.

S. B. Smithson and coworkers recently attempted to derive models of deep crustal structure and petrology through interpretation of seismic reflection profiles. Their approach has been to generate theoretical profiles by assigning seismic velocities to rocks in known structures and then comparing those to real profiles. One model, presented by Smithson, P. N. Shive, and S. K. Brown (Fig. 3), shows many of the petrologic features anticipated by studies of xenoliths and the Ivrea Zone. In this model the middle crust consists of migmatites, granitic gneiss, and augen gneiss. Lower crustal levels consist of heterogeneous assemblages of granulite facies rocks of overall intermediate composition surrounding numerous igneous bodies.

Conclusion. The current picture of the continental crust is that it is heterogeneous on all scales, with rock composition changing both laterally and vertically, and with metamorphic grade increasing downward in most cases. The middle and lower crusts are variable in composition on a regional and global scale, implying that crustal evolution is a complex process and probably differs from region to region. The present state of crustal research does not permit calculation of average middle and lower crustal composition or construction of a single petrologic model to describe the continental crust of the Earth. Recognition of additional exposed cross sections, coupled with xenolith investigations and geophysical exploration, will ultimately lead to an understanding of the complexities of crustal petrology and evolution.

For background information *see* GEOPHYSICAL EXPLORATION; PETROLOGY in the McGraw-Hill Encyclopedia of Science and Technology.

[DAVID M. FOUNTAIN]

Bibliography: R. W. Kay and S. M. Kay, *Rev. Geophys. Space Phys.*, 19(2):271–297, 1981; T. R. McGetchin and L. T. Silver, *J. Geophys. Res.*, 77(35):7022–7037, 1972; K. R. Mehnert, *Neues. Mineral Abh.*, Part 2, 125:156–199, 1975; S. Schilt et al., *Rev. Geophys. Space Phys.*, 17(2):354–368, 1979; S. B. Smithson, P. N. Shive, and S. K. Brown, Seismic velocity, reflections, and structure of the crystalline crust, in J. G. Heacock (ed.), *The Earth's Crust*, Amer. Geophys. Union Monogr. no. 20, pp. 254–270, 1977.

Photographic materials

In recent years the cost of silver halide–bearing photographic materials has been severely affected by the unprecedented price fluctuations of silver. Businesses like the printing industry that are heavily dependent on silver products must seriously consider alternative nonsilver photographic processes. Systems that are available or under development include those based on diazo, vesicular, photopolymer, metal etch, photodelamination, and electrophotographic processes. These systems are expected to have a major impact on industrial photography in the 1980s.

The major disadvantage of nonsilver systems is that they are much less sensitive to light than silver-imaging systems are (see table). Although it is not practical to use nonsilver systems as camera films to photograph original copy, they can be used in applications requiring the copying of an original while it is held in intimate contact with the unexposed light-sensitive film. Powerful light sources with energy output in the ultraviolet region must be used for exposure because of the relatively low light sensitivity of most nonsilver films. Uses for such a process include reproduction of engineering drawings, reproduction of microfilms, production of phototools in printed circuit board manufacture, and image assembly in the production of printing plates.

Diazo. Diazo processes have been used extensively in a variety of commercial photographic papers and films. The process depends on the light-induced decomposition of diazonium salts and subsequent dye formation in an alkaline environment. In practice, a diazo coating on a film base is exposed to a high-intensity light source followed by development with aqueous ammonia to give a colored dye image in unexposed areas. Exposure and development can be done automatically in a simple, low-cost, tabletop processor. The process is positive-working, so that a copy of the original is produced. Different-colored dyes account for the variety of products seen in the marketplace. Diazo films are used extensively in microfilm applications, in printed circuit board manufacture, and to a small extent as graphic arts films in the printing industry. Advantages of these films are lower cost, excellent resolution, room light handling, and rapid processing. Disadvantages include limited shelf life of the unexposed films, long exposure times, the ammonia smell, lack of dot etching, and long plate exposures in critical printing applications.

Vesicular. One of the earliest nonsilver films to be used commercially was a vesicular film. In this product, exposure decomposes a diazonium compound, producing nitrogen gas which, when the film is heated during development, expands within the polymeric coating creating a light-scattering bubble image. Unexposed areas in the film are fixed by a final overall exposure, producing nitrogen gas, which slowly diffuses out of the film.

Although light-scattering images produced by this process have been used extensively in microfilm applications for projection viewing, little use has been made of them in the graphic arts. The process is completely dry, but the utility of the system appears to be limited by the characteristics of the image.

Photopolymer. Light-induced polymerization can be used to change properties such as solubility, adhesion, or tackiness. The process has been used successfully in printing plates, proofing products, and photoresists. Solubility changes by photopolymerization have been used to produce nonsilver graphic arts films.

A commercially available photopolymer contact film has a lightly matted black photopolymer coating which hardens on exposure; it is processed by being wet with dilute alkali, rinsed with water to remove unexposed coating, and then dried (illustration *a*).

In graphic arts applications the advantages demonstrated by this film include room light handling, very rapid processing, environmental compatibility, excellent dimensional stability, and wide exposure and development latitude. Although it does not perform equivalently to silver products when exposed through the base or out of contact, many of the use requirements of silver products are provided in this

Approximate energy requirements of photographic systems

Exposure index	Energy, ergs/cm^2	Amplification factor	System	Resolution, l/mm
	0.0001			
10^4		100–1,000	Fast silver recording films	50
	0.001			
10^3		1	Human eye	10
	0.01			
10^2		100–1,000	Common silver camera film	50–200
	0.1	10	Direct-writing silver	50
10		1,000–1,000,000	High-contrast films	200–800
			Litho camera films	200–800
	1	10–100	High-speed dry silver	200–800
		10	Electrostatics	8–10
1		10	Silver print-out proof	100
	10	10–100	Conventional dry silver	50–200
0.1				
	100			
0.01				
	1,000	1,000–10,000	High-resolution silver	1,000–2,000
10^{-3}				
	10,000	10–1,000	Photopolymer	20–100
10^{-4}				
	100,000	10–100	Photochromics	100–2,000
		10–100	Bichromated gelatin	100–1,000
10^{-5}				
	1,000,000			
10^{-6}		10–100	Vesicular	200–500
	10,000,000			
10^{-7}		10–100	Diazo	1,000–2,000
	100,000,000			

nonsilver system. Currently available films are negative-working—that is, a positive original copies as a negative image—but positive-working films are under development.

Metal etch. A number of companies have announced the development of photopolymer-resist, metal-etch films for the graphic arts. These developments are closely related to the photopolymer wash-off technology discussed above, as well as to the photoresist technology used in metal etching. In one approach polyester base is covered with a vacuum-deposited layer of bismuth metal and overcoated with a photoresist layer. The photoresist layer is similar in composition to presensitized printing plates. Both positive- and negative-working systems have been demonstrated. In the negative-working process, exposure hardens the photoresist layer in image areas (illustration *b*).

The film is then processed by rinsing with alkali solution to remove the unexposed polymer, etching with acid (ferric chloride) to remove bismuth in clear areas, and washing and drying. The bismuth layer is black and opaque and becomes the halftone image when used in graphic arts.

The preferred composition in an alternate approach has an aluminum or aluminum-iron metallic layer instead of bismuth. This could have an advantage of allowing the removal of the photoresist layer and the metal in one step in the alkaline bath.

These products are still at the development stage, and therefore little is known of their actual performance.

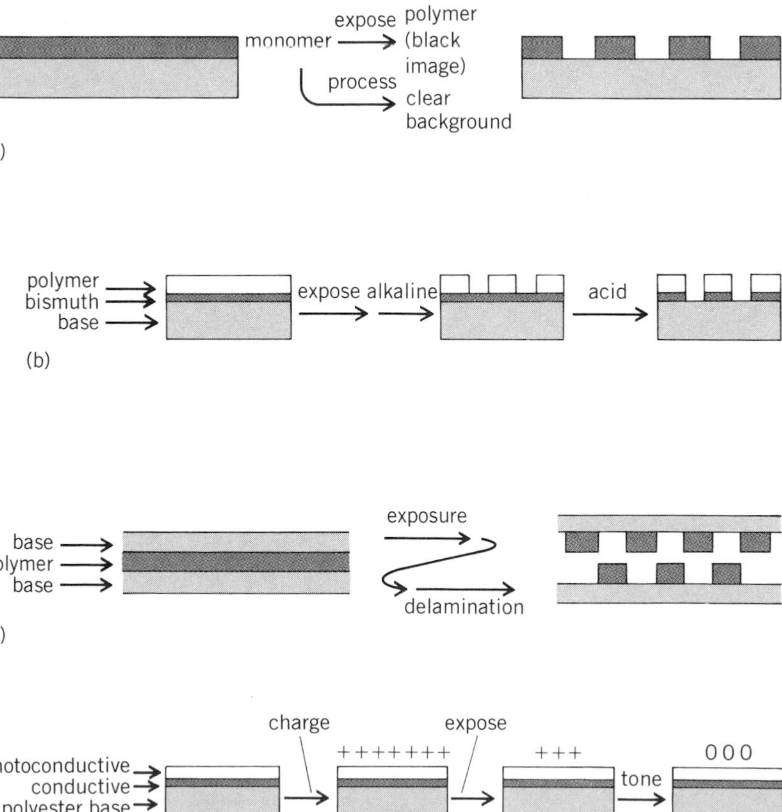

Nonsilver photographic processes. (*a*) Photopolymer. (*b*) Metal etch. (*c*) Delamination. (*d*) Electrostatic.

Delamination. In another approach, photopolymer technology is being applied to the development of multilayer structures which can be exposed and then delaminated to produce both positive and negative images in a completely dry process. The effect of the exposure is to change the adhesion of the photosensitive layer so that the interface at which delamination occurs changes as a function of exposure (illustration *c*). Products have been developed that can be peeled apart at room temperature, while others require hot temperatures to effect image separation. One product currently being tested as a graphic arts film consists of a thin metallized film which, after exposure, is immersed in steam or hot water prior to peeling. Although products based on this technology are in use today for contact line reproduction, they have not yet been successfully used in general graphic arts applications. Limitations seem to be image resolution, inconsistencies caused by the critical adhesion balance required, and operator dependence on the peeling process.

Nevertheless, the dry processing, general convenience, absence of processing chemicals, and pollution-free characteristics of delamination development make this an attractive process. Research to improve this process for use as graphic arts films is under way in the laboratories of several companies.

Electrophotography. This process has been applied extensively on the fringes of the graphic arts and most successfully in office copy, color copy, color proofing, and short-run printing or duplication. The process generally involves applying an electrostatic charge to the surface of a photoconductive material, exposing the image in order to dissipate the surface charge in the exposed areas, and applying toner to the residual charged areas to produce an image (illustration *d*). The final image is obtained by fusing the toners onto the support or by transfer and fusing to some other substrate.

Positive- and negative-working processes exist today, but the positive process predominates. Electrostatic systems are attractive because they are spinoffs from known technology, the rapid dry processing, and potential tie-in to electronics. However, these advantages may be offset by the likely high investment cost and, to date, inferior image quality.

In the near future the acceptability of nonsilver systems for contacting applications will increase because of the advantages of freedom from silver price fluctuations, room light handling, fast access, and simple processing. For camera work, however, complete replacement of silver is unlikely and hybrid silver-nonsilver systems are a possible compromise.

For background information *see* PHOTOGRAPHIC MATERIALS in the McGraw-Hill Encyclopedia of Science and Technology.

[JOHN L. W. JONES]

Bibliography: A. B. Cohen, *Amer. Printer Lithographer*, November 1979; J. Kosar, *Light-Sensitive Systems: Chemistry and Application of Non-Silver Halide Photographic Processes*, 1965; O. C. Rodenberg and L. E. Controls, *J. Appl. Photographic Engineering*, 5(4):244, 1979.

Plant growth

Plants, unlike most animals, increase in volume during most of their lifetimes. Roots proliferate, leaves originate and expand, shoots extend, and branches may develop to produce an ever-changing form.

Growing regions formed from changing cells. The growing regions of higher plants include the apical (tip) regions of roots and shoots and the vascular cambium, a ring of cells which in woody plants produces growth in girth. These growing regions often appear unchanging, or slowly changing, in time. A photomicrograph of a root tip 8 h after seed germination, for instance, may appear similar to a photomicrograph of a similar root tip 16 h after germination, but the cellular elements which compose the growing structure are displaced through it and experience continuing change. The root tip is composed of a procession of cellular elements, each of which may divide, enlarge, and undergo the biochemical and cytological changes associated with maturation.

The hooked stem found in seedlings of dicotyledonous (broad-leaved) plants is a heuristically useful example of an unchanging structure composed of changing elements. When a dicotyledonous seed germinates in the soil, the shoot grows upward with a curved stem which is thought to protect the plant apex from being damaged by sharp particles during soil penetration. After the stem emerges from the soil, light usually stimulates opening of the hook and subsequent straight growth. If the seedling is maintained in dim light, however, the hook can be retained during growth. (Bean sprouts purchased commercially have maintained their hooked stems for many hours of growth.) A simple marking experiment reveals the active nature of the process of hook maintenance: A mark placed on the apical side of the hook migrates through the bent structure and appears on the basal, straight portion in perhaps 8 h of growth. The mark is attached to a cell on the stem. The marking experiment implies that the hook is composed of a procession of cells, each of which curves and then straightens. At any one time the hook has elements which are increasing in curvature (on the apical side) and other elements which are decreasing in curvature (on the basal side). The result of this activity is a structure which at any fixed distance from the plant apex has approximately constant curvature but whose elements are continually changing in curvature.

The root tip and seedling hook are typical growth organs in that they are formed from changing cellular elements. This situation is reminiscent of a waterfall or the wake of a boat, structures which themselves do not change but whose material (real) elements keep changing. In fact, concepts and equations from fluid dynamics prove useful in analyzing plant development. In particular, flow-field kinematics, a description of motion and deformation of particles without reference to underlying forces or causes, can be used to solve many problems in plant growth.

Local and material specifications. The examples of the root tip and the seedling hook illustrate an important distinction made by fluid dynamicists who differentiate the local or eulerian specification from the material or lagrangian specification of a variable. This distinction corresponds in botanical terms to the difference between a site-specific and a cell-specific pattern. In the examples cited above, it can be said that local curvature change in the seedling hook is negligible but the material curvature changes are large. Similarly, the protein content encountered 4 mm from the root tip may be changing only slowly, while there is a large decrease in protein concentration experienced by the group of cells which are moving to a location more than 4 mm from the tip. In plants, if the apex is chosen as origin of the coordinate system, local changes are often small and cellular changes large.

The choice of apex as origin may be difficult to visualize, for cells must be imagined as moving back from the root or shoot tip while the intuitive idea is that the tip moves away from the nongrowing region. The advantage of choosing the tip as origin is similar to the advantage gained by choosing the tip of a boat as the (moving) origin in studies of boat wakes. The steady, or time-independent, nature of the pattern becomes obvious, and it can be seen that the pattern behind the boat is formed from water particles which change in a predictable way. Choosing the tip as apex also makes botanical sense because cell files originate near the apex. In this (moving) reference frame, developmental age can be correlated with position. The history of a cell can be inferred by looking at more apical cells, while the fate of the cell can be guessed by examination of cells further from the apex.

Cellular rates of change. Recognition that the cells continually change position relative to the plant tip leads to an appreciation that both the spatial pattern and movement during growth must enter into the calculation of cellular rates of change. The material or cell-specific rate of change of $\boldsymbol{a}$ can be related to the spatial pattern via an expression known as the material derivative of $\boldsymbol{a}$, symbolized $D\boldsymbol{a}/Dt$: [Eq. (1)], where t is time, and $\boldsymbol{v}$ is velocity

$$\frac{D\boldsymbol{a}}{Dt} = \frac{\partial \boldsymbol{a}}{\partial t} + \boldsymbol{v} \cdot \nabla \boldsymbol{a} \qquad (1)$$

Material derivative (cell-specific)	Local derivative (site-specific)	Convective change (displacive)

of displacement from a material point on the plant. This equation implies that cell specific rates of change of $\boldsymbol{a}$ can be evaluated if data are obtained on the spatial distribution of both $\boldsymbol{a}$ and growth velocity.

Quantitative growth descriptions. Physically, plant growth may be specified by a field of strain rate tensors. If tiny spheres were painted throughout a growing tissue, the spheres would deform during growth into larger ellipsoids. On a growing leaf, if growth in thickness is negligible, the problem reduces to the deformation of surface circles of unit area into ellipses. The major axis of each ellipse points in the direction of local maximum stretch rate, while the minor axis points in the direction of minimum stretch. The area of the ellipse gives the magnitude of the local strain rate. Such an analysis was performed on the cocklebur leaf at the half-expanded stage. The results indicated that for this leaf, growth is rather isotropic (similar in all directions) but heterogeneous (different in different regions), with less growth (local strain) at the tip and progressively more at the base of the leaf, which was growing 50% per day. On much of the growing root, the problem is one-dimensional. The growth rate distribution can be visualized as the rate at which tiny line segments painted on the root would lengthen. Mathematically, the strain rate in the root is given by the velocity gradient, which has been evaluated with an elegant photographic technique. Corn root growth analysis revealed that while the root as a whole grows at 2 mm per hour, only the apical centimenter is stretching. The maximum of about 40% per hour strain rate occurs at about 4 mm from the tip.

Physiological implications. Many physiological problems can be solved if information on the spatial distribution of growth velocity is acquired with information on the spatial distribution of another variable of interest. One application of growth kinematics is the derivation of growth rate distributions which produce observed curvatures. In the seedling hook, if A represents a point on the apical side of the hook, and B represents a point on the basal side, then the strain rate must increase from outside to inside at A, where tissue is increasing in curvature, but must decrease from outside to inside at B, where tissue is straightening. The difference in stretch rate between the inside and the outside can be related quantitatively to the local and displacive curvature change experienced by the moving tissue elements by Eq. (2). Here M is the strain rate at the

$$M(s, o) - M(s, i) = \frac{\partial}{\partial t}[\ln(1 + Kw)] + u\frac{\partial}{\partial s}[\ln(1 + Kw)] \qquad (2)$$

location s units from the tip on the outside (o) or inside (i) of the hook, t is time, K is local curvature, u is velocity of displacement from the tip, and w is the width of the stem.

Another application is the calculation of local production rates in growing tissues. The continuity equation, which is a statement of the conservation of mass, indicates that the rate at which a substance is produced locally (by biosynthesis, or transport and deposition) equals the observed rate of concentration change, plus an amount (the existing concentration times the local stretch rate) to compensate for growth dilution, plus a third term (the concentration gradient times the local growth velocity) which represents the convective derivative of the substance. To calculate a net deposition rate, one must measure the spatial distribution of the substance plus the spatial distribution of the growth velocity. These two sets of data can be used with their spatial derivatives to compute the net production rate. Thus, if growth data are collected at the same time as biochemical or cytological data, net synthesis rates can be computed.

The recent availability of low-cost interactive computer graphics equipment implies that the growth data is now relatively easy to analyze. Much progress is being made in evaluation of material (cell-specific) rates of change and local production rates in growing tissues.

For background information *see* PLANT GROWTH in the McGraw-Hill Encyclopedia of Science and Technology.

[WENDY K. SILK]

Bibliography: R. O. Erickson and W. K. Silk, *Sci. Amer.* 242(5):134-151, 1980; W. K. Silk and R. O. Erickson, *J. Theoret. Biol.*, 76:481-501, 1979.

Plant hormones

The complex processes leading to the growth and development of higher plants are controlled by a variety of plant hormones, the best studied of which are the auxins, cytokinins, gibberellins, abscisic acid, and ethylene. Other compounds with hormonelike properties occur in plants, and they also may function to regulate developmental processes in plants. There are several peculiarities of plant hormones, one being that different hormones may control the same biological process, for example, cell elongation, in different plants or even in different tissues or developmental stages of the same plant. In addition, unlike most animal hormones, all the major plant hormones are compounds of relatively low molecular weight (<500 daltons).

With the exception of ethylene, plant hormones occur naturally in the form of the free, active hormone and also as chemically bonded conjugates with little or no biological activity (see table). The process of conjugate formation and release may be the mechanism by which plants exert fine control over the concentrations of these important regulatory compounds (see illustration). For example, light, which inhibits the rate of cell elongation in dark-grown grass seedlings, causes a change in the ratio of free-to-conjugated auxin. Since the level of free auxin controls growth in this plant tissue, it is this shift to more conjugated and lower amounts of free hormone which is responsible for the lower rate of cell elongation.

At least in the case of auxins, and probably gibberellins and cytokinins as well, the formation of conjugates is involved in transport of the hormone from one place to another within the plant. It is possible that the compound to which the hormone is linked (usually a sugar or amino acid) serves as a type of biological "ZIP Code" which determines the final destination of the hormone once it enters the transport system. In addition to targeting the destination of the hormone, conjugation also renders the compound immune to many of the enzymes which would normally degrade the free hormone. Thus, conjugation prevents further metabolism until the hormone has reached its site of action. As yet, there is no direct knowledge of the chemical mechanism by means of which the hormone promotes growth once it reaches its site of action. This detailed knowledge awaits further study.

Cloning of higher plants. The cloning of plants, that is, the production of many plants of the same genetic composition from individual cells or plant parts, has a long history in agricultural practice. The propagation of deciduous trees and vines by cuttings dates back to antiquity and is the most common method of cloning biological material in use today. Farmers in Afghanistan and Dutch gardeners used seeds of grain to aid in the rooting of such cuttings, taking advantage of a rich natural source of the auxin, indole-3-acetic acid. In modern nursery practice the synthetic auxins, indole-3-butyric acid and naphthalene acetic acid, are used in a similar manner, and commercial preparations of these plant hormones are available to the amateur gardener at most nursery and gardening stores.

Growth of individual plant cells in culture is usually dependent on added auxins and cytokinins, and manipulations involving changes in type and amount of these two classes of hormones are the basis for nearly all plant tissue culture. Practical uses of plant tissue culture include embryo culture and aseptic seed germination techniques for the development of hybrid varieties, especially in the orchid industry. Since the early 1950s plant scientists have utilized cultures derived from the apex of plant shoots to produce virus-free plants, and there is interest in Europe and Asia in the use of tissue culture for the production of secondary products for medicinal purposes.

Currently over 300 different species of plants,

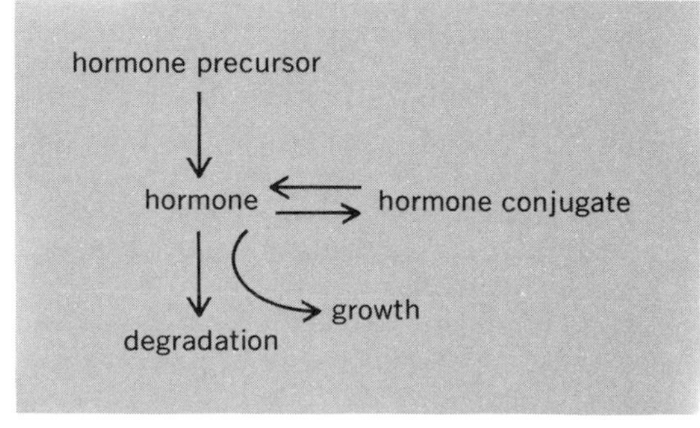

Scheme for the metabolic control of the level of plant hormones.

Some naturally occurring plant hormones and their conjugates

Hormone	Hormone conjugate
Auxin Indole-3-acetic acid 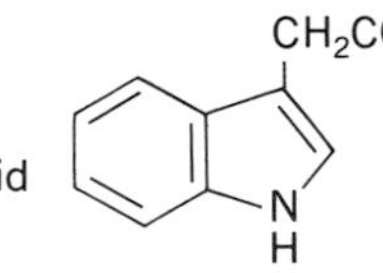 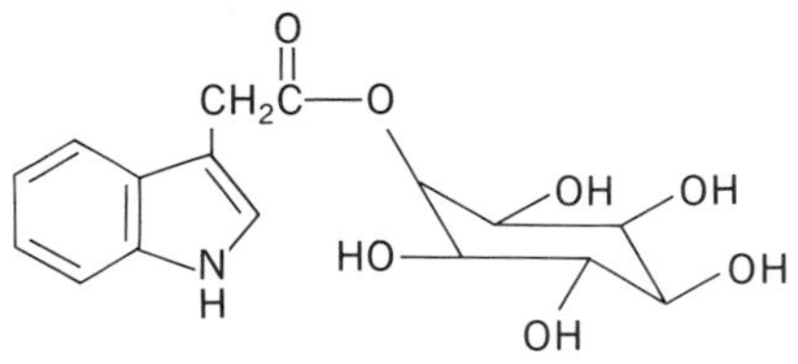 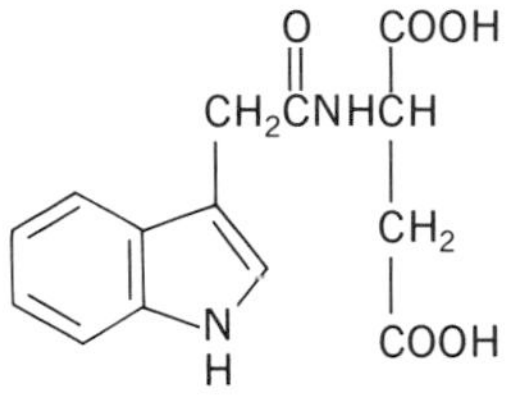	
Cytokinin Zeatin 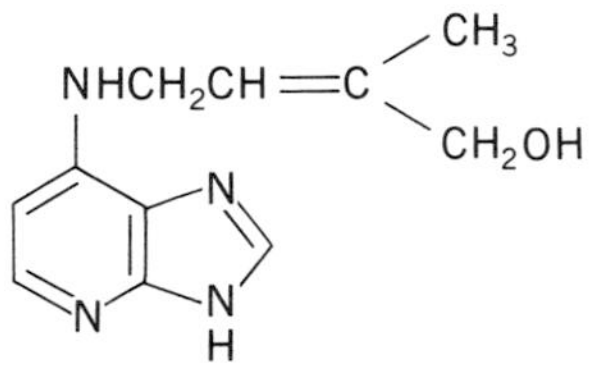 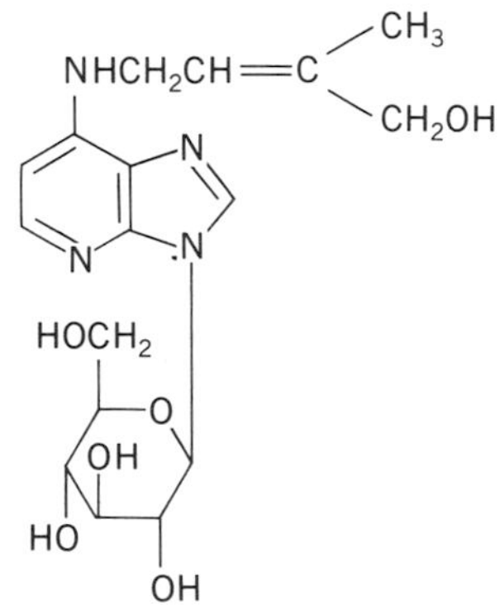	
Gibberellin GA_3 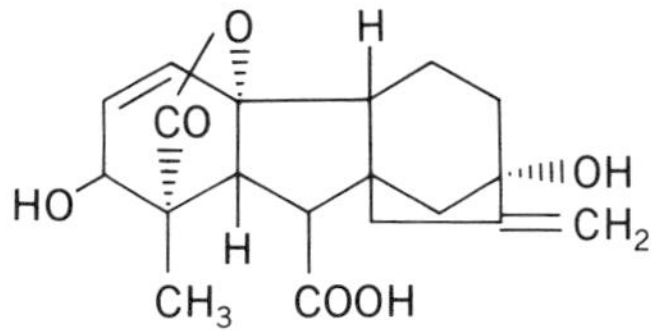 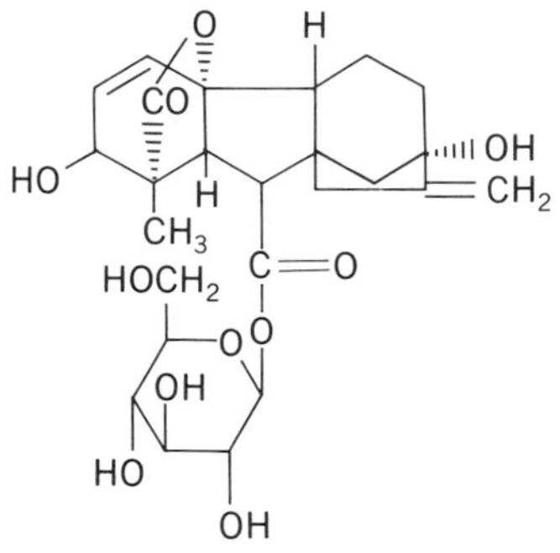	
Abscisic acid 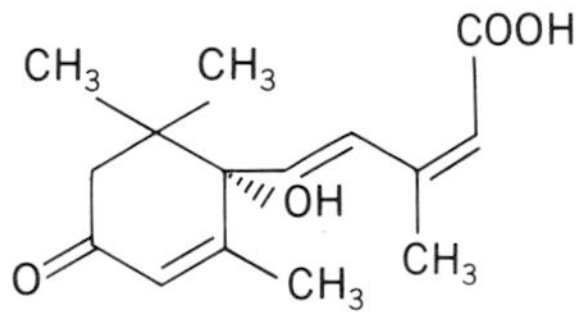 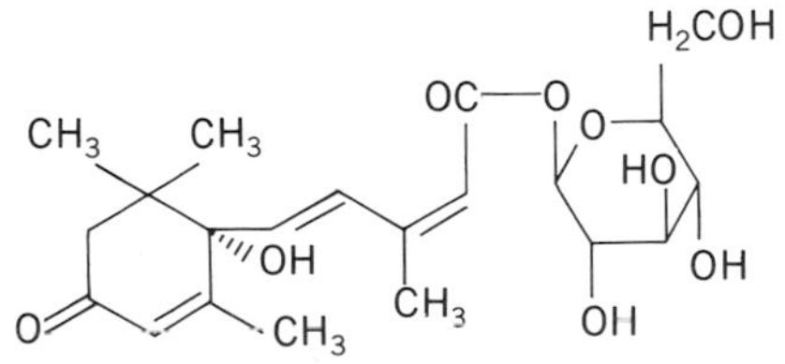	

ranging from ferns to forest trees, are propagated clonally via tissue culture. Under development are methods for the production of "clonal seeds" by somatic cell embryogenesis. From a single culture flask, thousands of embryos can be produced. These embryos could be packaged in a protective coat of nutrients and hormones for later planting in the soil (a kind of manufactured seed), or the germinating embryos could be planted into the soil directly from the test tube. These methods show promise for situations in which rapid introduction of new and better varieties of plants is desired, and they are especially interesting to the forest and orchard industries, where classical breeding programs have required years or even decades for the introduction of a new variety.

Small pieces of many plants can be cultured on nutrient media containing sugars, mineral salts, vitamins, and suitable plant hormones (auxins and cytokinins). These tissues grow in a random fashion, forming a clump of cells known as a callus. The callus can be divided, and each piece will form more callus when placed on fresh media. With most tissue lines, such manipulations can be carried out indefinitely. With some plant species, the amorphous callus can be induced to form recognizable tissues. For example, with low levels of auxin and high amounts of cytokinins, bud formation can be induced; the obverse leads to root formation. Thus, regeneration of the entire plant from callus can be accomplished by suitable hormonal manipulation. However, many agronomically important plant species have not yet been regenerated in this manner, and more experimentation is necessary.

Formation of new plants from callus and tissue pieces is an important advance, but to take full advantage of the genetic systems elegantly developed for microorganisms, it is necessary to develop methods for handling single plant cells. This has been accomplished by growing plant cells in rapidly shaken cultures, where some species will proliferate as discrete, single cells. The best cell culture of this type, the sycamore maple (*Acer pseudoplatanus*), has been maintained as a single-cell suspension culture for almost 2 decades. Other plant species, however, do not form cell suspensions and instead grow in small, hard, peanutlike clumps which make them unsuitable for genetic investigations. By careful manipulation of the hormonal content of the media, changes in growth habit are possible. Most promising has been the recent introduction of the use of conjugated forms of plant hormones in plant cultures. Amino acid conjugates of the natural auxin indole-3-acetic acid has produced suspension cultures from species which had proved to be impossible to work with when grown with free hormone or with the synthetic growth regulators. Some tissue cultures are able to form their own hormone conjugates, and the efficacy with which they carry out this process may be the basis for ease of tissue regeneration.

Mechanism of cell elongation. Plants are able to grow by two basic processes: cell division, which increases the number of cells within the plant, and cell elongation, which results in increased cell length. Cell elongation is one of the best-studied processes controlled by plant hormones and is the primary process resulting in early seedling growth; during this stage it is regulated by the auxin level within the tissue. The mechanism by which auxins elicit cell elongation has been studied for over a century but is not yet understood. When the tip of a growing seedling (thought to be the source of auxin) is removed, cell elongation stops and does not continue unless auxin is supplied to the tissue. Within minutes after auxin is added to the plant, growth resumes. The speed with which this response occurs appears to be too fast to be explained by production of new cell proteins, and thus the mechanism is best explained by a hormone-induced change in the rate of existing metabolic reactions.

Growth, including both cell division and cell elongation, is a complex process involving changes in a multitude of biochemical events. Some of the rapid changes observed following auxin treatment are increased protoplasmic streaming, changes in the cell wall that allow cell elongation, and a lowering of the extracellular pH. Auxin treatment of seedling plant tissue results in changes in the physical characteristics of the plant cell wall so that it can elongate in response to the force generated by the plant cell's own turgor pressure. The biochemical processes which cause these changes in the physical properties of the wall are largely unknown, but the changes are the result of enzyme activities associated with the cell wall. Most evidence indicates that glucose polymers (like cellulose, but of a different chemical linkage from cellulose) are degraded during cell elongation and that this is related to the autolytic degradation seen in isolated cell wall preparations. Enzymes which are active in producing these changes in cell wall preparations are now being isolated and characterized. Recent evidence confirms these conclusions, since it has been shown that antibodies produced against cell wall proteins inhibit both autolytic degradation of isolated cell walls and the process of cell elongation.

Agricultural applications. The most widespread use of growth-regulating chemicals is for weed and vegetation control with synthetic auxins such as 2,4-dichlorophenoxyacetic acid (2,4-D) and 2,4,5-trichlorophenoxyacetic acid (2,4,5-T) and their homologs. In addition, synthetic auxins are used for floral thinning of overproductive orchard trees and also to prevent preharvest fruit drop. As already discussed, auxins are used widely for the propagation of woody cuttings because they stimulate rooting. The synthetic auxins also prevent early bud growth and are used to prolong the storage life of potatoes and also to reduce frost damage in grape by delaying bud break until the danger of cold conditions has passed.

Spraying fruit trees with gibberellins in the fall delays spring flowering and is used in regions where early frosts might occur. Gibberellins increase the number of female inflorescences in hops, resulting in higher yields, and they are also used to stimulate the malting process in barley. Gibberellins and, in some species, auxins are effective in aiding in fruit set, thus allowing fruit development even from flowers which are not pollinated.

Cytokinins are less often used than the other hormones in practical application. Sprays of cytokinins stimulate resting buds to grow in some crops, while gibberellins accomplish the same task in other species. Cytokinins also stimulate seed germination by overcoming natural seed dormancy mechanisms. Some varieties of apple show improved size and shape when treated with cytokinins; however, the effect is not consistent with different varieties.

Ethylene has been found useful for increasing the growth of figs and also for speeding fruit ripening. The compound ethephon releases ethylene and is used to hasten ripening of berries, tomatoes, and melons and also to aid in the harvest of olives by accelerating fruit drop.

Many other commercially available compounds mimic some of the effects of the natural plant growth regulators. As more is learned about the naturally occurring plant growth regulators, the hormones, devising new and more effective ways to control plant growth may become possible.

For background information *see* ABSCISIC ACID; AUXIN; BREEDING (PLANT); CYTOKININS; GIBBERELLIN; PLANT GROWTH; PLANT HORMONES in the McGraw-Hill Encyclopedia of Science and Technology.

[ROBERT BANDURSKI; JERRY D. COHEN]

Bibliography: N. B. Mandava (ed.), *Plant Growth Substances*, 1979; F. Skogg (ed.), *Plant Growth Substances*, 1979, 1980; R. J. Weaver, *Plant Growth Substances in Agriculture*, 1972.

Plant pathology

Plant diseases are caused by a variety of viruses and pathogens, such as fungi, bacteria, nematodes, and higher plants, and by a host of abiotic agents such as nutrient-deficient soil, acid rain, and air pollution. Three areas of study important to the understanding of plant diseases are discussed in this article: host-parasite interactions, the effects of plant exudates on the functioning of pathogens, and infectious hypovirulence.

HOST-PARASITE INTERACTIONS

An important subfield of plant pathology is the biochemistry of host-parasite interactions. Research in this subfield is based on the rationale that all disease processes are controlled by molecules in the host interacting with those of the pathogen. Understanding interactions at the biochemical level will aid in determining how plants resist plant pathogens, and ultimately will improve the crop productivity of the world. This approach parallels that taken by medical professionals who wish to understand and control human disease by studying the molecular interactions of the disease process.

In plants, however, the response to an invading parasite is not the same as that occurring in human or other animal systems. For instance, plants do not have an immune system. How, then, are pathogens specific for certain hosts? How do pathogens of plants cause disease and how do plants defend against the invading pathogen? The answers to some of these questions have come with new and expanding studies on the biochemical mechanisms controlling host-parasite interactions.

Phytotoxins. Many plant pathogens, especially fungi and bacteria, release phytotoxins that are in part responsible for some of the effects of plant disease. Some of the most biologically interesting toxins specifically affect the same plants or varieties of plants that are attacked by the pathogen that produces them (host-specific toxins). The host specificity of the toxin produced by the fungus seems to be related to the presence of receptor sites in the sensitive host. These receptor sites are proteins found in or on the cells of the plant host. When the fungal toxin contacts the receptor site protein, a series of events is triggered that leads to cellular death; eventually the entire leaf or plant may die.

Host-specific toxins are now known from at least 15 plant-pathogenic fungi, including *Helminthosporium maydis*, the causal agent of southern corn leaf blight that devastated the 1970 United States corn crop. This blight was one of the most destructive and costly plant diseases occurring over a short time. The main destruction was caused by the toxins produced by the fungus in its sensitive host. The host was corn carrying Texas male sterile (TMS) cytoplasm. This cytoplasm simultaneously carried sensitivity to the *H. maydis* toxins and male sterility. Switching to (N) cytoplasm corn resulted in resistance to *H. maydis* and avoidance of subsequent epidemics of southern corn leaf blight.

Besides the host-specific toxins, plant parasites produce a variety of phytotoxins that do not possess host specificity. These compounds belong to such diverse chemical classes as peptides (amino acid–derived), phenolics, terpenoids, glycosides, combinations of these classes, and many others. Thus, they have no structural features in common. They have diverse modes of action, are found in plants infected by the parasite that produces them, and usually cause one or more symptoms of the disease. Some plant diseases in which one or more toxins are involved in disease expression are Dutch elm disease, halo blight of beans, scald of barley, alfalfa wilt, Victoria blight of oats, stem canker of tomato, and eyespot disease of sugarcane.

Phytoalexins. Resistance to a toxin appears to be a passive response (lack of a toxin receptor) of the plant. However, a common response in plants, especially in the legumes, to invasion by a pathogen is the synthesis of phytoalexins. Phytoalexins are low-molecular-weight antibiotics capable of inhibiting the growth of microorganisms, including plant pathogens. About a hundred different phytoalexins have been isolated and characterized. A single plant species is capable of producing several structurally related phytoalexins, but the mechanisms by which phytoalexins inhibit the growth of pathogens have not been determined. Healthy plants do not normally contain detectable amounts of phytoalexins. Plants synthesize these compounds at the sites where invasion of the tissues is occurring. The cells producing the phytoalexins do so in response to the presence of certain molecules in the invading pathogen. These molecules have been called elicitors.

One such elicitor is a complex carbohydrate present in the cell walls of a plant-pathogenic fungus. Not only will this substance induce phytoalexin formation in the plant, but small fragments of the complex carbohydrate also possess elicitor activity. Applying an elicitor molecule from a pathogen prior to invasion by a pathogen may confer disease resistance by virtue of the formation of phytoalexins. The use of these compounds, or their related synthetic analogs, as a means of conferring protection on plants is currently in its infancy, but holds promise for the future.

The study of plant host-parasite interactions at the biochemical level provides new approaches for pest control such as use of specific phytotoxins as tools for screening plants for resistance to disease, use of phytotoxins as molecular models in developing new herbicides (weed control), and application of elicitors or elicitorlike molecules in conferring disease resistance to plants.

[GARY STROBEL]

EFFECT OF PLANT EXUDATES ON PATHOGENS

Many plant pathogens—the fungi, nematodes, and other organisms that cause plant disease—are profoundly affected by chemicals that the plant releases. This happens before the pathogen penetrates the plant, and often increases both the vigor of the attack and the severity of the resulting disease.

Plants release many kinds of chemicals into the environment as they grow. This release may occur with normal, healthy tissue, but more often is associated with injuries to plant tissue. All parts of plants are subject to injury by forces of humans and nature. Roots as well as foliage are constantly subjected to stresses and strains that result in microscopic ruptures in the surface layer of cells. Then chemicals from the interior leak out of the plant tissue into water drops on leaf surfaces, or into the water films that surround the roots. Materials that diffuse from inside the plant to the environment are known as plant exudates.

Chemicals also diffuse through unwounded surfaces, particularly where the walls of the epidermal cells meet the cuticle. There are also some special structures such as nectaries and hydathodes through which materials may be exuded; these are important in the functioning of specific plant pathogens. These structures also serve as avenues of entrance for pathogens into plants. For example, the bacteria which cause fire blight of apple and pears are carried by insects attracted to the sugars in the nectar of the flowers. Once deposited in this rich nutrient solution, the bacteria increase in number and penetrate from the nectaries into the fast-growing susceptible tissue of the young shoots. Here they cause lethal necroses. The bacteria causing cabbage black rot, *Xanthomas campestris*, and the bean anthracnose fungus, *Colletotrichum lindemuthianum*, both enter their hosts in drops of guttation fluid exuded from the hydathodes. These drops are later drawn back into the leaf as its dries, allowing the pathogens to enter the leaves and to grow and spread through adjacent vascular tissue.

The interaction between plant exudates and plant pathogens normally occurs in one of two general sites on the plant: the leaves and the root system. Since these two structures are very different and exist in an entirely different environment, they will be considered separately. First, however, it is necessary to look at plant pathogens to see what they do that is affected by plant exudates.

Plant exudates and airborne pathogen. Plant pathogens are microscopic organisms that can cause disease only if they are able to enter the plant tissue. Plant pathogens often have difficulty in penetrating the outer epidermal layers of the plant leaves, stems, fruit, and roots. It is in this process of penetration that plant exudates are of most importance.

Spores of some leaf-infecting fungi grow much better in drops of water placed on leaf surfaces than in drops placed on a glass slide. When water is applied to the leaf surfaces as by rain or dew, or if leaves are dipped in water, substances inside the plant diffuse out. Analysis of leaf leachings has shown that a wide range of organic and inorganic compounds, including sugars and most of the amino and organic acids, have been found to diffuse from plants in this way. Compounds of this kind stimulate the germination of spores of plant-pathogenic fungi such as *Botrytis cinerea*, which causes gray mold in many plants, and provide the energy that the fungus needs to penetrate the host tissue. Another example is the fungus causing anthracnose of pepper, which does not infect green fruit but on red fruit it forms the appressoria that are essential for penetration of the plant. Exudates from the red tissue contain sucrose which is stimulatory and amino acids, which inhibit infection in this instance. The effect of the sugar is predominant, however, and is responsible for the fact that infection occurs. The amounts and kinds of substances leaching into water from plant tissue increase with age of the foliage and are greatest as senescence approaches.

The nature of the fungus on the leaf surface also affects the nature of compounds produced by plants. Some germinating spores produce substances (elicitors) that stimulate a reaction of the host tissue in which phytoalexins are produced. These are very inhibitory to many fungi and prevent further growth of the fungus. For this reason, they are an important mechanism of disease resistance. Certain organisms known to be pathogenic in such tissues are unaffected by these substances, however, and are consequently able to develop and cause disease.

Root exudates and soil-borne pathogens. Roots grow in a complex environment out of sight and thus "out of mind." Plant exudates are extremely important in this environment, and those from seeds and roots have been studied to a much greater extent than those from foliage. In order to understand the effect of plant exudates on plant pathogens in the soil, it is necessary to understand the world in which they function. The soil in which plants normally grow contains vast numbers of microorganisms. In 1 g of soil, there may be several million bacteria, a similar number of actinomycetes, thousands of fungal spores, and hundreds of nematodes. Fungi and nematodes are the most common kind of plant pathogens in soil. Fungi persist in the soil in a variety of forms (propagules), including spores and mycelium, that are adapted to survive in the company of all other soil microorganisms. The spores may consist of single cells or of many cells. They range widely in size, but most are 0.01 to 0.001 the diameter of the root tip that moves through the soil. They are generally dark in color and have thickened walls that ward off attacks of other microorganisms. These microorganisms in the soil are dependent for energy on organic matter derived from plants that can carry on photosynthesis. Thus, soil microorganisms depend for their life on plant tissue or plant exudates or on excreta or remains of animals that, in turn, were dependent on higher plants. There are so many microorganisms and there is such intense competition in the soil that every bit of nutrient energy made available is quickly consumed and only those organisms able to attack the most resistant residues can remain active for long. In the absence of a readily available energy source, most soil microorganisms exist in a metabolically quiescent state ready to grow when a new substrate becomes available.

Consequently, when nutrients are added to soil, there is a flurry of activity as each microorganism competes for a share. Seeds and roots are major sources of nutrients in the soil, and they serve as a focus of microbial activity.

Spermasphere. Seeds placed in soil have minute amounts of nutrients in their coats, and more leaches from the seed as it takes up water from the soil and swells. As these nutrients diffuse from the seed, there is created a zone of increased microbial activity known as the spermasphere that may extend several millimeters from the seed coat. Most of the microorganisms that are stimulated to grow in this zone are saprophytes that use the nutrients available. A few, such as the fungus *Pythium*, however, are plant-pathogenic. When stimulated to grow in the spermasphere, they gain the energy needed to grow to and penetrate the seeds and the developing seedlings. Plant pathogens that develop in the spermasphere are most dangerous in wet soils and at low temperatures where plants develop slowly. Stimulated by exudates from the seeds, they multiply rapidly and colonize the seed. As the soil moisture increases, more soluble sugars are released from the seeds. Under such conditions, the extent of seed rot is proportional to the amount of carbohydrates lost from the seed. The dimensions of the spermasphere, that is, the distance that the effect of the seed exudates reaches from the seed itself into the soil, is illustrated by results of a study with the fungus *Fusarium*. Within 16–24 h after planting bean seeds in a moist soil, the chlamydospores of *Fusarium* were stimulated to germinate in a zone approximately 1 mm wide surrounding the seed. This is a zone 200 times greater than the diameter of the spore. Within 3 days, spores in an 18–20-mm-wide zone of the seed were affected by the exudates. It is to protect seeds in this situation that farmers often treat seeds with a chemical coating that provides a barrier of fungicidal material. This delays the development of the plant pathogens until the seedling is resistant to infection.

Rhizosphere. Roots that develop after the seed germinates start as a simple structure known as the radicle and ultimately develop into a complex root system by which the plant absorbs all the water and nutrients it requires. The growing part of this system consists of primary and secondary roots that ramify throughout the soil mass wherever water and nutrients are available to be absorbed and transported back to the growing shoot. Like the seeds from which they started, the growing roots release materials into the soil. As they do so, a zone of increased microbial activity develops around them. This zone of soil in which the plant affects the level of microbial activity was designated as the rhizosphere by Hiltner in 1904. A rhizosphere exists around every root of a plant, and the cumulative effect from all roots is easily detectable. The zone around the individual root is of the most significance to plant pathogens.

What is the origin of the root exudates that create the rhizosphere? Root growth, resulting from the division and elongation of cells located just behind the root cap, forces the cap onward through the soil. Despite the lubricating action of mucigel excreted by root cap cells, large shearing forces develop as the root cap is forced through the soil. Substantial amounts of organic debris, partially embedded in the mucigel, are left in the soil as cells are crushed and torn away. In a growing root tip, these cells are continually replaced with new cells which, in turn, are lost as root growth continues. More materials that may serve as substrates for the microbial community diffuse from the thin-walled immature cells elongating behind the root tip. As root hairs form behind the zone of elongation and later die, more nutrients are released into the soil. Further back along the root, where secondary roots break through the cortex of the primary roots, leakage again occurs and nutrients are made available that stimulate microbial growth.

Any plant pathogen or other soil microorganism in the path of the root tip comes under the influence of the root that develops and matures at that site. From it the microorganism derives energy needed to develop, and if it is a plant pathogen, to penetrate the root tissue. Root exudates consist of a variety of materials, including soluble sugars and amino acids, proteins, and virtually all other kinds of molecules present in the protoplast of cells. All are used by one or more of the organisms present in the soil.

Reaction of fungi to root exudates. Typical of many of the fungi that infect plant roots are members of the genus *Fusarium*. These fungi reproduce on roots, and some species grow on other substrates in the soil. They produce spores that may germinate immediately or they may form thick-walled resistant chlamydospores that persist for long periods in the soil. Root exudates provide the energy required for these spores to germinate and infect root tissue. This is a nonspecific stimulation that occurs around the roots of susceptible and nonsusceptible plants alike; it affects all microorganisms able to utilize the nutrient material from the root. The black root rot fungus, *Thielaviopsis basicola*, is a more fastidious fungus, however. It germinates in the vicinity of susceptible roots and responds not to amino acids and sugars but to long-chain fatty acids and lecithin.

Amount of root exudates. Many factors affect the amount of exudate given off by a root. Soil water content is an important factor in the production of exudates by roots and seeds, and in their utilization by plant pathogens and other microorganisms. In water-saturated soils, a reduction in oxygen levels increases the permeability of seeds and roots. The resulting increased leakage of metabolites, in turn, stimulates increased root pathogen activity.

Water is also the vehicle that carries energy-rich exudates from the root or seed into the soil. At moderate levels of soil moisture, fungal spores can germinate and the hyphae can grow through the soil to the root. Repeated wetting and drying of the soil results in a substantial increase in root exudation. Soil particles may shift during the drying cycle, wounding root hairs in the process.

Resistance of soils to root growth also affects root exudation. The fine particles of a moist clay loam soil slip aside easily and cause little injury as the root tips are forced through the soil by the growing root. The much larger particles of a sandy soil do not yield easily, and so the cells of the advancing root cap are bruised and torn away and the roots branch more frequently. The amount of organic matter in the rhizosphere increases dramatically, and the infection of roots by pathogens increases because of resultant stimulation.

Specificity of root exudates. The tomato plant is highly susceptible to the vascular wilt fungus *Verticillium*, whereas wheat is resistant. Root exudates from tomato have a much greater stimulatory effect on microsclerotia of this fungus than do those of wheat. This appears to be due to a quantitative difference in amount of exudates rather than a specific stimulant unique to the tomato plant.

There is only one plant-pathogenic fungus that appears to be stimulated solely by plants susceptible to its attack. The onion white rot fungus, *Sclerotium cepivorum*, forms small black spherical resting propagules in and around infected plants in the soil. A very low percentage of the sclerotia are able to germinate in natural soil. This does not indicate that they are dead or dormant, for they germinate readily if they are separated from soil and placed on an agar medium. This inhibition in soil changes dramatically when roots of a plant belonging to the genus *Allium* (includes onion, garlic, and leek) develop in the soil. The sclerotia in the rhizosphere of these plants germinate in large numbers and infect the root. What happens to trigger this reaction? How are roots of this single genus of plants able to induce the germination of the sclerotia of the pathogen and in a real sense be responsible for their own demise? The roots of this genus contain a nonvolatile water-soluble precursor which, when acted on by soil microorganisms, releases volatile metabolites that stimulate sclerotial germination. The active materials appear to be *n*-propyl and allysulfides and disulfides. The precursors of these are probably *n*-propyl and allyl cystines that are water-soluble and metabolized by bacteria in the soil. The specific response suggests that volatile compounds in some way remove a restraint on an endogenous metabolic pathway responsible for the inhibition of sclerotial germination in the soil.

As described, the survival of many plant-pathogenic fungi depends entirely on the chance that a susceptible root will pass their way, stimulate their germination, and provide a substrate within which the fungi can grow and increase free of competition with all of the soil microorganisms. This is true also for most nematodes that attack plants. Nematodes are small worms (eelworms) larger than fungi and bacteria, but still too small to be seen without a low-power microscope. Some nematodes persist in the soil in the form of cysts that contain large numbers of eggs. These nematodes are particularly dependent on root exudates in order to function. The eggs in the nematode cysts remain dormant in the soil for long periods of time. The exudates of roots of certain plants overcome this dormancy and stimulate the nematodes to develop through the larval stage and break out into the soil. They continue to respond to this chemical by moving through the soil to the source root (chemotaxis), and infection follows. Hundreds of chemicals have been tested in an effort to determine the nature of the active agent in the exudate. The compound sought is apparently active in extremely low levels and breaks down rapidly, and has not been identified. It was thought for some time that this was a specific mechanism by which pathogenic nematodes located roots of a susceptible host plant. But when plant breeders developed cultivars of potatoes resistant to the potato cyst nematode, they found these plants all stimulated the nematodes in the cysts. Moreover, some nonhost plants have been found to stimulate egg hatch whereas certain host plants have no effect on cysts of nematodes that are highly pathogenic to them.

Roots infected by nematodes of the genus *Meloidogyne* develop knotlike swellings. Such roots produce an abundance of root exudates that stimulate the development of the fungus *Rhizoctonia*. The root is then invaded more vigorously by mycelium of *Rhizoctonia* than it otherwise would. The result is more severe root damage than if either pathogen were acting by itself. Such an interaction is called synergism. A similar reaction was discovered when it was noted that bean and clover plants infected by certain viruses were more subject to a root rotting fungus, *Fusarium solani*, than when the virus was absent. When soil in which the virus-infected bean plants were growing was leached repeatedly with water, the root rot was not increased. This showed that something outside the root was responsible and that an increase in susceptibility of root tissue was not the mechanism involved. An assay of the leachates revealed an increased exudation of amino acids from roots of virus-infected plants. These amino acids, in turn, were shown to stimulate germination and development of spores of the root rotting fungus. This increased exudation occurred approximately 3 days after plants were infected with the virus and continued for about 12 days before it returned to normal levels.

Finally, there is yet another kind of plant pathogen able to function because of exudates of plant roots. Plants of the genus *Striga*, commonly known as witch weed, produce several hundred thousand seeds that remain dormant in the soil for many years. *Striga* is a serious problem in the tropics and semitropics when it is present, for when a plant approaches that produces the chemical strigol the dormant seed are induced to germinate. *Striga* then grows into the root and produces haustoria, by which it obtains nutrients. This obligately parasitic plant pathogen causes a great reduction in yield of plants such as corn, sorghum, rice, and sugarcane.

Plant exudates are diverse in nature and arise as a result of wounds to plant surfaces, and of natural leakage of organic materials generally from immature tissues of roots. On leaves and particularly on and around roots, plant exudates have a dramatic effect as they provide the nutrient energy that propagules of plant pathogens require to compete successfully in the rhizosphere and penetrate into plant tissue.

[J. E. MITCHELL]

INFECTIOUS HYPOVIRULENCE

Hypovirulent means less virulent than normal, and the phenomenon offers possibilities for control of some diseases. Either the genetic makeup of the pathogen or an infectious agent can be responsible for hypovirulence. An aggressive infectious hypovirulent agent could control a disease by infecting the entire normal pathogen population and reducing the disease to a minor irritation. Recently, infectious hypovirulence has received considerable attention in plant pathology because it might provide a way to control the chestnut blight fungus, *Endothia parasitica*. Infectious hypovirulent agents have been found in isolates of the chestnut blight fungus, and these agents may be helping to control the disease in Europe. Early research on the possibility of control of blight on the American chestnut in the eastern United States reveals much about the nature and limitations of infectious hypovirulence.

Chestnut blight was introduced into the United States from China and Japan in the early 1900s. The American chestnut was highly susceptible to the disease, and the blight fungus rapidly spread over the range of the species. The disease spread 10–20 mi (16–32 km) per year, no individuals were resistant to the disease, and efforts to eradicate or even slow its spread were not successful. By 1950 the disease was present throughout the natural range of American chestnut, and the host had been reduced from a premier timber species to a short-lived understory species.

Chestnut blight was first reported on European chestnut in Italy in 1938. This host was highly susceptible and the disease spread rapidly, but by 1953 some chestnut stands were reported to be recovering. Later, when abnormal isolates of the fungus were recovered from healing cankers, these isolates were considered the cause of healing. Less virulent than normal isolates, they were called hypovirulent. Some blight cankers initiated by normal isolates could be controlled by inoculating the canker margin with hypovirulent isolates. The hypovirulent isolates appeared to exclude the normal isolates. Subsequently, infection of isolates in culture demonstrated that the hypovirulent agents were infectious.

Infectious hypovirulent agents associated with plant diseases have thus far been viruses or viruslike particles (VLP). In the case of *E. parasitica*, the presence of double-stranded ribonucleic acid (dsRNA) has been associated with hypovirulence, and the dsRNA is presumed to be of viral origin.

Isolates. In some plant pathogens, infectious hypovirulence has been limited to a few isolates of the fungus, but in *E. parasitica* many isolates are infected. Normal isolates are readily infected only if they are of the same vegetative compatibility group. Infection occurs when hyphae of the two isolates contact each other, fuse, and exchange cytoplasm.

In chestnut blight, the role of infectious hypovirulence in relation to the progress of the disease remains to be established. The hope, of course, is that a virulent isolate is infected and becomes hypovirulent as a result of that infection. Another possibility, however, is that the fungus usually becomes infected only after it has been weakened by aging. This phenomenon, called senescence, is used by some scientists to explain why certain fungi lose their capacity for sexual reproduction and pathogenicity. Under unfavorable host or environmental conditions, it is hypothesized that the senescence syndrome becomes common. Accordingly, European chestnut may be recovering because the disease conditions are in some way less favorable for *E. parasitica* than are conditions with American chestnut in the United States.

Cankers. A French scientist suggested that the frequency of recovery of isolates with infectious hypovirulence from a canker was related to the amount of stem swelling and the depth of the canker. Thus, the more superficial the canker the more hypovirulent isolates present. Unfortunately no objective data have been presented to support this concept. In the United States, swollen, superficial cankers were reported by early workers as one of a variety of symptoms (Fig. 1). Recently, cankers on American chestnuts have occasionally yielded isolates with infectious hypovirulence, but the relationship of canker healing to abundance of hypovirulent isolates has not been confirmed.

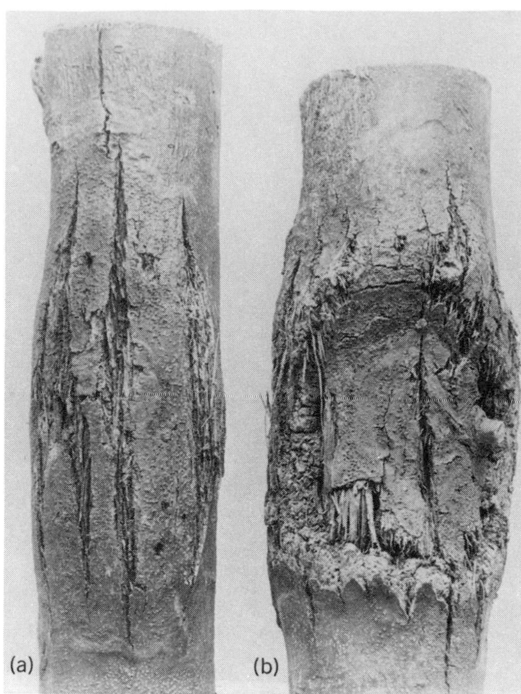

Fig. 1. Chestnut blight cankers on American chestnut illustrating *(a)* the swollen, superficial type of canker and *(b)* the sunken "normal" canker.

If superficial, swollen cankers are caused by isolates with infectious hypovirulence, inoculation of chestnut trees with these isolates should produce this type of canker. American isolates of *E. parasitica* with infectious hypovirulence are so debilitated by the infectious agent that they are essentially avirulent (without virulence) and no cankers are produced. Some European isolates with infectious hypovirulence are virulent, but no studies have reported the formation of swollen, superficial cankers. Therefore there is no confirmation that isolates with infectious hypovirulence cause these cankers.

Spread of isolates. Individual blight cankers on American chestnut trees have been controlled by inoculating the canker margin with isolates from the same vegetative compatibility group having infectious hypovirulence. This is one step in confirming the utility of infectious hypovirulence. However, the hypovirulent isolates must survive and spread to adjacent trees to ensure survival of the chestnut. In France, repeated annual inoculations with hypovirulent isolates are sometimes necessary to ensure spread of 1–2 m per year from the inoculated trees. In the United States, there has been no confirmation of spread of the isolates with infectious hypovirulence. A rate of spread of 1–2 m per year would be 16,000 times slower than the 10–20 mi (16–32 km) per year reported for the normal isolates in the United States and would seem to be noncompetitive. Fungus spores are the usual means of disease spread. Conidia, the asexual spores of *E. parasitica*, have only a small amount of cytoplasm present. Therefore cultures started from single conidia from isolates with infectious hypovirulence may be either normal or hypovirulent. No information is available about cultures from ascospores, the sexual spore stage, but the current hypothesis is that the sexual stage "cleans up" viral infections and would be less likely to retain the infection. The French believe infectious hypovirulence is spread in their orchards by ants feeding on the fungus mycelium. Insect vectors might circumvent the problem of the infectious factors in spores. The rate of spread reported from France supports the hypothesis of ants as vectors.

Survival is critical for any plant pathogen. The chestnut blight fungus has survived for decades in isolated locations to enable it to infect new sprouts or new host material. For biological control, isolates with infectious hypovirulence must survive as well as normal isolates. Evidence so far suggests that infectious hypovirulence decreases survival.

Occurrence. The widespread occurrence of *E. parasitica* isolates with the infectious element is intriguing. Infectious hypovirulence has been reported in other plant pathogens, including *Helminthosporium victoriae*, *Rhizoctonia solani*, and *Geumannomyces graminis*, but it was detected only in single or occasional isolates. The limited occurrence of infectious hypovirulence in these species did not provide opportunities for following the infection process among isolates. The presence of infectious hypovirulence in an isolate of *E. parasitica* changes its growth habit in agar culture in color, rate, or relative amount of aerial mycelium (Fig. 2). When the hypovirulence property infects a normal isolate, growth of the isolate is changed in a readily recognizable fashion (Fig. 2).

Infection. Infection depends upon vegetative compatibility or heterokaryon formation. In heterokaryon formation two hyphae contact each other, fuse, and exchange nuclei with the result that the hyphae have

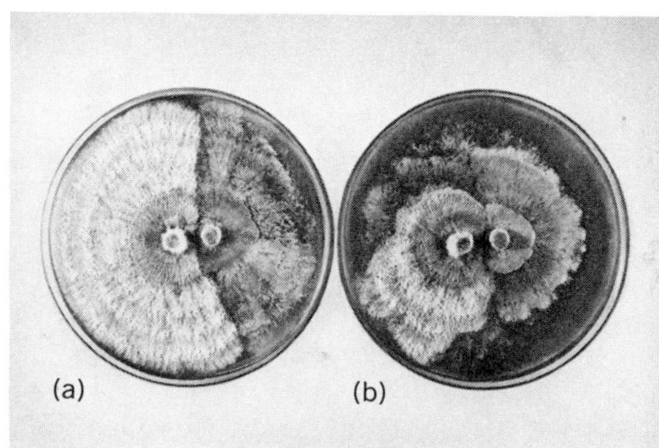

Fig. 2. Cultures of *Endothia parasitica* showing the growth habit of normal isolates and isolates with infectious hypovirulence. *(a)* Normal isolate with fluffy, light mycelium (left) paired with incompatible, hypovirulent isolate with a suppressed mycelium (right). *(b)* Infection and conversion of the growth habit of a normal isolate (left) by a compatible hypovirulent isolate. The normal isolate starts growing as on the left, but following infection by the hypovirulence factor subsequent growth is suppressed.

two genetically different nuclei. In infection with the hypovirulence factor only cytoplasm has to be exchanged, so that fusion may be brief, that is, interrupted by an incompatibility factor, but the normal cytoplasm is infected. Vegetative compatibility is an inherited trait and is generally recognized as a barrier to rapid spread of mycoviruses, VLPs and dsRNA. However, in *E. parasitica* infections occur 15–20% of the time even among apparently incompatible combinations. Infectious hypovirulence in *E. parasitica* should be a useful tool for studying the exchange of cytoplasmic material among isolates from different vegetative compatibility groups since the effect of exchange (infection) is so dramatic in culture.

Control techniques. Proof that infectious hypovirulence provides natural control of chestnut blight on either American or European chestnut trees would be an important step in the utilization of this phenomenon. This unique system of biological control would have applications in many other diseases that have defied conventional control techniques. The widespread occurrence of the infectious hypovirulence property in association with *E. parasitica* contrasts with scientists' failure to establish hypovirulent populations. If nature's successful methods are discovered, scientists may be able to utilize this phenomeon.

For background information *see* PLANT PATHOLOGY in the McGraw-Hill Encyclopedia of Science and Technology.

[E. G. KUHLMAN]

Bibliography: S. L. Anagnostakis and P. R. Day, *Phytopathology*, 69:1226–1229, 1979; G. W. Bowen and C. Theodorau, Growth of ectomycorrhizal fungi around seeds and roots, in G. C. Marks and T. T. Kozlowski (eds.), *Ectomycorrhizae: Their Ecology and Physiology*, pp. 107–150, 1973; M. G. Hale, L. D. Moore, and G. J. Griffen, Root exudates and exudation, in Y. R. Dommerques and S. V. Krupa (eds.), *Interactions between Non-Pathogenic Microorganisms and Plants*, pp. 163–203, 1973; R. A. Jaynes and J. E. Elliston, *Phytopathology*, 70:453–456, 1980; E. G. Kuhlman, in J. G. Horsfall and E. B. Cowling (eds.), *Plant Disease: An Advanced Treatise*, vol. 5: *Hypovirulence and Hyperparasitism*, 1980; W. L. MacDonald et al. (eds.), *Proceedings of the American Chestnut Symposium*, 1978; J. E. Mitchell, The effect of roots on the activity of soil-borne plant pathogens, in *Encyclopedia of Plant Physiology*, new series, vol. 4: *Physiological Plant Pathology*, pp. 104–128, 1976; A. P. Rovira and C. B. Davey, Biology of the rhizosphere, in E. W. Carlson (ed.), *The Plant Root and Its Environment*, pp. 153–204, 1974.

Pogonophora

Pogonophora are wormlike marine animals which build chitinous tubes. They are found in the deeper parts of the oceans, except for a few small species which reach up onto the continental shelf at high latitudes where the bottom water remains cold all year. In most pogonophores, the greater part of the tube and the animal inside lie buried in the mud of the sea floor, but members of the specialized subgroup Vestimentifera are found attached to rocks, usually near hydrothermal vents. All known Pogonophora lack an internal digestive system, and in this they differ from most other free-living metazoans. A species of polychaete, some oligochaetes, a bivalve mollusk, and certain brachiopods lack a gut or have a reduced gut in the adult stage, but these are rare and specialized examples when compared with the whole phylum Pogonophora. Recent studies suggest that bacteria in the cells of Pogonophora could be chemoautotrophic and might provide a supply of organic matter to their host. Such symbiosis with bacteria, and uptake of dissolved organic matter, are the two hypotheses currently favored to explain the mode of nutrition in Pogonophora.

Hypotheses. The first example of a pogonophore was discovered in 1914, and since then about 200 species have been described. There has been much discussion about the likely method of feeding in the absence of an internal digestive system (see table).

Particles. The first hypothesis, that feeding might still take place on particles even without a gut, was inferred from studies of the morphology of the group. It seemed a likely method for the multitentaculate species, and agreed with more general studies which seemed to show a preponderance of suspension feeders in the deep sea. So far it has not been possible to experiment on these large multitentaculate pogonophores, owing to the difficulty of collecting them alive. The hypothesis of particulate feeding has been rejected for the small species, since histological and histochemical evidence cannot be found for the external digestion, which would be necessary if particles were to be utilized.

Pinocytosis. The second hypothesis, direct epidermal uptake (or cellular ingestion) of small particles, remains a possibility for the large species, particularly the Vestimentifera, which expose their tentacles to the surrounding water. Experiments show that the epidermal cells of the small species can take in large molecules such as ferritin and ^{14}C-labeled proteins. However, the nutritional value of this process is doubtful since the small pogonophores appear to stay inside their tubes, through which there is little or no circulation of water.

Dissolved organic matter. The third hypothesis, use of dissolved organic matter, is quite supportable since pogonophores, as well as several other soft-bodied marine invertebrates, can absorb amino acids and other organic solutes through their skin from very dilute solutions. Experiments have been made on four of the smaller species. One of them, *Siboglinum ekmani*, more completely studied than the rest, would be able to meet its minimum food requirements (or maintenance metabolism), as estimated from oxygen consumption, by uptake of amino acids, glucose, fatty acids, and hydroxy acids from concentrations similar to those existing in

Summary of the hypotheses of feeding methods in pogonophores

Method	Hypothesis	Evidence
Particulate	Particles collected by the tentacles are digested externally. Soluble products are absorbed by the epidermis.	Deduced from morphology. No evidence for external digestion in small species.
Pinocytosis	Very small particles are taken into epidermal cells by pinocytosis.	Ferritin and ^{14}C-labeled proteins are incorporated into epidermal cells in small species.
Dissolved	Dissolved organic compounds are absorbed by epidermal cells.	^{14}C-labeled amino acids, sugars, fatty acids, and hydroxy acids are taken up by the epidermis and reach all parts of the body.
Chemoautotrophy	Bacteria inside cells use energy from oxidation of sulfide to fix CO_2 into organic compounds, some of which pass to the host.	Bacteria and enzymes for oxidation of sulfide and CO_2 fixation by the Calvin cycle found in Vestimentifera. Bacteria and some Calvin cycle enzymes are now found in small species also.

the mud in which it lives. Dissolved organic matter taken up this way may enter the amino acid pool (glycine), be converted to other compounds (glucose, alanine), or undergo metabolization to CO_2 and other volatile compounds (amino acids, glucose, fatty acids, hydroxy acids). From this it can be presumed that the pogonophore is feeding under the experimental conditions. However, the rates at which the compounds are taken in through the skin are not significantly greater than the highest rates for epidermal uptake by polychaete worms, which have a functioning alimentary system. Furthermore, it is not yet clear whether dissolved organic matter can provide enough food for growth and breeding, as well as for maintenance metabolism.

Chemoautotrophic bacteria. The fourth hypothesis is the most recently developed, though a suggestion of association with microorganisms was put forward in 1957. New ideas came from studies of the specialized animals which were discovered in 1977 living around hydrothermal vents in rift zones in the Pacific. The vents have a source of organic production independent of surface photosynthesis, through activity of chemoautotrophic bacteria which use energy derived from oxidation of sulfide to fix carbon dioxide and inorganic nitrogen. The vent water is rich in hydrogen sulfide and contains large numbers of sulfide bacteria. Most of the animals living around the vents are assumed to filter off and digest the bacteria. However, one of the most common animals around some of the vents is the very large vestimentiferan pogonophore *Riftia*, and the question has again arisen as to how such an animal could feed without an internal digestive system. Some of the clues found are: (1) particles of sulfur are present within the tissues of *Riftia*; (2) tissue extracts contain enzymes associated with oxidation of sulfide and fixation of CO_2 by the Calvin cycle; (3) numerous gram-negative bacteria are found in the same tissues; and (4) stable carbon isotope ratios in the tissues indicate that *Riftia* has a different food source from the filter-feeding clams and mussels at the same sites, and that this food source also differs from that of oceanic animals not associated with hydrothermal vents. From these findings it looks as if organic matter is produced by chemoautotrophic bacteria inside *Riftia* and some of it is passed to the animal. The energy for CO_2 fixation may come from oxidation of sulfide, and the pogonophores may help to supply oxygen for this process.

Carbon and energy sources. Recently, similar methods have been applied to the small pogonophores from the continental slope and the Norwegian fiords: some of them contain enzymes associated with fixation of CO_2 by the Calvin cycle, and gram-negative bacteria are present in the tissues at the hinder end of the body, where most Calvin cycle enzyme activity is found. Analysis of the stable carbon isotope ratios of one species shows the lowest proportion of ^{13}C so far reported for any marine animal, which suggests that dissolved organic matter from the mud is not a major source of food. It appears that internal chemoautotrophic bacteria are again supplying organic matter to the animal, but that they are using a carbon source which is already low in ^{13}C. This carbon source could be biogenic (respired) CO_2, or methane produced by anaerobic bacterial decomposition outside the animal. The energy source for CO_2 fixation is not yet established. The hind end of the animal, where the bacteria are, lies deep in the mud, and the hemoglobin-rich blood may supply oxygen for the chemoautotrophic reactions since there is a special arrangement of blood sinuses close to the bacteria-containing cells. Although chemoautotrophy seems likely on the basis of these findings, measurement of total CO_2 fixation in one species shows that carbon supplied in this way would be less than that provided by uptake of dissolved organic matter, though together the two methods of nutrition would provide more than the minimum carbon budget for the pogonophore.

Trophic diversity. A plausible interpretation of recent results is that the gutless Pogonophora may have evolved in a partly anoxic environment in association with a range of chemoautotrophic bacteria which are adapted to different sources of carbon and different supplies of energy. At one extreme there is the giant vestimentiferan *Riftia*, living in the sulfide-rich waters of the hydrothermal vents where the endosymbiotic bacteria may supply all the organic matter needed by the host. At the other extreme is the very small species *S. ekmani*, which can meet its minimum metabolic needs from uptake of dissolved organic matter, which has little or no capacity for fixation of CO_2 by the Calvin cycle, and in which the apparently small number of bacteria may function more as a detoxification mechanism. In between are *S. atlanticum* from the continental slope and *S. fiordicum* from the shallower parts of the fiords, both of which may combine uptake of dissolved organic matter with partial reliance on chemoautotrophic symbionts. The recent discoveries, first in the hydrothermal vents and now in more "normal" habitats, open the way to future studies of this fascinating relationship between marine invertebrates and chemoautotrophic microorganisms.

For background information *see* DIGESTION (INVERTEBRATE); FEEDING MECHANISMS (INVERTEBRATE); POGONOPHORA in the McGraw-Hill Encyclopedia of Science and Technology.

[ALAN J. SOUTHWARD; EVE SOUTHWARD]

Bibliography: H. Felbeck, *Science*, 213:336–338, 1981; M. L. Jones, *Proc. Biol. Soc. Wash.*, 93:1295–1313, 1981; A. J. Southward et al., *Nature, Lond.*, 293:616–620, 1981; A. J. Southward and E. C. Southward, Dissolved organic matter and the nutrition of the Pogonophora: A reassessment based on recent studies of their morphology and biology, *Kieler Meeresforsch.*, 1981.

Powder coating

Powder coating is a method of industrial finishing which uses dry powder resin rather than traditional solvent-based paints.

Powder coatings are used extensively in metal finishing on appliances, automotive applications, and a wide variety of metal parts. Because of the efficiencies of material utilization, energy conservation, and pollution and waste advantages, powder coatings are an important finishing alternative for the future.

Recent legislation regarding clean air, clean water, and hazardous waste makes powder coating very desirable. Because no organic solvents are used, there are little or no evaporative or baking by-products to be exhausted into the atmosphere. With traditional solvent-based systems, significant amounts of solvent-laden air must be exhausted from the spray application booth and oven. Many of these solvents are photochemically reactive, and the amount of exhaust is closely regulated via the Clean Air Act.

In powder coating, air is the only vehicle carrying the dry powder in the spray booth. Any oversprayed air-powder mixture can be filtered. This is so effective that the filtered air can be exhausted directly back into the plant, and the powder can be reused. Increased efficiency is achieved through both reduction of energy required for makeup air to replace large exhausts to the outside and in simple coating material utilization. In solvent-based painting, the overspray is trapped in dry filters or waterwash collection booths and must be disposed of. Similarly, because of the absence of organic solvents in powder coating, the ovens used do not require large exhausts for safety reasons, an additional energy-conserving feature.

Processes. Basically, all powder coating processes heat-fuse the powder to the substrate into a continuous coating. There are two primary application methods: fluidized-bed coating and electrostatic application. Application specifics such as the required film thickness, type of material required, part configuration, and materials of construction dictate the type of process used.

Fluidized-bed coating. This was the first major application technique used for applying powder in the United States. It consists of a tank separated into a plenum chamber and working chamber by a porous member too fine to allow the powder to pass, but adequate to give uniform passage of air (Fig. 1).

Powder is placed in the working chamber, and the plenum chamber is pressurized to 2–3 SCFM [standard cubic foot per minute = 1 ft^3 (0.0283 m^3) at 60°F (15.6°C) and 1 atm (101.325 kPa) pressure] per square foot (0.2–0.3 standard cubic meter per square meter) of porous plate area. The air diffuses into the powder in the working chamber, buoying it and increasing its volume by 50–75%. The powder is at ambient temperature, but actually looks like a boiling liquid.

The part to be coated is heated slightly higher

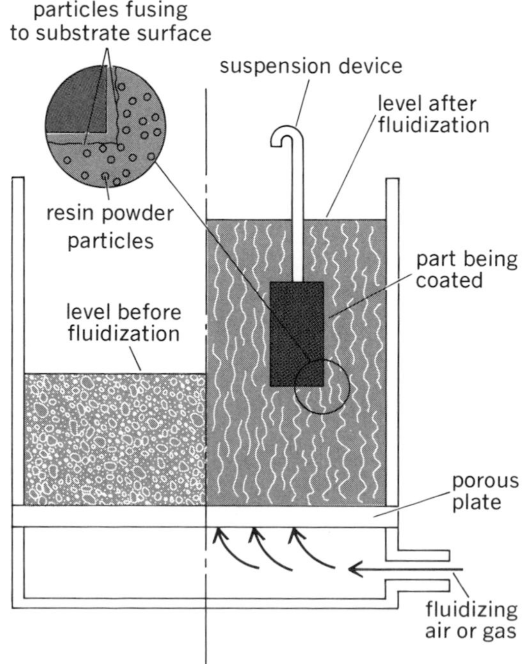

Fig. 1. Typical fluidized bed before (left) and after (right) fluidization.

than the fusion point of the coating material being used, and is immersed into the powder, which fuses onto the part and coalesces to form a smooth, even film. On parts with small mass, such as wire goods, additional heating may be required for complete fusion.

The process lends itself to coating intricately shaped parts with good uniformity in areas that are difficult to reach otherwise. An automated fluid-bed system provides excellent reliability.

The system is generally utilized for functional coatings of 10–15 mils (250–280 μm). About 8 mils (200 μm) is the minimum thickness obtainable, since 60–70% of the thickness is deposited in the first second of dip time. The total thickness is dependent upon part temperature and dip time in the powder.

Typical end uses are metal circuit boards, pole-mount and transformer covers, bus-bars, metal furniture, and fabricated wire. These applications require heavier film thicknesses.

Electrostatic powder coating. This type of coating can be achieved via the spray or electrostatic fluidized-bed technique. In both techniques the principle is basically the same. A high-voltage, low-amperage charge is applied to the powder. When these particles are directed to a grounded conductive part at room temperature, they cling to it because of the electrostatic potential. The part is then transferred to the baking oven, where the powder melts, flows, and cures to form a smooth, hard, inert finish. These processes are capable of applying film thicknesses ranging from 1.0 to 15 mils (25 to 380 μm). For this reason, most traditional painting applications use electrostatic techniques.

The amount of powder applied to a part is a function of the material being used and the design and operation of the system. Any powder that does not become attached to the part is collected as overspray and recycled to the feed hoppers for reuse. This efficiency makes material utilization about 98%.

Collection is accomplished by filtering the powder-air mixture via one of a number of ways using cloth or paper filters. The filtered air is clean enough to be exhausted back into the working area.

Spray operations can be achieved through directed spray guns or through a disk positioned in an omega conveyor loop (a conveyer with a disk applicator). Both gun and disk operations are similar to conventional liquid spray and disk operations.

The electrostatic fluidized bed incorporates the two processes of fluidized-bed and electrostatic application (Fig. 2). An electrode system is placed in the working chamber of a fluidized-bed tank. This grid imposes a charge on the powder which is attracted to any grounded part passing over the bed.

The process is especially applicable to continuous coating of sheet, screen wire, and two-dimensional objects. Since the effective throwing height over the bed is only 2–3 in. (50–75 mm), this system is not used for parts with a greater height.

Pherostatic systems have been introduced which utilize a combination bed to coat more complex objects such as capacitor cans and three-dimensional objects (Fig. 3). Coatings in the 0.7–3.0-mil (18–75-μm) range can be applied with relatively high line speeds.

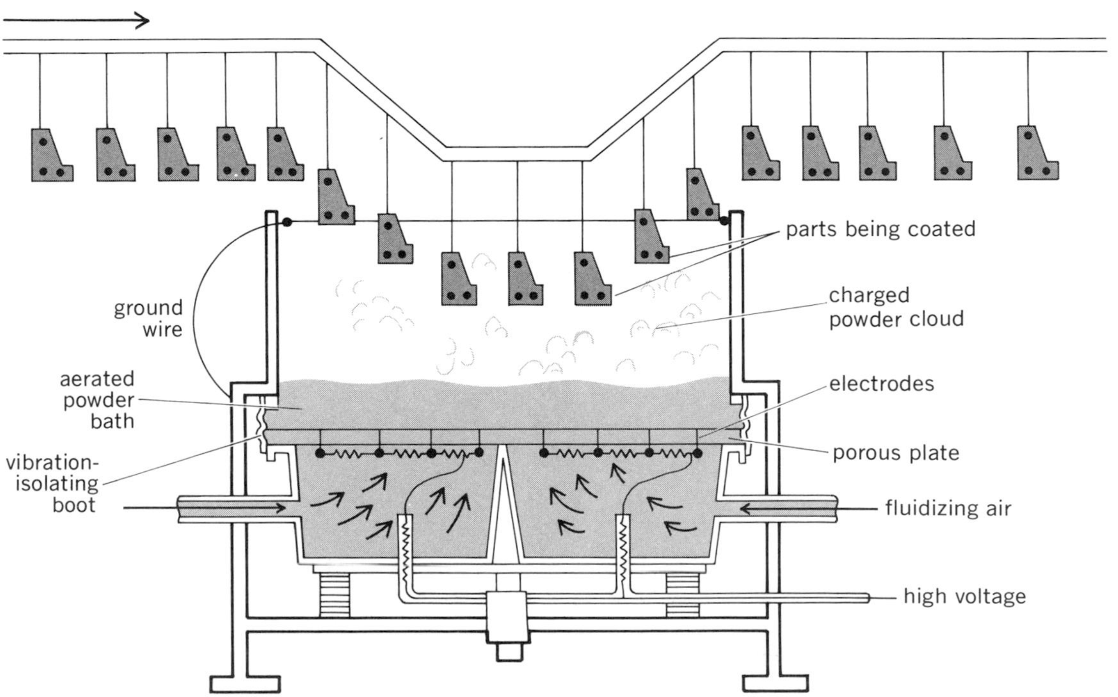

Fig. 2. Typical electrostatic fluidized-bed coating system.

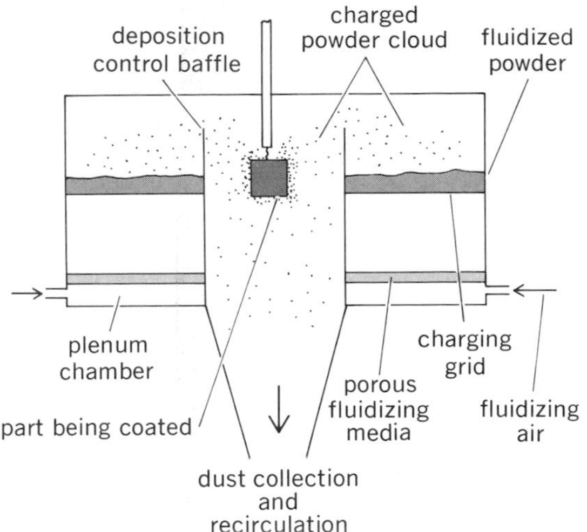

Fig. 3. Typical pherostatic system.

Materials. Most powder materials are manufactured by melt extrusion techniques. Ingredients of resin, curing agent, pigments, and modifiers, which are generally in solid form, are preblended, thoroughly melted, and homogeneously mixed by means of extrusion. The resultant flake or chip is then pulverized into the preferred particle size, depending upon application technique (see the table). Materials can be either thermosetting or thermoplastic. While electrostatic spray is the predominant method of applying thermosetting materials, fluidized-bed coating is the predominant method of applying thermoplastics. Both methods of application can be used with either type, however.

Thermosetting. These resins will chemically cross-link or cure, forming a coating film that becomes relatively inert and will not remelt. Such formulations typically provide very good adhesion to metal as a one-coat system (no primer). Thermosetting materials are used primarily in paint replacement applications where film thicknesses average 1.0–2.0 mils (25–50 μm). Thermosetting epoxies find use in thicker film applications such as pipe coating and electrical insulation. Generic types of thermosetting material include epoxy, polyester, epoxy-polyester hybrid, and acrylic.

Typical particle size distribution for powder coating

Particle size, μm	Percent greater than
Electrostatic spray powder	
100	0
80	5
60	15
40	40
20	70
0	100
Fluidized-bed powder	
200	5
150	20
100	50
50	80
0	100

Epoxy is applied when mechanical strength and corrosion resistance are required. It will chalk or change color when exposed to the ultraviolet component of sunlight, making it undesirable for exterior applications. Seventy percent of all thermosetting powder coating used today is epoxy.

Polyester resins are classified into two types. The more common type is cured with a isocyanate cross-linker, and provides a smooth, decorative coating in the 1.0–2.5-mil (25–64-μm) range, with resistance to color change and good chalking. This produces a material combining the good adhesion and flexibility properties of epoxy and weathering resistance. The second type is cross-linked with a very-low-molecular-weight glycidal or epoxy functional curing agent and exhibits very good color and exterior durability. This type is mainly utilized where good edge coverage and thick film (above 2.0 mils or 50 μm) mechanicals are required.

The epoxy-polyester hybrid utilizes the chemical reaction of certain epoxy and polyester resins, and should be considered part of the epoxy family from a performance standpoint. The strong tendency for poor ultraviolet resistance remains even when large percentages of polyester are used. The hybrids are softer than epoxies and tend to have better electrostatic application characteristics.

Acrylics exhibit good exterior durability combined with excellent detergent and alkali resistance. They have a higher degree of surface hardness and scratch resistance than polyesters. They display less flexibility than polyesters, but in most cases will pass impacts which would otherwise deform the metal and ruin a fabricated piece.

Thermoplastic. These resins will not chemically cross-link. They are nonreactive and will remelt. In many cases a primer is required in order to achieve the desired adhesion to metal. Most thermoplastic materials are used in thicker film applications. Generic types include polyvinyl chloride, nylon, and polyethylene-polypropylene.

Polyvinyl chloride's initial application used liquid vinyl coatings for items such as dishwasher baskets. Vinyls have now found use in the metal furniture industry and as a low-voltage electrical insulating coating. In order to achieve good adhesion, a primer must be utilized.

Nylon coatings are noted for excellent abrasion resistance and their low coefficient of friction. Good adhesion requires the use of a primer. Nylon coating is effectively used in mechanical design considerations. When the strength of metal is required in conjunction with high abrasion and bearing requirements, nylon coatings offer a solution.

Polyethylene-polypropylene provides very good chemical resistance, and tends to be harder and tougher than polyethylene. Although these materials require a primer for good adhesion, some manufacturers offer so-called self-adhering grades.

For background information *see* METAL COATINGS; POLYMER; SURFACE COATING in the McGraw-Hill Encyclopedia of Science and Technology.

[RONALD FARRELL]

Power factor controllers

The power factor controller (PFC) is a solid-state electronic device that reduces excessive energy waste in ac induction motors. Its development was the result of a request by solar power engineers to reduce the energy consumption of motors used in a solar-heated and solar-cooled demonstration house.

The significance of the power factor controller lies in the fact that the nearly a billion induction motors used daily in homes and industry of the United States alone consume the equivalent of approximately 7,500,000 barrels ($1.2 \times 10^6 m^3$) of oil daily. Obviously then, each percent reduction in power consumption results in a savings of the equivalent of 75,000 barrels (12,000 m^3) of oil daily. The power factor controller is applicable to both single-phase and three-phase induction motors. Since it is connected to the power lines of the motor and requires no modification to the motor itself, it may be applied to existing motors as well as to new installations.

Induction motors have the outstanding advantages of being simple, rugged, and reliable, and of having a relatively low initial cost. They are long-lasting and essentially maintenance-free and have a speed characteristic which is relatively insensitive to changes in loading or in applied voltage. It is this feature, however, which causes the induction motor to waste excessive amounts of energy when operated, as most are, at less than full rated load.

Causes of induction motor waste. All the electric power wasted within a motor is caused by current

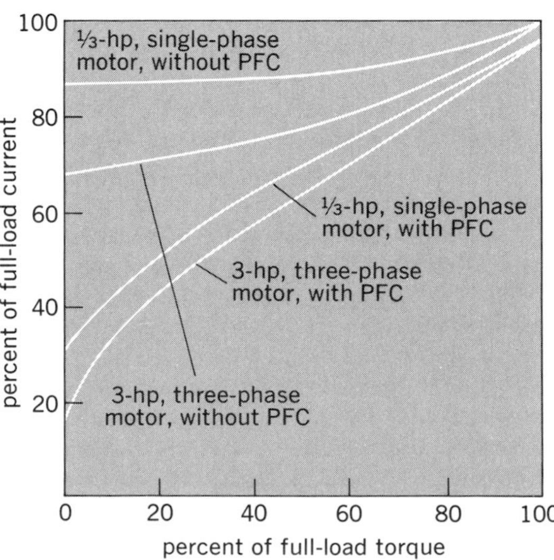

Fig. 1. Current versus load in typical 1/3-hp (250-W) single-phase and 3-hp (2.2-kW) three-phase motors, with and without a power factor controller.

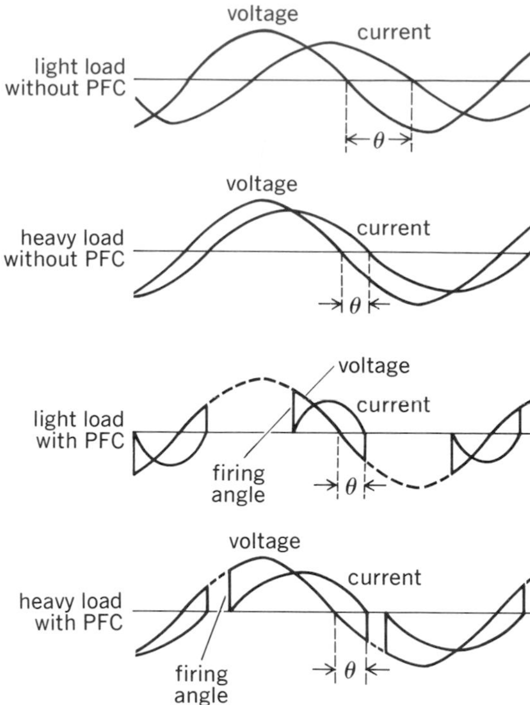

Fig. 2. Voltage and current waveforms of motor with and without a power factor controller.

flow and is the result of copper losses, hysteresis losses, and eddy current losses. Some waste is unavoidable. However, current that flows in excess of that required to satisfy a given load results in excessive power waste. It is this excessive waste which is minimized by the power factor controller.

Unlike the dc motor, which changes its speed in response to a change in load or applied voltage, the induction motor is a constant-speed machine. Its speed is determined by the frequency of the applied voltage and, within limits, is essentially insensitive to changes in loading or amplitude of the voltage. If the load is removed from a dc motor or if the applied voltage is increased, its speed will increase. The increased speed results in an increase in counter electromotive force (emf), which opposes the applied voltage and reduces current flow. This mechanism inherently regulates the motor current (and hence motor losses) to only that required to satisfy the load.

The induction motor, however, cannot appreciably change its speed and hence cannot always be optimally loaded or supplied with the optimum voltage because of the following considerations: (1) The motor is usually designed with a safety margin in output torque to allow for low line voltage conditions, but rarely is subjected to low voltage. (2) The power distribution system usually delivers higher than the minimum required voltage to allow for variations due to line losses and changes in loading on the system. (3) Motors are available only in incremental horsepower ratings ($^1/_2$, 1, $1^1/_2$, 2, 3, 5,

$7^1/_2$, 10, 15, and so forth; 1 hp = 0.75 kW). (4) Because many types of loads are uncertain or unpredictable and because the incremental difference in price for the next-larger horsepower rating is small, it is standard practice to select a larger-size motor than is ultimately required. (5) Many loads are cyclic, and the motor runs at idle or very light load much of the time. Examples of such applications are elevators, escalators, saw mills, rock crushers, punch presses, industrial sewing machines, and machine tools such as lathes, drill presses, and grinders.

The induction motor therefore is usually subjected to a line voltage significantly greater than that required to satisfy loading conditions. Since the motor cannot appreciably change its speed or counter emf, the high line voltage (high with respect to a light load) causes current to flow in excess of that required by the load. This results in excessive power waste which is dissipated as heat in the motor. The curves in Fig. 1 indicate that the idling and light load current in small motors, both single-and three-phase, remains high relative to the current at rated load. These motors waste about the same amount of power whether idling or fully loaded. The curves also show the reduction in current typically attainable through use of the power factor controller.

Operation. An understanding of the characteristics of induction motors is helpful in explaining the operation of the power factor controller. The power input to the motor is equal to the product of the applied voltage E, the current I, and the cosine of the phase angle θ between the voltage and current ($EI \cos \theta$). Cosine θ is defined as the power factor of a motor, and θ is the parameter which is held constant by the power factor controller.

An idling motor is highly inductive, resulting in a large phase lag θ between voltage and current as indicated in Fig. 2. This phase lag decreases as the motor is loaded and becomes less inductive. Typically, the phase lag in a well-constructed motor will vary from 30° under heavy load to 80° at no load. If the applied voltage could be sufficiently reduced at no load, the current would decrease and the phase lag would once again decrease to 30° before the motor speed decreased appreciably. It is this task which the power factor controller performs. The device varies the applied voltage as a function of the load to maintain the phase angle at a constant predetermined value.

The power factor controller senses the line voltage and current and produces a voltage proportional to the phase angle between the two [$V(\theta)$ in Fig. 3]. This voltage is summed with a reference or command voltage (θ command) that is indicative of a desired phase angle. The difference in these two voltages is an error voltage which, after conditioning, controls the turn-on point or firing angle of a solid-state switch (triac) that is in series with the motor (Fig. 3). The triac is turned on for a portion of each half-cycle to vary the applied voltage to the motor. Only the firing angle or turn-on point of the

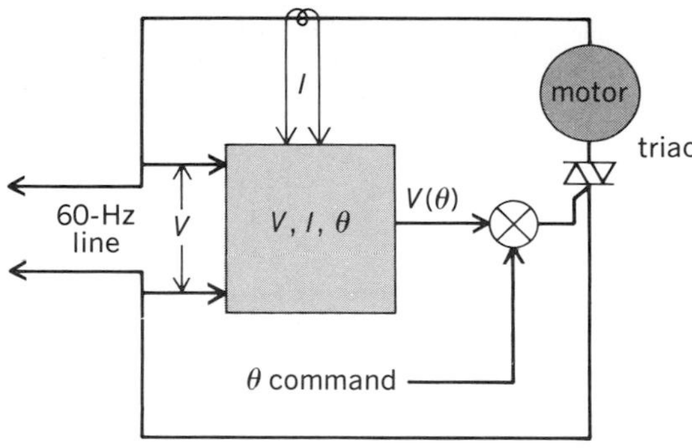

Fig. 3. Block diagram of a power factor controller.

triac is controlled. Once the device is turned on, it will remain on until the current goes to zero. The time between the zero crossing of the voltage and the zero value of the current is the phase angle (θ in Figs. 2 and 3) sensed by the system.

Under idling or lightly loaded conditions, the triac is turned on late in each half-cycle to apply a low effective voltage to the motor. An increase in load is sensed by a decrease in phase angle. The resulting increased error voltage, after amplification and conditioning, advances the firing angle or turn-on point of the triac to increase the voltage applied to the motor. By varying the applied voltage as a function of the load, the phase angle is forced to remain constant at the commanded value regardless of changes in load. Thus by controlling the magnitude of the applied voltage to only that necessary to overcome torques attendant to less than fully loaded conditions, the wasted power associated with full line voltage is reduced. For application to three-phase motors, a triac and its associated signal conditioning are required for each phase.

Factors affecting savings. The amount of energy savings resulting from the use of the power factor controller depends primarily on three factors: the quality of the motor's construction, the magnitude of the line voltage, and the loading on the motor. All these factors are depicted in the curves of Fig. 4, which indicate the power consumption versus load of typical motors with and without the power factor controller. The better-constructed motors, indicated by Fig. 4*b*, contain more iron, iron with better magnetic properties, thinner laminations, more copper, and a smaller air gap than the more economically constructed motors. The purchase cost of these motors is greater, but they waste less power proportional to their size, and hence less savings can be realized by application of the power factor controller. The curves also show the effect of line voltage on power consumption and savings. The line voltage will vary from one utility company to another, it will vary as the loading on the distribution system changes and with respect to the location of a motor within a distribution system. The high line voltage curves of Fig. 4 indicate that a small change in voltage can make a significant difference in power

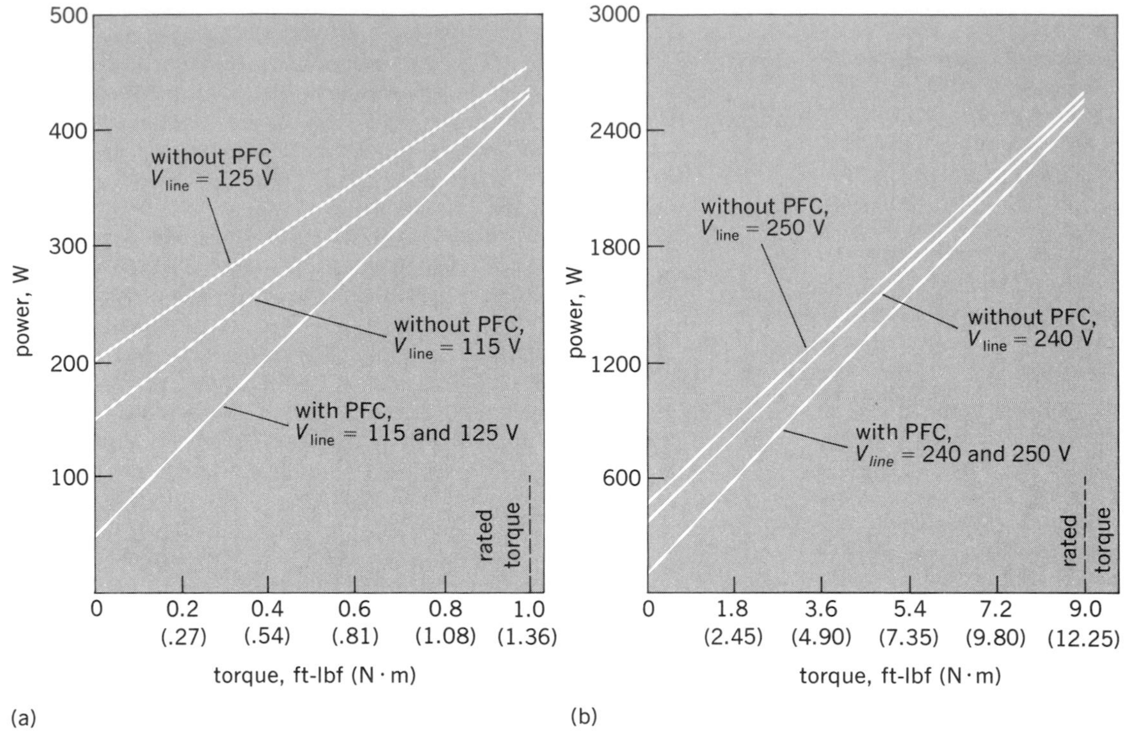

Fig. 4. Power consumption of motors as a function of load and line voltage (V_{line}), with and without a power factor controller, showing savings realized with a PFC. (a) Typical 1/3-hp (250-W) single-phase motor. (*b*) Typical 3-hp (2.2-kW) three-phase motor.

consumption. The most significant factor affecting savings is the loading on a motor. The lighter the load, the greater the potential for reducing the power consumption. As indicated by Fig. 4, savings may range from nothing to a few percent at rated load and up to 50–70% in an idling or lightly loaded motor. It is in applications with light or cyclic loads such as machine tools where the greatest benefits may be derived by use of the power factor controller.

Other benefits. In addition to the energy-saving feature, there are other benefits to be gained through use of the power factor controller. Reduced power waste results in lower operating temperature. Theoretically, each 10°C that the temperature is reduced will double the life of a motor. In many applications, use of the power factor controller lowers the operating temperature 10–20°C.

Power losses in utility company distribution systems range from 7 to 10%. A reduction in current flow in motors through use of the power factor controller results in reduced current flow throughout the entire system, and hence in less wasted power in that system. Also, the capacity of a distribution system is fixed by the losses (wasted power) in the system. Since the power factor controller reduces wasted power, its use results in increased system capacity. For each percent that wasted power is decreased, the utility can add a like percent of customers with no increase in capital investment.

Large users of motors operating with light or cyclic loads are often charged a penalty for poor power factor. Capacitors, which are used to correct the power factor and avoid this charge, reduce wasted power in distribution systems but have virtually no effect on motor power. Use of the power factor controller would minimize the need for capacitors and reduce energy waste in both the motor and the distribution system.

The power factor controller can reduce air-conditioning requirements and costs. In a textile mill with 3700 sewing machines, tests indicated that the power factor controller could reduce energy consumption by 33%, or by approximately 60 W on each machine. The total heat output into the building due to excessive wasted power is 222 kW (3700 × 60 W). The air conditioning required to neutralize this heat load could be eliminated by use of the power factor controller.

Finally, the cost effectiveness of the power factor controller is enhanced if it serves double duty as an on-off starter for the motor. Because of the many advantages of electronic control systems, there is a trend toward replacing electromechanical motor starters with solid-state starters. When the power factor controller is installed in original equipment as part of a solid-state starter, its energy-saving feature can be incorporated for only a small fraction of the total cost.

For background information *see* INDUCTION MOTOR; POWER FACTOR; SEMICONDUCTOR RECTIFIER in the McGraw-Hill Encyclopedia of Science and Technology.

[FRANK J. NOLA]

Bibliography: NASA, *Power Factor Controller*, MFS 23280, March 1979; F. Nola, Article 4, *IEEE Textile Industry Technical Conference*, May 7, 1980; U.S. Department of Energy, *Energy Efficiency and Electric Motors*, HCP/M50217-01, May 18, 1978.

Precambrian

The chronology of terrestrial evolution was originally erected through studies of hard, organic exoskeletons which first appeared at the base of the Cambrian, about 570 million years (m.y.) ago. The immense duration and the complexity of the preceding "lifeless" Precambrian Eon became appreciated only after the advent of isotopic dating techniques, which placed the formation of Earth at about 4650 m.y. ago. Consequently, the Precambrian, usually subdivided into the Archean (more than 2500 m.y. ago) and the younger Proterozoic (2500 to 570 m.y.), covers the initial 87% of the terrestrial lifespan. The Precambrian witnessed the emergence and growth of continents, the development of the oceans, the emergence and diversification of life, and the transformation of the originally inhospitable atmosphere of volcanic gases into a breathable atmosphere with abundant oxygen. In addition, Precambrian shields contain the bulk of economically recoverable global reserves of iron, nickel, chromium, gold, and uranium as well as significant amounts of other metals. Frequently the ore-generating periods were caused by unique circumstances typical of a given aging stage of the Earth and never repeated since. To interpret the complex Precambrian record from fragmentary evidence of its physical, chemical, and biological impact on the planet Earth is a complicated task, and only the outline of some of the major issues will be covered in this article.

Hadean gap. Hadean, the pre-rock period, is the interval between the formation of the Earth and about 3750 m.y. ago, the age of the oldest presently known rocks from the Isua area of western Greenland. The Moon, in contrast to the Earth, retains its geologic record mostly from this time interval. This was also the time of intense meteoric bombardment throughout the inner solar system, abundantly illustrated by lunar craters, and it is unlikely that the early Earth could have escaped the impacting. Yet, despite an intensive search for vestiges of such bombardment or for any rocks of this age, the gap has not yet been bridged. *See* LIFE, ORIGIN OF.

The convective thermal dissipation of collisional, gravitational, and radioactive energy, although declining exponentially with time, remains the driving force of present terrestrial geologic processes, while in the smaller Moon thermal convection ceased 3000 to 4000 m.y. years ago. The perpetuity of geologic processes on the Earth results in continuous cyclic formation and destruction of rocks, with progressively older rocks having a statistically higher chance of being obliterated. Indeed, the area extent of progressively older continental rocks diminishes exponentially with increasing geologic age. Nevertheless, the complete absence of pre-Isua rocks is enigmatic. This absence implies very high mantle-crust recycling rates during the earliest stages of terrestrial evolution, in accord with the high heat flow typical of that time. However, it is not clear why the accompanying magmatism and plutonism failed to generate global continental crust, or if they did so, how such a buoyant crust could have been recycled so efficiently back into the dense terrestrial mantle.

Formation of continental crust. Continents, the rafts of lighter material (ρ = 2.7 g/cm^3) floating on the denser (ρ = 3.3 g/cm^3) mantle, were derived through fractional melting of mantle silicates. Tectonic provinces of the continents, as revealed by radiometric dating, follow a concentric pattern, with the oldest nuclei usually in the centers of continents and progressively younger provinces situated toward the periphery. Consideration of Rb/Sr, U-Th/Pb, and particularly of the recently developed Sm/Nd isotopic methods indicates that this pattern is a consequence of lateral growth of continents over geologic time. The initial nucleation, about 3750 m.y. ago, was followed at first by relatively sluggish episodic growth until about 3000 m.y. ago. The bulk of the continents appears to have been formed between approximately 3000 and 2000 m.y. ago, but their growth likely continues, at much slower rates, even today.

The present-day continents have an irregularly stratified structure, with the upper-continental crust (about 15 km thick) containing relatively less dense granitoid bodies which segregated as silicate melts at 16-km depths or more. Such melts can form either from young volcanic and sedimentary precursors of mature island arcs or from older, continentally derived, thick sedimentary (geosynclinal) accumulations. Granitic melts of the latter type contain more radiogenic strontium isotopes and can thus be distinguished from the former group. The proportion of such continental granitic rocks was relatively minor in the Archean but increased markedly during the subsequent Proterozoic and Phanerozoic times. This suggests that prior to about 2000 m.y. ago, continents were growing mostly at the expense of the oceanic crust (and mantle), but subsequently the cyclic processes of continental destruction and regeneration gained in significance. Recycling, because of its repetitive nature, causes not only more efficient segregation of incompatible elements into the upper continental crust (and thus more fractionated chemical composition) but also tectonic stabilization of large continental blocks (cratons). This has both theoretical and economic implications.

From the theoretical point of view, it is not clear whether the present mode of sea-floor spreading and plate tectonics operated prior to the late Proterozoic, which is the age of the oldest known ophiolites (from the Arabian Peninsula). Most of the Proterozoic orogenies are intracratonic and not, as demanded by plate tectonics, peripheral. The older, Archean record suggests an even more peculiar tectonic regime, with numerous small convection cells in the mantle, in stark contrast to the long, arcuate orogenic belts of the Phanerozoic. Thus the post-Hadean evolution of the Earth was likely characterized by three discrete types of tectonic regimes, which perhaps imply a change in the geometry of heat-dissipating convective cells in the mantle.

Economically, the less evolved Archean upper continental crust, with its numerous ultramafic bodies, has a high frequency of chromium and nickel ores. In contrast, the subsequent development of cratons enabled formation of important sedimentary ores, for example, gold and uranium from Witwatersrand in South Africa or Blind River in Canada, banded iron ore formations on all continents, and copper deposits from Zambia. Cratonization, however, was only one of the preconditions of ore formation, and some additional necessary attributes will be listed briefly below.

Oceans. The generally invoked large-scale melting of the early mantle should have been accompanied by its degassing and thus by early formation of most of the present-day atmosphere and hydrosphere. However, the continuous slow degassing of primordial helium into today's oceans shows that the mantle is not yet completely degassed. It is not clear therefore whether the bulk of the global atmosphere and hydrosphere accumulated "instantaneously" or was degassed more gradually, principally prior to the early Proterozoic. For the globe of constant size, with early continents absent or small, the primordial oceans should have been shallower and of global extent. The subsequent growth of continents, because of isostatic considerations, should have led to areal shrinking and deepening of the oceans.

Archean stromatolites, approximately 3400 m.y. old, from the Warrawoona Group in the Pilbara Block, Western Australia. Stromatolites are domal, laminated, calcareous (in this case later silicified) sedimentary structures formed in shallow-water environments under the influence of a mat or assemblage of sediment-binding blue-green algae or bacteria. (*Courtesy D. R. Lowe, Louisiana State University*)

The present-day chemistry of sea water is controlled by: continental river influx; basalt–sea-water interaction in hydrothermal cells on spreading centers (for example, midoceanic ridges); and efflux via, and interaction with, sediments. Today, continental river influx is the largest flux, but in the Archean it was considerably smaller and basalt–sea-water interaction was enhanced due to higher geothermal gradients. Consequently, hydrothermal and volcanic exhalations into the early oceans were a major factor buffering sea-water chemistry, with the transition toward a modern kinetic steady state accomplished mostly during late Archean–early Proterozoic times. These exhalations were a direct source of, for example, copper and zinc in the ubiquitous massive sulfide deposits of the Archean, and of iron in the coeval volcanogenic (Algoma-type) iron formations.

Atmosphere and life. There is evidence (see illustration) for the existence of life, probably anaerobic and photosynthetic bacteria, since at least 3500 m.y. ago. Disputed direct evidence and circumstantial evidence, based on carbon isotopes in carbonates and graphites, indicate that life may be at least as old as the oldest terrestrial rocks. Furthermore, carbon isotope data do not preclude that the early biomass could have been considerable and it could have photosynthetically produced some oxygen. More research is needed, but the realization of the great antiquity and of possible conservatism of life is one of the most stunning advances of recent months. Furthermore, the distinct possibility of the presence of oxygen producing photosynthetic biota since about 2800 m.y. ago forces a rethinking of the models of atmospheric evolution. The existence of detrital uraninite (a mineral soluble in oxygenated waters) in the previously mentioned uranium deposits of South Africa and Canada and other geological evidence suggest low P_{O_2} of the pre-2000-m.y. atmosphere-hydrosphere system. The subsequent increase to a significant percentage of present-day oxygen levels was considered to be the consequence of the emergence and proliferation of the oxygen-producing photosynthetic biomass in the early Proterozoic. If, however, oxygen-producing biologic systems are more ancient, an efficient sink for the generated oxygen is required. Such a sink could have been the prolific pumping of sea water through, and reaction with, basalts at spreading centers. The well-publicized Galapagos black smokers, submarine vents spewing black clouds of H_2S and minerals for endemic biota, are an active example of such oxygen-consuming hydrothermal cells. The exponential decline in geothermal gradient, coupled with the coeval growth of continents at about 3000–2000 m.y. ago, resulted in the diminished capacity of this sink and in the gradual accumulation of oxygen in the atmosphere. Simultaneously, upwelling of the deep unoxygenated ocean waters (containing dissolved Fe^{2+}) onto the newly generated continental shelves culminated in precipitation of the enormous banded iron formations containing the bulk of the global reserves of iron. Thorough oxygenation of the atmosphere-hydrosphere system eventually led to a termination of banded iron formations and detrital uranium deposits, for example, with subsequent steady states producing new sets of circumstances conducive to formation of other types of economic deposits.

Synopsis. The Precambrian Eon of Earth history approximates the stages of childhood, adolescence, and early adulthood in human life. The early, precocious growth period, which produced the bulk of the global hydrosphere, atmosphere, lithosphere, and biosphere, culminated about 2000 m.y. ago in a system somewhat analogous to the present-day one. Although the evolution was not terminated at this time, the subsequent "early adulthood" was controlled mostly by cyclic phenomena. The above-outlined terrestrial evolution was ultimately a consequence of the changing mantle heat regime and interaction of this inorganic cause with advancing life.

For background information *see* LIFE, ORIGIN OF; PLATE TECTONICS; PRECAMBRIAN; STROMATOLITE in the McGraw-Hill Encyclopedia of Science and Technology.

[JÁN VEIZER]

Bibliography: A. Kröner (ed.), *Precambrian Plate Tectonics*, 1981; J. W. Schopf (ed.), *Origin and Evolution of the Earth's Earliest Biosphere: An Interdisciplinary Study*, 1982; J. Veizer and S. Jansen, *J. Geol.*, 87:341–370, 1979; B. F. Windley, *The Evolving Continents*, 1977.

Printing

Development of automatic controls for gravure and web offset printing presses through the application of distributed intelligence systems is expected to revolutionize control of the presses and enhance management capabilities. The result will be improved and more consistent printing quality, greater productivity, and reduction of waste, energy, and worker-power requirements. This will drastically change an industry whose craft orientation has relied on human interpretation and hand adjustments.

Systems. Microprocessors grouped in a press-side data terminal will be interconnected in a network or hierarchy that starts with control of one machine and leads to precise management of the entire plant. It will provide fully integrated plant control not only over the pressroom, but over film-preparation, plate, and cylinder departments, and the bindery as well.

All major press manufacturers in the United States and Europe have separate developments under way. The Presfax system was developed by the Harris Corporation; the George Hantscho Company has introduced the POMS (Press Operation and Maintenance System); the Goss Waste Program from the MGD Division of Rockwell International has been in successful operation at several major newspapers; and the Baker Perkins Company in England and several press builders in Germany also have systems under development.

Many manufacturers of peripheral equipment have developed support units that are compatible with the systems of press manufacturers. Among the most prominent are Datamat (Butler Automatic, Inc.) and Microsplice (Microtrol, Inc.) for controlling the roll stand and measuring the paper being fed into the press. At the delivery end, developments by Baldwin Automatic and the emTECH Corporation solve the problem of count control and provide input to integrated data collection systems.

A system representative of the most advanced developments now available is the Hurletron Print Management System 700 from Hurletron Altair. It can be installed on web offset and gravure presses of all manufacturers and will accept input from both the roll stand and delivery controls made by the other suppliers.

Hurletron Print Management System 700. The primary level of control is initiated through sensors or measuring devices mounted in the press. Its incremental motion resolvers or encoders measure cylinder or web movement at accuracies of 0.001 in. (25 μm), or better, over cylinder diameters of 8–60 in. (20–150 cm). The web scanner (or eye) permits scanning of copy or register marks up to 1 in. (2.5 cm) away from the web. It is panchromatic and has the ability to generate more than one type of output. It can simultaneously be used as a probe for both circumferential and transverse control.

These sensors connect to microprocessor controls, which close the loop by driving stepping or DC servo motors at rates limited only by the application or job. The fact that the controllers are program-controlled and not hard-wired allows tuning of the stimulus-response characteristics to a refinement never before possible. For instance, error averaging may be selected and its output weighted by process speed and the dynamic nature of the error. These calculations set the rate of compensation response which is applied to each repeat or turn of the cylinder. These devices control circumferential and transverse register, lateral web guiding, and backup and cutoff control. The latest development has been a monitor for color-to-color register.

The next level of control is a distributed intelligence system: a press-side data entry terminal (Hurletron Predat 746). One controller is dedicated to each press and has access to every control sensor mounted in the press. The status of each sensor—its operating mode and dynamic activity—is available for monitoring and analysis on a video display. Operator interaction with the Predat is through English language displays (or menus) and simple keyboard responses. An extensive dictionary, tailored to each application, requires only yes/no, accept/reject, and simple button pushing on a telephone-type touch pad.

The parameters for press control are established by the pressman who communicates with press sensors by selections from the menu and inputs on the touch pad. Tolerances are selected for web tension, lateral movement, backup accuracy, cutoff and fold accuracy, and register variation. Additional functions may be included. In fact, any value that can be measured can be sensed, and any adjustment that can be driven can be controlled. The Predat has the capacity for 256 separate sensor inputs—128 analog and 128 digital. Logical extension of controls includes automatic press start-up and shut-down routines, folder compensation set-ups and oven temperature control.

Inputs from Datamat or Microsplice units will provide complete records of the paper used. Roll identification, weight, and yield are essential records. The input of wrapper, slab, and core weights, the identification of roll defects, and the reason for rejection are required by the "War on Waste." From the data entered, basic weight can be calculated and complete consumption totals can be relayed to accounting.

The emTECH count control system provides accurate counts of good signatures produced and a breakdown of the signatures wasted identified by cause. The combination of roll data, good signature count, and the record of press operation from the Predat provides a complete record of press operation and essential management information.

At the press, the Predat transmits its control data to a dedicated computer in a location remote from the hostile pressroom environment. In the Hurletron system it is called the DSAM (Data Storage and Analysis Module). It has the capacity to handle data from four presses simultaneously. The unit contains several microprocessors operating in a multiprocessing environment. The large video display has a full keyboard and rotating magnetic storage device for program retrieval and data storage.

This device accomplishes dynamic monitoring of each process loop or individual units within a loop. As an example, a continuous display of the register variation over the preceding 10 revolutions is available on request. The DSAM can recall from memory the press conditions that existed for the 30 s prior to an unscheduled stop. This new and powerful diagnostic tool determines the cause of repetitive occurrences such as web breaks.

The printer at the DSAM produces management reports with timing, accuracy, and detail never before available. The reports generated, like the system itself, are hierarchical. Top management reports are brief summaries of plant operations and indices reflecting the productive effectiveness of the plant. The primary indices are average net production speeds and earned hour percentages. Equally important are indices of average makeready time, run interval, and delay intervals. At the plant manager level, reports are in adequate detail to identify the problem areas and to compare current operations with historical records for the preceding month and year. At the supervisory level, reports are in complete detail for diagnostic investigation of problem areas.

Information in the DSAM is transmitted to the company's main frame computer for use in payroll, job costing, and inventory control. Hurletron's Print Management System 700 is representative of the new generation of press control which allows effective management of the printing process.

For background information *see* MICROPROCESSOR; PRINTING in the McGraw-Hill Encyclopedia of Science and Technology.

[GORDON BARRETT]

Problem solving (psychology)

Problem solving consists of an attempt to convert some initial state of a task (say, a chessboard before the first move is made) into a final or goal state (checkmate) through a series of actions or problem-solving steps. In some problems such as chess, both the initial and final state are well known (these are sometimes referred to as well-structured problems). In other cases, either the initial state or the final state, or both, may be poorly understood (these are referred to as ill-structured problems). If the necessary problem-solving steps are also known ahead of time, problem solving consists of finding the right steps to apply at the right time. If some of the problem-solving steps are not known, the problem-solving process entails creating or inventing new ways to convert one state of a task into another.

By observing problem-solving activity through the technique of analyzing "thinking aloud" protocols, it has been possible to determine the way in which a given problem is solved. Such analyses reveal that for individuals who are relatively inexperienced at a task (novices), the activity of problem solving is often one of trial-and-error search for steps to relate initial and final states of a task. Such a search, often referred to as a means-ends analysis, can require large amounts of time if the problem is complex, and it has been successful primarily in games (such as tic-tac-toe) and relatively simple problems. For experts, the analysis of "thinking aloud" protocols reveals that the activity of problem solving involves the use of problem-specific methods and techniques for solution (often referred to collectively as heuristics). These heuristics are acquired through experience and represent the basis for expertise in a specific domain of tasks.

Problem solving in physics. One area of investigation in the last several years has been the solution of relatively well-structured (textbook) physics problems. Undergraduates as well as graduate students and professors have been observed solving such physics problems through the technique of protocol analysis. Typical analyses of problem solving identify two types of problem-solving steps: representing the problem situation, and applying the principles of physics to the representation of a problem in order to generate a solution. Representation involves translating the statement of the problem into a standard form that is accessible to the principles of physics. For many problems in physics, this translation may entail identifying the problem as an instance of some standard or prototype problem (for example, the problem of the harmonic oscillator, free fall at constant gravity, or block on a frictionless plane) that has known solution methods. More generally, translation consists of equating aspects of the problem with idealized objects such as massless rods, frictionless surfaces, and point masses, whose behavior can be represented fairly clearly by a few principles of physics. Of course, a given representation can be inappropriate or flatly wrong, and even a good representation may not be tractable (such as the classic three-body problem in physics).

Once a representation is created, problem solving proceeds by finding principles of physics that will generate a solution. In some instances, single principles (say, uniform linear motion) will be sufficient to derive a solution, while in other cases combinations of principles (rotational as well as linear motion) are necessary. In order to be applied efficiently either singly or in combinations, principles of physics need to be selected so that they fit the assumptions and data of a given problem. When several principles are involved, the coordination of their application becomes important. Prototypical problems in physics are those for which single principles suffice, or ones for which the coordination of principles is taught explicitly in introductory textbooks (for example, the simultaneous use of momentum and kinetic energy conservation in an elastic two-body collision). In more advanced texts, procedures are identified for handling additional complexities, and algorithms are derived for dealing with general cases (such as lagrangian and hamiltonian equations in advanced mechanics). Problems that require a novel coordination of principles for their solution, even though all the requisite principles may be known, provide opportunities for a level of problem-solving activity sometimes referred to as heuristic thinking.

When problem solving fails because relevant prototypes are either not known or have not been learned, specific heuristics or shortcuts must be employed. In mechanics problems, for example, the goal of problem solving is typically to infer the behavior of a physical system over long spans of time, given only a description of its initial state. One heuristic for accomplishing this goal is to begin with the initial state of the system (such as a body at rest) and piece out the time course of system behavior by applying iteratively the laws that say what happens in the next instant of time (this process is often referred to as simulation). A second heuristic for specifying the behavior of a physical system is to integrate the differential laws of motion to yield an expression that describes the state of the system at any point in time as a function of its initial state. Because mathematical complexities often make either of these heuristics difficult to carry out in practice, additional heuristics are required. One such heuristic involves describing steady-state solutions of the system (the places where the laws of motion state that nothing changes). Another heuristic involves describing asymptotic behavior of the system (what happens to it in the long run). Yet another heuristic is to specify a property of the system (such as energy or momentum) that is invariant over time, even though there is variation in other aspects of the system. These heuristics (and others like them) are typically learned and applied by individuals in specific problem-solving tasks. Indeed, the power of heuristics as problem-solving tools derives from their development in the successful solution of particular types of problems.

The use of heuristic thinking is often associated with level of expertise (that is, it is used more often by experts than novices). Such an association does not mean, however, that individuals at more advanced stages of training and experience do only heuristic thinking, or that this is their preferred method of solving problems. Like other cognitive activities, problem solving requires the expenditure of mental effort. Because individuals at all levels of expertise seem to prefer to expend the least amount of effort that will do the job, experts as well as novices generally solve problems by relating them to known prototypes. It is only when truly unfamiliar or difficult problems are encountered that the expert resorts to the more demanding skills of heuristic thinking, which are for the most part unavailable to the novice.

Medical problem solving. Many of the features just described for solving problems in physics can also be found in the behavior of individuals attempting to solve the more poorly structured problems of medicine. Like problem solving in physics, problem solving in medicine has been studied intensively for a number of years by using techniques for analyzing "thinking aloud" protocols. Whereas in physics equal amounts of effort are often expended on problem representation and solution generation, in medicine one aspect of problem solving deals almost exclusively with finding a good representation (the problem of medical diagnosis).

Solving a diagnostic problem in medicine consists of an application of hypothetico-deductive reasoning in which the basic units are questions, cues, and diagnostic hypotheses. Diagnostic problem solving typically begins the elicitation of a small number of cues via questions (usually through the taking of a patient history). These cues suggest a limited number of diagnostic hypotheses which, in turn, suggest new revised hypotheses. Eventually, hypotheses are evaluated with respect to their relative ability to account for cues. Physicians of all levels of skill appear to generate hypotheses in similar quantities, use those hypotheses to guide collection and interpretation of additional patient data, and evaluate the relative ability of the hypotheses to account for those data. Differences between experts and novices (and among members of each group) are found at a more detailed qualitative level in the content of physician knowledge and the features of a case.

At a detailed level, the activity of individual problem solvers is guided by representations of the problem situation. Images of "what the problem is" enable forward searching strategies for the collection of additional information, as well as heuristics for the generation and evaluation of possible categories within which to interpret this information. Current models of problem solvers in medicine focus upon the structure of memory. Such models propose a large number of recurring patterns or prototypical combinations of data (for example, combinations of patient signs and symptoms) that are familiar and can be recognized. The organization of these prototypes is hierarchical, with the more general patterns at the top and the most specific ones at the bottom. For the expert, access to a particular pattern or prototype occurs at multiple levels in the hierarchy (top, middle, and bottom) and is a function of other patterns (pattern-to-pattern links) and data (data-to-pattern links).

Patterns in memory are established in the form of expectations to be matched against external data, so that a particular configuration of data can be interpreted as an instance of a given pattern or prototype. Unfamiliar experiences are interpreted by building new prototypes. This occurs either through a form of data-driven processing in which commonalities are abstracted from recurring patterns or cues, or through a more conceptually driven form of processing in which higher-level rules are used to derive patterns that can then be matched with unfamiliar data.

Although there are similarities between problem solving in areas as diverse as chess, physics, and medicine, the secret of problem solving lies in mastery of large amounts of task-specific knowledge. It is this knowledge, gained through experience and stored in memory, that accounts for the major differences between experts and novices in a field and between experts or between novices. Teaching problem-solving skills has been a topic of interest to psychologists and educators alike. The biggest payoff from instruction in problem solving seems to be simply learning more about what task specific knowledge needs to be learned as the skills of problem solving are practiced.

For background information *see* PROBLEM SOLVING (PSYCHOLOGY) in the McGraw-Hill Encyclopedia of Science and Technology.

[PAUL JOHNSON]

Bibliography: D. Connelly and P. E. Johnson, *Human Pathol.*, 11:412–419, 1980; A. S. Elstein, L. S. Shulman, and S. A. Sprafka, *Medical Problem Solving*, 1978; J. Larkin et al., *Science*, 208:1335–1342, 1980; Larkin, J. H. Teaching problem solving in physics: The psychological laboratory and the practical classroom, in D. T. Tuma and F. Reif (eds.), *Problem Solving and Education*, 1980; A. Newell and H. A. Simon, *Human Problem Solving*, 1972.

Prolactin

Prolactin is one of the several protein hormones of the anterior pituitary gland. It is closely related chemically to another hormone of the gland, growth hormone, and to a similar hormone produced by the placenta, placental lactogen. Prolactin is an unusual hormone in several respects. It is the most versatile hormone of the pituitary in terms of the number and diversity of its actions among the various vertebrate groups.

Osmoregulation. Prolactin's most prominent role in bony fishes and amphibians involves regulation of water and electrolyte balance when they are in fresh water. This osmoregulatory role of the hormone involves actions on various organs, including the gills, skin, gut, kidney, and urinary bladder. The net effect that prolactin has is to help the animals retain NaCl while reducing water influx from the dilute external environment. The hormone also facilitates water elimination, at least in some bony fishes in fresh water.

Prolactin regulates water and electrolyte balance in reptiles, birds, and mammals also, and in these groups the hormone's effects are exerted on the gut and kidney. The hormone also stimulates the functions of some specialized structures found in birds that operate to eliminate excess NaCl that the animals may ingest. These specialized organs, called nasal or orbital salt glands, enable some birds, such as ducks or sea gulls, to drink sea water.

Although the osmoregulatory effects of prolactin are not as prominent among tetrapods as in teleosts or aquatic amphibians, the hormone may be important for regulating salt and water balance in embryonic and fetal mammals, which are immersed in amniotic fluid. Thus, during the aquatic (intrauterine) phase of mammalian development, prolactin has osmoregulatory functions similar to those found in the fresh-water-dwelling lower vertebrates. Placental lactogen, the hormone related to prolactin, may also be important for osmoregulation in mammalian embryos and fetuses.

Growth and development. A second general function of prolactin among the vertebrates involves the control of growth and development. This action is particularly prominent in larval amphibians where the hormone promotes growth of structures that are specialized to facilitate adaptation to the aquatic habitat, such as the gill and the tail. Prolactin also antagonizes the metamorphosis-inducing actions of thyroid hormones in larval amphibians. Thus, it favors the growth and retention of larval structures and the continuation of aquatic existance of amphibians generally while aiding their capacities to osmoregulate in the hypoosmotic medium.

Prolactin promotes growth or development of specialized organs in virtually all vertebrate classes. Examples include the seminal vesicles of catfishes and rodents, the mammary glands of mammals, the mucus–secreting skin glands in fishes and amphibians, and the crop sac of pigeons and doves.

Interaction with steroids. Many of the physiological functions of prolactin involve effects on integumentary structures. These include skin glands in fishes, amphibians, and mammals. In the mammals the oil-secreting (sebaceous) glands are included. In addition, the hormone acts synergistically or antagonistically with adrenal and gonadal steroids on a variety of organs. Synergistic effects with androgens are seen in the growth-promoting effects of prolactin on male sex accessory organs, such as the prostate and seminal vesicles. Similarly, prolactin works in concert with ovarian steroids to promote mammary growth, and it acts synergistically with adrenal steroids to stimulate milk secretion. On the other hand, prolactin and cortisol have opposing actions in the regulation of water and electrolyte balance in teleosts.

Reproductive functions. Although prolactin is concerned primarily with water and electrolyte balance in fishes and aquatic amphibians, the majority of its functions in the homeothermic vertebrates are related to reproduction. Accordingly, it was suggested that the osmoregulatory actions of the hormone represent its more primitive functions and the reproductive effects evolved later. However, prolactin is also involved in the control of some reproductive functions even in the cold-blooded vertebrates.

Birds. In some species of birds prolactin inhibits the functions of the gonads after the mating phase of the reproductive cycle. At the same time it promotes the development of parental behavior so that the parent birds will incubate their eggs and care for their hatchlings. In concert with these effects, prolactin promotes the development of structures which are specifically concerned with the incubation of eggs. These structures are called incubation patches, which develop on the lower abdomen of some species. These areas of skin become defeathered and highly vascularized. Thus, they facilitate the transfer of heat from parent to egg during the incubation period.

Among pigeons and doves, the crop sac, which is a pouch in the esophagus, is used to feed the hatchlings. The cells lining the crop sac undergo marked proliferation, and they are sloughed off into the interior of the crop. These cells, called crop milk, are fed to the young along with food that the parent pigeon has ingested. The production of this crop milk is controlled by prolactin, and its function is analogous to that of milk secretion in mammals because it provides nourishment from the body of a parent to the growing young.

Mammals. The reproductive functions of prolactin in mammals are similar to those found in birds. Its most widely known function is that of promoting mammary gland growth and milk secretion. In addition, prolactin can have both gonad-stimulating and gonad-inhibiting functions among mammals. Low concentrations of the hormone appear to be essential for the production of androgens (male sex hormones, such as testosterone) by the testis. In bringing about this effect prolactin sensitizes the Leydig cells of the testis to the stimulating action of luteinizing hormone, a pituitary hormone that stimulates gonad functions. Conversely, when prolactin levels are high, the hormone becomes inhibitory to the functions of the testis because it inhibits the secretion of the pituitary hormones (the gonadotrophins) that stimulate the production of sperm and testosterone. This inhibitory effect of prolactin is accomplished by an action on part of the brain called the hypothalamus. The hypothalamus produces neurohormones, and some of these regulate the secretion of the hormones of the anterior pituitary gland. The hypothalamic neurohormone that regulates the secretion of the gonadotrophins is called gonadotrophin-releasing hormone. When prolactin levels in blood are high, the hormone inhibits the secretion of the gonadotrophin-releasing hormone by the hypothalamus. This effect reduces the secretion of the gonadotrophins, which results in decreased testosterone secretion and reduced sperm production.

In females, prolactin is important for the production of progesterone by the corpus luteum of the ovary, and its role here seems to be that of maintaining the sensitivity of the corpus luteum cells to the stimulatory action of luteinizing hormone. This function is analogous to that on the Leydig cells in the testis. Prolactin also participates in the regulation of ovarian follicular development, but in an inhibitory manner. When prolactin levels are high, follicular growth is inhibited because the hormone reduces the ability of the follicles to produce androgens. Reduced androgen production decreases the follicle's ability to produce estrogens. Since estrogens promote follicular growth by a direct action on the cells of the follicle, inhibition of follicular growth occurs.

Inhibition of follicular estrogen production by high blood prolactin levels reduces follicular growth and prevents ovulation by an extraovarian mechanism also. The final phases of follicular growth and subsequent ovulation require that the pituitary gland secrete increasing amounts of luteinizing hormone. This increase is brought about by rapidly rising levels of estrogen in the blood. Since high blood prolactin levels reduce follicular estrogen synthesis, the rise in luteinizing hormone is prevented and follicular growth and ovulation are inhibited. In females, prolactin also inhibits gonad function by an action on the hypothalamus similar to that in males wherein it depresses synthesis and secretion of the gonadotrophin-releasing hormone.

Pathology. Studies in recent years have uncovered a number of pathological conditions associated with abnormally high blood levels of prolactin in humans. This condition, called hyperprolactinemia, is associated frequently with breast growth (gynocomastia) and milk secretion (galactorrhea) in men. Hyperprolactinemia in nonpregnant women can also result in breast development accompanied by galactorrhea. In women, hyperprolactinemia is frequently accompanied by a suspension of menstrual cycles, or amenorrhea, and in men high levels of the hormone can cause infertility and impotence. In both sexes the infertility is due to inhibition of gonadotrophin secretion because of reduced secretion of gonadotrophin-releasing hormone. Hyperprolactinemia is frequently caused by small tumors of the pituitary gland called microadenomas. These tumors can usually be treated successfully by surgery and sometimes by drug treatment.

Contraceptive function. The inhibitory effects of hyperprolactinemia on secretion of gonadotrophin-releasing hormone and on the developing follicle operate in concert to inhibit fertility in women. This action of prolactin is probably the most effective means of limiting population control in developing countries. During lactation, high blood prolactin levels are maintained by the suckling infant. As a result, women may experience a period of natural infertility of between 1 and 3 years, depending on the frequency and intensity of suckling. Thus, prolactin has been called nature's contraceptive.

It is evident from the foregoing that prolactin is important for various kinds of physiological regulation among vertebrates. The hormone is vitally important for the provision of nourishment (in the form of milk) to the vast majority of human infants throughout the world. When scientists learn to duplicate prolactin's antifertility actions in women, the result may be a more satisfactory means of birth control than those presently available.

For background information *see* ADENOHYPOPHYSIS HORMONE; OSMOREGULATORY MECHANISMS; SALT GLAND in the McGraw-Hill Encyclopedia of Science and Technology.

[CHARLES S. NICOLL]

Bibliography: A. Bartke, *Fed. Proc.*, 39:2577–2581, 1980; W. C. Clarke and H. A. Bern, *Hormonal Proteins and Polypeptides*, 8:105–197, 1978; P. W. Howie and A. S. McNeilly, *Eur. J. Clin. Invest.*, 9:237–238, 1979; C. S. Nicoll, *Fed. Proc.*, 39:2563–2566, 1980; C. S. Nicoll, Physiological actions of prolactin, in E. Knobil and W. H. Sawyer (eds.), *Handbook of Physiology*, sec. 7: *Endocrinology*, vol. 4, pt. 2, 1974; M. Wallis, *J. Mol. Evol.*, 17:10–18, 1981.

Proton

The nucleus of an atom is made of protons and neutrons. These two particles have very similar properties; for example, each has a mass of about 10^{-27} kg and a radius of about 10^{-15} m. Their main difference is that the neutron carries no electric charge. Although it was believed for many years that they were elementary particles, it is now known that they have quite a complicated internal structure. The experimental study of this structure, and its theoretical interpretation, has been important in the great advances that have occurred in particle physics since the late 1960s.

Also, a completely new kind of experiment with protons is beginning; it is inspired by attempts to unify the strong, electromagnetic and weak interactions in a single quantum field theory. The simplest such grand unified theory predicts that the proton is not stable, but decays with a half-life of about 10^{31} years. The search for proton decay is now under way.

Structure. Early in the 20th century Lord Rutherford discovered the existence of the atomic nucleus by firing a beam of alpha particles at a thin gold foil. He found that some of the alpha particles were deflected through a wide angle, and interpreted this as being caused by small hard centers inside the gold atoms. Modern experiments that explore the structure of the proton and neutron are similar in principle to the Rutherford experiment. Suitable projectile particles are fired at them and are found to be deflected through a wide angle, thus indicating that there is something small and hard inside.

The first experiments of this kind, at the Stanford Linear Accelerator Center, used electrons as the projectile particles. An electron interacts with the target by exchanging a photon with it (Fig. 1). According to simple relativistic kinematics, together with Planck's law, $E = h\nu$, where E and ν are the energy and frequency of a photon and h is Planck's constant, the wavelength of the photon decreases as the energy and the angle through which the electron scatters increase. Short wavelengths are needed if the short-distance details of the target's structure are to be probed. Present-day experiments are able to resolve details whose linear scale is about 1/100 of the proton radius. The radius of the small hard objects that have been discovered inside the proton and neutron is not greater than this, and it may well be very much less.

According to Planck's law, the short-wavelength photons are highly energetic. Einstein's relation $E = mc^2$, where E and m are the energy and mass of a system and c is the speed of light, allows this energy to materialize in the form of matter. That is, when the photon hits the target, the target breaks up, and a large number of new particles are produced, mostly pions (π mesons). But, rather mysteriously, among the target fragments there is no direct sign of the small hard scattering centers that cause the wide-angle scattering of the electron. Their presence, and properties, can be inferred only indirectly.

Quarks. This is not to say that there is any doubt at all that the scattering centers do exist. There is

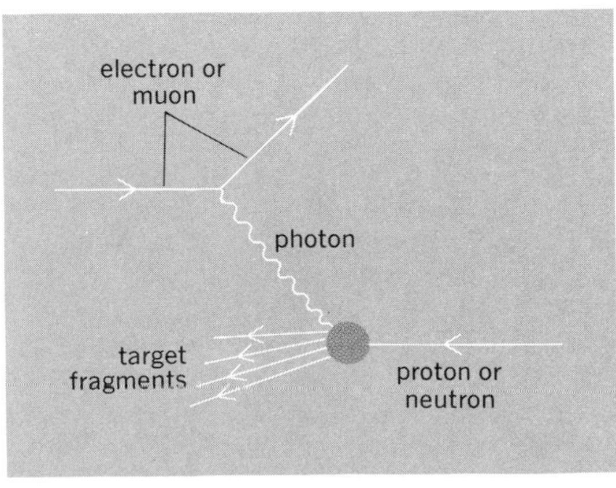

Fig. 1. Scattering of an electron or muon on a proton or neutron target through photon exchange. The target breaks up in the process.

now a huge quantity of indirect evidence, from a large variety of different kinds of experiments, that the small hard objects are quarks. The idea that protons, neutrons, and other particles might be made of quarks was introduced in 1963 by Murray Gell-Mann and George Zweig in order to explain the spectrum of the very large number (several hundred) of fundamental particles that have been discovered. According to Gell-Mann and Zweig, a striking feature of quarks is that, in units of the proton charge e, they have fractional charge, either $\frac{2}{3}e$ or $-\frac{1}{3}e$. In 1928 Paul Dirac predicted that corresponding to each particle there is an antiparticle of opposite charge, so there are antiquarks having charge $-\frac{2}{3}e$ or $\frac{1}{3}e$. In the simplest picture, a pion consists of a quark and an antiquark (and hence has integral charge), while protons and neutrons consist of three quarks, $(\frac{2}{3}e, \frac{2}{3}e, -\frac{1}{3}e)$ and $(\frac{2}{3}e, -\frac{1}{3}e, -\frac{1}{3}e)$.

In some very careful experiments at Stanford University, which are essentially a sophisticated version of the Millikan oil drop experiment, William Fairbank and collaborators seemed to find particles of the right fractional charge to be identified as quarks or antiquarks. However, at present their results are not understood, as no other experiments have been able to obtain corroboration of them. Most theorists have come to believe that, as a matter of fundamental principle, quarks and antiquarks cannot exist as free particles, but are permanently confined within their parent protons, neutrons, pions, and so forth. *See* QUARKS.

Further experiments. The electron-scattering experiments have been superseded by muon-scattering experiments, because beams of energetic muons are easier to produce. These experiments were done first at Fermilab, but now the most accurate muon-scattering experiments are those performed at the European laboratory CERN. Like the electron, the muon interacts electromagnetically with the proton or neutron target by exchanging a photon with it. Fermilab and CERN also do neutrino-scattering experiments. The principle of these is similar, except that the interaction is not electromagnetic but weak. (The weak interaction is responsible for neutron decay, for example.) So the information obtained from neutrino scattering is a little different from that obtained from electron or muon scattering. The former complements the latter and helps to confirm the quark interpretation. Further verification comes from proton-proton scattering experiments, electron-positron annihilation experiments, and various others.

Detailed quark-gluon structure. The various experiments find that actually the three-quark picture of the proton or neutron is a little too simple. It is true that, in a Lorentz frame where the proton or neutron is fast-moving, three quarks are found to have a significant fraction of its momentum. However, there are also a very large number of objects that have small momentum. These include a large number of quark-antiquark pairs. Each of these pairs is electrically neutral and so would not interact with a long-wavelength photon, but a small-wavelength photon can resolve the pair and interacts with the charged quark and antiquark separately. It is found that about half the total momentum of the proton or neutron is carried by the charged quarks and antiquarks (including the original three quarks of the simple model).

Evidence is growing that the other half of the momentum is carried by objects known as gluons. Just as the electromagnetic force between charged particles is known to be transmitted by photon exchange, it is believed that the strong force that holds the quarks together inside the proton or neutron is transmitted by gluon exchange. The quantum field theory of the electromagnetic force is called quantum electrodynamics, while that of the strong force is called quantum chromodynamics. This is because, while the photon couples to the electric charge of a particle, the strong force couples to an attribute of the quarks called *color*. (This is a technical term that has nothing to do with color in the ordinary sense.)

Color confinement. If several quarks of different colors are bound together, they can make states that are uncolored. It is apparently only these uncolored states that can exist as free particles, for example, the proton or neutron. Separate quarks, being colored, are not observed as free particles, so the quark-confinement problem is really a color-confinement problem—according to the present theoretical thinking.

When the short-wavelength photon in the muon-scattering experiment hits a quark in the proton or neutron, the quark tries to escape. As it separates from the rest of the target particle, a color-electric field is set up, and it is assumed that the structure of the theory is such that the energy of this field increases with the separation. When it becomes large enough, quark-antiquark pairs can be created in the field (again according to the relation $E = mc^2$). The quark that is trying to escape then combines with one of the antiquarks, to form one of the pions that is observed in the experiment.

Before this happens, the quark can radiate gluons, just as electrons radiate photons. Experimental evidence that gluon radiation does occur is slowly accumulating. Much more work needs to be done to verify all these ideas and explore them further. [PETER LANDSHOFF]

Stability. Recent theoretical speculations have led to predictions of proton decay which are now in the process of being tested. After radioactivity was discovered in 1896, it was thought for some time to be a property of heavy elements only (such as uranium); these elements were shown to decay in many steps of varying half-life, by emission of alpha or beta rays, to stable elements. Later it was found that a few of the lighter elements (for example, ^{40}K) also had radioactive isotopes. Still later it was discovered that radioactive isotopes of all elements can be artificially produced. All radioactive isotopes ultimately decay in one or more steps to stable ones. Whether these isotopes are absolutely stable depends on the fate of the proton.

It is not possible to determine how long an isolated proton has existed as a proton. It may have been "born" that way in the big bang, or it may be the descendant of one or more radioactive baryons; for example, it may have arisen in the decay of a neutron ($n \rightarrow p + e^- + \bar{\nu}_e$) or a lambda ($\Lambda^0 \rightarrow p + \pi^-$), and so forth, thus conserving baryon number, no matter how long the decay chain is. According to the quark hypothesis, the proton is made of three quarks, two u (up) quarks, and one d (down) quark; the neutron is made of two d quarks and one u quark; and the Λ^0 of a u, a d, and an s (strange) quark. Thus, neutron decay corresponds to a transition $d \rightarrow u$ and Λ^0 decay to a transition $s \rightarrow u$. In such transitions, as well as in all other known particle interactions, the number of quarks is always conserved. Their net number does not change: a quark can change into another quark, or quark plus antiquark pairs can be produced. It is estimated that there are more than 10^{80} quarks in the universe, but it is not known whether they will exist forever.

Until a few years ago it was widely believed that protons are absolutely stable. This belief was first explicitly expressed by H. Weyl, and later by E. C. G. Stückelberg and E. P. Wigner. Wigner suggested that a conservation law for the number of heavy particles (protons and neutrons) might be responsible for the stability of the protons in the same way as the conservation law for charges is responsible for the stability of the electron. Without a conservation law for heavy particles, the proton could disintegrate, through the emission of a light quantum, into a positron, just as the electron could disintegrate, into a light quantum and a neutrino, were it not for the conservation law for the electric charge.

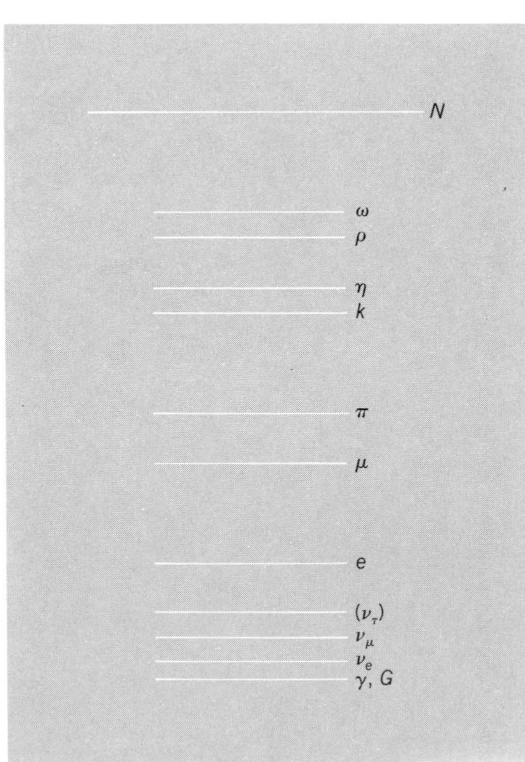

Fig. 2. Schematic energy-level diagram of particles of mass lower than that of a nucleon. (*From M. Goldhaber, Status of conserved quantum numbers, in D. B. Cline and F. E. Mills, eds., Unification of Elementary Forces and Gauge Theories, pp. 531–548, Harwood Academic Publishers, 1978*)

Until recently it was generally believed that there are three fundamental absolute conservation laws (other than those for the dynamic quantities energy, momentum, and angular momentum):

1. Total electric charge (ΣQ) is conserved. ($Q = -1$ for a negative electron, $+1$ for a positron or proton, and so forth). *See* ELECTRIC CHARGE CONSERVATION.

2. Total baryon number (ΣB) is conserved. ($B = +1$ for a baryon, for example, a proton, and -1 for an antibaryon.)

3. Total lepton number (ΣL) is conserved. (L is a number assigned to weakly interacting particles, for example, electrons, muons, and neutrinos; $L = +1$ for an electron, negative muon, or neutrino, -1 for a positron or antineutrino, and so forth.)

In 1954 M. Goldhaber, together with F. Reines and C. Cowan, Jr., started experimental tests of the assumption of proton stability. Such tests have been refined over the years so that it is now known that protons live longer than 10^{30} years. Questioning the proton's stability is not the same as predicting its decay. But the question stimulated some theoreticians to consider theories in which a proton can disappear. Some of these theories are testable with present-day techniques and have in turn stimulated renewed experimental efforts.

Proton decay modes. It will be assumed that in any proton decay, electric charge, mass-energy, momentum, and angular momentum would be conserved. Since there are no baryons lighter than the proton, B would not be conserved, and since the proton is a fermion (spin 1/2), at least one fermion (or, more generally, an odd number of fermions) would have to be emitted in order for angular momentum to be conserved. The only lighter fermions are leptons, e, μ, ν, . . . ; thus at least one lepton would be created for example, e^+; that is, L would also not be conserved. The proton could decay only if its electric charge were exactly equal to that of, say, a positron. Experiment shows that these charges do not differ by more than 10^{-20} of a unit of charge (electron charge).

Figure 2 is a schematic energy-level diagram of all particle families of mass lower than a nucleon mass. Thus one or more of these particles may appear in proton decay, provided their total masses add up to less than a proton mass. The photon (γ) and the graviton (G) are believed to have mass zero, and so may some or all of the neutrinos, though at present only upper limits are known for their masses. The mesons shown decay rapidly into leptons or photons. The positrons annihilate when they meet negative electrons. Including all particle and antiparticle members of these families (their number is given in parentheses), there are 24 possible different particles among the potential decay products of a proton: $\gamma(1)$, $G(1)$, $\nu_e(2)$, $\nu_\mu(2)$, $\nu_\tau(2)$, $e(2)$, $\mu(2)$, $\pi(3)$, $k(4)$, $\eta(1)$, $\rho(3)$, $\omega(1)$. *See* NEUTRINO.

Thus in the decay of the proton at least two particles would be expected to be emitted (to conserve energy and momentum) and at least one particle of positive charge and at least one lepton (these two properties could be combined, for example, in a positron). But since more than two particles could be emitted, the decay modes might be very complicated. Some of the contemplated experiments therefore concentrate on some simple possible decay modes (for example, $p \rightarrow \pi^0 + e^+$).

Unifying theories. The 1979 Nobel Prize in Physics was awarded to Sheldon Glashow, Abdus Salam, and Steven Weinberg for their success in unifying theories of electromagnetic and weak forces. For some time it had been known that these forces are not constant; they change with energy. The new formulation, which makes use of the mathematical language of group theory, implies that the two types of forces should become comparable in strength when tested by probes with an energy equivalent to about 100 proton masses, the approximate masses of new particles (called $W\pm$ and Z^0 bosons) which mediate transitions between quarks. Accelerators with sufficient energy to produce such particles are now being built. In weak interactions virtual transitions are postulated in which a quark emits a W^- while changing into another quark, for example, $d \rightarrow u$; the W^- then changes into an electron and antineutrino. This is supposed to be the process underlying beta decay of the neutron, a process which exhibits neatly the three fundamental conservation laws of electric charge, baryon number or quark number, and lepton number.

A few years ago theoreticians started to consider larger mathematical groups to unify the strong, weak, and electromagnetic forces. This led to predictions of a mechanism by which the proton might decay. Some of these theories predict that the strong, weak, and electromagnetic forces become comparable at a very large energy, named the unification mass, equal to $\sim 10^{15}$ proton masses. At that energy two u quarks can combine into a boson X whose mass equals the unification mass. This boson can decay into $e^+ + \bar{d}$. Inside a proton this process can happen via a virtual transition of two u's into an X which would then decay into $e^+ + \bar{d}$, where the $\bar{d}$ combines with the leftover d to form a π^0, thus leading to the reaction $p \rightarrow \pi^0 + e^+$, which would be a prominent decay mode of the proton. Many other possible decay modes have been calculated, for example, $p \rightarrow \pi^+ + \bar{\nu}_e$. The lifetime of the proton is predicted to be about $10^{31\pm1}$ years, only a little longer than the existing experimental limit.

Experimental tests. Many experiments to test these theories are on the way. The nucleon decay detectors must contain well above 10^{32} nucleons in order that one may hope to observe a few events per year, thus they must weigh more than 100 metric tons. Detectors of about 100 to 10,000 metric tons are being readied for this important test. To reduce the effect of cosmic radiation, such experiments are best done underground.

One of these experiments uses a water Cerenkov detector of nearly 10,000 metric tons' weight; it is being installed in a salt mine near Cleveland, OH. A reaction such as $p \rightarrow \pi^0 + e^-$ leads to the production of Cerenkov radiation which can be detected by photomultipliers. The experiment is designed to confirm or falsify the theoretical predictions.

An Indian-Japanese collaboration has recently reported tentative evidence for a possible proton decay. Their experiment is located in the deep Kolar Goldfield mine in South India. In a 150-ton detector three events were seen, which are, however, incomplete as far as demonstrating energy and momentum conservation. Further experiments are needed to draw definite conclusions.

Cosmological implications. If, as some believe, the universe continues to expand indefinitely, and protons do decay, there will be left chiefly low-energy photons and neutrinos, together with gravitons, all particles of zero or very small mass. Thus most mass would asymptotically be transformed predominantly into kinetic energy. This transformation would take at least 10^{20} times the present age of the observable universe.

For background information *see* ELEMENTARY PARTICLE; FUNDAMENTAL INTERACTIONS; GLUONS; LEPTON; QUANTUM CHROMODYNAMICS; QUARKS in the McGraw-Hill Encyclopedia of Science and Technology. [MAURICE GOLDHABER]

Bibliography: G. Feinberg and M. Goldhaber, *Proc. Nat. Acad. Sci.*, 45:1301–1312, 1959; H. Georgi and S. L. Glashow, *Phys. Today*, pp. 30–39, September 1980; M. Goldhaber, P. Langacker, and R. Slansky, *Science*, 210:851–860, 1980; G. t' Hooft, *Sci. Amer.*, 242 (6):90–114, June 1980; J. Iliopoulos, *Contemp. Phys.*, 21:159–183, March/April 1980; M. Jacob and P. V. Landshoff, *Sci. Amer.*, 242(3):66–75, March 1980; S. Weinberg, *Sci. Amer.*, 244(6):64–75, June 1981.

Pyroxene

Pyroxene minerals occur in a wide variety of rocks in the Earth's crust and in its mantle to a depth of several hundred kilometers. Pyroxenes are almost ubiquitous in lunar rocks and in stony meteorites, and it is very likely that they will be found on the other terrestrial planets. Recent advances in pyroxene mineralogy include: (1) determination of crystal structures at high temperatures or high pressures; (2) discovery of new space groups for pyroxene structures; (3) elucidation of intergrowth relationships by means of high-resolution transmission electron microscopy; (4) chemical analyses of a large

Major chemical subdivisions of pyroxenes*

Subdivision and members	Formula	Space group
Magnesium–iron pyroxenes		
Enstatite	$Mg_2Si_2O_6$	*Pbca, P2₁/c, Pbcn*‡
Ferrosilite	$Fe_2^{2+}Si_2O_6$	*Pbca, P2₁/c*
Orthopyroxene	$(Mg,Fe^{2+})_2Si_2O_6$	*Pbca*
Pigeonite	$(Mg,Fe^{2+},Ca)_2Si_2O_6$	*P2₁/c, C2/c*
Calcium pyroxenes		
Augite	$(Ca,R^{2+})(R^{2+},R^{3+},Ti^{4+})(Si,Al)_2O_6$†	*C2/c*
Diopside	$CaMgSi_2O_6$	*C2/c*
Hedenbergite	$CaFe^{2+}Si_2O_6$	*C2/c*
Johannsenite	$CaMnSi_2O_6$	*C2/c*
Calcium–sodium pyroxenes		
Omphacite	$(Ca,Na)(R^{2+},Al)Si_2O_6$	*C2/c, P2/n, P2*
Aegirine-augite	$(Ca,Na)(R^{2+},Fe^{3+})Si_2O_6$	*C2/c*
Sodium pyroxenes		
Jadeite	$NaAlSi_2O_6$	*C2/c*
Acmite	$NaFe^{3+}Si_2O_6$	*C2/c*
Ureyite	$NaCr^{3+}Si_2O_6$	*C2/c*
Lithium pyroxenes		
Spodumene	$LiAlSi_2O_6$	*C2* (~*C2/c*)

*After M. Cameron and J. J. Papike, Crystal chemistry of silicate pyroxenes, in C. T. Prewitt (ed.), *Reviews in Mineralogy*, vol. 7, chap. 2, Mineralogical Society of America, 1980.

†$R^{2+} = Mn^{2+}, Fe^{2+}$, Mg; $R^{3+} = Fe^{3+}, Cr^{3+}$, Al.

‡Multiple entries indicate polymorphs having identical composition.

number of pyroxenes from lunar highlands and mare (basaltic) rocks; and (5) major progress in determining the phase-equilibrium relations and thermodynamic properties of pyroxenes.

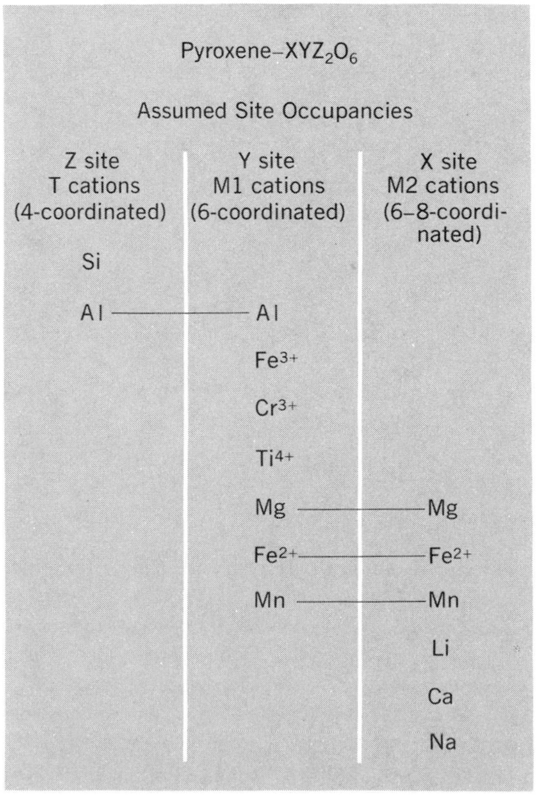

Pyroxene–XYZ_2O_6

Assumed Site Occupancies

Z site T cations (4-coordinated)	Y site M1 cations (6-coordinated)	X site M2 cations (6–8-coordinated)
Si		
Al	Al	
	Fe^{3+}	
	Cr^{3+}	
	Ti^{4+}	
	Mg	Mg
	Fe^{2+}	Fe^{2+}
	Mn	Mn
		Li
		Ca
		Na

Fig. 1. Distribution of common cations in the rock-forming pyroxenes. (*From M. Cameron and J. J. Papike, Crystal chemistry of silicate pyroxenes, in C. T. Prewitt, ed., Reviews in Mineralogy, vol. 7, chap. 2, Mineralogical Society of America, 1980*)

Crystallography and crystal chemistry. All pyroxenes consist of SiO_4 tetrahedra that are linked in single $(SiO_3)_\infty$ chains. In the three-dimensional structure, layers of chains are separated by layers of octahedra that contain mainly divalent cations (monovalent, trivalent, and quadrivalent cations occur as minor constituents of many natural pyroxenes and as major components of less common types). M1 octahedra are regular in shape and accommodate smaller cations; M2 octahedra tend to be larger and less regular in shape. The table lists the cations that are commonly found in the three types of sites. In addition to the four major space groups (orthorhombic: *Pbca, Pbcn;* monoclinic: *C2/c, P/2₁/c*) that have long been known, *P2* and *P2/n* space groups have recently been reported for omphacites. A *P2₁ca* space group has also been reported for a very slowly cooled lunar pyroxene. The table gives the space groups for a variety of pyroxenes.

In the past dozen years, major advances have been made in the determination of detailed crystal structures of pyroxenes at high temperatures or at high pressures. This information has been of considerable importance in aiding the understanding of the behavior of pyroxenes in the Earth's interior. Although healing produces expansion and increased pressure results in compression, in detail the effects are not exact opposites. As a first approximation, the SiO_4 tetrahedra change very little in size: compression and thermal expansion are accomplished mainly by changes in the angles between tetrahedra along the SiO_3 chains and in the size and shape of the octahedral sites (M1, M2). An important aspect of the high-temperature studies is the recognition that the *P2₁/c* pyroxenes (usually low-calcium) convert upon heating to the *C2/c* structure typical of high-calcium clinopyroxenes. Because the structures are the same at high temperatures, the high- and low-calcium clinopyroxenes can be treated as obeying a single thermodynamic equation of state at their temperatures of formation.

Chemistry. Although the major chemical variations of pyroxenes have long been known (Fig. 1), the past decade has seen a quantum increase in chemical data for pyroxenes from the Moon, meteorites, and terrestrial midocean-ridge basalts. These pyroxenes carry a chemical signature of the planet in which they formed. For example, many pyroxenes from lunar mare basalts are much more iron-rich than their terrestrial counterparts.

Phase equilibria. Major advances have been made in the determination of the phase equilibrium relations of pyroxenes. The complicated polymorphic relations of $Mg_2Si_2O_6$ are still not well understood, however. Pyroxenes of intermediate calcium contents [$0.15 < Ca/(Ca + Mg + Fe) < 0.35$ in Fig. 2] are generally not stable at the temperatures at which most rocks form; instead there occur two coexisting pyroxenes—high-calcium (augites) and low-calcium (pigeonites or orthopyroxenes or both). Potentially the compositions of these coexisting pyroxenes should serve as a geothermometer—that is, as an indication of the temperature of formation of the rock in which they occur—but some problems remain in applying the phase-equilibrium data to natural occurrences. Figure 3 illustrates the principle of the geothermometer; iron has been omitted for simplicity. At 1300°C, the two pyroxenes indicated (Fig. 3*a*) should form in equilibrium if the bulk composition is suitable. If a similar pair of pyroxenes were found in a rock, the temperature of formation would be taken as 1300°C. In reality, the effects of Fe and of other elements must be taken into account to apply the geothermometer to most rocks. Although these effects can sometimes be handled graphically, it is best to handle them via a thermodynamic model.

The content of Al in pyroxenes is a complex function of both temperature and pressure. In theory, if the temperature can be estimated independently, the Al content can often be used to estimate the pressure at which the pyroxenes (and the rocks containing them) crystallized. Unfortunately, there is still considerable disagreement among various experimenters as to the equilibria of aluminous pyroxenes at high pressures, in large part because of difficulties in obtaining and measuring the equilibrium partitioning of Al. Major progress is being made in several laboratories, and the ambiguities should be largely overcome in the next several years. Development of thermodynamic solution models will aid in applying the experimental results to natural occurrences.

High-resolution transmission electron microscopy is providing much information on microtextures of pyroxenes. Ultimately such observations should aid greatly in applying experimental results to the interpretation of the cooling histories of natural pyroxenes.

Application of phase-equilibrium results to natural pyroxenes presumes that the latter reflect equilibrium; this is not always the case. In particular, quickly cooled pyroxenes from lava flows tend not to crystallize at equilibrium. For example, many pyroxenes from lunar basalts have Al contents corresponding to high pressures of crystallization, even though all other indications are that the basalts crystallized at the lunar surface.

Thermodynamics. Progress on the thermodynamics of pyroxenes is being made on two fronts. Calorimetric measurements made on well-crystallized synthetic pyroxenes are adding greatly to knowledge

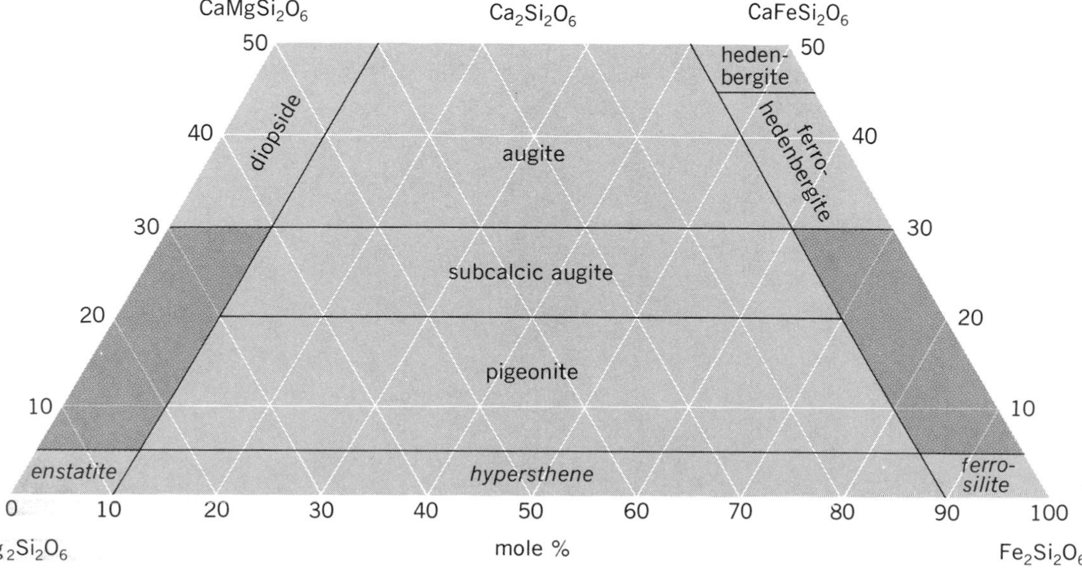

Fig. 2. Simplified nomenclature for the Ca-Mg-Fe pyroxenes. Enstatite, hypersthene, and ferrosilite are orthopyroxenes; other minerals are clinopyroxenes. Shaded areas indicate compositions that rarely, if ever, occur in nature.

of the free energies of formation of end-member pyroxenes. Calorimetric studies of solid-solution series (for example, $Mg_2Si_2O_6$–$Fe_2Si_2O_6$) yield information on departures from ideality, and aid in formulating thermodynamic solution models. The results of phase-equilibrium studies are also being used to develop such models. Most progress has been made for the $Mg_2Si_2O_6$–$CaMgSi_2O_6$ system; Fig. 3 was calculated from such a thermodynamic model. The Gibbs free energy of mixing for the clinopyroxenes in this system can be expressed by Eqs. (1)–(3). R

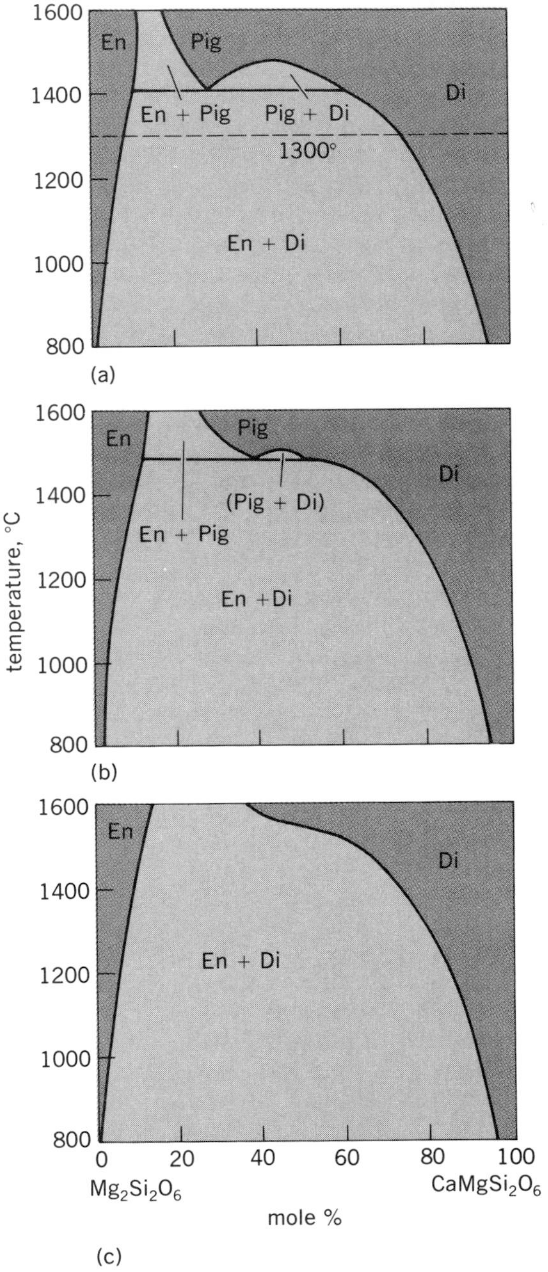

$$G(\text{mixing}) = G(\text{configurational}) + G(\text{excess}) \quad (1)$$

where

$$G(\text{configurational}) = RT(X_1 \ln X_1 + X_2 \ln X_2) \quad (2)$$

and

$$G(\text{excess}) = X_1X_2[X_1(31216 - 0.0061P) + X_2(25484 + 0.0812P)] \quad (3)$$

is the gas constant (8.3143 joules/kelvin-mole); T is the temperature in kelvins; P is the pressure in bars (1 bar = 10^5 Pa); X_1 is the mole fraction of $Mg_2Si_2O_6$, and X_2 the mole fraction of $CaMgSi_2O_6$; G is expressed in joules.

Because Fe and Mg partition between the M1 and M2 sites, thermodynamic expressions for pyroxenes containing those elements must include effects due to ordering.

Physical properties. Recent advances in measuring the physical properties of pyroxenes have come both in the determination of compressibilities and thermal expansions and in the determination of seismic (acoustic) velocities. These velocities have been traditionally measured by attaching transducers to crystals in various orientations, a method that requires crystals approximately 1 cm on edge. Many pyroxenes of geophysical interest have been unavailable in the requisite size or degree of purity, and a major improvement has been the development of a technique that uses Brillouin scattering to measure the velocities. This method can be used on crystals as small as 100 μm on a side, a decrease of a factor of 10^6 in the volume required.

Future developments. Crystal-structure determinations at simultaneous high pressures and temperatures are needed to aid in understanding the behavior of pyroxenes in the Earth's interior. Transmission electron microscopy has already provided much information on microtextures of natural and synthetic pyroxenes, and promises more. The next few years should see major advances in the development of thermodynamic models and their use for geothermometry and geobarometry.

For background information *see* BASALT; GEOLOGIC THERMOMETRY; HIGH-PRESSURE MINERAL SYNTHESIS; METEORITE; PYROXENE in the McGraw-Hill Encyclopedia of Science and Technology.

[DONALD H. LINDSLEY]

Bibliography: W. A. Deer, R. A. Howie, and J. Zussman, *Rock-Forming Minerals*, vol. 2a: *Single-Chain Silicates*, 1978; D. H. Lindsley, J. E. Grover, and P. M. Davidson, in S. K. Saxena (ed.), *Advances in Physical Geochemistry*, 1981; C. T. Prewitt (ed.), *Reviews in Mineralogy*, vol. 7: *Pyroxenes*, 1980.

Fig. 3. Phase diagrams calculated for the $Mg_2Si_2O_6$–$CaMgSi_2O_6$ system. The thermodynamic model used in the calculations fits available phase-equilibrium data from 1 atm (100 kPa) to 40 kilobars (4 GPa). Shaded areas indicate where one pyroxene is stable; unshaded areas where two pyroxenes are stable. (*a*) 10 kilobars (1 GPa). Data points indicate the compositions of orthopyroxene (En = enstatite) and clinopyroxene (Di = diopside) that could coexist at 1300°C, and indicate the potential of coexisting pyroxenes to serve as a geothermometer. (*b*) 20 kilobars (2 GPa). (*c*) 30 kilobars (3 GPa). The field for pigeonite (Pig) as a discrete phase dies out as pressure exceeds approximately 21 kilobars (2.1 GPa).

Quantum solids

At very low temperatures all classical motion, such as thermal vibration, is frozen out, and matter either becomes inert and uninteresting to the physicist or begins to show the effects of quantum mechanics on a macroscopic scale. In solid ^{3}He at temperatures within 0.001°C of absolute zero, such effects lead to an ordering of the nuclear spins into a periodic antiferromagnetic arrangement throughout the lattice. Recent nuclear magnetic resonance (NMR) experiments have supplied the first information on the spin arrangement in the antiferromagnetic state. The results show the structure to be much more complicated than previously thought, and provide renewed stimulus for further research into the quantum tunneling processes involved.

Atomic exchange. The classical picture of a solid at low temperature is of a set of atoms almost at rest on the lattice sites, with only very small vibrations due to whatever thermal energy is left. In solid helium, by contrast, even at the absolute zero of temperature, the atomic mass is so small and the interatomic force so weak that the atoms are forever in a state of violent zero-point motion. Each atom makes large excursions from its lattice site, until a collision with a neighbor sends it back. One dramatic consequence is that the solid tends to shake itself apart to a liquid at arbitrarily low temperatures: a pressure of 34 atm (3.4 MPa) must be applied just to form a solid. A more subtle consequence is that sometimes in this solid the zero-point motion of neighboring atoms is such that they exchange lattice sites. Exchange happens very rarely—about once in every 10,000 to-and-fro motions—but this is sufficient to lead to an ordering of the nuclear spins at a temperature much higher than nuclear magnetic ordering temperatures of conventional solids, where the ordering is produced by the tiny direct magnetic interaction between the nuclear moments.

The way that an exchange of spatial coordinates can lead to an effective spin interaction is known from electron physics. Consider two hydrogen atoms approaching each other. If the electron spins are antiparallel, both electrons may go into a molecular orbital surrounding the two nuclei, and the two atoms bind strongly. On the other hand, if the spins are parallel, the Pauli exclusion principle forbids them from both occupying this binding orbital, and the two hydrogen atoms repel. Any direct magnetic interaction between the spins is quite unimportant, yet the energy depends strongly on the orientation of the spins.

In the same way, if neighboring atoms in solid ^{3}He exchange places, the energy depends on the orientation of the nuclear spins. If, as might be expected, pairwise exchange between nearest neighbors is most important, a very simple expression for the energy E is given below, where $\vec{s}_i$ is the direc-

$$E = J \sum_{i,n} \vec{s}_i \cdot \vec{s}_{i+n}$$

tion of the spin on the ith site, and the sum over n runs over the eight nearest neighbors with which particle i can change places. The positive constant J sets the scale of the energy dependence and is related to the frequency of the exchange process ν through $J = h\nu$, where h is Planck's constant.

Although the origin of the energies is a difference in spatial wave functions, it is easy to see that the above equation predicts a definite arrangement of the spins as the lowest energy state. The equation shows a tendency for a spin and its neighbor to be antiparallel ($\vec{s}_i \cdot \vec{s}_{i+n} = -1$). In the body-centered cubic structure of solid ^{3}He, this implies a ground state in which all the corner sites have nuclear spin pointing "up" along some arbitrary direction, and all the center sites have spin "down" along this direction. Investigating the ground state provides a good test of the validity of the simple energy expression above.

NMR experiments. If a magnetic field is applied to magnetic moments, such as those associated with the ^{3}He nucleus, they will precess about the field—the same motion as a spinning top in the gravitational field. Furthermore, the frequency of precession, known as the Larmor frequency, is characteristic of the particular species. Observing this resonant frequency becomes a useful tool, because interactions of the magnetic moments with each other or with other species may shift the frequency from the Larmor value.

The use of NMR as a method of diagnosing the antiferromagnetic ordering in solid ^{3}He follows from a simple observation. In an NMR experiment all the ^{3}He spins tend to precess together. Such a motion does not change the exchange interaction, which depends only on the angles between pairs of spins. Consequently, the large exchange interaction, which determines the ordering of the ground state but is unknown at present, does not lead to any shifts from the Larmor frequency. However, the much smaller nuclear magnetic dipole interaction can give shifts, and is easily calculated for any spin arrangement. The idea, then, is to investigate the nuclear magnetic dipole energy with NMR and to use this to characterize the ground state.

Results. In these NMR experiments large shifts from the Larmor frequency were observed. This in itself is an important observation, and conclusively eliminates the equation presented above as a possible description of solid ^{3}He. This is because the spin arrangement in the ground state of this equation retains the cubic symmetry of the lattice. In their work on the dielectric properties of matter, H. A. Lorentz and L. Lorenz showed that any cubic array of dipoles leads to no net dipole energy, and so in solid ^{3}He to no shift of the NMR frequency. In fact, this single observation eliminated nearly all of the previously proposed spin arrangements for the antiferromagnetic phase of solid ^{3}He.

A careful analysis of the full NMR spectrum (precession frequency as a function of magnetic field) led to a new suggestion for the ground-state spin arrangement (see illustration). The arrangement consists of sets of planes of atoms parallel to one

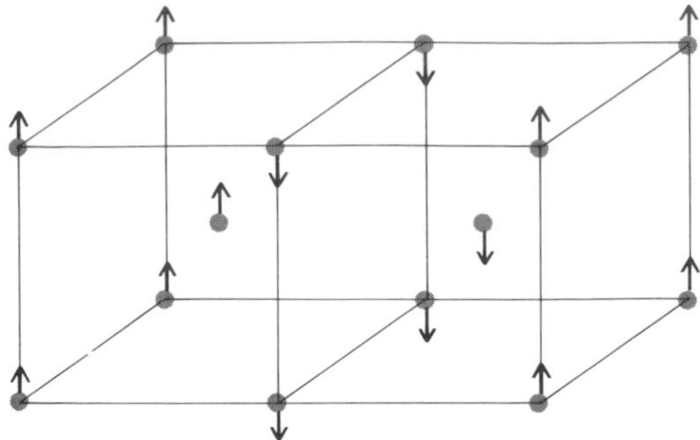

The up-up-down-down structure proposed for solid ^{3}He, showing the simplest building block for constructing the periodic arrangement throughout the lattice. The arrows indicate the direction of the nuclear spins at each lattice site.

pair of the cube faces. On each plane, the nuclear spins are parallel (ferromagnetically aligned). The planes then follow an antiferromagnetic sequence, with a four-plane repeat length: up-up-down-down, where up and down are with respect to some arbitrary direction which is constant throughout the crystal. In fact, all that can be deduced from the NMR data are the rather unusual symmetries of the state: one pair of cube faces is picked out of three originally equivalent pairs, and then all the spins are specified by a single direction (the up-down direction referred to before). Any further assignment of the microscopic structure (for example, a four-plane repeat distance, rather than six) involves a rather imprecise quantitative comparison between the magnitudes of the predicted and measured shifts. However, in the 2 years since the NMR experiments were done, no state that gives a better fit to the data has been put forward.

Implications for atomic exchange. If, for the moment, the proposed arrangement in the antiferromagnetic state is accepted, consider what this implies for atomic exchange in solid ^{3}He. It is apparent that the proposed state suggests a competition between ferromagnetism (pairs of planes of parallel spins) and antiferromagnetism (alternate pairs of planes with spins in opposite directions). It is known from general arguments that exchange processes of an odd number of particles promote ferromagnetism, and of an even number, antiferromagnetism. A simple model allowing exchange of rings of three particles and four particles, rather than the pairs described by the above energy equation, has been shown to lead to the proposed ground state. Although much work remains to be done to test this model, it is the best candidate at present to replace the original description.

The new experimental results have shown that the consequences of atomic exchange in solid ^{3}He are much richer than previously thought. This presents a challenge to the physicist: to derive these results from a first-principles treatment. Only then can it be confidently said that the subtle interplay of single-atom quantum effects and the many-particle interactions of condensed matter are truly understood in this system.

For background information *see* ANTIFERROMAGNETISM; MAGNETIC RESONANCE; QUANTUM CHEMISTRY; QUANTUM SOLIDS in the McGraw-Hill Encyclopedia of Science and Technology.

[M. C. CROSS]

Bibliography: E. D. Adams et al., *Phys. Rev. Lett.*, 44:789–792, 1980; D. D. Osheroff, M. C. Cross, and D. S. Fisher, *Phys. Rev. Lett.*, 44:792–795, 1980; Proceedings of the 16th International Low Temperature Conference, *Physica B+C*, to be published.

Quarks

Quarks are subnuclear, strongly interacting constituents of matter carrying fractional electric charge, ±1/3 or ±2/3 the charge on an electron. Quarks are now firmly established and universally accepted as the fundamental constituents of hadronic matter: protons, neutrons, mesons, strange particles, and so forth. Thus all known matter is now understood to be composed of quarks, charged leptons (electrons, muons, and the tau lepton), and neutrinos, together with the field quanta corresponding to the various interactions: the photon (electromagnetism), the intermediate vector bosons (the weak interaction), and eight gluons (the strong interaction), plus the graviton (gravity). The quarks, charged leptons, and neutrinos may be grouped as shown in the table. Quantum chromodynamics (QCD) is a theoretical model which successfully relates many of the observed properties of hadrons and their interactions by introducing a new quantum number which may take on three values. In analogy with the three primary colors, this new quantum number has been (somewhat arbitrarily) labeled color; thus each quark is thought to exist in three states: red, blue, or green. Known hadronic particles are all presumed to be white, or color-neutral, either through a combination of all three quarks—red, blue, and green—in the particles (baryons) or through a quark and an antiquark of the same color being bound together (mesons). The straightforward QCD theory does not permit the existence of hadronic states containing one or two quarks with a net color.

Older quark searches. Since the inception of the quark model in 1961, quarks have been sought by experimental physicists almost everywhere: in cosmic rays, with high-energy particle accelerators,

The fundamental particles*

Family	q	I	II	III
Charged lepton	$-e$	e	μ	τ
Neutrinos	0	ν_e	ν_μ	(ν_τ)†
Quarks‡	$2/3e$	u	c	(t)†
	$-1/3e$	d	s	b

*For each particle there is also an antiparticle.
†The ν_τ and t have not yet been observed.
‡Each quark exists in three color states.

and in stable matter. The negative results of these many experiments stimulated the theoretical physics community to develop QCD with its confinement properties. No free quarks have been found with particle accelerators in which the searches were sensitive to masses of up to 25 GeV/c^2, nor have they been found in cosmic rays among the reaction products of primary cosmic rays.

A wide variety of searches for samples of stable matter with fractional electric charge has been carried out using air, sea water, lunar rock, and many other minerals and materials. Results have been generally negative, down to concentrations corresponding to 1 quark per 10^{22} nucleons, and in some cases, much less.

Fairbank's experiment. One search has consistently reported positive results, however. William M. Fairbank and collaborators have results from a long series of experiments using superconducting niobium spheres, each of about 90 micrograms, with consistent values of $\pm 1/3$ electron charge ($\pm 1/3e$) observed (see illustration). There have been 40 measurements made on 13 spheres, and the experiments have become progressively cleaner. Several aspects of the data are noteworthy. The measurements have been repeated on five of the small niobium balls, and in three of them it has been found that the charge changes by $\pm 1/3e$ between measurements. Most of the measurements have given integral values of charge (no quarks), suggesting that indeed the mass range is on the edge of sufficient sensitivity to barely find quarks. On the other hand, the ease with which some of the samples gain and lose quarks would suggest that they are either quite numerous or that they reside on the surface and are easily gained or lost. Fairbank's experiments employ a cryogenic system in which the niobium is in a superconducting state such that it is a perfect diamagnet, that is, with a magnetic permeability of zero. This permits the niobium balls to be suspended in a magnetic field where they can be delicately manipulated by an electric field in order to measure their charge. To make the samples good superconductors, the niobium must be very pure, and the metal spheres must be annealed so that each is a single crystal.

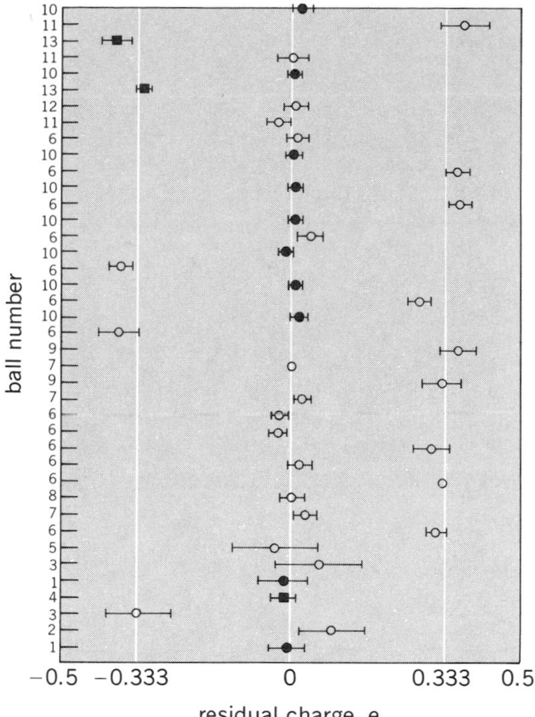

key:
$R = 140\ \mu m$ $R = 116\ \mu m$ $R = 98\ \mu m$

The residual charges measured by Fairbank's group on 13 small niobium spheres, with the measurements chronologically ordered from bottom to top. The number of the ball is noted on the left for each measurement. Symbols indicate ball radius R.

Other experiments. G. Morpurgo and colleagues have conducted a series of experiments seeking quarks in stable matter over at least as long a period as Fairbank has. These experiments have also employed magnetic levitation but with ferromagnetic samples at room temperature. In the latest reports no evidence has been found for fractional charge in over 3 milligrams of iron samples. Both Morpurgo and Fairbank have encountered and solved a host of exceedingly subtle and sophisticated experimental problems to reach their present state of certainty. As a measure of the problems in these experiments, the ratio of the applied electric force on each sphere to its weight is about 10^{-8}.

In a totally different experiment, R. Bland and colleagues have constructed a modern version of the classical Millikan oil drop experiment, using a laser, photomultipliers, and computer logic to automate the charge measurements; it has thus been possible to process a very large number of droplets in a reasonable time. These researchers have used mercury as the sample material, and have reported results based on a total of 100 micrograms; no evidence of quarks has been observed.

K. O. H. Ziock carried out magnetic levitation experiments using iron samples several years ago, with ultimately inconclusive results. It is hoped that measurements can be resumed with improved apparatus. Others are planning similar experiments.

The obviously ambiguous situation created by Fairbank's results has stimulated a number of other new searches for quarks in stable matter. R. Hagstrom is building a sensitive analog to the old shot tower. Using ink jet printing technology, he plans to generate a stream of nearly identical tiny droplets of triethylene glycol at the top of an evacuated pipe which is several meters high and contains a pair of vertical electrodes. A strong horizontal electric field in the tower will produce a displacement of a charged droplet reaching the bottom of the tower proportional to the charge-to-mass ratio of the droplet. It will be possible to process milligrams of material in a reasonably short time and hence achieve a sensitivity comparable to that of Fairbank and

Morpurgo, and a wide variety of liquids, both as pure substances and as solutions, can be used. Two other groups are building similar droplet-tower experiments.

All these experiments seek evidence of quarks in bulk sample experiments, without attempting to remove or concentrate the possible quarks. A large number of other experiments seek to evaporate, ionize, electrically concentrate, or otherwise extract quarks. Thus earlier experiments used a molecular-beam electrometer system and an ion-beam mass spectrometer; both experiments set upper limits of no more than 1 quark per 10^{22} nucleons in the samples studied. However, the number of atoms actually studied was much less than 10^{22}, and the limits quoted depend upon the assumptions made concerning the properties of quarks in a host sample. A negative result can prove either that there are no quarks or that the assumptions are wrong.

Recently multimillion-volt electrostatic accelerators have come to be used as very sensitive mass spectrometers for trace nucleus separation with applications in radioisotope dating. These same installations can be used for quark searches, and such searches have already begun. Other searches are planned using dye laser spectroscopy, solid-state devices, and other ingenious methods. Many of these searches involve separation or enrichment assumptions, however, and a negative result would seem therefore less significant than it would with the magnetic levitation, Millikan, or falling-droplet types of experiments.

Interpretation. It seems at first difficult to see how quarks could exist in stable matter and not be found with particle accelerators or in cosmic rays. In fact, it is straightforward to relate quark concentrations in the Earth's crust to cosmic-ray production, and these calculations verify such skepticism. There are three relevant remarks here. First, Fairbank reports his experimental results as the observation of fractional charge; whatever is observed may be some phenomenon quite different from the quarks of QCD. Second, quarks might have been made by cosmic rays in the material from which the Earth condensed billions of years ago; thus the concentration seen by Fairbank may be greater than what one would deduce from cosmic-ray fluxes. Third, free quarks could be a result of processes occurring during the big bang, that primordial explosion which is presumed by cosmologists to be the beginning of the universe as it is now known. Matter and energy at the exceptional densities which characterized the early instants of expansion might have permitted broken confinement, and some small fraction of the quarks, for example, 10^{-20} or so, might have remained free in the subsequent expansion. Cosmic-ray experiments would not have detected primordial quarks if their abundance were less than about 10^{-11} per nucleon.

If free, physically isolated quarks exist, the theory of QCD must be altered. Modifications to permit broken confinement are possible, but in each case they appear cumbersome and add complications, such as new, electrically neutral quarks.

It would be easier to accept Fairbank's results as evidence for free quarks if Morpurgo and others had less convincing negative results, or if the quarks did not seem to come and go so readily in Fairbank's measurements. In any event, the Fairbank result has stimulated a broader, more intelligent spectrum of quark searches in stable matter than had occurred prior to his reports.

George Zweig has explored the properties of quarks in matter from the perspective of a physical chemist, and has tried to understand where atoms containing quarked nuclei might be found. Atoms have been parameterized by their ionization potential, electron affinity, and ion radius, and attempts made to deduce by interpolation the properties of quarked atoms of various species. It has been concluded that an atom containing a quark would behave very differently from a normal atom of the same element. Thus, if quarks are in Fairbank's niobium sample, they are almost certainly not on niobium. Zweig hopes that his studies will help to guide experimentalists to the most promising quark ores; he further argues that many of the earlier quark enrichment schemes (of experiments in the 1960s and early 1970s) would have effectively rejected quarks from the sample.

It is certain that the puzzle of the free quarks will remain lively and exciting until Fairbank's result is either independently verified or definitively refuted. In either case, the search for free quarks forms a fascinating chapter in the current evolution of the rapidly increasing comprehension of the fundamental structure of matter.

For background information *see* QUANTUM CHROMODYNAMICS; QUARKS in the McGraw-Hill Encyclopedia of Science and Technology.

[LAWRENCE W. JONES]

Bibliography: L. W. Jones, *Rev. Mod. Phys.*, 49:717–752, 1977; G. S. LaRue, J. D. Phillips, and W. F. Fairbank, *Phys. Rev. Lett.*, 46:967–970, 1981; M. Marinelli and G. Morpurgo, *Phys. Lett.*, 98B:465–468, 1981.

Railroad engineering

Throughout the world a new era of rail passenger systems has emerged. The Shinkansen "Bullet" trains of Japan (Fig. 1), the all-new very-high-speed (TGV) Paris-Lyon line in France (Fig. 2), and the Direttissima of Italy are models of modern railroad technology. Achievements in these countries have raised the state of the art in line with the most vital requirements of rail passenger transport: frequency, safety, speed, and comfort.

Modern systems. The new systems are highly coordinated, bringing together the assets of tremendous organizations. The passenger railroad itself is supported by heavy industry, heavy construction, generations of progressive research and development, automation and electronic data processing, advanced traffic control and systems management, and modern communications. The Japanese Shinkansen, only a single 663-mi (1067-km) corridor, has been operating since 1964 at high speed. It has carried more than 1,700,000,000 passengers without a casualty, and is now being extended to a 2100-mi (3380-km) interisland system.

Although national railroads have the reputation for being deficit operations, these new high-speed passenger systems, as measured during more than a decade of Japanese experience, show evidence of being profitable. The revenues of the Shinkansen, after the first 2 years of operation, have exceeded expenses, and the operating ratio of that system is around 60%. The total contribution of the Shinkansen to the passenger revenue of the Japanese National Railways is 36%. This is most significant in a total system that produced more than 194,183,-800,000 revenue passenger miles (312,500,000,-000 passenger kilometers) in 1979. These trains produce heavy traffic because of their frequency, safety, speed, and reliability. They operate up to nine full trains, of 16 cars each (1342–1483 passenger capacity), per hour between Tokyo and Osaka (320 mi or 515 km) at a scheduled average speed of 101 mph (162 km/h), and up to two trains per hour between Tokyo and Hakata (663 mi or 1067 km) at a scheduled average speed of 93 mph (150 km/h). Business travel represents 62.5% of all ridership during weekdays.

The French Paris-Lyon system and the Italian Direttissima line are following the Japanese example. Other countries are planning similar systems. In each case a dilemma had to be resolved. The choices involved either upgrading the existing conventional railway system to a greater or lesser extent or developing a totally new system. The Japanese, French, and Italian lines plus an all-new Soviet line from Moscow to the south are all built on entirely new roadbeds, have eliminated highway grade crossings, and are dedicated entirely to high-speed passenger service. On the other hand, the British, German, and Canadian developments, as well as the Northeast Corridor Improvement Program in the United States, are being improved and upgraded on existing roadbeds and have not eliminated all crossings at grade.

Fig. 1. Japanese Shinkansen "Bullet" train. (*Japanese National Railways*)

Fig. 2. French National Railroads TGV train. (*French National Railroads*)

Roadbeds. The speed regime of these new railroads, 125–155 mph (200–250 km/h), dictates the necessity for a totally new roadbed because of the demand for a more perfect geometry and higher tolerances. High speeds demand straight lines. The curvatures, gradients, and other physical characteristics which were acceptable in an earlier, less hurried era have become handicaps. Furthermore, since the 50–70-mph (80–113-km/h) speed and the unmatched door-to-door advantage of the automobile reduce elapsed time appreciably, railway competition is forced into the over-100-mph (160-km/h) zone to regain its share of the market by delivering passengers to their destination first. Journey time is perceived in terms of the round trip, so frequency of service must be high. Frequency can save hours; speed can save minutes.

Passenger trains designed to run at 125–155 mph (200–250 km/h) or more create uncompromising safety demands that require high technology and superior maintenance performance. A fundamental requirement is a roadbed designed with absolutely minimum curvatures. For example, the curve radius on the newer Shinkansen route segments is never less than 4000 m. At the same time, the grade of the Shinkansen does not exceed 15 ft/1000 ft (15 m/1000 m) or 1.5% gradient. Such strict parameters demand an almost straight, all-new roadbed which results in a very high utilization of viaducts, bridges, and tunnels when the terrain presents natural obstacles. These conditions are now accepted as worldwide standards for high-speed railways.

The new railways are more nearly straight and level than those being replaced. With the reduction of circuitry, distances are reduced, enhancing the high-speed advantage and further reducing elapsed time. The Shinkansen south of Okayama, a 250-mi (400-km) segment, is 56% in tunnels and another 8% on bridges. This is the equivalent of having almost 60% of the mileage between New York City and Washington in tunnels. In all, the Japanese National Railway system has 1350 mi (2173 km) of bridges and 1120 mi (1802 km) of tunnels.

On the other hand, in France railway design has taken a different course with the Paris-Lyon Très Grande Vitesse (TGV; very high speed) route. In order to hold down the initial cost of construction the engineers have, where possible, taken advantage of the flat or rolling terrain to lay the all-new roadbed on the ground, letting it rise and fall with natural contours. It has the general appearance of a slowly curving motorway, sometimes curving as it rises and falls roller-coaster fashion. However at 13 m (43 ft) width it is less than half the width of a four-lane highway, and has at least three times the passenger carrying capacity per hour. Further, it has no tunnels.

Avoiding tunnels has cut costs and has removed the possibility of aerodynamic problems that would have been encountered when two very-high-speed trains passed underground. The Japanese trains, cruising at a slightly lower speed, are slightly pressurized and there is no noticeable aerodynamic problem. Most passengers scarcely notice the event in the tunnels.

This important decision on the part of the French engineers requires the acceptance of gradients to 35/1000 (3.5%). The accepted International Standard is 15/1000 (1.5%). A TGV trainset climbing at 260 km/h (162 mph; on full power will run for 3.5 km (2.17 mi) and rise 122 m (400 ft) vertically as its top speed falls to an acceptable 220 km/h (137 mph). In other areas land-fill embankments 32 m (105 ft) high and deep cuts, the deepest at 40 m (131 ft), have been built to keep the rise and fall of the roadbed within the 35/1000 gradient limits.

While the original design permitted a 30% reduction in initial construction costs, the operating costs of the TGV system are high because of the power required to maintain high speeds on the steep gradients. It is anticipated that high fuel costs eventually will require a modification of the design of the roadbed, with the construction of some tunnels and smoothing out of the excessive gradients.

When a decision has been made to build an all-new, high-speed rail passenger system, railroad planners have found it necessary to restrict the system to passenger trains and to exclude freight. The Japanese included the concept of a dual system in their initial planning and development process until actual tests proved that high-speed, high-frequency freight operations would be too dangerous.

Good track structure resembles sculpture, and railroading may be considered an art. The Shinkansen line was conceived and designed by precision engineers to join Tokyo and the great cities of Japan with ribbons of steel, fitting them to the terrain, controlling the contours, the gradient, and the curvature thus forming the final geometry so that speeding trains under controlled conditions could safely meet planned schedule requirements with wheel loads at equilibrium at all times. All-new high-speed systems are now being built on concrete slab track, similar to much of the earlier construction on the Shinkansen line. This keeps maintenance costs to a minimum and assures achievement of the highest standards. Track designed to produce the balance required for structural equilibrium is also physiologically comfortable, an important consideration in any passenger system.

Maintenance. After the conception and design of a three-dimensional structure, it must be built and maintained within the limits of design. Long before the final system had taken shape, Japanese engineers had placed an electric-track integrated inspection train in operation. Beginning with the first test segment of track south of Tokyo and continuing over each new segment, this highly instrumented electronic train, made of seven multipurpose inspection cars, runs the same schedule as the commercial trains to inspect, detect, and measure irregularities of both track and electrical components. As a mark of its excellent performance, this measurement train records all track irregularities to the extent that when the track is out of its planned design alignment by 1.5 mm or more per meter, maintenance equipment is dispatched to that point the same day. With the Tokyo-Hakata line in full operation, this track-measurement train is run at least once a week on the Hikari high-speed schedule. At other times this train, carrying an inertial gyroscopic unit monitored by a computer, is used as an aid in construction to check the new track sections being built to the north of Tokyo on the new Shinkansen routes planned for early operation. The use of the electric-track integrated inspection train is of prime importance in this high-technology rail system.

Rolling stock. After the so-called route-roadbed-track and the dedicated-passenger-system decisions have been made, the next phase of the planning process for these new railroads is design of rolling stock. Locomotives are eliminated, and electricity is chosen as the power source. The overriding consideration of safety in the 125–155-mph (200–250-km/h) speed range requires that rolling stock meet high standards.

Experience has shown that the speed ceiling above which passenger trains hauled by heavy locomotives must be abandoned is in the vicinity of 120 mph (193 km/h). Above this speed, it is necessary to use self-propelled sets with driving axles distributed so as to reduce the per-axle mass, the truck mass, and the nonsuspended mass. There must be compatibility between track design and rolling stock. Examples are the superior technology of the French electric-traction Très Grande Vitesse (Fig. 2) on the very-high-speed Paris-Lyon route, and the Japanese Shinkansen equipment with its perfect safety record. Both were designed specifically for those systems. The Shinkansen employs small computer-controlled electric motors on each axle in the train, that is, 64 motors.

Management. There has been considerable technological commonality in the course of these developments as well as a similarity in the structuring of organization and management. These railroads all have had direct access to excellent in-house research and development facilities. Technological innovations are generated for the British Railways by a staff of 900 located at the Railway Technical Centre laboratory facilities in Derby. In France railway technology is developed by all members of the Federation of Railway Industries in close collaboration with the French National Railroads. The Railway Technical Research Institute of Japan, established in 1907, has a staff of 700 engineers alone.

Similarly, they all have heavy construction and heavy maintenance subsidiaries. British Rail Engineering Ltd., a subsidiary of the British Railway Board, is the largest engineering complex of its kind in the United Kingdom, with 13 major plants (employing over 35,000 people) strategically placed throughout the country. Its principal role is to overhaul and repair all locomotives, carriages, and wagons (freight cars) in the British Railway fleet and to manufacture new locomotives and rolling stock. The Japanese National Railway and the French National Railroads have similar large complexes. One of the most important elements of modern railway development in Japan has been the Japan Railway Construction Corporation. This heavy-construction organization has been responsible for much of the work on the new high-speed lines. Work is well along on the world's longest undersea tunnel, about 33 mi (53 km), between the main island of Honshu and the northern island, Hokkaido. Furthermore, among the 20 longest center-span bridges in the world, 10 of them will be Japanese railroad bridges, including the world's longest.

The Italians have tackled heavy-construction projects on their new high-speed line. The line from Paris to Lyon in France, an all-new construction project, has been used to set a new rail speed record.

Signaling and communications. The high-speed operations and all-new construction have created a demand for new signaling and communications. Higher speeds, higher frequencies of service, and reduced headway depend heavily upon automatic controlling systems. The new rolling stock is designed with modern braking systems able to stop the train in about the same distance as the slower and heavier units being replaced; yet high speed and short headway intervals place extreme demands upon signal systems to assure proper safety and traffic control, especially during emergencies. This is accomplished by depending upon electronic data processing, not only for information and reservations functions but for automatic traffic control and emergency requirements of high-speed, high-volume operations.

Operations. There are a number of reasons for countries to move swiftly into the development of high-speed corridors. First, these high-speed systems are profitable. In the densely populated regions of Japan, up to 1,000,000 passengers a day ride the corridor. In 1980 the average was 344,000 per day. This produces a positive cash flow which is used to support the costly outlying passenger routes and special passenger services that the government stipulates must be low-cost. Second, these corridor operations are frequent and safe. Passengers find them more convenient and comfortable than a long automobile trip. In many areas they actually produce better door-to-door time than airline schedules do. Third, they are environmentally acceptable. These new corridors operate on electricity that is clean and up to 30% more fuel-efficient than diesel power. Perhaps the most important reason for corridor development from the point of view of the railroad is that it offers an opportunity to revitalize the entire system and to upgrade the total infrastructure. Generation-long declines in ridership and revenues are being reversed by the application of high technology. This revitalizes not only the railroads but also the huge national infrastructures that support the railroads. Good transportation itself generates commerce, and trains revitalize cities.

For background information *see* RAILROAD CONTROL SYSTEMS; RAILROAD ENGINEERING in the McGraw-Hill Encyclopedia of Science and Technology.

[L. FLETCHER PROUTY]

Satellite navigation systems

The Navstar Global Positioning System (GPS) is bringing a revolutionary capability for worldwide real-time navigation. The natural advantages of a satellite-beacon system using highly precise atomic clocks will provide a level of accuracy in self-positioning and navigation never before possible. Sophisticated signal processing allows GPS receivers to compare signals from four satellites to provide accurate three-dimensional readouts of position, velocity, and time. In addition, satellite signals will be continuously available anywhere on Earth and in any weather. Although this system is being developed for military use, GPS has extensive civilian applications.

The capabilities of GPS will foster oil exploration far from shore-based beacons and reduce the hazards and cost of tanker and shipping operations. The ability of GPS to provide accurate altitude and velocity information in addition will enhance aircraft and helicopter operations. For example, GPS can provide for precision landing approaches, low-altitude helicopter activities, and operations off established flight lanes. GPS will also allow the worldwide synchronization of clocks to within fractions of a microsecond.

GPS operation. The Navstar Global Positioning System is a U.S. Department of Defense satellite-based navigation system being developed under the direction of the Air Force. Planned to become fully operational in the mid-1980s, GPS will utilize rubidium and cesium atomic clocks for precision system timing, L-band spread-spectrum digitally encoded signals for accuracy and interference rejection, chip circuitry for reduced receiver size and costs, and the NASA space shuttle for orbit emplacement of the satellites.

GPS is planned to have a constellation of 18 satellites in circular 12-h (20,190 km altitude) orbits in inclined orbital planes. This constellation will ensure a favorable user-satellite geometry necessary for accurate and continuous worldwide navigation.

The GPS signals are broadcast on two L-band downlink frequencies (L1 and L2) at 1575.42 and 1227.60 MHz. These signals are encoded with two unique biphase-modulated pseudorandom-noise codes on orthogonal carriers—a short, 1023-bit code with a clock rate of 1.023×10^6 bits per second, called the coarse/acquisition or C/A code, and a very long, fine-grained code at 10.23 MHz called the precise or P code. The C/A code is used to quickly acquire the signal and can be used for moderate-accuracy navigation. The P code is used for high-precision timing and navigation.

These L1 and L2 GPS signals also include a 1500-bit navigation data message containing parameters describing the satellite ephemeris and satellite clock corrections. These data are modulo-2-added to the ranging codes at a 50-bit-per-second rate prior to carrier modulation. The ephemeris- and clock-state parameters allow the user to calculate the position of each satellite at the time of signal transmission.

The dual-carrier-frequency transmission allows accurate determination of and compensation for variable ionospheric delays. The effect of ionospheric disturbances, which plague other navigation systems, can therefore be eliminated by using dual-channel receivers.

The primary measurement performed by GPS receivers is the apparent propagation time of the coded signal transmitted from each of several satellites to the user. This measurement is accomplished by determining the time delay between the code sequence received from a particular satellite and its precise copy (local replica) generated within the receiver by using cross-correlation techniques. This delay, multiplied by the speed of light, is known as pseudorange because, in addition to the actual range delay, it also contains an unknown phase offset between GPS time and the receiver clock. However, by making this same type of measurement on signals from four satellites, four simultaneous equations can be solved for the three coordinates of user position and the receiver-clock offset.

The accuracy of this measurement depends on the code used. Correlation between the received code and the local replica can be resolved to within one chip or code period. Thus, the higher-frequency P code allows a more precise measurement of the time offset between the received code and its local replica.

User velocity is determined by measuring the rate of change of pseudorange (range rate). This can be accomplished by measuring the Doppler shift of the carrier frequency due to relative user-satellite motion. From a knowledge of the satellite motion provided via the navigation data message, user velocity can then be computed.

The present GPS constellation consists of 6 of the planned 18 satellites. These are distributed in two planes inclined at 63° to the Equator. Today this constellation provides a capability for accurate positioning (better than 1 m) and navigation (10 m rms). However, this capability is limited to the few hours that four satellites are in the users' view. This present phase of the GPS program supports various programs of test and evaluation, and is expected to continue.

Types of users and receivers. GPS has undergone extensive testing and performance validation by the Department of Defense. The accuracy obtained using the C/A code has proved to be much greater than anticipated. There has been concern over making GPS capabilities available to a potential enemy. In recognition of this problem the Department of Defense has advised that future satellites will have the capability to restrict the P code to only military and authorized users and to degrade the performance of the C/A code to levels based on national security considerations. It is projected that the C/A code performance (initially) will provide only 200 m accuracy, CEP (circular error probable). However, this issue will be periodically reviewed with the intent to provide the highest-accuracy performance commensurate with the national security and national utility.

Thus, it appears that there will be at least two classes of civilian GPS users: those for whom a navigational accuracy of 200 m is sufficient and those who need higher-precision navigation.

The first group might include cargo ships, fishing vessels, and sailboats, and light planes, helicopters, and spacecraft. Members of the first group can use the GPS L1 C/A code for normal-mode navigation, and would utilize simple, L1 C/A only receivers. Members of the second group could use the differential navigation technique discussed later.

Normal mode navigation. Although a full navigation solution (three-dimensional position and time) requires pseudorange measurements from only four satellites, a typical navigation receiver will sequentially acquire and track all satellites in view. This provides some additional margin in the event that some satellites are obscured. The sequence rate is largely a function of the user vehicle dynamics. A GPS receiver in a corporate jet might sequence through all visible satellites to provide a new navigation solution every 3 or 4 s, while a receiver on a fishing traveler could put out solutions at a more leisurely pace. The user in this mode is completely independent of ground- or shore-based control. Navigation solutions are presented to each user in a familiar format, yet all navigation solutions are tied to a common reference system.

Differential navigation technique. Navigators who require the high-precision capabilities of GPS can use a technique called relative or differential navigation. This technique is commonly used with other navigation systems, such as loran C. Similar to relative positioning, this technique allows cancellation of errors common to each observer, for example, errors due to the satellite clock variations (natural and artificial), the ephemeris, and propagation delays. Differential navigation is suitable for use in precision approach and landing of aircraft, for harbor approach and entry of marine vessels, and for geophysical surveys in remote locations.

Here, one GPS receiver is located at a well-surveyed point: an entrance to a harbor, the end of a landing strip, an oil rig in the far north (see illustration). This base station receiver determines its present position by using GPS, compares it with the accepted position, and data-links the resultant errors to a mobile GPS receiver. This receiver then uses this information, plus its received GPS data, to remove the same errors from the solution of its position. The errors are typically highly correlated for all users in the vicinity of the base station. Thus, the error correction improves as the user approaches the base station.

Accuracy improvements have been examined with simulations and actual tests. The differential technique typically improves the (degraded C/A) horizontal accuracy from 250 to 5 m and the altitude accuracy from 250 to 9 m. Accuracies achievable with the nondegraded C/A signal are even better. Differential navigation using only the L1 C/A code could provide higher-accuracy navigation than use of the L1 P code in the nondifferential mode described above.

As mentioned above, ionospheric effects can be eliminated with the use of a dual-channel receiver. However, civilian receivers are likely to be low-cost single-channel receivers. The use of the differential GPS technique removes this dual-frequency requirement and provides for a relatively accurate compensation for ionospheric delays.

The one major limiting factor of this technique is its use of a data link between the base station and the user vehicle. Its use results in added cost, and a restriction on use of differential GPS to a localized area. However, as is important in precision approaches to an airport or harbor, the accuracy of differential GPS increases as the user approaches the most critical part of the journey—the landing or harbor entry.

Unresolved issues. GPS is clearly seen by many communities of users to be the first truly universal navigation system. However, for the civilian community, at least three main issues are yet to be resolved:

Cost. From the users' perspective, the cost of the receiver equipment must be brought to levels competitive with currently used systems. From the government's perspective, the source of revenue required to operate and maintain the system has yet to be established.

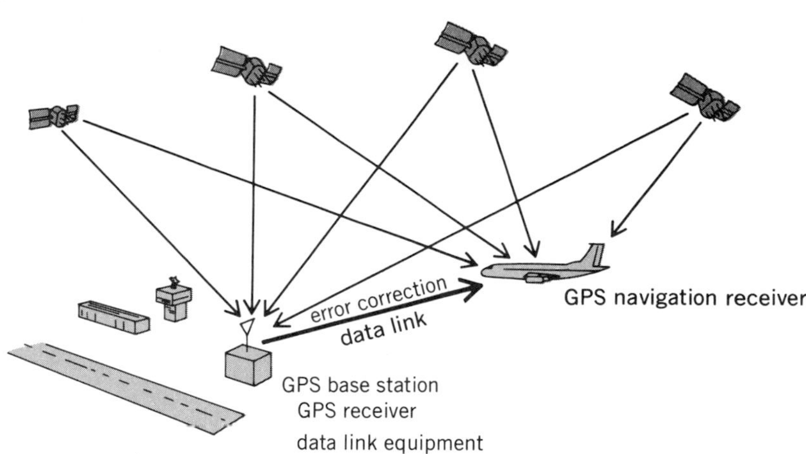

Differential navigation with GPS.

Management. Experience with the first satellite navigation system, Transit, has demonstrated the need for civilian user involvement in the management of the system, to assure its dependability and responsiveness to user needs.

Accuracy. Denial of accuracy, and selective-availability questions and policy must be resolved before large groups of users will commit themselves to GPS. The degraded accuracy of the present baseline concept excludes several large user groups from its use.

These questions are thorny but will have to be settled before the GPS system can be utilized to its full potential.

For background information *see* SATELLITE NAVIGATION SYSTEMS in the McGraw-Hill Encyclopedia of Science and Technology.

[PAUL D. PERREAULT]

Bibliography: J. Beser and B. W. Parkinson, *The Application of AVSTAR Differential GPS to Civil Helicopter Operations,* NASA Contractor Rep. CR166169, June 1981; G. P. Dinneen, *Statement before House Subcommittee on Transportation, Aviation and Communications of the Committee on Science and Technology,* February 21, 1980; P. D. Perreault, *Microwave Syst. News,* 11(1):61–93, 1981; J. J. Spilker, Jr., *Navigation,* 25(2):121–146, 1978.

Saturn

Saturn is the sixth planet from the Sun and the most distant body in the solar system known to the ancients. Because it comes no closer to Earth than about 1.25×10^9 km, Saturn has largely defied the efforts of astronomers to understand its nature and evolution despite centuries of telescopic observation. The planet's exquisite ring system, first noted by Galileo Galilei in 1610, baffled early astronomers for decades until Christiaan Huygens correctly deduced its true nature in the mid-17th century. Among the satellites that circle Saturn is Titan, a planet-sized body discovered by Huygens that is larger than Mercury.

In 1977, armed with this and other basic data, the National Aeronautics and Space Administration (NASA) launched two Voyager spacecraft toward the outer solar system. The twin probes passed Jupiter in 1979 and used the strong Jovian gravity to propel them on toward Saturn. *Voyager 1* reached Saturn in late 1980, passing only 6500 km from Titan on November 11 and 126,000 km from the planet's swirling gold-hued cloud tops a day later. *Voyager 2* arrived on August 25, 1981, coming closer to the planet by some 40,000 km. These two encounters vastly increased knowledge of the Saturnian system, providing thousands of high-quality pictures and other data that will form the basis of further research for years to come.

The planet. Saturn is composed almost entirely of hydrogen and helium atoms, mixed in a ratio of 9:1. Through a telescope, the planet does not display the vivid color exhibited by Jupiter, probably because pigments in the Saturnian clouds are formed at a slower rate than in Jupiter's or are more thoroughly mixed throughout the atmosphere. Images from *Voyager 1* show little in the way of atmospheric markings on Saturn, but *Voyager 2* recorded banding (Fig. 1), shear zones, ovals, and other features similar to (but smaller than) markings on Jupiter.

Winds. Winds blow at extremely high speeds on Saturn. Near the equator, the Voyagers measured velocities of about 500 m/s (1800 km/h), blowing primarily toward the east. At higher latitudes, the winds change direction and form alternating eastward- and westward-moving currents. This pattern is remarkably symmetric north and south of the equator, leading some scientists to suggest that the winds do not arise in the upper atmosphere but are actually driven from a source deep within the interior. Because Saturn has no solid surface, its wind velocity is determined relative to the rotation of its interior. *Voyager 1* detected a radio pulse, believed to arise from the interior, with a period of 10 h 39 min 24 s. This is what scientists now use as the length of one Saturnian day.

Fig. 1. Saturn from *Voyager 2* at a distance of 21 × 10^6 km, showing light and dark atmospheric bands and ring spokes.

Temperatures. Saturn circles the Sun in $29^1/_2$ years, and the Voyagers passed by during northern hemisphere spring. When *Voyager 2* flew behind Saturn, its radio beam penetrated the atmosphere and found a temperature minimum of 82 K, although this increased to 143 K at the deepest levels probed—about 1200 mb or 120 kPa (surface pressure on Earth is about 1000 mb or 100 kPa). Temperatures in the upper atmosphere near Saturn's north pole were some 10 K colder than at more temperate latitudes, a difference which may be seasonal in nature. Auroral emissions were also observed near the poles and across the planet's sunlit face.

Magnetosphere. Saturn also has a magnetosphere, driven by an intrinsic magnetic field. The field axis is offset from the rotation axis by only 1°—much less than other planets' magnetic fields. The magnetosphere extends about 10^6 km from Saturn, where it encounters the solar wind. Changes in the wind have a corresponding effect on the shape and extent of the magnetosphere, which contains neutral atoms and a plasma consisting mostly of hydrogen ions, oxygen ions, and electrons. The plasma accumulates near the magnetic equator, forming a highly charged sheet of material. Some ions have been accelerated to high velocities which correspond to temperatures of hundreds of millions of kelvins. But the plasma is too tenuous to affect spacecraft as they pass through it.

Ring system. Perhaps the greatest surprises and the most perplexing puzzles discovered by the Voyager spacecraft are found in the majestic system of rings that gird Saturn about its equator. Through ground-based telescopes, astronomers have long recognized three prinipal bands that are distinct in character. The outermost pair, A and B, are separated by a clearing called the Cassini Division. Ring C, closest to the planet, is dimmer than the other two and more difficult to observe. A still fainter ring, D, exists between C and the planet, but in Voyager images it is very faint, and cannot be seen from Earth. Another faint ring, F, lies just outside the A ring and was discovered by *Pioneer 11* in 1979. It was better observed by the Voyagers, which also discovered two others, E and G, still farther from the planet.

Voyager 1 found a great deal of unexpected structure in the classical trio of A, B, and C. One explanation was that the structure might be unresolved ringlets and gaps, created by tiny satellites immersed within the system and sweeping out bands of particles as they orbited Saturn. But the improved views afforded by *Voyager 2* revealed no sign of such moonlets, although the ring structure was found to be even more complex than that seen during the first flyby (Fig. 2). Another *Voyager 2* in-

Fig. 2. Detailed structure of Saturn's B ring, imaged by *Voyager 2*. Finest details are about 15 km across.

strument monitored the light from a star passing behind the rings. This star-occultation experiment could resolve ring structures only 300 m wide, and it showed that few clear gaps exist anywhere in the system. Moreover, abrupt changes recorded at the rings' edges indicate a thickness in those locations totaling only about 200 m—even though the classical ring trio is 275,000 km across. In other words, a phonograph record would have to be more than 4 km across to match the A ring's ratio of diameter to thickness.

Of all Saturn's rings, B is the most opaque and shows the most structure (Fig. 2). Its many ringlets and gaps appear to be variations in the abundance and proximity of ring material. These features probably arise from density waves or other stationary wave forms. Density waves are created by subtle gravitational effects from Saturn's satellites. They propagate outward from the positions within the rings where particles orbit in harmony, or resonance, with the satellites. For example, one resonant location occurs near the outer edge of the B ring, just inside the Cassini Division. There, a particle makes two orbits around Saturn in the same time taken by the satellite Mimas to complete one orbit. Mimas interacts with such particles and perturbs them, forming an eccentric ringlet even though surrounding rings are circular.

Eccentric ringlets are also found in almost every case where clear gaps appear. All seem to show variations in brightness due to the clumping, kinking, or other disruption of ring material. Two separate, discontinuous ringlets were found in an A-ring gap called the Encke Division. In *Voyager 2*'s best images, at least one of those ringlets has multiple strands. Some scientists believe the only plausible explanation for clear gaps and the irregular material within them is the presence of adjacent satellites that have escaped detection by the Voyager cameras. Others put more trust in wave theory or in the subtle resonances with satellites that are not yet fully understood.

Amid the many unexpected phenomena found by the Voyager spacecraft, a class of B-ring features called spokes continues to baffle researchers (Figs. 1 and 3). The spokes are dark markings that cut across the B ring's bright face in a 14,000-km-wide band corresponding to the most opaque region of the entire ring system. Voyager images of some spokes, narrow ones that have an alignment radial to the planet, suggest that the features formed just before being photographed. Broader, less radial spokes appear to have formed earlier than their narrow counterparts, and in some cases new spokes seem to be reprinted over older ones.

As the spacecraft departed, they turned back toward Saturn and recorded images in which the spokes appear brighter than their surroundings, indicating that they scattered sunlight toward the cameras better than adjacent ring material did. This suggests that the spokes consist of very fine particles. One theory holds that this fine material, once charged by solar ultraviolet radiation, becomes levitated above the ring plane through interactions with Saturn's magnetic field. But the spokes have also been detected underneath the ring plane, on its unilluminated side, so some process other than photoionization may be at work.

The F ring is probably the most puzzling feature observed by the spacecraft. Photographs by *Voyager*

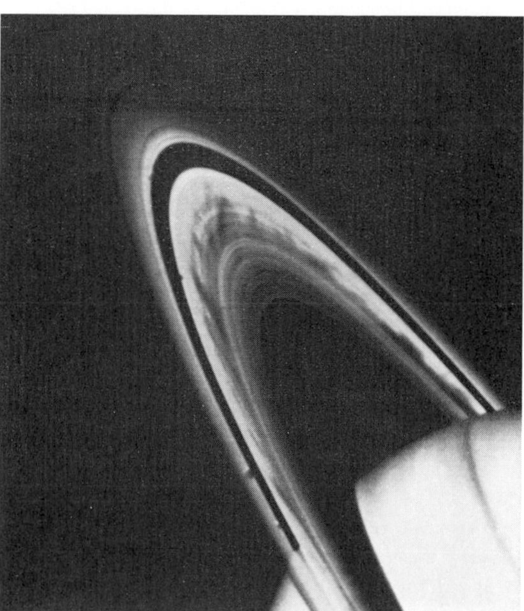

Fig. 3. Saturn's rings from *Voyager 2*, clearly showing ring spokes.

Fig. 4. Saturn's satellite Enceladus, imaged by *Voyager 2*.

1 show three separate strands that appear twisted or braided. *Voyager 2* found five distinct strands in a section that had no apparent braiding. During the stellar occultation, the brightest component resolved into at least 10 strands. Twists in the F ring are believed to arise from gravitational perturbations induced by two small satellites, 1980 S 26 and 1980 S 27, discovered by *Voyager 1*. Clumps in the ring occur with a roughly uniform spacing of 9000 km, coinciding with the arc of F-ring material that would slide past the outer of these shepherding satellites while the inner satellite completed one circuit along its faster orbit. By analogy, one theory holds that similar mechanisms create the irregular ringlets seen in gaps elsewhere in the main ring system.

Satellites. Before *Pioneer 11*'s encounter with Saturn in 1979, the planet was believed to have 9 (perhaps 10) satellites. At least 17 Saturnian satellites are now known, 3 of these discovered by *Voyager 1*. No new satellites have yet been confirmed in *Voyager 2* data, although electromagnetic "shadows" seen by the cosmic-ray instrument provide evidence for more material—possibly an eighteenth satellite—in the same orbit occupied by Mimas.

The innermost satellite, 1980 S 28, was discovered by *Voyager 1* and is about 30 km in diameter. It orbits just outside the A ring and may actually provide gravitational perturbations that act to constrain the ring's outer edge. Farther out are 1980 S 27 (100 km across) and 1980 S 26 (90 km across), the two shepherding satellites mentioned above. These satellites have temporary designations given them by the International Astronomical Union; in time, they will acquire names similar to those of other Saturnian moons.

Three other satellites observed closely by the Voyager spacecraft were recently discovered by terrestrial observers. One, 1980 S 6, shares the orbit of Dione and travels about 60° ahead of it. Two others circle Saturn in the same orbit as Tethys, one ahead and the other behind it. They are designated 1980 S 13 and 1980 S 25, respectively, and are comparable in diameter to the 30-km 1980 S 6.

Mimas, Enceladus, Tethys, Dione, and Rhea are roughly spherical in shape and probably consist mostly of water ice; their diameters are 390, 510, 1050, 1120, and 1530 km, respectively. The ice makes all five highly reflective; Enceladus reflects almost 100% of the light striking it. Unlike the other four satellites, which have heavily cratered surfaces, Enceladus exhibits some regions with little or no cratering (Fig. 4). This suggests a relatively fresh surface, and other evidence indicates that Enceladus is by far the most geologically active satellite in the Saturnian system. In fact, some scientists believe it is still undergoing change. Mimas bears a large impact crater roughly one-third the diameter of the entire satellite. Tethys, a much larger satellite, also has such a crater with a diameter greater than Mimas itself, as well as a huge rift that may have formed during expansion of Tethys' crust. The surfaces of Dione and Rhea may have been partially coated with material from their interiors.

Iapetus, Phoebe, and Hyperion (respectively 1440, 200, and 300 km across), best observed by *Voyager 2*, have somewhat darker surfaces that may contain near-black organic compounds formed in the bitter cold of deep space. Iapetus itself is almost totally black on one hemisphere and about 10 times brighter on the other—a bizarre dichotomy that may result from the satellite's capture of dark debris, falling toward Saturn from Phoebe, on the hemi-

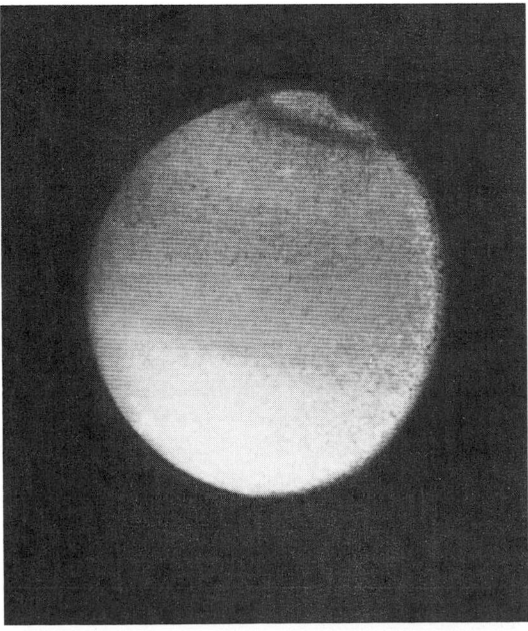

Fig. 5. Saturn's satellite Titan, imaged by *Voyager 2*. Lack of features is due to opaque haze shrouding the surface.

sphere facing forward along its orbit.

Titan, Saturn's largest satellite, is 5150 km in diameter—large enough to attract and retain a substantial atmosphere composed primarily of nitrogen and methane. The satellite's surface is hidden from view by an opaque photochemical haze suspended in the atmosphere (Fig. 5), which was also discovered to contain traces of hydrogen cyanide (HCN) and other organic compounds. Thus, from the terrestrial perspective, Titan may well be the most interesting object in the solar system. Its surface temperature is about 95 K and its pressure 1600 mb (160 kPa). These conditions are close to the point at which methane condenses, so the existence on Titan of liquid-methane rain, snow, lakes, or oceans cannot be ruled out. However, the low temperature undoubtedly inhibits more complex organic activity.

Future encounters. Although *Voyager 1* is now on a course taking it well above the ecliptic plane and out of the solar system, *Voyager 2* is expected to fly by Uranus in January 1986, and Neptune in September 1989. Uranus also has a ring system, though of a markedly different character than Saturn's. If the spacecraft remains operational until the end of this decade, the United States will have successfully visited every planet in the solar system except Pluto.

For background information *see* SATURN in the McGraw-Hill Encyclopedia of Science and Technology. [J. KELLY BEATTY]

Bibliography: J. K. Beatty, *Sky Telesc.*, 62(5):430–444, 1981; J. K. Beatty et al. (eds.), *The New Solar System*, 1981; R. Berry and R. Burnham, *Astronomy*, 9(11):6–30, 1981; J. A. Wood, *The Solar System*, 1979.

Sea anemone

Coelenterates (for example, hydroids, jellyfish, sea anemones, and corals) have long been considered primitive or simple animals meriting little examination, but now scientists are learning the full repertoire of their behavior. The physiological and morphological bases of this behavior are beginning to be characterized, but varied hypotheses are still being formulated and tested. Close examination has shown coelenterates to be capable of a variety of fairly complex behaviors, including escape responses, symbiont selection, and aggression, as well as the long-recognized feeding and locomotory behaviors. The sea anemone's acrorhagial response has recently attracted attention because of its interface with the fields of ecology, population genetics, and immunology as well as animal behavior-ethology. Because the acrorhagial response is associated with a relatively simple animal, it has great comparative value in generating and testing hypotheses of phenomena that also occur in more complex animals or population interactions.

The acrorhagus and aggression. All sea anemones possess tentacles which are generally used in feeding. In addition, some species have small round structures at the base of the tentacles called acrorhagi (Fig. 1). As in all coelenterates, such structures consist of three tissue layers: the interior endoderm, the largely acellular mesoglea, and the exterior ectoderm. Each tissue layer is considered to be one cell thick, but a certain degree of stratification can occur. In the sea anemone, body parts such as the acrorhagi and tentacles are hollow (the interior is continuous with the digestive cavity). The sea anemone is in effect like a rubber glove filled with water, and by moving water and using various muscles (hydrostatic skeleton), the animal can inflate or contract the column, tentacles, or acrorhagi.

Fig. 1. A sea anemone in the midst of an acrorhagial response. The large arrows mark unexpanded and expanded acrorhagi. The expanded acrorhagus has just been withdrawn, leaving ectoderm (denoted by the smaller arrow) on the target anemone.

Acrorhagi were first described in the early 19th century. It was soon realized that the acrorhagus with its very dense concentration of nematocysts (stinging capsules) must have a special function, but its nature was controversial and remained a mystery. Based on morphological evidence, the various uses ascribed to the acrorhagus included photoreception, sensitivity to touch, and defense. It was only in 1954 that E. Abel recognized the significance of and described acrorhagial response behavior. With slight variations the same basic behavior has been observed in all the acrorhagi-bearing sea anemones examined, primarily several species of *Anthopleura* in North American and *Actinia* in Europe and Australia. When an acrorhagi-bearing anemone touches certain conspecifics, or some other coelenterates, the acrorhagi in the area of contact swell and elongate (Fig. 1). The expanded acrorhagi are placed on the target animal. When the expanded acrorhagus touches the target animal, the dense concentration of nematocysts in the tip (Fig. 2) fire and the nematocyst threads penetrate the target tissue. Following discharge of the nematocysts, the acrorhagus is withdrawn and the application process is repeated. Pieces of acrorhagial ectoderm anchored by the fired nematocysts peel off the withdrawing acrorhagus and remain, adhering to the target (Fig. 3). The ectoderm lost by the attacking

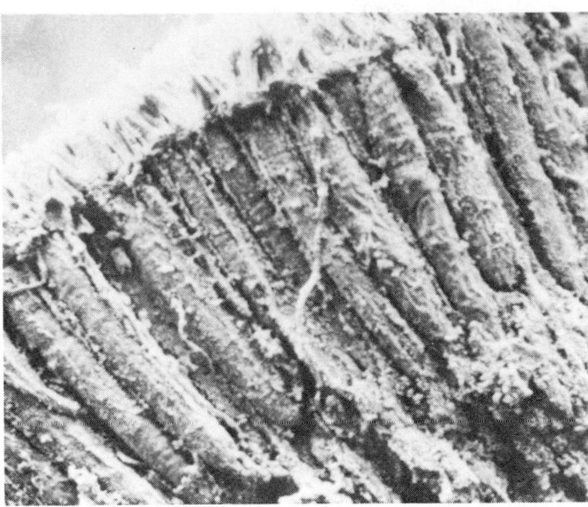

Fig. 2. Scanning electron micrograph of a fracture through the ectoderm of an acrorhagus tip showing the dense concentration of undischarged nematocysts (stinging capsules).

anemone is later regenerated. Within as short a time as 20 min the tissues of the target animal receiving the nematocyst discharge exhibit signs of necrosis. In addition, over a period that can last for several days, the target animal may keep the afflicted tentacle or other body area contracted and even receive other mechanical or physiological damage from the attached ectoderm. The "loser" in some manner avoids further contact or leaves the area.

Both the initiation of the acrorhagial behavior and the acrorhagi nematocyst firing are very specific, that is, responsive only to certain conspecifics or other coelenterates but not to inert objects of metal or glass or noncoelenterate prey or predators. Even when a different species of sea anemone could serve as either prey or acrorhagial target, the two responses are mutually exclusive; predation and the acrorhagial response never occur during the same encounter. Acrorhagial or tentacular contact with an appropriate target is required to initiate the application behavior. A single, brief touch can be sufficient to induce a response within as short a time as 20 s. The resulting acrorhagial application is directed toward the area of stimulation. When the stimulus is moved around the anemone, the zone of acrorhagial expansion and the direction of application follow the stimulus.

The acrorhagial response, a directed behavior that causes injury to a target animal, fulfills the criteria for aggressive behavior and therefore establishes sea anemones as the simplest animals in which aggressive behavior has been demonstrated. The acrorhagial aggressive behavior has been examined on several levels and from corresponding perspectives: (1) the cellular and chemical level—determining what interactions of the relatively limited cell types and their products in these anemones provide the cytological and physiological basis for the response; (2) the organismic level—analysis of the parameters of the response and what such aggressive behavior may mean to the individual sea anemone; (3) the population level—a multifaceted exploration of the social and ecological consequences of aggressive interactions, both interspecifically (xenogeneic) and intraspecifically (allogeneic), with concomitant consideration of the population genetics of this system of histoincompatibility.

Cellular and chemical basis of response. The acrorhagial response must be thought of in two major parts: the recognition of an appropriate target

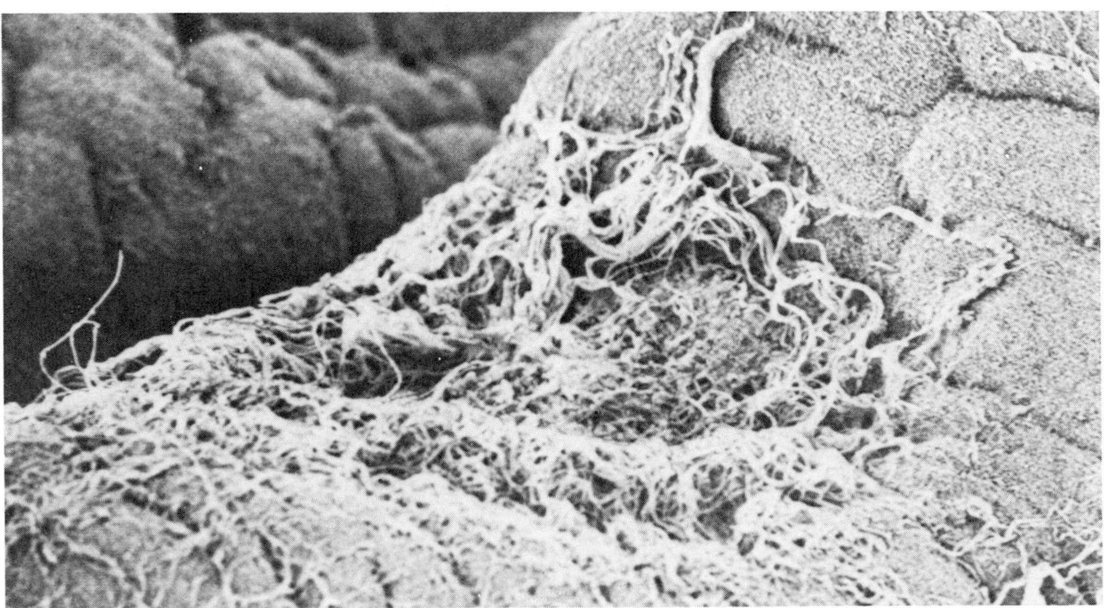

Fig. 3. Scanning electron micrograph showing a piece of acrorhagial ectoderm adhering to a target anemone tentacle. It is held in place by the many fired nematocyst threads, large numbers of which penetrate the tissue and release toxin.

and the effector phase. There is certainly a finely tuned specificity capable of allogeneic as well as xenogeneic discrimination associated with all the receptors of the acrorhagial response. Nothing has been definitively established about the chemical nature of the receptors or the signal transduction, although recently it has been proposed that the acrorhagial response might be immunologic in nature. With its exquisite specificity and cytotoxic effects, the acrorhagial response is certainly an example of self/not-self recognition (specific recognition of foreignness). While self/not-self recognition is the cornerstone of immunology, it should be understood that not all self/not-self phenomena are immunologic. The existence of a behavioral effector component in the acrorhagial response does not preclude the use of a recognition system similar to that of other coelenterate self/not-self or immunologic systems. However, until more is known about acrorhagial recognition mechanisms, receptors, and molecular pathways, the acrorhagial response cannot be considered immunologic and investigation of a common underlying recognition mechanism among various coelenterate self/not-self phenomena remains for future studies.

Three separate components constitute the effector phase of the acrorhagial response: acrorhagi expansion, application behavior, and the acrorhagial nematocyst discharge and associated ectodermal tissue loss. The initiation of expansion and application is controlled by receptors located in the acrorhagi and tentacles, and the receptors for nematocyst discharge are restricted to the acrorhagus tip and may well be the cnidocytes (cells containing the nematocysts) themselves. There is definitely coordination between the three components, including a requirement for acrorhagi to be expanded before firing their nematocysts. However, how that integration is achieved is only now being critically examined. Although recordings have been made of the electrical pulses associated with acrorhagial activity using a suction electrode (which integrates electrical activity in the area of the electrode), cytological and ultrastructural studies have indicated that the neurons of the acrorhagus exist as a diffuse nerve plexus that does not appear as well developed as in other regions of the anemone. Nonnervous electrical coupling/conduction (neuroidal or myoidal conduction) has been demonstrated in another class of coelenterate. In addition, other methods of integration, such as chemical and mechanical, must be considered. In fact, a system of mechanical coupling between acrorhagial expansion and activation (threshold lowering) of the receptors, associated with acrorhagial nematocyst discharge, has been proposed.

Because earlier researchers studying coelenterate behavior emphasized the independence of parts rather than the whole coelenterate acting as a unit, concepts of separate, essentially isolated conducting systems for each behavior developed. Even as recently as 1961 the existence of a nervous system in coelenterates was still being debated. It is now suggested that the equivalent of a central nervous system in dispersed form exists in coelenterates. In the light of present understanding, more considerations of behavioral integration and underlying mechanisms are being developed to present an accurate picture of the capabilities of these animals, which are regarded as models of the simplest nervous system.

Influencing factors. The primary factor in initiation of an acrorhagial response is the presence of a nondiffusible, eliciting factor of unknown chemical composition bound to the surface of ectodermal and endodermal (and probably all) cells of a target animal. In at least some anemone species, the response is stronger when elicited by an allogeneic rather than a xenogeneic stimulus. The genetic differences of the interacting anemones presumably result in quantitatively or qualitatively different recognition factors. Present data suggest that, in common with other coelenterate self/not-self recognition systems, acrorhagial recognition is the result of a complex genetic system such as the mammalian histocompatibility system. Use of asexually derived anemones (clones) presents the investigator with a large number of genetically identical animals (equivalent to mouse inbred strains) that potentially allow dissection of the genetic influence and, if in the future controlled breeding is feasible, will permit detailed genetic studies. If two interacting anemones have the same genetic makeup (genotype), there is no acrorhagial response. The specific differences seem to determine which one (or both) will attack and the severity of the attack. Other factors have been implicated in affecting the response and its outcome: (1) residence—in some studies the winner was almost invariably the anemone that was stationary at the time of contact; (2) prior interactions—even 2 h after an interaction the response-eliciting threshold is lowered; and (3) size—larger *Actinia equina* seemed to have a lower threshold and therefore won more encounters. There does not appear to be any acclimation process in the acrorhagial response; that is, anemones with the same genotype separated for a long time and then introduced are compatible, and incompatible individuals remain incompatible even after continued exposure. Without a doubt other factors also influence the acrorhagial response, perhaps to various degrees depending on the genetic differences of the anemones.

Population level. Acrorhagi-bearing sea anemones have two basic life styles, living as solitary individuals or as groups of anemones that appear to have the same genotype as a result of asexual reproduction (clone mates). The acrorhagial response is certainly used among allogeneic individuals, and its use is a major cause of individual spacing patterns (distribution). However, the response may serve different functions. In asexually grouped anemones, aggression may be used in straight competition for space, and a clonal division of labor has been demonstrated; that is, warrior anemones on the edge of the groups have more and larger acrorhagi and lack gonads, whereas the reverse holds true for clone mates in the center of the groups. In more solitary systems, aggression may serve to protect an area for the settlement of young. It has been hypothesized that one of the major results of acrorhagial interactions (combined with other ecological factors) could be maintenance of a heterogeneous gene pool.

The more such simple animals as sea anemones are critically examined, the more it is realized how complex they really are. In a truer perspective, they might be more accurately classified as relatively simple. As such, acrorhagi-bearing sea anemones provide investigators in different fields with excellent experimental animals for investigation of problems that are difficult to approach in more complex animals.

For background information *see* COELENTERATA in the McGraw-Hill Encyclopedia of Science and Technology.

[CHARLES H. BIGGER]

Bibliography: C. H. Bigger, *Biol. Bull.*, 159:117–134, 1980; R. C. Brace, J. Pavey, and D. L. C. Quicke, *Anim. Behav.*, 27:553–561, 1979; L. Francis, *Biol. Bull.*, 150:361–376, 1976; R. Lubbock and G. A. B. Shelton, *Nature*, 289:59–60, 1981.

Sedimentary rocks

The scarcity of dolomite, $CaMg(CO_3)_2$, in recent marine carbonate sediments cannot be explained simply by the thermodynamic properties of dolomite; its formation seems to be somehow inhibited by sea water. Lately, extensive dolomite formation was discovered in various Holocene-Pleistocene organic-rich marine sediments. Laboratory experiments on the chemical controls affecting the dolomitization of calcite, $CaCO_3$, and aragonite, $CaCO_3$, indicate that the most important inhibitor of dolomite formation in sea water is dissolved sulfate (SO_4^{2-}). Dolomitization of calcite is already inhibited at SO_4^{2-} concentrations of approximately 5–7% of sea water value (28 millimolar). Aragonite dolomitization, however, although strongly retarded at these low dissolved SO_4^{2-} values, is inhibited only at SO_4^{2-} concentration values of approximately 50% of sea water value. Thus, favorable sites for dolomite formation in the marine environment are those where dissolved sulfate concentrations are low. The most effective processes of sulfate removal from, or its dilution in, marine pore fluids are microbial reduction in organic-rich sediments and mixing of sea water with large amounts of fresh water. The relative paucity of dolomite-rich carbonate rocks (dolostones) formed during the last $1-1.5 \times 10^8$ years, as compared with their abundance in older sediments, could be explained on the basis of changes in the depositional environments and in the carbonate mineralogy of the primary calcareous sediments.

Background. Carbonate rocks are very common in the geological column. Among the minerals composing these rocks, three are prominent: calcite, aragonite, and dolomite. The original name of dolomite was dolomie, in honor of the French geologist D. G. Dolomieu. Although dolomite is the thermodynamically stable mineral in sea water, in most carbonate sediments formed in the present ocean, calcite and aragonite are widespread, but dolomite is exceedingly rare; the formation of dolomite seems to be inhibited by sea water.

During the last 30 years, for both scientific and economic reasons, many attempts were made to explain the scarcity of Recent dolomites. Carbonate rocks composed mainly of dolomite generally have high porosities and are therefore important reservoir rocks for oil. Most sedimentologists assumed that the formation of dolomite is mainly controlled by the dissolved magnesium/calcium (Mg^{2+}/Ca^{2+}) ratio in sea water. The molar Mg^{2+}/Ca^{2+} ratio in sea water is 5.3; it was therefore assumed that dolomite formation requires a still higher ratio. This explanation, however, had to be abandoned when, mainly during the last several years, dolomite was observed as an important constituent of many modern marine organic-rich sediments; for example, in the California Borderlands, Gulf of California, Japan Trench, Cariaco Basin off the coast of Venezuela, and the Solar Lake in Israel.

Dolomite forms either as a primary precipitate, as in reaction (1), or as a replacement of $CaCO_3$, as in reactions (2) and (3), which proceeds through a solution-precipitation mechanism.

$$Ca^{2+} + Mg^{2+} + 2CO_3^{2-} \rightarrow CaMg(CO_3)_{2\ (solid)} \quad (1)$$

$$2CaCO_{3\ (solid)} + Mg^{2+} \rightarrow CaMg(CO_3)_{2\ (solid)} + Ca^{2+} \quad (2)$$

$$CaCO_{3\ (solid)} + Mg^{2+} + CO_3^{2-} \rightarrow CaMg(CO_3)_{2\ (solid)} \quad (3)$$

Generally, young dolomite is not strictly stoichiometric, contains more Ca^{2+} than Mg^{2+}, and/or is not perfectly ordered. Such a mineral is called protodolomite. With time or rising temperature, it transforms to ideal dolomite, with a 1:1 molar $CaCO_3/MgCO_3$ ratio and an ordered structure with respect to calcium and magnesium.

Experiments conducted recently in the laboratory

Experimental transformations of aragonite and calcite into dolomite

Reaction	Temperature, °C	Duration
Aragonite + Mg^{2+} → dolomite	200	~3 days
Aragonite + Mg^{2+} → dolomite	150	35–45 days
Calcite + Mg^{2+} → dolomite	200	12–14 days
Calcite + Mg^{2+} → dolomite	150	6–8 months

have shown that the dolomitization of $CaCO_3$ is controlled primarily not by the molar Mg^{2+}/Ca^{2+} ratio but by the dissolved sulfate concentration which inhibits its formation. The mechanism of the inhibiting effect of SO_4^{2-} is as yet unknown. Probably, SO_4^{2-} inhibits or prevents the nucleation of dolomite or the solution of the metastable $CaCO_3$ minerals. Dolomitization proceeds through protodolomite. Dissolved SO_4^{2-} also retards the rate of protodolomite transformation to dolomite.

Experimental work. The lowest temperature at which the calcite-dolomite phase boundary in an aqueous solution with sea water ionic strength has been determined experimentally is 170–200°C. In these experiments, the calcite-dolomite phase boundary was found to lie at a molar Mg^{2+}/Ca^{2+} ratio between 0.51 and 1.06, and the dolomite-magnesite boundary between 2.11 and 3.70. Sea water at ambient temperature has a Mg^{2+}/Ca^{2+} molar ratio of 5.3.

Both calcite and aragonite, in contact with sulfate-free $MgCl_2 + CaCl_2 + NaCl$ solutions of sea water ionic strength and with molar Mg^{2+}/Ca^{2+} ratios within the stability field of dolomite, transform into dolomite within a few days to a few months, depending on the temperature and the nature of the $CaCO_3$ mineral. Examples are given in the table. In each experiment, the formation of protodolomite preceded that of dolomite. At 200°C aragonite transformed to protodolomite in 1 day.

By adding 1 mM SO_4^{2-} to the above solution, aragonite transformed to dolomite in 5 to 6 days, but calcite only in 40 to 45 days.

The presence of protodolomite or dolomite seeds enhances $CaCO_3$ dolomitization in both the absence and presence of dissolved SO_4^{2-}.

The effect of dissolved SO_4^{2-} on the transformation of protodolomite to dolomite was determined at 150°C. Protodolomite was reacted in a solution containing $MgCl_2$, $CaCl_2$, and NaCl of sea water ionic strength, within the stability field of dolomite, with and without dissolved SO_4^{2-}. X-ray diffraction analysis showed that after 6 weeks in the absence of dissolved SO_4^{2-}, the experimental product was an almost ideal dolomite, but in the presence of 2 mM dissolved SO_4^{2-}, the original protodolomite was practically unchanged.

On the basis of scanning electron microscopy of the solids and oxygen isotope analysis of the starting materials (solid $CaCO_3$)—the solid reaction products (dolomite) as well as of the experimental solutions—it was demonstrated that dolomitization of $CaCO_3$ proceeds through a solution-precipitation mechanism. Illustrations *a* and *b* show dolomite rhombs synthesized from a Maastrichtian coccolith chalk from the North Sea and from synthetic aragonite, respectively. The reaction products consisted only of well-developed dolomite crystals, and no traces of the original textures were present. The oxygen isotope value of the chalk was +27.86 parts per mil (‰) relative to standard mean ocean water (SMOW), that of the solution −14.7‰, and that of the dolomite produced from the chalk at 200°C (illustration *a*) −1.90‰.

Dolomite rhombs formed by experiment. (*a*) Dolomitization of coccolith chalk from the North Sea, at 200°C, in a solution of 0.08 M $MgCl_2$ + 0.06 M $CaCl_2$ + 0.28 M NaCl. Duration of experiment: 4 weeks; porosity of chalk: 37.2%. (*b*) Dolomitization of synthetic aragonite, at 200°C, in a solution of 0.08 M $MgCl_2$ + 0.06 M $CaCl_2$ + 0.28 M NaCl + 0.001 M Na_2SO_4. Duration of experiment: 4 days.

Geochemical implications. One of the main results of these experiments is the proof that calcite and aragonite transform relatively easily into dolomite, not only in sulfate-free or -poor artificial sea water, but even in artificial solutions at 100–200°C with a molar Mg^{2+}/Ca^{2+} ratio of only 1–2, a ratio much lower than the sea water ratio of 5.3. The important condition for dolomite formation is therefore not a high Mg^{2+}/Ca^{2+} ratio but a low SO_4^{2-} content. Sea water generally contains 28 mM SO_4^{2-}. In some particular environments, however, such as in pore fluids within organic-rich marine sediments, SO_4^{2-} is quickly removed by microbial reduction. This happens, for example, in continental margin sediments with high rates of deposition, as in the Gulf of California.

Other sedimentary low-sulfate environments favorable for the transformation of $CaCO_3$ into dolomite are limestones exposed on land or within supratidal zones through which mixed fresh and sea water circulates. Some of the first discoveries of Recent dolomites were made in the sebkhas of Abu Dhabi in the Persian Gulf, where precipitation of gypsum ($CaSO_4 \cdot 2H_2O$) and anhydrite ($CaSO_4$) together with bacterial sulfate reduction is common. It was believed that dolomite formation in this environment resulted from a very high Mg^{2+}/Ca^{2+} ratio caused by the removal of Ca^{2+} in the evaporites. It now turns out that the most important operative factors for dolomite formation in this environment are the SO_4^{2-} concentration lower than that of sea water, the $CaCO_3$ mineral species for dolomitization which is aragonite, and high surface temperatures which may reach 45°C—and not necessarily the high Mg^{2+}/Ca^{2+} ratio. Protodolomite is the first mineral to form. The rate of its transformation to dolomite, the thermodynamically stable mineral, is also strongly controlled by dissolved SO_4^{2-} concentrations.

In environments free of or low in sulfate, dolomite formation may be limited by lack of supply of any of the following constituents: Ca^{2+}, Mg^{2+}, HCO_3^-, or solid $CaCO_3$. Sea water of the open ocean is characterized by the relationship $Mg^{2+} > Ca^{2+} > HCO_3^-$. The concentration of HCO_3^- is only 2.5 milliequivalents, and this constituent would therefore be the limiting factor in the formation of primary dolomite. In contrast, in organic-rich sediments, HCO_3^- is continuously produced by bacterial sulfate reduction, methanogenesis, or fermentation. In this environment, dissolved Ca^{2+} will be the controlling factor for the formation of primary dolomite. Finally, in areas of extensive silica diagenesis, or by the interaction of basalt with sea water, Mg^{2+} may be so strongly depleted that this element may become the controlling factor for primary dolomite precipitation during late diagenesis. Most dolomite in the geological record is, however, formed by the replacement of solid $CaCO_3$ [reactions (2) and (3)] and not by direct precipitation from sea water [primary dolomite reaction (1)].

The relative proportions of dolomitic and calcitic carbonate rocks vary greatly with geologic age. Whereas in Precambrian and Early Paleozoic carbonates dolomite is by far dominant, younger carbonate sediments, and particularly Cenozoic ones, are largely calcitic. This difference does not reflect changes in the sea water Mg^{2+}/Ca^{2+} ratio, but can satisfactorily be explained by a change in the nature of the sedimentary material and a different path of diagenetic evolution.

Until the Cretaceous, carbonate sedimentation took place mainly in shallow shelf areas where aragonite and high-Mg calcite are very important carbonate mineral species to form, and are rather rapidly dolomitized even in aqueous solutions containing dissolved sulfate up to a concentration approximately one-quarter that of sea water. Shallow-water carbonates are also more likely than deep-sea carbonates to be exposed to mixed fresh and sea water percolation, and to contain more organic matter (for example, to be associated with algal mats and bacterial sulfate reduction).

In Mid to Late Mesozoic, marine carbonate sedimentation shifted to the pelagic environment by the appearance of coccolithophorids and foraminifera, and the main carbonate mineral species was low-Mg calcite. Since Late Cretaceous, the fertility of calcareous microplankton increased, and the ocean became gradually more undersaturated with respect to the carbonate minerals. Therefore, more of the unstable carbonate minerals, aragonite and high-Mg calcite, dissolved on the shelf before burial, and the ratio of low-Mg calcite to aragonite and high-Mg calcite increased. Of all $CaCO_3$ minerals, low-Mg calcite is the most resistant to dolomitization. The presence of only 5–7% of the normal sea water concentration of SO_4^{2-} is enough to inhibit dolomitization of low-Mg calcite.

For background information *see* DIAGENESIS; DOLOMITE; DOLOMITE ROCK; SEDIMENTARY ROCKS in the McGraw-Hill Encyclopedia of Science and Technology. [M. KASTNER; P. A. BAKER]

Bibliography: P. A. Baker and M. Kastner, *Science*, 213(4504):214–216, 1981; R. G. C. Bathurst, *Carbonate Sediments and Their Diagenesis*, 2d ed., 1975; W. H. Berger, in C. Emiliani (ed.), *The Sea*, vol. 7, 1981; R. M. Garrels and F. T. Mackenzie, *Evolution of Sedimentary Rocks*, 1971; J. A. McKenzie, *J. Geol.*, 89:185–198, 1981; D. H. Zenger, J. B. Dunham, and R. L. Ethington, *Concepts and Models of Dolomitization*, Soc. Econ. Paleontol. Mineral. Spec. Publ. no. 28, 1980.

Sickle cell disease

Recent advances in the understanding of protein and nucleic acid structure and function have provided new insights into sickle cell disease. In particular, a rational framework for understanding the pathophysiology of sickle cell anemia has been provided by new knowledge of the physical state of hemoglobin in the human red cell by more sophisticated methods for measuring the degree of hemoglobin aggregation or polymerization in cells. New techniques of gene analysis have also permitted the antenatal diagnosis of sickle cell anemia by direct analysis of gene structure.

Molecular defect. Sickle cell disease is caused by the presence of an abnormal hemoglobin, sickle cell hemoglobin (hemoglobin S, HbS). The genetic defect in sickle cell disease is a single nucleotide change from GAG to GTG at the sixth codon of the β-globin gene. The nucleotide change results in a single amino acid change at position six of the β-globin chain with the substitution of a valine for a glutamic acid. The sickle β-globin (β^S) chains are then incorporated into intact hemoglobin tetramers. The normal major adult hemoglobin (hemoglobin A, HbA) contains two α and two β chains to form an $\alpha_2\beta_2$ tetramer. Sickle cell hemoglobin is $\alpha_2\beta_2^S$.

Genetics. Each individual has two genetic loci (genes) for β-globin. Normal individuals have two β^A-globin genes. Individuals with sickle cell trait have one β^A-globin gene and one β^S-globin gene. Sickle cell trait, which affects approximately 7% of blacks in the United States, is an asymptomatic condition. Persons with two β^S genes have sickle cell anemia. This is the most severe type of sickle cell disease, and the major form which produces clinical problems. Other variants of sickle cell disease are those in which one gene is a β^S gene and the other β gene is another abnormal β gene, usually either a β thalassemia gene or a β^C gene (another β-globin gene abnormality).

Pathophysiology. The single amino acid change in the β-globin chain in β^S-globin is responsible for the unique and pathologic quality of sickle hemoglobin: its preferential gelation or aggregation under conditions of low oxygen tension (deoxygenation). Hemoglobin S, when deoxygenated, tends to aggregate and form polymers of hemoglobin which are more insoluble than normal hemoglobin. Recent analyses of these polymers indicate that they are composed of coils or helices of 14–16 HbS tetramers. The change in physical-chemical state of HbS is due to the fact that the substitution of a valine for a glutamic acid on the surface of the HbS molecule leads to abnormal intermolecular contacts of HbS molecules with each other, and the formation of aggregates and polymers. For example, the β6 valines of the HbS tetramer are thought to interact abnormally with β85 phenylalanines and β88 leucines of adjacent tetramers. The presence of polymer increases with increasing deoxygenation of the cells and eventually leads to deformity of the cell's shape. This deformity, when viewed by microscopic examination of the deoxygenated red cells, resembles a sickle and gives the disease its name.

The formation of HbS polymers has been extensively investigated recently by techniques such as nuclear magnetic resonance to determine the polymer content in cells. The primary event in this process appears to be nucleation or joining of hemoglobin S molecules by intermolecular contacts such as those described above. Some polymers exist even in relatively fully oxygenated HbS cells. As the red cells move from arterial to capillary and to prevenous circulation, oxygen tension decreases, and there is increasing HbS polymerization. This polymerization can be shown to be accelerated by heating HbS solutions, providing an acidic pH, or increasing the hemoglobin concentration of the deoxygenated HbS solutions. All of these effects are demonstrated in patients with sickle cell disease, whose sickling becomes worse when fever, acidosis, or dehydration is present.

The effect of adding other hemoglobins, such as fetal hemoglobin and hemoglobin A to an HbS solution at a concentration known to form a gel, is used to define the ability of these other hemoglobins to interfere with HbS aggregation. For example, hemoglobin F or fetal hemoglobin (HbF) will prevent the gelation of a hemoglobin S solution if present in significant amounts. This is reflected in clinical medicine by the finding that if sufficient amounts of HbF are present in cells, the severity of HbS aggregation and sickling in these cells is diminished. In certain populations of HbS individuals, especially of Saudi Arabian extraction who have >20% hemoglobin F, the presence of sickle cell anemia does not lead to its usual pathologic consequences because of the ameliorating effect of HbF.

Pathologic process. In individuals with sickle cell anemia, the aggregation and polymerization of HbS has two major effects: stasis or slowing of blood flow, and anemia. There is an increased viscosity of the blood as gelation or aggregation of HbS molecules occurs, and the movement of sickled red cells through the microcirculation (capillaries and other small vessels) is slowed dramatically. This in turn leads to increasing local deoxygenation within the small blood vessels in regions of the body where this process occurs. Deoxygenation itself is created by this slowed blood flow and leads to further aggregation of sickle hemoglobin and increasing sickling. This process produces a so-called vicious cycle, of stasis–deoxygenation–more stasis, and is responsible for the painful episodes which accompany sickle cell anemia, so-called painful crises. In addition, the slowed and decreased blood flow to vital tissues can result in tissue damage by ischemia and infarction of organs such as the liver, kidney, and heart.

The red cell containing mainly HbS in sickle cell disease is also abnormal in its physical state and has a markedly decreased life-span in the body. While normal red blood cells can survive approximately 120 days, those of patients with sickle cell anemia have a markedly reduced survival to 10–20% of this length of time due to premature sequestration of the abnormal cells in the liver, or in the spleen. The shortened life-span of the red cells results in the patient with sickle cell anemia having a chronic hemolytic anemic state with hemoglobin levels of approximately 40–80% of normal. Increased bone marrow production of red cells provides the compensation for the markedly decreased survival of sickle red cells in sickle cell disease. The increased turnover of red blood cells may lead to deficiency of folic acid in individuals with sickle cell anemia, and therapy with this compound is routinely advised.

The role of the red cell membrane in sickle cell anemia remains controversial. It is unclear whether the red cell membrane is specifically altered in sickle cell disease or is simply affected by the presence of the HbS polymer in the red cells. Similarly, a variety of electrolyte disturbances of the red cell, including abnormal calcium transport, may simply be secondary to the presence of HbS. Another phenomenon seen in sickle cell disease is the presence of irreversibly sickled cells: cells which do not return to normal shape when they reenter the arterial circulation and are appropriately reoxygenated. Although these cells are undoubtedly more frequent in patients with sickle cell anemia than in those with other sickle cell disorders, it is unclear whether the number of these cells has any significance with respect to the severity of sickle cell anemia.

Diagnosis. Sickle cell anemia can be distinguished from sickle cell trait and other sickle syndromes by hemoglobin electrophoresis. Hemoglobins S and A separate clearly on electrophoresis. The presence of large amounts of HbA indicates the presence of sickle cell trait. Hemoglobin D can be confused with hemoglobin S since they both migrate in the same position on hemoglobin electrophoresis routinely performed at pH 8.6. Hemoglobin D does not have any pathologic effects. Hemoglobins S and D can, however, be separated by agar gel electrophoresis at pH 6. If the β thalassemia gene is present in addition to the sickle cell gene, it can be detected by other methods. A variety of tests based on the decreased solubility of sickle hemoglobin as compared with normal hemoglobin are available for screening for the presence of some sickle cell hemoglobin; however, these tests do not distinguish between sickle cell trait, sickle cell anemia, or its variants.

Gene diagnosis of the sickle syndromes is also available due to recent advances. Since all of the body's cells contain the same genetic material in DNA, the β-globin genes of an individual can be analyzed from any tissue; usually peripheral blood cells are used, since the white blood cells contain nuclei and have adequate amounts of DNA. The DNA isolated from these cells is cleaved by specific enzymes called restriction enzymes which cut DNA only at defined nucleotide sequences. The change in nucleotide sequence of sickle DNA as compared with normal DNA results in different sizes of DNA containing the β^S- and β^A-globin genes. These differently sized pieces are detected by a radioactive nucleic acid probe which specifically recognizes β-globin DNA sequences. By using this procedure, it is now possible to detect and to distinguish sickle cell anemia, sickle cell trait, and normal individuals, by DNA analysis in antenatal diagnosis. First, fetal cells are obtained by removal of fluid from the amniotic fluid sac surrounding the fetus (amniocentesis). The DNA of these cells is then analyzed, and can be used to determine if a fetus with sickle cell anemia is present in a pregnancy at risk. Fetal blood can also be analyzed for the presence of sickle cell hemoglobin, but this procedure is more complex and involves greater risk.

New approaches to therapy. There are several levels at which sickle cell disease might be attacked. At present, there is no satisfactory treatment for sickle cell anemia except for supportive therapy, including hydration, treatment of infection, and analgesia for the sickle cell painful crises, and blood transfusions when the anemia becomes severe. Since the concentration of hemoglobin in the red cell is an important factor in the development of sickle cell gelation, therapy directed at diluting the hemoglobin concentration in the cells of patients with sickle cell disease has been tried by lowering the serum sodium concentration in the red cells. Although this approach is theoretically of great interest, it is difficult to attain adequate degrees of hyponatremia, and concomitant hemodilution to accomplish the desired effect.

Another approach to the therapy of sickle cell disease is to interfere with the intermolecular interactions between adjacent HbS tetramers which are associated with the formation of sickle aggregates and polymers. Many compounds, including small peptides, alkylating agents, and aspirin and its derivatives, have been identified as potentially useful in interfering with this process; however, none of these agents as yet has been shown to be an effective antisickling agent in practice. There is great hope that a suitable agent will become available from these or other compounds in the near future as more information becomes known about the specific intermolecular reactions which promote sickling and how to interfere with them. Another potential approach to therapy would be either to correct the nucleotide defect in the sickle gene by some genetic manipulation or to increase fetal hemoglobin production. This latter effect would result in decreased hemoglobin S polymerization due to the inhibitory effect of fetal hemoglobin on this process. Addition of normal β^A-globin gene to sickle cells is another theoretical approach to therapy.

For background information *see* HEMOGLOBIN; HUMAN GENETICS in the McGraw-Hill Encyclopedia of Science and Technology. [ARTHUR BANK]

Bibliography: A. Bank, J. G. Mears, and F. Ramirez, *Science*, 207:486–493, 1980; I. M. Klotz, D. N. Haney, and L. C. King, *Science*, 213:724–731, 1981; C. T. Noguchi and A. N. Schechter, The intracellular polymerization of sickle hemoglobin and its relevance to sickle cell disease, *Blood*, in press.

Sign language

The term sign language has been applied loosely to various uses made of different kinds of gesturing. Most interesting to science, however, and best qualified to be called sign languages are the languages of deaf populations. These languages differ from more familiar languages by being signed (or gestured) instead of spoken. Much of the confusion that arises when the term sign language is used stems from the circumstances that some deaf people also use gestures, especially of the hands, to represent the letters (finger spelling) or the words (signed English, signed Swedish, etc.) of the written canon of some spoken language. Once these other language representations, used mainly for hearing-deaf communication, are set aside, the remaining system of regular interaction among deaf persons themselves may be seen as a language that is signed—a sign language in the strict use of the term.

History of American Sign Language. Sign languages of deaf groups were legislated against by educators of the deaf in their congress at Milan in 1880. During the following 7 decades, many states and most schools prohibited their use with and exposure to young deaf children. Sign languages were thus proscribed in the pious but futile hope that, deprived of them, deaf children would succeed in learning to speak well and to understand all that was said in their presence by looking at speakers' faces. The result was to deprive sign languages of respectability, but by no means to eliminate them.

By 1956, when scientific study of sign languages began, most schools kept signing out of elementary school classrooms by the simple expedient of hiring hearing teachers ignorant of signing and conditioned to remain uninterested in what the children were signing to one another. Naturally, deaf children everywhere did sign, inventing their own signs if a deaf community's signed language was unavailable to them. In a few American schools the children's signing was used educationally in a limited way: hearing and deaf teachers in the vocational, not the academic, divisions of the large state schools were permitted to sign.

The scientific status of sign languages changed radically in 1960 with the publication of *Sign Language Structure* as Occasional Paper 8 of *Studies in Linguistics*, and its review in *Exceptional Children. A Dictionary of American Sign Language* followed in 1965; in it were listed 3,000 signs of American Sign Language (ASL or "Ameslan") in notational symbols for their simultaneous aspects (what act, its actions, its location). Previous and subsequent "sign language dictionaries" list English words and describe the sign in words or pictures as the supposed equivalent of the word. Linguistic, sociolinguistic, and psycholinguistic study of sign languages grew rapidly in the 1970s; a dozen books on ASL alone appeared between 1972 and 1981. Sign languages from around the world have also been reported on and examined in more or less detail: Polynesian, Japanese, Taiwanese, New Guinean, Australian (of hearing aborigines), Indian, Israeli, African, Russian, French, Italian, Spanish, Danish, Norwegian, Swedish, Finnish, Caribbean, early American, Central American.

American Sign Language structure. ASL forms its words from aspects of bodily actions visible at the same time, and thus is unlike spoken words, formed of successive vocal sounds. Arm-hand actions form part of the semantic base, but facial and other changes are often integral parts of a lexical unit; and these nonmanual actions are regularly used to form sentences. To illustrate: "A man was looking for you last Friday. He came back this afternoon." With no change from normal head posture and facial expression, these make two ASL sentences. But if the signer's head is tilted and a changed facial expression is held while the first is signed (returning to normal with the second), the sentence is complex in ASL and means "The man who was looking for you last Friday came back this afternoon."

Signers also use space to signal syntactic relations. Many ASL verbs are varied in their manual spatial dynamic enactment to give information about their arguments (subject, object, indirect object, and so on). Hence, like other much-inflected languages, ASL does not use fixed order of words alone to denote phrase relations. Adverb-adjective or adverb-verb phrases in English often translate in one, inflected, sign. Arm-hand configuration and action provide the phrase head, while nonmanual and dynamic features supply modifications; for example, the sign meaning "sick" can easily be performed to mean chronically, suddenly, or figuratively ill; or the sign for "drive," to indicate "drive slowly," "drive while exhausted," and so forth.

Social and artistic aspects. An unexpected change in social status of deaf people accompanied the new scientific status of signed languages. With the first hint of respectability—sign languages could be and were discussed in scientific papers—the Communicative Skills Division of the National Association of the Deaf launched a nationwide program of sign language teaching. Supported by government agencies, this program introduced many hearing persons, some of them teachers of deaf children, to signs. This program also gave employment to deaf persons whose sign language knowledge and teaching abilities had before been grievously under- or unemployed.

The new respectability and attraction of American Sign Language have had effects also on the arts. Beginning as a little theater movement at Gallaudet College under the direction of Frederick Hughes, a deaf man, and developed professionally by George Detmold, sign language drama in the 1960s achieved far wider acclaim with the founding of the National Theater of the Deaf and a first full-length play in ASL, *Sign Me Alice* by Gilbert Eastman. Some appreciation of the irony in these changes comes when it is recalled that only a dozen years before the play *Children of a Lesser God*, half of which was in sign language, won the New York Drama Critics' Circle Award, an organization devoted to oral education of the deaf and to stamping out sign language mounted a letter-writing campaign to keep the signing National Theater of the Deaf off network television. Now, of course, the presence, usually in a small insert picture, on the television screen of a sign language interpreter at events like presidential inaugurations and papal visits is commonplace. Only slightly less visible than sign language drama is sign language dance, an art form that combines modern dance movement with a kind of enlargement and enhancement of sign language signs and phrases. Originated by two deaf dancers,

Carolyn Bateman and Colette Foley, it was developed by Peter Wisher, a pupil of Martha Graham, and its works are not so much choreographed as translated out of sign language texts into artistic movement which can be "read" as well as enjoyed.

These social and artistic aspects of American Sign Language are necessary foundations for any consideration of the scientific and psychological significance of signed languages. Sign language figures in recently renewed interest in the origins of language, in some of the more successful attempts to see how much language capacity other primates may have, and in relations of language functions to the brain and behavioral sciences. The basic reason, however, for scientific interest in signed languages is not their curiosity value but their deep involvement with the culture of the deaf community and so their equal status with spoken language as a way in which human beings make use of their innate human language capacity. *See* APE LANGUAGE.

Psychological aspects. One of the most obvious implications for the psychology of sign language use is the information about language processing it affords. Speech and hearing as language transmission and reception channels have been much studied, as have short-term memory, slips of the tongue, pause phenomena, and processing times. Now information from sign language motor activity and vision, as well as language channels, furnishes information on the same matters from a different source. Researchers are currently adding substantially to psycholinguistic knowledge.

Evidence from language ontogeny, the development in stages of various language skills and functions, is also being increased by considering not just the acquisition of sign language by deaf children of deaf signing parents but also the acquisition of spoken language and sign language ("bilingualism") by deaf or hearing children with one deaf and one hearing parent and by hearing children of deaf parents. The whole matter of language acquisition and deafness is complex, as these few permutations suggest, and it has been studied from various theoretical viewpoints.

Social psychology now takes notice of signing instead of speaking as a factor, and studies of teaching, interpreting, enculturation, institutions, and life histories, when sign language is a means of interpersonal communication, have begun to appear.

Brain function. One of the most interesting new areas of research sign languages is that of brain control centers—not just hemispherical specialization but areas of the brain long known from studies of stroke and trauma patients to be vital to language functioning. Doreen Kimura, a leader in this research, has found that older theories or summations of experiments hold only if sign language is ignored. That is, localization of language control on the left (for most) and motor control on the right grossly oversimplifies and so misrepresents. Instead, Kimura finds, the sequencing and simultaneous production of muscle actions that put one or more body parts into changing positions relative to others are what the left hemisphere specializes in. This description of course fits both the articulation of speech and the manual, brachial, facial, and other articulations of sign language.

Kimura's treatment of the matter of brain function and language function opens up new possibilities for research. Because left hemisphere and frontal lobe areas involved in speech and sign language expression are also implicated in facial expression and indications of affect, future research may well look at total biological functioning, with language and cognition included. Fragmentation of human interaction into "verbal" and "nonverbal" has hampered both linguistic and psychological research. Sign language, with its nonvocal output and use of vision instead of audition, can lead the concerned sciences back to considering mind and nature as a necessary unity, or at least to looking at language functioning in a far less narrowly restricted scenario.

Sensory systems. *The Senses Considered as Perceptual Systems* by James J. Gibson offers an explanation of the remarkable changes both in the social milieu of deaf people and in scientific contemplation of their sign languages. Unlike a narrowly conceived Skinnerian scheme, Gibson's conception sees the senses as active perceptual systems managed by the organism to bring in information about its environment. Since information is difference, each perceptual system is controlled systematically to maximize its information yield. In this view, because what each system detects is different from the information at the moment from other sensory systems, the whole is equal to more than the sum of its parts. But if that kind of logic seems suspect, Boolean algebra instead may be used to summarize the Gibsonian view: the sum of information—sensory information—that the organism has to act on at any instant is *one*. All the information the senses gather is the whole of what the organism has of this kind to work with.

Blind persons and deaf persons work with one less sensory system in operation than does the rest of the population; yet their behavior and their general success in functioning in the human world should make it clear that for them, too, the sum of all their functioning senses bring in is the same unit—one, the total information. And if this example does not suffice, the uncanny knowledge of the world which deaf-blind persons show they have by their behavior is convincing evidence that they too construct a whole, not a partial, world from what they perceive.

This kind of significance attaches to the relatively new study of signed languages. Previous generations and eras treated deaf persons as defective, that is, as partial human beings. Medical sciences considered deaf persons as ill and in need of remedies or therapy, without realizing that certain kinds of deafness are no more curable than blue eyes or five fingers on a hand. Taking their cue from this, the psychological sciences in the recent past looked on deaf persons as beings without language, when in fact it was the beholder who was without insight. Discoveries about sign languages and their working have turned psychologists to serious consideration of deaf persons as persons and as persons whose total sensory input at any moment is unity. As a result, ideas about language, intelligence, learning, cognition, and communication are being revised to take account of the new information that sign languages transmit and yield.

For background information *see* COGNITION; INFORMATION PROCESSING; INTELLIGENCE; LEARNING, NEURAL MECHANISMS OF; PSYCHOLINGUISTICS in the McGraw-Hill Encyclopedia of Science and Technology. [WILLIAM C. STOKOE]

Bibliography: J. J. Gibson, *The Senses Considered as Perceptual Systems*. 1966; D. Kimura, Neuromotor mechanisms in the evolution of human communication; in H. D. Steklin and M. J. Raleigh (eds.), *Neurobiology of Social Communication in Primates*, pp. 197–219, 1979; *Sign Language Studies*, W. C. Stokoe (ed.), quarterly, 1973 onward; W. Stokoe, D. Casterline, and C. Croneberg (eds.), *A Dictionary of American Sign Language*, rev. ed., 1976.

Soil

Recent research involving soil problems has yielded information concerning the effects of trace-metal complexation by sewage sludge in natural waters, the metal uptake by plants following applications of sewage sludge to soils, and the effects of biodegradation of soil organic matter on drained organic soils.

Trace-metal complexation by sewage sludge. The disposal of municipal sewage sludge in natural waters and on land brings with it the possibility that the concentrations of trace metals in the environment may increase to potentially hazardous levels. The fates of trace metals such as Ni, Cu, Cd, and Pb in surface and subsurface waters mixed with sewage sludge pose a complicated scientific problem that involves complexation, adsorption, coagulation, oxidation-reduction, and biological phenomena, with both equilibrium and kinetic features. With respect to the prediction of trace-metal solubility, mobility, and bioavailability, however, it is believed that much can be learned from studies of metal complexation reactions with the water-soluble, organic fraction of sewage sludge.

Fulvic acid fraction. The fulvic acid fraction of sewage sludge is considered to include the extractable, organic compounds that are most reactive toward trace-metal cations in soluble forms. This fraction is obtained from a sludge by extraction with 0.5 *M* NaOH, followed by addition of 6 *M* HCl until the pH value is 1.0. All precipitated material is discarded, and the clear, yellow supernatant solution is retained for purification. Purified sludge fulvic acid is a heterogeneous mixture of organic compounds exhibiting a wide range in chemical composition and functional group character. It has an average mole mass of 250 daltons and contains about 40% C, 7% H, 3% N, 8% S, and 42% O. Its principal components are known to include carbohydrates, amino acids, and detergents (anionic surfactants) in both unaltered and biodegraded forms.

Acidic functional groups. Given the complicated nature of fulvic acid, a description of its reactions with trace metals at the most detailed level would seem impossible to achieve in the time frame of interest to environmental management. Instead, it has been found that much information can be had from research focused only on the chemistry of the acidic functional groups in fulvic acid. These functional groups have been studied by spectroscopic and electrochemical methods. The principal moieties are carboxyl (COOH), amine (NH_m, $m = 0, 1, 2$), phenolic hydroxyl (ϕ-OH), and sulfonic (SO_3H) groups. Very little is known about the molecular structural framework into which these acidic functional groups are bonded in sewage sludge fulvic acid. However, electrochemical studies have shown that these groups fall into four distinct classes with respect to their proton dissociation reactions. This fact is illustrated in Fig. 1, which is a graph of the number of moles of protons bound per kilogram of fulvic acid extracted from Rialto sewage sludge (material which has been collected from a water treatment plant near Rialto in southern California) versus

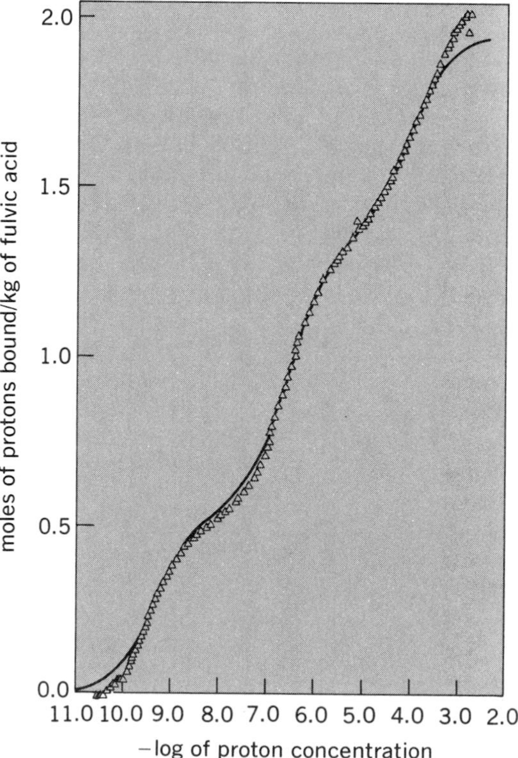

Fig. 1. Experimental and predicted values for proton binding in a sample of fulvic acid from Rialto (California) sewage sludge. The three distinct inflection points indicate three classes of proton dissociation behavior. Fulvic acid content is 1.0 kg/m³ at 25°C.

Values of the parameters n_i and K_i for trace-metal complexes with sewage sludge fulvic acid

Trace metal	n_1	log K_1	n_2	log K_2
Cu	0.096	3.88	0.78	2.11
Cd	0.050	3.04	1.06	2.27
Pb	0.135	4.22	1.35	2.62

the proton concentration in moles per cubic decimeter. This graph shows three distinct inflection points and, therefore, three classes of proton dissociation behavior. The line through the data points is Eq. (1), where n_H is the number of moles of pro-

$$n_H = \sum_{i=1}^{3} \frac{n_i\, K_i[H^+]}{1 + K_i[H^+]} \quad (1)$$

tons bound per kilogram, n_i is the maximum number of moles of protons per kilogram bound in class i, K_i is the proton association constant for class i, and $[H^+]$ is the proton concentration in moles per cubic decimeter.

For the data in Fig. 1, $n_1 = 0.614$ mol/kg, $n_2 = 0.797$ mol/kg, $n_3 = 0.552$ mol/kg, log $K_1 = 4.06$, log $K_2 = 6.55$, and log $K_3 = 9.30$. Similar values of these parameters have been found for proton binding by other sludge fulvic acids. It is believed that class 1 represents mostly carboxyl groups and class 3 is mostly phenolic hydroxyl groups. However, no class is likely to contain a single type of acidic functional group in just one molecular chemical environment.

Trace-metal complexes. The concept underlying the use of Eq. (1) is that, despite its heterogeneous nature at the molecular level, sludge fulvic acid taken as a whole possesses average properties, in respect to ion binding, that are classifiable and describable with only a few macroscopic parameters. Thus, according to Fig. 1, the acidic functional groups in fulvic acid fall into only three classes with regard to proton binding, and each class can be described by a certain maximum number of moles of bound protons and a single proton-association constant. This same approach has been applied successfully to describe the complexes of sludge fulvic acid with metals such as Cu(II), Cd(II), and Pb(II). Figure 2, a graph very similar to that in Fig. 1 except that it pertains to metal binding instead of proton binding, was also derived from work with fulvic acid extracted from Rialto sewage sludge. In this case, well-defined inflection points are not apparent in the graphs for Cd and Pb binding, but the close-fitting lines through the data points represent Eq. (2), where n_M is the number of moles of the bivalent

$$n_M = \sum_{i=1}^{2} \frac{n_i K_i[M^{2+}]}{1 + K_i[M^{2+}]} \quad (2)$$

trace metal cation M^{2+} complexed per mole of fulvic acid, n_i is the maximum number of moles of metal bound in class i, K_i is the metal complex stability constant for class i, and $[M^{2+}]$ is the molar concentration of M^{2+}.

For the trace-metal cations, only two classes of complex with sludge fulvic acid are found. It is believed that one of these (that with the smaller value of K) corresponds to all three classes of proton complex shown in Fig. 1. The other corresponds to a strongly acidic (log $K \approx 1$) class of functional groups that titrate below pH 2 and, therefore, are not shown in Fig. 1. This strongly acidic class may comprise primarily sulfonic groups in the detergent components of sludge fulvic acid. The values of the parameters n_i and K_i for complexes of fulvic acid with Cu(II), Cd(II), and Pb(II) are listed in the table. These data are of direct use in that they can be input into computer models of trace-metal chemistry in natural waters and in soil to help predict the solubility of the metals as affected by sewage sludge. Calculations of this type indicate that, in soil water, sewage sludge fulvic acid complexes account for 20–50% of the total soluble concentration of metals such as Ni, Cu, Zn, Cd, and Pb.

[GARRISON SPOSITO]

Sewage sludge applications on cropland. Increased emphasis has been placed on applying sewage sludges on agricultural cropland because of environmental or economic constraints associated with alternative sludge management methods such as incineration and ocean dumping. Even though sewage sludges contain all the essential major (nitrogen, phosphorus, and potassium) and minor (zinc and copper, for example) nutrients required for growth of agricultural crops, the majority of sewage sludges also contain some undesirable heavy metals such as nickel, cadmium, and lead. Extensive research has

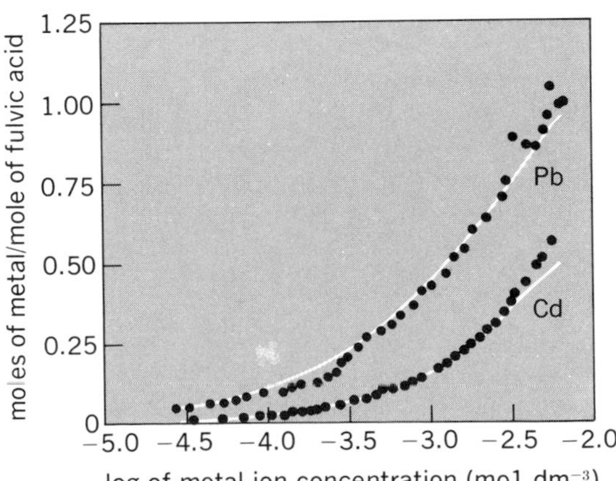

Fig. 2. Experimental and predicted values for binding of Cd and Pb by fulvic acid from Rialto sewage sludge. Fulvic acid content is 0.5 kg/m^3 at 25°C, and pH = 5.

been conducted during the past 10 years to evaluate the accumulation of heavy metals by crops grown on soils treated with sewage sludge. With respect to human health, cadmium in crops has received the most attention. This research has indicated that significant increases in the concentrations of zinc and cadmium in crops can occur following sewage sludge applications to soils. Only small amounts of lead, nickel, and copper applied to soils in sewage sludges are accumulated in growing crops. These studies have also shown that cadmium uptake by crops can be minimized by limiting the amount applied each year, by growing crops that exclude cadmium, and by maintaining soils at a near-neutral pH. Regulations have been enacted recently by the U.S. Environmental Protection Agency to prevent excessive contamination of crops with cadmium following application of sewage sludge on agricultural cropland.

Metals in sludges. Sewage sludges are the solid materials generated during several different stages of biological treatment of wastewater by municipal sewage treatment plants. The solids collected at various stages are normally subjected to further treatment by chemical, physical, or biological processes. The heavy metals ultimately found in sewage sludges enter the sewage treatment plant from industrial, urban runoff, and domestic sources. Trace elements such as cadmium, lead, zinc, copper, nickel, chromium, and arsenic tend to form sparingly soluble salts or are adsorbed on the surface of organic and inorganic solids. As a result, these trace elements are concentrated in the sewage solids relative to the amount present in the sewage entering the treatment plant. A variety of other trace elements, including mercury, chromium, cobalt, molybdenum, selenium, and vanadium, can also be found in sewage sludges. These trace elements are not normally of concern in land application of sewage sludge because of either low concentrations found in the vast majority of sludges or the fact that plant uptake is minimal in soils treated with sewage sludges, as has been shown by experimental studies.

Metal concentrations in plants. The metals of primary concern when considering sludge applications on agricultural cropland are lead, zinc, copper, nickel, and cadmium. The reason underlying concern for lead applications on soils is the potential for ingestion of lead by animals grazing on pastures or forages treated with sewage sludges and also ingestion or inhalation of soil and dust by infants. Extensive studies have demonstrated that growing crops on sludge-treated soils does not significantly increase lead concentrations in plant materials. Zinc, copper, and nickel are important because of the potential for decreased plant yields (phytotoxicity) if they accumulate to high concentrations in soils. Both zinc and copper are essential micronutrients for all plants, and therefore sewage sludge applications can be beneficial for soils deficient in these metals. Of greatest concern is cadmium, a naturally occurring metal which is present in all soils and crops. Cadmium is taken up by all plants, but the concentration can be increased by applying sewage sludges containing cadmium. Cadmium is accumulated in the kidney of humans and animals and, after a long exposure period (30–50 years), a subclinical disease called proteinuria can develop. The currently allowable rates of cadmium addition to cropland will not decrease crop yields, although cadmium can be toxic to plants at extremely large rates of addition. The current regulations for cadmium applications on cropland were designed to minimize increases in the cadmium content of the human diet.

Plants are quite variable with respect to metal uptake from soils containing either naturally occurring metals or metals originating from sewage sludge applications. The metal content of crops in a variety of locations has been determined, and these data have shown that the trace-metal concentration in a crop can be quite variable, depending on the location in which the crop is grown. A variety of plants have been grown on soils treated with sewage sludge under experimental conditions. When crops were grown on soils not treated with sewage sludge, the concentrations of cadmium were 0.03 mg/kg in peas and 0.62 mg/kg in lettuce. When crops were grown on the same soil treated with sewage sludge, the cadmium concentration in peas was 0.04 mg/kg while lettuce contained 2.67 mg/kg. This example illustrates two important points. First, all crops currently entering the human diet contain heavy metals such as cadmium. Second, the response of different crops to a single application of sewage sludge will be quite different. Plants containing the greatest concentration of metals include the leafy vegetables such as lettuce, carrot greens, beet tops, and Swiss chard. When the various parts of a plant are analyzed for zinc, cadmium, and other metals, it has been shown that the leaf and stalk contain significantly greater concentrations of metals than the grain. It has also been shown that grain crops differ in accumulation of metals. In general, concentrations of cadmium will decrease in this order: soybeans > wheat > corn. In commercial agriculture, a variety of cultivars are used in different regions of the United States to optimize crop yields. It has been shown that the range for metal content in different cultivars of the same crop may be as large as the increase found after applying sewage sludge. Metal uptake by crops can be controlled by selecting crop species and cultivars of a given species that tend to accumulate minimal amounts of metals. These data all indicate that if the effect of sewage sludge on the metal content of crops is going to be evaluated, it is essential that the same cultivar of a crop species be grown to make a valid comparison.

Soil parameters. The properties of the soil treated with sewage sludge will also influence the metal concentrations found in the crop grown. The parameter of greatest significance is soil pH. As soil pH is decreased (soil becomes more acidic), the availability of metals to plants increases quite dramatically. Several studies have been conducted to evaluate the effect of limestone applications on metal uptake by crops grown on soils treated with sewage sludge. For a variety of small grains and leafy vegetables, the concentration of cadmium can be decreased by a factor of 2 to 10 by a single limestone application. An additional soil parameter which is currently being used to determine appropriate total metal additions to soils is soil cation exchange capacity. Cation exchange capacity is a measure of the net negative charge associated with the soil particles. Even though the metals applied to soils in sewage sludge are not present as exchangeable cations, cation exchange capacity appears to be related to other soil constituents which tend to minimize the availability of metals to plants. The Environmental Protection Agency, along with some state regulatory agencies, has developed regulations in which the total amount of cadmium that can be applied to soils is related to the soil cation exchange capacity.

Cadmium limits. Current regulations concerning the application of sewage sludge on soils are primarily concerned with minimizing the adverse effect resulting from additions of cadmium to soils. The regulations developed by the EPA limit the amount of cadmium applied each year to 0.5 kg/ha if leafy vegetables, root crops, or tobacco are grown. If other human food-chain crops are grown, the current annual cadmium limit is 2 kg/ha. This annual application limit will be decreased to 0.5 kg/ha in 1987. Both of the annual limitations require that soil pH be maintained at 6.5 or above whenever sewage sludges are applied. The total amount of cadmium that can be applied to soils over a period of years is based upon soil pH and cation exchange capacity. If the soil has a natural pH of 6.5 or below, then only 5 kg/ha of cadmium will be allowed over the lifetime of the sludge application site. If the soil is naturally pH 6.5 or above, or if liming will be employed to maintain the soil at pH 6.5 or above, then the total amount of cadmium that can be applied is scaled to the cation exchange capacity of the soil. Soil cation exchange capacities have been separated into three categories: less than 5, 5 to 15, and greater than 15 milliequivalents/100 g with cumulative cadmium limits of 5, 10, and 20 kg/ha, respectively. An additional alternative is available for sites where all crops grown are used for only animal feed. For these types of sites, there are no annual or cumulative limitations on the amounts of cadmium applied, but the soil pH must be maintained at 6.5 or above to minimize cadmium uptake by plants, and all crops produced must be utilized for animal feed.

Summary. Sewage sludges contain the essential nutrients required for fertilization of agricultural crops. Because sludges also contain some undesirable heavy metals (primarily cadmium), it is essential that accurate sludge composition data be available prior to applying sewage sludges on land used for growing agricultural crops. Crops showing the least amount of metal uptake include corn and other cereal grains. Leafy vegetables accumulate the greatest concentrations of cadmium. In summary, sewage sludges can definitely be applied to agricultural soils to attain crop yields essentially the same as those obtained by applying commercial fertilizer materials; however, a greater level of management is required to ensure that soil pH is maintained at 6.5 or above (to minimize metal uptake) and to ascertain that neither annual nor cumulative limitations on cadmium additions are exceeded. Current guidelines concerning total additions of zinc, lead, nickel, and copper to soils should maintain soil productivity and allow the growth of any crop in the future.

[LEE E. SOMMERS]

Soil organic matter oxidation in organic soils. Organic soils (Histosols) are those containing high percentages of colloidal organic matter, and are of considerable ecological and economical importance to humans. The organic matter content must be greater than 20% and may be nearly 100% of the soil mass. In contrast, soils which contain less organic matter are classified as mineral soils.

Histosol characteristics. Histosols can be divided into two types based upon their water content: swamps, marshes, and bogs, which are characteristic of nascent organic soils and are water-saturated; and drained, cultivated Histosols. Both soil types occur more frequently in temperate regions which have experienced glaciation, but they may be found in tropical and subtropical regions, such as in the Florida Everglades. Histosols provide many esthetic and practical uses. The swampy soils not only serve as a habitat for the specialized plant and animal communities characteristic of these ecosystems, but they are also water reservoirs and natural water purification systems. Bog deposits have been mined for biomass sources for energy production. When drained, these soils, which then resemble potting material for the growth of house plants, provide an ideal medium for the cultivation of sugarcane, vegetables, pasture grasses, and minor crops, such as blueberries and mint. The productivity of drained Histosols is exemplified by the Florida Everglades Agricultural Area, where farm income for 1979 exceeded $500,000,000. This productivity is threatened by the biodegradation of the soil organic matter. When the soils are drained, the microbial oxidation of the accumulated organic matter to gaseous and water-soluble products is accelerated. Thus, this nonrenewable resource is lost. The processes involved in this biodegradation of the soil organic matter are the subject of this section.

Effects of drainage. Before Histosols are drained, organic matter accumulates, partly as a result of the oxygen-limiting conditions imposed by the water-saturated environment. Following draining of the soil, this limitation is removed, and the aerobic oxidation of the accumulated organic matter by the soil microbial community is stimulated. Thus, instead of being accumulated, organic matter is lost from the ecosystem. In mineral soils, where organic matter is a small part of the total soil matrix (generally less than 4%), the physical manifestations of organic matter loss are minimal. But, in Histosols, the result of the oxidation of the soil organic matter can be quite dramatic. The oxidation of organic matter to gaseous products results in a loss of soil mass and volume. Thus, the surface elevation declines. This is termed soil subsidence. In the Everglades Agricultural Area, the soil is subsiding at about 3 cm/year. The organic matter lost annually required about 50 years to accumulate under the previously waterlogged conditions. Subsidence rates of 2.7, 1.75, and 7.6 cm/year have been measured in southern Quebec, western Netherlands, and the California Delta, respectively. Other factors, such as compaction and erosion, contribute to this loss of soil elevation, but generally, microbial oxidation of soil organic matter is the major cause of subsidence.

Oxidation processes. With initial drainage of a Histosol, a wide variety of biochemicals is made available for the microbial community to oxidize for carbon and energy. These compounds include the substituents of the partially decayed plant and animal remains (proteins, carbohydrates, lignin, and so on) and the humified soil components characteristic of colloidal soil organic matter. The more readily oxidizable plant and animal remains are catabolized first. Thus, after the initial period of biological activity, a more biodegradation-resistant component of soil organic matter remains. When the swamp is initially drained, the soil organic matter, which at that time retains some of the original plant structure, is referred to as peat. As the plant structure is lost and the oxidation state and degradation resistance of the biomass increase (that is, the humification of the substrate increases), muck is formed. Although muck contains readily decomposable substituents which compose the soil microbial community, it is primarily made up of well-humified plant and animal remains.

This oxidation of the soil organic matter in drained Histosols is characterized by several properties. First, the important reactions are aerobic; that is, they require oxygen to occur. This provides more energy to the growing microbe and results in a more rapid turnover of the available substrate than occurs in oxygen-depleted ecosystems. Second, the most rapid reactions involve the oxidation of the readily degradable substrates. These include the carbohydrates and proteins of recently incorporated plant and animal remains, as well as the same type of compounds in the soil microbial biomass. These substrates are oxidized rapidly in the soil environment because they provide an abundant supply of energy for the growing cell with a minimum energy investment by the microbe to recover the stored energy. In cropped Histosols, the majority of the microbial respiration measured involves oxidation of this type of substrate. Third, a portion of the respiration of the microbial community in Histosols is provided by the oxidation of well-humified substrates. These substrates are found in the soil in three main groups: humin, fulvic acid, and humic acid. These three classes of compounds are delineated on the basis of their solubility in water, acid, and alkali, and are characterized by a high content of aromatic ring–containing compounds. These humified substrates are random polymers of aromatic ring compounds linked by acid, alcohol, and ether bonds. Readily degradable compounds are said to be humified when they become chemically bonded to one of these aromatic polymers.

Stability of humified compounds. Several factors contribute to the relative biological stability of the humified compounds: (1) The initial reactions in the oxidation of the aromatic ring–containing nucleus require the direct incorporation of molecular oxygen. In soil microsites containing high biological activity or water, oxygen may become limiting. (2) The aromatic polymers contain ether linkages. This type of bonding requires large amounts of energy for the microbe to cleave it. (3) The polymers and the attached humified compounds present a bulky structure for the soil microbe and its enzymes to approach and degrade; that is, the enzyme is prevented by the physical structure of the soil organic matter from reaching that portion of the molecule where its activity is expressed. This is defined as steric hindrance. (4) The chemical nature of the substrate is complex. Because these humic substances are random polymers, they contain many different types of chemical linkages. Thus, a microbe needs several types of enzymes to degrade the substrate completely to carbon dioxide and water. Few, if any, bacteria or fungi have all of the requisite enzymes. Thus, each microbial species can only partially degrade the humified substrates. (5) Humic acid and humin are water-insoluble. For a microbe to oxidize a biochemical, it must be water-soluble. Thus, the microbe must produce an enzyme or chemical which makes the substrate water-soluble before degradation of the compound can occur.

Complexity of conditions. These facts stress the complexity of the biotransformations involved in the oxidation of the soil organic matter of Histosols. These reactions are very similar to those involved in the oxidation of organic matter in mineral soils except that the magnitude of the reactions is accentuated by the greater concentrations of soil organic matter in drained Histosols. Although understanding of the enzymological reactions involved in the oxidation of the soil organic matter of Histosols and the factors limiting this activity have increased in the last few years, the only practical means of controlling the subsidence of these soils remains to return as much of the soil profile as possible to the water-saturated conditions existent prior to draining.

For background information *see* SOIL; SOIL CHEMISTRY in the McGraw-Hill Encyclopedia of Science and Technology. [ROBERT L. TATE, III]

Bibliography: Council for Agricultural Science and Technology, Ames, IA, *Report no. 83*, 1980;

P. H. Given and C. H. Dickinson, *Soil Biochem.*, 3:123–212, 1975; *Sludge-Health Risks of Land Application*, Ann Arbor Science Publishers, 1980; *Soils for Management of Organic Wastes and Waste Waters*, American Society of Agronomy, Madison, 1977; J. C. Stephens, *J. Irrig. Drain. Div. Amer. Soc. Civil Eng.*, 95:285–305, 1969; R. L. Tate, III, *Adv. Microbial Ecol.*, 4:169–201, 1980; R. L. Tate, III, *Appl. Environ. Microbiol.*, 37:1085–1090, 1979; U.S. Environmental Protection Agency, *Fed. Reg.*, 44:53438–53468, 1979.

Soil–water relations

There are a number of approaches for handling the problems associated with characterizing the movement of water in soil. While the similitude analysis concept dates back to the 1950s, only recently has its practical value become evident. In addition, there has been a great deal of work involving improved field measurements and procedures which can yield reliable parameters to represent water or solute transport.

Similitude scaling. The movement of moisture in soils constitutes a particularly difficult category of so-called transport phenomena. The two coefficients describing capacity and conductance are dependent—enormously and hysteretically—on soil wetness. In addition, the driving force contains an extra (gravity) term. Consequently, the application of similitude analysis to the scaling of soil moisture behavior is unusually difficult. Nevertheless, it is especially useful. The basic analysis described below dates back to 1956, but in the last 5 years a new awareness of its practical potential has been emerging.

Similitude analysis concept. Results obtained from dimensional analysis can also be obtained using similitude analysis, decidedly the more powerful approach for many problems, including soil moisture behavior. The analysis consists of deducing the formulation (in terms of basic differential equations and boundary/initial conditions) that underlie a given problem into the simplest possible reduced forms. To this end, each action variable (such as pressure, velocity, and the four coordinates of position and time) is combined with qualities that are known as system parameters (such as viscosity, surface tension, and a characteristic geometrical length and/or time) to form dimensionless terms called reduced variables. Any solution obtained for such a reduced formulation thus represents (or telescopes into this single reduced solution) an unlimited set of different physical solutions, each representing differences in the physical values of the system parameters. Even when an actual analytic solution for the given reduced formulation is not easily calculated, the scaling relationships implicit in the reduced variables are still applicable (to the extent that the formulation itself is applicable), which justifies the employment of physical scale-modeling techniques.

Similitude applied to soil moisture. Two microscopic differential equations are developed. Within individual pores of a coarse soil, the shape of the air-water interface is governed by a differential formulation of surface tension forces, given by Eq. (1), where p is the air-water pressure difference,

$$p/\langle 1/R\rangle = 2\sigma \tag{1}$$

$\langle 1/R\rangle$ is the Gauss mean curvature of the interface, and σ is the surface tension. The detailed shapes of the pores along with the contact angle (γ) provide the boundary formulation for such systems. E. E. Miller and R. D. Miller argued that the multivalued solutions of this formulation are intrinsically hysteretic (time-scale invariant). Flow within the water-filled spaces is subject to the differential formulation of viscous flow (the viscous truncation of the standard Navier-Stokes formulation), Eq. (2),

$$(\mathbf{f} - \nabla p_w') = \eta\nabla^2\mathbf{v}' \tag{2}$$

where $\mathbf{f} = \rho\mathbf{g}$ is the gravitational body force (per unit volume), and where at points within the water p_w' *is the pressure*, $\mathbf{v}'$ is the water velocity, and η is the viscosity. The primes denote a microscopic viewpoint. The boundary condition, that flow velocities go to zero as solid surfaces are approached, can reasonably be extended to the air-to-water interfaces. An immediate consequence is that the flow pattern $\mathbf{v}(\mathbf{r}')$ (where $\mathbf{r}'$ is the position vector) varies in direct proportion to the driving force, $(\mathbf{f} - \nabla p')$. This is simply the microscopic version of Darcy's law, deduced from elementary similitude.

Microscale similitude. For microscopic similitude of the shapes of air-water interfaces and of flow patterns $\mathbf{v}'(\mathbf{r}')$, the geometry of pore walls must everywhere be similar in the sense of similar triangles. Two porous media exhibiting such a geometric similitude, as illustrated in Fig. 1, are termed similar media. In this figure the air-water interfaces also exhibit similitude, a condition described as being in similar states. Such microscopic similitude is highly academic. However, just as a bucketful of soil can be stirred up and then two samples removed

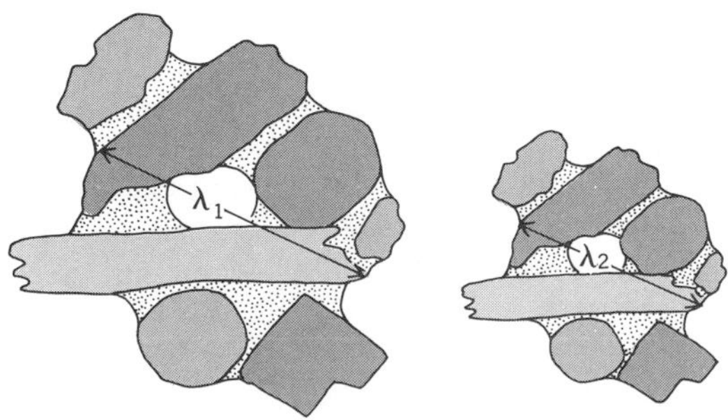

Fig. 1. Porous media exhibiting geometric similitude. The two characteristic lengths, λ_1 and λ_2, connect corresponding points in the scale-modeled media. (*From E. E. Miller and R. D. Miller, Physical theory for capillary flow phenomena, J. Appl. Phys., 27:324–332, 1956*)

that are considered identical, even though every microscopic detail is different, so also can this similitude concept be extended to a statistical or macroscopic criterion. The characteristic microscopic length (λ) is used for such comparisons. If $\{\mathbf{r}_1/\lambda_1\}$ is a certain position on the solid surface of soil sample number 1, then $\{\mathbf{r}_2/\lambda_2\}$ is exactly the corresponding position on the solid surface of soil sample number 2. (This was ensured in the preparation of Fig. 1 by simply preparing a photographic enlargement of medium 1 and then changing the subscript for λ.) Speaking generally, the reduced position vector, $\mathbf{r}'_*$ (using an asterisk subscript to denote reduced quantities) is thus defined by formulation (3). When reduced, formulation (1) for interface shape then becomes Eq. (4). Here $p_* \equiv \{(\lambda/\sigma)p\}$ and $\langle 1/R \rangle_* \equiv \{\lambda \langle 1/R \rangle\}$, with braces denoting reduced groupings.

$$\mathbf{r}'_* \equiv \{\mathbf{r}'/\lambda\} \tag{3}$$

$$p_*/\langle 1/R \rangle_* = 2 \tag{4}$$

Mesoscopic scale. Central to analysis of soil moisture is the concept of seeking an intermediate or mesoscale formulation. For this mesoscale, the differential element of space contains so many individual pores that it is feasible to employ averages over many pores for such properties as wetness (θ) and average velocity or flux density ($\mathbf{v}$). (Note that no prime is needed for macroscopic flux density.) At the same time, these differential elements must also be kept small enough when compared to the whole system to justify employing the methods of calculus for obtaining system solutions. Applying mesoscale averaging to Eq. (4) yields a hysteretic relationship between water volume content (θ) and the air-water pressure difference (p). This relationship is called a hysteresis function ($\theta_H(p_*)$), where θ is intrinsically a reduced quantity and need not be starred. Thus the wetness-characteristics of any set of similar media can all be telescoped into one functional form ($\theta_H(p_*)$).

Mesoscale averaging of the proportionality of $\mathbf{v}'$ to driving force in Eq. (2) yields Darcy's law as a simple consequence of similitude, Eq. (5). Here $\mathbf{v}$ is

$$\mathbf{v} = K_H(p)(\mathbf{f} - \nabla p) \tag{5}$$

$\mathbf{v}'$ averaged over all local (mesoscale) space—air, water or solid; and ∇p has replaced ∇p_w, because the pressure of air has been assumed essentially constant within a soil. K is just the conductivity or Darcy K. Mesoscale averaging of the reduced form of Eq. (1) also gives formulation (6), so that $K_{*H}(p_*)$

$$K_* \equiv \{(\eta/\lambda^2)K\} \tag{6}$$

is a telescoped representation of all $K_H(p)$ functions for any set of similar media. The usual conservation-of-matter condition also applies to soil water, Eq. (7).

$$\partial\theta/\partial t = -\operatorname{div} \mathbf{v} \tag{7}$$

Macroscale: similitude of darcian solutions. The final stage of similitude analysis is reached in the consideration of solutions of reduced versions of Eqs. (5) and (7) for whole flow systems. This introduces another system parameter, a macroscopic characteristic length (L), describing the basic size of whole systems. To reduce Eqs. (5) and (7), this term (L) must be used in relating the operators ∇ and div (both reciprocal lengths) to the system size, as shown by formulations (8). When Eqs (4)–(8) are combined, no ambiguity is found in distributing the system parameters among the action variables, **f**, **v**, **r** (macroscale position vector, no prime), and t, shown by formulations (9). The reduced differential equations given by Eqs. (10) have exactly the same form as the physical versions.

$$\nabla_* \equiv \{L\nabla\} \qquad \operatorname{div}_* \equiv \{L \operatorname{div}\} \tag{8}$$

$$\mathbf{f}_* \equiv \{(L\lambda/\sigma)\mathbf{f}\} \tag{9a}$$

$$\mathbf{v}_* \equiv \{(\eta/\sigma)(L/\lambda)\mathbf{v}\} \tag{9b}$$

$$\mathbf{r}_* \equiv \{(1/L)\mathbf{r}\} \tag{9c}$$

$$t_* \equiv \{(\sigma/\eta)(\lambda/L^2)t\} \tag{9d}$$

$$\begin{aligned} \mathbf{v}_* &= K_{*H}(p_*)(\mathbf{f}_* - \nabla_* p_*) \\ \partial\theta/\partial t_* &= \operatorname{div}_* \mathbf{v}_* \end{aligned} \tag{10}$$

Limitations and applications. In 1959, it was

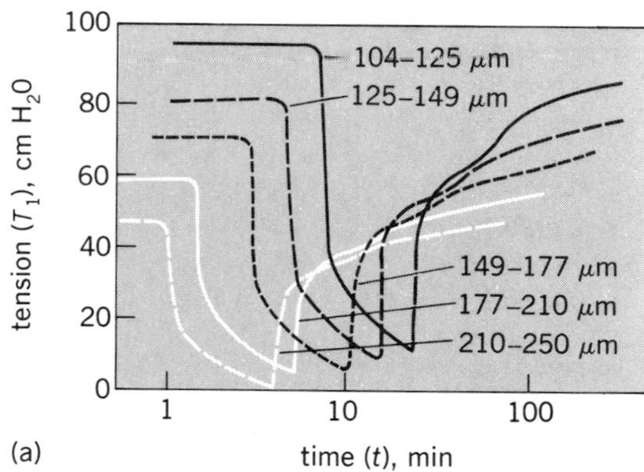

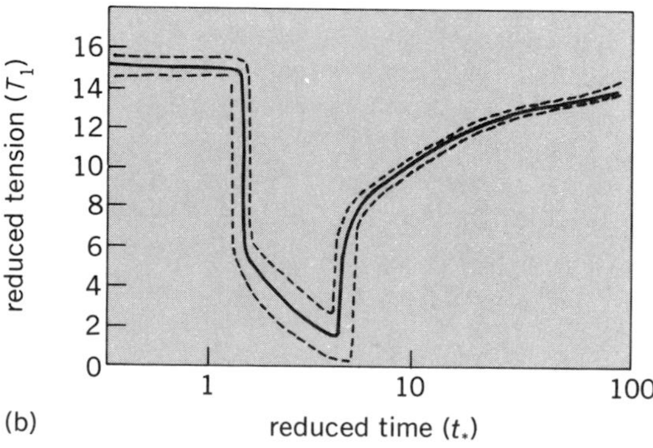

Fig. 2. Similar sands subjected to identical reduced boundary conditions involving infiltration, then redistribution. Tension versus time at an internal point is shown in (*a*) physical variables and (*b*) reduced variables. (*From G. E. Wilkinson and A. Klute, Some tests of the similar media concept of capillary flow: II Flow systems data, Soil Sci. Soc. Amer. Proc., 23:434–437, 1959*)

shown that similitude scaling was borne out experimentally so long as the active pores were large enough to be governed by classical physics [Eqs. (1) and (2)], but that it began to deviate noticeably when the operative water channels were down to a few molecules thick, as in swelling clay soils (Fig. 2).

Finer pores require larger systems. For soils, the fluid parameters and gravity are relatively constant, so the greatest interest is in comparing reasonably similar media having different textures (different λ's). As shown by formulations (9), the reduced body forces can be the same for two soil systems only if they have the same product of the conductivity term and the characteristic microscopic length ($L\lambda$). Hence when the particles are finer, the system must be larger. In this case, the time scale varies as $1/\lambda^3$. Thus when 2 in. (5 cm) of rain is imbibed by a finer soil, the system can be compared to one in which 1 in. (2.5 cm) is imbibed by a soil that is two times coarser, remembering that the redistribution processes are then eight times faster.

There are implications for dissimilar media. The micro length (λ) can be defined mesoscopically in many ways, for example, by specifying that $\{(\eta/\lambda^2)K\} = 1$ at saturation. Whatever the choice, the range of characteristics and behavior of different soils is greatly decreased when reduced variables are used, a little more effectively for some choices of definition of texture than for others. This reveals the term λ as an obvious first step in parameterizing soil moisture characteristics, a very important stage for such practical applications as watershed modeling. As an added advantage, recent extensive experiments by A. W. Warrick and colleagues suggest that the spatial variability of soils in any one field may be rather well represented as simply a spectrum of similar media in which only the value of λ varies.

[EDWARD E. MILLER]

Water movement through soil. Recently a number of new approaches to the problem of representing water and solute flow through soil have been introduced. This new outburst of creativity stems in large part from the repeated failures suffered by laboratory-based models when they were applied to experimental measurements of material transport through natural field soil. Since attempts at prediction of the movement of chemicals through soil now form a significant part of the myriad environmental impact reports written each year for various proposed land-use operations, researchers are looking as never before at the differences between transport through isotropic, homogeneous soil and transport through structured field soil. The latter medium, certain to contain aggregates, cracks, holes, boulders, and various other selective pathways and obstacles, tends to disperse surface-applied solutes both laterally and vertically, even when the application is uniform across the soil surface. In an effort to take such variations into account, recent soil water research, both in measurement and in modeling, is directed more toward statistical approaches and probabilistic inferences, rather than to deterministic representations of material transport in space and time.

Until recently, one-dimensional models of water and chemical transport were commonly applied to entire fields. These models, largely based on material movement through laboratory soil columns, tended to work well in the field only when applied to relatively small plots. In larger fields involving transitions across soil horizons or in horizontal areas encompassing large variability in soil packing and aggregation, such models, particularly of solute transport, required enormously large values of effective solute dispersion in order to describe the large spreading and dilution of the field-averaged concentrations of surface-applied chemicals.

Realizing that the failure of the one-dimensional theory was largely due to the variations in soil geometry, researchers have recently concentrated on obtaining better procedures for designing and making measurements in the field and on obtaining a set of parameters to represent water or chemical transport which would characterize the degree of variability encountered in the field. Several representative studies published in the last few years and discussed below summarize the main efforts of these new research studies.

Variograms and kriging. T. M. Burgess and R. Webster have described the use of two statistical procedures, the variogram and kriging, to maximize the information content of a given set of soil samples. These techniques, developed over the years by mining geologists and others, are used to establish a minimum distance within which measurements of a given property of interest are correlated, and then to find the most accurate means of interpolating between measurements. The variogram is used to find the correlation length and to describe the extent of the correlation within this length as a function of the distance between points. Once the variogram has been obtained, kriging employs the relationship between variance and distance to design an optimum method of interpolation which uses all of the points within the correlation length to determine the value of a new point. The theory also provides a method of calculating an estimation of variance at every point, which may be useful in designing a sampling network for making subsequent measurements.

Scaling theory. A. Warrick and A. Amoozegar-Fard used the concept of scaling theory to simulate water and solute movement through a hypothetical heterogeneous field. Scaling theory represents soil heterogeneity as a relative magnification of the same unit structure of the poruous medium. The water transport and retention properties of one region may be calculated from those of another if the magnification factor is known. Therefore, only a single local length scale is needed to define the relative heterogeneity of a given region, and only a reference set of transport coefficients and the distribution of scaling factors are needed to characterize the entire field. These authors modeled water and solute transport by assuming that all transport took place in the vertical direction and that transport within a given region could be characterized by the appropriate scaling factor. From a distribution of scaling factors, and a dimensionless, scaled form of the transport equation, they were able to represent an entire hypothetical field by making a single computer calculation.

Chemical transport through aggregated soil. P. S. C. Rao and coworkers conducted a combined laboratory and theoretical study to look at the influence of soil aggregates on solute transport. Their model combined effects of solute transfer by moving soil water in the interaggregate region, with solute diffusion into and out of aggregates which were wet but contained stagnant water. They were able to derive an expression for the mass transfer coefficient describing the rate of movement of solutes into and out of the aggregate in terms of experimentally accessible soil parameters, and observed good agreement between predicted and measured values of chemical transport in a laboratory study of solute leaching in aggregated soil. This model is partly deterministic and partly statistical in the sense that a distribution of aggregate sizes is assumed in the calculation. It represents an attempt to account physically for some of the features of a heterogeneous porous medium which cause deviations between ideal and observed behavior.

Stochastic model of solute transport. A theory for solute transport in the field proposed by G. Dagan and E. Bresler combined the effects of variable water application across the field, scaled variations in soil hydraulic properties, and idealized solute transport by piston flow (that is, no diffusion) in a statistical model, to predict horizontally averaged soil solute concentration as a function of depth and also deviations from average behavior. The appearance of a surface-applied chemical at a given depth and time was represented only in terms of a probability distribution, rather than through a deterministic model. All of the variability of the field and the irrigation system was incorporated in a joint probability distribution. These authors produced several scenarios of downward salt movement as a function of variations in surface application and in soil properties predicting lateral variations in the depth of the solute front as well as predicting the probability that ponding would occur at any given point on the surface.

Summary. The approaches outlined above are typical of the new direction in soil water research, with an emphasis on solving practical problems in natural field situations through a combination of deterministic and statistical methods. Regardless of the particular approach taken, models currently proposed for use in the field recognize the limitations posed by spatial variations in soil geometry, and seek to characterize not only the mean but the extreme behavior of materials flowing through soil.

For background information *see* CALCULUS OF VECTORS; SOIL in the McGraw-Hill Encyclopedia of Science and Technology.

[W. A. JURY]

Bibliography: T. M. Burgess and R. Webster, *J. Soil Science*, 31:315–331, 1980; G. Dagan and E. Bresler, *Soil Sci. Soc. Amer. J.*, 43:461–466, 1979; D. E. Elrick, J. H. Scandrett, and E. E. Miller, *Soil Sci. Soc. Amer. Proc.*, 23:329–332, 1959; D. Hillel, *Applications of Soil Physics*, vol. 2, 1980; E. E. Miller and R. D. Miller, *J. Appl. Phys.*, 27:324–332, 1956; P. S. C. Rao et al., *Soil Sci. Soc. Amer. J.*, 44:1139–1146, 1980; A. W. Warrick and A. Amoozegar-Fard, *Water Resour. Res.*, 15:1116–1120, 1979; A. W. Warrick, G. J. Mullen, and D. R. Nielsen, *Water Resour. Res.*, 13:355–362, 1977; G. E. Wilkinson and A. Klute, *Soil Sci. Soc. Amer. Proc.*, 23:434–437, 1959.

Solar greenhouse

The greenhouse is one of the most energy-intensive structures used in food and flower production today. Thin, transparent covering materials necessary to provide optimum lighting conditions during daylight are also transparent to escaping heat during night hours.

All greenhouses are solar collectors. They are specifically designed to capture solar energy and make it available to plants for photosynthesis. In most cases, this collection process also results in collection of excess heat which must be vented from the structure during daylight hours in order to maintain temperatures desired for optimum plant growth. During the past few years a considerable amount of research has been directed toward collection, storage, and use of this excess heat to supply some of the nighttime energy needs of the greenhouse. Greenhouses which have been so modified are referred to as solar greenhouses.

The solar greenhouse has five major components; a solar collector; a heat storage system; a heat transfer and control system; a backup heater; and some type of heat conservation system (Fig. 1).

Collector. Solar collectors are designed to capture incoming short-wave radiation from the Sun and convert it into longer-wave heat energy. The most commonly used type of collector is the covered flat plate unit. The cover is constructed of material transparent to the shorter-wave visible light and relatively opaque to the longer-wave infrared heat energy. Behind the cover plate is an absorber designed to capture incoming solar energy with a minimum amount of reflection and convert it into heat energy. The space between the cover plate and the absorber is used to contain the heat transfer fluid (typically air).

The greenhouse itself is the primary collector in most solar greenhouses. The glass or plastic cover

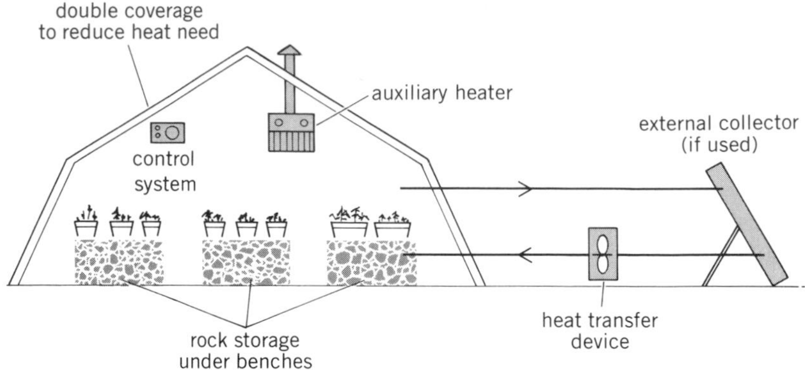

Fig. 1. Basic components of a solar greenhouse heating/storage system.

serves as the transparent cover, the ground and plant surfaces as the absorber, and the air within the building as the heat transfer medium. Because it must serve primarily as a plant growth unit, the greenhouse is not as efficient a collector as a separately designed unit. It is highly unlikely that energy captured from the greenhouse alone will reduce total annual supplemental requirements by more than 25%.

Separate external collectors using water or air as the heat exchange medium can be sized to provide a significant amount of the supplemental heat required. These units normally operate at collection efficiencies of 40–50%. Specific sizing depends on calculated greenhouse needs, storage type and temperature, and anticipated sunlight for the particular area.

It should be noted that there is a wide variation in cost among different types of external collectors. These can range from as low as $30 per square meter up to $400 per square meter, or more. As a result, most designers agree that external collectors offer only limited potential for economic return until fuel costs reach a level much higher than at present.

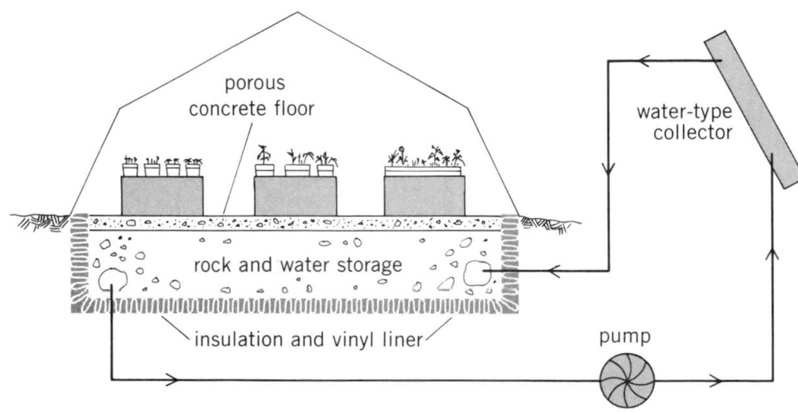

Fig. 2. Schematic of the solar collection and storage system developed by Rutgers University.

Heat storage. The greatest need for supplemental heat in the greenhouse usually occurs between 3 and 7 A.M., when outside temperatures are at their minimum level and residual heat from plants and soil is essentially depleted. In order to make solar energy available during this time period, some type of heat storage is required. Crushed rock, water, and combinations of the two are most commonly used.

Small hobby greenhouses can combine passive solar with water storage of heat by placing containers of water within the greenhouse in locations where they will be heated directly by the sun. These containers then give off their stored heat at night directly to the plant growth space. This system is not practical in larger greenhouses because of the loss of floor space required for containers.

Greenhouses with raised benches can incorporate rock storage directly below the bench area. Rocks are contained in solid enclosures, and warm air from near the ridge of the greenhouse is circulated through the rock whenever building temperatures are too high. At night, room air is circulated through the rock to pick up the stored heat and add it to the room. Rocks are capable of storing approximately 1300 kilojoules per cubic meter per degree Celsius. A normal working range for greenhouse rock storage would be 10–15°C. Storages are usually designed to allow for airflow of 0.01 m^3 per second^2 per 10 m^2 of collection surface.

A combination water-rock storage system developed by Rutgers University shows considerable potential for commercial greenhouse applications (Fig. 2). A solid vinyl plastic liner is placed over a layer of insulation beneath the greenhouse floor. This is filled with a layer of rock and capped with a porous concrete floor. The storage unit is filled with water which serves as a heat transfer fluid carrying heat from low-cost solar collectors adjacent to the structure. On mild nights, the greenhouse is heated directly by heat transfer from the floor. When weather is colder, warm water is pumped from the storage and used to provide additional heat.

Heat transfer and controls. Active collection and storage systems require the use of a heat-carrying fluid (usually water or air), a pumping device, and controls. The amount of fluid required and the design of pumping systems is based on the size of collector, size of storage, anticipated heat collection, and the method of extracting heat from storage. Three controls commonly used in solar systems are thermostats, differential thermostats, and power louvers.

Thermostat. This is an on-off switch which is activated by a temperature change. Normal thermostats turn a switch on when temperature falls below a certain point. These are used to turn on heating systems which extract heat from storage. Reverse-acting thermostats turn a switch on when temperature rises. These are used to move heat from a collector to storage or to operate ventilating fans designed to provide cooling.

Differential thermostat. This is a logic-based device designed to check temperatures at two places simultaneously and turn on a switch when one temperature exceeds the other. Differential thermostats are almost always used to control fluid flow through remote collector-storage units. They turn on the solar system whenever the collector temperature exceeds storage temperature.

Power louvers. These are electrically controlled dampers which can be used to direct airflow to move heat into or out of storage. Power louvers are normally controlled by thermostats.

Backup heaters. No matter how large or how well designed a solar system is, there is always a high probability of encountering several cloudy days in a row which will exhaust the heat storage system. This means a backup heating system must be supplied in order to minimize the possibility of crop loss.

Backup heaters must be designed to carry the maximum anticipated heat load for the structure. Because of this, solar systems are frequently considered additions to heating systems rather than replacements.

Conservation measures. The economical application of solar energy to greenhouse heating requires the use of energy conservation techniques which effectively lower the need for supplemental heat and thus reduce the size of the solar system needed. There are four general conservation methods.

Double glazing. Two layers of cover material (glass, plastic, or a combination) effectively reduce conductive heat loss from the greenhouse. Research has shown a total heat savings of 25 to 50% when comparing double coverage with the traditional single glazing systems. Additional layers will save even more heat; however, they may reduce light transmittance to the point where crop growth is impaired.

Night covers. These are flexible layers of insulating fabric which are pulled over the plant growth canopy at night. These act both to reduce the volume of space heated and to provide an additional layer of heat-resistant material between the crop and outside conditions. Heat savings of nearly 60% can be realized with tight-fitting covers. This reduction in energy requirements makes the use of solar heating much more feasible in many units.

Insulated walls. Several new greenhouse designs utilize conventionally built insulated north walls to replace the traditional glazing. This greatly reduces the heat loss through this section. The overall effect on energy usage will depend on size of the greenhouse. Commercial-sized units are not improved as much by this technique as are smaller hobby-type units.

Lower temperatures. Heat loss is directly affected by the temperature difference between inside and outside. If crops will tolerate lower night temperatures, a considerable amount of energy can be saved. Research is currently under way to determine how much temperature setback can be tolerated by various crops.

For background information *see* HEAT TRANSFER; SOLAR ENERGY; SOLAR HEATING AND COOLING in the McGraw-Hill Encyclopedia of Science and Technology.

[RICHARD E. PHILLIPS]

Bibliography: *Agricultural Energy*, vol. 2: *Biomass Energy and Crop Production*, American Society of Agricultural Engineers, pp. 501–580, 1981; D. S. Ross et al., *Energy Conservation and Solar Heating for Greenhouses*, Northeast Regional Agricultural Engineering Service, Cornell University, 1978.

Solar-powered aircraft

Solar-powered flight, using only the energy of the Sun converted directly into electricity, became possible only recently. Development of photovoltaic solar cells and advances in lightweight construction using artificial plastics have made it feasible. Although today's photovoltaic cells convert only a little more than one-tenth of the Sun's available energy into electric power, a solar collection array the size of an aircraft's wing can now provide sufficient electricity to achieve flight, as demonstrated recently by the world's first piloted solar-powered aircraft, the *Solar Challenger*.

Solar cells. A solar cell is a wafer-thin slice of dark-blue silicon crystal, chemically treated to produce a positive and negative face. Typically it is about 0.015 in. (0.38 mm) thick, measures 2–5 in.^{2} (13–33 cm^2), and is extremely brittle, like glass. When the cell is exposed to sunlight, dc electricity is generated by the photovoltaic effect, which causes electrons to flow from the illuminated face. Such cells were first developed in the mid-1950s, primarily for the space program.

Each cell produces approximately 0.5 V and 0.2 A/in.^{2} (0.03 A/cm^2). Connecting them in series and parallel produces the desired working voltage and current. Although even good cells are only 14% efficient, 1 ft^2 (0.09 m^2) can produce about 10 W. However, the sunlight must be perpendicular to the cell for maximum efficiency. A lower angle of incident light, haze or clouds, dust or shading, or a build-up of heat will each reduce the power output. All these factors have to be considered when designing a solar-powered aircraft. The only advantages to using solar cells aboard aircraft are that the higher a solar plane flies, the clearer the air and the cooler the cells become, both factors improving performance.

Development. Initial development of solar-powered flight took the form of models and drones. The first successful solar-powered flight took place on November 4, 1974, with a 32-ft (9.8-m) pilotless drone (technically not a model). Designed by Roland Boucher and Robert Boucher of Venice, CA, the *Sunrise I*, built of balsa wood, weighed 26 lb (12 kg), and was designed for the U.S. Air Force. The first flight of *Sunrise II*, a lighter version, took place on September 12, 1975. Four thousand cells totaling 40 ft^2 (3.7 m^2) produced 600 W of electricity. Both aircraft were launched with an elastic line catapult before commencing solar flight.

These two drones were prototypes for a proposed research vehicle theoretically capable of gaining altitude to 100,000 ft (30 km) on solar energy during the day, and gliding at night while remaining above the clouds. Although one drone reached 17,000 ft (5.2 km) by remote control, the proposed 100 ft-span (30-m) final version was never commissioned.

In Europe, model flyer F. Militky made the first model-size solar-powered flight for manufacturer H. Graupner. Militky had earlier made the first piloted electric flight on October 21, 1973. He used a 39-ft-span (12-m) converted HB-3 motor glider. It weighed 978 lb (444 kg) and was fitted with a 13-hp (10-kW) electric motor and batteries. His radio-controlled *Solaris*, spanning 7 ft (2.1 m), used 96 cells to power the electric motor.

This pioneer model paved the way for other German models, notably those of H. Shenk, who made a 28-min flight on September 16, 1977; the *Utopie* by R. Stuck in France; and a series of solar-powered models designed by H. Wakamatsu for Toshiba and Mabuchi in Japan, culminating with a design called *Helios*.

Only a handful of solar-powered models worldwide have flown to date. With an average span of 6–9 ft (1.8–2.7 m), and using approximately 100–150 cells, these 2–3-lb (0.9–1.4-kg) radio-controlled models have, nevertheless, provided useful information on the practical difficulties involved with solar power.

Two further developments led toward piloted solar flight. In England, F. To's *Solar One* was flown on December 19, 1978. Piloted by designer D. Williams, the 68-ft-span (21-m) glider-type aircraft used 24 Nicad batteries charged by 750 solar cells mounted on the inboard area of the wings. Subsequent flights by K. Stewart on June 13, 1978, reached an altitude of 40 ft (12 m) during a flight of less than 1 mile (1.6 km). In the United States, L. Mauro made a flight in April 1978 at Roubidoux with a converted biplane hang glider, *Solar Riser*, using the same principle.

Neither airplane, however, could truly be called solar-powered. In each case flight depended primarily on electric battery capacity which could have been charged from any source before flight. By carrying their generating equipment on the wings, they had the distinction of being self-sufficient electric aircraft. In essence, each required too much power and yet had too few cells for the solar flight.

Design considerations. To achieve efficient flight, aircraft wings ideally must be smooth and streamlined. Fitting cells onto this smooth surface can produce aerodynamic problems, disturbing the boundary layer of airflow. Yet, to install them under transparent covering reduces power output and causes the cells to overheat due to the greenhouse effect. With the cells mounted on the wings of an aircraft, the exact orientation relative to the Sun is constantly changing as the plane alters heading in turns, resulting in fluctuations of power generated.

The most important factor in solar-powered flight is power-to-weight ratio. In flight, aircraft are continuously defying the pull of gravity; thus, minimizing the additional weight of solar cells, wiring, and electric motor is very important. Furthermore, the area of the solar cell array and, hence, available power is limited by the size of the aircraft's wings. Thus, highly efficient design, lowweight, and energy conservation are required in order to make flight possible.

Piloted solar flight. P. MacCready applied his experience with ultralightweight human-powered aircraft to the field of solar-powered flight. Winner of two Kremer prizes with his Gossamer aircraft, *Condor* and *Albatross*, he had already mastered lightweight, low-energy designs. His human-powered aircraft typically had a 96-ft span (29-m), yet weighed only 70 lbs (32 kg). A structural frame of thin-walled aluminum tube, and later carbon-fiber tube, externally wire-braced, with expanded polystyrene components and a Mylar plastic film covering only 0.0005 in. (12 μm) thick, accounted for this low weight. The aircraft were designed to fly on a minute amount of power, the $^1/_4$ to $^1/_3$ hp (200 to 250 W) produced by a human, compared to the 60 to 100 hp (45 to 75 kW) of conventional light aircraft.

Gossamer Penguin. MacCready soon realized that the limited human power factor could easily be replaced by a tiny electric motor, and that the little electricity needed to fly his designs could easily be supplied directly by photovoltaic cells without any battery storage capacity. A spare airframe, the 72-ft-span (22-m) *Gossamer Penguin*, was converted, by mounting a 50-ft^2 (4.6-m^2) panel of 3920 cells. This first aircraft could fly only in the calm air soon after dawn, and therefore the solar panel had to be mounted vertically above the wings to face the rising Sun low on the horizon. On May 18, 1980, MacCready's 13-year-old son, Marshall, made the first piloted solar-powered flight of a few hundred feet. Later a woman pilot, J. Brown, flew almost 2 mi (3 km) during a 14-min flight to establish a new record on August 7, 1980.

Solar Challenger. The *Penguin* was not completely satisfactory, requiring tow-assisted takeoffs, but it provided important operating experience for the *Solar Challenger*, the first aircraft designed purposely for piloted solar-powered flight in turbulent conditions, and capable of long-distance flight.

The *Challenger* project was sponsored by the DuPont Corporation, manufacturer of many of the high-strength lightweight materials used for the aircraft's construction. The 47-ft-span (14-m) cantilever wing is constructed around a single tubular main spar fabricated from carbon fiber, Nomex honeycomb, and Kevlar bonded with epoxy, a sandwich combining light weight with great strength and requiring no bracing wires despite a weight of just 55 lb (25 kg). Similarly, throughout the rest of the airframe meticulous attention was paid to saving weight on each component, producing a total weight of 125 lb (56 kg) less cells, motor, and instruments.

Aerodynamically the flying surfaces utilize computer-designed airfoils for maximum lift and minimum drag to fly with the least waste of energy. By aircraft standards the airfoils are unique, with the top surfaces perfectly flat for the rear 85%, seemingly upside down compared to conventional sections. This allows all the photovoltaic cells mounted directly on the upper surface of the wing and stabilizer to have an identical angular orientation, thereby producing a similar output, which is very important for coupling the combined energy of the cells.

The *Challenger* succeeds because it is aerodynamically clean and efficient, producing minimal disturbance to the air through which it flies, and thereby conserving energy. The construction of the aircraft achieves this at minimum weight, again reducing power requirements and maximizing the energy available for flight. Care was taken in design to ensure that shadows from propeller arc or vertical fin never fall across the cells, thereby reducing power. Finally, maximum use is made of the horizontal flying surfaces to incorporate sufficient cells to achieve flight.

The 16,128 cells of the *Challenger* typically produce 2400 W, depending on Sun condition. However, because the cells do not work in shade, when the Sun goes behind a cloud the *Challenger* reverts to being a glider, altitude then being its store of energy.

The propulsion unit is built around a permanent-magnet dc motor, rated at 5 hp (3.7 kW). To improve efficiency, a large-diameter, slow-revving propeller is used, driven through a 23:1 two-stage gear reduction. This allows the electric motor to operate efficiently at 7500 revolutions per minute (rpm) while the propeller rotates at about 300 rpm. The propeller also features variable pitch, operated by the pilot in flight, which allows the thrust to be controlled on an aircraft which otherwise has no throttle. Since no fuel is consumed, and there is no noise or pollution, there is no reason not to fly at full power, except for landing or slow flight.

Initial flight trials during the winter months of 1980–1981 showed surprisingly high performance, despite this being the worst time of year for solar flight, with the Sun relatively low in the sky. The first flight, on November 20, 1980, at El Mirage, CA, lasted 2 min 50 s (see illustration), but subsequent flights in Arizona lasted up to 1 h 55 min with an altitude record of 3500 ft (1070 m) recorded.

After further refinements to the *Challenger*, completed in spring 1981, the aircraft set new records of 8 h 22 min flight duration and 14,300 ft (4.4 km) altitude during test flights at Shafter Airport near Bakersfield, CA. These tests preceded a long-distance flight attempt in Europe, crossing the English Channel.

On July 7, 1981, the *Solar Challenger* took off from Cormeilles-en-Vexin airfield northwest of Paris. Headed toward England, the solar plane climbed to 11,000 ft (3.4 km) before leaving the French coast at Le Tréport. After a flight of 165 mi (265 km) lasting 5 h 23 min the *Solar Challenger* piloted by S. Ptacek landed at its destination airport, RAF Manstan, near Ramsgate in Kent.

World's first piloted solar-powered flight at El Mirage, CA, by *Solar Challenger* on November 20, 1980. (© *Martyn Cowley, 1980*)

Prospects. With these impressive performances by this prototype solar-powered aircraft, the future potential seems to point toward specialist research planes for weather monitoring or aerial survey, where duration and low-running cost rather than ability to fly fast or carry cargo are important. Reduction in cost of solar cells may also herald a new sport aircraft for the future.

For background information *see* AIRPLANE; GLIDER; SOLAR CELL in the McGraw-Hill Encyclopedia of Science and Technology.

[MARTYN B. COWLEY]

Bibliography: M. B. Cowley, *AeroModeller*, 46(545):280–287, June 1981; M. B. Cowley, *Flight Int.*, 119(3782):1865–1868, June 13, 1981; P. F. Kluge, *Smithsonian*, 2(2):70–76, February 1981; C. Phinizy, *Sports Illus.*, 54(8):34–40, February 1981.

Space flight

The period from September 1980 to November 1981 was marked by two spectacular successes in the United States Space Program: the Voyager missions to Saturn, and two orbital flights by the shuttle *Columbia.* The Soviet Union launched a total of 92 missions during this reporting period. Significantly, 47 of those missions were identifiable as military missions. The balance consisted of 16 civil communications missions; 16 earth resources or related missions; 8 crewed space-program launches; 1 interplantary mission; 2 radar calibration missions; and 2 missions which were unidentifiable. Major missions are listed in the table.

Space missions, September 1980–November 1981

Payload name	Launch date	Payload country or organization	Purpose and comments
GOES-D	Sept. 9, 1980	United States	Weather observation
Soyuz 38	Sept. 18, 1980	Soviet Union	Crewed
Progress 11	Sept. 28, 1980	Soviet Union	Uncrewed resupply spacecraft to *Salyut 6*
FLTSATCOM-D	Oct. 30, 1980	United States	Part of worldwide Armed Forces communications system
SBS-A	Nov. 15, 1980	SBS	Satellite Business Systems advanced communications satellite
Soyuz T-3	Nov. 27, 1980	Soviet Union	Civil, crewed
INTELSAT V-A	Dec. 6, 1980	INTELSAT	Largest commercial, international communications satellite
Cosmos 1241	Jan. 21, 1981	Soviet Union	Target for antisatellite interceptor
Progress 12	Jan. 24, 1981	Soviet Union	Uncrewed resupply spacecraft to *Salyut 6*
Cosmos 1243	Feb. 1, 1981	Soviet Union	Target for antisatellite interceptor
COMSTAR D-4	Feb. 21, 1981	COMSAT General	Domestic communications satellite
Soyuz T-4	Mar. 12, 1981	Soviet Union	Crewed
Cosmos 1258	Mar. 14, 1981	Soviet Union	Interceptor, "killer" satellite; destroyed *Cosmos 1241*
Soyuz 39	Mar. 22, 1981	Soviet Union	Crewed
STS 1	Apr. 12, 1981	United States	First reusable shuttle spacecraft, *Columbia*
Cosmos 1267	Apr. 25, 1981	Soviet Union	Docked with *Soyuz 6* on June 19, 1981
Soyuz 40	May 14, 1981	Soviet Union	Crewed
NAVY NOVA-1	May 15, 1981	United States	Navy satellite (transmit)
GOES-E	May 22, 1981	United States	Meteorological satellite
INTELSAT V-B	May 23, 1981	INTELSAT	Second in series of largest commercial, international communications satellites
NOAA-C	June 23, 1981	United States	Environmental monitoring satellite
Dynamic Explorers A and *B*	Aug. 3, 1981	United States	Sun-Earth energy studies
FLTSATCOM-E	Aug. 6, 1981	United States	Part of worldwide Armed Forces communications system
SBS-B	Sept. 24, 1981	SBS	Satellite Business Systems advanced communications satellite
Solar Mesopheric Explorer	Oct. 6, 1981	United States	Atmospheric research
Venera 13 (Venus 13)	Oct. 29, 1981	Soviet Union	Exploration of Venus (expected to arrive March 1982; possible soft landing)
STS 2	Nov. 12, 1981	United States	Second launch of world's first reusable, crewed spacecraft, *Columbia*
SATCOM III	Nov. 19, 1981	RCA	Third in series of SATCOM communications satellites

Voyager mission. Two uncrewed spacecraft, *Voyagers 1* and *2*, completed highly successful fly-through encounters of the Jovian system during 1979, and went on to rendezvous with Saturn and its system of rings and satellites on November 12, 1980, and August 24, 1981, respectively. *Voyager 2* will continue on to encounter Uranus in January 1986, and Neptune in August 1989.

The historic Voyager encounters with Jupiter provided startling new information about the Jovian system. Jupiter's atmosphere is complex, with layers of colorful clouds above a deep atmosphere of hydrogen and helium. The Great Red Spot, large enough to hold several Earths, is a tremendous atmospheric storm that rotates counterclockwise, with one revolution every 6 days.

One of many spectacular discoveries was that of active volcanoes on Io, Jupiter's innermost Galilean satellite. Additional discoveries concerning Jupiter included a thin ring of material around the planet and two new satellites. Continuing on to Saturn, *Voyager 1* discovered three new satellites, and its photographs revealed that Saturn's rings have a highly complex structure with hundreds of individual features. *Voyager 2* furthered the exploration of Saturn by photographing the topside of the rings and Saturn's northern hemisphere. Storms as huge as the entire United States were found to sweep through the higher latitudes of Saturn, producing a jet stream shooting through the cloud tops at a speed of 500 km/h.

Voyager 2 created some concern when a malfunction caused one of its movable platforms to jam, pointing the cameras randomly into space; however, numerous command signals from Earth started the platform moving again.

The spacecraft provided a relatively close look at the most unusual of Saturn's satellites, Hyperion. *Voyager 2* passed within 500,000 km of the hockey puck–shaped satellite. Scientists determined that the satellite had a squat cylindrical shape, 220 mi (350 km) across and 130 mi (210 km) thick. The strange body puzzled scientists because its long axis points approximately 45° away from the center of Saturn. Saturn's gravity would normally be expected to align Hyperion's axis with the huge planet's center. It is believed that the oddly shaped satellite may have collided with another body, about 10^8 years ago, knocking it from its normal attitude.

Both spacecraft are continuing their outward journey through the cosmos. *Voyager 1*, traveling above the elliptic plane, will not encounter another planet. Meanwhile, Uranus and Neptune await the arrival of *Voyager 2*. *See* SATURN.

Space shuttle. On April 12, 1981, the space shuttle, *Columbia*, was launched for its first flight into orbit. Astronauts John W. Young and Robert L. Crippen crewed the history-making flight. After 36 orbits, *Columbia* returned to Earth on April 14, touching down precisely on target at Edwards Air Force Base, CA.

Both Young and Crippen praised the performance of the shuttle. The steering during the ascent phase was termed excellent. Performance of the vehicle during reentry was outstanding and demonstrated the accuracy of its control system. The protective tiles on the spacecraft suffered only minimal damage during the ascent phase. Both the liquid and solid propellant systems provided the expected efficiency, while the orbital performance of *Columbia* exceeded expectations.

The second flight of *Columbia* began on November 12, after delays. The first delay was due to thermal tile adhesive damage because of a fuel spill on the launch pad during preflight servicing. Replacing the dislodged tiles took weeks. Originally scheduled for a new launch date on November 4, the mission was delayed an additional 8 days when an on-board fuel cell malfunctioned at T minus 31 seconds in the final countdown. During that delay, a contamination of the oil used in the hydraulic control system was also discovered.

Following the correction of these problems, the shuttle was launched on its second flight. After 54 h in orbit, the mission was recalled, due to fuel-cell failure.

In spite of the fact that *Columbia* did not remain in orbit for its scheduled 5-day mission, the crew—astronauts Richard Truly and Joseph Engle—completed an estimated 90% of the planned experiments. Most important of these was the testing of a 50-ft (15-m) arm located in the spacecraft bay. That arm is to be used to place satellites in orbit and to retrieve others from orbit for repair. That test was fully successful.

Columbia returned to Earth in such excellent condition that the flight director was quoted as saying that after replacement of the faulty fuel cell and refueling, the spacecraft was theoretically ready for another mission. The third flight of the shuttle was tentatively planned for early 1982. *See* SPACE SHUTTLE.

Soviet crewed space program. Some space scientists outside the Soviet bloc believe that the Soviet Union reached a plateau in the development of space stations in the latter months of 1981. This was refected by a hiatus in crewed launches during that period. It is believed that a more-cylindrical-type Salyut spacecraft which would allow modules to be attached at either end, or possibly at the sides, will be developed and used in a new phase of that program. The advanced Salyut would have the same mass as the earlier spacecraft and would utilize the same launch vehicle. The later version may also feature return capsule capability which, it is believed, was demonstrated by the test launches of *Cosmos 929* (1977) and *Cosmos 1267* (1981) which docked with *Salyut 6*.

The advanced space station is expected to be expandable, not in the classic circular-structure form popularized in science fiction, but in a manner which one space expert said would resemble a shish kebab. This approach would overcome engineering problems which construction of a wheel-shaped space station would present.

Some space engineers believe that the advanced Salyut will have a design life of as long as 5 years and will be crewed continuously during its lifetime. The station's crew size could reach as many as 12 cosmonauts, depending on the size of the vehicle used for launch. The present Soviet station could accommodate six persons, transporting them to the Salyut for a relatively short period of time.

While the advanced Salyut space station was expected to begin in the first half of 1982, space scientists feel that the program's first significant step was taken with the launching of *Cosmos 1267* on April 25, 1981, and its subsequent docking on June 19, 1981. It is further expected that the advanced space station's orbit may be pushed higher than the previously used 51°, possibly as high as 70°. This could allow for a greater Earth surface area surveillance for both civil and military purposes.

Earth-Sun relations. The United States launched the *Solar Mesospheric Explorer* satellite and *Dynamic Explorers A* and *B* to investigate Earth-Sun relations.

Solar Mesopheric Explorer. The *Solar Mesospheric Explorer* is an atmospheric satellite, placed in polar orbit on October 6, 1981. That orbit is circular, and Sun-synchronous (3 A.M.–3 P.M.), 336 mi (540 km) above the Earth. A Sun-synchronous orbit allows the spacecraft to observe any spot on Earth at the same relative local time of day. In its orbit, the satellite studies reactions between sunlight, ozone, and other chemicals in the atmosphere and how concentrations of ozone are transported within the atmosphere in the region from 19 to 55 mi (30 to 90 km) altitude.

Five scientific instruments simultaneously monitor ozone and minor atmospheric-constituent quantities, water-vapor abundance and temperature, and the amount of incoming solar radiation to determine the role each plays in ozone production and distribution throughout the mesophere.

The mesosphere is that region of the atmosphere between 19 and 50 mi (30 and 80 km) altitude—above the stratosphere and below the ionosphere. Radiation of short wavelength in that region causes a variety of photochemical reactions; the most notable is a layer of ozone that effectively absorbs solar ultraviolet radiation and causes a warm layer at about 18 mi (29 km) altitude. The scientific data from this spacecraft will help to develop models for ozone production and depletion on a global scale.

Five instruments perform the collection and monitoring of data: an ultraviolet ozone spectrometer to measure ozone (O_3) between 25 and 43 mi (40 and 70 km) altitude; a 1.27-micrometer spectrometer to measure the ozone between 31 and 56 mi (50 and 90 km) altitude and hydroxyl (OH) between 37 and 56 mi (60 and 90 km) altitude; a nitrogen dioxide spectrometer, designed to measure nitrogen dioxide (NO_2) between 12 and 25 mi (20 and 40 km) altitude; a four-channel, infrared radiometer to measure the temperature and pressure between 12 and 43 mi (20 and 70 km) altitude, and water vapor and ozone between 19 and 40 mi (30 and 65 km) altitude; and a solar ultraviolet monitor to measure incoming radiation from the Sun.

The spin-stabilized satellite is made up of two major modules: the observatory module and the spacecraft module. The body is a cylinder 69 in. (1.7 m) tall and 50 in. (1.25 m) in diameter. The solar array is a disk located on the booster end of the cylinder. That array is 88 in. (2.2 m) in diameter.

Dynamic Explorers A and B. Two scientific satellites were launched simultaneously on August 3, 1981. *Dynamic Explorer A* was placed in a highly elliptical polar orbit, ranging from 420 to 15,500 mi (675 to 24,945 km). *Dynamic Explorer B* was injected into a significantly lower polar orbit, ranging from 190 to 808 mi (306 to 1300 km) altitude.

The Dynamic Explorer spacecraft, each weighing approximately 1100 lb (500 kg), work in harmony to provide an understanding of the processes by which energy from the Sun flows through interplanetary space, entering the region controlled by the forces from the magnetic field which surrounds the Earth. This causes a reaction that results in the aurora (Northern Lights), disrupts radio transmissions, and may influence basic weather patterns.

The Dynamic Explorer satellites provide specific knowledge about the interaction of energy, electric currents, electric fields, and plasma (ionized atomic particles) between the magnetosphere, the ionosphere, and the atmosphere.

Detecting and monitoring Earth resources. The United States launched three satellites to monitor the Earth's atmosphere, oceans, and land areas.

GOES-D. The Geostationary Operational Environmental Satellite *GOES-D*, launched on September 9, 1980, was the first United States satellite with the capability of near-continuous monitoring of atmospheric water vapor and temperatures. It provides data for the western half of the United States, Canada, and the eastern Pacific Ocean. The spacecraft's monitors provide improved sea surface-temperature information, which is of increasing importance to the fishing and marine transportation industries, weather forecasters, and other interests. Information relayed back to Earth by *GOES-D* provides great assistance to weathercasters in predicting severe local hurricanes, storms, and other short-term weather phenomena.

The spacecraft is 85 in. (216 cm) in diameter. Its overall height is 11.5 ft (350 cm). At launch, *GOES-D* weighed 1840 lb (836 kg) and 874 lb (397 kg) after firing the apogee motor which placed it in synchronous orbit.

The drum-shaped base of the satellite spins and houses the Earth, Sun, and x-ray sensors, energetic particle monitor, and axial and radial thrusters, the latter providing control and altitude maintenance. The entire side of the drum surface is covered with solar panels. A boom is situated at the edge of the drum and supports a magnetometer. Protruding from the base of *GOES-D* at the center is the bottom of the primary instrument, the visible infrared spin-scan radiometric atmospheric sounder (VAS). Mounted atop the VAS housing are the S-band, high-gain antenna; the UHF antenna; and the S-band omnidirectional antenna. The antennas are mechanically despun by an electric motor so that they point at the Earth while the satellite is spinning for stability.

In addition to the VAS, the spacecraft payload includes the space environmental monitor (SEM). SEM includes a magnetometer, a solar x-ray telescope, and an energy particle monitor, designed to provide direct quantitative measurements of the important effects of solar activity for use in real-time solar forecasting and subsequent research. On board is also a data collection system (DCS), providing communications relay from data collection platforms on land, at sea , and in the air to the command and data acquisition station (CDA); the system also enables interrogation of the CDA by the satellite.

The telemetry, tracking, and command (TTC) subsystem uses S-band frequencies to transmit wideband video to the CDA, for relay of "stretched" data from the CDA via satellite to the national weather service, and for transmission of weather facsimile data to local ground stations equipped to receive S-band automatic-picture-transmission data. The subsystem employs UHF for transmissions from the DCS platforms to the spacecraft and then, via S-band to the CDA, and uses the VHF and S-band for commanding the spacecraft and for transmitting the space-environment-monitoring data.

The data collected by *GOES-D* is vital to the analysis of storms and the short-lived weather phenomena they produce. Weather forecasters and other scientists use the data recorded by the spacecraft to study severe storms and storm-spawned phenomena such as hail, flash floods, and tornadoes.

GOES-E. *GOES-E*, launched on May 22, 1981, is the second in a series of three improved GOES spacecraft. It is used to monitor the eastern half of the United States and Canada, all of Central and South America, and much of the Atlantic Ocean. *GOES-E* watches for hurricane development and movement in the Caribbean Sea, locates Gulf Stream system currents for marine interests, warns

Florida citrus growers of approaching crop-killing frosts, and provides government and private weather forecasters with a variety of information crucial to accurate forecasting of the weather.

NOAA-C. This satellite, designated *NOAA-C* at launch but listed as *NOAA 7* after achieving orbit on June 23, 1981, carried aloft the most versatile scanning radiometer ever used in an environmental spacecraft, gathering visual and infrared imagery and measurements in five spectral channels. This permits more accurate evaluation of sea surface temperatures, as well as of the temperatures of land, ice, surface water, and clouds under the satellite's polar orbital path.

Specifically, *NOAA-C* provides improved data-handling capabilities through: high resolution, day and night cloud cover observations on both a local and global scale; high resolution observations of sea surface temperatures; improved observation of vertical temperatures and water vapor profiles in the troposphere and lower stratosphere on a global basis; observations of vertical temperature profiles in the middle and upper stratosphere on a global basis; operational flight of a high-capacity data collection/relay and platform location system; and observations of electron and proton flux densities and total particle energy densities in the near-earth space environment.

The satellites in the NOAA series have the capability of receiving, processing, and retransmitting data from free-floating balloons, buoys, and remote automatic weather stations located around the world. They can also track the free-floating weather buoys.

Space communications. Communications satellites were launched in the Fleet SATCOM (FLTSATCOM), Satellite Business Systems (SBS), INTELSAT V, and COMSTAR series.

FLTSATCOM D and E. Launched on October 30, 1980, and August 6, 1981, respectively, these craft (Fig. 1) are advanced Earth satellites designed to provide extensive communications capability for the U.S. Navy and the U.S. Air Force. The spacecraft each provide 23 UHF and 1 SHF communications channels. Services provided are: fleet broadcast, channel 1; fleet relay, channels 2 through 10; Air Force narrow band, channels 11 through 22; a DOD wide band, channel 23; and an additional telemetry tracking and command (TT and C) S-band channel.

The spacecraft each consist of two major hexagonal elements, a payload module, and a spacecraft module. Twelve panels enclose the payload and spacecraft modules, and the majority of the electronic equipment is mounted on these panels.

Contained within the spacecraft modules are the Earth sensors; apogee kick motor; altitude and velocity controls; telemetry, tracking, and command subsystems; electrical power distribution; and the solar array which, before emplacement in orbit, is folded around the spacecraft module.

During the transfer orbit operations and through apogee kick motor burns, the spacecraft is stabilized by spinning. After kick motor burn, the spacecraft is despun, and solar arrays and the UHF antenna are deployed. Once in geosynchronous orbit, the Sun and Earth are acquired through sensors and the spacecraft is pointed at the Earth's center by yaw jets and a reaction wheel. The craft is next rotated by a clocked drive which keeps the solar array normal to the orbit plane. The satellite is then rotated at a uniform rate to keep its solar array pointed at the Sun. A redundant monopropellant hydrazine thruster system provides spacecraft control and velocity maneuvers. The solar arrays, along with batteries, interface with a command electrical power bus which distributes primary power to the subsystem equipment converters.

Satellite Business Systems. *SBS-A* and *SBS-B* (Fig. 2), launched on November 15, 1980, and September 24, 1981, respectively, are 85-in.-di-

Fig. 1. A FLTSATCOM communications satellite being encapsulated in its protective nose cone. (*NASA*)

Fig. 2. *SBS-A* communications satellite being prepared for launch. (*NASA*)

Fig. 3. *INTELSAT V* communications satellite. (*INTELSAT*)

ameter (216-cm) spacecraft. After deployment in geosynchronous orbit, the telescoping solar panel cylinder is extended, and the communications antenna is raised, giving the satellites an overall height of 21 ft 8 in. (6.6 m).

SBS-A and *-B* are spin-stabilized spacecraft utilizing two concentric, cylindrical solar panels. When the lower panel is deployed downward, the combination provides a 914-W, end-of-life power required for communications. The power system uses improved solar cells, providing 20 mW/cm^2. Two nickel cadmium batteries power the craft during eclipse periods.

The satellite's repeater has 10 channels, each 43 MHz wide, and each using a 20-W multicollector traveling-wave tube. A signal strength of more than 43 dBW (decibels above 1 W) is devloped by the single large shaped beam reflector. This is for use in the primary eastern United States coverage zone. Redundant receivers, using solid-state microwave integrated-circuit techniques, provide service over the most densely populated regions of the United States.

The operational spacecraft provides integrated, all-digital transmission of telephone, electronic mail, computer, and video-teleconferencing to business and industrial clients. The SBS domestic satellite system provides fully switched private networks for business, government agencies, and other organizations with large and varied communications requirements.

SBS is a consortium consisting of IBM, COMSAT General, and Aetna Insurance companies. Launch services are provided by NASA on a reimbursable basis.

INTELSAT V-A and V-B. The *INTELSAT V-A* spacecraft (Fig. 3), launched on December 6, 1980, and *INTELSAT V-B*, launched on May 23, 1981, each weighed 4521 lb (1928 kg) at launch and 2286 lb (1037 kg) after apogee motor firing. Their height is 21 ft (6.4 m) and their diameter is 21 ft (6.4 m). After full deployment, the diameter is 22.25 ft (6.8 m). The main body is 5.4 by 5.8 ft (1.66 by 1.77 m).

INTELSAT V is a modularly constructed, three-axis body, stabilized by Earth and Sun sensors and a momentum wheel. Its winglike, Sun-oriented solar array panels produce a total of 1241 W of electrical power. The spacecraft has an average capacity of 12,000 simultaneous, two-way telephone circuits and two television channels. It has six communications antennas in two global coverage horns, two hemispherical/zone offset-fed reflectors and two offset-fed spot beam reflectors. This geosynchronous satellite has almost double the communications capacity of earlier INTELSAT series spacecraft. The *INTELSAT V* was built in the United States, using components built by France, the United Kingdom, Japan, West Germany, and Italy, and was launched by NASA on a cost-reimbursable basis.

The INTELSAT global satellite system comprises two segments. The space segment, consisting of satellites owned by the International INTELSAT Satellite Organization, is composed of more than 100 countries. The satellites are positioned over the Atlantic, Pacific, and Indian ocean regions. The ground segment consists of Earth stations owned by telecommunications entities in the countries in which they are located.

The 190 Earth station antennas, operating with the satellites, provide more than 840 pathways, or direct communications links, between more than 130 countries, territories, and possessions.

COMSTAR D-4. This craft was launched into geosynchronous orbit on February 21, 1981, to provide domestic satellite service. It has a capacity of 18,000 circuits, or the equivalent of 18,000 simultaneous telephone conversations. Included in its instrumentation are 24 transponders, each having a bandwidth of approximately 34 MHz with center frequencies spaced at 20-MHz intervals within 500-MHz-wide bands at 4 and 6 GHz. To reuse frequencies, polarized signals are employed. Twelve transponders receive and transmit horizontally, and twelve vertically. This technique allows the doubling of the capacity of the satellite's system without mutual interference. The *COMSTAR D-4* also carries a millimeter-wave experimental package with which communications tests are conducted for possible use in future satellite systems.

The satellite is 8 ft (2.4 m) in diameter and 20 ft (6.1 m) in height. At launch, it weighed 3348 lb (1517 kg) and, after apogee motor firing, weighed 1746 lb (791 kg) in orbit.

For background information *see* APPLICATIONS SATELLITES; COMMUNICATIONS SATELLITE; METEOROLOGICAL SATELLITES; SPACE FLIGHT in the McGraw-Hill Encyclopedia of Science and Technology.

[LARRY G. HASTINGS]

Bibliography: *COMSAT Guide to INTELSAT, MARISAT and COMSTAR*, pp. 13, 42, 1981; NASA, *A Record of NASA Space Missions, 1958–1980*, 1981; NASA, *Satellite Handbook*, 1981; NASA, *The Voyager Report*, D-23835A, pp. 762–775, 1981; NASA releases nos. 80-137, 80-140, 80-158, 80-199A, 80-10187, 81-27, 81-50, 81-55, 81-60, 81-93, 81-106, 81-130, 81-114, 81-172.

Space shuttle

A new era of transportation has commenced with the launch of the reusable space shuttle.

In March 1981, the orbital vehicle *Columbia* (*OV 102*) successfully completed integration testing at Kennedy Space Center. Installation of the thermal protection system had been completed with roll-out in November 1980 to the Vertical Assembly Building for mating with the other elements of the shuttle vehicle. The development flight test program of the Space Transportation System (STS) consists of four orbital flights, of which the first two, *STS 1* and *STS 2*, were launched in 1981.

STS 1. The terminal countdown for the initial attempt to launch *STS 1* was conducted April 10, 1981. The countdown proceeded normally until T minus 20 min when the orbiter general-purpose computers (GPCs) were scheduled for transition from the ground mode to the flight mode. The launch was held for the maximum time and scrubbed when the four primary GPCs would not provide the correct timing for the backup flight system GPC. Analysis and testing indicated that the primary set of GPCs had provided incorrect timing to the backup flight system at initialization.

Ascent. Lift-off from the Kennedy Space Center, Launch Pad 39A, occurred at 07:00:03.9 EST on April 12, 1981, with John W. Young and Robert L. Crippen as crew members. With both solid-fuel rocket boosters (SRBs) and all three main engines delivering sea-level rated thrust, the space shuttle vehicle lifted vertically off the launch pad until the tower structure was safely cleared. Next, a pitchover and roll maneuver placed the *Columbia* on its correct launch azimuth heading to achieve the desired orbital inclination of 40.3°. The vehicle was flown through the high-dynamic-pressure region, during which programmed main engine throttling down to a 65% power level and elevon deflection schedules were successfully implemented to maintain acceptable structural load margins during first-stage ascent. The SRBs burned out approximately 1 s early and separated 2 min 11 s after lift-off, with all separation constraints satisfied with comfortable margins. The first-stage trajectory lofted high, resulting in SRB burnout approximately 3000 m higher than predicted.

During second-stage flight, all orbiter subsystems functioned normally. The flash evaporator cooling effectively maintained desired cabin environmental conditions and provided more than adequate heat rejection from critical components. Main engine throttling successfully limited structural loading to 3 *g*, as guidance brought the vehicle to the desired cutoff conditions before providing a main engine cutoff (MECO) signal to terminate main engine thrust. Main engine cutoff was achieved at about 8.5 min, targeted for an altitude of 60 nautical miles (n mi; 111 km) and an inertial velocity of 25,668 ft/s (7824 m/s).

External tank separation was successfully accomplished on schedule, and postseparation maneuvers provided comfortable clearances from the spent tank. Maneuvers to the desired insertion burn attitude were accomplished using reaction control system (RCS) jets, and the dual orbiter maneuvering system (OMS) engine insertion burn was accomplished, as scheduled, to acquire a 57 × 133 n mi (106 × 246 km) orbit. The subsequent circularization burn, using the two OMS engines, placed the orbiter in a 133 n mi (246 km) circular orbit. These burns were completed by 46 min into the flight.

Orbital operations. On-orbit operations began after the vehicle orbit was circularized at 133 n mi (246 km). The initial preparation for orbital operations consisted of reconfiguring the data processing system software from ascent to on-orbit functions and the successful performance of the other operations critical to staying on orbit. Included in the critical operations were payload bay door latch and door drive tests ending with the doors opened and the radiators activated and deployed. Crew optical alignment sight calibration, inertial measurement unit alignments, and fuel-cell power-plant purges were also performed. Subsystem status was evaluated by the crew and the mission control center at 3 h 16 min into the flight, and a "go" given for continued on-orbit operations.

The crew removed their ejection escape suits and configured the vehicle for on-orbit operation. A wastewater dump was performed. Following a meal, the crew executed two orbital maneuvering system burns, each burn using a different engine in the prescribed crossfeed mode to raise and circularize the vehicle orbit at 148 n mi (274 km). Part one of the flight control system checkout was performed, using auxiliary power unit (APU) no. 1 to verify vehicle aerosurface drive capability. An aft-reaction control system thruster test was performed on those thrusters not utilized during ascent. Vacuum inerting of the main propulsion system, inertial measurement unit alignment and accelerometer calibration, and a supply-water dump were conducted prior to the next crew meal. Following the meal, the inertial measurement units were again aligned, and the crew optical alignment sight was calibrated and verified. The fuel-cell power plants were again purged in preparation for the first crew sleep period.

The second day began with another inertial measurement unit alignment and fuel-cell power-plant purge. The initial RCS test firing was successfully conducted prior to a vehicle gravity gradient free-drift evaluation period of two orbital revolutions. During this time, the inertial measurement unit (IMU) accelerometers were calibrated. The crew then conducted part two of the flight control system checkout. Following the gravity gradient operations, a series of RCS test firings were conducted, including a fourth (unplanned) firing to evaluate vehicle maneuverability. The payload bay doors were closed but not latched prior to the last test firing and were reopened at its conclusion. Later the crew again aligned the inertial measurement units, purged the fuel-cell power plants, and conducted an ejection escape-suit donning and seat ingress exercise. Vehicle passive thermal control attitude tests, namely, barbecue (vehicle rotated at a constant rate to maintain temperature balance), RCS manual rotation, and attitude hold, were conducted following suit doffing using both the primary and vernier RCS thrusters. These tests were conducted using various digital autopilot deadband control limits. The remainder of the day was spent conducting crew optical alignment sight calibration and performing housekeeping tasks such as wastewater and supply-water dumps, meal preparation, carbon dioxide–absorber replacement, IMU alignment, and fuel-cell power-plant purging.

Following the second crew sleep period, the day began with similar housekeeping chores being conducted prior to the crew initiating more than 6 h of deorbit preparation activities. Those activities included repeating part two of the flight control–system checkout, vehicle reconfiguration for entry, ejection escape-suit donning, closing of the payload bay doors, the final IMU alignments, final fuel-cell power-plant purges, and two orbits of mid-body thermal conditioning. All pre-deorbit activities were then complete.

Descent. The orbiter traversed the entry phase from deorbit to touchdown and rollout on runway 23 at Edwards Air Force Base, CA, with a near-nominal descent. Data indicate that the descent events occurred properly and as scheduled. Performance parameters were very near the planned values.

The deorbit firing occurred at approximately 53.5 h into the mission during the 36th revolution, with targeting for a range from the entry interface of about 4400 n mi (8100 km). A two-OMS firing was performed having velocity residuals less than 2 ft/s (0.6 m/s), thereby not requiring trimming. During the subsequent 25-min coast to the 400,000-ft (120-km) entry interface, *Columbia* was maneuvered to the proper attitude for entry, and the final entry preparations were performed. The range-to-go and path angle at the entry interface were nominal. About 4 min later the communications and tracking blackout occurred during the high-speed portion of entry. Navigation was updated on exiting the blackout, and the entry/terminal area energy management (TAEM) interface was reached at Mach 2.5 and 80,000 ft (24 km) altitude, 6 min from touchdown.

The trajectory parameters observed before and after blackout very nearly matched the preflight predictions, thus suggesting nominal performance during the blackout region and trajectory dispersions within allowable tolerances. Because of an initial cross-range of approximately 350 n mi (650 km), to the landing site to the right of the trajectory plane, the initial bank direction was also toward the right.

Three bank reversals were commanded at the planned 19,000, 10,000, and 5000 ft/s (5.8, 3.0, and 1.5 km/s) velocities. The crew indicated that the first roll reversal had a slight overshoot and that each subsequent reversal exhibited better performance.

The guidance system executed the internal switching properly from phase to phase. The switch from entry to TAEM guidance occurred at 2500 ft/s (760 m/s) with minimum transients.

S-band communications were established at approximately 10,700 ft/s (3260 m/s), and TACAN acquisition and update occurred at approximately 6500 ft/s (1980 m/s). The rudder was activated at Mach 3.5, air data from the probes became available at Mach 2.5, and the fuselage vents were opened at Mach 2.4.

The crew selected control-stick steering as planned for the last two roll reversals with no stability problems. The crew commented that the switching from auto to manual to auto to manual was exceptionally smooth.

As planned, the crew again selected the control-stick steering mode just prior to reaching the heading alignment circle and used this mode for the remainder of the flight. The heading alignment circle was traversed with small errors and a near-nominal, manually controlled landing was made. The gear was lowered at 270 knots (139 m/s), touchdown occurred approximately 5000 ft (1500 m) down the runway at approximately 195 knots (100 m/s), the nose was lowered to the runway at approximately 165 knots (85 m/s), and the orbiter was stopped approximately 16,000 ft (4900 m) from the runway threshold.

Evaluation. Observation of the orbiter showed that all the critical thermal protection system tiles and reinforced carbon-carbon material sections were intact. Several noncritical tiles were lost during launch. Even so, thermocouple data indicated top surface temperatures below predicted values with no burn-throughs in the several missing tile locations. The tile on the underside of the vehicle performed better than expected with backface structural temperatures lower than expected.

Failure of a tape recorder resulted in the loss of thermal entry data during blackout which hampered the evaluation of the orbiter's thermal characteristics. However, evaluation of the data before and after blackout and the postflight inspection of the thermal protection system confirmed that the orbiter design was adequate for continuing the flight test program.

The most significant *STS 1* problem was the overpressure condition of the aft heat shield after SRB ignition which was significantly larger than predicted. The overpressure resulted from a pressure wave that was generated by the SRB exhaust plume rebounding from the launch pad. This was observed in measurements on the orbiter heat shield and on the upper surface of the body flap which read 2 psi (14 kPa) above ambient or four times the predicted value. All other systems' performances were as expected with only minor problems, none of which affected mission operations.

STS 2. After the ferry flight from Edwards Air Force Base, turnaround operations, and a facility modification, which used water in fabric troughs to absorb the pressure wave, thus reducing the SRB plume rebounding, *Columbia* was ready for its second launch on November 4, 1981.

Launch and ascent. The countdown proceeded normally until T minus 9 min when the ground launch sequencer (GLS) stopped the count for a violation of the liquid oxygen mass quantity redline (limit). The automatic sequencer resumed the countdown approximately 2 min later, when the liquid oxygen mass quantity redline was cleared.

The three orbiter APUs were started on time and in sequence, and a "go" was given on all three units even though the lubrication oil outlet pressures on APUs 1 and 3 were higher than anticipated. The countdown continued normally until T minus 31 s, when the GLS halted the count due to a violation of the orbiter oxygen tank pressure limits (800 psia or 5.52 MPa absolute pressure). A real-time decision had been made to lower the oxygen tank pressure limits to 775 psia (5.34 MPa absolute pressure) and continue the count, but the GLS operator was unable to clear the limits.

The orbiter APUs were turned off, and planning was begun for a recycle at T minus 9 min. After further analysis of the higher-than-expected lubrication oil outlet pressures on both the orbiter APUs 1 and 3, it was determined to scrub the launch since there were no APU test data available for mission duty cycles with a possible clogged filter and contaminated oil. The oil filters and the oil were changed on APUs 1 and 3, and the countdown was restarted for *STS 2*. A problem in an orbiter data link between the computer and the communications system resulted in a 2-h launch slip with *STS 2* being launched at 10:10 A.M. EST on November 12, 1981. The ascent phase of the flight was uneventful, with all systems performing as planned.

Orbital operations. Following the OMS-2 circularization firing, on-orbit operations were initiated, including reconfiguring the data-processing system. The payload bay doors were opened, theodolite measurements of structural alignment were made, and the radiators configured for heat rejection. After both onboard and ground assessment of orbiter status, a "go-for-orbit operations" was given by the flight director by the designated time of 3 h 15 min into the mission. Following the go-for-orbit decision, the flight crew removed their ejection escape suits and continued reconfiguring the orbiter subsystems for orbit operations.

Approximately 2½ h into the mission a caution and warning light indicated that the water produced by the fuel cell had a high pH reading (indicating that potassium hydroxide—the fuel-cell electrolyte—had exited the fuel cell). Since other measurements showed proper performance, it was initially concluded that this indication was a transducer malfunction. However, 2 h later the fuel-cell performance deteriorated, and the fuel cell was deactivated.

Due to the fuel-cell difficulty, and because a critical IMU alignment had not yet been completed, the OMS-3A/B and OMS-4 burns were delayed for one orbit. The delay was to allow an orderly completion of tasks and an assessment of the failure impact. The OMS-3A and -3B burns were made on orbit 6. At 8 h 33 min 20 s after lift-off, the OMS-4 firing was made which increased and circularized the vehicle orbit at 143.5 n mi (265.8 km). From this point, the flight crew followed a modified time line, completing most of the scheduled flight tests and all of the necessary subsystem activities. Meanwhile, the OSTA-1 (Office of Space and Terrestrial Applications) pallet experiments initiated data collection. The crew then began their first sleep cycle.

Although the orbiter could have successfully completed all scheduled tests and flown the originally planned 5 days, the NASA mission management team elected the more conservative approach of adopting a modified mission plan of about 54.5 h. The modified mission would satisfy more than 90% of the higher-priority flight tests.

Three high-priority tests originally scheduled for the first flight day were rescheduled to the second day, beginning with the remote manipulator system group 1 tests. These were highlighted by spectacular television pictures of the manipulator operation. During the balance of the second flight day, other flight-test objectives were satisfied by the RCS leak detection test, vernier RCS heaters tests, a modified hydraulic circulation pump test, suit donning, and other tests. Meanwhile further OSTA-1 data takes were successfully accomplished.

Following the second crew sleep period, housekeeping chores were conducted prior to the crew initiating more than 6 h of deorbit preparation activities. Those activities included the flight control system checkout, vehicle reconfiguration for entry, ejection escape-suit donning, closing of the payload bay doors, the final IMU alignments, final fuel-cell power-plant purges, and two orbits of mid-body thermal conditioning. This completed the pre-deorbit firing (burn) preparation activities.

Descent. The orbiter successfully landed on lakebed runway 23 at Edwards Air Force Base, completing the modified *STS 2* mission. Deorbit was from revolution 36, requiring an entry cross-range to the runway of about 70 n mi (130 km). Entry phase ranging and heating were near nominal considering the minimal cross-range requirement. The terminal area and approach/land phases experienced severe wind conditions. High-altitude tailwinds (becoming crosswinds and headwinds on final approach) resulted in a low-energy trajectory which touched down slightly short of the targeted point on runway 23. All entry flight test objectives were met.

The retrograde deorbit burn was extended by firing in a slightly out-of-plane attitude. This technique expended the excess OMS propellant available, and together with a null-type firing of the forward RCS yaw thrusters, produced the desired center of gravity for entry. The burn was both accurately planned and executed. No state vector update was required during the postburn pass over Guam.

The deorbit firing and ensuing entry were satisfactorily accomplished on two fuel cells by following preplanned electrical power management procedures. The deorbit firing was controlled using only two cathode-ray-tube (CRT) displays. The third CRT display was powered up after closing the vent doors, just prior to the entry interface. Navigation aids such as the TACANs and MSBLS (microwave scanning beam landing system) were kept powered off until needed after exiting the radio-frequency blackout. Nonessential equipment such as some of the development flight instrumentation (DFI) transducers, the DFI transmitter, and numerous heaters were kept powered off throughout the entry.

APU 3 was started a few minutes prior to the deorbit burn firing, and APU 1 and 2 started about 13 min after the burn as planned. A hydraulic thermal conditioning test, consisting of cycling the aerodynamic surfaces for 5 min to circulate warm fluid through the actuators, was accomplished according to plan starting 13 min before the entry interface.

The entry interface targeting resulted in a flight range to the runway of about 4470 n mi (8280 km), indicating a near-nominal entry trajectory for a 70-n mi (130-km) cross-range deorbit. Bank-reverse-bank ranging maneuvers occurred somewhat earlier than originally planned, but they were consistent with the cross-range requirement. Preliminary data indicate that boundary-layer transition was later than on *STS 1*. Some regions of the OMS and fuselage side appeared hotter than *STS 1*. Postflight examination revealed some discoloration of the flexible reusable surface insulation in these areas. All of the 25 maneuvers programmed for the acquisition of aerodynamic data during entry were accomplished, although the last maneuver for flutter evaluation had to be cut short in order to start the heading alignment circle. The radiator heat sink test was performed as planned. The radiators were precooled on orbit prior to closing the payload bay doors.

The transition from the entry phase to TAEM took place without incident, although a bank-reversal was required to acquire the heading alignment circle. The remaining range-to-go at the TAEM interface was a nominal 60 n mi (117 km), but the position was displaced about 4 n mi south (cross-range) compared with the predicted position. The TAEM trajectory was complicated by strong wind conditions which continued through touchdown. The winds were generally from the southwest and reached 99th percentile levels. The vehicle entered the heading alignment circle at about an altitude of 40,000 ft (12 km). Flight control was assumed to counter the severe crosswinds expected during the turn. The crew maneuvered the orbiter to avoid the downwind slide and the increased *g* levels characteristic of the auto system. The maximum normal load factor observed during the turnaround was 1.9 *g*.

The orbiter remained in TAEM guidance until the 5000-ft (1.5-km) altitude. At this point the guidance properly defaulted to the approach-land mode. Quick-look data show that the orbiter was in the auto guidance mode well below the nominal 2000-ft (600-m) altitude for preflare initiation, indicating an auto-controlled flare path closely matching that expected by the crew. The vehicle touched down short of the nominal landing point, an expected result of the low-energy heading alignment circle and the severe headwind on final approach. The wind at touchdown was varying between 18 and 24 knots (9 and 12 m/s) and coming from about 30° left of the runway heading. The crosswind component is expected to at least partially satisfy the test objective of landing and controlling the rollout in a crosswind.

Evaluation. The *STS 2* flight significantly contributed to the certification data base which will verify the space shuttle as an operational vehicle. The only *STS 2* problem of significance was the fuel cell. Postflight analysis has shown the cause to be contamination, probably introduced during the manufacturing process, which blocked an aspirator, a device which removes water from the fuel cell. This resulted in "flooding" of 2 of the 64 cells composing the fuel cell, which in turn led to the observed loss of voltage. The remaining fuel cells have been inspected and found to be without contamination.

The *STS 2* mission inaugurated the era of the reusable shuttle vehicle with the refurbished *Columbia* making its second space flight. Following the flight, astronauts Joe H. Engle and Richard H. Truly acclaimed *Orbiter 102* a solid vehicle. The November 12–14 mission was of less duration than originally planned, but the high-priority flight-test objectives were successfully completed. Tests of the remote manipulator system, provided by Canada, were very successful, as was the collection of Earth observation data by the experiment (OSTA-1) pallet. Since the majority of the planned *STS 2* flight tests were accomplished, only minor modifications to *STS 3* and *4* flight planning will be necessary.

STS 2 also demonstrated important designed-in operational capabilities by the continuing of all major flight operations in the presence of a significant subsystem failure. Sophisticated payload operations were among the activities carried on without being restricted by the two (versus three) fuel-cell operations. Two additional development test flights are to be flown in 1982, with the first operational flight planned for late in the year.

For background information *see* SPACE FLIGHT; SPACE SHUTTLE in the McGraw-Hill Encyclopedia of Science and Technology.

[AARON COHEN]

Stem

The form or morphology of an organism is the result of the number, shape, and arrangement of its cells. Plant cells are produced in meristems, which are more or less localized regions of cell divisions. Plants possess meristems throughout their lives and thus are continually growing and changing their form. Since meristems occur at the apices of all shoots and roots and their branches, their number in a single plant may be large. Furthermore, many plants have additional extensive nonapical meristems concerned with further primary and secondary increase in thickness of the plant stem. The most familiar of these lateral meristems is the vascular cambium which produces the wood in forest trees. Another such lateral meristem—the primary thickening meristem—exists in all monocotyledons with wide stems and short internodes. The existence of the primary thickening meristem in the stem of monocotyledonous plants has been recognized for over a hundred years and has been observed in most monocotyledonous species, from small herbaceous lilies

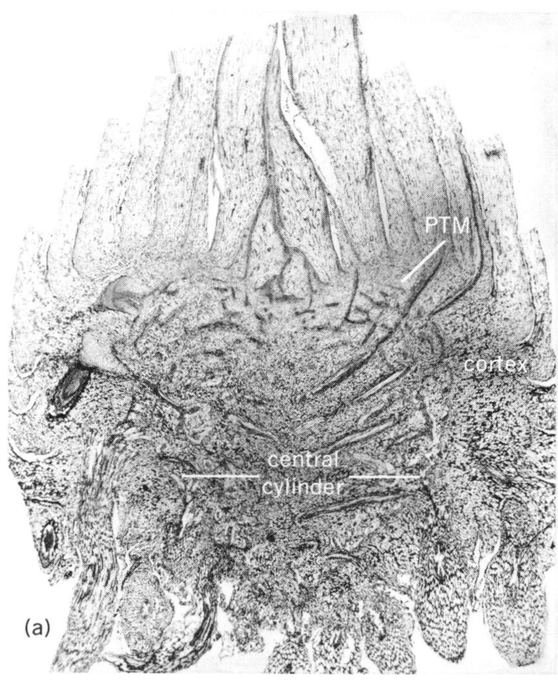

Fig. 1 Histology of the onion stem. (*a*) Median longitudinal section of a 7-month-old stem. (*b*) Apical meristem and subjacent tissue of same stem. (*c*) Enlargement of primary thickening meristem (PTM) from *b*. (*From D. A. DeMason, Function and development of the primary thickening meristem in the monocotyledon, Allium cepa L., Bot. Gaz., 140:51–66, 1979*)

to large arborescent palms. The primary thickening meristem has been distinguished as being a cambiallike zone consisting of tangentially flattened cells. Recent research has expanded knowledge of the extent, position, and functioning of this meristem in monocotyledons.

Monocotyledon anatomy. Two major groups exist among the flowering plants: the dicotyledons and the monocotyledons, distinguished by the number of seed leaves, or cotyledons, they possess in the embryo. Monocotyledons are distinguished from dicotyledons on the basis of numerous other structural characters, including differences in stem anatomy, leaf morphology, and floral structure. Generally, monocotyledons have a more complex stem anatomy than dicotyledons. The primary stem of dicotyledons is divided into two regions, the outer cortex and the inner pith. These regions are separated by a single ring of vascular bundles and eventually by the vascular cambium. Monocotyledonous stems also consist of two tissue regions, the cortex and the central cylinder. These are often not easily distinguished because monocotyledons possess numerous vascular bundles which occur in all areas of the central cylinder.

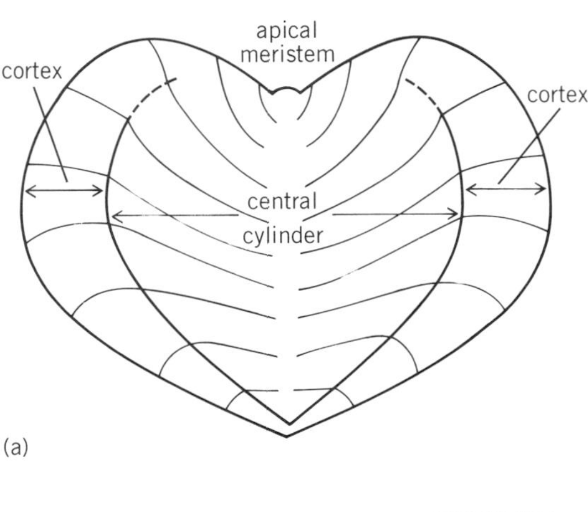

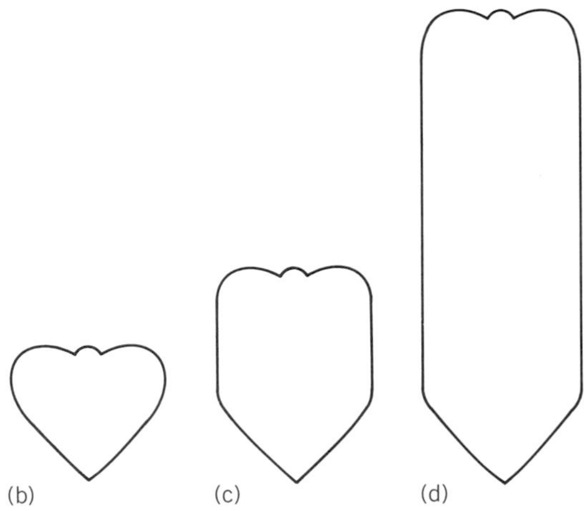

Fig. 2. Stem structure and development. (*a*) Pattern of cell alignment in the vegetative stem of onion. The lines indicate the course of cell files traced from photograph of 7-month-old stem in Fig. 1 (*from D. A. DeMason, Function and development of the primary thickening meristem in the monocotyledon Allium cepa L., Bot. Gaz., 140:51–66, 1979*). (*b–d*) Schematic outlines of median longitudinal sections through a large monocotyledon during various stages of stem growth. (*b*) Shape produced by primary thickening growth. (*c*) Shape produced during early maintenance growth. (*d*) Shape produced after extensive maintenance growth.

The top of the stem or crown of monocotyledons that have extensive stem thickening exhibits some distinctive morphological and histological features. Because of extensive stem thickening at the distal end of the shoot, the apical meristem and the youngest leaf primordia become sunken below surrounding stem tissues (Fig. 1*a*, and *b*). Related to this, the crown has a very distinctive pattern of cell alignment. In longitudinal section, the cells of the stem are arranged in files which run from the leaf bases through to the center of the central cylinder. These files of cells remain organized during growth; however, their angle from the vertical varies at different levels of the stem (Fig. 1*b* and 2*a*). In transverse sections, the cell files form concentric circles at the top of the crown. The files in these monocotyledons can be thought of as being arranged like a stack of saucers; the saucers at the base of the stack have less-inclined sides than those at the top of the stack.

The base of the stem in monocotyledons with extensive stem thickening is obconic in shape, that is, it resembles an inverted cone. In large monocotyledons, the stems can be divided into two phases of stem growth: phase I termed primary thickening, and phase II termed maintenance growth. In phase I the diameter of successive internodes increases as the plant grows in height to produce the obconic base. After the plant reaches its mature diameter through the process of primary thickening, it may enter phase II, and increase in height without further increase in diameter (Fig. 2*b–d*). Onions and, probably, other small monocotyledons cease vegetative stem growth during primary thickening and never undergo maintenance growth.

Primary thickening meristem. The extensive stem thickening which occurs in monocotyledons with wide stems and short internodes is thought to be the result of activities of the primary thickening meristem, which has been viewed in the past as a narrow, compact meristematic zone of tangentially flattened cells. Early anatomical work had been concerned only with descriptive histology. From this narrow perspective, botanists had envisioned the primary thickening meristem as comparable and similar to the only other major lateral meristem in seed plants—the vascular cambium. Traditionally, the vascular cambium has been envisioned as being uniseriate and producing tissues bidirectionally; therefore, so has the primary thickening meristem. Experimental techniques have been used recently in the stem of onion (*Allium cepa*) to locate the primary thickening meristem. Onion was selected for these studies because: the anatomy of the crown shows features typical of monocotyledons with broad stem axes, including files of cells and a primary thick-

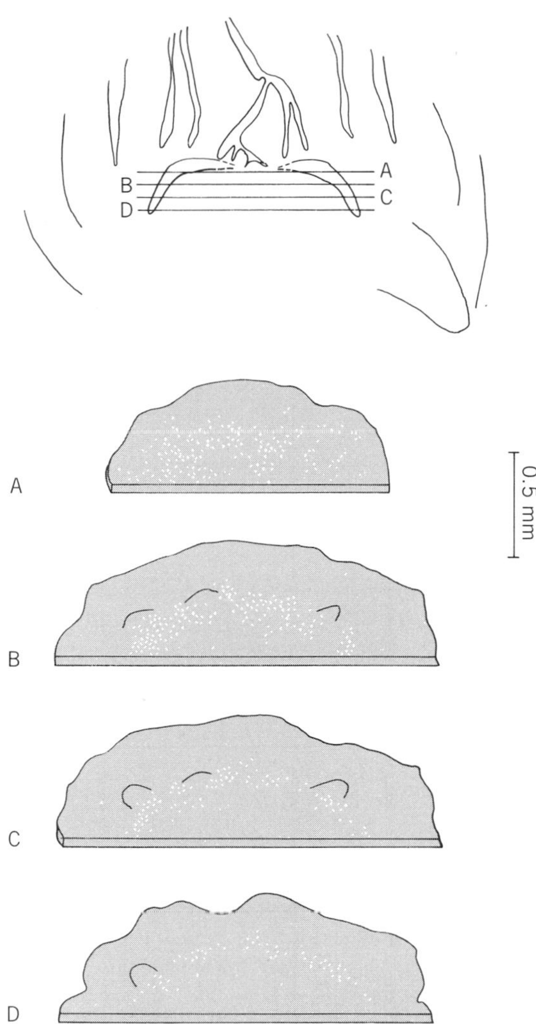

Fig. 3. Localization of primary thickening meristem. At top, outline drawing of 2-month-old plant shows positions of levels A–D. Crescent-shaped areas in crown indicate position of primary thickening meristem from protein staining with aniline blue-black. Levels A–D show results of three-dimensional analysis of labeled 3-month-old plant. Dots indicate the positions of labeled nuclei. *(From D. A. DeMason, Localization of cell division activity in the primary thickening meristem in Allium cepa L., Amer. J. Bot. 67:393–399,1980)*

ening meristem; the stem shows marked increase in stem girth during development; and the stem is sufficiently small so that sections include the entire stem, and measurements and comparisons can be made between plants of all ages.

Cells within a meristem are usually very active metabolically and contain greater amounts of protein and RNA than mature parenchyma cells; thus specific stains for protein (aniline blue-black) and RNA (azure B) were used on sections of the crown of onion. Also, onion plants were placed in a nutrient solution with tritiated thymidine, which is incorporated into replicating DNA—a process necessary and prerequisite to cell division. Nuclei that incorporated the radioactive base were localized by autoradiography, and then the resulting autoradiographs were analyzed three dimensionally.

These techniques reinforce one another and allow a more accurate visualization of the extent of the primary thickening meristem in onion. The primary thickening meristem appears as a diffuse meristem. It is broad at its distalmost region (near the apical meristem), but narrows basipetally before ceasing meristematic activities even more basipetally in the stem (Fig. 3). It is obviously not a uniseriate layer and probably does not produce tissues bidirectionally. It should be viewed as a diffuse meristematic zone that matures centrifugally in the central cylinder and centripetally in the cortex.

Histological studies and development. To understand the mechanism of stem thickening and the function of the primary thickening meristem, it is important to study changes in stem anatomy when the stem actually changes diameter, as in seedling development, bud development, or the transition to flowering. This has been done in onion. Onion plants were grown from seed and sampled at regular intervals during vegetative and floral development, fixed, and processed for light microscopy. Observations were made on the extent of cell alignment, the existence of the narrow zone of tangentially flattened cells, and the angle of cell alignment at a chosen level within the stem of the cell files in both the cortex and the central cylinder.

Among other results, it was found that: cell files are initiated quite soon after germination (in 2-week-old seedlings); the number of cells in a cell file increases until the primary thickening meristem ceases activity; the angle of the cell files changes during development and maturation; and the primary thickening meristem exists only in the vegetative stem, not in the inflorescence. These observations allow speculation on both the mechanism of stem thickening and the function of the primary thickening meristem in monocotyledons.

Mechanism of thickening. The mechanism of primary thickening of the stem axis of onion appears to involve three processes: production of new cell files; enlargement of cell files by cell division in the primary thickening meristem and subsequent cell enlargement; and reorientation of the cell files. One can use these processes to describe the phenomena seen at the whole plant level during primary thickening and maintenance growth. The histology of the crown during both primary thickening and maintenance growth are identical. All three important histological characteristics are present: sunken apical meristem, primary thickening meristem, and anticlinal cell files arranged at various angles. The difference between these two phases is that during primary thickening the stem increases in diameter as it grows in height so that successive internodes are wider than the previous ones; during maintenance growth the stem diameter does not change through time. One would predict that, during primary thickening, successively produced cell files would increase in length and number of cells per file, whereas during maintenance growth, successive cell files would maintain a constant length or number of cells per file. Further studies are necessary to test this hypothesis.

Conclusions. To date, all evidence confirms the fact that the primary thickening meristem is appropriately named and does function in the process of stem thickening in monocotyledons with short, squat stem axes. The evidence is as follows: the increase in number of cells per file ceases when cells in the region of the primary thickening meristem mature; the primary thickening meristem is associated with growth in width of the vegetative axis but not growth in length of the infloresence axis; and cell division occurs in the primary thickening meristem. The primary thickening meristem differs functionally from the vascular cambium in a number of ways: it is associated with primary growth only; it is a wide and diffuse meristem and does not function as a uniseriate layer; and it produces cell files at a vertical angle which reorient only secondarily to a horizontal angle.

For background information *see* APICAL MERISTEM; LATERAL MERISTEM; LILIOPSIDA; STEM in the McGraw-Hill Enclopedia of Science and Technology.

[DARLEEN A. DEMASON]

Bibliography: D. A. DeMason, *Amer. J, Bot.*, 66:347–350, 1979, and 67:393–399, 1980; D. A. DeMason, *Bot. Gaz.*,140:51–66, 1979.

Superconducting detectors

Very sensitive receivers for electromagnetic wavelengths close to 1 mm are being developed that use superconducting tunnel junctions for the nonlinear element. The junction response depends on tunneling of single electrons (more precisely, quasiparticles) through an oxide barrier between two superconductors. The tunnel junctions used have been extensively developed for Josephson-effect digital logic. The sensitivity of these new quasiparticle receivers is expected to approach fundamental physical limits. Applications for receivers with improved sensitivity in this wavelength range include radio astronomy and satellite communications.

The ultimate sensitivity of heterodyne mixer receivers of electromagnetic waves is set by the Heisenberg uncertainty principle. Zero-point fluctuations in the number of photons present in the device lead to a minimum noise temperature which is equal to the photon energy expressed as a temperature. Although heterodyne receivers at certain infrared wavelengths have approached this quantum noise limit, current millimeter- and submillimeter-wave receivers have noise temperatures that are one to two orders of magnitude larger than this quantum noise. Recent development of microwave mixers which use the nonlinear single-particle tunneling currents in superconducting tunnel junctions suggest that heterodyne receivers employing these mixers can approach the limit set by the quantum noise.

Josephson receivers. For a number of years, Josephson junctions have been explored as nonlinear elements for millimeter-wave detectors, mixers, and amplifiers. The Josephson effect is the tunneling of correlated pairs of electrons through an insulating barrier between two superconductors. The Josephson tunnel currents are highly nonlinear and behave as a quantum-mechanical oscillator with a frequency proportional to the applied voltage. These pair tunneling effects have been employed in direct detectors, heterodyne mixers with internal and external local oscillators, and various kinds of parametric amplifiers. Much interesting physics has been done and a variety of devices invented and tested. Unfortunately, the noise temperatures predicted and achieved have not been sufficiently low to justify practical applications.

Quasiparticle mixers. Recent interest has shifted to the tunneling of single electrons (quasiparticles) across Josephson junctions when electron pairs are broken by a sufficiently large voltage bias applied to the junction. In the first application of this phenomenon to microwave devices, a super-Schottky diode was developed in which quasiparticles tunnel from a superconductor through a Schottky barrier into a heavily doped semiconductor. Classical mixing and direct detection with very low noise were observed, but applications at millimeter wavelengths were hindered by the series resistance in the semiconductor.

Because of the intensive work on evaporated-film superconductor-insulator-superconductor (SIS) tunnel junctions intended for use in digital computers, these junctions became available with impedances and *RC* relaxation times adequate for operation at near-millimeter wavelengths. The first heterodyne experiments were done at 36 and 115 GHz. These experiments showed low-noise mixing with conversion loss, but without series resistance so that higher frequency operation was possible.

When computations based on the quantum theory of mixer operation were done for the ideal SIS junction, they gave a prediction of conversion gain along with low noise. Conversion gain is important because it reduces the problem of noise in the intermediate-frequency amplifier. In order to observe this quantum effect, it is necessary to use nearly ideal SIS junctions or high microwave frequencies, or both.

Experimental gain in excess of 4 dB has recently been observed in a single sideband of an SIS quasiparticle heterodyne mixer receiver operated at 36 GHz. This mixer had a single sideband input noise temperature less than 10 K, which is to be compared with the uncertainty principle limit of 2.3 K. Negative resistance effects that accompany gain have also been seen at 115 GHz. These results suggest that millimeter-wave receivers with noise approaching the quantum limit can be built by using SIS quasiparticle technology.

At higher frequencies, Josephson tunneling phenomena will be troublesome. However, mixing with gain and low noise at submillimeter wavelengths is also predicted in superconductor-insulator-normal (SIN) metal junctions that do not suffer from pair tunneling. The series resistance problem associated with the super-Schottky diode should not limit these SIN devices because the normal conductor can be a thin normal metal film backed with superconductor. The superconducting energy gap, which is about 800 GHz in lead junctions, should provide an upper frequency limit to efficient mixer performance.

Applications to astronomy. A complete 115 GHz SIS quasiparticle heterodyne receiver with a system noise temperature of 300 K has been developed and used for observations of interstellar carbon monoxide (CO) at the Owens Valley (California) Radio Observatory. The performance of this receiver does not yet reflect the full potential of the new technology, but its rapid development does illustrate the optimism of the radio astronomy community and the practicality of the SIS devices for use on a telescope.

Although improvements in presently available receivers are needed, quantum noise-limited receivers are not generally required for ground-based astronomy. Atmospheric emission at near-millimeter wavelengths provides an effective noise limit that is significantly larger than the quantum limit, even from high mountain sites. Experiments from aircraft, balloons, and satellites, however, can profit from the full theoretical sensitivity.

A number of features make SIS and SIN quasiparticle mixers very attractive for use in very-low-noise heterodyne receivers at near-millimeter wavelengths. The junctions are rugged and reliable. They are planar thin-film devices, well adapted to stripline technology. Power dissipation in the cryogen is very low. The local oscillator power required is very low. Series arrays of junctions can be used to match impedances or to increase the saturation threshold. Several types of harmonic mixers also appear practical.

Quasiparticle detectors. In addition to low-noise mixing with gain, the quantum theory of photon-assisted tunneling predicts that a tunnel junction can be operated as a direct detector of microwave photons with a responsivity of one tunneling electron for each microwave photon absorbed. A detector has been made at 36 GHz with a quantum efficiency of essentially unity and a noise equivalent power of 10^{-16} W/$\sqrt{\text{Hz}}$. The detector sensitivity is limited by inverse-frequency noise at low frequencies. When operated with high-frequency modulation, the SIS quasiparticle detector is shot-noise-limited and is the most sensitive broadband receiver available at millimeter wavelengths. Good performance is anticipated at submillimeter wavelengths where direct detection can have significant advantages over heterodyne receivers.

For background information *see* JOSEPHSON EFFECT; MICROWAVE SOLID-STATE DEVICES; RADIO ASTRONOMY; RADIO RECEIVER; SUPERCONDUCTING DEVICES; UNCERTAINTY PRINCIPLE in the McGraw-Hill Encyclopedia of Science and Technology.

[P. L. RICHARDS]

Bibliography: W. R. McGrath et al., *Appl. Phys. Lett.*, 39:655, 1981; T. G. Phillips et al., *IEEE Trans. Magnet.*, MAG-17:684–689, 1981; P. L. Richards and T.-M. Shen, *IEEE Trans. Electr. Devices*, ED-27:1909–1920, 1980; J. R. Tucker, *IEEE J. Quant. Electron.*, QE-15:1234–1258, 1979; J. R. Tucker, *Appl. Phys. Lett.*, 36:477–479, 1980.

Supergravity

The word supergravity refers to the latest attempts at constructing a unified field theory. The notion of a unified theory of all interactions (forces) responsible for natural phenomena has fascinated physicists for over a century. The first, and only, completely successful unified theory was constructed by James Clerk Maxwell, in which the up-to-then unrelated electric and magnetic phenomena were unified in his electrodynamics. Early in this century, the problem of embedding electrodynamics and gravitation into a unified theory was pursued by A. Einstein, H. Weyl, Th. Kaluza, O. Klein, and others. Later, with the discovery of weak and strong nuclear forces, it became clear that the unification problem was more complex than had been anticipated. Moreover, experimental discoveries in nuclear and particle physics and the emergence of internal symmetry groups provided the hint that a more attainable goal on the way to full unification was to first unify electromagnetic, weak, and possibly strong forces. If successful, the next step would be unification with gravity into a superunified theory. Supergravity models are special superunified theories. To put the aims and the achievements of these models in a proper perspective, it will be helpful to briefly recount the earlier stages of unification.

Electroweak theory. The second stage of unification concerns the unification of electromagnetic and weak interactions, using Maxwell's theory as a guide. This was accomplished by S. L. Glashow, A. Salam, J. C. Ward, S. Weinberg, and others, making use of the non-Abelian gauge theories invented by C. N. Yang and R. L. Mills, and of spontaneous symmetry breaking pioneered by Y. Nambu and G. Goldstone. The symmetry of Maxwell's theory is very similar to spatial rotations about an axis, rotating the vector potentials while leaving the electric and magnetic fields unchanged. It is a local invariance because the rotations about a fixed axis can be made by different amounts at different points in space-time. Thus, Maxwell's theory is invariant under a one-parameter group of transformations U(1). In Yang-Mills theory this local invariance was generalized to theories with larger symmetry groups such as the three-dimensional rotation group SO(3) $\simeq$ SU(2) which has three parameters. The number of parameters of the local symmetry (gauge) group is also equal to the number of 4-vector potentials in the gauge theory based on that group. A detailed analysis of weak and electromagnetic forces shows that their description requires four 4-vector potentials (gauge fields), so that the gauge group must be a four-parameter group. In fact, it is the product SU(2) · U(1). Of the four gauge fields, one transmits electromagnetic force which is long-range, indicating that the corresponding quanta (photons) are massless. The other three transmit short-range weak forces of the type responsible for the neutron β-decay, indicating that the quanta of short-range forces are massive. Instead of the ad hoc introduction of masses, which destroys the gauge invariance, a more satisfactory method of making weak forces short-range is by the spontaneous breaking of the SU(2) · U(1) symmetry down to $U_Y(1)$ of Maxwell's theory, thus ensuring that the photon remains massless.

Grand unified theories. In the third stage of unification, electroweak and strong forces are regarded as different components of a more general force which mediates the interactions of particles in a grand unified model. Strong forces are responsible for the interactions of hadrons and for keeping quarks confined inside hadrons. They are described by eight massless 4-vector potentials (gluons), the corresponding eight-parameter group being SU(3). This local symmetry is called color, and the corresponding theory quantum chromodynamics (QCD). Thus the gauge group of a grand unified theory must include SU(3) · SU(2) · U(1) as a subsymmetry. The most dramatic prediction of these theories is the decay of proton, first pointed out by J. C. Pati and A. P. Salam. *See* PROTON.

Supergravity and supersymmetry. The last, and the most ambitious, stage of unification deals with the possibility of combining grand unified and gravitation theories into a superunified theory, also known as supergravity. To achieve this, use is made of the dual role played by local symmetry groups. On the one hand, they describe the behavior of forces. On the other hand, they classify the elementary particles (fields) of the theory into multiplets: spin-zero fields in one multiplet, spin-$^1/_2$ fields in another multiplet, and so forth, but never fermions and bosons in one irreducible multiplet. This last restriction used to be a major obstacle on the way to superunification. This is because, of all the elementary particles, only the quanta of gravitional field (gravitons) have spin 2, so that a multiplet of elementary particles including the graviton must of necessity involve particles of different spin. But then by an internal symmetry transformation, which is by definition distinct from space-time (Lorentz) transformations, one can "rotate" particles of different spin into one another, thus altering their space-time transformation properties. This apparent paradox can be circumvented if both the internal symmetry and Lorentz transformations are part of a larger group which also includes the spin-changing transformations. This is how supersymmetry makes its appearance in supergravity theories.

Local supersymmetry transformations in field theory were first discussed by Y. A. Golfand and E. P. Likhtman. Under supersymmetry transformations, fermions and bosons mix. To visualize these transformations, it is useful to compare and contrast them to the transformations in the familiar three-dimensional euclidean space. In such a space, the coordinates (x_1, x_2, x_3) of a point are given by real numbers, so that they commute: $x_1x_2 = x_2x_1$, and so forth. Unlike real numbers, there is another class of numbers, known as Grassmann numbers, the elements of which do not commute: $\theta_1\theta_2 = -\theta_2\theta_1$. In particular, $\theta_1^2 = \theta_2^2 = 0$. Since their square yields a real number, zero, they may be thought of as square roots of real numbers in the same sense that the Dirac equation is the square root of the Klein-Gordon equation. The numbers θ_1 and θ_2 may be thought of as components of a two-dimensional (anticommuting) spinor $\theta = (\theta_1,\theta_2)$ in a spinor space. Now an even more general space can be envisioned, a superspace, in which some of the coordinates are x-like (commuting) and some are θ-like (anticommuting). Translations in superspace are examples of supersymmetry transformations. When $\vec{x}$ is changed by an amount $\vec{a}$, θ does not change. But when θ changes by an amount $\varepsilon = (\varepsilon_1,\varepsilon_2)$, $\vec{x}$ also changes according to the rule given below,

$$\theta \rightarrow \theta + \varepsilon$$
$$\vec{x} \rightarrow \vec{x} + \varepsilon^{+}\vec{\sigma}\,\theta$$

where ε^+ is the adjoint of ε, and $\vec{\sigma}$ consists of 2 × 2 Pauli matrices $(\sigma_1,\sigma_2,\sigma_3)$. Thus, by putting together a 3-vector $\vec{x}$ and a two-component spinor θ, it is possible to construct a supersymmetric multiplet $(\vec{x},\theta)$ on which supersymmetry transformations can be defined. These transformations can also be defined by putting together multiplets of fields of different spin to form a supermultiplet. It is precisely this feature which is indispensable in the construction of supergravity theories.

Supergravity models. From the beginning, the development of superunified theories proceeded along two distinct but related directions: explicit supergravity models in four-dimensional space-time; general theories in superspace aimed at first constructing all the possibilities allowed by the invariances and then specializing to those with desirable particle spectra. The two approaches are complementary to each other since each has advantages which the other lacks.

In the first explicit supergravity model, the supermultiplet consisted of a spin-2 particle, the graviton, and a spin-$^3/_2$ particle, called gravitino. In a supersymmetric (super-Lie) group, internal and space-time symmetry transformations are bridged by θ-like transformations discussed above. As a result, the form of the internal symmetry group is determined by the particle content of a supermultiplet, which in turn is fixed by the requirement that it con-

One-graviton supermultiplets of particles with spin equal to or less than 2

SO(N) or SU(N) internal symmetry	Number of particles of a given spin in a supermultiplet				
	Spin 0	Spin 1/2	Spin 1	Spin 3/2	Spin 2
$N = 0$	—	—	—	—	1
$N = 1$	—	—	—	1	1
$N = 2$	—	—	1	2	1
$N = 3$	—	1	3	3	1
$N = 4$	2	4	6	4	1
$N = 5$	10	11	10	5	1
$N = 6$	30	26	16	6	1
$N = 7$	70	56	28	7	1
$N = 8$	70	56	28	8	1

tain only one spin-2 particle, the graviton, and no particles with spin higher than 2. Then, depending on whether the gravitational sector is Einstein's theory or Weyl's theory, the internal symmetry groups turn out to be, respectively, SO(N) or SU(N), where $N = 1, \ldots, 8$. The SU(N) series are known as conformal supergravity theories. The spectrum of particles for each of these possibilities is given in the table. The largest internal symmetry group in the SO(N) series is thus SO(8), which does not contain SU(3) · SU(2) · U(1) of the grand unified theories as a subgroup. This means that it cannot accommodate the known quarks, leptons, gluons, and weak gauge bosons in a single supermultiplet. Opinions vary as to whether SO(8) should be regarded as a global symmetry or a local one. If it is only a global symmetry, a model has recently been constructed which has the spectrum of SO(8) and additional symmetries including a local SU(8) symmetry.

Superunification in superspace. One problem with supergravity models in four dimensional space-time is that the transformation laws of the fields in a supermultiplet change from one model to the next, so that, in contrast to the Yang-Mills theory, the knowledge of known models is not very helpful in constructing others. This is because, as discussed above, the space-time coordinates $x^\mu = (t, x)$ by themselves do not form a supersymmetric multiplet, and in the absence of their partners $\theta = (\theta_1, \theta_2)$, the supersymmetry transformations enter the picture in an indirect manner. A superspace, on the other hand, has both x-like and θ-like coordinates, so that it is naturally suited for implementing local supersymmetry transformations of general supergravity theories. General theories in superspace and their physical interpretations have been studied from various points of view, and there is no consensus as to which set of criteria are the most reasonable ones. In one of the superspace approaches, it can be shown that a supergravity theory, like pure gravity, is a nonlinear realization of a gauge symmetry: The fields transform linearly in the internal symmetry (Yang-Mills) sector but nonlinearly in the space-time (gravity) sector. Then, SO(N) and SU(N) groups, respectively, appear as local symmetries, and it is possible to derive model-independent transformation laws for every field by suitably generalizing Yang-Mills transformations. Using these it is also possible to write down all the invariants that are needed for the construction of lagrangians.

In the same way that in a relativistic quantum theory one is often confronted with unphysical degrees of freedom such as timelike oscillations, which must be eliminated by imposing gauge conditions, superunified theories in superspace contain unphysical degrees of freedom which must be eliminated by imposing a suitable set of gauge conditions. At present this problem is under study, and no systematic method of dealing with it has been found.

Outlook. One of the initial motivations for constructing supergravity models was the hope of obtaining finite quantum theories of gravity coupled to matter. Arbitrary couplings of gravity to matter lead to quantum theories in which the amplitudes for various processes are infinite, and these infinities (divergences) cannot be isolated and eliminated from the theory in a process known as renormalization. In supergravity models constructed so far, the degree of renormalization is improved in the sense that these gravity-coupled-to-matter theories are as renormalizable as pure gravity. But the overall picture is not as hopeful as it was once thought. The reason for this can be seen from the point of view of the nonlinear realizations. Local supersymmetry transformations are nonlinear. Just as general coordinate invariance does not improve the renormalizability of pure gravity by completely eliminating the available invariants, local supersymmetry transformations do not appear to limit the construction of invariants to a finite number, an essential criterion for the elimination of divergences. This does not mean, however, that superunified theories are useless and that they must be abandoned. Even if these obstacles cannot be overcome within the framework of the currently popular models, the rich structure of these theories is likely to provide a point of departure for the construction of other physically acceptable theories.

For background information *see* FUNDAMENTAL INTERACTIONS; GRAVITATION; QUANTUM CHROMODY-

NAMICS; RELATIVISTIC QUANTUM THEORY; RENORMALIZATION; SPINOR; SYMMETRY LAWS (PHYSICS) in the McGraw-Hill Encyclopedia of Science and Technology. [FREYDOON MANSOURI]

Bibliography: D. Z. Freedman and P. van Nieuwenhuizen, *Sci. Amer.*, 238(2):126–143, February 1978; H. Georgi and S. L. Glashow, *Phys. Today*, 33(9):30–39, September 1980; G. B. Lubkin, *Phys. Today*, 30(6):17–19, June 1977; A. Salam, *Rev. Mod. Phys.*, 52(3):525–538, July 1980.

Superionic conduction

Some inorganic crystalline solids, such as sodium beta alumina and Li_xTiS_2, have ionic conductivities comparable to the ionic conductivity of aqueous sodium chloride (NaCl). These superionic conductors offer the potential for revolutionizing some electrochemical devices, particularly batteries. Materials like sodium beta alumina have high ionic conductivities and very low electronic conductivities. These materials may find use as porous separators between the solid or liquid electrodes of a battery. The sodium sulfur battery is a good example. In this cell the sodium beta alumina acts as a separator between the molten sodium and sulfur electrodes. Li_xTiS_2 is an example of a fast ionic conductor that also has a high electronic conductivity, on the order of 10^6 $(\Omega\text{-m})^{-1}$. Materials of the Li_xTiS_2 type may make very good battery electrodes because they can rapidly exchange both ions to the elctrolyte and electrons to the external circuit. There have been many recent advances and discoveries in mixed ionic-electronic conductors like Li_xTiS_2.

The conventional wisdom of a few years ago, that diffusion in solids should be slow, is no longer supported by experiment. The first observations of high ionic conductivity were so startling that the term superionic was applied. Today it is widely recognized that there are no fundamental reasons to exclude extremely fast ionic conduction in solids; therefore the special designation of superionic is being dropped in favor of the view that there is a continuum of behavior from ionic insulators to excellent ionic conductors. There still remains some debate about the possible existence of a unique superionic phase in which new conduction mechanisms prevail.

Ionic conduction in layered compounds. Structure and bonding play an essential role in the ionic conductivity of compounds like Li_xTiS_2. In these compounds the crystals have a layered structure as shown in Fig. 1. In each TiS_2 (titanium disulfide) layer there are strong covalent bonds, making each layer like a macroscopic molecular sheet. These sheets are held together loosely by relatively weak bonding. This two-dimensional structure lends itself to chemical reactions that spread the layers apart and incorporate an ion between layers. These intercalation ("going between") reactions are illustrated by the $Li-TiS_2$ reaction in Fig. 1. The intercalated Li (lithium) ions resemble ions adsorbed on a surface; they are free to move from site to site, and an arbitrary number of ions (less than the total number of sites) can be adsorbed on this internal surface. There are many defects on the two-dimensional Li lattice, and ideal conditions for rapid diffusion exist.

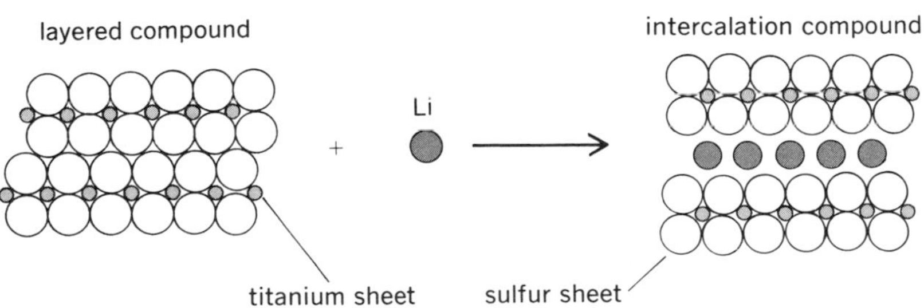

Fig. 1. Structure of TiS_2 and its intercalation reaction with Li.

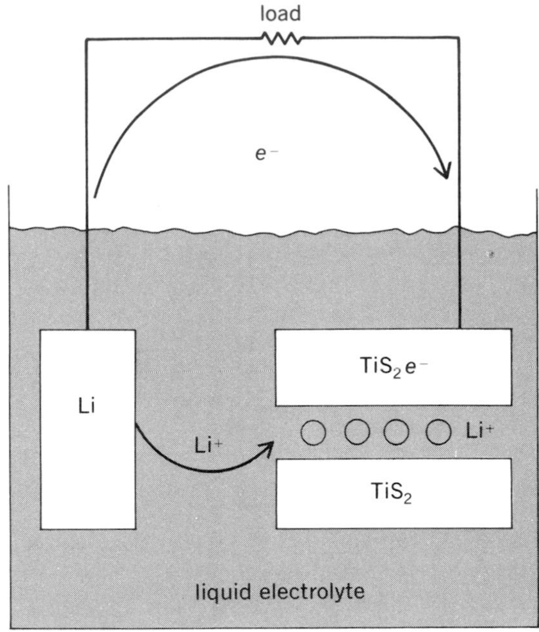

Fig. 2. Li/TiS_2 battery.

Battery applications. Intercalation compounds have been studied for their superconductive, lubricating, and catalytic properties, but here their electrochemical reactions will be chiefly discussed. In 1976, it was pointed out that the Li intercalation reaction can be of use as a battery cathode. The electrochemical cell is schematically illustrated in Fig. 2. When the battery is discharged, Li^+ ions are transported from the Li anode and through the liquid electrolyte, and are intercalated into the TiS_2. Simultaneously an electron passes through the external circuit and is donated to the TiS_2 layers. The electron maintains charge neutrality in the Li_xTiS_2 and becomes one of the conduction electrons. The composition can be changed continuously from TiS_2 to Li_1TiS_2, and as the composition changes, the cell voltage drops as shown in Fig. 3.

The Li/TiS_2 cell has many desirable features. It is reversible by reversing the current and deintercalating the Li. It has a high energy density of four to five that of the lead-acid battery. It operates at room temperature, and it can reside in the charged (deintercalated) condition indefinitely. The materials are fairly inexpensive and abundant. Development of this cell is continuing in several laboratories where there are needs for rechargeable, reliable cells of high energy density. The original expectation that this cell would be a viable automobile power battery has been dampened by the absence of a suitable electrolyte that can carry the very large current densities required for automotive power. Nonetheless, intercalation batteries are recognized to be a qualitative advance in battery technology because they incorporate the intercalation concepts that may be applicable to many other battery couples.

Study of molecular bonding and structural ordering. Beyond possible battery applications, the electrochemistry of intercalation has proved to be a valuable means for learning about bonding in layered compounds and about phase transitions in two dimensions. When Li is intercalated into TiS_2, it loses its valence electron to the TiS_2 layers and becomes a Li^+ ion. This ionization provides the energy that drives the reaction. As more Li are intercalated, they repel each other, making it harder to insert more Li. This repulsion causes the cell voltage to drop as shown in Fig. 3. The cell voltage is the difference in electrochemical potential between Li^0 metal in the anode and Li^+ ions in TiS_2; it is the Gibbs free energy of reaction. All the thermodynamic parameters for the reaction can be determined from electrochemical measurements. For example, the temperature derivative of the cell voltage is the temperature derivative of the Gibbs free energy or the entropy of reaction. A complete set of such thermodynamic parameters has been measured on Li_xTiS_2, with the result that subtle anomalies were discovered at several values of x. These are just visible in Fig. 3 as the plateau in voltage versus x near $x = 0.25$. These features have been associated with structural ordering of the intercalated Li^+.

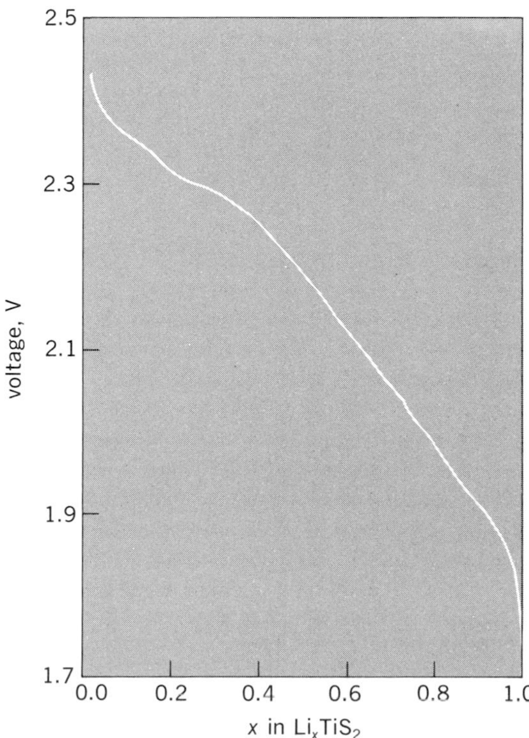

Fig. 3. Voltage-composition relation for the Li/TiS_2 cell. The plateau near $x = 1/4$ is caused by the structural ordering of intercalated Li.

The intercalated Li^+ ions in Li_xTiS_2 reside in sites between the sulfur layers formed by the close packing of six sulfur ions. These octahedral sites lie on the triangular TiS_2 lattice. Thus, the Li^+ are confined to positions on a two-dimensional triangular array. They are more or less free to move around on this lattice, but the electrostatic repulsion of one ion for another causes them to move as far apart from one another as possible. The result is that the ions form simple, ordered, two-dimensional arrays whenever the composition is such that all near-neighbor Li ions can be equidistant. These ordered compositions are $x = 1, 1/3, 1/4, 1/7, 1/9, \ldots$. The TiS_2 lattice is insensitive to the Li^+ ordering. This means that Li_xTiS_2 is a good model material to use in studying the ordering of ions confined to a noninteractive two-dimensional lattice, or a lattice gas. The two-dimensional lattice gas is an old, classical problem in solid-state physics that has relevance for both structural ordering and magnetic ordering. This model has been used to reproduce the features of Fig. 3 and some of the other electrochemical results.

The electrochemistry that provides the energetics of reaction can be combined with other physical measurements such as nuclear magnetic resonance, magnetic susceptibility, and electrical resistivity to extract information about the molecular bonding in Li_xTiS_2. For example, the interatomic potential energies have been determined, and from a knowledge of the electrostatic potential energy between the layers as a function of position, the diffusivity of the Li^+ has been calculated with good agreement between theory and experiment. Li_xTiS_2 has proved to be a highly successful model system in which electrochemistry has been combined with other physics to yield relevant thermodynamic and kinetic information.

Other ionic conductors. Experience with intercalation electrochemistry has led to exploration in many other novel ionic conductors. All these mixed electronic-ionic conductors have some form of surface ionic conduction. One example is the one-dimensional compound $NbSe_3$. In this compound there are long chains of $NbSe_3$ (niobium triselenide) molecules. The chains are weakly bound together. As in the intercalation reaction, Li or other alkali metals can react on the internal surfaces provided by these molecular chains. The reaction appears to be reversible by deintercalation. The common feature between the one-dimensional and two-dimensional materials is that these are solids having molecular sheets or chains with strong internal bonds but weak bonds between molecules. Ionic reactions take place on the surface of the molecules.

The ultimate continuation of the concept of a molecular solid would be to the amorphous molecular solid in which there are isolated molecules weakly bound to one another. Amorphous MoS_3 (molybdenum trisulfide) supplies an example of such a solid. It has no long-range crystal structure and can reversibly react with Li as a highly effective electrode. This material has an enormous surface area. The reaction with Li is not yet fully understood, but it appears to involve transport of ions on the internal surface of this highly dispersed solid, thus carrying the concept of internal surface to the molecular limit.

For background information *see* CHEMICAL THERMODYNAMICS; DIFFUSION IN SOLIDS; ELECTROCHEMISTRY; IONIC CRYSTALS; PHASE TRANSITIONS in the McGraw-Hill Encyclopedia of Science and Technology. [ARTHUR HOWARD THOMPSON]

Bibliography: G. C. Farrington and J. L. Briant, *Science*, 204:1371–1379, 1979; R. A. Huggins and A. Rabenau, *Mater. Res. Bull.*, 13:1315–1325, 1978; A. H. Thompson, *J., Electrochem. Soc.*, 126:608–616, 1979; M. S. Whittingham, *Progr. Solid-State Chem.*, 12:41, 1978.

Superplasticity

Superplastic behavior in metallic materials has created considerable interest because of the extremely large elongations obtained under low loads. This unusual metallic state has opened new avenues to metal forming, and the potentialities have stimulated considerable research on new alloys for commercial applications. The possibility of choosing an alloy that is normally high-strength, treating it to obtain the superplastic condition, and then deforming it into complicated parts with relative ease is very attractive. Moreover, after the part has been formed while superplastic, the original alloy properties can be restored by a simple heat treatment.

The exploitation of superplasticity in aluminum-base alloys, however, has lagged behind advances with other alloy systems. This situation had come about because of the difficulty of producing aluminum alloys with the ultrafine grain size required for this effect, and the inability to prevent excessive grain growth at the high temperatures used during superplastic forming. Recently there have been concerted efforts to overcome these difficulties in aluminum alloys. As a result, a dramatic increase has occurred in the number of commercial superplastic aluminum alloys, their applications, proprietary processes, and patents reported in the technical literature from 1978 to the present.

The basic requirements for superplasticity consist primarily of three microstructural attributes of the metal grains—they must be small, equiaxed, and stable. The desired microstructural conditions include: a metallic matrix with an ultrafine grain size (of the order of 10 μm or less); grains that are essentially equiaxed (to enable grains to slide and rotate); a well-dispersed second phase (in order to maintain the very small matrix grain size); a balance between the hardness of the second phase and the matrix (to control cavitation during superplastic forming); a preponderance of suitable high-angle interfaces (to promote grain boundary sliding, or GBS, during superplastic deformation); mobility of both grain and interphase boundaries (so as to dissipate stress concentrations that build up during grain boundary sliding); and the ability of boundaries or interfaces to resist tensile separation (which limits the amount of elongation obtainable).

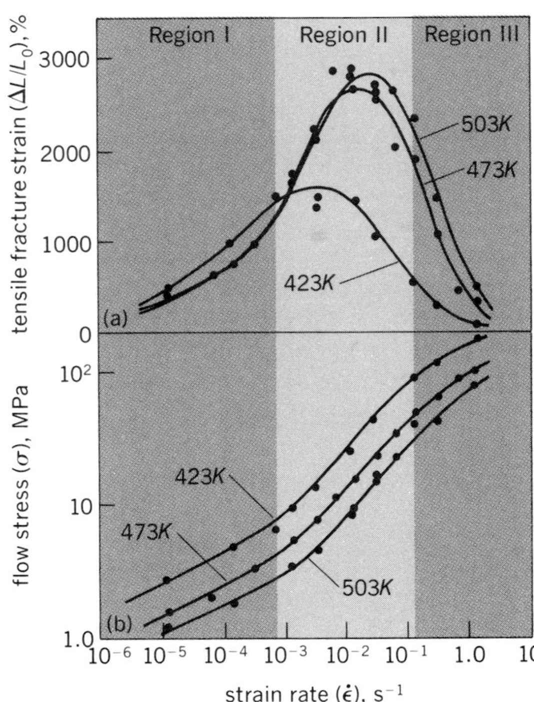

Fig. 1. Typical elongation and flow stress curves for Al–78%Zn alloy, with grain size 2.5 μm. (*a*) Tensile fracture strain versus strain rate. (*b*) Flow stress versus strain rate. (*From F. A. Mohamed, M. M. I. Ahmed, and T. G. Langdon, Factors influencing ductility in the superplastic Zn-22 pct Al eutectoid, Met. Trans. A, 8A:933–938, 1977*)

Mechanical requirements. These generally include test temperatures of $0.4T_M$ (where T_M is the absolute melting temperature) or higher, low strain rates (in the vicinity of about 10^{-3} s^{-1}), and corresponding low stresses (of a few thousand pounds per square inch or less; 1000 pounds per square inch = 6.9 MPa). Typical elongation and flow stress curves for the superplastic Al-Zn eutectoid alloy are shown in Fig. 1. The sigmoidal-shaped ln σ–ln $\dot{\epsilon}$ curves are divided into three regions (I, II, and III) according to their slopes and relative stress levels. Specific deformation mechanisms are identified with each region—superplasticity occurs in region II—and there is a sharp drop in the ductilities achieved in regions I and III.

The three regions in Fig. 1 can be represented empirically by Eq. (1), relating stress σ and strain

$$\sigma = K\dot{\epsilon}^m \tag{1}$$

rate $\dot{\epsilon}$, where m is the strain rate sensitivity index and K is a constant. The maximum value of m lies typically in the range 0.5 to 0.8 and usually coincides with the maximum elongation.

The curves in Fig. 1 apply to an alloy of constant grain size (here, d = 2.5 μm). When grain size is plotted as a variable versus the normalized stress, σ/G or τ/G, where G is the shear modulus and τ is the shear stress, the diagram looks like that shown in Fig. 2, called a deformation mechanism map. As stress is increased (along a constant grain size of d = 2.5×10^{-4} cm), it can be seen that the alloy would undergo superplastic deformation (region II) at stress values of τ/G of about 10^{-4}. Deformation mechanism maps hold great promise for both theoretical and practical purposes, since they afford a broad overview of all deformation mechanisms, possible interactions, and the temperature and stress dependence in various regions, as well as predictive capabilities for optimum forming conditions.

The temperature dependence of the strain rate at high temperatures can be expressed very generally by Eq. (2), where the stress σ is raised to the power

$$\dot{\epsilon} = \frac{AGb}{kT}\left(\frac{b}{d}\right)^p\left(\frac{\sigma}{G}\right)^n D_0 \exp(-Q/RT) \tag{2}$$

n. The grain size d appears reciprocally, while the diffusional contribution enters in the $D_0 \exp(-Q/RT)$ term. A is a dimensionless constant, b is the Burgers vector, k is Boltzmann's constant, T is the absolute temperature, D_0 is a frequency factor, Q is the activation energy, and R is the gas constant; p and n are constants with absolute values generally lying between 0 and 4. Under conditions of steady-state strain rate $\dot{\epsilon}_s$ and a constant grain size, Eq. (2) can be written in the dimensionless form as shown in Eq. (3), where D_b is the grain boundary diffusivity.

$$\dot{\epsilon}_s kT/D_b Gb = \text{constant } (\sigma/G)^n \tag{3}$$

A plot of Eq. (3) is shown in Fig. 3 for the Al-Zn eutectoid alloy with three different grain sizes. The three regions in this plot are identified with Coble creep, superplasticity, and dislocation climb–controlled creep as the stress increases from low to high, respectively. Figure 3 is not a schematic plot, but embodies a quantitative, theoretical treatment based on a model of superplasticity controlled by grain boundary sliding with dislocation accommodation, as well as experimental data points. The plot reveals the operative regions of the various rate-controlling mechanisms over wide ranges of temperature, stress, and strain rate, and also the influence of grain size. Predictions made with the appropriate rate-controlling mechanism compare well with experimental results.

Mechanisms. At the present time it is generally agreed that grain boundary sliding is the primary operative mechanism during superplastic flow. However, sliding cannot occur continuously on all boundaries without an additional process or processes to permit accommodation of grain shape

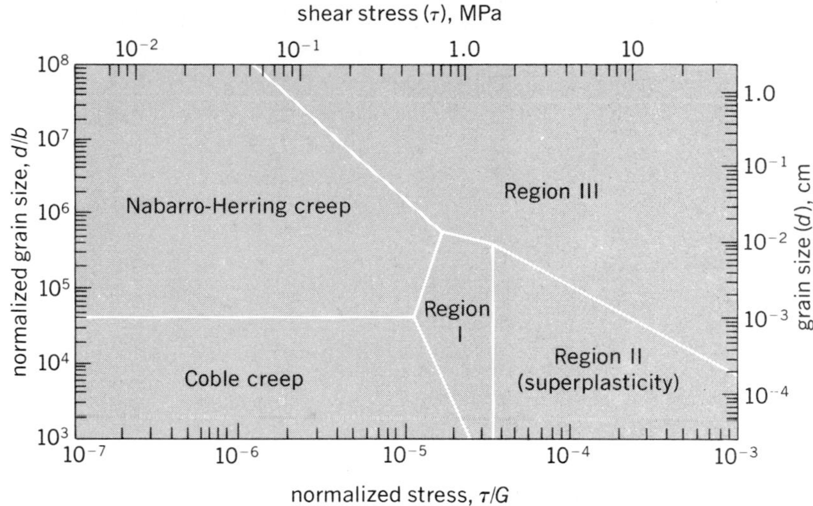

Fig. 2. Deformation mechanism map of normalized grain size versus normalized shear stress for Al–78%Zn alloy at 503 K. (*From F. A. Mohamed and T. G. Langdon, Deformation mechanism maps for superplastic materials, Scripta Met., 10:759–762, copyright 1976 by Pergamon Press, Ltd.*)

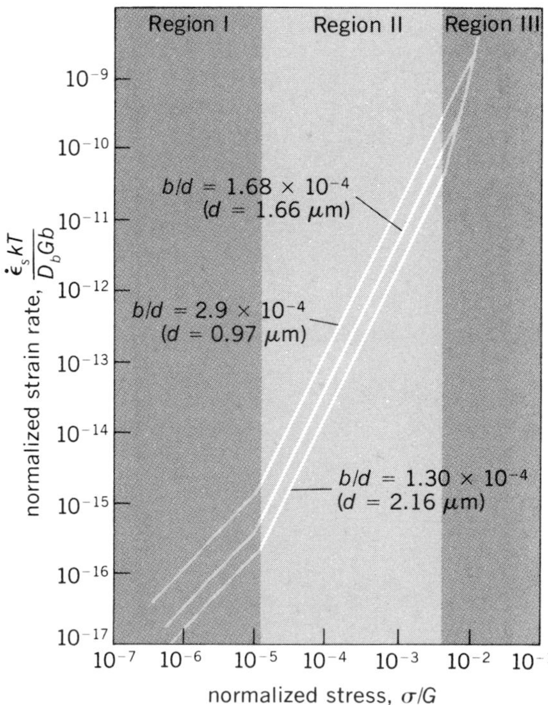

Fig. 3. Comparison of the operative creep mechanisms in Al–78%Zn alloy at 423 K. Region I refers to Coble creep, region II to superplasticity, and region III to dislocation climb. (*From S. C. Misro and A. K. Mukherjee, The relative contribution of Coble creep, dislocation climb and superplasticity at elevated temperatures, in J. C. M. Li and A. K. Mukherjee, eds., Rate Processes in Plastic Deformation of Materials, pp. 434–458, American Society for Metals, 1975*)

changes, contiguity of sliding grains, and relaxation of stress concentrations at grain boundary ledges and triple junctions. Several mechanisms have been proposed as possible accommodation processes. These are diffusional flow, crystallographic slip, grain boundary migration, and recrystallization. Theoretical treatments, as well as evidence supporting these various accommodation processes, have been offered, but the situation has not yet been completely clarified.

Aluminum alloys. Much of the early work on superplasticity was conducted on eutectoid or eutectic alloys, such as the Al-78%Zn or Al-33%Cu alloys, respectively. These alloys possess a large volume of second phase (about 50%) that stabilizes the grain size. Unfortunately, such alloys have inferior mechanical properties and are not suitable for structural materials. Consequently, recent alloys development has concentrated on aluminum alloys with compositions closer to those of commercial aluminum alloys, and with much smaller amounts of second phase.

Two main classes of superplastic aluminum alloys have emerged lately. These are the Al-Zn-Mg alloys and the Al-Cu-base alloys. Examples of typical compositions are given in the table. Zirconium additions are used frequently because of the marked effect this element has on inhibiting recrystallization, and because the zirconium-containing particles ($ZrAl_3$) are quite stable at high temperatures.

New procedures have been developed to produce a very fine grain size in aluminum alloys, with microstructures that are designed to prevent excessive grain growth during superplastic deformation. Thermomechanical processing can give a very fine grain size by means of suitable hot or cold working at selected temperatures, either before or after quenching or aging. For example, in one procedure, the alloy is first overaged to generate large precipitates (for example, $CuAl_2$ in Al-Cu-X alloys), then rolled to produce significant deformation around these precipitates, and finally, recrystallized in order to obtain a fine grain size which can be stabilized at elevated temperatures by a fine precipitate (for example, $ZrAl_3$).

Typical superplastic aluminum alloys

Alloy type	Specific alloys
Al-Zn-Mg	Al–8Zn–1Mg–0.5Zr
	Al–10.7Zn–0.9Mg–0.4Zr
	Al–9Zn–0.9Mg–0.3Zr–0.09Cr
	Al–7.6Zn–2.75Mg–2.3Cu–0.15Zr (V96Ts)
	Al–5.8Zn–2.2Mg–1.6Cu (7475)
Al-Cu	Al–6Cu–0.5Zr (Supral)
	Al–6.4Cu–0.6Mn–0.13Ti–0.05Zr–0.19Fe (D20)
	Al–6.3Cu–0.28Mn–0.06Ti (01201)
Al-Ca	Al–5Ca–5Zn (Alcan 0850)
	Al–7.6Ca
Al-Mg	Al–6Mg–0.37Zr–0.07Cr–0.16Mn
	Al–6Mg (AMg6), and with Zr (AMg61)
Al-Si	Al–14.6Si–4.9Mg
	Al–12Si–4Cu
Al-Zn	Al-78Zn plus 0.95Cu–0.045Mg–0.025Ca (HSZ)
Al–Mn	Al–0.9Mn–0.4Zr

Other methods for obtaining a very fine grain size involve rapid quenching of suitable alloys from the liquid state in order to generate finely dispersed particles. The resulting casting, either with or without subsequent deformation, is thereby endowed with superplastic properties. Another approach uses rapid solidification processing to obtain metal powders with superior uniformity and fine-scale structure. When consolidated, the alloy has a more homogeneous, fine-scale microstructure than conventional ingot material. Other processes that impart the necessary fine structure are spray casting, splat quenching (Al-8Fe, 2024), and melt spinning (MS 7075 ribbon).

Applications of aluminum alloys. Recently reported applications of superplastic aluminum alloys that utilize their unique properties are numerous, including pierce drawing (AMg6), deep drawing (modified for superplastic behavior), precision or isothermal forging (HSZ), and hot pressing for ribbed stiffeners (AMg6). In other novel applications, alloys of Al with 20 or 30 atomic % Ge were successfully reduced to wire from castings, after a ductile structure of uniformly dispersed Ge spheroids was achieved. Also, gallium has been added to dilute aluminum alloys to encourage grain boundary sliding and superplastic behavior.

It can be seen from the above that superplasticity has matured into a major process for industrial exploitation. The current theoretical problems are relatively minor and should be essentially solved in the near future.

For background information *see* ALLOY; ALUMINUM; SUPERPLASTICITY in the McGraw-Hill Encyclopedia of Science and Technology.

[ERVIN E. UNDERWOOD]

Bibliography: A. Arieli and A. K. Mukherjee, *Met. Trans. A*, 13A:717–732, May 1982; J. W. Edington, *Metals Technol.*, 3:138–153, 1976; D. M. R. Taplin, G. L. Dunlop, and T. G. Langdon, *Annu. Rev. Mater. Sci.*, 9:151–189, 1979; B. M. Watts et al., *Metal Sci.*, vol. 11, pts. I and II, pp. 189–197, 198–206, 1976.

Symmetry laws (physics)

Symmetry considerations have played an important role in the development of all branches of physics. During the 1960s and 1970s several examples of a new type of symmetry, called dynamic symmetry, were discovered. This, in turn, has led to a new and deeper understanding of the properties of matter.

Geometric symmetries. For many years, the only examples of symmetries known in nature were of the geometric type. This type occurs, for example, in crystals and molecules. Within a crystal (or a molecule), the constituent parts (atoms) arrange themselves in a very definite geometric structure. Symmetries of this type can be easily visualized, since they are related to properties of matter in ordinary space. For example, the twisted molecule H_3C-CCl_3, composed of three hydrogen (H) atoms, two carbon (C) atoms, and three chlorine (Cl) atoms (shown in Fig. 1), has a geometric symmetry. A mathematical description of symmetries is provided by the theory of group transformations. Since by rotating the structure of Fig. 1 through a vertical axis by angles that are multiples of 120°, one obtains the same figure, one says that this structure is invariant under the group of transformations composed of all rotations by angles that are multiples of 120°. In mathematical language, this group of transformations is called C_3.

Dynamic symmetries. In the late 1930s physicists realized that other types of symmetries could exist in nature, related not to the geometric structure of the constituents of matter but to the laws which govern the dynamic behavior of these constituents. These are called dynamic symmetries. An example is provided by the hydrogen atom. Here the proton and the electron are bound together by the electromagnetic interaction. The laws governing the dynamic behavior of the proton and the electron under the action of the electromagnetic interaction have a dynamic symmetry, known as O(4) symmetry, from the name of the group of transformations which describes it. As a result, the structure of the hydrogen atom shows striking regularities. Although these regularities had been known for many years, it was only in the 1930s that physicists realized that they were related to the occurrence of a dynamic symmetry.

The first application of the idea of dynamic symmetries to complex systems is due to E. Wigner, who in 1936 suggested that symmetries of this type could be found in the spectra of atomic nuclei. While geometric symmetries are related to groups of transformations which form a discrete set, dynamic symmetries appear to be related to groups of transformations which form a continuous set. The foundations of the mathematical theory of this type of groups had been laid down by S. Lie and others in the 1890s.

Wigner and G. Racah developed this theory further in the late 1930s and applied it to several physical systems. The Wigner model, known as the SU(4) model, from the name of the group of transformations which describes its properties, was followed by other models based on the same ideas. Another model, which received considerable attention in the late 1950s, was developed by J. P. Elliott to describe the collective spectra of light atomic nuclei. However, it was not until the late 1960s that the idea of dynamic symmetries received full recognition.

Dynamic symmetries in elementary particle physics. The importance of dynamic symmetries was recognized in 1961, when M. Gell-Mann and Y. Ne'eman independently proposed a model to describe properties of subnuclear particles, such as protons or neutrons (known as hadrons). This model, as well as other similar models, is constructed as follows. One begins by assuming that the physical system under study is made of some basic building blocks. In the model of Gell-Mann and Ne'eman there are three building blocks, introduced by Gell-Mann and G. Zweig and called quarks. Since the group of transformations which describes the properties of these three building blocks is called SU(3), this model is sometimes referred to as the SU(3) model. Next, one writes down the mathematical equations governing the dynamic behavior of these building blocks. When a dynamic symmetry is present, these equations can be solved in a closed, analytic form and lead to regularities in the properties of the system under study. For a quantum-mechanical system, such as an atom, an atomic nucleus, or a subnuclear particle, the most important property is the set of its allowed energies. This is called the energy spectrum. Thus, when a dynamic symmetry is present, the energy spectrum must follow simple rules. In the Gell-Mann–Ne'eman SU(3) model, these rules are referred to as Gell-Mann–Okubo mass formulas. An example of an energy spectrum displaying the Gell-Mann–Ne'eman symmetry, that of the baryon decuplet, is shown in Fig. 2. Baryons are one of the two possible varieties of hadrons. Both the observed (Fig. 2*a*) and the predicted spectra (Fig. 2*b*) display striking regularities. Within the accuracy of the Figure, the observed and predicted spectra are indistinguishable. The allowed energy levels are labelled by the baryon names, Ω, Ξ*, and so forth.

The Gell-Mann–Ne'eman model was subsequently expanded to include more building blocks, following the discovery in 1974 of a new subnuclear particle, known as J/ϕ. At present, five quarks appear to be needed to describe the experimental energy spectra of subnuclear particles, and perhaps more will be necessary in the future.

Dynamic symmetries in nuclear physics. Another area of physics in which the idea of dynamic symmetries has played an important role in the last few years is the description of collective properties of atomic nuclei. In 1974, A. Arima and F. Iachello suggested that collective properties of atomic nuclei with an even number of protons and neutrons could be built out of six building blocks, called interact-

SYMMETRY LAWS (PHYSICS)

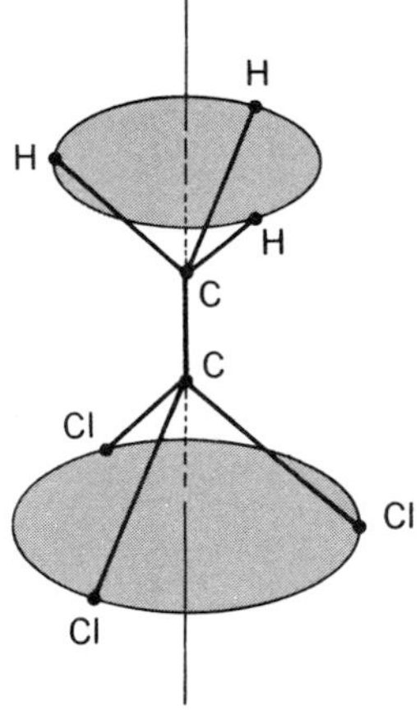

Fig. 1. A geometric symmetry: the twisted H_3C-CCl_3 molecule. This molecule has C_3 symmetry.

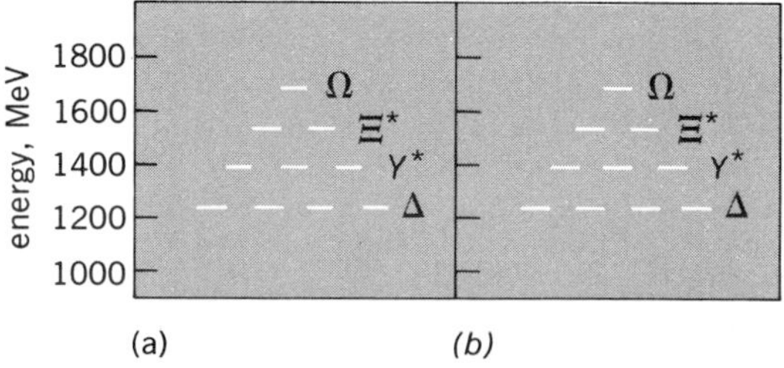

Fig. 2. A dynamic symmetry in elementary particle physics: the energy spectrum of the baryon decuplet. *(a)* Observed spectrum. *(b)* Predicted spectrum.

ing bosons. Since the group of transformations which describes the properties of these six building blocks is called SU(6), this model is called SU(6). As in the case of the Gell-Mann–Ne'eman SU(3) model of hadrons, dynamic symmetries arise in atomic nuclei whenever the six building blocks combine in a way which leads to an energy spectrum with simple structure. However, unlike hadrons, in which only one form of symmetry appears to be realized in practice, nuclei appear to have three forms of symmetry. This situation is similar to that encountered in the study of the geometric symmetries of a crystal or of a molecule. The atoms in a crystal can arrange themselves in many different ways (forms), each characterized by a geometric symmetry (cubic, hexagonal, and so forth). Similarly, the interacting bosons in an atomic nucleus can arrange themselves in different ways, each characterized by a dynamic symmetry. An example of a dynamic symmetry in nuclei is shown in Fig. 3. As in the previous case, the only way the occurrence of a dynamic symmetry can be graphically shown is by displaying the spectrum of its allowed energies. Once more, striking regularities occur in both the observed (Fig. 3*a*) and the predicted (Fig. 3*b*) spectra. Contrary to the example shown in Fig. 2, slight differences occur here between the observed and predicted spectra. The allowed energy levels are labeled by the values of their intrinsic spin, 0, 2, 4, and so forth.

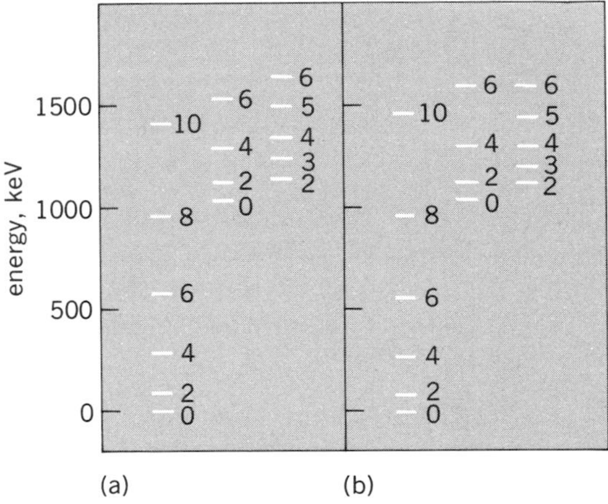

Fig. 3. A dynamic symmetry in nuclear physics: the energy spectrum of the nucleus gadolinium-156, composed of 64 protons and 92 neutrons. *(a)* Observed spectrum. *(b)* Predicted spectrum.

Here, too, it was subsequently realized that the six building blocks introduced by Arima and Iachello were not sufficient to describe all observed phenomena. This scheme has therefore also been extended to include more building blocks.

The idea of dynamic symmetries has also been used recently in other fields of physics, such as atomic and molecular physics.

Supersymmetries. As the idea of dynamic symmetries spread from one field of physics to another, generalizing the original type of geometric symmetry, physicists attempted a further generalization of this concept. Physicists believe that all building blocks of matter, be they subnuclear particles or atomic nuclei or atoms, have an intrinsic property called spin. When the intrinsic spin is an integer, the building blocks are called bosons; when the building blocks have half-integer spin, they are called fermions. The symmetries described above apply to physical systems which are composed either of fermions or of bosons alone. However, in 1974, D. V. Volkov and V. P. Akulov and, independently, J. Wess and B. Zumino suggested that there could exist in nature dynamic symmetries which would apply to mixed systems of bosons and fermions. These symmetries have been called supersymmetries. Ordinary dynamic symmetries are described mathematically, as mentioned above, by the theory of groups of transformations, known as Lie groups. Supersymmetries are described by the theory of new types of groups of transformations, known as graded Lie groups. The theory of graded Lie groups was pioneered by F. A. Berezin and V. G. Kac, and is still in the process of being fully developed. Although the search for supersymmetries in nature has been concentrated mostly in the field of elementary particle and gravitational physics, the first experimental example of supersymmetries in physics may have actually been found in 1980 by A. B. Balantekin, I. Bars, and Iachello in the spectra of heavy atomic nuclei. The full extent and implications of this new type of symmetry, in both elementary particle and nuclear physics are, however, not completely known.

In conclusion, it has become clear in the last decade that the concept of dynamic symmetries appears to be applicable in many areas of physics, and it is not an isolated accident occurring in a particular field. Furthermore, the generalization of these ideas to increasingly complex structures appears to provide a powerful guiding principle for analyzing and studying these structures.

For background information *see* ELEMENTARY PARTICLE; NUCLEAR STRUCTURE; QUARKS; SYMMETRY LAWS (PHYSICS) in the McGraw-Hill Encyclopedia of Science and Technology.

[FRANCESCO IACHELLO]

Bibliography: M. Gell-Mann and Y. Ne'eman, *The Eightfold Way*, 1964; F. Iachello (ed.), *Interacting Bosons in Nuclear Physics*, 1979; F. Iachello, *New Sci.*, 85(1193):390–391, February 1980; Y. Nambu, *Sci. Amer.*, 235(5):48–60, November 1976.

Taphonomy

Taphonomy is the science of the preservation of fossils; its domain begins with the death of organisms and includes the events that befall their remains until they become part of rocks. Taphonomy covers all events during this transition from the biosphere to the lithosphere, including mode of death, scavenging, surface weathering, transport of organic remains by animals, wind, water, reerosion of buried material, modification of organic remains as tools by hominids, and differential dissolution of mineral tissues within a range of chemical composition. Organic remains bear evidence of their preservational history in their degree of completeness, in damage patterns, in their orientation with respect to sediments and to other organic remains, in mineralization features, and in their abundance in certain sedimentary environments.

Taphonomy is a recent addition to paleobiology, with most work having occurred during the last 25 years. Taphonomic studies have influenced field collecting techniques by showing the importance of the body-part composition of fossil assemblages as indicative of specific preservational processes. Previously often only the most taxonomically diagnostic fossils were considered worthy of collection. Taphonomy has introduced new quantitative techniques into the analysis of fossil assemblages, including mathematical models of disarticulation and scattering and simulations of fossil assemblages with different preservational probabilities. In addition, by demonstrating the discrepancies between the information content of fossil assemblages and of modern communities, taphonomic studies have directed paleoecologists to consider models of community reconstruction appropriate to the fossil record.

The central issue of taphonomic studies is to elucidate how biological information has been lost from the original living systems. Every fossil assemblage is the result of historically unique living assemblages that have survived certain biological and physical degradative processes. Figure 1 illustrates the overall conceptual framework. Organic remains pass through several stages, each represented by a level and by many possible processes at each transition. With each transition, the amount of material diminishes by orders of magnitude. It is feasible to study only a small part of this sequence at once. One approach is to study the fossil assemblages to discern characteristic features associated with different sedimentary environments. Most vertebrate fossils from the Miocene Siwalik sediments of Pakistan are found in deposits representative of river channels; in the early Eocene Willwood formation of Wyoming, most vertebrate fossils are found in deposits interpreted as floodplain soil horizons. Soil accumulations are more likely to reflect original life composition of vertebrate assemblages than are fluvial accumulations. An example of the former is illustrated by Fig. 2. The species diversity of the surface skeletal assemblage has been compared to that of the living mammals of Amboseli Park, Kenya. Skeletal remains of mammals smaller than 50 kg underrepresented living species diversity. Bones entrained by moving water can be considered sedimentary particles like sand, silt, and clay, but the shapes of most bones are more irregular than the shapes of inorganic particles. Transport behavior of bones can be predicted from a modification of the equations describing sediment transport. Experiments documenting the transport and burial of bones in flumes (experimental waterways) and rivers permit comparison of predicted and observed behavior of different skeletal elements in moving water. For the experiments to date, there is an approximate match between predicted and observed behavior; the effects of shape are more difficult to quantify than those of density or volume.

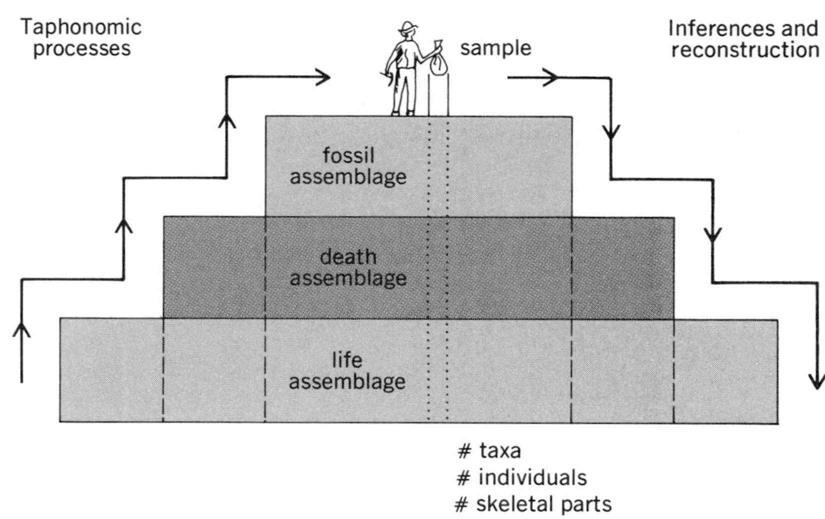

Fig. 1. A conceptual framework for taphonomy. Organic remains originate in living assemblages and become transformed into death assemblages, fossil assemblages, and collected fossil samples by taphonomic processes. At each transition, both material and information are lost from the original living assemblage. (*Courtesy of C. Badgley*)

Differences related to taxonomy. Most taphonomic research up to this time has focused on vertebrate fossils, much of it in connection with the study of early human evolution. The three main tax-

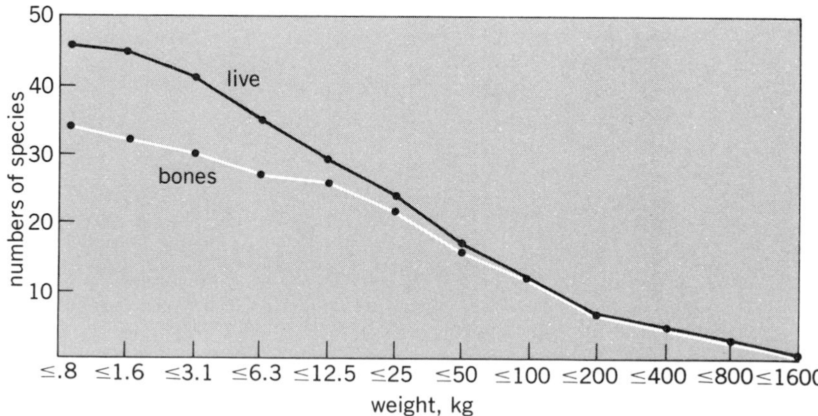

Fig. 2. Plot of numbers of live wild species and those represented by bones in the Amboseli surface assemblage versus body-size categories. The bone sample consists of over 20,000 identifiable pieces from approximately 1200 carcasses. (*From A. K. Behrensmeyer and A. P. Hill, eds., Fossils in the Making: Vertebrate Taphonomy and Paleoecology, University of Chicago Press, 1980*)

onomic groups of fossils—vertebrates, invertebrates, and plants—pose fundamentally different taphonomic problems. It is the exception that fossils from any two of these groups are preserved together in significant quantity. Most invertebrates with fossilizable hard parts (shells, tests) live in shallow seas. With the exception of fish, most vertebrates and plants are preserved in terrestrial deposits, and those environments chemically favorable to the preservation of bones and teeth are generally inimical to the preservation of plant tissues. Thus, specific reconstructions of past living assemblages are usually based on only one of the three taxonomic groups. Each group is distinct in other taphonomic features as well.

Vertebrates. The typical vertebrate body has over 200 bones. Since terrestrial vertebrates are mainly fossilized in fluvial systems, considerable effort has been devoted to distinguishing river-transported from untransported bone assemblages. Both experiments with bones in flumes and rivers and observations of fossil assemblages in different lithologies (sedimentary rock types) indicate that moving water sorts and winnows the complete vertebrate skeleton into distinctive dispersal groups of skeletal elements. Jaws, teeth, and skulls are most resistant to fluvial transport; vertebrae are transported most readily.

Many animals act as taphonomic agents. Predators and scavengers disassemble and destroy skeletons of their prey. Large vertebrates trample surface bones into soft sediment. Some rodents and insects collect bones. Different human groups use distinctive butchery techniques evidenced by skeletal-part composition and surface damage patterns on bones. Modern taphonomic studies have documented examples of these processes as a basis for distinguishing agents of accumulation of fossil assemblages. Much of this work arose in the context of research on hominid origins in eastern and southern Africa. Criteria for distinguishing hominid accumulations from those created by other animals include skeletal element representation, fracture patterns at both the macroscopic and microscopic scales, presence or absence of etching due to stomach acids, and presence or absence of artifacts.

Invertebrates. Organic remains are lost to the fossil record either by lack of preservation or by transportation of material away from the original living system. Soft-bodied marine organisms generally constitute over half of marine communities; while they leave no direct remains, they often leave indirect remains in the form of trace fossils (tracks, burrows, boreholes). Most community reconstructions of invertebrate fossils are based on in-place deposition, where the main avenue of information loss is nonpreservation. Consequently, there has been more emphasis on determining how modern assemblages of invertebrates would look if fossilized than for vertebrate assemblages. For example, it has been estimated that 23% of the species in a modern oyster community could potentially become fossilized. It then becomes an ecological issue to interpret a community on the basis of one-quarter representation.

Plants. Living plants shed separate parts which are often adapted for transport (pollen, fruits, seeds). While there has been little taphonomic work on plant megafossils (nonmicroscopic plant parts), there has been considerable analysis of modern and fossil pollen (palynology). Palynologists have mapped the abundance and distribution of pollen from modern land surfaces in relation to the source species and to climatic features, such as mean annual temperature and precipitation values. Since both the starting points and end points of this system can be observed directly, the environmental indicators can be described mathematically by transfer functions. These have been used to reconstruct temperature and precipitation regimes through the Quaternary on the basis of pollen records. The depositional environments in which plant megafossils are preserved do not indicate significant transport away from the site of origin. Taphonomic work leading to vegetation reconstruction of Tertiary floras suggests that leaf and stem litter layers record rather closely the original spatial heterogeneity of the vegetation. The various plant organs (wood, leaves, flowers, roots, seeds, fruits, pollen) that constitute the living plant do not tend to be preserved together, so it is difficult to reconstruct whole organisms.

Relationship to paleoecology. Current paleontological issues reflect the influence of modern population and community ecology: these issues include the nature of speciation, the origin of new adaptations, changes in species diversity over geologic time, and changes in community organization as major taxonomic replacements occur. As paleontological studies acquire a more ecological perspective, there is increased concern with reconstructing past environments in detail comparable to that of the present. These details include taxonomic diversity, relative abundance of taxa, structural diversity of habitats, and microhabitat preferences of selected taxa. The extent to which these features are discernible in fossil assemblages depends on their taphonomic histories. For example, a fossil assemblage representing catastrophic mortality contains a record of population structure. Thus, the fossils themselves represent two sequential phenomena: the individuals, populations, and communities of the once-living systems and the survivors of complex preservation histories. Taphonomy and paleoecology are closely allied, taphonomy representing the sampling processes that have acted upon past ecosystems. Most taphonomic processes are not ecological in character. Taphonomic reconstructions rely on the guidance of well-developed ecological models to recreate information of paleoecological significance.

For background information *see* FOSSIL; FOSSIL SEEDS AND FRUITS; PALEOBOTANY; PALEOECOLOGY; PALEONTOLOGY; PALYNOLOGY; TRACE FOSSILS in the McGraw-Hill Encyclopedia of Science and Technology. [CATHERINE BADGLEY]

Bibliography: A. K. Behrensmeyer and A. P. Hill

(eds.), *Fossils in the Making: Vertebrate Taphonomy and Paleoecology*, 1980; D. P. Gifford, *Advances in Archaeological Method and Theory*, vol. 4, pp. 365–438, 1981; V. A. Krasilov, *Paleoecology of Terrestrial Plants*, 1975; D. Lawrence, *Geol. Soc. Amer. Bull.*, 79:1315–1330, 1968.

Termite

Termites exist in rigid totalitarian societies in which morphologically specialized individuals execute specific tasks: the king and queen reproduce, the workers forage and build, and the soldiers defend. Com-

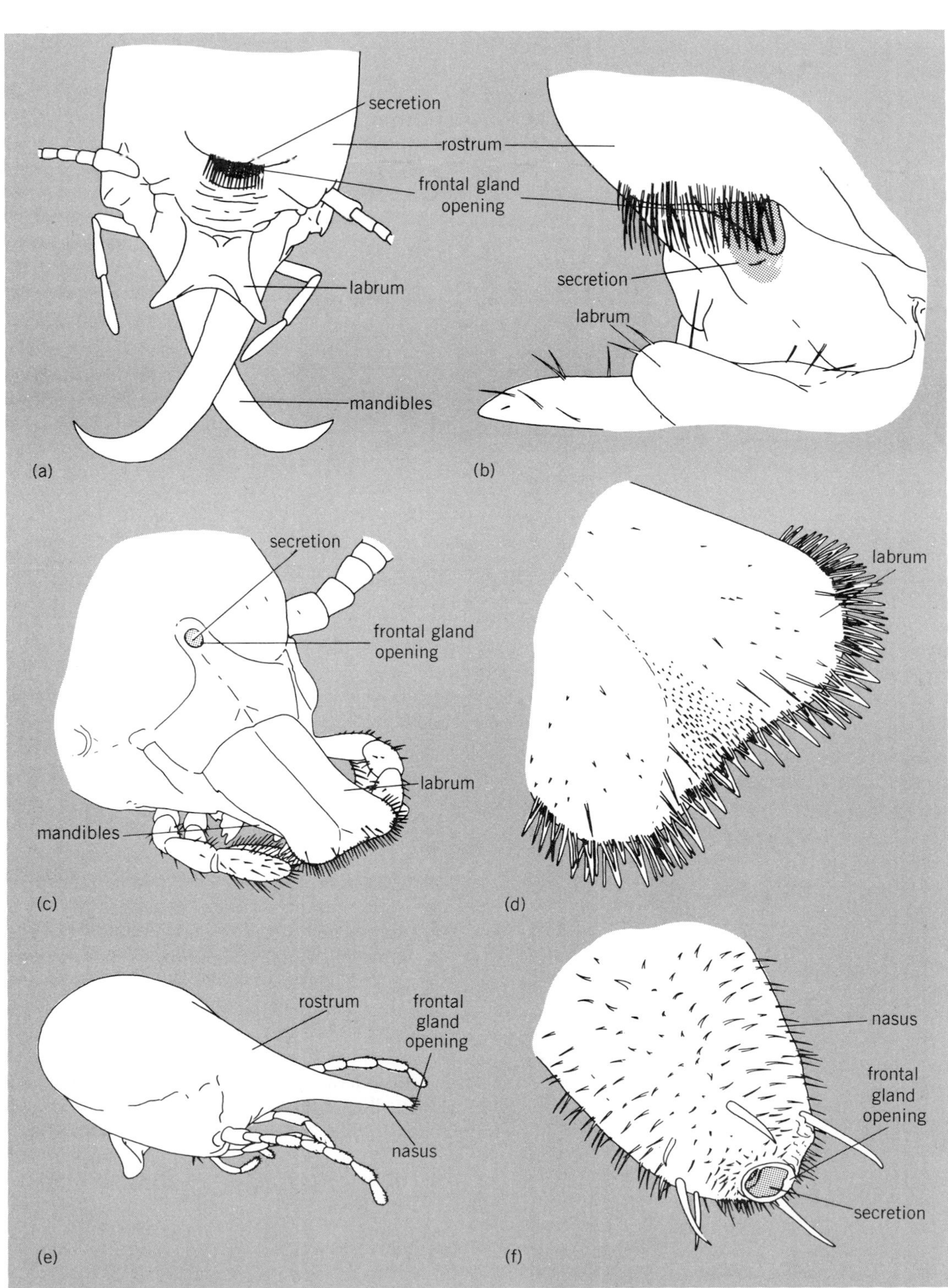

Fig. 1. Soldier heads (left) and magnified chemical "delivery systems" (right) of termite soldiers: (*a, b*) *Cubitermes umbratus* (biting-antihealant). (*c, d*) *Schedorhinotermes lamanianus* (poison brush). (*e, f*) *Trinervitermes bettonianus* (glue squirter). (*From G. D. Prestwich, Termite chemical defense: New natural products and chemosystematics, Sociobiology, 4:127–137, 1979*)

munication among individuals of each society is based on the exchange of chemical signals, either by smell (olfaction using antennas) or by taste (contact chemoreception). Building, food finding, nest mate recognition, trail following, growth, alarm, and defense all involve specific chemical cues for these blind, deaf, and dumb creatures.

A colony in danger of attack by aggressors (varying in size from ants to aardvarks) mobilizes a suicide squad of soldiers. Some African termite soldiers have razor-sharp jaws which slice through the bodies of their attackers. Some South American soldiers have huge squarish heads with mandibles longer than their bodies; when the termites are disturbed, the jaws snap with tremendous force and a loud click, producing irregular gashes in the enemy. Some American dry-wood soldiers use their heads like drain plugs, guarding the entrance into the colony with a sort of dangerous trapdoor. An unusual Malaysian termite soldier is a living bomb; when it explodes, it inundates its attacker with a thick, yellow, pungent liquid. Similarly, soldierless worker termites in Africa and South America literally explode in an ultimate self-sacrifice, bathing the attacker in a malodorous and foul-tasting mixture of feces and gut contents.

In addition to these unusual defensive tactics, a large number of termite species have evolved unique forms of chemical warfare. There are three types: biting, with the simultaneous addition of an oily or toxic antihealing chemical to the wound; brushing, with an enlarged paintbrushlike upper lip that applies a contact poison to the surface of the attacker; and glue squirting, in which an irritating, viscous entangling agent is sprayed onto potential predators (Fig. 1).

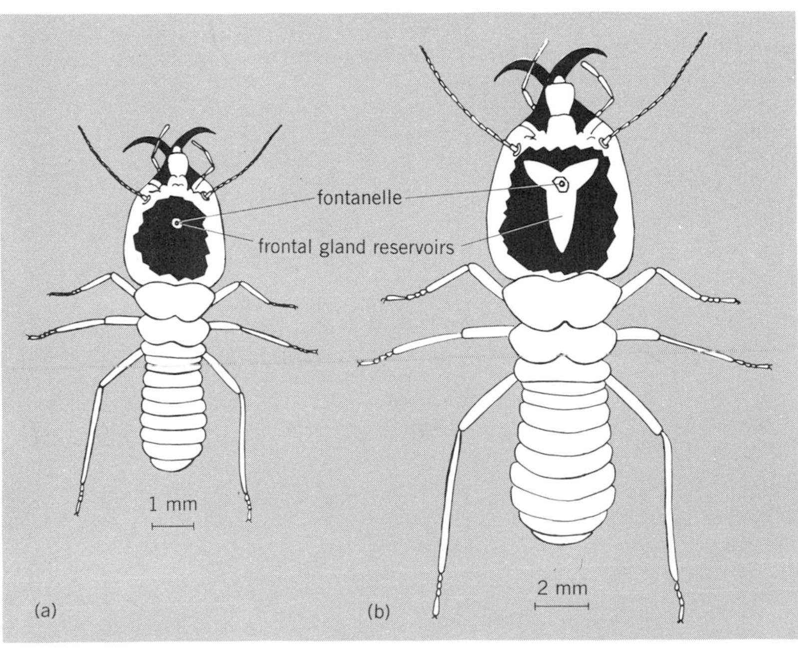

Fig. 2. *Macrotermes michaelseni* (*a*) minor and (*b*) major soldiers. Frontal gland reservoirs are shown within a cutaway cuticle of the head; the glands open through the fontanelle. (*From G. D. Prestwich et al., Soldier frontal glands of the termite Macrotermes subhyalinus: Morphology, chemical composition, and use in defense, J. Chem. Ecol., 3(5):579–590, 1977*)

Colony defense by the African mound-building termite *Macrotermes* is the job of a sterile female caste. Small soldiers escort workers during food gathering and mound repair, and large soldiers prevent attackers from reaching the inner hive containing young larvae and the royal pair. These soldiers (Fig. 2) have evolved hard shieldlike heads and sharp swordlike mandibles for defense. The chemical secretion of the major soldier is contained in a large frontal gland which constitutes 10% of the termite's dry weight. In this species, a greasy mixture, similar to paraffin wax, of long-chain alkanes and alkenes is coated onto wounds made by the mandibles, and ants thus attacked slowly die by loss of hemolymph. The antihealant property of this simple defense secretion is presumed to result from the ability of this waxy material to masquerade as cuticle, which prevents the initiation of coagulation and resclerotization required for normal wound healing in insects. In *Cubitermes* (Fig. 1), a humivorous African termite, G. D. Prestwich and collaborators discovered that diterpene (C_{20}) hydrocarbons possessing unusual molecular structures serve as antihealants.

The more advanced soldier termites of the subfamily Rhinotermitinae employ a paintbrush technique to apply a lipophilic (oil-soluble) contact poison on the cuticle of attacking arthropods. The primary morphological trends in this subfamily are the reduction of the mandibles and the elongation of the labrum (the upper lip with its "paintbrush") of the small soldiers (Fig. 1). Concomitantly, specific biochemical machinery has evolved in these termites to synthesize, store, and detoxify highly insectidal compounds. One soldier may contain as much as 35% of its dry weight as defense secretion, enough to kill thousands of ants. As the labral brush evolved to be longer, the chemical secretion evolved to become increasingly toxic. As the toxicity increased, the efficiency and substrate specificity of the detoxication system may also have increased.

Three genera of the "poison paintbrush" termites have been investigated. The most primitive is *Prorhinotermes*, from Florida, whose soldiers still have mandibles and which possesses an unusual nitroolefin. *Schedorhinotermes*, from Africa, produces a mixture of vinyl ketones, and the most advanced Guyanese termites in the genus *Rhinotermes* have recently been shown to possess highly reactive β-ketoaldehydes. These contact poisons are all electrophilic, meaning that they can react rapidly and often lethally with electron-rich biological molecules in enemies. However, each of these toxin-producing species detoxifies its electrophilic poison by simple reductive and oxidative processes common in fatty acid metabolism.

The most diverse and abundant group of termites in the tropics—over 500 of the 2000 described species worldwide—are those species with glue-squirting soldiers. These highly evolved termites of the subfamily Nasutitermitinae eject a viscous, irritating glue from a nozzlelike frontal gland contained in an elongation of their foreheads (called the nasus). Prestwich's group determined the molecular structures of many of the terpenoid components of the pine-resin-like secretion of these chemical bazookas. The first of these unusual carbon skeletons was named "trinervitanes." Studies have been undertaken to determine how these soldiers assemble their bizarre glue molecules from simple precursors like acetic acid, and how this defense secretion has evolved over tens of millions of years. Keys to the morphological and chemical evolution of defense in nasute termites can be found in the New World, where many South American rain forest termites have soldiers with ice tong mandibles and nozzlelike heads. These evolutionary relics, however, can neither make terpenoid glues nor eject the primitive secretion that they do manufacture.

[GLENN D. PRESTWICH]

Bibliography: G. D. Prestwich, *J. Chem. Ecol.*, 5:459–480, 1979; G. D. Prestwich, *Nat. Geogr.*, 153:532–547, 1978; G. D. Prestwich, *Sociobiology*, 4:127–137, 1979; G. D. Prestwich et al., *J. Amer. Chem. Soc.*, 102:6825–6828, 1980.

Tooth

Teeth are organs composed of three distinct mineralized tissues. The intimate integration of mineral and organic components was demonstrated over 100 years ago with optical techniques. Recent studies of tissue structures using electron microscopy, crystal-chemical characterization of the mineral phase (a calcium phosphate, hydroxyapatite, in all three tissues), and biochemical analyses of the protein matrix materials have elucidated the exquisite tissue-specific cellular controls. Modulation of crystallite size, composition, and distribution is crucial to the function of the tissues and the organs.

Growth and development. The adult human normally possesses 32 permanent teeth, 16 in the upper jaw (maxilla) and 16 in the lower jaw (mandible). The first or baby teeth, 10 in each jaw, are deciduous: after they erupt one at a time into the oral cavity, they exfoliate as permanent teeth move through the gum to take their places. Each tooth is a separate organ composed of three distinctive mineralized tissues: enamel, dentin, and cementum. Figure 1 illustrates the gestation, development, and wear (attrition) of a single-root tooth.

At about 6 weeks, or when the fetus is about 11 mm in length, ectoderm spreads a band of epithelial tissue across the future dental arch in what will become the oral cavity. At specific nodes on the arch, beginning anteriorly and moving posteriorly, the proliferating epithelium buds into the mesenchyme (Fig. 1*a*), producing a bell-shaped structure (Fig. 1*b* and *c*). The structure is composed of the enamel organ enclosing the dental papilla, and both are surrounded by the dental sac. The papilla produces the dentin and pulp, and the sac gives rise to cementum and the periodontal ligament.

The inner cells of the enamel organ induce the papilla mesenchymal cells to differentiate into odontoblasts. Elongate enamel-producing cells, the ameloblasts, oppose odontoblasts across a basement membrane, the future dentin-enamel junction. Both cell types move away from the junction as they lay down an organic matrix which subsequently mineralizes. Tooth roots do not develop until enamel-dentin proliferation has defined the entire crown. Enamel grows in thickness but does not extend down the root. Cementum envelops the dentin of the

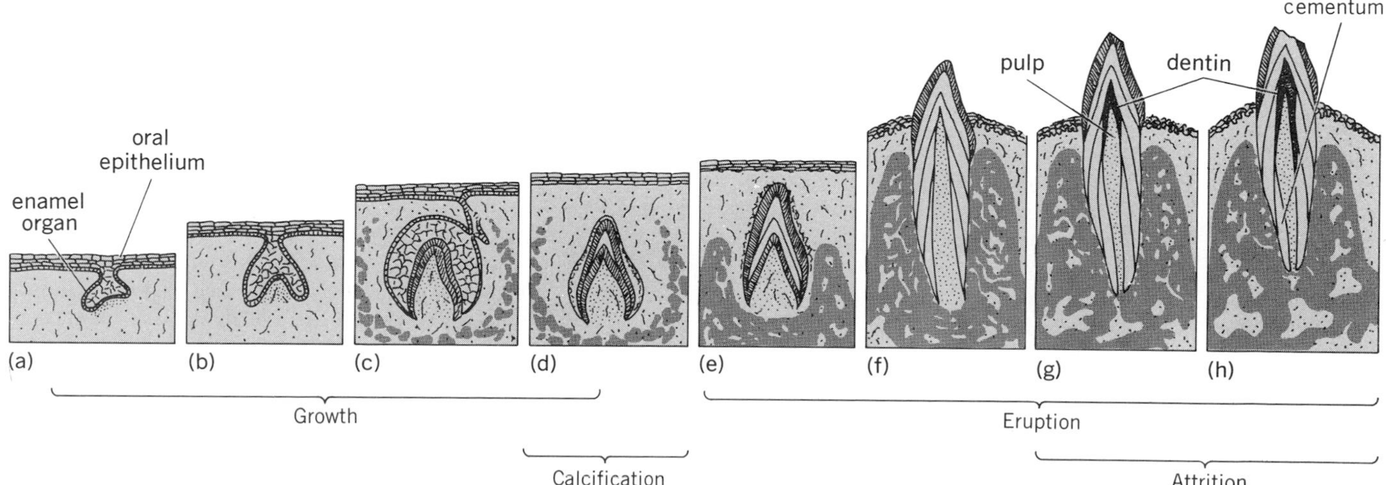

Fig. 1. Diagram of the cycle of a tooth from initiation to degeneration: (*a*) initiation, (*b*) proliferation, (*c*) morphodifferentiation, (*d*) calcification, (*e*) before emergence, (*f*) after emergence, and (*g*, *h*) degeneration. The dark blotches (*d–h*) represent the alveolar bone tissue of the jaw. (*From S. N. Bhaskar, ed., Orban's Oral History and Embryology, 9th ed., C. V. Mosby, 1980*)

root and butts against the enamel (Fig. 1*d–f*). The growth of multiple-root teeth (molars) parallels that of the single-root teeth (canines and incisors).

By the end of the second fetal month, 20 deciduous teeth are in progress. Subsequent extension of the dental lamina and arch provides space for the additional teeth, the permanent molars. The first permanent molar initiates in the gum behind the premolars during the fourth fetal month but does not erupt for 6 years. The second molar initiates during the first year after birth, erupting around 12 years, and the third molar, or wisdom tooth, begins after 4 years of age. Often third molars become impacted. They are perfectly formed but do not move upward to their proper position at the rear of the tooth row. Excision, though not mandatory, is recommended since sideward migration of this tooth can cause dissolution of bone or the roots of adjacent teeth.

As enamel, dentin, and cementum are developing, so does the surrounding bone tissue of the jaw (Fig. 1*d–h*). The inorganic material in all four tissues is a calcium phosphate crystallizing in the apatite mineral group. Hydroxyapatite, $Ca_5(PO_4)_3$-OH, is the closest chemical and structural approximation to the biologic minerals. There are, however, critical differences between these tissues: the relative amount of mineral, the size and disposition of the crystallite, the composition and organization of the organic matrix, the location and activities of cells, the vascularity, the developmental history, and the structures, as well as the distinctive functions of these dental elements.

Enamel. Enamel is the most highly mineralized tissue in the body. About 96% by weight and 87% by volume of the tissue is mineral. The remainder is protein, an asymmetric molecule with a composition resembling that of keratin, and water.

Sections of enamel submitted to electron microscopy allow single crystals of hydroxyapatite to be observed. They range in size up to 50 nanometers in cross section and 1000 nm in length, their size depending on the age of the tissue. The crystals nucleate and grow, and are oriented, in part, in and on the very small amount of protein matrix extruded by the ameloblast. The protein-crystal complex is packaged into three-dimensional structures called rods or prisms. The cross section of a prism is roughly 5 × 9 micrometers (μm), and the length traverses the enamel layer extending from the dentin-enamel junction to the surface of the tooth.

Prisms are extracellular structures secreted by at least two ameloblasts. Individual prisms are rarely straight but interdigitate and intertwine as they are produced. When enamel is cut perpendicular to the length of the prisms, parallel to the surface and the dentin-enamel junction, a regular keyhole pattern, typical of *Homo sapiens*, is observed with the electron microscope. If longitudinal sections are examined at lower resolution, in transmitted light with an optical microscope, variations in prism direction and interprism mineralization create alternating dark and light areas, called Hunter-Shreger bands. Appositional growth lines, called striae of Retzuis and equivalent to tree rings, can also be observed. They are the expressions of cyclical variations in structure and mineralization during rhythmic growth. Metabolic disturbances may also be recorded in the tissue. For example, the enamel of deciduous teeth commences before and continues after birth. A discontinuity is often observed in the sequence of striae coinciding with birth: the natal line.

Occasionally the prism pattern may be obliterated and replaced by a dense, highly mineralized layer. Growth striae, however, continue to be readily detected as ridges and grooves on the protected surfaces between teeth, especially on newly erupted deciduous teeth. With the scanning electron microscope up to 30 grooves per millimeter have been counted near the cementum-enamel junction and 10 on the cusp, the apex of the tooth. Apposition rates of enamel are mentioned below.

Newly erupted teeth (Fig. 1*f*) are covered with a cuticle that is rapidly worn away during mastication and replaced by a pellicle, a coating of proteins from salivary gland exudates. It is in and on the pellicle that colonies of microorganisms flourish, cause decay, and form bacterial plaque.

The enamel organ and the amelobasts disappear by the time a tooth erupts. The enamel is a semiporous layer when first formed and permits relatively large molecules (I^-, urea) to penetrate. By the time the maximum enamel thickness is attained (about 2.5 mm at the cusp) and the mineralization is complete, only surface exchange is possible. With continued use the enamel wears thin (Fig. 1*g*). At specific areas of contact between teeth on the upper and lower jaws it may actually disappear (Fig. 1*h*).

Dentin. Dentin is quite distinct from enamel. The most obvious differences are in mineral/matrix ratio and the size of crystallites. At maturity, dentin is less than 65% mineral by volume. The matrix material is another asymmetric protein, called collagen. The crystallites are orders of magnitude smaller, the maximum roughly 10 × 100 nm, but usually it is finer-grained. It is virtually impossible, even employing electron microscopy, to measure individual crystallites. In this respect dentin most closely resembles cementum and bone.

There are no prisms in dentin. Odontoblasts secrete predentin, an organic matrix composed of fine collagen fibrils and proteoglycans, sugar-protein molecules. After a brief interval, knots of mineral appear in the matrix and in time coalesce into fully mineralized dentin. As the odontoblast continues its production of matrix, the cell body migrates away from the mineralizing front toward the center of the tooth, the pulp chamber (Fig. 1), and site of neural tissue and the blood supply. Retreating from the dentin-enamel junction, the odontoblast leaves behind cytoplasmic processes in the mineralized tissue. Occasionally a process will extend into the enamel, but in the dentin each process is distinctly located in a channel or tubule.

The mineralized tissue in dentin, therefore, is intertubular. Each tubule is defined by a peritubular halo that in maturity becomes more densely mineralized than adjacent dentin. The cytoplasmic processes in the tubules, measuring up to 25 nm in diameter, may retreat, leaving behind extracellular fluid. With age or in response to disease or trauma, the channels may become occluded by fine mineral particles. The mineral matter in dentin, then, is a complex composite of hydroxyapatite crystallites generated at different stages and at several sites during the history of the tooth.

The appositional growth of dentin may also produce striae (lines of von Ebner) representing daily increments. To distinguish the lines in adult permanent teeth, it is often necessary to etch the tissue before optical examination. However, it was discovered, inadvertently, in fact, that there are alternative markers, specific for mineralized tissues. The tetracycline drugs are one example. On ingestion, either orally or by injection, a fraction of the circulating tetracycline molecule selectively attaches to actively mineralizing sites, such as the mineralizing front in dentin, and becomes permanently bound.

Figure 2, which shows a section through a third molar, demonstrates the sequestering of many sequential doses of tetracycline. The bands define the mineralization history of this tooth. The width of most bands at points of minimum distortion is 35 μm. Assuming that a band corresponds to the normal period of ingestion of tetracycline, 12 days, the rate of growth of dentin in this tooth is slightly less than 3 μm per day. The tooth was photographed with ultraviolet light since the tetracycline molecule fluoresces under ultraviolet radiation and remains UV-active after incorporation in living tissues. If the drug dose is high or sustained, it is possible that attachment of tetracycline will inhibit mineralization and disrupt the normal tooth structure; this is one reason why this family of drugs is not prescribed for children and adolescents.

The gestation period for the dental organs to eruption is 20 years. Once a tooth has formed, there is very little possibility of its changing shape or size. There are, however, changes in the level of mineralization, either normal or pathogenic (caries).

Cementum. As the enamel and dentin complete the crown, the cementoblasts differentiate from the connective tissue cells of the dental sac and commence forming the third dental tissue. A very thin layer grading from 20 μm at the dentin-enamel junction to 200 μm at the root tip, cementum is not grossly different from dentin but performs an exceedingly important specialized function. It is the site for attachment of the tooth into its bony socket.

Cementoblasts produce mineralized tissue in a fashion comparable to that of dentin except that the cells do not migrate. They become engulfed in the tissue, taking on a stellate appearance (cementocytes) with anastomosed processes. In addition to possessing a mineralizing collagenous matrix, the cells extrude collagen fibrils. The fibrils aggregate into ropes (Sharpy's fibers), one end of which is embedded in the mineralized tissue and the other end can be traced into the periodontal pocket. The fibrils are ligaments, or guy wires which suspend the organ and attach to the alveolar bone. A ligament is all "soft" tissue, that is, without mineral. The pocket contains connective tissue, cells, blood, and nerve endings. The alveolar bone is a portion of the mandible or maxilla, the jaw bones.

Cementum is distinct among the dental tissues in being vascular. Like bone, it does undergo some resorption and redeposition, especially when submitted to trauma such as when teeth move in their sockets (naturally or under orthodontic treatment). In general, the forces applied do not cause degeneration of the tissue but elevate the turnover within the affected tissue so that there is a continual steady state, with formation of new ligamental attachments. This site, the alveolar bone, is the most rapidly turning-over bone tissue in the body. The relationship between tooth ligament and bone is delicately balanced, a system of tissues capable of rapid response to local pressures. Cementum, the link between two mineralized organs, is a critical tissue.

Hydroxyapatite. Hydroxyapatite, the mineral of teeth, is most suitable for the biological needs of vertebrate tissues. First, the compound is highly insoluble and extremely stable over wide chemical conditions from acid to alkaline pH. Both Ca and P

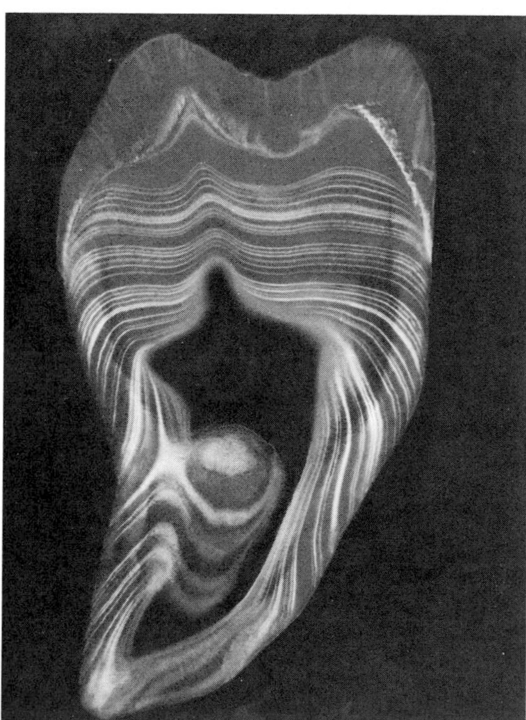

Fig. 2. Section through third molar photographed with ultraviolet light. The bands, fluorescent marks of tetracycline incorporation, illustrate the growth history of the tooth. Enamel cap, dentin body, and cementum are not visible. (*From H. C. W. Skinner and J. Nalbandian, Tetracyclines and mineralized tissues: Review and perspectives, Yale J. Biol. Med., 48:377–397, 1975*)

are essential elements in metabolism. Mineralized tissues are thought to function as a storehouse of elements as well as the framework for muscular attachment. The apatite crystal structure accepts substitution of many other elements: Na, K, Pb, and Sr for Ca; SiO_4, SO_4, and CO_3 for PO_4; and F for OH with minimal structural alterations. F replacing OH produces fluorapatite, an even more stable crystallographic configuration, for example. The hardness of hydroxypapatite is well above that of $CaCO_3$, the other inorganic compound commonly found in biologic materials (shells).

Ubiquitous in nature and the predominant phosphate mineral, hydroxyapatite usually occurs as fine-grained aggregates, but single crystals up to 60 ft (18 m) in length can be found in some geologic environments.

The fine-grained hydroxyapatite found in cementum and dentin maintains the structural integrity of these tissues, cements the ligaments, and, when required, can be resorbed and redeposited (alveolar bone). The size is influenced by the composition of the fluid from which these crystallites are precipitated, which in turn is controlled by the cells. Enamel directly subjected to enormous physical and chemical stresses in the oral cavity is virtually all hydroxyapatite of larger grain size. This means smaller surface area for the crystallites, a further reduction in the solubility of the mineral phase, obviously a distinct functional advantage for the enamel layer.

Research prospects. To design a replacement for teeth, structures which will perform under the conditions of chemical and physical stress and maintain viability, remains a challenge. Dental researchers today are elucidating precise compositions of the organic and inorganic components, background necessary before adequate alternatives, that is, substitute materials, can be devised. Perhaps the most promising research activities, however, are those preventing decay and loss of permanent teeth: basic research in nutrition, genetic abnormalities, growth, development, and fluoridation.

For background information *see* DENTITION; TOOTH in the McGraw-Hill Encyclopedia of Science and Technology.

[H. CATHERINE W. SKINNER]

Bibliography: S. N. Bhashka (ed.), *Orban's Oral Histology and Embryology*, 9th ed., 1980; A. E. W. Miles (ed.), *Structural and Chemical Organization of Teeth*, 1967; H. C. W. Skinner and J. Nalbandian, *Yale J. Biol. Med.*, 48:377–397, 1975.

Toxic shock syndrome

For more than 50 years illnesses have been described in the medical literature which were indistinguishable from toxic shock syndrome, but it was not until 1978 that this severe illness was recognized, by James Todd, as a distinct clinical entity. Todd had seen seven children with strikingly similar illnesses during a 2-year period. The children, four girls and three boys, had had a sudden onset of high fever, low blood pressure, sore throat, nausea and vomiting, diarrhea, liver and kidney failure, sunburnlike rash, and peeling of the skin during convalescence. All the patients had recovered without any long-term effects, but the cause of their illnesses remained a mystery.

This newly recognized disease remained in obscurity until early 1980, when the Centers for Disease Control (CDC) began receiving reports from physicians of a new and unusual illness. The cases reported to CDC were almost identical to the cases described by Todd but involved adult women almost exclusively. By May 1980 a total of 55 cases of toxic shock syndrome had been reported to CDC from eight states. These cases demonstrated that toxic shock syndrome was seen primarily in previously healthy young women and occurred most often during their menstrual periods. These findings were announced by CDC and led to a number of studies which demonstrated that the use of tampons was a major risk factor associated with the development of toxic shock syndrome and that the bacterium *Staphylococcus aureus* could be cultured from the vagina from most patients with the syndrome. These studies gained enormous attention in the lay press, partly because of the sensational occurrence of death in previously healthy people and perhaps to an even greater extent because of the potential impact on the large number of American women who use tampons, estimated to be more than 50,000,000.

Epidemiology. Toxic shock syndrome is primarily but not exclusively a disease of young menstruating women. Of the more than 1500 cases reported to CDC, 98% have occurred in women, and 95% of these women have had onset of illness during their menstrual period. Cases in men and nonmenstruating women (7% of all cases) have been associated with a focal staphylococcal infection, such as an infected wound, an abscess, osteomyelitis, or pneumonia. Several cases have been seen in women during the postpartum period and are currently thought to be related to a vaginal infection with *S. aureus*.

The age of female patients has ranged from 6 to 61 years, with a mean age of 23 years. More than one-third of the cases have occurred in women 15–19 years of age (see illustration). Although cases of toxic shock syndrome have been recognized in blacks, Asians, Hispanics, and American Indians, most reported cases (98%) have occurred in whites.

Although the number of cases of toxic shock syndrome reported to CDC each month is changing, estimates for the incidence of toxic shock syndrome in the United States during 1980 range between 3 and 15 cases per 100,000 menstruating women per year. Documented cases have also been reported in Canada, Great Britain, Sweden, the Nethcrlands, Germany, Australia, and New Zealand.

Studies of women with menstrually associated toxic-shock syndrome have shown that almost all were wearing tampons at the time of illness as compared with the estimated 70–80% of healthy women who use tampons during their menstrual periods.

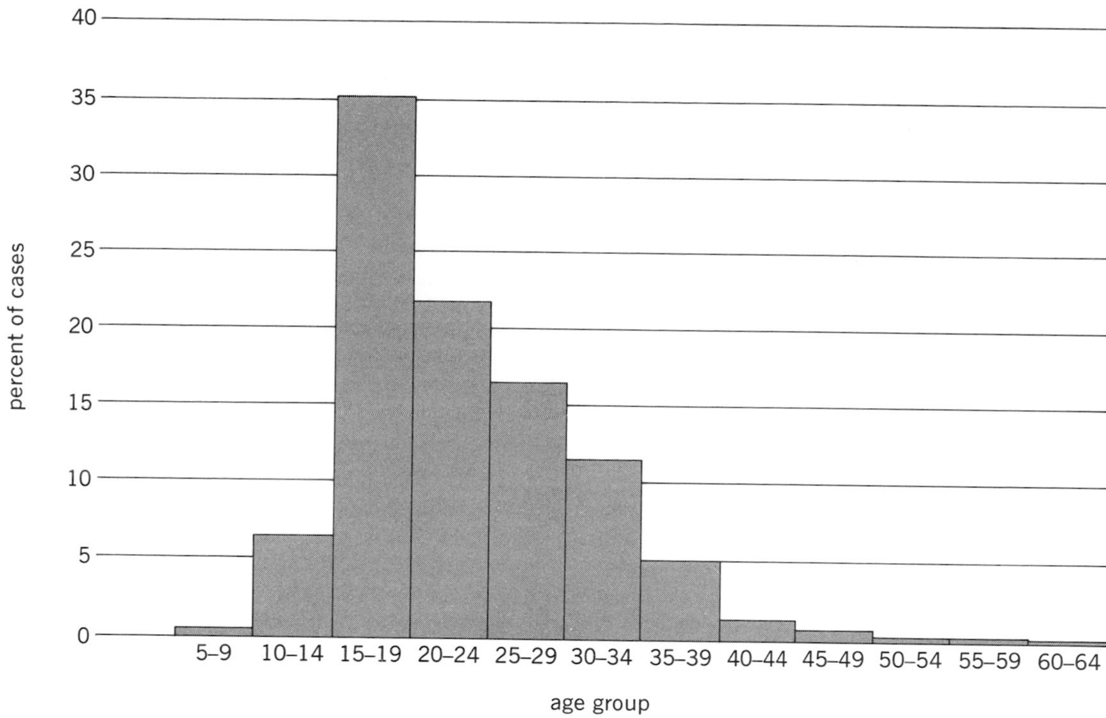

Cases of toxic shock syndrome, by age group, in the United States.

Studies have also shown that a greater percentage of women with toxic shock syndrome used tampons continuously, that is, all day and all night, than did their healthy counterparts. No differences were found between women with toxic shock syndrome and other women in the amount or duration of menstrual flow or in the frequency of tampon change. Women who used superabsorbent tampons changed them as often as did women who used regular tampons. Physical activity was not a risk factor in the development of toxic shock syndrome, and the type of exercise (in particular, swimming) was not associated with acquisition of illness. Women who became ill early in their menstrual cycle (by the fourth day) did not differ significantly in any studied factors from those who became ill later in the menstrual cycle (day 5 or later).

Information supplied by tampon manufacturers revealed that before 1977 all tampon products were made of rayon or a blend of rayon and cotton. Since 1977, 44% of tampon products, representing 65% of the estimated market, have contained more-absorbent synthetic materials, including polyacrylate fibers, carboxymethylcellulose, high-absorbency rayon-cellulose, and polyester foam. There were no statistically significant differences between cases and controls in absorbency of tampon used, use of deodorized tampons, or use of brands containing high-absorbency materials. However, subsequent investigations at CDC demonstrated that while cases of toxic shock syndrome had occurred in users of all brands and styles of tampons, the users of a single brand of tampon, Rely (Proctor & Gamble), were at an increased risk of developing toxic shock syndrome. That brand of tampon was removed from the market in late September 1980, and with its removal a decrease in the number of cases reported to CDC was noted.

Clinical picture. In a typical menstrually associated case, a previously healthy young woman develops fever, chills, nausea and vomiting, diarrhea, and muscle tenderness during her menstrual period. Watery, nonbloody diarrhea, high fever, and other symptoms persist or worsen over 24 to 48 hours, usually accompanied by headache, sore throat, and abdominal pain. A sunburnlike rash which blanches with pressure is usually evident at this point, most often on the trunk and face. In the early stages a diagnosis of viral gastroenteritis is often suspected, and nonspecific treatment at home can result in progressive dehydration, hypotension, and increasing lethargy, mental confusion, and coma. Patients who survive show peeling of the skin 7–14 days after the acute illness. There may be fine, flaking of the skin over the face and trunk or impressive peeling of thick layers of skin over the palms, soles, fingers, and toes.

Approximately one-third of patients with menstrually related toxic shock syndrome have experienced one or more recurrences during subsequent menstrual periods. Most patients who have had recurrences have experienced them within the first two or three menstrual periods following their initial episode, but recurrences have occurred as much as a year or more later. Recurrences are generally milder than the original episode, but can be more severe.

Failure to treat the initial episode with an effective antistaphylococcal antibiotic and continued use of tampons predispose to recurrences.

Differential diagnosis. Toxic shock syndrome can be confused with a variety of other severe illnesses, depending on which signs and symptoms predominate. Patients with toxic shock syndrome who have had a prominent rash and fever have been suspected of having scarlet fever, Rocky Mountain spotted fever, leptospirosis, meningococcemia, and Kawasaki disease; virtually all reported cases of adult Kawasaki disease have, in fact, been toxic shock syndrome. Patients in whom vomiting, diarrhea, and abdominal pain are prominent symptoms have often been diagnosed as having viral gastroenteritis, appendicitis, or pelvic inflammatory disease. A number of these patients have undergone exploratory abdominal or pelvic surgery.

The similarity of toxic shock syndrome to other toxin-induced illnesses such as scarlet fever and staphylococcal food poisoning and the almost uniform absence of staphylococci in the bloodstream suggest that toxic shock syndrome is caused by one or more staphylococcal toxins. A specific toxin has been identified which is produced by virtually all *S. aureus* isolated from toxic shock syndrome cases. But as yet the toxin has not been found in the blood of patients with toxic shock syndrome, nor has the toxin been shown to produce a similar illness in animals. The role of this toxin or other toxins in the development of the widespread effects seen in toxic shock syndrome remains to be elucidated.

Course of illness and prognosis. Patients suspected of having toxic shock syndrome are given a thorough physical examination, including a careful vaginal examination for women. The use of tampons should be discontinued. In patients in whom a nonvaginal source of infection with *S. aureus* is suspected, a thorough search for the site of infection should be undertaken. The treatment of toxic shock syndrome is primarily supportive. Most patients experience severe fluid losses as a result of vomiting, diarrhea, and sweating, and rapid intravenous fluid replacement is critical. Additional supportive measures which are occasionally necessary to treat the complications of toxic shock syndrome include renal dialysis and ventilatory assistance. Most patients respond dramatically to fluid replacement and antibiotics, although blood pressure may take 2 or 3 days to return to normal. Patients usually can begin taking fluids and medications orally within 2 or 3 days and can often be discharged within a week after admission to the hospital.

Over 90% of the severely ill patients survive. The majority of patients who recover have no apparent long-term effects, although some patients have suffered the secondary effects of profound shock. Many patients report hair loss and nail changes 1 to 3 months after their acute illness, but these changes are self-limited. Several patients have undergone normal pregnancies following toxic shock syndrome.

The well-documented association of toxic shock syndrome with the use of all vaginal tampons remains unexplained. Mechanical blockage of menstrual fluids, absorption of cervical secretions (which might restrict the growth of pathogenic microorganisms), alterations in the normal vaginal flora, and tampon-induced vaginal ulcerations and abrasions have all been postulated as explanations, but there is little or no evidence that they have an etiologic role. Despite testing of many unused tampons of different brands, intrinsic contamination of tampons with *S. aureus* has never been documented. A few menstrually associated cases have been observed in nontampon users, so an explanation of the role played by tampons must also allow for the establishment of similar conditions in the absence of tampons.

Women who have had toxic shock syndrome should not use tampons until they have had a follow-up vaginal culture to determine whether *S. aureus* has been eradicated from the vagina. However, routine vaginal cultures from women who have not had toxic shock syndrome should be discouraged because there is currently no way of determining which of the 8–10% of healthy women who carry *S. aureus* in the vagina are at risk of developing toxic shock syndrome. Women who have not had toxic shock syndrome should be advised that the risk of developing it is small, even among tampon users. These women can almost entirely eliminate their risk of developing toxic shock syndrome by not using tampons at all, or they can reduce their small risk by using tampons intermittently during the menstrual period (in particular, by discontinuing their use at night); but there is no evidence that changing tampons more frequently than usual will prevent the development of toxic shock syndrome. Whether the use of less-absorbent tampons decreases the user's risk of developing toxic shock syndrome is unknown at this time. [BRUCE B. DAN]

Bibliography: Centers for Disease Control, *Morbid Mortal Weekly Rep.*, 29:441–445, 1980; K. N. Shands et al., *New Engl. J. Med.*, 303:1436–1442, 1980; F. A. Stevens, *J. Amer. Med. Ass.*, 88:1957–1958, 1927; J. Todd et al., *Lancet*, 2:1116–1118, 1978.

Transistor

The first microwave gallium arsenide field-effect transistor (GaAs FET) was developed in 1967. Even the early devices showed promise for use as low-noise, high-frequency amplifiers. Recent advances in crystal growth and lithographic definition, and better designs have resulted in dramatic improvements in the capabilities of these devices. The low noise figures of the latest devices (less than 2 dB at 12 GHz) rival those of simple parametric amplifiers, and may make direct, low-cost, satellite television receivers possible. The broad bandwidth and moderate power (a few watts at 10 GHz) capabilities of these devices make them suitable for traveling-wave tube (TWT) replacements in some electronic warfare (EW) systems.

Operating principles. The structure of the early microwave FETs is shown in Fig. 1. The device is made by epitaxially growing a moderately doped n-type (10^{16}–10^{17} donors per cubic centimeter) gallium arsenide layer 0.1 to 0.5 μm thick on a semi-insulating gallium arsenide substrate about 400 μm thick. This grown region forms the channel of the device. The material around the channel region is removed by mesa etching to isolate it. Two ohmic contacts are made to the channel, a source and a drain (Fig. 1). Current between the source and drain is carried by free electrons in the channel region. The n-type gallium arsenide is chosen for the channel region because its electron-saturated drift velocity is nearly twice that of silicon. This permits electrons to transit across the device in the minimum time, giving the highest possible operating frequency. A third contact is made to the channel, between the source and drain. This is a Schottky barrier rectifying contact, usually made with aluminum or titanium (with other metals overlaid), and is called the gate. When the gate is biased negatively with respect to the source, no appreciable gate current flows, because of the rectifying nature of this contact. However, the negative gate bias causes a depletion region to form in the channel (Fig. 1). When sufficient negative gate bias is applied, the depletion region will extend down to the semi-insulating layer and block the carrier flow between the source and drain; this is called pinchoff. This type of device operation is termed the depletion mode. The maximum frequency of oscillation for the device is called f_{max}. Experimentally this has been shown to depend primarily on the gate length L, as given by Eq. (1), where B is a constant ($B \cong 3 \times$

$$f_{max} \cong \frac{B}{L} \text{ Hz} \qquad (1)$$

10^{10} for L in micrometers). Hence for $L = 1$ μm, $f_{max} = 30$GHz.

The amount of power the device can produce depends on breakdown voltage and gate width W. Typical small signal devices have W/L ratios of 300, and power devices use ratios up to 2400. The noise

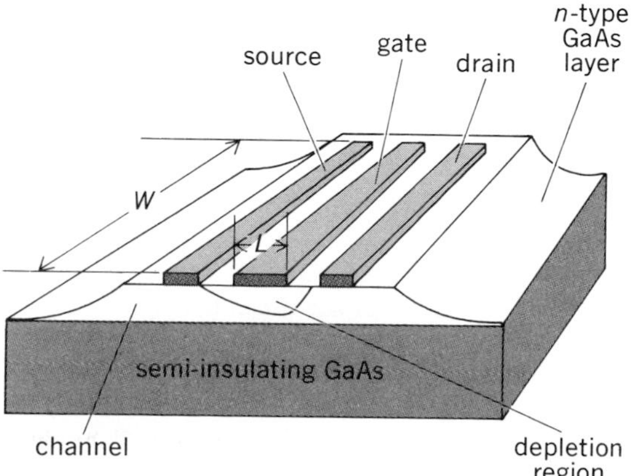

Fig. 1. Structure of early microwave field-effect transistor using a single epitaxial layer.

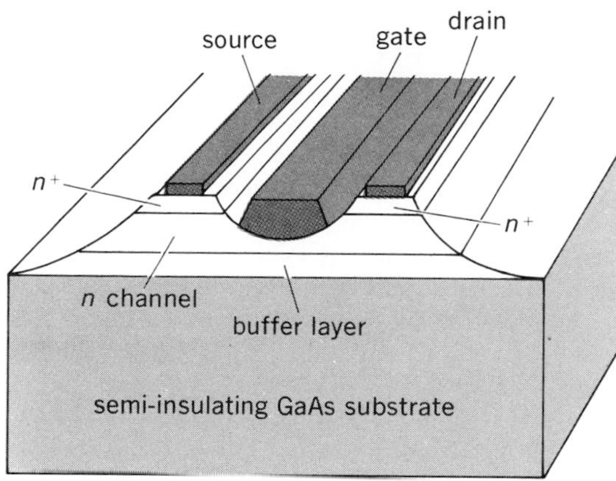

Fig. 2. Structure of a modern low-noise, high-frequency field-effect transistor showing three epitaxial layers.

performance of these devices has been examined, and it has been found that the noise figure F_{min} can be expressed as Eq. (2), where k is a constant, f is

$$F_{min} = 1 + kfL\sqrt{g_{mo}(R_g + R_s)} \qquad (2)$$

the frequency, L is the gate length, g_{mo} is the transconductance (gain), and R_g and R_s are the gate and source series resistance. Equations (1) and (2) show that performance improvements are in general expected with decreasing gage lengths, L.

Improvements. Recent advances in device modeling, crystal growth techniques, and submicrometer lithography (using electron-beam exposure) have produced dramatic improvements in upper frequency limits and noise performance of GaAs FETs. Figure 2 shows the cross section of a modern recessed gate microwave GaAs FET. Three epitaxially grown layers are now used on the semi-insulating substrate. First, an undoped, high-resistivity buffer layer is grown on the substrate to prevent device degradation due to substrate defects. Next, the channel layer is grown, and finally an N^+ (heavily doped) region is grown on top of the channel to make good low-resistance contacts to the source and drain regions. An opening is chemically etched through the N^+ layer down into the channel region, which separates the source and drain regions and exposes the channel for gate metallization. The curved shape of the channel top makes the effective length of the gate narrower than its actual length (because the channel is thinner in the middle of the gate and pinches off there first). It is now known that this structure also reduces R_s and R_g. The use of the n^+ regions and the curved recessed gate was identified as the primary way to make low noise FETs.

A 0.35 dB noise figure has been achieved at 4 GHz in a similar device using an $L = 0.3$ μm gate length. Short-gate-length devices have also demonstrated very high operating frequencies. An FET gain of 5 dB, with 100 mW of power output at 30 GHz, has been achieved with a 0.6 μm gate-length device and a 57-GHz oscillator has been made from a device with a 0.5-μm-gate length. At lower frequencies, near 10 GHz, power outputs of approximately 500 mW per millimeter of gate width are produced.

Prospects. When the gate length becomes short enough, so-called ballistic transport of the electrons occurs. It has been suggested that for electron paths less than 0.5 μm it is possible for the electrons to overshoot their saturated drift velocities. Experimental devices are currently being built to test this theory. In a different technique to reduce electron transit time, the free electrons are provided by a very thin heavily doped region adjacent to the channel. Thus, the electrons in the channel move more freely because of the lack of doping atom collisions. Development is also being done on new materials such as InGaAs (indium gallium arsenide), which has been shown to have an electron-saturated drift velocity more than twice that of gallium arsenide in an FET structure. GaAs FETs have also been fabricated by ion implantation directly into the semi-insulating substrate (or buffer layer). This has the advantage of deleting the mesa etch step and could result in higher-reliability "planar" FETs. *See* MICROSCOPIC SOLID STRUCTURES.

The semi-insulating substrate on which the FETs are made is also a low-loss microwave dielectric material. This means that microwave circuit elements such as transmission lines, inductors, and capacitors can be made on the substrate surface by lithographic patterning of the metal at the same time the FET is being metallized. Thus, a complete circuit including active and passive devices can be made in a very small area. This type of structure is called a monolithic microwave integrated circuit (MMIC). MMICs may become fairly inexpensive circuits in large-volume production, thus making direct satellite television receivers economically feasible.

For background information *see* MICROWAVE SOLID-STATE DEVICES; TRANSISTOR in the McGraw-Hill Encyclopedia of Science and Technology.

[R. J. CHAFFIN]

Bibliography: S. Bandy et al., *Appl. Phys. Lett.*, 38(10):817–819, May 15, 1981; R. Dingle et al., *Appl. Phys. Lett.*, 33(7):665–667, October 1978; H. Fukui et al., *IEEE Trans. Elec. Dev.*, ED-27(6):1034–1037, June 1980; W. Hooper, and W. Lehrer, *Proc. IEEE*, 55(7):1237–1238, July 1967; M. Shur and L. Eastman, *IEEE Elec. Dev. Lett.*, EDL-1(8):147–148, August 1980.

Trans-Neptunian planet

The search for a trans-Neptunian planet can best be explained by considering the search for the planet Pluto. When Pluto was discovered, it was thought to be the source of the gravitational force which was influencing the motions of Uranus and Neptune. When it was shown conclusively in 1978 that this planet was not large enough to exert this force, the search for another planet had to be continued. The current search consists of two phases which have been greatly aided by the technological advances of the past 10 years. The computational phase uses sophisticated computer programs which are run on high-speed computers with large memories. This phase involves computing an orbit and comparing it against thousands of observations which have been made over more than 360 years. The observational phase has been greatly aided by advances made in solid-state electronics. These allow the observations to be done more accurately and make it possible to see faint objects more clearly.

Early searches. The search for a trans-Neptunian planet began in the late 19th century. It had become apparent that Uranus and Neptune were being disturbed in their orbits. The first serious attempt came in 1877, when David Todd, an American astronomer, did a graphical analysis of the perturbations (deviations from a predicted orbit) of Uranus. No planet was found, however. In analyzing the differences between the computed and observed orbits for comets, William Pickering of Harvard computed orbits for several hypothetical planets. These differences can, in general, be periodic or secular, or a combination of both. A periodic difference is one in which the size of the numbers increases and decreases, following a wavelike pattern. A secular difference is one in which the size of the numbers steadily increases or decreases with time. As the result of his analysis, Pickering announced a prediction for a planet "o." This gave new impetus to an intensive exploration of space beyond Neptune under the direction of Percival Lowell. He conducted two searches for a missing planet during 1905–1909 and 1910–1915. In addition to mathematical computations, Lowell made a photographic search along a narrow band of sky around the ecliptic (the plane in space formed by the Earth's orbit around the Sun). He spent much of his own money and time in his two searches, and employed more than a dozen assistants at one time or another. He computed two possible orbits for a planet he called "x." Lowell died in 1916 without knowing that in the spring of 1915, as he readied his work for publication, faint images of a new planet beyond Neptune were recorded on two sets of the "x" search plates. The images went unnoticed for 15 years.

In 1929 Clyde Tombaugh began another photographic search for the missing planet using the new 13-in. (33-cm) refracting telescope at Lowell Observatory in Flagstaff, AZ. He found it (Pluto) in February 1930 by using a machine called a blink comparator to look at two photographic plates taken on January 23 and 29. This is a device in which one looks at two plates through an eyepiece. A light source is made to shine alternately on each plate in rapid succession. If the plates are taken far enough apart in time, one can see a nearby object seem to blink on and off as it moves between two positions in the sky while the background stars remain fixed. This procedure is very arduous, since there are also some stars which vary in brightness over short periods of time. Sometimes a third plate must be taken to resolve the problem.

Pluto's failure to explain perturbations. Although Pluto was near its predicted position, a great controversy existed over the fact that it seemed too small to account for the perturbations (small changes) in the orbits of Uranus and Neptune. Lowell had predicted a mass of 6.6 times the Earth's mass, and Pickering had predicted a factor of 2.0. Other determinations by other astronomers gave smaller masses. As the years went by, the mass determinations became smaller and smaller, and it seemed highly unlikely that Pluto could produce the observed effects. In July 1978 J. Christy of the U.S. Naval Observatory discovered a moon (Charon) orbiting Pluto while he was analyzing plates taken with the 61-in. (155-cm) astrometric reflecting telescope at Flagstaff. This permitted the first direct determination of Pluto's mass (0.0023 times that of the Earth). This value made it definite that Pluto was too small to affect the motions of Uranus and Neptune. There had to be some other force acting in the outer solar system. It had also been calculated that six long-period comets, which have passed through the solar system more than once, show unaccounted forces acting perpendicular to their plane of orbit around the Sun. Nongravitational forces on comets do not act in this direction, but all these comets go out beyond Neptune.

Current search. A team of astronomers at the U.S. Naval Observatory has been gathering all the observations which have been made of the outer planets (Jupiter, Saturn, Uranus, Neptune, and Pluto) in an attempt to determine their orbits more accurately. In the case of Neptune, which was discovered in 1846, an observation made by Galileo in 1613 has been found. This observation may prove valuable in analyzing Neptune's orbit. It allows for an observational set to span almost three orbital periods of the planet. The current procedure for analyzing the outer-planet orbits at the Naval Observatory is the following: A preliminary orbit is determined by using an orbital element integrator. This is a complex computer program which computes an orbit by using orbital elements, which are quantities describing its size, shape, orientation in space, and relation to the Sun. The program calculates a position for the planet and then uses this position to calculate another, and so on. A set of differences in the positions is then formed between the computed orbit and the actual observations. These differences are analyzed by using an algebraic manipulator. This is a set of many computer routines which allow one to manipulate algebraic expressions stored in the computer's memory. The results of this analysis are corrections to the orbital elements which defined the preliminary orbit. This procedure has already yielded a better determination of Pluto's orbit. It has also been used to compute a more accurate period for the revolution of Charon around Pluto. It is hoped that this process will yield, in addition to better-computed orbits for the outer planets, a direction to the origin of the unknown force acting on Uranus and Neptune. Once a direction is predicted, a photographic search can be conducted to see whether another planet can be found. It is still uncertain, however, whether the orbits of the outer planets can be determined accurately enough to allow this direction to be determined.

However, one must be cautious not to jump to the conclusion that an unknown force must be produced by another planet. Urbain Leverrier, the French mathematician who successfully discovered Neptune by studying the irregularities in the orbit of Uranus, spent 15 years studying the motion of Mercury in an attempt to find a missing planet closer to the Sun. His efforts to find this planet, which was later given the name Vulcan, failed. It was not until 50 years later that Albert Einstein gave the correct explanation. Newton's law of gravity needed a small correction which was provided by Einstein's theory of relativity. Therefore, it is not certain that the observed irregularities in the motions of the outer planets are produced by an unknown planet.

Results of preliminary investigation. After Tombaugh discovered Pluto, he once again took up the search for another planet. Over the years he examined 90,000,000 star images from 50° south of the ecliptic to the North Pole. He photographed two-thirds of the sky but did not find another planet. Nonetheless, the result of a preliminary investigation by T. C. Van Flandern and R. A. Harrington of the Naval Observatory is that a single undiscovered planet could entirely explain the observed irregularities in the orbits of the outer planets. They have also shown that the satellites of Neptune were strongly disturbed in the distant past by a close approach of a planet-sized body. In fact, they concluded that Pluto and Charon could both be escaped moons of Neptune. The body could have passed Neptune close enough to pull these moons out of their orbits. They did not entirely escape the solar system, however. Gravitational forces finally stabilized their motions in such a way that Charon now orbits Pluto, which in turn orbits the Sun. This would also explain the fact that Pluto is the only planet that crosses the orbit of another planet. It last crossed Neptune's orbit in January 1979. It will reach perihelion (the point in its orbit that is closest to the Sun) in September 1989. Neptune is now farther from the Sun than Pluto is. In March 1999 Pluto will again cross Neptune's orbit and will return to being the farthest known planet from the Sun. There is no danger of collision of the two planets, since Pluto's orbit is inclined in such a way with respect to Neptune's orbit that there is never a close approach. Harrington and Van Flandern calculated that a planet in the size range of two to five times the mass of the Earth was required, and that this mass could explain the observed irregularities in the planets' orbits if it were currently located 50 to 100 times farther from the Sun than the Earth is (Neptune is 30 times as far from the Sun as the Earth is).

Advantages over previous searches. The present

search has some distinct advantages over the searches conducted by Lowell. More than 70 years of additional observations are available on which to compute orbits of the outer planets. The larger number of observations allows smaller errors to be assigned to the predicted orbits. (In general, the errors should decrease as the square root of the number of observations.) The observations themselves are also becoming more accurate as increasingly sophisticated electronic equipment is being developed. It is also possible to compute orbits much faster than Lowell and his dozen assistants did. What had taken them months can now be done in a few hours. The space probes which have been launched in the last few years have given better determinations of the masses of Uranus and Neptune. Improved information on the masses should result in better orbit determinations.

A photographic search will be able to use the high resolution resulting from forming images using charge-coupled devices (CCDs). These devices store the photons of light coming into a telescope as electric charges on an electronic chip. With the proper computer programs, an image can be formed which resembles a photograph. Much time will still be required to analyze these electronic photographs as well as standard photographs, since the missing planet will be very faint indeed. The initial search will use standard photographs from ground-based telescopes, and if a promising object is observed it will probably be examined more closely with CCDs abord the Space Telescope. *See* ELECTRONIC CAMERA.

For background information *see* CELESTIAL MECHANICS; PLUTO; SATELLITE ASTRONOMY; SOLAR SYSTEM in the McGraw-Hill Encyclopedia of Science and Technology.

[KENNETH F. PULKKINEN]

Bibliography: R. S. Harrington and T. C. Van Flandern, *Icarus*, 39:131–136, 1979; W. G. Hoyt, *Planet X and Pluto*, 1980; P. K. Seidelmann et al., *Icarus*, 44:19–28, 1980; T. C. Van Flandern and K. F. Pulkkinen, *Astrophys. J.*, Suppl. 41:391–411, November 1979.

Ultrasonics

Very-high-frequency sound can be transmitted through the skin into the body, producing images of internal soft tissue organs to display both normal anatomy and pathologic processes. This noninvasive means of imaging has gained wide acceptance and application in clinical medicine because sound waves in the intensity and frequency used have not been shown to harm the body. Consequently, sonographic equipment has undergone rapid development over the past 10 years, and images of considerable clarity can now be produced. Through the use of rapid processing computers, static and dynamic imaging modes are available for producing large-area stationary images or smaller-area images that display motion. Ultrasound is used in many medical specialties and can detect diseases of the heart, abdomen, extremities, eyes, blood vessels, and the unborn fetus.

Ultrasound, by definition, is sound at frequencies greater than 20,000 Hz; diagnostic ultrasound uses frequencies between 2 and 10 MHz. Sound at these frequencies is transmitted through the body in the form of longitudinal pressure-wave fronts of compressed molecules. The sound waves reflect off tissue interfaces at the surface of and inside organs. The reflection process is determined by very slight changes in the densities of body tissues. The returning reflected echoes are detected by the equipment and converted into images. The component responsible for both sending and receiving the sound pulses is the transducer. The sound waves are actually produced by a waferlike ceramic disk in the face of the transducer that is applied to the surface of the skin with a coupling water-soluble gel. The gel also provides a lubricating medium, allowing the operator to slide the transducer across the skin surface during the examination. When the ceramic disk is subjected to a rapidly oscillating current, it bulges and contracts slightly. This phenomenon is referred to as the piezoelectric effect. This disk motion pushes and pulls on the adjacent skin surface, propagating the sound wave into the body in the form of wave fronts of compression and rarefaction of the molecules in the sound-beam path. All of these processes are very subtle and cannot be perceived by the individual being insonated. The returning reflected echoes from inside the body are detected by the transducer through the inverse process from transmission and are converted back into electrical pulses. The stronger returning echoes create proportionally stronger electrical pulses. The transducer accomplishes this by sending intermittent short pulses of sound, then listening for returning echoes. The distance and direction of the returning echoes are simultaneously registered by the equipment, and the electrical pulses are given a shade of gray corresponding to their strength and assigned a position as a dot on a television screen representing their position in the body. The images are created from numerous such dots in up to 32 shades of gray and are displayed on a black background. The stronger echoes are depicted in light gray shades, while the weaker echoes are seen as darker areas on the image. Regions that contain fluid register no echoes and are black on the image (Figs. 1 and 2).

Imaging. Static scanners accumulate information over a few seconds and display it as a composite picture, while dynamic scanners continuously display returning information in the form of images produced at frame rates up to 60 per second. This, in effect, produces a motion picture composed of continuous sound images. The single images in both dynamic and static imaging are planar and represent sections of the body around 1 cm in thickness. Thus, the body is examined as a series of sectional images, much like a loaf of bread that is composed of multiple slices. These images can be produced in any direction, and a precise knowledge of human anatomy is necessary to interpret the information. Both static and dynamic imaging have their particular applications, but they are usually used in a complementary fashion on each patient. The most recent technological developments have been directed toward improving dynamic imaging because of its increased operating speed and additional diagnostic information gained through watching body motions. Both types of equipment produce permanent records on photographic or x-ray film and can be interfaced with a video tape or video disk data storage and processing systems since the data are digitalized by the computer when an image is produced (Fig. 3).

Limitations and applications. There are some limitations to the use of diagnostic ultrasound. In the frequencies used, it cannot penetrate regions of the body that contain air, such as the lungs or abdominal intestine. Also, bone causes considerable sound-beam reflection, and images of bone are not of useful quality.

Though ultrasound equipment has applications crossing the lines of many medical specialty areas, it is usually used by radiologists who are physicians trained in medical imaging procedures. The two most frequent applications of diagnostic ultrasound are the detection of intraabdominal disorders and examining the fetus and gravid uterus in pregnancy.

Intraabdominal disorders. Abdominal sonographic diagnosis can be applied to all of the intraabdominal organs with the exception of the intestine. Sonography has become the primary diagnostic imaging modality of the liver and biliary system. Gallstones can be detected with an accuracy of around 95%. Tumors or abscesses in the liver can be identified, and the images are frequently used to guide the insertion of a biopsy needle to obtain a small piece of tissue for pathologic analysis. In this way, a diagnosis can be obtained without the need for surgery. The pancreas, kidneys, spleen, and pelvic organs can also be examined. A variety of abnormalities, such as fluid-containing cysts, purulent abscesses, clotted blood collections, and tumors, can be identified.

Fetal scanning. One of the most exciting areas of recent ultrasound development has been in the field of dynamic scanning of the fetus in the uterus. The developing human embryo can first be identified at the gestational age of 5 weeks, when it is approximately 8 mm in length. The growth and development can be followed from this point through the entire pregnancy. The fetal organs become large enough to examine individually between 16 and 20 weeks' gestation. Congenital anomalies and fetal structural abnormalities can frequently be identified, and ultrasound is used along with needle aspiration of the amniotic fluid surrounding the fetus (amniocentesis) in the early diagnosis of fetal abnormalities. The placental position is readily identified to exclude the possibility of its blocking the birth canal (placenta previa). Future developments will include the ability to monitor fetal physiologic functions, such as umbilical vein blood flow to the fetus and fetal cardiac functions. Ultrasound imaging can be used in the delivery room to assist the obstetrician in identifying the fetal position or possible retained placental components postdelivery.

Cardiac imaging. Ultrasound imaging of the heart and its valves is accomplished with dynamic scanning equipment. The transducers are small enough to be placed against the anterior chest wall to project the sound beam between the ribs. Only areas

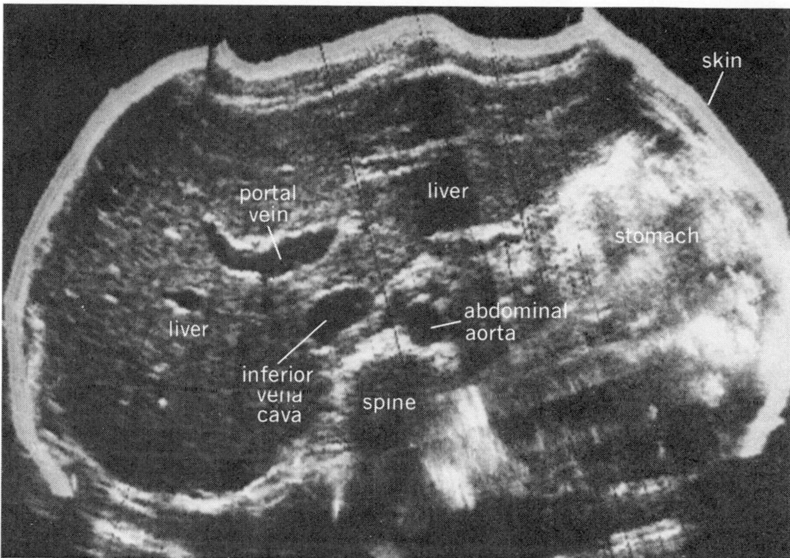

Fig. 1. Static image through the upper abdomen. In this cross section, the anterior surface is at the top and the right side of the subject is at the left. The white band surrounding the exterior of the image is the skin surface.

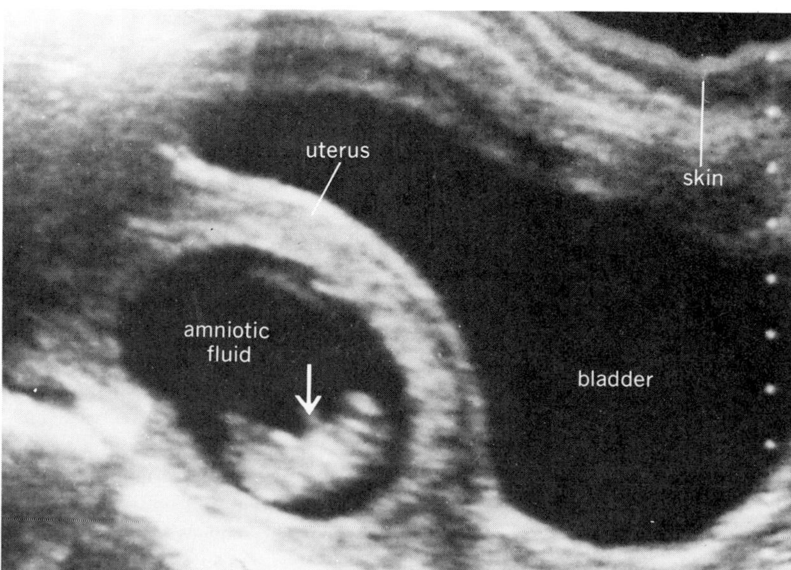

Fig. 2. Static scan through the midline of the pelvis along the length of the gravid uterus containing 7-week-old fetus measuring 3 cm in length. The arrow points to a partially formed arm.

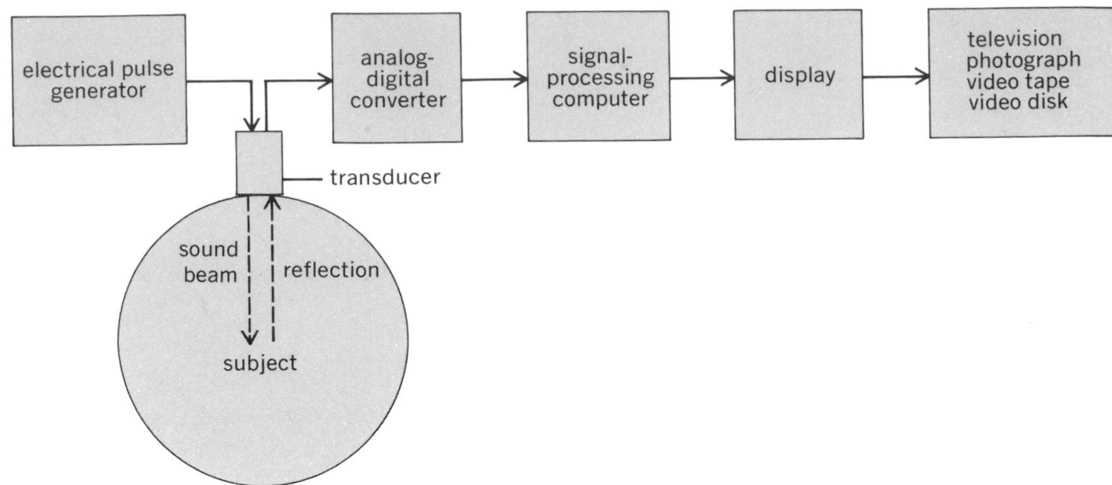

Fig. 3. Components of a standard diagnostic ultrasound system.

where the heart contacts the chest wall can be imaged since the air in the lungs reflects the beam totally. With an image production rate of 30–60 frames per second, the motion of the heart beating can be analyzed. The myocardial contractility, chamber volume, and valve motion for each of the four chambers and valves can be recorded on video tape and reviewed for diagnostic purposes. Disease states that result in abnormal motion of the valves or functional performance of the muscular walls of the chambers are most readily appreciated. The coronary arteries are not seen well enough to provide useful information. The cardiac applications have been extended with new smaller equipment into the area of pediatric heart disease and heart malformations of newborns. Future developments are likely to include computer graphics and three-dimensional images created from the two-dimensional sectional images available with present equipment.

Vascular sonography. Vascular sonography includes the use of both dynamic imaging equipment and nonimaging techniques to measure the velocity of blood flow by Doppler analysis. The Doppler effect is created when a sound beam is directed into a blood vessel at an angle. The blood cells in the moving bloodstream act as reflectors, and the frequency of the returning sound waves is proportional to the velocity of blood flow in the blood vessel. Dual-element transducers can create a picture of a blood vessel while a Doppler beam is steered into the lumen of the vessel to register flow. These techniques are used to evaluate vascular occlusions and stenoses (partial occlusions) in patients with suspected vascular disease. This rapid and accurate technique can be performed without patient discomfort. The previously predominant technique was vascular angiography, which is uncomfortable, uses x-rays, has greater risk to the patient, and is similar to a minor surgical procedure. Present equipment can examine vessels near the skin surface, and the greatest applications have been in the neck on the arteries to the brain in stroke victims and arteries and veins in the extremities. New developments will produce equipment capable of measuring blood flow in the deeper abdominal vessels and the fetus.

Other developments. High-frequency small transducers have been developed to examine small superficial areas of the body such as the eye, thyroid gland, and scrotum. The infant brain can be imaged through the anterior fontanel (the anterior soft spot) in newborns suspected of brain hemorrhage or tumor. Ultrasonic scanning of the breast for detection of tumors is in the developmental state and shows promise. Future developments in diagnostic ultrasound will involve further use of computers for more rapid and complete data processing and enhanced image resolution. More compact equipment will make portable studies of critically ill patients easier. There will be continued emphasis on the development of dynamic imaging because of its speed and overall greater information content.

For background information *see* BIOMEDICAL ENGINEERING; ULTRASONICS in the McGraw-Hill Encyclopedia of Science and Technology.

[JACK H. HIRSCH]

Bibliography: J. L. Rose and B. B. Goldberg, *Basic Physics and Diagnostic Ultrasound*, 1979; R. E. Sabbagha, *Diagnostic Ultrasound Applied to Obstetrics and Gynecology*, 1980; D. A. Sarti and F. W. Sample, *Diagnostic Ultrasound Text and Cases*, 1980.

Ultrastable materials

As technology achieves ever-increasing precision in areas such as giant telescopes, laser gyros, diamond turning machines, and space applications, requirements for ultrastable materials become more demanding. In response to these demands a number of materials with extremely low thermal expansivity have recently been developed. In addition, even when thermal expansivity is exactly zero, requirements for exceptional temporal stability are increasing (for example, no creep or relaxation of dimensions with time when the temperature is held perfectly constant). Finally, many space applications call for all the above plus high stiffness-to-density ratio in order to minimize overall weight, as in giant space-borne radio antennas (Fig. 1).

Stable materials, newly developed for the above applications, can be divided into two classes. The first class is characterized by minimal thermal expansivity over a wide range of temperatures and minimal temporal instability (creep). Examples of this class include quartz (fused silica), TiO_2-doped fused silica (Corning 7971, ULE), Cervit or Zerodur (glass ceramics), and the invars (iron-nickel-cobalt alloys). The second class, intended for strong, stable, lightweight structures often for space applications, is characterized by all the above requirements, plus low density and high stiffness in order to minimize overall weight. Examples of this class include graphite/epoxy (oriented fibers of graphite held together by epoxy), graphite/glass (oriented fibers of graphite in a glass matrix), and fiber-metal composites (for example, graphite/magnesium). In place of graphite, fibers of many other strong materials such as silicon carbide or aluminum oxide have also been used. The underlying idea behind these composites is to tailor-make many desired characteristics by utilizing pairs of materials with opposite thermal expansivities, at the same time benefiting from high thermal conductivity and high stiffness-to-density ratio.

Fig. 1. Lockheed's 30-ft (9.1-m) furlable antenna, constructed from graphite/aluminum composite material. (*Lockheed*)

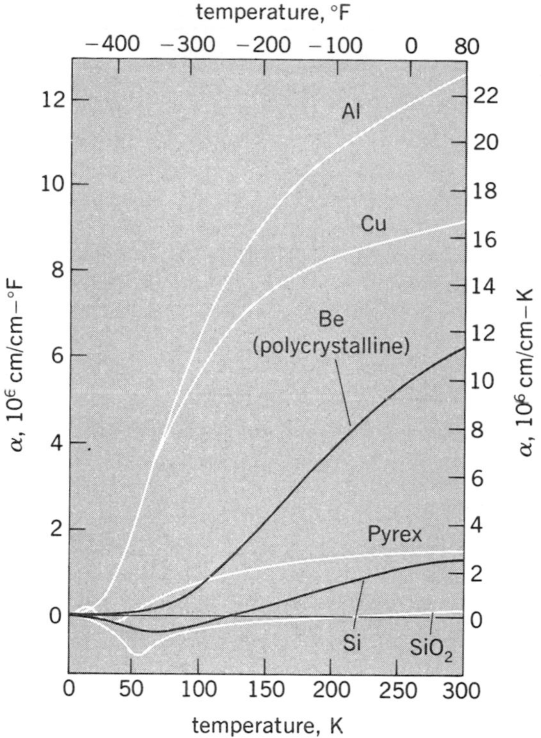

Fig. 2. Thermal expansion coefficient α versus temperature for several common materials.

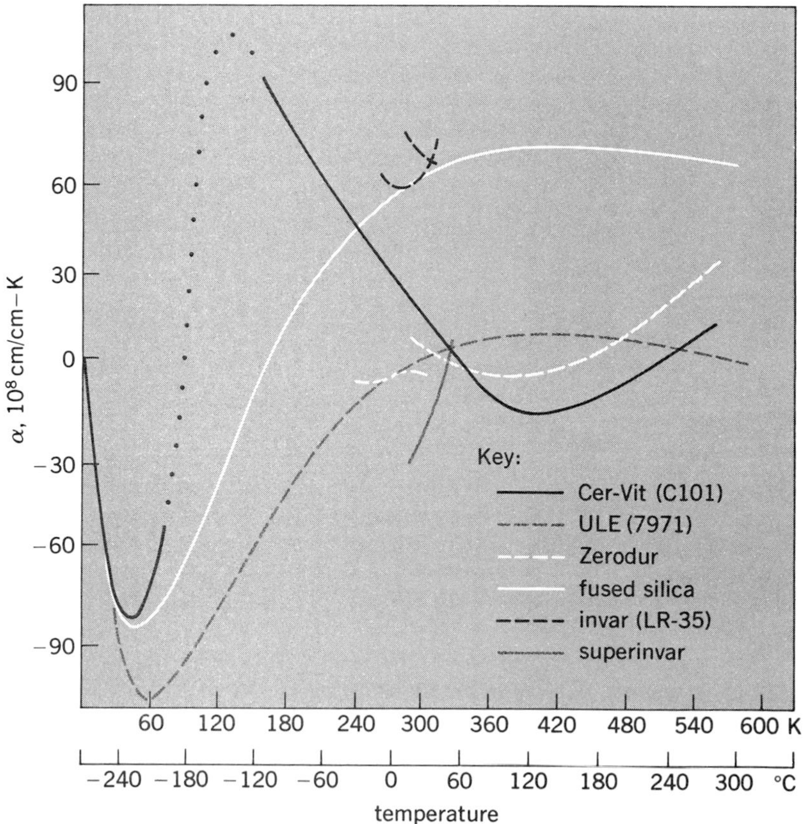

Fig. 3. Thermal expansion coefficient α versus temperature for several ultralow-expansivity materials.

Average daily length change of several materials maintained at constant temperature

Material	$\Delta L/L$, parts per billion
Corning 7971 ULE	−0.17
Corning 7940 fused silica	−0.51
Cervit (Owens Illinois C101)	+0.50
Zerodur (Heraeus-Schott)	0
Homosil fused silica (Heraeus-Schott)	−0.56
Invar LR-35 (Universal Cyclops)	+5.64
Superinvar (Simonds Saw & Steel)	0

In the following discussion, the thermal and temporal expansion characteristics of materials in the first class are described. The second class is then discussed briefly; however, the field is too new, and the engineering trade-offs too numerous, to do more than introduce these materials.

Thermal expansion. Thermal expansion is usually measured by recording the length change ΔL in a sample of length L when the temperature changes an amount ΔT. The thermal expansion coefficient is defined as $\alpha \equiv (1/L)(\Delta L/\Delta T)$. To provide some appreciation of the low-thermal-expansion materials which have recently been developed, Fig. 2 shows the expansivities of a number of common materials compared with quartz (SiO_2), which has a relatively small dimensional change with temperature, and Pyrex (Corning type 7740), which is a low-expansivity glass useful for telescope mirrors (such as Palomar's 200-in. or 5-m reflector) and for cooking ware. In contrast, Fig. 3 shows some of the new lower-expansivity materials which have been engineered. Concerning the composition of the materials in Fig. 3, fused silica = quartz = SiO_2. ULE is Corning's type 7971, TiO_2-doped SiO_2. Cervit 101 and Zerodur are glass ceramics, originally manufactured by Owens-Illinois and Schott respectively. Invar and Superinvar are metal alloys of iron/nickel and iron/nickel/cobalt respectively. Most of the materials shown were developed for applications in which low expansivity was required only near room temperature (300 K). However, with space applications becoming more numerous, it is becoming important to know the expansivity at a variety of temperatures, including those near absolute zero.

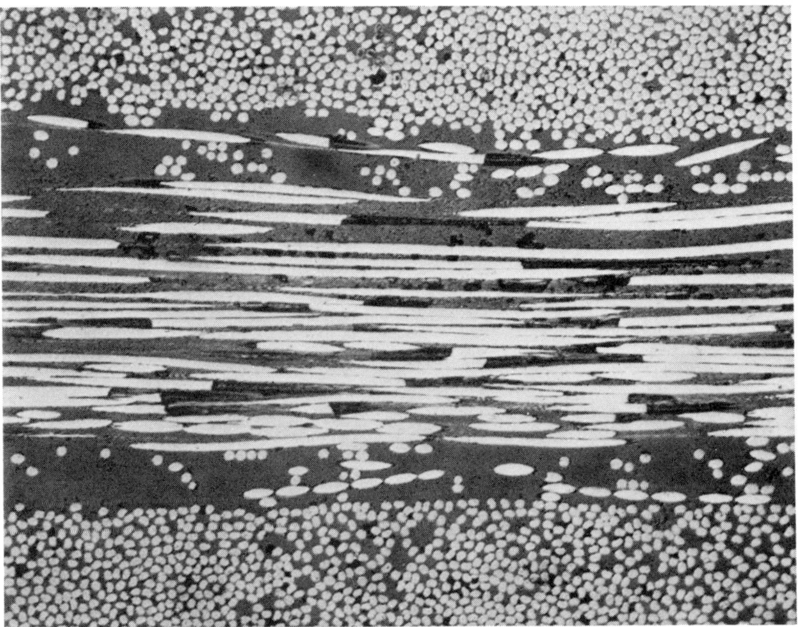

Fig. 4. Photomicrograph of graphite/glass composite. Graphite fibers (white; about 50 micrometers in diameter) are in two orientations. Some are seen end-on; others are seen on-edge. (*United Technologies Research Center*)

Temporal stability. Only recently has it been possible to measure reliably, in a reasonably short time, dimensional changes smaller than a part per million per year. The need for such measurements was underscored when it was discovered that samples of fused silica held at constant temperature changed dimension as much as 1 part per million over a 3-year period. This led to further investigations at constant temperature, utilizing an iodine-stabilized laser, the results of which are summarized in the table. Zerodur and Superinvar were as stable as could be observed (less than 1 part per billion per day), while fused silica, Cervit, and ULE showed small dimensional changes with time. The alarming changes in LR 35 Invar may have been the result of improper heat treatment, although similar results were later obtained using different samples. Superinvar, for all of its desirable characteristics, is difficult to reproduce exactly, and there is a need and a good likelihood of success for improving the Invar manufactured in the future.

Composite materials. The thermal expansivity of the composite materials can be tailored to the needs of a structural design, as can several other parameters. Because there are so many variables involved (for example, fiber diameters, volume fractions, and fiber orientations), there is no basic curve showing expansivity versus temperature. The thermal expansion coefficient can be made to cross zero and remain near zero about as well as in the case of the best materials shown in Fig. 3. Figure 4 shows a highly magnified picture of 90° crossed graphite fibers in a glass matrix, and Fig. 5 shows articles which have been made from this material.

Of the three types of composites mentioned above, graphite/epoxy has the main advantage of economy of manufacture; the graphite/metals, their high stiffness-to-density ratio and high thermal conductivity; and graphite/glass, its stability and low out-of-plane thermal expansion. All composites are

Fig. 5. Articles made from graphite fiber–reinforced glass by hot-forming and machining with a single-point hardened steel tool. (*United Technologies Research Center*)

designed to have low expansion in one or two dimensions. Expansion in the hard-to-control third direction (called out-of-plane) simply depends on the highest-expanding material used, since in this direction there is no cancellation of opposing expansivities (Fig. 4).

The question of stability with respect to time and environmental conditions (vacuum, humidity, temperature cycling) remains to be further investigated with the ultraprecise accuracy now available.

For background information *see* COMPOSITE MATERIAL; CREEP (MATERIALS); GLASS AND GLASS PRODUCTS; GYROSCOPE; OPTICAL TELESCOPE in the McGraw-Hill Encyclopedia of Science and Technology.

[STEPHEN F. JACOBS]

Bibliography: F. Aronowitz, The laser gyro, in M. Ross (ed.), *Laser Applications*, ch. 3, 1971; F. Aronowitz, *Proc. SPIE*, 157:2–6, 1978; J. W. Berthold, III, S. F. Jacobs, and M. A. Norton, *Appl. Opt.*, 15:1898–1899, 1976; J. W. Berthold, III, S. F. Jacobs, and M. A. Norton, *Metrologia*, 13:9–16, 1977; Large new-technology optical telescopes proposed, *Phys. Today*, 34(8):17–19, August 1981; T. Saito (ed.), *Proc. SPIE*, vol. 159: *Precision Machining of Optics*, 1978; *SAMPE Proc.*, National Symposium, 1979.

Water pollution

A research group at the Massachusetts Institute of Technology has completed the first phase of the development of a computer model for assisting in oil spill cleanup operations. The project, known as Oil Spill Cleanup—An Economic and Regulatory Model, began in July 1979 with support from the MIT Sea Grant Program, and has received additional support from the Commonwealth of Massachusetts, the Doherty Foundation, the JBF Scientific Corporation, the National Oceanic and Atmospheric Administration (NOAA), the U.S. Coast Guard (USCG), the U.S. Navy, and the Texaco Corporation. The goal of the project is to examine existing and alternative systems for oil spill cleanup in the United States and create a computerized tool which will provide the user with the ability to analyze complex decisions regarding oil spill cleanup. It is the first model which attempts to integrate all parts of a spill response system and explicitly incorporate analytical descriptors of system performance as well as decision-making techniques. It can be used in a variety of ways, including planning for the long-term response needs of a region, assisting the so-called on-scene coordinators in cleaning up a specific spill, evaluating the environmental and economic damages of a spill versus the cost of cleanup, and helping government and industry to evaluate their options in responding to oil spills.

Oil spill cleanup. Oil pollution of the oceans has already proved to be a complex problem for many industrialized nations around the world. Oil spills occur on a day-to-day basis from a variety of sources: operational discharges from tankers, vessel collisions, offshore blowouts, pipeline ruptures, and others. When an oil spill occurs, emergency action must be taken so that pollution of the sea and the coastline is kept under control. Part of such action involves the dispatching of specialized cleanup equipment to the spill site in order to contain and recover the spilled oil. Equipment may be a combination of booms, which are protective barriers that help prevent the uncontrollable spreading of the oil; skimmers, which are devices that pump and recover the oil; barges, which can be used to transport the recovered oil to a disposal site; and a variety of other devices. In addition, dispersants can be used in order to drive the oil into the water column, and sorbents in order to absorb the slick. The real-world decision maker (in the case of the Coast Guard, the on-scene coordinator) is faced with an extremely complex problem. It is necessary not only to address such issues as availability of equipment, performance degradation with bad weather, and uncertain movement of the oil slick, but also to balance the potentially high and uncertain cost of pollution damage with the similarly high cost of cleanup operations. The problem becomes even more complex for the strategic planner, who must recommend optimal stockpiles of cleanup equipment required to respond to spills which may occur in the future, or for the policy maker, who must assess all the technical, logistical, environmental, socioeconomic, and legal implications of each decision. All parts of the problem are interdependent, and arriving at sound decisions depends on whether the effect of each part of the problem on the whole operation can be accurately evaluated.

The problem addressed by an oil spill pollution response study involves determination of the kind of emergency system needed, within the confines of a specific budget, to respond to oil spill accidents in an optimal manner. "Optimal manner" is subject to many different interpretations, such as a response to any particular spill that is as fast as possible; keeping to a minimum total cost due to oil spills; designing cleanup equipment that operates in the most cost-effective manner; or as set by President

Carter in his March 1977 message to Congress, a response within 6 hours to a spill of 100,000 tons. It is clear that some of these goals are in conflict with one another, some are not well defined, and some, such as the 6-h response time, have turned out to be impossible to implement within the existing budgetary constraints. The goal of the MIT model is to design a system in such a way so as to minimize the expected long-term costs averted if the proposed system is implemented instead of the existing one. "Costs averted" is defined as the net difference of the damages averted by the new system less the additional investment and operating costs of that system. A similar approach is currently discussed in the new Superfund legislation, which specifies that any cleanup operation has to be justified on a cost-benefit basis.

Structure of the model. The overall structure of the model is depicted in the illustration. There are three types of elements: inputs, submodels, and outputs. Inputs provide the decision maker with all the information necessary for the evaluation of the decisions at hand. The submodels are computer algorithms where all the calculations, assessments, and trade-offs are performed. Finally, outputs are decisions, recommendations, and other issues relevant to the specific problem addressed.

Spill incidence model. This is the so-called generator of oil spills. These spills are simulated according to probability distributions concerning their volume, frequency of occurrence, and geographical location. Such probability distributions can be derived by using several methods from data related to past oil spills, for example, the USCG's Pollution Incidence Reporting System (PIRS). In the model, such information is provided from the historical spill data part of the input.

Strategic model. This submodel evaluates planning decisions in response to oil spills that may occur in the future and for a given geographical area. Such strategic decisions involve issues such as locations, quantities, and types of equipment that should be stockpiled in order to cope with the full range of potential oil spills that may occur over a specified period of time. The strategic problem, a special case of what is known in the mathematical programming and operations research literature as the facility location problem, has main applications in urban emergency service systems. The objective function for the problem in question is to minimize the expected total costs from oil spills over the period of consideration, costs which consist of fixed investment costs, spill-specific cleanup costs, and damage costs. Inputs for the strategic model come from the spill incidence model (spill probabilities), from the equipment performance and cost data part of the input, and from the damage assessment model (damage costs from potential spills). The baseline cleanup system part of the input describes the current system for responding to oil spills, against which any system proposed by the strategic model can be compared.

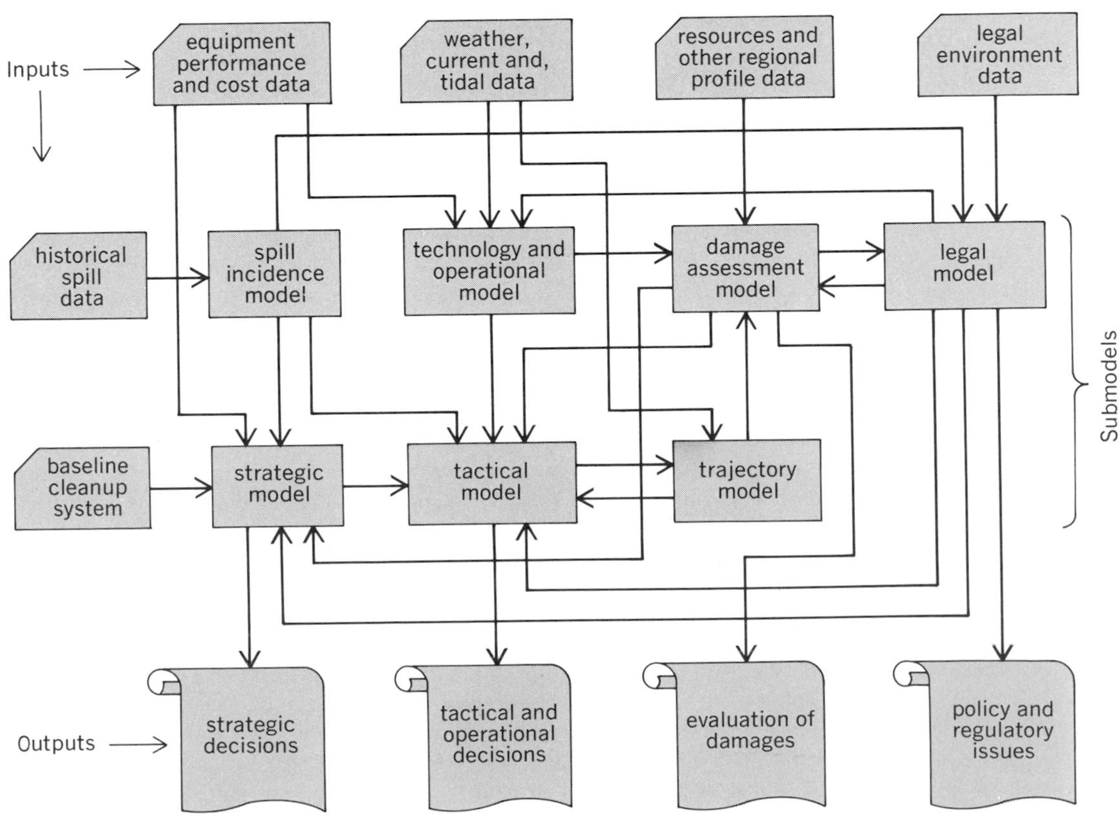

Structure of the oil spill cleanup model.

Tactical model. This addresses decisions to be made upon the occurrence of a specific spill. Such decisions involve the aggregate level of oil recovery capability (that is, gallons of oil recovered per hour) necessary to control the spill. Another consideration is the necessity of dynamic adjustments in time of this level throughout the duration of the spill due to changes in outflow rate and weather conditions (weather, current, and tidal data part of the input). In addition, the tactical model is linked with the technology and operational model to determine what specific sets of cleanup equipment (booms, skimmers, dispersants, and so on) should be dispatched to the spill site from the stockpile of such equipment located in the vicinity of the spill. The latter model provides more detail on actions that should be undertaken at the spill site, taking into account issues such as efficiency and geometric characteristics of booms and skimmers, efficiency of dispersant application, as well as the relevant costs of cleanup. The tactical problem can be modeled as a dynamic programming optimization problem.

Tactical and operational decisions critically depend on what happens, or may happen, to the oil once it enters the marine environment, and vice versa. Two components of the overall model are linked with the tactical model for that purpose: the trajectory model and the damage assessment model.

Trajectory model. This model tracks the movement and spreading of the oil on the sea surface. The processes which act on the oil once it is discharged into the water are numerous. Evaporation, natural dispersion, drift, emulsification, biodegradation, photooxidation, and sedimentation are the most important. Each of these depends on the type of oil as well as the general environmental conditions (weather, current, and tidal data). For instance, while a gasoline spill will evaporate completely after a few hours, a heavy crude spill will tend to form oil-water emulsions that can last for a very long period of time. Also, if the sea state is above a certain level, oil will be totally dispersed into the water. Oil drift is generally assumed to follow the so-called three percent law, according to which the velocity of the centroid of the spill is the vector sum of the current, tidal, and 3% of the prevailing wind velocities.

Damage assessment model. This takes into account the movement of the oil furnished by the trajectory model and evaluates damages to resources impacted as the oil moves through offshore and coastal areas. The resources and other regional profile data part of the input is a region-specific inventory of environmentally and economically sensitive resources, tabulated in a grid format. This submodel evaluates potential damages to marine fisheries, organisms, tourism, and other categories. It is linked to both the tactical and the strategic models.

Legal model. This model takes the legal environment data as inputs. The legal model addresses issues such as the impact of various state, Federal, or international regulations on delay for initiating cleanup efforts, on liability and compensation, and on other policy and regulatory matters. The use of chemical dispersants, an issue widely debated by the U.S. Environmental Protection Agency (EPA), is one of the legal concerns of the model.

Other features. Modularity is one of the important features of the overall model. This will allow replacement of specific submodels in case further progress is made in the corresponding area. The computer code is written in FORTRAN-IV and is implemented in an interactive mode on a VAX/VMS (virtual memory system) computer.

Uses of the model. There are many ways the MIT model can be used. Evaluation and optimization of tactical and operational decisions in an individual spill is one. The usefulness of such a model lies not so much in its ability to solve a complex problem as in its ability to address important "hypothetical" questions. For instance, what would happen if additional quantities of equipment were stockpiled in the geographical area under study? What would happen if cleanup equipment performance deteriorated because of bad weather? What would be the impact on damages and cleanup costs of a policy that excluded dispersants or prohibited the use of U.S. Navy cleanup equipment in a non-Navy spill? Although the oil spill cleanup problem is abundant with data whose values are not well established, extensive sensitivity analysis on those values is the only way to establish the degree of importance of the particular parameter in relation to the overall problem. In that respect, the ability to answer hypothetical questions is valuable in providing feedback to subsequent efforts in areas where data collection is important.

Similar questions can be also addressed in conjunction with other components of the model, such as the strategic and damage assessment levels. Potential users can be Federal regulatory and policy agencies (USCG, EPA, NOAA, U.S. Navy), state agencies, oil companies, spill cleanup cooperatives and contractors, and international bodies such as the International Tanker Owners Pollution Federation or the Intergovernmental Maritime Consultative Organization (IMCO).

For background information *see* WATER POLLUTION in the McGraw-Hill Encyclopedia of Science and Technology. [HARILAOS N. PSARAFTIS]

Bibliography: T. Audunson, The fate and weathering of surface oil from the Bravo blowout, *Mar. Environ. Res.*, vol. 3, 1980; J. W. Devanney and R. J. Stewart, Bayesian analysis of oil spill statistics, *Mar. Technol.*, vol. 11, no. 4, 1974; W. Mead and P. Sorenson, The economic cost of the Santa Barbara oil spill, *Proceedings of the Santa Barbara Oil Symposium*, 1970; J. D. Nyhart, H. N. Psaraftis, and W. S. Laird, The legal environment component of an oil spill cleanup model, *Oil Spill Conference*, Atlanta, 1981; H. N. Psaraftis, A. V. Baird, and J. D. Nyhart, National response capability to oil spills: A systems approach, *OCEANS '80 Conference (IEEE)*, Seattle, 1980.

Wind tunnel

Recent developments in the technology of wind tunnels include the planning and construction of three major new test facilities as elements of a National Aeronautical Facilities Program, and the construction of magnetic suspension and balance systems that suspend wind tunnel models without mechanical support.

National Aeronautical Facilities Program. The wind tunnel is a simulation device used for research to conduct investigations under closely controlled conditions, at lower cost and with less risk than would be involved in flight testing, and for the development of new aircraft to aid in arriving at the best aerodynamic design prior to actual construction. A high degree of confidence in test results is important because of the high cost of resolving problems encountered first in flight or finding flight performance to be less than that predicted. In most simulations, compromises must be made which restrict the range of applicability. Facilities designed to test different components or components in different flow regimes, therefore, differ in character. Historically, when operation in new flight regimes was anticipated, new facilities were required—most notably as flight speed increased.

In the mid 1960s, NASA and the Department of Defense jointly undertook a review of their aerodynamic test facilities and then projected testing requirements to identify serious deficiencies in test capabilities for future aircraft developments. This in-depth review with active participation by the aviation community has led to the identification of three major new aerodynamic test facilities required for future aircraft and engine developments, and they are now under construction as elements of a National Aeronautical Facilities Program. These facilities will be used for research and for the development of both commercial and military aircraft.

80 × 120 ft subsonic wind tunnel. Deficiencies in existing wind tunnels resulting in an inability to obtain highly reliable force and moment data on vertical/short takeoff and landing (V/STOL) aircraft in hover and transition to forward flight and on the dynamics loads of helicopters at high forward flight speeds led to the development of this new facility. V/STOL aircraft operating in these modes are characterized by a large downward deflection of the airstream. The presence of wind tunnel walls constrains the deflected airflow and induces a modification of the pressures acting on the test article. For normal aircraft, tunnel wall corrections are modest and can be predicted with acceptable accuracy. For the highly deflected flows of V/STOL aircraft in these operating modes, the wall effects are larger and the accuracy of prediction is poorer. For powered-lift V/STOL aircraft, the propulsion system jet exhaust or propeller slipstream strongly influences the flow over the wing, fuselage, and tail surfaces. Simulation of these influences using scale models is subject to question, and the greatest confidence in test results requires a full-scale vehicle with an operating propulsion system.

High-forward-speed flight of helicopters increases the unsteady dynamic loads on the rotor resulting from the different relative velocity encountered by the advancing and retreating blades on opposite sides of the rotor disk. As helicopters operate at higher forward-flight speeds, control of these loads and indeed safe operation becomes more difficult. Full-scale wind-tunnel testing is the key to the successful design of higher-speed helicopters because the scaling of the flow over the blades cannot be done with confidence. The most economic means for providing these new test capabilities was determined to be through modifications and additions to the existing 40 × 80 ft (12 × 24 m) subsonic wind tunnel at the NASA Ames Research Center, Moffett Field, CA (Fig. 1). The original set of six 6000-hp (4.5-MW) drive motors is being replaced by higher-power units, increasing the drive horsepower from 36,000 to 135,000 (27 to 101 MW). When operating with the existing test section, the test velocity will be increased from 200 to 300 knots (100 to 150 m/s), providing for test of full-scale helicopters at high forward-flight speeds. The quieter new wind tunnel drive system will also enhance the quality of acoustic measurements on helicopters and V/STOL aircraft.

The new facility will incorporate adjustable turning vanes and louvers upstream of the drive motors and air exit louvers and guides downstream. Use of these devices will permit the drive system to induce airflow through a new 80 × 120 ft (24 × 37 m) test section and exhaust that flow through the air exit, resulting in a nonreturn operating mode. The maximum velocity in the new test section will be 100

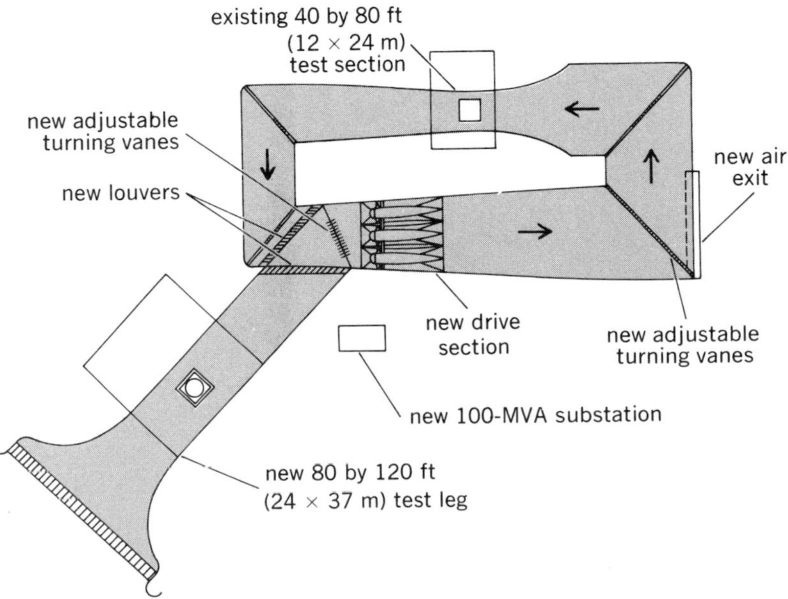

Fig. 1. Modification to the NASA Ames Research Center 40 × 80 ft (12 × 24 m) subsonic wind tunnel. (*NASA Ames Research Center*)

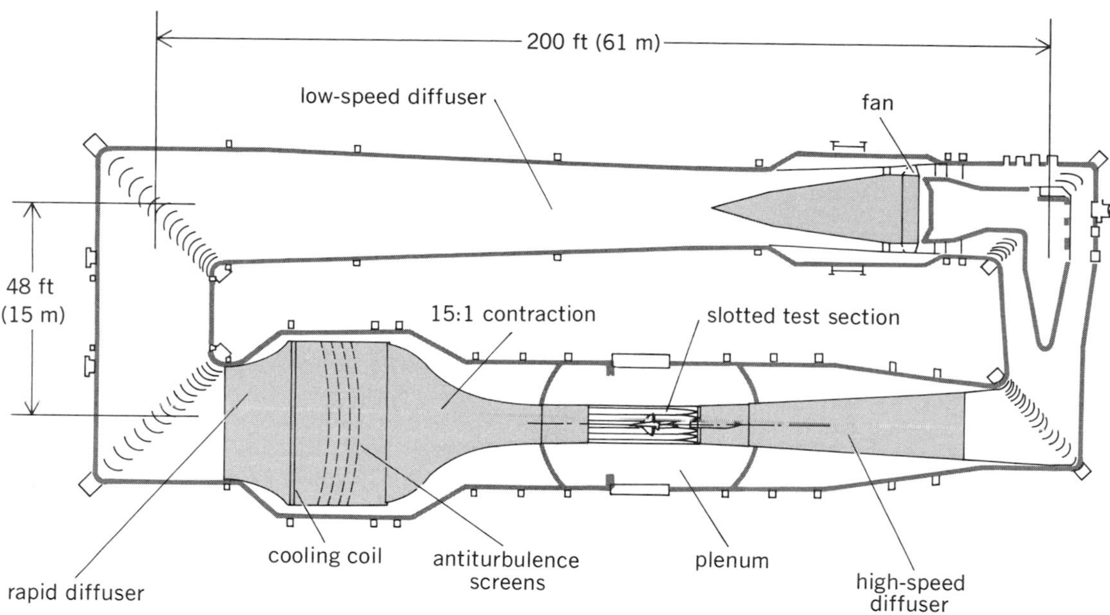

Fig. 2. Plan of tunnel circuit at the National Transonic Facility. (*From R. R. Howell, The National Transonic Facility. Status and Operational Planning, paper presented at the AAIA 11th Aerodynamic Testing Conference, Colorado Springs, March 18–20, 1980, AAIA 80–0415 pp. 1–9, 1980*)

knots (50 m/s), sufficient for testing V/STOL aircraft in hover and through the critical transitional flight regime.

National transonic facility. A major deficiency in existing transonic test facilities has been identified as an inability to provide high test Reynolds numbers representative of flight at transonic speeds. The National Transonic Facility (NTF) is being constructed at the NASA Langley Research Center, Hampton, VA, to overcome this deficiency.

Reynolds number is a scaling factor in the sense that two steady fluid flows are similar if the Reynolds number is the same for both. It represents the ratio of inertial forces to viscous forces in the flow, and is defined as $R = \rho Vl/\mu$, where ρ is the mass density, V the flow velocity, l a characteristic length, and μ the coefficient of viscosity. It is a very important parameter in the modeling of flows about flight vehicles because the viscous surface flow is extremely important in determining the resultant forces and moments on the test article. Particular problems have been encountered in testing at transonic speeds where local embedded or recompression shock waves occur near the surface of the vehicle and influence the coupling of the viscous surface flow with the external flow field.

Increased test Reynolds number can be achieved in a number of ways by varying separately or in combination the factors in the Reynolds number equation. The concept that is embodied in the National Transonic Facility is to influence ρ, V, and μ by greatly reducing the fluid temperature. This is done by expanding liquid nitrogen directly into the tunnel to cool the test gas. Lowering the fluid temperature results in a large increase in ρ and a large decrease in μ, which more than offset a reduction in V for the same test Mach number that occurs because the speed of sound is reduced. The reduction of the stagnation temperature from 322 K (120°F) to 100 K (−280°F) increases the test Reynolds number by a factor of approximately 5. A major advantage of this approach to high test Reynolds numbers is that the reduction of the velocity V compensates for the increase in density ρ, resulting in essentially the same model loads at five times the test Reynolds number (since the loads are proportional to ρV^2). The wind tunnel drive power required, which is proportional to ρV^3, is reduced by about 50%. Additional energy, however, must be expended to provide the cryogenic nitrogen, which may be thought of as an energy storage medium. An extremely valuable feature of the National Transonic Facility results from the ability to vary the stagnation temperature. This feature provides the capability to conduct Reynolds number studies at constant Mach number and dynamic pressure (hence, constant model distortion under load), aeroelastic studies at constant Reynolds number and Mach number (by varying dynamic pressure), and Mach number studies at constant Reynolds number and dynamic pressure. This is a new and unique capability not found in existing large wind tunnels.

A plan view of the National Transonic Facility is shown in Fig. 2. Not shown are the drive motor building and the nitrogen storage facilities. Isolation valves (not shown) are incorporated so that the test section and the plenum that surrounds it may be isolated, purged, and recharged with fresh air, allowing entrance of crews to make model or instrumentation changes without purging the entire cir-

Table 1. Operating characteristics of the National Transonic Facility

Characteristic	Value
Mach number	0.1–1.2
Reynolds number	3,000,000–120,000,000
Pressure	1–8.85 atm (101–897 kPa)
Temperature	78–339 K
Test section size	2.5 m square
Length	7.62 m
Drive horsepower	126,000 (94 MW)

cuit. The circuit is heavily insulated to minimize heat flow from the outside to the cold test medium.The operating characteristics are given in Table 1.

Aeropropulsion Systems Test Facility. The design and development of aircraft engines also require a realistic simulation of the true operating environment to assure achievement of the desired performance in a timely and cost-effective manner. To accomplish this, air density and temperature and flight Mach number must be simulated, as well as the true flow field which the engine will encounter in service. The local flow field in the region of the engine air inlet is influenced by the presence of certain aircraft elements, particularly those forward of the inlet location, and by flow changes resulting from maneuver or the encounter of gusts. The effects of these time-dependent variations are particularly severe in the case of combat aircraft capable of rapid acceleration and maneuver.

The Aeropropulsion Systems Test Facility (ASTF), being constructed at the U.S. Air Force's Arnold Engineering Development Center in Tullahoma, TN, will provide such simulation of flight conditions for current and advanced supersonic turbojet and large subsonic turbofan propulsion systems. The facility will provide for direct-connect testing in two separate test cells, and provisions are made for incorporating free-jet test installations. Direct-connect testing simulates the airflow conditions at the engine compressor face. Steady-state pressure distortion patterns, discrete-frequency pressure fluctuations, or turbulent flow patterns can be simulated in this mode of testing. The facility is designed to accomodate free-jet adjustable-attitude nozzles located ahead of an aircraft propulsion system air inlet to simulate the external flow field at various flight attitudes. This test mode requires two to three times the airflow required by the engine, and is used for simulation of a range of flight conditions for a supersonic aircraft propulsion system. A model of the Aeropropulsion Systems Test Facility is shown in Fig. 3. The air supply building houses the air compressor equipment. The compressed air is then either cooled or heated to provide the proper altitude airflow conditions at the model location, and is introduced to the engine in one of the test cells. An exhaust gas cooler is required because the exhaust flow is heated by the engine, and the exhauster building equipment provides suction to simulate the appropriate altitude ambient pressure. The operating characteristics of the facility are given in Table 2.

Table 2. Operating characteristics of the Aeropropulsion Systems Test Facility

Characteristic	Value
Airflow	1450 lb/s (658 kg/s)
Air temperature	200–822 K
Refrigeration capacity	23,000 tons (81 MW)
Installed horsepower	611,000 (456 MW)
Test cell dimensions	
Diameter	8.53 m
Length	25.91 m

[J. LLOYD JONES]

Magnetic suspension and balance systems. Magnetic fields can be used to suspend wind tunnel models without support interference while measuring air loads which can be scaled to free-flight values. The scaling laws that relate aerodynamic forces and moments acting on scale models in wind tunnels to those aerodynamic forces and moments acting on the full-scale flight vehicle are reasonably well known. But actual application of these laws is limited by several practical restrictions, one of which is aerodynamic interference from the mechanical support itself.

Choice of nonmechanical support. The basic principle of developing interference-free support of the model is relatively simple in concept—action at a distance. The two most obvious possibilities of this kind are electrostatic or magnetostatic fields. A third choice could be either an acoustic or electromagnetic radiation field. However, at the present time the level of technical development of radiation generation devices is too low to allow practical employment of radiation suspension. Although laser pulses can be directed at micrometer-sized particles so that the particle is suspended against gravity for very short periods of time, the particle does not survive the experiment. For this reason, electromagnetic radiation does not seem to be a practical source for the suspending force for reasonable-sized models and for extended periods of time. The lack of a suitable source causes the same conclusion to be reached about the use of acoustic radiation. Thus electric or magnetic fields seem to be more attractive for this application. Electric fields and gradients of electric fields can be generated that are strong enough to suspend extended models, but the use of these fields is subject to limitations of a different kind. First is the lack of practical quantities of materials that can be charged permanently, as the electric analog of the permanent magnet does not possess comparable properties in terms of strength or lifetime. Further, many materials, including working fluids suitable for use in wind tunnels, tend to break down or change their nature in the presence of strong electric field gradients. Consequently, the only practical source of long-range

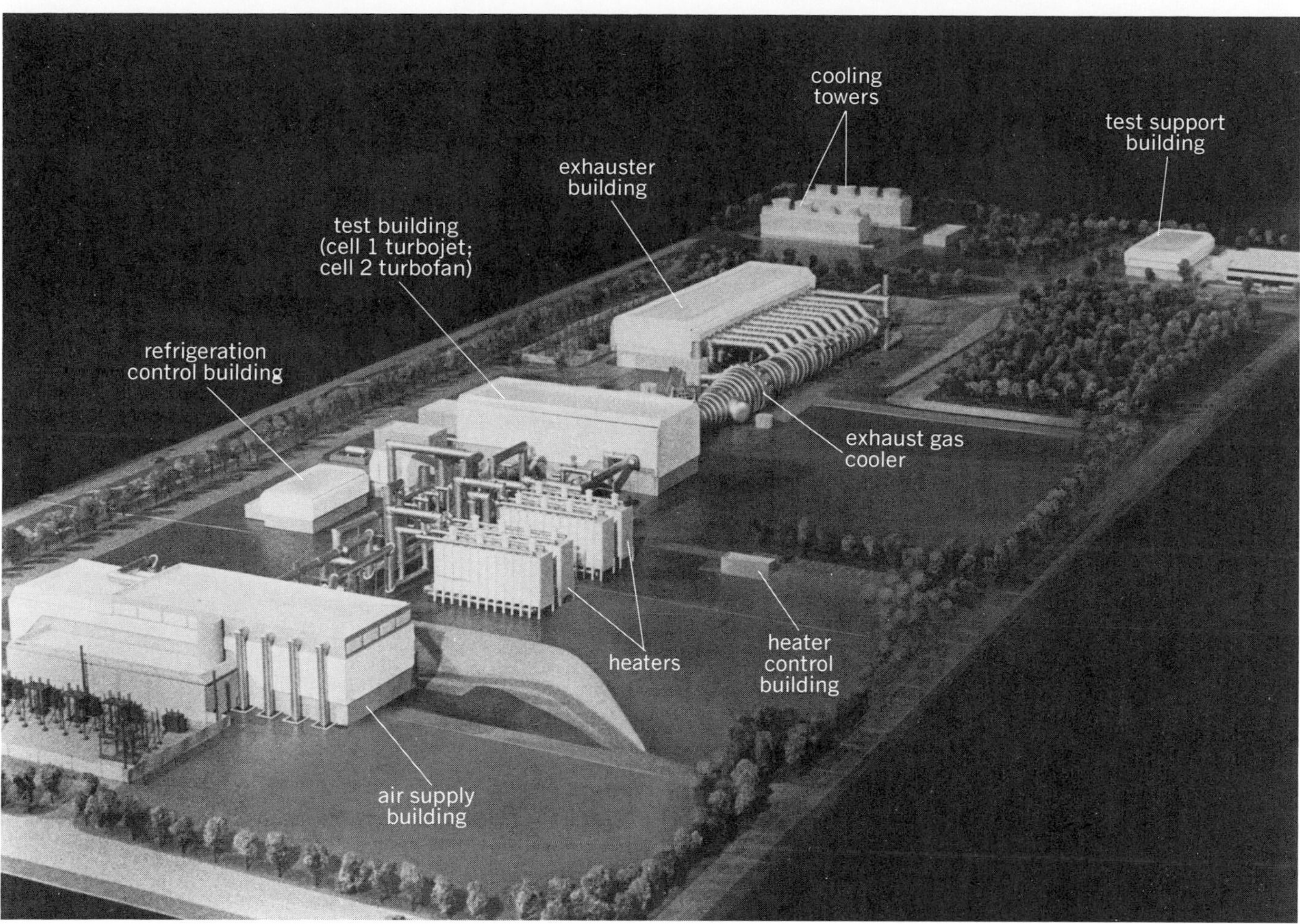

Fig. 3. Model of the Aeropropulsion Systems Test Facility. (*From W. R. Bates, The Aeropropulsion Systems Test Facility, paper presented at the AIAA/SAE 13th Propulsion Conference, Orlando, FL, July 11–13, 1977, AIAA 77–913, 1977*)

forces able to suspend models being tested in wind tunnels is magnetic.

Diamagnetic and paramagnetic suspension. There are two different ways that magnetic fields interact with matter. If the nature of the matter is such that it tends to exclude magnetic field lines, for whatever reason, the material is said to be diamagnetic. It is easy to envision a scheme to suspend a diamagnetic model. In one such configuration an intense magnetic field is developed such that there is a local minimum field in a fixed volume of space. This minimum exists over an area that is more or less the size of the object to be suspended. The diamagnetic model can be "floated" or suspended against gravity in the cup-shaped region where the local minimum exists. The stronger fields act like a wall to restrict lateral displacement of the diamagnetic model. Diamagnetic suspensions are useful only at temperatures where superconductivity exists, since there are no materials whose diamagnetic constant is sufficiently large at room temperature. In other words, superconductors act very nearly as a perfect diamagnetic material. Superconducting bodies allow circulating or eddy currents to flow at the body's boundary, and these eddy currents in turn induce a magnetic field that cancels the applied field on the inside of the body. Hence the applied field seems to be excluded from the body's interior. This line of reasoning suggests that one can use alternating currents to induce eddy currents in the model and thereby support it against gravity and the aerodynamic forces and moments. Unfortunately, unless the model is made of super-conducting materials, the heat generated by the alternating eddy currents cannot be dissipated effectively, so ultimately the model will melt. Applying this principle, E. C. Okress has developed vacuum furnaces in which material to be melted is suspended by the diamagnetic effect produced by the eddy currents. In this way the melt is produced without contamination from the furnace walls.

There is another class of materials called paramagnetic, which act to attract the applied magnetic field. Hence the internal magnetic field is more intense than the external magnetic field. A special class of paramagnetic material is called ferromagnetic. This class of material, which includes iron and nickel, acts almost like the antithesis of the perfect diamagnetic material. In the appropriate magnetic field, ferromagnetic materials can be magnetized to relatively high strengths. Some of these substances can also retain the induced magnetism for very long time periods, and these are used to make permanent magnets. Ferromagnetic materials acting in conjunction with independently powered electromagnets constitute the basis for the magnetic suspension systems that are in use in wind tunnels today.

The first practical magnetic suspension and balance for use with wind tunnels was constructed in 1957 at ONERA (Office National d'Etudes et de Recherces) in Paris by M. Tournier and P. Laurenceau. Their results encouraged experiments in England and the United States.

Position sensing and control. A support-interference-free model suspension apparatus needs magnetic and other elements to operate successfully. Since no material support exists, a model position sensor is needed. The position sensor serves two needs. The first need is to define the orientation of the model with respect to the airstream. This is vital because the aerodynamic forces and moments are very sensitive to model orientation. The second need is equally important. The model suspension apparatus based on ferromagnetic materials and electromagnets is basically unstable. That is to say, even if the magnetic forces and moments acting on the model exactly balance the gravity and aerodynamic forces and moments, the resulting equilibrium is unstable to any kind of change. Any perturbation in the electric currents in the coils, in the airstream, or in the sensed position will cause the model to move away from the equilibrium state. The model's motion away from the desired location and orientation causes the position sensor to generate an error signal. This is used to change the current in the appropriate electromagnet. In this way the model can be accelerated back toward the equilibrium position. Determining just the proper change in the current flowing in the electromagnet is necessary but can be difficult. Consequently, another important element of the magnetic suspension apparatus is the automatic control. A cyclic or closed-loop arrangement is shown in Fig. 4. It is useful to describe the operation of this control.

For the sake of convenience, assume that a signal proportional to position and orientation is compared to the desired or set position and orientation signal. The difference between the actual and desired signal is called the error signal. It is amplified and conditioned in the automatic controller, and the modified signal is fed to the appropriate power supply, which provides a modified current flow in the electromagnet coils. This flow in the coil induces a modified magnetic field that in turn causes a change in the net force and moment on the model. The changed force and moment field act in direction and magnitude to cause a motion that tends to reduce the error signal to zero. This process is repeated continuously, with the result that the model is maintained at the desired position and orientation within allowable error. Of course, the actual pro-

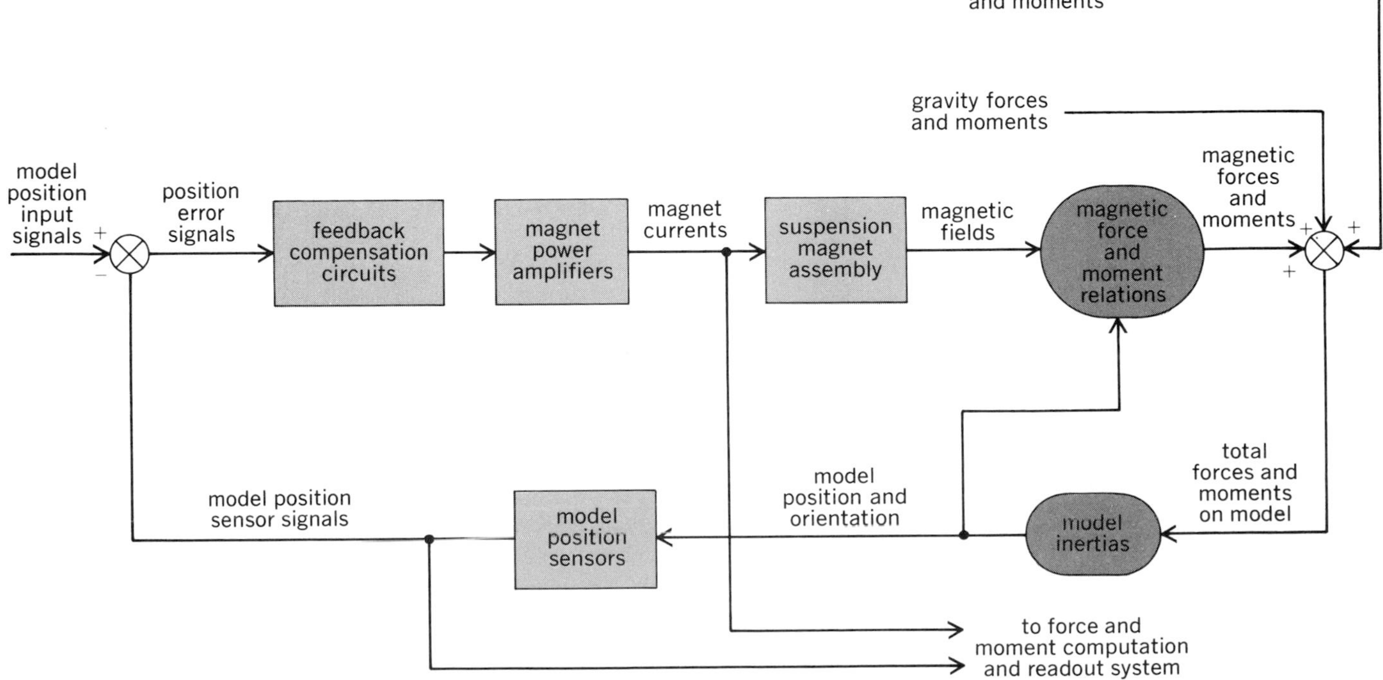

Fig. 4. Closed-loop automatic control system for magnetic suspension of a wind tunnel model.

cess is complicated because the model can pitch, roll, and yaw, and it can move laterally, vertically, and longitudinally. Thus the sensor must be able to sense six simultaneous components of motion.

The forces and moments acting on the model in the wind stream are deduced from measurements of the currents flowing in each of the coils. In principle, the relation between coil currents and the applied forces and moments can be found theoretically. In practice, it is much easier to derive the relation between the coil currents and the forces and moments through a calibration procedure.

Coil design. At the present time, nearly all of the operational systems use coils of copper wire. In the future, superconducting wire may be used for those applications where the special properties of superconductors can be used to advantage either to reduce the amount of current consumed or to reduce the volume of a particular coil. The design of coils is now understood to the extent that designers of magnetic suspension and balance systems have a great deal of flexibility and can tailor the design quite closely to the intended use.

Model design. The models should be made either of a permanently magnetized material like ferrite or samarium cobalt or of a material that has as low a magnetic retentivity as possible, such as pure, soft iron. If the latter material is selected, the model may be totally constructed from it or the model may have a core in the shape of an ellipsoid of revolution. This shape of the core has the property that it can be magnetized uniformly; thus it is also a useful shape for permanently magnetized cores. In either case, these material choices simplify the calibration procedures by providing stable repeatable calibration constants. The choice of materials is dependent on the particular application.

Sensing equipment. The nature of the position sensor is based upon the use of partially interrupted light or other electromagnetic beams, or upon the use of an additional unsteady magnetic field that senses the position of the model through coupling between several coils. In the latter scheme the model acts like the core in a differential transformer. In either scheme the model position is sensed to less than a thousandth of an inch or a few hundredths of a millimeter. Similarly the angular orientation of the model can be measured to between 0.01 and 0.02° on a regular basis. This precision in the information about the position and orientation of the model, when coupled with the precise instruments that are now available to measure electric currents, provides this apparatus with the potential to determine aerodynamic data very accurately. Several very promising experiments have recently been conducted that suggest that magnetic suspension and balance systems will be of great value in the future for measurement of unsteady aerodynamic characteristics.

For background information *see* CONTROL SYSTEMS; MAGNETISM; REYNOLDS NUMBER; TRANSONIC FLIGHT; WIND TUNNEL in the McGraw-Hill Encyclopedia of Science and Technology.

[EUGENE E. COVERT]

Bibliography: D. D. Baals (ed.), *High Reynolds Number Research*, NASA CP-2009, 1977; W. R. Bates and Z. B. Edwards, *The Aeropropulsion Systems Test Facility*, AIAA Pap. no. 77–913, 1977; E. E. Covert et al., Magnetic balance and suspension systems for use with wind tunnels, *Process in Aerospace Sciences*, vol. 14, pp. 27–107, 1973; C. W. Haldeman et al., *Application of Superconducting Coils to the NASA Prototype Magnetic Balance*, NASA Contractor Rep. 165660, January 1981; C. W. Haldeman and E. E. Covert, New techniques for production of spinning and coning motion with magnetically suspended wind tunnel models, *Proceedings of the 8th International Congress on Instrumentation in Aerospace Simulation Facilities*, 1979; W. B. Igoe, *Characteristics and Status of the U.S. National Transonic Facility*, AGARD Lec. Ser. 111, Cryogenic Wind Tunnels, 1980; K. W. Mort, P. T. Soderman, W. T. Eckert, *AIAA J. Aircraft*, 16(8):571–575, August 1979; M. Tournier and P. Laurenceau, *La Recherche Aeronaut.*, no. 59, pp. 21–27, July/August 1957.

McGRAW-HILL YEARBOOK OF SCIENCE AND TECHNOLOGY

List of Contributors

List of Contributors

A

Alford, Dr. N. McN. *Department of Metallurgy and Science of Materials, Oxford University, England.* CEMENT—coauthored.

Altamuro, Vincent M. *President, Management Research Consultants, Yonkers, NY.* AUTOFACTURING—feature.

Axel, Dr. Peter. *Department of Physics, University of Illinois.* PARTICLE ACCELERATOR.

B

Badgley, Dr. Catherine E. *Department of Anthropology, Peabody Museum, Harvard University.* TAPHONOMY.

Baker, Dr. Paul A. *Department of Geology, Duke University.* SEDIMENTARY ROCKS—coauthored.

Balas, Prof. Mark J. *Electrical, Computer, and Systems Engineering Department, Rensselaer Polytechnic Institute.* CONTROL SYSTEMS.

Bandurski, Dr. Robert S. *Botany and Plant Pathology Department, Michigan State University.* PLANT HORMONES—coauthored.

Bank, Dr. Arthur. *Professor and Acting Chairman, Department of Human Genetics, and Director, Division of Hematology, Columbia University.* SICKLE CELL DISEASE.

Barger, Prof. Vernon. *Department of Physics, University of Wisconsin.* NEUTRINO.

Barrett, Dr. Gordon S. *EMTECH Corporation, Caldwell, ID.* PRINTING.

Beatty, Dr. J. Kelly. *"Sky & Telescope," Sky Publishing Corporation, Cambridge, MA.* SATURN.

Beene, Dr. James R. *Physics Division, Oak Ridge National Laboratory.* NUCLEAR ORIENTATION—coauthored.

Bemis, Dr. Curtis E., Jr. *Physics Division, Oak Ridge National Laboratory.* NUCLEAR ORIENTATION—coauthored.

Bigger, Dr. Charles H. *Department of Microbiology and Immunology, School of Medicine and Dental Research Institute, University of California, Los Angeles.* SEA ANEMONE.

Boucot, Dr. Arthur J. *Department of Geology, Oregon State University.* BRACHIOPODA.

Bousfield, Dr. Edward L. *Senior Scientist, National Museum of Natural Sciences, Ottawa, Ontario, Canada.* AMPHIPODA.

Braslow, Albert L. *Newport News, VA.* LAMINAR FLOW.

Breazeale, Prof. Mack A. *Department of Physics, University of Tennessee.* NONLINEAR ACOUSTICS.

Buell, Heinz. *Kearfott Division, The Singer Company, Wayne, NJ.* DOPPLER NAVIGATORS.

Buffum, Harvey E. *Director, Operations Technology, Boeing Commercial Airplane Company, Seattle, WA.* AIRCRAFT MANUFACTURING PRODUCTIVITY.

Burns, David M. *Director, Climate Project, American Association for the Advancement of Science, Washington, DC.* THE PLANETARY GREENHOUSE—feature.

Burns, Dr. Roger G. *Professor of Mineralogy and Geochemistry, Department of Earth and Planetary Sciences, Massachusetts Institute of Technology.* MARINE MINING.

C

Carroll, Robert G., Jr. *Celanese Corporation, Charlotte, NC.* GEOTEXTILES.

Chaffin, Dr. R. J. *Sandia National Laboratories, Albuquerque, NM.* TRANSISTOR.

Chestnut, Dr. Charles, III. *School of Medicine, University of Washington.* OSTEOPOROSIS.

Clarke, Dr. Malcolm R. *The Laboratory, Marine Biological Association of the United Kingdom, Plymouth, England.* CEPHALOPODA.

Cohen, Aaron. *NASA Lyndon B. Johnson Space Center, Houston, TX.* SPACE SHUTTLE.

Cohen, Dr. David. *National Magnet Laboratory, Massachusetts Institute of Technology.* BIOMAGNETICS.

Cohen, Dr. Jerry D. *Plant Biochemist, Beltsville Agricultural Research Center, U.S. Department of Agriculture.* PLANT HORMONES—coauthored.

Coleman, Dr. J. M. *Coastal Studies Institute, Louisiana State University.* MASS WASTING—coauthored.

Conlon, Dr. T. W. *Nuclear Physics Division, AERE, Harwell, Oxfordshire, England.* ACTIVATION ANALYSIS.

Covert, Prof. Eugene E. *Department of Aeronautics and Astronautics, Massachusetts Institute of Technology.* WIND TUNNEL—in part.

Cowley, Martyn B. *Simi Valley, CA.* SOLAR-POWERED AIRCRAFT.

Crandell, Dr. Dwight R. *Geological Survey, U.S. Department of the Interior, Denver, CO.* VOLCANO—feature; in part.

Cronin, Dr. Thomas W. *Department of Biology, Yale University.* CRAB—coauthored.

Cross, M. C. *Bell Telephone Laboratories, Murray Hill, NJ.* QUANTUM SOLIDS.

D

Dan, Dr. Bruce. *Bacterial Diseases Division, Center for Disease Control, Atlanta, GA.* TOXIC SHOCK SYNDROME.

Datta, Dr. Vijay K. *Gibbs & Hill Engineers, New York, NY.* COGENERATION.

Davies, Prof. D. R. *The John Innes Institute, Norwich, England.* BREEDING (PLANT)—in part.

Dayanandan, Dr. P. *Division of Biological Sciences, University of Michigan.* GRAVITROPISM—coauthored.

Dehmelt, Dr. Hans. *Department of Physics, University of Washington.* ELECTRON.

Delwiche, Dr. C. C. *Department of Land, Air, and Water Resources, College of Agricultural and Environmental Sciences, University of California, Davis.* NITROGEN CYCLE—in part.

DeMason, Dr. Darleen A. *Department of Botany and Plant Sciences, University of California, Riverside.* STEM.

Deuser, Dr. W. G. *Department of Chemistry, Woods Hole Oceanographic Institution.* MARINE SEDIMENTS—in part.

Dodd, J. L. *Natural Resource Ecology Laboratory, Colorado State University.* GRASSLANDS—coauthored.

Double, Dr. D. D. *Department of Metallurgy and Science of Materials, Oxford University, England.* CEMENT—coauthored.

Dunn, Dr. S. Thomas. *President, Dunn Technology Incorporated, Vista, CA.* LASER IMAGING.

Duxbury, Dr. John. *Department of Agronomy, Cornell University.* NITROGEN CYCLE—in part.

E

EerNisse, Dr. Errol P. *President, Quartex, Inc., Salt Lake City, UT.* ACCELEROMETER.

El-Sayed, Prof. Sayed Z. *Department of Oceanography and Geosciences, Texas A&M University.* KRILL.

F

Farrell, Ronald. *General Manager, Powder Coatings Region, Glidden Chemical Coatings, SCM Corporation, Charlotte, NC.* POWDER COATING.

Felker, Dr. Peter. *Caesar Kleberg Wildlife Research Institute, Texas A&I University.* LEGUMINOUS TREES.

Ferry, Dr. David K. *Department of Electrical Engineering, Colorado State University.* MICROSCOPIC SOLID STRUCTURES.

Fink, Lester H. *Systems Engineering for Power Incorporated, Oakton, VA.* ELECTRIC POWER SYSTEMS.

Forward, Dr. Richard B., Jr. *Marine Laboratory, Duke University.* CRAB—coauthored.

Fountain, Dr. David M. *Department of Geology, University of Montana.* PETROLOGY.

Fudenberg, Dr. H. Hugh. *Professor and Chairman, Department of Basic and Clinical Immunology and Microbiology, Medical University of South Carolina, Charleston.* IMMUNOLOGY—coauthored.

G

Gibbs, Prof. Hyatt. *Optical Sciences Center, University of Arizona.* OPTICAL BISTABILITY.

Giunipero, Dr. Larry C. *Management Department, College of Business, Florida State University.* MATERIAL REQUIREMENTS PLANNING.

Goette, Dr. Eckart E. *Midrex Corporation, Charlotte, NC.* IRON METALLURGY.

Goldhaber, Dr. Maurice. *Brookhaven National Laboratory, Upton, NY.* PROTON—in part.

Green, Dr. David E. *Institute for Enzyme Research, University of Wisconsin.* CELL MEMBRANES.

Grivell, Prof. L. A. *Afdeling Moleculaire Biologie, Laboratorium voor Biochemie, Universiteit van Amsterdam, Netherlands.* MITOCHONDRIA.

Grot, Dr. Walther G. *Plastics Products and Resins Department, E. I. du Pont de Nemours & Company, Wilmington, DE.* ION EXCHANGE.

Grover, Kenneth M. *President, GSA International Corporation, Katonah, NY.* HYDROELECTRIC POWER.

Guilbault, Dr. George G. *Department of Chemistry, University of New Orleans.* ENVIRONMENTAL CHEMISTRY—coauthored.

H

Harlan, Prof. Jack R. *Department of Agronomy, University of Illinois.* BREEDING (PLANT)—in part.

Hastings, Larry G. *U.S. Department of Energy, Washington, DC.* SPACE FLIGHT.

Hayes, William C. *Editor in Chief, "Electrical World," McGraw-Hill, New York, NY.* ELECTRICAL UTILITY INDUSTRY.

Heeger, Dr. Alan J. *Department of Physics, University of Pennsylvania.* CONDUCTING POLYMERS—coauthored.

Hirsch, Dr. Jack H. *Ultrasound Laboratory, University Hospital, University of Washington.* ULTRASONICS.

Hodgson, Dr. Harlow J. *Agricultural Experiment Station, University of Wisconsin.* ENERGY SOURCES.

Hopwood, Prof. D. A. *Head, Department of Genetics, The John Innes Institute, Norwich, England.* BACTERIAL GENETICS.

Houghton, Dr. Donald C. *Department of Pathology, Medical School, University of Oregon.* LUPUS ERYTHEMATOSUS.

Hsu, Dr. Kenneth J. *Geological Institute, Swiss Federal Institute of Technology, Zurich.* EXTINCTION (BIOLOGY).

I

Iachello, Prof. Francesco. *Department of Physics, Yale University.* SYMMETRY LAWS (PHYSICS).

J

Jacobs, Prof. Stephen F. *Optical Sciences Center, University of Arizona.* ULTRASTABLE MATERIALS.

Janata, Dr. Jiri. *Department of Bioengineering, University of Utah.* ION-SELECTIVE MEMBRANES AND ELECTRODES.

Johnson, Dr. Paul. *Center for Research in Human Learning, University of Minnesota.* PROBLEM SOLVING (PSYCHOLOGY).

Jones, J. Lloyd. *Los Altos, CA.* WIND TUNNEL—in part.

Jones, Dr. John L. W. *Technical Service Manager, Photo Products Department, E. I. du Pont de Nemours & Company, Wilmington, DE.* PHOTOGRAPHIC MATERIALS.

Jones, Prof. Lawrence W. *The Harrison M. Randall Laboratory of Physics, University of Michigan.* QUARKS.

Jones, Prof. R. Alan. *School of Chemical Sciences, University of East Anglia, Norwich, England.* CATALYSIS.

Jury, Dr. William A. *Department of Soils and Agronomy, University of California, Riverside.* SOIL-WATER RELATIONS—in part.

K

Karwacki, Francis A. *Communication Navigation Technology Directorate, Naval Air Development Center, Warminster, PA.* GYROSCOPE.

Kastner, Dr. Miriam. *Scripps Institution of Oceanography, La Jolla, CA.* SEDIMENTARY ROCKS—coauthored.

Kaufman, Dr. Peter B. *Division of Biological Sciences, University of Michigan.* GRAVITROPISM—coauthored.

Kay, Dr. Richard F. *Department of Anatomy, School of Medicine, Duke University.* RAMAPITHECINES AND HUMAN ORIGINS—feature.

Klein, Prof. Richard M. *Department of Botany, University of Vermont.* ACID RAIN; BONSAI.

Klitgord, Dr. Kim D. *Geological Survey, U.S. Department of the Interior, Woods Hole, MA.* MARINE GEOLOGY.

Kristof, Emory. *National Geographic Society, Washington, DC.* ELECTRONIC CAMERA.

Kuhlman, Dr. E. G. *Forestry Sciences Laboratory, U.S. Department of Agriculture, Research Triangle Park, NC.* PLANT PATHOLOGY—in part.

Kvenvolden, Dr. Keith A. *Geological Survey, U.S. Department of the Interior, Menlo Park, CA.* MARINE SEDIMENTS—in part.

L

Landshoff, Dr. Peter. *Department of Applied Mathematics and Theoretical Physics, University of Cambridge, England.* PROTON—in part.

Lindsley, Dr. Donald H. *Department of Earth and Space Sciences, State University of New York, Stony Brook.* PYROXENE.

Lines, Dr. Malcolm E. *Bell Telephone Laboratories, Murray Hill, NJ*. MAGNETIC FERROELECTRICS.

Lord, Dr. Elizabeth. *Department of Botany and Plant Sciences, University of California, Riverside*. FLOWER.

Lorian, Dr. Victor. *Chairman, Department of Pathology, Division of Microbiology and Epidemiology, Bronx-Lebanon Hospital Center; Professor of Laboratory Medicine, Albert Einstein College of Medicine*. ANTIBIOTIC.

Lucchesi, Dr. Claude. *Department of Chemistry, Northwestern University*. ANALYTICAL CHEMISTRY.

Lumpkin, Thomas. *Consultative Group on International Agricultural Research, Washington, DC*. FERN—coauthored.

M

MacDiarmid, Prof. Alan G. *Department of Chemistry, University of Pennsylvania*. CONDUCTING POLYMERS—coauthored.

Macdonald, Prof. Kenneth C. *Department of Geological Sciences, University of California, Santa Barbara*. OCEANIC HOT SPRINGS—feature.

McFarlane, Robert. *Philips Laboratories, Briarcliff, NY*. COMPUTER STORAGE TECHNOLOGY.

Mallet, Dr. V. N. *Chemistry Department, Université de Moncton, New Brunswick, Canada*. ENVIRONMENTAL CHEMISTRY—coauthored.

Mansouri, Dr. Freydoon. *Department of Physics, Yale University*. SUPERGRAVITY.

Meagher, Dr. E. P. *Department of Geological Sciences, University of British Columbia, Vancouver*. MINERALOGY.

Meyer, Dr. Henry O. A. *Department of Geosciences, Purdue University*. DIAMOND.

Miller, Dr. Edward E. *Departments of Physics and Soil Science, University of Wisconsin*. SOIL-WATER RELATIONS—in part.

Mitchell, Dr. J. E. *Department of Plant Pathology, University of Wisconsin*. PLANT PATHOLOGY—in part.

Mitchell, Dr. Rodger. *Department of Zoology, Ohio State University*. AGRICULTURAL SYSTEMS.

Mullineaux, Dr. Donal R. *Engineering Geology Branch, Geological Survey, U.S. Department of the Interior, Denver, CO*. VOLCANO—feature; in part.

N

Neu, Dr. Harold C. *Professor of Medicine and Pharmacology, and Chief, Division of Infectious Diseases, College of Physicians and Surgeons, Columbia University*. CEPHALOSPORINS.

Nicoll, Dr. Charles S. *Departments of Physiology and Anatomy, University of California, Berkeley*. PROLACTIN.

Nola, Frank J. *NASA Marshall Space Flight Center, Huntsville, AL*. POWER FACTOR CONTROLLERS.

Norman, Prof. Eric B. *Department of Physics, University of Washington*. ELECTRIC CHARGE CONSERVATION.

Nosanow, Prof. Lewis H. *Office of the Provost, University of Chicago*. HYDROGEN, SPIN-POLARIZED ATOMIC.

O

Ostrom, Dr. John H. *Division of Vertebrate Paleontology, Peabody Museum of Natural History, Yale University*. DINOSAUR.

P

Papendick, Dr. Robert I. *Research Leader, Land Management and Water Conservation Unit, Science and Education Administration, U.S. Department of Agriculture, Pullman, WA*. ORGANIC FARMING—feature; coauthored.

Parmater, John Q. *Structural Dynamics Research Corporation, Milford, OH*. COMPUTER-AIDED ENGINEERING.

Parr, Dr. James F. *Chief and Supervisory Microbiologist, Biological Waste Management and Organic Resources Laboratory, Beltsville Agricultural Research Center, U.S. Department of Agriculture*. ORGANIC FARMING—feature; coauthored.

Parton, W. J. *Natural Resource Ecology Laboratory, Colorado State University*. GRASSLANDS—coauthored.

Perreault, Dr. Paul D. *Program Manager, GPS Precision Systems, Stanford Telecommunications, Inc., Sunnyvale, CA*. SATELLITE NAVIGATION SYSTEMS.

Phillips, Dr. Richard E. *Professor of Agricultural Engineering, University of Missouri*. SOLAR GREENHOUSE.

Phillips, Dr. Tommy L. *Department of Botany, School of Life Sciences, University of Illinois*. CARBONIFEROUS.

Plucknett, Prof. Donald L. *Scientific Advisor, Consultative Group on International Agricultural Research, Washington, DC*. FERN—coauthored.

Pobell, Dr. Frank. *Institute for Solid State Physics, Nuclear Research Laboratory, Jülich, West Germany*. MICROKELVIN TEMPERATURES.

Pond, Dr. David M. *Tennessee Research Laboratories, Eastman Kodak Company, Kingsport*. CHEMISTRY IN A PETROLEUMLESS WORLD—feature; coauthored.

Ponnamperuma, Dr. Cyril. *Laboratory of Chemical Evolution, University of Maryland*. LIFE, ORIGIN OF.

Poole, Dr. Colin F. *Department of Chemistry, Wayne State University*. GAS CHROMATOGRAPHY.

Prestwich, Dr. Glenn D. *Department of Chemistry, State University of New York, Stony Brook*. TERMITE.

Prior, Dr. David B. *Coastal Studies Institute, Louisiana State University*. MASS WASTING—coauthored.

Prouty, Dr. L. Fletcher. *Formerly, Senior Director, Corporate Communications, National Railroad Passenger Corporation, Washington, DC*. RAILROAD ENGINEERING.

Psaraftis, Prof. Harilaos N. *Ocean Engineering Department, Massachusetts Institute of Technology*. WATER POLLUTION.

Pulkkinen, Kenneth F. *Nautical Almanac Office, U.S. Naval Observatory, Washington, DC*. TRANS-NEPTUNIAN PLANET.

Purser, Kenneth. *President, General Ionex Corporation, Newburyport, MA*. MASS SPECTROMETRY—in part.

R

Rankin, Dr. Mary Ann. *Marine Biological Laboratories, Woods Hole, MA*. INSECT PHYSIOLOGY.

Rawlins, Dr. Stephen L. *National Program Staff Specialist, Beltsville Agricultural Research Center, U.S. Department of Agriculture*. AGRICULTURE.

Richards, Prof. Paul L. *Department of Physics, University of California, Berkeley*. SUPERCONDUCTING DETECTORS.

Richardson, Dr. C. A. *University Marine Biological Station, Millport, Isle of Cumbrae, Scotland*. BIVALVIA.

Rubin, Dr. Sally Jo. *Microbiology Laboratory, St. Francis Hospital, Hartford, CT*. CLINICAL BACTERIOLOGY.

Rumbaugh, Dr. Duane M. *Department of Psychology, Georgia State University*. APE LANGUAGE—coauthored.

S

Saaty, Prof. Thomas L. *Graduate School of Business, University of Pittsburgh*. ANALYTIC HIERARCHIES.

Salans, Dr. Lester. *Associate Director, Diabetes Branch, National Institute of Arthritis, Metabolism and Digestive Diseases, National Institutes of Health, Bethesda, MD*. DIABETES.

Salmon, Dr. Stuart. *General Electric Manufacturing Technology Laboratory, Cincinnati, OH*. GRINDING.

Savage-Rumbaugh, Dr. E. Sue. *Yerkes Regional Primate Research Center, Emory University.* APE LANGUAGE—coauthored.

Schmidt, Dr. George T. *Program Manager, Guidance and Navigation Advanced Programs Division, The Charles Stark Draper Laboratory, Inc., Cambridge, MA.* MISSILE GUIDANCE.

Seager, Dr. C. H. *Sandia National Laboratories, Albuquerque, NM.* GRAIN BOUNDARIES.

Seidel, Prof. George E., Jr. *Animal Reproduction Laboratory, Colorado State University.* BREEDING (ANIMAL).

Shadomy, Dr. Smith. *Department of Medicine, Medical College of Virginia.* ANTIFUNGAL AGENTS.

Shapiro, Dr. Mark H. *Department of Physics, California Institute of Technology.* EARTHQUAKE.

Silk, Dr. Wendy K. *Department of Land, Air, and Water Resources, University of California, Davis.* PLANT GROWTH.

Singh, Dr. R. Paul. *Departments of Agricultural Engineering and Food Science and Technology, University of California, Davis.* FOOD MANUFACTURING.

Skinner, Dr. H. Catherine W. *Jonathan Edwards College, Yale University.* TOOTH.

Sommers, Dr. Lee E. *Department of Agronomy, Purdue University.* SOIL—in part.

Southward, Dr. Alan J. *The Laboratory, Marine Biological Association, Plymouth, England.* POGONOPHORA—coauthored.

Southward, Eve. *The Laboratory, Marine Biological Association, Plymouth, England.* POGONOPHORA—coauthored.

Sposito, Dr. Garrison. *Department of Soil and Environmental Sciences, College of Natural and Agricultural Sciences, University of California, Riverside.* SOIL—in part.

Starke, Prof. Edgard A., Jr. *Department of Chemical Engineering, Georgia Institute of Technology.* ALLOY.

Stegeman, Prof. George I. *Optical Sciences Center, University of Arizona.* NONLINEAR INTEGRATED OPTICS.

Stewart, Dr. William E., III. *Lutz, FL.* INTERFERON.

Stokoe, Dr. William C. *Department of Linguistics, Gallaudet College.* SIGN LANGUAGE.

Strobel, Dr. Gary A. *Department of Plant Pathology, Montana State University.* PLANT PATHOLOGY—in part.

Sugarman, Dr. Barrett. *Department of Medicine, Veterans Administration Medical Center, Houston, TX.* BACTERIA.

Sugiyama, Dr. Hiroshi. *Food Research Institute, University of Wisconsin.* INFANT BOTULISM.

T

Tanaka, Prof. Toyoichi. *Department of Physics, Massachusetts Institute of Technology.* GEL.

Tate, Dr. Robert L., III. *Department of Soils and Crops, Cook College, Rutgers University.* SOIL—in part.

Terrell, Dr. James. *Physics Division, Space Sciences Office, Los Alamos National Laboratory.* GAMMA-RAY ASTRONOMY.

Thompson, Dr. Arthur Howard. *Exxon Laboratories, Princeton, NJ.* SUPERIONIC CONDUCTION.

Toon, Dr. Owen B. *NASA Ames Research Center, Moffet Field, CA.* CLIMATIC CHANGE.

Tracy, Dr. Robert J. *Department of Geology and Geophysics, Yale University.* METAMORPHIC PETROLOGY.

Turnbull, Dr. David. *Department of Applied Physics, Harvard University.* METASTABLE SOLID PHASES.

U

Underwood, Prof. Ervin E. *Department of Chemical Engineering, Georgia Institute of Technology.* SUPERPLASTICITY.

V

Veizer, Dr. Jan. *Department of Geology, University of Ottawa, Ontario, Canada.* PRECAMBRIAN.

Vinson, Dr. S. Bradleigh. *Department of Entomology, Texas A&M University.* ECOLOGICAL INTERACTIONS.

Virella, Dr. G. *Department of Basic and Clinical Immunology and Microbiology, Medical University of South Carolina, Charleston.* IMMUNOLOGY—coauthored.

W

Walsh, Dr. William J. *Energy and Environment Systems Division, Argonne National Laboratory.* AUTOMOTIVE VEHICLE.

Webb, Everette L. *Director of Engineering, Boeing Commercial Airplane Company, Seattle, WA.* AIRPLANE.

Whittaker, Dr. A. Greenville. *(Retired) Aerospace Corporation, El Segundo, CA.* CARBYNES.

Wilkins, Dr. Charles L. *Department of Chemistry, University of Nebraska.* MASS SPECTROMETRY—in part.

Winchester, Dr. John W. *Department of Oceanography, Florida State University.* ATMOSPHERIC SAMPLING.

Wolff, Dr. Jack E. *Senior Chemical Engineer, Tennessee Research Laboratories, Eastman Kodak Company, Kingsport.* CHEMISTRY IN A PETROLEUMLESS WORLD—feature; coauthored.

Wolken, Dr. George T. S., Jr. *Physical Sciences Section, Battelle Memorial Institute, Columbus, OH.* COMPUTATIONAL CHEMISTRY.

Woodmansee, Prof. R. G. *Natural Resource Ecology Laboratory, Colorado State University.* GRASSLANDS—coauthored.

Wüst, Dr. Jürg. *Institute of Medical Microbiology, University of Zurich, Switzerland.* ANTIMICROBIAL AGENTS.

Z

Zuberer, Dr. David A. *Department of Soil and Crop Sciences, Texas A&M University.* NITROGEN FIXATION.

McGRAW-HILL YEARBOOK OF SCIENCE AND TECHNOLOGY

Index

Index

Asterisks indicate page references to article titles.

A

B

C

F

G

H

I

J

K

N

O

P

Q

R

T

U

V

W

X

Y

Z

DOES NOT CIRCULATE